Textbook of
Highway and Traffic Engineering

Textbook of
Highway and Traffic Engineering

Subhash C Saxena

BSc (civil engg), MASc (highway and traffic engg), FASCE, FIE, FZwIE, MIRC

Visiting Professor, Yagyavalkya Institute of Technology (YIT), Jaipur

Former

Principal
Maharishi Arvind Institute of Engineering and Technology, Jaipur

Dean
Faculty of Engineering, National University of Science and Technology, Bulawayo, Zimbabwe

Professor of Civil Engineering
NUST and UZ (Zimbabwe), UDSM (Tanzania), UM (Malaysia)
University of Sulaimania (Iraq)
SVNIT (Surat)

Lecturer in Civil Engineering
Indian Institute of Technology, Roorkee

Author of

*Tunnel Engineering, Textbook of Railway Engineering, Traffic Planning and Design,
Airport Engineering: Planning and Design*

CBSPD

CBS Publishers & Distributors Pvt Ltd

New Delhi • Bengaluru • Chennai • Kochi • Kolkata • Lucknow • Mumbai
Hyderabad • Jharkhand • Nagpur • Patna • Pune • Uttarakhand

Textbook of
Highway and Traffic Engineering

ISBN: 978-81-239-2417-5

Copyright © Author and Publisher

First Edition: 2014
  Reprint: 2017, 2020, 2023

Published by Satish Kumar Jain and produced by Varun Jain for

CBS Publishers & Distributors Pvt Ltd
4819/XI Prahlad Street, 24 Ansari Road, Daryaganj, New Delhi 110 002, India
Ph: 011-23289259, 23266861   Website: www.cbspd.com
                                     e-mail: delhi@cbspd.com

Corporate Office: 204 FIE, Industrial Area, Patparganj, Delhi 110 092
Ph: 011-4934 4934   Fax: 011-4934 4935   e-mail: publishing@cbspd.com; publicity@cbspd.com

Branches

- **Bengaluru:** Seema House 2975, 17th Cross, K.R. Road, Banasankari 2nd Stage, Bengaluru 560 070, Karnataka, India
  Ph: +91-80-26771678/79   Fax: +91-80-26771680   e-mail: bangalore@cbspd.com
- **Chennai:** 7, Subbaraya Street, Shenoy Nagar, Chennai 600 030, Tamil Nadu, India
  Ph: +91-44-26680620, 26681266   Fax: +91-44-42032115   e-mail: chennai@cbspd.com
- **Kochi:** 42/1325, 1326, Power House Road, Opp KSEB, Power House, Ernakulam 682 018, Kerala, India
  Ph: +91-484-4059061-65   Fax: +91-484-4059065   e-mail: kochi@cbspd.com
- **Kolkata:** 147, Hind Ceramics Compound, 1st Floor, Nilgunj Road, Belghoria, Kolkata-700056, West Bengal, India
  Ph: 033-25633055, 033-25633056   e-mail: kolkata@cbspd.com
- **Lucknow:** Basement, Khushnuma Complex, 7-Meerabai Marg (Behind Jawahar Bhawan) Lucknow 226001, India
  Ph: 0522-4000032   e-mail: tiwari.lucknow@cbspd.com
- **Mumbai:** PWD Shed. Gala no. 25/26, Ramchandra Bhatt Marg, Next to JJ Hospital Gate no. 2, Opp. Union Bank of India
  Noorbaug Mumbai-400009, Maharashtra, India
  Ph: 022-66661880/89   e-mail: mumbai@cbspd.com

Representatives

- **Hyderabad** 0-9885175004 • **Jharkhand** 0-9811541605 • **Nagpur** 0-9421945513
- **Patna** 0-9334159340 • **Pune** 0-9923910676 • **Uttarakhand** 0-9716462459

Printed at Sanjay Printer, Sahibabad (UP), India

to

---

my wife Aruna
son Sundeep
daughters Shalini and Shivani

Preface

My students, fellow professionals and colleagues in India and abroad who have read my previous books *Textbook of Railway Engineering* and *Airport Engineering: Planning and Design* have been requesting me to write a book on 'highway and traffic engineering', for a long time. This book is an outcome of their suggestions.

The basic objectives of writing this textbook are:

- To provide a basic book on highway and traffic engineering for students preparing for diploma, undergraduate BE (civil), and postgraduate ME (highway and traffic/transportation) degrees of different universities and technical institutes.
- To develop comprehensive understanding of highway and traffic engineering aspects for professionals, practising engineers and planners engaged in planning, design and building of highways.
- To create ability in consultants and experts to consider and evaluate spectrum of alternatives available in highway and traffic engineering problems, systematically.
- To stimulate learning and research abilities for engineering teachers and research scholars in the area of highway and traffic engineering.
- To meet the requirements of candidates preparing for UPSC, PSCs, Institution of Engineers, and other competitive examinations in highway and traffic engineering.

Highway and traffic engineers of today are expected not only to be conversant with traditional engineering skills but also have decision-making and analytical skills which are technologically achievable, socially acceptable and environmentally feasible. The subject matter presented in this text is up-to-date, taking into account technological advances in a style meeting the needs and interests of the students. This book is not a manual, and for detailed guidelines, specifications, recommendations and procedures, reference to Indian Road Congress (IRC) publications is made frequently. Readers should consult the most up-to-date version of IRC publications when engaged in actual planning, design, construction, operation and maintenance of highways/roads.

The text presented in this book introduces the subject beginning with the pivotal role of highway transportation to chapters on specific areas of highway planning, route selection, geometric and pavement design, construction, maintenance, quality control, traffic operation and safety aspects. It concludes with chapters on environmental and economic considerations. Worked examples, diagrams, tables, etc. are integral part of the main text for developing the creative problem-solving approach and learn theory and its applications. Theoretical and objective type questions are included at the end for revising the theory and preparing for various university and competitive examinations. Most of the material presented in the book is the outgrowth of author's lectures for both undergraduate and postgraduate students on highway and traffic engineering, at universities in different countries.

The subject of highway and traffic engineering is dynamic and flow of technology is so voluminous that there will be a need for future updating of the contents of the book. It is hoped that initial efforts will justify further improvements in subsequent editions. Reactions, comments, suggestions, corrections and feedback from the readers are most welcome for improvement and enhancement of the book.

Subhash C Saxena

Acknowledgments

The concept and planning of the contents of this *Textbook of Highway and Traffic Engineering* was developed jointly with Prof (Dr) Rajat Rastogi, of IIT-Roorkee. I extend my sincere gratitude to him for his contributions, and writing initial chapters for this book.

Special acknowledgments are due to Indian Road Congress (IRC), Indian Standard Institute (ISI), Australian Road Research Board (ARRB), Transport and Road Research Laboratory, UK (TRRL), Federal Aviation Agency, USA, and various authors of books and papers, for use of data developed and published by them, in this text. I apologize for any unintentional imprecision in my referencing or interpretation of their work.

I appreciate the interaction with my colleagues Mr Tarun Rawat, MD, Theme Engineering Services, Jaipur, and Mr Vishnu Sharma, Head, Civil Engineering, Yagyavalkya Institute of Technology, Sitapura, Jaipur, that proved very useful.

I am thankful to Mr YN Arjuna, Senior Director, CBS Publishers & Distributors, for producing such a nice text out of my manuscript written in very poor handwriting.

Finally, I want to thank my family — wife Aruna, son and daughter-in-law Sundeep and Ritu, and daughters and sons-in-law Shalini and Mohit, Shivani and Vishesh — for their patience and continuous encouragement.

I express my deep appreciations to Ms Pallavi Saxena, software engineer, for her continuous support and proofreading work. Arjun, Anvita and Aadi assisted me in using computer facilities. My love and thanks to them.

Subhash C Saxena

Contents

1. Introduction to Transportation   1

2. Highway Terminology   38

Introduction of Technical Terms Commonly used in Highway and Traffic Engineering 38

3. Highway Development and Planning   56

15. Highway Construction and Stabilisation   602

16. Pavement Maintenance, Evaluation and Strengthening   662

List of Tables

List of Figures

Abbreviations (Acronyms)

AADT	Annual Average Daily Traffic
AASHTO	American Association of State Highway and Transportation Officials
AC	Asphaltic Concrete
ACV	Air Cushioned Vehicles
ADB	Asian Development Bank
ADT	Average Daily Traffic
ARAI	Automotive Research Association of India
ARRB	Australian Road Research Board
ASTM	American Society of Testing Materials
ATC	Area Traffic Control
BC	Bituminous Concrete
BCR	Benefit Cost Ratio
BM	Bituminous Macadam
BOOT	Build, Own, Operate and Transfer
BOT	Build, Operate and Transfer
BRTS	Bus Rapid Transport System
BS	British Standards
BSI	Botanical Survey of India
BUSG	Built-up Spray Grout
CBD	Central Business District
CBR	California Bearing Ratio
Cd	Candela
CDMA	Compacted Density of Mixed Aggregates
CIDA	Canadian International Development Agency
CIRT	Central Institute of Road Transport
CM	Crusher Macadam
CNG	Compressed Natural Gas
CO	Carbon Monoxide
CPCB	Central Pollution Control Board
CPM	Critical Path Method
CRCP	Continuously Reinforced Cement Concrete
CRF	Central Road Fund
CRRI	Central Road Research Institute
CSIR	Council of Scientific and Industrial Research
CVPD	Commercial Vehicles Per Day
dB	Decibels
DBM	Dense Bituminous Macadam
DBOT	Design, Build, Operate and Transfer
DHV	Design Hourly Volume

DPR	Detailed Project Report
DST	Department of Science and Technology
EA	Environmental Audit
EDM	Electronic Distance Measurement
EIA	Environment Impact Assessment
EPA	Environmental Protection Agency
EPC	Environment Protection Council
ESA	Equivalent Single Axle
ESCI	Engineering Staff College of India
FAA	Federal Aviation Agency, USA
FRC	Fibre Reinforced Concrete
g	Acceleration due to gravity
GDP	Gross Domestic Product
GEMS	Global Environmental Monitoring System
GIS	Geographic Information System
GL	Ground Level
GNP	Gross National Product
GPS	Global Positioning System
GSB	Granular Sub-base
HC	Hydro Carbon
HCV	Heavy Commercial Vehicle
HDM	Highway Design and Maintenance
HFL	Highest Flood Level
HHV	Highest Hourly Volume
HMSO	Her Majesty Stationary Office, UK
HPS	High Pressure Sodium
HRB	Highway Research Board
HWY	Highway
ICB	International Competitive Bidding
ILO	International Labour Organisation
IRAP	Integrated Rural Accessibility Planning
IRC	Indian Road Congress
IRR	Internal Rate of Return
IS	Indian Standards
ISO	International Standard Organisation
IT	Information Technology
ITS	Intelligent Transport System
IVR	Interactive Voice Response
LDD	Luminaire Dirt Depreciation
LIMV	Linear Induction Motor Vehicle
LPS	Low Pressure Sodium
MAGLEV	Magnetic Levitation
Max	Maximum
MDR	Major District Road
MES	Military Engineering Services
MOEF	Ministry of Environment and Forest

MORD	Ministry of Rural Development
MORTH	Ministry of Rural Transport and Highways
MOST	Ministry of Surface Transport
MOU	Memorandum of Understanding
NAASRA	National Association of Australian State Road Authority
NABARD	National Bank for Agriculture and Rural Development
NBWL	National Board for Wild Life
NCAT	National Centre for Automotive Testing
NCEP	National Committee on Environment Planning and Coordination
NEERI	National Environmental Engineering Research Institute
NH	National Highway
NHAI	National Highway Authority of India
NHDP	National Highway Development Project
NICMAR	National Institute of Construction Management and Research
NITHE	National Institute for Training of Highway Engineers
NMT	Non Motorized Transport
NOC	No Objection Certificate
NPV	Net Present Value
NREGA	National Rural Employment Guarantee Act, 2005
NREGP	National Rural Employment Guarantee Programme
NREP	National Rural Employment Programme
NRRC	National Rural Road Corporation
NRRDA	National Rural Road Development Agency
NRTP	National Road Transport Policy
NTPC	National Transport Policy Committee
ODR	Other District Road
OMC	Optimum Moisture Content
OMMS	On-line Management and Monitoring System
PC	Prime Coat, Premix Carpet
PCA	Portland Cement Association
PCB	Pollution Control Board
PCI	Pavement Condition Index
PCU	Passenger Car Unit
PERT	Program Evaluation and Review Techniques
PI	Plasticity Index
PIU	Project Implementation Unit
PMGSY	Pradhan Mantri Gram Sadak Yojna
PPP	Public Private Participation
PRT	Personalised Rapid Transit
PSI	Present Serviceability Index
PSR	Present Serviceability Rating
PSV	Polished Stone Value
PWD	Public Works Department
R&D	Research and Development
r	Correlation coefficient
RCC	Reinforced Cement Concrete

REAAA	Road Engineering Association of Asia and Australia
RIDF	Rural Infrastructure Development Fund
RTIM	Road Transport Investment Model
ROB	Railway Over Bridge
ROW	Right of Way
RUIDP	Rajasthan Urban Infrastructure Development Project
SCOOT	Split Cycle Offset Optimization Technique
SCOPE	Scientific Committee on Problems of the Environment
SDBC	Semi-dense Bituminous Concrete
SGB	Specific Gravity of Bitumen
SGM	Specific Gravity of the Mix
SGMA	Specific Gravity of Mixed Aggregates
SH	State Highways
SI	Standard International
SIDA	Swedish International Development Agency
SP	Special Publication
SPCB	State Pollution Control Board
SVD	Selective Vehicle Detection
TACV	Tracked Air Cushioned Vehicles
TOFC	Trailer On Flat Car
TOPAZ	Technique of Optimum Placement of Activities in Zones
TOPICS	Traffic Operation Plan for Increased Capacity and Safety
TQM	Total Quality Management
TRB	Transport Research Board, USA
TRRL	Transport and Road Research Laboratory, UK
TSM	Traffic System Management
UMTA	Unified Metropolitan Transport Authority
UNCED	United Nations Conference on Environment and Development
UNEP	United Nations Environment Programme
UNESCO	United Nations Educational, Scientific and Cultural Organisation
USEPA	United States Environment Protection Agency
UTC	Urban Traffic Control
UTP	Urban Transport Planning
UTPP	Urban Transport Planning Process
VFB	Voids Filled with Bitumen
VIM	Voids in Compacted Mix
VMA	Voids in Mix Aggregates
VR	Village Road
VRDE	Vehicle Research and Development Establishment
WB	World Bank
WBM	Water Bound Macadam
WC	Wearing Course
WCED	World Commission on Environment and Development
WHO	World Health Organisation
WMM	Wet Mix Macadam
X	Cross-section

Symbols for Units

(a) SI units

Sl. No.	Quantity	Name of unit	Symbol
1.	Length	Metre	m
		Millimetre	mm
		Kilometre	km
		Micrometre or micron	μm
2.	Area	Square metre	m^2
		Square millimetre	mm^2
3.	Volume	Cubic metre	m^3
		Litre	Lit
4.	Mass	Kilogram	kg
		Gram	gm
		Ton	t
5.	Density	Kilogram per cubic metre	kg/m^3
		ton per cubic metre	t/m^3
6.	Time	Second	s
		Minute	min
		Hour	h
		Day	d
		Year	a
7.	Temperature	Degree Celsius	°C
8.	Velocity	Metre per second	m/s
		Kilometre per hour	km/h
9.	Force	Newton	N
		Kilonewton	KN
		Meganewton	MN
10.	Moment	Newton metre	N.m
		Kilonewton metre	KN.m
		Meganewton metre	MN.m
11.	Pressure/stress	Pascal (newton per square metre)	
		Megapascal	Pa (N/m^2)Mpa (MN/m^2 or N/mm^2)

(b) Non-SI units

Quantity	Unit name	Symbol	Relationship to SI unit
Area	Hectare	ha	$1 \text{ ha} = 10000 \text{ m}^2$
Energy	Kilowatt-hour	kWh	$1 \text{ kWh} = 3.6 \text{ MJ}$
Mass	Metric ton or tonne	t	$1 \text{ t} = 1000 \text{ kg}$
Plane angle	Degree (of arc)	0	$1° = 0.017453 \text{ rad}$
Speed of rotation	Revolution per minute	r/min	
Temperature interval	Degree celsius	°C	$1°C = 1 \text{ K}$
Time	Minute	Min	$1 \text{ min} = 60 \text{ s}$
	Hour	h	$1 \text{ h} = 3600 \text{ s}$
	Day (mean solar)	d	$1 \text{ d} = 86400 \text{ s}$
	Year (calendar)	a	$1 \text{ a} = 31536000 \text{ s}$
Velocity	Kilometre per hour	km/h	$1 \text{ km/h} = 0.278 \text{ m/s}$
Volume	Litre	l	$1 \text{ l} = 0.001 \text{ m}^3$

Conversion Factors

(a) Length measures

	FPS Units				Metric		
	Inches	Feet	Yards	Miles	Centimeters	Meters	Kilometers
Inches	1	0.08333	0.027778	–	2.54	0.0254	–
Feet	12	1	0.33333	–	30.48	0.3048	–
Yard	36	3	1	0.000568	91.44	0.9144	0.000914
Mile	63,360	5,280	1,760	1	160,934	1609.34	1.6093
Centimeters	0.393701	0.0328	0.010936	–	1	0.01	–
Meters	39.3701	3.28084	1.09361	0.00062	100	1	0.001
Kilometers	39370.1	3280.8	1,093.6	0.62137	100,000	1000	1

(b) Area measures

	Sq. inch	Sq. foot	Sq. yard	Sq. mile	Sq. centimeter	Sq. meter	Sq. kilometer
Sq. inch	1	0.00694	0.000772	–	6.452	0.0006451	–
Sq. foot	144	1	0.111	–	929	0.092903	–
Sq. yard	1.296	9	1	0.00003228	8361	0.83613	–
Sq. mile	–	–	3097.600	1	–	2590000	2.590
Sq. centimeter	0.1550	–	–	–	1	–	–
Sq. meter	1,550.00	10.764	1,19599	–	10,000	1	–
Sq. kilometer	–	–	–	0.386	10^{10}	1,000,000	1

(c) Volume measures

Cubic foot	= 1,728 cubic inch, 28316.91 cubic cm, 28.3161 litres
Litre	= 61.026 cubic inch, 0.035316 cubic ft, 1000 cubic cm, 0.001 cubic m
Cubic meter	= 35,315 cubic ft, 1,000,000 cubic cm, 1,0000 litre

(d) Weight measures

Kilogram (kg)	= 1000 gms, 2.20462 lbs, 0.001 metric tone
Pound (lbs)	= 0.453592 kilograms, 453.592 gms, 0.0004536 metric tone

(e) Other units

1 micron	= 1/1,000,000 meter = 1/1000 millimeter = 1/25,000 inch
1 sq. mile	= 640 acres
1 acre	= 0.4047 hectares, 4840 sq yards, 4047 m^2
1 hectare	= 2.47105 acres
1 HP	= 550 ft lbs/sec, 33,000 ft. lbs/min, 745.70 watts, 0.7457 kilo watts
1 ton (2000 lbs)	= 0.9072 tonnes = 907.2 kg
1 statute mile (5280 feet)	= 1.6090 km = 1609 m
1 nautical mile (6080 ft)	= 1,8531 km = 1853 m
1 degree fahrenheit	= (VF-32) 5/9°C
1 gallon (US liquid)	= 0.003785 m^3
1 ft/second	= 0.3048 m/s

Introduction to Transportation

1.0 PREAMBLE

Transportation can be defined as an act, process or instance of moving, conveying or transferring an object or a person from one place to another using suitable source of energy. Going to office from home is an act of moving a person, whereas, taking raw material from a manufacturer or retailer to the construction site is the act of moving an object. There is a certain process involved in moving coal from storage area to furnaces in power plants or moving persons to higher altitudes using cables and suspended cabins. These are examples of conveying man or material between two distant places when normal movement becomes impossible or quite difficult. Similarly, any instance of moving the object or the person from one to another type of mode of travel, say for example, from train to bus at an intermediate stoppage, is a classic example of transferring the object or the person.

All these movements are either guided by scientific laws or are restricted by physical features of the geography of the area in which that movement takes place. Whatsoever is the type of the movement, a source of energy is always required to execute that movement. This source of energy may be derived either from a mechanical device or by converting potential energy to kinetic energy or from human or animal energy. Movements using an IC engine, sliding along a slope and walking or moving in an animal drawn cart are examples of such uses of energy, respectively. It means transportation involves an application of science and mathematics by which the properties of matter and the sources of energy in nature are made useful to move man and material from one place to another. The frictions caused by the physical features of the geographical area had spurred technological inventions and innovations that can help in overcoming those frictions so as to achieve the desired objectives of moving man and materials. The frictions caused by rivers and mountains are overcome using bridge and tunnel technologies, respectively. Along with movement, the safety of the item being moved needs to be ensured. This may be in terms of provision of safe storage along the route; availability of supporting mechanism from below to maintain movement at a predefined level; and placement of a controlling system that facilitate movement without collisions, theft and damage. Efficiency is another important aspect related to the movement or transportation. Any movement should be performed within a stipulated time so that the time and quality utility can be attached to that movement. Similarly, the movement should be performed between places of interest to provide place utility. At the same time, any system or mechanism developed, formulated or invented to achieve the above objectives should be convenient to use for the user. Otherwise, the usefulness of the transportation system will be lost.

At this point, it is worthwhile to note a definition that can comprise of all the points of considerations as discussed before to delineate the breadth of transportation engineering.

Transportation engineering can be defined as a useful combination of physical facilities, objects and devices moving/operating in a medium, and the controlling devices or systems, that can provide a safe, efficient and convenient movement of objects or persons between desired locations; that can overcome the frictions posed by the geographical features of the area; and that can provide time, place and quality utility to the one which is moved using that combination.

There are certain terms used in this definition. These terms signify certain functions that are inherent in any movement. Physical facilities provide guidance and support to the objects that are moved in a medium using those facilities. They also point towards the connectivity and accessibility available between the places of interest. The objects are devices moving in a medium basically indicate towards the mobility imparted to the objects or persons who wished to be moved between places using the connectivity available through physical facilities. The controlling devices ensure the safe and efficient movement, and attach utility to that movement.

1.1 FIELDS OF TRANSPORTATION ENGINEERING

Transportation engineering is a combination of different fields, which are unified by the use of the scientific methods and certain basic fundamentals. Primarily, field of transportation can be broadly classified as:

   i. Transportation economists
   ii. Transportation lawyers
   iii. Transportation engineers
   iv. Transportation planners and
   v. Other transportation professionals.

Transportation economists, lawyers and planners are out of scope of this book, and hence are simply defined to give an idea of the transportation domain in which they work.

Transportation economists generally deal with the economics related to transportation projects, project finances and budgeting, economic viability and feasibility of the projects.

Transportation lawyers deal with the legal issues involved in the planning and construction of transportation facilities, their operation and maintenance. They need to examine and study the legalities involved in the implementation of proposed rules and regulations and have to study in advance the legal possibilities which can stop or delay the implementation of transportation projects.

Transportation planners make estimates of future requirements to improve the existing facilities, plan new facilities, transit systems, re-route the traffic, make rearrangements within the system, take policy implementation decisions, etc. They may work as:

- Consulting firms
- System planners
- Large scale engineering design analysts
- State/Provincial highway/Railway planners
- Regional planners
- System operation planners

- Traffic planners
- System operators
- System optimization analyst/managers
- System regulators
- System maintenance schedulers, etc.

A transportation engineer remains involved with works that are related to transportation. The various fields in which a transportation engineer works are:

- Highway and airfield pavements
- Bridge design
- Design of canals, docks, railways, port facility, intermodal terminal facility, etc.
- Mechanical designs of vehicles, design and working of engines, aeronautics, marine vessels, naval architects, etc.
- Communication, safety (control of flow), etc.

Transportation engineering, therefore, is an extremely broad field encompassing different kinds of professional activity. Broadly, from transportation perspective it can be classified as:

- Transportation system engineers or system planners
  - Mainly working in the area of design and planning of complete transportation system for a region, and includes different modes, technologies, and their interrelationships.
- Transportation system component engineers
  - Mainly work on the design of components and procedures of their use.

The divisions of transportation engineering field is shown in Fig. 1.1.

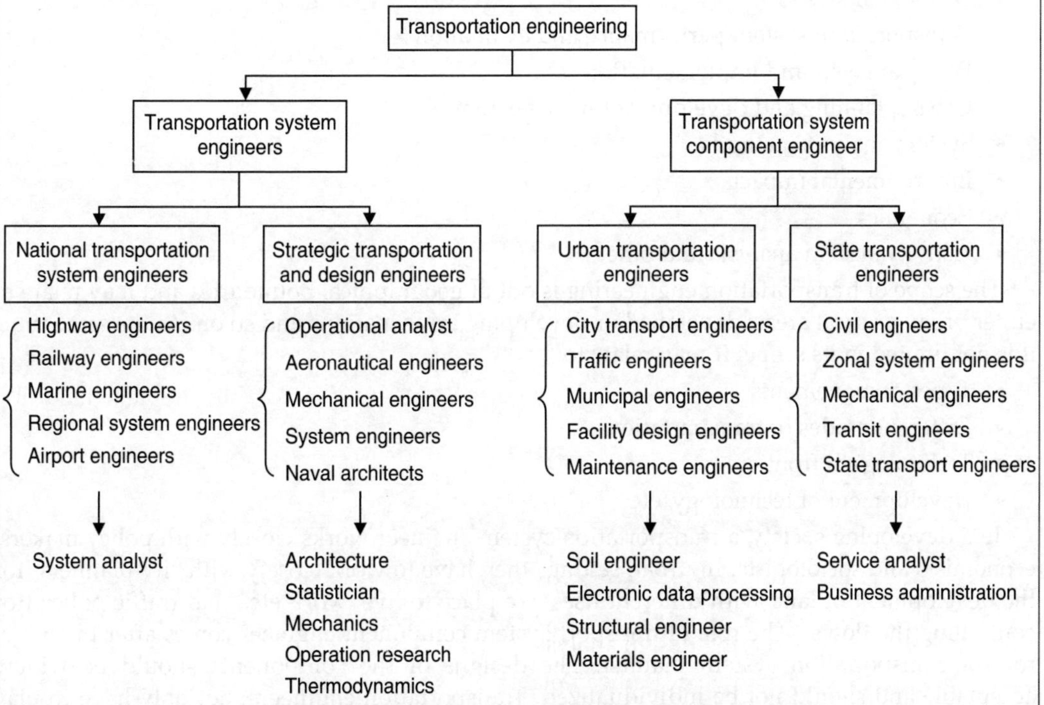

```
                        Transportation engineering
                 ┌──────────────────┴──────────────────┐
      Transportation system                  Transportation system
          engineers                            component engineer
     ┌──────────┴──────────┐              ┌──────────┴──────────┐
National transportation   Strategic transportation   Urban transportation   State transportation
  system engineers        and design engineers          engineers              engineers
```

National transportation system engineers	Strategic transportation and design engineers	Urban transportation engineers	State transportation engineers
Highway engineers	Operational analyst	City transport engineers	Civil engineers
Railway engineers	Aeronautical engineers	Traffic engineers	Zonal system engineers
Marine engineers	Mechanical engineers	Municipal engineers	Mechanical engineers
Regional system engineers	System engineers	Facility design engineers	Transit engineers
Airport engineers	Naval architects	Maintenance engineers	State transport engineers

System analyst	Architecture	Soil engineer	Service analyst
	Statistician	Electronic data processing	Business administration
	Mechanics	Structural engineer	
	Operation research	Materials engineer	
	Thermodynamics		

Fig. 1.1: Fields of transportation engineering

There are large number of specialists working in different fields who assist in the planning, design and implementation of transportation facilities. A short list of such application specialities is given below:

- Highway engineering
- Freight transportation
- Marine transportation
- Transportation management
- Traffic engineering
- Urban transportation planning
- Developing country transportation planning
- Rail transport
- Port development and planning
- Airport planning
- Trucking
- Transportation regulations
- Transportation engineering
- Transportation and economic development
- Transportation economics
- National transportation policy
- Transportation environmental analysis and others.

Similar to above, there are methodological specialities who work in the area of forecasting and projection of demands. Such specialities are:

- Demand analysis
- Transportation system performance and evaluation
- Policy analysis and implementation
- Urban planning and development management
- System analysis methods
- Environmental impacts
- Economics
- Activity system analysis and others.

The scope of transportation engineering is out of geographical boundaries and may relate to either urban or rural areas; developed or developing areas, nations and so on. At the same time, it is not limited in its scope. It may relate to:

- Types of movements
- Different modes of transportation
- Technological frontiers
- Development of technology, etc.

In a developing society, a transportation system engineer works closely with policy makers, economists and sociologists. In urban settings they have to work closely with city planners (for the development of land with different uses like place to live, work etc.) and traffic police (for controlling the flows). The role of transport system component engineer comes after the initial role of transportation system planner. The designs of the components should be widely acceptable and should not be individualized. Transportation engineers, not only have to play their role in government or public sector but also in private sector, e.g. truck lines, airlines, inland and ocean water transportation, container companies.

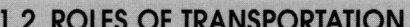

1.2 ROLES OF TRANSPORTATION

Transportation plays different roles in the society, like:

- Economic role
- Social role
- Political role, and
- Environment role.

These are discussed in detail in the following successive sections.

1.2.1 Economic Role of Transport

This can be further discussed under following headings:

- Place, time and quality utility
- Destination choice
- Long v/s short distance movement
- Location of an activity
- Extent of freight transport.

Place, time and quality utility of goods

One of the main objectives of transportation is to attach utility to the movement of objects or persons. This may be in terms of place utility, time utility or quality utility.

Place utility means a commodity having good chances of consumption at a place other than the place of production. It is decided based on the inter-relation of cost at the place of manufacturing, cost at the place of consumption and paying willingness of buyer. Cost at the place of consumption is taken as a function of threshold transport cost (i.e. fixed charges of loading a vehicle, billing, documentation, etc.) and product of unit charge of transport and distance between production and consumption places. This is shown in Fig. 1.2.

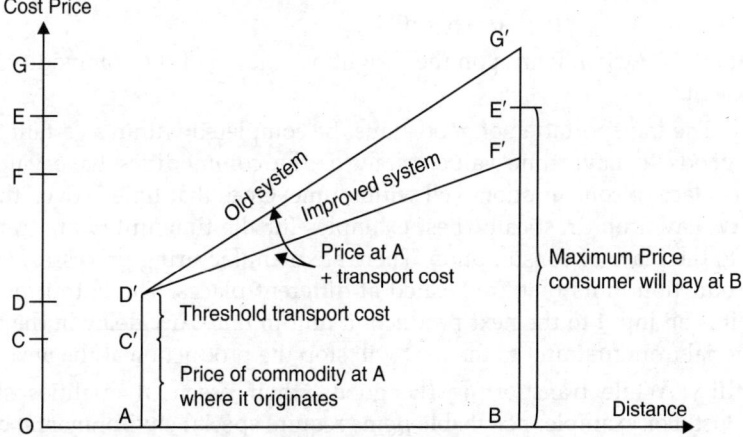

Fig. 1.2: Place utility

In Fig. 1.2, place 'A' is the production centre and place 'B' is the consumption centre. 'OC' denotes the cost of production of the commodity at place 'A' and 'CD' is the threshold cost, as defined before. Therefore, 'OD' is the total cost at the place of manufacturing. In the old transport system, the cost of transportation is 'DG'. This makes the total cost of the commodity at place 'B' equal to 'OG', which is more than the willing price 'OE' consumer may pay at place

'B'. With high transport cost, there will not be any buyer at place B. When system improves the transport cost reduces and is equal to 'DF', thus bringing the total cost of the commodity at place 'B' equal to 'OF'. This is less than the willing price the consumer may pay for the commodity at place 'B'. Therefore, the commodity will be consumed at the place 'B', thus giving it a place utility. The difference in the total cost and the willing price, i.e. 'EF' can be interpreted in different ways. This may allow the increase in the price of the commodity at place 'A' by an amount equal to F'E' (manufacturer's view point) and still the commodity will be purchased at the place 'B'. This may also provide surplus money to the user equivalent to 'EF' that can be used for the purchase of another commodity, which could not be purchased before due to shortage of money available in hand or it may be used to purchase more quantity of the same commodity, may be for stocking or satisfying the actual needs, which were left unsatisfied before. The decision regarding use of excess money in hand will depend upon the users' or consumer's view point.

If the price of commodity is greater than 'OE' at the place of consumption, then the value of commodity at the place 'A' would become zero as there will be no consumer for that commodity at place 'B'.

Mathematically:

Cost of transport (P) = Fixed cost + variable cost

Fixed cost = cost of manufacturing + threshold cost

$$= OC + CD = \alpha$$

Fixed cost is not dependent on the distance between place of manufacture and place of consumption.

Variable cost = Unit operating cost × distance moved

$$= \beta \times d$$

The unit operating cost will be dependent upon the cost of fuel, wears caused during operation, cost of lubricants, and wages paid to the operators.

Therefore, the cost of transportation will be written as:

$$P = \alpha + \beta \times d$$

Here the ratio '/' is dependent upon the weight or volume of the commodity that is moved through the system.

Time utility: The transportation of goods must be completed within a certain time period in order for the goods to have time value. Many of the commodities have value if they are delivered to the place of consumption well within time. Once that time is over, the value of the that commodity may drop drastically. Best examples for the time utility are transportation of raw materials to the place of consumption. There are manufacturing processes, which operates using stage production units that are located at different places. The output of the one stage production unit is an input to the next production unit in line. Any delay in the transportation of the raw material from first unit to the next will stop the production at the next unit in line.

Quality utility: While transporting the goods, their essential qualities should not be diminished or lost. For example: perishable items require special environmental conditions like controlled temperature, pressure or humidity during transport to reduce their natural deterioration. Such items will lose their utility if delivered under deteriorated condition. Sometimes, processed material is transported instead of the raw material, which reduces the weight of the commodity, thus making transportation in bulk possible and easy. Damage to goods during shipment or loading or unloading is another point having economic implications. This may need special packing requirements or use of a specific type of vehicle and is dependent on the condition of the pathway.

Destination choice and transport cost

Destination choice also plays an important role in arriving at transportation cost of materials. The destinations, which are nearby or are connected with other places having continuous movement of traffic between them, will cost less in transportation than those destinations, which are in a remote area or are not having traffic on regular basis or are less likely to be reached. This will increase the cost of transport between such places. The taxes to be paid at the time of crossing the boundaries between states or nations also become a part of the threshold cost.

Further, there are fixed charges for certain minimum weight of the cargo to be moved. With the increase in the weight of the cargo to be shipped, the total cost of transport will not increase proportionately. It will be a loss to the shipper if the shipment is falling below the threshold limit of weight to be transported. In such cases, the cost to be paid by the shipper may be reduced by sharing the space in vehicle with other shippers. Maximum cargo that can be shipped is defined by the vehicle cargo holding capacity. Similarly, maximum weight to be shipped is defined based on the characteristics of the supporting system of the vehicle.

Sometimes, special schemes are implemented to generate revenues and to provide some benefit to the shippers. Golden card scheme allowing extra load above the restricted value is one such scheme. In this scheme the shipper pays extra money for getting golden card and is allowed to ship load more than the restricted load that can be transported using a type of a vehicle.

Use of different transport technology

Technological innovations have resulted in large number of transport technologies. Different technologies are suitable for movement in different distance bands or for movement in varying terrain conditions. Technologies like turbo, jets, rockets, etc. can be used only for travel to long distances say across continents, whereas, technologies based on animal power or human power can be used only for short distance travels. Use of technology can also reduce the time of travel between two distant locations. The distance which were used to be covered in days using runners or animals in old good period can now be covered in some hours only. There is also a difference in the total tonnage hauled between places in the modern times. In good old days, it was possible only by employing a large numbers of labourers or animals to deliver large quantities, but today even a single movement of train can do the same work with higher efficiency and less number of persons employed for the same. This has become more resource efficient. As per one study, the cost per ton-mile has decreased from human back-horse-cart-truck-rail system. This indicates the change from slow movement to faster movement, from smaller loads to bigger loads, from small distances to large distances, and from energy consuming to energy efficient modes of transportation.

Long-distance v/s short distance traffic

There is a compositional change in the movements categorized by distance travelled. Two distinct movements are person movements and freight movements. Large number of person movements can be categorized either in short distance movements or in medium distance movements. Short distance movements are mostly carried out using personal vehicles or public transport facilities available in the area. Medium distance movements are made possible using public transport like bus or train. Few person movements are observed for long and very long distances, which are generally covered using air transportation facilities. Even if such facilities are not available at the point of start of journey, they are reached by using other modes of transportation. Freight movement is the one, which has economic considerations. Large amount of freight is moved from one part of the nation to another part, may be due to location of place of manufacturing and consumptions. Not necessarily, these can be hauled directly to the desired places using a single transportation technology. At times, it requires a mix of technologies, working in coordination with each other, to complete the movement. The shift of freight may be between different types of modes within a system or between different systems.

One example for such a requirement may be the commodities transported via sea routes to other nations. Once these commodities reach the harbor or port of a nation, they are unloaded and transferred to transport modes like trucks and railways for further distribution in different parts of the nation.

Changes in location of an activity

The provision of transportation facilities in an area(s) may cause a change in the activity pattern of that area. Lets us look at an economic scenario. Certain commodity is manufactured at one place and is supplied to another place where it is consumed. After certain period of time another manufacturer sets up a manufacturing unit to produce the same commodity at a distance farther away compared to distance at which the first unit is from place of consumption. Owning to either reduced cost of production or reduced cost of transportation or relaxations in taxes by the government, the second unit supplies the same commodity at lower price than the first unit. This will cause a shift in the consumption pattern of that commodity, the one produced by second unit being consumed more than that of first unit. Depending upon the price difference, this will make a change in the activity pattern, sooner or later. The first unit may get closed down, if improvements to bring down the price of the commodity are not taken or are not made competitive to those offered by second unit. One drastic change, as an after effect, will be the change in the employment pattern in that area.

Figure 1.3 presents a similar activity scenario. Place 'A' is the place of manufacturing a commodity that is consumed at place 'B'. The cost of production at place 'A' is 'AC'. The total cost of production is 'AD', which includes threshold cost 'C'D'. The cost of transportation between place 'A' and 'B' is 'D'E', thus bringing the price cost at place 'B' to 'OE'. Say, this is less than the price cost that a consumer can afford at place 'B' and hence, is consumed. After certain time period, a new manufacturing unit is set at place 'K' to produce the same com-modity. Its distance from the place of consumption, i.e. 'KB' is greater than that of first unit, i.e.

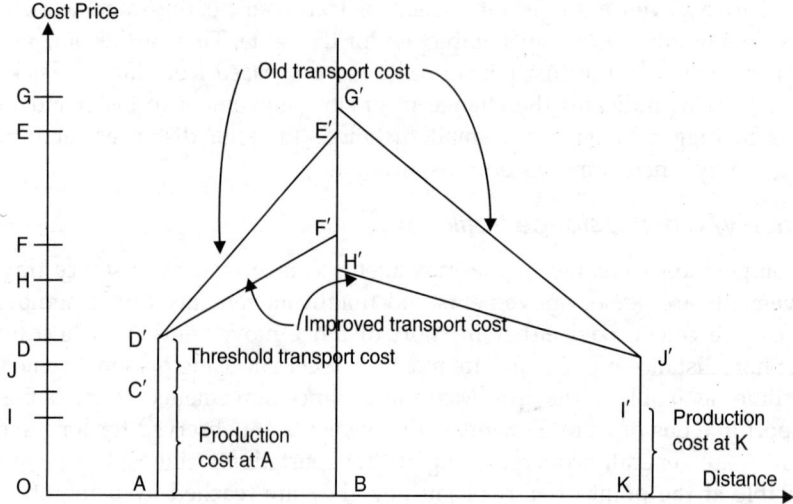

Fig. 1.3: Effect of location of an activity

'AB'. The cost of production at location 'K' is lower than that at 'A'. The threshold cost being same at both the locations, the total cost of production at place 'K' comes to be lower than that at place 'A'. Due to longer distance for supply, the price cost of commodity produced at place 'K' comes out to be higher (OG) than that from place 'A' (OE) and hence, will not be consumed at place 'B'. After some time a new transportation technology is introduced, which brings down the transportation cost between the places, but the reduction in transportation cost is higher between places 'K' and 'B', as compared to between places 'B' and 'A'. This brings the reduction in price cost of the commodity at place 'B', the new value being 'OF' for commodity produced at place 'A' and 'OH' for that produced at place 'K'. As OF > OH, commodity produced at place 'K' will be consumed more than that produced at place 'A'. This transportation improvement may even cause stopping of the production at place 'A'. This change will affect the pattern of production and the pattern of settlement, and the shipment will change accordingly. The change may cause localization or concentration of developmental activities and loss of employment at other places. Therefore, the balancing of economic activities between places is required.

With the changed economic scenario certain income effects may be visualized:

- Saving of money while same living standard is maintained
- More purchase of same commodity
- Spending of saved money on some other commodity
- May cause reduction in time spent on working also

Some more examples depicting the effects of the transportation infrastructure improvements are:

- Constructions of bypass may increase the cost of land available on both the sides of the road; and may induce ribbon development, unauthorized constructions, etc.
- Construction of flyover may reduce delays to crossing traffic, reduce operating cost and travel time, but may cause loss to market activities, which were flourishing on the sides of the road before the construction of the flyover due to coming within ramp areas.

Extent of freight transport

It helps in increasing the range of goods available for consumption, thus improving upon the living standard of the users. Cost of extraction versus transportation, e.g. oil, decides whether a commodity can be consumed at a place away from the place of extraction, or the price of the commodity or its quality. The opportunities of freight transportation may induce the shift in the location of production, employment, etc.

1.2.2 Social Role of Transport

This can be further discussed under following headings:

- Formation of settlements
- Size and pattern of settlements
- Ease of travel
- Long distance travel

Formation of settlements

Formation of settlement helped the end of nomadic life. It made the transportation of food and fuel possible, as well as the storage of food for emergencies. The saving in time that was spent in searching food and shelter during nomadic period, due to better transportation resulted in giving time to other activities. Most of the initial settlements were located near water bodies, especially flowing water, because of ease of transportation. This caused the development of terminal facilities near such trans-shipment points and warehousing facilities to store the material either to be transported through water route or received at that terminal from other locations. Human being could devote more time to professions or activities like farming, cattle rearing, manufacturing, etc. Higher level of interaction started between such settlements, bringing cultural and temporal changes in societies. Availability of material not available at a place otherwise due to transportation routes improved the style of living. Increasing level of technological up gradations and development of transportation systems has lowered the travel times causing horizontal swelling of cities.

Size and pattern of settlement

Size of settlement is governed by population and per capita requirement. It should be able to sustain the population residing therein. If the distance of supply to a point is doubled then due to circular area of influence, the area of land supporting the settlement gets quadrupled. This is known as 'Lardner's law of Squares'. If the distance related transportation cost is reduced then area of possible production for the city increases. The law is depicted in Fig. 1.4.

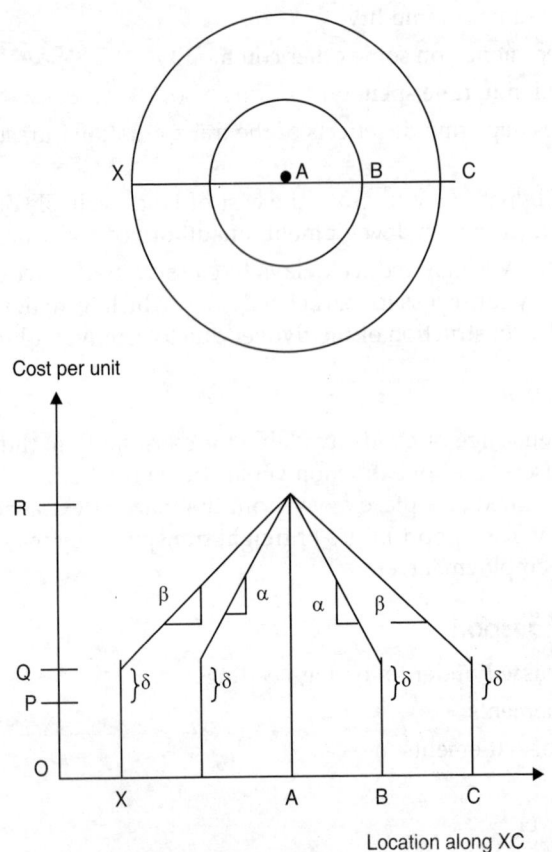

Fig. 1.4: Location and size effect of the city

Where, OR = Affordable cost at 'A'

OP = Production cost of commodity

δ = Threshold cost

α = Unit rate of transportation, defining the extent of settlement, i.e. AB

β = Unit rate of settlement after improvement $(< \alpha)$

Technological developments have reduced the transportation costs in recent years. But the availability and affordability of motorized modes has induced the sprawling of cities. The effect of reduction in the unit transportation cost is the increase in the size of the city within affordable prices. Figure 1.4 shows that under the constraint of affordable price at location 'A' and the existing unit cost of transportation 'α' the city can expand upto location 'B'. But with the transportation improvements, the unit cost of transportation reduces to 'β', thereby increasing the distance that can be traveled within the money available to the user. Hence, the city now can expand to location 'C'.

Another effect is the increase in spendable money that will be available with the household, if no shifting to outer locations takes place. This will result in taking up those activities that could not be taken up before due to unavailability of money. The high population growth has been observed at the two extremes, i.e. high incoming groups and low-income groups. If the size of the household is big, then it may affect the spending power of the households belonging to low-income groups, especially on transport.

In the initial periods of formation of settlements, the central area acts as a core area of activities. Most of the activities remain concentrated in a small area termed as CBD (Central Business District). Dense settlement takes place near CBD area thus increasing the value of land in that area. Similar settlement patterns can be observed along the through routes that connect CBD with other locations in a city. Therefore, the land value increases along the transport networks (ribbon developments, linear developments, etc) also, thus creating higher densities near the facilities. With time, the central areas or areas near the transport facilities become congested not fit to live in due to less air circulation and increasing pollutions, and cause outer flux of population to open areas. This becomes possible again because of lower transportation costs due to improved transportation systems.

The city size has its effect on the development of activity centers also. Single central core causes the concentration of activities, resulting in large number of trips made to that area. This causes congestion, pollution and delays. In its place, the development of number of small activity centers throughout the city can cater to the needs of local areas, thus reducing the need to travel long distances within city, as well as, to a specific location like CBD. The major activity centers may cater to specialized needs and can act as employment hubs.

Dispersal of activities and related developments away from transit routes due to availability of access modes also creates its effect on the settlement patterns. The size of the population and spatial form of urban areas are matters of social choice, with technology limiting the range of choices but not specifying them.

The size of the city and the settlement pattern also affects the personal movement. This needs to be studied with respect to the trade-off between travel time and travel cost. The accessibility and affordability is to be maintained along with mobility. The use of walk mode reduces the extent of the city to around 5 km, so that any location is reachable within half an hour. Animal drawn vehicles or trams increase the extent of the cities to around 13 km. Mostly these are affordable to the masses. The automobile dependence with the technological innovations has increased the city size to around 45 km.

Ease of travel

The technology has helped in easing out the travel burden and has brought about the following changes in the society, settlement and travel patterns-

- Choice of place of residence with respect to the place of work
- Depletion of population from older areas/central areas, causing change in population density pattern
- Dispersal of business and activities in a city or outside the city, like relocation of major businesses namely, wholesale fruit markets, transport providers, etc.
- Increase in traffic congestion due to increase in per capita trips and increase in automobile ownership
- Dangerous levels of noise and air pollution, creating problem of dispersal of pollutants and health problems
- Improvement in accessibility and connectivity along with mobility
- Extreme reliance on automobile causing
  - Disparity in travel opportunities among population groups
  - Automobile oriented development
  - Overlooking of the needs of disabled, elderly and children
- Requirement of certain level of income to use new form of transport, like travel in air-conditioned services
- Use of vehicle, as a symbol of status in the society rather than as a need of travel, reducing social interactions and creating barriers among masses

Long-distance travel

The technological improvements have made it feasible to travel long distances within shorter time periods. The various effects of long-distance travel can be listed as follows:

- Vacation, weekend trip over long distances become possible for masses using affordable modes, for example travel to resorts or amusement parks, nearby hill stations, etc.
- Improvement in level of understanding due to mixing or coming in contact of persons from different cultures, traditions, etc.
- Availability of industrial products at long distances bringing changes and improvements in patterns of living
- Faster travel has increased business trips over long distances within same time period bringing in economic revolution
- Improvement in rural economy and living due to better connectivity, bringing them out of isolation and at times inducing migration to urban areas.

1.2.3 Political Role of Transport

The following aspects need to be discussed:

- Rule of an area
- Political choices in transport
- Financing of transport

Rule of an Area

Transportation has enabled governance of vast area feasible and easy. In ancient period, passing on of the information about the unrest or the problems faced by the people of an empire used to take many days. It was done either by employing runners, or by using drummers located at audible distances passing-on coded information or sending a person on a horse back. Similarly, sending assistance to affected areas from central seat of governance used to take many days. The problems were acute in the case of boarder areas having strategic importance and facing

continuous threats from invaders. The day to day policing of an area have also become possible due to availability of good transportation networks and means of trans-portation. Various political aspects that can be taken care of by transportation are:

- Passing of information related to good governance of an area to all parts of the region or nation governed by a government
- Meeting out the needs of the area and providing assistance to the masses during emergencies
- Use of communication technology to pass-on the information at a faster rate to make possible the equitable governance of different areas, like central or state government decisions, gazette notifications, information on internal or external security of areas, etc.
- Smooth functioning of a representative form of government, as there is no requirement for the government to send a representative physically to get the orders or directions from the higher leadership and wait till the representative returns with the required details.
- Strategic requirements, like movement of defense personal and machinery to the locations which are otherwise inaccessible, like Siachin Glacier; maintaining a continuous supply of required material for their sustenance; providing connectivity to remote but strategically important areas, etc.
- Policing of internal areas, ensuring day to day safety of residents

Political choices in transport

The choice of nation's transport system is necessarily a political one. Some examples of communication in the ancient times as pointed before are through drums, horses, messengers, Pony Express system, etc. In the later period, road networks were developed by Romans for the movement of troops in their administered area. The ancient civilizations used maritime intercity transport system as is clear from their location of settlement. Various types of movements of persons and goods require provision of systems that can satisfy the needs of intra-town, intra-urban, intercity, rural-urban movements. Mobility hierarchy has been provided to provide connectivity between different parts of the country after getting approval from governing political wings. One such example is of Rural Highways, having hierarchical system comprising of Expressways, National highways, State highways, Major District roads and Rural roads. Recent examples of political choices are construction of roads under Pradhan Mantri Gram Sadak Yojna (PMGSY), North-south and East-west freight transportation corridors, Golden quadrilateral, rail links upto higher altitudes and strategic locations like Udhampur and Srinagar, year round ferry service in Ganga river between Allahabad and Howrah, etc.

Similarly, mobility hierarchy can be discussed for urban areas. It includes freeways, arterial roads, sub-arterial roads, collector streets and local streets. The decision of providing a private owned vehicle network or a public transport network in a city is also dependent on political choices. Certain examples of such choices are metro or suburban rail (MR) or light rail transit (LRT) systems, Sky bus system, Bus Rapid Transit (BRT) system, construction of flyovers, etc.

Financing of transport

Financing of transport system is one of the biggest tasks. Mostly it is the responsibility of the central government but the priorities changes with the change in the government. Local bodies have to be provided with sufficient technical knowhow so as to maintain roads and facilities falling under their jurisdictions. One recent example of such initiative is National Rural Employment Guarantee Assurance (NREGA) scheme of central government which envisages the passing of technical knowhow related to road construction, maintenance and quality assurance to local persons through training. Policies need to be implemented wherein the user pays for the provision of facilities and their maintenance, like Road pricing, Tolls, etc, so that the money collected is utilized efficiently in maintaining the old facilities or for providing new ones. Private participation, e.g. BOT, etc. and Public-Private-Participation (PPP) are some such schemes to pool resources for transport infrastructure development. Fixing of priorities and phasing of activities are done so that works can be taken up within available resources. Economic viability of transportation projects should also be analyzed before the provision of facilities.

1.2.4 Environmental Role of Transport

The following aspects need to be discussed:

- Pollution
- Energy consumption
- Land and aesthetics
- Safety

Pollution

Transportation sector is the main polluting sector. Almost seventy percent of the pollution is due to this sector. This may be of any type like air pollution, water pollution, noise pollution, waste, land pollution, etc. The pollution can be defined as unwanted by-product, which deteriorates environment and causes health hazards. The contamination of air from transportation sector usually takes place due to internal/external combustion engines, burning of fossil fuels, etc. One study has observed that a large number of police personals in Bangkok deputed at the roadside to control traffic are affected with respiratory diseases. The pollution level on the road has been found equivalent to smoking of many cigarettes in a day. The gravity of air pollution is associated with the natural dispersal of pollutants or use of cutting-edge technology to reduce concentration of pollutants. These may help in maintaining environmental quality. This is also dependent upon type of urban development, wind velocity and its direction, and the rate of dispersal of pollutants. This process is time dependent. Certain emission control devices are available that can be installed on the vehicles to reduce the harmful emissions. New technology vehicles come with energy efficiency norms and thus affect the environment at reduced level. The fuels are modified, e.g. premium oil and lubricants, Xtra mile oil, etc. that are supposed to improve engine efficiency and reduce pollution. The search for alternate fuels is on, some examples being hybrid fuels, hydrogen fuel, battery operated engines, etc. Some of such vehicles are already under design testing phase, as well as in operation.

Another pollution caused by transportation is noise pollution. Noise is defined as an unwanted sound. It causes physical and psychological effects like deafness, sleep disturbances, startles, loss of concentration, etc. There are noise-sensitive areas like areas with hospitals and schools, where noise control should be implemented. Noise abatement techniques are in use, whereby the noise can be tackled either at the source of the noise itself or by interrupting the path of the transmission of noise disturbances. Some examples are noise barriers, use of acoustic material, etc.

One important source of disturbance or land uses along the side of a facility is vibration induced due to the movement of vehicles on that facility. The transmission of vibrations from railway networks or rapid transit lines causes disturbances to certain activities that require concentration. The effect of vibrations can be reduced by the use of dampeners or by relocating the transit routes or by developing thick vegetation between the source and the affected.

Pollution of water due to transportation is generally associated with the inland water transportation routes or distance sea routes. This is mostly caused due to the release of oil,

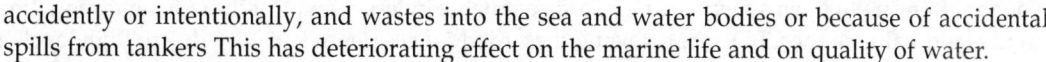

accidently or intentionally, and wastes into the sea and water bodies or because of accidental spills from tankers This has deteriorating effect on the marine life and on quality of water.

Pollution of ground happens due to the discharge of waste onto ground, or dumping loose cargo or time-barred products on vast lands. This makes the land as good as barren with no productive value. The metal deposits and fluid discharges at the time of breaking or processing of such waste render the land further unusable.

Energy consumption

One of the negative effects of transportation is the consumption of energy in different forms. Fuel consumption in the transportation sector is one of the concerns because of the cost of fuel, which is to be imported from the oil rich nations and the depleting nature of that fossil fuel. This causes the outward flow of money, putting pressure on foreign exchange reserves and makes the nations dependent on others. If the comparison is made between different types of travel modes and travel systems then it becomes clear that the consumption of material during vehicle manufacture and operation of motorized vehicles is much higher than those of non-motorized nature. This is another burden on nation's limited resources.

Another point of concern is the loss of energy during movement through the congested networks due to delays and forced stops. Many studies have estimated such losses generally in terms of the loss of productive man-hours. These are found to be much larger in size, at times amounting to some percent of nation's GDP. The distribution of resources should be done equitably without being biased towards certain class of system users. One example of unequitable allocation of resources is the provision of high class facilities for small proportion of car users in developing countries. This is again mostly done at the expense of facilities for mass transport systems or non-motorized modes.

Land and aesthetics

The way in which the transportation system is designed and implemented in a city defines whether it is going to have good effect or a bad effect on the society and the environment. As far as norms are concerned, the land area required for the circulation purpose should ideally be as high as 25 to 30 percent, whereas, it is around 13 percent in India. Such circulation networks should be continuous and wide as far as possible. This will reduce the expenses on wear and tear of vehicles, improve fuel efficiency, reduce polluting emissions, reduce delays and loss of productive time and disperse emissions at a faster rate. This will certainly improve the aesthetics of the network also.

Another aspect to look at is the transportation hierarchy of networks systems. Freeways, arterials, interchanges, etc. may come out to be good or bad depending upon their designs and the ultimate satisfaction of the desired goals. As far as possible, the crossing traffic should be separated out at different levels. The designs should also be user friendly, with no confusion at all while making any maneuver. This will improve the movements, reduce the delays caused due to confusing states and reduce the hazardous situations on the road. A proper or improper design of the system may also induce changes in the residential and business patterns. One such example already discussed is that of the loss of business to the shops located at the sides of the road due to the construction of flyover at that road.

The change in the value of land and its use with the provision of transportation network is another effect of transportation. The change in the value of land may render the area highly desirable or undesirable. Certain examples to this effect are the construction of bypass, CBD areas, etc. Due to congestion a person may like to move out from the CBD area. If this trend persists then one may find an increase in the land value at the outskirts of the city, being most sought after due to open space and more area available in lesser amount initially, and reduction in the land value in the CBD area for being discarded for residential activity due to high level of congestion and pollution. Such changes may also bring in the displacement of facilities. If this happens, it may result in the eating up of additional land available, which otherwise would have been used for some other desired activities, like agriculture. It also has implications related to sustenance of that area. Big transportation projects may also result in displacement of households and their rehabilitation. As far as possible, a good rehabilitation scheme should be implemented so that least legal proceedings get initiated.

The transportation infrastructure can also be used as a structure of tourist attraction. It all depends upon its designing and maintaining the aesthetics. Certain examples are already there, like Howrah bridge, Ram Jhula and Laxman Jhula (swing) at Rishikesh, Sydney harbor link, Worli-Bandra and Worli-Nariman Point sea-links under development in Mumbai, etc.

Safety

One of the disturbing effects of transport is the loss of life and injury to transport users. A large number of accidents are reported involving different types of vehicles. Similarly, a large numbers of in-vehicle and out-of-vehicle users get affected due to accidents. Special mention may be made of the accidents involving freight transport. There is a need to improve the system on different accounts like,

- Periodic inspection of transport vehicle and equipment, e.g. rail and air transport
- Issuing license to operators (for vehicle and route, both) after proper examination of their capabilities
- Use of technological improvements in designs and operations, e.g. expressways, use of instrumental aids, etc.
- Implementing safety measures, in general and specific to conduits
- Stringent measures before issuing driving license to drivers
- Higher enforcement of rules and regulations
- Introducing rapid accident prevention and curing system on highways and urban roads.

1.3 INTEGRATING AND COORDINATING ROLE OF TRANSPORTATION

Transportation has an integrating and coordinating function at all the three steps of economy, i.e. production, distribution and consumption. These are shown in Fig. 1.5.

The process of production requires raw material that is made available through transportation networks and varied means of transportation. The mobility of manpower needed for controlling the manufacturing processes at the plant or for the distribution of produce and manufactured goods becomes feasible only through the transportation means and networks. Availability of items at the place of consumption is again controlled through transportation systems. Apart from these, transportation plays an important role in the governance of an area, in the movement of defense personal and equipment to the border areas, in speedier communication between distant places (apart from information technology) and in generating revenues from use of transportation infrastructure projects. Similarly, markets get directly connected to the production places and consumption places. It also plays its role in the energy sector. Consumption of fuel and other forms of energy is one such aspect. High consumption of fossil fuel and burden placed due to its import has brought forth the need to search alternate fuels that are sustainable, available easily and cheap.

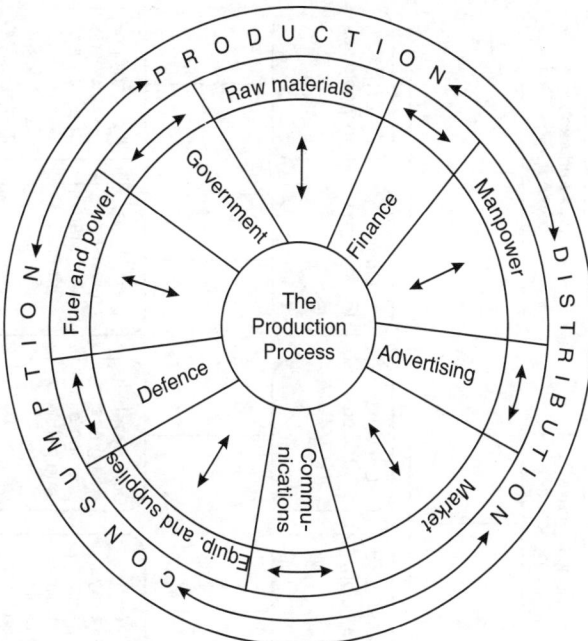

Fig. 1.5: Economic role of transportation

1.4 TRANSPORTATION TECHNOLOGIES

A good numbers of transportation modes are available throughout the world. Each is suitable for different travel distances. They can be categorized based on either the supporting medium, technology used for propulsion and the ownership. At times, certain transportation modes are fitted with technologies that make the use of different supporting medium feasible. A detailed classification chart is given in Fig. 1.6. The major transportation technologies are defined in successive paragraphs below.

1.4.1 Prepared Surface-based Technology

The prepared surface-based technology uses a surface that is prepared to take loads of the vehicles that are expected to move above it. The surface is usually prepared to specifications such that it has sufficient strength to support the vehicle movements without failure. There can be two types of technologies involved in such a system. They are:

- Rail-based technology
- Steel-rimmed or pneumatic wheel-based technology

Rail-based technology

Rail-based technology can be defined as a combination of flanged wheel with a rail. The rail may be either conventionally rigid or flexible. Rails provide lateral guidance and support from the bottom. The mobility is achieved from the engine mounted on a carrier. A good number of modes or systems constitute this category. They are intercity railways, rapid transits, street cars (light rail systems), some designs of monorails, automobile carriers, and tracked air cushioned vehicles (TACVs). The rolling guidance is achieved through the rail and flange combination. This technology has single degree of freedom.

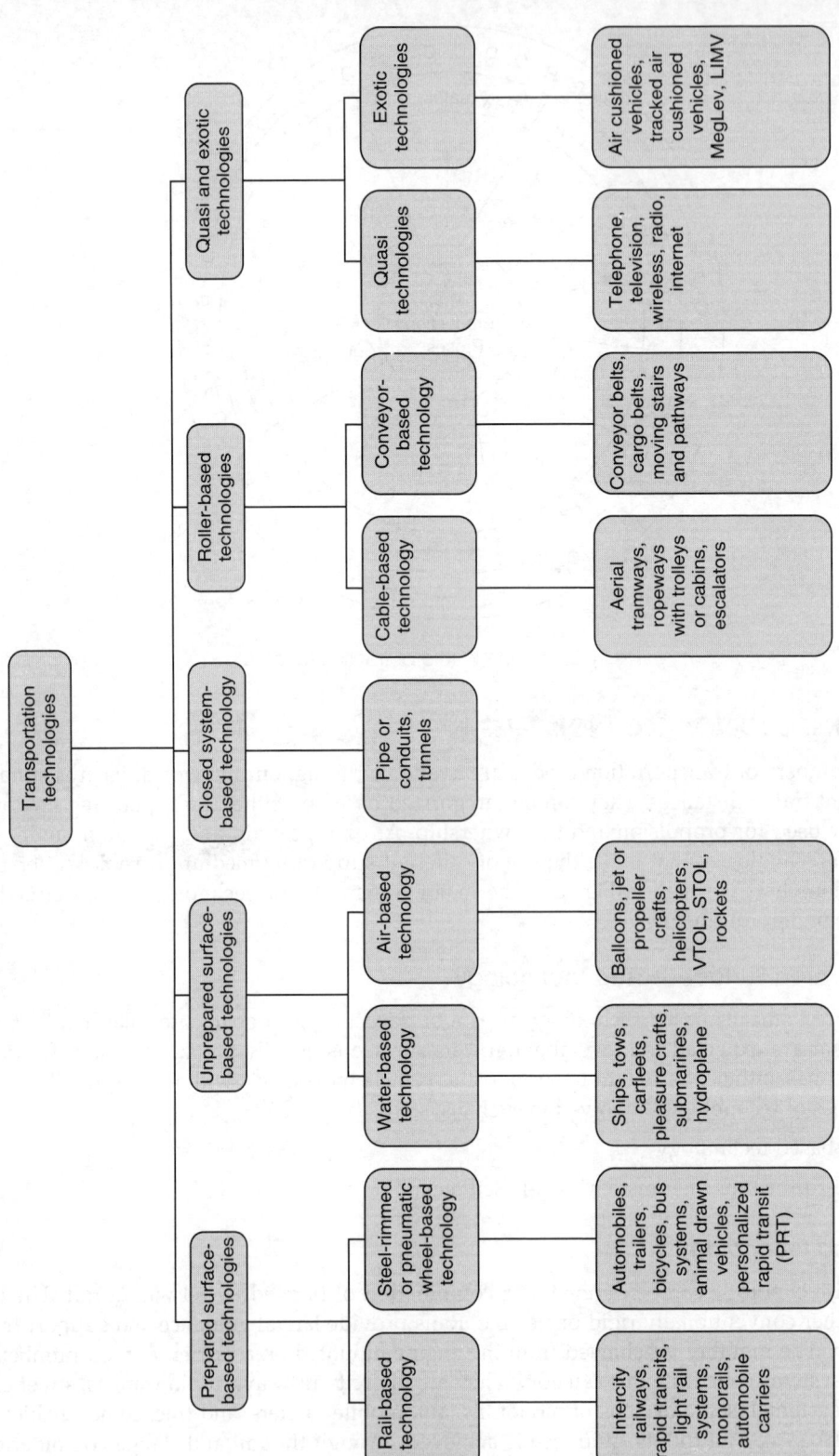

Fig. 1.6: Classification of transport technologies

Steel-rimmed or pneumatic wheel-based technology

The steel-rimmed or pneumatic wheel-based technology uses a flexible or a rigid supporting surface, which is prepared for a smooth and safe movement of steel rimmed or pneumatic wheel vehicles. Examples in this category are automobiles, trucks, buses, tractor-trailers, bicycles, motor bikes, bus systems, animal drawn vehicles, and some of the personalized rapid transit (PRT) people carriers. Mobility in this technology is achieved through the propulsive unit mounted on a vehicle base. The steering system of the vehicle provides maneuverability to the vehicle. This technology has double degree of freedom.

1.4.2 Unprepared Surface-based Technology

The unprepared surface-based technology does not require a physical supporting system that keeps the vehicle in position during its movement. The support is derived from the medium in which a vehicle moves. The exclusion of physical supporting system makes these technologies more hazard-prone. Two of such technologies mostly in use are:

- Water-based technology
- Air-based technology

Water-based technology

Water-based technologies are those technologies, which use water as a medium for the movement of vehicles. The vehicles derive support from the fluid characteristics of the water to remain in position. Such characteristics are buoyant force, draft and block coefficient. Similarly, certain vehicular characteristics like metacentric height, angle of heel and tilt, etc. also act towards the stability of the vehicle. These vehicles consist of engine on board for their mobility. Such technologies are, in general, free from grade resistances. Water bodies used for movement may be channels, rivers, lakes, sea and oceans. Some of the examples of vehicle types using water-based technology are ships, tows, carfleets, pleasure crafts, etc. Technological advancements have made it possible for the vehicles to move through the medium, such as in the case of submarines, or to use two different mediums such as by hydroplanes. These vehicles have two to three degree of freedom.

Air-based technology

Air-based technology uses air space at a height more than the nominal height above the ground for the movement of specially designed vehicles. These vehicles are aerodynamic in design and uses on-board engine to produce propulsive power or thrust for their forward motion. The support is derived from the medium in the form of lifts and drags using components installed on the wings, whereas, the stability along the three axes is achieved using components installed at the tail and wings of the airplane. The long chord of the wing defines the load carrying capacity of the airplane. The air-based technology has three degree of freedom. Examples falling under this technology are balloons, jet or propeller crafts, helicopters, vertical take-off and landing crafts (VTOL), Short take-off and landing crafts (STOL), rockets, etc. These vehicles are able in negotiating any gradients.

1.4.3 Closed System-based Technology

The closed system-based technology allows movement through enclosed surfaces built artificially. The movement is through conduits. Such technology can be used for both, passenger and freight movements. If designed and constructed in rightful manner, it provides a safe and protected passage for their movement. These technologies are mostly used to traverse certain topographical sections, which are otherwise difficult to traverse. Tunnel is an example for passenger movement through such sections. The bed below provides support for the movement and enclosed sides provide guidance for movement. Pipes are the example of transportation of freight. Liquid is used to support the freight as a floating substance and pumping pressure or gravity is used as a propulsion medium. So it provides safe containment, as well as, a roadway. Some of the examples of freight transportation are gas, heat, sewage, petroleum products, etc.

1.4.4 Roller Supported Technology

Roller supported technology is utilized to transport both, the passengers and the freight. Mostly, it is used for freight transportation. Use for passenger transportation is purely for recreational or pleasure trips. Examples of such movements are roller skates. Two such technologies generally in use are,

- Cable-based technology
- Conveyor-based technology

Cable-based technology

Cable-based technology is used primarily for moving products of mines or relatively short distances over a terrain that is too rough for the economical construction and operation of other technologies of transport. These are based on moving suspended cars on cables tightened or slackened at either of the two ends. Rollers are used for such movements. There are specialized designs like aerial tramways, ropeways, trolleys, cabins, etc. that are used to transport passengers or freight over rough terrain.

Conveyor-based technology

Conveyor-based technology is in use again mostly for moving cargo or freight over short distances but for gentle gradients. In this technology, flat conveyors with projected sides are moved using motorized rollers at its bottom. These systems are mostly in use for transporting granular particles of bulks, and therefore, cannot be operated at steep gradients. Such examples are transporting coal in power plants to turbines, mine products to ground levels, etc. In the case of passenger transportation, exclusive applications using conveyor principle are cargo belts at air terminals, escalators or moving stairs, moving pathways, etc.

1.4.5 Quasi and Exotic technologies

Transportation substitutes or alternatives are defined as quasi transport technologies. These technologies reduce the need of moving between places with the desired intention. Some of such technologies are telephone, television, transmission by wire or radio, etc.

Exotic technologies are mostly those technologies that are basically design concepts under pilot testing phases, or the tested concepts that are ready for use. These technologies have found limited implementation due to the cost involved in it. Mostly, they are in use for limited distances on trial basis. Some of the examples of such technologies are air cushioned vehicles (ACV), tracked air cushioned vehicles (TACV), linear introduction motor vehicle (LIMV), magnetic levitation system (MEGLEV), very high speed rail-based systems, etc.

1.5 COMPARISON OF MAJOR TECHNOLOGIES

A comparison of major transportation technologies is given in Table 1.1.

Table 1.1: Comparison of major technologies

S.No	Factor	Prepared Surface Technology		Unprepared Surface Technology	
		Rail-based	Pneumatic wheel-based	Water-based	Air-based
01	Distance coverage	Moderate to long	Small to moderate	long	Moderate to long
02	Speed of movement	Medium	Low to medium	Low	High
03	Hauling capacity	High	Low	High	Medium
04	Type of traffic handled	Freight and passengers	Freight and passengers	Freight and passengers	Passengers and freight
05	Choice of modes	Limited	Varied	Few	Few
06	Fuel consumption (as ratio w.r.t. railway)	1.0	4.0	5.2	25.0
07	Accidents: Numbers Intensity	 Less Medium	 Large Low	 Less High	 Less High
08	Degree of freedom	One	Two	Two to three	Three
09	Unit of transport	Assemblage	Single or with trailer	single	single
10	Flexibility: Route flexibility Operational flexibility	 Limited Limited	 High High	 Nil Limited	 Limited Nil
11	Dependability	Good to excellent	Fair to good	Fair	Fair
12	Gradients traversed	Steeper	Gentle	Nil	Steepest
13	Noise range	70-80 dBA	70-85 dBA	55-68 dBA	75-95 dBA
14	Storage requirements	Manageable	High and unmanageable	Not an issue	Limited and available
15	Guidance system	Rail-flange guidance	Lateral wheel guidance	Electronic guidance	Electronic guidance
16	Support systems	Combination of rail, sleepers and ballast, superelevation design, suspension system of vehicle	Pavement crust thickness, super-elevation design, suspension system of vehicle	Buoyancy, depth of channel, draft and block coefficient	Lift, burble angle, long chord of wings
17	Stability of vehicle	Derailment, overturning	Skidding or slipping, overturning	Metacentric height, size and velocity of wave	Stability w.r.t. lateral, longitudinal and vertical axis

(Contd.)

Table 1.1: Comparison of major technologies (*Contd.*)

S.No	Factor	Prepared Surface Technology		Unprepared Surface Technology	
		Rail-based	*Pneumatic wheel-based*	*Water-based*	*Air-based*
18	Resistances	Due to track profile, and starting and acceleration, Train, and Wind resistances	Air, Rolling, Grade, Curvature, Transmission loss resistance	Skin friction, Streamline, Eddy-current, Wave and Air resistance	Drag resistance, Wing area, Angle of attack
19	Accessibility	Upto medium	High	Limited	Low
20	Cost of: Physical Features Maintenance Terminals	Moderate Moderate Moderate	Moderate Moderate Low	Nil High High	Nil High High

1.6 COMPONENTS OF TRANSPORTATION SYSTEM

The various components of a system are:

1. **Object:** Object can be defined as any item that can be moved in the system using system characteristics. It may be a passenger or freight.
2. **Path:** Path is defined as a location in space along which the object flows or moves.
3. **Vehicle:** Vehicle is the one, which gives the object mobility on a particular type of path employed, and which can be propelled on that path. It may also serve to protect the object from damage.
4. **Container:** It is the device into or onto which the objects to be transferred are placed in order to facilitate the movement. It does not itself posses either mobility or the capability of propelling itself on the path or both. It protects the object and facilitates loading and unloading.
5. **Way-Link:** Way links are the paths in which the flow is constrained to follow a particular route, as in the case of a railway track, highway, pipes, aircrafts, etc.
6. **Way Intersections:** Flows of two or more links can be merged together at intersections, and a single flow can be separated to follow two or more distinct paths at intersections.
7. **Terminal:** Terminals are the points through which traffic is transferred from one vehicle or container to another within the system or between different systems. These are the facilities which accept the objects to be moved into the system or get them out of the system at the end of the journey. It also facilitates the storage of the vehicle when not in use.
8. **Operations Plan:** It is essential that the terminals be operated in a manner that the traffic flowing through them can be accommodated, that vehicles are available to accept the traffic, that the traffic is routed via the proper links and the intersections through the system to the final destination of the traffic. Therefore, it can be defined as a set of the procedures by which the co-ordination of these activities is achieved. The main objective of these procedures remains the safe and efficient movement of the objects.

9. **Maintenance of subsystem:** It is a system, which is primarily treated as a function of cost and management associated with each of the physical components identified earlier. This helps in maintaining the system so that all the components can work together in a coordinated way.

10. **Information and Control subsystem:** It is similarly treated in the context of components and operations where they apply. These provide the safety measures not only for an individual component but also for whole of the system. These may take any shape like devices, information, rules and regulations, design guidelines, etc.

The interrelationship of all the above components is shown in the flow chart given in Fig. 1.7 below. The components related to major transport systems are given in Table 1.2.

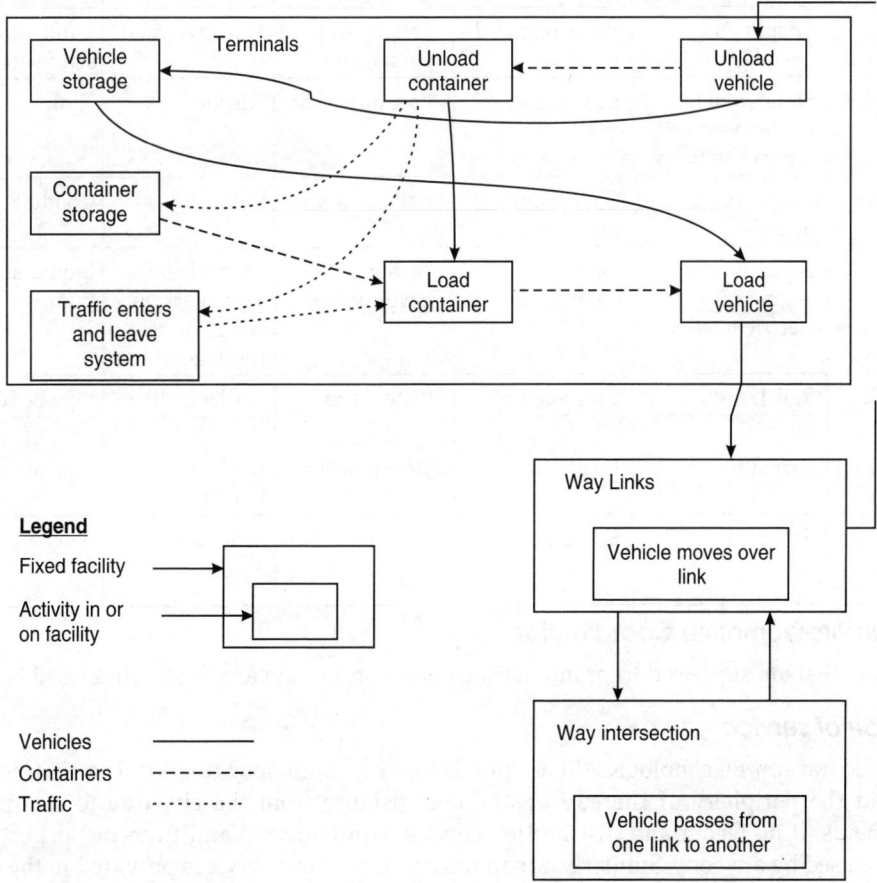

Fig. 1.7: Interrelationship between components of a transportation system

1.7 TRANSPORTATION COORDINATION

An ideal transportation system is the one which facilitates the movement of an object between consigner's door and consignee's door in a single haul. But it is quite difficult to have such a system. Sometimes, such an ideal mode cannot be realized. Each of the transport mode or a system possesses certain inherent technical and economic advantages and disadvantages. For example, a good bus raid transport (BRT) system can be designed and operated in a city. But it is not possible to connect each and every house with its network. Therefore, a person has to walk to a bus stand to get into BRT system. In such cases, the inherent advantages of a system can best be realized or disadvantages overcome, only if two or more of the transport systems are combined together to provide a joint or coordinated transport service. Another example can be the provision of feeder services to main traffic corridors like urban railways or BRTs.

Transport coordination, in this way, can bring in faster and more dependable services and at the same time may economize the transportation. This will be beneficial for both, the passengers as well as to shippers.

Table 1.2: Components of major transportation systems

Component type	Rail-based	Pneumatic wheel-based	Water-based	Air-based	Closed system
Traffic	Freight and passenger	Passenger and freight	Freight and passenger	Passenger and frieght	Fluid and semi-solids
Terminal	Shipper and consignee docks, railway stations	Bus station	Port or harbor	Airport	Tanks
Container	Closed or open Box	Closed or open Box	Cabin, deck	Cabin	Conduit
Vehicle	Separate IC engine or electric traction	On-board IC engines	Turbines on a lower deck	Jet engines or propellers on wings or below fuselage	Liquid and pumps
Way link	Rail Track	Carriageway	Water strip	Air lane	Pipe or conduit
Way intersection	Turnout	Road junction	Strip junction	Air lane junction	Pipe junction
Operations plan	Schedule	Schedule	Schedule	Schedule	Batching

1.7.1 Factors Promoting Coordination

The factors that are supposed to promote transport coordination are briefly discussed below.

Extension of service

Most of the transport technologies do not provide service from door-to-door. The air services are located at the periphery of the city or at some distance from the city due to its operation requirements. If no feeder and distribution service is provided to and from the airport then it cannot be used by anybody. Similarly, urban rail services or bus services provided in the city ply on a predefined route. To improve their ridership, they require connecting services to bring or take back home the potential passengers. The flexibility of truck services is utilized by railways to extend their services to the door of shipper. In all these examples, it is clear that a combination of different services can help in expanding the area of the influence of a major service. Another example of extension of service can be the provision of separate platforms for meter gauge and broad gauge rail services at the same terminal, thus extending the service of railways in low traffic areas.

Economy

The coordination of services between transport modes or within a transport system can bring in financial economy or land use economy. Sometimes, it is possible to provide a single terminal

facility that operates two or more systems together or consist of facilities that are common between different transport systems. This type of coordination can make financial economy feasible. Example for such coordination may be operating railway system and road transport system from the same terminal. The platforms for trains and buses can be provided on the two opposite sides of the terminal building and all administrative units, utilities and facilities can be located within the terminal building. Similarly, it is possible to use same land area for different types of services or systems. It reduces the cost of acquisition of costly land, thus economizing the provision of a service. There can be many examples in this case also, the most seen being the joint use of bridge by railways and road-vehicles at two levels, or road bridge with a pipeline fitted on a side of the road, thus reducing the need of constructing a separate pier for pipeline.

Convenience

The transport coordination facilitates the concentration of objects at one place through a major transportation system and the dispersal of the same using another transportation system. The bulk shipment carriers like railways and waterways accumulate large quantities of freight at a single place may be in the form of stock piles, tanks, warehouses, and the same is distributed to different places of consumption in small lots using road transporters like truckers. The same is true for farm products or dairy products. A simple example of news paper can also be looked in. It is dispatched in bulks from printing press using big automobiles to distant places and is later distributed to the readers by vendors moving on bicycles. In terms of passenger movement, the example of providing railway and road-based system at one place makes it convenient for those passengers who do not have railway services in their area but do have bus services. So they feel themselves free from the stresses of shifting luggage to long distances before getting another forward service.

Speed of transfer

Speed of travel do matter in long distance travel, may be inter-city or inter-country or inter-continent. It is difficult to use the same mode for all travel distances with the same travel efficiency. The coordination helps in achieving the speed efficiency. In case of certain perishable objects, the speed of movement becomes a major factor even if the travel distances are not too long. To get to speedier service, one needs to use another service. For example, to use air service one has to use own travel mode or public mode available to access it.

1.7.2 Limitations of Coordination

Transport coordination should be exercised only when it makes a real contribution to the overall economy and efficiency of movement. Merging of two distinct facilities or two facilities under use by different systems but located at acceptable distance from each other makes no contribution to either efficiency or economy. While planning new facilities, the provision of different transport systems at one location should be emphasized. Coordination should also be examined and properly planned from the view point of frequent interchanges, rehandling of equipment and loading and unloading of equipment. Each handling or interchange increases the possibility of loss or damage to the objects to be transported. Therefore, interchange facilities should be designed based on minimum requirement of handling of the objects and equipment and need of reverse movements.

1.7.3 Types of Transport Coordination

Transport coordination, as discussed before, can be carried out between different systems of transportation, i.e. inter-system coordination, or between different players within the system, i.e.

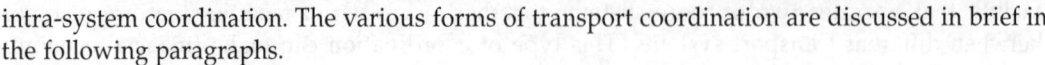

intra-system coordination. The various forms of transport coordination are discussed in brief in the following paragraphs.

Joint use of a terminal

Terminals are used jointly both for inter-system coordination and intra-system coordination. In most of the metro or metropolitan cities, terminal railway stations host facilities for both inter-city travel and intra-city travel. One such example can be of Mumbai Central Railway station, which hosts platforms for inter-city trains as well as for intra-city (Suburban Railway) local trains. This is a good example of intra-system coordination. Another example of intra-system coordination is of bus stations provided in most of the big cities. They provide bus services that ply between big cities, may be non-stop or with limited stops and at the same time also provide services that provide connectivity to interior location of that area, thus making it feasible to move between interior areas to big cities located at long distances. An example of inter-system coordination can be of ferry terminals that are generally accessed through pneumatic wheeled vehicles. So it is a combination of water transport and land transport by road. The same is with railway stations, airports, etc. There are examples wherein union depot is provided for rail, bus and air transport for lake and river transport.

Such coordination is advantageous as it improves convenience, transfer, land-use economy, and allows provision of better facilities for passengers or freight, which otherwise could not be provided due to fund shortages. The coordination helps in reducing the number of services required and frequency of individual shipper, saving of mileages, and avoids the duplication of facilities thus saving resources. But such constructions require large piece of land often at central locations of a city that may consume good proportion of money resources and may cause congestion on roads around the terminal especially during rush hours. Many of the facilities will remain unused and unproductive during lean periods. The individual carriers may also find it disadvantageous as they lose their identity and economic advantage once enjoyed from an individual location over other carriers, and have to spend big amount on spaces used for their operation and administration. It also makes it difficult to apportion user costs and charges among different transport systems using that terminal.

Coordination of schedules

Most of the transport systems or carriers operate with fixed schedules. These schedules are mostly well publicized or published for wider circulation. It is, therefore, possible to rearrange the schedules of different carriers or transport systems or carriers within a transport system for their mutual benefits. Arranging of the schedules of one carrier to connect with the schedules of another carrier can improve upon the ridership of both the carriers. The requisite to such an operation is to have either a common terminal for two different systems or a feeder or bridge line service for a major carrier. Some of the examples of such a coordination can be provision of platforms for long distance trains and local trains at the same terminal; the matching of schedules of two trains reaching a junction station from two different directions and moving forward as a link train; and provision of bus feeder services and distribution services according to the departure and arrival of trains at a railway station. At times, the commuters themselves fix the schedule of their movements according to the schedules of main mode to be taken for long distance commuting.

The coordination of schedules makes the movement of objects or passengers feasible between locations not connected directly. It relieves them of mental and physical stresses of reaching unconnected locations. The intermediate connectivity improves the turnover of the operators. It helps the commuters in reaching distant places in minimum possible time. But, such coordination has its disadvantages also. A commuter has to move within time frame. One has to catch a feeder bus service at a pre-specified time to get the main service say a train. In case of a link train, if one

train gets late in its running then other train has to wait for it. This causes unnecessary delay to the passengers of the first train. Variable schedules of feeder and distributor services have to be formed to adjust the flow on main corridors or with main carriers.

Interchange of equipment

The coordination with interchange of equipment involves the movement of the equipment and the content carried in it freely between two different carriers. Such interchanges save rehandling costs and loss of time in transferring freight. Disadvantages of this coordination includes the difficulties of getting the equipment returned, maintenance required for foreign or off-line equipment, determination and collection of user costs and empty return of equipment causing loss of revenue.

The concept of containerization fits in the above definition of coordination. Containers may be simple boxes that are set on truck bodies, or on flat cars, or in gondola cars, or in ship's holds or on demountable truck bodies or chassis for land movement. The standard containers can be hauled by largest aircrafts only, if air transfer is required. Smaller units that fit the curves of a fuselage are already in use. Individual freight items loaded in the container need not be rehandled from the time that container is closed and sealed by the shipper until the seal is broken by the consignee.

Joint use of right-of-way

Provision of transport systems on dedicated but separate right-of-way has caused shifting of productive land to these facilities, the loss of other productive activities and an increase in the cost of providing these facilities. Due to these reasons, the possibility in the joint use of right-of-way started receiving adequate attention during the last decade or so. In such coordination, two or more than two transport systems use the same right-of-way for their operation. The combinations may involve only passenger movements or freight movements or a mix of the two. There are many examples to define such coordination. At many places, especially over big rivers, bridges are constructed with double decks or more, with each deck being dedicated to one system of transportation. It may be rail at one level, road-based vehicles at next level and another level for peddlers and pedestrians. Another example is of using the same deck of the bridge for the movement of vehicles, as well as, to provide support to pipelines. Sometimes, they are provided below ground level either under footpaths or central medians. But in such cases, the pipelines should be buried deep enough to avoid breakage and disastrous fire in the event of an accident. Many a times, the central median of expressways, freeways or superhighways is used as a right-of-way for railways or rapid transit systems, or for pipelines or systems like skybus.

Such coordinated provisions help in reducing the cost of construction of the facilities, optimal use of land space occupied, and possibility of providing interchange facilities on route and eliminates the monotony of movement when covering longer distances. The major problem with this system lies in the maintenance schedules of individual transport systems, which may cause hindrance to the other transport systems, and may disrupt the operations during events of hazardous nature.

Piggy-back system

This term is used to describe a type of coordination that has come into common use, especially on railways and road-based systems. This is also known as 'trailer-on-flat car' (TOFC) service system. Five systems are generally recognized under this category. They are,

- Vehicles of common highway carriers are hauled by railways. In this system the shipper deals with highway carrier who then deals with the railways.
- Only trailers owned by railways are carried by railways.
- Anyone's trailer is carried by railways
- An intermediate or forwarding agency or broker secures the freight, loads it onto its trailer and onto its own flat car and turn it over to the railways to haul.
- First plan with joint railway and highway rates for haul with provision of highway carriers.

This coordination provides door-to-door service. Much of the expenses and nuisance of highway haul, like traffic congestion, personal problems, traffic rule violations, accident hazards, delays, restrictions on weight and size, limitations on vehicle size, etc. either get avoided or are markedly reduced in nature. Though such a coordination relieves the shipper from the stresses of managing the haul, there are always chances of a breakup between partners, i.e. railways and highway operators that can stop the system from working.

Ferry system

This type of coordination can be defined as the movement of especially highway vehicles by car floats or ferries across the lakes and channels. The water carrier provides a container service, whereby the container, equivalent to highway trailer without a vehicle is transferred to and from the ship to the highway for land movement and vice versa. This type of coordination provides connectivity between places across water bodies when no other type of connectivity is provided. The main problem remains about the size of the ferry fitting the size of the vehicle that is to be ferried across the water body. Further, the system becomes unoperational during high tides, floods, cyclones, etc.

1.8 POPULATION AND MOTOR VEHICLES GROWTH

The population of India is increasing continuously at a high rate. At present, the total population of India is only second to that of China and at the present rate of population increase it is slated to surpass the population of China in next 25 to 30 years. The population statistics of India are given in Table 1.3.

Table 1.3: Population history

Year	Total Population	Decadal Growth
1950	357,000,000	—
1960	443,000,000	24.09%
1970	553,000,000	24.83%
1980	684,000,000	23.69%
1990	838,141,000	22.53%
2000	1,004,591,054	19.86%
2005	1,095,054,669	—
2007	1,129,866,154	12.47%*

*With respect to 2000 population

The population growth of the country has seen the decreasing trend continuously since 1960. The decadal growth rate has decreased by 4.23% in the last 50 years. During the current decade it is increasing at a rate of around 1.7% annually. The population demographics by age show an increasing trend in the age group of 15-34 years and for age group 65 years and above. It indicates that the life expectancy is increasing in the country. The total population of the country is expected to reach 1300 million by the year 2020. The statistics are given in Fig. 1.8. The increasing size of the population will obviously put more pressure on the transportation infrastructure. Heavy investment will be needed to overcome the expected shortage of transportation needs.

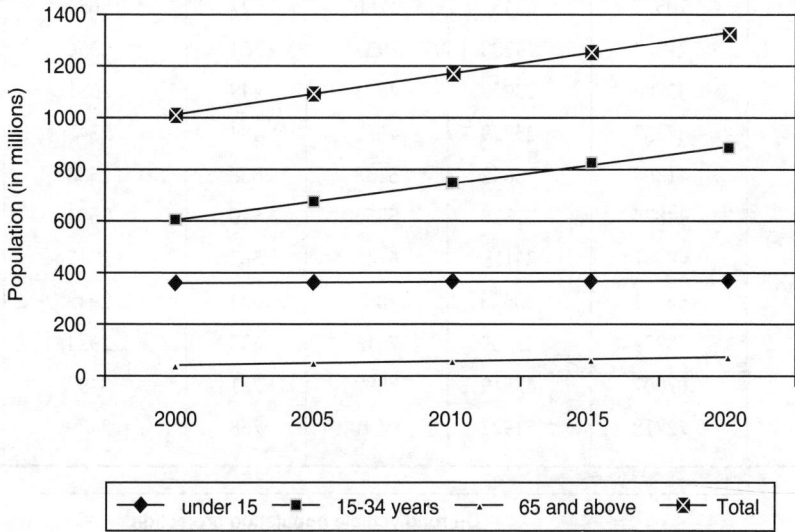

Fig. 1.8: Population projection for India

Another influencing factor affecting transportation is the number of vehicles that get added every year. The data of registered motor vehicles for the country is given in Table 1.4. The population of motor vehicles has increased 237 times between 1951 and 2004. Compared to this, the increase in population between 1950 and 2005 is 206.7%. Starting from 1976 and till 1991, for each successive five year period the rise has been almost 100 percent. Next higher increase in vehicle registration has been observed between 1991 and 1996, and 2002-2003. This high increase in vehicle registration has its effect on the travel needs of private vehicle owners, as well as, on the needs on public transit users.

Table 1.4: Total number of registered motor vehicles in India, 1951-2004, in thousands

Year (As on 31st March)	All Vehicles	Two Wheelers	Cars, Jeeps and Taxis	Buses	Goods Vehicles	Others
1	2	3	4	5	6	7
1951	306	27	159	34	82	4
1956	426	41	203	47	119	16
1961	665	88	310	57	168	42
1966	1099	226	456	73	259	85

(Contd.)

Table 1.4: Total number of registered motor vehicles in India, 1951-2004, in thousands (*Contd.*)

Year (As on 31st March)	All Vehicles	Two Wheelers	Cars, Jeeps and Taxis	Buses	Goods Vehicles	Others
1	2	3	4	5	6	7
1971	1865	576	682	94	343	170
1976	2700	1057	779	115	351	398
1981	5391	2618	1160	162	554	897
1986	10577	6245	1780	227	863	1462
1991	21374	14200	2954	331	1356	2533
1996	33786	23252	4204	449	2031	3850
1997	37332	25729	4672	484	2343	4104
1998	41368	28642	5138	538	2536	4514
1999	44875	31328	5556	540	2554	4897
2000	48857	34118	6143	562	2715	5319
2001	54991	38556	7058	634	2948	5795
2002	58924	41581	7613	635	2974	6121
2003	67007	47519	8599	721	3492	6676
2004	72718	51922	9451	768	3479	6828

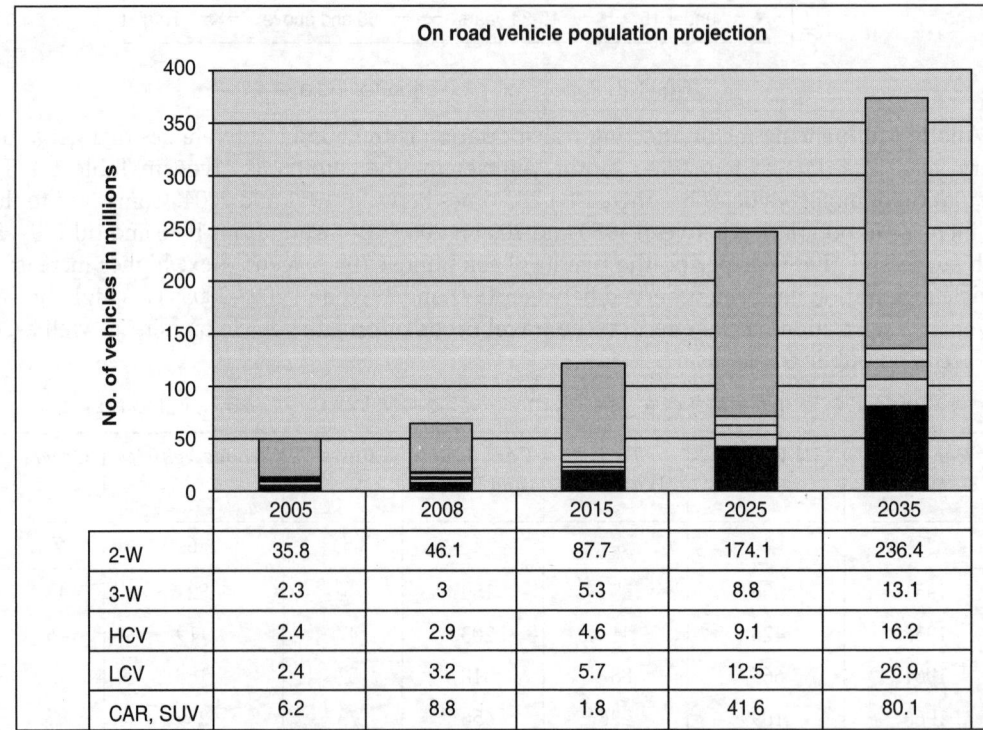

On road vehicle population projection	2005	2008	2015	2025	2035
2-W	35.8	46.1	87.7	174.1	236.4
3-W	2.3	3	5.3	8.8	13.1
HCV	2.4	2.9	4.6	9.1	16.2
LCV	2.4	3.2	5.7	12.5	26.9
CAR, SUV	6.2	8.8	1.8	41.6	80.1

Fig. 1.9: Projection of vehicle population for India

The vehicle population projection estimates are shown in Fig. 1.9. The estimates indicate exponential rise after 2008. This rise is quite high in the motorized two-wheeler segment and is picking big strides in car or SUV segment. The vehicle population is expected to reach 175 million by the year 2020. The vehicle ownership is expected to rise to 0.131 from 0.045 between 2005 and 2020.

1.9 GROWTH OF HIGHWAYS IN INDIA

A look at the budget outlay for transportation during five year plans indicates a decrease in the total share with the change of plans since 1951. The first three plans saw almost one-fifth of the total outlay spent on transportation. After that it has been observed to be fluctuating between 16 and 12%. Even from the committed outlay, a major share has gone to railways. The expenditure on roads has been 6.7% during the first five year plan and that has reduced to 3% during eighth plan. The budget outlay for transportation sector as a whole and road transport in specific is given in Table 1.5 for the period between 1951 and 1997. Table 1.6 gives the outlay for the ninth and tenth plan upto 2005.

Table 1.5: Budget outlay for transportation sector during five year plans (Rs. in crores)

Five Year Plan	Outlay for Transportation sector	Outlay for Road Transport	Outlay for Roads
First Plan, 1951-1956	434	12	135
Second Plan, 1956-1961	1100	18	224
Third Plan, 1961-1966	1983	27	440
Fourth Plan, 1969-1974	2522	128	862
Fifth Plan, 1974-1979	5543	503	1701
Sixth Plan, 1980-1985	13962	1276	3807
Seventh Plan, 1985-1990	29457	2151	6335
Eighth Plan, 1992-1997	56142	3850	13210

Table No. 1.6: Outlay (Centre and States) for road transport sector during IX plan and X Plan (Rs. in crores)

YEAR	Outlay			Revised Outlay		
	Centre	States	Total	Centre	States	Total
1	2	3	4	5	6	7
IX Plan (1997-2002)	60.00	7355.26	7415.26	42.78	5026.15	5068.93
X Plan (2002-07)	210.00	9206.90	9416.90	...	...	...
2002-03	30.00	2209.49	2239.49	29.75	1837.20	1866.95
2003-04	40.00	1361.32	1401.32	35.48	1577.47	1612.95
2004-05	44.00	1498.50	1542.50	35.48	1299.11	1334.59

Source: Planning Commission, Govt. of India.

For the financial year 2006-07, the actual expenditure of the Department of Road Transport and Highways has been 22807.86 crore (Appropriation Accounts 2006-07).

India, having one of the largest road networks of 3.314 million km, consists of Rural roads and Highways namely National Highways, Expressways, State Highways, Major District Roads,

Other District Roads and Village Roads with following length distribution (MOSRT&H Annual Report 2007-08):

National Highways/Expressways	66754 km
State Highways	128000 km
Major and other District Roads	470000 km
Rural Roads	2650000 km

The road network formed by the above categorized roads is assisted by the roads that are constructed and maintained by different authorities like Zila Parishads, Panchayat Samities or Village Panchayats, Urban authorities like Municipalties, and specific agencies like Port trusts, Forest departments, Irrigation department, Electric supply department, etc. Roads under national schemes like Pradhan Mantri Gram Sadak Yojna (PMGSY), Jawahar Rojgar Yojna (JRY), National Rural Employment Guarantee Scheme (NREG), etc. are providing connectivity at the grass root level, as well as, generating employment in rural areas. Road network figures of upto March 31, 2002 are given in Table 1.7.

Table 1.7: Total and surfaced road length by categories in India as on 31st March (kms)

India/Category	Total/ Surfaced	1998	1999	2000	2001(P)	2002(P)
1	2	3	4	5	6	7
A. HIGHWAYS	T	1822874	1888261	1953843	1967005	1981409
	S	1079020	1124064	1146232	1166209	1175353
a - PWD Roads	T	826515	859851	915487	925838	921248
	S	714803	748288	784056	798057	796910
(i) National Highway	T	38517	49585	52010	57737	58112
	S	38354	49368	51952	57679	58006
(ii) State Highway	T	136489	137950	132797	132100	137711
	S	134304	135679	130592	129862	135546
(iii) Other PWD Roads	T*	651509	672316	730680	736001	725425
	S*	542145	563241	601512	610516	603358
b-Panchayat Raj Roads	T	996359	1028410	103856	1041167	1060161
	S	364217	375776	362176	368152	378443
(i) Zilla Parishad Roads	T	451574	456666	483137	487604	499462
	S	247959	250995	272899	275786	283832
(ii) Village Panchayat Roads	T	445353	425486	408524	406150	412595
	S	84374	69485	53705	55675	57338
(iii) CD/Panchayat Samiti Roads	T	99432	146258	146695	147413	148104
	S	31884	55296	35572	36691	37273

(Contd.)

Table 1.7: Total and surfaced road length by categories in India as on 31st March (kms) (*Contd.*)

India/Category	Total/ Surfaced	1998	1999	2000	2001(P)	2002(P)
1	2	3	4	5	6	7
B. URBAN ROADS	T	236055	237866	248408	252001	250122
	S	178877	180558	188325	191797	190102
(i) Municipal Roads	T	212635	214475	224983	228607	226706
	S	157458	159169	166936	170437	168719
(ii) MES Roads	T	11921	11883	11941	11905	11918
	S	11763	11725	11806	11770	11783
(iii) Railway Roads	T	10282	10282	10322	10319	10325
	S	8464	8464	8505	8504	8510
(iv) Major Port Roads	T	832	841	680	687	689
	S	810	818	666	673	675
(v) Minor Port Roads	T	385	385	482	483	484
	S	382	382	412	413	415
C. PROJECT ROADS	T	269427	270523	213827	223665	225116
	S	50523	50758	56041	56541	55034
(i) Forest Department	T	161954	162508	119701	129205	130346
	S	9662	9666	13317	13456	13916
(ii) Irrigation Department	T	73803	74017	61247	61475	61627
	S	19145	19051	21382	21637	19496
(iii) Electricity Department	T	4583	4657	4349	4356	4369
	S	3978	4049	3960	3961	3984
(iv) Sugar Cane Authority	T	22741	22972	23121	23240	23319
	S	12062	12293	12373	12491	12570
(v) Coal Mines Authority	T	3909	3923	2955	2928	2985
	S	3478	3493	2797	2771	2819
(vi) Steel Authority	T	2437	2446	2454	2461	2470
	S	2198	2206	2212	2225	2249
D. RURAL ROADS (PMGSY)**	T	—	—	—	3996	26697
	S	—	—	—	—	—
INDIA	T	2328356	2396650	2416078	2446667	2483344

(*Contd.*)

Table 1.7: Total and surfaced road length by categories in India as on 31st March (kms) *(Contd.)*

India/Category	Total/Surfaced	1998	1999	2000	2001(P)	2002(P)
1	*2*	*3*	*4*	*5*	*6*	
		(3228356)#	(3296650)#	(3316078)#	(3346667)#	(3383344)#
	S	1308420	1355380	1390598	1414547	1420489
		(1491622)#	(1538582)#	(1573800)#	(1597749)#	(1603691)#

(P) Provisional.

\# Includes rural roads constructed under Jawahar Rojgar Yojna as on 31.3.1996.

* Reconciled figures for the years 1998 and 1999.

The increase in the road lengths since 1951 has been substantial even though still not upto the mark. Table 1.8 gives the total road length during different years.

Table 1.8: Growth of roads during different years

Year	Total road length (km)
1951	3,99,943
1961	7,08,122
1971	9,17,880
1981	14,85,421
1991	20,16,594
2001	33,46,667

Different types of road works were taken up during the tenth plan. These included widening works, from single lane to two lane or from two lane to four lane system, strengthening of roads, construction of bridges and railway over-bridges (ROBs) and bypasses, improvement of junctions or shoulders, etc. The outlay of different works during the tenth plan is given in Table 1.9.

Table 1.9: Details of achievement of physical targets during the tenth plan period
(upto Dec. 2006) (km or No)

Sl. No.	Name of Scheme/ project/programme	Unit	2006-07				Total %age of Achievement during the 2006-07
			Overall Target during the 2006-07*	*Target*	*Achieve-ment*	*Shortfall and reasons therefore*	
1	Widening to 2-lanes	km.	1157.00	705.81	708.62	Land Acquisition	100.39
2	Widening to 4-lane	km.	1323.00#	559.35	334.71	ROB	59.84
3	Strengthening	km.	534.00	357.40	655.87	Clearance, Environment and Forest	183.51
4	Improvement of Riding Quality	km.	2087.00	1323.97	1006.13		75.99
5	Construction of Bypasses	No.	11	7	1	Clearance, Law and Order Problems, etc.	14.28

(Contd.)

Table 1.9: Details of achievement of physical targets during the tenth plan period
(upto Dec. 2006) (km or No) (*Contd.*)

Sl. No.	Name of Scheme/ project/programme	Unit	2006-07				Total %age of Achievement during the 2006-07
			Overall Target during the 2006-07*	Target	Achieve-ment	Shortfall and reasons therefore	
6	Major Bridges	No.					
7	Minor Bridges	No.	144	84	55		65.47
8	ROB/RUB No.	No.					
9	Others (Paved Shoulders, Junctions Improvement, Road Safety, Toll Plaza, etc.)	-	Nil	Nil	Nil		Nil

Note: * Includes the Annual Targets for State PWDs, NHAI and BRO

Includes 2.5 Km. length of widening to 8-lane in Delhi.

1.9.1 Growth of National Highways

The development of national highways was considered as main artery of the road networks. Major surge was experienced during fourth plan, ninth plan and tenth plan. The growth of the national highways during different five year plans is given in Table 1.10.

Table 1.10: Addition to NH length during different plan periods

Plan/Period	NH Length (km)
As on 01.04.1947	21378
Pre Plan period (1947-51)	22193
First Plan (1951-56)	22193
Second Plan (1956-61)	23707
Third Plan (1961-66)	23886
Interim Plan (1966-69)	23938
Fourth Plan (1969-74)	28757
Fifth Plan (1974-79)	28977
Interim Period (1979-80)	29023
Sixth Plan (1980-85)	31980
Seventh Plan (1985-90)	33612
Interim Period (1990-92)	33689
Eighth Plan (1992-97)	34298
Ninth Plan (1997-2002)	58112
Tenth Plan (2002-2007)	67120
Eleventh Plan (2007-2012)	67284 – 530 denotified = 66754

The National Highways have been classified depending upon the carriageway width of the highway. Generally, a lane has a width of 3.75 meters in case of a single lane and 3.5 meters per lane in case of multilane National Highways. The percentage of National Highways in terms of width is as under:

Single Lane/Intermediate lane   18350 km (27%)

Double lane   39079 km (59%)

Four Lane/Six lane/Eight Lane   9325 km (14%)

Still large proportion of national highway length is either single lane or intermediate lane wide. This is below the standard norms of national highways, according to which it should be double lane wide. This is going to be the emphasis in the eleventh and twelfth five year plan.

The Government of India has constituted National Highway Development Project (NHDP) to oversee the construction, maintenance and upgradation of national highways in the country. Some of the projects with high significance to infrastructure development namely Golden Quadrilateral, North-south and East-west corridors for movement of freight, addition of lanes to the existing system, efficient movement through the removal of bottlenecks by constructing bypasses, etc. have been taken up under NHDP. These works are taken up in phased manner. The lengths to be constructed, the targets achieved and the final date of completion (expected) are given in Table 1.11.

Table 1.11: Overall status of different phases of NHDP as on February 2008

Phase	Total Length (in km)	Length Completed in km	Likely date of Completion
I GQ, EW-NS corridors, Port connectivity and others	7,498	7035	97% of GQ will be completed by Mar - 08
II 4/6-laning North South-East West Corridor, Others	6,647	1123	Dec - 2009
III Upgradation, 4/6-laning	12,109	330	Dec - 2013
IV 2-laning with paved shoulders	20,000	-	Dec - 2015 (as per financing plan)
V 6-laning of GQ and High density corridor	6,500	NIL	Dec - 2012
VI Expressways	1000	NIL	Dec - 2015
VII Ring Roads, Bypasses and flyovers and other structures	700 km of ring roads/bypass + flyovers, etc.	NIL	Dec - 2014

The road length density of national highways in different states and union territories with respect to the area of the state or union territory and with respect to the population (census 2001) are given in Appendix 'A'. Similarly, the list of various national highways, by their number, route and length between terminating locations, categorized by states and union territories are given in Appendix 'B'.

1.10 TRANSPORTATION ADMINISTRATION IN INDIA

The hierarchical structure of administrative setup in India related to transportation sector is given in Fig. 1.10.

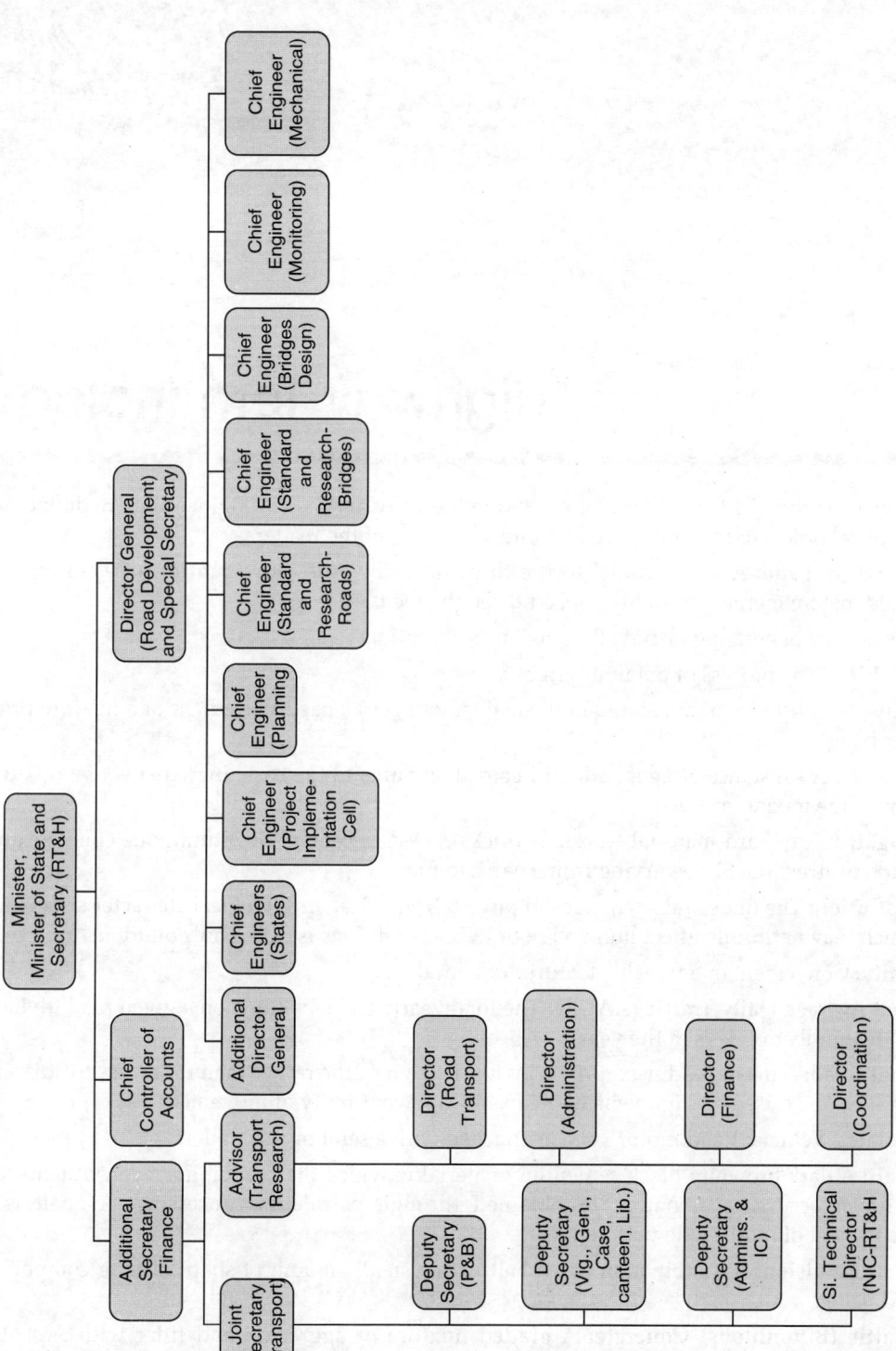

Fig. 1.10: Administrative setup of highway administration at center in India

Highway Terminology

In this chapter, the terms commonly used in highway and traffic engineering are defined in alphabetical order for remembrance and understanding of the readers.

Acceleration Lane: A lane parallel to the through traffic lane, provided for accelerating the speed, before merging with high speed through lane traffic.

Access: A way of entering or travelling towards a location.

Acquisition: The process of obtaining right of way.

Additive: A substance or agent added in small amounts to a basic ingredient of a mixture prior to mixing.

Admixture: A substance or agent added in small amounts to the basic ingredients of a mixture during the mixing process.

Aggregates: Any hard material (rocks or bricks), used in cement or bituminous concrete and road construction. Sizes varying from coarse to fine.

Air Pollution: The undesirable changes in physical, chemical, or biological characteristics of air which may harmfully affect human life or living conditions is called air pollution.

Amenity: Convenient or agreeable features of a locality.

Annual Average Daily Traffic (AADT): The total yearly traffic volume on a given road divided by the number of days in the year.

Arterial Road: A major road used primarily for through traffic rather than for access to adjacent land, characterised by high vehicular capacity and continuity of movement.

Articulated Vehicle: Road motor vehicle attached with a semi or full trailer.

Asphalt: A dark brown to black cementing material in which the predominate constituents are bitumen occurring in nature or obtained through petroleum processing. Asphalt is a constituent of most crude petroleums.

Asphalt Emulsion: A combination of asphalt with a small amount of soap-forming compound and water.

Asphaltic (Bituminous) Concrete: A graded mixture of aggregate and filler with asphalt/bitumen, placed hot or cold and rolled for making a road.

At-grade Intersections: At-grade intersections are those where roads meet at the same level, i.e. there is no vertical separation between two roads.

Attracted Traffic: When a new road is constructed, or old road is improved, in addition to its own traffic some traffic is attracted from the nearby road. This traffic is called attracted traffic.

Auto Rickshaw: A motor operated vehicle having three wheels less than 450 mm in diameter.

Auxiliary Lanes: Lanes other than main driving lanes of a roadway, such as turning lanes, truck lanes, passing lanes, parking lane, etc.

Average Daily Traffic (ADT): The average two-way daily traffic volume for a specified day.

Average Spot Speed: The arithmetic mean of the speeds of all traffic or component thereof, at a specified point.

Backfill: Material used to replace or the act of replacing material removed during construction. Also may denote material place or the act of placing material adjacent to structures.

Barricade: A device which provides a visual indicator of a hazardous location, or a desired path motorist should take.

Barrier: A device which provides a physical limitation through which a vehicle would not normally pass.

Base Course: The base course in the main load carrying course within the pavement, placed on a sub-base or sub grade and to support surface course.

Base Year: The year to which the data is related.

Basic Capacity: The maximum number of passenger cars that can pass a given point on a lane or roadway during one hour under the most nearly ideal roadway and traffic conditions that can be attained.

Bearing Strength (Bearing Capacity): The measure of the ability of a pavement to sustain the applied load.

Benefit/Cost Ratio (B/C Ratio): A ratio used to compare the benefits vs the cost of proposed alternatives.

Berm: The shoulder of a paved road or ditch.

Bicycle: or cycle is a two-wheeled road vehicle fitted with pedals and using human energy as its source of moving.

Binder Course: A plant mix of graded aggregate (generally open graded) and bituminous material which constitutes the tower layer of the surface course.

Binder: Bituminous cement or modified asphalt cement that binds the aggregate particles into a dense mass.

Bituminous Concrete: A designed combination of dense graded mineral aggregate, filler and bituminous cement mixed in a central plant, laid and compacted while hot.

Bleeding of Water: Excess water pumped through surface cracks under action of traffic.

Bottleneck: A highway section with reduced capacity that experiences operational problems such as congestion.

Boulevard: An improved strip of land between the roadway and the sidewalk or between two opposing roadways.

Braking Distance: The distance travelled by a vehicle after application of brakes.

Bridge: A structure which provides a roadway or walkway for the passage of vehicles and pedestrians across an obstruction, gap, water course, or facility and which is typically greater than 3 m in span.

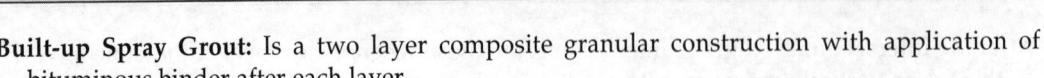

Built-up Spray Grout: Is a two layer composite granular construction with application of bituminous binder after each layer.

Bump: Localised upward displacement of the pavement which may be caused by the absorption of moisture in swelling soils or load induced plastic deformation pushing upwards materials not directly under the load.

Bus: A motor vehicle designed for the transportation of more than 10 persons.

Bus Bays: These are special areas made along the road side for use of the buses.

Bus Rapid Transit: A modern, high-tech, rail-like bus system, with rapid, convenient, frequent service and quick collection of fares.

Buslane: A street or highway lane intended exclusively or primarily for buses either all day or during specified periods.

Bypass: An alternative road constructed to enable through traffic to avoid congested urban areas or other obstructions to movement.

Caisson: A wood, metal or concrete casing sunk or constructed below ground or water level.

California Bearing Ratio (CBR): The bearing ratio of soil determined by comparing the penetration load of the soil to that of a standard material. The method covers evaluation of the relative quality of subgrade soil but is applicable to sub-base and some base-course material.

Camber: Camber is raising the centreline of a road pavement higher than its edges, in order to provide rapid drainage of surface water.

Candle Power: Luminous intensity expressed in candles.

Candela: Candela is the unit of luminous intensity. It is defined as luminance of a black body at a temperature of freezing of platinum.

Capacity of Road: Capacity is the maximum number of vehicles that can pass over a given section of a lane or highway in one direction (both direction for a 2 lane or 3 lane highway) during a given period of time (usually one hour) under prevailing roadway and traffic conditions.

Carpool: An arrangement in which a group of people share the use and possibly the cost of a car in travelling to and from pre-arranged destinations together.

Carriageway: That portion of the road devoted particularly to the use of vehicles, inclusive of shoulders and auxiliary lanes.

Catch Drain: Catch drains are located on the high side of cutting on the top of batter to intercept the flow of surface water and upper soil seepage water.

Causeway: A carriageway across a water course or across tidal water, specially constructed to resist the effects of submergence.

Cement Concrete: Cement concrete is a composite mixture of portland cement, water and inert aggregates.

Central Business District (CBD): The downtown retail trade and commercial area of a city.

Centroid: An assumed point in a zone that represents the origin or destination of all trips to or from the zone. It may not be the geographical central of the zone/area.

Channelisation: Channelisation involves use of islands at the intersections, to guide and protect the traffic and pedestrians.

Channelised Intersection: At-grade intersection in which traffic is directed into definite paths by islands.

Chippings: Fragments of single sized material of nominal size between 3 to 25 mm.

Clay: Fine grained soil that exhibits plasticity over a range of water content, and that exhibits considerable strength when dry. Also that portion of the soil finer than 2 mm.

Climbing Lane: An auxiliary lane introduced at the beginning of a sustained upgrade in the direction of traffic flow, to be used by slow moving vehicles such as trucks and buses.

Cloverleaf Interchange: An 8 ramp interchange resembling a cloverleaf, in which 4 ramps lie in a diamond roadway configuration and 4 ramps lie in a cloverleaf configuration. This type of interchange does not require a traffic signal.

Coal Tar: A by-product in the manufacture of gas from coal. It is obtained by destructive distillation of coal and is used for road works.

Coefficient of Friction: Coefficient of friction is the ratio of the friction force at and along an interface to the force normal to the interface.

Collision: An incident resulting in property damage, personal injury or death and involving the loss of control and/or the striking of one or more vehicles with another vehicle, a person, an animal or an inanimate object.

Colour Sequence: A pre-determined order of traffic signal indications within a cycle.

Compaction: Compaction is the process of soil densification by which the air voids of a soil are reduced. It involves application of loads through rolling, kneading, tamping, rodding or vibratory actions at specified moisture content

Comprehension: The ability of drivers to understand the meaning of a sign message including any symbols or abbreviations.

Consolidation: Consolidation is the process by which the moisture content and volume of the soil reduces under long term sustained loads.

Construction Joint: A joint made necessary by a prolonged interruption in the placing of concrete.

Contour: Graphical plot consisting of a smooth curve, statistically regressed through points of equal levels.

Contract: Contract is a voluntary but legally enforceable agreement between two parties in which one party, for a consideration, offers to do something for the other.

Contraction Joint: A joint at the ends of a rigid pavement/slab to control the location of transverse cracks.

Contrast: Contrast refers to difference in colour or in brightness which allows a target, such as a sign message or symbol to be seen against the sign background.

Controlled Access Road: A highway for through traffic with access from abutting properties or joining roads is controlled.

Controlled Intersection: An intersection where traffic approching from any or all directions is regulated by some form of traffic control device. Where regulatory controls such as give way signs, stop signs, police and traffic signals may be used.

Cordon Line: An imaginary line enclosing the whole areas of study.

Corridor: A strip of land between two terminal within which traffic, topography, environment and other characteristics are evaluated for transportation purposes.

Courses: Pavement courses are layers which are placed and compacted and as an entity to form a course.

Crack: A fissure or open seam not necessarily extending through the body of a material.

Creep: Time dependent deformation of a material under sustained load.

Critical Length of Grade: The combination of gradient and length of grade that will cause a designated vehicle to operate at some predetermined minimum speed.

Critical Path Method (CPM): A scheduling technique which uses single completion time estimates for each activity to construct the sequence of work activities which will govern the completion date.

Cross Fall: The fall given to the surface of any part of the roadway, at right angles to road length.

Cross-Section: The transverse profile of a road showing horizontal and vertical dimensions.

Crown: The highest point in cross-section of a curved road surface, near or at the centre point.

Crude Oil: Unrefined petroleum.

Culvert: A structure which is designed to provide an opening under a roadway, railway or side entrance for the passage of water, livestock or pedestrians.

Curve: A horizontal or vertical deviation in the roadway.

Cuttings: That portion of the site where road formation has been excavated below the ground level.

Cycle (Traffic Signal): When referring to traffic signal, cycle describes one complete sequence of signal indication.

Cycle Length: The time required for complete sequence of traffic signal indications (R-RY-G&Y) is called cycle length or time cycle.

Cycle-track: In urban areas, where number of bicycles or motor cycles is large, a cycle-track (lane) is provided separately for such traffic.

Dead Load: The weight of all material supported by the structure and not subject to movement.

Deceleration Lane: A lane which provides space for left turning vehicles to slow down separating from through traffic, before making a left turn.

Decibel (dB): A unit of measure of sound level. The number of decibels is calculated as ten times the base 10 logarithm of the square of the ratio of the mean square sound pressure and the reference mean square sound pressure of 20 mPa, – the threshold of human hearing.

Delay: This is the time lost while traffic is impeded by some element over which the driver has no control.

Dense Graded Aggregates: A well graded aggregate so proportioned as to contain a relative small percentage of voids.

Density: The number of vehicles per kilometer on the travelled way at a given time instant.

Density of Traffic: Density of traffic is the number of vehicles occupying a unit length of the road. (vehicles/kilometer)

Design Capacity: The flow that the designer assumes in order to design all associated traffic facilities. Maximum number of vehicles that can pass over a lane or a roadway during one hour without operating condition falling below a particular level.

Design Hourly Volume (DHV): The amount of traffic a transportation facility (road) is designed to carry in one hour.

Design Noise Level: The maximum traffic noise allowed that has an acceptable impact on human activities.

Design Speed: The highest speed at which a vehicle can travel safely on the relevant section of road in good weather and low traffic. It is used in design.

Design Vehicle: A selected motor vehicle whose weight, dimensions and operating characteristics are used to establish highway design controls to accommodate vehicles of a designated type.

Design Volume: Volume of traffic determined for use in design, representing traffic expected to use a highway. It is an hourly volume.

Destination: The zone in which a trip terminates (ends)

Detour: An alternative road which traffic may use while another road is temporarily closed.

Developed Traffic: Due to the development of the adjacent area of land, more traffic continues to develop on a new or improved road, for many years. It is called developed traffic.

Diamond Interchange: A 4-leg interchange, generally having more than or e highway grade separation, with direct connections for the major right turning movements.

Distortion: Irregular deformation of the pavement which may be the result of differential settlement of the lower pavement layers.

Distress: Excessive cracking or deformation

Diverging: The dividing of a single stream of traffic into separate streams.

Divided Highway: A highway, street or road with opposing traffic, separated by a median.

Donation: The voluntary conveyance of private property to public ownership and use, without compensation to the owner.

Dowel: A load transfer element usually consisting of a plain round steel bar.

Driveway: Vehicle access provision between the edges of a carriageway and the property boundary.

Driver: Any person who drives a vehicle or rides to drive a slow vehicle.

Durability: Durability is a measure of a rocks (or aggregates) ability to resist repeated loading, wearing, and weathering (or degradation).

Elasticity: The property of a material that permits it, after loading to return to its original unloaded condition.

Embankment: A barrier comprised of earth and constructed above the natural ground surface to carry a road.

Encroachment: A structure or object placed in, under or over a highway right of way.

Engineered Soil: Placed soil of known geotechnical properties.

Environment: The totality of man's surroundings social, physical, natural, etc.

Environmental Impact Study: An analysis of the environmental impacts that proposed transportation projects could potentially have.

Expansion Joint: A joint located to provide for expansion of a rigid concrete slab, without damage to itself, adjacent slabs or structures.

Expressway: A highway for through traffic with full or partial control of access and generally with grade separation at intersections.

Expropriation: Acquiring property for highway purposes by the right of eminent domain.

Fatal Accident: An accident in which one or more persons are killed.

Faulting: Relative movement of slabs in a cement concrete pavement due to differential deformation.

Feeder Streets: These streets allow traffic to circulate within neighbourhood areas, taking motorist to major arterial streets.

Filler: A fine mineral powder added to bituminous mixes. It may be stone dust, cement, slag dust, coal dust, etc.

Fixed Delay: Delay caused by traffic controls.

Fixed Time (Pre timed) Signals: Signals in which the sequence Red - Red yellow - Green - Yellow, etc. as they appear on each approach of an intersection is fixed and repeat after a fixed interval in seconds.

Flag Stone: A flat and relatively thin slab of natural or artificial stone for pavement, used for walking purposes.

Flared Intersection: An unchannelised intersection or a divided highway intersection without islands other than medians, where the travelled way of any intersection leg is widened or an auxiliary lane added.

Flash Point: The lowest temperature at which the vapour of a substance takes fire, but does not continue to burn, under specified conditions of the test.

Flashing Lights: Flashing lights warn the approaching drivers, of certain critical locations like railway crossing, obstructions on road, etc. to slow down and proceed with caution.

Flexible Pavement: A pavement structure that maintains intimate contact with and distributes load to the subgrade, due to its flexibility.

Flow: The number of vehicles passing a point on a road per hour.

Fluxing: Softening of hard bitumen or asphalt to desired consistency by mixing with certain oils. The product is called flux oil.

Fog Seal: A thin application of bituminous material without cover aggregates.

Footpath: Portion of road set aside for use of pedestrians.

Forecasting: The process of determining the future value of land use, and number of trips, traffic, population, etc.

Foundation: That part of a structure or bridge that transfers load to the soil, or rock. It supports the structure.

Freeway: A highway for through traffic with full control of access and with grade separation at all intersections.

Frontage Road: A road adjacent and typically running parallel to a highway allowing access to nearby properties.

Frost Heave: A seasonal upthrust of the ground or pavement caused by the formation of ice layers or lenses in frost susceptible soils.

Gap: The time interval between the arrival at a point of one vehicle and the arrival at the same point of the next relevant vehicle.

Generated Traffic: In addition to the normal growth of the traffic on a road, year after year, the new or improved road creates new traffic for itself, called generated traffic.

Geometric Design: The dimensional design of a road within the surrounding topography, including all surficial dimensions, but excluding dimensions relating to thickness and structure, i.e. arrangement of visible elements of a road such as alignment, grades, sight distance, slopes, etc.

Glare: Glare is luminance greater than that to which the eye is accustomed.

Gradation: Gradation is the particle size distribution of a soil. Properties like internal friction, void content, wear resistance and permeability depend on the distribution of particle sizes.

Grade: The slope of the roadway segment expressed in percent.

Grade Separated Intersections: (Flyovers) locations where one or more crossing conflicts are separated through an overpass or underpass.

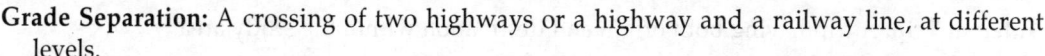

Grade Separation: A crossing of two highways or a highway and a railway line, at different levels.

Gradient: Gradient (grad-line) is the rate of rise and fall along the length of the road.

Grading: Particle size distribution recorded as percentage of the various grain sizes present in an aggregate as determined by sieve analysis.

Gravel: Rounded or water worn stones of irregular shape and size occurring in natural deposits with or without finer material.

Grit: Fine sharp edged aggregates, coarse sand, or broken stones, found in natural state.

Gross Vehicle Weight: The maximum total weight of a traffic vehicle.

Grubbing: Uprooting and removing of the stumps and roots of small trees, hadges from the site of work.

Guard Rail: A safety feature element intended to redirect an errant vehicle away from the approach embankment.

Gutter: An open drain constructed along the sides of a carriageway, to carry away the water drained from the surface of the pavement.

Hair Pin Bend: It is a curve (bend) in alignment of roads, resulting in reversal of direction of flow of traffic. Common on hill roads.

Hardness: Hardness relates to how difficult it is to crush and compact the material. It is related largely to the strength and fracture toughness of the material.

Head Way: The time interval between consecutive vehicles passing a given point measured from front-to-front of the vehicle.

High Occupancy Vehicle Lane (HOV Lane): A lane reserved for vehicles travelling with two or more passengers.

Highway: An important road for vehicular traffic.

Home Based Trip: A trip which has one end of the trip at home.

Horizon Year: The future year for which forecasting is done.

Hump: A device on the road which is intended to slow motorists down, but not deter their passage.

Ignition Point: The temperature at which the vapour of a substance takes fire and continues to burn.

Illumination: The unit flux incident on a surface per unit area (e.g. of pavement). Its unit is Lux (Lx) and is equivalent to Lx/m^2.

Information Signs: These signs inform and advise road users of direction, distances, destinations, routes, points of interest, location of services, etc.

Infrastructure: Permanent, fixed structures such as roadway, railways, etc.

Intangible Benefits: Those benifits which can not be converted into money value directly such as benefits of pleasure, easiness and comfort in driving, decreased commodity prices, etc.

Intelligent Transportation System (ITS): ITS applies current and emerging technologies in information processing, communication, and electronics, to solve transportation problems.

Interchange: A system of inter connecting roadway in conjunction with one or more grade separations, providing for the movement of traffic between two or more roadways on different levels.

Inter Green Period: The period between one phase losing right of way and next phase gaining right of way is called inter green period, in traffic signals.

Interchange Ramp: A turning roadway at an interchange for travel between intersection legs.

Internal Trip: A trip having both origin and destination within the study area

Intersection: An intersection (or junction) is the location at which two different traffic streams cross or merge on different roads. The area common to two or more highways that come together at an angle.

Intersection Angle: The angle between two intersection legs.

Intersection Leg: Anyone of the highway radiating from and forming part of an intersection.

Interzonal Trip: A trip between two different zones.

Island: A defined area between traffic lanes for control of vehicle movements or for pedestrian refuge within an intersection, a median or an outer separation is considered an island.

Joint: A designed vertical plane of separation or weakness in a pavement.

Land Use: The purpose or activity for which land or the structure on the land is being used. Land uses may be commercial, residential, officials, industrial, educational, recreational, etc.

Lantern: A housing for one or more lamps, together with any reflector, diffuser or other enclosure associated with the lamp.

Lay-by: The local widening of a carriageway to enable vehicles to draw off the road for temporary parking or stoppage, without obstructing the traffic flow.

Layers: Layers are those portions of course which are placed and compacted as an entity.

Leveling Course: The layer of material placed on an existing surface to eliminate irregularities prior to placing an overlaying course.

Levels of Service: Levels of design or maintenance that correlate with the expected and desired speed and use of the road.

Light: Radiant or luminous energy in the visible spectrum (wave lengths of 380 – 780 nm)

Limited Access: A highway that has some access restriction, but does not have fully controlled access.

Line of Sight: Direct path from the source to receiver without any intervening objects or topography.

Liquid Limit: Water content corresponding to the boundary between the liquid and plastic states.

Live Load: A load imposed by vehicles, pedestrians, equipment or components subject to movement, other than collision load.

Loadtransfer Device: A mechanical means designed to carry loads across a joint.

Loam: A mixture of sand, silt and/or clay with organic matter.

Local Street: A street or road primarily for access to residence, business or other abutting property.

Longitudinal Joint: A joint normally placed between traffic lanes to control longitudinal cracking.

Loop: A one way turning roadway that curves about 270° degrees to the left to accommodate a right-turning movement. It may include provision for a right turn at a terminal to accommodate another turning movement.

Luminance: The luminous intensity per unit area measured at a point on a projected surface and in a given direction.

Luminous Flux: The light being emitted by a light source (e.g. by a lantern) or received (e.g. by pavement). It is measured in Lumen (*Lm*). This is the radiant power given by light source.

Luminous Intensity: Luminous flux emitted per unit solid angle.

Lux: Lux is the unit of luminous flux. It is the flux emitted in a solid angle of one steradian by a source having a uniform point source of one candela

Macadam: Macadam is a word used to describe an open graded pavement course of angular aggregate, to which smaller sized stone is added during construction, with attention being given to working these smaller stones into the upper layer of the in-situ course. It is named after its inventor.

Maintenance: All works, which are required for upkeep of a road or its associated elements or both, so as to prevent deterioration of quality and efficiency, after its construction.

Mall: An area for walking permitting movement in various directions.

Markings: Markings are applied to pavements to guide, warn or regulate traffic.

Mass Transit: Services provided for passengers and their incidental baggage, on a route, following a fixed schedule and fares.

Mathematical Model: A set of equations relating quantifiable characteristics of the system being modelled.

Medians: Medians are used to separate opposite traffic streams

Median Opening: A gap in a median provided for crossing and turning traffic.

Median Speed: The speed represented by a middle value when all speeds are arranged in ascending order. It is 50th percentile speed.

Median Strip: A dividing strip in the middle of a roadway.

Merging: The converging of separate streams of traffic into a single stream.

Mesh: The square opening of a sieve.

Minimum Turning Radius: The radius of the minimum turning path of the outside of the outer front tire of a vehicle.

Mobilization: Preparatory work, such as movement of personnel, equipment, supplies and incidentals to the project site, which must be performed prior to beginning actual project work.

Modal Speed: The speed which occurs most frequently.

Model: A quantitative description of the behaviour of a system developed to predict its performance in new situation and/or illuminate the workings of the system under existing conditions.

Moisture Content: The ratio, expressed as percentage, of the mass of water in a material to the dry mass of the material.

Motor Vehicle: Mechanically propelled road vehicle fitted with an engine.

Negotiation: The process by which property is sought to be acquired for highway purposes through discussion, conference, and final agreement upon the terms of a voluntary transfer of such property.

Network Planning: A broad term used to describe the representation of a construction job as a network of events.

Noise: Any unwanted sound.

Non Destructive Testing: The in-situ determination of the physical properties of a component without impairment of or removal of any material.

Non Home Based Trip: The trip which does not end or start from home.

Nuclear Gauge: Instruments used to measure in-place density, moisture content or asphalt content through the measurement of nuclear emissions.

Odometer: A distance measuring device in a motor car.

Off-Peak Period: Non-rush period of the day when travel activity is generally low.

One Way Street: Those roads on which movement of vehicles in allowed only in one direction.

Open Graded Aggregates: A well graded aggregate mix containing little or no fines, with a relatively large percentage of voids.

Operating Speed: Operating speed is the speed that drivers judge to be possible under prevailing traffic conditions.

Operational Delay: Delay caused by interference between components of traffic.

Optimum Moisture Content (OMC): The water content at which a soil can be compacted to a maximum dry density by a given compactive effort.

Origin: The zone in which a trip originates (begins)

Overpass: A bridge that passes over a freeway or roadway.

Overall Speed: Overall speed is the average speed for a journey, including any stops, and delay.

Overall Travel Time: The time of travel including stops and delays except those of the travelled way.

Overtaking (Passing) Sight Distance: The minimum distance open to the view of the driver of a vehicle, intending to overtake a slow moving vehicle ahead with safety against the traffic of opposite direction.

Overtaking Zone: Zones/lengths where overtaking by the vehicles can be done safely.

Parapet: Parapet are low walls, or railings or a combination of both which are located along the outside edge of bridge decks. They are designed to prevent vehicles from running off the sides of the bridge.

Para-transit: Transport modes which are neither personal nor of mass transportation system such as car pooling, dial-a bus/taxi, etc.

Parking Accumulation: The total number of vehicles parked at a given time.

Parking Duration: The average time spent in a parking space

Parking Lane: Parking lane is an auxiliary lane provided along urban roads to allow curb parking.

Parkways: An arterial highway for non-commercial traffic, with full or partial control of access and usually located within a park or a ribbon of parklike development.

Passenger: Any person, other than a driver, who is in or on a vehicle.

Passenger Car: A motor vehicle designed for transportation of not more than 6 persons.

Passing Sight Distance: Maximum sight distance on 2 lane highway that must be available to the driver of one vehicle to pass another vehicle safely and comfortably, without interfering with the speed of an oncoming vehicle, travelling at the design speed.

Pattern: A pattern is an association of units forming a recognisably district landscape.

Pavement Markings: Are lines, symbols, words, numerals, messages or other devices set in the pavement or applied, or attached to the pavement or curb, to control, warn, guide, or inform the road users.

Pavement Structure: Pavement structure provides a surface of acceptable strength and rising quality, with adequate skid resistance, favourable light reflecting characteristics and low noise. It is combination of sub-base, base course and surface course placed on subgrade.

Paving: Laying of hot bituminous mix with a paving machine on the highway surface.

Peak Period: A time period or periods when travel activity is at its heaviest.

Pedestrian Actuated Signals: Signals provided on busy urban streets for crossing of the pedestrian. They temporarily show a red light for the traffic to stop, permitting the pedestrian to cross the road.

Pedestrian: Any person other than a driver or passenger on road is called a pedestrian.

PERT(Program Evaluation and Review Technique): A scheduling technique which uses a range of possible completion times. PERT is more applicable to development and CPM to construction control.

Phasing of Traffic Signals: The procedure by which the traffic streams are separated at an intersection is known as phasing.

Plastic Limit: Water content corresponding to the boundary between the plastic and the semisolid states.

Plasticity: The ability of a material to deform permanently without losing its strength.

Point Load: Point loads are generated when loads are unevenly distributed and are concentrated on a very small area.

Pot Holes: A steep sided bowl-shaped cavity caused by either loss of surfacing and basecourse erosion or advanced cracking under traffic or severe weather.

Pre-stressed Concrete: Reinforced concrete in which internal stresses and deformation are initially induced, of such magnitude and distribution that the subsequent stresses and deformations resulting from dead and live loads are counteracted to a desired degree.

Prime Court: An application of a low viscosity liquid bituminous material to coat and bind mineral particles preparatory to placing a base or surface course

Private Road: A road not available for general public use.

Progressive System of Signals: A signal system in which signal lights are so controlled that a given stream of traffic gets a continuous movement (green light), at a planned rate of speed, which may vary in different parts of the system.

Projected Land Use: The anticipated way a specific area will be utilised in the future. (residential, commercial, industrial, etc).

Pumping: Vertical movement of the base course moisture and suspended fines to the surface (usually through joints or cracks). Common in cement concrete and stabilised pavement.

Quality Assurance: Planned and systematic action necessary to provide confidence that a product or service will satisfy given requirements for quality.

Quality Control: Operational, process control techniques or activities that are performed or conducted to fullfil contract requirement for material or equipment quality.

Radial Highway: An arterial highway leading to or from an urban centre.

Radial Road: A road or highway radiating from the centre of an urban area.

Railings: Railings are placed for safety of people, along the roads in city areas and at other dangerous places such as intersections, etc.

Ramp: An interconnecting road of a grade separated inter-sections, i.e. providing connection between different levels.

Random Sampling: Procedure for obtaining non-biased, representative samples.

Rapid Transit: Mass transit service, without interference or many stops, provided through fast buses, minibuses or railways.

Reconstruction: Works which alter the pavement width and/or alignment.

Reflection Crack: A crack appearing in a resurface or overlay caused by movement at joints or cracks in underlying base or surface.

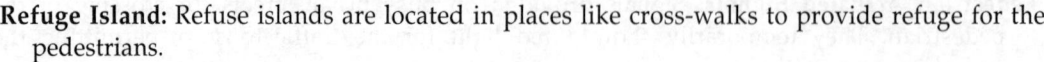

Refuge Island: Refuse islands are located in places like cross-walks to provide refuge for the pedestrians.

Regulatory Signs: These signs indicate legal restrictions and failure to comply with them is an offence.

Rehabilitation: Works carried out on the existing alignment and width to restore or improve pavement strength.

Reinforcement: Steel imbedded in a rigid concrete slab to resist tensile stresses and detrimental opening of cracks.

Repeatability: The within laboratory precision of a test result. Test done again in the same laboratory.

Reprocessing: The renewal of an existing surface by scarifying, remixing with or without additional material and relaying.

Re-producability: The between laboratory precision of a test result. It is the difference between the results from different laboratories using same procedure.

Resurfacing: The placing of one or more new courses on an existing surface of a road.

Retaining Wall: A wall built to hold back earth or water.

Reversible One Way Street: Those streets in which the street is one-way street in one direction for one part of the day and one-way in the opposing direction at another period of the day.

Reverse Curve: A curve consisting of two arcs of the same or different radii curving in opposite directions and having a common tangent or transition curve at their point of junction.

Right of Way: The entire strip or area of land for highway purposes.

Rigid Pavement: A pavement structure that distributes load to the subgrade having its surface course of portland cement concrete slab of relatively high bending resistance.

Ring Road: A circumferential road built around an urban area to enable free flow of traffic.

Road Bed: The graded portion of a highway within top and side slopes, prepared as a foundation for the pavement structure and shoulder.

Road: Road is a line of communication between places for use of foot passengers, riders and vehicles.

Road Furniture: Lamp posts, telephone posts, signs, trees, etc. called road furniture, should be located to minimise danger to the vehicles, which may be using the shoulder.

Roadside: A term denoting the area adjoining the outer edge of the roadway.

Roadway: The portion of a highway, including shoulders for vehicular use. A divided highway has two or more roadway.

Rotary Intersection (Roundabout): A rotary or roundabout is a channelised intersection where traffic moves clockwise around a central island.

Roughness: Roughness of road surface is a measure of its unevenness.

Roundabouts (Rotary): Roundabouts provide a channelised intersection where the traffic moves clockwise around a central island.

Route: Route is the way taken in getting from starting point to destination.

Rubble: Pieces of stone or broken brick of irregular size and shape.

Running Speed: The speed over a specified section of a highway, being the total distance divided by overall travel time (time of travel inclusive of stops and delays)

Running Time: The time during which vehicle is in motion.

Rut: A groove or depression formed in a surface layer longitudinal to the road, by the wheels of travelling vehicles.

Rutting: Longitudinal wheel track depressions.

Sag: The depression formed by the junction of two falling gradients.

Sand Asphalt: A mixture of sand and asphalt, either plant mixed or road mixed.

Saturation Flow: Saturation flow of an approach is the maximum flow/hour (in pcu), that can pass through an intersection, from one approach, without impedance by signals.

Scheduling: The allocation of resources to tasks whose time sequence is fixed but whose cost is time dependent.

Scoring: Localised marks (wheel imprints), which may be caused by prolonged parking.

Screed: A strip of wood or metal, used as a guide for a template or straight edge, for finishing surface to required profile.

Seal Coat: A thin treatment consisting of bituminous material, usually with cover aggregates, applied to a surface course.

Segregation: The separation of aggregates by size resulting in a non-uniform material.

Semitrailer: A vehicle designed for carrying persons or property and drawn by a truck tractor on which part of its weight and load rests.

Separator: A device for dividing traffic streams. Medians are a type of separator.

Service Life: The length of time that a facility is expected to provide a specified service.

Service Road: (Frontage road) A subsidiary carriageway constructed between the main carriageway and the road boundary to serve properties abutting on the road and connected only at selected points with the main carriageway.

Setback Line: The line outside the right of way, established by public authority, on the highway side of which the erection of buildings or other permanent improvements is controlled.

Settlement: A general lowering of the road surface which may be due to such factors as change in subgrade water content, inadequate construction compaction or heavy/frequent axle loads.

Shingle: Consists of coarse rounded material may be stones, pebbles, etc. greater in size than gravel and smaller than boulder.

Shoulders: Shoulders are strips between the outer edge of the pavement and inner edge of the drain for accommodating stopped vehicles, etc.

Side Track: A track adjacent to the carriageway to permit the passage of traffic while the carriageway is temporarily closed.

Side Walk: An exterior pathway with a prepared surface (concrete, bituminous, brick stone, etc) intended for pedestrian use.

Sight Distance: Sight distance is the distance at which driver can see a specified object ahead of him, in clear weather and visibility conditions.

Skidding: Skidding occurs when path travelled along the road surface is more than circumferential movement of the wheel, i.e. when the wheels slide without revolving.

Skip Graded Aggregate: Aggregate possessing disproportionate distribution of successive particle sizes.

Slipping: Slipping occurs when driving wheels of a vehicle revolve more than the longitudinal movement. Slipping occurs on wet, icy, slippery surfaces.

Slump: Measurement related to the workability of concrete.

Slurry Seal: A seal coat consisting of a semi-fluid mixture of bituminous emulsion and fine aggregates.

Sound Barrier: A wall or other barrier that separates a highway from residential or commercial areas adjacent to the highway with the purpose of reducing roadway noise.

Space Mean Speed: Space mean speed is the arithmetic mean of the simultaneous speed of each vehicle in a given short length of lane.

Space-headway: Space headway is the distance from head to head of successive vehicles.

Spacing: The distance between consecutive vehicles measured front to front.

Spalling: Separation or removal of a portion of the surface concrete.

Speed: Speed is the rate of movement of traffic, i.e. distance covered in unit time (kilometer/ hour).

Speed-change Lane: An auxiliary lane, including tappered areas, primarily for the acceleration or deceleration of vehicles entering or leaving the through traffic lanes.

Spot Speed: Spot speed is the instantaneous speed of a vehicle at a point in time.

Stabilisation: Stabilisation of soils by the use of additives is the process by which the addition of a little percent of material such as lime, cement, bitumen, to a soil produces large increases in the properties of that soil.

Stack Interchange: An interchange with several flyover ramps.

Stage Construction: A process by which a road is open to traffic at various stages of its construction history.

Steradian: A steradian is the unit measure of a solid angle. It is equal to the solid angle subtended at the centre of a sphere by unit area of its surface. The whole space surrounding a point subtends a solid angle.

Street: A street is a town or village road that has (mainly) continuous houses on one side or both.

Stripping: Stripping of the smaller surface aggregate from the bitumen film.

Study Area: The area covered for the purpose of data collection and planning of a transportation study.

Sub-Base Course: Sub-base course is a course whose prime purpose is to protect and separate the base course from the subgrade. It supports base course.

Subgrade: Subgrade is the natural or fill formation on which the pavement structure is constructed. It supports load transmitted by the pavement.

Subsidence: A localised rather abrupt lowering of the road surface which may result from poor drainage, compressible soils or the collapse of underground cavities.

Subway: An underground passage or tunnel to permit the movement of traffic pedestrians.

Summit: The peak formed by the junction of two rising gradients.

Surface Course: Surface course of a pavement is to carry the local effects caused by wheel loads and withstand the atmospheric environment.

Surface Treatment: One or more applications of bituminous material and cover aggregates of thin plant mix on an old pavement or any element of a new pavement structure.

Tack Coat: An application of bituminous material to an existing surface to provide bond with a superimposed course.

Tandem Axle Load: The total load transmitted by two or more consecutive axles.

Tangible Benefits: Those benefits which can be converted into money value such as reduction in road user costs, travel time costs, accident costs, etc.

Tar: Tar is a by-product of the carbonisation of coal at temperatures over 600 °C. It was used in place of bitumen earlier.

Template: A full-sized mould, pattern or frame shaped to serve as a guide in forming or testing contour or shape.

Temporary Signs: These signs warn of temporary hazardous or deleterious conditions.

Terrain: Terrain is topography, hydrology, drainage, geology, pedology and vegetation of an area.

Thermoplasticity: The ability of a material to become plastic as its temperature is raised.

Thirtienth Highest Hourly Volume (30 HV): The hourly volume in both directions of travel that is exceeded by 29 hourly volumes during a designated year.

Through Trip: A trip having both origin and destination outside the study area.

Tie Bar: A deformed steel bar or connector imbedded in the concrete across a joint to prevent separation of abutting slabs.

Time Headway: The time interval between head to head of successive vehicles passing a given point measured from front to front of the vehicles.

Time Mean Speed: The time mean speed is the arithmetic mean of spot speeds of a short line of traffic measured over a given time interval, as they pass the same point.

Title: The evidence of a person's right to property or the right itself.

Toll Road/Bridge: A road or bridge open to traffic only upon payment of a toll or fee.

Traffic Calming: The process of designing streets or adding design elements to tame fast traffic and address unsafe traffic conditions.

Traffic Control Device: Any sign, signal, marking or installation placed or erected under public authority, for the purpose of regulating, warning, or guiding traffic.

Traffic Density: The number of vehicles using the road per hour during peak periods.

Traffic Engineering: The phase of engineering which deals with the planning and geometric design of streets, highways and abutting lands, and with traffic operations thereon, as their use is related to the safe, convenient and economic transportation of persons and goods.

Traffic Islands: Traffic islands are raised constructed areas of pointed portions within the pavement, which are used to guide and control of vehicular traffic or for pedestrian refuge.

Traffic lane: The portion of the travelled way for movement of a single line of vehicles.

Traffic Modelling: Using a computer programme to analyse the prominent ways people travel.

Traffic Sign: Traffic sign is a device mounted on a fixed or portable support, whereby a specific message is conveyed by means of symbols or words for purpose of regulating, warning, guiding or informing the traffic.

Traffic Signals: Traffic signals are devices which by means of changing coloured lights, regulate the movements of traffic, assigning use of intersection first to one stream of traffic, then to the other.

Trailer: A vehicle designed for carrying persons or property and drawn by a motor vehicle which carries no part of the weight and load of the trailer.

Travel Time: The time required to travel between two points, including the terminal time at both ends of the trip.

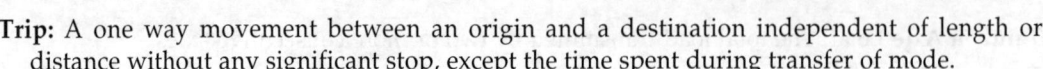

Trip: A one way movement between an origin and a destination independent of length or distance without any significant stop, except the time spent during transfer of mode.

Trip Ends: Either end of the trip where trip originates or where trip ends.

Trip Mode: The means of transport by which a trip is made, e.g. on-foot trip, car trip, bus trip, taxi-trip, etc.

Trip Time: The time of the day at which a particular trip is made, e.g. morning trip, night trip, peak-hour trip, etc.

Truck Tractor: A motor vehicle designed for drawing other vehicles but not for a load other than a part of the weight of the vehicle and load drawn.

Turf: A type of grass or herbs.

Turn Over: Average number of times a parking space is used by different vehicles during a given period of time.

Uncontrolled Intersection: Intersection where no regulatory devices are used.

Underpass: This is the location where a roadway, railroad, or other feature passes under the subject highway.

Uninterrupted Traffic Flow: Vehicles are free to move and are not required to stop by any cause external to the traffic stream.

Urban Area: A population cluster of more than 1000 people.

Utilities: Transmission and distribution lines, pipes, cables, and other associated equipment used for public services such as lighting, heating, gas, oil, water, sewage, cable-vision, data communication, telephone, etc. are utilities.

Vehicle Actuated Signals: These signals change the length of green signal interval, in accordance with the actual traffic volume on the particular approach of the intersection.

Viaduct: An elevated roadway structure which carries a road.

Viscosity: A measure of the resistance to flow, and is one method of measuring the consistency of bitumen.

Volume of Traffic: Volume of traffic is the number of vehicles passing a given point during a specified period of time (vehicles/hour).

Water Bound Macadam (WBM): The surface layer of a road in which road metal has been consolidated with water and earthly material or rock particle, called water bound Macadam.

Warning Signs: These signs inform road user of unexpected or hazardous situations on or adjacent to the road.

Warping Joint: A joint in which flexure is permitted but separation and vertical displacement of abutting rigid concrete slabs are prevented by metal ties and mechanical or aggregate interlock.

Warrant: A criterion used to determine whether the construction of a traffic facility or the institution of a traffic control device can be justified.

Water Table: The upper surface of the zone of saturation of the soil where the groundwater is not confined by an overlying impermeable formation.

Wave: Wave on the pavement surface with crests at least 60 cm apart, may be due to deep differential subgrade deformation.

Wearing Course: The top or surface course of a bituminous pavement over which the traffic moves.

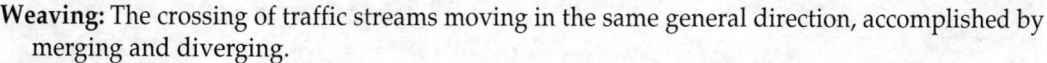

Weaving: The crossing of traffic streams moving in the same general direction, accomplished by merging and diverging.

Weaving Length: The length of the carriageway between adjacent radial routes around a traffic roundabout.

Well Graded Aggregate: An aggregate mix possessing proportionate distribution of successive particle sizes.

Whip-off: The loss of cover aggregates from a freshly laid surface-dressing due to traffic.

Zero Air Voids Curve: Curve showing the zero air voids density as a function of water content.

Zones: For the purpose of data collection and aggregation, the area under study within the cordon line is subdivided into smaller areas called zones.

Zoning: The division of an area into districts and the public regulation of the character and intensity of use of the land and improvements thereon.

Highway Development and Planning

The first forms of road transport were either by using animals like horses and oxen or by humans themselves carrying goods on backs and heads over pre-identified tracks. In the Stone Age humans did not need constructed tracks in open country. The first improvements probably would have been that of clearing trees and big stones from the identified paths. The earliest records of such paths have been found around some springs near Jericho and date from about 6000 BC. These paths or tracks were later flattened or widened to accommodate increase in human and animal traffic. Some of these tracks finally got developed into extensive networks, allowing communications, trade and governance over wide areas. The Incan Empire in South America and the Iroquois Confederation in North America, neither of which had the wheel, are examples of effective use of such paths.

The authentic information on early period roads is available in literature and appears to start since 5000 BC onwards. Brief description of these is given in the following sections.

3.0.1 Harappan Roads

Street paving has been found from the first human settlements around 4000 BC in cities of the Indus Valley Civilization on the Indian subcontinent, such as Harappa and Mohenjodaro. Harappan are admired by world historians for their advance town planning and scientific drainage system. During Harappan the roads were 9 feet to 34 feet wide and main roads used to cross at right angles. They probably believed that blowing winds would automatically clean the roads which used to cross at right angle. The turn in the alleys was semicircular and not at right angle. The houses were made after the straight roads were constructed. In most of the houses the windows were not present on the ground floor for maintaining privacy and the main door of the house was not on the main road but on the side streets. These roads were brick paved. According to Dr R S Bisht, former Joint Director General of Archeological Survey of India, Harappans never allowed vehicular traffic inside their cities which explains the condition of their roads which remained as they were for a long time.

3.0.2 Wheeled Transport and Stone Roads

Wheels appear to have been developed in ancient Sumer in Mesopotamia around 5000 BC, perhaps originally for the making of pottery. The first simple two-wheel carts, apparently developed from travois, appear to have been used in Mesopotamia and northern Iran in about 3000 BC and two-wheel chariots appeared in about 2800 BC. Heavy four-wheeled wagons were developed about 2500 BC. Two-wheeled chariots with spoked wheels appear to have been developed around 2000 BC by the Andronovo culture in southern Siberia and Central Asia.

Wheeled-transport created the need for better roads. In urban areas it began to be worthwhile to build stone-paved streets and, in fact, the first paved streets appear to have been built in Ur in 4000 BC. Corduroy roads were built in Glastonbury, England in 3300 BC. The Minoans on the island of Crete built a 30-mile (50-kilometre) road from Gortyna on the south coast over the mountains at an elevation of about 4,300 feet (1,300 metres) to Knossos on the north coast. Construction of layers of stone, the roadway took account of the necessity of drainage by a crown throughout its length and even gutters along certain sections. The pavement, which was about 12 feet (360 centimetres) wide, consisted of sandstone bound by a clay-gypsum mortar. The surface of the central portion consisted of two rows of basalt slabs 2 inches (50 millimetres) thick. The centre of the roadway seems to have been used for foot traffic and the edges for animals and carts. It is the oldest existing paved road. There is evidence of the building of special roads for religious purposes and transport about 800 BC in Greece. These ceremonial roads were paved with shaped stones and contained wheel ruts about 55 inches (140 centimetres) apart. The Porta Rosa was the main street of Elea dating Hellenistic age (IVth-IIIrd century BC). It connected the northern quarter with the southern quarter. The road was 5 meters wide and had an incline of 18% in the steepest part. It was paved with limestone blocks, girders cut in square blocks, and with a small gutter on one side for the drainage of rain water.

3.0.3 Road along Amber Route

A few remnants of these roads survive today. They were constructed by laying two or three strings of logs in the direction of the road on a bed of branches and boughs up to 20 feet (6 meters) wide. This layer was then covered with a layer of transverse logs 9 to 12 feet in length laid side by side. In the best log roads, every fifth or sixth log was fastened to the underlying subsoil with pegs. There is evidence that the older log roads were built prior to 1500 BC. They were maintained in a level state by being covered with sand and gravel or sod.

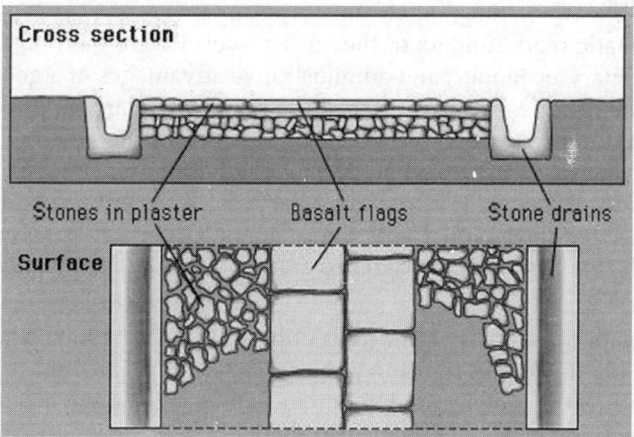

Fig. 3.1: Ancient Cretan road

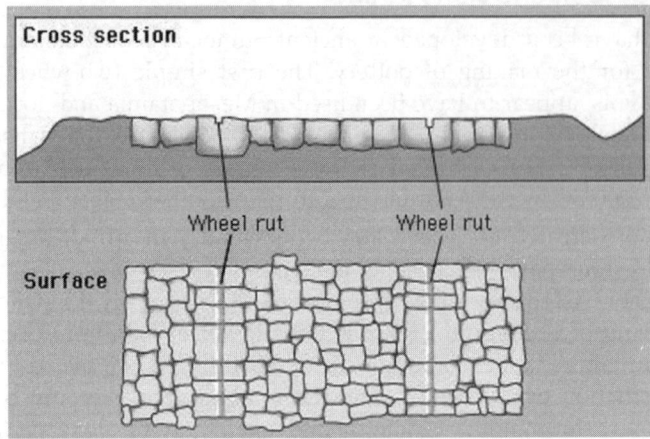

Fig. 3.2: Ancient Greek road

3.0.4 Persian and Babylon Road (Royal Road)

The earliest long-distance road was a 1,500-mile route between the Persian Gulf and the Mediterranean Sea. It came into some use about 3500 BC, but it was operated in an organized way only from about 1200 BC by the Assyrians, who used it to join Susa, near the Persian Gulf, to the Mediterranean ports of Smyrna (Izmir) and Ephesus. The route was duplicated between 550 and 486 BC by the great Persian kings Cyrus II and Darius I. Starting from Susa, it passed northwestward to Arbela, proceeded westward through Nineveh to Harran, a major road junction and caravan centre and then continued to twin termini at Smyrna and Ephesus. The Greek historian Herodotus, writing about 475 BC, put the time for the journey from Susa to Ephesus in 93 days, although royal riders traversed the route in 20 days and mail couriers could travel 2,699 km in seven days.

In Babylon about 615 BC the Chaldeans connected the city's temples to the royal palaces with the processional way, a major road in which burned bricks and carefully shaped stones were laid in bituminous mortar.

3.0.5 Roman Roads

The greatest systematic road builders of the ancient world were the Romans, who were very conscious of the military, economic, and administrative advantages of a good road system. The Romans drew their expertise mainly from the Etruscans particularly in cement technology and street paving–though they probably also learned skills from the Greeks (masonry), Cretans, Carthaginians (pavement structure), Phoenicians, and Egyptians (surveying). The Romans began their road-making task in 334 BC and by the peak of the empire, had built nearly 53,000 miles of road connecting their capital with the frontiers of their far-flung empire. Twenty-nine great military roads, the *viae militares*, radiated from Rome. The most famous of these was the Appian Way.

The typical Roman road was bold in conception and construction. Where possible, it was built in a straight line from one sighting point to the next, regardless of obstacles, and was carried over marshes, lakes, ravines, and mountains. In its highest stage of development, it was constructed by excavating parallel trenches about 40 feet apart to provide longitudinal drainage–a hallmark of Roman road engineering. The foundation was then raised about three feet above ground level, employing material taken from the drains and from the adjacent

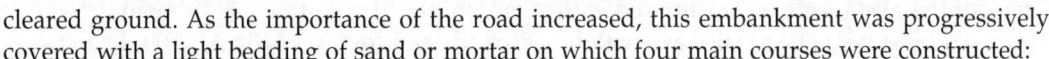

cleared ground. As the importance of the road increased, this embankment was progressively covered with a light bedding of sand or mortar on which four main courses were constructed:

1. The *statumen* layer 10 to 24 inches (250 to 600 millimetres) thick, composed of stones at least 2 inches in size
2. The *rudus*, a 9-inch-thick layer of concrete made from stones under 2 inches in size
3. The *nucleus* layer, about 12 inches thick, using concrete made from small gravel and coarse sand, and, for very important roads and
4. The *summum dorsum*, a wearing surface of large stone slabs at least 6 inches deep.

The total thickness thus varied from 3 to 6 feet. The width of the Appian Way in its ultimate development was 35 feet. The two-way, heavily crowned central carriageway was 15 feet wide. On each side it was flanked by curbs 2 feet wide and 18 inches high and paralleled by one-way side lanes 7 feet wide. This massive Roman road section, adopted about 300 BC, set the standard of practice for the next 2,000 years.

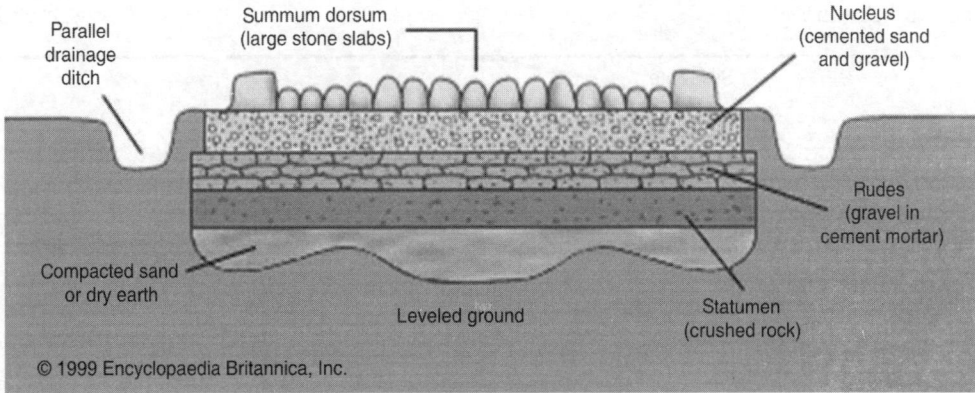

Fig. 3.3: Ancient Roman road

3.0.6 Roads in India

Street paving seems to have been common in the towns in India at the beginning of the common era, and the principles of drainage were well known. The crowning of the roadway and the use of ditches and gutters were common in the towns. Northern and western India in the period 300 to 150 BC had a network of well-built roads. The rulers of the (4th century BC), which stretched from the Indus River to the Brahmaputra River and from the Himalayas to the Vindhya Range, generally recognized that the unity of a great empire depended on the quality of its roads. The Great Royal Road of the Mauryans began at the Himalayan border, ran through Taxila (near modern Rawalpindi, Pakistan), crossed the five streams of the Punjab, proceeded by way of Jumna to Prayag (now Allahabad, India), and continued to the mouth of the Ganges River. "Ministry of Public Works" was responsible for construction, marking, and maintenance of the roads and rest houses and for the smooth running of ferries. Evidence from archaeological and historical sources indicates that by AD 75 several methods of road construction were known in India. These included the brick pavement, the stone slab pavement, a kind of concrete as a foundation course or as an actual road surface, and the principles of grouting (filling crevices) with gypsum, lime, or bituminous mortar.

3.0.7 Roads in China

China had a road system that paralleled the Persian Royal Road and the Roman road network in time and purpose. Its major development began under Emperor Shihuangdi about 220 BC.

Many of the roads were wide, surfaced with stone, and lined with trees; steep mountains were traversed by stone-paved stairways with broad treads and low steps. By AD 700 the network had grown to some 25,000 miles (about 40,000 kilometers). Traces of a key route near Xi'an are still visible.

3.1 HISTORY OF MODERN ROADS

In Europe, gradual technological improvements in the 17th and 18th centuries saw increase in commercial travel. This created an incessant demand for better roads. In 1585, an Italian engineer Guido Toglietta wrote a thoughtful treatise on a pavement system using broken stone that represented a marked advance on the heavy Roman style. In 1607, Thomas Procter published the first English-language book on roads. The first highway engineering school in Europe, the School of Bridges and Highways, was founded in Paris in 1747. Up to this time roads had been built, with minor modifications, to the heavy Roman cross section, but in the last half of the 18th century the fathers of modern road building and road maintenance appeared in France and Britain.

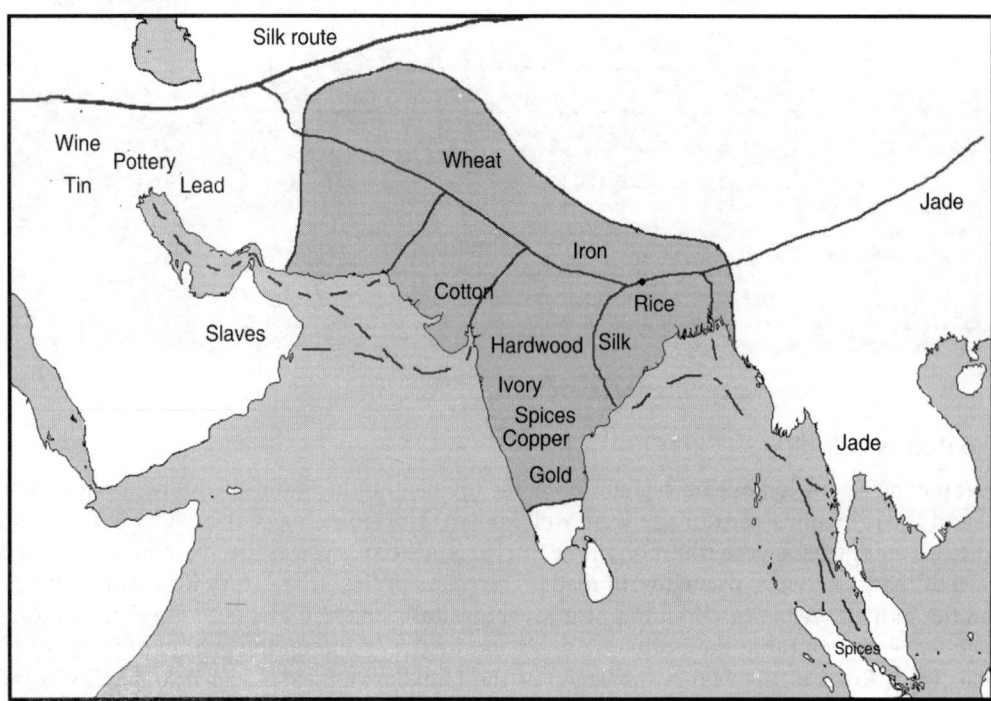

Fig. 3.4: Trade-routes during Mauryan empire (4th – 2nd BC)

3.1.1 French Road Construction

Pierre-Marre-Jerome Tresaguet, an engineer from an engineering family, became in 1764 Engineer of Bridges and Roads at Limoges and in 1775 Inspector General of Roads and Bridges for France. In that year he developed an entirely new type of relatively light road surface, based on the theory that the underlying natural formation, rather than the pavement, should support the load. His standard cross section was 18 feet wide and consisted of an eight-inch-thick course of uniform foundation stones laid edgewise on the natural formation and covered by a two-inch layer of walnut-sized broken stone. This second layer was topped with a one-inch layer of

smaller gravel or broken stone. The purpose of these layers was to transfer the weight of the road and its traffic to the ground, while protecting the ground from deformation by spreading the weight evenly. The upper running surface provided a smooth surface for vehicles, while protecting the large stones of the lower layers. In order to maintain surface levels, Tresaguet's pavement was placed in an excavated trench – a technique that made drainage a difficult problem.

3.1.2 Scottish Road Construction

Thomas Telford, born of poor parents in Dumfriesshire, Scotland, in 1757, was apprenticed as a stone mason. Intelligent and ambitious, Telford progressed to designing bridges and building roads. Under his supervision 1,500 km of roads and 1,000 bridges were built in Scotland between 1802 and 1822. He placed great emphasis on two features: (1) maintaining a level roadway with a maximum gradient of 1 in 30 and (2) building a stone surface capable of carrying the heaviest anticipated loads. His roadways were 18 feet wide and built in three courses:

1. A lower layer, seven inches thick, consisting of good-quality foundation stone carefully placed by hand (this was known as the Telford base)
2. A middle layer, also seven inches thick, consisting of broken stone of two-inch maximum size, and
3. A top layer of gravel or broken stone up to one inch thick.

3.1.3 English Road Construction

The greatest advance came from John Loudon McAdam, born in 1756 at Ayr in Scotland. McAdam began his road-building career in 1787 but reached major heights after 1804, when he was appointed General Surveyor for Bristol, then the most important port city in England. The roads leading to Bristol were in poor condition, and in 1816 McAdam took control of the Bristol Turnpike. There he showed that traffic could be supported by a relatively thin layer of small, single-sized, angular pieces of broken stone placed and compacted on a well-drained natural formation and covered by an impermeable surface of smaller stones. Drainage was essential to the success of McAdam's method, and he required the pavement to be elevated above the surrounding surface. He had noticed in his observations that coaches with narrow, iron-tyred wheels and moving at relatively high speed were causing significant damage to roads, but that areas of small broken stones were most resistant to damage, while the areas that had large surface stones degraded fastest. His solution was to create roads with three layers of stones laid on a crowned subgrade with side ditches for drainage. The first two layers consisted of angular hand-broken aggregate, maximum size 3 inches (75 mm), to a total depth of about 8 inches (200 mm). The third layer was about 2 inches (50 mm) thick with a maximum aggregate size of 1inch (25 mm). Each layer would be compacted with a heavy roller, causing the angular stones to lock together with their neighbours. In practice, his roads proved to be twice as strong as Telford's roads. The principles of the "macadam" road are still used today.

Blind Jack Metcalf (1717-1810) built about 300 km (180 miles) of turnpike road between 1753 and 1810, mainly in Lancashire, Derbyshire, Cheshire and Yorkshire. He understood the importance of good drainage and surfaced his roads with "a compact layer of small, broken stones with sharp edges", rather than the naturally rounded stones traditionally used in European road building. British turnpike builders began to realize the importance of selecting clean stones for surfacing, and excluding vegetable material and clay to make better lasting roads.

3.2 DEVELOPMENT OF MODERN PAVED ROADS

Various systems had been developed over centuries to reduce washways, bogging and dust in cities, including cobblestones and wooden paving. Tar-bound macadam (tarmac) was applied to macadam roads towards the end of the 19th century in cities such as Paris. In the early 20th century tarmac and concrete paving were extended into the countryside. Incidentally, bicyclists were among the early campaigners on what was called the Good Roads Movement. Bicycling was an extremely popular recreation among the middle and upper classes in the late 19th century and was more fun on paved roads.

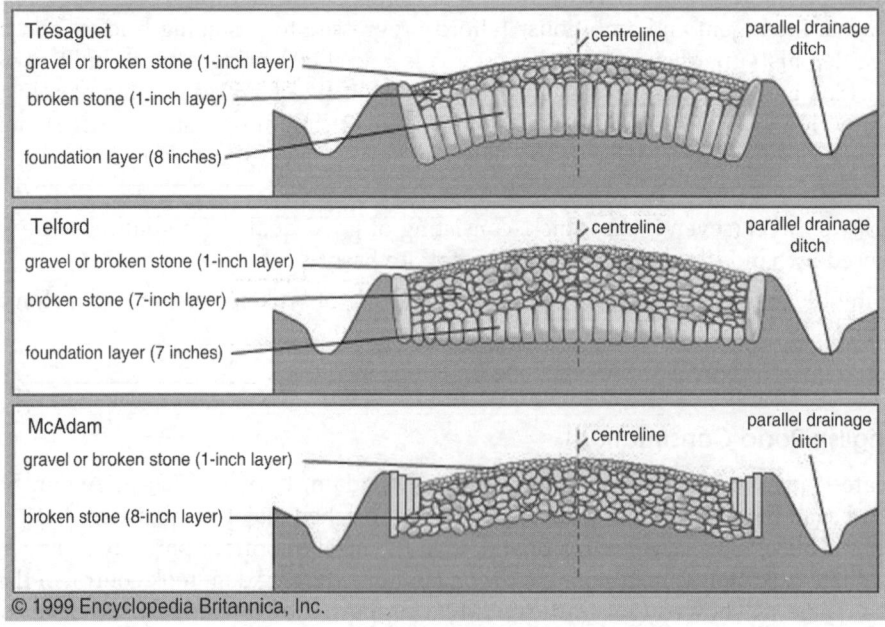

Fig. 3.5: Road construction in modern era

The developments have lead to the two broad types of road constructions, namely flexible pavements and rigid pavements, consisting of different layers. The profiles have also got standardised with time. These are shown in Figs. 3.6 and 3.7.

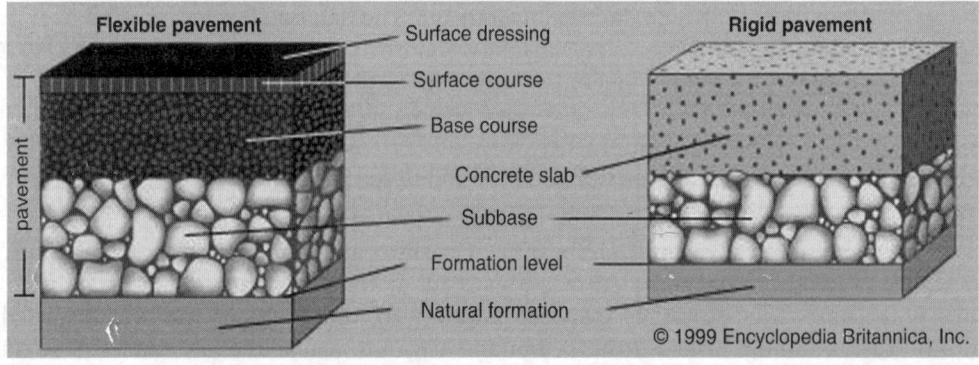

Fig. 3.6: Modern pavement structure

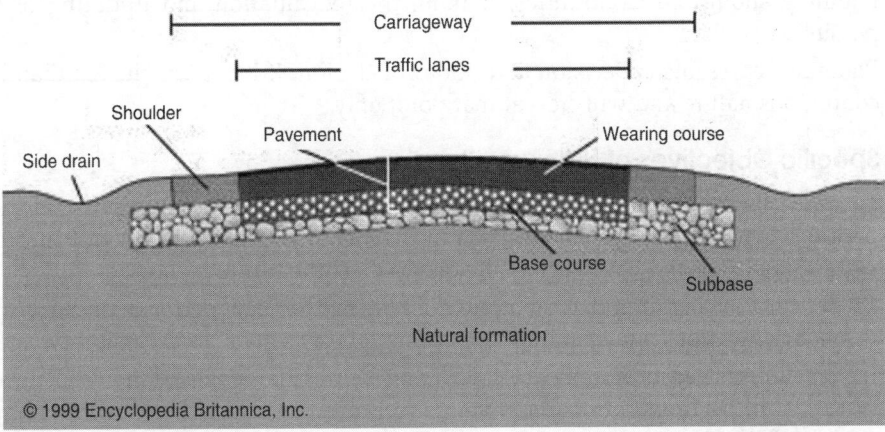

Fig. 3.7: Modern roadway cross-section

3.3 HIGHWAY PLANNING

Highways are growing since their conception. They have become big in terms of their numbers, classification and lengths. These can be developed systematically and in a coordinated way by doing proper planning. Planning, in general, can be defined as an organized, rationally conducted process of collection, analysis and presentation of facts about all facets of a system. In a similar way, Highway planning can be defined as a scientific approach related with collection and analysis of information of sectors like population and vehicular growth, land-uses and economic activities, government developmental policies and socio-economic trends in an area, that provides a first-hand account of existing facilities; allows examination of road network capacities; helps in identifying future facilities and financing systems for their implementation; and gives opportunities to innovate during the implementation of decisions taken. This also allows taking up the works in a phased manner giving them relative priority so that the financial resources do not feel the burden of implementation.

In general, the characteristics of highway planning may be summarized as A3CD. These are defined below.

1. Planning should **A**nticipate the needs that will be created over the planning horizon based on existing facilities, probable growths of influencing attributes and under impacts of developmental policies.

2. Planning should be **C**omprehensive in nature. It should take into consideration all the influencing attributes in their present form and the likelihood of their impacts in future. This may require:

   a. Complete information on economic, population and land use elements,

   b. Estimates of future demands related to movement of objects through public or private networks,

   c. Inclusion of terminal facilities and traffic control systems in the planning, and

   d. The information on activities, those are existing or likely to take place in the whole influence area over the horizon period where the forces of development may create an effect even in future.

3. Planning should be **C**ooperative, implying an agreement between agencies involved in planning and implementation of decisions. These agencies may be purely government, semi-government or private in nature.

4. Planning should be Continuous, making the reevaluation and updating of the plans possible.

5. Planning represents a Decision and not a conclusion. If it is a conclusion than it loses its continuous nature and will stop at that point only.

3.3.1 Specific Objectives of Highway Planning

1. To know the limitations of the road system, existing as well as planned, analyze the time period when the old system needs improvement; identify the ways of improving the system; and plan and construct a new one to meet the future needs.

2. To develop an integrated road network system that can perform primary as well as secondary functions, and can improve upon mobility as well as accessibility.

3. To develop phased programme for different hierarchial levels of the road development system, starting from national and state level to urban and village level, within the constraints of available resources.

4. To provide economic activity and efficiency through reduced costs of construction and operation, and optimal development of road network.

5. To identify the sources of money needed for implementation and optimize expenditure and revenues to be collected through taxation, tolls, road user pricing policies, etc.

6. To have the most suitable type of road system out of the available funds.

7. To establish priorities of construction and maintenance of the road system.

8. To reduce social and environmental costs arising out of road system development due to different reasons like accidents, congestion and environmental degradation.

3.3.2 Need of Highway Planning

The planning of highways results in the following merits:

a. Lost ground demand: The backlog of construction, which has been postponed during the past so many years due to lack of resources, may be material, money, manpower or equipment can be taken up and incorporated in the new plans.

b. Replacement of facilities: With time the old facilities wear out and needs either maintenance or complete replacement. This becomes possible when taking up highway planning.

c. Expansion of facilities: The expansion of facilities is important from the view point of increasing vehicles on road. If no management measures work then it is to be taken up. Planning helps in identifying such requirements.

d. Quality of service: The users of facilities always demand higher level of service than that is provided. The use of higher standards in construction and implementation of periodic maintenance can be taken care of while planning.

e. Elimination of duplication: The duplication of resources in the provision of facilities due to inadequate or unplanned works is not desired and this can be eliminated through proper planning.

3.4 HIGHWAY DEVELOPMENT AND PLANNING IN INDIA

3.4.1 Pre-British Period

The construction of roads has been an organized function of the state from times immemorial in our country. Archaeological findings of road and city planning at Mohanjodaro and Harappa have been dated to 3000 BC. Kautilya's Arthashastra, one of the immortal woks on government functions and politics, written as early as 300 B.C. speaks of officers of the state in-charge of finance, public works and royal correspondence. According to this Shastra, duties of a king included construction of roads for good communication. The system was co-operative in nature as is clear from the record that states :

"Whoever stays away from any kind of co-operative construction shall send his servants and bullocks to carry on his work and shall have a share in the expenditure but no claims to profit."

This ancient book contains details of layouts of villages, townships, forts, width of roads, chariot road, royal roads, roads leading to military stations, gardens, graves and forests, burial ground, etc. which remind us of the fact that ideas of modern town planning are not really modern, as thought by us. The construction of structures with massive stone blocks would not have been possible without the provision of a good road network. Emphasis was given to road construction and on provision of a good road network during all big dynasties without which the governance of such a big area would not have been possible. It is true for Mauryan, Gupta and other Hindu dynasties as well as to the Mugal period. During Suri period a road was constructed running through the width of the country from far east to far west. It is still available is named as 'Grand Trunk Road'.

3.4.2 British Period

The same emphasis remained as such during the British period also. In the aftermath of industrial revolution in Europe in 18th century, the need for construction of roads, railways and irrigation works etc. came to the forefront. The works of construction of railways were executed by different companies, and the public works like, roads, buildings and irrigation were entrusted to the charge of military board form strategic and governance reasons. This arrangement continued from year 1773 to 1858. In the year 1849, a department for public works was created. It was immediately entrusted with the improvement of Grand Trunk Road to Peshawar including construction of about 100 bridges on it. The roads from Kalka to Shimla and Chini to Sutlej were also completed by the year 1854. PWDs worked under the charge of a chief engineer who worked under the Lt. Governor of the province. A Secretary of the department of public works was appointed in the government of India for the first time in year 1854. During the years 1863-66, the department of public works in government of India was split in three separate branches to deal with military works, civil and irrigation and railways works. These branches were placed under the charge, of an under secretary each in the government of India in year 1867 and had an inspector general of works attached to each of them to co-ordinate the functions of each wing throughout the country. By the year 1870, the posts of under secretaries controlling these three branches were upgraded to those of deputy secretaries. In 1872, it was decided that the branches dealing with the military works should be transferred from the secretariat to the military department. This transfer was completed by the year 1890. In December 1911, a committee of experts was appointed by the secretary of state to advise the government with regard to the site of the new capital (in Delhi) and its layout. Sir Edvin Lutyens, an eminent and world famous architect, was chosen to be the architect and designer of the new capital city. Work started in December 1913 with a slow pace up to 1920, but took a faster track afterwards.

3.4.2.1 Road Development Committee, 1927

At a meeting of council of state, held on 9th of February 1927, a resolution of road development was unanimously adopted. The Government of India in the department of commerce by their resolution No. 489-T(1), dated the 3rd November 1927 appointed a committee from the

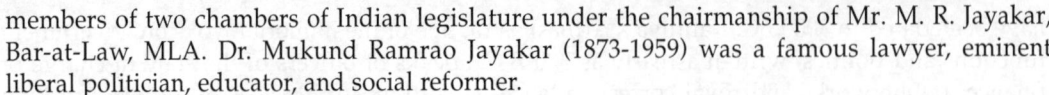

members of two chambers of Indian legislature under the chairmanship of Mr. M. R. Jayakar, Bar-at-Law, MLA. Dr. Mukund Ramrao Jayakar (1873-1959) was a famous lawyer, eminent liberal politician, educator, and social reformer.

The terms of reference of the committee were:

i. To examine the desirability of developing the road system of India and in particular the means by which such development could most suitability be financed; and

ii. To consider, with due regard to the distribution of central and provincial functions, whether it is desirable that steps should be taken for the co-ordination of road development and research in road construction, by the formation of Central Road Board or otherwise.

The committee suggested that the development of the road system of India is desirable for the general welfare of the country as a whole, and in particular:

a. For the better marketing of agricultural produce;

b. For the social and political progress of the rural population, which will be advanced by the increased use of motor transport;

c. As a complement to railway development.

The committee also looked into the aspect of financing for road development in India and highlighted that road development in India is passing beyond the financial capacity of local governments and local bodies. No increase in the expenditure on roads from existing revenues was recommended. Additional taxation on motor transport, over and above the existing, was suggested for purposes of road development. It included,

a. A duty on motor spirit

b. Vehicle taxation

c. License fees for vehicles plying for hire.

A rise in the duty on motor spirit to 6 annas per gallon was the suggested amount out of which 2 annas were to be spent on road development. This was coupled with the effort of reducing the price of petrol in inland towns. The additional revenues were to be controlled and distributed through the constitutional processes prescribed by the Government of India Act and maintained as a separate account. The division of this grant was suggested as:

a. One-sixth should be retained by the Government of India as a reserve

b. Out of the remaining five-sixths:

i. An apportionment should be made among the provinces in the ratio which the consumption of petrol in each province bears to the total consumption in India in each year.

ii. The balance, representing the consumption of petrol in minor provinces and administrations and Indian States, should be allotted as a lump sum to the Government of India.

The unexpanded amount was allowed to be carried over for expenditure in that province in the following year. The decision on vehicle taxes and license fees for vehicles plying for hire was left entirely on local governments, local legislatures and local bodies. A system of reciprocal exemptions was suggested so that each vehicle would be taxed only at the place of registration. The license fees was linked with local conditions and designated as a tool to limit the number of vehicles playing on any road so that it remains lower than the numbers that the road can economically carry.

Certain principles were suggested for financing the road development through loans. The development of village roads was given higher importance and it was hoped that they will receive more attention and larger grants from local governments and local bodies in future.

The following was suggested for the coordination of road development in India:

a. The appointment of a Central Road Board with executive powers, administering a separate road fund was not recommended.

b. A periodical road conference, consisting of the member of the governor general's executive council in charge as chairman, the members of the standing committee of the Indian legislature for roads, representatives of the departments of the Government of India concerned with roads, representatives of the local governments and, if so desired by them, of Indian states, to discuss subjects of common interests.

c. A separate road department in the Government of India was not recommended, but attaching a road engineer to the department that deals with roads was recommended.

d. The Government of India should reconsider the recommendations of previous committees that all matters relating to communications and transport should be dealt with by one department.

e. The Indian Railways Act and the Devolution Rules should be amended so as to enable the railway administration to contribute towards the construction and maintenance of feeder roads.

f. A contribution should be made from the army budget towards the cost of repairing damage done to roads by military transport; and when a road or bridge is required for military reasons to be of higher class than is necessary for civil purposes, part of the cost should be borne from the army budget.

g. Monopolies of public motor services on roads are undesirable.

h. Road tolls on all traffic should be abolished as soon as possible, except tolls on bridges where a definite service is provided to replace a ferry or a bad river crossing.

3.4.2.2 Central Road Fund, 1929, 2000, 2007

One of the recommendations of road development committee was to place duty on petrol and to keep a part of additional revenues for the road development programme. In its accordance the Central Government created a dedicated fund, called Central Road Fund through collection of cess from petrol and diesel. Presently, Rs. 2/- per litre is collected as cess on petrol and high speed diesel (HSD) oil. The fund is distributed for development and maintenance of national highways, state roads, rural roads and for provision of road over brides/under bridges and other safety features at un-manned railway crossings as provided in Central Road Fund Act, 2000. The fund will remain under the control of central government and will be utilized for the:

a. Development and maintenance of national highways,

b. Development of rural roads,

c. Development and maintenance of other state roads including roads of interstate and economic importance,

d. Construction of roads either under or over the railways by means of a bridge and erection of safety works at unmanned rail-road crossing, and

e. Disbursement in respect of such projects as may be prescribed.

The central government shall be responsible for the administration, management, coordination, timely utilization and release of funds, and sanction of schemes, formulation of fund allocation criteria and project sanction criteria.

Allocation of funds shall be done as follows:

a. 50 percent of cess on high speed diesel oil for the development of rural roads,

b. Rest 50 percent of cess of high speed diesel oil and 100 percent of cess on petrol as follows;

i. As such 57.5 percent of sum on the development and maintenance of national highways,

ii. As such 12.5 percent for the construction of road under or over the railway by means of a bridge and erection of safety works at un- manned rail-road crossing, and

iii. Balance 30 percent on the development and maintenance of roads other than national highways, out of which 10 percent (or 3 percent of total state road share) to be kept as central reserve for allocation on schemes of inter-state and economic importance as approved by central government.

The Government of India vide GSR-475(E) dated July 10, 2007 notified the rules for the disbursement of Central Road Fund in respect of specific projects, schemes and activities relating to development and maintenance of state roads including roads of inter-state connectivity and economic importance. These rules were called Central Road Fund (State Roads) Rules 2007. The types of works to be considered in this scheme shall comprise -

1. Construction of missing bridges, cross-drainage works, rehabilitation of bridges, widening of two lanes, strengthening of weak pavement sections;

2. Engineering aspects of road safety works covering improvements of traffic junctions, road markings, signaling, construction of subways and over-bridges, construction of parking lay-bays, bus sheds and the like;

3. Construction of bypasses, parallel service roads, along national highways and state highways, in built-up areas in exceptional cases, and

4. Development of connecting roads to national highways from rural roads as well as to tourist important places.

The projects that shall be considered as projects of economic importance on state roads and major district roads are-

1. Road either directly connecting to or leading to an important market centre, economic zone, industrial zone, agricultural region, tourist centre, and the like where significant economic activity is being undertaken,

2. Roads benefitting vulnerable sections of the society, such as, schedule castes, schedule tribes, ghat roads, and roads connecting ecologically sensitive areas,

3. Roads leading to centre of economic activities, schools, and education institutions,

4. Roads leading to socially important infrastructure, such as, cremation grounds, bathing ghats, orphanages, old age homes, and public utilities, and

5. Roads connecting the state highways, national highways, and link roads connecting the tourist destinations.

Roads of inter-state connectivity are fully financed and roads of economic importance are financed fifty percent.

3.4.2.3 Motor Vehicles Act, 1939, 1988

Transport wing of the Department of Road Transport and Highways is responsible for following legislations: -

1. Motor Vehicles Act, 1988.

2. Central Motor Vehicles Rules, 1989.

3. The Carriers Act, 1865.

The first enactment relating to motor vehicles in India was the Indian Motor Vehicles Act, 1914, which was subsequently replaced by the Motor Vehicles Act, 1939. The act of 1939 had been amended several times. In spite of several amendments it was felt necessary to bring out a

comprehensive legislation keeping in view the changes in the transport technology, pattern of passenger and freight movements, development of the road network in the country and particularly the improved techniques in the motor vehicles management.

A working group was, therefore, constituted in January, 1984 to review all the provisions of the act of 1939. The recommendations of the working group were discussed in the meeting of transport ministers of all states and union territories and the final recommendations were presented to the parliament as a bill. It was then implemented as Motor Vehicle Act, 1988.

Motor Vehicles Act, 1988 lays down the principles and procedures and the authorities responsible for the following:

a. Issue of driving licenses;

b. Issue of permits;

c. Grant of fitness certificates for the vehicles on roads;

d. Prescribing of emission and safety related norms for motor vehicles;

e. Norms for type approval in conformity of production of new motor vehicles;

f. Issues relating to compensation in case of motor vehicles accidents etc.

The Motor Vehicles Act, 1988 has so far been amended a number of times. These are,

(a) Amendment, 1994

This included:

i. Rationalization of the definition of the various categories of motor vehicles,

ii. Mandating of a minimum one year experience of driving a light motor vehicle before a person can be granted a license for transport vehicle and tightening of norms for drivers transporting dangerous or hazardous goods,

iii. Encouraging use of battery, CNG and solar energy as an auto fuel by exempting vehicles using such fuel from the requirement of permit or fixation of fare by the state government,

iv. Empowering central government to make rule for standardizing components in motor vehicles, and

v. Increasing the amount of compensation in the event of death from Rs. 25,000/- to Rs. 50,000/-in respect of no fault liability, etc.

(b) Amendment, 2000

This included:

i. Authorized use of LPG as an auto fuel,

ii. Buses used by educational institutions brought under the purview of permit regime, and

iii. Alterations made in transport vehicle without prior approval of the registering authority were barred.

(c) Amendment, 2001

i. Need to bring the buses playing on CNG within the purview of State Transport Authority in respect of fixation of fares and route permits.

(d) Amendment, 2004

These included:

i. Safety norms for various components of agricultural tractors such as power steering, lamps, light, parking light, etc.,

ii. Extension of Bharat Stage-II emission norms for four wheeled vehicles in Solapur and Lucknow from 1-6-2004,

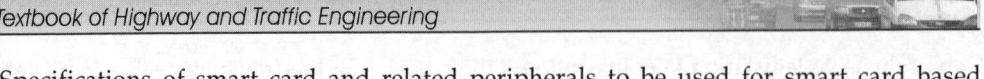

iii. Specifications of smart card and related peripherals to be used for smart card based driving licence (DL) and registration certificate (RC), and

iv. Introduction of Bharat Stage-III emission norms in 11 mega cities from 1-4-2005.

(e) Amendment, 2005

These included:

i. Updating of the list of dangerous and hazardous goods in the central motor vehicles rules,

ii. The terms "Battery Operated Vehicle" and "Power Tiller" have been defined,

iii. The emission norms, overall dimension and other related norms for "Power Tiller" have been prescribed,

iv. Time limits for various functions discharged by licensing and registering authorities and appellate authorities under the Motor Vehicles Acts/Rules have been specified,

v. It has been made mandatory for the manufacturer to supply a protective headgear conforming to BIS Standards at the time of sale of the two wheelers, subject to the exceptions under Section 129 of Motor Vehicles Act, 1988,

vi. Type approval rules for CNG/LPG vehicles have been rationalized,

vii. States have been empowered to prescribe special provisions such as fog lamp, power steering, defogging and demisting systems in transport vehicles plying in hill areas.

The vehicle emission norms, high security registration plates and issuing of smart card based license and registration certificate are other areas where progress has been made. The information about these is given in Appendix C, D and E respectively.

3.4.2.4 Road Development Plan, 1943-1963 (Nagpur Road Plan)

First long-term vision for road development in India was drafted in the conference of chief engineers of all the states and union territories that was held at Nagpur in the year 1943. The plan was prepared for a twenty year period of 1943 -1963.

Some of the major recommendations of Nagpur plan are:

1. Road Development Plan should consist of short-term measures to overcome the damages caused by war and a long-term planning for good communication throughout India.

2. Though road was a provincial subject, it was brought under centre to bring balanced and coordinated development.

3. Roads were classified based on their function as,

   a. 'National Highways', having national importance as for strategic, administrative and other purposes.

   b. 'Provincial and State Highways', which are other main roads of a province or a state,

   c. 'District Roads', taking traffic from the main roads to the interior of the districts. According to their importance, they are further classified as 'Major District Roads' and 'Minor District Roads.'

   d. 'Village Roads', which would link villages to road system.

4. There should be balanced development of all classes of roads.

5. National Highways should work as a framework for road development and centre should be responsible for their construction and maintenance.

6. All other roads should be designed and constructed by the highway department of a province or a state.

7. The national highways, provincial highways and major district roads must be provided with a durable hard pavement crust.

8. The other district roads and village roads should be provided with a properly engineered earth surface, with modifications as necessary.

9. All national highways, provincial highways and major district roads should be adequately bridged. The interruptions to submersible bridges and causeways should not generally be more than twelve hours at a time or more than six times a year.

Salient features of the Road Development Plan were:

1. The plan was formulated keeping in view the requirements of growing population and increasing vehicular traffic for the next twenty years.

2. The road network development was based on 'Star and Grid' pattern.

3. The area was classified based on the potential of agricultural production as agricultural area and non-agricultural area.

4. The connectivity by road was defined based on population residing in villages so that each village gets some length of the road inducing development. A wide range of population was considered for the categorization of villages.

5. The functionally classified roads were further grouped into two categories only with regard to calculating road lengths.

6. The target length of the roads was defined based on the size of the area that it will serve and was decided as 16km per 100 km sq of area.

7. The size of the grid for the first category of roads in agricultural area was decided as grid of 16 km side and for non-agricultural area it was of 64 km side.

8. An allowance of 15 percent of the road length calculated based on the formulae recommended by the committee should be provided for agricultural and industrial development that may take place during the next twenty years.

9. The existing length of railway track was also considered while computing the length of roads falling in the first category.

The formulae based on which the road lengths in the two categories were calculated were:

Kilometerage of national and
provincial highways and major = $A/8 + B/32 + 1.6\,N + 8T + D - R$
district roads

Where,

A = Agricultural area of the state or province concerned, in km sq

B = Non-agricultural area of the state or province concerned, in km sq

N = Number of towns and villages having population of 2000-5000

T = Number of towns and villages having population over 5000

D = An allowance for industrial and agricultural development during the next twenty years, approximately 15 percent of total road length calculated

R = Length of railway in the province or the state concerned, in km

According to the above formula, no village will be more than 8 km away from the main road in a highly developed agricultural area, with an average distance less than 4.8 km in many cases. In non-agricultural area, the corresponding distances will be 32 km and between 9.8 and 11.2 km, respectively.

Kilometerage of other district

roads and village roads = 0.32 V + 0.8 Q +1.6 R + 3.2 S + D

Where

V = Number of towns and villages having population sf 500 or less

Q = Number of towns and villages having population of 501-1000

R = Number of towns and villages having population of 1001-2000

S = Number of towns and villages having population of 2001-5000

D = An allowance for industrial and agricultural development during the next twenty years, approximately 15 percent of total road length calculated

The targets of this plan were achieved before the plan period in the year 1961. These are given in Table 3.1

Table 3.1: Nagpur Plan targets and achievements by 1961

S. No.	Category	Nagpur Plan Targets, km	Achievements by 1961, km
1.	(a) National Highways	26715	22636
	(b) National Trails	6680	—
2.	State Highways	86825	62052
3.	Major District Roads	80145	113483
4.	Other District Roads	133580	111961
5.	Village Roads	198755	388841
6.	Unclassified Roads	—	10149
	Total	532700	709122

3.4.3 The Road Transport Corporations Act, 1950

Subject to rules made under this act, a corporation shall consist of a chairman and such number of other members as the state government may think fit to appoint. Rules made under this act shall provide for the representation both of the central government and of the state government concerned in the corporation in such proportion as may be agreed to by both the governments. In case the capital of a corporation is raised by the issue of shares, provision shall also be made for the representation of such shareholders in the corporation.

The corporation should:

1. Exercise its power to provide or secure or promote the provision of an efficient, adequate, economical and properly coordinated system of road transport services in the state or part of the state for which it is established and in any extended area.

2. Provide for any ancillary service,

3. Provide for its employees suitable conditions of service including fair wages, establishment of provident fund, living accommodation, places for rest and recreation and other amenities,

4. Issue passes to its employees and other persons either free of cost or at concessional rates it may deem fit to impose,

5. Grant refund in respect of unused tickets and concessional passes.

Subject to the provisions of this act, the powers include-

a. To manufacture, purchase, maintain and repair rolling stock, vehicles, appliances, plant, equipment or any other thing required for the purpose of any of the activities of the corporation.

b. To acquire and hold such property, both movable and immovable, as the corporation may deem necessary for the purpose of the said activities, and to lease, sell or otherwise transfer any property held by it.

c. To prepare schemes for the acquisition of, and to acquire, either by agreement or compulsorily in accordance of the law of acquisition for the time being in force in the state concerned the whole or any part of any undertaking of any other person to the extent to which the activities thereof consist of the operation of road transport services in that state or in any extended area.

d. To purchase by agreement or to take on lease or under any form of tenancy any land and to erect such building as may be necessary for the purpose of carrying on its duties.

e. To authorize the disposal of scrap vehicles, old tyres, used oils, or any other stores of scrap value.

f. To enter into and perform all such contracts as may be necessary for the performance of its duties and the exercise of its powers under the act.

g. To purchase vehicles of such type as may be suitable for use in the road transport service operated by the corporation.

h. To purchase or otherwise secure by agreement vehicles, garages, sheds, office buildings, depots, land, workshops, equipment, tools, accessories to and spare parts for vehicles, or any other article owned or possessed by the owner of any other undertaking for use thereof by the corporation for the purposes of its undertaking.

i. To do anything for the purpose of advancing the skill of persons employed by the corporation or efficiency of the employees of the corporation or of the manner in which that equipment is operated, including the provision by the corporation, and the assistance by the corporation to others for the provision of facilities for training, education and research.

j. To enter into and carry out agreement with any person carrying on business as a carrier of passengers or goods providing for tile carriage of passengers or goods on behalf of the corporation by that other person at a through fare or freight.

k. To provide facilities for the consignment, storage and delivery of goods.

1. To enter into contracts for exhibition of posters and advertising boards on and in the vehicles and premises of the corporation and also for advertisement on tickets and other forms issued by the corporation to the public.

m. With the prior approval of the state government to do all other things to facilitate the proper carrying on of the business of the corporation.

3.4.4 National Highway Act, 1956

The National Highways Bill, 1956 having been passed by both the Houses of Parliament received the assent of the President on 11th September, 1956. It came into force on 15th day of April, 1957 as The National Highways Act, 1956. The act empowers the central government to declare each of the highway specified in schedule 2 as national highway, to declare any other highway as national highway, and to omit any highway from the schedule. The highway so omitted cease to be a national highway. It also provides power to the central government to acquire land for public purpose as required for building, maintenance, management and operation of a national highway or part thereof. It also gives powers to an authorized person to:

a. make any inspection, survey, measurement, valuation or enquiry

b. take levels

c. dig or bore into sub-soil

d. set out boundaries and intended lines of work

e. mark such levels, boundaries and lines placing màrks and cutting trenches or

f. do such other acts or things as may be laid down by rules made in this behalf by that government.

Such acquisitions have to be declared/published through gazette notification. The Act further gives power to the central government to take possession of the land as required for the development of national highways through depositing amount determined by competent authority for the land with the authority and the authority would issue a notice to that effect to the owner of the land. Forceful possession is allowed at the expiry of sixty days from the date of issue of notice to the owner.

Though the central government is responsible for the development and maintenance of all the national highways but by notification in the official gazette it may direct that any function regarding development and maintenance of any national highway shall also be exercisable by the government of the state within which the national highway is situated. The central government is also empowered to levy fees for services or benefits rendered in relation to the use of ferries, temporary bridges and tunnels on national highways and the use of sections of national highways. The central government can also enter in to an agreement with any person for the maintenance and development of whole or part of the national highway and the person will be entitled to collect and retain fees at a rate that is agreed upon between the two based on expenditure involved, interest on the capital invested, reasonable return, the volume of traffic and the period of such agreement.

Various amendments have been made to the above act from time to time. These are:

1. The National Highways (Amendment) Act, 1977 (30 of 1977).

2. The National Highways (Amendment) Act, 1992 (1 of 1993).

3. The National Highways (Amendment) Act, 1995 (26 of 1995).

3.4.5 Road Development Plan, 1961-1981 (Bombay Road plan)

The first twenty year Road Development Programme became deficient in many respects due to the changed economic, agricultural and industrial scenario. Further, the plan objectives or targets were achieved by the year 1961 and hence, there was a need to redraw the targets of road development that should be achieved in the future. Another twenty year road development plan was drafted by the roads wing of government of India, and the same after deliberations was adopted by the chief engineers of all states and union territories at their conference meeting held at Bombay in 1959. The plan has also been known as Bombay Plan.

The main objectives of the plan were:

1. Provision of good communication facilities in the rural areas is essential to check the increasing rate of urbanization and migration to urban areas.

2. The future road system should, besides taking care of agricultural and non- agricultural areas, also take into account the needs of the semi-developed and undeveloped areas, administrative headquarters, places of pilgrimage and cultural institutions, important

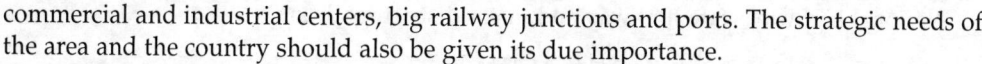

commercial and industrial centers, big railway junctions and ports. The strategic needs of the area and the country should also be given its due importance.

3. The road density should be increased to 32 km per 100 km sq area. This target was set taking into account the level of expected development and the need of the rural as well as urban areas.

4. Every village -
   a. In a developed area and agricultural area should be brought within 6.5 km of the metalled road and 2.5 km of any road.
   b. In a semi-developed area should be brought within 13 km of a metalled road and 5 km of any road.
   c. In an undeveloped area and uncultivable area should be brought within 20 km of a metalled road and 8 km of any road.

5. To achieve the overall road density target, the road density in classified areas should be 44 km per 100 km sq in developed and agricultural area, 19 km per 100 km sq in semi-developed area and 12 km per 100 km sq in undeveloped and uncultivable area.

6. Every town with population above 2000 in plains, above 1000 in semi-hills, and above 500 in hilly areas should be connected by a mettalled road.

7. An allowance of 100 percent should be made in hilly regions. Hills with altitude more than 2300 meters may be ignored in road length calculations.

8. A target length of 1600 km should be changed into expressway.

Other salient features of the road development plan were:

1. Specifications of works to be adopted should permit 'Stage Construction' in view of limited funds available.

2. Funds for road construction and maintenance should come not only from direct beneficiaries but also from indirect beneficiaries.

3. Classified village roads should be brought under village panchayat to whom state PWD should give necessary technical advice.

4. Highway department should have a cell to deal with formulation of standards related to traffic engineering, roads, etc.

5. A two year professional course should be given to fresh entrants to enable them to understand the requirements of the job.

6. Arterial roads should be fully bridged and made two-lane wide to withstand truck-trailer loads and movements.

7. Maintenance of roads should be accorded high priority

8. The length of railway track should be dealt independently of the road development.

9. An allowance of 5 percent should be given for future development and other unforeseen factors.

The road lengths of different categories of roads were calculated using the formulae given below:

$$\text{National highway (km)} = \left[\frac{A}{64} + \frac{B}{80} + \frac{C}{96}\right] + [32K + 8M] + D$$

$$\text{National highway and state highway (km)} = \left[\frac{A}{20} + \frac{B}{24} + \frac{C}{32}\right] + [48K + 24M + 11.2N + 1.6P] + D$$

National highway, state highway and major district road (km)

$$= \left[\frac{A}{8} + \frac{B}{16} + \frac{C}{24}\right] + \left[\begin{array}{c} 48K + 24M + 11.2N + \\ 9.6P + 6.4Q + 2.4R \end{array}\right] + D$$

National highway, state highway, major district road and minor district road (km)

$$= 3\left[\frac{A}{16} + \frac{B}{32} + \frac{C}{48}\right] + \left[\begin{array}{c} 48K + 24M + 11.2N + \\ 9.6P + 12.8Q + 4R + \\ 0.8S + 0.32T \end{array}\right] + D$$

National highway, state highway, major district road and minor district road and village roads (km)

$$= \left[\frac{A}{4} + \frac{B}{8} + \frac{C}{12}\right] + \left[\begin{array}{c} 48K + 24M + 11.2N + \\ 9.6P + 12.8Q + 5.9R + \\ 1.6S + 0.64T + 0.2V \end{array}\right] + D$$

Where

A = Developed and agricultural area, km sq

B = Semi-developed area, km sq

C = Undeveloped and uncultivable area, km sq

K = Number of towns with population over 100000

M = Number of towns with population between 100000 and 50000

N = Number of towns with population between 50000 and 20000

P = Number of towns with population between 20000 and 10000

Q = Number of towns with population between 10000 and 5000

R = Number of towns with population between 5000 and 2000

S = Number of villages with population between 2000 and 1000

T = Number of villages with population between 1000 and 500

V = Number of villages with population below 500

D = An allowance for future development and other unforeseen factors, taken approximately 5 percent of road length calculated.

The total cost of works involved in Bombay Plan was estimated at Rs. 520000 crores. The road length targets and the road length achieved by 1981 are given in Table 3.2. Though the over-all target was achieved by the year 1981, the targeted length of main road system, i.e. national highways, state highways and major district roads was not achieved. The expenditure was also short of the suggested outlay even without considering inflation.

In addition to the road lengths achieved by the year 1981, there was an additional road length of 309785 km of roads as,

a. Urban roads, within municipalities (123569 km)

b. Project roads, under the charge of forest department, irrigation department and other project authorities (186216 km)

3.4.6 Road Development Plan, 1981-2001 (Lucknow Road Plan)

The Chief Engineers of States and Union Territories, and Indian Roads Congress at their Lucknow conference formulated a long term road development plan for the country for the period 1981-2001.

Table 3.2: Bombay plan targets and achievements by 1981

S.No.	Category	Bombay Plan Targets, km	Achievements by 1981, km
1.	National Highways	51500	31737
2.	State Highways	112650	95491
3.	Major District Roads	241400	153000
4.	Rural roads:		
	(a) Other District Roads	289680	912684
	(b) Village Roads	362100	
	Total of 4	651780	
	Total	1057330	1192912

The goals and policies defined in this plan were:

1. Road network should be developed to provide accessibility to all villages with a population of above 500 by the turn of the century.

2. Roads should be built in less industrialized areas so as to attract the growth of industries.

3. Roads should be a major choice of construction programme to generate employment.

4. Long-term master plans should be prepared at all levels.

5. Transport requirements of small cities and towns should be studied.

6. The national highway grid should be increased to 100 km side by expanding it to 66000 km.

7. Expressways should be constructed on all major traffic corridors to provide speedy travel.

8. Road improvements for energy conservation should be taken up.

9. The environmental equality should be maintained and improved.

10. Road safety measures should be given importance to bring down road accidents.

11. The annual rate of growth of 7.5 percent for passenger traffic and 10 percent for freight traffic was recommended.

Other specific features of the plan were:

1. State Highways should be extended to serve district headquarters, sub-divisional head-quarters, major industrial centers, places of commercial interest, and places of tourist attraction, major agricultural market centers and ports.

2. Major District Roads should serve and connect all towns and villages with a population of 1500 and above, not connected directly by NH or SH.

3. Other District Roads should serve and connect villages with a population in range 1000 and 1500.

4. Village Roads should serve and directly connect all villages with a population of 500 - 1000.

5. The overall road density in the country should be increased to 82 km per 100 km sq area by the year 2001. The corresponding values of planned road densities are 40 km per 100 km sq area in hill areas of altitude up to 2100 meters and 15 km per 100 km sq area in hill areas for altitude above 2100 meters with respect to M.S.L.

6. There should be a road within a distance of 3.0 km in plains and 5.0 km in hilly terrain connecting all villages or group of villages with population less than 500.

7. A new road classification system was introduced for the calculation of road lengths with specific functional capabilities. This was Primary Road system, Secondary Road system and Tertiary Road system.

8. Expressways with a total length of 2000 km should be developed based on traffic requirements.

9. By 2001, 72 percent of NH and 25 percent of existing SH should be double- lane. Whole of the primary network and all existing secondary network should be fully blacktopped. Certain part of secondary network should also have roads with intermediate lane width.

10. The widening of pavements should be adopted in stages.

11. Ribbon developments should be checked by suitable legislation and regulation.

12. The primary and secondary network should be fully bridged with high level bridges for NH and at least submersible bridges for whole SH and 50 percent of MDR. Causeways may be allowed for MDRs if economy dictates.

13. New pavement design methodology should be evolved relating pavement thickness with number of repetitions of standard axles.

14. Road improvements can bring about 15 percent savings in vehicle operating costs.

15. Maintenance of roads should receive attention.

16. A separate coordinating agency (Rural Engineering Organisation) in states for rural roads was recommended.

17. Sources of highway financing should be explored.

18. A good management information system is needed to handle data of highway department.

The road lengths of different categories were calculated using the formulae given below:

A. Primary System:

    a. Expressways = 2000 km

    b. National Highway, km = Area of state, km sq / 50

B. Secondary System:

    a. State Highway, km   = Area of state, km sq / 25

                  or = 62.5 × number of towns in state

                      – area of state, km/sq/50

    b. Major District Road, km = Area of state, km sq / 12.5

                  or = 90 × Number of towns in the state

C. Tertiary System:

    a. Total Road Length, km = 4.74 × Number of villages and towns in the state

    b. Rural Road Length, km = Total Road length – Sum road length of National Highways, State Highways and Major District Roads

Rural road length is defined as the sum of road lengths falling in previous categories as Other District Roads and Village Roads. Based on the road density, the total road length to be constructed can be calculated. Another approach is to use grid pattern. Assuming a uniform distribution of habitations and knowing the number of villages and towns in the country, the average area per human settlement comes out to be 5.634 sq.km. Assuming the square shape of settlement, the side of one square works out to be $\sqrt{5.634} = 2.37$ km. Thus the total road length share of each village would be 4.74 km. This is used in calculation of total road length above. The road lengths targeted and achieved by 2001 are given in Table 3.3.

Table 3.3: Lucknow plan targets and achievements by 2001

S. No.	Category	Lucknow Plan Targets, km	Achievements by km 2001, km
1.	National Highways	66000	57700
2.	State Highways	145000	124300
3.	Major District Roads	300000	2994000
4.	Rural Roads	2189000	
	Total	2700000	3176000

3.4.7 The National Highways Authority of India Act, 1988

For the development, maintenance an d management of national highways the National Highways Act, 1956 (48 of 1956) was enacted. Under the provisions of this act the Central Government had to face certain difficulties in developing and maintaining the national highways mentioned in the schedule of the said act and it was felt necessary to constitute a separate authority with statutory powers for the development, maintenance and management of national highways. To achieve this objective the 'National Highways Authority of India Bill,' 1988 was introduced in the parliament. The National Highways Authority of India Bill, 1988 having been passed by both the houses of parliament received the assent of the president on 16th December, 1988. It came into force on 15th June, 1989 as The National Highways Authority of India Act, 1988 (68 of 1988).

According to the act,

a. For discharging the functions under the act, if any land is required by the authority then that land shall be deemed to be land needed for a public purpose and such land can be acquired under the provisions of the National Highways Act, 1956. The functions of the authority are laid explicitly and are listed under NHAI afterwards.

b. The central government has been empowered to provide any capital that may be required by the authority or pay to the authority by way of loans or grants such sums of money as it may consider necessary for the efficient discharge of the functions by the authority.

c. For anything and for any damage caused or likely to be caused by anything which is in good faith done or intended to be done under the act or the rules or regulations made there under, no suit, prosecution or other legal proceeding shall lie against the authority or any member or officer or employee of the authority.

d. Any person authorised by the authority may, at all reasonable times, enter upon any land or premises and make any inspection, survey, measurement, valuation or enquiry, take levels, dig or bore into sub-soil, set out boundaries and intended lines of work, mark such levels, boundaries and lines by placing marks and cutting trenches, or do such other acts or things as may be prescribed. No such person shall enter any boundary or any enclosed court or garden attached to a dwelling house without the consent of the occupier

The Central Government, through gazette notification, may temporarily divest the authority of the management of any National Highway or under emergency supersede the authority for a period as specified in notification but not exceeding one year. The authority is also bound by the directions as given by the Central Government from time to time on questions of policy.

3.4.8 Pradhan Mantri Gram Sadak Yojna (PMGSY), 2000

Government launched the 'Pradhan Mantri Gram Sadak Yojana' on 25th December, 2000 to provide all-weather access to unconnected habitations. The Pradhan Mantri Gram Sadak Yojana

(PMGSY) is a 100% Centrally Sponsored Scheme. 50% of the cess on High Speed Diesel (HSD) is earmarked for this programme. The primary objective of the PMGSY is to provide connectivity, by way of an all-weather road (with necessary culverts and cross-drainage structures, which is operable throughout the year), to the eligible unconnected habitations in the rural areas, in such a way that all unconnected habitations with a population of 1000 persons and above are covered in three years (2000-2003) and all unconnected habitations with a population of 500 persons and above by the end of the tenth plan period (2007). In respect of the hill states (North-East, Sikkim, Himachal Pradesh, Jammu and Kashmir, Uttaranchal) and the desert areas (as identified in the Desert Development program) as well as the tribal (Schedule V) areas, the objective would be to connect habitations with a population of 250 persons and above. The PMGSY will permit the upgradation (to prescribed standards) of the existing roads in those districts where all the eligible habitations of the designated population size have been provided all-weather road connectivity. However, it must be noted that upgradation is not central to the programme and cannot exceed 20% of the state's allocation as long as eligible unconnected habitations in the state still exist. In upgradation works, priority should be given to through routes of the rural core network which carry more traffic.

3.4.8.1 Salient features

1. The unit for this programme is a habitation and not a revenue village or a panchayat. An unconnected habitation is one with a population of designated size located at a distance of at least 500 meters or more (1.5 km of path distance in case of hills) from an all-weather road or a connected habitation.

2. The population, as recorded in the census 2001, shall be the basis for determining the population size of the habitation. The population of all habitations within a radius of 500 meters (1.5 km. of path distance in case of hills) may be clubbed together for the purpose of determining the population size.

3. The selected habitations become part of a core network. A Core Network is that minimal network of roads (routes) that is essential to provide basic access to essential social and economic services to all eligible habitations in the selected areas through at least a single all-weather road connectivity. The routes are defined as through routes and link routes.

   a. Through routes are the ones which collect traffic from several link roads or a long chain of habitations and lead it to marketing centers either directly or through the higher category roads, i.e. the district roads or the state or national highway. Through routes arise from the confluence of two or more link routes and emerge on to a major road or to a market centre.

   b. Link routes are the roads connecting a single habitation or a group of habitations to through routes or district roads leading to market centers. Link routes generally have dead ends terminating on a habitation.

4. Preference should be given to those roads which serve a larger population. Habitations with population 1000+ should be considered first, 500+ habitations next and 250+ habitations at the last.

5. The PMGSY envisages only single road connectivity to be provided.

6. The programme considers road construction under two schemes, new connectivity and upgradation. New connectivity may involve 'New Construction' where the link to the habitation is missing and additionally, if required, 'Upgradation' where an intermediate link in its present condition cannot function as an all-weather road. Upgradation, when permitted would typically involve building the base and surface courses of an existing road to desired technical specifications and/or improving the geometrics of the road, as required in accordance with traffic condition.

7. PMGSY does not permit repairs to black-topped or cement roads, even if the surface condition is bad.

8. The role of different agencies including the intermediate panchayat, the district panchayat as well as the state level standing committee in the formulation of plan and deciding the core network is outlined. In the identification of the core network, the priorities of elected representatives, including MPs and MLAs, are also expected to be duly taken into account and given full consideration.

9. The preparation of core network is based on fixing the priority of connectivity to a habitation giving due consideration to the set of socio-economic/infrastructure variables best suited for the district, categorising them and by according relative weightages to them.

10. Block level master plan includes drawing up of the existing road network, identification of unconnected habitations and preparing the plan required to connect these unconnected habitations. This is termed as 'Comprehensive New-Connectivity Priority List (CNCPL)' and is placed before intermediate panchayat for approval and sent to the members of parliament and MLAs for their comment. The next step is to place it before district panchayat and state level agency for approval.

11. States may, each year, distribute the state's allocation among the districts giving 80 percent on the basis of road length required for providing connectivity to unconnected habitations and 20 percent on the basis of road length requiring upgradation under the PMGSY.

12. In addition to the allocation to the states, a special allocation of up to 5% of the annual allocation from the rural roads share of the diesel cess will be made for:
    a. Districts sharing borders with Pakistan and China (in coordination with Ministry of Home Affairs)
    b. Districts sharing borders with Myanmar, Bangladesh and Nepal (in coordination with Ministry of Home Affairs)
    c. Left Wing extremists areas in the district identified by the Ministry of Home Affairs
    d. Extremely backward districts (as identified by the planning commission) which can be categorised as special problem areas
    e. Research and Development Projects and Innovations

13. In order to manage the rural road network for upgradation and maintenance planning all states will carry out, every 2 years, a pavement condition survey of all through routes resulting in a 'Pavement Condition Index (PCI)' between 1 and 5.

14. Once no new-connectivity is left in a district, the district will prepare a 'Comprehensive Upgradation Priority List (CUPL)'. The priority will be set as follows:
    a. **Priority-I** will be through routes which are constructed as WBM roads.
    b. **Priority-II** are other fair weather through routes or gravel through routes or through routes with missing links or lacking cross drainage.
    c. **Priority-III** will be other through routes which are at the end of their design life, whose PC1 is 2 or less, i.e., are 'Poor' or 'Very Poor'.

15. In case of upgradation, the Annual Average Daily Traffic (AADT) estimate is made, which is the basis for the prioritisation as well as the design. An axle load survey may also be carried out, on selective basis, on the roads where heavy traffic is expected with wide variations in the axle load spectrum.

16. The proposals so submitted by district panchayats shall be verified by 'State Technical Agencies (STAs)' approved by National Rural Roads Development Agency (NRRDA). STAs can also make ground check before verifying the proposals.

17. A three-tier Quality Control mechanism is envisaged under the Pradhan Mantri Gram Sadak Yojana. The State Governments would be responsible for the first two tiers of the quality control structure. The PIU will be the first tier. As the second tier of the quality control structure, periodic inspections of works will be carried out by quality control units, set up/engaged by the state government, independent of the executive engineers/ PIUs. These officers/agencies may be called 'State Quality Monitors (SQM).' As the third tier of the quality control structure, the NRRDA will engage independent monitors (individuals/agency) for inspection, at random, of the road works under the programme. These persons may be designated as 'National Quality Monitors (NQM).'

18. Proper use of communication systems is in place to bring in efficiency. The district 'Project Implementation Units (PIUs)' place whole of the information on CNCPL on internet through 'On-Line Management and Monitoring System (OMMS)' software for global viewing. The STAs after verification of proposals enter their decision through the same software. In this regard, there should be National Informatics Officer (NIO) and State Informatics officer (SIO).

19. The proposal should be accompanied by certain certificates as:

    a. A certificate that land is available. It must be noted that the PMGSY does not provide funds for land acquisition. No objection of land owners whose land will be acquired should also accompany the proposal.

    b. A certificate that funds are available for the project submitted under state/district/ block grant.

    c. A certificate that no forest clearance is required and where so it is required a certificate to that reason.

    d. No objection of panchayat and interview of persons living in that habitat regarding proposing of cart track as a new-connectivity. This can be done by organizing a simple, non-formal transect walk by the assistant engineer at the time of preparation of DPRs.

    e. Confirmation from CNCPL/CUPL

20. Some of the proposals are funded under World Bank scheme and are so designated as World Bank projects. The additional mandatory requirements in such projects is filling up of Environmental Protection Form and going by the conditions laid in it.

21. The PIU will ensure the following in preparing the Detailed Project Reports (DPRs):

    a. The rural roads constructed under the Pradhan Mantri Gram Sadak Yojana must meet the technical specifications and geometric design standards given in the 'Rural Roads Manual of the IRC (IRC:SP20:2002)' and also, where required, the 'Hill Roads Manual (IRC:SP:48).'

    b. Normally rural roads would need to be designed to carry upto 45 commercial vehicles per day (CVPD) only. In the case of new construction for eligible habitations of population below 1000 the design may be for traffic below 15 CVPD.

    c. In case of new construction to connect habitations with population below 500 where the projected traffic growth is likely to be very low, the carriageway may further be restricted to 3.0 m.

    d. Where the road passes through a habitation, the road in the built-up area and for 50 meters on either side may be appropriately designed preferably as a cement road or with paved stones, besides being provided with side drains.

    e. Wherever local materials, including fly ash, are available, they should be prescribed subject to adherence to technical norms and relevant codes of practice.

    f. The rural roads constructed under PMGSY must have proper embankment/drainage.

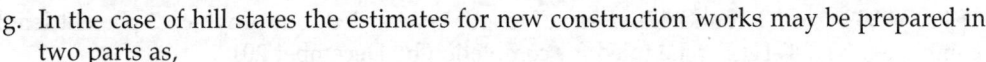

g. In the case of hill states the estimates for new construction works may be prepared in two parts as,

    i. The first stage will consist of formation cutting, slope stabilization, protection works and drainage works.

    ii. The second stage will include the WBM layers and bituminous surface course.

23. The maintenance component includes a 5-year maintenanec contract, as:

a. In case of link routes (new construction) the component shall comprise of 5 year routine maintenance

b. In case of associated rural through routes not requiring upgradation, the component shall include 5-year routine maintenance including one renewal as per cycle

c. In case of through routes taken up for upgradation, 5-year routine maintenance and a renewal at the end of the period

Table 3.4 gives an overall account of new connectivities under PMGSY programme.

Table 3.4: PMGSY programme: new connectivity

Habitation Population Group	Number of rural unconnected Habitation	Length Required (km)	Estimated cost (Rs billion)
1000+	60.030	138.888	784.18
500-999	79.208	160.754	
250-499*	39.530	75,690	
Total	178,768	375,332	784.18

*Only in hill states, desert and tribal areas as per PMGSY eligibility.

3.4.9 National Highway Development Projects, 2000 onwards

The National Highways have a total length of 66,590 km to serve as the arterial network of the country. The development of National Highways is the responsibility of the Government Of India. The Government of India has launched major initiatives to upgrade and strengthen National Highways through various phases of National Highways Development project (NHDP), which are briefly as under:

NHDP Phase I: NHDP Phase I was approved by Cabinet Committee on Economic Affairs (CCEA) in December 2000 at an estimated cost of Rs. 30,000 crore comprises mostly of GQ (5,846 km) and NS-EW Corridor (981km), port connectivity (356 km) and others (315 km). About 93% of NHDP-I project has been completed. Around 12% is done through PPP route on BOT (Toll) [6.0%] and BOT (Annuity) [6.0%] mode.

NHDP Phase II: NHDP Phase I1 was approved by CCEA in December 2003 at an estimated cost of Rs.34,339 crore (2002 prices) comprises mostly NS-EW Corridor (6,161 km) and other National Highways of 486 km length, the total length being 6,647 km. The total length of Phase I1 is 6,647 km. Around 24 percent work is through PPP on BOT (Toll) [11 percent] and BOT (Annuity) [13 percent]. 87 percent of length is awarded out of which around 12 percent work has completed. NHDP-I1 was scheduled for completion by Dec. 2009.

NHDP Phase III: Government approved on 5.3.2005 upgradation and 4 laning of 4,035 km of National Highways on BOT basis at an estimated cost of Rs. 22,207 crores (2004 prices). Government approved in April 2007 upgradation and 4 laning at 8074 km at an estimated cost

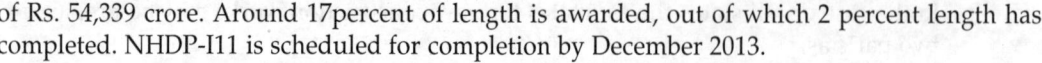

of Rs. 54,339 crore. Around 17percent of length is awarded, out of which 2 percent length has completed. NHDP-I11 is scheduled for completion by December 2013.

NHDP Phase IV: There is a proposal under consideration for widening of 20,000 km of existing single/intermediate/two lane highways to two lane with paved shoulders at an estimated cost of Rs. 27,800 crore through PPP route on BOT (Toll)/BOT (Annuity) basis.

NHDP Phase V: CCEA has approved on 5.10.2006 six laning of 6,500 km of existing 4 lane highways under NHDP Phase V (on DBFO basis). Six laning of 6,500 km includes 5,700 km of GQ and other stretches. A length of 148 km awarded. NHDP-V was scheduled for completion by December 2012.

NHDP Phase VI: CCEA has approved on November 2006 for 1000 km of expressways at an estimated cost of Rs. 16680 crores. NHDP-VI is scheduled for completion by December 2015.

NHDP Phase VII: CCEA has approved on December 2007 for 700 km of ring roads, bypasses and flyovers and selected stretches at an estimated cost of Rs. 16680 crores through PPP route on BOT (Toll) mode. NHDP-VII is scheduled for completion by December 2014.

3.4.10 Control of National Highways (Land and Traffic) Act, 2002

In order to provide for control of land within the national highways, right of way and traffic moving on the national highways and for the removal of unauthorised occupation of the land within the national highways the 'Control of National Highways (Land and Traffic) Bill' was introduced in the parliament. The Control of National Highways (Land and Traffic) Bill having been passed by both the houses of parliament received the assent of the president on 14th January, 2003.

According to the Act:

1. The Central Government has to establish one or more highway administrations to exercise powers and discharge functions under the act, as well as, one or more 'National Highway Tribunals' specifying the limits of the highway within which, or the length of highway, on which, such tribunal/tribunals may exercise jurisdiction for entertaining and deciding the appeals filed before it/them. The tribunal is to be headed by a person from Indian legal services having qualified to be a judge of a high court.

2. No person can occupy any highway land or discharge any material through drain on such land without written permission of the highway administration.

3. The highway administration is empowered to take actions, impose penalty on persons, remove constructions or unauthorized occupations as deemed necessary in the interest of traffic on highway and the safety of structures forming part of the highway, and to recover the cost or expenditure incurred, if any, from the person concerned.

4. No person shall have right of access to a highway either through any vehicle or on foot by a group of five or more persons except permitted by the highway administration.

5. The highway administration shall have the power in the interest of the safety and convenience of the traffic to refuse, regulate or divert any proposed or existing access to the highway.

6. If at any time the highway administration is satisfied on the information received by it or otherwise that any Highway within its jurisdiction or any part thereof is or has become congested or unsafe for vehicular or pedestrian traffic by reason of damage or otherwise, it may, either close the highway or such part thereof to all traffic or to any class of traffic, or regulate the number and speed of vehicles to be used on the highway or such part thereof, as the case may be, in the manner as it may deem fit.

7. If the highway administration is satisfied that it is necessary in the interest of public safety or convenience, or because of the nature of any road or bridge so to do, it may prohibit or restrict the use of any highway or part thereof by a class or classes of traffic either generally or on specified occasion or time as specified in the notification and when such prohibition or restriction is imposed, the highway administration shall cause such traffic sign to be placed or erected at suitable places for the convenience of the traffic as may be prescribed.

8. No person can construct, install, shift, repair, alter or carry any poles, pillars, advertisement towers, transformers, cable wire, pipe, drain, sewer, canal, railway line, tramway, telephone boxes, repeater station, street, path or passage of any kind on highway land or across, under or over any highway without prior written permission of the highway administration.

3.5 ROAD DEVELOPMENT PLAN VISION: 2021

The current road development plan was approved by the council of Indian Roads Congress at its meeting held at Kolkata on November 4th, 2000. The plan has clearly pointed out the deficiencies and needs, and made recommendations accordingly. These are discussed in the following successive sections.

3.5.1 Deficiencies Observed

1. Targets of NH and SH could not be achieved by the end of the Lucknow Plan. The targets achieved were around 85 percent.

2. Major part of the NH and SH network was expected to be double-lane by the end of the 1981-2001 plan. But by the end of the plan, only 34 percent was double-lane, 1 percent was four-lane and 65 percent was single-lane.

3. Still 40 percent of the villages were yet to be connected with all-weather roads.

4. A long leap needs to be taken in commercial vehicle kilometerage. It has to be increased from 250-300 km per day to 500-600 km per day, as in the case of developed countries.

5. At present, NH is only 2 percent of total road network and SH is 4 percent of the total road network, but both carries 40 percent of the total road traffic.

6. There is shortage of funds for maintenance by around 40 percent. The replacement cost of the existing network stands at Rs 500,000 crores.

7. There is a mismatch in road accidents and population statistics at global level. The share of fatalities on roads is 9 percent with respect to world statistics, whereas, the population share stands at only 1 percent.

8. A very small amount (0.20 percent) of automobile sales is spent on research and development needs against 3-5 percent spent in developed countries.

9. The increase in fuel consumption due to congestion is 10-15 percent per year. The loss due to poor condition of roads is estimated as Rs 25000 crores per year.

10. The network is also deficient in terms of load taking capacity. Only 10-20 percent of NH and SH network can take up legal axle load of 10.2t.

11. Overloading of vehicles is another problem and it causes premature failure of the pavement. For a pavement with 10 years of design life, the life left is 6.5 years due to 10 percent overloading and 3.5 years with 30 percent overloading.

12. The economic loss due to lack of modern NH system is estimated as Rs 10000 crore per year.

3.5.2 Needs Projected

1. The specialization gaps of National Highway Authority of India (NHAI) relating to legal, contract, administration, transport planning, land management, social aspects, etc. should be filled for better planning and implementation of projects.

2. The Institutional Development Strategy (IDS) studies should be carried out at state level and the institutional reforms so identified should be implemented.

3. The available resources should be effectively integrated through Rural Engineering Organizations (REO).

4. Training should be imparted to working personals at all levels through National Institute of Training for Highway Engineers (NITHE), Central Road Research Institute (CRRI), National Institute of Construction Management and Research (NICMAR), etc.

5. The International Competitive Bidding (ICBs) Procedures and FIDIC conditions of contract should be incorporated for highway projects.

6. There is a need to switch over from quality control to quality assurance.

7. The pre-construction activities like land acquisition, resettlement, plantation, environmental clearance, removal of utilities, etc. should be attended before the award of the contract.

8. New firms and small size consulting firms may be encouraged to participate.

9. The equipment industry needs modernization.

10. Strategies that conserve fuel need to be adopted.

11. Road safety measures should be included in the design, construction and operation phases of the project.

12. A healthy Decision Support System (DSS) and an efficient Dispute Resolution Mechanism (DRM) should be placed.

13. The policies that can improve environment should be included in the plans.

3.5.3 Recommendations of the Plan

1. A comprehensive highway act to take care of efficient land and traffic management, effective control on ribbon development and prevention of encroachment should be introduced.

2. Additional levy of Rs. 1.00 per liter of petrol and diesel should be fixed to support construction of village roads.

3. Toll should be collected through toll plazas located on improved facilities. The design should give adequate consideration to minimum spacing, integration with upgrading project and service roads in urban and rural areas for local traffic.

4. Thrust areas, which should be given more importance are:
   a. Pavement design methodologies
   b. Pavement evaluation methodologies
   c. Performance evaluation or fatigue analysis when using additives/waste construction material
   d. Highway cost allocation studies based on user charge principle for cost recovery
   e. Guidelines on transport planning for small and medium sized towns
   f. Guidelines for interchanges on intercity roads or in urban areas

5. Technology upgradation in maintenance operations is essential.

6. The Bharat Stage-I1 and Euro-IV equivalent norms for vehicle emission levels should be implemented.

7. An integrated multi-model system of transport should be developed and designs should consider multi-axle vehicles.

8. Mass transit system should be provided in metro cities.

9. Inter-modal movement should be promoted through Inland Container Depots (ICDs) and Inland Freight Stations (IFS).

10. The capacity of NH should be augmented by taking following measures:

    a. Providing minimum 2-lane carriageway with hard shoulders over whole length

    b. Upgrading at least half of the network to 4 or 6 lane by 2021

    c. Constructing Golden Quadrilateral by 2003

    d. Constructing North-South corridor (Srinagar to Kanyakumari) and East-West corridor (Silchar to Porbandar) by 2009

    e. Strengthening of wear pavements and rehabilitation of bridges, construction of bypass, railway overbridges, etc.

11. The capacity of SH network can be augmented by:

    a. Providing minimum 2-lane on whole SH network by 2021

    b. Upgardation to 4-lane system of 10000 km

12. The capacity of MDR network can be augmented by upgrading 40 percent of the network to minimum 2-lane carriageway by 2021.

13. Villages with population above 1000, between 500 and 1000, and less than or equal to 500 should be connected with all-weather roads by 2003, 2007 and 2010, respectively.

14. In case of urban networks, the following actions should be taken up to improve the traffic movements:

    a. Formulating an Integrated Land use-Transport plan

    b. Strengthening of public transport system

    c. Provision of non-motorized transport system

    d. Implementing traffic management measures

    e. Provision of transport nagars, bus terminals, etc.

    f. Construction of expressways, flyovers, cycle tracks, bypass, etc.

15. A minimum 10 percent of the outlay should be kept as special allocation for North-East region. Similarly, special attention should be given to island areas like Lakshdweep, Andaman and Nicobar islands.

16. Consideration should be given to the provision of wayside facilities like petrol pumps, toilets, drinking water, snack bars, dhabas, repair shops, rest rooms, etc. as part of the modernization.

17. IS0 9000 system should be adopted in road construction industry.

3.5.4 Salient Features

1. Expressways of length 10000 km should be constructed by 2021.

2. The expansion of NH network should be limited to 22000 km in next twenty years. The SH network should be expanded to 160000 km and MDR network to 320000 km.

3. Road density achieved at the end of the plan should be 1.00 km per sq.km of area.

4. The estimated traffic movement on these networks would be of the order of 3000 vpk for passenger movements and 800 vpk for freight movements.

5. The finance requirements for the plan period are estimated as:

   a. Rs. 100000 crores for expressways

   b. Rs. 33000 crores for NH network

   c. Rs. 16500 crores for bridges and bypasses

   d. Rs. 200000 crores for SH network

   e. Rs. 100000 crores for MDR network

   f. Maintenance cost as Rs. 5000 crores per year

   g. Rs. 140000 crores for 4 and 6-lanning

   h. Rs. 33000 crore for upgradation to 2-lane system

6. In plains, the connectivity to village should be 100 percent for population 1500 and above and 50 percent for population 1000-1500. In case of deserts, coastal areas and tribal or hill areas, this may be 500 and above, and 200-500, respectively.

7. Roads should be available within 3.0 km in plains and 5.0 km in hills.

8. Environmental clearance is needed if the cost of the project is Rs. 50 crore or more.

9. Rehabilitation plan should be prepared if the people displaced due to a project are 1000 or more.

Further information contained in the plan is given in Appendix G

3.6 NATIONAL TRANSPORT POLICY, 2005

3.6.1 Background

The support required to maintain the level of economic activity has increased over the years and has stressed the transportation system available in an urban area quite significantly. This has resulted in the following:

1. Access to jobs, education, recreation and similar activities has become increasingly time consuming causing loss to the tune of billions of man hours in traffic. The primary reason for this has been the explosive growth in the number of motor vehicles, coupled with limitations on the amount of road space that can be provided.

2. The increase in cost of travel has made access to livelihoods, particularly for the poor, far more difficult. This is largely because of cheaper non-motorised modes like cycling and walking becoming extremely risky due to sharing of the same right of way with motorized modes, and the same modes becoming impossible to use because of sprawl of cities and increase in travel distances.

3. The accident rates have increased by more than 2 times during 1981 and 2001. The number of persons killed in road accidents has also gone up by around 2.5 times and it includes a high share of cyclists, pedestrians or pavement dwellers. This again has impacted the poor more severely.

4. Rapid motor vehicle growth has also caused severe air pollution, adversely affecting the health of the people and their quality of life.

The improvements in road transportation sector need to be looked at two segments:

1. Transport services that serve the public or commercial customers directly, and

2. Transport infrastructure that is used by the transport service providers

It is further envisaged that the passenger and freight transportation by road will increase in the future due to:

1. Substantial investment in national highway network
2. Freight movement by road offering optimal solution to the cost of transport, logistics, and inventories
3. Higher demand for inland transport for moving cargo from production centers to the gateway ports and
4. Accelerated urbanization creating additional demand for transportation.

Such an endeavor calls for a conducive road transport regime geared to meet requirements of faster mobility, safety, access to social and economic services and minimizing the impact of negative externalities (e.g. pollution, accidents, etc.).

3.6.2 Need of a National Policy

A central policy is considered necessary as:

1. Several key agencies that would play an important role in urban transport planning work under the central government, with no accountability to the state government
2. Several acts and rules, which have important implications in dealing with urban transport issues, are administered by the central government
3. A need exists to guide state level action plans within an overall framework
4. A need exists to guide central financial assistance
5. A need exists to build capacity for urban transport planning as also develop it as a professional practice
6. A need exists to take up coordinated capacity building, research and information dissemination to raise the overall level of awareness and skills

3.6.3 Objectives of National Transport Policy

The endeavor of the National Road Transport Policy (NRTP) should be to promote modern, energy efficient and environment friendly road transport with following objectives:

1. Incorporating urban transportation as an important parameter at the urban planning stage rather than being a consequential requirement.
2. Encouraging integrated land use and transport planning so that travel distances are minimized and access to livelihood, education, and other social needs, especially for the marginal segments of the urban population is improved.
3. Improving access of business to markets and the various factors of production.
4. Bringing about a more equitable allocation of road space with people, rather than vehicles, as its main focus.
5. Investing in transport systems that encourage greater use of public transport and non-motorized modes instead of personal motor vehicles.
6. Establishing regulatory mechanisms that allow a level playing field for all operators of transport services.
7. Introducing intelligent transport systems for traffic management.
8. Increasing effectiveness of regulatory and enforcement mechanisms.
9. Addressing concerns on road safety and trauma response.
10. Reducing pollution levels through changes in traveling practices, better enforcement, stricter norms, technological improvements, etc.

11. Building capacity (institutional and manpower) to plan for sustainable urban transport and establishing knowledge management system that would service the needs of all urban transport professionals, such as planners, researchers, teachers, students, etc.

12. Promoting the use of cleaner technologies.

13. Raising finances, through innovative mechanisms that tap land as a resource, for investments in urban transport infrastructure.

14. Associating the private sector in activities where their strengths can be beneficially tapped.

15. Taking up pilot project that demonstrate the potential of possible best practices in sustainable urban transport.

3.6.4 Policy Statement

1. In the sphere of road infrastructure government will endeavor to augment road capacity by promoting:
   a. Corridor Management Plan for major highways to address the concerns of ribbon development, encroachments, uncontrolled access and safety.
   b. Way-side amenities, parking space etc, along highways under Public Private Partnership (PPP) scheme.
   c. PPP units at national and state level to disseminate information on PPP and render advisory services.

2. Government will strive to strengthen road maintenance activities:
   a. Through Pavement Management System and Bridge Management System for proper upkeep of road network.
   b. By ensuring adequate availability of funds for maintenance of the nationally financed road infrastructure like highways and locally owned assets as well. To this end, a fixed share of Central Road Fund (say 10%) will be earmarked for maintenance of road network. States may consider allocating 10 per cent of Motor Vehicle Tax and VAT collected on sale of vehicles to this dedicated Fund.

3. Strengthening public road transport to facilitate access to essential socio-economic services and provide affordable mobility through a package of measures which amongst others would include:
   a. Provide fiscal incentives to Public Transport and facilitate harmonization of Motor Vehicle Tax among State/UTs.
   b. Liberalize issue of stage carriage permits.
   c. Use of information technology to enhance efficiency and productivity of public bus transport in particular.
   d. Provision of assistance/funds to strengthen public transport system in States/UTs through centrally sponsored scheme.
   e. Product differentiation in bus passenger services to serve different class of passengers to wean away commuters from personalized mode and raise fare box collection.
   f. Review of Road Transportation Act, 1950 in view of the changed economic environment with greater role for market forces. Government will encourage SRTUs to explore alternate institutional models which offer greater autonomy and flexibility from the point of view of commercial operation.

4. Providing affordable mobility to people who do not have access to personalized mode and reside in rural and remote areas. To this end, government will endeavor to encourage participation of private operators on nonviable rural/remote routes through:

a. Auctioning of combination of such routes to private operator(s) so as to enable them to compensate their losses on account of operation of non viable routes

b. Offering non viable routes to bidder asking for lowest subsidy/financial support

c. Subjecting non viable routes to lower rates of taxation or permit fees and allowing alternate competing modes of passenger road transport.

5. Setting up independent Road Transport Regulator at State level with following functions:

a. Fix price band for different kinds of services in an objective and transpar>nt manner.

b. Ensure service coverage across all regions and provide mechanism for compensation for non remunerative routes.

c. Benchmark quality/service standards for road passenger services.

d. Promote competition to curb anti competitive practices.

e. Address/adjudicate operational issues like access to terminals/common infrastructure.

6. Facilitating smooth and seamless flow of freight movement by road across States/UTs so as to foster single barrier free domestic market through:

a. Use of Intelligent Transport System (ITS) for automation of the commercial and regulatory documentation that has to accompany commercial vehicles and goods and electronic payment of various charges without stopping the vehicle enroute etc.

b. Devising "Green Channel" facility to high value/export cargo with single destination.

c. Adoption of "Single Window Clearance System" for all authorized charges/clearances both at origin and at check posts.

d. Rationalization of discretionary powers vested with officials of various departments, consistent with requirements of national security, law and order and important legal requirements.

e. Creation of a web-based database of the vehicles having national permits and adopting e-payment scheme for payment of taxes of various states so as to reduce the scope of revenue leakage, delayed remittance and malpractices plaguing the national permit system.

7. Making concerted efforts to curb the menace of vehicle overloading through strict monitoring and enforcement of provisions relating to permissible weight.

8. Reducing the scope of human error in road accidents and to meet the human resource requirement and development of road transport sector through:

a. Establishing model driver training institutions across the country with requisite infrastructure support.

b. Placing a mechanism for inspection of motor driving training schools to ensure compliance with prescribed standards.

c. Making refresher training course from certified/model schools mandatory before renewal of license for drivers of transport vehicles.

d. Identifying and financing industrial training institutions for imparting motor driving training as 'Trade' so as to create a pool of competent driver training instructors.

e. Making evaluation and issuance of driving license IT based so as to reduce scope for subjectivity and extraneous considerations.

9. Making public the institutional responsibilities of stake holders and measures to promote road safety, traffic management and post-accident care through:

a. Setting up of the National Road Safety and Traffic Management Board (NRS and TMB) with regulatory, advisory, capacity building and research functions under an act of the parliament along with State Road Safety and Traffic Management Board in the states on the lines similar to NRS and TMB.

b. Use of PPP in rescue, evacuation and trauma care of accident victims for effective delivery of emergency relief services.

c. Setting up truck terminals outside the cities. This will result in quicker turnaround time, speedy loading/unloading of freight and reduction in traffic congestion.

10. Fostering a sustainable road transport which is efficient in terms of resource use (fuel, road space, etc), puts less strain on environment and is affordable. This will become possible by:

    a. Use of bio-fuels

    b. Making provision for Non-motorized Transport (NMT) infrastructure on all roads, segregating motorized and non-motorized transport etc.

    c. Enhancing the effectiveness of Inspection and Certification (I and C) process, linking registration/insurance of vehicles to I and C, and covering both safety and emission norms. This can be achieved through PPP.

    d. Creating dedicated resource pool through levy of 'Green Cess' by states on the older vehicles, which could be used to ensure compliance to environmental regulations.

11. Promoting research and development in transport sector through:

    a. Use of intelligent transport system for addressing the problems of transport sector.

    b. Use of modern technology in construction and maintenance of road infrastructure as well as for rolling stock.

    c. Closer collaboration with academia and industry to promote research and development in the sector and strengthening of the existing institutions involved in road transport research.

12. Assisting States/UTs to enhance data collection, accessibility to facilitate planning, policy formulation and monitoring by:

    a. Making vehicle registration IT based and creating a centralized registry/depository of all information on motorized vehicles. All the regional transport authorities should be computerized and networked.

    b. Putting in place mechanism to generate economic data on road transport sector on a regular basis through electronic format.

Some of the other salient features are:

1. Internalizing the features of sustainable transport systems in the formulation of an integrated master plan giving due attention to channelize the future growth of a city around a preplanned transport network rather than develop a transport system after uncontrolled sprawl has taken place.

2. All urban development and planning bodies in the state should have a in-house transport planner as well as representation from transport authorities in their management.

3. Support to the extent of 50 percent of the cost involved in developing integrated land use and transport plans will be given by Government of India to the implementation bodies.

4. The Government of India would select a few sample cities for carrying out fully supported pilot studies.

5. The focus of the principles of road space allocation should be moving people and not moving vehicles. Accordingly, the road space should be allocated on more equitable basis by reserving lanes and corridors exclusively for public transport and non-motorized modes of travel.

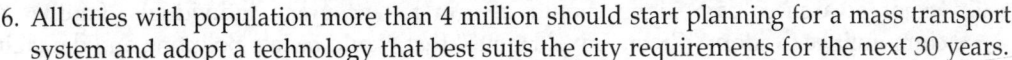

6. All cities with population more than 4 million should start planning for a mass transport system and adopt a technology that best suits the city requirements for the next 30 years.

7. A serious relation should be developed between user cost, operating cost and the services provided by the operators or service providers. Higher cost of public transportation affects the poor, who are the main users of these transport facilities, and lower fares result in compromise in services or loss in operating cost.

8. A shift to public transport system from private vehicle users need to be stimulated through provision of better services at a specified premium that can subsidize the transport for the poor.

9. A knowledge centre would be established to provide necessary information required for taking the right technological decisions for a specific city when planning for a public transport system for that city.

10. The government would finance or support the mass transit systems in the form of equity participation and/or viability gap funding for the provision of infrastructure provided the users (direct and indirect beneficiaries) pay for the operating costs and rolling stock.

11. Non-motorized traffic should be segregated from motorized traffic to improve traffic flow as well as the safety of users of non-motorized modes. Facilities like parking places, rest areas, drinking water, etc. should be provided along the NMT corridors. Users of the facility should be contacted before providing any such facility.

12. Provision and construction of cycle tracks and pedestrian paths should be given priority under the National Urban Renewal Mission (NURM).

13. The area plans for congested areas should be formulated and implemented with an appropriate mix of modes of transport including exclusive zones for non-motorized modes.

14. The use of personal motor vehicle should be discouraged. This may be done through market mechanisms such as higher fuel taxes, higher parking fees, reduced availability of earning spaces, longer time taken in travel vis-a-vis public transport, etc. or by implementing congestion pricing schemes.

15. Parking on right of way should be prohibited through appropriate legislation. Less land space should be used as a parking space in urban areas. Multi-level parking complexes should be a mandatory requirement in city centers. Graded scale of parking fee should be levied to recover the economic cost of land used in parking.

16. The freight and passenger movements should be separated from each other to improve passenger or users' safety and smooth movement of freight.

17. A Unified Metropolitan Transport Authority (UMTA) should be set up in all million plus cities, to facilitate more coordinated planning and implementation of urban transport programmes and projects and an integrated management of urban transport systems.

18. Capacity building should be addressed at two levels – institutional and individual. Institutional capacity means creating a pool of knowledge and a knowledge management center. This responsibility can be discharged by The Institute of Urban Transport (IUT), Ministry of Urban Development. As for individual capacity, the training and skill development of the public officers and other public functionaries should be taken up. Academic programrnes in urban transport would also be strengthened.

19. The research and development in the area of cleaner technologies need to be encouraged.

20. The commercial utilization of land resources is recommended to raise additional resources.

21. Public private partnership should be encouraged in identified activities related to construction, operation and maintenance of facilities through well structured procurement contracts.

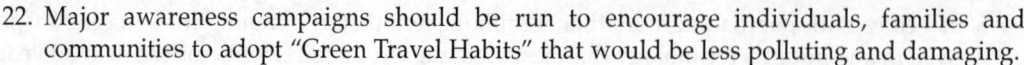

22. Major awareness campaigns should be run to encourage individuals, families and communities to adopt "Green Travel Habits" that would be less polluting and damaging.

3.7 RURAL ROAD DEVELOPMENT PLAN VISION: 2025

3.7.1 Historical Background

In the seventies, the Government of India launched the Minimum Needs Programme (MNP) with the objective of bringing the rural population into the mainstream of national development. This programme covered various activities including rural roads and rural housing. The MNP was later merged with the Basic Minimum Services (BMS) programme formulated in 1996, which had similar Ministry of Rural Development objectives. Various employment generation programmes were started from time to time, like National Rural Employment Programme (NREP), Rural Landless Employment Generation Programme (RLEGP), the Integrated Rural Development Programme (IRDP), the Jarvahar Rozgar Yojana (JRY), the Jawahar Gram Samrudhi Yojana (JGSY) and Swarnajayanti Gram Swarozgar Yojana (SGSY), all of which aimed at the creation of rural economic infrastructure with employment generation as a major objective. The JGSY, the Employment Assurance Programme (EAS), and the Food for Work Programme were revamped and merged under the new Sampoorna Gramin Rozgar Yojana (SGRY) in 2001. Rural infrastructure creation, including rural roads, received encouragement under all the above programmes. Funds from the Rural Infrastructure Development Fund (RIDF) were channelised into several small rural infrastructure projects through the help of the NABARD. Recently, Government of India, with the twin objective of employment assurance and poverty reduction in rural areas, has launched another scheme of National Rural Employment Guarantee Programme (NREGP).

The village connectivity achieved under MNP and BMS programme upto the year 2000 are given in Tables 3.5 and 3.6.

Table 3.5: MNP norms for connectivity of village with rural roads (census data of 1981)

Type of Area	Population of village for 100 per cent connectivity	Population of village for 50 per cent connectivity
Plains	1500 and above	1000 – 1500
Desert, Coastal, Tribal	1000 and above	500 – 1000
Hills	500 and above	200 – 500

Table 3.6: Connectivity of villages with roads achieved upto year 2000

Population category	Total No. of villages	No. of villages connected by 1980	No. of villages connected by 1985	No. of villages connected by 1990	No. of villages connected by 1995	No. of villages connected by 2000
1500 and above	71623	37950 (53%)	49495 (69%)	59722 (83%)	65704 (92%)	70000 (98%)
1000-1500	58229	21970 (38%)	28732 (49%)	35362 (61%)	44120 (76%)	50000 (86%)
Less than 1000	459465	107324 (23%)	142020 (31%)	166311 (36%)	173837 (38%)	200000 (43%)
1000 Total	589317	167244 (28%)	220247 (37%)	261395 (44%)	2836661 (48%)	320000 (54%)

Note: Figure within brackets give the percentage of villages in each population category to the total number of villages in that category. The basis for population is 1981 census.

Source: Planning Commission and MORTH Road Development Plan vision 2012 (published by IRC in 2001).

3.7.2 National Committee on Rural Infrastructure

Continuing the momentum gathered, the Government set up a National Committee on Rural Infrastructure in 2004, under the chairmanship of the Prime Minister, with Ministers of Agriculture, Shipping and Road Transport, Power, Rural Development, Water Resources, Panchayati Raj, Communications and IT, Non-Conventional Energy Sources and Deputy Chairman and four Members of the Planning Commission as Members, with the objective of improving rural infrastructure in a time-bound manner through initiating policies, effecting internal prioritization and developing innovative financing arrangements.

3.7.3 Connectivity Vision

The proposed plan looks beyond PMGSY and gives due consideration to the requirement of connectivity to habitations with lower population thresholds not presently covered under the PMGSY. Existing roads and tracks are also to be brought to a fully serviceable condition. The following vision for new connectivity is recommended under the programme:

- Habitations with population above 1000 (500 in case of hill, NE states, deserts and tribal areas) by year 2009-10
- Habitations with population above 500 (250 in case of hill, NE states, deserts and tribal areas) by year 2014-15
- Habitation with population above 250 by year 2021-22

Rural roads serve the accessibility function. They feed traffic into and receive traffic from the secondary system, (State Highways and Major District Roads: SH and MDR), which in turn is supported by and supports the primary system (National Highways: NH). The secondary system contributes both to the rural economy and to the industrial development. They combine the mobility and access function. The road transport system can function efficiently only if all the three groups of roads are developed harmoniously and are integrated into one another. No single class of road can function efficiently if its linkages with the other are deficient.

Further, the rural roads should be constructed not only to improve accessibility but also to act as a facilitator to:

- Promote and sustain agricultural growth
- Improve basic health and hygiene
- Provide access to schools and other educational opportunities
- Provide access to economic opportunities
- Create employment opportunities
- Enhance democratic processes and bring people into national mainstream
- Enhance local skills
- Reduce vulnerability and poverty
- Act as infrastructure multiplier

Planning should take into account the market centers or Rural Business Hubs (RBH), which are centers of activities for marketing of agricultural produce and inputs, servicing of agricultural implements, health, higher education, postal, banking services, etc. They are generally growth centers located on higher category of roads or at the confluence of roads emanating from a number of habitations. Maximum distance between a village and RBH would normally be 15-20 km. Planning should take into consideration that an RBH may not be fully developed at present but a big village may have such potential in future. Similarly, many through routes have the potential to become Major District Roads (MDR) over the next 20 years. Route alignment, design and construction standards must keep this point in view in order to reduce the need for avoidable investments when such roads become MDR on the basis of traffic volumes etc.

The annual investments required for construction, upgrading and maintenance during the period 2007-2025 would be as shown in Table 3.7.

Table 3.7: Phasing of investment for development and maintenance of rural roads

Average Annual Investment (Rs. Million)

Period	Construction + upgradation	Maintenance	Total
2007 – 12	75000	40500	115500
2012 – 17	120000	54000	174000
2017 – 22	152000	84500	236500
2022 – 25	182000	106000	288000

As per phasing of new construction and upgrading of rural roads, the yearly targets as envisaged during various plan periods for construction, upgradation and maintenance are given in Tables 3.8 and 3.9.

Table 3.8: Target of annual length for construction and upgradation

Plan Period	Construction (km)	Upgradation (km)	Total (km)
2007 – 12	26000 (1600 gravel)	11200	37200 (1600 gravel)
2012 – 17	40000 (4300 gravel)	23000	63000 (4300 gravel)
2017 – 22	44000 (10000 gravel)	43000	87000 (10000 gravel)
2022 - 25		73000	73000

Table 3.9: Target of annual length for periodic maintenance

Plan period	Length (km)
2007 – 12	50,000
2012 – 17	60,000
2017 – 22	110,000
2022 - 25	140,000

3.7.4 Salient Features

1. Rural roads are low volume facilities basically serving as an access function. The design speed and level of service expected are low. The design standards should be in harmony with such expectations.

2. Geometric standards, particularly gradients, are difficult to change later, and hence should be selected carefully with the future requirements in view.

3. A design period of 10 years is considered adequate, with rehabilitation being planned based on road condition.

4. A roadway width of 7.5m and carriageway width of 3.75m is suggested in general for rural roads. In case the traffic volume is less than 100 motorised vehicles per day, this may be reduced to 6.0m and 3.0m respectively.

5. Subject to consideration of rainfall and traffic, blacktopping may be restricted to through routes and to link routes that link bigger villages say with population above 1000. For

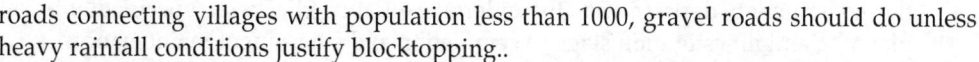

roads connecting villages with population less than 1000, gravel roads should do unless heavy rainfall conditions justify blocktopping..

6. The initial cost is an important consideration. Many roads particularly through routes will in due course carry fairly substantial traffic but it is preferable to optimize costs by stage construction in tune with traffic growth. Such stages may be gravel road and bituminous road in plain terrain, and bridle path (2m wide), jeepable road (3m wide) and motorable road (5m wide formation) in hilly areas.

7. Drainage and erosion control measures and protection works against landslides and rock fall etc should form part of the initial design.

8. In the case of rural roads, it is permissible to put up with minor interruptions to traffic. The road can be submerged upto 6 times a year, the period not exceeding 24 hours at a time.

9. The rural road network should be evolved keeping in view the bridging needs of the people. Construction of bridges shall be deemed to be an integral part of rural roads network in order to ensure all weather connectivity. At intermediate locations, suitable suspension foot-bridges may be provided for use by pedestrians and pet animals.

10. Standard specifications, designs and quantities should be prepared for different aspects of road construction and especially for bridges.

11. Use of local materials and skills and labour-intensive technology should be preferred. It should be ensured that the construction technology is 'Intermediate' and not 'Equipment-Intensive'. The construction approaches and equipment should be made tractor oriented as this is the one vehicle available in villages and can be used throughout the year in every terrain. The technique should result in lower cost and higher employment opportunities.

12. Heating of bitumen and aggregates, producing mix, transporting mix and using it at an appropriate and correct temperature is a big problem in rural areas. It is therefore, suggested to use cold-mix technology using emulsions in surface dressing and carpeting.

13. Land compensation policies should be formulated by state governments that avoid competing demands and encourage land donations. Community should be involved both for acquiring land and its use for road development.

14. GIS-based rural planning and development road connectivity and network planning should be encouraged.

15. Innovative ways of funding rural roads projects need to be adopted apart from financial assistance received from government in the form of allocations of funds or Rural Infrastructure Development Funds (RIDF) managed by National Bank for Agriculture and Rural Development (NABARD). At present, only around 40 percent of the road taxes collected is spent on roads. Such innovative measures may include levying marketing fee and cess on agriculture produce, stamp duty on land transactions, vehicle fee, etc.

16. Mobile maintenance units should be incorporated for maintenance of rural roads. Though costly, they can provide maintenance of around 200 km road length. Therefore, these can be adopted depending upon the density of rural roads.

17. A healthy equipment leasing mechanism should be developed to ease out the financial burden on contractors. The equipment industry also needs to focus on production of low-end technology machines, which can be used cost-effectively by the local contractors in construction and maintenance of rural roads.

18. A quality assurance system should be developed covering all the aspects of rural roads. Over a period of time, the sector should move towards Total Quality Management (TQM) as practiced universally. A well-trained quality appreciation unit should be set up within

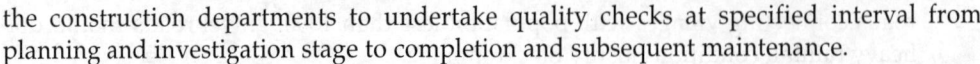

the construction departments to undertake quality checks at specified interval from planning and investigation stage to completion and subsequent maintenance.

19. With the rapid expansion of rural road network and general prosperity that will take place in the rural areas in the coming years, enforcement of motor vehicle rules should receive attention. Panchayati Institutions can play effective role in supplementing the police efforts in increasing the awareness for observance of traffic rules. The head of implementation units of rural roads may be included in the district level safety committees. Similarly, State Level Quality Coordinator of the SRRDA may be nominated to the State Level Safety Council.

20. The environmental aspects need consideration at the time of planning, design, construction and maintenance of rural roads. Their likely impact and the mitigation measures proposed should be spelt out in the project report. Vriksha Rojgar Yojna should be extended to plant trees along the rural roads. An All India Code of Practice for Environmental Protection should be developed in line with 'Environmental Code of Practice' as adopted for World Bank assisted rural roads projects.

21. There are no regulatory provisions in the country to preserve and manage the rural road assets. There is a need to introduce 'Rural Road Management Act', the provision of dedicated funds for maintenance, higher role of Panchayati Raj Institutions in road management, constructing maintainable roads, etc. on the similar lines as 'Road Maintenance Initiative' (RMI) formulated by World Bank and adopted in several countries of-African continent.

22. A change in institutional setup is needed. It is necessary to create specialized Rural Road Engineering Unit (Divisions) so as to ensure the necessary technical expertise and management quality. Such rural roads divisions would be under a separate Rural Roads Wing of the PWD/Rural Development Department and should be headed by a full time chief engineer.

The plan gives adequate importance to social and economic development of villages through the provision of rural road connectivity. These are discussed in the following successive paragraphs.

1. Rural roads should be used as an entry point for poverty alleviation since lack of access is accepted universally as a fundamental factor in continuation of poverty. A study (Fan, Hazell and Thorat, 1999) carried out by the International Food Policy Research Institute on linkages between government expenditure and poverty in rural India has revealed that an investment of Rs. 10 million in roads lifts 1650 poor persons above the poverty line. The relationship between the two is shown in the Fig. 3.8.

2. The provision of good rural roads changes the characteristics of rural transport. People tend to travel more, the ownership of vehicle increases, and the cost of travel and transport comes down.

3. Land donation efforts have several important social and economic impacts, which may need to be addressed, generally in a more planned way than if compensation is paid through an acquisition procedure. Impacts can include the following:
   • Loss of frontage for houses/shops on existing alignment
   • Loss of agricultural land
   • Loss of income
   • Landlessness in rare cases
   • Fragmentation of holding (in rare cases, where improvement in geometrics require realignment)

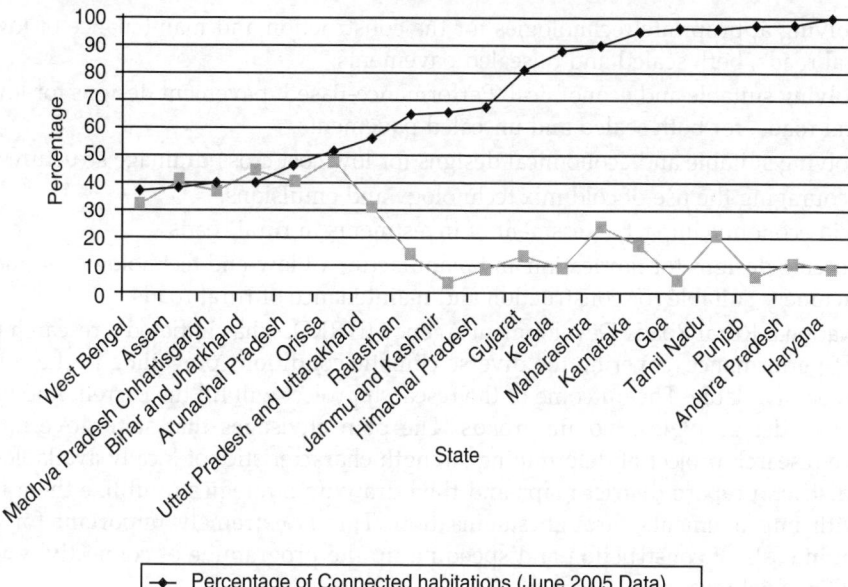

Fig. 3.8: Relationship between connectivity and rural poverty

4. An expenditure of Rs. 1 crore in rural roads is likely to create 40,000 man-days of employment (taking an average wage rate of Rs. 75 per day). The employment in terms of mandays likely to be generated by the development of rural roads is given in Table 3.10.

Table 3.10: Employment potential of rural roads

Period	*Annual Investment*	*Employment Potential (Number of Mandays)*
2007 – 12	Rs. 11550 crore	460 million
2012 – 17	Rs. 17400 crore	700 million
2017 – 22	Rs. 23650 crore	950 million
2022 - 25	Rs. 28800 crore	1150 million

3.7.5 Emerging Areas for Research and Development

Some of the more important thrust areas for medium term R&D work and long-term R&D work requiring immediate attention are:

- Critical appraisal of design and construction practices being adopted for low volume roads around the world.
- Review of the existing geometric design standards pertaining to low volume rural roads.
- Evolving low cost drainage and erosion control measures for low volume roads and preparing a comprehensive manual
- Identification of sources of locally available materials for road construction at district level and determining the strength and other characteristics of such materials.
- Developing stabilisation techniques for improving performance of locally available softer materials.

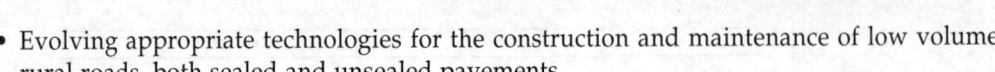

- Evolving appropriate technologies for the construction and maintenance of low volume rural roads, both sealed and unsealed pavements.
- Evolving suitable and economical 'Performance-Based' pavement designs for low volume rural roads, for both sealed and unsealed pavements.
- Evolving suitable and economical designs for low cost cross-drainage structures
- Encouraging the use of cold mix technology and emulsions.
- Socio-economic impact assessment of investments in rural roads
- Evolving designs for fabrication and manufacture of low-end technology and inexpensive machinery suitable for construction and maintenance of rural roads.

The National Rural Roads Development Agency (NRRDA) has initiated a research project on pavement performance, covering the diverse climatic conditions prevailing in the country and the soil types available. The outcome of the research project will fill the current knowledge gap in pavement design of low volume roads. The plan envisages that state governments will undertake research project of determining strength characteristics of locally available materials in each district, prepare district maps and then draw up strategies to utilize them in various layers with improvements through stabilisation. This is extremely important for achieving reduction in costs of construction and speeding up the programme of connectivity within the available financial resources.

The rural road planning procedures need refinement. The approaches used should be supportive to the on-going programme rather than obstructive in nature. One such approach is 'Integrated Rural Accessibility Planning' (IRAP) developed by International Labour Organisation (ILO), which looks directly on rural road networks and the distribution of socio-economic facilities and services. The approach is implemented in two states of India, Rajasthan and Orissa. The approach is given in Table 3.11.

Table 3.11: Integrated rural accessibility planning – an ILO tool

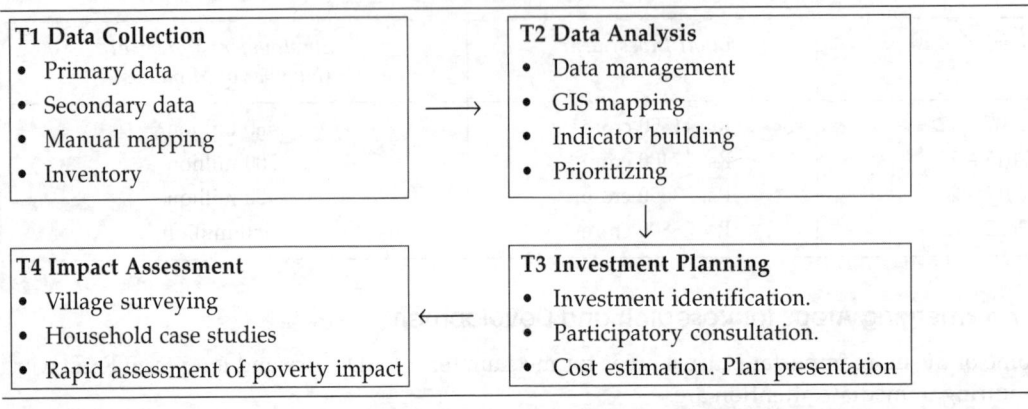

T1 Data Collection	T2 Data Analysis
• Primary data	• Data management
• Secondary data	• GIS mapping
• Manual mapping	• Indicator building
• Inventory	• Prioritizing

T4 Impact Assessment	T3 Investment Planning
• Village surveying	• Investment identification.
• Household case studies	• Participatory consultation.
• Rapid assessment of poverty impact	• Cost estimation. Plan presentation

Any developmental programme requires the assessment of its success through measurements in terms of objectively assessed indicators, on physical achievements, performance levels and the impact of the programme on the society. The indicators and strategy to measure and assess performance of rural roads are listed in Table 3.12.

3.8 HIGHWAY ADMINISTRATION AND ORGANIZATIONS

Number of organization are working in the area of road development in the country. They are concentrating on planning, construction, maintenance, operation, specifications and guideline, training and manpower development, vehicle testing, evaluation and guidelines, transport management and so on. Some of these are discussed in the following successive sub-sections.

Table 3.12: Transport indicators for rural transport sector

Sl. No.	Indicator	Content	Collaborators	Available to be initiated	Strategy for Data Collection Storage and Analysis
1.	Network	Physical length, length by type, length by ownership, surfaced, unsurfaced, condition, etc.	PWDs, REOs Panchayati Raj Departments and others.	Mostly available except condition which needs to be assessed periodically.	All concerned road agencies can be asked to collect the data, store and consolidate. The departments may be required to carryout the inventory once in 2 years and update the data base with changes in the status. Can be done immediately. The information is readily available but only requires reconciliation and updation.
2.	Traffic	Growth, frequency, private, public, motorized vehicles, light commercial vehicles, heavy goods vehicles and non-motorized vehicles	Transport department, public sector transport operators	Not available except in aggregate terms.	The breakup of traffic plying on rural roads is not explicitly available. Such data is to be deduced based on traffic census on rural roads. Special care is to be taken in respect of non-motorized vehicles since, as of now, their registration is not compulsory. Methodology to be developed and needs orientation.
3.	Passenger travel and freight movement patterns	Frequency, purpose, cost, time, passenger km by mode, commodity tonne/km by type	Public transport operators, public works department, freight operators	Not available	To be assessed by organizing origin-destination surveys. Requires design of formats, training of enumerators, working out logistics, etc. This can be done on sample basis once in five years as it requires elaborate arrangements for data collection and analysis.
4.	Public transport	Number of seat km provided, seat km used, fare structure, trip frequency	Public transport undertaking	Available but requires filtering out for rural transport	State Transport Undertaking (STUs) are to filter out the data for rural operations from their records. Can be readily obtained where STUs are operating. Otherwise requires special surveys.

(Contd.)

Table 3.12: Transport indicators for rural transport sector (*Contd.*)

SI. No.	Indicator	Content	Collaborators	Available to be initiated	Strategy for Data Collection Storage and Analysis
5.	Accidents	Vehicle type, persons involved, fatality, grievous injury, minor injuries, property damage.	Traffic police	Mostly available but needs streamlining	At present data recording is not scientific, Requires better organized way of data collection and analysis, Numbers can be obtained with immediate effect since compilation is done regularly. However, detailed instruction for database management are required.
6.	Revenue indicators	Registration and other taxes	Transport excise departments	Available	Needs filtering out data
7.	Financial indicators	Capital investment for new construction, upgradation, renewal, routine maintenance.	PWDs, REOs, Panchayati Raj department and others.	Partly available	Each department can easily work out the details yearly. This facilitates planning and operationalization of the work programmes for each financial year.
8.	Coverage indicators	Habitation population connected, access to schools, dispensaries and other services	District administration and Panchayati Raj Institutions	Partly available	This data can be readily compiled by PIUs of PMGSY programme in association with other organization like Panchayats.
9.	Outcome indicators	Growth of income, employment, increased enrolment in schools in rural areas, reduction in child mortality, reduction in maternal mortality (in tune with MDGs) including productions and productivity	District administration	Partly available, but requires restructuring	Requires design of formats for data collection, storage and analysis. Many of the parameters can be at panchayat level to be consolidated at Block/District/ State and National level. Indicators like income, employment can be at district/state level. Steps can be initiated by the SRRDAs in each state.

3.8.1 Ministry of Road Transport and Highways, MORTH

An apex organisation under the Central Government, is entrusted with the task of formulating and administering, in consultation with other central ministries/departments, state governments/ UT Administrations, organisations and individuals, policies for road transport, national highways and transport research with a view of increasing the mobility and efficiency of the road transport system in the country.

The department has two wings: **Roads wing** and **Transport wing.**

Roads wing deals with development and maintenance of national highway in the country. The main responsibilities are:

- Planning, development and maintenance of national highways in the country.
- Extends technical and financial support to state governments for the development of state roads and the roads of inter-state connectivity and economic importance.
- Evolves standard specifications for roads and bridges in the country.
- Serves as a repository of technical knowledge on roads and bridges.

Transport wing deals with matters of road transport. The wing is responsible for:

- Motor vehicle legislation,
- Administration of the Motor Vehicles Act, 1988
- Taxation of motor vehicles,
- Compulsory insurance of motor vehicles,
- Administration of the Road Transport Corporations Act, 1950,
- Promotion of transport co-operatives in the field of motor transport.
- Evolves road safety standards in the form of a national policy on road safety and by preparing and implementing the annual road safety plan.
- Collects, compiles and analyses road accident statistics and takes steps for developing a road safety culture in the country by involving the members of public and organising various awareness campaigns.
- Provides grants-in-aid to non-governmental organisations in accordance with the laid down guidelines.

Administrative Setup

- Minister and Minister of State
- Secretary (RT and H)
- Additional secretary and Financial Advisor
- Chief Controller of Accounts
- Director General (Road Development) and Special Secretary
- Additional Director General (Road Development)
- Chief Engineers (States, UTs, North-East States, Project Implementation Cell, Planning, Standard and Research [Roads], Standard and Research [Bridges], Bridge design, Monitoring (Mechanical)
- Joint Secretary (Transport and Administration) and Chief Vigilance Officer
- Advisor (Transport Research)
- Director (Road Transport and IT)
- Director (Administration)
- Director (Finance)

- Director (Coordination and Parliament)
- Deputy Secretary (P and B)
- Deputy Secretary (Vigilance, Gen., Cash, Canteen and Lib.)
- Deputy Secretary (Administration and IC)
- Sr. Technical Director (NIC-RT and H Division)

3.8.2 Central Public Works Department, 1911

Central Public Works Department was set up in December 1911 on the proclamation of change of the capital from Calcutta to Delhi. The charge of execution of the work was entrusted to imperial Delhi committee, which had chief commissioner of Delhi as president and Chief Engineer as Enginner-Member. The works of the Capital Project were in the charge of the Chief Engineer, a Superintending engineer (Civil), a Superintending Engineer (Electrical and Mechanical) and one Executive Engineer. Sardar Bahadur, Shri Teja was the first Indian Chief Engineer of the CPWD. CPWD came in existence on 1st April, 1930 to look after the vast office and residential campus of the Central Secretariat and allied offices. At that time, the department had a cadre of only two permanent circle (civil), i.e. circle I and II and six divisions for the works at Delhi, Simla, Dehra Dun, Ajmer and Indore. With the increase in the workload, the number of circles and divisions increased from time to time. Gradually by the end of year 1940, the CPWD was entrusted with all the centrally financed civil works.

3.8.3 National Rural Roads Development Agency, 2000 (NRRDA)

The Ministry of Rural Development has set up the National Rural Roads Development Agency (NRRDA) to provide operational and management support to the PMGSY Programme. The NRRDA provides support, *inter alia*, on the following:

   i. Designs and specifications and cost norms.
  ii. Technical agencies
 iii. District rural roads plans and core network.
  iv. Scrutiny of project proposals
   v. Quality monitoring
  vi. Monitoring of progress, including online monitoring
 vii. Research and development
viii. Human resource development
  ix. Communication

3.8.4 National Highway Authority of India (NHAI), 1988

The National Highways Authority of India was constituted by an act of Parliament, the National Highways Authority of India Act, 1988. It is responsible for the development, maintenance and management of national highways entrusted to it. It consists of a chairman, five full-time members and four part-time members who are appointed by the central government.

The functions of the authority are:

1. Survey, develop, maintain and manage highways vested in, or entrusted to it;
2. Construct offices or workshops and establish and maintain hotels, motels, restaurants, and rest-rooms at or near the highways vested in, or entrusted to, it;
3. Construct residential buildings and townships for its employees;
4. Regulate and control the plying of vehicles on the highways vested in, or entrusted to it for the proper management thereof;

5. Develop and provide consultancy and construction services in India and abroad and carry on research activities in relation to the development, maintenance and management of highways or any facilities thereat;

6. Provide such facilities and amenities for the users of the highways vested in, or entrusted to, it as are, in the opinion of the Authority, necessary for the smooth flow of traffic on such highways;

7. Form one or more companies under the Companies Act, 1956 (1 of 1956) to further the efficient discharge of the functions imposed on it by this act;

8. Engage, or entrust any of its functions to, any person on such terms and conditions as may be prescribed;

9. Advise the central government on matters relating to highways;

10. Assist, on such terms and conditions as may be mutually agreed upon, any state government in the formulation and implementation of schemes for highway development;

11. Collect fees on behalf of the central government for services or benefits rendered under section 7 of the National Highways Act, 1956 (48 of 1956), as amended from time to time, and such other fees on behalf of the state governments on such terms and conditions as may be specified by such state governments; and

12. Take all such steps as may be necessary or convenient for, or may be incidental to, the exercise of any power or the discharge of any function conferred or imposed on it by this act.

The organizational structure or NHAI is given in Figure 3.9.

3.8.5 Central Road Research Institute (CRRI), 1948

The Institute, established in 1948, is a national research laboratory under the Council of Scientific and Industrial Research New Delhi India. The institute is an ISO 9001 certified organisation for providing services in road and transport research, highway engineering, pavement design and maintenance, traffic and transport planning, geotechnical and bridge engineering are the major areas. Research level and technical services in roads, traffic, environmental and road safety aspects, airfield pavement, landslide mitigation are executed professionally by the Institute for both public and private sector customers in India and overseas. The Institute has large number (over 200) of highly qualified and experienced scientific and technical staff uniformly distributed in different areas. The laboratory testing facilities and field investigation equipments are the latest and state of the art. Computers are fully networked and IT facilities are efficiently managed. Institute conducts training and refresher courses on different contemporary topics. The library and documentation facilities in the institute are excellent.

The various divisions of CRRI are:

- Bridges and Structures
- Geotechnical Engineering
- Human Resource Planning
- Pavement Engineering and Materials
- Road Development Planning and Management
- Research and Development Support Services
- Traffic and Transportation Planning

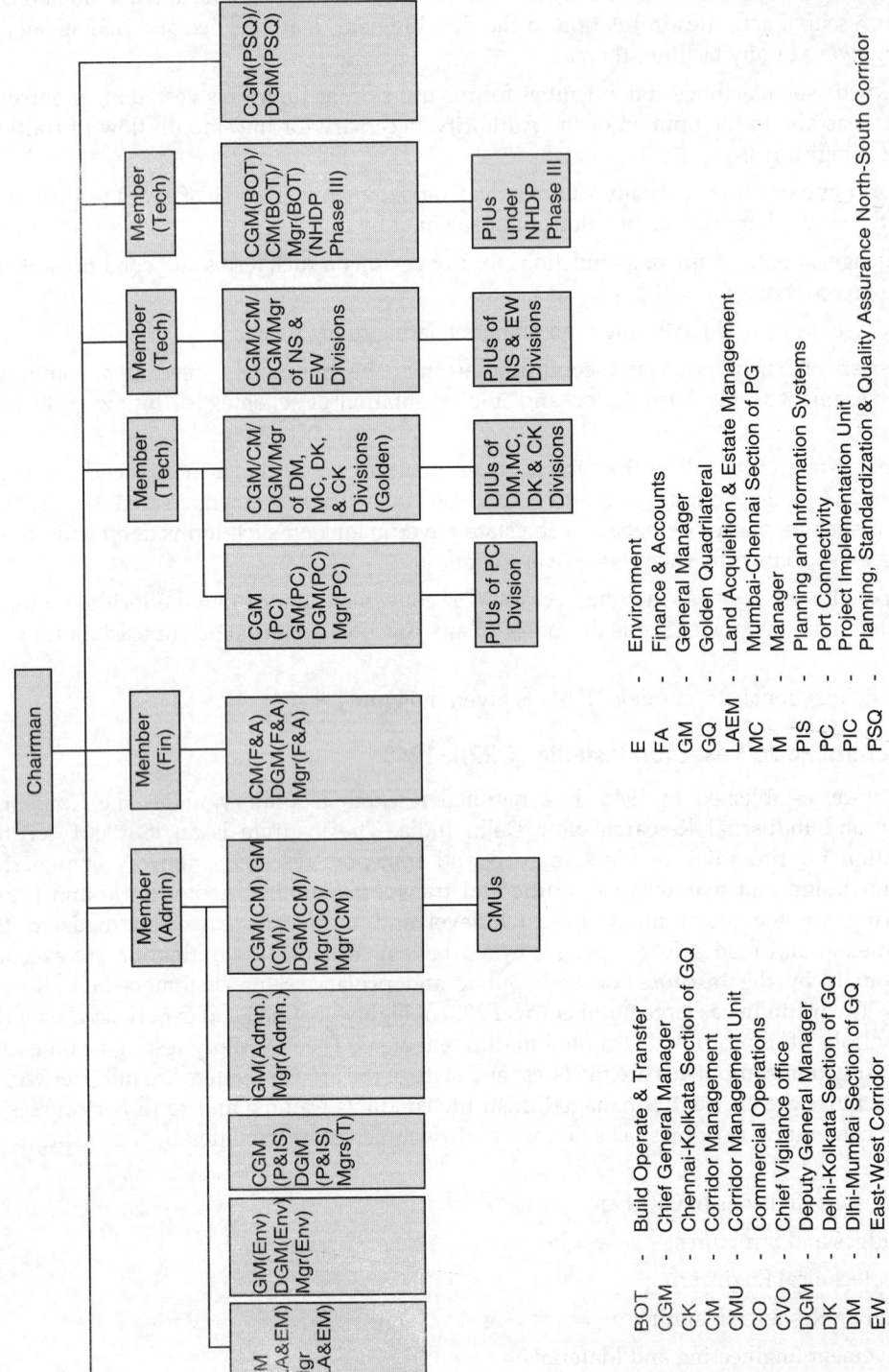

BOT - Build Operate & Transfer
CGM - Chief General Manager
CK - Chennai-Kolkata Section of GQ
CM - Corridor Management
CMU - Corridor Management Unit
CO - Commercial Operations
CVO - Chief Vigilance office
DGM - Deputy General Manager
DK - Delhi-Kolkata Section of GQ
DM - Dlhi-Mumbai Section of GQ
EW - East-West Corridor

E - Environment
FA - Finance & Accounts
GM - General Manager
GQ - Golden Quadrilateral
LAEM - Land Acquieltion & Estate Management
MC - Mumbai-Chennai Section of PG
M - Manager
PIS - Planning and Information Systems
PC - Port Connectivity
PIC - Project Implementation Unit
PSQ - Planning, Standardization & Quality Assurance North-South Corridor

Fig. 3.9: Organizational structure of the national highway authority of India

3.8.6 Automobile Research Association of Indian (ARAI), 1966

The Automobile Research Association of India (ARAI) has been playing a crucial role in assuring safe, less polluting and more efficient vehicles since its establishment in 1966 at Pune. ARAI provides technical expertise in R&D, testing, certification, homologation and framing of vehicle regulations.

ARAI is a co-operative industrial research association established by the automotive industry with the Ministry of Industries, Government of India. It works in harmony and complete confidence with its members, customers and the Government of India to offer the finest services, which earned for itself IS0 9001, IS0 14001, OHSAS 18001 and NABL accreditations.

ARAI is well-equipped with state-of-the-art infra-structural facilities and highly qualified manpower. ARAI offers comprehensive R&D services in the fields of engine development, alternate fuels, NVH-noise, vibration and harshness, computer-aided engineering, structural dynamics, automotive electronics and materials. ARAI offers expert services in testing, certification and homologation of complete vehicles, engines, systems and components. It covers the area of vehicle evaluation, emission, safety, materials, EMI/EMC etc.

3.8.7 Vehicle Research and Development Establishment (VRDE), 1965

The history of Vehicles Research and Development Establishment dates back to 1929, with the inception of Chief Inspectorate of Mechanical Transport (CIMT) at Chaklala (now in Pakistan). In 1947, the establishment was shifted to Ahmednagar and was renamed as Technical Development Establishment (Vehicles), popularly known as TDE [V]. In 1962, the engineering wing was separated, with the establishment of an independent establishment at Pune, named as R&DE (Engrs). Further in 1965, the activities were bifurcated between R&D and Inspection and two separate establishments, viz. Vehicles Research and Development Establishment (VRDE) and Controllerate of Inspection Vehicles (CIV), now known as Controllerate of Quality Assurance Vehicles (CQAV), came into existence.

A detachment of VRDE was created at Avadi, Madras in 1966, to assist in a production of tanks at Heavy Vehicle Factory, Avadi which eventually led to the formation of CVRDE and further bifurcation of roles. Starting with design modifications and technical evaluation of vehicles, VRDE has graduated over the years into an organization capable of undertaking innovative development of total vehicles incorporating latest technologies. The National Centre for Automotive Testing (NCAT), a separate division of VRDE, provides one stop solution to all vehicles testing and evaluation requirements of defence services as well as automotive industry.

Following are enlisted in its charter:

- Research, design, development and technical trials and evaluation of all types of light tracked vehicles upto 25 ton class for combat and specialist roles.
- Research, design, development and technical trials and evaluation of all types of wheeled vehicles including transporters, trailers and various types of specialist role vehicles.
- To act as a nodal point for interaction and transfer of technology between various defence organizations and public/private industries on matters related to light tracked and wheeled vehicles.
- To carryout roadworthiness, fuel efficiency and pollution tests on vehicles and type testing of automotive engines on behalf of government, semi-government and other agencies and issue certificates.

3.8.8 Central Institute of Road Transport, 1967

The Central Institute of Road Transport (CIRT) was established in 1967 on the joint initiative of the Association of State Road Transport Undertakings (ASRTU) and the then Ministry of

Shipping and Transport, Government of India. CIRT is committed to improve the efficiency and productivity of the transport sector, with particular emphasis on State Transport Undertakings.

CIRT offers management development programmes covering general management, transport operations and maintenance engineering. The programmes are meant for practising managers in STUs, other organisations operating transport services besides road transport officials. In addition, the Institute undertakes consultancy and research assignments on transport policy, transportation planning, traffic management, maintenance management, materials management, human resource management and management information systems. CIRT has a sophisticated automobile component testing laboratory, recognised by the Bureau of Indian Standards for testing a wide range of automobile components. The institute publications provide a window on the transport industry and are avidly sought by the Planning Commission, World Bank and several other leading national and international organisations. CIRT publishes Indian Journal of Transport Management quarterly.

The objectives of CIRT are:

- To promote and provide for the study of the principles and practices of organisation and management in various spheres of public transport
- To conduct research on critical issues in transport development and to undertake consultancy assignments to improve organizational effectiveness and operational efficiency
- To impart training and education to all categories of managers and professionals engaged in furthering the cause of public transport
- To help the government and its agencies in evolving integrated policies and legislative framework to enable a balanced growth of the transport sector
- To be the clearing house of ideas, information and data pertaining to transport industry with specific reference to safety, environment and productivity
- To lay down standards, specifications and norms for various materials used in heavy vehicles through performance' evaluation, material analysis and quality monitoring
- To undertake research for improvement and technological upgradation of vehicles used for passenger transportation together with their components and assemblies; and
- To engage in any other activity as may be conducive or incidental to the attainment of the above objectives

3.8.9 Indian Roads Congress, 1934

The Indian Roads Congress (IRC) is a premier technical body of highway engineers in the country. The IRC was set up in December, 1934 on the recommendations of the Indian Road Development Committee best known as Jayakar Committee. As the activities of the IRC expanded, it was formally registered as a Society in 1937 under the Societies Registration Act of 1860. The Congress provides a national forum for sharing of knowledge and pooling of experience on the entire range of subjects dealing with the construction and maintenance of roads and bridges, including technology, equipment, research, planning, finance, taxation, organisation and all connected policy issues. In more specific terms the objectives of the congress are:

- To promote and encourage the science and practice of building and maintenance of roads;
- To provide a channel for the expression of collective opinion of its members regarding roads;
- To promote the use of standard specifications and to propose specifications;
- To advise regarding education, experiment and research connected with roads;

- To hold periodical meetings, to discuss technical questions regarding roads;
- To suggest legislation for the development, improvement and protection of roads;
- To suggest improved methods of administration, planning, design, construction, operation, use and maintenance of roads;
- To establish, furnish and maintain libraries and museums for furthering the science of road making;
- To publish, or arrange for the publication of proceedings, journals, periodicals and other literature for the promotion of the objects of the society;
- To accept subscriptions, subsidies, donations, endowments and gifts in furtherance of the objects of the society;
- To invest and deal with the funds of the society or entrusted to the society, to acquire and hold any movable or immovable property, and to borrow or raise money for the furtherance of the objects of the society and to sell, lease, exchange, or otherwise deal with the same;
- To grant pay, prizes honoraria, or scholarships (including traveling scholarships) for meritorious work in furtherance of the object of the Society; and
- To do all such other lawful things as may be, incidental or conducive to the attainment of the above objects

IRC publishes material on wide aspects related to roads and traffic as follows:

1. Journal of Indian Roads Congress (Quarterly)
2. Indian Highways (Monthly)
3. Highway Research Bulletins
4. Highway Research Record (Annual)
5. IRC specifications, standards, design codes covering various aspects of roads, bridges, materials, construction practices, technology, management, etc.
6. Special Publications in respect of selected topics
7. Ministry of Road Transport and Highways specifications, pocket books, standard designs for inter-sections, standard drawings and plans for highways, bridges or various spans and carriageway width (Reinforced Cement Concrete, Pre-stressed cement Concrete Composite Construction, etc)
8. IRC Seminar Publications
9. Highway Research Board Publications
10. State-of-the-Art-Reports on important emerging topics
11. Miscellaneous Publications

3.8.10 Highway Research Board

The primary aim is to serve as national center for road research with the following objectives:

- To ascertain the nature and extent of research required.
- To correlate research information from various organizations in India and abroad with a view to exchanging publications and information on roads.
- To co-ordinate and conduct correlation services.
- To sponsor basic research through universities and research organizations.
- To collect and disseminate results of research.
- Any other matter related to road research.

3.8.11 National Institute for Training of Highway Engineers (NITHE),1983

NITHE was registered as a Society in January 1983. It was operating from Delhi up to September 2001 and has now shifted to its own campus at NOIDA, UP. It was funded by the Ministry of Shipping, Road Transport and Highways (MOSRT&H) under World Bank Loan Assistance. It organises training in Highway Engineering for Central/State Governments, Public and Private sectors and Colombo Plan and SAARC countries. The trainings are foundation courses for newly recruited engineers, specialised areas of highway engineering, highway management and administration programmes, refresher courses at all the levels, management development programmes, orientation courses and strategic training.

The objectives of the institute are:

- To impart training to engineers at entry level and during service at different levels of central and state governments, public and private sectors.

- To help highway engineers build up character and develop an all round personality as a part of human resource development.

- To assist various organisations in developing their training institutes and training of their faculty.

- To promote co-operation and foster exchange of knowledge, ideas and experience in the sphere of highway engineering among engineers in India and abroad.

3.8.12 National Institute of Construction Management and Research (NICMAR)

National Institute of Construction Management and Research (NICMAR) is a leading educational Institute established by the Indian construction industry. NICMAR is an autonomous, non-government, non-profit academic body, incorporated in India on September 1983 as a 'Society' and a public charitable 'Trust'. It is recognised by Government of India as a Scientific and Industrial Research Organisation (SIRO). Major academic programmes of the institute were started with technical assistance from UNDP and academic inputs from the faculty of Loughbourgh and Indian Institute of Management, Ahmedabad. The institute runs post graduate, graduate and graduate certificate programmes in varied areas.

The post graduate programmes running are in advanced construction Management, Project Engineering and Management, and Real Estate and Urban Infrastructure Management. Two-year post graduate programmes are in Construction Management, Project Management, and Infrastructure Development and Management. One-year graduate programmes are in Construction Project Management, Building Maintenance Management, Construction Safety Management, highway project management, oil and gas pipeline project management, quantity surveying, construction contracts management, and construction business management. Graduate certificate programmes (six-month duration) are in construction contracts management, construction quality management, construction safety management, site personal management, and construction equipment management.

3.8.13 The Engineering Staff College of India (ESCI)

The Engineering Staff College of India (ESCI) is a forum of The Institution of Engineers (India), the country's premier professional organization and has completed twenty six years in service to the nation in imparting continuing education for the engineers and managers in engineering profession as well as providing consultancy services to the industry and government agencies. ESCI is certified for quality under ISO 900 : 2000. It is an autonomous institution recognised by AICTE, and has formal linkages with a large number of eminent engineers, managers and

academicians who serve as an adjunct faculty and consultants to ESCI. Their experience is best utilised to structure and deliver well-focused continuing education programmes of current topical interest. In addition to the open programmes, ESCI also offers customised programmes to suit the requirements of any individual organization. Such in-house programmes are provided successfully for various State Government departments, State Electricity Boards, the Nuclear Power Corporation, NTPC, NMDC, Singareni Collieries, DFID -Government of United Kingdom, Bhutan Power Corporation, Nepal Electricity Authority and many other industrial and Government entities. Unlike most other training institutions, ESCI is thus able to deliver courses and services of specific relevance and focus to the client organizations. On an average about 6000 engineers and managers are given training at ESCI every year.

3.9 HIGHWAY HIERARCHY

The highways are classified based on the function they serve in the country. According to the area they are provided, they can be termed as Rural Highways and Urban Roads. The Road Development Plan Vision 2021 categorises the rural highways in three groups as follows:

A. Primary System: The primary roads are the main arteries through which the traffic will move across country. This forms the main grid or network. The roads in this category are categorized further as,

    a. Expressways: These are routes of superior type having facilities like controlled access, grade separation at crossing, adequate road signing, parking and recreational centers, traffic and medical aid posts, petrol filling stations and telephone booths. Slow moving traffic is not allowed on these roads. Higher speeds, comfort and safety are possible. Considerable savings in operating costs and travel time is also possible.

    b. National Highways: National Highways are the main highways running through the length and breadth of the country, connecting ports, foreign highways, capitals of states, and strategically important areas. They should run throughout the year uninterruptedly and be of high grade construction. The central government is responsible for the construction and maintenance of these highways. The following criteria is followed to declare a new route as National Highway:

        i. Roads which run through the length and breadth of the country

        ii. Roads connecting adjacent countries

        iii. Roads connecting the national capital with state capitals and road connecting mutually the state capitals

        iv. Roads connecting major ports, large industrial centers or tourist centers

        v. Roads meeting very important strategic requirements

        vi. Arterial roads which enable sizeable reduction in travel distances and achieve substantial economy thereby

        vii. Roads which help opening up large tracts of backward areas and hilly regions.

B. Secondary System: Secondary roads assist the traffic to move from lower level to the higher level of service. These basically collect the traffic from districts and rural areas and transfer that to the main grid or function in the reverse direction. This category of roads are also divided into two as,

    a. State Highways: The state highways are main arteries of traffic within a state. They connect the district headquarters and important cities, with each other and with the National Highways or highways of other adjoining states. They are of same standards as National Highways. The criteria adopted in developing SH network is:

    i. They should serve all district headquarters and sub-divisional headquarters

    ii. They should serve major industrial centers

    iii. They should connect places of commercial interest

    iv. They should serve places of tourist attraction

    v. They should serve major agricultural market centers

    vi. They should connect important ports in the state.

b. **Major District Roads:** Major district roads are roads traversing each district, serving areas of production and markets, connecting these with each other or with national or state highways or railways or important navigational routes. They should be able to take traffic to the heart of rural areas throughout the year with only minor interruptions. The following criteria may be kept in view in developing the MDR network:

    It should connect all towns and villages above a population of 1500 which are not connected by NH or SH.

C. **Tertiary Roads:** Tertiary Roads collect the traffic from the heart of the country and transfer them to higher level of services. These basically provide accessibility to masses and villages. These are now defined as rural roads, which previously were defined as other district roads and village roads. These roads connect the villages or group of villages or small towns with each other and to the nearest district road, national or state highway or railway or navigational routes.

3.10 INVESTEMENT IN HIGHWAY SECTOR

The investment in highway sector has remained a major component as the development of roads has been given due importance from the very start of planning. This is evident from the data given in Table 3.13. It has taken a big leap during ninth and tenth plan.

Further, the progress of road network during different decades in terms of road lengths under different categories and the improvement in accessibility of villages is depicted in Table 3.14. As on March 2006, around 60 percent villages have been connected by the roads.

Table 3.13: Investment in roads

Period	*Investment (Rs. Crore)*
First Plan (1951 – 56)	135
Second Plan (1956 – 61)	224
Third Plan (1961 – 66)	440
Period 1966 – 69	309
Fourth Plan (1969 – 74)	862
Fifth Plan (1974 – 79)	1701
Sixth Plan (1980 – 85)	3807
Seventh Plan (1985 – 90)	6335
Period 1990 – 92	3779
Eighth Plan (1992 – 97)	13210
Ninth Plan (1997 – 2002)	39331
Tenth Plan (2002 – 07)	59490

Source: Planning Commission and IRC Road Development Plan Vision 2021

Table 3.14: Progress of road network ('000 km)

	1950 - 51	*1960 - 61*	*1970 - 71*	*1980 - 81*	*1990 - 91*	*2000 - 01*	*2005 – 06*
Total length	400	515	915	1485	2327	3176	3316
Of which surfaced length	156	234	398	684	1090	1600	1700
National highways	22	23	24	32	34	58	67
State highways	45	62	70	95	127	124	132
Major District Roads and Rural Roads	333	429	821	1358	2166	2994	3117
Percentage of village with population above 1000 connected with all-weather roads	32%	36%	40%	46%	73%	90%	92%
Overall village accessibility	20%	22%	25%	28%	44%	54%	60%

Source: Basic Road Statistics, Planning, Commission and Road Development Plan Vision: 2012

ILLUSTRATIVE EXAMPLE

Example: Area of an state is 20,000 km^2 of which 5000 km^2 is agricultural. The number of towns and villages and their population is given below:

Population	*Number of villages/towns*
5000 above	15
2001 – 5000	35
1001 – 2000	100
.501 – 1000	200
Less than 500	600

The length of the railway track in the state is 300 kms: using Nagpur plan formulae calculate, for the state the following:

i. Road length of NH, PH and MDR
ii. Road length of ODR and VR
iii. Total road length per 100 km^2 of area.

Solution: As per Nagpur plan, the road lengths are given by following formula:

i. Length of NH, PH and MDR $(L_1) = \dfrac{A}{8} + \dfrac{B}{32} + 1.6\,N + 8T + D - R$ (kms)

ii. Length of ODR and VR $(L_2) = 0.32\,V + 0.8Q + 1.6\,R + 3.2S + D$ (kms)

Here

A = Agricultural area = 5000 km^2
B = Non – Agricultural area = 20,000 – 5,000 = 15,000 km^2
$N = 35, T = 15, D = 15\%$ road length
$R = 300$

$$\text{Road length (l)} = \left[\frac{5000}{8} + \frac{15000}{32} + 1.6 \times 35 + 8 \times 15 \right]$$

$$= 1269$$

Length of NH PH and MDR $(L_1) = 1269 + 15\%\ (l) - 300$

$$= 1269 + .15 \times 1269 - 300 = \textbf{1160 Kms.}$$

For

$L_2 : V = 600, Q = 200, R = 100, S = 35, D = 15\%$

Road lenth $(L_2) = 0.32 \times 600 + 0.8 \times 200 + 1.6 \times 100 + 3.2 \times 3.5$

$$= 192 + 160 + 160 + 112 = 624$$

Length of ODR and VR $= 624 + .15 \times 624$

$$= 624 + .15 \times 624 = 624 + 94 = \textbf{718 kms}$$

Total length $= L_1 + L_2 = 1160 + 718 = 1878$

iii. length per 100 km^2 area $= \dfrac{1878}{20,000} \times 100 = \textbf{9.39 km}$

Highway Alignment and Location Surveys

4.0 PREAMBLE

All road projects related to the new constructions start with the finalization of alignment of the desired road. In case of road improvements, sometimes the work relates to the improvement in road alignment with the objective of making traffic movements more easy and smooth. In general, the alignment can be defined as follows,

Alignment of a road is a location in space depicted by a line, drawn straight or curved or as a combination of both, between two desired places whose locations are fixed in space, and that fulfills the desired objectives by fitting in well over the political and physical map of the area.

The direction and position given to the centerline of the road on ground is its alignment. The line plotted between the two places of interest shows the centreline of the road. The movement of this centreline in 'X-Y' coordinate system and 'X-Z' coordinate system defines the profile of the road in the horizontal direction and vertical direction. It, therefore, means that a road profile esssentially consists of two types of profiles, namely horizontal profile and vertical profile. These are defined further below and are discussed in detail in the chapter on highway geometric.

a. **Horizontal Profile or alignment:** Horizontal profile or alignment defines the layout of a road in plan, connecting the two desired places and passing through the geographical and physical features of the area. The plotting of this plan requires a topographical map of the area of concern. The profile is deviated in Y-direction with respect to the X-direction based on the physical features that come along the X-direction as an obstruction to movement. Such deviation requires incorporation of curves to facilitate a change in the direction of movement. These curves are known as 'horizontal curves'. The latitudes and longitudes shown on the map help in the measurement of distance travelled between the two desired places.

b. **Vertical Profile or alignment:** Vertical profile or alignment of the road defines the layout of the road in elevation. It is not necessary that the ground between the two desired places that need to be connected by road is leveled and has constant elevation above the mean sea level (MSL). The change in the elevations of the intervening ground between the two places and the relative elevation of the two places is indicated by the contour maps drawn for that area. Contours are freely moving lines drawn by joining points representing same elevation above MSL. This necessitates the change in the profile in 'X-Z' direction. This is accomplished by joining the two points having different elevations with a straight line. Such a profile in 'X-Z' direction is defined as 'gradient'. If there is a requirement of joining two gradients that are moving in opposite directions, then they are connected using curves in vertical profile, and these are defined as 'vertical curves'.

While fixing an alignment, horizontally and vertically, there are number of important factors that need to be kept into consideration. At the same time, an ideal alignment should have certain features and these should become the part of the proposal as far as possible. These features are discussed under the next section.

4.1 IDEAL ALIGNMENT

An ideal alignment is the one that satisfies the economic, social, geographical, environmental and political constraints in best of the way; needs minimal resources for its construction, operation and maintenance; maximizes user's safety and comfort; and provides proper connectivity with equal emphasis to accessibility and mobility. An ideal alignment in this sense should fulfill the following:

a. **Shortest Path:** The line profile marked on the topographical and contour map between the two places of desire should be straight and connect them directly, thus representing the shortest distance that needs to be travelled between the two places. Such connectivity satisfies the economic constraints and social constraints in terms of resources required for its construction and the distance travelled. But this may not be the possibility always. It may happen that the intervening area consists of natural or manmade obstructions that would necessitate deviation in the direction of alignment, and thus making the alignment not the shortest one. Possibility of shifting the obstruction instead of alignment and the comparative economic feasibility should be examined before taking any decision.

b. **Easy Profile:** The profile is rated easy if it requires minimal resources for the construction, operation and maintenance of the road and maintains the comfort and safety of the user while traversing it. As far as possible, the profile should pass through the terrain that is not tough and do not requires highly specialized and sophisticated construction technology and skills. Otherwise, these will increase the cost of construction. If the profile comprises of a good number of deviations in horizontal or vertical plane (i.e. curves and gradients), then it increases the cost of operation in terms of consumption of fuel and wear and tear of the vehicle engine and parts. At times, the selection of a profile is such that it needs maintenance at much faster rate, i.e. periodically. The users' perspective is taken into consideration in the form of the ease of driving along the profile and the level of comfort achieved during travelling along it.

c. **Safety along Profile:** The safety along the profile is to be ensured during both the phases, construction of the road and operations thereafter on it. The safety concern is related to both, the human being and the equipment. During the construction phase, it is the manpower and machinery employed, and during operations, they are users and vehicles. The safety should be ensured by proper selection of the alignment, physical design features of the road and provision of safety information and devices along the profile. Another safety aspect is related to the surrounding area on both sides of the road. The profile should pass through an area that is regarded as safe for the travelers.

d. **Purpose:** The alignment should fulfill the purpose with which it is selected. The purpose may be purely social, i.e. providing connectivity between two places, improving transportation network in an area; economics based like providing a freight corridor, linking places of production and consumption, improving access to industrial areas, harnessing of resources of an area; or political in nature, say constructing a road with strategic reasons, satisfying the demand on an area, development of a specific area. The intended purpose should not be defeated at any cost.

e. **Aesthetics:** As far as possible, the profile should pass through an area that gives pleasant feelings to the travelers moving along that profile. It improves level of comfort during ride and removes monotony. The profile should be such that the road fully integrates with the surrounding landscape of the area.

f. **Preservation of Ecology:** The big transportation infrastructure projects require big area for their provision. The intervening area between the two places usually passes through rural areas or forest areas. Provision of an alignment through such area has a possibility of disturbing the ecology and bio-diversity of that area. The fixation of the alignment should be such that minimum disturbance is caused to the local flora and fauna due to either construction of the road or operation of vehicles on the same. An environmental impact assessment study of the impacts of the road should be made to ensure that the adverse effects are kept to the minimum.

The above mentioned points are the guiding principles for the selection of a road profile. Apart from these there are factors, which need to be taken into consideration while deciding upon the profile of the road. These are discussed in the following section.

4.2 FACTORS CONTROLLING THE ALIGNMENT

There are factors, which are part of the abovementioned points. These are discussed below under two categories, general factors and specific factors. Specific factors are listed related to hilly areas, desert area, different types of soils and water logged areas.

4.2.1 General Factors

a. **Cost-Economics of the project:** The cost of the project is dependent upon the selection of the profile and the terrain through which it passes. If the terrain is quite tough and construction of the road requires specialized techniques and equipment than the cost of the road project would be very high, whereas, if the design features are kept within acceptable limits or guidelines, then the cost would be much lower than the former case. The economics of the project depends upon the source from where the money will come for the construction of the road and on the operation and maintenance of the facility provided. If the operating cost is quite high, very few will like to use the facility, whereas, if it is beneficial in terms of savings made in travel time or improvement in the quality of riding then more persons will like to use the facility. The facility in such a case may also fetch revenues in terms of tolls that can be charged from the users. It is, therefore, important to look at two factors that govern the cost-economics, i.e. cost-benefit ratio and internal rate of return that will make the repayment of the capital cost feasible within a certain time frame. Another way of achieving economics is to select the profile that balances the cutting and filling, as shown in Fig. 4.1

b. **Obligatory Points:** It is usually difficult to finalize a profile of the road that is satisfying the desired elements as mentioned in the previous section and is economical under all considerations. Deviations in the profile are made due to many factors, out of which one is the obligatory points that may or should not be traversed. Passing through or away of the obligatory points is decided based on the characteristics of the obligatory point into consideration. There are certain points that need to be kept into mind before selecting the final profile. These are:

i. The location should steer clear of obstructions like places of worships, archeological and historical monuments, cemeteries, burning ghats, edges of properties, areas prone to subsidence due to mining, and as far as possible, the need to re-cross the railway line, public utilities say water lines, overhead lines, and public facilities like hospitals, schools, play grounds, or dense and ecologically important wooded areas, built-up areas, areas prone to flooding, etc. should be avoided. The profile should not pass through the marshy land, lowlying lands, areas with poor drainage, areas with poor drainage material, costly land, lake, pond, etc.

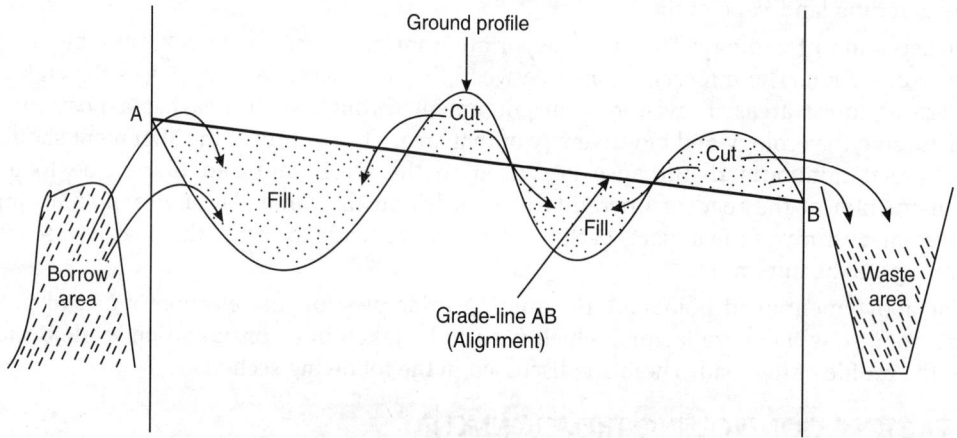

Fig. 4.1: Balancing of cut and fill material

ii. The profile should be located along the edges of the properties so as to avoid interference to the activities carried out on that property and to minimize the possibility of crossing of the road by the local population. It should pass through the important cities and towns, mountain passes, locations with better type of soil, sites for tunnels and locations for bridges across the rivers. As far as possible, efforts should be made to locate the bridge where -

1. The river is straight both on the upstream and downstream side
2. The location is sufficiently away from the influence of the confluence of tributaries
3. The channel is well defined and narrow
4. The banks are high, rocky/firm and well defined above the HFL

iii. It should be close to the source of embankment and pavement materials so as to reduce the distance of haulage and the cost of transportation. The profile should cross the river normal to its direction from view point of cost minimization. As far as possible, areas likely to be unstable due to toe-erosion by river should be avoided.

c. **Physical Features of the Alignment:** While an alignment is selected, one of the important decisions that are usually taken simultaneously is the level of the road facility that is to be constructed or provided between the two desired locations. This has its effects on certain aspects like the land that is to be acquired for the construction of the road and for its future expansion. This further depends upon the geometric guidelines available for different types of road facilities, like national highway, state highway, district or rural roads. Important parameters considered are number of traffic lanes, width of carriageway and shoulders, road land width, radius of horizontal curves, rate of rise or fall of the profile in longitudinal

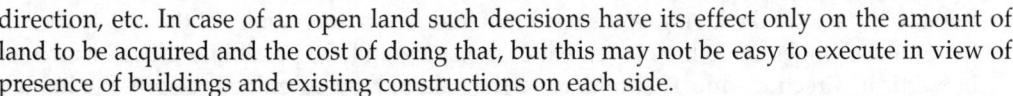

direction, etc. In case of an open land such decisions have its effect only on the amount of land to be acquired and the cost of doing that, but this may not be easy to execute in view of presence of buildings and existing constructions on each side.

d. **Traffic Projections and Forecasting:** The decision attributes listed in the previous sub-section related to physical features are also dependent upon the traffic that is supposed to move on the facility for which profile is selected. Such dependent attributes are number of traffic lanes, design of intersections, design of pavement and economic appraisal of the project. The volume of traffic will depend upon the importance of the two places that need to be connected through the link under consideration and the towns and cities along the profile that would also be connected additionally. Population residing within certain km distance on either side of the profile is expected to contribute to the traffic on this link. Initially, the movement of freight traffic is considerably more than the passenger traffic between the newly connected places. As a thumb rule, the traffic is expected to vary as a square of the population that is served by the proposed link or profile. In case of modifications needed to the already existing alignment, the new traffic count will become the basis of changes in the alignment. The estimated traffic is projected over a certain period that is decided based on the type of project, importance of the road, availability of the finances and other related factors. In case, the alignment is designated as trunk route, the period may span between 10 and 20 years, whereas, for lower category roads it may be 5 to 10 years.

e. **Political Will and Obligations:** The decision makers in most of the governing systems are the politicians in power. The ideology and policies of a government defines the areas of importance for that government. The planned outlay dedicated to the cause decides the amount of work that can be executed in a given time frame. Fortunately, the trans-portation and communication have been given due importance over the planning years and hence, there is substantial growth in the two sub-sectors. Sometimes, the population of an area demands for certain links between the desired places or for the upgradation of the existing links, and these are sanctioned out of political obligations of the politicians towards the area they represent in the parliament or the legislative council. It may happen that such obligations may fall short of economic feasibilities.

f. **Hydrological Aspects:** Natural drainage pattern of an area through which the profile or alignment is expected to cross is one of the parameters that need to be given due importance. Any alignment that crosses the drainage profile repeatedly will require fast periodic maintenance culminating in costs. Rivers and ponds are important hydrological features that need to be traversed by the alignment. In case of rivers, a suitable bridge location, not having meandering section and minimum possible in width, need to be identified for the construction of bridges. During fixing of the alignment by the side of the river, the direction of flow of the river and High Flood Level (HFL) records for past 50 years shall be kept in view. As far as possible, the alignment should move along the watershed line. In case of ponds and lakes, the need of bridges should be avoided by going along its periphery thus keeping the cost of project to the minimum. Ground Water Table (GWT), its depth and variation in depth with time and seasons also affects the life of the road constructed on the selected profile.

g. **Rock-mass and Soil Characteristics:** In most of the cases, the roads are constructed in embankment. In some of the cases as the conditions necessitate it may be in cutting also or may remain at level with the surrounding ground level. In all such cases, it is necessary to ensure that the soil mass allows speedier drainage of water percolating from the top surface or seeping from the sides. The soil mass should be of better quality. The same is the case with rock mass, if the road passes through such area. It should not be fissured and should provide a stable surface for the laying of the road material. The susceptibility of the soil or rock mass to volume change due to the movement of moisture, and the possibility of water movement in vertical direction due to capillary action should be examined before finalizing the alignment of the road. The strength of the supporting mass is another important factor that needs consideration as whole of the load is finally transferred to the bottom-most layer, i.e. soil or rock mass. It should be easy to work with as well as to compact.

h. **Environmental and Ecological Considerations:** The location of the road profile should be such that it fully integrates with the surrounding landscape of the area, with minimum of the interference caused to it. In this respect, it would be necessary to study the environmental impact of the road construction and operation and to examine the adverse effects that it may cause to the environment and ecology of the area. In cases where the cutting of the wooded area is unavoidable the road profile should be aligned on a curve so as to preserve an unbroken background. The widening of the road becomes necessary with the increase in the traffic and it leads to the cutting of the trees on one or either side of the profile or road. This should be compensated by an equal afforestation along the profile or in additional areas at suitable locations. These aspects should be kept in view even at the time of project preparation. The aim should remain to cause minimum disturbances to the flora and fauna, habitat of species and the interference with the paths followed by animals during transitions. Adequate and secure crossings should be located with this reason along the profile. The main ecological problems associated with hill roads are geological disturbances, degradation of land, soil erosion, loss of forestry and vegetation, aesthetic degradation, siltation of water reservoirs, etc.

It is rare that all the above factors are satisfied by a single alignment and therefore highway engineer should make a judgement is selecting the feasible factors to minimize total cost to possible extent.

4.2.2 Specific Factors

a. **Alignment in hilly areas**

i. **Accessibility:** The terrain of the area makes it difficult to construct many roads in the hilly areas. It is, therefore, advisable to orient the profile in a way that it benefits a good number of habitations, irrespective of size, and not necessarily only by direct connectivity. Bridal paths leading to the main road may also be considered during the selection of the profile. Another point is the choice of gradient enabling the access to points at different elevations. The profile should attain the ruling gradient for most of its length, thus making it possible for even small vehicles and non-motorized movements. Steep gradients and other inaccessible areas should be avoided as far as possible.

ii. **Stability of Rock and Soil Mass:** The stability of rock and soil mass in hilly areas is one of the most important factors that need consideration. It is important to examine the characteristics of these masses with respect to stability, strength, moisture movement, dip of the strata, faults and folds, if any, unstable hilly features, areas having frequent landslides or settlement problems and up slope benched agricultural fields with potential for standing water should be avoided as far as possible. The road profile should be selected on the strata having dip away from the valley or having mass that has stabilized over the years. Precautions should be taken in new hills and in areas having uncompacted mass that can move with a trigger. Careful study should be done of the geological maps of the area.

iii. **Drainage Pattern:** Drainage of the area and its pattern over the adjoining land is another important area that decides the stability and need for maintenance. Locating profile along the river valley sides allows using the inherent advantage of use of gentle gradients, but requires a large number of cross-drainage structures and protection works against erosion. Side drains should run along the whole length of the profile and catch drains should be located wherever a transverse valley reaches the profile level. It should be kept in mind that cross-drainage structures increase the cost of the project tremendously and hence should be kept as minimum as possible. As far as possible, the natural drainage of the area should not be disturbed.

iv. **Resisting Length and Geometric:** The resisting length is calculated based on the concept of work done in moving the loads along the profile and takes into consideration the horizontal length between the places to be connected by the profile, the difference in elevation of the two places under consideration, and the sum of ineffective rise and fall in excess of floating gradients. This factor indirectly indicates towards the extra consumption of energy and resources that would remain present due to the adjustments done in the selection of profile. Geometric requirements of the roads in hilly areas are different than those in plain areas. Specifically, consideration should be given to the carriageway width, width of road land, shoulders depending upon the location, the radius of curves and the gradients.

v. **Land Features:** The physical features of the area govern the selection of the alignment more empathetically in hilly areas as compared to plain areas. In crossing mountain ridges, the location should be such that the road preferably crosses the ridge at their lowest elevation. Such locations also need hair-pin bends. These bends should be as minimum as possible and wherever unavoidable should be located on stable and gentle hill slopes. The areas liable to snow drift should be avoided. The alignment should pass through such side of the hill that received plenty of sunlight and is not falling in shade. In certain cases, traversing along the hill may not be an economical solution and it may become expedient to negotiate high mountain ranges through tunnels. As far as possible, attempt should be made to avoid the following features

1. Unstable hill features and areas having perennial landslide or settlement problems
2. Areas subject to seepage or flow from springs, hydel channels, subterranean channels, etc.
3. Steep hill sides
4. Area subject to flooding or water logging
5. Areas liable to snow drift or avalanches
6. Locations involving unnecessary and expansive destruction of wooded areas

b. **Alignment in Deserts**

i. Soil Characteristics: The soil mass usually available in the desert areas is sand. Sand is a non-cohesive mass and the gradation of the material defines the portability of that due to natural forces like heavy wind. Locations where sand is loose and unstable should be avoided as this makes the pavement surface highly accident and abrasion prone due to presence of fine slipping material that also works as an abrasive charge under load applications. Preference should be given to areas having coarse sand than to areas having fine wind blowing sand. The locations along the ridges having vegetation make a good choice for alignment. Drainage is not a problem in sand.

ii. Sand Dunes: Sand dunes have the tendency to shift with heavy winds. These also have a unique pattern of movements. As far as possible, the profile of the road should run parallel to the sand dunes. If required the sand dunes should be crossed without disturbing their existing profile, which may be stable in nature. In locating a road profile in an area having longitudinal sand dunes, the best location is always at the top of a ridge or in the inter-dunal space. Location along the face of the longitudinal dunes should be avoided.

c. **Alignment in Problematic Soils**

Different types of soils are available in different parts of the country. Certain types of soils like expansive soils, saline soils, and marine clays cause problems for the roads constructed over them. It is therefore required to given due consideration to the following:

 i. The areas with expansive soils show problem of strength when moist. Such soils should be stabilized, especially by mixing lime in puverised soil, to gain strength.

 ii. Locations where large salt deposits occur should be avoided. In case it is required to pass profile of the road through medium or highly saline soils, precaution should be taken to divert the water away from the road bed. Further, if the saline soil is wet, then embankment of good quality soil should be laid over the saline soil.

 iii. Marine clay is soft and compressible. On the application of load it has tendency to settle. In case road profile has to pass through such area, the ground improvement methods should be adopted. If the site is under influence of rise and fall of water, the subgrade should be 1.0 m above the highest water table.

d. **Alignment in Water Logged Areas**

 i. Water has detrimental effect on the life of the material used in the pavement crust, as well as on the adhesion of bitumen with the aggregates as used in the surface or wearing course. Therefore, if the alignment passes through water logged area then the height of the embankment should be adequately above the level of standing water, if any.

 ii. Other preventive measures include the laying of capillary cut-offs or blanket below pavement so as to improve drainage facility.

4.3 ROUTE LOCATION PROCESS

The route location process may be divided into the stages:

 i. 'Broad-Band' Location approach which is the selection of a general location or corridor from a broad band or area of interest between into points between origin and destination, or some intermediate points. These points may be determined by physical, socio-economic or political considerations. The distance between two points should be large enough to encompass all general locations between the origin and the destination. The width of the broad band or corridor will depend upon level of details available and may vary from 500 m to 5 km or so.

 ii. 'Narrow-Band' location is selection of a preliminary alignment from the optimum corridor. The width of this preliminary alignment may be 5 to 10 times the width of final alignment. The narrow band selection (preliminary alignment) is there subjected to final design process during which the precise combination of alignment and grade is decided. Figure 4.2 shows the two processes diagrammatically.

At the broad-band approach three corridors are selected and corridor 2 is selected as the optimum corridor. At the narrow band approach the optimum corridor 2 is enlarged, the data is refined and two alignments are considered, one of which will be selected for final location and design.

At times, the alignment which is shortest may be costtier than other alignments in initial cost. All requirements can not be fully justified simultaneously. Deviation from shortest route may be necessary due to following points through which road must pass or must not pass.

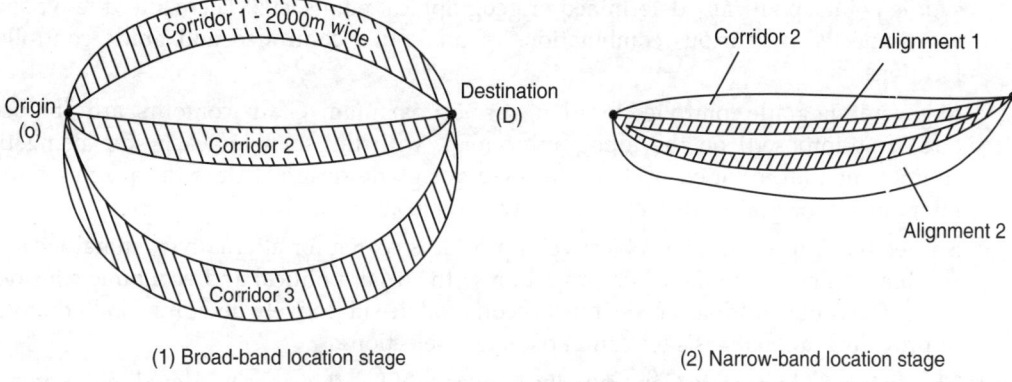

(1) Broad-band location stage	(2) Narrow-band location stage

Fig. 4.2: Route location process

Point through which road alignment has to pass:

* Important towns and cities
* Places of social, commercial, political and defence importance
* Site of location of major bridges, river-crossing
* Passing the hill or saddles
* If tunnels are unavoidable, road alignment should pass through tunnel site.
* Politically dictated locations
* Other geographical control points.

Point through which road alignment should not pass:

* Where land acquisition is very costly.
* Religious places like temples, churches, mosques tomb etc.
* Area liable to flooding (water logged areas), marshy areas
* Other areas protected by law.

4.3.1 Search Techniques for Alignment

Route location process, is the process of searching the suitable alignment for roads. Following are some of the basic and modern techniques.

a. **Upgrading:** Most basic technique is upgrading the existing location, such as non-motorable tracks, footpaths. Many modern highways have evolved in this way from horse and wagon paths.

Where travel ways have been already established to accommodate non-motorized traffic, they are usually well located to minimize drainage, to follow ridge line, etc and they could be used as primitive locations wherever possible. These paths are nearly in conformity with the topography of the terrain, and a direct horizontal alignment consistent with minimum earthwork.

b. **Line Projection:** Line projection is the process of fixing the alignment of a highway by drawing the proposed centerline on a topographic map, using this to prepare a ground profile, and planning the grade line. The objective is to obtain best combination of alignment and grade consistent with the terrain and established geometric design standards. It can be used as a general indicator either for the broad-band or narrow-band stage.

Line projection is a trial error process, using topographic maps. The first step is to mark critical terrain control points, through which alignment must pass (such as river crossing, saddle points, politically determined or geographical points, etc.) There after these points are connected by various combinations of non-grade controlled and grade controlled locations.

In generating grade controlled location, various maximum 'Grade-contours' are 'Stepped' out (superimposed) on the topographic map, with the span for each step being the contour-interval, divided by the percentage of grade desired. Thus for a 5 m contour interval and 5% grade, the step interval would be 5.0/.05 or 100m.

c. **Subjective Approach:** The subjective approach is search for alternative general location combining upgrading, and line projection with social, political and economic consideration. The route location process must accommodate subjectively determined alternatives, to provide a rational basis for comparison and selection.

d. **The Network Approach:** Conceptually, networks of location are considered in the process of search and selection. A network is comprised of links and nodes, representing route segments (with regard to route location) and junctions the route segments would be defined by road characteristics such as curvature, gradient, surface type, etc. The junctions merely represent points of branching and interconnection to neighbouring links. Any route alignment would then be a sub-set of these links and nodes, interconnected to form a path, through network, between points of interest.

Usefulness of network approach in the route location process is that it can facilitate disaggregate analysis, and approximate construction, maintenance and vehicle operating cost can be evaluated on a link by link basis. Total cost of route alignment would be sum of the costs of the links comprising the alignment between origin and destination.

The number of alternatives can be infinite and with the help of computer-assisted procedures, the evaluation becomes easy and fast.

Computer aided route location models are available incorporating most of the parameters, used in different countries such as CARROLS (Computer Aided Rural Road Location System), RSM (Route Selection Model) developed by TRRL, U.K, GCARS (Generalised Computer Aided Route Selection System), developed in U.S.A., OPTLOC (Optimal Rural Highway Location System) was developed in Australia.

4.4 SURVEYS FOR FINALIZING ROUTE LOCATION

4.4.1 Stages in Project Preparation

Stages involved in the preparation of the project are:

a. Prefeasibility study

b. Feasibility study/preliminary project report preparation

c. Detailed engineering report and plan of construction .

A pre-feasibility report is prepared to provide information regarding the broad features of the project, finances involved and the probable returns. It is based on reconnaissance survey, wherein information on the present status of the profile or the existing road, deficiency or distress identification, development potential environmental impact and traffic data is collected, approximate cost estimated and economic analysis performed. Feasibility study is intended to establish whether the proposal is acceptable in terms of soundness of engineering design and expected benefits from the investments involved. It forms a basis for an investment decision or of Administrative Approval (AA). Herein, the traffic projections are done based on

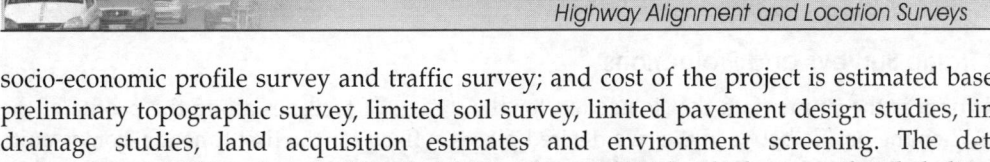

socio-economic profile survey and traffic survey; and cost of the project is estimated based on preliminary topographic survey, limited soil survey, limited pavement design studies, limited drainage studies, land acquisition estimates and environment screening. The detailed engineering report consists of a Detailed Project Report (DPR), which covers detailed alignment surveys, final location survey, soil and material surveys, pavement design studies, drainage studies, environmental impact and management plan, detailed drawings, estimates and implementation schedules. This becomes the basis for the Technical Approval and Financial Sanction (TA and FS).

Figure 4.3 gives the stages in project preparation.

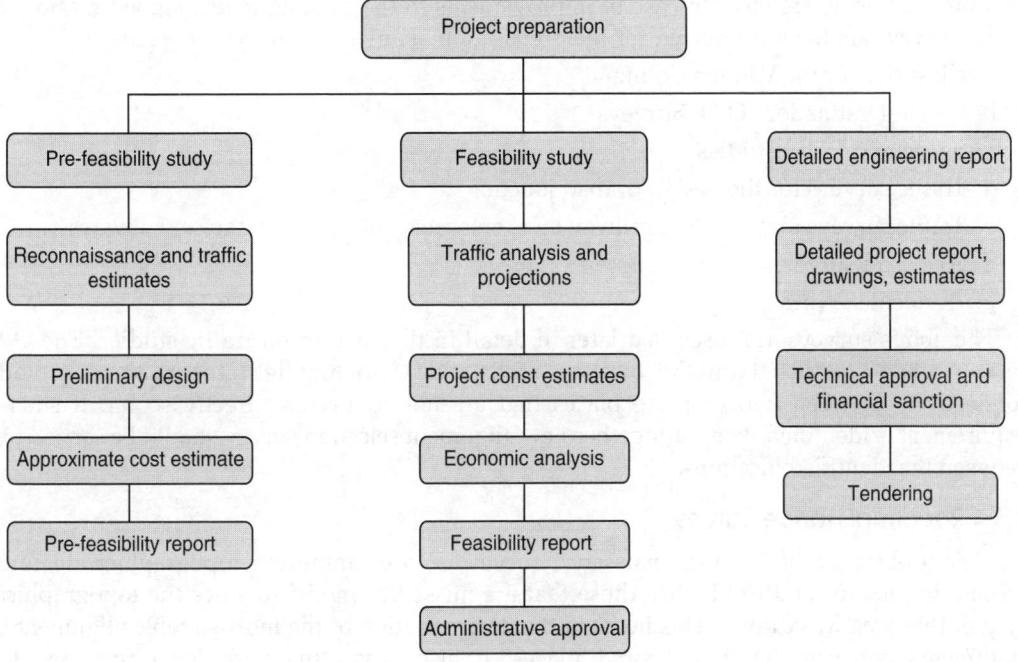

Fig. 4.3: Stages in project preparation

4.4.2 Surveys of Socio-economic Profile

Socio-economic characteristics of an area define the economic activities that may take place in the area. These activities translate into traffic that may move on the selected alignment. This data therefore needs to be gathered at two levels, one at region or state level and other in project influence area. The state or region level data helps in generating the relationship between traffic growth and different economic scenarios, whereas, the project level data would represent the development that would take place in near vicinity of the alignment in future and the additional traffic that would be generated out of that development. The type of data that should be collected includes information on population, culture, land use, land value, institutions, transportation, historical sites, utilities and services, community boundaries and movements, employment potential, tax base and dynamic changes expected to take place in the area. Data should be collected for last ten years and include aspects related to population, economic activities and transport system. The historical data should be correlated to determine the growth trends and elasticity of traffic with respect to economic activities and Gross Domestic Product (GDP). These should be further used to get the growth rates and trends for the future.

4.4.3 Traffic Surveys and Projections

Traffic data is the prime data that is needed for the design of facilities along the profile of the road. When the profile between the two desired places is fixed for the first time it becomes quite difficult to estimate the traffic that would use the proposed link. One way of doing it is based on past experiences in similar conditions of alignment selection. The socio-economic profile of the area, the economic activities in that area and the possible growth rate can provide some idea of the traffic that would be coming on the proposed link. The extent of area expected to produce traffic on the link is taken as around 15 km initially on both the sides of the link and it is assumed that it spreads to 25 km in due course of time. Therefore, an extensive socio-economic and land use survey need to be carried out in the catchment area of the profile if the road alignment is newly selected. In case of improvements to the existing roads, a good number of traffic surveys are to be conducted for the preparation of any road project. These are:

a. Classified Traffic Volume Counts

b. Origin-Destination (O-D) Surveys

c. Speed and Delay studies

d. Traffic surveys for the design of road junction

e. Traffic surveys for replacing railway level crossings with over-bridges or subways

f. Axle load surveys

g. Accident surveys

The above surveys are discussed later in detail in the chapter on traffic studies. The O-D survey when conducted on the existing roads may also highlight the need to provide connectivity between two or more places that are not connected directly so far. If such a requirement is identified then a altogether new alignment selection survey should be carried out between the identified locations.

4.4.4 Reconnaissance Survey

The main objective of the reconnaissance survey is to examine the topographical features present in the area and to identify those that are missed or modified since the topographical map of that area was drawn. This helps in the determination of the most suitable alignment or alignments between the two desired places. It also paves the path for future detailed investigations. This survey actually helps in identifying the deviations that need to be made to the alignment(s) drawn based on map-study done in the office. The survey is carried out in the following sequence:

a. Study of maps

b. Aerial Reconnaissance, if required and feasible

c. Ground Survey

The first stage in reconnaissance survey is map study. Different types of maps like, topographical sheets, agricultural, geological, metrological, soil profile, drainage pattern maps, as well as aerial photographs are available. Topographical sheets are available with Survey of Indian (SOI) on the scale of 1:25,000 1:50,000 and 1:250,000. State maps are available on scale of 1:1,000,000. Plastic relief maps, if available, can provide information on ridges, valleys, peaks, etc with contour information (3-D) for a region. These are drawn on scale of 1:15,000,000. Vegetation maps from Forest Survey of Indian (FSOI) are available on scale 1:25,000,000. Similarly, maps from National Bureau of Soil Survey and Land Use Planning (NBSS & LUP) and Geological Survey of India (GSI) are also available. After going through all these maps and giving due consideration to all controlling factors, the feasible alignments are determined. Aerial photographs, if available, on a scale of 1:50,000 can be enlarged five times to provide details of

geological, geomorphological, groundwater, environmental attributes of interest. Small Format Aerial Photographs (SFAP) can also be used with the same purpose. The aerial photographs can further reduce the alternate alignments determined based on study of other maps. Aerial photographs are available from National Remote Sensing Agency (NRSA), Hyderabad, Air Survey Company, Kolkatta and the Indian Air Force (IAF).

Aerial reconnaissance provides a bird view of the determined alignments and help in identifying features that may cause deviation in alignment or rejection of any alignment. It may also indicate towards the need of ground reconnaissance of certain sections of the determined alignments. It also helps in making observations regarding the correctness of obligatory points as given in the map, correctness of control points as marked in the map, existence of other control points like, major saddles or passes, river crossings, slide or slip areas, marshy areas, camping sites, rocky areas and vegetation.

Ground reconnaissance is carried out by walking or riding along the determined alignments and required information is collected along the way for the evaluation of the alignment. Reference pegs are left along the alignment for further survey that needs to be conducted under preliminary or detailed surveys. Equipment usually carried during this survey are compass, Abney level, Pedometer, Aneroid barometer, Clinometers, Ghat tracer, etc that are used for collecting data at bends and slopes. In difficult terrain, Global Positioning System (GPS) and Differential GPS (DGPS) can also be used.

The final result of the reconnaissance survev is the determination of alternate alignments. It has following objectives:

- To acquire the knowledge of physical features of the area like rivers, valleys, cultivated land, forests, hill, existing roads, canals, etc. and the topography (plain, rolling, hilly). Confirm features indicated on map study
- To collect geological information regarding:
  - Nature of soil, and strength characteristic
  - Surface formation of ground
  - Dip of the existing rock
  - Hill shapes
  - Drainage conditions and floods, etc.
- To collect information regarding availability of construction materials, labour, and sources of water, etc for proposed alternative alignments.
- To see the number, width and type of rivers, and streams crossing the proposed alignment for determining bridge sites and other requirements.
- To locate the control points through which alignment should pass, and also those along which alignment should not pass, and determine length of each alternative.
- To decide maximum gradient and curvature for proposed alignment, like sharp curves, reverse curves, hair-pin bends, etc.
- To judge the requirements for land acquisition and available right of way, built up areas and other land uses and land values.
- To assess extent and type of plants, trees and other vegetation, and scenic spots.
- To locate if there are existing animal tracks, non-motorable tracks or foot paths.
- To collect climatic information regarding rain fall, temperatures, snow, wind, fog, storms, ground water table, etc.
- To note towns, villages, etc which are connected, population they serve, and their economic status and mode of transport.

- To locate railway and road crossings, places of religious importance, other structures like school, grave yards, etc.
- To note other ecological, historical, aesthetical factors which may help in further study.

4.4.5 Preliminary Survey

The purpose of conducting preliminary survey is to collect more information on ground along the alignments that are determined at the end of reconnaissance survey. It is an instrument based traverse survey conducted to collect topographical details of physical features of the land on either side of the alignment(s) along with taking longitudinal section and cross-sections at regular intervals. The data collected at this stage becomes the basis for the final determination of the alignment of the road. Apart from the topographical details, information on nature of soil or rock strata, availability of construction material, drainage, metrological factors, sub-soil and flood water levels, etc is also collected to assist in taking final decisions.

The survey procedure includes running of a traverse along the centre line of the determined alignment. The lengths and intermediate angles between the straight lines are measured carefully. Equipments used are theodolite with Electronic Distance Measurement (EDM), GPS and level. EDM or Total Station is used for the measurement of distances and angles, GPS is used for taking coordinates of necessary points of the traverse and level is used for fly-leveling at an interval of 50 m along longitudinal profile and for cross-sectional details at 100 to 200 m in plain terrain, 50 m in rolling terrain and up to 25 m in hilly terrain. Bench marks are made at regular intervals and connected to GTS datum. Sufficient area should be covered on both the sides of the alignment. Physical features, such as, buildings, monuments, burial grounds, cremation grounds, places of worship, posts, pipelines, existing roads, railway lines, stream/river/canal crossings, cross-drainage structures, etc that are likely to affect the alignment should be located by means of offsets measured from the traverse line. In case of existing alignments, measurements of existing carriageway, roadway, location, radii of horizontal curves should be taken. In addition to these, the nature and extent of grades, ridges and valleys, and vertical curves should be covered in the alignment passing through rolling or hilly terrain.

The data so collected is plotted to produce plans and longitudinal and transverse sections of the alignment. The contours should be traced at critical locations like, hair-pin bends, sharp curves, bridge crossings, etc to examine the suitability of the alignment. The whole exercise of preliminary survey prepares ground for the submission of feasibility report.

Finally drawings are prepared for each alignment, providing following information:

- Length of alternative routes
- Various possible gradients, along the route
- Quantity of earthwork
- Maximum height and length of embankment and cutting.
- Characteristics of rivers to be crossed such as H.F.L, LW.L, flow direction, stream profile, cost of boring, difficulties in alignment etc.
- Geological information like soil type, rocks and slips, etc.
- Details of crossings of other roads, canals, railway line etc.
- Availability of materials, labour, water, food and equipment etc.
- Details of various affected properties with ownerships for land acquisition.
- The climatic conditions of routes traversed.

In the end, a comparative study of various results is made, with relative merits and demerits. The route which is most economical and best from all consideration of an ideal alignment, is selected and plotted on the map.

4.4.6 Feasibility Report

Feasibility report is prepared for the connectivity between the two desired locations may be through different alternative alignments. It is prepared after carrying out engineering surveys and investigations with sufficient accuracy and details along the alternative alignments. It provides complete information on the need and scope of the project, socio-economic profile of alignment influence area, methodology adopted, types of surveys and investigations conducted, engineering details, design standards adopted, reasonably accurate estimate of costs, sources of funding, budgetary provisions, and the implementation programme including prequalification, bidding, construction supervision and contract management. The economic analysis, including cost-benefit assessment, should be based on current costs and sensitivity analysis should be presented for different economic scenarios. The report should accompany the following drawings:

a. Location map

b. Plans showing various alternate alignments considered and the selected alignment

c. Typical cross-sections showing pavement details

d. Drawings for cross-drainage and other structures

e. Road junction drawings

f. Strip plan

g. Preliminary land acquisition plans

The feasibility report should be presented in three volumes as, Volume -I -Main text and Appendices, Volume -II-Design Report and Volume -III -Drawings.

4.4.7 Final Location Survey (Detailed Survey)

Once the discussions on the feasibility report are over and the alignment is finally selected out of the various alternate alignments a final location survey is conducted for the selected alignment. Before carrying out the final location survey certain points need to be given consideration. They are:

a. Few alternate alignments of the centreline of the selected alignment are drawn and studied with respect to the engineering, economic and aesthetic requirements.

b. A trial gradeline is drawn taking into account the control points already set during preliminary survey and the vertical curves are designed.

c. The horizontal alignment of the profile is examined and adjustments are made: The horizontal and transition curves are designed with respect to visibility and the final centreline is marked on the map.

The purpose of the final location survey is to lay the centreline of the road in the field and to collect information that is required for working drawings. The accuracy maintained in this survey is of the highest quality. The centreline of the road or alignment is transferred to the ground by means of continuous transit survey and placing hubs or pegs along that. Double reversal method should be adopted at all horizontal intersection points (HIP) and intermediate points of transit (POT)on long tangents. At curve, the beginning of spiral curve (BS), beginning of circular curve (BC), end of circular curve (EC) and end of spiral curve (ES) should be fixed and referenced. The angle of intersection of road with road or road with railway track should be measured. Such points should be serially numbered and identified by coordinates on final plan drawings. These should not be disturbed or altered during construction. Final centreline should be staked at an interval of 50 m in plain and rolling terrain and 20 m in hilly terrain for taking levels. In case of existing roads paint mark with button headed steer nails may be used. Distance should be measured continuously following the horizontal curve. To establish firm control on elevations, design and constructions, permanent bench marks should be established at regular interval of 2 km and temporary bench marks at an interval of 250 m. Single datum should be used as far as possible.

Levels should be taken along the centreline of the alignment at all staked stations. Cross-sections should be taken at 50-100 m intervals in plain terrain, 50-75 m interval in rolling terrain and 20 m in hilly terrain. Cross-sections should also be taken at all critical points like crossings and deviations namely, beginning and end of spiral and circular curves, middle of circular curve, etc. These are extended up to right of way limit on either side of the centre line with levels shown at every 2-5 m interval. Centreline profile should extend up to 200 m beyond the limits of the project for proper connection of grades at both ends. For intersecting alignments, it should be up to 150 m. at railway level crossings, the level of top of the rails, and in the case of subways, the level of the roof should be noted.

4.5 MATERIAL SURVEYS

In general, material surveys are required to:

a. determine the nature and physical characteristics of soil and soil profile
b. determine the salt content in soil
c. classify the earthwork involved in various categories like rock excavation, earthwork in hard soil, etc.
d. gather information regarding sub-soil water level and flooding, etc
e. locate sources of aggregates, ascertain their suitability and availability for use
f. locate source of good quality water for use in different construction works

Information on above aspects is collected from the concerned authorities like PWDs, ground water commission, etc and from the available information like geological maps, soil maps, etc. Information from similar constructions or constructions requiring similar construction material may prove more useful. The investigations to be carried out will depend on the construction activity in which that material is to be used. Soil is used as subgrade material or for construction of embankments. The investigations depend upon the height of the embankment and are as follows:

a. Gradation test, for soil classification and sand content
b. Atterburg limits (liquid limit and plasticity index)
c. Density and optimum moisture content
d. Deleterious constituents (for salt infected areas)
e. Shear parameters and consolidation properties
f. Stability analysis for slopes and embankments
g. Identification of land slide or rock fall prone areas
h. California Bearing Ratio (CBR) test or 'K' value test for pavement crust design

In case of aggregates (naturally occurring or artificially prepared from rock mass) for use in pavement courses, suitable quarries as sources, amount likely to be available from each are identified, and physical and strength characteristics are determined. The materials that can be used are stone aggregates, murum, gravel, kankar, etc. Ceratin waste products of heavy industries like slag, fly ash, pond ash, etc can also be used in the construction of embankments and pavements. Project preparation should aim at maximum utilization of these substitute materials in the construction process. This would help in environmental and economical development. Possibility of getting overburnt bricks and brick bats to be used as brick

aggregates in the near vicinity of the project should also be envisaged and recorded. The material so available should be tested based on quarry source. Samples should be representative as far as possible. Three specimens per sample should be tested for each type of material met with. The tests required to be conducted on aggregate material are:

a. Gradation of the material
b. Aggregate abrasion value or aggregate impact value test
c. Flakiness and elongation test
d. Specific tests like soundness test, stone polishing value test, water absorption test, etc

4.6 SURVEYS FOR ROAD IMPROVEMENTS

Improvements of existing roads usually include strengthening or widening of road and construction of new facilities along the road. Such measures are taken to improve the traffic condition over the existing facility. The justification and scope of measures are dependent upon the road inventory and condition surveys. In case road registers and bridge registers are maintained by the concerned PWD, then these should be consulted and made use of. In case these sources are not available, a road inventory should be prepared and fresh condition survey should be carried out. Road inventory data includes data like type of terrain, land use, formation width, type, width and condition of carriageway, type, width and condition of shoulders, height of embankment, depth of submergence, etc. The pavement or road characteristics that are considered are:

a. Riding quality
b. Pavement width
c. Vertical profile
d. Horizontal curvature

The riding quality is measured by a bump integrator or roughometer. In case these instruments are not available, a International Roughness Index can be estimated based on World Bank guidelines and correlated with 5th Wheel Bump Integrator value using relation given in HDM-I11 volume-I. The roughness estimation scale for paved roads with surfacing or surface treatments varies between 0 and 12, representing high ride comfort at over 120 km/h and necessitating reduction in speed below 50 km/h, respectively. In case of unpaved roads with gravel or earth surface, the scale varies between 0 and 24, representing fine blanded surface of gravel or soil with excellent longitudinal and transverse profile and a poor earth surface with frequent moderate defects and depressions, respectively. The amount of undulations, surface depressions, potholes, corrugations, etc keep increasing as the number of roughness increases. The surface condition of the pavement is assessed by a team of experts, who rate the condition of the pavement on a five point scale based on amount of depressions, ruts, potholes, cracks, etc. The average value of all the experts is used as a "Pavement Condition Index" and recorded for use.

For pavement widening projects, the information needed is the subgrade soil strength and traffic projections or forecast for horizon or design year. The investigations carried out in such cases are:

a. California Bearing Ratio (CBR) test or 'K' value test indicating strength of soil subgrade
b. 'K' value test for materials used in pavement layers
c. Traffic census and growth rates for future traffic projections and predictions

For pavement strengthening projects, information on subgrade soil strength and pavement thickness and composition is needed. Deflection characteristics of the pavement are also required. Other information collected is related to pavement edge drop, condition of embankment, road side drain, etc. In case, the need of construction of additional lanes for widening of the road is known prior to the project preparation, the embankment and the pavement should be constructed eccentric with respect to the total land available. The subsequent construction may then be undertaken symmetrically with respect to the centre line of the land. In case of old roads, the provision may be made either symmetrically on both sides of the road or on one side of the road. The points to be considered in such case are:

a. The availability of land and convenience of additional acquisition if buildings or constructions are present on each side

b. The width of new construction and facility of compaction equipment to operate

c. Technical convenience for construction of additional structures and necessary protection works

d. Technical convenience for locations of additional two-lane carriageway preferably on upstream side of the flow of water

The additional investigations carried out, apart from the above listed, are:

a. Benkelman beam deflection data for existing roads

b. Pavement condition survey

c. Riding quality survey

The pavement width can be measured easily. The vertical profile and horizontal curvature are measured quickly by car-mounted instruments or can be evaluated from topographical survey.

In case of bridges, the inventory information includes data like, location of bridge, name of river and type of crossing, length/span of bridge, average vertical clearance, type of foundation, sub-structure and super-structure of bridge, type and condition of deck, carriageway, footway and railings, High Flood Level (HFL), thickness of slab or girder, type of protection works, etc.

4.7 DRAINAGE STUDIES

The water may come on the road in different forms like precipitation falling on the road, surface runoff, seepage water moving through sub-terranean channels or moisture rising by capillary action. Drainage studies have the following principal objectives:

a. Fixing the grade line of the road

b. Design of pavement

c. Design of the surface/sub-surface drainage system

Main components of the drainage investigation are determination of HFL and ponded water level, depth of water table, range of tidal levels and amount of surface run-off. Special investigations for cut sections are also carried out for roads in hilly and rolling terrain.

HFL is determined based on the information available with irrigation department, water marks left on trees and structures, and by gathering information locally. The height of embankment should be fixed based on HFL. If the embankment blocks the natural drainage paths then adequate number of openings shall be provided. The design HFL should be based on return period of flood and decided based on the importance of the structure.

In case water stagnates by the roadsides, the level of standing water should also be taken into consideration.

The knowledge of depth of water table helps in fixing the subgrade level, deciding the thickness of pavement, provision of capillary cutoffs or design of intercepting drains, etc. It may be measured at open wells along the alignment or at holes specially bored for the purpose. The

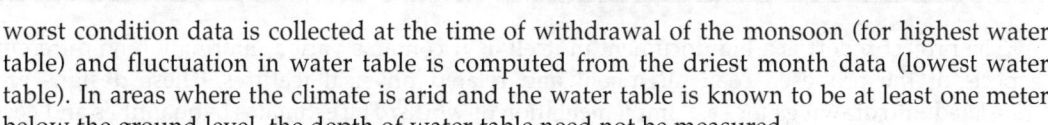

worst condition data is collected at the time of withdrawal of the monsoon (for highest water table) and fluctuation in water table is computed from the driest month data (lowest water table). In areas where the climate is arid and the water table is known to be at least one meter below the ground level, the depth of water table need not be measured.

The surface water is usually carried from the pavement through side drains. Following investigations need to be carried out for the design of these drains:

a. Study of ground contours of the land adjacent to the road for determining the catchment contributing to the flow

b. Determination of the surface characteristics of the catchment area like type of soil, vegetation, slopes, etc

c. Study of ground contours for locating the outfall points.

For cross-drainage structures, the decisions are taken at two levels, first to select the site and second to collect data for the design of cross-drainage (CD) structure. Following points should be given consideration in the selection of site:

a. The site should be on straight reach of the stream sufficiently below bends

b. The location should be as far away from the confluence of the large tributaries as possible

c. The bands should be well defined

d. As far as possible the site should enable a straight alignment and a square crossing.

The essential data needed for the design of new CD structure are:

a. Identification of catchment area

b. Taking three cross-sections of the stream, one at the selected site, one upstream and another downstream of the site. Approximate distance upstream or downstream of the selected site varies between 150 m and 160 m depending upon the size of the catchment area.

c. The maximum HFL

d. Longitudinal section of the channel showing levels of bed, low water level and HFL

e. Velocity of the flood, and

f. Trial pits, upto the depth of rock or firm mass and in case of their absence up to twice of the maximum depth of existing or anticipated scour line.

For existing drainage structure requiring improvement, the data required are:

a. Type of structure and details of span, vent height, etc

b. Existing width of carriageway

c. Condition of foundation, sub-structure, super-structure, etc

d. Load carrying capacity of the structure

e. Adequacy/inadequacy of waterway, signs of silting or blocking of ventway, over-topping of structure, scour level, patterns, etc.

4.8 PREPARATION OF DETAILED PROJECT REPORTS (DPRs)

The finalized alignment of the proposed road along with the reports of different investigations, related designs of facilities, and plans and drawings should be presented in a proper form for the consideration of competent authority. The complete project document should include:

a. The report

b. Estimates

c. Drawings

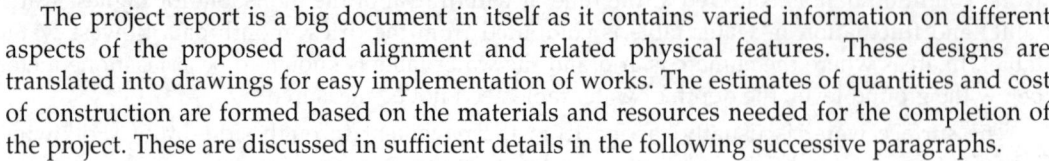

The project report is a big document in itself as it contains varied information on different aspects of the proposed road alignment and related physical features. These designs are translated into drawings for easy implementation of works. The estimates of quantities and cost of construction are formed based on the materials and resources needed for the completion of the project. These are discussed in sufficient details in the following successive paragraphs.

4.8.1 Project Report

The project report consists of the following sections:

a. **Executive Summary:** It is the summary of whole report and includes salient features of the proposal. This should be clear and straight forward. In case, the work is taken up under a scheme, the documents related to that scheme should form the part of the executive summary.

b. **Introduction:** This should include the name of the work, broad scope of the work, history, geographic, climatic and demographic details like economic activity, population served, transport facilities, drainage problems, topographical features, geological features, metrological data, and the justification of the project proposal.

c. **Socio-economic Profile:** The profile should include the socio-economic data of the project influence area as well as the region.

d. **Social and Environmental considerations:** This should include the social impacts like generation of employment, improvement in lifestyle and daily wage rate, health impacts, relocation requirements, resettlement and rehabilitation needs, and beneficial and detrimental impacts to the environment like different types of pollutions, loss of flora and fauna, ecological disturbances, erosion of productive soil cover, interruption to available drainage pattern, changes in landscapes, etc. The possible measures to minimize such detrimental impacts should also be discussed. The Ministry of Environment and Forests has issued environmental guidelines for highway projects.

e. **Traffic Surveys and Traffic Forecasts:** The traffic survey data, traffic growth computed for different horizon years, design traffic with respect to the design elements like pavement width, pavement crust design, intersections, crossings, etc should be included in this section of the report. In case of road under improvement or upgradation, the accident data should also be included, especially the list of accident prone locations.

f. **Engineering Surveys and Investigations:** This section should include information of the alternate alignments considered and the one selected based on the reasons. The merits of that alignment and the developmental effect it would bring in the area should be highlighted. The details of the selected alignment like longitudinal profile data, topographical and geological features, obligatory points, production and consumption centers, soil profile along the alignment, the formation levels and ground levels, the earth work involved, cross-sectional details at regular intervals, etc should also form the part of this section. Further, the decisions taken regarding the total road land, width of carriageway, width of shoulders, acquisition of structures alongside roadway, nature of cuttings or high banks, nature of gradients, radii of curves, etc should be noted. In case of existing roads, additional information includes road inventory, road condition survey, geo-technical investigations, bridge inventory if any, bridge condition survey, etc.

g. **Material, Labour and Equipment Survey:** The results of material survey like the availability of type of road construction material in near vicinity of the alignment, number of quarries for stone aggregates available and their transportation lead, the type of soil cover available locally, the physical and engineering properties of the stone aggregates and soil, the location and size of borrow pits, cost of transportation or bitumen or cement,

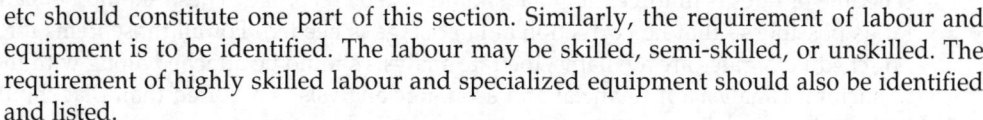

etc should constitute one part of this section. Similarly, the requirement of labour and equipment is to be identified. The labour may be skilled, semi-skilled, or unskilled. The requirement of highly skilled labour and specialized equipment should also be identified and listed.

h. **Design Standards and Specifications:** A complete list of standards that are used in the selection of alignment, fixing the physical features of the alignment, design of the facilities, preparation of estimates, preparation of abstract of cost, etc. should be given. Wherever, some deviation is made it should be highlighted. In general, such guidelines include specifications for fixing road geometric, specifications for road construction materials, specifications for materials used in layers of the pavement, specification related to special provisions to be taken for roads constructed in special areas, and standards related to road furniture, and design guidelines related to design of horizontal and vertical curves, sight distances, set-back distances, transition curves, super-elevation design, design of pavement, design of culverts and bridges, design of drainage structures and protection structures, etc.

i. **Geometric Designs:** Here all the standard values related to various geometric features of the road that should be provided along the alignment are listed. These includes type of terrain, different cross-sectional elements like lane width, carriageway width, width of shoulders, right-of-way width, road land, side slopes and drains, and camber (based on terrain classification), design of horizontal curves and super-elevation, length of transition curves, extra-widening at curves, provision of extra-widening, transition curve length and super-elevation at curve, choice of gradients, length of vertical curves, design of intersections if any, etc. Each curve should be identified by a number and drawing detail should follow the design for proper implementation. Fixing of grade line with respect to HFL and transverse section at regular interval showing ground level, formation level and final road (pavement) level should also constitute this part. In case of existing roads, the deficiencies in road design, remedial measures in problematic areas like landslide prone areas, need for increasing the carriageway width, etc. should be included.

j. **Pavement Design:** The soil investigation data and traffic counts and projections are utilized in the design of pavements. The decisions include the type of pavement construction (bituminous or cement-concrete), design of pavement crust, i.e. depth of pavement based on traffic load (million standard axles) and soil strength (CBR or 'K' value), number of layers, type of material to be used in those layers, specification of such material, design of wearing surface or slab, etc. All the calculations along with drawings should be given. In case of existing roads, the Benkelman beam data is used for the design of overlays on the existing pavement surface. Roads are usually designed in embankment. In such cases, the height of embankment, the slope of sides, stability of slopes, the use of waste material in the construction of the embankment, etc becomes the decision parameters.

k. **Drainage Facilities:** The hydrological data investigation becomes the part of this section. The decision parameters include HFL, water table, ponded water table, seepage flows, closure periods in case of canals, etc. The design features should include surface and sub-surface drainage measures like longitudinal side drains, catch water drains, longitudinal-transverse sub drains, blanket courses, etc. Design features of small cross-drainage structures like culverts should be enclosed. This includes design calculations for waterway, deck level, HFL, protection structures, etc. Design calculations and drawings should be attached.

1. **Rates and Cost Estimates:** The cost estimates are prepared based on schedule of rates published from time to time by regional PWDs. In case any correction to the last published

schedule of rates is in force then that should also be recorded. These should be realistic as far as possible so that no correction at later stage is needed. For all those items for which no schedule of rates are available, the rate analysis should be attached along with the basis of such rate analysis. If financial and economic analysis is required than that should also be included after the estimates.

m. **Construction Programme:** The executing agency of the work should be identified. The proposed period of completion with phased programme of execution of different works should be highlighted. This may be in the form of construction schedule defined by bar chart or Critical Path Method (CPM). Further included is the budgetary requirement at different stages of the construction.

n. **Conclusions and Recommendations:** These should form the last part of the report.

4.8.2 Estimates

This is an important section of the report and the timely completion of the project depends upon the accuracy of estimates to a great extent. The estimate consists of:

a. **General abstract of cost:** This gives the total cost of the project with the general breakup under major heads like land acquisition, site clearance, earthwork, sub- bases and bases, bituminous/cement concrete pavements, cross-drainage works, other structures, miscellaneous items, percentage charge for contingencies, work- charged establishments, quality control, cost of shifting utilities like electric lines, telephone poles, underground cables, gas lines, sewers, water pipes, cost of arboriculture, cost of removal of trees and compensatory afforestation.

b. **Detailed estimate of major heads:** This should consist of -
   i. Abstract of cost
   ii. Estimate of quantities
   iii. Analysis of rates for items not covered by the relevant schedule of rates
   iv. Quarry/material source charts.

4.8.3 Drawings

Project drawings should depict the proposed works in relation to the existing features. These should be drawn in uniform manner with regard to size, scales and the details to be incorporated. Following drawings should be included in the detailed report:

a. Locality map cum site plan
b. Land acquisition plans
c. Plan and longitudinal sections
d. Typical cross-sections
e. Detailed cross-sections
f. Drawings for cross-drainage works
g. Road junction intersection drawings
h. Drawings for retaining walls and other structures
i. Drawings for wayside amenities
j. Location of various road signs, markings, etc.

Following scales are commonly used for drawings:
Key map—1: 250, 000
Index map—1: 50, 000

Land acquisition plan—1: 2000 to 1 : 8000

Horizontal plan—1: 2500   on hill roads—1: 1000

Vertical plan—1 : 250   on hill roads—1 : 100

Cross-section elements (intervals)

> Plain area—50–100 m
> Hilly area—20 m   } Scale : 1 : 100
> Rolling area—50–75 m

Cross-Drainge structures – 1 : 50

Road Junction/Intersections – 1 : 500 to 1 : 600

Pavement Materials and Testing

5.0 PREAMBLE

An understanding of the soils on which highways/roads are founded and other materials used in construction of pavements is important. Use of locally available material is required to obtain economically constructed facilities. Different materials used in pavement structure are soils, aggregates which affect stability and durability, and binding materials such as bitumen and cement. Aggregates form 70 to 80 percent of the road structure and their characteristics and quality plays vital role in life and economy of a road. Suitability of locally available aggregates for use in different courses a pavement structure is determined by conducting certain tests. Aggregates are held in position by binding materials like bitumen and cement to provide strong and stable structure. Physical tests are conducted on bituminous binders and cements to determine suitability. Soils are foundation material for highways as a subgrade, and soil-aggregate mixes are used in pavement layers. Soils form bulk of material in embankments. Figure 5.1 shows use of these materials in layers of a flexible pavement.

This chapter summarises characteristics and qualities of these materials, determined through various tests to establish their suitability to perform intended function in the pavement structure. Details of test procedure and laboratory equipment required are specified in standard testing manuals, and readers should follow them when required.

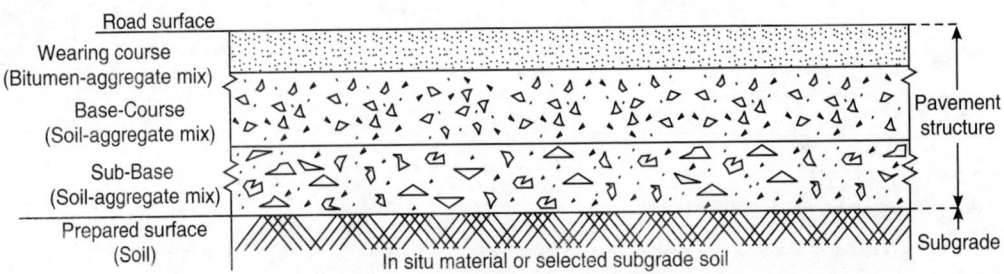

Fig. 5.1: Material used in flexible pavements

The basic requirements for pavement materials are as follows:

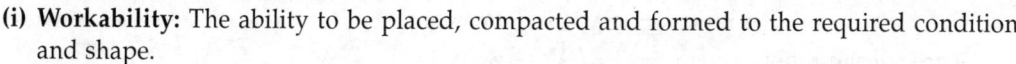

(i) Workability: The ability to be placed, compacted and formed to the required condition and shape.

(ii) Economy: The material must be available and workable at an acceptable cost.

(iii) Strength: The ability to resist loads without unacceptable deformation or crushing in service.

(iv) Durability: The ability to maintain its desired characteristics with time.

(v) Volume stability: The ability to resist significant changes in volume as conditions such as moisture content change.

(vi) Wear resistance: The ability to resist erosion, abrasion and polishing and provide frictional resistance in surface course.

(vii) Impermeability: The ability to resist moisture penetration through surface course.

Tests are performed to detect whether or not materials meet above requirements.

5.1 SOIL CHARACTERISTICS

Soils play most important role in highway construction and engineer should have a thorough understanding of soils and how they behave.

Soils are composed of solid, liquid and gaseous components as shown in Fig. 5.2.

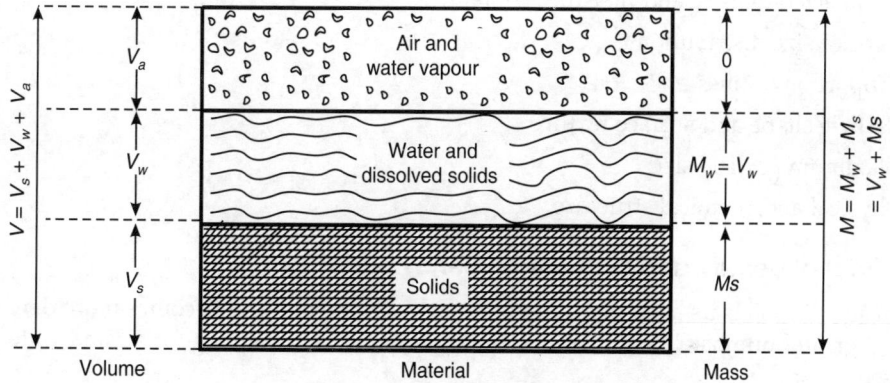

Fig. 5.2: Component materials in soil mass

Following terms can be derived from simple quantities shown in above figure.

$$Ms = \text{Mass of solids in the sample is determined by overdrying the sample to constant mass.}$$

$$Mw = \text{Mass of water} = \text{Total mass } M - \text{Mass of solids } Ms$$
$$= M - Ms = V_w.$$

$$Vs = \text{Volume of solids}$$
$$V_w = \text{Volume of water and dissolved solids}$$
$$V_a = \text{Volume of air voids}$$

Therefore:

Bulk density	d	$= M/V$
Dry density	d_d	$= Ms/V$
Solid density	d_s	$= Ms/Vs$
Moisture (water) content	w	$= M_w/M_s \times 100$ percent
total void ratio	e	$= (V_a + V_w)/V_s$

air void ratio $a = V_a/V \times 100$ percent

Degree of saturation $\qquad S_r = V_w/(V_a + V_w)$

Porosity $\qquad\qquad\qquad \eta = (V_a + V_w)/V \times 100$ percent.

The total void ratio 'e' is equivalent to the term VMA (voids in mineral aggregate) which is used in bituminous mix design.

Following relationships between above terms can also be derived.

(i) $\qquad\qquad$ Porosity $\eta = \dfrac{e}{1+e} \times 100$ percent

(ii) $\qquad\qquad$ Dry density $d_d = \dfrac{P}{1+w}$

(iii) $\qquad\qquad$ Dry density $d_d = \dfrac{1-a}{\dfrac{1}{d_s}+w}$

(iv) $\qquad\qquad$ Bulk density $d = d_d + S_r\,\eta$,

Those properties which affect behaviour of materials (soils) depend upon factors like:

(i) Compacted density and moisture content

(ii) Particle size distribution

(iii) Proportion of fines

(iv) Particle shape and surface texture

(v) Maximum particle size

(vi) Physical and chemical structure.

5.1.1 Effect of Density and Moisture on Behaviour of Soil

The density of a soil is its mass per unit volume. As a particular soil becomes more dense, it will contain a greater number of particles and the (pore) volume remaining for air and water will be decreased.

For coarse-grained soils, increased density and to a lesser degree decreased moisture content improve the physical properties of soil which are of primary importance in construction of pavements. Strength is increased, consolidation under loading and the rate of water movement through the soil are decreased.

Clays have high affinity for water. Clayey soils expand when come in contact with water. This results in uneven road surface and often failure of the roadway.

5.1.2 Compaction and its effect on Strength of Soils

Compaction is the process by which the air voids ratio of a soil is reduced. This is usually done by mechanical means such as rolling.

Consolidation is the process by which the moisture content and volume of the soil reduce, under long term sustained load.

A well compacted layer of material will suffer little permament or elastic deformation under subsequent traffic. In addition, compaction also raises the strength and stifness of the material and makes the entry of water more difficult, decreases permeability, increases the long term layer strength and inhibits consolidation. After selecting the pavement material, it is compacted to some 'Critical (optimum) Density' at which it will behave in the intended way.

The resistance of a material to compaction and in service deformation depends upon the internal friction and cohesion. Internal friction is influenced by the shape and surface texture of the coarse particles. It increases markedly as density increases and is little influenced by moisture content.

Cohesion is influenced by the properties of fine particles. Cohesion arises both from the molecular forces between very small particle and from tension in the soil water. It is greatly influenced by moisture content, decreases as moisture content increases to reach zero when the material becomes liquid. Cohesion increases with density.

From a compaction viewpoint, pavement materials can be of two types:

(i) Those from which water can be readily squeezed out during construction. The moisture content at which these are placed is not critical.

(ii) Those from which water is not readily squeezed out during construction (typically clays). The contained water takes some of the compaction pressure (The water pressure is called pore pressure). For such soils, there is an optimum moisture content (OMC) at which the maximum density will be achieved for a particular compactive effort.

5.1.3 Grading of Soils

Particle size is determined for coarse-grained soils by seiving a soil sample through standard seives and for fine-grained soils by sedimentation principle and a hydrometer. The test results are plotted as the percentage of the soil sample passing a given size to produce a particle size distribution curve.

Grading is the particle size distribution of a soil. Classification of particulate material by grading is important in pavement engineering as many relevant properties such as internal friction, void content, wear resistance and permeability depend on the distribution of particle sizes. A well graded distribution is one which permits each particle to fit into the voids created by inter-particle contact of the larger sizes. Poorly gaded mixes are limited in their compaction potential as it is impossible to produce a close-packed geometric arrangement in such cases. Gaps and voids always exist within the pavement. A uniformly graded soil is one with a predominance of single sized particles. A gap graded mix has at least one size range of particles missing. An open graded mix is a form of uniform grading which contains only small amounts of fine aggregates, (particles below 1 mm)

In order to achieve a desired particles size distribution, it is necessary to mix various aggregate fractions together.

5.1.4 The Atterberg Limits (Consistency Limits)

Pavement materials (soils) consist of 3 phases, gaseous, liquid and solid. As water is added to a particulate material, the internal cohesion drops as interparticle interlock is diminished. Eventually the mixture becomes a liquid with no effective particle interlock. It will then flow under its own weight. The moisture content at which this occurs is known as "Liquid Limit" (in percent).

A mixture just below the liquid limit will be in the plastic stage in which it can be permanently deformed under load without loosing its strength, i.e. It can be moulded and rolled into threads. For example clays can readily be made plastic by moistening and kneading. The moisture content at which the material becomes too dry to be plastic is called "Plastic limit". In many fine grained soils the plastic limit is a little below optimum moisture content (OMC).

The further removal of moisture will cause volume changes with in the soil until a moisture content called shrinkage limit is reached. Below which no volume changes occur. These three limits are called consistency or Atterberg limits. The plasticity index (PI) is the difference between plastic limit (PL) and liquid liquid (LL). It is thus a measure of the moisture range over which the soil will remains plastic and relate mainly to the activity of the clay component of the material. PI greater than 10 indicates clayey soil. Materials with low plasticity indices make the best subgrade and base course and a maximum value of 6 is a typical limit.

5.1.5 Classification of Soils (Soil Types)

Soils composed of mineral particles formed by the physical weathering of rocks are usually called 'Granular Soils', whereas soils formed from chemical weathering are usually called 'Clays'. Subdivision of soils is illustrated in Fig. 5.3.

Soil names in this figure depend upon particle size and range from boulders to fine clay. Clayey soils are prone to swell and shrink considerably with changes in moisture content. The presence of some clay in a material is desirable, as in small amounts it will increase strength and wear resistance. High plasticity indexes indicate the presence of clay, however a clay content of over 20% will mean that clay properties will dominate. Table 5.1 shows Casagrande soil classification

5.2 SOIL TESTS AND THEIR SIGNIFICANCE

Significance, synopsis of test method and typical test results of common soil tests on highway materials and soils are described in following paragraphs:

1. Particle size distribution test

   (Mechanical and Hydrometer procedure)
2. Specific gravity test for soils.
3. Consistency test and Indices (Atterberg limits)
4. Moisture and density test (Proctor's compaction test)
5. Field density by sand replacement test
6. California bearing ratio test
7. North Dakota cone test
8. Dynamic Cone Penetration Test (DCPT)
9. Plate bearing test
10. Shear test
11. Unconfined compression test
12. Consolidation test.

These tests are conducted as per Indian Standards (IS), or American Society of Testing Materials (ASTM) standards, and/or British Standards (BS), as indicated. For detailed procedures, equipment, and application, readers should consult a textbook of soil mechanics, as it is beyond the scope of this text.

5.2.1 Particle Size Distribution Test

(Mechanical and Hydrometer Procedure)

(IS: 2720 (iv) and IS: 1498-1970, ASTM: D422)

(a) Significance of the test

The mechanical analysis of a soil is the determination of the percent of individual grain size present in the sample. The results of the test are of most value when used for classification purposes.

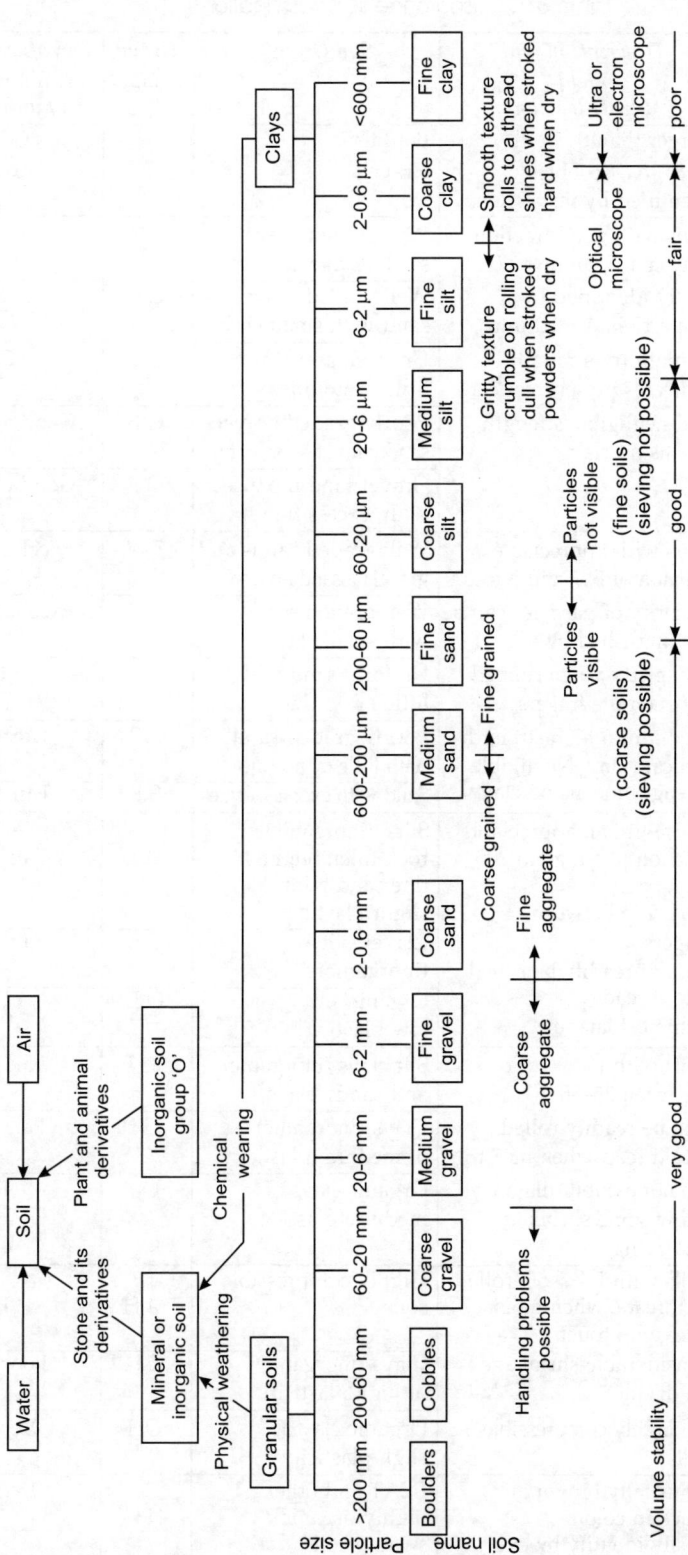

Fig. 5.3: Classification of soils

Table 5.1: Casagrande soil classification

Major Division		Description and Field Identification	Sub Group	Group Symbol	Value as road foundation in unfrost area
	Boulders and Pebbles	Larger than 8" in dia or 8" to 3" cobbles Identified by inspection	Boulder Gravels	–	Good to excellent
COARSE GRAINED SOILS	GRAVEL and GRAVELLY SOILS	Soils with good fraction between 3" and No. 7 B.S. easily identified by inspection. A medium to high dry strength indicates presence of clay and negligible strength means no clay	Well graded gravel sand mixture no fines	G W	Excellent
			Well graded gravel, sand with small clay	G C	Excellent
			Uniform gravel with little or no fines	G U	Good
			Poorly graded gravel sand with little or no fine	G P	Good to excellent
			Gravel sand mixture with excess fine	G F	Good to excellent
	SANDS and SANDY SOILS	Soils with appreciable fraction of BS7 and No. 200 majority of particles can be distinguished by eye. Feel gritty when rubbed between the fingers Medium to high strength indicate clay. Negligible strength means no clay	Well graded sands and graveling sand no fine	S W	Excellent to good
			Well graded sand with small clay	S C	Excellent to good
			Uniform sand with little or no fine	S U	Fair
			Poorly graded sand with little or no fine	S P	Fair to good
			Sand with excess of fines	S F	Fair to good
FINE GRAINED SOILS Containing little or no coarse grained material	FINE GRAINED SOILS HAVING LOW PLASTICITY (Silts)	Soils with an appreciable fraction passing No 200 LL < 35 No gritty between fingers Cannot readily be rolled into thread Exhibit dilatancy	Silts (inorganic) rock filler. Slightly fine sand with slight plasticity	M L	Fair to poor
			Clayey-silts (inorganic)	C L	Fair to poor
			Organic silt low plasticity	O L	Poor
	FINE GRAINED SOILS HAVING MEDIUM PLASTICITY	Soils with LL between 35–50 Can be readily rolled into thread when moist Do not exhibit dilatancy Show some shrinkage on drying	Silt clays (inorganic) and sandy clays	M I	Fair to poor
			Clays (inorganic) of medium plasticity	C I	Fair to poor
			Organic clays of medium plasticity	O I	Poor
	FINE GRAINED SOILS HAVING HIGH PLASTICITY	Soils with L L > 50, rolled in threads when moist Greasy to touch considerable shrinkage on drying All highly compressible soils	Highly compressible soils	M H	Poor
			Clays (inorganic) of high plasticity	C H	Poor to very poor
			Organic clay of sligh plasticity	OH	Very poor
FIBROUS ORGANIC SOILS WITH VERY HIGH COM-PRESSIBILITY		Unusually brown or black in colour, Easily identify by eye	PEAT and other highly organic swamp soils	Pt	Extremely poor

It is found that larger the grain size, the better are engineering properties. Detrimental capillary or frost damage are not a problem with coarse (sandy) material, whereas it can be very damaging with fine-grained silts and clays.

For soil stabilisation, use is frequently made of grain size analysis for mix design and control. Percent of cement to be used in soil-cement mixtures, is estimated on the basis of grain size. For mechanical stabilisation, results of gradation test are used to determine the size and percent of aggregates and fines.

At times, the degree of permeability (measure of the amount of water that will flow through a material) is estimated on the basis of grain size. Sands are more permeable than silts and silts are more permeable than clays.

(b) Synopsis of Test Method

The mechanical analysis consists of two parts:

(i) The determination of the amount of coarse material by sieve analysis.

(ii) The analysis of fine-grained fraction using a hydrometer (sedimentation method), using Stoke's law.

The sieve analysis is a simple test consisting of sieving a measured quantity of material through successive smaller sieves. The weight retained on each sieve is expressed as percentage of the total sample.

The hydrometer analysis is conducted on a sample of the material that passes No: 200 sieve. The test is based on the principle that the soil can be dispersed uniformly through a liquid. The specific gravity of the soil-liquid mixture is then measured at various time intervals. Stoke's law is used to compute the rate of setting of the various sizes, i.e. larger grains settle more rapidly than smaller grains. The computations include corrections for temperature, viscosity of the liquid and the specific gravity of soil particles. The results are first expressed as percent of the sample used in the hydrometer and then converted to percentages of total soil sample if there is a coarse grained fraction.

(c) Typical Test Results/Comments

Identification as per particle size is

Gravel   60–2.0 mm

Sand   2.0–.06 mm, Coarse: 0.6–2.0 mm, Medium: .2–.6 mm, Fine: .06–.2 mm

Silt   .06–.002 mm, Medium: .006–.02 mm, Fine: .002–.006 mm

Clay   <.002 mm

Same can be represented on the log-log paper graphically, as shown in Fig. 5.4.

The results of the test are of great value in earthwork, embankment and mechanical stabilisation of soil, for designing soil aggregated mixes, in making correlations between grain size distribution of soil and general soil behaviour as a subgrade material, its permeability characteristics, bearing capacity and other properties. Field identification of soils can be made approximately as shown in Table 5.2.

The grain size distribution curve gives the exact idea regarding gradation of the soil. It can be identified whether the soil is well graded, uniformly graded or poorly graded. Often specifications call for checking gradation or grain size distribution as criterion for selecting the soil.

5.2.2 Specific Gravity Test for Soils

(IS: 2720-III) (ASTM-D854)

(a) Significance of the Test

The specific gravity of a soil is the ratio of the weight in air of a given volume of soil particles at a stated temperature to the weight in air of an equal volume of distilled water.

Table 5.2: Field identification of soils

Soil type	Visual appearance	Resistance of dry piece of soil against crushing between fingures	Moist soil ribboned between thumb and fingers
Sand	Individual grains seen, Flows freely when dry	None	Can not be ribboned
Loam	Uniform mixture of sand, silt and clay	Low	Can not be ribboned
Silt	Soft flour like feel when dry	Medium	Has a tendency to ribbon with broken appearance. Feels smooth
Clay	Hard lumps when dry	High	Forms long, thin flexible ribbons, smooth greasy feel and sticks to fingers
Organic soils	Generally dark in colour, have distinctive smell, some fibrous materials.	Low	–

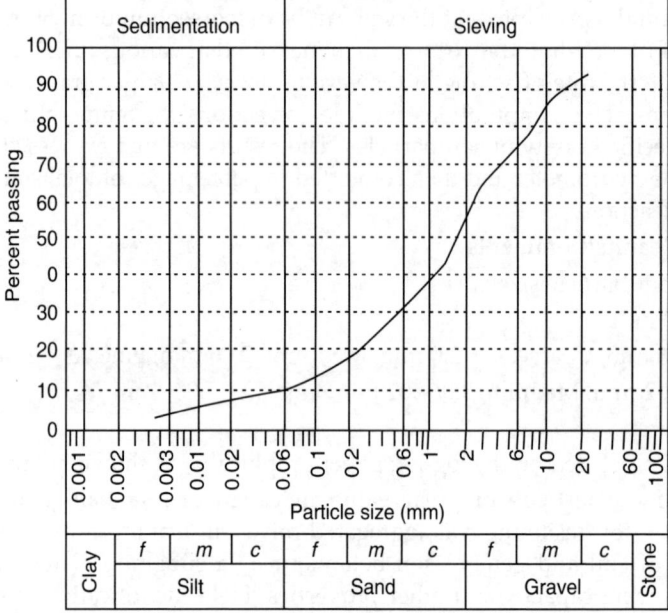

Fig. 5.4: Typical grading curve

Specific gravity is used frequently in computations of many laboratory test results, settlement, degree of saturation and void ratio calculations.

(b) Synopsis of Test method

The prescribed weight of the sample (all passing the sieve No: 4 or the No: 10 sieve depending upon purpose of test) is placed carefully in a calibrated pycnometer. Distilled water is added to fill the flask about three fourth full. The entrapped air in the soil then is removed by partial vacum or by boiling. The calibrated pycnometer then is filled with distilled water and weighed.

The specific gravity is computed, using the determined weights and temperature corrections. Figure 5.5 shows the equipment for determination of specific gravity of soils.

(c) Typical Test Results/Comments

The specific gravities of soils range, from below 2.0 for organic or porous particle soils to over 3.0 for soils containing heavy minerals. Most soils, however, have specific gravities in the range of 2.65 to 2.85.

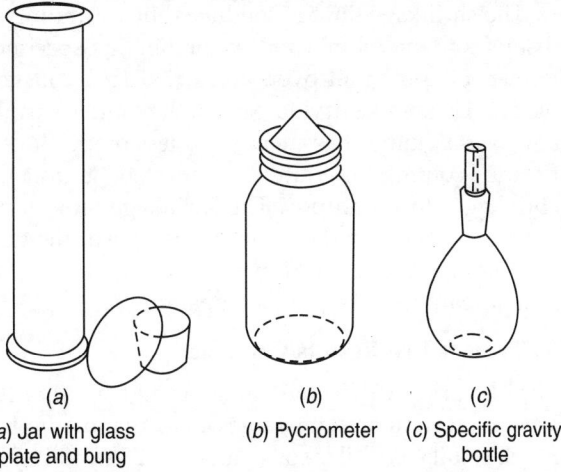

(a)   (b)   (c)

(a) Jar with glass   (b) Pycnometer   (c) Specific gravity
plate and bung   bottle

Fig. 5.5: Equipment for determination of specific gravity of soils

Accurate results depend upon care in obtaining weight and temperature measurements. A small error may be quite significant in the results. Calibration of pycnometers, complete removal of entrapped air and drying of the samples should be done with precision.

5.2.3 Consistency Test and Indices (Atterberg Limits)

(IS: 2720-V) (ASTM: D423, D424, D427)

(a) Significance of the Test

Consistency means the property of a soil which is manifested by its resistance to flow. The consistency tests are: the liquid limit, the plastic limit, and the shrinkage limit. Plasticity index is then calculated. Engineering properties of soils vary with amount of water present. The results of consistency tests are expressed as moisture content. They are arbitrarily used to differentiate between the various states of the material.

The 'Liquid Limit' (LL) is the moisture content at which the soil changes from the liquid to the plastic state.

The 'Plastic Limit' (PL) is the border between the plastic and semi-solid state.

The 'Shrinkage Limit' (SL) is the semi-solid from the solid state. The 'Plasticity Index' (PI) is the arthematic difference between the liquid limit and plastic limit, i.e. it is the range of moisture content over which the soil is in the plastic state.

Generally, soils with high liquid limits are clays with poor engineering properties. A low plasticity index indicates a granular soil with little or no cohesion and plasticity. LL and PI are used as quality measuring tests for pavement materials, in order to exclude those granular material with too many fine-grained particles.

Required plasticity and plasticity index values are specified for filler material, in bases and sub-bases.

(b) Synopsis of Test Method

The liquid limit test consists of molding a soil put in a brass cup, cutting a groove in the pot with a special cutting tool and dropping the cup onto a solid base from a constant height. The liquid limit is that moisture at which the groove closes for a length of 1.25 cm under 25 impacts.

The plastic limit test consists of rolling a wet soil sample into a thin thread, with hand. The procedure is repeated until the sample crumbles when the diameter of the thread is equal to 3 mm (1/8"). The moisture content of the soil in this latter condition is the plastic limit of the soil.

The shrinkage limit is conducted by saturating a soil sample, placing the material in a small dish of known volume and weighing. The specimen is then placed in an oven and dried to a constant weight. During the drying period, the sample shrinks and loses volume at a rate more or less proportionate to the volume of water evaporated, untill the shrinkage stops abruptly. The shrinkage limit is the moisture content of the saturated sample at the time the shrinkage ceases. Figure 5.6 shows the liquid limit apparatus (Casagrande apparatus).

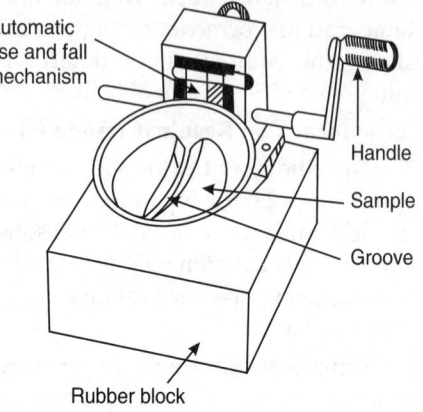

Fig. 5.6: Liquid limit apparatus (Casagrande apparatus)

(c) Typical Test Results/Comments

The LL varies widely and values as high as 80 to 100 are not uncommon with values 40 to 60 more typical for clay soils. For silty soils values of 25 to 50 can be expected. Liquid limit test will not produce a result for sandy soils as they are termed "Non-plastic".

The PL of silts and clays will not vary too widely and will range from 5 to 30. Normally silty soils will have the lower plastic limit. For the shrinkage limit, clays may range in values from 6 to 14, with silty materials most frequently showing values beween 15 and 30. Pure sand will show no decrease in volume during the drying period.

The plasticity index can be as high as 70 to 80 for the very plastic clays. Commonly, clays will have PI between 20 and 40. The silty material normally range in PI between 10 and 20.

For quality evaluation, bases, sub-bases and subgrades are sometimes restricted to those materials with liquid limit of 25 or less and a maximum PI of 6, i.e. a predominantly granular material. IRC requirements for WBM are the following:

Plasticity index,	*Soil Type*	Requirements of screening and binding material in WBM		
–	sand			
5–15	Sand-clay	**Screenings:**	LL: max: 20%	
15–25	Sand-clay		PI: max: 6	
25–40	Silty clay		Passing 75 μm max: 10%	
> 40	Heavy clay	**Binding material:** P.I.: max: 6 (Base)		
			max: 4–9 (Surfacing)	

For the liquid limit the most common source of error are:

(i) inaccurate height of drop of cup.

(ii) A worn up cup due to scratching with grooving tool.

(iii) Too thick a soil pot.

(iv) The rate of dropping the cup.

(v) The human error in deciding when groove has closed.

For controlling the earthwork or embankment construction plasticity index of the soil is determined to judge the quality of work.

For filler material in stone soling, sand or any other granular matter having a plasticity index of not more than 6 is used.

5.2.4 Moisture and Density Test (Proctor Compaction Test)

(IS-2720 – VII) (ASTM: D698)

(a) Significance of the Test

The moisture-density test helps in field compaction of soils so as to develop best engineering properties of the material. Standard test as conducted in the laboratory, uses a constant compactive effort, assuming that it is similar in magnitude to the weight, impact and action of the average construction equipment. As might be anticipated, a greater compactive effort will bring an increase in density.

Certain amount of water is needed in order to get densities desired. Water acts as a lubricant, however too much water tends to force the particles apart and higher densities can not be achieved. This test also delineates how much water should be used during compaction. Maximum density can be obtained at optimum moisture content, using best practical conditions of compaction in the field.

This test is useful in subgrade, sub-base, base course construction and compaction for a road.

(b) Synopsis of Test Method

The total sample is permitted to dry. From this, a sample of the material passing the No: 4 sieve is selected. This soil is then compacted in three layers into a metal cylinder mold of known volume. A metal rammer is dropped from a height of 30 cms onto the soil in the mold. A total of 25 bellows per layer is used. The weight of the soil in the mold is determined, and with the known volume of the mold, the density is computed by dividing the weight by the volume. A moisture content determination is made on the sample in the mold. The soil is then removed from the mold, pulverized, an increment of water mixed into the sample and the compaction procedure repeated. The test continues untill the weight of the compacted soil sample in the mold is less than that obtained in the preceeding step. Figure 5.7 shows the compaction mould and hand hold hammer for compaction test.

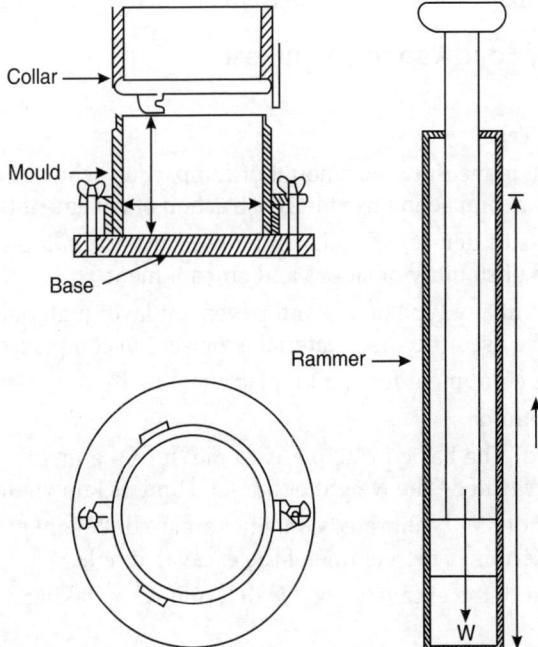

Fig. 5.7: Compaction mould and hand hammer for compaction test

(c) Typical Test Results/Comments

Moisture content verses the density plot is made. Calculations are then made of the density of the soil grain only, i.e. excluding the weight of water. This density is also plotted verses the moisture content and is termed as the 'dry-density' curve. The resulting plots are curved lines showing higher densities with increased moisture content up to some peak and then lower densities with increased moisture content.

The density of the peak of the dry-density curve is called the "Maximum Density" and the moisture content at that point is termed as "Optimum Moisture Content".

The following ranges of values might be anticipated for standard moisture-density test:

Clays:	Maximum density	—	90–105 lbs/cub.ft (1.44–1.68 g/cc)
	optimum m.c	—	20–30%
Silty clays:	Maximum density	—	100–115 lbs/cub.ft (1.6–1.84 g/cc)
	optimum m.c	—	15 to 25%
Sandy clays:	Maximum density	—	110–135 lbs/cub. ft (1.76–2.16 g/cc)
	optimum m.c	—	8 to 15%.

For modified procedure using an increased compactive effort maximum densities of 10–20 lbs/c.ft can be anticipated with optimum moisture content of 3 to 10 percent.

For sandy or gravelly soil with no fines, there is no significant change in density with the use of water.

Many compaction specifications require that a percent of the maximum density be achieved. This percentage varies from 95 to 100% for granular materials and 90–95% for the fine grained soils and clays. In the earth work, it is responsibility of the field engineer to ensure that density assumed in design is achieved at the expected moisture content. He should tests samples for moisture and density and take appropriate corrective measures.

In laboratory, precautions must be taken in mixing of the water into the soil thoroughly. In taking the moisture sample, care should be taken to obtain a representative sample.

5.2.5 Field Density by Sand Replacement Test

(IS: 2720 part-XXVIII)

(a) Significance of the Test

It is a quality control test to measure the amount of compaction achieved during the construction. It is a field control test for compaction used in construction of pavement layers, and subgrade soil.

Determination of in-situ density of soil, can be used in calculating over burden pressure of soil deposits in analysis of stability of slopes and embankments.

This tert can be used in any type of soil and pavement layer materials, unlike the core cutter method which can not be used if coarse material is present in compacted layer.

Figure 5.8 shows the equipment for sand replacement method.

(b) Synopsis of test Method

This is a simple method. The basic principle is to measure the in-situ volume of a hole, from which the material is excavated. The weight of the sand and its known density gives the volume.

A sand pouring cylinder and calibrated container are used. Weight of the sand filling the hole and density of sand determines the volume of the excavated hole.

The moisture content of the excavated soil is determined by taking a sample of soil from it.

(c) Typical test Results

This test determines field density, or (in-place density), moisture content and dry density of a compacted layer, and is used to check the amount of compaction attained in the field.

Sand used for the test should be clean, dry and uniformly graded. The depth of the excavated hole, size of pouring cylinder, and calibration container used depend upon the grain size of the material/soil, whose density is to be determined in the field.

5.2.6 California Bearing Ratio (CBR) Test

(IS: 2720-XVI) (ASTM: D1883)

(a) Significance of the Test

CBR test is a comparative measure of the shearing resistance of a soil. It is used with empirical curves to design flexible pavements. It is one of the most common test, developed in early 1940 by U.S. corps of engineers.

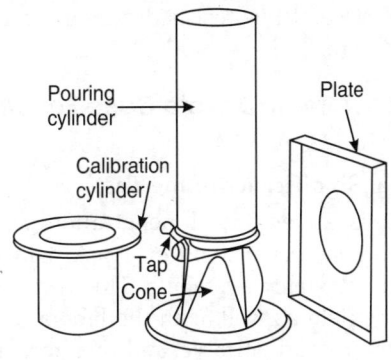

Fig. 5.8: Equipment for sand replacement method

CBR values of base, sub-base and subgrade are determined to design the total structure of flexible pavement. It is one of the most commonly used methods to evaluate the strength of subgrade soil for design of pavement thickness.

(b) Synopsis of Test Method

This test consists of measuring the load required to cause a plunger of standard size (50 mm dia) to penetrate a soil specimen at a specified rate (1.25 mm/min). CBR is the load required to force the piston into the soil a certain depth, expressed as a percentage of the load, required to force the piston the same depth into a standard sample of crushed stone. Penetration loads for the crushed stones have been standardised. The resulting ratio is known as 'California Bearing Ratio" abbreviated as CBR.

Penetration depth of 2.5 mm and 5 mm are commonly used in design. Procedures are prescribed for laboratory compacted specimen of swelling, non-swelling and granular materials. There is a standard procedure for performing inplace field tests and tests upon undisturbed samples obtained from the field. Soil must all pass 19 mm sieve. Soaked CBR is appropriate for worst conditions in the field.

It is assumed that the crushed rock with CBR of 100 would require no cover other than a wearing course. Justification of comparing clay to a crushed rock is that the test has a record of relative reliability and usefulness. Though the test results can not be related accurately with fundamental properties of the material, this test is very useful in design of flexible pavements. Figure 5.9 shows equipment for CBR test.

(c) Typical Test Results/Comments

CBR is an empirical value used in design of flexible pavement. Typical values are:

Soil Type	Approx: Soaked CBR
Sand	8–25
Sand clay	4–8
Silty clay	3–5
Heavy clay	1–3

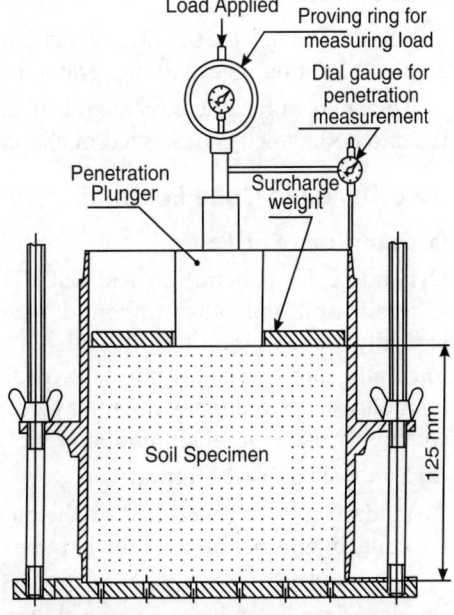

Fig. 5.9: Equipment for CBR test

Use of this test is described in chapter 14 on pavement design.

5.2.7 North Dakota Cone Penetration Test

(ISI: 2720 part XXXII)

(a) Significance of the Test

It is an empirical penetration test developed by North Dakota Department of U.S.A. which is used for designing the flexible pavements. This test can be performed both in the field as well as in the laboratory. Its equipment is very simple and portable, and therefore usable for quick controls in the field.

One design method for flexible pavements has been developed using this method by North Dakota State of United States of America. Pavement thickness requirements are worked out using an empirical procedure and design chart using this method.

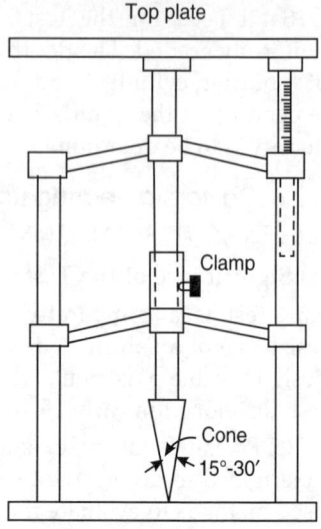

Fig. 5.10: North Dakota cone test

(b) Synopsis of Test Method

The cone test apparatus consists of a vertical member with a sharp steel cone of angle 15° 30' attached to one end. This can move up and down by a supporting frame, which is also provided by a locking device. On the top of the vertical shaft there is a top plate on which load can be placed. The penetration of the cone under the load can be measured by a graduated scale.

Penetration of this cone is noted under different loadings on the plate starting from 5 kg. Emprical formulae are then used to calculate bearing capacity of the soil/material, to be used in emprical design charts for flexible pavement design. Figure 5.10 shows typical north Dakota cone apparatus.

Typical Test Results

This test calculates the bearing capacity of soil (kg/cm^2), which does not contain coarse particles and thus limiting its use to fine grained soils, like silt and clays.

Thickness of flexible pavement is obtained by a formulae or chart, developed empirically for this method, which is described in chapter 14, on pavement design.

5.2.8 Dynamic Cone Penetration Test (DCPT)

(a) Significance of Test

Dynamic cone penetration test (DCPT) is done for rapid determination of the strength of subgrade soil and other unbound pavement layers. It can be performed in the field. It is basically a penetration test, in which a standard cone is driven in the mass with the standard force and rate of penetration is recorded. This test can also be conducted to determine the boundaries between pavement layers with different strengths and their thicknesses. It is commonly used for rehabilitation of bituminous pavement for design purposes.

(b) Synopsis of Test Method

A standard cone is penetrated by the blows of a standard hammer. DCP value is penetration of the cone, in mm per blow of the hammer.

The standard cone is made of steel with an angle of 60°, having the diameter of 20 mm. The standard drop hammer is of 8 kg weight which slides over a 16 mm dia steel rod with a fall of 57.5 cms. The measurements can be taken up to a depth of 1.2 m, with the extension rod. Penetration of the cone is measured on a graduated scale, at each blow of the weight. A plot is made between penetration and number of blows. Figure 5.11 shows typical Dynamic Cone Penetrometer (DCP).

(c) Typical Test Results

Penetration per blow is inversely proportional to the strength of the material. For a stronger material DCP value is lesser, i.e. penetration (mm) per blow will be less.

5.2.9 Plate Bearing Test

(IS: 1888–82) (ASTM: D1195)

(a) Significance of Test

The plate bearing test can be used to measure the strength of any component of the pavement structure. It can be performed on the surface of the subgrade, top of the sub-base, top of the base course or surface of the finished pavement.

This test is often used for thickness design of highway and airport pavements, using:

(a) For highways:
  – 30 cm diameter bearing plate
  – 5 mm deflection
  – 10 repetition of load
(b) For airport runways:
  – 75 cm diameter bearing plate
  – 12.5 cm. deflection
  – 10 repetitions of load.

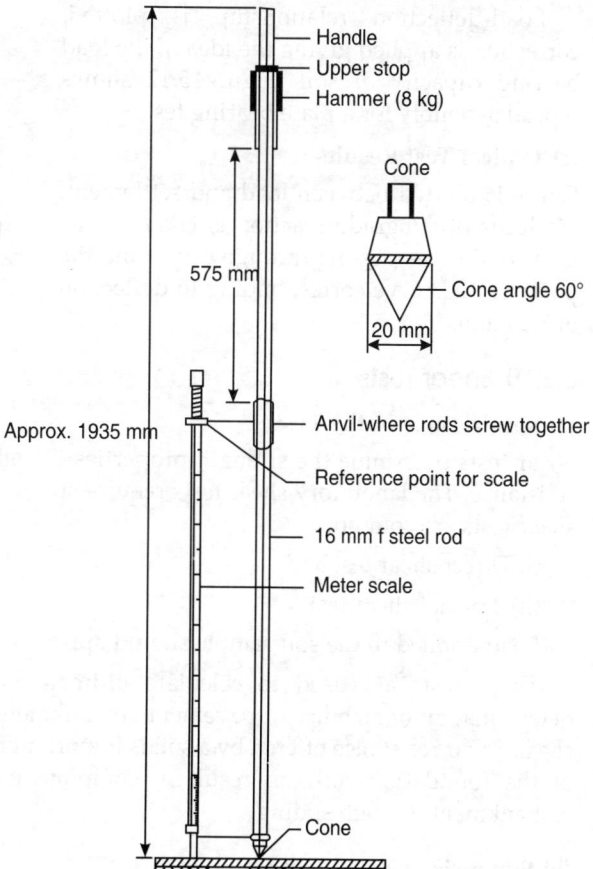

Fig. 5.11: Typical dynamic cone penetrometer (DCP)

Plate is used to calculate 'K' value (Modulus of subgrade reaction) in calculation of load stresses in Westergard's method of rigid pavement design.

(b) Synopsis of Test Method

Equipment for plate bearing test includes a loading device, a hydraulic jack assembly, bearing plates, and the necessary dial gauges, etc. loading device can be a truck, an anchored frame or other structure.

Bearing plates are of steel, circular in shape not less than 2.5 cm thick with diameter from 15 cm to 75 cm.

Plate of selected diameter is carefully centred under the applied load and the average vertical movement of the bearing plate is measured by dial guages, placed as prescribed.

Load is applied increasingly, maintaining a specified rate and then gradually released to note rebound deflection. There is standard procedure of performing this test depending upon the purpose.

Load-deflection relationship is plotted, correction is applied giving the idea of the load bearing capacity of soil. Figure 5.12 shows typical assembly for a plate bearing test.

(c) Typical Test Results

Curve is plotted between load and settlement. Modulus of subgrade reaction is taken as the slope of the line passing through origin and the point on the curve corresponding to deflection of 1.25 mm.

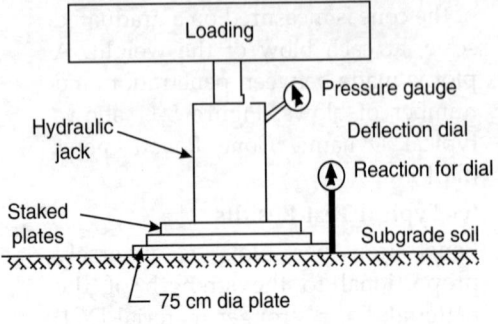

Fig. 5.12: Typical plate load test in the field

5.2.10 Shear Tests

(a) Significance of the Test

Shear tests determine the strength properties of soil mass, i.e. cohesion and angle of shearing resistance. The laboratory shear tests require specimen, similar to field conditions. Two type of shear tests are common:

(i) Direct shear test and

(ii) Triaxial shear test

Load applied to the soil sample should approximate the actual field loading conditions.

These tests are used to calculate ultimate bearing capacity of the soil, used in the determination of stability of pavement embankment and slopes, to prevent their failing under shear. Shear resistance offered by a soil is important factor in many highway problems. Shearing of the foundation soil can result in complete pavement disintegration and failing of an embankment through sliding.

(b) Synopsis of Test Methods

(i) Direct Shear Test

Specimen used in the direct shear test may be saturated prior to testing if it is desired to represent the most critical conditions. The failure of the test specimen is caused along a pre-determined plane of soil and both the shear resistance and normal stresses are measured directly. Figure 5.13 shows shear box test apparatus.

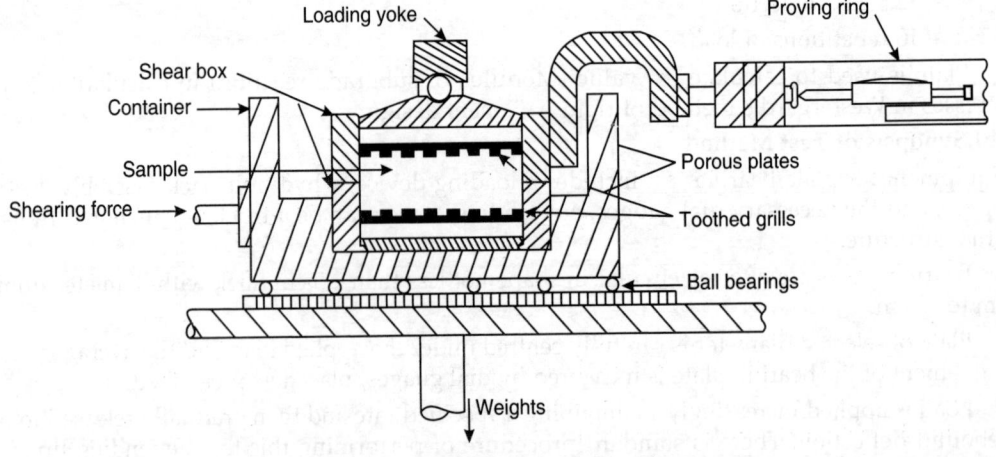

Fig. 5.13: Shear box test apparatus

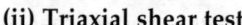

(ii) Triaxial shear test

In this test, the cylindrical specimen is subjected to three compressive stresses right angles to each other and one of these stresses is increased until the specimen fails in shear. It differs from direct shear test in that the plane of shear failure is not pre-determined. This test is considered better than the direct shear test. Figure 5.14 shows Tri-axial test equipment.

(c) Typical test results

For both the shear tests, load-versus-displacement curves are plotted and a critical point on each curve is determined. This point is related to the allowable displacement, the slope of the load-displacement curve and other factors. Cohesion (C) and angle of internal friction(ϕ) are found, from the curve.

Values of C and ϕ are considered constant for both laboratory and field conditions and the actual shearing resistance can then be calculated.

Typical values for friction and cohesion are as follows:

Sandy soils: ϕ = 28° to 45°

C = 0 to 2.06 mn/m^2

Clay soils: ϕ = 0° to 15°

C = 0.7 to 13.8 mn/m^2

5.2.11 Unconfined Compressive Strength Test

(a) Significance of the Test: The unconfined compressive strength test of a natural soil, is basically a shear test, as it is essentially a triaxial shear test with zero lateral pressure.

This test may be used to compare suitability of the soil for treatment with a given additive and to compare different mixtures, to select the additive content to be used in construction. Unconfined compressive strength data is mainly used for control purposes.

(b) Synopsis of Test Method: This test can only be carried out on cohesive soils or soils stabilised with an additive which binds the particles together.

A cylindrical specimen is prepared either by static or dynamic compaction method. It is then put to compression test, where the load is applied at a pre-determined uniform rate of deformation. The maximum load applied by the compressive machine is recorded as the unconfined compressive strength.

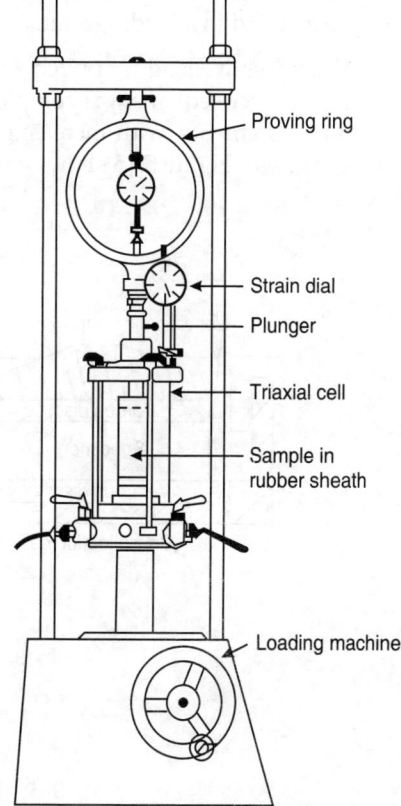

Fig. 5.14: Tri-axial test equipment

(labels on figure, top to bottom:) Proving ring · Strain dial · Plunger · Triaxial cell · Sample in rubber sheath · Loading machine

(c) Typical Test Results: Unconfined ultimate compressive strength is generally used in soil stabilisation for highway works. It is also possible to plot a load-versus-deflection curve. For a reasonably rigid material, the curve may be a straight line over a long range, indicating fairly elastic material.

5.2.12 Consolidation Test

(a) Significance of the Test

This test determines in an accelerated manner both the rate of settlement and the total amount of settlement to be expected under the total load. These values are important for analyses of highway embankment settlements and stability of slopes during construction.

Road pavements can be constructed safely on the embankment, when it is known that the bulk settlement is complete.

Consolidation test can also be used to determine how rapidly the height of an embankment can be increased during construction without a shear failure is produced in the soil.

(b) Synopsis of Test Method

Prediction of the settlement of a structure by use of a laboratory consolidation test requires that the sample used should be as nearly identical and representative of field soil mass as possible.

The soil sample is cut and trimmed so as to fit into a special metal ring provided for the test. Porous discs are placed on top and beneath the specimen and the assembled sample discs and ring are placed in a loading unit.

A compressive load is applied and changes in the thickness of the sample are read at set time intervals. After settlement is complete (say after 24 hours) the applied unit load is noted. 4 to 6 loading increments, with increasing load, depending upon the actual loading expected in the field are used. Figure 5.15 shows equipment for consolidation test.

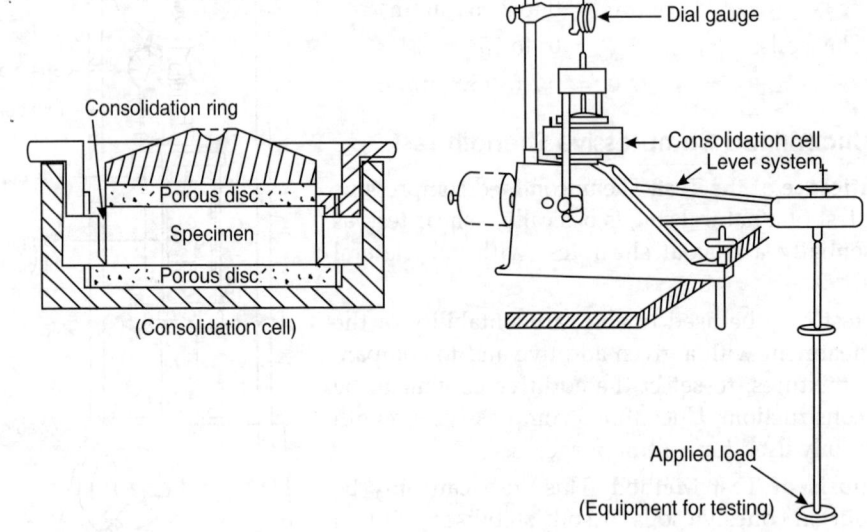

Fig. 5.15: Equipment for consolidation test

(c) Typical Test Results

Two principal values obtained from the consolidation test are the compressive index and coefficient of consolidation. These values are calculated from the test data and are used to estimate the rate, and total settlement under a given load condition.

The values for the coefficient of consolidation are generally:

Silty clays:	0.2 to 2.0 cm^2/sec
Clays:	0.02 to 0.1 cm^2/sec

The coefficient is used to estimate the amount of settlement for a given period of time under a given increament of load.

The settlement for the time period is compared with the total settlement for the load, and the ratio of the two, usually expressed as a percentage is called the degree of consolidation.

5.3 CHARACTERISTICS OF ROAD AGGREGATES

5.3.1 Types of Road Aggregates

(a) Natural aggregates (rocks): Aggregate is the term used to describe stone pieces used for road making. Igneous rocks are mainly crystalline and formed during the cooling of molten rock. Sedimentary rocks are formed from water deposition of granular (insoluble) soil, or air deposition of fine granular soil, through the weathering of other rocks. Sedimentary rocks are mechanically weak, they have bedding planes. They are not used commonly in road construction. Metamorphic rocks are formed when igneous and sedimentary rocks are subjected to heat and/or pressure of such magnitude that new minerals and textures are formed. They also are not of use in road making.

(b) Artificial aggregates: Artificial rocks such as slag, flyash are aften used in construction of roads. Slags are mainly from various metallurgical processes, particularly from iron and steel making and are commonly used artificial rock. Flyash comes from burning of coal in power stations. It is a material which is inert in itself but reacts with lime and water to form cemented products.

(c) Gravels: Gravels are naturally occurring particulate rocks and an excellent road making material.

Aggregates are sometimes obtained from deposits of naturally occurring and adequately sized and shaped rocks, usually gravels. Aggregates are also obtained by crushing rocks using mechanically driven hammers or gyratory crushers. Screening is done to remove over size material and produce gradings suitable for use. Aggregates are classified by their size as:

Coarse fraction: Material retained on a 4.75 mm sieve

Fine fraction: Material passing a 4.75 mm sieve

Coarse aggregate: Material retained on a 2.36 mm sieve

Fine aggregate: Material passing the 2.36 mm sieve but retained on a 425 μm sieve

Binder: Material passing the 425 μm sieve

Filler: Material passing a 75 μm sieve.

Coarse aggregates should be cubical and angular in shape rather than flaky and elongated. They should be durable – durability is a measure of a rock's (aggregate's) ability to resist repeated loading, wearing and weathering (or degradation). Figure 5.16. shows types of road making aggregates.

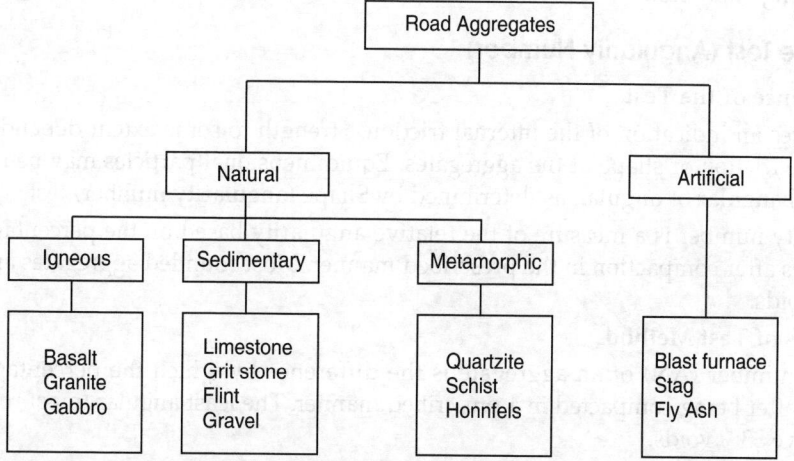

Fig. 5.16: Types of road aggregate

5.3.2 Functions and Characteristics of Road Aggregates

Aggregates in a road pavement perform following basic functions:

- Support the main stresses occurring within the pavement
- Resist wear due to abrasion of traffic
- Resist weathering action

Their properties and performance depend on both, the inherent properties of the individual particles and the means by which they are held together, e.g. interlocking, cementation, etc.

Ideal characteristics for aggregates used in bituminous mixes are:

- Strength and toughness
- Ability to crush into chunky particles free from flakes and pieces that are unduly thin or elongated
- Low porosity
- Hydrophobic characteristics
- Particle size and gradation according to the type of construction.

5.4 AGGREGATE TESTS AND THEIR SIGNIFICANCE

Significance, test methods in short and typical test results of the following basic tests on aggregates are described in following paragraphs.

1. Shape test (Angularity number)
2. Flakiness and elongation test
3. Specific gravity test
4. Water absorption test
5. Gradation test (Sieve Analysis)
6. Aggregate crushing value test
7. Aggregate impact test
8. Los Angles abrasion test
9. Deval abrasion test
10. Dorry abrasion test
11. Soundness test (Durability test)
12. Stripping value test.

5.4.1 Shape Test (Angularity Number)

(a) Significance of the Test

This test gives an indication of the internal friction. Strength to some extent depends upon the inter particle friction or shape of the aggregates. Equidimensional particles may be round, sub-rounded, subangular or angular, as determined by Shape (angularity number) test.

Angularity number is a measure of the relative angularity based on the percentage of voids in aggregates after compaction in the prescribed manner. Most rounded aggregates are found to have 33% voids.

(b) Synopsis of Test Method

Angularity number (AN) of an aggregate is the difference by which the percentage of voids exceeds 33, after being compacted in a prescribed manner. The least angular (most rounded) are found to have 33% voids.

Angularity number: AN = (Percent volume of voids) – 33

$$= (100 - \text{percent volume of voids} - 33$$
$$= (67 - \text{percent volume of voids})$$

Aggregate sample is compacted in the cylinder in 3 layers, using 100 blows. Cylinder is then weighed with compacted aggregate in it.

Angularity Number $AN = 67 - \dfrac{W_a}{G_a.W_w} \times 100$

Where

W_a = wt of dry agg. to fill cylinder in 3 layers

W_w = wt of water to fill cylinder

G_a = specific gravity of aggregates.

(c) Typical Test Results/Comments

Angularity number is expressed as a whole number

0 –   Rounded aggregate

3 – 6- Subrounded aggregate

8 – 11- Angular aggregate.

Crushed aggregates with higher angularity number are preferred.

5.4.2 Flakiness and Elongation Test

(*IS: 2386-I*)

(a) Significance of the Test

Flaky and Elongated stone pieces (aggregates) produce unstable and weak mix. They are usually unsuitable for road works.

"Flaky" are those having least dimension (thickness) smaller than 0.6 times the mean dimension of aggregate. (passing and retained certain sieves)

"Elongated" are those having greatest dimension (length) greater than 1.8 times of the mean dimension of aggregate (passing and retained on certain sieves)

Flaky aggregates get crushed under traffic. It is difficult to get edges of flaky particles coated with bitumen in premixes Elongated particles have poor strength and create problem during compaction.

This test eliminates aggregates unsuitable for bituminous roads.

(b) Synopsis of Test method

200 pieces of aggregates are taken and they are made to pass through the gauge of suitable opening, depending upon sizes of aggregate (passing and retained) e.g Size of aggregate passing 50 mm and retained 40 mm:

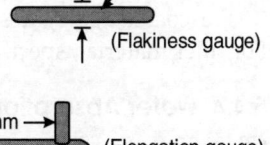

27 mm

(Flakiness gauge)

81 mm

(Elongation gauge)

Mean dimension is $\dfrac{50 + 40}{2}$ = 45m :

For flakiness: Thickness gauge is 0.6 × 45 = 27 mm

For elongation: greatest dimension   = 45 mm

Clear distance of length gauge   = 1.8 × 45 = 81 mm

Weight of flaky and elongated material, passing the respective gauges is taken accurately and expressed as percentage of total weight, of aggregates.

(c) Typical Test Results/Comments

Flakiness index of more than 15–25 is not desired. As per IRC recommendations, for WBM

Road layer	Flakiness index
Sub-base	–
Base with black topping	Max: 15
Surfacing	Max: 15

The limits for elongation index are same as for flaky aggregates.

5.4.3 Specific Gravity of Aggregates

(IS: 2386-II) (ASTM: C127)

(a) Significance of the Test

Specific gravity is the ratio of the mass of a given volume of aggregates to the mass of equal volume of water at a specified temperature.

The specific gravity of aggregates is important from the stand point of mixture calculations such as void contents in mixes.

Specific gravity could be expressed as "Apparent Specific Gravity" and "Bulk Specific Gracity"

$$\text{App: sp. gravity} = \frac{W}{(V_S + V_i)d}$$

$$\text{Bulk sp. gravity} = \frac{W}{V \cdot d}$$

Where d = unit weight of water.

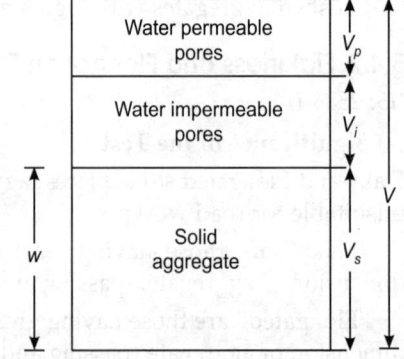

It is useful in:

– Billing and transportation purposes
– Proportioning of aggregates in bituminous mixtures by weight.

Coarse and fine fractions have different specific gravities.

(b) Synopsis of Test Method

The test is carried out using a pycnometer. Fine Aggregates are immersed in water for 24 hours, then water is drained and aggregates are saturated surface dried. Aggregates are then put in pycnometer and filled with distilled water and weighed. Aggregates are then ovendried and weighed.

For coarse aggregates, the wire bucket method is used for determination of specific gravity. For filler material, specific gravity bottle is used.

5.4.4 Water Absorption Test

(IS: 2386-III) (ASTM: C-127)

(a) Significance of the Test

This test shows the water absorption capacity of aggregates indicating the amount of binder required to satisfy the absorption of the aggregates in bituminous surfacing.

(b) Synopsis of Test Method

The aggregate sample is soaked in distilled water for 24 hours. After surface drying it is weighed. Then it is ovendried and weighed again. The difference between the weights of the saturated and oven dried sample as a percentage of the latter indicates the water absorption.

(c) Typical Test Results/Comments

Water absorption is high for porous aggregates. They require extra binder in a bituminous mix.

0.5% extra binder is recommended for every 1 percent water absorption in excess of 2 percent.

5.4.5 Gradation Test (Sieve Analysis)

(IS: 2386) (ASTM: C-136)

(a) Significance of the Test

This test establishes total percent passing through a particular size of sieve opening. It is useful in

- Controlling the aggregate specifications for construction of road.
- Determining the suitability of aggregate mix for a specific use

(b) Synopsis of Test Method

A weighed quantity of thoroughly dried aggregate is shaken over a set of sieves having selected sizes of square openings. One having the largest opening is rested on the top and successive smaller openings are placed beneath. A pan is placed below the bottom sieve to collect all material passing through it. Shaking is normally done with a mechanical sieve shaker.

The weight of material retained on each sieve size is determined and expressed as a percent of the weight of the original or total sample. These datas may be plotted on a chart/diagram.

(c) Typical Test Results

Table 5.3: Size requirements for mineral aggregates

Coarse Aggregate

S.No.	Standard size of aggregates	Designation of sieve through which the the aggregates shall wholly pass	Designation of sieve on which the aggregates shall wholly be retained
1	75 mm	106 mm	63 mm
2	63 mm	90 mm	53 mm
3	45 mm	53 mm	26.5 mm
4	26.5 mm	45 mm	22.4 mm
5	22.4 m	26.5 mm	13.2 mm
6	13.2 mm	22.4 mm	11.2 mm
7	11.2 mm	13.2 mm	6.7 mm
8	6.7 mm	11.2 mm	2.8 mm

5.4.6 Aggregate Crushing Value Test

(IS: 2386-IV)

(a) Significance of the Test

Road aggregates get crushed, by movement of roller during construction or due to heavy wheel loads. They must therefore be strong enough to support these loads. Crushing value test measures the strength. It gives a relative measure of the resistance of an aggregate to crushing under gradually applied compressive load. The weight of the fines produced in this process expressed as a percentage of the total weight represent the crushing value. Aggregates with lower crushing value indicate longer service life to the road and therefore more economical in performance. Aggregates used in road construction must be strong enough to withstand crushing under roller and traffic.

(b) Synopsis of Test Method

Aggregates (P 12.5–R 10 mm) are dried and filled in a cylinder in 3 layers, each tamped 25 times. Aggregates are then transferred to cylinder of the appratus, resting on base plate in 3 layers each layer tamped 25 times as before.

Load through the piston plunger is applied uniformly at rate of 4 ton/minute, untill 40 tons is reached. Crushed material passing sieve 2.36 mm is weighed.

$$\text{Aggregate crushing value } = \frac{\text{Wt. of crushed material passing 2.36 mm}}{\text{Total weight of dry sample}} \times 100$$

Figure 5.17 shows typical test equipment.

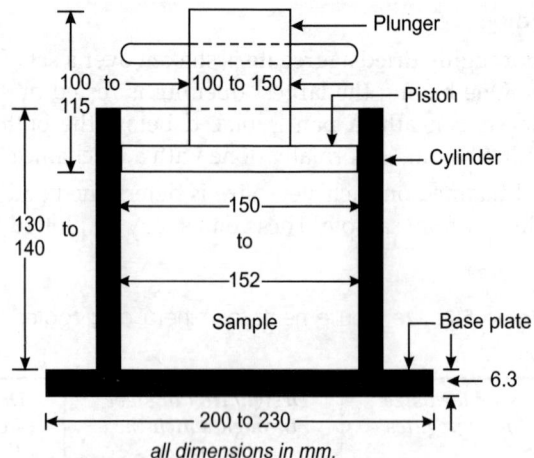

all dimensions in mm.

Fig. 5.17: Typical crushing test equipment

(c) Typical Test Results/Comments

In general, aggregate crushing value should not exceed 30 to 45%. Suitability of aggregates is judged, depending upon the proposed use in the payment layer, and for different types of road construction as given below:

Type of Road Construction	*Aggregate crushing value not to exceed*
1. Flexible Pavements	
(a) Soling	50
(b) Water Bound Macadam	40
(c) Bituminous Macadam	40
(d) Bituminous surface dressing or thin pre-mix carpet	30
(e) Dense mix carpet	30
2. Rigid Pavements	
(a) Other than wearing course	45
(b) Surface or wearing course	30

5.4.7 Aggregate Impact Test

(IS: 2386-IV)

(a) Significance of Test

Aggregates in pavements are also subjected to impacts/blows of wheels, specially at irregularities in roads resulting in their breaking down into smaller pieces.

Aggregate impact test helps in determing the resistance of stones to fracture or breakdown under shock and impact. It measures 'Toughness' of the aggregates, i.e. resistance to compaction with roller and resistance to impact and crushing action of traffic.

Aggregates should have sufficient toughness to resist their disintegration due to impact, or shocks.

(b) Synopsis of Test Method

Aggregates 1 kg in weight (P 12.5-R10 mm) are dried in oven. 1/2 kg is filled in standard cylindrical measure in 3 layers, tamping each layer 25 times.

Material is then transfered to cup of the impact testing machine compacted 25 strokes of tamping road. Hammer of machine is then allowed to fall freely 38 cm height, 15 blows (times).

Crushed material is passed through 2.36 mm sieve and weighed. Test is repeated for a second observation. Figure 5.18 shows aggregate impact test equipment.

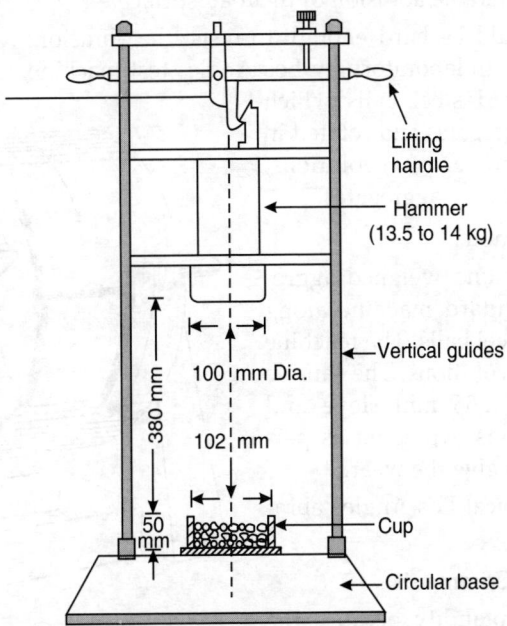

Fig. 5.18: Aggregate impact test equipment

$$\text{Agg. impact value (AIV)} = \frac{\text{Weight of material passing sieve 2.36 mm}}{\text{Total weight of dry sample}} \times 100$$

(c) Typical Test Results/Comments

Max. AIV allowed is 50 (dry condition), 60 (wet condition)

Aggregate impact values:

   (a) Very strong aggregate: < 10%
   (b) Strong aggregate: 10–20%
   (c) Satisfactory for road surfacing: 20–30%
   (d) Weak for road surfacing: > 30%

Maximum allowable value is 35 for bituminous courses. For built up spray grout, the aggregate impact value is max: 40.

Requirement of Coarse aggregate for WMB

Coarse aggreagate	Aggregate impact value
Sub base	max: 50
Base with bit. topping	max: 40
Surfacing	max: 30.

5.4.8 Los Angle's Abrasion Test

(IS: 2386-IV) (ASTM: C131)

(a) Significance of the Test

The aggregates to be used for wearing course of pavement, must have certain hardness against abrasive action. The soil particles between the pneumatic tyres of the moving vehicles and the road surfaces causes abrasion of road aggregates. The steel reamed wheels of animal drawn vehicles also cause considerable abrasion of the road surface.

Road aggregates should be hard enough to resist the abrasion. Resistance to abrasion of aggregates is determined in laboratory by Los-Angles test machine, which produces abrasive action by use of standard steel balls which when mixed with the aggregates and rotated in a drum for specified number of revolutions, causes impact and abrasion on aggregates.

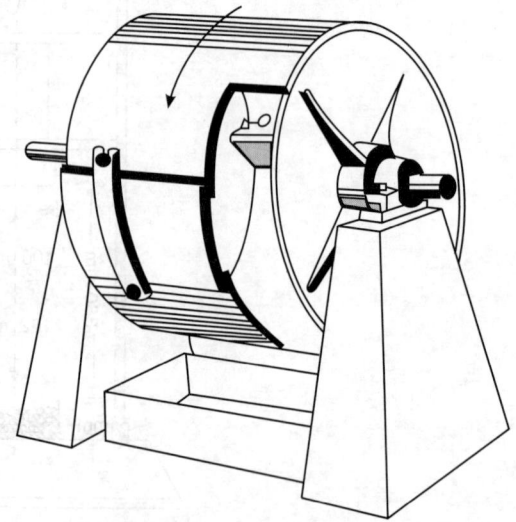

(b) Synopsis of Test Method

A dry sample of washed and weighed aggregates are put into a standard machine along with certain number of steel balls. The machine is then rotated for 500 revolutions. The sample is then sieved through 2.36 mm sieve and weight passing the sieve is expressed as percentage of total weight to give the wear.

Figure 5.19 shows typical Los Angles abrasion test machine.

(c) Typical Test Results/Comments

This test has more acceptability because the resistance to abrasion and impact is determined simultaneously. The maximum permissible values, specified by IRC for different road constructions are:

Fig. 5.19: Typical Los Angle's abrasion test equipment

S.No.	Type of pavement layer	Maximum permissible abrasion value%
1.	Water bound Macadam (Sub-base):	60
2.	Water bound Macadam (Base with bituminous surfacing):	50
3.	Bituminous bound Macadam:	50
4.	WBM surface course:	40
5.	Bituminous penetration Macadam:	40
6.	Bituminous surface dressing, cement concrete surface course:	35
7.	Bituminous concrete surface course:	30

5.4.9 Deval Abrasion Test

(a) Significance of the Test

This test is not very common these days. It also measures the abrasion of coarse aggregates. In this test abrasion as well as impact takes place due to the steel balls used as abrasive charge.

Aggregates used in pavements specially in wearing course must be hard enough to bear abrasive action of the traffic.

Abrasive action in this test is created by cast iron or steel balls.

(b) Synopsis of Test Method

Deval abrasion test machine consists of a hollow cast iron cylinder whose one end is closed, and other end is open which is used for putting the aggregate sample and six steel/cast iron spheres as abrasive charge.

The cylinder is mounted on a shaft at an angle of 30° with axis of rotation.

Clean and dried sample of aggregates, with chosen gradings, is weighed and placed in the Deval abrasion machine with 6 steel/cast iron spheres and rotated at a speed of 30 to 35 rpm for 10,000 rotations.

After rotations are complete the material is removed from the machine and is sieved on a 1.70 mm IS sieve. The material retained on the sieve is washed, dried and weighed.

Figure 5.20 shows typical Deval abrasion equipment.

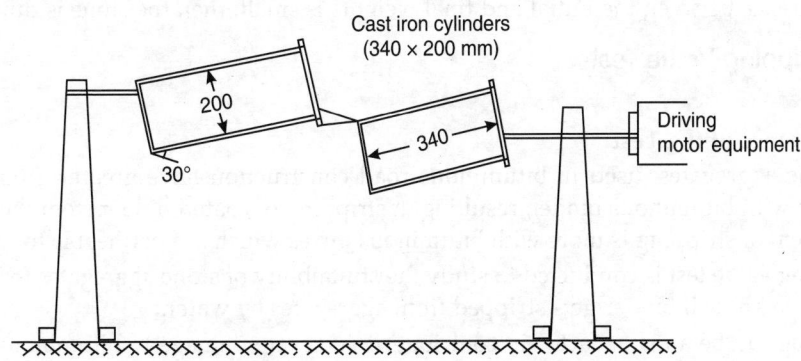

Fig. 5.20: Typical Deval abrasion test equipment

(c) Typical Test Results

The weight loss due to abrasion is the difference between the original weight of the test sample and the weight of the material retained on the 1.7 mm IS sieve. It is expressed as percentage.

This test is being replaced by Los Angle's abrasion test described earlier.

5.4.10 Dorry Abrasion Test

(a) Significance of the Test

This test was commonly used in U.K. for testing the abrasion. In this test a flat circular disc of iron or steel is rotated in horizontal plane at a speed of 30 rpm. A fixed weight of 2 kg is used to press the test sample down on the disc. Through a device standard abrasion sand is poured continuously on the disc in front of the test sample at a rate of 680 to 900 gm. per minut.

Like other abrasion test, the value of the loss of weight, of the sample aggregates is noted after passing through standard B.S. seives.

This test is simple, but preparation of test sample is time consuming.

(b) Synopsis of Test Method

Clean aggregates of specified size passing 12.5 mm and retained on 8.3 mm sieve, about 33 cm^3 are used. The sample in tray is placed on the abrasive machine and loaded. Standard Abrasive sand is fed continuously on the disc of the machine which is rotated at speed of 28-30 rpm.

After 500 revolutions, the test samples are removed and weighed.

(c) Typical Test Results

The aggregate abrasion values is expressed as the percentage loss in weight due to abrasion. Its use was initially common in U.K., but now a days it is replaced by Los Angles abrasion test.

5.4.11 Soundness Test (Durability Test)

(*IS: 2386–V*)

(a) Significance of the Test

This test is conducted to determine the resistance of the stone against sun, wind, rain etc. It indicates the susceptibility of aggregates to disintegration under atmosphere.

(b) Synopsis of Test Method

A sample of stone of definite size (4 cm cube) is kept in a 14% solution of NO_2SO_4 (Sodium sulphate) for 2 hours at 20°C. Then it is dried at 100°C. The process is repeated 12 times. Each time, the weight is determined any difference in weight is noted.

(c) Typical Test Results/Comments

If the difference between the initial and final weights is small, then the stone is durable.

5.4.12 Stripping Value Test

(*ISI-6241*)

(a) Significance of the Test

Some stone aggregates, used in bituminous road construction have greater affinity towards water than with bituminous binder, resulting in stripping of coated binder from the aggregates. This problem of stripping is more with bituminous mixes which are permeable to water.

Stripping value test is conducted to study the stuitaibility of stone aggregates for bituminous roads, and to know if binder gets stripped from aggregates by water.

Stripping can be a major factor in contributing to severe damage to bituminous pavements such as rutting, shoving, raveling, cracking, pot holes etc.

(b) Synopsis of Test Method

This is a static immersion test. Dry and clean aggregates passing 20 mm IS sieve and retained on 12.5 mm sieve are heated up to 150°. These aggregates are mixed with bitumen, thoroughly till they are completely coated.

The mixture is then transfered to a water beaker and allowed to cool at room temperature, for about 2 hours. Distilled water is then added and after 24 hours, it is estimated visually how much stripping of bitumen from coated aggregates has taken place.

(c) Typical Test Results

Stripping value is the ratio of the uncovered area observed visually to the total area of aggregates, expressed as percentage. Siliceous type aggregates, have low stripping values.

Due to visual assessment it has poor reproducibility. Anti-stripping agents are used, to reduce stripping. IRC recommends maximum stripping value as 25% for aggregates to be used in bituminous construction.

5.5 TYPE AND CHARACTERISTICS OF BITUMINOUS MATERIALS (BINDERS)

Bituminous road construction of different types, use one of the following bituminous material:

  (i) Paving Bitumen (penetration grade)

  (ii) Cut back Bitumen (liquid bitumen)

 (iii) Emulsions of Bitumen (liquid bitumen)

 (iv) Tar

  (v) Road oils

 (vi) Primers

(i) Paving Bitumen

Bitumen is a residual material obtained in the process of distillation of petroleum. It is a black or brown thermo-viscous material. It posses the characteristics of adhesion to road aggregates and its viscosity varies with temperature. Bitumen is insoluble in water. It completely dissolves in cabon disulphide but is non-volatile and resistant to most acids, alkalis and salts. Bitumen is composed mainly of hydrocarbons and their derivatives. Most of the bitumen in our country is obtained from imported crude petroleum.

Paving bitumens are rich in aromatic and napthenic hydrocarbons with paraffins. Bitumen is a thermoplastic viscoelastic adhesive, whose deformation response is dependent on both its temperature and the rate at which it is loaded. Its properties are

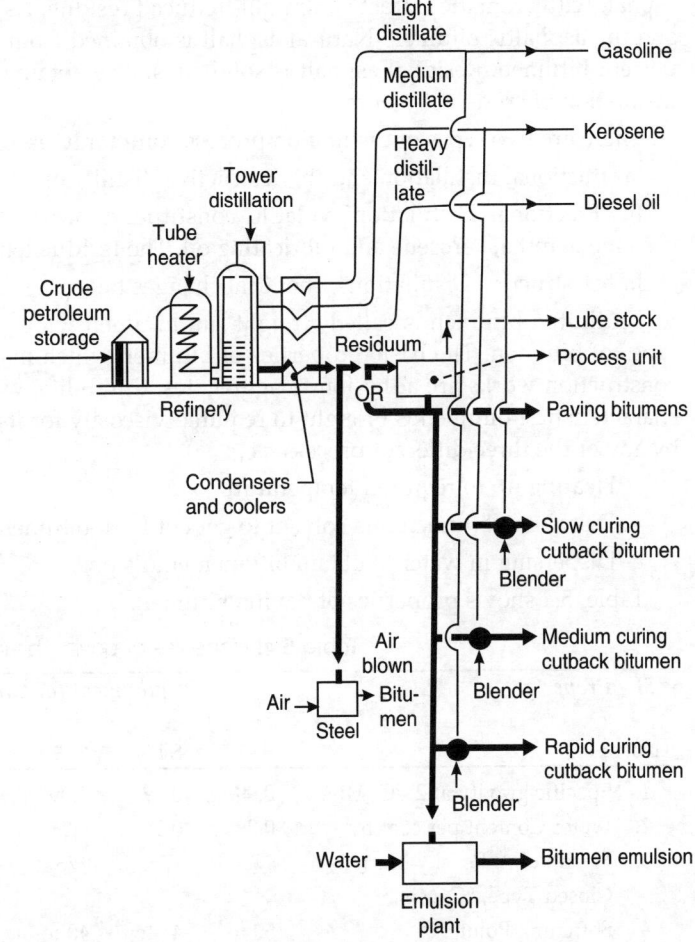

Fig. 5.21: Manufacturing process of bitumen

cheapness, workability, strength, durability, imperviousness and adhesion. It works as a binder for holding aggregates together in road construction, forming a stiff and impervious material.

Bitumen exists in solution as a natural constituent of most petroleums and usually is obtained as residue from the distillation of crude petroleum. It is therefore also known as 'Residual Bitumen'. In India it is obtained from imported petroleum crude. Bitumen is produced at Digbai refinery, Haldia refinery and Madras refinery. Figure 5.21 shows the production process of bitumen.

Fractionating removes products such as petrol and remaining long residue is removed by distillation. Three main processes normally used for separation and production of Bitumen are distillation, distillation plus air blowing, and distillation followed by blending with solvent.

In process of straight reduction, crude petroleum is heated and then flashed into fractionating column. The lighter fractions are separated leaving behind the long residue containing heavy oils and bitumen. By vacum distillation most of the heavy oil fraction is removed, and 80/100 penetration bitumen is obtained as short residue. This is a simple, straight and efficient process for obtaining bitumen.

In Propane deasphalting process Bitumen is a by-product of propane deasphalting and fractionation process. In this process paving bitumen is obtained by blending precipitated asphalt with aromatic extract or straight- reduced residue. Asphalt is a mixture of inert material and the asphaltic bitumen. Natural asphalt is obtained from lakes and it may contain 40 to 80 percent bitumen. Residual asphalt is solid substance obtained as base material after fractional distillation of crude petroleum.

There are two types of distillation process commonly used:

(a) fractional distillation and (b) destructive distillation.

In 'Fractional Distillation' volatile constituents are obtained as successive fractions of gasoline, naptha, kerosene and lubricating oil. The residue left is bitumen.

In "Destructive Distillation", chemical changes take place under extreme heat and pressure.

Undiluted bitumen is called straight run bitumen which is further treated to obtain other types of bitumen. The basic properties of bitumen which make this material suitable for road construction works are adhesion to aggregates, durability, water proofing, and thermoviscous characteristics. Bitumen is brought to required viscosity for its application in road construction, by any of the three different process, i.e.

 – Heating up to required temperature
 – Dissolving bitumen in a solvent to get cut-back bitumen
 – Dispersing in water to obtain bitumen emulsions.

Table 5.4 shows properties of paving bitumen.

Table 5.4: Properties of paving bitumen

Sl. No.	Property	Requirement for Grades					Reference to method of test
		S35	S45	S65	S90	S200	
1.	Specific gravity at 27°C, Min.	0.99	0.99	0.99	0.98	0.97	IS: 1202–1978
2.	Water Content per cent by	0.2	0.2	0.2	0.2	0.2	IS: 1211–1978
3.	Flash Point, Pensky Martens Closed Type, °C, Min.	175	175	175	175	175	IS: 1209–1978 (Method A)
4.	Softening Point, °C	50 to 65	45 to 60	40 to 55	35 to 50	30 to 45	IS: 1205–1978
5.	Penetration, at 25°C, 100g,	30 to 40	40 to 50	60 to 70	80 to 100	175 to 225	IS: 1203–1978
6.	Ductility at 27°C in cm, Min. (Ductility requirements of Digboi bitumens are given in brackets.)	50 (10)	75 (12)	75 (15)	75 (15)	– (15)	IS: 1208–1978
7.	(a) Loss on heating, per cent by weight, Max.	1	1	1	1	2	IS: 1212–1978
	(b) Penetration of residue expressed as percentage of item 5, Min.	60	60	60	60	60	IS: 1203–1978
8.	Matter soluble in carbon disulphide/trichloroethylene, per cent by wt. Min.	99	99	99	99	99	IS: 1216–1978

Bituminous binder should meet the requirements of viscocity, temperature susceptibility and stripping as per IS: 73–1992.

Paving Bitumen 30/70, 80/100 are commonly, used for pre-mix carpet, built up grout, penetration macadam, bituminous macadam, asphaltic concrete etc. depending upon climatic conditions of the area. For surface dressing work 80/100 bitumen is more common. For areas with hot climate, a harder paving bitumen 30/40 is used. 30/40 bitumen is also advantageous for roads with large number of repitition of wheel loads like expressways, urban roads etc.

Bitumen with Additives (Rubberised-Bitumen)/Modified Bitumen

There is a growing trend to use rubber and other polymers in bitumen. Latex, strap rubber have been tried. The behaviour of rubberised bitumen is intermediate between that of rubber and of bitumen. Both rubber and bitumen are viscoelstic, rubbers are less sensitive to the effect of temperature. Bitumen treated with modifiers are known as 'Modified Bitumen'.

Epoxy resins have been used to increase stiffness but it is not good in bonding properties. Ethyl Vinyl Acetate (EVA) has also been tried. It is not yet possible to disperse rubber into tar or bitumen and obtain a mixture that is stable for road works.

There are some observations where it is found that aggregate-bituminous mixture containing rubber are more resilient than without, and are better for providing cushion to traffic vibrations. Bitumen also reduces surface cracking due to temperature. Rheological properties of bituminous binder are considerably improved by addition of rubber/resins. The important contribution of addition of rubber is that the change in viscosity with temperature is significantly reduced. The softening point of the bitumen is increased and brittle point is lowered. Crumb rubber modified bitumens (CRMBs) are being tried in road works.

The cost of rubberised bitumen is increased but the same could be off-set by the improved tenacity, resistance to deformation, susceptibility to temperature changes, resistance to stripping by action of water, and longer life of surfacing. These improved properties of the binder may be used in locations like bus stops, steep gradients, roads having high traffic, traffic with accelerating and decelerating situations. Research for use of rubber in bitumen are in progress both in laboratory and field, all over the world.

Indian Road Congress, based on the performance of pavements constructed by modified bitumen has published guidelines details of which are included in its publication IRC: SP-53 "Guidelines on use of Polymer and Rubber Modified Bitumen in Road construction".

Advantages of using modified bitumen are:
- Improved performance and durable roads
- Lower susceptibility to temperature
- Higher resistance to deformation
- Greater life of mixes
- Better adhesion between bitumen and aggregates
- Better resistance and lesser cracking.

Performance of modified bitumen is dependent on strict control of temperature during construction. Viscosity temperature relationships of modified binders help in knowing the temperatures at which specified viscosity can be achieved for road construction operations.

(ii) Cut-back Bitumen

As bitumen is not volatile, it is mixed with volatile solvents to obtain a solution which can be applied at low temperatures. Volatile diluent reduces the viscosity of the bitumen temporarily. The new compound is called 'Cut-Back' bitumen. Cut back bitumen are more workable and they can retain their low viscosity for a much longer time than practical with a heated bitumen. The effect of solvent diminishes with time, as solvent get evaporated, and bitumen develops the binding property.

Different type of solvents are used to produce short, medium, or long term effects, and they are called

(i) Rapid curing cutback (RC): Solvents are naptha or petrol.

(ii) Medium curing cutback (MC): Solvents are kerosene, light diesel oil.

(iii) Slow curing cut back (SC): Solvents are diesel oil, furnace oil etc.

The above classification is based on the rate of curing or hardening after the application of cut-back bitumen.

Suffix of numeral 0, 1, 2, 3, 4, and 5 to above RC, MC, and RC categories of cut back bitumen indicate the thicker or more viscous form of cutback. RC-0, RC-1 contain high proportion of solvents as compared to RC-4, RC-5 etc. The number also indicates the viscosity levels at specified temperatures and setting characteristics.

Rapid Curing cut-back (RC) are used for surface dressing in cold weather, while medium curing (MC) are suitable for pre-mix with less quantity of fine aggregate. Slow Curing (SC) are better in pre-mix containing large quantity of fine aggregates.

Table 5.5 shows details and properties of cut backs, and primers.

Table 5.5: Properties of cutback bitumen

S.N.	Type of cutback Bitumen	Penetration value of Base Bitumen	Type of Solvent used	Amount of Solvent percent by volume of cutback-bitumen	Viscosity range of cut back bitumen cst
1.	Rapid curing (RC)	80/100	Naptha	15-45	70-6000
2.	Medium curing (MC)	80/100	Kerosene	15–45	30–6000
3.	Slow curing (SC)	80/100	Heavy Distillate	0–50	70–6000
4.	Bituminous Primer	80/100	Diesel oil furnace oil	50–70	15–4000

(iii) Emulsions of Bitumen

An emulsion is a mixture of two immiscible liquids one of which is dispersed in the other in the form of very fine particles. Emulsion is composed of bitumen, water and an emulsifying agent, such that bitumen is mechanically broken into fine globules. A colloid mill is generally used for the preparation of an emulsion.

The emulsifying agents consist of soaps resinous bodies and it is absorbed on the surface of each globule of bitumen forming a film, which prevents the bitumen globules from coagulating.

Two types of bitumen emulsions are used, viz anionic and cationic. In the anionic emulsions, the emusifier used is generally fatty acid metallic soap. The water is alkaline and bitumen particles are negatively charged. Anionic bitumen are suitable for use with calcarcous aggregates like limestone. In the cationic bitumen emulsion, the emulsifier used belongs to long chain of amine or quarternary ammonium salt class. Water is acidic and bitumen particles are positive charged. Cationic bitumen are suitable for use with siliceous aggregates like granite, sand stone.

Bitumen emulsions can also have three different setting characteristics i.e. rapid setting, medium setting and slow setting. The rapid setting (RC) are suitable for surface dressing and patch work. The medium setting (MC) are used for pre-mix. The slow curing emulsion are used as slurry seal, etc. Emulsion are commonly used for maintenance and repair works of the road,

and soil stabilisation. Emulsions can also be used in wet weather and rains. Tables 5.6 and 5.7 show properties of emulsions.

Table 5.6: Properties of anionic bitumen emulsion

Sl. No.	Property	Requirement for		
		Rapid setting type	*Medium setting type*	*Slow setting type*
1.	Viscosity by Standard Saybolt Furol Viscometer, seconds at 25°C	20-100	20-100	20-100
2.	Water content, per cent by weight		Not more than 45	
3.	Settlement, 5 days, per cent, Max.	3	3	3
4.	Demulsibility, 35 ml of 0.02 N Calcium chloride, per cent, Min.	60	–	
5.	Demulsibility, 50 ml of 0.1 N Calcium chloride, per cent, Min.	–	30	–
6.	Miscibility in water, appreciable coagulation in two hours	–	Nil	–
7.	Modified miscibility with water, difference of bitumen content, Max.	–	–	4.5
8.	Cement mixing test, per cent, Max.	–	–	2.0
9.	Sieve test, per cent, Max.	0.10	0.10	0.50
10.	Particle charge	Negative	Negative	Negative

Table 5.7: Properties of cationic bitumen emulsion

Sl. No.	Property	Requirement for		
		Rapid setting type	*Medium setting type*	*Slow setting type*
1.	Residue on 600-micron IS Sieve, per cent by mass, Max.	0.05	0.05	0.05
2.	Binder content, per cent by mass, Min.	57	57	57
3.	Viscosity by Saybolt Furol Viscometer, seconds			
	at 25°	–	–	20–100
	at 50°C	50–400	50–400	–
4.	Coagulation of emulsion at low temperature	Nil	Nil	Nil
5.	Storage stability after 24 hr, per cent	1	1	1
6.	Particle charge	Positive	Positive	Positive
7.	Stability to mixing with coarse aggregate (per cent coagulation)	20–80	< 40	< 5
8.	Stability to mixing with cement (per cent coagulation)	–	–	< 2
9.	Miscibility with water	Nil	Nil	Nil

(Contd.)

Table 5.7: Properties of cationic bitumen emulsion (*Contd.*)

Sl. No.	Property	Requirement for		
		Rapid setting type	*Medium setting type*	*Slow setting type*
10.	Tests on residue			
	(a) Residue by evaporation, per cent, Min.	57	57	57
	(b) Penetration at 25°C 100g, 5 sec, in 1/10 mm.	60–210	60–210	60–210
	(c) Ductility at 27°C, in cm, Min.	100	100	100
	(d) Solubility:			
	– in carbon disulphide	99	99	99
	– in trichloroethylene	97.5	97.5	97.5

Emulsions are used as binder for surface dressing works, for sealing, filling cracks in pavements, for pre-coating of aggregates and for tack coats, etc. Rapid setting are good for surface dressing and liquid seal. Cationic emulsions can be used with wet aggregates. Medium setting emulsions are used for premix open graded patching and slow setting are used for slurry seals.

(iv) Road tar

Road tar is obtained from the coal tar which is a by-product in the process of destructive distillation of coal and wood. Bitumen has now replaced tar for construction of roads. Road tar also is a thermoviscous material, viscosity varying with temperature. Road tars have better adhesion quality, but are more susceptible to temperature than bitumen. They get oxidised sooner than bitumen and thus have shorter life, and less durability.

Road tars are classified in five grades RT-1 to RT-5, depending upon their viscosity and use. RT-1 is low viscosity tar used for surface dressing in cold weather. RT-2 and RT-3 are commonly used for surface dressing, pre-mixing and carpeting, etc. RT-4 and RT-5 are good for bituminous macadam or grouting, etc. Basic properties of 5 types of road tars are given in Table 5.8

Table 5.8: Properties of road tars

S.No.	Characteristics	RT-1	RT-2	RT-3	RT-4	RT-5
1.	Viscosity					
	(a) Temperature °C	35	40	45	55	65
	(b) Seconds	30–55	30–55	35–60	40–60	–
2.	Softening point of residue	48	50	52	54	56

Figure 5.22 outlines the manufacture process of road tars.

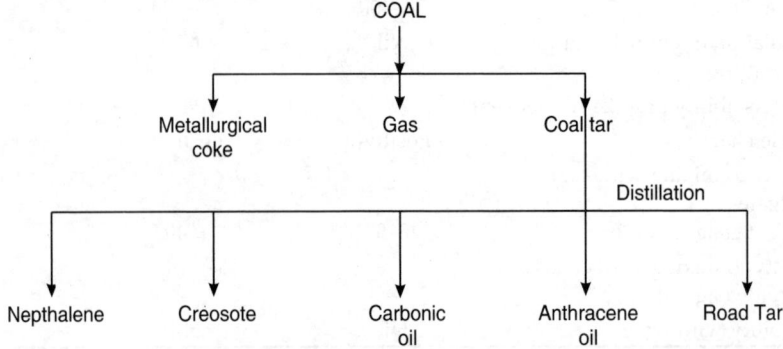

Fig. 5.22: Outline of road tar manufacturing

Tables 5.9 and 5.10 show selection of bitumen and tar for road works.

Table 5.9: Selection of bituminous binder for road works

S. No.	Type of binder	Application
1.	30/40 Penetration Grade	Penetration macadam in normal summer temperatures and in plains. Hot mix asphaltic concrete in normal summer temperatures and in plains. Bituminous macadam (hot mix) in normal summer temperatures and in plains.
2.	60/70 Penetration Grade	Penetration macadam in winter and at high altitudes. Hot mix asphaltic concrete in normal summer temperatures and in plains. Hot mix bituminous macadam in summer climate and in plains.
3.	80/100 Penetration Grade	Surface dressing, seal coat and premixed chipping carpet in normal summer temperatures and in plains. Penetration macadam in summer temperatures and in plains. Hot mix asphaltic concrete and bituminous macadam in summer temperatures and in plains. Hot mix asphaltic concrete and bituminous macadam in winter temperatures and at high altitudes.
4.	180/200 Penetration Grade	Surface dressing, seal coat and premixed chipping carpet at winter temperatures and at high altitudes. Penetration macadam at winter temperatures and at high altitudes. Hot mix asphaltic concrete and macadam at winter temperatures and at high altitudes.
5.	Cut-back RC-3	Surface dressing and seal coats at normal summer temperatures and in plains. Surface dressing and seal coats at winter temperatures and at high altitudes. Cold-mix macadam and bituminous carpets at summer and winter temperatures, both in plains and in high altitudes.
6.	Emulsions	Cold weather work. Remote area work. Wet condition of aggregates. Maintenance and patching work. Sealing of fine cracks. Prime coat and tack coat.
7.	Medium-curing and slow-curing cutbacks	Priming.

Table 5.10: Selections of tar for road works

Grade	Uses
RT-1	Surface dressing for very cold weather conditions and at high elevation on hill roads.
RT-2	Surface dressing under normal climatic conditions.
RT-3	(A) Surface dressing and renewal coats. (B) Precoating chippings; light chipping carpet.
RT-4	Premix tar macadam.
RT-5	Grouting.

(v) Road oils

Raod oils are slow curing liquid asphalt. The are also petroleum distillate from which volatile fractions are removed. They have slow setting property. They are also obtained by blending bitumen with solvents, like slow curing (SC) cut-backs.

Road oils are commonly used on earth road as dust preventive.

(vi) Primers

Primers penetrate into the road surface, to bind the particles. It could be a road oil, or a cut-back asphalt or a low viscosity road tar.

Primers stick to road aggregates and are of low viscosity. Viscosity of the primers is adjusted by using different percentages of bitumen or tar in the solvent.

Bituminous primers differ from cut-back bitumen and primers that in cut-back bitumen, distillate evaporates, but in primer it is absorbed by the surface on which the primer is sprayed. The viscosity of the primer is to be adjusted to the porosity of the surface on which it is sprayed. Primers are sprayed on stabilised soil and water bound macadam base-courses, before putting the bituminous course. It plugs the capillary voids in the base course and binds dust and loose particles, and promotes adhesion between base course and bituminous surfacing.

5.6 BITUMEN TESTS AND THEIR SIGNIFICANCE

Significance, test procedure in short and typical test results for common bitumen tests are described in following paragraph:

1. Penetration test
2. Specific Gravity test
3. Stripping test
4. Flash and Fire point test
5. Ductility test
6. Softening Point test
7. Viscosity test
8. Distillation test (Composition Test)
9. Water Content test
10. Loss on heating test
11. Ash Content test
12. Solubility test
13. Marshall Stability test
14. Determination of Bitumen Content test
15. Float-test

For performing tests on bituminous binders it is necessary to take proper care in collection of samples, and to follow the testing procedure cafefully. IS: 1201-1978- a publication of Indian

Standards should be followed to know the details of various types of sampling apparatus, sampling procedures, labelling, storing, packaging of samples.

Above tests are used for paving bitumen, cut-back bitumen and bitumen emulsions, as mentioned.

5.6.1 Penetration Test

(IS: 1203) (ASTM: D5)

(a) Significance of the Test

It is a consistency test, which determines the hardness or softness of bitumen (resistance to flow). Bitumen should have consistency needed to enable it to function satisfactorily as a cementing material under the traffic, climate, foundation and other conditions, to which bituminous surface will be exposed:

This test is used for classification of bitumen grade and for evaluating consistency of bituminous materials. It is being replaced by viscosity test, in many present day specifications.

The penetration value as such has identifying character and bears no quality statement. The lower the value of penetration, the harder or more viscous is the bitumen.

(b) Synopsis of Test Method

The test consists of measuring depth in tenths of millimeter (0.1 mm) to which a standard loaded needle will penetrate vertically in 5 seconds. The sample is maintained at 25°C. The penetrometer consists of a needle assembly with a total weight of 100 gms and a device of releasing and locking it in any position. Lower the penetration, harder is bitumen and vice-versa. Figure 5.23 shows typical needle penetration test equipment.

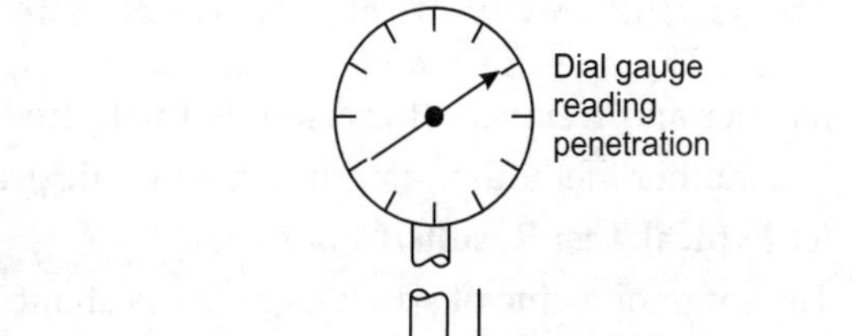

(c) Typical Test Results/Comments

Penetration grade bitumen are known to reduce in penetration with age and they develop cracking tendencies. Generally high penetration grades are used in cold climates while low grades in hot climates. The grading of bitumen helps to assess its suitability for use in different climatic conditions and types of construction. For 30/40 Penetration grade the penetration value at 25°C $\left(\text{in } \dfrac{1}{100} \text{ cm}\right)$ should be between 20 to 40.

Fig. 5.23: Penetration test apparatus

The test trails are continued untill at least 3 penetration values are within tolerance limits. Results of 3 independent observations should not differ the mean by more than following limits.

Penetration	0–80	80–225	above 225
Repeatability	4%	5%	7%

For bituminous macadam and penetration macadam, IRC suggests bitumen grades 30/40, 60/70, and 80/100. Higher penetration grades like 180/200 are used in cold regions, so that

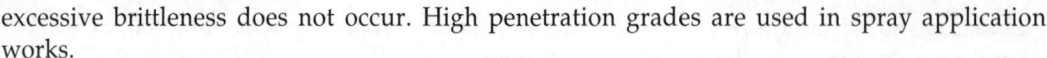

excessive brittleness does not occur. High penetration grades are used in spray application works.

5.6.2 Specific Gravity Test

(IS: 1202) (ASTM: D70)

(a) Significance of the Test

It is desirable to know specific gravity of the bitumen used, for making volume corrections when volume measurements are made at high temperatures. It is also used as one of the factor in determination of voids in compacted bituminous paving mixes.

Specific gravity of a bituminous binder is the ratio of weight of given volume of bitumen to the weight of an equal volume of water both at specified temperature. Bitumen expands when heated and contracts when cooled.

– Volume and weight relationship is useful for billing and transportation.
– It differentiates between different type of binders.
– Used in determination of percentage voids in bituminous mix.

(b) Synopsis of Test Method

The specific gravity of bitumen is determined by the pycnometer method/specific gravity bottle. The specific gravity bottle (50 ml capacity) is filled with bitumen in the prescribed manner and weighed. It is also filled with freshly boiled distilled water and weighed.

Another method of determining specific gravity is called "Displacement Method."

(c) Typical Test Results/Comments

The common value of specific gravity is about 1.025. Specific gravity value is normally supplied by manufacturers. The specific gravity of refinery penetration grade bitumen normally lies between 1.00 and 1.05.

5.6.3 Stripping Test

(IS: 6241) (ASTM: D1664)

(a) Significance of the Test

During the life time of bituminous surface, the bitumen should posses and retain good affinity or adhesion for the aggregate with which it is combined, inspite of the destructive agencies of traffic, water and climate. The affinity of bitumen for aggregates is usually considered in relation to the possibility of water separating the bituminous coating from the aggregates. This test is conducted to determine the effects of moisture upon the adhesion of the bituminous film to the surface particles of the aggregates. It is of significance to ascertain the suitability of both the materials, i.e. bitumen (binder) and aggregates, because particular binder may be satisfactory with one type of aggregates and unsatisfactory with another.

(b) Synopsis of Test Method

The test involves 200 gms of aggregates passing 20 mm size and retained on 12.5 mm sieve, are mixed with 5 percent binder by weight, heated to 160°C. The aggregates are also heated to 150°C prior to mixing. After complete coating, the mixture is transferred to 500 ml beaker and allowed to cool at room temperature. Distilled water is then added to immerse the coated aggregates. The beaker is covered and kept at 40°C. After expiry of 24 hours, it is cooled to room temperature and the extent of stripping is estimated visually.

(c) Typical Test Results/Comments

The visual assesment of stripping leads to poor reproducibility. Maximum allowable stripping value is 25%. A higher value indicates possibility of loss of adhesion between aggregates and

bitumen in the presence of water. The more the stripping value, the poorer are the aggregates from point of new of adhesion.

To check stripping tendency anti-stripping agent may be added as per specification of the type of agent.

5.6.4 Flash and Fire Point Test

(IS: 1209) (ASTM: D92)

(a) Significance of the Test

Bitumen must be heated to quite high temperatures to make it fluid enough for satisfactory use. The 'Flash Point' of bitumen indicates the temperature to which the material may be safely heated without danger of instantaneous flash, in the presence of an open flame.

This temperature, however is usually well below that at which the material will burn. That temperature is called 'Fire Point'. The bitumen should be as safe as possible to handle at the highest temperatures to which it will be subjected during construction and maintenance operations. This is important for safety reasons. At high temperatures, bituminous materials emit hydrocarbon vapours which are susceptible to catch fire. Heating tempeatures of bituminous materials should be restricted to avoid hazardous conditions. Flash point and fire point tests are used to determine the temperatures to which bituminous materials can safely be heated.

(b) Synopsis of Test Method

Flash point is measured in cleveland open cap. A brass cup is partly filled with bitumen and heated at prescribed rate. A small flame is played over the surface of the sample periodically, and temperature at which sufficient vapours are released to produce an instantaneous flash is designated as the 'Flash point'. The test continues until the vapours remain burning for 5 seconds. This is "Fire Point". Figure 5.24 shows typical Flash Point test.

(c) Typical Test Result/Comments

The results should not differ from the mean by more than 3°C. By addition of small quantities of silicon, the flash point of a bitumen can be raised. It is primarily a safety test though may also be considered as indirect reflection of binder volatility.

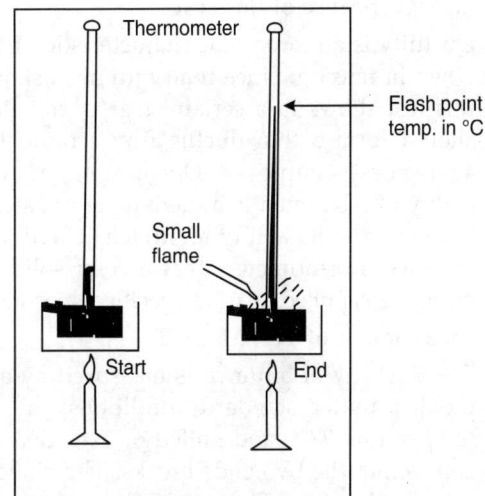

Fig. 5.24: Flash point test

Table 5.11: Working temperatures of bituminous material (IRC)

Type of construction	Temperature at application/ mixing		Temperature of mix at laying °C
	Bitumen °C	Aggregate	
Tack Coat Surface dressing 80/100 Pen	163–177	dry	–
Penetration Macadam 80/100 Pen	163–177	do	–
30/40 Pen	177–190	do	–
Buit up spray grout 80/100	163–177	do	–
Open graded mixes: 80/100	150–165	125–150	110–135
60/70	165–180	do	do

(Contd.)

Table 5.11: Working temperatures of bituminous material (IRC) (*Contd.*)

Type of construction	Temperature at application/ mixing		Temperature of mix at laying °C
	Bitumen °C	Aggregate	
Dense/semidense mixes 80/100	150–177	155–163	121–163
60/70	165–177	do	do
30/40	175–190	do	do

Difference in temperature between the binder and aggregates at the time of mixing shall not exceed 25°C in case of open-graded mixes (e.g. BM) and 14°C for dense semi-dense mixes.

The safe limit for heating of bitumen is normally 50°C below the flash point—the temperature up to which the binder can be safely heated.

The flash point of most penetration grade bitumen lies in the range 245–335°C, while rapid curing cut-backs may flash at temperatures as low as 27°C. medium–curing cut-backs usually flash between 52°–99°C, while slow curing ones have flash points above 110°C.

5.6.5 Ductility Test

(*IS: 1208*) (*ASTM: D113*)

(a) Significance of the Test

Ductility is an important characteristic of bitumen. A ductile material is one which elongates when in tension. Since road bitumens should be elastic to overcome the surface movements, this test shows in a certain degree, the elasticity of the material. It is necessary that binder should form a thin ductile film around the aggregates so that physical interlocking of the aggregates is improved. Ductility might also be considered to have some relationship to the ability of a bitumen to adhere to aggregates. This test gives a measure of adhesive property of bitumen and its ability to stretch. The ductility of a bituminous material is expressed as the distance in centimeters that a semi solid briquette will elongate before breaking. When the briquette is pulled apart at specified speed and temperature.

(b) Synopsis of Test Method

The ductility of bitumen is measured by an extension type of test. A briquette of bitumen is molded under standard conditions and dimensions. It is then brought to standard test temperature 27°C and pulled or extended at specified rate of speed cm/minute till the thread connecting the two ends breaks. The elongation in cms at which thread of material breaks is called as Ductility. Figure 5.25 shows typical ductility test.

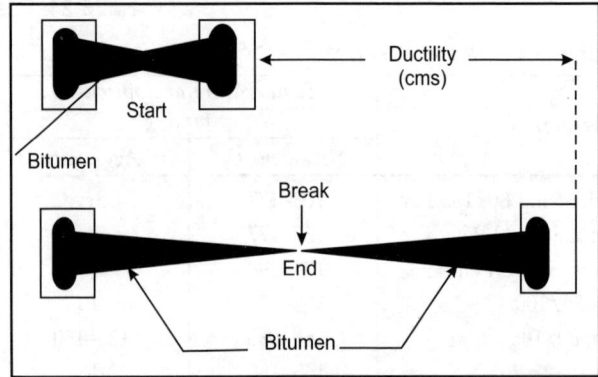

Fig. 5.25: Ductility test

(c) Typical Test Results/Comments

At least two normal tests are conducted for each sample and average value reported. The results should not differ from the mean by:

(a) Repeatability = 5%

(b) Reproducibility = 10%

The most common causes of low ductility in bitumen are presence of excess quantities of wax or the use of considerable air blowing in refining process.

Ductility at 27°C minimum should be 10 cm for 30/40 grade.

For 80/100 bitumen the ductility value usually is 75 cm and above. Presence of wax content in bitumen reduces its ductility value. For a waxy bitumen of 80/100 grade the minimum ductility value specified is only 15 cms. Bitumen with low ductility value may get cracked especially in cold weather. This test is used to obtain bitumen least likely to harden in service.

Table 5.12 shows ISI requirements for values of minimum ductility for various grades of bitumen.

Table 5.12: Requirements of ductility value

Source of Paving Bitumen	Penetration Grade	Minimum Ductility value (cms)
Assam petroleum	A 25	5
	A 35	10
	A 45	12
	A 65, A 90, A 200	15
Other sources	S 35	50
	S 45, S 65, S 90	75

The ductility test is actually a measure of the internal cohesion of a bitumen. Bitumen possesing high ductility are normally cementitious and adhere well to aggregates. Bitumen possessing high ductility are usually highly susceptible to temperature changes, while low ones are not. The lack of ductility does not necessarily indicate poor quality of bitumen.

Bitumen of low susceptibility and low ductility are highly desirable as crack fillers in roadways.

5.6.6 Softening Point Test

(IS: 1205) and *(IS: 334-1982)* *(ASTM: D2398)*

(a) Significance of the Test

The softening point of bitumen is the temperature at which the substance attains a particular degree of softening. Different grades of bitumen soften at different temperatures. It is used to characterise the harder materials, showing the temperature at which they can reach arbitrary degree of softening. The bitumens do not have a definite melting point, but instead the transition from solid to liquid is slow and gradual. Determination of softening point helps to know the temperature up to which a bituminous binder should be heated to have sufficient fluidity for applications in road use. This test is another measure of consistency, and is also called 'Ring and Ball Test".

This test also indicates the atmospheric temperature at which bleeding of bitumen will commence in bituminous surfacing.

(b) Synopsis of Test Method

The softening point is determined by the 'Ring and Ball' test method. In this test the heated bitumen is poured into a brass ring of specified dimensions. The sample thus prepared is suspended in a water bath and a steel ball of specified dimension and weight is placed in the centre of the sample. The bath is heated at a controlled rate and the temperature at the instant the steel ball reaches the bottom of the glass vessel is recorded. This temperature is called softening point of bitumen. Figure 5.26 shows typical ring and ball softening point test.

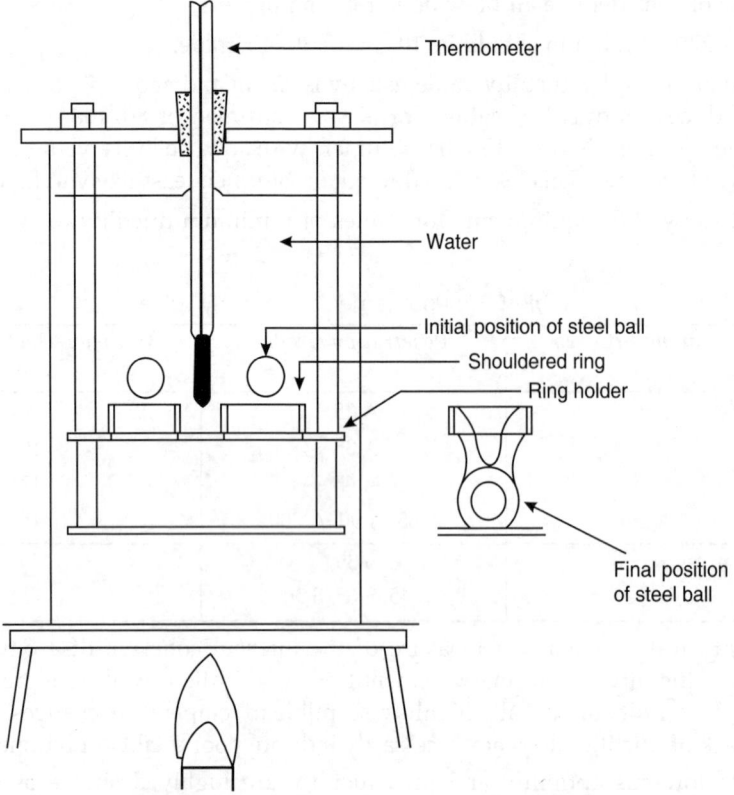

Fig. 5.26: Apparatus for ring-and-ball test

(c) Typical Test Results/Comments

The mean of the temperature readings is reported as softening point. The test results should not differ by the mean, by more than following:

Softening point	Repeatability	Reproducibility
Below 30°C	2°C	4°C
30°C – 80°C	1°C	2°C
Above 80°C	2°C	4°C

Softening point of 30/40 grade should be between 50°C to 90°C, and 80/100 between 35°C to 50°C as per Indian standards.

It is better to use the bitumen at a particular work which has softening point higher by 5°C to 10°C than maximum atmospheric temperature, to ensure that they will not flow during service. Higher the softening point, the lesser the temperature susceptibility. Bitumen with higher softening point may be preferred in warmer places.

5.6.7 Viscosity Test

(Is: 1206 part III) (ASTM: D-88)

(a) Significance of the Test

Viscosity of a fluid is the property by virtue of which it offers resistance to flow. Higher the viscosity, slower will be the movement of liquid. Viscosity test is used to control the consistency which in turn determines the grade of bitumen. It is important for establishing the relationship between temperature and viscosity. Viscosity is more scientific parameter to measure consistency. Consistency/viscosity is resistance of a material to flow. Since the binders vary from thin liquids to semi-solids, and the temperatures for processing at construction site are quite different from those to which road surfacing subjected, it is important to evaluate consistency of bituminous binders, for their suitability under different climate, temperature and load conditions. The viscosity affects the ability of the binder to spread, move into and fill up the voids between aggregates. It plays important role in coating of aggregates.

If the surface between two parallel surfaces is filled with a liquid, and one of the surface is moved parallel to other a force(F) which resists the movement is developed.

i.e. $$F = \eta \frac{A\,v}{d} \quad \text{where} \quad \eta = \text{Coefficient of viscosity}$$

$$A = \text{Area}$$
$$v = \text{relative velocity}$$
$$F = \text{Force}$$

Viscosity (η) is the force in dynes offered by the liquid to the movement of a surface of 1 sq. cm, moving with a velocity of 1 cm/second, at a distance of 1 cm from another fixed surface. Figure 5.27 shows schematic diagram of viscosity measurement.

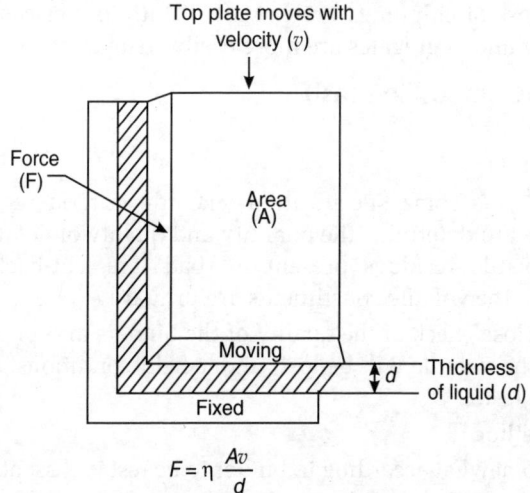

$$F = \eta \frac{Av}{d}$$

Fig. 5.27: Schematic diagram of viscosity measurement

The binders are classified into viscosity ranges according to the time, (in seconds), required for a given amount of binder to flow through specified orifice at the bottom of the container. The viscosity of bituminous binders fall vary rapidly as the temperature rises. Different methods are used for determination of viscosity.

(b) Synopsis of Test Method

There are several types of viscometers in use: In the Saybolt Furol viscosity test, a given volume of material is heated in a standard 'tube' in the bottom of which there is an orifice of specified shape and dimensions. After a particular temperature is reached the stopper of the orifice is

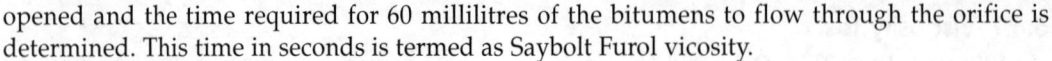

opened and the time required for 60 millilitres of the bitumens to flow through the orifice is determined. This time in seconds is termed as Saybolt Furol vicosity.

Other methods to evaluate the viscosity of a material are based on:

– The flow of a body through a liquid.

– The rotation of one of two coaxial cylinders when space between is filled with a liquid.

(c) Typical Test Results/Comments

A minimum viscosity at certain temperature 60°C (140°F) or 135°C (275°F) is prescribed in specifications, for easy mixing and laydown at those temperatures. A complete viscosity-temperature curve of the bitumen being used will be more useful, to provide optimum fluidities for mixing, spraying and other operations.

**Viscosity Requirements and Quantity of
Bituminous Binder (IRC)**

Type of Surface	Kinematic viscosity of Primer at 60°C (CENTISTOKES)	Quantity per 10 sq. m (kg)
Low Porosity	30–60	6 to 9
Medium Porosity	70–140	9 to 12
High Porosity	250–500	12 to 15

Binders having very low viscosity can be advantageously used in exceptionally cold weather conditions. High viscosity binders have to be heated before their application. If a binder with too low viscosity is pre-mixed with an aggregate, it may flow off the stone while enroute from the mixing plant. Conversely, if the viscosity is too high, the mixture may be unworkable by the time it reaches the site. If too low a viscosity is used for surface dressing purposes, the result may be "Bleeding" or loss of chippings under traffic. With low-viscosity application temperatures can be kept lower and aggregates are more easily coated.

5.6.8 Distillation Test (Composition Test)

(ASTM: 402) (IS: 1213-1978)

(a) Significance of the Test

It is a composition test, as some specifications include criteria regarding composition of bitumen. This test is used to determine the quantity and quality of volatile constituents and the amount of the non-volatile residues present in road tars, cut-back bitumen and binder emulsions. In emulsions, the volatile constituents are primarily water.

This test enable the close check of the quality of the binders and provide useful information at which temperature, the volatile will be lost, under field conditions, and the type of residue left from the distillation test.

(b) Synopsis of Test Method

Test procedure differs somewhat according to binder to be tested. Essentially, all involve heating a specified quantity of binder, at a specified rate and then determining the amount of distillate removed at prescribed temperatures.

(c) Typical Test Results/Comments

This test offers a close check on the quality of the binder.

The moisture content of road tars must not exceed 0.5 percent, water content in excess of this will lead to undesirable foaming if tar is heated beyond 100°C.

This test provides information on the type of volatiles in the binder and on the rate at which these volatile will be lost under field conditions. The residue left from the distillation test can also provide useful information.

5.6.9 Water Content Test (Composition Test)

(IS: 1211-1978) (IS: 73-1961)

(a) Significance of the Test

Although water content can be determined during the distillation test sometimes it is desirable to find only the moisture in the binder, without carrying out the whole distillation procedure.

It is a useful test if the binder is to be heated above 100°C, because then 'foaming' will occur. Bituminous binder should only contain extremely low moisture content, about 0.2 percent. If water content is more than specified limit it may cause cracking and frothing during heating of bitumen.

(b) Synopsis of Test Method

In this test specified amount of binder is mixed with a pre-determined amount of solvent and then heat is applied. This causes the water to separate and its volume is expressed as percentage by weight of the original sample.

The petroleum spirit is used as the solvent for bitumen and coal-tar solvant for road tars, with which it is immiscible.

5.6.10 Loss of Heating Test (Composition Test)

(IS: 1212) (ASTM: D-6)

(a) Significance of the Test

This test is somewhat representative of heating conditions is the field storge drums. This test is very useful for bitumen. It is an indicator of the volatile content and the hardening of the binders.

Often the penetration test is carried out on the residue of the loss of heating test. Specifications usually require that maximum loss of weight on heating and the maximum drop in penetration do not exceed certain limits. The loss on heating should be with in the limits as prescribed in IS: 73-1961.

(b) Synopsis of Test Method

The sample of bitumen (50 gm) is heated for 5 hours at 165°C. After cooling, the loss of weight is expressed as a percentage of the original weight.

Penetration test is done on residue of the loss on heating test and expressed as percentage of original penetration, before heating.

(c) Typical Test Results/Comments

The minimum penetration of the residue ranges from 60 to 75 percent of the original penetration. Loss on heating maximum should be 3%

This test is of use only as a general indication of volatile content under specified conditions of test.

5.6.11 ASH Content Test (Composition Test)

(ASTM: D 482)

(a) Significance of the Test

The ash content of a bitumen is the percentage by weight of inorganic residue left after ignition of the sample. This test is carried out on penetration grade and cut-back bitumen. It ensures that there are not undesirable amounts of mineral matter in bitumen.

The ash test is useful in determining the composition of the binder.

(b) Synopsis of Test Method

The weighed amount of sample of bitumen is gently heated untill it begins to burn at specified temperature and then it is fired untill the ash is free from carbon. The ash content of a bitumen is the percentage by weight of the inorganic residue left after ignition of the sample.

5.6.12 Solubility Test (Composition Test)

(IS: 1216) (ASTM: 2042) (IS: 73-1961)

(a) Significance of the Test

The solubility test is a measure of the 'Purity' of bitumen. The portion of the bitumen that is soluble in trichloroethylene represents the active cementing constituents. Bitumen manufactured from petroleum should be almost entirely soluble in solvents such as carbon disulfide or carbon tetrachloride also.

The test shows how much actually binder is available for use in the road surfacing, and indicates the degree of purity and freedom from contamination.

(b) Synopsis of Test Method

Solubility is determined by dissolving the bitumen in the given solvent and separating the soluble and insoluble, portions by filtering. After filtering the solution, the residue retained is determined and the percentage of the soluble material is calculated. Because of fire hazard associated with carbon disulfide it is substituted by carbon tetrachloride in which it is equally soluble.

(c) Typical Test Results/Comments

As a protection against contamination or the presence of impurities most specifications require 99% solubility in carbon tetrachloride for bitumens derived from the refining of petroleum.

Solubility in CS_2 should be 99% minimum

5.6.13 Marshal Stability Test

(ASTM: D-1559 (8)

(a) Significance of the Test

Marshall test method is a most common and convinient method of bituminous mix design. In designing of bituminous mixes, this test determines the two important properties of strength and flexibility.

Strength is measured through 'Marshall Stability' of the mix and the flexibility is measured in terms of 'Flow Value'. This test determines the "Optimum Binder Content" for the aggregate mix type and the traffic intensity.

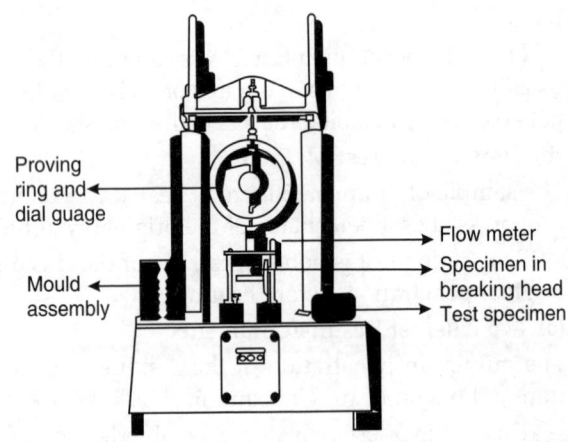

Fig. 5.28: Marshall stability testing equipment

(b) Synopsis of Test Method

After selection of the type and grading of the aggregates to be used in the bituminous mix, Marshall test specimen are prepared using different ranges of bitumen content. For each bitumen content two samples are prepared and tested. There is a standard procedure for making samples, using mould and hammer, and compaction procedure.

Test specimen are put in the Marshal machine and stability and flow values are recorded, for various samples. Curves are then plotted to show maximum stability and corresponding optimum bitumen content. There may be correction to be applied to stability value as per thickness of compacted specimen. 5 samples are then prepared using 0.5% Bitumen content above the optimum value for two and 0.5% bitumen content below the optimum and one with optimum bitumen.

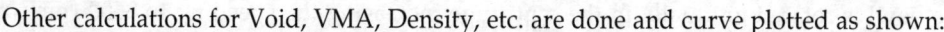

Other calculations for Void, VMA, Density, etc. are done and curve plotted as shown:

Figure 5.28 shows Marshal stability testing equipment. Figure 5.29 shows typical relationship between Marshall test parameters. Table 5.13 lists requirements of bituminous concrete mix.

(c) Typical Test Results/Comments

Table 5.13: Requirements of bituminous concrete mix

S. No.	Parameter	Value
1.	Marshal stability (ASTM: D 1559) on specimen compacted by 50 blows on each side.	340 kg. minimum
2.	Marshall flow (0.25 mm)	8–16
3.	Percent voids in the mix	3–5
4.	Percent voids in mineral aggregates filled bitumen	75–85
5.	Binder content percent by weight of total mix	5–7

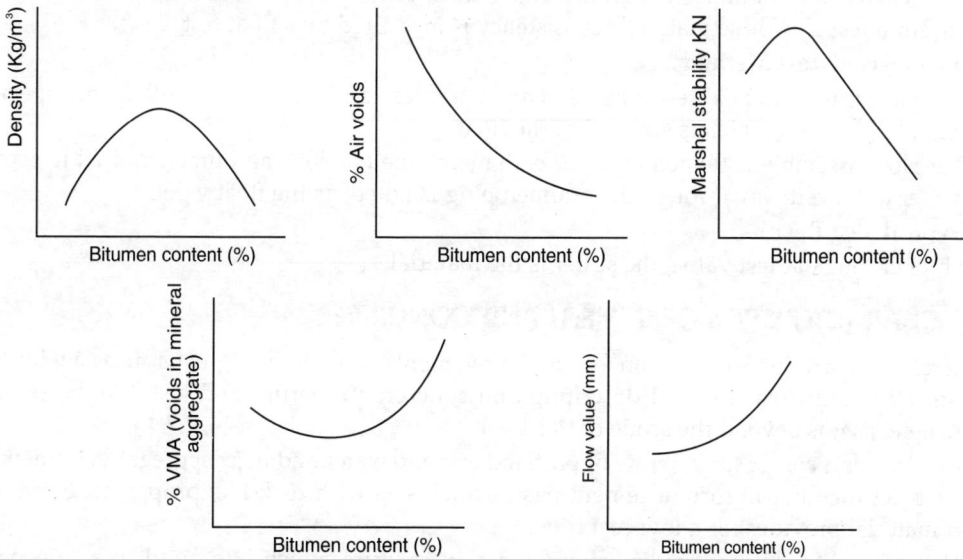

Fig. 5.29: Typical relationship between Marshall test parameters

5.6.14 Determination of Bitumen Content by Centrifuge Extractor

(a) Significance of the Test

This test is a field test to exercise quality control and ensure that the specified amount of bitumen has been used. The centrifuge extractor is used for the quantitative determination of bitumen in hot-mixed paving mixtures and samples of existing pavement.

The bitumen content is calculated by difference of the weight of the extract aggregates, moisture content and ash from the weight of the sample taken for the test.

(b) Synopsis of Test Method

The centrifuge extractor consists of a revolving bowl inside a housing. The bowl is provided with a cover plate and an outlet. The system is operated manually.

A 1000 gm. sample of asphalt mix, is broken and heated to about 115°C. It is put in the bowl and weighed. Benzene or trichoroethane is added to cover the sample and allowed to soak it for one hour.

The bowl is rotated gradually in the centrifuge increasing the speed up to 3600 rpm, untill the solvent ceases to flow from the outlet. After stopping the centrifuge 200 ml of trichloroethane or benzene is added and rotated again (not less than 3 washings) untill the extract is no longer cloudy and is fairly light in colour.

Filter is removed from the bowl and loose particles from the filter are brushed in the bowl. The sample and filter is dried to constant weight in an oven at 98°C to 105°C and weight of the filter and bowl with dry aggregates is obtained.

For the three different samples, the bitumen content is calculated.

(c) Typical Test Results/Interpretation of Results

The test results indicate the quantity of bitumen that has been used in a bituminous mix, which can be compared with the optimum required, for best quality of work.

The performance of the road is affected if lesser or more quantity of bitumen is used.

5.6.15 Float Test

(a) Significance of the Test

There are several bituminous materials, whose consistency can not be measured by viscosity or penetration test. For these materials consistency is measured by a float test.

(b) Synopsis of Test Method

The apparatus used in float test consists of an aluminium and a brass collar filled with specimen material to be tested which is screwed to the float.

The float assembly is floated in water bath maintained at 50°C and time required in seconds for water to force it way through the bitumen plug is noted, as the float value.

(c) Typical Test Results

The higher the float test value, the stiffer is the material.

5.7 CHARACTERISTICS OF CEMENT AND CONCRETE

Concrete is used in construction of rigid pavements, and it has greater uses in building construction industry. Detailed description on concrete properties and tests can be searched elsewhere, as it is beyond the scope of this book.

Concrete is a composite mixture of portland cement, water and inert aggregates. Cement and water react together to form a cement paste which sets with rock-like properties. Sand as an inert material provides for a low cost filler.

Cement is basically burnt limestone and clay mixture which is ground to a fine power. Gypsum is added at final stage to retard the hydration process. Cement's chief chemical constituents are lime (CaO), and Silica (SiO_2). Water reacts with cement to produce calcium silicates and aluminates.

Aggregates used in cement should be inert, which do not react with the alkali in the cement.

The tendency of concrete to segregate during placement, and possibility of air voids, should be controlled through proper quality control. Vibrators control these problems.

Curing is important after placing the concrete. Keeping the concrete moist and surface water absorption low permits strength gains resulting from continued hydration.

The mechanical properties of concrete are determined by unconfined compressive strength after 28 days. Unconfined compressive test is performed on concrete samples of cylindrical shape. Test are made for concrete road construction to determine consistency, air content and strength. Commonly used tests are:

- Slump test
- Air content test
- Strength tests (Flexural strength).

Geometric Design of Highways

6.0 PREAMBLE

Geometric design of highways includes all the visible features of a roadway like alignment, curves, sight distances and cross-section elements, etc. There are several types of roads and highways. At one extreme there are high volume, high speed facilities (expressways/freeways) carrying through traffic, with no attempt to serve the nearby property or local traffic, and on the other hand there are local roads (urban roads), or city streets, carrying low volume of traffic or low speed, and mixed type of traffic resulting into congestion and crawling speeds. Traffic and geometric design requirements for different types of roads vary from facility to facility, in view of the level of service expected.

The basic aim of geometric road design is to produce roads which are both operationally efficient and safe. Capacity often dominates urban designs, whereas safety and adequacy are usually key factors in rural road design.

Geometric design is concerned with relating the physical elements of the highway to the requirements of the drivers and the vehicles. It mainly consists of those elements which are visible features of a roadway, and does not include the structural design of the facility.

Standards and controls in geometric design of roads are necessary to:
- Maintain a degree of uniformity
- Proper utilisation of road funds in view of future growth.
- Provide satisfactory and standard designs, which are easily understood and followed.
- Ensure safety and efficiency of operation through uniformity and consistency.

Indian standards for geometric design of rural and urban roads (highways) are recommended by Indian Road Congress IRC-73, which are suitably modified from time to time in view of latest engineering practices and research.

6.1 BASIC CONSIDERATION IN GEOMETRIC DESIGN

A basic principle of the geometric design of roads is that the appearance of the road should clearly indicate to the driver the speed and path that his vehicle should adopt in order to

proceed with comfort, economy and safety. If this is not possible, additional devices such as signing, marking, signalling and lighting, etc. should be used, to guide the driver.

In preparing the geometric design of a new highway or redesigning an old one, the following basic aspects should be considered:

(i) The design must be adequate for the estimated future traffic volume, e.g. the design speed, design volume, design vehicle, capacity, should be selected properly. It should not become obsolescent in near future, before design period.

(ii) The design must be safe for driving and should have confidence of majority of drivers.

(iii) The design must be consistent and must avoid sudden changes in alignment, grade, sight distance, etc. The driving must be comfortable and enjoying.

(iv) The design must be complete. It must include the necessary roadside treatment and provide essential traffic control devices such as marking, sign, lighting, intersections, etc.

(v) The design must be as economical as possible, relative to initial costs and maintenance cost.

(vi) The design should be esthetically pleasing to the user and who live along the highway

(vii) It should be beneficial to social and community values of the adjacent area.

(viii) The design should be ecologically harmless. It should not degrade the environment.

(ix) The design should be in conformity with standards proposed for different elements. Rectifications in design at a later date are very costly and difficult.

6.2 FACTORS INFLUENCING GEOMETRIC DESIGN ELEMENTS

The principal factors influencing the selection of geometric design aspects of a road are the following:

1. Classification of roads (highway type)
2. Type of terrain
3. Design period
4. Future traffic
5. Design hourly volume of traffic
6. Directional distribution of traffic
7. Composition of traffic
8. Design vehicle
9. Design speed
10. Design capacity

The geometric features of a highway can not be changed or corrected at a later stage even if it may be possible to improve certain elements (remove deficiencies) it requires heavy expenditures. It is therefore important that geometric design requirements should be considered very carefully right in the beginning.

6.2.1 Classification of Roads

Depending upon their relative importance and service provided roads are classified in different categories. Minimum geometric design standards are governed by specific transportation service provided by the road.

Roads are classified based on different criteria such as:

A. According to location

(a) Rural (Non-urban) roads

(i) National Highways (NH): Main highways running through the length and breadth of the country connecting major parts, foreign highways, state capitals, large industrial and tourist centres, etc.

(ii) State Highways (SH): Arterial routes of a state linking district headquarters and important cities within the state and connecting them with National Highways or highways of the neighbouring states.

(iii) Major District Roads (MDR): Important roads within a district serving areas of production and markets, and connecting these with each other or with the main highways.

(iv) Other District Roads (ODR): Roads serving rural areas of production and providing them with outlet to market centre, taluka/tehsil headquarters, block development headquarters or other main roads.

(v) Village Roads (VR): Roads connecting villages or group of villages with each other and to the nearest road of a higher category.

(vi) Hill Roads (HR): Roads located in hilly areas or on hills or mountains.

(b) Urban roads

Urban roads are classified in following categories:

(i) Expressways: These include freeways and parkways and provide rapid and efficient movement of large volume of through traffic between areas and across the urban area. They are not intended to provide land access service. Expressways are divided highways with high geometric design standards using full or partial control of access. Recommended land width is 50-60 m.

(ii) Major arterial streets: They provide principal network for through traffic flow between areas and across the city. Significant intra-urban travel between major suburban centres takes place on this network. They are generally divided highways with full or partial control of access. Recommended lane width is same as for expressway.

(iii) Sub-arterial streets: These are functionally similar to arterial streets, but have lower level of mobility and service, as compared to major arterial streets. Control of access may not be rigid. Recommended land width is 30-40 m.

(iv) Collector streets: The purpose of collector street is to collect traffic from local streets and feed it to the arterial streets. They provide traffic movement between arterial and local streets with access to abutting property. They are located in residential, commercial and industrial areas. Recommended lane width is 20-30 m.

(v) Local streets: These streets provide direct access to the abutting land and are for local traffic movement. Depending upon the activity on the adjoining land they may be residential, commercial, or industrial local streets. Parking and pedestrian movement is generally allowed on these roads. Recommended land width for local streets is 10-20 m.

In residential areas urban roads/city road must serve following purposes:

- Provide access to the arterial road networks.
- Provide access to public transport
- Permit easy interchange between travel modes

- Provide access to private property
- Accommodate vehicular traffic in an equitable manner
- Permit movement of necessary large vehicle, e.g. garbage trucks, etc.
- Accommodate pedestrian and cyclists
- Minimise local travel
- Discourage unnecessary traffic particularly through traffic
- Provide safety of operation and use
- Provide space for local social activities, like children playgrounds
- Provide space for the reticulation services
- Be pleasant in appearance and harmonise with the environment
- Provide necessary parking
- Discourage speeding
- Encourage obedience of the traffic laws
- Ribbon development should be controlled

B. According to traffic

(i) Based on traffic density

- Heavy traffic roads
- Medium traffic roads
- Low traffic roads, etc.
- One, two, three ... lane roads, depending upon traffic volume.

(ii) Based on character of traffic

- Mixed traffic roads
- Commercial traffic roads
- Exclusive bus road/lane etc.

(iii) Based on speed of traffic

- Road having design speed of 70 Km/h, 80 Km/h, etc.
- Other high speed roads
- Expressways: Complete separation from opposing stream, no crossing road, specially designed interchanges, no pedestrian or slow vehicles, high speeds, moderate gradient, good shoulders and facilities along the road.
- Freeways: Road having all the above features for free and high speed movement of vehicles.
- Parkways: Roads located in park areas, where traffic volumes are low.

C. According to tonnage (load)

This classification is useful for structural design. Typical classification according to tonnage is given in Table 6.1.

Different countries have their own method of naming various types of roads.

Table 6.1: Classification based on tonnage (IRC)

Class	Tonnage per day	Commercial vehicle per day
A	0–100	0–15
B	100–400	15–45
C	400–1500	45–150
D	1500–4000	150–1500
E	4000–120,000	1500–4500
F	More than 120,000	More than 4500

D. According to control of access

The condition where the right of owners or occupants of abutting land or other persons to access, light, air or view in connection with a highway is fully or partially controlled by the authorities. Control of access ensures safe and efficient travel, by controlling ribbon development along the highway. This helps in eliminating congestion, reducing speed and level of service.

There may be two types of access controls:

(i) Full control of access

Control of access is exercised to give preference to through traffic by providing access connections with selected public roads only and by prohibiting crossings at grade direct private driveway connections.

(ii) Partial control of access

Partial control of access is exercised to give preference to through traffic to a degree that, in addition to access connections with selected public roads, there may be some crossings at grade and some private driveway connections.

With control of access, the capacity of road does not diminish with time and capacity is maintained throughout its life.

The degree of access control depends upon the level of service required, frequency of accidents, traffic pattern and composition; vehicle operating costs, travel times, land uses, public opinion and legal considerations. Adequate provisions are made in a legislation to regulate access to arterial/rural roads.

Control of access can be an urban roads (expressways, arterial highways, sub-arterial streets, collector streets, local street) and rural roads (national highways, state highways, major district road)

Location, type, length and other features of a controlled access road depends upon the local needs. In the long run control access highways are most economical in time savings, safety, comfort and relaxation.

Roads with full or partial control of access are often called "freeway" or "parkways".

Expressway: An expressway is a divided highway for through traffic, with full or partial control of access and generally with grade separation at intersections.

Freeway: A freeway is an expressway with full control of access. The term refers to freedom of movement without traffic signals. All crossings are separated by bridges, pedestrians are excluded.

Parkway: It is a type of highway provided for non-commercial traffic with full or partial control of access and usually located in parklike development.

Some roads are charged for their use for certain period of time, to recover the costs involved in their construction to high standards. They are called 'Toll-Highways.'

Toll Highways: It is a major highway, which is urgently needed and is economically justified, but can not be constructed by usual resources. Some tax is collected by users to pay for its construction, to recover the cost in certain period of time.

E. According to use during the year

(i) **All weather roads:** Roads which can be used throughout the year, irrespective of weather conditions.

(ii) **Fair-weather roads:** There are roads which can not be used during rainy season due to water streams crossing the road or quality of road which gets damaged and becomes impassable during rainy season. It is reconstructed/improved after rains every year.

F. According to type of surfacing

(i) **Paved Roads:** Roads which have bituminous or other type of hard surface on the top, like water bound macadam (WBM) or stabilized roads/concrete roads.

(ii) **Unpaved Roads:** Roads which do not have hard pavement on the surface like earth roads, gravel roads, etc.

Road which are provided with bituminous course on the top are also called "black topped road" or "flexible pavement" and those made by cement concrete are called "rigid pavement".

6.2.2 Terrain Type (Topography)

Geometric design standards are influenced by the terrain in which road is located, in view of economics. The terrain is classified on the basis of cross slopes of the area in the following four categories:

(i) **Steep terrain:** The terrain where cross slopes are greater than 60 percent.

(ii) **Mountainous terrain:** The terrain where cross slope are between 25 to 60%.

(iii) **Rolling terrain:** The terrain with cross slopes from 10 to 25 percent

(iv) **Plain (level) terrain:** The terrain with cross slopes less than 10 percent (0-10%).

6.2.3 Design Period (Lifetime)

Geometric design elements must take care of the traffic likely to use the road for next 15 to 25 years, called design period. A design period of 20 years is commonly used for rural roads due to following reasons:

(i) Reasonable estimates of traffic are not possible for longer periods than 20 years. Attracted, generated, development and normal traffic growth is predicted for this period.

(ii) Reasonably good pavement lasts for 20 years, and it is desired that geometrics should not fail, before pavement fails structurally.

(iii) Public investments should generally to paid off in 20 years.

A design period of 15-20 years is adopted for arterial and sub-arterial roads and 10-15 years for collector and local streets in urban areas.

6.2.4 Future Traffic

A road is designed to accommodate the future traffic (normal, attracted, generated, and developed traffic) that occurs within its life time — 20 years.

Attracted traffic: When a new road is constructed or an old road is improved, in addition to its own traffic some traffic is attracted from the nearby roads. This traffic is called attracted traffic.

Generated traffic: In addition to the normal growth of the traffic on a road, year after year, the new or improved road creates new traffic for itself which may be made up of:

- New trips which were not there before
- Trip which were previously made by buses or other public transport
- Trips which were previously made on other roads.

This traffic is called generated traffic and generates within first 1 to 2 years, a new road is opened depending upon quality of new road. It may vary from 5 to 25% from place to place. Judgement based on previous experience on similar situations is used to predict this traffic.

Developed traffic: Due to development of the adjacent area or land, because of new or improved highway, more traffic continues to develop for many years, after a road is constructed.

The traffic resulting from such development must be accounted in estimating future traffic volume, based on future land use forecast.

Traffic projection factor considering normal growth, attracted, generated and developed traffic, is worked out, which is multiplied by current traffic to obtain future volumes of the traffic for design year.

6.2.5 Design Hourly Volume (DHV)

Traffic volume is usually expressed in terms of annual average daily traffic (AADT) and is estimated from traffic counts and forecasts for the end of design period of the road. Average daily traffic (ADT) is not of much use in geometric design of highways, because it does not indicate significant variation in traffic, occurring during various months of the year, days of the week and hours of the day. Traffic volume on certain day may be considerably higher than an average day.

Traffic volumes expressed on the hourly basis, reflect better operating conditions and are used in geometric design. Traffic volumes on any road vary during different hours of the day and hourly volumes change throughout the year. It must therefore be determined which of these hourly volumes should be used in design. If maximum peak hour volume is used in design, it would be wasteful over design as traffic will be rarely so large. If average hour traffic volumes are used for design, it will result in inadequate design as volumes will exceed too often. For selecting design hourly volumes a curve, showing hourly traffic volumes as percentage of ADT is plotted as shown in Fig. 6.1.

Based on several observations, the conclusion is drawn that 30th highest hourly volume of the year (30th HV) should be used in design. It is exceeded 29 times during the year and is about 15-16% of ADT, for American Conditions. 8-10% AADT volume, is assumed to be the 30th HV for Indian conditions based on several studies conducted on Indian Highways.

Characteristics of 30th HV is that there are many hours in which traffic volume is not much less than 30th HV and as a percentage of ADT, it varies only slightly from year to year.

There are unusual traffic fluctuation on seasonal or recreational roads during weekends or during few months only. On such roads higher traffic volumes occur during the few weeks only and substantially lower volumes occur during most of the year. In such situations, economic conditions dictate use of lower volumes than 30th HV on these roads–may be used 80th HV or 100th HV as shown in Fig. 6.1.

The design hourly volume (DHV) should therefore be 30th HV for the future chosen year for design, but on highly seasonal (resort) highways 80th to 100th hourly volumes may be used as design hourly volumes.

Unlike rural roads, the peak hour volumes occur on urban roads during morning and evening. Peak hour traffic on urban roads may be 8-10 percent of total daily traffic and it is in one direction during morning and reverse direction in the evening. Urban roads are therefore designed on the basis of busier peak hour traffic among morning and evening, based on traffic volume study for 16 hours (from 6 am to 10 pm) to capture peak hour flows.

6.2.6 Directional Distribution of Traffic: (D)

Design hourly volume (DHV) is the total traffic in both directions of travel on the road. The knowledge of hourly traffic in each direction of travel is essential for geometric design. For the same average daily traffic (ADT) or design hourly volume (DHV), if there is high percentage of traffic in one direction during peak hours, wider pavement will be required.

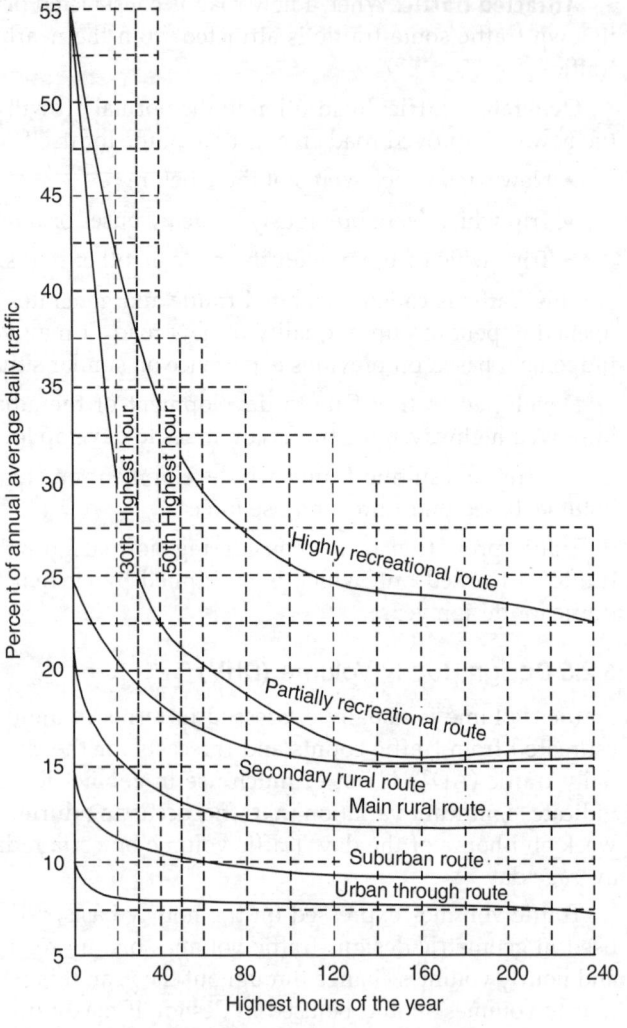

Fig. 6.1: Highest hourly volumes as percentage of ADT

Traffic distribution (D) by direction during peak hours is generally the same from year to year, so the observed directional distribution may be assumed to apply to the DHV for future year.

$$DHV \text{ in one direction} = K \times D \times ADT$$

$$\text{where } K = \text{\%age ratio of } \frac{DHV}{ADT}$$

$$ADT = \text{Two way average daily traffic}$$

$$D = \text{percentage of traffic in predominant direction during design hour.}$$

6.2.7 Composition of Traffic

Vehicles on the road are of different sizes, weights and operating characteristics. The overall effect on traffic operation of a truck may be equivalent to 3-5 passenger cars, depending upon the conditions of road (sight distance, gradient, etc.).

Larger proportion of trucks and buses means greater traffic load and more highway capacity is required.

For geometric design of a highway, it is essential to know percentage of trucks, buses (commercial vehicles). Reasonable estimate are therefore made regarding the percentage of trucks (T), during peak hours through traffic field studies for the design year. Traffic volume is often converted into passenger car units (PCU). Table 6.2 shows passenger car equivalents of different vehicles used in India.

Table 6.2: Passenger car equivalents

Serial Number	Type of Vehicle	Urban Road	PCU value for the analysis of				
			Uncontrolled intersections	Signalized intersections	Roads in Rural Areas		
					Single lane	Two-lane	Multilane
1.	Passenger car	1.0	1.0	1.0	1.0	1.0	1.0
2.	Bus or Truck	3.0	4.6	3.3	4.0	4.0	4.5
3.	Multi-axle truck	-	-	4.0	4.0	6.0	6.0
4.	LCV	2.0	1.5	2.2	2.4	2.6	2.8
5.	Tractor	4.0	2.5	-	3.0	4.0	3.0
6.	Tractor trailer	5.0	3.0	-	4.6	6.0	4.5
7.	3-wheeler	0.7	1.0	0.8	1.2	1.4	1.6
8.	Scooter/motorbike	0.4	0.5	0.5	0.3	0.4	0.5
9.	Pedal cycle	0.5	0.3	0.3	0.5	0.6	0.6
10.	Rickshaw	1.8	2.0	2.0	1.5	2.0	2.5
11.	Horse cart	4.5	4.0	3.8	4.0	5.5	4.0
12.	Bullock cart	6.0	8.0	-	7.0	8.0	7.0

6.2.8 Design Vehicle

A selected motor vehicle, the weight, dimensions, and operating characteristics of which are used in highway design is called 'Design Vehicle'.

The design vehicle should be the largest which represents the significant percentage in the traffic stream for the design year.

At locations where trucks are very few in number 'Passenger Car' may be used as design vehicle.

For designing a road, on which largest volume is that of trucks, the design requirements of WB-40 vehicle are used and check should be made that it will be possible for WB-50 (largest vehicle) to negotiate curves by lowering speed, particularly where pavement is curbed. AASHTO has standard vehicle classes as: Passenger Cars (P) Single unit truck (SU), Single unit bus (BUS), Semi-trailers (WB-40, WB-50) and full trailer combination (WB-60). In India, vehicles are standardized as Single unit trucks (length: 11 m), Semi-trailers (length: 16 m), truck trailer combination (length: 18 m), Single unit bus (length: 12 m). Maximum width of vehicle is 2.5 m, and maximum height is 4.75 m. As per Indian Road Congress the maximum weight of single axle vehicle is 10.2 tonnes and tandem axle vehicle is 18 tonnes.

Table 6.3 shows dimensions and turning radii of Indian Design Vehicles.

Table 6.3: Dimensions and turning radii of design vehicles

S.No.	Vehicle Type	Overall Width (m)	Overall Length (m)	Overhang Front (m)	Minimum Turning	
					Rear (m)	Radius (m)
1.	Passenger Car (P)	1.4 – 2.1	3 – 5.74	0.9	1.5	7.3
2.	Single Unit Truck (S.U.)	2.58	9	1.2	1.8	12.8
3.	Semi Trailer and Single unit Bus (WB-12 m)	2.58	15.0	1.2	1.8	12.2
4.	Large Semi-Trailer (WB-15 m)	2.58	16.7	0.9	0.6	13.71
5.	Large Semi-Truck Trailer (WB-18 m)	2.58	19.7	0.6	0.9	18.2

Major roads and high speed highways are designed for largest vehicles while road in the city or other unimportant roads are designed for passenger cars. Figures 6.2 to 6.5 show turning radii of some of the AASHTO vehicles.

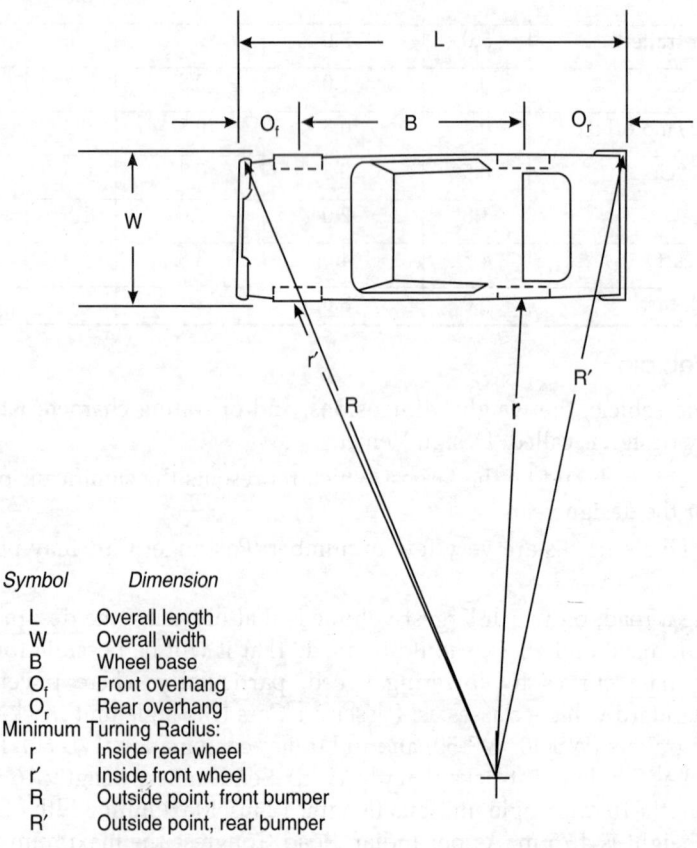

Symbol	Dimension
L	Overall length
W	Overall width
B	Wheel base
O_f	Front overhang
O_r	Rear overhang
Minimum Turning Radius:	
r	Inside rear wheel
r′	Inside front wheel
R	Outside point, front bumper
R′	Outside point, rear bumper

Fig. 6.2: Symbols for dimensions of design vehicle

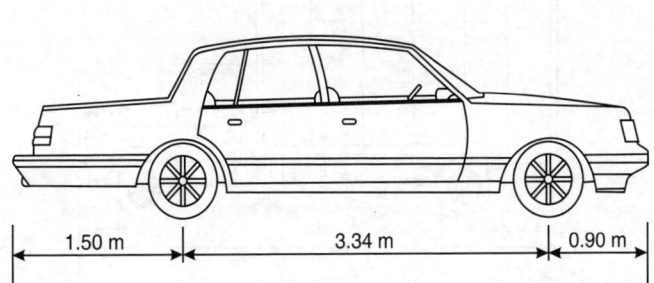

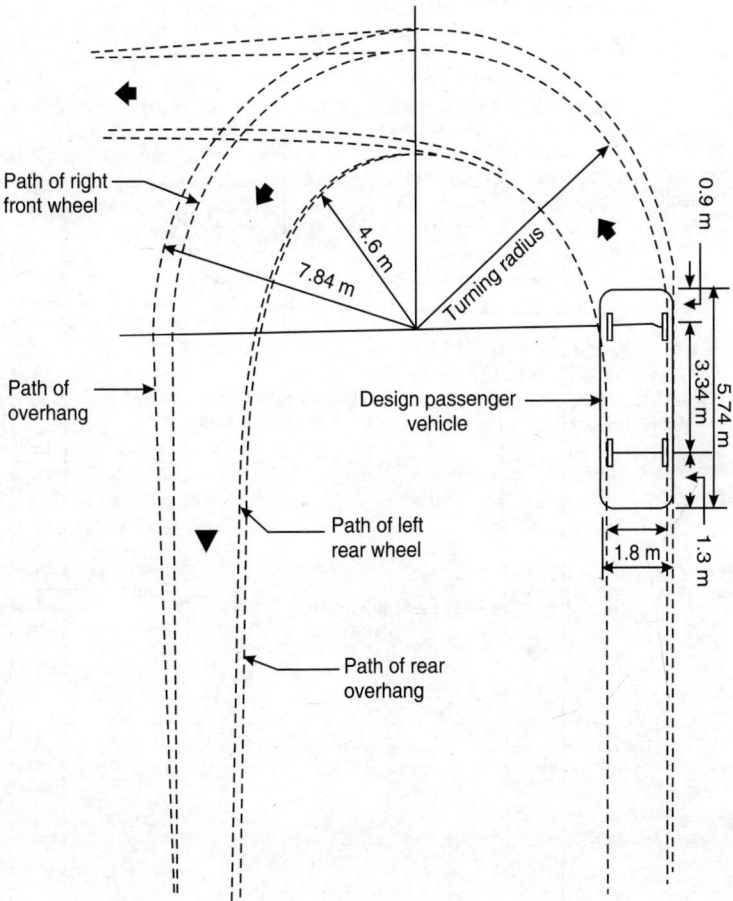

Fig. 6.3: Minimum turning path for passenger car design vehicle (P) IRC

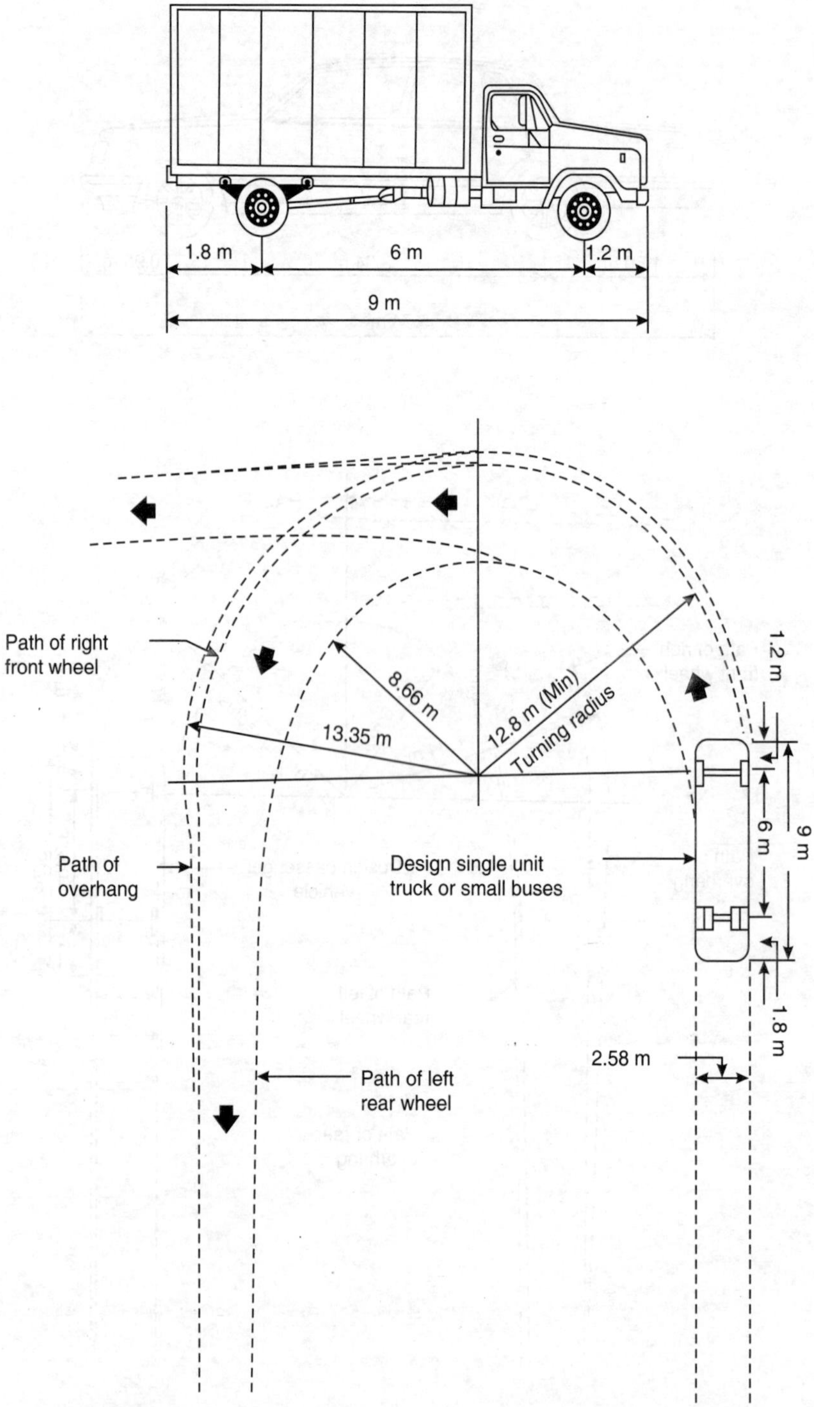

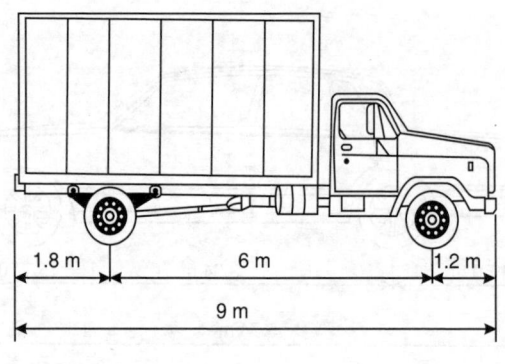

1.8 m   6 m   1.2 m

9 m

Path of right
front wheel

8.66 m

13.35 m

12.8 m (Min)
Turning radius

1.2 m

6 m

9 m

Path of
overhang

Design single unit
truck or small buses

1.8 m

2.58 m

Path of left
rear wheel

Fig. 6.4: Minimum turning path for single unit truck design vehicle (SU) IRC

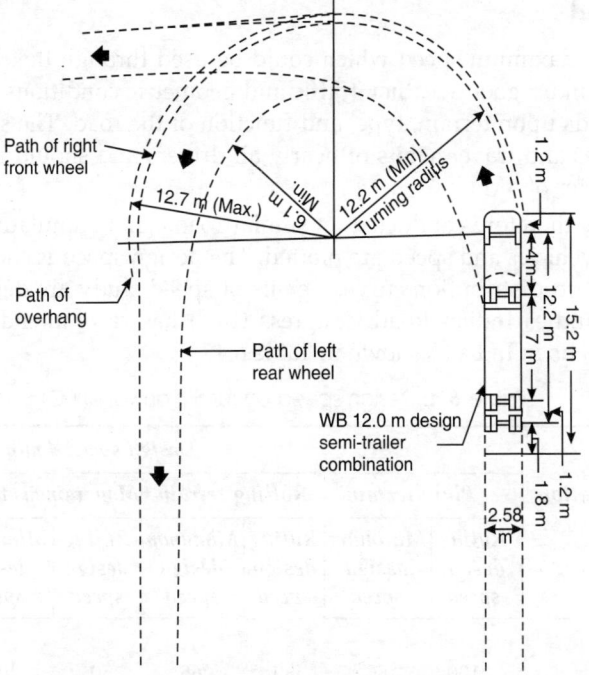

(a) Minimum turning path for semi-trailer (WB-12.0) design vehicle

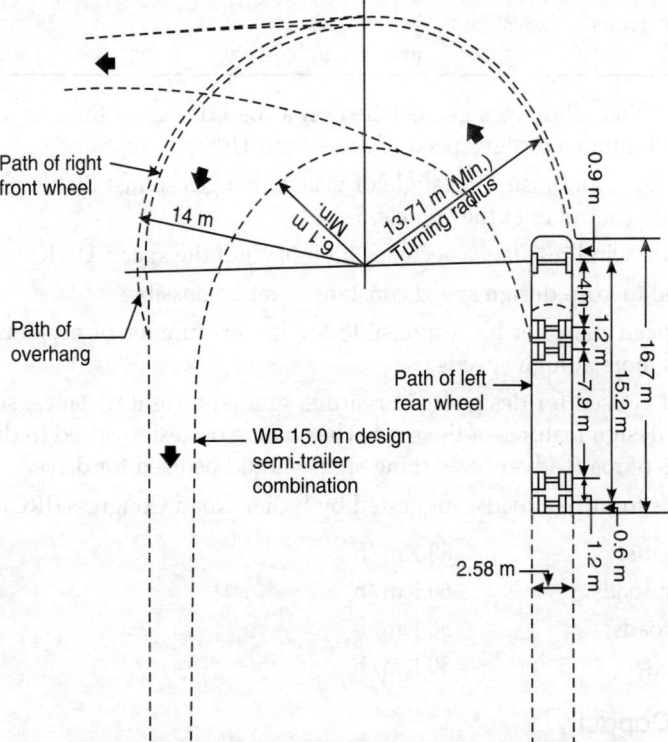

(b) Minimum turning path for semi-trailer (WB-15.0) design vehicle

Fig. 6.5: Minimum turning path for semi-trailers

6.2.9 Design Speed

Design speed is the maximum speed which could be used through the section of a road, with safety and comfort under good weather, traffic and geometric conditions of the road. Choice of design speed depends upon terrain, type, and function of the road. The speed taken for design should fit the desires and travel habits of nearly all drivers and should be suitable for type of highway and character of terrain.

Speed studies are therefore conducted on similar roads, and cumulative distribution curves between percent of vehicles and speed are plotted. The design speed is chosen as 98th percentile speed from such speed distribution curve. Details of speed study are described in chapter–10. Based on several studies Indian Road Congress (IRC) has recommended design speeds for various classes of roads in India as shown in Table 6.4.

Table 6.4: Design speed on rural highways (IRC)

S.No.	Road classification	Design speed, km/h							
		Plain terrain		Rolling terrain		Mountainous terrain		Steep terrain	
		Ruling design speed	Minimum design speed	Ruling design speed	Minimum design speed	Ruling design speed	Minimum design speed	Ruling design speed	Minimum design speed
1.	National and State Highways	100	80	80	65	50	40	40	30
2.	Major district roads	80	65	65	50	40	30	30	20
3.	Other district roads	65	50	50	40	30	25	25	20
4.	Village roads	50	40	40	35	25	20	25	20

It is observed that all drivers get satisfied on a road designed for a speed of 110 Kph. The reasons for not adopting a design speed of more than 110 kph are:

- As we go faster and faster the angle of vision changes so fast that beyond 110 Kph one is likely to lose control over the vehicle.
- Noise due to wind resistance becomes high beyond the speed 110 Kph.
- It is desired to keep design speed constant as far as possible
- Though speed may not be responsible for larger number of accidents, the severity of accident is more at high speeds.

Design speed is used for designing curvature, gradient, sight distance, super elevation and other geometric design features of the road. The change of design speed (reduction), if required on some sections of road, advance warning signs should be used for drivers.

Design speeds for urban roads, suggested by Indian Road Congress (IRC) are:

Arterial roads:	80 Km/h
Subarterial roads:	60 Km/h
Collector roads:	50 Km/h
Local streets:	30 Km/h

6.2.10 Design Capacity

Capacity of a road is the maximum number of vehicles that a road/highway can accommodate. It indicates the effectiveness (efficiency) of the road and its facilities in serving the traffic.

If V is the speed of vehicles in km per hour and

H is the average distance between centre to centre of vehicle (headway) m,

Capacity C is given by $C = \dfrac{1000\,V}{H}$

The clear spacing between the vehicles should be equal to stopping sight distance, for safety.

Therefore H = Length of vehicle + Stopping sight distance

$$(L) + (SSD)$$

$$\text{Capacity } C \;=\; \frac{1000\,V}{L+SSD}.$$

Such ideal conditions when vehicle move with the spacing equal to SSD are rarely met. The above equation gives the capacity value which can be termed as "Theoretical Capacity" or "Basic Capacity" of a road, and can be defined as "Maximum number of vehicles that can pass through a given point on a road, during one hour, under most nearly ideal roadway and traffic conditions, which can possibly be attained."

For design purposes, IRC has recommended capacity of different types of roads as shown in Table 6.5. Capacity standards help in rational evaluation of the improvements needed for better operation of traffic.

Table 6.5: Capacity of different types of roads (IRC)

S. No.	Type of road	Capacity (Passenger car units per day in both directions
1.	Single-lane roads having a 3.75 m wide carriageway with normal earthen shoulders	1,000
2.	Single-lane roads having a 3.75 m wide carriageway with adequately designed hard shoulders 1.0 m wide	2,500
3.	Two-lane roads having a 7 m wide carriageway with normal earthen shoulders	10,000
4.	Roads of intermediate width, i.e. having a carriageway of 5.5 metres with normal earthen shoulders	5,000

Note: Capacity of highways having a dual carriageway will depend on factors like the directional split of traffic, degree of access control, composition of traffic etc. Depending on the actual conditions, capacity of a 4-lane divided highway could be upto 20,000-30,000 pcus.

Detail description on capacity of highway and levels of service is included in the chapter–9.

The test of design adequacy is made by comparing design volume with design capacity, where capacity is not equal to or greater than expected volume, some other alternate must be provided. Roads are designed for design hourly volumes (DHV) for some future year (20 years), volumes greater than those estimated may occur or estimated volumes may occur sooner than expected, exceeding the design capacity, resulting in lower type of service than planned.

Stage construction technique is used to accommodate current volume and increased volumes over years.

6.3 ELEMENTS OF GEOMETRIC DESIGN

The geometric design elements of the road depend upon several factors, involving relationship between dimensions and layout of the road and needs of vehicles and road users.

Following are the basic element considered in geometric design of highways:

1. Cross section elements including sight distances.
2. Horizontal alignment (radius of curve, transition curves, super-elevation)
3. Vertical alignment (gradient, its length, sag and valley curves)

6.3.1 Cross-Sectional Elements

Features which are accommodated in the total width of the land on either side of a road are called cross-sectional elements as shown in Fig. 6.6. The width reserved and other elements of highway cross-section are:

(a) **Right of Way (Permanent road land width):** Right of way (ROW) is the area on the two sides of the road up to road land boundary, reserved for road purposes. It should be sufficient for the construction of the road and its elements. It may be necessary to allow for future widening and foreseeable minor realignment and also to provide for services such as telephone lines, electric poles, etc. The use of this land is controlled by highway authorities. It depends upon several factors such as type of road, drainage requirements, landscaping, planting, cost of land and other local constraints.

(b) **Building line:** In order to prevent ribbon development along the two sides of the road, building lines are established, so that buildings may not be constructed in future. It limits future buildings in relation to a road. Building activity is not allowed within a prescribed distance from road.

(c) **Control line:** Control line is a line, in addition to building line to exercise control on the nature of building activity for a further distance beyond the building line. IRC recommendation for building and control lines are shown in Table 6.6 and Fig. 6.6.

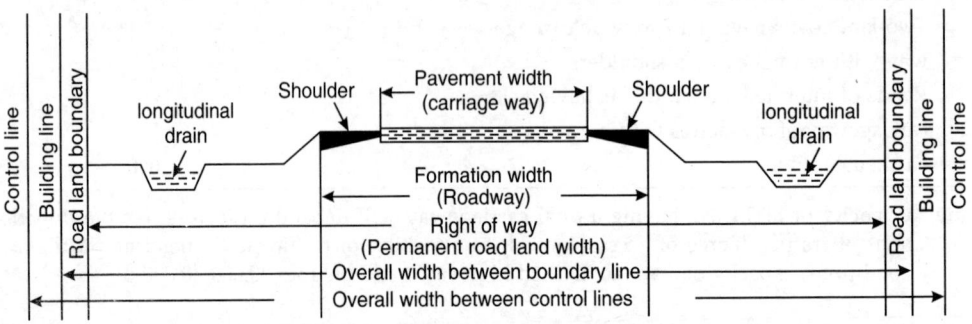

Fig. 6.6: Cross-section terminology

(d) **Formation width:** Formation width is the width of the pavement plus the width of two shoulders one on either side. It depends upon the expected traffic. Due to economy formation width on hilly terrain is minimised to possible extent.

(e) **Pavement width (Carriageway width):** Pavement width is the paved width of the road on which vehicles move. It depends upon the width of a traffic lane and the number of lanes. The width of the pavement set aside for single stream of vehicle operation is called a 'Lane'.

Table 6.6: Recommended standards for building and control lines and land widths (IRC)

S.No.	Road classification	Plain and rolling terrain							Mountainous and steep terrain			
		Open areas (Rural)				Built up areas (Urban)			Open areas (Rural)		Built up areas (Urban)	
		Overall width between Building lines	Overall width between control lines	Land width (normal)	Land width (range)	Distance between Building line and road boundary (set back)	Land width (normal)	Land width (range)	Distance between building line and road boundary (set back)	Land width	Distance between building line and road boundary (set back)	Land width
		(m)	(m)	(m)	(m)	(m)	(m)	(m)	(m)	(m)	(m)	(m)
1.	National and State Highways	80	150	45	30-60	3-6	30	30-60	3-5	24	3-5	20
2.	Major district roads	50	100	25	25-30	3-5	20	15-25	3-5	18	3-5	15
3.	Other district roads	25-30	35	15	15-25	3-5	15	15-20	3-5	15	3-5	12
4.	Village roads	25	30	12	12-18	3-5	10	10-15	3-5	9	3-5	9

In India, single lane roads are 3.75 m wide. Table 6.8 shows typical standards of pavement width.

Table 6.7: Width of pavement (carriageway)

Lane description	Width (m)
Single lane	3.75
Double lane without raised curbs	7.00
Double lane with raised curbs	7.5
Multilanes-width per lane	3.5
Less important two lane road	5.5
Village roads (low volume roads)	3.00

Number of lanes are decided as per traffic volumes for the design year.

(f) **Shoulders:** Shoulders are the strips between the outer edge of the pavement and inner edge of the drain. Shoulders may be turfed (a special type of grass), or made of gravel and crushed stones, or bituminous type. They should be well compacted to support occasional use by vehicles, usable in all weather. Surface colour should be different from pavement for easy distinction.

The purpose of providing shoulders are:

- Parking of vehicles
- Removal of out-of-order vehicles from pavement.
- Putting traffic sign
- Space for people to walk
- Providing lateral stability and adding structural strength to the pavement
- Helping in overtaking and crossing of vehicles safely and conveniently.
- Increasing safety, quality of service and comfort
- Improving sight distance, specially on curves
- Better appearance of highway
- Improving capacity and speed by developing a sense of openness to the drivers.
- Protecting edge of carriageway from breaking.

The width of the shoulders should be adequate to prevent tendency of drivers to drive closer to the centreline of the road. It should be enough to allow vehicles to stop on shoulders without obstruction to lane traffic. Indian practice is to use 2.5 m wide shoulders on two-lane rural roads.

Highway capacity is considerably influenced by the type and width of shoulder. Substandard design of shoulders reduces the capacity of a road. The capacity of a 2 lane road is expected to reduce by 30 percent when no proper shoulders are provided.

(g) **Median Strip (Traffic Separators):** For separating the traffic streams 'Medians' are provided. The longitudinal separation between lanes/traffic streams may be in form of open land, road makings or concrete type.

The purposes of providing medians are:

- Separate opposing direction of traffic streams.
- Separate slow and fast moving traffic in same direction
- Minimise head light glare from opposing traffic

- Help in movement of traffic by channelising traffic into separate streams at intersections.
- Provide refuge area in case of an emergency

Width of the median depends upon available land. Minimum desirable width of median is 5 m, but it can be 3 m if space is not available, in rural areas. On urban roads median width may vary from 1.2 m to 5 m. On bridges and viaducts the width of the median may be from 1.2 to 1.5 m. Medians may be paved flush, curbed and crowned, depressed, or roads at separate levels with natural ground as median, as shown in Fig. 6.7. A special study should be made to determine the type of median most suitable for a particular condition.

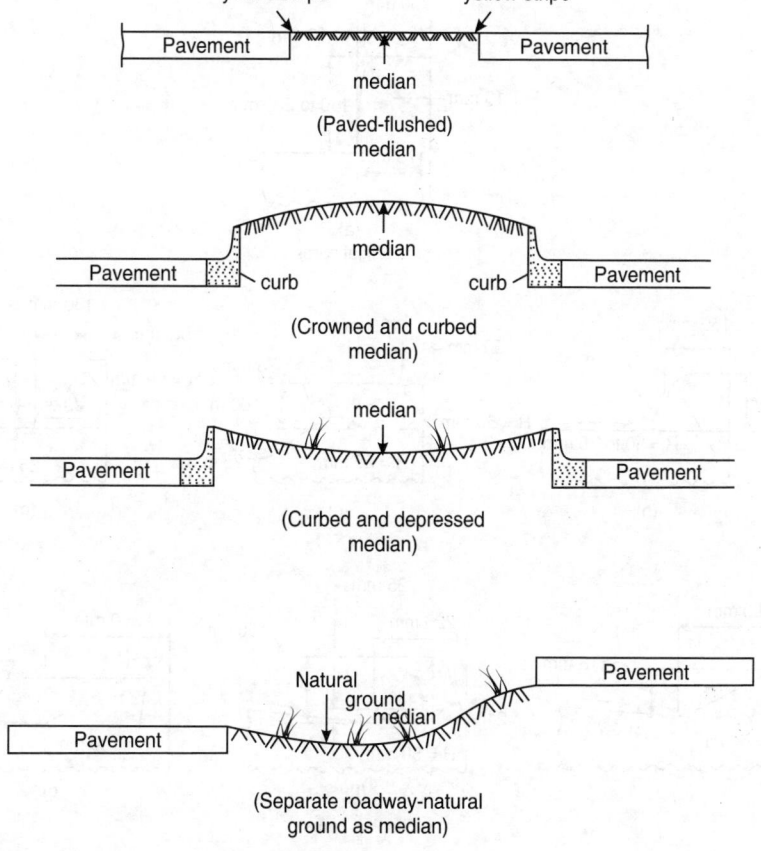

Fig. 6.7: Types of medians

(h) **Curbs:** Curbs are dividing lines between carriageway and shoulders (footpaths). It separates footpaths, islands, parking spaces, etc, clearly defining the edge to vehicle operators.

Purposes of providing curbs are:

- Strengthen the edges of the pavements and prevent lateral spread.
- Improve the appearance by acting as longitudinal edge marking.
- Clearly show the pavement edge.
- Facilitate in providing drainage of water.

- Provide safety and protect pavement edges, specially on urban roads.
- Manage any difference in level between footpath and carriageway.
- Work as a traffic barrier.

Curbs are classified into three categories depending upon their height.

(i) **Low (mountable) curbs:** These curbs are 7 to 9 cm high with top sloped so that vehicles can climb.

(ii) **Semi-mountable curbs:** These curbs are 15 to 20 cm high and do not allow the vehicles to climb easily, however in emergency vehicles can climb with difficulty.

(iii) **Barrier curbs:** These are 23-45 cm high. Vehicles can not climb over it. These are used on bridges. Typical curbs are shown in Fig. 6.8.

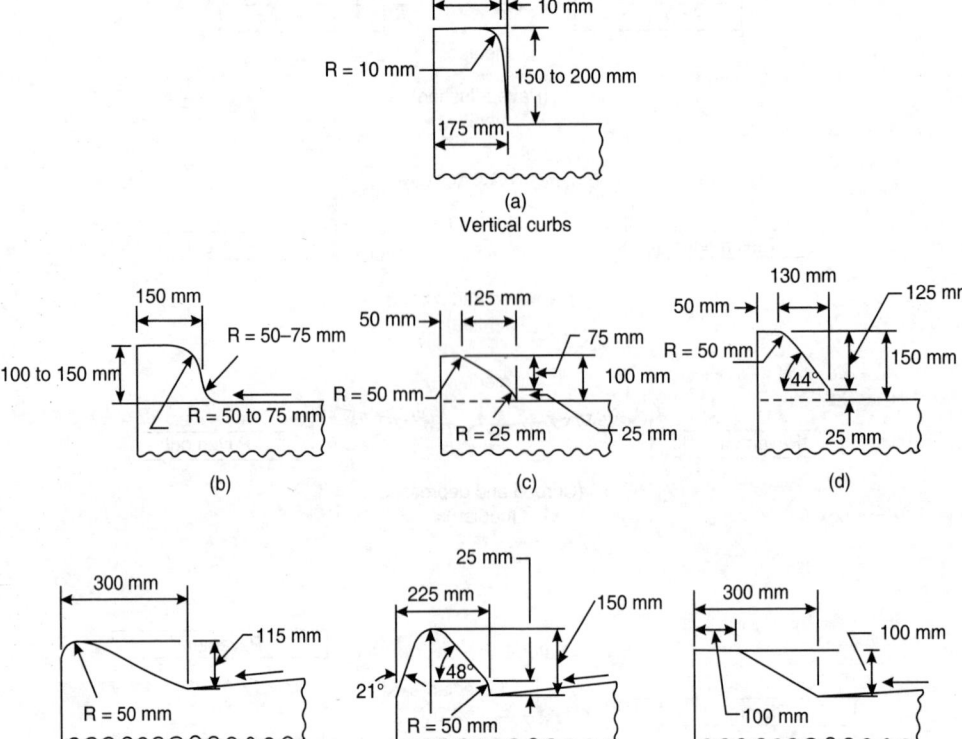

Fig. 6.8: Typical highway curbs

(i) **Camber (cross-slope):** Cross-slope is the slope of the surface of the pavement measured normal to the centreline of the road. It is provided to drain off water from the road surface. The central highest point is called 'Crown'. The cross-slopes on straight roads should be as flat as required for drainage, depending upon the type of the road.

Table 6.8 shows the cross-slopes recommended by IRC. In areas having high intensity of rainfall steeper values and in areas of low intensity of rainfall lower values than shown in the table can be used.

On hilly roads, depending upon the side conditions, unidirectional cross-slopes are also used. Unidirectional cross-slopes are also adopted on divided roads, having a median, carriageway sloping towards outer edge. Undivided roads are crowned in the middle.

Table 6.8: Cross-slopes for different types of road surfaces (IRC)

S.No.	Type of Surface	Cross-slope m/m	percent
1.	High type bituminous surface or cement concrete.	1 in 60 to 1 in 50	1.7 to 2.0
2.	Thin bituminous surfacing	1 in 50 to 1 in 40	2.0 - 2.5
3.	Water bound macadam or gravel	1 in 40 to 1 in 30	2.5 - 3.0
4.	Earth	1 in 30 to 1 in 25	3.0 - 4.0

The cross-slope for earth shoulders should be at least 0.5 percent steeper than the slope of the pavement subject to a minium of 3 percent. Shoulders should normally have same cross-fall as the pavement on superelevated sections.

(j) **Footpaths (Sidewalks):** Footpaths are provided along the curbline in urban roads. They are raised strips constructed on both sides of the roads for providing protection to pedestrians. Minimum width of footpaths should be 1.5 m. Higher widths are used depending upon number of people walking on it.

Pedestrian sidewalks ensure safety of pedestrians. They should be sloped properly to drain away the rain water. Normal slopes are 1 in 40 to 1 in 30.

(k) **Cross-walks (Zebra crossings):** Pedestrians should cross the road only at authorised places for safety, minimising hazard with vehicular traffic. Pedestrian cross-walks should therefore be provided at places where conflict exists between vehicular traffic and pedestrians at mid-blocks near school, hospitals, etc and also at intersection of roads in urban areas.

To be effective pedestrian cross-walks are clearly marked. The width of cross-walks depends upon the number of pedestrians at a particular location.

(l) **Pedestrian Refuge islands:** For safety of pedestrians, refuge islands are provided on the roads which have wide carriageway. Pedestrians who can not cross the whole width of the road at one time they can take refuge at such islands.

Central islands should be so designed and placed that they do not obstruct the movement of vehicular traffic. They should be well lighted, having reflectorised sides so that they are visible during the night time.

(m) **Cycle tracks:** Separate cycle tracks may be required where cycle traffic is high. Where per hour cycle traffic is 400 or more and vehicular traffic is 100-200 vehicles per hour. If motor vehicles are more than 200 per hour a lesser volume of cycle traffic say 100 cycles/hr may justify a separate cycle track.

Separate cycle track should have a minimum lane width of 1 m and should be provided on both sides of the road. Wider cycle tracks, with good riding surface and lighting are needed for safety.

(n) **Parking lanes:** Parking lanes are provided along urban roads to allow curb parking. Parking lanes should be allowed only in one direction parallel to main road. Clearance of 1 m to 1.5 m from parked vehicles to edge to traffic lane is necessary. Parking lane should be 3 to 3.7 m wide. Conversion of parking lane to through traffic lane should be possible in future if necessary, to increase the capacity of the road.

(o) **Speed change lanes:** Drivers have to adjust their speed while entering or leaving an intersection. On entering the speed is reduced to safe value at which the intersection can be negotiated and speed is increased when leaving the intersection to match the highway speed. To avoid traffic disruption and hazard, speed change lanes are generally provided on high type facilities (highways, expressways, etc). These lanes increase the capacity of the intersection.

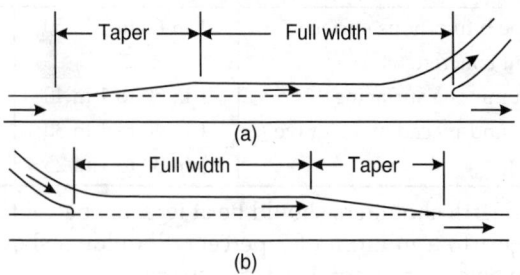

Fig. 6.9: (a) Deceleration lane and (b) Acceleration lane

(p) **Right-Turn lanes:** Storage lanes for right turning traffic enable the vehicles to slow down to the necessary turning speed without impeding the flow of through traffic and waiting at locations where signals are provided.

(q) **Bus-bays:** Bus-bays are special areas made along the road side for storing the buses. Typical bus storage bays are shown in Fig. 6.10. The minimum desirable length of a bus bay for 40 to 45 seat bus (app. 11 m long) requires 25 m for approach side, 19 m for departure side and 32 m for mid block single bus zone. Location of bus bays should be carefully selected to avoid creating hazardous situation and minimise the obstruction to other traffic.

(r) **Railing and Guard stones:** Railings are placed for safety of people, along the roads in city areas and at other dangerous places such as intersections, steep grades, sharp curves etc.

Guard stones are used on curves on rural roads. Guard rails or guard stones should be installed before the highway is opened to traffic and should be highly visible to be fully effective.

(s) **Driveways:** Driveways are short links of road (detours) which connect important highway facilities like service stations, restaurants, fuel pump, etc.

(t) **Lateral clearance:** The distance between extreme edges of the carriageway from any structure on side is called lateral clearance. For safety purposes the lateral clearances should be as large as possible, however the minimum clearance recommended for national highways (NH) and state highways (SH) is 2.5 m to 2 m, major other district roads (MDR and ODR) is 1.5 to 2 m, and village road (VR) is 1 m to 1.5 m.

(u) **Road furniture:** Lamp post, telephone post, sign boards, sign stones, trees, etc are called road furniture. They should be placed to minimise danger to the vehicles which may be using the shoulder.

(v) **Batter slopes:** Cuttings may be stable in solid rocks, where a slope of 2.5 to 1 may be used. For materials other than rock, cut slopes may vary between 1.5 to 1 or 2 to 1. The materials in fill may not stand above a slope of 1.5 to 1, unless suitably faced.

Flat batter slopes reduce the severity of accidents in which vehicles run off a road losing control on steep embankment or running into a drain. A slope of 4 horizontal to 1 vertical (4 to 1) is sufficient to reduce such incidents.

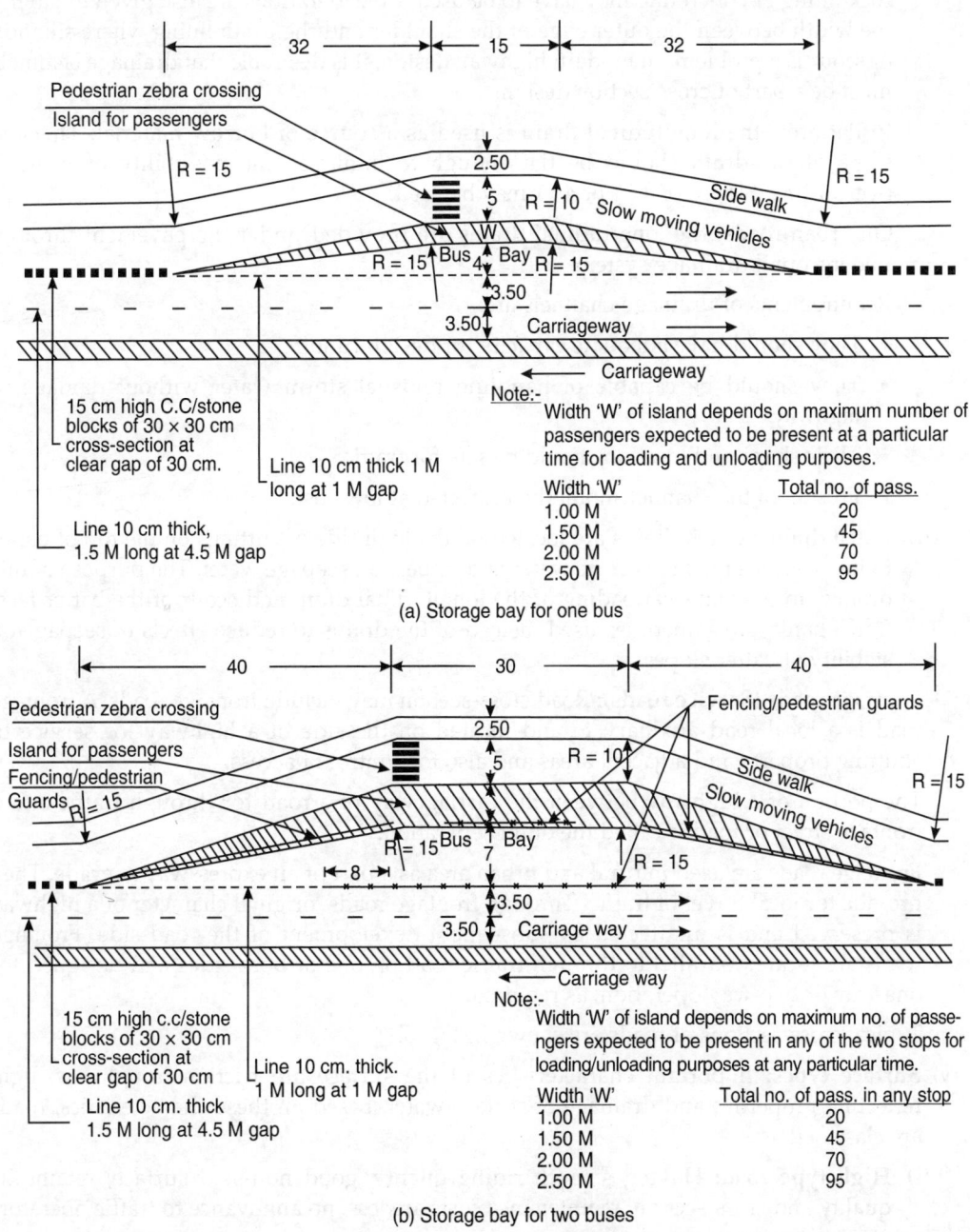

(a) Storage bay for one bus

Note:-
Width 'W' of island depends on maximum number of passengers expected to be present at a particular time for loading and unloading purposes.

Width 'W'	Total no. of pass.
1.00 M	20
1.50 M	45
2.00 M	70
2.50 M	95

(b) Storage bay for two buses

Note:-
Width 'W' of island depends on maximum no. of passengers expected to be present in any of the two stops for loading/unloading purposes at any particular time.

Width 'W'	Total no. of pass. in any stop
1.00 M	20
1.50 M	45
2.00 M	70
2.50 M	95

Fig. 6.10: Typical bus-bays (IRC)

(w) Drains

(i) Longitudinal drains: Drainage is required to discharge rain water from road surface to the drain. Longitudinal drains are located on the outside of shoulders. Sometimes subsoil drain is provided at the edge of the pavement. When scour is likely to occur because of the nature of material or longitudinal gradings some protection by grassing, rock lining or concreting may have to be used. Considerations are also given to paving the width between the outer edge of the shoulder and the drain lining where siltation or scour is a problem. In modern highway design, it is desirable that drainage channels must be a part of cross-section design.

In flat area, the longitudinal drain is used as a source of borrow material. The side slope of the drain should be flat enough to minimise the possibility of vehicles overturning. Slopes of 3 : 1 or 4 : 1 may be used.

On urban roads, the longitudinal drains are provided under the pavement through underground drainage system.

Requirements of drainage channels are:

- They should have adequate capacity for design runoff.
- They should be capable of handling unusual storm water without damage to highway.
- Their shape and location should be safe for traffic.
- Erosion of the channel should be protected at low cost.

(ii) **Catch drains:** Catch drains are located on the high side of cuttings on the top of batter to intercept the flow of surface water and upper soil seepage water. The purpose of this drain is to prevent overloading of the longitudinal drain and scour of the batter face. Catch banks are sometimes used instead of the drains to reduce effects of seepage on stability of batter slopes.

(x) **Frontage roads (Service roads):** Road cross-section may include frontage roads. A frontage road is a local road auxiliary to and located on the side of a highway for service to abutting property and adjacent areas and also for control of access.

The portion of a highway between the carriageway of a road for through traffic and a frontage street or road is called the outer separation.

Frontage roads are used in rural and urban area as a part of an expressway at grade. They provide for local travel of traffic. Through frontage roads, original character of a highway is preserved and is unaffected by subsequent development of the road side. Frontage roads are located parallel to through traffic road on one or both sides and designed for one-way or two-way operation, as required.

Typical cross-sections of roads are shown in Fig. 6.11

(y) **Surface types:** Important characteristics of the surface are friction, roughness, light reflecting properties and drainage of surface water. Based on these characteristics, roads are classified as:

(i) **High type road:** Having smooth riding quality, good non-skid surface, retains its quality and cross-section, maintenance costs are less, no annoyance to traffic operators etc. such as cement roads, bituminous concrete roads.

(ii) **Intermediate type road:** Cheaper than high type roads, having relatively lower strength, lower traffic volumes, slightly inferior riding surface than high type roads, such as water bound macadam, bituminous surfaces.

(iii) **Low type road:** Quality and cross-section not retained. High maintenance costs, not liked by operators, cheap in initial cost, such as earth roads, gravel roads, etc. Geometric design standards vary with type of road.

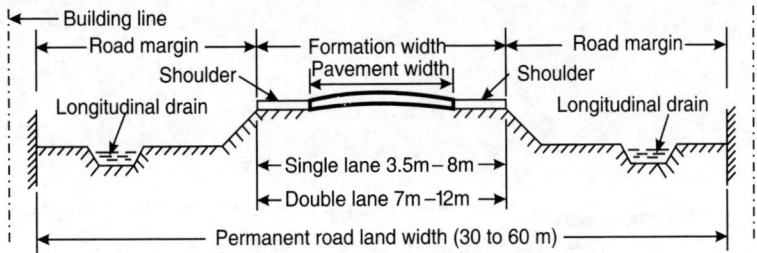

1.   Cross-section of a rural road on embankment

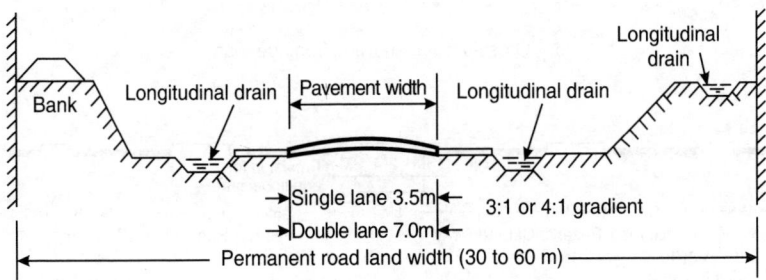

2.   Cross-section of a highway in cutting

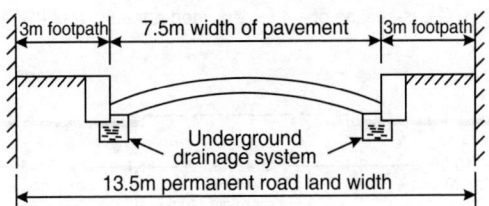

3.   Cross-section of a 2-lane urban road

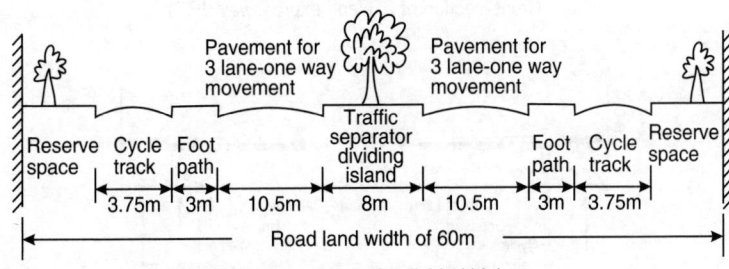

4.   Cross-section of a divided highway

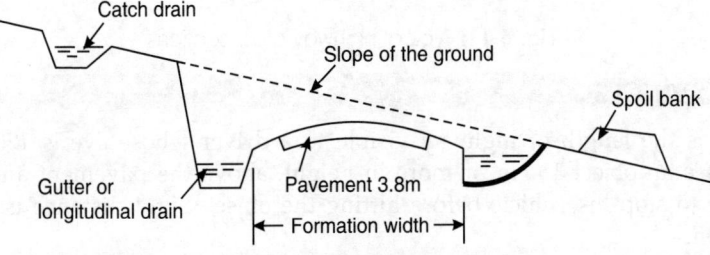

5.   Cross-section of a road in hilly area

Fig. 6.11 (*Contd...*)

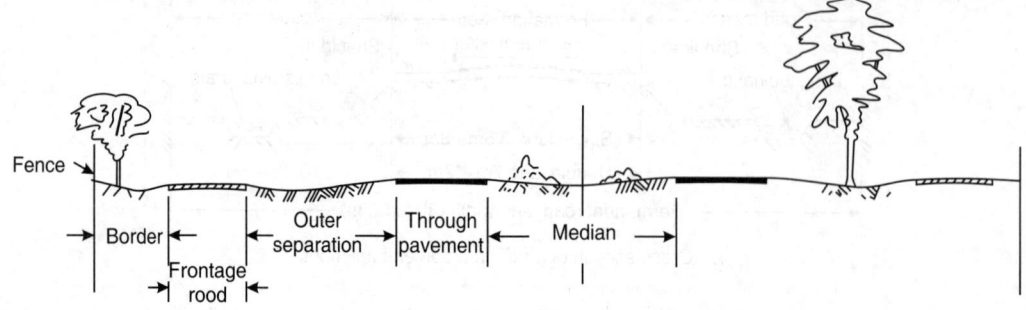

6. Cross-section showing frontage road

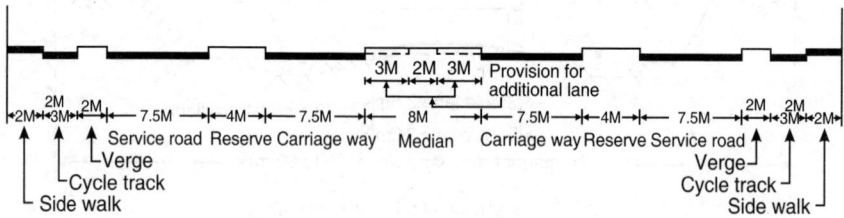

7. Cross-section of a 4 lane divided arterial street (IRC)

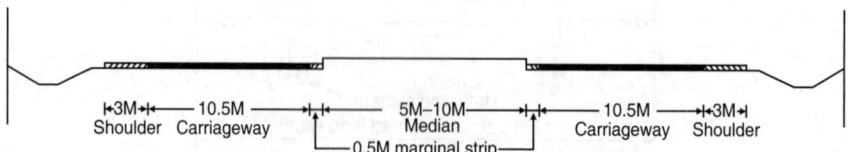

8. Cross-section of a 6 lane expressway (IRC)

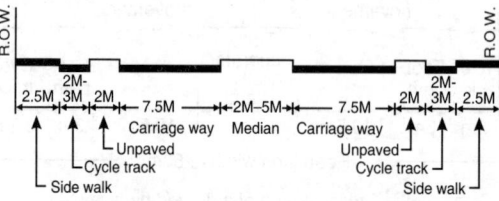

9. Cross-section of a 4 lane divided sub-arterial street (IRC)

Fig. 6.11: Typical highway cross-sections

6.4 SIGHT DISTANCES

Sight distance is the length of highway visible to a driver whose eye is 120 cm above the pavement, to see an object 15 cm or more in height, above the pavement ahead of him. He should be able to stop his vehicle before hitting the object. Sight distance is needed for the following reasons:

(i) To give driver enough time to avoid an obstruction or oncoming vehicle.

(ii) To give two cars coming is opposite directions, the opportunity to stop before colliding.

(iii) To allow drivers to safely overtake an vehicle or abort overtaking.

(iv) To provide vehicles stopped at intersections with sufficient sight distance for them to cross other road safely.

Sight distances are classified as:

(a) Stopping sight distance

(b) Passing (overtaking) sight distance

(c) Intermediate sight distance

6.4.1 Stopping Sight Distance

Large objects, fallen trees, animals etc. on the road require that vehicle must stop before striking them. Such hazards should be visible at a distance large enough that driver can stop the vehicle. Stopping sight distance is defined as "The minimum distance required within which a vehicle moving at a design speed can be stopped before reaching the object ahead". It is measured between two points one 1.2 m above the pavement (Drivers eye) and other 0.15 m height representing the object.

Stopping sight distance depends upon following factors:

- perception and reaction ability of driver
- efficiency of the brakes of the vehicle
- frictional resistance between road surface and tyres of the vehicle.
- speed on which vehicle is moving (design speed) and
- whether road is on straight, upgrade or downgrade.

Analysis of stopping sight distance

Stopping sight distance is made up of two elements:

1. Distance travelled by the driver after seeing the object, before applying the brakes. This is be called 'Perception and Reaction Distance' or simply 'Reaction Distance' and
2. Distance travelled after application of the brakes before stopping. This is called 'Braking Distance'

Stopping sight Distance		Perception and	+	Braking
(SSD)	=	Reaction Distance		Distance

1. Reaction distance:

The perception and reaction time (t), depends upon driver's own ability, based on PIEV theory and other physical and psychological factors, however on an average it is taken as 2.5 seconds in calculations of SSD (It may vary from 0.5 seconds to 3-4 seconds depending upon complexity of the situation.

Perception and reaction distance = Speed × time $(V \times t)$

If V = speed in km. p. hr.

t = reaction time in seconds.

$$\text{Reaction Distance} = \frac{V \times 1000}{60 \times 60} \times t = 0.28 \, Vt$$

2. Braking distance: (L)

The distance travelled by the vehicle after applying the brakes can be calculated by equating the kinetic energy developed to work done in stopping the vehicle.

Workdone in stopping the vehicle = Kinetic energy

If

L = Braking distance

f = Coefficient of longitudinal friction between road and tyres) normally taken as 0.4. (It however varies from 0.40 to 0.35)

W = Total weight of the vehicle

P = Maximum frictional force developed

g = Acceleration due to gravity

Maximum frictional force P = Coefficient of friction × Weight

$$= f \times W$$

Work done in stopping the

vehicle against friction = Force × Distance

$$= P \times L = f \times W \times L \tag{i}$$

The kinetic energy for the design speed 'V' $= \dfrac{1}{2} mV^2$

$$\text{K.E.} = \frac{1}{2} \frac{W}{g} V^2 \tag{ii}$$

equating (i) and (ii)

$$f W L = \frac{1}{2} \frac{W}{g} V^2$$

$$L = \frac{V^2}{2gf} \quad \text{Converting } V \text{ in m/sec}$$

$$= \frac{(0.28\,V)^2}{2 \times 9.8 \times 0.4}$$

$$= .01\,V^2$$

Stopping sight distance = Reaction Distance + Braking Distance

$$0.28\,Vt + .01\,V^2$$

where

V = speed in km. p. hr.

t = Reaction time of drivers.

Braking distance on grade

When ascending the gradient → Component of gravity helps in braking.

When descending the gradient → Component of gravity opposes braking.

The equation of braking distance: $\dfrac{V^2}{2g\,(f \pm n)}$ where n = percent gradient.

(For grade = 3%, n = 0.03)

$$\left. \begin{array}{l} \text{Upgrade} = \dfrac{V^2}{2g\,(f+n)} \\[4mm] \text{Downgrade} = \dfrac{V^2}{2g\,(f-n)} \end{array} \right]$$

Coefficient of friction varies from 0.40 to 20 km/hr to 0.35 at 100 km/hour.

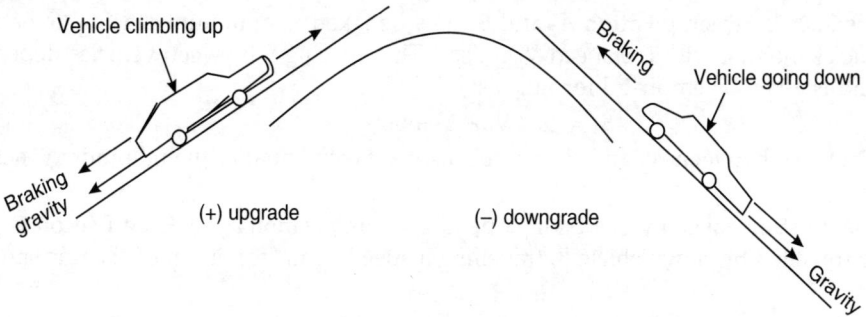

Fig. 6.12: Braking distance on grades

6.4.2 Passing (Overtaking) Sight Distance

Overtaking or passing sight distance is the minimum distance open to the view of the driver of a vehicle, intending to overtake a slow moving vehicle ahead with safety against the traffic of opposite direction.

Analysis of Overtaking Sight Distance. Refer to Fig. 6.13.

A. The vehicle intending to overtake, A_1, A_2, A_3 are its positions at different intervals, of time.

B. The vehicle to be overtaken B_1 and B_2 are its positions.

C. Vehicles coming from opposite direction, C_1 and C_2 are its positions.

Overtaking sight distance is thus divided into 3 parts, d_1, d_2 and d_3.

d_1 = Distance travelled by A, to reduce the speed equal to B, till it gets chance to overtake. A_1 and A_2 are its 2 positions in time.

d_2 = Distance travelled by vehicle A from position A_2 to A_3 (vehicle A shifts to adjoining lane, overtakes B, and comes back to original lane, at A_3, ahead of B.

d_3 = From the position A started overtaking and reached A_3, the vehicle C in opposing lane moves from position C_1 to C_2, a distance of d_3.

For calculating the values of d_1, d_2 and d_3, certain assumptions are to be made.

Distance d_1. The vehicle A, moving at the designed speed V m/sec, is forced to reduce the speed to that of slow moving vehicle B, and follows B till there is an opportunity for safe overtaking operation.

If V_b is speed (m/sec.) of slow vehicle B, and t is the reaction time of driver in seconds, then

$$d_1 = V_b \times t$$
$$V_b = \text{m/sec speed of } B$$
$$t = \text{Seconds, reaction time}$$

on the average the value of t is taken as 2 seconds.

Distance d_2. From position A_2, vehicle A starts accelerating, shifts to adjoining lane, overtakes vehicle B and comes back to original lane ahead of B, in position A_3. (Distance d_2)

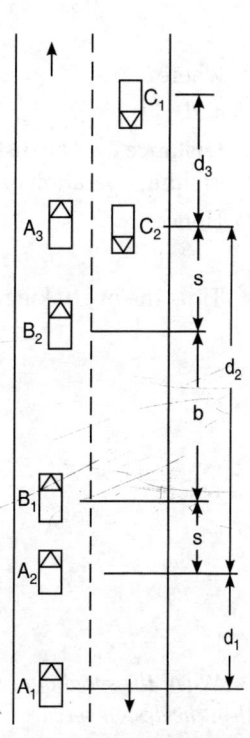

Fig. 6.13: Overtaking manoeuvre

Distance between A_2 and A_3 (d_2) can be analysed in 3 parts:

(i) **Distance between position A_2 and B_1** may be taken as minimum spacing 'S' between the two vehicles moving with the speed V_b m/sec. This spacing S between vehicles depends upon speed and is given by empirical formula as:

$$S = (0.7\,V_b + 6) \text{ metres}$$

(ii) **Distance between A_3 and B_2** may be also be taken equal to the minimum spacing 'S' as above.

(iii) Let the time taken by overtaking operation from position A_2 to A_3 be T seconds. Then the distance travelled by slow vehicle B, travelling a speed V_b m/sec. in time T is b, as shown :

$$b = V_b \times T \text{ metres} \tag{iii}$$

So the total distance $d_2 = S + b + S = (b + 2S)$ metres.

Overtaking Time 'T'. The overtaking time 'T' depends upon the speed of the vehicle B and acceleration of overtaking vehicle 'A'.

Using the general formula, for distance d_2 travelled by a uniformly accelerating body, with initial speed of V_b m/sec and 'a' the acceleration, m/sec^2

$$d_2 = V_b T + \frac{1}{2}\,aT^2 = (b + 2S) \tag{iv}$$

As in (iii) $\qquad\qquad b = V_b T$, and so from (iv)

$$2S = \frac{1}{2}\,aT^2$$

or $\qquad\qquad\qquad\qquad T = \sqrt{\dfrac{4S}{a}} \qquad \text{seconds}$

where $\qquad\qquad\qquad\qquad S = 0.7\,V_b + 6$

and $\qquad\qquad\qquad\qquad d_2 = (V_b T + 2S) \text{ metres.} \tag{vi}$

Distance d_3. The distance travelled by vehicle C moving at design speed V m/sec, during the overtaking operation of vehicle A i.e., during time T, is d_3 between positions C_1 to C_2.

Hence $\qquad\qquad\qquad\qquad d_2 = V \times T \text{ metres.}$ V = design speed

$$T = \text{Overtaking time as above.}$$

Thus the overtaking sight distance $= d_1 + d_2 + d_3$.

$$d_1 = V_b t$$
$$d_2 = V_b T + 2S$$
$$d_3 = VT$$
$$OSD = V_b t + V_b T + 2S + VT$$

where $\qquad\qquad\qquad\qquad V_b = \text{Speed of overtaken vehicle m/s}$

$\qquad\qquad\qquad\qquad\qquad t = \text{Reaction time in seconds}$

$\qquad\qquad\qquad\qquad\qquad T = \text{Overtaking time in secs.}$

$\qquad\qquad\qquad\qquad\qquad S = \text{Minimum spacing between vehicles, m.}$

$\qquad\qquad\qquad\qquad\qquad V = \text{Design speed of road, m/sec.}$

When the speed of overtaken vehicle (V_b) is not given, it is assumed 15 to 16 Km/h (4.5 m/sec), less than the design speed.

Overtaking sight distance is measured between two points both 1.2 m above carriageway, representing the positions of the drivers.

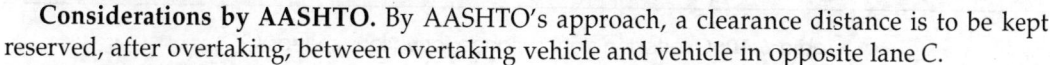

Considerations by AASHTO. By AASHTO's approach, a clearance distance is to be kept reserved, after overtaking, between overtaking vehicle and vehicle in opposite lane C.

This is given by d_4

$$d_4 = \frac{2}{3} d_2.$$

Therefore OSD (PSD) by AASHTO = $d_1 + d_2 + d_3 + d_4$

On two-way road, minimum OSD = $d_1 + d_2 + d_3 + d_4$

On one-way road (no opposing vehicle), OSD = $d_1 + d_2$.

Overtaking Zones. On the roads, where safe overtaking distance cannot be provided all along, overtaking opportunities for vehicles are provided at frequent intervals. These zones, where overtaking can be done safely, are called "Overtaking Zones".

Sign post is put sufficient distance in advance to indicate the starting of overtaking zone and also at the end of over-taking zone.

Sign post is placed 1 OSD ahead of beginning of over-taking zone and 1 OSD, before zone ends, as shown in Fig. 6.14.

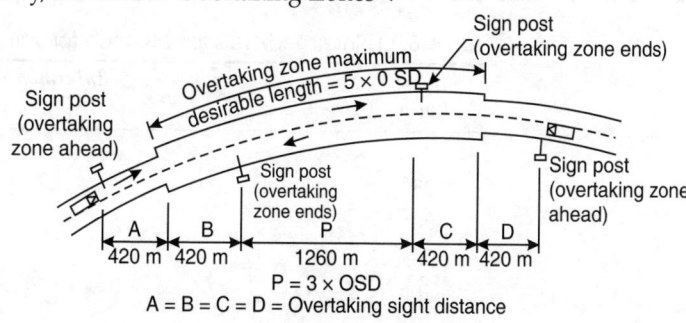

Fig. 6.14: Overtaking zone

6.4.3 Intermediate Sight Distance

Intermediate sight distance is twice the safe stopping distance. Intermediate sight distance is allowed where it is not possible to provide full passing sight distance. It provides reasonable opportunities to drivers to overtake, by observing certain cautions, while overtaking.

IRC recommendations for stopping, overtaking and intermediate sight distances are given in Tables 6.9 to 6.11. Figure 6.15 shows stopping, intermediate and passing sight distances.

Table 6.9: Stopping sight distance for various speeds (IRC)

Speed	Perception and brake reaction		Braking		Safe stopping sight distance (metres)	
V	Time, t	Distance (metres)	Coefficient of longitudinal	Distance (metres)	Calculated	Rounded off values
(km/h)	(sec.)	$d_1 = 0.278\ Vt$	friction (f)	$d_2 = \dfrac{v^2}{254 f}$	$d_1 + d_2$	for design
2.0	2.5	14	0.40	4	18	20
25	2.5	18	0.40	6	24	25
30	2.5	21	0.40	9	30	30
40	2.5	28	0.38	17	45	45
50	2.5	35	0.37	27	62	60
60	2.5	42	0.36	39	81	80
65	2.5	45	0.36	46	91	90
80	2.5	56	0.35	72	118	120
100	2.5	70	0.35	112	182	180

Table 6.10: Overtaking sight distance for various speeds (IRC)

Speed km/h	Time component, seconds			Safe overtaking sight distance
	For overtaking manoeuvre	For opposing vehicle	Total	
40	9	6	15	165
50	10	7	17	235
60	10.8	7.2	18	300
65	11.5	7.5	19	340
80	12.5	8.5	21	470
100	14	9	23	640

Table 6.11: Intermediate sight distance for various speeds (IRC)

Speed km/h	Intermediate sight distance (metres)
20	40
25	50
30	60
35	80
40	90
50	120
60	160
65	180
80	240
100	360

6.4.4 Application of Sight Distances

Single/Double lane roads: Safe stopping distance is the minimum basic distance to be provided. Whereever possible overtaking sight distance should be provided. Where it is not feasible intermediate sight distance may be provided. Proper overtaking zones should be marked or shown by signs.

Multilane roads: Divided highways with 4 or more lanes should only be designed for a safe stopping distance. Undivided highways with 4 lanes there are sufficient opportunities for overtaking within half of the carriageway. Such roads should be designed for safe stopping distance.

On obstructed horizontal curves: If there are obstruction on horizontal curves to obtain minimum sight distance these obstruction should be removed to provide lateral clearance. For such obstruction a set-back distance must be used to clear the site.

6.4.5 Headlight Sight Distance at Valley Curves

During night travel the design must ensure that the roadway ahead is illuminated by the vehicle headlights on a sufficient length on valley curves to enable drivers to apply brake to stop when necessary. This distance is called "Headlight Sight Distance". It should be atleast equal to the safe stopping sight distance.

For headlight sight distance on valley curves, the height of the headlights above road surface is assumed 0.75 m and the useful beam of headlight is up to one degree upward from the grade of the road. The height of object is considered nil.

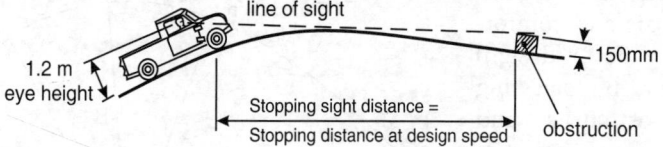

(a) Stopping sight distance on vertical alignment

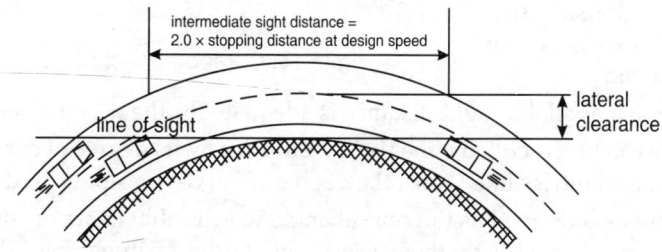

(b) Intermediate sight distance on horizontal alignments

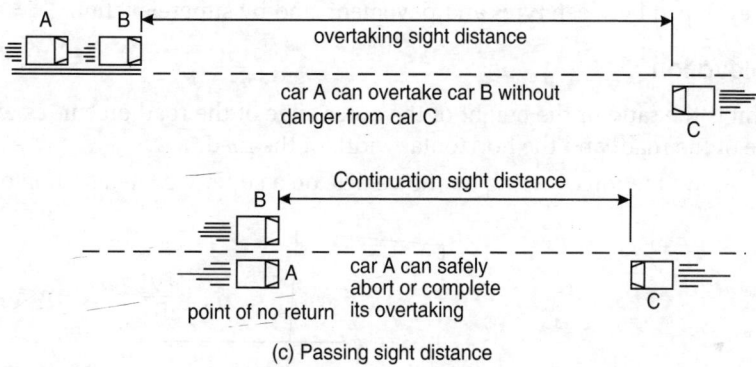

(c) Passing sight distance

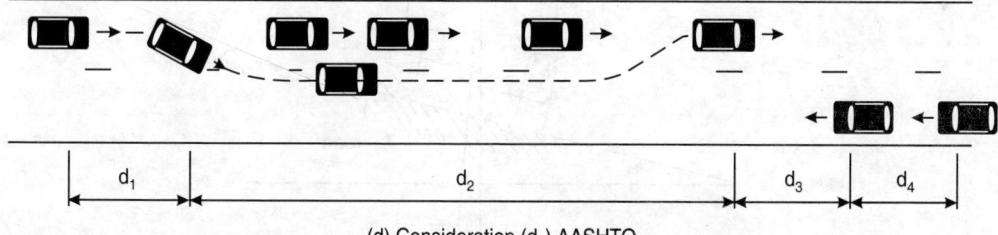

(d) Consideration (d_4) AASHTO.

Fig. 6.15: Concept of sight distances

6.5 HORIZONTAL ALIGNMENT

Design standards used for a road should be kept uniform, throughout, by proper design of horizontal alignment. The horizontal alignment should be as direct as possible, conforming with the topography of the terrain, minimising the earth work and providing good drainage. Horizontal alignment of a road is usually a series of straights (tangents) and circular curves connected by transition curves. (Fig. 6.16). The curves are segments of circle connected to the straight by transition curves.

The purpose is to design a highway that is safe and comfortable for travel at uniform (design) speed throughout, by selecting suitable radius of curve, and length of transition curves and tangents. Minimum curve design is based on two factors:

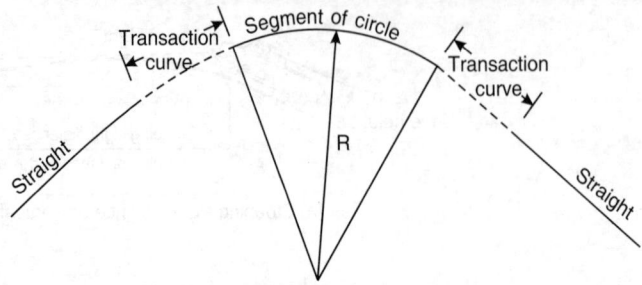

Fig. 6.16: Components of horizontal alignment

(a) ensuring that the side friction is not excessive for design speed and

(b) making sure that available sight distance is adequate for the design speed.

A circular curve can be described either by its radius or by its degree of curvature. India uses radius while in some countries like U.S.A. the degree of curve concept is used.

The degree of curve is defined as the central angle which a 100 m arc of curvature subtends. As a vehicle traverses a curved path, there acts an outward centrifugal force through the centre of the gravity of the vehicle, which tries to topple or turn down the vehicle. For a given radius and speed a resisting force is required to maintain the vehicle in its path. It is provided by the side friction developed between tyres and pavement, and by superelevation.

6.5.1 Superelevation

Super elevation is the ratio of the height of the outer edge of the road on curves with respect to the inner edge of the road, and the horizontal width of the road.

Figure 6.17 shows the forces acting on the vehicle on a super elevated section of a road.

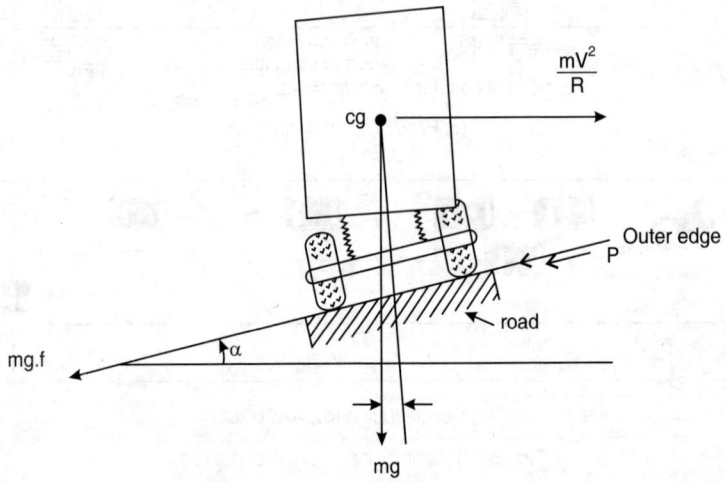

Fig. 6.17: Force acting on a vehicle on superelevated road

In the figure let:

$$m = \text{mass of the vehicle} \left(\frac{W}{g}\right), \text{where } W \text{ is weight of vehicle.}$$

$$e = \text{super elevation, (for small angle is } \tan \alpha), \text{ m/m}$$

$$v = \text{speed of the vehicle m/sec.}$$
$$f = \text{coefficient of lateral friction}$$
$$R = \text{radius of the curve m,}$$
$$V = \text{speed in km/hr.}$$
$$g = \text{acceleration due to gravity } (g = 9.81 \text{ m/sec}^2)$$
$$P = \text{side friction force resisting the centrifugal force } (Nf)$$
$$Cg = \text{Centre of gravity}$$
$$N = \text{Normal force}$$

The centrifugal force $= \dfrac{mV^2}{R}$

For equilibrium

$$\frac{mV^2}{R} \cos \alpha = mg \sin \alpha + P$$
$$= mg \sin \alpha + Nf$$

Resolving forces, N is given by $N = mg \cos \alpha + \dfrac{mV^2}{R} \sin \alpha$

Therefore $\dfrac{mV^2}{R} \cos \alpha = mg \sin \alpha + f \left(mg \cos \alpha \, \dfrac{mV^2}{R} \sin \alpha \right)$ dividing by $mg \cos \alpha$

$$\frac{V^2}{gR} = \tan \alpha + f + \frac{fV^2}{gR} \tan \alpha$$

the term $\dfrac{fV^2}{gR} \tan \alpha$ is very small and can be neglected and $\tan \alpha = e$ the super-elevation.

$$\frac{V^2}{gR} = e + f$$

If V is in km. p. hour, $g = 9.81$ and R is in metres.

$$\frac{V^2}{127\,R} = e + f, \quad \text{or} \quad e = \frac{V^2}{127\,R} \quad \text{when friction is neglected i.e. } f = 0.$$

This equation establishes the basic relationship between speed of the vehicle (V in km/hr), the radius of curve (R in m), Super-elevation (e) and the coefficient of lateral friction (f).

If coefficient of friction (f) is neglected or it equals to zero, the whole counter balancing force is provided by the superelevation. There is a limitation imposed by the steepness of e. In specifying a permissible maximum value of superelevation considerations must be given first to slow moving vehicles.

The value of the coefficient of the friction f for a treaded tyre on an icy pavements is about 0.1 and this sets an upper limit to superelevation for such condition.

Figures 6.18 and 6.19 show superelevation on curves without and with transition curve.

If limiting value of f is known, the minimum radius of curve can be calculated for any given speed.

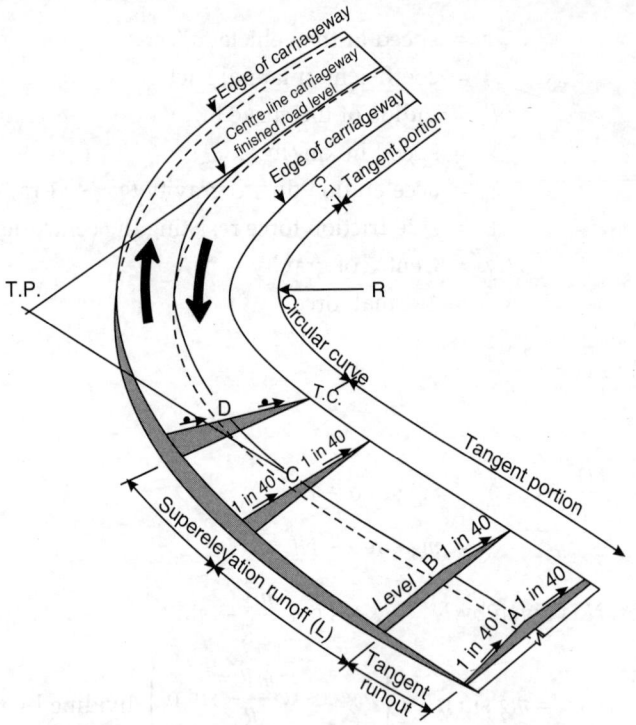

Fig. 6.18: Superelevation on curve without transition curve

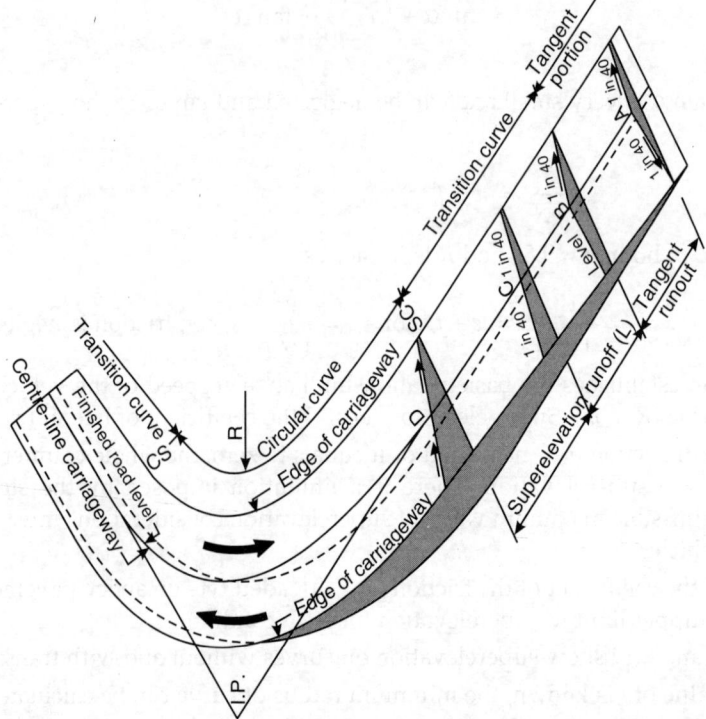

Fig. 6.19: Superelevation on curve with transition curve

Maximum superelevation. The superelevation is chosen primarily on the basis of safety, comfort and appearance. Following factors should be considered:

- The design speed of the curve is taken as the speed at which 85th percentile of drivers are expected to negotiate it.
- The stability of laden commercial vehicles.
- The difference between inner and outer formation levels.
- The length available to introduce the necessary superelevation.

Use of maximum superelevation is made when the radius of curves approaches a maximum value for a particular speed. This happens when there are constraints on increasing the radius of an individual curve.

The maximum superelevation ranges from 0.12 to 0.10 (12 to 10 percent) in mountainous terrain, to 0.06 to 0.07 (6 to 7 percent) in flat terrain.

Minimum superelevation. At low and intermediate ranges of design speeds, it will be desirable to superelevate all curves at least to a value equal to the normal cross slope on straights, unless the radius of individual curve is so large that it can be regarded as straight for the particular speed.

The value of side friction coefficient 'f' is more for low speed varying from 0.4 at 50 Kph to 0.11 at 130 Kph on paved roads.

6.5.2 Methods of Applying Superelevation

There are 3 methods of applying superelevation as shown in Fig. 6.20.

(a) Rotation about centre line: This gradually lowers inner edge raising the upper edge, keeping the same level of centre line.

(b) Rotation about inside edge: This raises the centre and outer edge.

(c) Rotation about outside edge: This depresses the centre and the inner edge.

Method (a) is commonly used as it leads to minimum displacement of the carriageway edge.

Method (b) is preferable where drainage problems may rise and

Method (c) leaves outer edge free.

First of all cambered, the outer half of the camber is gradually raised to make the road level and thereafter, one of the above method is used to provided needed super elevation.

As per IRC recommendations for mix traffic of slow and fast vehicles superelevation is calculated by assuming that centrifugal force corresponding to three-fourth the design speed is balanced by super-elevation and rest one fourth by side friction using the formula shown in Fig. 6.21.

$$e = \frac{V^2}{225\,R}$$

Where e = superelevation in metres/metre

       V = Speed in km/hr and

       R = Radius of curve in m

6.5.3 Minimum Radius of Horizontal Curve

The minimum radius of horizontal curve for a given design speed can be calculated by the formula:

$$R_{minimum} = \frac{(V_{design})^2}{127\,(e_{max} + f_{max})}$$

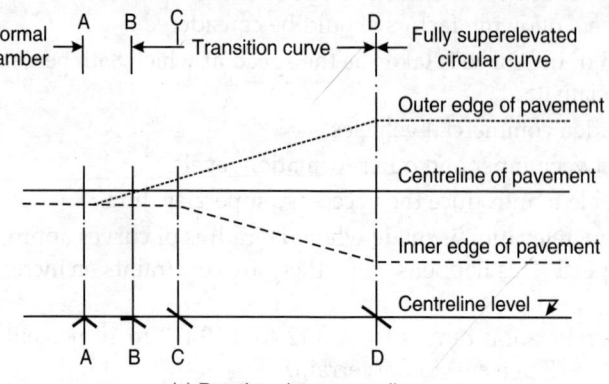

(a) Rotation about centre line

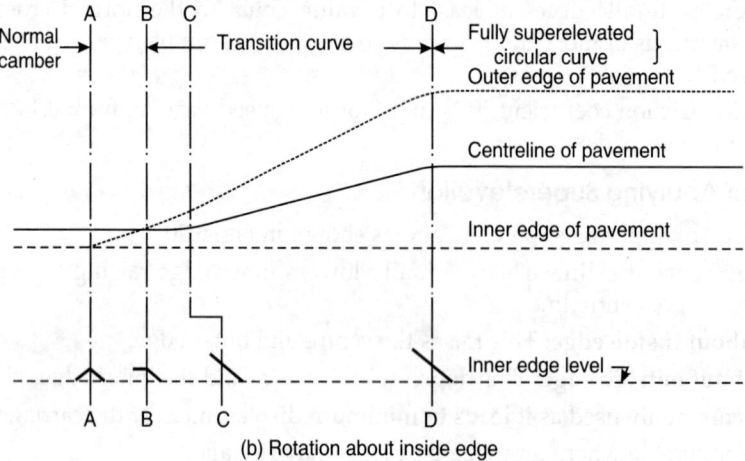

(b) Rotation about inside edge

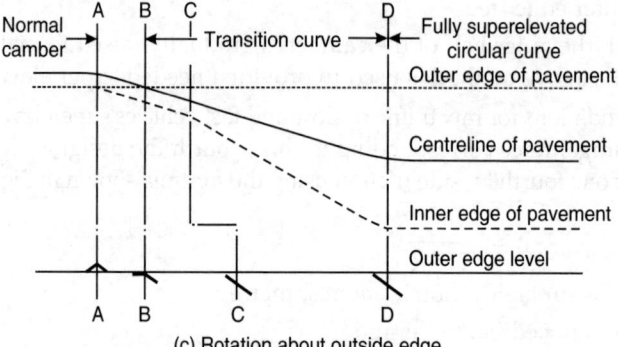

(c) Rotation about outside edge

LEDEND

Cross section at AA–normal camber
Cross section at BB–adverse camber removed
Cross section at CC–superelevation equal to camber
Cross section at DD–full superelevation achieved

NOTE:–

The rate of change of superelevation (longitudinal slope of edge compared to centreline) should be minimum 1 in 150 for roads in plain and rolling terrain and 1 in 60 in mountainous and steep terrain the actual rate used will determine the distances AB,BC and CD

Fig. 6.20: Different methods of attaining superelevation (IRC)

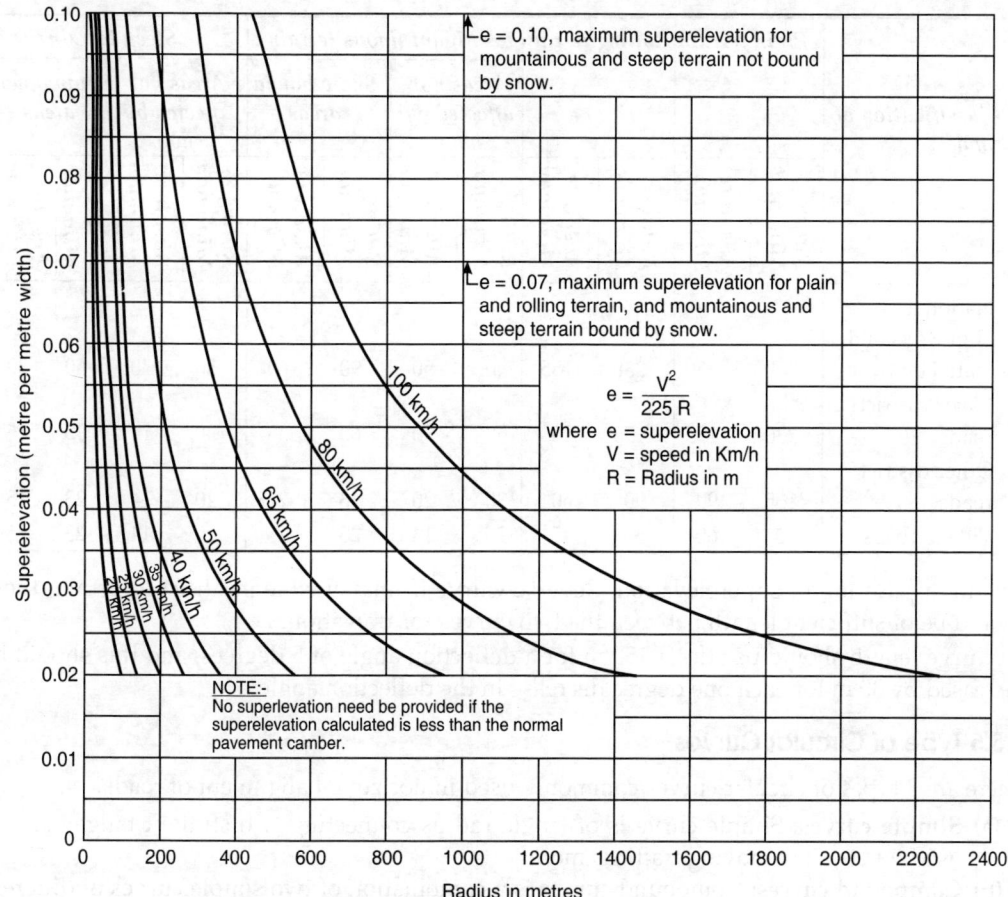

The chart contains the following labels:

e = 0.10, maximum superelevation for mountainous and steep terrain not bound by snow.

e = 0.07, maximum superelevation for plain and rolling terrain, and mountainous and steep terrain bound by snow.

$$e = \frac{V^2}{225\,R}$$

where e = superelevation
       V = speed in Km/h
       R = Radius in m

NOTE:-
No superelevation need be provided if the superelevation calculated is less than the normal pavement camber.

Superelevation (metre per metre width) — vertical axis

Radius in metres — horizontal axis

Curves labelled: 100 km/h, 80 km/h, 65 km/h, 50 km/h, 40 km/h, 35 km/h, 30 km/h, 25 km/h, 20 km/h

Fig. 6.21: Superelevation for different speeds and radii of curve

where V_d = design speed in km/hr.
             R_m = minimum radius of curve
             e_{max} = maximum value of superelevation
             f_{max} = maximum value of side friction factor.

If we assume value of $f = 0.15$

In plain and rolling area, and snow bound area ($e = 7\%$) and $R = 0.0357\,V^2$

In hilly area not bound by snow ($e = 10\%$), and $R = 0.0315\,V^2$

6.5.4 Ruling Minimum Radius of Curve

For ruling minimum radius of horizontal curve it is assumed that future speeds shall be higher than the present speeds. The design speed is therefore increased by 16 Km/h for use in the above formula for calculating the minimum radius of horizontal curve. Table 6.12 shows minimum and ruling radius of horizontal curves for different terrain and roads.

To break monotony and to avoid glare from headlights the use of long straight road is not recommended. Some artificial curvature may be introduced by gently deflecting the alignment to the left and right. Abrupt reversal of alignment is also undesirable. Curves should be of sufficient length to avoid the appearance of a kink as this not only makes driving easier but also

Table 6.12: Minimum radii of horizontal curves for different terrain conditions   (*metres*)

Classification of road	Plain terrain		Rolling terrain		Mountainous terrain				Steep terrain			
					Areas not affected by snow		Snow bound areas		Areas not affected by snow		Snow bound areas	
	Ruling Minimum	Absolute Minimum	Ruling Minimum	Absolute Minimum	Ruling Minimum	Absolute Minimum	Ruling Minimum	Absolute Minimum	Ruling Minimum	Absolute Minimum	Ruling Minimum	Absolute Minimum
1. National Highways and State Highways	360	230	230	155	80	50	90	60	50	30	60	33
2. Major district roads	230	155	155	90	50	30	60	33	30	14	33	15
3. Other district roads	155	90	90	60	30	20	33	23	20	14	23	15
4. Village roads	90	60	60	45	20	14	23	15	20	14	23	15

aids in application of superelevation. Reverse Curve if unavoidable in difficult terrain, there should be of sufficient length between the two curves for transition.

Curve length should be atleast 150 m for a deflection angle of 5 degrees, and this should be increased by 30 m for each one degree decrease in the deflection angle.

6.5.5 Type of Circular Curves

There are 3 types of circular curves commonly used in horizontal alignment of roads.

(a) **Simple curves:** Simple curve is of single radius connecting two straight tangents. It is used where slow moving traffic is more.

(b) **Compound curves:** Compound curves is a combination of two simple curves of different radii, in same direction joining together at a point. Where simple curve is not possible to use, due to some obstructions, compound curves are used. Compound curves are advantageous in placing the road to fit the ground. This is true in mountainous region where 2, 3, or 4 simple curves of different radii may be needed.

(c) **Reverse curves:** Reverse curves are compound curves having two different arcs of a circular curve either of same or different radii, but in different (opposite) directions, with a common tangent. These curves are dangerous to negotiate, uncomfortable due to sudden change of direction and difficult for provision of superelevation. Their use is limited to very difficult or unavoidable situations.

Figure 6.22 shows different type of circular curves.

Where the radii of curve is such that the superelevation required is less than the camber, the normal camber section of the road is continued on the curve. Table 6.13 shows radii beyond which super elevation is not required for different cambers and design speeds.

6.6 EXTRA WIDENING ON CURVES

On horizontal curves it has been found that drivers require wider pavements due to following two reasons:

(i) Driver have difficulty in steering the vehicle due to rigid wheel base of the vehicle. The rear wheels do not follow the path of the front wheels. The inner rear wheel becomes off the

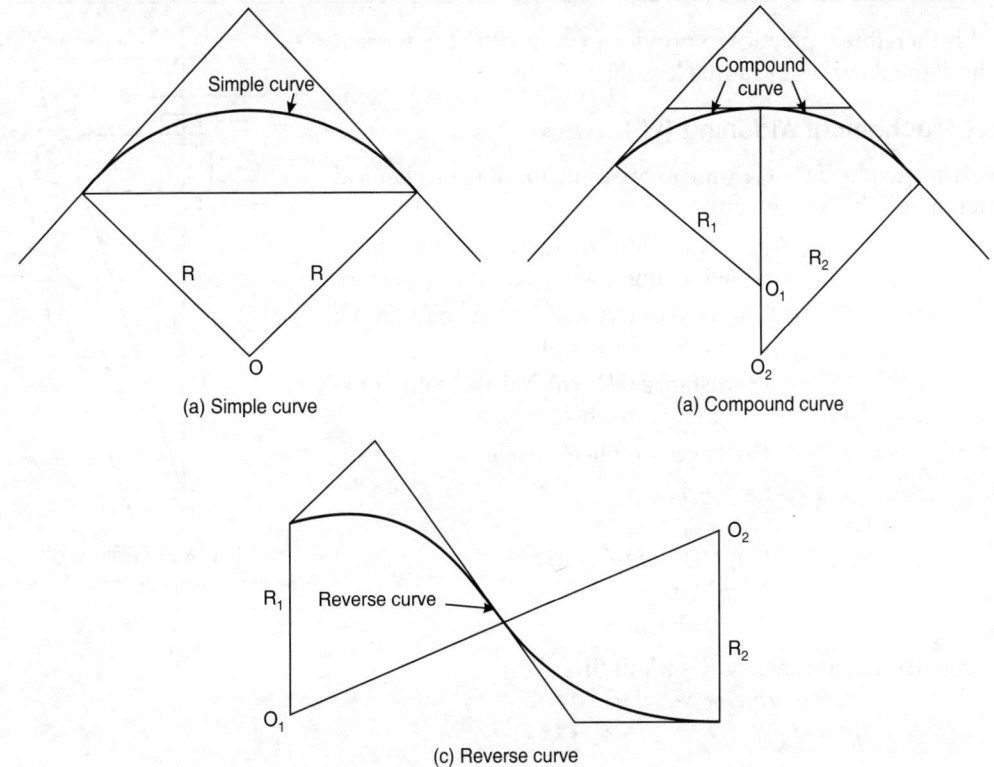

(a) Simple curve

(a) Compound curve

(c) Reverse curve

Fig. 6.22: Type of circular curves

Table 6.13: Radii beyond which superelevation is not essential for different cambers and speeds (IRC)

Design Speed (K.P.H.)	Radius in metres beyond which no super-elevation is needed for values of camber of				
	4%	3%	2.5%	2%	1.7%
20	50	60	70	90	100
25	70	90	110	140	150
30	100	130	160	200	240
35	140	18	220	270	320
40	180	240	280	350	420
50	280	370	450	550	650
65	470	620	750	950	1100
80	700	950	1100	1400	1700
100	1100	1500	1800	2200	2600

pavement due to this off-tracking on curves. Widening needed due to this mechanical reason is called "Mechanical Widening of Road on Curves".

(ii) Psychologically drivers need greater clearance on curves than on straight roads, as they shy away from the edges of the carriageway while traversing a curve. Widening due to Psychological reason is called 'Psychological Widening of Road on Curves'.

It is therefore a practice to provide extra width of pavement on horizontal curves of radius less than 300 m.

6.6.1 Mechanical Widening (W₁)

Referring to the Fig. 6.23 an expression for the mechanical widening can be worked out.

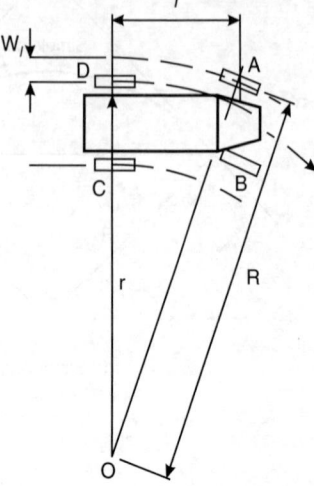

W_1 = extra width required for mechanical widening

R = distance OA which is the path of the outer front wheel in m.

r = distance OD which is the path of the outer rear wheel in m.

l = length of wheel base in m.

$r = R - W_1$

(i)

From the triangle ODA, $OD^2 = OA^2 - AD^2$

Fig. 6.23: Extra width requirements on circular curves

or

$r^2 = R^2 - L^2$

(ii)

From (i) substituting $r = R - W_1$ in (ii)

$$(R - W_1)^2 = R^2 - L^2$$

or

$$R^2 - 2RW_1 + W_1^2 = R^2 - L^2$$

$$W_1(2R - W) = L^2$$

∴

$$W_1 = \frac{L^2}{2R - W}$$

As an approximation the term $(R - W)$ can be replaced by mean radius of curve R_c.

$$= \frac{L^2}{2R_c} \text{ approx.}$$

(iii)

If the number of traffic lanes is n, the equation for mechanical widening becomes $\dfrac{nL^2}{2R_c}$.

6.6.2 Psychological Widening (W₂)

The empirical formula used for additional psychological widening is:

$$W_2 = \frac{V}{9.5\sqrt{R}}$$

where

W_2 = Psychological widening in m.

V = Design speed in km./hr.

R_c = Radius of curve m

Thus total widening required on a horizontal curve is given by

$$W = W_1 + W_2 = \frac{nL^2}{2R_c} + \frac{V}{9.5\sqrt{R_c}} \qquad \text{...(iv)}$$

where n = number of lanes.

On two lane roads and wider roads total widening is provided on horizontal curve, however on single lane roads, outer wheels can make use of the shoulder, only mechanical component of widening is provide.

Table 6.14 shows the extra width of pavement to be provided on horizontal curves.

Table 6.14: Extra width of pavement on horizontal curve (IRC)

S.No.	Radius of curve in m	Extra width in m	
		Two lane	Single lane
1.	upto 20	1.5	0.9
2.	21 to 40	1.5	0.6
3.	41 to 60	1.2	0.6
4.	61 to 100	0.9	Nil
5.	101 to 300	0.6	Nil
6.	Above 300	Nil	Nil

The widening is started from the beginning of the transition curve and is uniformly increased so that full widening is achieved till the end of the transition curve. Extra width should continue over the full length of the circular curve. Extra widening is provided on both side of the carriageway except on hill roads, having radius less than 50 m, it may be provided in full on inside of the curve.

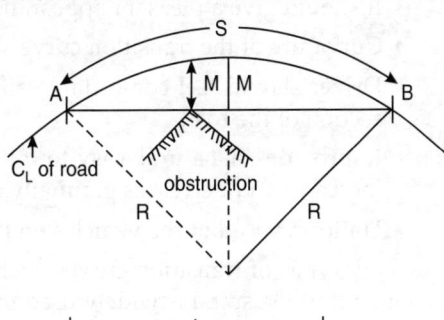

6.6.3 Sight Distance on Horizontal Curve

Another important element of horizontal alignment is sight distance across inside of the curve.

Where there are sight obstructions such as walls, cut slopes, buildings, etc. certain minimum off set or set back, from the obstruction, to the centre line of road must be available, to avoid vision obstruc-tion of driver, on curve.

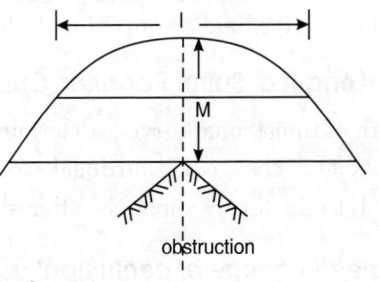

This set back depends upon:

(i) Sight distance desired

(ii) Length of curve.

Case I: When length of curve is more than S.D. $L > S$

$$S = \text{Sight distance}$$
$$M = \text{Set back}$$
$$R = \text{Radius of curve}$$

$$M = \frac{S^2}{8R}$$

Case II: When length of curve is less than S.D. $L < S$

$$M = \frac{L(2S - L)}{8R}$$

6.7 TRANSITION CURVES

A vehicle can not instantaneously change from a straight path to a circular one of a constant radius, and it is usual practice to employ a transition curve changing the radius from infinity at the start to that of the circular one at the end. Curves connecting straight lengths and circles are known as 'Simple Transition Curves' and those connecting two circular curves are known as 'Compound Transition Curves' Transition curves are also called "Easement Curves."

A transition length is also necessary to gradually increase the super elevation from zero at tangent point to full value at circular curve. It also presents a visually pleasing line without a disjoint at the tangent point. Gradual increase in extra widening of pavements from zero at tangent point to full value at circular curve is also provided through transition curve.

A transition curve is provided on either side of the circular curve.

Requirements of an ideal transition curve are:

- Transition curve should be tangential to straight section of the road. Its radius of curvature should decrease gradually.
- The length of transition curve should be enough to provide for required super-elevation gradually.
- It should give a pleasant appearance.
- Curvature of the transition curve should confirm with the curvature of circular curve.
- Driver should feel comfortable and safe while steering from normal position to curved section of the road.
- It provides a natural easy-to-follow path for the motorists so that centrifugal force increases and decreases gradually as vehicle enters and leaves the circular curve.
- Uniform speed of the vehicles on the curve are maintained.

Many types of transition curves such as the spiral, the lemniscate, and cubic parabola are common, but the spiral is widely used by highway engineers due to the ease with which it can be set out in the field compared with the other two curves.

6.7.1 Length of Spiral Transition Curve

Length of transitional curve is determined to satisfy the following two conditions:

(i) Rate of change of centrifugal acceleration should be safe and

(ii) Introduction of super-elevation should be at a reasonable rate.

1. Rate of change of centrifugal acceleration basis

Radial acceleration is given by $\dfrac{V^2}{R}$

where
V = speed in meter/sec

R = Radius of curve in meter

At the beginning of spiral transition curve centrifugal acceleration is 0 and radius is infinity and at the end of spiral transition curve centrifugal acceleration is maximum and radius is maximum curvature of R_m as shown in Fig. 6.24.

Let
L_S = length of spiral transition curve in meters.

V = Design speed in meter/second

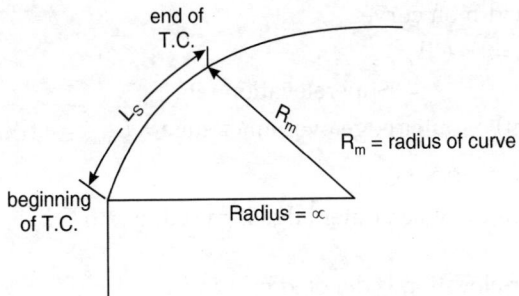

Fig. 6.24: Spiral transition and circular curve

If t is time in seconds, that a vehicle moving with a design speed takes, over the length L_S of transition curve

Then
$$t = \frac{L_S}{V}$$

The maximum centrifugal acceleration, to be introduced in time t is $= \dfrac{V^2}{R}$

∴ Rate of change of centrifugal acceleration $C = \dfrac{\text{Maximum Centrifugal acceleration}}{\text{time}} = \dfrac{\frac{V^2}{R}}{\frac{L_S}{V}}$

$$= \frac{V^2}{R} \times \frac{V}{L_S} = \frac{V^3}{L_S R}$$

Value of C depends upon the speed and varies inversely with radius.
IRC has recommended the value of C for design speed V as

$$C = \frac{80}{75 + V} \text{ m/sec}^3$$

Maximum and minimum value of C are 0.8 to 0.5
Knowing value of C

$$L_S = \frac{V^3}{CR} \text{ when } V \text{ is m/sec.}$$

$$\boxed{= \frac{V^3}{46.5\, CR} \text{ when } V \text{ is km/hour.}}$$

where
R = Radius of circular curve in meters
C = Allowable rate of change of centrifugal acceleration in m/sec^3.

2. Rate of Introduction of superelevation basis

The outer edge of the highway should be raised, at the following desirable rates.
   (i) In open area = 1 in 150
   (ii) In build up (urban area) = 1 in 100
   (iii) In hilly area = 1 in 60.

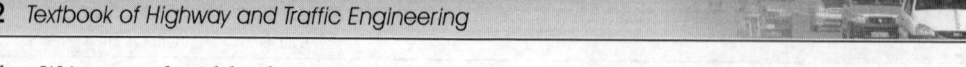

If W is normal width of pavement

W_e is the extra width on curve

Total width on curve $= W + W_e$

$$e = \text{Superelevation rate}$$

Maximum rise (height) of outer edge over inner edge $= E_{max} = e\,(W + W_e)$

(a) If pavement is rotated about C.L.

The maximum amount by which outer edge is raised, w.r.to C.L. is $E_{max}/2$

if rate of change of super-elevation is denoted by 1 in N.

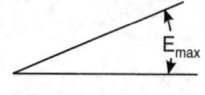

(where N = 150, 100, 60, for plain, built up, hilly area)

Length of spiral transition curve $L_S = \dfrac{E_{max}}{2} \times N$

$$= \dfrac{e\,(W + W_e)}{2} \times N$$

S.E. Rotation about C.L.	$L_S = \dfrac{eN}{2}\,(W + W_e)$

(b) If pavement is rotated about inner edge

The maximum amount by which outer edge is raised with respect to inner edge is E_{max}.

Length of spiral transistion curve $L_S = E_{max} \times N$

$$= e\,(W + W_e) \times N$$

S.E. Rotation about inner edge.	$L_S = eN\,(W + W_e)$

Higher of the value of L_S, obtained by above two methods is used in design. Values of L_S should be checked by minimum length required by IRC and highest of the three should be used.

Minimum length of transition curve for different speeds and curve radii as recommended by IRC are given in Table 6.15, and are based on the empirical formulae:

(i) For plain and rolling terrain: $L_S = \dfrac{2.7\,V^2}{R}$

(ii) For mountainous steep terrain: $L_S = \dfrac{V^2}{R}$

where V = speed in km. p. hr.

R = radius of curve in m.

6.7.2 Providing Spiral Transistion Curve

Spiral transition curve BJC and DGE, as shown in Fig. 6.25 are provided on both ends of circular curve CD, of radius R. This curve gets shifted towards inside and this shift S is given by formula

$$S = \dfrac{L_S^{\,2}}{24R} \qquad \text{length } BJ \text{ and } JC \text{ are approx. equal to } \dfrac{L_S}{2}.$$

Where L_S = length of spiral transition curve.

R = Radius of the circular curve.

Table 6.15: Minimum transistion lengths for different speeds and curve radii (IRC)

Plain and rolling terrain

Curve radius R (metres)	Design speed (KPH)					
	100	80	65	50	40	35
	Transistion length-metres					
45					NA	70
60				NA	75	55
90				75	50	40
100			NA	70	45	35
150			80	45	30	25
170			70	40	25	20
200		NA	60	35	25	20
240		90	50	30	20	NR
300	NA	75	40	25	NR	
360	130	60	35	20		
400	115	55	30	20		
500	95	45	25	NR		
600	80	35	20			
700	70	35	20			
800	60	30	NR			
900	55	30				
1000	50	30				
1200	40	NR				
1500	35					
1800	30					
2000	NR					

Mountainous and steep terrain

Curve radius R (metres)	Design speed (KPH)				
	50	40	30	25	20
	Transistion length-metres				
14				NA	30
20				35	20
25			NA	25	20
30			30	25	15
40		NA	25	20	15
50		40	20	15	15
55		40	20	15	15
70	NR	30	15	15	15
80	55	25	15	15	NR
90	45	25	15	15	
100	45	20	15	15	
125	35	15	15	NR	
150	30	15	15		
170	25	15	NR		
200	20	15			
250	15	15			
300	15	NR			
400	15				
500	NR				

NA-Not applicable.

NR-Transistion not required.

6.7.3 Double Spiral Curve (Fig. 6.26)

If there is no intermediate circular arc, the two spirals make a 'Double Spiral Curve' which is transitional throughout. It theoretically will require changing super-elevation throughout, which is objectionable. However superelevation could be made constant, around the middle portion of the curve, since the actual deviation of the spiral from a circle is very small here.

The middle fifth of the curve (if total length of curve is 500 m, the 100 m length of centre) is very near an arc of circle, in which the contact super elevation can be provided.

6.7.4 Unequal Spirals (Fig. 6.27)

Sometimes the spirals used with the existing circular curve, at the two ends of the curve may be of unequal length, without introducing complications.

The intermediate circular curve, remains fixed in position and spirals are fixed on either end of the curve.

6.7.5 Spiraled Reverse Curve (Fig. 6.28)

Sometimes when it is required to improve the alignment in every possible way, with out changing radically the existing layout, an spiral, may be introduced, between the two reverse curves. Such a curve is called 'Spiraled Reverse Curve'.

The intervening spiral (increase in length of intervening tangent), improves the new alignment and makes it short.

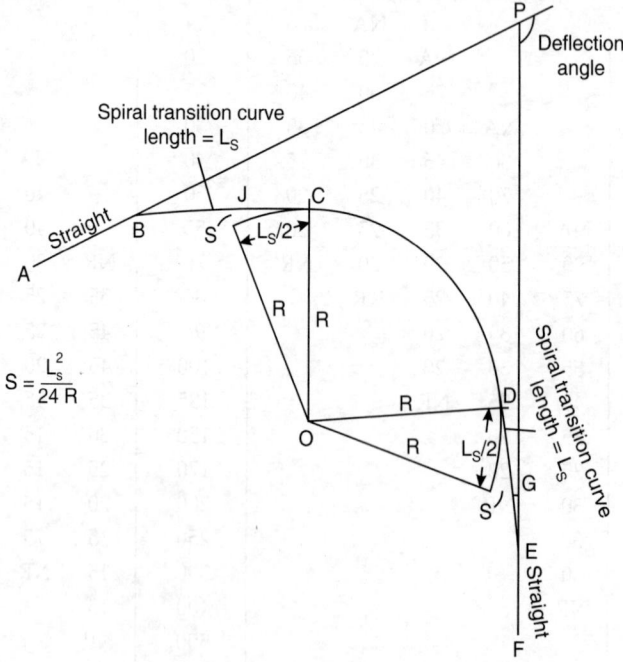

Fig. 6.25: Spiral transistion curve

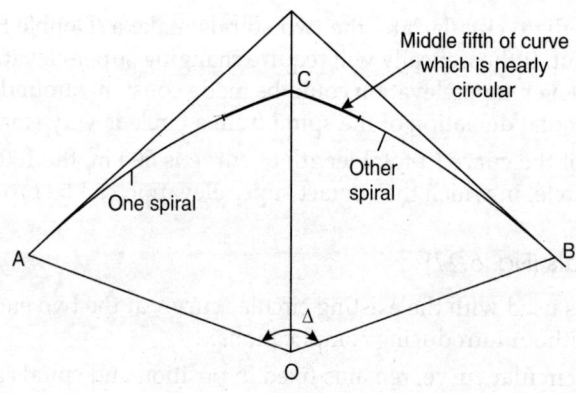

Fig. 6.26: Double spiral curve

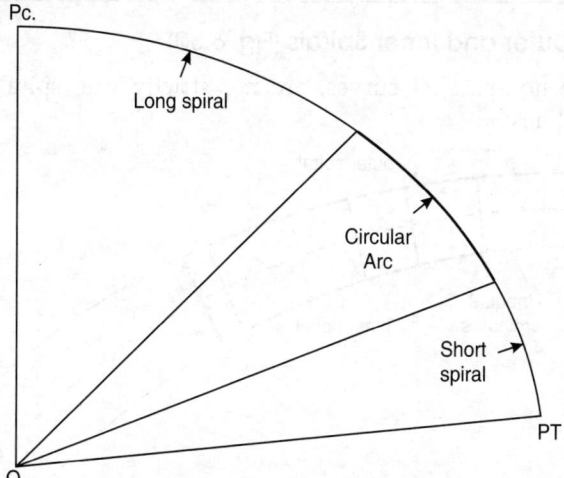

Fig. 6.27: Unequal spiral curve

6.7.6 Parallel Spirals (Fig. 6.29)

The two edges (outer and inner) are some-times provided (designed) as two separate spirals. The outer spiral has different variable radius and is parallel to inner spiral.

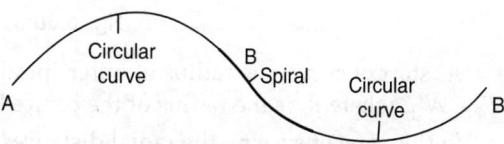

Fig. 6.28: Spiral reverse curve

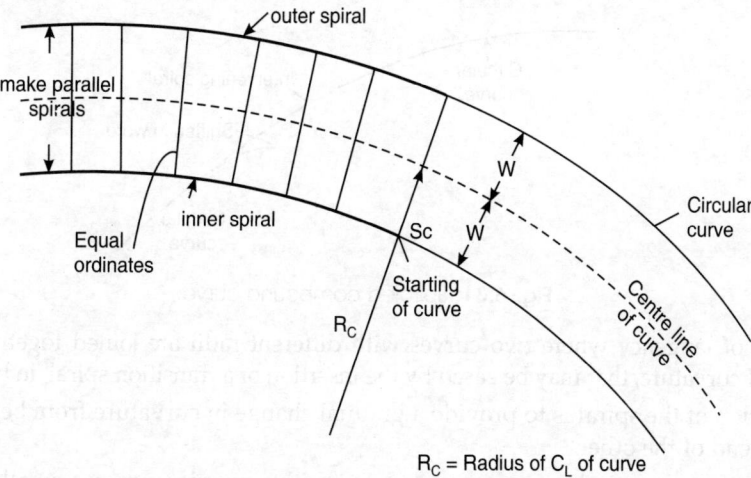

R_C = Radius of C_L of curve

Fig. 6.29: Parallel spirals

At starting curve, of radius R_c, the radius of outer spiral R_0 will be $R_c + W$ and of inner spiral will be $R_c - W$. Where

$\qquad R_c$ = Radius of the centre line of the circular curve

$\qquad W$ = radial distance from centre line of spiral to outer (or inner) parallel spiral.

In this design, the radial distances remain constant to make the two spirals parallel.

6.7.7 Non-Parallel Outer and Inner Spirals (Fig. 6.30)

The outer and inner non-parallel curves, are not strictly true spirals, but they represent satisfactory reference lines.

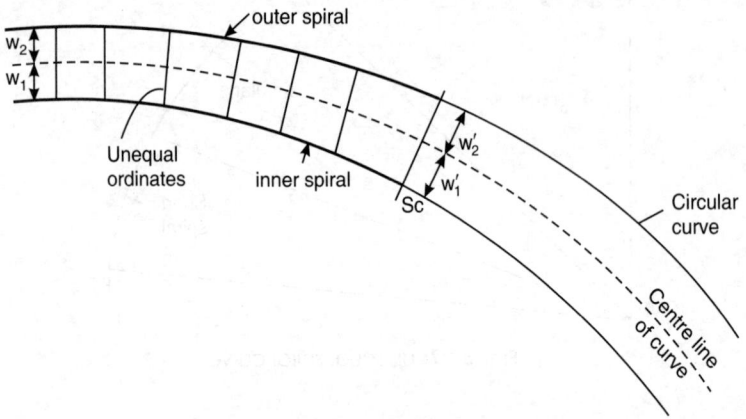

Fig. 6.30: Non-parallel spirals

At start of curve, the radius of outer spiral will be $R_c + W'_2$ and radius of inner spiral will be $R_c - W'_1$, where R_c is the radius of the centre line of the circular curve.

In this type of system, the radial distances are variable, from point to point of two spirals.

6.7.8 Spiraled Compound Curves (Fig. 6.31)

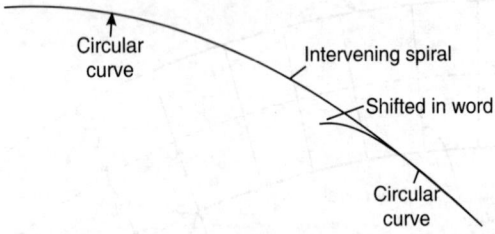

Fig. 6.31: Spiraled compound curve

At the point of tangency where two curves with different radii are joined together, there is an abruptness of curvature, that may be eased by the insertion of a transition spiral, in between.

The function of the spiral is to provide a gradual change in curvature from beginning of one curve to the end of the other.

As in the case of a spiral connecting a tangent with a circular curve, a small radial shift, is necessary in order to allow proper clearance for the spiral.

This shift or separation of two branches of the compound curve, is effected by moving either the sharper curve radially inward or the flatter curve radially outward.

6.7.9 Two Centred Compound Curve with 3 Spirals (Fig. 6.32)

TS = Point of change from tangent to spiral.

SC = Point of change from spiral to circle.

CS = Point of change from circle to spiral.

PCC = Point of compound curvature.

ST = Point of change from spiral to tangent.

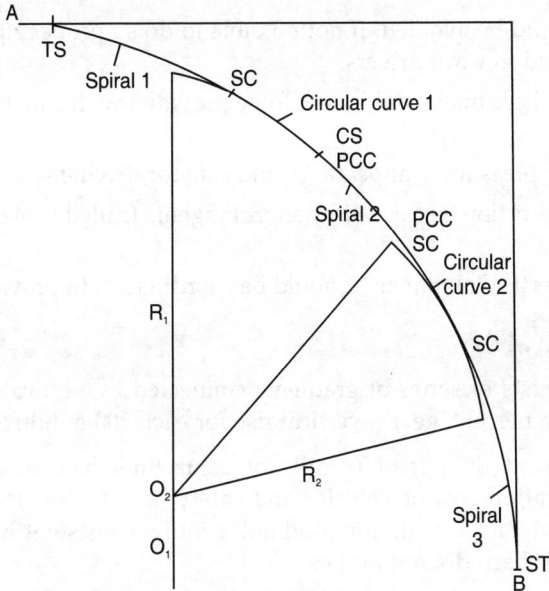

Fig. 6.32: Two central compound curve with 3 spirals

6.7.10 Grade Compensation on Curves

Due to both gradient and curve at same location, there will be increased resistance and it is necessary to ease the gradient at horizontal curves. The gradient is decreased to compensate for the extra tractive effort involved at such sharp curves, specially on hill roads.

This reduction in gradient at horizontal curve is called "Grade Compensation". It is calculated from the formula:

$$\text{Grade compensation in percent} = \frac{30 + R}{R}$$

subject to a maximum value of $75/R$.

where R = Radius of the curve in m.

IRC suggests that grade compensation is not required for gradients flatter than 4 percent, as such gradients need not be compensated (eased) beyond 4 percent.

6.7.11 Points to be Kept in View for Horizontal Alignment

For a good horizontal alignment following points should be kept in view:

- As large as possible radius of curve should be used, for a given design speed.
- Alignment should fit the natural topography and contours
- As far as possible number of curves should be the minimum required
- Abrupt turns are unsafe and should be avoided
- Curves should be atleast 150 m long for a deflection angle of 5° and for each degree decrease of the angle, minimum length should be increased by 30 m.

- Formation of kinks should be avoided.
- Alignment should be as direct as possible.
- Sharp curves should be avoided. If not possible to do so proper signs ahead of the curve should be provided to warn drivers.
- Reverse curves, where unavoidable should be provided with suitable and long transistion curves.
- Curves should be pleasing in appearance and safe for driving.
- Curves in same direction, separated by short tangents (called broken-back curves) should be avoided.
- Horizontal and vertical alignments should be coordinated, to provide harmony.

6.8 VERTICAL ALIGNMENT

Vertical alignment consists of series of gradients connected by vertical curves. Gradients are normally expressed as a percentage, i.e. vertical rise for each 100 m horizontal distance.

The vertical alignment of a road (profile of centreline) has great influence upon the construction cost, operation cost of vehicles and safety of vehicles. The objective of vertical alignment is to provide a smooth longitudinal profile consistent with the surrounding countryside and has no sharp discontinuities.

Grades should not change very frequently because of their detrimental effect upon speed, capacity and safety.

Vertical profile should be well coordinated with the horizontal alignment. When grades change (intersect) a vertical curve is provided to avoid abrupt change in the profile, causing sudden impacts, and unsafe conditions. A parabolic curve is used in vertical alignment. These curves are convex when two grades meet at "Summit" and concave when they met at a 'Sag' . Accordingly they are known as Summit (crest) or Sag (valley) curves.

Figure 6.33 shows summit and valley curves.

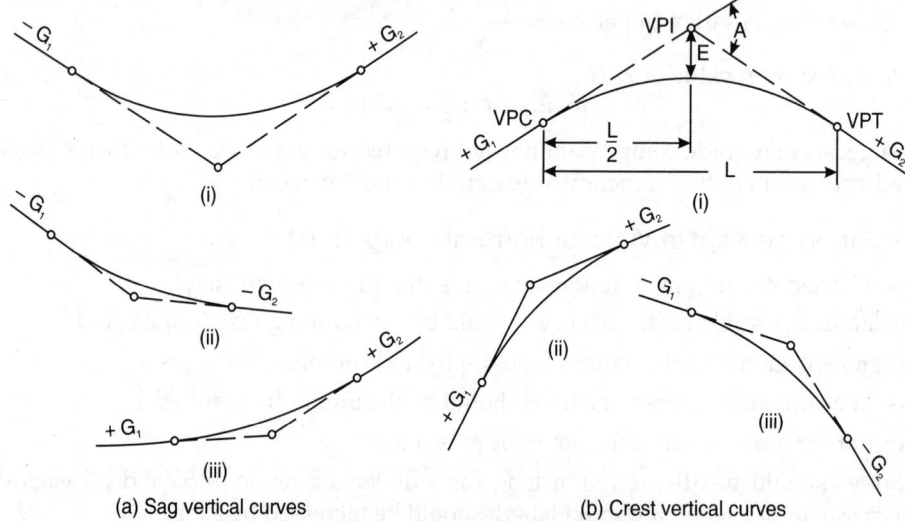

(a) Sag vertical curves   (b) Crest vertical curves

Fig. 6.33: Types of vertical curves

The vertical alignment must provide adequate sight distance over all crests and must not present any sudden hidden changes in alignment to the driver. Short sag curves and short gradients between two crests or sags should be avoided. A smooth profile which changes gradually is preferred to a line with numerous breaks and short segments. At crest curves, the minimum length may be fixed by sight distance. At sag curves, the length may be fixed approximately by comfort related to vertical acceleration, appearance, or drainage and head lights. The effect of changes in horizontal alignment on the smoothness of vertical profile should be carefully considered.

6.8.1 Gradient (Grad line)

Grad line (gradient) is the rate of rise and fall along the length of the road. It is not practical to construct roads with flat gradients all the time to permit all vehicles to operate at high speeds. It is therefore necessary to adopt modest design standards. The effect of gradient is more pronounced on trucks than on passenger cars, and on animal drawn vehicles.

Gradients need to be considered from the standpoint of both length and steepness, and the speed at which heavy vehicles enter the gradient. They are chosen to balance the construction cost against the operating cost of heavy vehicles ascending them.

Design of grad line should consider the following requirements:

- The heavy vehicles of lower power range must be able to climb the grade and also slow vehicles.
- Quality of service considerations as grades cause the need for speed variations, gear changes and braking for vehicles.
- Grades cause speed disparities between vehicle types, leading to increased queueing and overtaking requirements.
- Grades should be smooth and economical (the cut and fill may be balanced).
- Long lengths of grades are desirable for easiness of driving and safety.
- Adequate sight distances should be available both at summit and valley curves.
- Special considerations should be given to drainage problem on valley curves.

6.8.2 Maximum Gradient

The maximum gradient (ruling, limiting and exceptional) depends upon type of topography, terrain and the design speed. Table 6.16 shows IRC recommended gradients for different terrains.

Table 6.16: Gradient for roads in different terrain (IRC)

S.No.	Terrain	Ruling gradient	Limiting gradients	Exceptional gradient
1.	Plain and Rolling	3.3% (1 in 30)	5% (1 in 20)	6.7% (1 in 15)
2.	Mountainous terrain and steep terrain having 3000 m above M.S.L.	5% (1 in 20)	6% (1 in 16.7)	7% (1 in 14.3)
3.	Steep terrain upto 3000 m height height above above M.S.L.	6% (1 in 16.7)	7% (1 in 14.3)	8% (1 in 12.5)

Gradients upto ruling gradients are used as a matter of course in design. However in special situations such as roads carrying high volume of slow moving vehicles it may be desirable to adopt a flatter gradient of 2%.

'Limiting Gradient' is used where the topography compels to do so to save cost. In such cases the length of continuous grade steeper than the ruling gradient should be as short as possible.

'Exceptional Gradients', are meant to be used only in very difficult situations and for short lengths not exceeding 100 m, usually in mountainous and steep terrain.

6.8.3. Minimum Gradient

On curved pavements, longitudinal gradient is provided to facilitate surface drainage. A value of 0.5% is used for this purpose, however on unlined drainage ditches a gradient of 1 percent may be used. Normally the camber of the road is sufficient to take care of surface drainage, however sometimes specially in cut sections it may be necessary to introduce a slight longitudinal gradient for drainage in side ditches.

6.8.4. Truck Climbing Lane/Critical Length of Grade

Critical length of a grade is the maximum length of an upgrade upon which a loaded truck can operate without an unreasonable reduction in speed. Extra truck climbing lanes are added for trucks on upgrade if their percentage is high.

A climbing lane should be provided where:

- Critical length of the grade is exceeded to drop the speed of a truck to about 40 kph.
- Percentage of trucks in the traffic stream is high.
- Traffic volume is high in relation to capacity.
- Traffic congestion is caused by a long or steep upgrade.
- Sight distance restrictions may call for an auxiliary lane or lanes to allow overtaking manoeuvres, even on moderate grades.
- The cost of the climbing lane should be justified by the savings in time and fuel to traffic.

6.9 VERTICAL CURVES

A parabolic curve is usually used to connect gradients in the profile alignment. Convex curves meet at 'Summit' and concave curves meet at a 'Sag' or "Valley". Both are designed as parabolic curves. Parabolic curves can be laid out easily and they provide comfortable transition from one grade to another. Vertical curves are usually not necessary when the total grade change from one tangent to the other does not exceed 0.5 percent (1 in 200).

6.9.1 Summit Curves

Length of summit curves

The length of summit curves is governed by sight distance requirements, using following formula:

(a) Situation when sight distance is less than length of curve: ($S < L$)

$$L = \frac{AS^2}{\sqrt{\left(2h_1 + 2\sqrt{2h_2}\right)^2}} \qquad \text{(i)}$$

(b) Situation when sight distance is more than length of curve: ($S > L$)

$$L = 2S - \frac{2\left(\sqrt{h_1} + \sqrt{h_2}\right)^2}{A} \qquad \text{(ii)}$$

Where:
S = Sight distance in m
L = length of vertical curve in m
h_1 = Height of eye above carriageway
h_2 = height of object above carriageway
A = algebraic difference in slopes

As $h_1 = 1.2\ m$ and $h_2 = 0.15\ m$, the above equations for safe stopping sight distance become:

when $S < L$ $\qquad\qquad L = \dfrac{AS^2}{4.4}$ $\qquad\qquad$ (iii)

when $S > L$ $\qquad\qquad L = 2S - \dfrac{4.4}{A}$ $\qquad\qquad$ (iv)

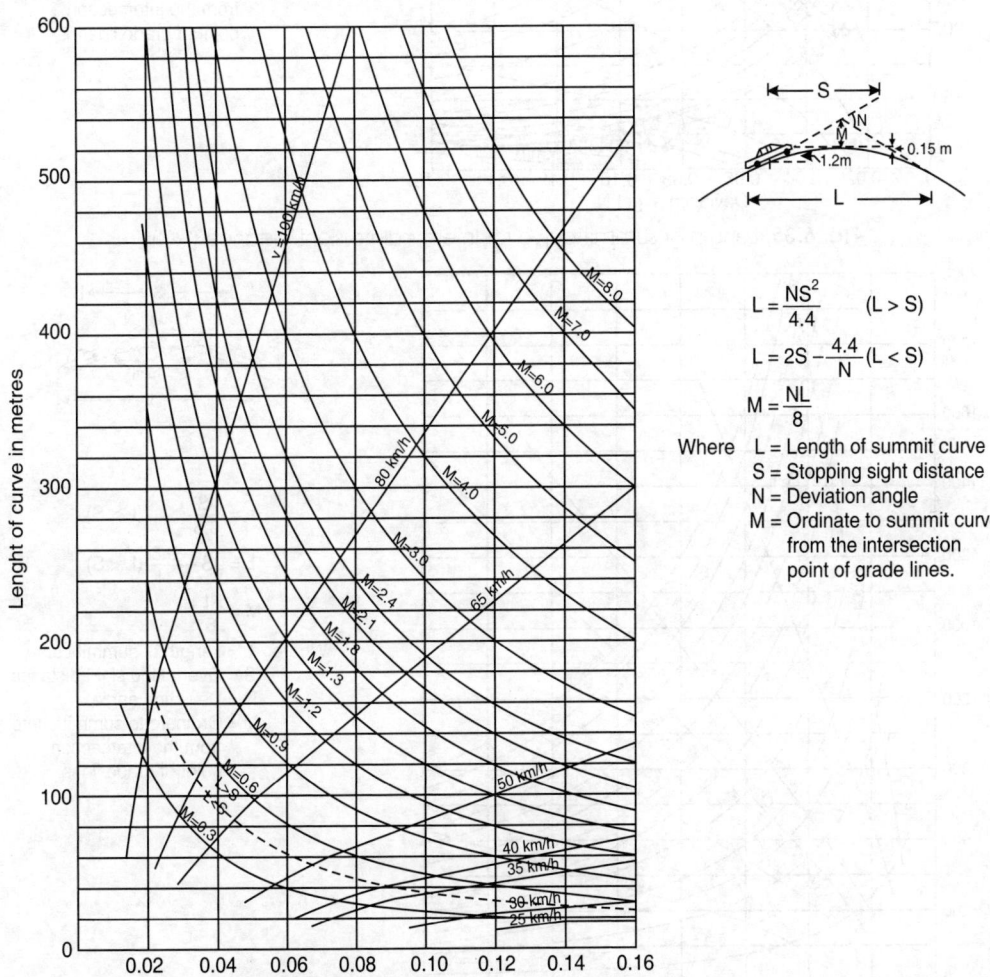

$$L = \frac{NS^2}{4.4} \quad (L > S)$$

$$L = 2S - \frac{4.4}{N}(L < S)$$

$$M = \frac{NL}{8}$$

Where L = Length of summit curve
S = Stopping sight distance
N = Deviation angle
M = Ordinate to summit curve
from the intersection
point of grade lines.

Fig. 6.34: Length of summit curve for stopping sight distance (IRC)

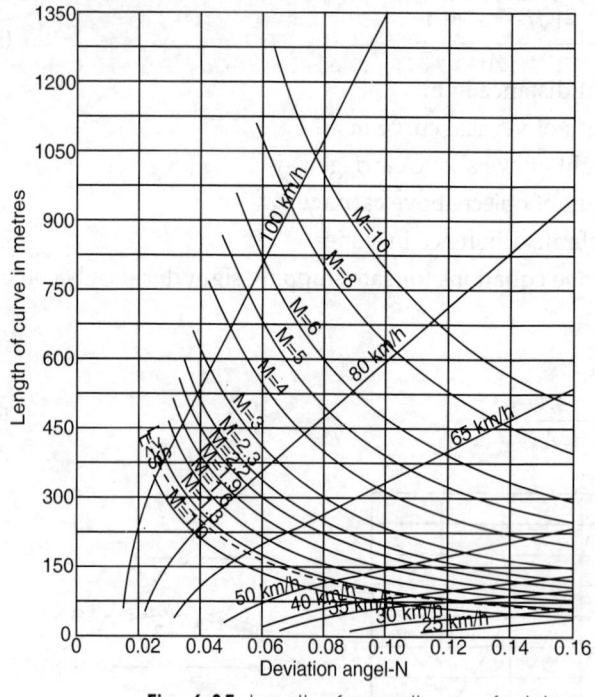

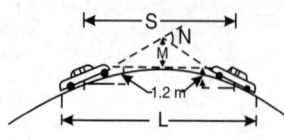

$$L = \frac{NS^2}{9.6} \quad (L > S)$$

$$L = 2S - \frac{9.6}{N} \quad (L < S)$$

$$M = \frac{NL}{8}$$

Where   L = Length of summit curve
S = Intermediate sight distance
N = Deviation angle
M = Ordinate to summit curve
from the intersection
point of grade lines.

Fig. 6.35: Length of summit curve for intermediate sight distance (IRC)

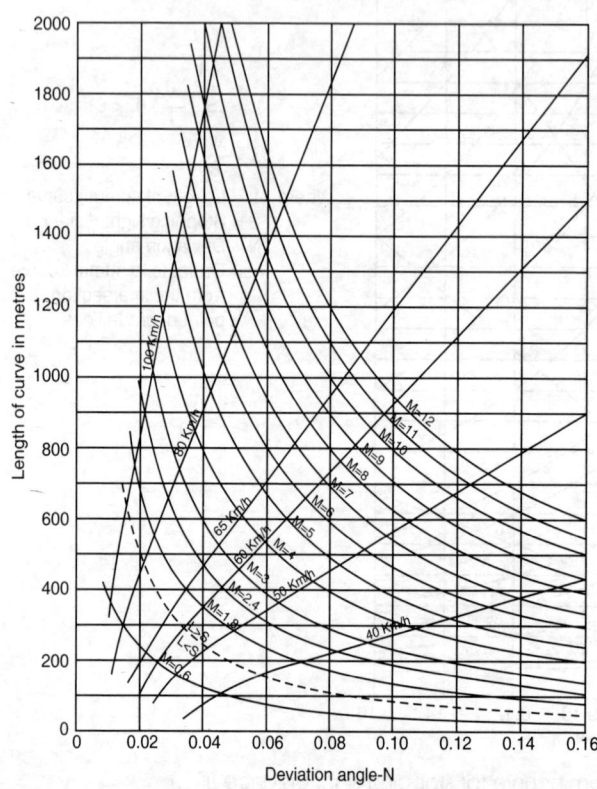

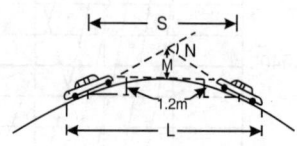

$$L = \frac{NS^2}{9.6} \quad (L > S)$$

$$L = 2S - \frac{9.6}{N} \quad (L < S)$$

$$M = \frac{NL}{8}$$

Where   L = Length of summit curve
S = Overtaking sight distance
N = Deviation angle
M = Ordinate to summit curve
from the intersection
point of grade lines.

Fig. 6.36: Length of summit curve for overtaking sight distance (IRC)

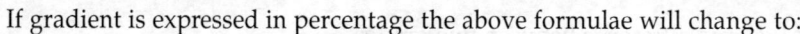

If gradient is expressed in percentage the above formulae will change to:

$$\text{when} \quad S < L, \qquad L = \frac{AS^2}{440} \qquad\qquad\qquad\qquad (v)$$

$$S > L, \qquad L = 2S - \frac{440}{A} \qquad\qquad\qquad\qquad (vi)$$

For intermediate or overtaking sight distance

$$h_1 = h_2 = 1.2\ m$$

The above equations will change to:

$$S < L \qquad L = \frac{AS^2}{9.6} \qquad\qquad\qquad\qquad (vii)$$

$$S > L \qquad L = 2S - \frac{9.6}{A} \qquad\qquad\qquad\qquad (viii)$$

Length of summit curve for stopping sight distance, intermediate sight distance and overtaking sight distance can be read from Figs. 6.34 to 6.36. Minimum length of vertical curve are given in Table 6.17.

Table 6.17: Minimum length of vertical curves (IRC)

Design Speed km/hr	Maximum grade change (percent), not requiring a vertical curve	Minimum length of vertical curve (m)
35	1.5	15
40	1.2	20
50	1.0	30
65	0.8	40
80	0.6	50
100	0.5	60

6.9.2 Valley Curves

Length of valley curves

The length of valley curves is governed by two criteria. Headlight beam distance should be equal to stopping sight distance, and the allowable rate of change of centrifugal acceleration of .6 m/sec^2 should be used for rider's comfort.

1. Stopping sight distance criteria:

(a) When length of curve is more than required sight distance: $(L > S)$

$$L = \frac{AS^2}{1.5 + 0.035\ S}$$

(b) When length of curve is less than the required sight distance: $(L < S)$

$$L = 2S - \frac{1.5 + 0.035\ S}{N}$$

Where:

A = Algebraic difference between two grades

L = Length of vertical curve m

S = Stopping sight distance m

Length of valley curve for various grade difference can be read from Fig. 6.37.

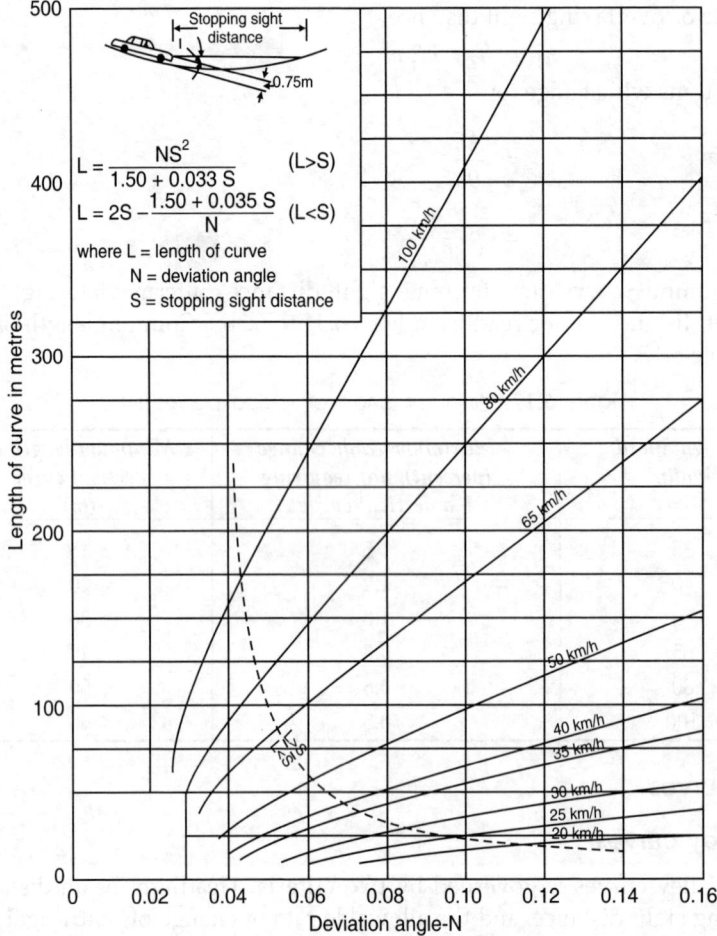

Fig. 6.37: Length of valley curve (IRC)

2. Allowable centrifugal acceleration criteria (Rider's comfort):

The radial acceleration is given by:

$$\alpha = \frac{V^2}{R} \qquad (i)$$

where

α = is radial acceleration m/sec^2

V = speed in m/sec

R = radius of curve in m

if V is in Km/h $\qquad \left(\text{As } V_{kmph} = \dfrac{V}{3.6} \ m/sec \right)$

$$\alpha = \frac{V^2}{13 R} \tag{ii}$$

Radial acceleration changes from zero in the beginning to α in length of $\dfrac{L}{2}$ (where L = length of sag curve).

The rate of change of radial acceleration $C = \dfrac{\dfrac{V^2}{13 R}}{\dfrac{L \times 3.6}{2 V}}$

$$= \frac{V^3}{23.4 \ LR} \ m/sec^3 \tag{iii}$$

Recommended value of C is 0.6 m/sec³, for rider's comfort.

$$0.6 = \frac{V^3}{23.4 \ LR} \tag{iv}$$

$$L = \frac{V^3}{14 R} \tag{v}$$

or $\qquad\qquad LR = \dfrac{V^3}{14} \tag{vi}$

For a transistion valley curve of cubic parabola, it is a property that:

$$L = (2 \ RLA)^{\frac{1}{2}} \tag{vii}$$

where $\qquad A$ = algebraic difference in grades expressed in decimal.

Putting value of LR from (vi) in (vii)

$$L = \left(2 \cdot \frac{V^3}{14} \cdot A \right)^{\frac{1}{2}}$$

$$= \frac{V^{3/2} \cdot A^{1/2}}{2.65} \tag{viii}$$

Usually first criteria of head light distance is higher and governs the design of valley (sag) curves.

6.9.3 General Points to be Kept in View for Vertical Alignment and its Combination with Horizontal Alignment

- Intersections on grades should be avoided. If unavoidable lattest gradients should be used.
- Grade line should not change abruptly, and should be in conformity with the terrain.
- Driving on grade line used should be safe, smooth and pleasant.

- Enough sight distances should be available. The hidden type of profile is dangerous.
- Broken-Back curves should be avoided. Broken back is when two vertical curves in the same direction are not provided with enough tangent length.
- In place of long grades short intervals of flatter grades should be used.
- Grad-line should be aesthetically pleasant. A balance between curvature and grades should be established.
- Combination of vertical and horizontal alignment should be done properly to ensure safety, comfort and uniformity.
- Sharp horizontal curves when combined with a summit curve, need special attention. Such curves should be avoided at or near the valley point of sag curve.
- Adequate sight distance should be available as far as possible. Proper signs should be erected at critical situations.
- Guard rails/railings should be used, at dangerous sections of vertical/horizontal curves and their combinations.

6.10 SAFETY AND GEOMETRIC DESIGN

Safety factors, must be taken into considerations during initial design and construction. High Design standards, generally result in low accidents.

Based on researches, following points need considerations:

- An error in understanding or judgment or a wrong action of driver, may result in an accident. Highways should be designed, so that a driver needs to make only one decision at a time. Standardisation plays an important role, so he knows what to expect on a certain highways.
- Accidents increase as the traffic volume increases. The lane width therefore directly affects the accident rate.
- Type of shoulders, its width, lateral clearances to roadside, must be considered jointly with pavement widths, in elevation of highway safety.
- Control of access, is one of the most important factor which reduces accidents to 40 to 50% or more depending upon type of road and area (urban or rural).
- Speed is also a contributing factor to accident. Generally at low speeds, accidens are less and non severe type. High speeds, are certainly responsible for more severe and fatal accidents. The safest speed for any highway depends upon traffic volumes, weather conditions, roadside development, cross-traffic and other factors.
- A divided highway with wide medians and wide shoulders, results in less number of accidents.
- There is considerable variation in accident rate with sharpness of curves and frequency of curves and grades.
- The accident rate on 3 lane highways is one third higher than accidents on 2 lane or 4 lane, highways for the same volume conditions.
- Roadside objects, such as culvert head walls, bridge abutments, utility poles, signs, trees, etc. must be controlled to some degree in design, construction and maintenance of highway to ensure safety.
- Uniform types and high quality of signs, markings, and traffic control devices, must be used for safety.

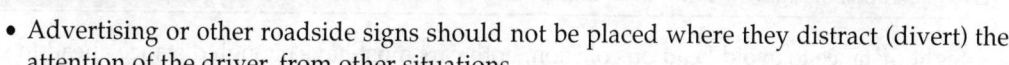

- Advertising or other roadside signs should not be placed where they distract (divert) the attention of the driver, from other situations.
- Increasing percentage of cross-traffic increase the accidents, at intersections. Improvement in safety by channelisation, refuge islands, sidewalks for pedestrain, control devices, signals, must be taken into account, in geometric design of roads.
- Considerations should be given to adequate lighting arrangement, type of bulbs, height of poles, glare, etc. so that night accidents are prevented.
- Steep side slopes are hazardous, slopes of 4 : 1 are considered safe. Steeper should avoided or guard rails provided, for safety, if steeper slopes are used.
- It should be recognised that accidents will happen and should be designed as safe as possible all elements should be designed as if "Accident Will Happen", so how design of the particular element will prevent it.

ILLUSTRATIVE EXAMPLES

Example 1. Calculate the safe stopping sight distance for design speed of 50 Km/h for:
  (a) Two-way traffic in a 2 lane road
  (b) Two-way traffic in a single lane road
Assume coefficient of friction as 0.37 and reaction time of drivers as 2.5 seconds.
Solution:

$$V = 50 \text{ Km/h} = 0.28 \times 50 = 14 \text{ m/sec.}$$
$$t = 2.5 \text{ seconds}, g = 9.8, \mu = 0.37$$

Stopping distance
$$= 14 \times 2.5 + \frac{14^2}{2 \times 9.8 \times 0.37}$$
$$= 35 + 27 = \textbf{62 m.}$$

  (a) 62 m.
  (b) Two-way in single lane = $2 \times 62 = \textbf{124 m.}$

Example 2. Calculate the minimum sight distance required to avoid a head on collision of two cars approaching from the opposite direction, if both the cars are speeding at 90 Km/h Assume total perception and break reaction time of 2 seconds, coefficient of friction of 0.8 and brake efficiency of 50%.
  Solution:

Stopping sight distance $Vt + \dfrac{V^2}{2g\mu}$.

$$V = 90 \text{ km. ph.}, V = 0.28 \times 90 = 25 \text{ m/sec}, t = 2 \text{ sec.}$$

As the break efficiency is 50%, the wheels will skid through 50% of braking distance and rotate through remaining distance.

The stopping distance for one of the cars

$$= Vt + \frac{V^2}{2g\mu \times \eta}$$

$$\eta = \text{efficiency of brakes.}$$

$$\text{S.D.} = 25 \times 2 + \frac{(25)^2}{2 \times 9.8 \times 0.8 \times 0.5} = 50 + 80 = 130.$$

Sight distance to avoid head on collision, both cars must have enough distance ahead to stop.

$$\text{S.D.} = 2 \times 130 = \textbf{260 m.}$$

Example 3. Find out the minimum non-passing sight distance on a highway at a descending grade of 2% using the following data:

    (i) Design speed = 80 Km/h

    (ii) Reaction time of drivers = 2.5 sec.

    (iii) Coefficient of friction between tyre and road surface = 0.35.

Solution:

On down grade $S = -2\% = -0.02$

$$\text{Stopping sight distance} = Vt + \frac{V^2}{2g\,(\mu - n)}$$

$$V = 80 \times 2.8 = 22.4 \text{ m/sec., } t = 2.5,$$

$$\text{S.D.} = 22.4 \times 2.5 + \frac{(22.4)^2}{2 \times 9.8\,(0.35 - .02)}$$

$$= \textbf{132 m.}$$

Example 4. The speed of overtaking and overtaken vehicles are 70 and 40 Km/h respectively on a two-way traffic road. If the acceleration of the overtaking vehicle is 3.6 Km/h per second,

    (a) Calculate safe overtaking sight distance.

    (b) How it is modified, considering AASHTO recommendations.

    (c) Overtaking sight distance for one-way operation on this road.

    (d) Draw a sketch of the overtaking zone and show the position of the sign posts.

Solution:

(a) Design speed (speed of A) = 70 Km/h

$$V_a = \frac{70 \times 1000}{60 \times 60} = 19.6 \text{ m/sec.}$$

Speed of slow vehicle (speed of B) = 40 Km/h

$$V_b = 11.2 \text{ m/sec.}$$

Acceleration $A = 3.6$ Km/h = 1.0 m/sec^2.

$$\therefore \qquad d_1 = V_b \times t \text{ (assume reaction time = 2 sec)}$$

$$= 11.2 \times 2 = 22.4 \text{ m.}$$

$$d_2 = V_b T + 2S$$

and

$$T = \sqrt{\frac{4S}{a}}$$

$$S = (0.7\,V_b + 6) \text{ m} = (0.7 \times 11.2 + 6) = 13.8 \text{ m}$$

$$\therefore \qquad T = \sqrt{\frac{4 \times 13.8}{1.0}} = 7.4 \text{ sec.}$$

$$d_2 = 11.2 \times 7.4 + 2 \times 13.8 \text{ m} = 110.5 \text{ m.}$$

$$d_3 = V \times T \qquad\qquad V = \text{design speed.}$$

$$= 19.6 \times 7.4$$

$$= 145 \text{ m.}$$

$\therefore$ Overtaking sight distance $= d_1 + d_2 + d_3$

$$= 22.4 + 110.5 + 145 = 277.9 \text{ m}$$

$$= \textbf{278 m.}$$

(b) By AASHTO consideration an additional clearance of $d_4 = \dfrac{2}{3} d_2$ should be added

$\therefore$ OSD by AASHTO $= 278 + \dfrac{2}{3} \times 110.5$

$$= 279 + 73.6$$

$$= 351.6 = \textbf{352 m}$$

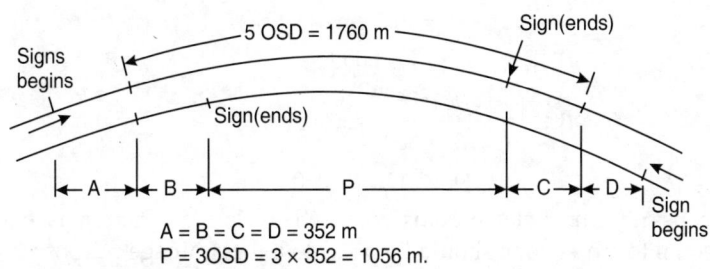

$$A = B = C = D = 352 \text{ m}$$
$$P = 3OSD = 3 \times 352 = 1056 \text{ m.}$$

(c) On one-way traffic OSD $\quad = d_1 + d_2 = 22.4 + 110.5 = 132.9 = \textbf{133 m.}$

(d) Sketch of overtaking zone is as above.

$\qquad$ Desirable length $= 5 \times \text{OSD} = 5 \times 352 \text{ m}$

$$= \textbf{1760 m.}$$

(Minimum could be 3 × OSD in restricted area).

$$A = B = C = D = 352 \text{ m}$$
$$P = 3 \text{ OSD} = 3 \times 352 = 1056 \text{ m.}$$

Example 5: For a cement concrete road, 7.5 m wide, having a design speed of 96 Km/h and assuming the coeff. of lateral friction as 0.15, calculate the following:

(i) Super elevation on a curve of 240 m radius.

(ii) Equilibrium S.E. for the condition when pressure on outer and inner wheels will be equal.

(iii) The superelevation required on a curve of 800 m radius.

(iv) Radius of the curve, for which no superelevation is required.

(v) If maximum superelevation is to be restricted to 1 in 15, for a 300 m radius curve, calculate the restricted speed.

(vi) Absolute minimum radius for allowable superelevation of 1 in 15.

(vii) The ruling minimum radius, assuming future design speed 16 Km/h higher.

Solution:

(i) $\qquad\qquad\qquad\qquad e + f = \dfrac{V^2}{127\,R} \quad \text{where} \qquad\qquad V = \text{speed Km/h}$

$\qquad\qquad\qquad\qquad\qquad\qquad\qquad\qquad\qquad\qquad\qquad\quad R = \text{Radius of curve in m}$

$\qquad\qquad\qquad\qquad\qquad\qquad\qquad\qquad\qquad\qquad\qquad\quad f = \text{Coeff. of friction}$

$\qquad\qquad\qquad\qquad\qquad\qquad\qquad\qquad\qquad\qquad\qquad\quad e = \text{Super elevation m/m}$

$$e + 0.15 = \frac{96 \times 96}{127 \times 240} = 0.30$$

$$e = 0.30 - .15 = .15$$

Height of outer edge over inner edge $= .15 \times 7.5 \times 100$

$$= \textbf{112.5 cms.}$$

(ii)   For equilibrium S.E. $f = 0$.

$$e = \frac{V^2}{127\,R} = 0.30.$$

Height of outer edge over inner edge $= .30 \times 7.5 \times 100$

$$= \textbf{225 cm.s}$$

(iii) $e = ?$ $R = 800$

$$e + 0.15 = \frac{96 \times 96}{127 \times 800} = .090$$

∴ $e = .090 - .15 = -.060\ (-ve)$

–Ve S.E. means no super-elevation is required as lateral friction is enough. However minimum S.E. equal to cross-slope should be provided, for drainage purposes.

For concrete road cross slope is 1 in 72

$$e = \frac{1}{72} = .014 \text{ Height of outer-edge over}$$

inner edge $= 7.5 \times .014 \times 100 = \textbf{10.5 cms.}$

(iv) When $e = 0, R = ?$

$$0 + 0.15 = \frac{96 \times 96}{127 \times R}$$

∴ $R - \frac{96 \times 96}{127 \times .15} = 484\text{ m}$

(v) $e = 1 \text{ in } 15 = 0.067,$
$$R = 300 \text{ find } V$$

$$.067 + 0.15 = \frac{V^2}{127 \times 300}$$

$$V = 91 \text{ Km/h}$$

Speed restriction $= 96 - 91 = 5$ Km/h

Design speed should be reduced to **91 Km/h.**

(vi) $e = 1 \text{ in } 15 = .067$
$$f = 0.15, \text{ find } R_{min.}$$

$$.067 + 0.15 = \frac{96 \times 96}{127 \times R_{min}}$$

$$R_{min.} = \frac{96 \times 96}{127\,(.067 + 0.15)}$$

$$= \textbf{334.5 m}$$

(vii) For ruling minimum radius it is assumed that speed in future will increase by 16 km. p.hr.

$$V_{future} = 96 + 16 = 112 \text{ Km/h}$$

$$\therefore \qquad R_{ruling} = \frac{112 \times 112}{127\,(.067 + 0.15)}$$

$$= \textbf{455 m}$$

Example 6. How much camber is to be provided (crown to be raised) on a road having width of 7 m with bituminous concrete surface located in heavy rainfall area?

Solution:

In heavy rainfall area the rate of camber is 1 in 50.

$$\text{Rise of crown} = \frac{1}{50} \times \frac{\text{Width}}{2} = \frac{1}{50} \times \frac{7}{2} = .07 \text{ m} = \textbf{70 mm.}$$

Example 7. What rate of super elevation should be provided on a horizontal curve of radius 90 m, if the design speed of the road is 50 Km/h. Take coeff. of friction as = 0.15. The width of road is 7 m, also calculate the height of outer edge to be raised over inner edge.

Solution:

Using the equation: $\qquad e + f = \dfrac{V^2}{127\,R}$

here $\qquad\qquad\qquad f = 0.15,\ V = 50,\ R = 90.$

$$e = \frac{50^2}{127 \times 90} - 0.15 = 0.22 - 0.15 = .07$$

$$= 7\%.$$

Height of outer edge to be raised $= e \times$ width of road

$$= 0.07 \times 7 = \textbf{0.49 m.}$$

Example 8. Calculate the maximum permissible speed on a horizontal curve of radius 200 m, if permissible super elevation is 0.07 and coeff. of friction is 0.15.

Solution:

$$e + f = \frac{V^2}{127\,R}$$

$$0.07 + 0.15 = \frac{V^2}{127 \times 200},\ V^2 = \sqrt{122 \times 2\cancel{00} \times 127} = \sqrt{22 \times 2 \times 127}$$

$$= \textbf{74.8 km. p. h.}$$

Example 9. Calculate the extra width required on a 2 lane curved road of radius 400 m. The design speed is 60 Km/h and length of wheel-base of vehicles is 6 m.

Solution:

Extra width is given by $\qquad W = \dfrac{n\,L^2}{2R} + \dfrac{V}{9.5\,\sqrt{R}}$

here $\qquad\qquad\qquad n = 2,\ L = 6,\ R = 400,\ V = 60,$

$$\therefore \qquad W = \frac{2 \times 6^2}{2 \times 400} + \frac{60}{9.5\,\sqrt{400}}$$

$$= \frac{36}{400} + \frac{60}{9.5 \times 20}$$

$$= .09 + .316 = .406 \approx .4 \text{ m}$$

Example 10. Ruling and minimum design speed for a highway is 100 and 80 Km/h. Calculate the radius of horizontal curve if superelevation is 0.07 and coefficient of lateral friction is 0.15.

Solution:

$$R = \frac{V^2}{127\,(e + f)}$$

Ruling minimum Radius $R_R = \dfrac{100^2}{127\,(.07 + 0.15)} = \mathbf{358\ m}$

$(V = 100)$

Absolute minimum Radius $R_M = \dfrac{80^2}{127\,(.07 + 0.15)} = \mathbf{229\ m}$

$(V = 80)$

Example 11. Determine the length of transistion curve for a national highway having width of 7 m wide and curve of 400 m radius. The design speed is 100 Km/h and value of c is 0.46. Use rate of change of acceleration criterion.

Solution:

(a) Rate of change of centrifugal acceleration basis:

Length of transistion curve $L_S = \dfrac{V^3}{C_R}$

here $V = 100 \text{ km/ph} = (.28 \times 100) \text{ m/sec. } e = 0.46, R = 400$

∴ $L = \dfrac{(.28 \times 100)^3}{0.46 \times 400} = \mathbf{119\ m}$

Example 12. Calculate the length of summit curve at the junction of an upward gradient of 1% and downgrade gradient of 2%. The stopping sight distance is 170 m on this highway. Assume sight distance is less than length of curve.

Solution:

Here S is less than L

$$L = \frac{NS^2}{440} = \frac{1 - (-2) \times 170 \times 170}{440}$$

$$= \frac{3 \times 170 \times 170}{440} = \mathbf{197\ m}$$

Assumption that sight distance is less than length of curve is correct. So length of summit curve is 197 m.

In case the assumption would have been wrong other formula when $S > L$ should have been tried.

Example 13. What should be the length of the summit curve for stopping distance of 180 m on a highway meeting on upward gradient of 1 in 200 and a downward gradient of 1 in 200. The height of driver and the object above the pavement are 1.2 m and 0.15 m respectively.

Solution:

(i) Assuming sight distance (S) is less than length of curve (L)

$$L = \frac{AS^2}{440} \qquad \text{where} \qquad A = \text{algebraic difference of slopes}$$

$$S = \text{stopping sight distance}$$

$$A = \frac{1}{200} = 0.5 - \left(-\frac{1}{200}\right) = (0.5 + 0.5)$$

$$= \frac{(0.5 + 0.5)\, 180 \times 180}{440}$$

$$= 73.6 \text{ this is less than } S\,(180)$$

therefore other formula should be tried.

(ii) Sight distance (S) is more than length of the curve.

$$L = 2S - \frac{440}{N}$$

$$= 2 \times 180 - \frac{440}{(0.5 + 0.5)}$$

$$= -80$$

This condition is not satisfied.

This means that grade change is small and no vertical curve is needed.

Example 14. Design a valley curve at the junction of a downward gradient of 1 in 30 and a level stretch from headlight consideration. The stopping sight distance is 180 m.

Solution:

(i) **Assume $L > S$:**

$$L = \frac{AS^2}{1.5 + 0.035\,S} \qquad A = 1 \text{ in } 30 = .033$$

$$= \frac{.033 \times 180^2}{1.5 + .035 \times 180}$$

$$= 137 \text{ m}$$

In this case L (137) is not greater than S (180) so other formula should be tried.

(ii) **Assume $L < S$:**

$$L = 2S - \frac{1.5 + 0.035\,S}{A}$$

$$= 2S - \frac{1.5 + 0.035\,S}{A}$$

$$= 2 \times 180 - \frac{1.5 + 0.035 \times 180}{0.033}$$

$$= \mathbf{123\ m}$$

The length of valley curve should be 123 m.

Example 15: A road has a total width of 7.5 including extra widening on curve and design speed of 60 Km/h. Calculate the length of transition curve and its shift on this curve of 200 m radius. Allowable superelevation is 1 in 150 and pavement is rotated about centreline.

Solution:

Calculating length of transaction curve by the two methods:

(i) **By rate of change of centrifugal acceleration criteria:**

$$L_S = \frac{V^3}{CR} \qquad \text{where} \qquad R = \text{radius of curve in m}$$

$$V = \text{speed in m/sec.}$$

$$C = \text{Allowable rate of change of centrifugal acceleration.}$$

and

$$C = \frac{80}{75 + V} \text{ m/sec}^3$$

$$= \frac{80}{75 + 60} = \frac{80}{135} = 0.6 \text{ m/sec}^3$$

$$L_S = \frac{V^3}{CR} = \frac{(.28 \times 60)^3}{0.6 \times 200} = \textbf{39.5 m}$$

(ii) **By rate of introduction of super-elevation criteria:**

$$\text{Superelevation rate} = \frac{V^2}{225\,R} = \frac{60^2}{225 \times 200} = 0.08.$$

As this value is greater than maximum allowable rate of .07 we have to limit this value to 0.07 and check for friction value to prevent skidding.

$$e + f = \frac{V^2}{127\,R}$$

$$f = \frac{V^2}{127\,R} - e$$

$$= \frac{60^2}{127 \times 200} - 0.07$$

$$= .141 - .07 = .071$$

As this value is less than allowable friction value of 0.15 super elevation rate of 0.07 is safe for this design speed.

Total width of pavement = 7.5 m.

Total rise of outer edge of pavement with respect to centre line

$$= \frac{eW}{2} = \frac{.07 \times 7.5}{2}$$

$$= \textbf{0.26 m}$$

here $\qquad N = 150.$

$$\therefore \qquad L_S = \frac{eN}{2} \cdot W = \frac{0.07 \times 150 \times 7.5}{2}$$

$$= 0.26 \times 150$$

$$= \textbf{39 m}$$

(iii) **Minimum value of L_S as per IRC**

$$L_S = \frac{2.7\,V^2}{R} = \frac{2.7 \times 60^2}{200}$$

$$= \textbf{48.6 m}$$

Higher value o the three L_S (i.e. 39.5, 39 and 48.6 m) should be used as design value. L_S = 48.6 m. **Ans.**

Shift:

The shift is given by

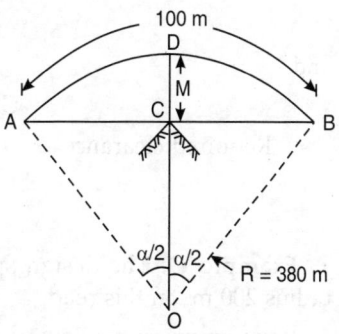

$$S = \frac{L_S^{\,2}}{24\,R} \qquad\qquad L = \text{Length of transition curve}$$

$$R = \text{Radius of curve}$$

$$= \frac{48.6^2}{24 \times 200}$$

$$= \textbf{0.49 m}$$

Example 16: What should be the clearance required from the centre line, on the inside of the horizontal curve so as to provide for:

(i) Stopping sight distance of 100 meters

(ii) Safe over taking sight distance of 350 meters.

The radius of horizontal curve is 380 m and length of curve is 225 m, on this highway.

Solution:

Case I:

$$\text{SSD} < \text{L}.$$

$$(100 \text{ m} < 225 \text{ m})$$

Required clearance 'M' = CD

$$\begin{aligned} CD &= OD - OC & OD &= R \\ &= R - R\cos\alpha/2 & OC &= R\cos\alpha/2 \\ &= R\,(1 - \cos\alpha/2) \end{aligned}$$

for finding α.

$$\frac{\alpha}{100} = \frac{180°}{\pi R}$$

$$\frac{\alpha}{100} = \frac{180°}{\pi \times 380} \qquad\qquad \therefore \quad \frac{\alpha}{2} = \frac{180°}{2\pi \times 380}$$

$$\therefore \qquad CD = 380\,(1 - \cos 7°.54°) \qquad\qquad = 7.54°$$

$$= 380\,(1 - .991) = \textbf{3.284 m}$$

Case II:

$$\text{SSO} > \text{L}$$

Required clearance = $DE = CD + CE$

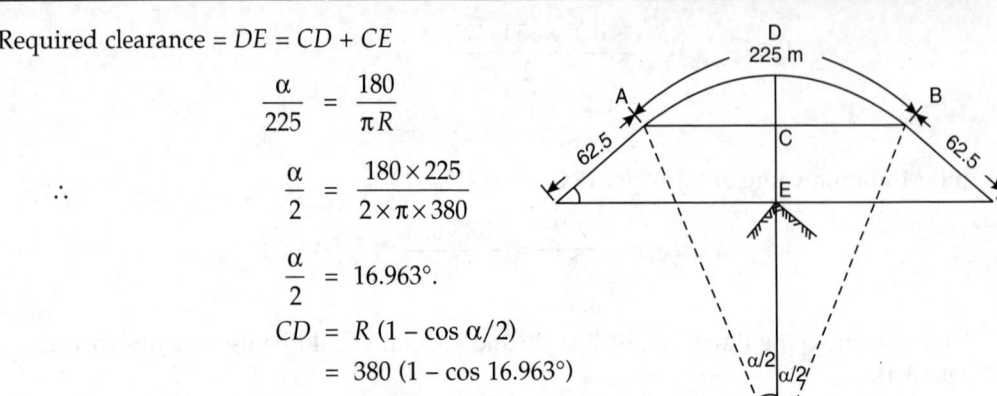

$$\frac{\alpha}{225} = \frac{180}{\pi R}$$

$$\therefore \qquad \frac{\alpha}{2} = \frac{180 \times 225}{2 \times \pi \times 380}$$

$$\frac{\alpha}{2} = 16.963°.$$

$$CD = R\,(1 - \cos \alpha/2)$$
$$= 380\,(1 - \cos 16.963°)$$
$$= 16.53 \text{ m}$$

and

$$CE = 62.5 \sin \alpha/2 = 62.5 \sin 16.963° = 18.23$$

$\therefore$ Required clearance $\quad DE = CD + CE$
$$= 16.53 + 18.23$$
$$= \textbf{34.76 m.}$$

Example 17: The design speed of a highway is 80 Km/h. There is a horizontal curve of radius 200 m, on this road.

If maximum superelevation of 1 in 15 is not to be exceeded, calculate maximum allowable speed on this curve.

Also determine the extra widening required and length of the spiral transistion curve, using following data.

Length of wheel base of largest vehicle = 6.1 m

Pavement width = 7.2 m.

Number of lanes = 2.

Rate of introduction of superelevation = 1 in 200.

Solution:

(a) Maximum allowable speed = $\mu = 0.15$, $e = .067$

$$e + \mu = \frac{V^2}{127\,R}$$

$$.067 + 0.15 = \frac{(V_{\text{allowable}})^2}{127 \times 200} \qquad\qquad \therefore V_a^2 = 0.217 \times 127 \times 200$$

$$V_a = \textbf{74 km. ph}$$

(b) Extra widening $(W_e) = \dfrac{NL^2}{2R_c} + \dfrac{0.1V}{\sqrt{R_c}}$

$$= \frac{2 \times (6.1)^2}{2 \times 200} + \frac{0.1 \times 74}{\sqrt{200}} = .186 + .551$$

$$= \textbf{.737 m}$$

Total width = $W + W_e = 7.2 + .737$

$$= 7.937 \text{ m}$$

(c) Length of transistion curve (L_S)

(a) **Rate of S.E. Basis:**

Assuming rotation of S.E. about C.L. = $\dfrac{e\,(W + W_e)}{2} N$

$$= \frac{.067}{2} \times 200$$

$$= \textbf{53.19 m}$$

(b) **Rate of change of centrifugal acceleration basis:**

Rate of change of centrifugal acceleration C.

$$C = \frac{73}{V + 64}$$

for speed between 33 to 96 Km/h

$$= \frac{73}{74 + 64}$$

$$= 0.529 \text{ m/sec.}$$

$\therefore$ $$L_S = \frac{V^3}{46.5\,CR}$$

$$= \frac{(74)^3}{46.5 \times .529 \times 200}$$

$$= 82.36 \text{ m}$$

Higher of the two is used in design.

$$L_S = \textbf{82.36 m.}$$

Road User and Vehicle Characteristics

7.0 PREAMBLE

Traffic involves the travel of discrete units (pedestrians, cyclist and vehicles of different type) having characteristics of independence, randomness, and human control. Road users mainly driver's behaviour guided by human factors influences all aspects of motoring. The safe and efficient operation of road traffic system ultimately depends upon the complex interaction between following four elements (Fig. 7.1).

 i. The road users (drivers and pedestrians)

 ii. The vehicle

 iii. The road or highway

 iv. The environment.

The degree of complexity of traffic behaviour depends upon variability and degree of interaction between these elements and on the physical constraints of time and space. The road user's performance and constraints to various stimuli varies over a wide range from individual to individual, depending upon psychological and physical abilities. The driver's task is to receive and process inputs, make choices about appropriate actions, execute the actions and observe their effects.

Roads or highways design standards are controlled by the physical dimensions, maneuverability, accelerating, braking characteristics of the vehicles. The pavement design is dependent upon axle load and spacing. The road transportation system is a series of interaction between the components: vehicle, road and driver. The driver brings varying degree of fatigue, attention, experience and training, health, etc. on to the driving scene. The vehicle he drives has its own set of characteristics and defects.

The environment in which the road is located, play a vital role in considering and planning several elements of road geometric design and safety. For example selection of speed, depends very much on environment and terrain. Natural environment, snow, rainfall, fog, play an

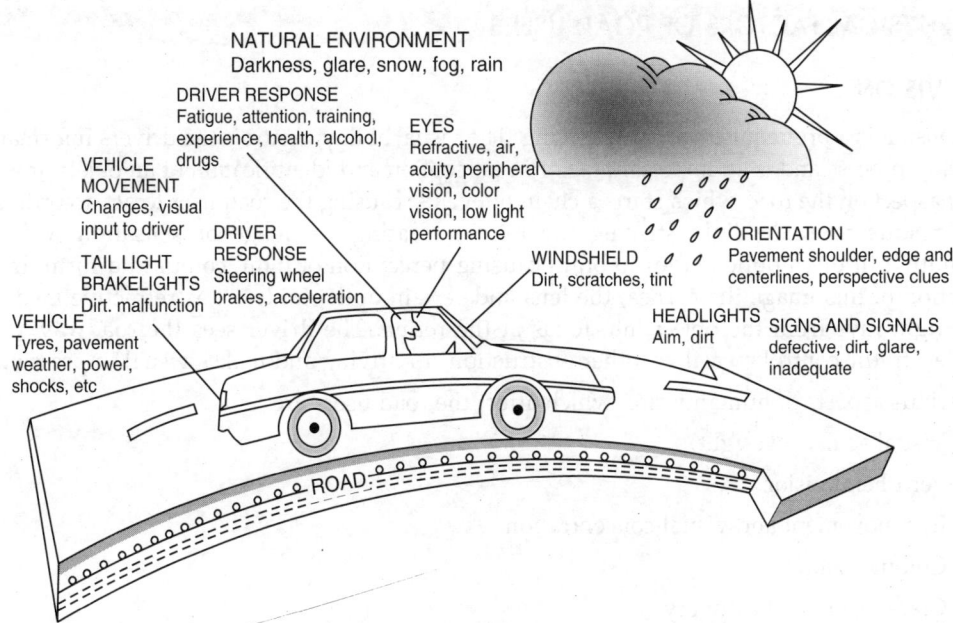

Fig: 7.1: Basic elements in traffic operation

important role in safety of road operation. Study and knowledge of human and vehicle characteristics is therefore very important in planning and design of highway facility, which has to operate in different environmental conditions throughout its life.

7.1 THE HUMAN BEHAVIOUR

Human factors and driver's behaviour affect efficient operation of road traffic system. The human performance varies so widely that considerations are given to a typical characteristics rather than using the average values. For example a traffic signal timed to permit an average pedestrian to cross the road safely might cause a severe hazard to an elderly person or others whose capability falls below those of average pedestrian.

Driver's role is to receive and process mostly visual inputs from his operating environment. There are many aspects of driver's behaviour involved in driving which must be studied and understood in predicting both 'Normal' and 'Abnormal' behaviour which might lead to accidents.

For practical solution to any highway and traffic problem, study of human psychology is important. The engineer has two approaches to plan traffic measures:

i. To design the highway facility as per defined human behaviour or

ii. Adjust the human behaviour as per fixed highway facilities through controls, regulations and restrictions.

In reality, both the above approaches are adopted in planning and design of highway facilities. Understanding of road user performance, his mental limitations and capabilities are therefore critical in traffic control, design and operational measures.

Human behaviour is most complex and least understood. It is affected by both internal and external factors. The internal factors can be classified into two categories

i. **Physical factors:** Vision, hearing, strength, judgement power and reaction time.

ii. **Psychological factors:** Motivation, intelligence, learning, emotions, individual differences and PIEV time.

7.2 PHYSICAL FACTORS OF ROAD USERS

7.2.1 VISION

Good vision is a pre-requisite for safe driving. The visual ability, controls the drivers information receiving process and is an important factor in perception and identification of an object. It is the visual aspect on the road which starts a chain of events, causing the road user to act accordingly. Sight results from the light striking the retina, creating an image or sensation which is transmitted through optic nerves to brain causing perception of light, colour and form. In the formation of this image the cornea, the lens and certain optic fluids act as refractive media to bring light rays from the object into focus at the retina. The driver sees the roadway, other vehicles, traffic control measures, other obstructions to driving and makes visual impressions.

Various aspects of human vision which affect the road users are:

i. Visual acuity (eyesight)

ii. Peripheral vision

iii. Eye movement and visual concentration

iv. Colour vision

v. Glare vision and recovery

vi. Perception of space (dynamic judgment of time and space)

7.2.2 Visual Acuity (Eyesight)

Visual acuity is the indication of the ability of the eye to see fine details. It relates to the field of clearest vision under average illumination. Poor visual acuity will directly influence driver's ability to read traffic signs and will increase his time to respond to them. The visual acuity varies markedly between individuals with the level of illumination, the length of viewing and the eccentricity of viewing.

The most clear vision is within a narrow cone of 3° to 5° as shown in Fig. 7.2.

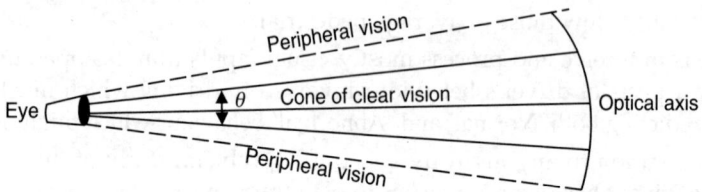

Fig. 7.2: Cone of clear and peripheral vision

Cone of vision (horizontal plane)θ	Clarity of vision
3° to 5°	Very clear vision
5° to 7°	Clear vision
Up to 12°	Satisfactory vision
Beyond 12°	Blurred vision

The value of θ — the cone of vision in vertical plane is 1/2 to 2/3 of horizontal plane of vision. Visual acuity diminishes with age. When placing traffic signs and symbols, etc it must be kept in mind that beyond 12° cone the vision becomes blurred.

7.2.3 Peripheral Vision

Peripheral vision is the field of view within which an individual can see objects, but without clear details or colour.

As the speed of vehicle increases, the angle of peripheral vision decreases. It is about 100° at speed of 30 Km/h but get reduced to 40° as speed increases to 90 Km/h. One moves his eyes or head to bring the object into his cone of clear vision and there is no conclusive evidence to show that road users with restricted peripheral vision are accident prone.

7.2.4 Eye Movement and Visual Concentration

While shifting his eye with movement of traffic to obtain clear vision of the object, the eye makes six different type of movements:

i. Fixing of eye at the object (takes 0.1 to 0.3 sec)
ii. Jumping of eye from one point of fixation to next takes 0.125 to 0.235 sec.
iii. Eye follows moving elements of the stream.
iv. Both the eyes shift (converge or diverge) together. It takes 0.3 to 0.5 sec.
v. Eye compensates for the movement of head
vi. Eye often moves involuntarily in response to noise or other extraordinary situations.

For clear vision these eye movements must occur constantly. Every man has an area of natural vision concentration where he pays maximum visual attentions, therefore all traffic control devices and facilities should be located in this area of vision, for better comprehension without moving their eyes and head greatly.

7.2.5 Colour Vision

Different colours have different wavelengths and thus their relative visibility changes. At very low illumination all colour sensation disappears. Optimum colour schemes for signals, i.e. red, yellow and green and markings, etc. are worked out keeping colour vision characteristics of the road users in view.

7.2.6 Glare Vision and Recovery

Adaptability to light changes is important factor of vision. Eye takes more time to adapt when going from light to dark and adjusts much faster while going from darkness to light. Glare recovery time is about 6 seconds or more when going from light to dark and about 3 seconds when going from dark to light. Tunnel lighting, head lights and intermittent highway lighting are the situations in which eye is subjected to rapid and severe changes in illumination. While designing light systems at such locations, this human characteristics of glare recovery is taken into account, by introducing changes gradually.

7.2.7 Perception of Space

Perception of space refers to the ability of the eye to judge the space, distance and time. He is required to make judgement of his own behaviour with respect to the behaviour of others in traffic stream. There are different abilities of the road users to perceive space.

Overtaking operations, movement at intersections, etc. require judgement of space and speed. Rear end collisions on high speed are largely due to failure of speed judgement.

The various factors of vision described above get modified, due to several factors such as fatigue, disease, alcohol, season, weather, time of the day, altitude and desire, etc. Vision of human beings is thus an important aspect of highway and traffic engineering elements for safety and efficiency. It has been estimated that over 90% of driver's information is acquired visually.

7.2.8 Hearing, Strength, Judgement and Reaction Time

a. **Hearing:** Hearing is needed to detect warning sounds, but lack of hearing can be compensated by use of hearing aids. Deafness is therefore no disqualification for driving. There appears to be no relationship between poor hearing and driving accidents.

b. **Strength:** Physical strength is a factor which is of importance in driving heavy vehicles as drivers with poor strength may lag behind in their braking or steering action. However in present day vehicles with power brakes and power steering, do not depend upon strength characteristics.

c. **Judgement:** Judgement power is ability to judge coming situation. It depends upon experience and one's own judgement power. Before the actual response, it is the right judgement of situation, for steps to be taken to avoid the mishap. Road user's judgement power as a driver, pedestrians or cyclist is therefore important in traffic situations.

d. **Reaction time:** Road user's response to act on a particular situation takes time. This reaction time taken for responding depends upon the physical as well as psychological factors. Increase in reaction time increases with complexity of the situation. The temporary factor, which affect the reaction time are: *alcohol*, which slows responses and increases reaction time, *fatigue*, which produces drowsiness or sleep reducing judgement capacity, *illness*, producing a tendency of diverting attention and slower reaction and *anger*, which may create me-first, feeling, affecting judgement or reaction time. Climate, season, weather, time of the day, altitude and light characteristics may produce complex responses in a road users (drivers).

7.3 PSYCHOLOGICAL FACTORS OF ROAD USERS

The role of psychological factors in road user (driver) behaviour are proved to be more significant, than the physical factors described above. The driver's behaviour depends upon his motives, intelligence, learning, emotions, personal factors, etc. in performance of the driving task.

7.3.1 Motivation

Motivation is the purpose for which road user enters the road. It affects time-distance economy on one hand and safety or comfort on the other. Motive may be business, reading, recreation marketing, trip to hospital in emergency, etc. How much safety in how much time demands the decision by the road users in the normal traffic operation. In emergent motives such as trip to hospital, fire hazard, etc. the behaviour becomes abnormal.

7.3.2 Intelligence

Intelligence of the road user has bearing upon the decision he takes to adjust himself and his behaviour according to a particular traffic situation. It is usually seen that persons of superior intelligence are not as mentally attentive to the task of the road use as others. Conclusive correlation between intelligence and behaviour in traffic is yet to be established.

7.3.3 Learning Processes

Learning processes depend upon motivation, intelligence and other factors and it develops skills, habits and abilities in road users to respond properly to total environment of traffic operation. Through the past experience road user learns and continues to respond in learned pattern of behaviour. Learned responses are powerful factors which may be beneficial or harmful in traffic operation. Through education and publicity good learning can be achieved.

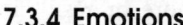

7.3.4 Emotions

Emotions strongly motivate the road user to inefficient, and random adjustment. Fear, anger, worry and other such emotions disorganize the reaction and behavior because intelligence and logic is lost. 'Me-first', 'Show-off', 'Racing' type feelings arise from emotions.

Personal differences vary among the road users. However, there are limits in the range and central tendencies can be established in traffic performance. Failure to recognize the wide range and variability of performance of road users may lead to conflicting and wrong conclusions.

7.4 PIEV TIME

Time required for response to a traffic situation such as putting on brakes, etc. depends upon the psychological processes of perception, intellection, emotion and volitions. This time factor of response called PIEV time deserves thoughtful consideration in analysis, interpretation and treatment of any traffic events. Reaction to an external situation involves a series of events related to human factors.

7.4.1 Perception

Sensations received by eyes, ears and body are transmitted to brain by nervous system. The time required for this purpose is called perception time. In all cases road user must perceive a complex and involved situation before he can react. As the complexity of the situation increases, the time required to perceive increases. Exact time required to perceive a traffic situation is highly variable and depends on the individual's psychological and physiological characteristics, and also on conditions to be perceived. The time required for this purpose is the 'Perception Time'.

7.4.2 Intellection (Identification)

Intellection is the forming of new thoughts and ideas. Perception may be simple knowing of a fact such as red or green signal, a hole in the road, etc. The time required for comparing, registering or regrouping (recall of old similar experiences) for identification and understanding is the 'Intellection Time'.

7.4.3 Emotion (Judgement)

Emotion traits of the individuals are closely linked with perception and intellection. Emotions may influence the final message, which is sent to the muscles to be carried out. Fear or anger may result is an abnormal behaviour leading to an accident. This involves the decision, making is which determination is made for right action (to stop, to apply brakes, to speed up, blow the horn, pass, etc.) Decision-making is the conversion of perceived information into some action upon the vehicle by the driver.

7.4.4 Volition (Reaction)

The will to take some act or produce some action is volition. While intellection process may suggest many answers, the final decision requires resolution of all impulses resulting in a volitional outgoing message which takes some definite action for execution of decision. In this process road users act in accordance with his own memories, prejudices, ideals, habits, desires and weaknesses.

The total time required to perceive and complete a reaction to a stimulus is the sum of the time necessary for perception, intellection, emotion and volition. It is called "PIEV" time. PIEV time may vary from 0.5 second for a simple problem to 3 to 4 seconds for a complex problem.

In an abnormal person, with certain deficiencies, PIEV time may be more or result in abnormal behaviour. This time also varies with age, fatigue, alcohol, etc. PIEV qualities of road user are taken into consideration in geometric and traffic design features of the roads. It is normal practice to use a value of 2.5 seconds for normal traffic situations, as reaction time. Perception and attention errors predominate in accidents, which could be avoided by alert and attentive driving.

In a moving vehicle, the driver has no time to see everything along the road or to think about what to do in all different traffic situations that may arise. Every driver therefore builds habits of selective seeing in traffic and of automatic responses to hazards along the roads.

Traffic engineer can plan form, size, shape, colour contrast and luminance, etc. of signs and other objects to guide and control road users, within human visual system.

7.5 BEHAVIOURAL ASPECTS AND PROVISIONS FOR PEDESTRIANS

Walking and cycling are known as soft transport modes. Walking is of course the fundamental human travel mode and is a component of almost all trips. Bus commuters walk to and from the bus stop, car owners and motor cyclists walk to their parked vehicles. Some people walk for health purposes and others have to walk as they have no other option. Pedestrian thus could be considered as most important traffic unit on a road. In India fatalities are highest amongst pedestrians using the road.

Pedestrians are the largest and major group of road users but they get lowest attention. The transport planning is directed towards motorized vehicles and bulk road space is kept for them, while negligible portion of road space is made available for pedestrians. The road traffic system is actually vehicle friendly as even the motor vehicle act does not recognize pedestrian as a basic traffic unit. Pedestrian as road users are the worst sufferers of noise and pollution created by vehicles. Most critical conditions arise at unsignalised intersections and mid block crossings where judgement of pedestrian could be impaired due to various reasons. Pedestrian, in the lack of sufficient footpaths have no alternative but to walk on the edge of the road, putting his life in danger.

7.5.1 Characteristics (Behavioural pattern) of Pedestrians

Following are some of the basic factors to be kept in mind while planning traffic and highway facilities for pedestrians:

- Pedestrians attempt to minimize travel distance even when this involves some traffic risk. They have a tendency to use shortest distance

- Pedestrian minimize travel effort and excess time

- Physical, mental, emotional and variability factors of pedestrians are similar to those of vehicle drivers but they become more compound due to the fact that many pedestrians lack the knowledge of the rules of the road and some are illiterate or unable to read. They feel that no traffic law is applicable to them

- Pedestrians walking speeds range from 1 to 1.5 m per second and their reaction times range from 4.0 to 5.0 seconds. For determining minimum phase for signals, this factor should be kept in view. Normal speed of 1.2 m/second is used in signal design

- Pedestrians specially old age people and infirm do not like foot-over bridges which require climbing up and down to 6 meters or so, for crossing a road.

- Pedestrians often cross the road in an impracticable manner and with over-optimistic estimate of a motorist's ability to either stop or avoid impact.

7.5.2 Provisions for Pedestrians

For pedestrians safety following provisions are to be used wherever possible:

- To reduce the problem of pedestrians-vehicular conflict, segregation of one form or the other, wherever warranted, should be adopted.
- Footpaths in good condition free from encroachments should be provided and maintained for pedestrians as a safe walkway on roads and bridges to save them from high and speedy traffic.
- Zebra crossings for pedestrians at all intersections, alongwith safety railings at the corner, are necessary for safe passage of pedestrians and must be well painted for visibility. The minimum width of pedestrians crossing should be 2 m.
- Pedestrian subways are necessary at heavy traffic locations and junctions. Guidelines stipulate provision of controlled pedestrian crossing when the hazard index determined as product of number of pedestrians (crossing) per hour and square of number of vehicles (PV^2) exceeds 10^8 for undivided carriageway and 2×10^8 for divided carriageway. Controlled crossing as mentioned above could be:
  - Automatic traffic signal
  - Push button pedestrian signal (pelican crossing)
  - Foot over bridge or
  - Pedestrian subway.
- Shops may be provided inside subways to take care of security aspects of pedestrians.
- Pedestrians should be educated about traffic rules and regulations
- A special traffic signal for blind pedestrians is operated by turning a special key in a lock mounted on the sign post. This causes the signal to change to red phase, 6 seconds after which a bell rings, indicating that it is safe to cross. The bell continues to ring for 20 seconds. The special signal lock key is distributed to all blind persons in the area.
- Pedestrians islands, crossings, markings, signs, control devices, etc. should be properly designed, maintained, and illuminated.

7.6 CHARACTERISTICS AND PROVISIONS FOR BICYCLE USERS

Bicycle as a vehicle is important on Indian roads. The length of a bicycle is about 1.9 m. Normal cycling speeds are 16 to 20 Km/h, but are sensitive to gradient. Bicycles are used by students and adults for journeys to work and other purposes. A drop in bicycle use is associated with a sharp rise in motor vehicle ownership and tendency to live away from home.

Advantages of the bicycles are:

- Its use demands only moderate effort
- It needs no fuel
- It has no emissions or noise
- Cycle offers door to door mobility
- There is no fixed time table for its use
- Require little parking space
- Can pass even in congested traffic
- It is relatively safer, more so when separated from fast traffic
- It is healthy to use cycling as it reduces calories
- It is convenient to cycle for short distances of 3 to 5 kms.

The disadvantages of the cycling are seen to be:

- Lack of safety, when uses the road with other fast traffic
- There is no weather protection in extreme cold, hot or during rains
- Even after locking, it is not safe if parked at an isolated place
- The distance which can be covered on a cycle is not very large
- Climbing a grade creates difficulties for the rider
- There are relatively few legal restrictions on cyclists
- Cyclists are cause of many accidents specially at intersections. Accident severity is also high.

7.6.1 Traffic Facilities for Bicycles

Almost all the urban roads are used by cyclists and they therefore require considerations at all stages. Special facilities (tracks) for cycle traffic are called bikeways.

Facilities provided exclusively for cyclist can be of the following types:

a. **Shared Paths:** Sometimes cycling is permitted on footpaths, sharing it with pedestrians. Bicycles are required to give way to the pedestrians. It provides low cost cycle paths.

b. **Shared Bikeways:** Shared bikeways use spare capacity of the road. They are relatively cheap. Cycle lane should be sufficiently wide to permit cyclist to travel away from the main traffic stream. Such facilities are generally provided on low volume city roads. Surface conditions of shared bikeways should be good, as many cyclists prefer to use main roads because they provide better cycling conditions.

c. **Restricted Bikeways (Bicycle lanes):** Restricted bikeways use signs, markings and physical barriers to denote an area of road for the exclusive use of cycles. This is possible on relatively wide road pavements with spare capacity.

d. **Exclusive Bikeways:** Exclusive bikeways are the special paths constructed for cycles. They are relatively expensive. Even after this complete segregation, intersections make them unsafe, requiring particular attention in design of intersections, where merging with other traffic occurs.

7.7 PHYSICAL CHARACTERISTICS OF VEHICLES

Physical characteristics of vehicles include:

a. Type of vehicles
b. Dimensions of vehicles
c. Other design features.

7.7.1 Type of Vehicles

Type of vehicles in traffic streams are usually heterogeneous having combination of fast moving and slow moving traffic.

Fast moving vehicles include passenger cars, trucks, truck-trailer combinations, buses, jeeps, street cars, mini-buses, vans, taxies, motor cycles, scooters, mopeds, tankers and other transport vehicles.

Slow moving vehicles are vehicles drawn by animals, such as bullock carts, horse drawn coaches, man-pulled rickshaws, cycles, pushed vehicles, etc.

Mixed traffic of slow and fast vehicles creates many operational and structural design problems. There are several makes and models of these vehicles.

American Association of State Highway and Transport Officials (AASHTO), has established six type of vehicles for use in geometric design. These are passenger car (P), single unit truck (SU), Single unit bus (BUS) and three types of trailer combinations having wheel base of 12 m, 15 m and 18 m (WB–12, WB–15 and WB–18).

Indian Road congress uses only commercial vehicles of 3 types, i.e. single unit truck, semi trailer and truck–trailer combination which govern the geometric design features of roads.

7.7.2 Dimension of Vehicle

Dimension of vehicles are of relevance for many traffic engineering purposes including geometric design of roads. All countries therefore have a limit for maximum allowable dimensions of the vehicle. The traffic lanes are limited to about 3.5 m in width and vertical clearance beneath structures is normally about 4.5 m. The type and structural design of pavement depends upon permissible wheel loads. Occasionally loads, dimensions of vehicles exceed the legal limits. Special permits are granted for these vehicles to use the road carefully. As wheel base increases the capacity and mass of the vehicle also increase.

Table 7.1 shows the vehicle characteristics affecting different elements of road geometrically and structurally. Table 7.2 shows typical maximum dimensions of motor vehicles.

7.7.3 Turning Radius

  i. **Low-speed turns:** The minimum turning radius for a vehicle at low speed (less than 16 km per hour) depends upon the wheel base and steering angle. The path followed and space required by a vehicle while making a sharpest possible turn is specially important in vehicle manoeuvring such as parking. Under-clearance of a vehicle is important in designing of vertical curves.

Table 7.1: Vehicle characteristics affecting road features

Vehicle characteristics	Affecting elements of road
Width	Lane width, width of shoulders, width of parking lots.
Length	Horizontal alignment, minimum turning radius, extra widening on curves, passing sight distance, road capacity, parking facilities.
Height	Clearance to be provided under bridges, subways, electric service lines, etc.
Weight	Pavement thickness, ruling and limiting gradients, design of road bridges.
Turning radius	Parking and turning manoeuvres, curves, intersection design, channelisation, driveways, etc.
Speed	Horizontal and vertical alignment, super elevation, limiting radius, sight distances, lane capacity, intersection design, skid resistance.
Power	Ruling and limiting gradient, speed of vehicle, various resistances, braking distance.
Acceleration characteristics	Time required to cross intersection, passing manoeuvers, gap acceptance
Braking characteristics	Stopping distance, overtaking sight distance, traffic capacity.
Headlights	Night operation, night accident.

Table 7.2: Typical maximum dimensions of motor vehicles

Particulars	Maximum Limits AASHTO	Maximum Limits IRC
Width	2.6 m (8.5 ft)	2.5m
Height	4.1 m (13.5ft)	Truck – 3.8–4.2 m Double decker bus – 4.75 m
Length a. Single unit track b. Bus c. Truck tractor and semi-trailer d. Other combination	12.2 m (40ft) 12.2 m (40ft) 16.75 m (55ft) 19.8 m (65ft)	11 m 12 m 16 m 18 m

For sharpest turn (90° or so) the minimum turning radius for vehicle P is 7.4 m, SU is 11.4 m, WB-40-50 is 12.2 m when vehicle turns at low speeds, the rear wheel tracks the front wheel on short radius curves. The difference between radii of rear and front wheel known as "Off Tracking" is dependent on the turning radius and the vehicle wheel base. For combination vehicles having more wheel bases amount of tracking is obtained by using the scale models.

Figures 6.3 to 6.5 show minimum turning path of road vehicles.

ii. **High speed turns:** At high speed turns (0.7 times of the design speed), the turning radius is controlled by amount of super elevation and the side friction factor between the tyres and pavements as given by the following formula.

$$R = \frac{V^2}{127(e+f)}$$

Where: R = Radius of curve (m)

V = Speed of the vehicle (km per hour)

f = Coefficient of side friction

e = Rate of super-elevation

The minimum turning path of the specific vehicle is particularly important in geometric design of roads.

7.7.4 Other Design Features

Various other features which must be considered in design of vehicles are:

- Head lights and driver's visibility determine the portion of the road lighted to driver's view. Driver should be able to see all sources of potential traffic conflict to avoid accident. Lighting characteristics of vehicles are of extreme importance in night journey.
- The corner posts to support top, wind shield wipers, hood, etc. are important factors in vehicle design so that vision is not obstructed. Figure 7.3 shows vision restrictions due to vehicle design.
- In order to identify a vehicle in darkness, tail lamps, stop lamps, side marker lamps and clearance lamp should be properly designed and provided.
- The dimensions of vehicle body such as driver's seats, location and placement of controls, steering wheel dimensions and positioning and other elements should be designed as per

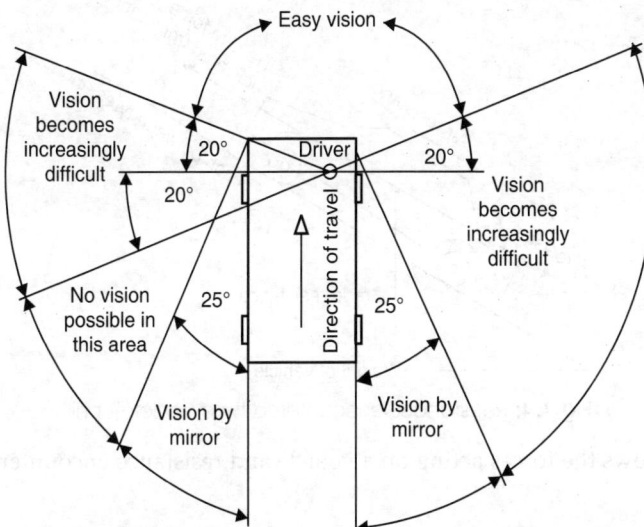

Fig. 7.3: Vision restriction due to vehicle design

human needs, comforts and safety. In the design of passenger cars there is a wide variation in these features among manufactures. Direction indicators should be properly placed, with flashing lights to draw attention of following drivers.

- The braking characteristics of vehicles are significant in road safety and skidding of vehicle when brakes are applied suddenly. Anti-skid and improved braking systems are introduced in new models of the vehicles.

- Emission, noise and carbon monoxide levels should be within human tolerance and safe limits prescribed, for all automobiles.

- Tyres are important feature of vehicles in skidding, braking, providing friction, etc. They should be wear, cutting, tearing resistant providing strength, flexibility, durability and high friction.

7.8 OPERATING CHARACTERISTICS OF VEHICLES

Operating characteristics of vehicle are necessary for highway and traffic engineers in geometric design of roads, economic analysis and over all performance.

7.8.1 Resistance and Power Requirements

The resistances to the motion to be encountered by a motor vehicle in operation can be divided in two categories.

   i. Those comprising of the forces which are always present. These are due to inertia, rolling and air resistance.

  ii. Second category of forces which arise in particular case such as on grades, for change of speed (braking and accelerating).

     The power is developed by the engine, to overcome these resistances to motion, for maintaining the required speed. The resistances to be encountered are following:

   a. Inertia force

   b. Rolling resistance

   c. Air resistance

   d. Grade resistance

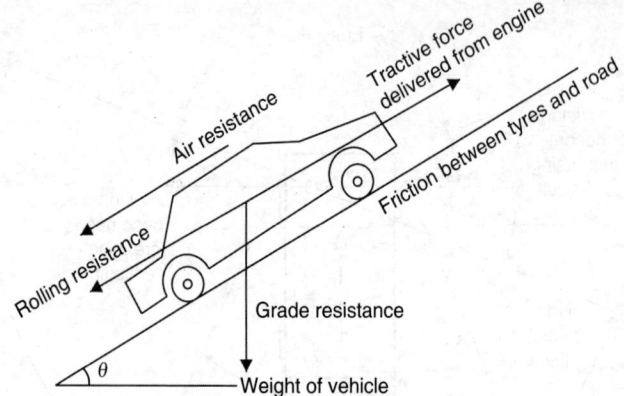

Fig 7.4: Resistances encountered by a vehicle uphill

Figure 7.4 shows the forces acting on a vehicle and resistance encountered when climbing up, on a road.

(a) Inertia Force (Fi)

It is the force required to bring a vehicle in motion. This force is given by the product of mass and acceleration.

$$Fi = m \times a$$

Fi = Inertia force

m = mass of the vehicle kg.

a = acceleration m/sec^2

(b) Rolling Resistance (R$_r$)

Once the vehicle is in motion, a rolling resistance comes into play due to:

 i. Surface roughness of the pavement

 ii. Irregularities of road surface

 iii. Friction in the moving parts of the vehicle.

Depending upon the type of road surface values of coefficient of rolling resistance are used, as shown in Table 7.3.

The rolling resistance is given by:

$$R_r = mfg$$

Where R_r = Rolling resistance N

m = mass of the vehicle kg.

f = Coefficient of rolling resistance

g = acceleration due to gravity m/sec^2

Rolling resistance is fairy constant up to speed of about 100 km. per hour. At higher speeds it increases as speed is increased further.

(c) Air Resistance (R$_a$)

The air resistance of a vehicle depends upon the following:

 i. Density of air

 ii. Speed of the vehicle

Table 7.3: Coefficient of rolling resistance (IRC)

Road Surface/Type	Coefficient of rolling resistance
Cement concrete and bituminous surface	0.01
Premix carpet (good condition)	0.16
Premix carpet (bad condition)	0.022
Gravel	0.046
Earth	0.055
Water bound macadam (good condition)	0.025
Water bound macadam (bad condition)	0.037

   iii. Frontal area of the vehicle

   iv. Direction of wind (stream flow of air which is displaced)

The air resistance is given by the formula:

$$R_a = KAV^2$$

Where R_a= Resistance due to air N

    A = Frontal area of vehicle Sq. m

    K = A constant-coefficient depending upon type of vehicle, varies from 0.36 to 0.48

    V = speed of the vehicle relative to air m/sec

At higher speeds, air resistance becomes very high being a function of square of the speed and is thus a large component in power consumption.

(d) Grade Resistance (R_g)

When a vehicle climbs an inclined plane, force is required to overcome the effect of gravity. The force developed by gravity is proportional to slope of the plane and weight of the vehicle. If θ is the angle of the slope, the component sin θ can be assumed to be equal to tan θ, so that for each one percent grade, the exerted force is equal to one percent of the weight of the vehicle.

$$R_g = \frac{mg \cdot i}{100}$$

Where R_g = force exerted to climb the grade

    m = mass of the vehicle.

    i = percent of slope

    g = acceleration due to gravity m/sec^2

When vehicle descends slope is negative. Negative force means no additional effort required to move downwards.

Extensive research is being conducted to develop accurate, economical methods of improving pavement coefficient of friction, by tyre design, design of surface course, etc. Improved automobile designs with more reliable and cheaper power (H.P) and reduced cost of maintenance and fuel are being developed from time to time.

Power Requirement of Vehicles

Vehicle power is required in starting, developing speed, accelerating, and climbing the grade. The total horse power required by a vehicle is the sum of the horse power required for rolling resistance, air-resistance and grade resistance.

$$
\begin{array}{l}
\text{Total H.P. required} \\
\text{(Power developed} \\
\text{by engine)}
\end{array}
=
\begin{array}{l}
\text{H.P. to overcome} \\
\text{Rolling resistance} \\
(R_r)
\end{array}
+
\begin{array}{l}
\text{H.P. to overcome} \\
\text{Air resistance} \\
(R_a)
\end{array}
+
\begin{array}{l}
\text{H.P. required for} \\
\text{grade resistance} \\
(R_g)
\end{array}
+
\begin{array}{l}
\text{H.P. required} \\
\text{for inertia} \\
\text{force for} \\
\text{acceleration} \\
(F_i)
\end{array}
$$

7.9 BRAKING AND ACCELERATING CHARACTERISTICS OF VEHICLES

7.9.1 Braking Ability of Vehicles

Because of sharp and rapid stops brakes must be applied to stop/slow down the vehicle as rapidly as possible in view of the traffic conditions and driver's judgement to prevent an accident. It is important to study the braking ability/slowing rate of the motor vehicles, for safety on roads, and skidding of vehicle after sudden application of brakes. Vehicle decelerates as brakes are applied. Only in emergency driver utilizes maximum deceleration to stop.

After application of the brakes the distance required to stop the vehicle is given by (neglecting resistances to normal motion)

$$ D_b = \frac{V^2}{2g(f \pm \tan\alpha)} $$

Where D_b = Braking distance, m

  V = speed of the vehicle Km/h.

  g = acceleration due to gravity 9.8 m/sec^2

  f = coefficient of friction between tyres and pavement

 $\tan\alpha$ = slope of gradient (+ for going up, – for going down)

On a straight road ($\tan\alpha = 0$), V in km per hour is will be given by:

$$ D_b = \frac{V^2}{254\,f} $$

7.9.2 Acceleration Characteristics

Acceleration data is used to determine the vehicle characteristics in situation such as

  i. The time to cross an intersection from a standing point

  ii. The distance required to pass other vehicles

  iii. The gap acceptance, etc.

Acceleration of a vehicle depends upon horse power, speed and grades. There are two theories of acceleration:

  a. Uniform acceleration,

  b. Non-uniform acceleration.

  a. **Uniform Acceleration Theory:** The basic equations of motions, reproduced below, are based on the assumption that acceleration of the vehicle remains constant throughout the journey with time:

$$V = V_0 + at \qquad \qquad \dots (i)$$

$$S = V_0 t + \frac{1}{2} at^2 \qquad \qquad \dots (ii)$$

$$S = \frac{V^2 - V_0^2}{2a} \qquad \qquad \dots (iii)$$

Where   V = Final speed of vehicle

V_0 = Initial speed of vehicle

a = Acceleration or retardation

t = Time

S = Distance travelled.

Above equations are based on uniform acceleration theory, but in actuality the acceleration is not uniform. The acceleration rate is maximum at low speeds. Actual acceleration rates must therefore be determined by tests. In a study of passenger cars, on the average, the instantaneous acceleration rates were found to be 4 Km/h/sec at speeds between 32–56 kph, decreasing to 3.2 km ph/sec at speed of 104 Km/h.

  b. Non-uniform Acceleration Theory: The acceleration of a vehicle is not constant, it varies inversely with speed. The acceleration behaviour of a vehicle may be described by the expression:

$$\frac{dV}{dt} = \alpha - \beta V \qquad \qquad \dots (i)$$

where

$$\frac{dV}{dt} = \text{Acceleration (rate of change of speed)}$$

$$\alpha, \beta = \text{Constants}$$

$$V = \text{Speed.}$$

Equation (i) can also be written as:

$$dt = \frac{dV}{\alpha - \beta V}$$

Integrating both sides: $\displaystyle\int_0^t dt = \int_{V_0}^{V} \frac{dV}{\alpha - \beta V}$

When $t = 0$, speed is V_0 – initial velocity

$t = t$, speed is V – final velocity.

For the above equation (i)

When speed = 0, acceleration is maximum = ∞

When acceleration = 0 speed is maximum = $\dfrac{\alpha}{\beta}$

By integrating, we get,

$$t = -\frac{1}{\beta} \left[\log(\alpha - \beta V) \right]_{V_0}^{V}$$

$$= -\frac{1}{\beta}\left[\log^{(\alpha-\beta V)} - \log^{(\alpha-\beta V_0)}\right]$$

$$= -\frac{1}{\beta}\log\frac{\alpha-\beta V}{\alpha-\beta V_0}$$

$$\log\frac{\alpha-\beta V}{\alpha-\beta V_0} = -\beta t$$

or $$\frac{\alpha-\beta V}{\alpha-\beta V_0} = e^{-\beta t}$$

$$\alpha - \beta V - \alpha e^{-\beta t} + \beta V_0 \times e^{-\beta t} = 0$$

divided by β

$$\frac{\alpha}{\beta} - V - \frac{\alpha}{\beta}e^{-\beta t} + V_0 e^{-\beta t} = 0$$

or $$V = \frac{\alpha}{\beta}(1-e^{-\beta t}) + V_0 e^{-\beta t} \qquad \qquad \dots (ii)$$

This is the expression for the speed of the vehicle at any time t. By substituting this value of V in equation (i), we can get the acceleration time relationship.

$$\frac{dV}{dt} = \alpha - \beta V = \alpha - \beta\left[\frac{\alpha}{\beta}(1-e^{-\beta t}) + V_0 e^{-\beta t}\right]$$

$$\frac{dV}{dt} = (\alpha - \beta V_0)e^{-\beta t} \qquad \qquad \dots (iii)$$

Since $V = \dfrac{dx}{dt}$, its integration provides the equation of distance, as a function of time, eqn. (*ii*)

$$V = \frac{dx}{dt} = \frac{\alpha}{\beta}(1-e^{-\beta t}) + V_0 e^{-\beta t}$$

$$x = \frac{\alpha}{\beta}t - \frac{\alpha}{\beta^2}(1-e^{-\beta t}) + \frac{V_0}{\beta}(1-e^{-\beta t}) \qquad \qquad \dots (iv)$$

Typical acceleration rates in normal use are:

Medium cars	3.8 km/h/sec.
Commercial vehicles	0.75–2 km/h/sec.
Sports cars	12–16 km/h /sec.

7.10 CHARACTERISTICS OF SLOW MOVING VEHICLES

On Indian roads, slow moving traffic consisting of animal drawn vehicles and man pulled cycle rickshaws, also shares the road space. Their dimension and speed vary considerably. Length of cycle rickshaws is about 2.60 m, while length of animal drawn (bullock, horse, donkey) carts vary from 4 m to 6 m. There is large variations in speeds. A bullock cart has slow speed of 3 to 4 kms per hour while horse drawn vehicles have a speed of 10 to 15 kms per hour. Hand carts pulled by manual labour, have much lower speeds 2 to 3 km per hour.

For providing access to various land uses, among non-motorised means of transportation, cycle rickshaws are pre dominant mode in many Indian cities. The advantages of cycle rickshaws are:

- Cheap and ideal for short distance travel.
- Non polluting and eco friendly
- No requirement for petrol or diesel.
- Cheap in initial cost
- Provide employment to large number of people
- Suitable to move on narrow city roads

Cycle rickshaws are used as a feeder service and for shopping, business and education purposes. The cycle rickshaws trips are usually 1.5 to 3 kms long on the average. Speed of cycle rickshaws are low varying from 5-8 kms per hour. Volume of cycle rickshaws has negative impact on speed of traffic flow on the roads, this increases congestion, cost and pollution.

Total elimination of cycle rickshaws may not be very easy in view of above advantages, however their use be restricted to identified internal roads of the city. Efforts are being made in this direction.

An alternative to cycle rickshaws, being tried is battery operated auto rickshaw. Battery operated auto rickshaws has a seating capacity of 6-8 while a cycle rickshaw has seating capacity of only 2 passengers. Speed of this mode is also greater as compared to cycle rickshaw, though initial cost of battery operated rickshaws is much more.

Because of their varying characteristics, slow speeds and heterogeneous character of slow moving vehicles, traffic speeds and capacity of roads are dropped, resulting in congestion and unsafe driving conditions. Frequent braking, slowing, starting and stopping creates confusion and irritation among drivers and vehicle operating costs are increased.

ILLUSTRATIVE EXAMPLES

Example 1. A vehicle is accelerating on a gradient of 1.5 percent (upwards) with a rate of 0.8 m/sec^2, from initial speed of 15 to 25 Km/h. Calculate the various resistances encountered by the vehicle and the horse power required, using the following data:

  i. Mass of the vehicle = 1500 kg
 ii. Coefficient for rolling resistance = 0.02
iii. Frontal area of vehicle = 3.5 m^2
 iv. Coefficient for air resistance = 0.40 kg/m^3

Solution:

$$\underset{\text{Total resistances encountered }R_T =}{} \underset{\substack{\text{Rolling} \\ \text{resistance} \\ \text{(Rr)}}}{} + \underset{\substack{\text{Air} \\ \text{resistance} \\ \text{(Ra)}}}{} + \underset{\substack{\text{Grade} \\ \text{resistance} \\ \text{(Rg)}}}{} + \underset{\substack{\text{Inertia} \\ \text{resistance} \\ \text{(Ri)}}}{}$$

$$= m.g.f + Ca.AV^2 + m.g.i. + m.a$$

here

$$mg = 1500 \times 9.81 = 14715 \text{ kg.}$$

$$V = \left(\frac{15+25}{2}\right) \times \frac{1}{3.6} = 5.55 \text{ m/sec}$$

 i. Rolling resistance (Rr) = m.g.f. = 14715 × 0.02 = **294.3 N**
ii. Air resistance (Ra) = Ca.A.V^2 = 0.4 × 3.5 × 5.55 × 5.55 = **43.12 N**

iii. Grade resistance (Rg) = mgi = $14715 \times \dfrac{1}{1.5 \times 100}$ = **98.1 N**

+ve for upgrade

iv. Inertia resistance (Ri) = m.a = 1500 × 0.8 = **1200 N**

+ve for accelerating

Total resistances	= 294.3 + 43.12 + 98.1 + 1200 = **1635.52**	
Power developed	= 1635.52 × 5.55	
	= 9077.136 W	

1 HP = 735 W

$$\therefore \qquad HP = \frac{9077.136}{735} = 12.34 \approx 14$$

Example 2. A driver traveling at 50 km per hour behind another vehicle decides to overtake it and presses the accelerator. The accelerating behaviour of the car is described by the following equation:

$$\frac{dV}{dt} = 1.22 - 0.015\,V$$

Where V is speed in m/sec and t is time in seconds. Determine:

a. The maximum speed of the vehicle
b. The maximum rate of acceleration
c. The rate at which the vehicle is accelerating after 3 second
d. The time it takes the vehicle to reach a speed of 120 Km/h.

Solution:

(a) The speed of the vehicle is maximum when acceleration is zero,

i.e.,
$$\frac{dV}{dt} = 0$$

$$0 = 1.22 - 0.015\,V$$

$$\therefore \qquad V = \frac{1.22}{0.015} = \textbf{81.33 m/sec (292 km p. hr.)}$$

(b) Acceleration is maximum when speed is zero (V = 0)

$$\frac{dV}{dt} = 1.22 - 0.015 \times 0 = 1.22 \text{ m/sec}^2 \text{ (4.39 Km/hr/sec)}$$

(c)
$$\frac{dV}{dt} = (\alpha - \beta V_0)\,e^{-\beta t}$$

here
$$V_0 = \frac{50 \times 1000}{3600} = 13.89 \text{ m/sec.}$$

$$= (1.22 - 0.015 \times 13.89)^{-0.015 \times 3}$$

$$= \textbf{0.967 m/sec}^2 \ \textbf{(3.48 km/hr/sec.)}$$

(d) V = 120 Km/h = $\dfrac{120}{3.6}$ = 33.33 m/sec.

$$e^{-\beta t} = \frac{\alpha - \beta V}{\alpha - \beta V_0} \qquad \text{or} \qquad t = -\frac{1}{\beta} C_N \left(\frac{\alpha - \beta V}{\alpha - \beta V_0} \right)$$

$$= -\frac{1}{0.015} \times (-0.34) = -\frac{1}{0.015} C_N \frac{1.22 - 0.015 \times 33.33}{1.22 - 0.015 \times 13.87}$$

$$= \textbf{22.67 seconds.}$$

Example 3. The driver of a vehicle traveling at 80 Km/h requires 8.5 m less to stop after applying the brakes up a grade, than when traveling down the same grade.

If the coefficient of friction is 0.55 calculate:

a. The percent of the gradient.

b. The braking distance on the down grade.

Solution:

(a) $D_b = \dfrac{V^2}{2g(f \pm \tan\alpha)}$ $\qquad$ Let $G = \tan \alpha = $ slope

$$\therefore \quad \frac{V^2}{2g(f-G)} - \frac{V^2}{2g(f+G)} = 8.5$$

$$\frac{V^2}{2g}\left[\frac{1}{(f-G)} - \frac{1}{(f+G)}\right] = 8.5 \qquad\qquad V = 80 \text{ Km/h.}$$

$$\frac{V^2}{2g}\left[\frac{f+G-f+G}{f^2-G^2}\right] = 8.5 \qquad\qquad = \frac{80 \times 1000}{3600} = 22.22 \text{ m/s.}$$

or $\qquad\qquad \dfrac{2G}{f^2-G^2} = \dfrac{8.5 \times 2 \times 9.81}{(22.22)^2} = 0.338$

$$2G = 0.338 f^2 - 0.338 G^2$$
$$0.338\,G^2 + 2G - 0.338\,(0.55)^2$$
$$0.338\,G2 + 2G - 0.102 = 0$$

$$G = -2 \pm \sqrt{\frac{4 + 4 \times (0.338) \times (0.102)}{0.676}} = \frac{-2 \pm 2.34}{0.676}$$

$$= -5.96 \text{ or } 0.050$$

Braking distance $\qquad\qquad \textbf{G = 5\%}$

(b) $\quad D_b$ (downgrade) $= \dfrac{22.22^2}{2 \times 9.81(0.55 - 0.05)} = \textbf{50.34 m.}$

Urban Traffic and Transportation Planning and Management

8.0 PREAMBLE

Transport planning is a systematic technique of understanding traffic and transportation characteristics, problems and requirements with the objective of producing safe, efficient and convenient transportation system which will meet current and future needs and preferences of an individual and the community and shall also promote social and economic development. Planning involves arrangement of activities within an environment which will yield benefits to both individuals and community.

The specific objective of planning are:

- To have most suitable type of transportation system out of available fund.
- To develop an integrated road transport system.
- To coordinate the total transport development with an emphasis on inter-modal concept.
- To optimise expenditures and revenues, ensuring that the benefits and costs of road transport programmes fall equitably on the community.
- To promote economic activity and efficiency through reduced cost of transportation.
- To plan for phased development programmes and stage construction techniques for promotion of urban and regional development.
- To know how to and when to improve old roads or construct new ones to meet future needs.
- To establish priorities of construction and maintenance of roads.
- To make plan for satisfying current and future needs
- To reduce accidents.
- To preserve and improve the environment.
- To review, revise and refine the planning as necessary.

A feature of all planning must be the ability to change, as transport planning is an integral part of urban planning.

The activities of people, by necessity become separated, and are spacially distributed at location dependent in terms of time, cost, convenience, social-economic conditions and some intrinsic site values.

Each land use, whether it be a school, factory, house or park is a generator of traffic. The traffic may be a pedestrian, cyclist or vehicular. Traffic flows will be developed either locally or nationally and will be distributed throughout the hours of the day, in varying proportions. A planned transport system allows better freedom to choose where to live, work or perform other activities and provides opportunities of selecting the mode of transport, route to be used for any activity. The various type activities are served with physical transport facilities such as roads, highways, railways, airports or harbours. Transport planning therefore primarily deals with the space and transport facilities related to human activities.

Figure 8.1 shows elements of transport system, with response to travel and Fig. 8.2 shows the hierarchy of travel choices

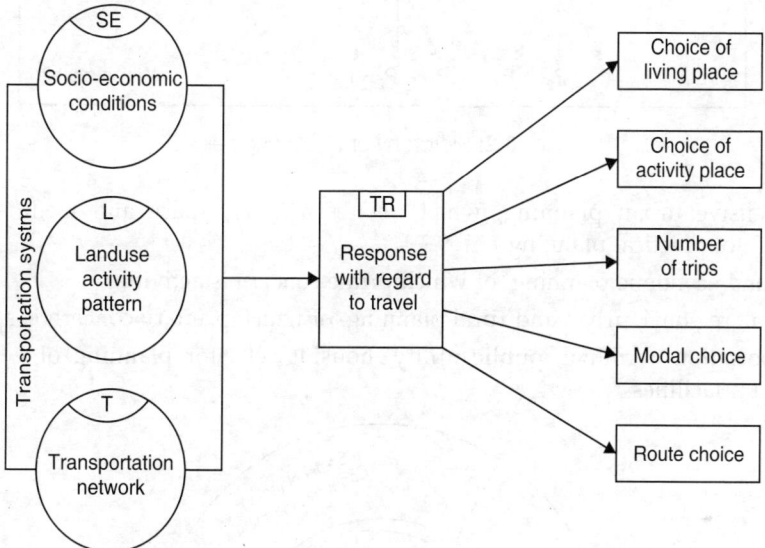

Fig. 8.1: Transportation system elements and travel response

Planned development of transportation system has positive impact on the society, resulting in reduced travel times, vehicle operating costs, accidents and fuel expenses.

8.1 URBAN TRANSPORT SYSTEM PLANNING

Traffic is growing faster on urban roads as cities and towns provide greater opportunities and comfortable living to workers in every sector of economy. The rapid movement of people either by mass transportation or in individual form of transport, has replaced the need to house workers in close proximity to their employment. This trend has encouraged further urban-isation causing not only traffic congestion and shortage of housing and land, but also creating difficulties in provision and administration of adequate quality level of services to the population. The growth and development of cities/towns called "Agglomerating Effect" is largely a legacy of past social, industrial and communicative pattern brought mainly by the needs to locate workers advantageously to the development and growth of industry. More often than not, the urban planners neglect transport planning. When system tends to fail and traffic problems increase, transport planning is sought.

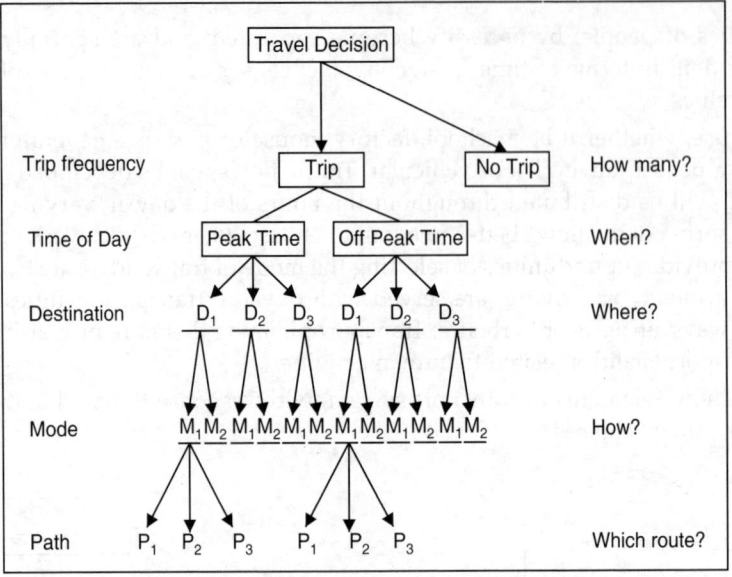

Fig. 8.2: Hierarchy of travel choices

Comprehensive urban planning, which relates activity, space and facilities, has three different kinds of practical planning:-

i. Social and economic planning, of which prime concern is activities.

ii. National, regional, urban and rural planning, of which primary concern is space, and

iii. Transportation, terminal, public utility, housing, etc. for planning of which primary concern is facilities.

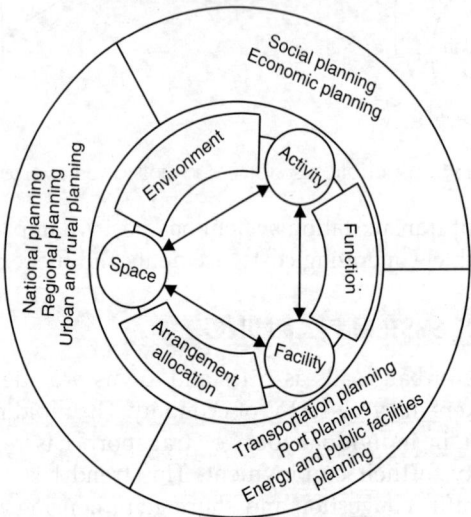

Fig. 8.3: Components of comprehensive urban planning

Figure 8.3 shows above components of comprehensive urban planning. Urban transportation system planning (UTP) is one important element of comprehensive urban planning- which is both a social and physical process, requiring high degree of interaction of professionals in

various fields. UTP intends to reach certain goals related to the development of urban complex. Transport planning proposes a system for transport which solves or reduces existing traffic and transportation problems, and provide safe and efficient movement of people and good, from one place to another. Transport planning aims at solving the traffic and transportation problems on short and long term basis.

Growth of urbanisation and population in towns/cities, rise in per capita income and development of economic activities and need to travel long distances has resulted in high ownership of private and public vehicles. With an unplanned or out dated planning and hapazard growth of cities, several urban transportation problems, are manifesting in high level of congestion, noise pollution, traffic fatalities and other severe traffic and transportation problems, like slowing down of average vehicles speed , wasteful consumption of fuel, etc.

8.2 URBAN TRAFFIC AND TRANSPORT PROBLEMS

Urban traffic in India consists of fast as well as slow animal drawn vehicles, man pulled rick-shaws, cycles, handcarts, etc. on the same carriage way, resulting in several type of problems. The most common problems and their ill effects are as under:

8.2.1 Congestion and Delay

As the road space has not been increased in proportions with the increase in number of vehicles, most of the urban roads are congested beyond tolerable limits. The main cause of congestion can be attributed to the inadequate capacity of roads, mixed traffic, rickshaws, auto-rickshaws, illegal parking, pedestrian movement, loading and unloading, encroachments on the road surface etc.

Increased congestion results in low operating speeds, delays, reduction in level of service, environmental and noise pollution, and other external and emotional effects. Heavy congestion occurs during peak hours on the urban and semi-urban roads. Buses are adversely affected by congestion as bus schedules are disrupted, affecting frequency of buses. The failure of transportation system is most poignantly felt in the problem of congestion on roads. Congestion increases the energy cost per kilometer , and vehicle operation costs.

8.2.2 Parking Problems

Increasing demand for parking of vehicles, limited parking supply, and absence of a parking policy are impediments to smooth flow of traffic specially in and around commercial areas and activity centers, of a city.

Parked vehicles occupy space intended for vehicular movement and reduce the width and capacity of the road resulting in increased congestion, pathetically low speeds, and accidents while parking and unparking. Parking generators do not provide their own parking space facilities, creating problem on the roads around them. Slow moving vehicles, e.g. rickshaws, auto-rickshaw make their parking on the road and near intersections, further complicating the problem. Non-availability of off-street parking, gives rise to traffic problem, which needs serious attention.

8.2.3 Problem of Encroachment on carriageway

Encroachments on urban roads and intersections, results in numerous circulation problems for the traffic. Commercial activities over spilling from adjacent stalls and shops, selling activities by hawkers at intersections, buses and mini buses loading and unloading on the main carriageway, workshops and repair shops for vehicles extended on the pavement, etc are the

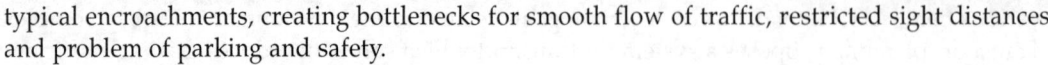

typical encroachments, creating bottlenecks for smooth flow of traffic, restricted sight distances and problem of parking and safety.

8.2.4 Lack of Pedestrian Facilities

Lack of footpaths and their inadequacy forces pedestrians to walk on the carriageway, exposing them to risks. Widening of roads to meet the demands of increasing number of vehicles, has reduced or removed the walking space for the pedestrian on either side of the road. Even where footpaths are provided they are encroached upon by various activities by the shopkeepers.

Pedestrian crossing facilities such as subways or foot over bridge are lacking on urban roads, forcing pedestrian to cross at unsuitable places along the road. The problem of pedestrian vehicle conflict are quite serious resulting in increasing pedestrian accidents every year.

8.2.5 Problem at Railway Level Crossing

Due to developing and increasing in size of cities, some of the railway level crossings are now located in the densely populated urban and residential areas of the city. With increase is number of trains per day and road vehicles, the delays at various level crossing are increasing, long queues of traffic are observed at these crossing.

8.2.6 Problem of City Transportation System (Buses)

The bus fleet consisting of public and private buses and mini buses is inadequate, unplanned, resulting in overloading as demand exceeds the supply. When an accident happens, the number of persons injured is usually high.

Lack of infrastructure facilities for buses, disorganised private sector operation of buses, lengthy and zigzag bus routes, inadequate supply of buses, lack of priority to buses on roads, inadequate and improperly located bus stops, etc. are main causes of problems of bus transportation. The preference for personalised mode of transport is due to inefficient public mass transportation system.

8.2.7 Problem of Heterogeneity of Traffic

Slow moving vehicles particularity cycles and man-pulled rickshaws in some cities/towns, are in significant proportions in traffic flow, along with vehicular traffic, on the same carriageway creating disorganised and unsafe driving conditions. Slow moving vehicles violate systematic file movement on roads. Most urban poor can not afford any private motorised transport at all and many can not even bear low fares on public transport, they are the forced to walk, or use cycles, with no separate right of way, suffering from pollution, safety and congestion problems.

Slow traffic occupies the road for longer time than fast moving traffic thus reducing the capacity of the road. They also reduce the overall speed on the road. Varying dimensions of traffic, different speeds, and character of heterogeneous traffic creates problems of capacity, safety and comfort.

8.2.8 Lack of Traffic Sense

Awareness towards traffic problems and responsibility as a citizen are very much lacking among the road users and drivers. Vehicle owners have tendency to park their vehicle anywhere they like on the main road, without any consideration to the problems it might create. Lack of traffic sense and knowledge of traffic rules creates several problems of road use. Not following lane driving rules and reckless driving aggravates the congestion and safety problem.

8.2.9 Inadequate Traffic Control and Management

In most of the cities traffic monitoring and control is lacking, which results in indisciplined behaviour by road users, encroachments, unauthorised parking, improper crossing and movement on roads, affecting the road use and safety.

8.2.10 Poor Maintenance of City Roads/Bridges

The conditions of the road network and riding quality is deplorable in many cities. Corrugations, potholes, rough riding surface are part of the high traffic roads. Bridges need widening along with road.

Due to inadequate funds allotted to maintenance, physical conditions of city roads and bridges result in poor riding quality and unsafe driving conditions. Large resources are required for sustainable development of urban transport, to keep up with the growth rate of traffic and utilization of roads.

8.2.11 Inadequate Drainage Facilities

Besides poor maintenance, drainage facilities are not properly designed on many roads. The existing drainage system is either inadequate placed at an improper level or not kept clean with the result that the storm water remains standing on the carriageway, leading to formation of cavities, potholes etc. Most of the towns/cities are developing fast with inadequate attention towards drainage plans. Presence of water on the paved area damages the pavement structure very fast and is problematic to users and road safety

8.2.12 Multiple Agencies Claiming Right on Roads

It is a common sight on urban roads that roads are cut for provision of electrical cables, telephone lines, sewer lines, etc, and not repaired for a long time. This results in unsafe driving conditions and accidents specially during the night time, as warning or informatory signs are rarely put on such locations. City witnesses many multiple agencies each claiming right of way for some purpose on the road.

If joints are not made completely watertight, resulting percolation creates further problems. It is the road and the transport system which get adversely affected due to operation of multiple agencies and lack of coordination between them.

8.2.13 Problem due to Advertisements

Advertisers, builders frequently place advertising hoardings, banners, posters etc close to the roads, intersections or at other dangerous locations, in urban areas with no concern for the traffic.

Attractive posters and sign boards, with bright lights in the night time draw attention of the drivers of fast moving vehicles causing accidents. It also disturbs the aesthetic value.

8.2.14 Unauthorised Land Uses

One of the major traffic problem is that residential buildings, are used for business and commercial activities like shops, hostel, institutes, training centers, etc. Many residential houses are converted to multistoried complexes or flats by the owner creating parking and other traffic problems.

8.2.15 Pollution and Noise Problem

Frequent stops and start operations due to congestion and poor road surface, create noise and emission of harmful gases, polluting the environment and affecting health of the people. The

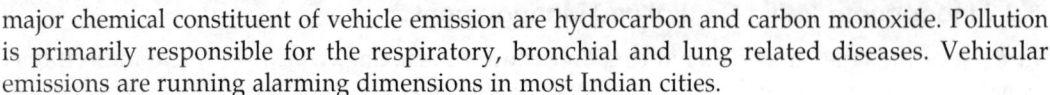

major chemical constituent of vehicle emission are hydrocarbon and carbon monoxide. Pollution is primarily responsible for the respiratory, bronchial and lung related diseases. Vehicular emissions are running alarming dimensions in most Indian cities.

8.2.16 Road Accidents and Safety

The rate of traffic related accidents involving death, injury and property damage is increasing mainly due to congested road, undisciplined drivers, heterogeneity of traffic and several other problems mentioned above.

Besides above problems, in some cities peculiar problems like stray cattles walking freely on roads, informal weekly markets on certain sections of the roads, violation of speeds, lack of coordination between different agencies responsible for constructing roads, managing traffic and other facilities, further aggrevates the problems of traffic safety. To avoid and control accidents transportation system must be planned in most conscious way.

For scientific urban transportation planning (UTP) it is important to have accurate, adequate and real time data and therefore a data collection, storing, managing and updating system must be established to get relevant and reliable data, in all cities and towns. Accurate data base will help in planning and solving the problems on short and long term basis.

8.2.17 Lack of Appreciation of Transportation Planning

Transport planning of a city can not be undertaken in isolation. Comprehensive planning for the city covering entire gamut of transportation vis-a-vis land use development should be undertaken. City grows haphazardly due to certain unaccounted factors and actual development is found to be different from the pattern envisioned in the master plan. No attention is paid to road infrastructure planning with development of the city, like sewer lines, water supply lines, telephone cables, etc. which will be using the right of way of the road.

8.2.18 Unauthorised dumping on Road

Construction material, garbage etc. is dumped by the roadside house owners on the road surface, shoulders and side drains, affecting the drainage, road-surface and flow of the traffic. There is no agency to penalise such offenders, at times municipalities themselves are responsible for such an act.

These traffic problems are too complicated, to be solved by any single isolated treatment. The road system and development on either side of the road get so saturated that rebuilding or widening is hardly possible. Traffic management measures, using traffic engineering techniques, enforcement of traffic laws and regulations, education and publicity, regulatory measures, can improve the present existing situation for sometime till a permanent solution of the problems is planned through comprehensive planning of land uses and transportations system on long term basis.

8.3 STEPS IN URBAN TRANSPORT PLANNING PROCESS

UTP begins by defining the problem and establishing the urban objectives and standards for best-use and balance of all modes of transportation. The problem of urban transportation is a complex one comprising many problems as such simple solutions of isolated improvement do not work. There is therefore a need, for comprehensive understanding of the problems and finding solutions to it on a very broad-base footing.

The basic approach and procedural stages for conducting an urban transport planning study involves following stages.

i. **General concept planning:** This defines the pattern of urban development in terms of broad social and economic goals.

ii. **Development or strategic planning:** This defines specific areas of land use, broad transport strategy, broad locations of transport facilities and future needs.

iii. **Comprehensive transport planning:** In this more specific transport plans are tested. The result defines the distribution of travel demand and the specific location of transport routes and corridors.

iv. **Highway project planning and route design:** Selected specific corridors or projects are detected.

The basic sequence of steps in transport planning are illustrated in Fig. 8.4 and briefly described as follows:

a. Inventory of existing conditions

b. Land use forecasts

c. Travel forecasts

d. Alternative plan preparation and testing

e. Objective and standards

f. Evaluation

g. Recommended plan and implementation

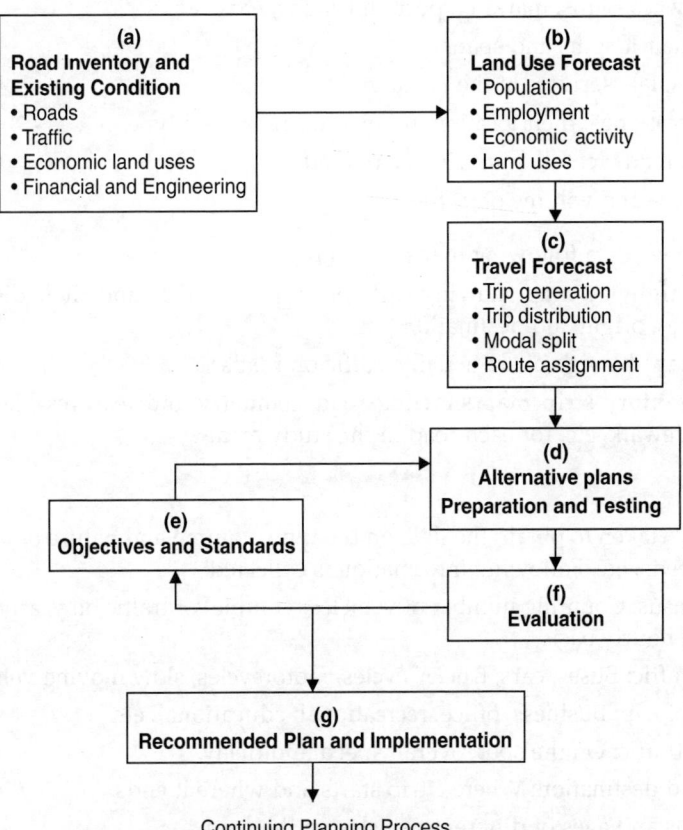

Fig. 8.4: Steps in transportation planning process

8.3.1 Inventories of Existing Conditions

For systematic urban transport planning, it is necessary to have a future vision, stated in form of aims and objectives, towards which the efforts will be devoted to achieve. Well defined and clearly understood objectives, consistent with the requirements and provisions provide guidelines in developing specific methodology for achieving long and short term plans

To understand the actual existing problems, to be addressed in planning, inventories and data collection is undertaken. Some of the required data on the road network, travel and traffic, economic, finance and engineering may be obtained through the concerned departments and agencies. Specially designed studies and surveys are conducted for collecting the data required for urban transport planning. Following basic data is required:

i. Road inventory data

This includes collection of data on existing roads such as:-

- Clear demarcation of various states, regions, or areas
- Man made and natural features adjacent to roads
- Designation and classification of road
- Road width, pavement width, surface type and condition
- Side walks and kerbs adjacent to roads
- Alignment and gradient of roads
- Safety devices, signs, marking, post, milestone, etc
- Bridges their lengths and spans
- Reservoirs, lakes, rivers in the vicinity
- Railway crossings, their location, number of tracks and type
- Nature of road, i.e. toll, municipal, city road
- Type, nature and volume of traffic.

Using the above data following maps are prepared:
- **General highway map** showing different type of roads and their distinctive features, their names, origin and destination.
- **Traffic map** showing average daily traffic on roads
- **Road inventory strip map** showing road geometry and features likes curves, grades, width, sidewalk, etc. for each road in the study area.

ii. Traffic Data

Studies are undertaken to obtain the data on the type, volume and nature of traffic, their origin-destination, speeds, etc. Following information is collected:

- Traffic census: Counting number of vehicles (volume) of traffic on yearly, monthly, weekly, daily and hourly basis.
- Type of traffic: Buses, cars, trucks, cycles, motorcycles, slow moving vehicles, etc.
- Purpose of trip: Business, office, recreational, educational, etc.
- Speed of traffic: On the spot, overall speed and delay.
- Origin and destination: Where a trip starts and where it ends.
- Conditions and ages of different vehicles on the road
- Accident data: To be collected from traffic police and other concerned agencies.

- Miscellaneous information such as axle loads, type of tires, overloading, etc.
- Parking practices and demands.
- Special data and information on bus passenger services, fares, their utilisation, etc.
- Data on commercial vehicles and their utilisation.

Various types of maps, diagrams, histograms are prepared to present above traffic data on volume, speed, O-D and parking, etc.

iii. Economic data

Economic data determines the economic status and development of the city/locality. It includes information on:

- Dispersal and location of the population
- Town/cities classified on population basis
- Location of existing industries and sources of material
- Anticipated future development of industries
- Classification of area as agricultural, industrial, etc.
- Location of marketing centers, their size and goods sold
- Number of people going for different works and their mode of transport
- Per capita income of the people living in the area

This information helps in planning the road system based on population demands, development, mobility and economic life of the people.

iv. Financial planning data

This data concerns with financial aspects such as allotment of funds, income from different roads and other sources of revenue. Basic data includes:

- **Statistics on income such as:**
  - Rate on import of motor vehicles and parts
  - Revenue from customs, excise duty on tire, etc.
  - Revenue from motor vehicle taxes
  - Registration fee of vehicle
  - Sale tax on motor vehicle
  - Income from other taxes, etc.
- **Motor vehicle use statistics such as:**
  - How ownership of vehicles is distributed among different types of people and areas
  - Age of various automobiles and weights carried by commercial vehicles
  - Annual travel, trip length, purpose of trip, road used, etc
  - Rate of fuel consumption by different vehicles and their age
  - Transport mode used by different of employee, etc
- **Statistics on road life such as:**
  - Facts regarding life of various roads, their construction and maintenance cost
  - Salvage value of various road elements
  - History of each section of the road, regarding maintenance and repair
  - Condition of pavements, etc

- **Statistics on fund allocation such as:**
    - Allocation of funds to different roads and regions.
    - Priority establishment keeping in view the political, financial obligations and traffic demands, etc.

v. Engineering data

The engineering data, to be collected includes.
- Topography of the area under study.
- Type of soils in the area
- Available materials of construction of roads.
- Methods of construction and maintenance
- Equipment availability for road construction
- Drainage problems in the area
- Availability of labour
- Road life studies, etc.

This type of data gathered/collected on the basic present day conditions is analysed and quantified as models by mathematical and statistical techniques. It is used to predict future demands for planning.

8.3.2 Land use Forecast

Above inventories describe the study area as it is at present. The next step is to produce land use predictions for the design year which is usually selected 15 to 20 years ahead. A land use plan for the design year is prepared on the basis of extrapolation of existing trends or on the levels of economic activity. Mathematical models such as intervening opportunities model are often used for landuse predictions, based on overall growth rate and available opportunities to grow.

The growth of land use of course is dependent upon transport facilities provided. Thus the process input will depend upon the process output and a proper solution of the system would need number of iterations. To start with some future transport facilities are assumed to predict land uses.

Land use models are often based on employment, service potential, population or economic activity. Such models first predict the population and residential areas for location of a service industry, consistent with observed travel behavior.

The economic activity type model for landuse prediction involve determination of residential locations on a competitive market basic. Models must be able to accommodate any policy effects and changes. Population, employment and income forecasts bases on the above land use plans are established for the design year. The alternative future land use patterns are also supposed in this stage. Then the estimates of future patterns of movement are derived, associated with particular set of landuse proposal.

8.3.3 Travel Forecast

Travel forecasting is a computer dependent mathematical process, based on present day observation, where by future travel pattern is predicted. It involves developing of mathematical formulae/models which correlate present day household structure, income of the people, car ownership by people, petrol sales etc in the study area, to produces present day travel pattern as observed/surveyed. These models in their present form, fit well the observed behaviour of the traffic, between different zone/areas of the city of study area.

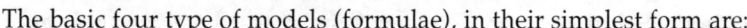

The basic four type of models (formulae), in their simplest form are:

i. Trip Generation models: determine how many trip the population, employment of study area are generated.

ii. Trip Distribution models: determine the origin and destination of each trip. Where generated trip will go.

iii. Model Split models: determine the mode of travel walk, car, bus, etc used by generated trips, in reaching from origin to destination.

iv. Route assignment model: determine the actual routes/roads used by the generated trips, between 0–D.

Using these models which are developed for present day trips, future trips-how they are generated, distributed, which mode and route they will follow can be determined by putting the future value of population, income, landuses, employment etc. in these models. The design year values of the parameters on which present models are based, are put in the models to estimate traffic pattern of the future (usually 20 years).

Figure 8.5 showing trip prediction model explains the four mathematical models listed above and how they answer several questions of traffic flow such as:

- How many trips are produced in zone 2
- How many trips are distributed to zone 3, from zone 1
- How many trips are there by mode 3 on route b, etc?

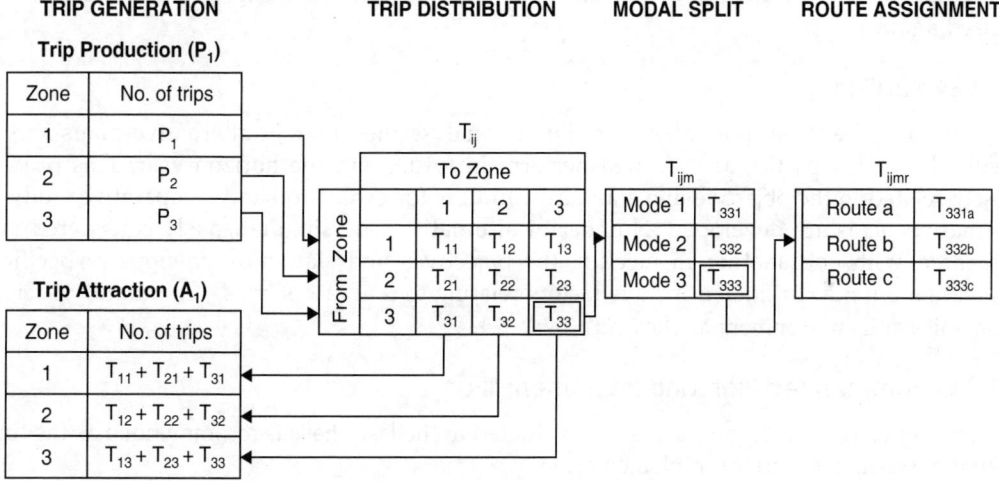

Fig. 8.5: Traffic prediction models

8.3.4 Alternative Plan Preparation and Testing

The next stage of urban planning process, after knowing the future travel pattern, involves preparation of alternative transport plans for accommodating the projected movements for the design year. (20 years) Total transport system of any city/area consists of many sub systems. A variety of landuse and transport alternatives are developed at this stage, to arrive at an optimal arrangement of transport plan. All the alternatives so developed should satisfy to a greater or lesser degree the previously established objectives and standards.

There is no general approach or methodology developed for developing/preparing the alternative solutions. Each alternative solution is tested for the extent to which it meets the stated objectives and standards. Solution generation is an activity that involves application of planning techniques towards suitably restoring a balance between demand and supply, meeting the stated strategic objectives. Alternative plans may involve TOPICS (Traffic Operation Plans for Increased Capacity and Safety) as well as TOPAZ (Techniques of Operation, Placement of Activities in Zones).

8.3.5 Objective and Standards

Infact objectives and standards are worked out in the beginning of the urban transport process (UTP). The objectives become the criteria against which, each alternative plan is evaluated. When an objective is well defined and understood, its achievement on long and short term basis is easily possible. The order and importance of the desired objectives should be listed. Objectives may belong to several groups as for example

- Contribution to socio-economic growth
- User's safety, mobility, comforts and convenience
- Efficiency, ease and cost of transport system
- Influence on urban environment, etc.

There will always be some constraints in achieving some of the stated objectives. Along with objectives, constraints in development of a transport plan such as finance, political, acceptable service to non car users etc. must also be worked out and taken into account. If a particular alternative plan does not meet the objectives to the desired extent it should be revised again.

8.3.6 Evaluation

This step in urban transport planning aims to analyse the different alternative plans and to predict how well particular plan will perform in future and the horizon year. This process closely related to the objective and standards. Indices for evaluation and comparative study of alternative plans are developed. Number of alternatives are simultaneously considered and compared with one another in meeting the objectives and standards previously specified. Evaluation criteria considered include factors relating to road users, roads, community, society, economics and environment as shown in Table 8.1.

8.3.7 Recommended Plan and Implementation

The most effective and economic plan as evaluated in the last phase is recommended as the final optimum solution (plan) for implementation.

Strategies for implementation of this final plan are then worked out. Time phasing for achieving the proposed transport network over the planning period may be done. There may be items/tasks to be implemented on short, intermediate and long term (i.e. 5, 10, 15, and 20 years) basis. Inputs necessary for phased implementation will include:

- Phase-wise investment plans
- Estimate of traffic for road sections, separately for each phase
- Cost Benefit analysis for justifying phased implementation.
- Co-operation and teamwork of financing and managing agencies etc.

It is also necessary that the plan and its phasing finds favour of the public in general.

Urban transport planning (UTP) is a continuous process and needs scrutiny, and updating from time to time.

Table 8.1: Objectives of highway planning

Objectives directly affecting the highway user are:
- to minimise motor vehicle operating cost
- to minimise motor vehicle travel time.
- to minimise traffic accidents and their cost and
- to maximise travel comfort convenience.

Objectives directly related to highway design are:
- to minimise the cost of highway construction, maintenance, and operation
- to maximise the quality of travel service and
- to provide high capacity facilities to be used by the maximum percentage of local area traffic.

Objectives related to community transportation are:
- to maximise coordination and integration of total transportation system of all modes
- to provide adequate transportation to all land areas and land uses in accordance with land-use plans: and
- to decrease congestion on roads used by public transport vehicles by diverting other vehicles to new highway.

Objectives related to community development factors are:
- to facilitate the achievement of the overall regional and community development plans:
- to preserve and enhance the community goals and aspirations:and
- to conserve open space.

Objectives related to community social factors are:
- to preserve historical areas, and sacred areas and objects:
- to minimise adverse effects on residents and establishments due to property acquisition and route location:
- to preserve the desired social characteristics of local areas and the community
- to provide for and preserve neighbourhood unity:
- to minimise air pollution:
- to minimise nuisance from noise:
- to preserve and enhance natural beauty and scenic attractions:
- to apply aesthetic principles in the design of the highway, to improve the appearance of the neighbourhood as much as possbile:
- to improve the opportunities for recreation: and
- to cause the minimum of disturbance during construction.

Objectives related to economic factors are:
- to minimise the adverse consequences to business, trade, and industry before, during and after construction:
- to maximise conservation of resources:
- to preserve land values: and
- to promote land use in accordance with economic development plans.

Thus after taking comprehensive view of existing urban form, existing transportation system, both for the city and the surrounding influence area, with necessary technical inputs of O-D, desire lines, spatial distribution due to future landuse developments plans for improving city's urban transportation system can be developed. Such plans should be periodically scrutinized and updated. After the master plan blue prints for ready execution and implementation mechanism should be worked out and put in place for implementation of priority schemes. From such comprehensive studies, traffic and transportation problems are handled in three broad stages:

i. Immediate improvement measures to optimise the existing facilities. They can be implemented immediately without land acquisition.

ii. Intermediate improvement measures which can be undertaken during next 5-10 years, after getting right of way and avaliability of resources.

iii. Long term improvement of traffic and transportation facilities to meet additional demands on long term basis (15-20 years), due to changed land uses and increased traffic demands.

These improvements include widening of roads and intersections, improvement of carriage way, providing parking and pedestrian facilities, imposing traffic regulations like one way street, no parking areas, speed limits, banning of certain turning movements and certain traffic, improvement in bus transportation, bus stops and routes, construction of flyovers, ROBS and plan for better mass transportation system with buses or trains (MRTS).

Figure 8.6 shows the study procedure adopted for transportation planning of Jaipur city on short and long term basis under Rajasthan Urban Improvement Development Project (RUIDP).

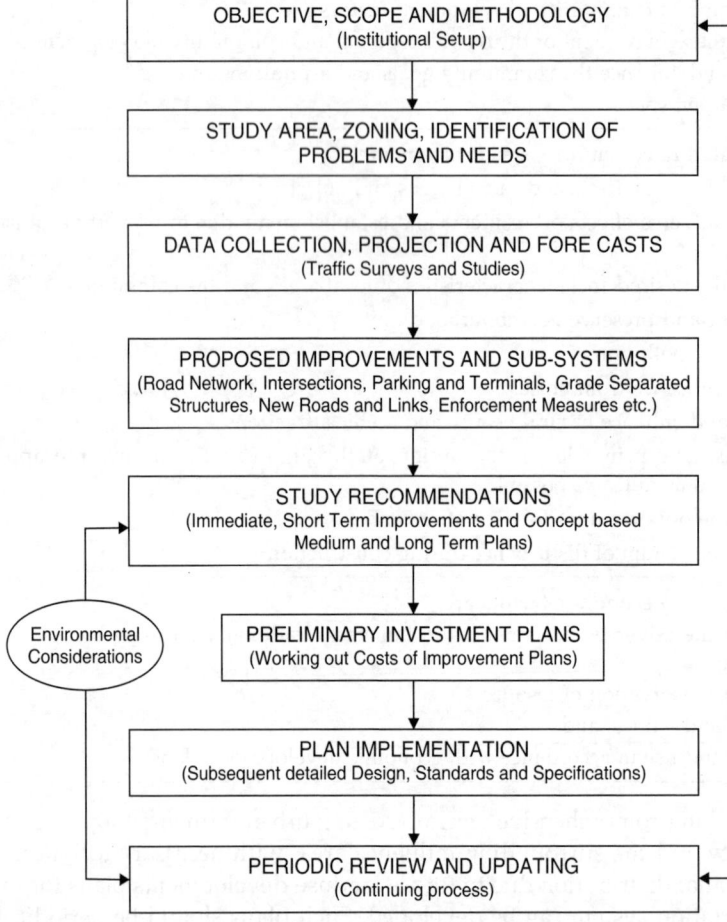

Fig. 8.6: Steps in traffic and transportation planning for Jaipur city (RUIDP)

8.4 LEVELS OF URBAN TRANSPORT PLANNING

Transport planning studies are carried out at a number of tiers/levels. Each level will have different objectives and decision taking criteria. Planning requirements and approach also changes from level to level, and so also the type of problems. Macro, meso and microplanning levels cover whole transportation requirements.

In an Australian study (Sydney), five levels of planning were identified, according to the type of plan. These levels are described below and shown in Figure 8.7

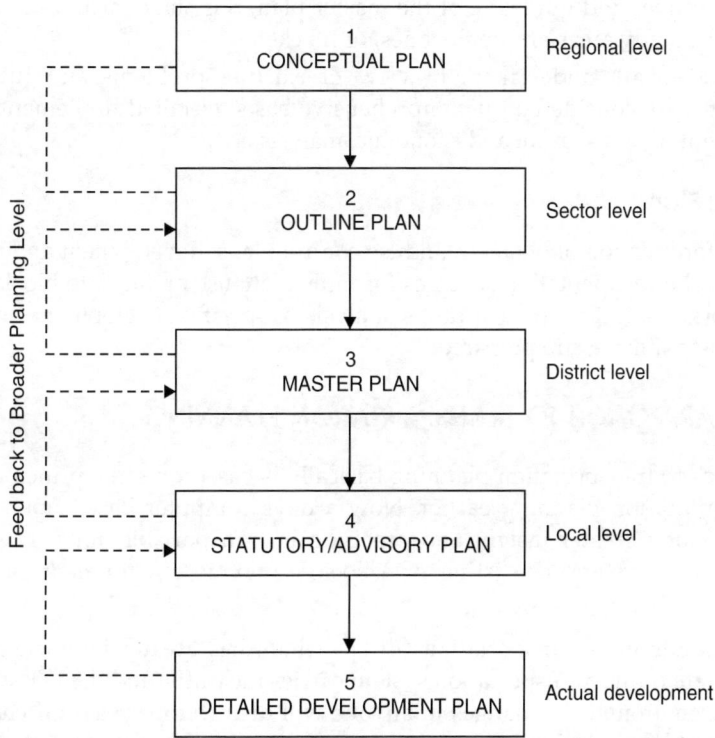

Fig. 8.7: Different levels of planning studies

8.4.1 Conceptual Plan

It is broadest level of planning where plan period is large 30 to 40 years. The area for planning and population is very large. Conceptual plan makes forecasts for landuses, population, employment and other activities, considering economic status of area.

It indicates, major roads and public transport system for the whole area, without analysis of right of way or costs.

8.4.2 Outline Plan

Keeping conceptual plan in view, outline plan is developed. Its plan period is smaller about 20 years. The study area and population is also lesser. It involves simplified forecasts of landuses, population, employment and other activities district by district. Benefit cost ratio is used for economic evaluation. It predicts major landuses, population, future traffic demands and models.

It shows major transport routes, reserves right of way, and undertakes costs into account.

8.4.3 Master Plan

With the objectives of outline plan, master plan for a shorter period 15-20 years is developed. Population, landuse, activities and employment are predicted on a zonal basis (whole area of study is divided into smaller zones). It predicts future demands for major and minor city roads. All routes are established, based on functional design studies and land uses. Detailed cost-benefit analysis is done.

8.4.4 Statutory Plan

Meeting the objectives and guideline of the master plan, this stage includes local planning for 5 to 10 years. The study areas are small zones, or blocks.

Detailed studies are undertaken, in view of existing problems and future needs. All significant routes are considered on comprehensive basis. Detailed implementation plans are worked out. Right of way control and economic analysis is done.

8.4.5 Detailed Plan

This considers formulation of detailed landuse control plans, development and re-development plans, design and implementation schemes for immediate use on block to block basis. Detailed designs are worked out. Benefit cost ratios analysis of specific development proposal is done, and detailed cost estimates are prepared.

8.5 SYSTEM APPROACH TO TRANSPORTATION PLANNING

System approach to transportation planning basically is based on same principles as described in urban transportation planning earlier. Now a days computer based operational research theory has provided greater insight and precision in transport planning. It considers urban transportation as a system, consisting of a socio-economic environment closely related to landuses.

Environment-Landuse-Transportation (ELT) system, has greater interaction among them. Landuses generate traffic, transportation system carries the traffic and the L-T system exists in a socio-economic environment. Changes in anyone will disturb the system. If changes are small system will adopt them and will attain the state of equilibrium. If these changes are large the system may become unstable and destroy itself, but even after its destruction, a new stage of equilibrium will be reached. The concept is that even if "Do Nothing" philosophy is adopted system being a close loop will still function, but the question is will that be a optimum or desired state of transportation system?

The System: System is defined as a set of components, that is organised in a way that action of the system towards specific goal and objectives, depends upon certain inputs, i.e. inter relationship between components is such that the system will respond to demands placed on it in a desirable manner.

The Environment: An environment may be defined as the set of all components outside a system which both influences the behaviour of the system and which in turn is influenced by the behaviour of the system. Thus the system is not alone, but it is in conjunction with the environment.

The concept of inter-relationship of System-Environment ensemble is illustrated in Figure 8.8. The systems planner is to design a system which achieves maximum integration (degree of fit) between system and its environment.

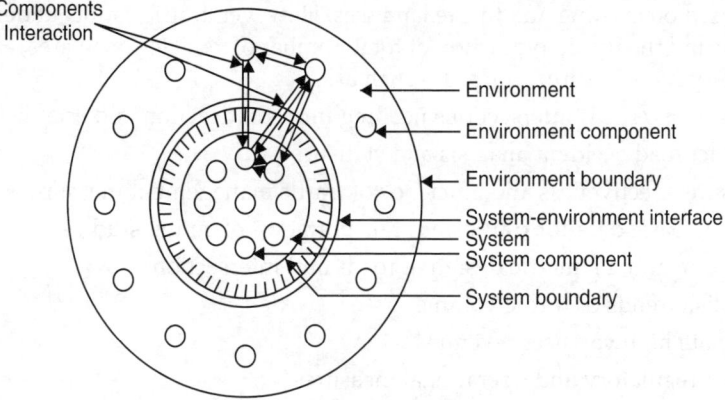

Fig. 8.8: System-environment ensemble

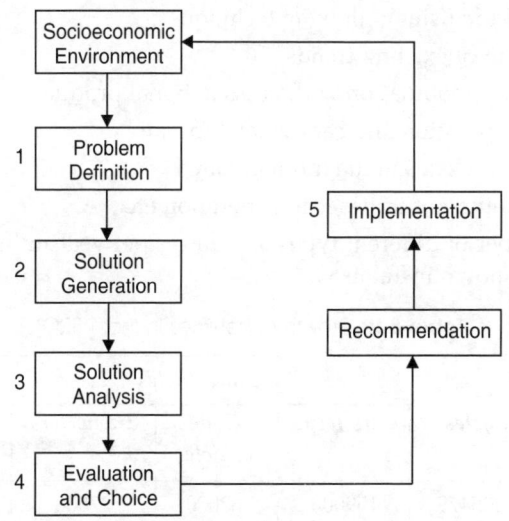

Fig. 8.9: Systems approach to transportation planning

General approach in systems planning methodology is similar to the planning process described earlier. The principle stages are:

 i. Problem definition
 ii. Solution generation
 iii. Solution analysis
 iv. Evaluation and choice
 v. Recommended plan
 vi. Implementation

The flow chart of these stages is shown in Fig. 8.9. Detailed description of these stages and development of various mathematical models is beyond the scope of this book. Readers may refer to any book on traffic and transportation planning for further details.

8.6 TRAFFIC FORECAST

Transportation planning forecasts the overall trips, their generation and distribution in the whole study area, for long and short term planning through mathematical and statistical techniques, based on existing and future landuses. However, traffic counts undertaken regularly or periodically and the future prediction of total number of vehicles in an area, at an intersection or on a road serves many direct purposes such as:

- To know areas/road/intersections needing immediate action and establish priorities
- To conduct road accident analysis and statistics
- To measure effectiveness and efficiency of traffic management measures.
- To provide basic data for traffic and transportation planning studies
- In structural and geometric design of roads and intersections
- To establish trends of traffic volume
- To estimate highway user revenues
- Justifying regulatory and operational measures
- In classification of vehicles typewise, and their composition.
- In design of traffic signals, etc.

Traffic forecasts are made using following techniques:-

  i. Simple extrapolation of existing trends.
  ii. Correlation of vehicle population with gross national product
  iii. Correlation with population and car ownership rate
  iv. Correlation will petrol sales in the influence area
  v. Mathematical modelling of traffic/trip generation etc.

Table 8.2 shows number of different types of vehicles per year in India from 1951. The trend of growth of vehicles is shown in Table 8.3.

Table 8.2: Growth of vehicles in India (IRC)

Year	Type of vehicles						
	Agricultural Tractors	Cycles	Cars/Jeeps	Good vehicles	2-wheelers	Buses	Auto Rickshaws
1951	8600	1205479	159000	82000	27000	34000	
1952	10281	1315068	166950	88560	24930	36380	
1953	12291	1479452	175298	95645	32079	38927	
1954	14694	1589041	184062	103296	34966	41651	
1955	17566	1753425	193265	111560	38113	44567	
1956	21000	1972603	203000	119000	41000	47000	
1957	22701	2136986	221270	127330	47970	48880	
1958	24540	2356164	241184	136243	56125	50835	
1959	26528	2575342	262891	145780	65666	52869	
1960	28677	2849315	286551	155985	76829	54983	500
1961	31000	3123288	310000	168000	88000	57000	1200
1962	34639	3452055	334800	183120	106480	59850	2100
1963	38706	3835616	361584	199601	128841	62843	3200

(Contd.)

Table 8.2: Growth of vehicles in India (IRC) (*Contd.*)

Year	Agricultural Tractors	Cycles	Cars/Jeeps	Good vehicles	2-wheelers	Buses	Auto Rickshaws
				Type of vehicles			
1964	43249	4164384	390511	217565	155897	65985	4700
1965	48327	4602740	421752	237146	188636	69284	6600
1966	54000	5095890	456000	259000	226000	73000	8800
1967	69288	5589041	492480	275440	273460	76650	11400
1968	80667	6136986	531878	291012	330887	80483	14400
1969	93915	6739726	574429	308473	400373	84507	18000
1970	109338	7397260	620383	326982	484451	88732	22200
1971	127295	8164384	682000	343000	576000	94000	27400
1972	148200	8986301	740000	364000	656000	100000	33800
1973	160985	9863014	709000	308000	734000	102000	41800
1974	174873	10849315	768000	323000	838000	105000	51700
1975	189959	11945205	766000	335000	946000	114000	63400
1976	206346	13150685	772000	345000	1055000	113000	75400
1977	224147	14465753	870050	376440	1412570	116620	87700
1978	269724	15890411	910653	395915	1615351	121453	100400
1979	271032	17620274	987235	436348	1885113	130275	113200
1980	293705	19232877	1049797	464736	2113853	137084	126400
1981	312632	21205479	1150265	546560	2615113	159567	149000
1982	323714	23287671	1224730	604670	3058030	171120	177000
1983	340602	25643836	1365086	666087	3645845	183045	211700
1984	357653	28219178	1433293	732463	4341459	196967	249900
1985	374709	31013699	1565861	811795	5147053	220885	295000
1986	391567	28590629	1737814	850985	6210843	224983	342900
1987	407980	33819724	1962114	968880	7695097	242150	396700
1988	435198	39320430	2246684	1097519	9249521	266008	457800
1989	464191	45055560	2433990	1161036	10906845	274858	531100
1990	591305	51043669	2638014	1218419	12543867	294701	614400
1991	635271	54646295	2893225	1334146	14124595	327284	681100
1992	682487	58506347	3140084	1498460	15571467	354350	734300
1993	706949	62651814	3260118	1582528	16922041	377168	798000
1994	781090	67277313	3370771	1773539	17815049	385976	870400
1995	839695	71757685	3760014	1919492	20690002	418775	952300

(*Contd.*)

Table 8.2: Growth of vehicles in India (IRC)

Year	Type of vehicles						
	Agricultural Tractors	Cycles	Cars/Jeeps	Good vehicles	2-wheelers	Buses	Auto Reckshaws
1996	908326	76730629	4104829	2014982	23097497	443732	1036200
1997	991565	82124156	4573743	2322000	25579373	478000	1127400
1998	1076285	87814352	5036844	2516000	28486993	536000	1226100
1999	1338933	98973009	5450737	2539000	31138830	537000	1337100
2000	1606278	111153489	6043406	2703000	33915802	553000	1458900
2001	1815020	125475559	6992000	2932000	38340817	633000	1597400

Table 8.3: Historical growth trend of vehicles in India (IRC)

S. No	Type of vehicle	1951-1980	1981-1990	1991-2001	1951-2001
1.	Cars	7.0	9.96	9.08	7.57
2.	Two wheelers	17.35	19.60	10.51	16.88
3.	Buses	4.70	7.03	6.50	5.86
4.	Goods vehicles	5.97	9.52	8.11	7.14
5.	Bicycles	9.90	9.52	8.22	9.60
6.	Agricultural tractors	12.97	6.18	10.84	10.40
7.	Auto Rickshaw	28.15	16.88	8.87	18.64

8.7 TRAFFIC SYSTEM MANAGEMENT (TSM) MEASURES

Traffic system management (TSM) measures are primarily concerned in the short/medium term solutions to improve the efficiency and safety of movements of pedestrian, vehicular and other traffic, on existing road system. TSM measures maximise the use of the existing system, without impairing environmental quality. These measures are cost effective. Managing of traffic is managing human beings as traffic is comprised of human beings. Best management is done suiting to human psychology as closely as possible.

Traffic management may involve alteration to the geometric layout, widening of roads, provision of control devices like signs, signals, marking, pedestrian crossing, etc and imposing restrictions like oneway streets, no turning movements, no entries etc. Traffic growth steadily erodes the benefits from traffic management measures.

Commonly used traffic system management techniques are:

8.7.1 One-way Streets System

One-way streets are those on which vehicles move in one direction only. One way movement reduces delays, conflicts, permitting higher speeds, as traffic is allowed to move only in one specified direction. They are useful in congested areas of city, where extensive measures are not possible.

Reversible one-way system can partially overcome capacity limitations of a network, during peak hours of traffic. Advantages and disadvantages of one-way street are included in Table 8.4. A grid pattern of rectangular roads is most suitable for one-way streets.

Detailed traffic studies are conducted, for implementation of one way system, and after studies are required to establish the usefulness of one-way system.

Table 8.4: Advantages and disadvantages of one-way streets

Advantages	*Disadvantages*
• Increase in capacity of roads and intersections	• Longer journey distances and increased traffic volume
• Accidents are reduced, eliminating head-on collisions	• Trouble for new (out side) drivers to locate alternative route
• Low cost control in imposing and maintenance	• Longer travel distances may involve increased fuel costs
• Traffic movement at intersections is simple as number of conflict points are reduced	• Loss of amenity for residents in areas of one-way streets and possible environment problems
• Higher operating speeds are possible	• Increased walking distances for bus passengers
• Additional pedestrian crossing can be used without serious disruption of traffic	• Buses may have to travel longer, and routes may change
• Reduction in journey times can be achieved	• Opposition from commercial intrest people and shopkeepers along one-way route
• Linking of traffic signals becomes easy	• Additional signs "One-Way" "No Entry" are required.
• Savings in accidents costs and lesser police control	• Higher speeds increase severity of accidents
• Traffic interference with parked vehicles is reduced as all vehicles face same direction.	• Hospital and school goer may suffer in some cases, so also emergency vehicles like ambulences and fire brigades.

8.7.2 Segregation of Traffic

For minimising traffic conflict and accident, the best technique is to provide separate traffic lanes for different slow and fast moving vehicles. Segregation of traffic may not be possible in many situation due to limited road width and the financial implications.

Segregation of traffic can be achieved through one way street approach, or by providing separate lanes/tracks for cycles, pedestrians and slow moving vehicles. Segregation can also be achieved by restricting entry of certain vehicles during particular hours of the day, e.g. trucks are usually not allowed on roads from 8 am to 8 pm in cities.

Method of segregation to be used (if possible) depends upon local traffic and system of roads in a locality. On roads having heavy vehicular traffic, slow moving vehicles like bullock carts, tongas, cycle-rickshaws are not allowed, for the whole day or certain hours of the day.

Exclusive bus lanes, bus transport system (BRTS), is used to provide separate lane for buses.

8.7.3 Restrictions on Turning Movements

i. **Right turns:** At certain intersections congestion is often caused during peak hours by right turning traffic, reducing capacity of the intersection. It may stop all movements at the intersection sometimes.

Right turning traffic is entirely banned during all or part of the day, depending upon the site conditions of streets and availability of alternative route. It should be observed that

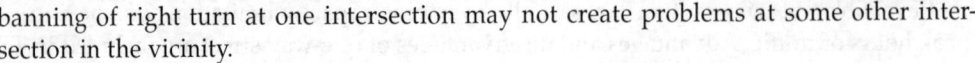

banning of right turn at one intersection may not create problems at some other intersection in the vicinity.

ii. **Left turns:** Left turns are rarely banned as they do not create any problem. Only in cases where number of pedestrians crossing the minor road is unusually heavy, left turn ban may be used, to avoid pedestrian vehicle conflict.

8.7.4 Reversible Traffic Flow

It is a common phenomenon these days that in the morning roads are filled to full capacity in one direction, while in the evening situation is reversed and capacity is reached in reversed directions. 'Inbound' and 'Outbound' traffic thus creates a ' Tidal Flow' situation during morning and evening but in different direction.

One solution to this problem is to allocate larger area of road (3-lanes) in one direction in the morning peak and less (1 lane) in other direction. It may be reversed during the evening peak.

The second method, if two or more routes between same origin and destination are available, one can be made completely one way for morning peak traffic and other made completely one-way to take care of evening peak.

8.7.5 Traffic Management for Public (Bus) Transport

Improvement of public transport services is very important in traffic management. More public transport use should be encouraged to reduce number of private vehicles on the road. Additional bus facilities may be provided for transferring private car/vehicle owner to use buses. Based on specific mass transit studies, following traffic management measures may be used.

- Use of privately sponsored buses and mini-buses.
- "Pick on Demand" type services may be introduced
- Bus stop locations should be examined, relative to origins and destinations of passengers, wrong location of stops, may create congestion on roads. Bus stops should not be located near intersection.
- "Park and Ride" type of integrated bus services which carry the persons to their destinations after parking at a central or remote place may be useful in cities.
- General improvement in bus services using vehicles designed with ease in boarding and comfortable ride.
- "Bus Only" lanes can be adopted on suitable roads. These are the lanes which are used exclusively by buses. Bus lane operation may be restricted to selected hours of the day.
- "Bus Priority Signals" are used ensuring that bus clears the signal ahead with minimum delay. Special loop detectors are used for this purpose. Emergency vehicles can also be equipped to receive priority. This can be achieved by advanced urban traffic control (UTC) systems such as SCOOT (Split Cycle Offset Optimisation Technique), and SVD (Selective Vehicle Detection).
- "Bus Bays" which are special areas on side of the road, where buses stop for boarding or alightening of passengers, are essential for passenger safety and free flow of traffic on roads.
- Smart card technology can be used for ticketing. This system offers advantages over the traditional paper-based system, in processing , reducing frauds, and using flexible fare structure.
- Information technology (IT) and Interactive Voice Response (IVR) application, help in public patronage and access to bus services.

8.7.6 Closing of Streets

In some situations number of side lanes join the main road, bringing and taking off the traffic, creating merging and diverging conflict and thus affecting the flow of the traffic on the main road. It may be advantageous in certain cases to close some to these lanes or make them one way, suiting to the situation.

8.7.7 Parking Management Measures

Control of curb (on-street) parking is necessary to reduce congestion specially during periods when movement of vehicles is the first priority. Parking of cars may be completely prohibited on narrow city roads in order to provide the movement of through traffic. Some measure to manage parking are:

- Time-limit parking, only during certain hours of the day parking be allowed on certain roads. Parking meters may be used
- Long-term parking should be encouraged at specified places, away from the road. "Park and Ride" type of system may be used if justified
- Parking should not be allowed near intersections
- Tougher action against illegal parking, including loading and unloading should be used
- Restrictions to parking in central congested areas.
- Imposition of high parking fees, to discourage parking in congested areas.

8.7.8 General Management Measures

Some general techniques of improving personal mobility, safety and environment with energy conservation are:

- Preferential treatment to high occupancy vehicles, car pooling, shared taxi, etc to be encouraged, like Singapore
- Segregate lane for 'Pool Cars' having more than two passengers as used in USA
- Use of staggered school hours/office hours may be worth while. Staggering of annual holidays, more variety in school hours in each locality may be used
- Supplementary extra charges may be introduced for vehicles which enter congested urban areas, except those of doctors, emergency vehicles, shopkeeper etc.
- Loading and unloading of goods vehicles should be permitted only at specified locations, during non-peak hours.
- Movement of cycle rickshaws and other slow traffic should be regulated properly and forcefully. They should be completely eliminated from certain roads.
- Strict enforcement of 'Speed Limit' is necessary on city roads. Speed limits should be displayed on all roads, at strategic locations.
- Heavy penalties should be imposed if stray cattle is found on the road.
- Safety belts must be made compulsory for safety on roads.
- Strict control of roadside hoardings, advertising boards is necessary. They should be properly and scientifically located.
- Illegal accesses, markets, stalls, which create congestion and unsafe travel conditions should be removed and site monitored to prevent their re-appearance.
- Intensive road safety propaganda, for proper road/lane use, knowledge of rules, and regulations, penalties etc. should be done with all possible ways like T.V., newspapers, lecturers, safety weeks, on road demonstrations, etc.

- Measures like weekend car scheme, auto control policy restricting vehicles of fixed plate numbers on fixed week days not allowed on road.
- Three-in-one regulation scheme, restricting entry during peak hours if car occupancy is less than 3 persons, etc.
- More discipline must be enforced in traffic movement, vehicles and pedestrians.
- Drivers must be made to conform to driving rules and courtesies. The defaulter should be immediately booked.
- For improvement of transportation in cities on day to day basis, a unified agency should be setup with comprehensive authority and representatives from various concerned organisations such as police (Traffic) municipal corporations, regional transport office (RTO), electricity department, public works department (PWD), forest and environment organisations etc. to minimise time lost in coordination in maintaining road in good condition all the time.
- In most of the cities there is no traffic engineering cell, not even a traffic engineer. How traffic can be managed properly without even a single trained person in the city? It is necessary to have a qualified expert to advise on traffic issues/problems
- Four-day a week working, keeping the same total hours, and through electronic technology enabling more working from home, can be considered.
- Building of roads to higher standards so that repairs are needed less often. All repair works should be done during night time.
- Adopt an "EASE" approach to traffic problems:

  E: Educate the masses, regarding their duties and road use

  A: Administrative reforms through integrating various independent interacting agencies into system for up keep of road-network.

  S: Systems-interactive, co-operative and supportive approach in maintaining and running good transportation system.

  E: Engineering measures and uses of latest technology in construction, maintenance, and operation of roads.

The alleviation of congestion, is possible through traffic system management (TSM) measures, if soundly based, permitting greater traffic flows to be accommodated, more safety on the existing road system. The traffic growth steadily erodes the benefits of TSM. Long term objectives should therefore be to find ways in which restraints can be imposed on traffic growth and use.

Particular attentions must also be given to enforcement and regulation, to obtain maximum benefits of TSM schemes. Traffic and transportation planning should be encouraged for exhaustive and comprehensive treatment of problems, creating suitable alternative long-term changes in urban layout, and alternative forms of transport system.

8.7.9 Legislative Measures

The main legislation governing the vehicles and drivers are various traffic acts, rules framed by police authorities, orders of state and local government, for regulation and control of mixed traffic. In India motor vehicle act, 1993, which was subjected to subsequent amendment, is complex, extensive and detailed.

Traffic laws and regulations impose a duty to comply on a driver and he offends when he fails in his duty. Penalties for a traffic offence are fines, imprisonment, disqualification from

driving, probation, etc. By observance of the traffic laws, safety, convenience, and easy of driving is achieved reducing delay and accidents.

Basic traffic rules for safety, which should be followed by pedestrian, cyclist, slow and animal drawn vehicles, and parked vehicles are enumerated below:-

- Speed limits should be followed by different vehicles under mixed traffic conditions. For light and medium traffic : 50 to 30 Km/h depending upon major and low traffic road to congested road in built up area. For heavy traffic: it should be 40 to 20 Km/h for similar roads.
- vehicle with projecting material shall be allowed in the town during night hours with red flags/red lights.
- abutting shopkeepers should not occupy the side walks meant for pedestrian movement.
- requiring that, wherever sidewalks are provided, the pedestrians should keep to the sidewalks.
- laying down that wherever pedestrian crossing are provided, pedestrians should not cross a carriageway at any place other than at the cross walk;
- laying down that wherever sub-ways or overbridges are provided for cross movement, pedestrians shall not use the carriageway for crossing.
- stipulating that not more than two bicycles shall remain abreast except at separate cycle tracks;
- prohibiting cyclists from using sidewalks;
- making it obligatory for cyclists to use separate cycle tracks wherever provided;
- prohibiting cyclists from being towed by any other vehicle;
- prohibiting cyclists from carrying any other person on the cycle;
- making it necessary for cyclists to keep to the extreme left of the carriageway where no separate cycle tracks are provided;
- making it compulsory for the cyclists to have a head lamp, back reflector, and rear mudguard painted white in the bottom portion.
- restricting the length of the load to 3 m and height to 2.5 m (measured from ground) and laying down that the width of load should not exceed the width of hand-cart by 0.3 m;
- restricting the use of hand-carts carrying long goods and pipes greater than 5 m length, unless accompanied by four attendants. Such projecting material should always be provided with red flags and during night hours with red lanterns or lights;
- stipulating that if the load exceeds 225 kgs, two able bodied attendant should be provided, and for lesser loads, one attendant should be provided;
- stipulating that hand-carts carrying loads in excess of 500 kg should be provided with pneumatic tyres and efficient brakes.
- restricting the height, length and width of load to safe limit, depending upon the size of the carriage;
- laying down that the animal drawn traffic should invariably keep to the extreme left of the carriageway having mixed traffic.
- prohibiting animals such as camels and elephants from being driven on the busy streets;
- regulating that the cattle in droves shall not be driven on busy streets at peak periods of traffic;

- laying down that cattle in droves shall be accompanied by sufficient number of attendants to completely keep them under control;
- requiring that a rider of a horse should keep to the extreme left of the carriageway as far as possible;
- prohibition of parking for a period exceeding 30 minute on busy streets during the whole day or during peak hours of traffic;
- designating parking places for different types of vehicles such as private cars, taxis, scooters, horse-drawn vehicles, cycles, auto rickshaws and cycle rickshaws;
- making it obligatory for traffic to halt in a street at a distance not greater than 0.6 m from the edge of the curb or pavement;
- restriction of stopping or parking of vehicles;
  - i. within an intersection area and 75 m on either side;
  - ii. on a pedestrian crossing or within 6 m on either side of it; and
  - iii. near corners, bends, rise or safety zones, and also entrances and driveways of properties;
- making it necessary for vehicles to park parallel to the kerb, unless angular parking is specially permitted;
- restricting parking on sidewalk and cycle tracks;
- restricting parking within 10 m of a bus stops; and imposition of parking fee where necessary;
- making it necessary for the vehicles halting for the purpose of loading and unloading to draw up parallel to and alongside the kerb or edge of the pavement and not blocking a cross street;
- restricting loading and unloading;
  - i. within an intersection area, and 75 m on either side;
  - ii. on a pedestrian crossing or within 6 m either side of it;
  - iii. near corners, bends, safetyzones, and also entrances and drive-ways of properties;
  - iv. on sidewalks and cycle tracks; and within 10 m of a bus stop.
- restricting the load projections on the sides to about 0.5 m from the longitudinal middle of the motor cycle, on the front to about 0.6 m from the front wheel and on the rear to about 1.0 m from the rear wheel;
- restricting the number of passengers to not more than two (one driver and one pillion rider) when no side car is attached and to not more than four when a side car is attached. In the case of learner drivers, no pillion rider or passenger in side-car is to be allowed;
- prohibiting scooters and motorcycles from towing any trailer (not being a side-car).

8.8 PEDESTRIAN AND CYCLIST MANAGEMENT MEASURES

On city roads in India, there is preponderance of pedestrian traffic due to socio-economic conditions and narrow streets. Most pedestrian management measures involve segregating the pedestrian from vehicular traffic. Following are the measures in use:-

8.8.1 Pedestrian Side Walks (Footpaths)

Pedestrian channelisation is done through provision of footpaths with guard-rails, so that pedestrian are kept away from the main road. Unfortunately there is no specific law which says that pedestrian must use the footpaths and not the carriageway. Footpaths should be wide

enough, smooth and flat. Pedestrian do not use the foot path if surface is bad or broken. Side walks should be provided on both sides of the road raised above the carriageway. Following points should be kept in view:

- Sidewalks should be well maintained and keep free from encroachments and hawkers.
- Sidewalks should be sloped adequately to drain water. A slope of 1 in 40 to 1 in 30 should be sufficient.
- Width of sidewalks depends upon the number of pedestrian. It varies from 1.5 m to 4 m.
- Pedestrian side walks should be close to the abutting buildings. Capacity of footpaths for different width is given in Table 8.5.

Table 8.5: Capacity of side-walks for different width (IRC)

Width of side-walk (metre)	Capacity in number of persons per hour	
	All in one direction	In both direction
1.50	1,200	800
2.00	2,400	1,600
2.50	3,600	2,400
3.00	4,800	3,200
4.00	6,000	4,000

8.8.2 Pedestrian crossing (cross-walks)

Crossing the carriageway at selected places reduces the risk of conflict with traffic.

Pedestrian crossing other than signals are of the two types:

- At grade zebra crossing and
- Completely segregated crossing subways or over bridges.

A zebra crossing is simply a portion of the road, that is reserved for use of pedestrian crossing the road. This strip is marked alternate black and white. Its success depends upon the drivers willingness to yield right of way to a pedestrian crossing, the road.

Segregated crossing completely avoid the possibility of conflict between the pedestrian and the vehicle. Foot over bridges for crossing require more time and energy to cross and therefore are not preferred by pedestrian, specially old, children and ladies.

Subways which are more expensive, meeting pedestrian desires, provide safer and swifter crossing, responding to requirement of pedestrian travel.

Pedestrian crossing should be:

- Provided at locations where pedestrian traffic accumulates, e.g. at school, hospitals or mid blocks.
- Provided at intersections for crossing the roads by pedestrians.
- Crossing the carriageway at right angles to minimise walking distance and exposure to vehicular traffic.
- Clearly marked and maintained, as required.
- Free from obstructions like signposts, lamp post, etc.
- At least 2 m to 3 m away from the stop line for the vehicles at intersections.

- Indicated by a sign if not controlled by a signal.
- Wide enough as required by the number of pedestrians crossing, but not less than the width of pedestrian sidewalks (1 m to 4 m).

Pedestrian crossing are either uncontrolled or controlled. Uncontrolled crossings are those which are marked by stands or paints, but not controlled by any signal, except, flashing beacons.

Controlled pedestrian cross-walks are those where movement of pedestrian and vehicles is regulated by traffic signals. Push button signals to be actuated by pedestrian are also used.

Figure 8.10 shows typical pedestrian crossings.

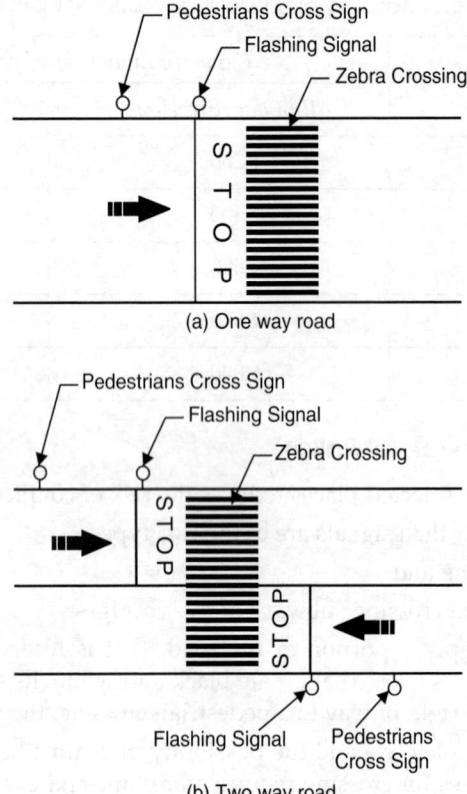

Fig. 8.10: Mid block zebra crossings

Grade Separated Pedestrian Facilities

8.8.3 Foot Over Bridges

A foot overbridge is a bridge over a road, away from intersection, provided for pedestrian to climb and cross the road. Foot over bridges have a disadvantage as one has to climb up almost two stories and also get down two stories in order to cross a road. It is practically not of much use for old, children, infirm and ladies.

In new bridges however automatically moving staircases are used for climbing and getting down making it usable for all type of people. Though the initial and maintenance costs of such foot over bridges is high, they are in use at many cities like Jaipur.

8.8.4 Subways

A subway is an underground tunnel across a carriage way on a straight section of a road or it may be a combination of underground tunnels connected through a pedestrian subway. Exact shape of the subway will depend upon the geometry, shape, size and angle of the intersection. Pedestrian subway adequately responds to most of the requirements of pedestrian travel. The minimum width of a pedestrian subway is 2.5 m and the vertical clearance should not be less than 2.5 m. Wider and higher subways are preferable.

Drainage is important for subways. A slope of 1 in 30 on each side should be used to remove water. Gullies should be provided to trap water entering from the roads or steps.

Capacity of a grade separated pedestrian facility is worked out as 50 persons per minute per meter on a level or upto 1 in 20 gradient and 35 persons per minute per meter width on steps or ramp over 1 in 20 gradient.

Indian Road Congress (IRC:103-1983) provides guidelines for controlled pedestrian crossing when the "Hazard index determined as a product of number of pedestrian crossing (P) and square of the number of vehicles (V^2) exceeds 10^8 for undivided carriage way and 2×10^8 for a divided carriageway", i.e. when hazard index $(PV^2) > 10^8$ undivided street and $(PV^2) > 2 \times 10^8$ for divided street.

Controlled crossing mentioned above could be automatic traffic signals, pedestrian push button signals (pelican crossings) foot over bridges or pedestrian subways.

Figures 8.11 and 8.12 shows typical layout of subways.

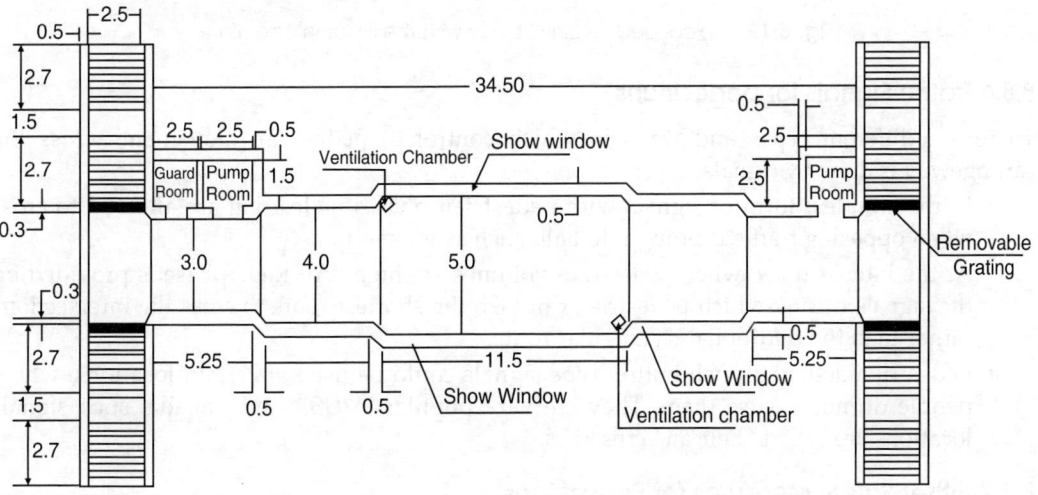

Fig. 8.11: Typical pedestrian subway at middle of the road

8.8.5 Grade Separation Across Railways

Road over bridges, i.e. grade separators are provided across existing railway crossings to reduce delay to the traffic and increase safety of operation.

It is recommended that a grade separator should be provide on an existing railway crossing if the product of ADT (fast vehicle only) and the number of trains per day exceeds 50,000 within the next 5 years.

For new constructions such a grade separation should be provided when the above figure is greater than 25,000.

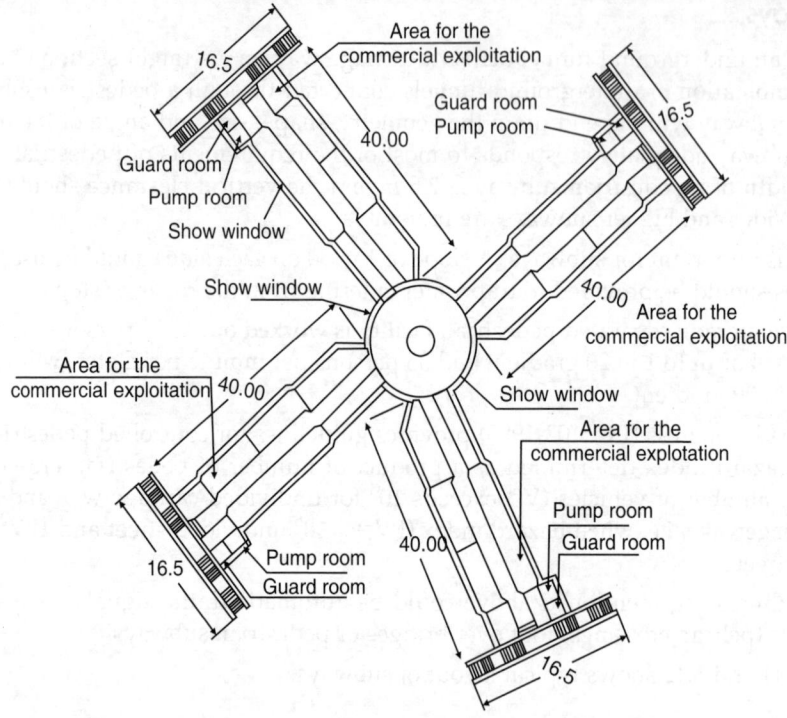

Fig. 8.12: Typical pedestrain subway with 4 exits at intersections

8.8.6 Traffic Signals for pedestrians

There are different types and ways in which control of pedestrian movement across the carriageway is done by signals.

- In the simplest form of signals with pedestrian phases pedestrian are allowed to cross when opposing traffic is brought to halt, such as intersections.
- At the intersections, when pedestrian volumes are high "All Red" phase is provided on the signals during which pedestrian can take the shortest route to cross the intersections rather than the traditional rectangular route.
- Pedestrian actuated push button type signals could be used at certain locations, where people do not misuse them. They are very popular in USA and Canada, such signals locations are called "Pelican" crossings.

8.8.7 Other Safety Measures for Pedestrians

- Education and instruction are important aspect of management programmes specially for children with limited ability to judge speed and distance, who should be adequately trained.
- Needs and characteristics of elderly and handicapped should be properly understood and taken care in crossing, while boarding, seating, and getting down a bus.
- Prohibiting hawkers, vendor, pillars from occupying foot paths.
- Shopkeeper who project their shops, block footpaths should not be allowed to do so.
- Pedestrian should obey traffic signals and cross the streets only at cross walks, subways, etc. wherever they are provided. Violation should be considered an offence.
- Certain shopping areas, street should be reserved for pedestrians only and no vehicular traffic should be allowed in those places.

8.8.8 Pedestrian Precincts

At certain locations in cities it is common to restrict vehicular traffic in certain areas, shopping places, markets, and on certain roads. On such locations only pedestrian movements, are allowed ensuring full safety and comfort. Such pedestrian precincts totally prohibit any vehicular traffic. On such locations, certain time either early in the morning or late in the night is allowed for delivery vans, garbage trucks, taxis or other essential vehicles to enter. Ambulances and emergencies vehicles may however be permitted.

8.8.9 Management of Cycle Traffic

Cycle traffic on Indian roads in and around towns/cities is quite significant. Cycle accidents due to conflict with fast moving traffic, can be avoided by providing separate cycle tracks along city roads.

As per IRC recommendations cycle tracks may be provided when peak hours cycle traffic is 400 or more on routes with motor vehicle traffic of 100-200 vehicles per hour. When motor vehicles are more than 200 per hour, separate cycle tracks may be justified, for a cycle traffic of only 100 per hour. Cycle tracks should be provided in both directions of travel.

The width of cycle tracks depends upon number of cycles to be accomodated, however the minimum width of a single lane tracks should be 1 m.

Cycle tracks should be well lighted and maintained in good condition. For drainage, proper gradient should be used.

8.9 INTELLIGENT TRANSPORT SYSTEMS (ITS)

Applications of information technology (IT), to traffic management systems, traffic safety and impact of traffic on environment, etc are now available. It is collectively known as "Intelligent Transport Systems" (ITS). ITS deals through most advanced communication and control technologies, for receiving and transmitting information on human, roads and automobiles. ITS creates ideal traffic system that will reduce congestion, accidents while saving energy and protecting environment.

What is ITS?

ITS as a key technology in improving road safety is common in Japan, USA, UK and Europe. Its popularity increased in mid 1990s.

Intelligent transport system covers a broad range of information technology and telecommunications for roads, road users and vehicles. It gives a fundamental solution to traffic related problems like traffic accidents, congestion, delays and pollution to improve transport efficiency, safety and sustainability. Table 8.6 covers the definition of ITS, given by different countries. The conceptual model of ITS is presented in Figure 8.13.

ITS can be used in 8 different aspects of highway and traffic management as shown in Table 8.7. The challenge for transportation professionals is to determine how and where ITS can realistically be used.

8.9.1 Advantages of ITS

Advantages of intelligent transport system, relating to safety, efficiency, economy and environment are summarised below:-

- Drivers of cars, trucks and buses avoid getting into crashes and keep them running off the road. ITS will maintain safe distance between vehicles and safe speed approaching danger spots.

Table 8.6: Definition of intelligent transport system

Source	Country/ Region	Definition
AUSTROADS	Australia	ITS covers the integrated application of modern computer, electronic information and communication technologies to improve all facets and all modes of transport and their linkages. ITS are essentially a diverse range of sophisticated tools that can reduce the environmental effects of transport, reduce congestion in a managed way, and improve transport efficiency safety and sustainability.
ITS Inc Australia		ITS is an umbrella term referring to the application of informed technology to transport operations in order to reduce operating costs, improve safety and maximize the capacity of existing infrastructure
ITS IncCanada	Canada	ITS is a broad range of diverse technologies applied to transportation to save lives, money and time. The range of technologies involved includes micro–electronics, communications and computer informatics and cuts across disciplines such as transportation engineering, telecommunication, computer science, etc.
ERTICO, Europe	Europe	ITS are the marriage of information and communication technologies with the vehicles and networks that move people and goods. 'Intelligent' because they bring extra knowledge to travelers and operators. ITS system help drivers navigate avoid traffic holdups and avoid collisions.
ITS, U.K.		ITS is a combination of Information Technologies and telecommunications, allowing the provision of on-line information applied to road, rail, air and sea transport to improve safety through the provision of on-line information to drivers in their vehicles and by equipping the vehicle with computerized systems which assist the driver.
ITS	Japan	ITS is a new transport system which is comprised of an advanced information and telecommunications network for users, roads and vehicles. ITS contributes much to solving problems such as traffic accidents and congestions.
VERTIS*		ITS offers a fundamental solution to various issues concerning transportation, which include traffic accidents, congestion and environmental pollution. ITS deals with these issues through the most advanced communication and control technologies. ITS receive and transmits information on humans, road and automobiles. VERTIS renamed as ITS Japan in June 2001
ITS Inc America	America	ITS is comprised of a number of technologies including information processing, telecommunications, control and electronics. Joining these technologies to transportation system to save lives, save time and save money and also for many transportation problems.
US Department of Transport		ITS are the integration of current and emerging technologies in fields such as information processing, communications, and electronics applied to solving surface transportation problems.

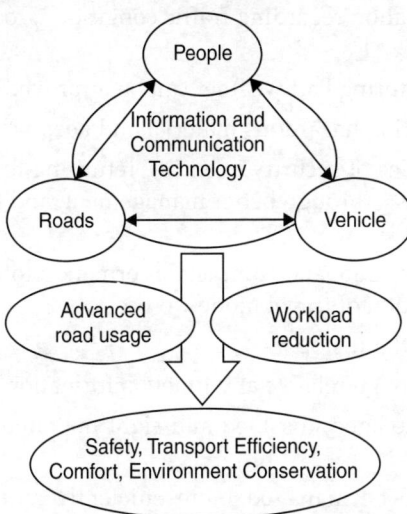

Fig. 8.13: Conceptual model of ITS

Table 8.7: Application of ITS – user's services

Advanced Traveler Information System	Pre–trip Information On-trip Driver Information On-trip Public Transport Information Personal Information Service Route Guidance and Navigation
Advanced Traffic Management System	Transportation Planning Support Traffic Control Incident Management Demand Management Policing/ Enforcing Traffic Regulations Infrastructure Maintenance Management Vehicle
Advanced Vehicle Control System	Vision Enhancement Automated Vehicle Operations Longitudinal Collision Avoidance Lateral Collision Avoidance Safety Readiness Pre-crash Restraint Deployment
Commercial Vehicle	Commercial Vehicle Pre-clearance, Commercial Vehicle Administrative Process Automated Roadside Safety Inspection Commercial Vehicle On-board Safety Monitoring Commercial Vehicle Fleet Management
Electronic Payment	Electronic Financial Transactions
Safety	Public Travel Security Safety Enhancement for vehicles. Intelligent Junctions
Public Transport	Public Transport Management Demand Responsive Transport Management Shared Transport Management
Emergency Management	Emergency Notification and Personal Security Emergency Vehicle Management Hazardous Materials and Incident Notification.

- ITS can provide information regarding traffic congestion, road accidents, road conditions, pedestrian crossing etc.
- Visibility in night and during bad weather can be improved for drivers.
- ITS can help in monitoring hazardous material and keep watch over transport facility.
- Coordination of activities of security agencies, returning to normal conditions as early a possible following a crisis through better management, and timely information is possible through ITS.
- ITS will provide fast, accurate and complete information to help travellers when to start a trip and what combination of travel modes to use.
- Drivers can pay toll without stop.
- Clearing of vehicles through reliable and timely information to the authorities will be easy.
- ITS will provide help to aged, disabled and sight impaired travellers, to get where they need.
- Transport system shall be managed more efficiently and safely, for motorist, cyclist, pedestrian and users of public transport.
- As travel will be faster, smoother, eliminating unnecessary travel and reduce congestion, environmental pollution will be reduced.
- Energy waste, wear and tear of vehicles and time of travellers will be saved.

8.9.2 ITS Technologies

Three types of ITS technologies include:-

   i. ITS for road/infrastructure:

      (a) Traffic management systems.

      (b) Traveller information systems.

      (c) Intersection traffic management systems.

      (d) Vulnerable road user/pedestrian protection systems.

  ii. ITS for vehicles:

      (a) Vehicle enhancement systems.

      (b) Collision avoidance systems.

      (c) Automated collision notification systems.

      (d) Driver status and performance monitoring systems.

 iii. ITS for road/infrastructure and vehicles:-

      (a) Drivers information systems.

      (b) Vision enhancement system.

      (c) Intelligent speed control system.

      (d) Collision avoidance system.

      (e) Commercial vehicle related technologies.

i. ITS for road/infrastructure

These ITS technologies are designed to optimise traffic flows, reduce delays, control pollution, reduce traffic conflict and reduce likelihood of crashes. Road side devices used for this purpose may be ramp meters, signalling systems, video cameras, radio-systems, optimised roadway lighting system, etc.

Traveller information systems, provide real time driving information and risk warning enabling the driver to react before accident occurs. Road side devices and in vehicle technologies contribute to an advanced traveller information. The devices used on the road may be signs providing weather, speed information, incident detection system, sensors, advisory radio and load condition detection sensor, etc.

Intersections collision avoidance systems are designed to identify approaching vehicle that have a high probability of violating the red light or stop sign, and warn the driver on crossing road to avoid a crash. Variable message signs or traffic signals warn other drivers of danger.

Venerable road user/pedestrian protections system creates safer intersections, increases time cycle for pedestrians to cross. Technologies include in pavement lighting, illuminated push buttons, pedestrian signals and automated detectors.

ii. ITS for vehicles

In vehicle vision enhancement services, through onboard systems that use infrared radiation, helps in reduced visibility during dusk, dark, rain and fog, and give drivers an enhanced view of what is ahead—pedestrian, animals and road side features.

Other systems are designed to help driver to judge proximity to other drivers or objects. This avoids rear end collisions, merging and lane change accidents, intersections conflicts, etc.

These systems obtain traffic information such as acceleration, relative speed and distance from other vehicles through sensor in the vehicle and then analyse likelihood of collision and warn.

Automated collision notification system, send out notification signals automatically when a crash occurs to emergency service providers, to get response faster. This saves from consequences of an accident to some extent.

Driver's status and performance monitoring systems keeps tabs on the drivers. The systems identifies dangerous drivers and provides suitable warning signals.

iii. ITS for road/infrastructure and vehicles

The combination of systems described above can actually improve performance and increase the benefits.

Driver information systems used to enhance drivers awareness of traffic conditions, weather information collected by roadside devices.

Improvement to roadway infrastructure such as reflective marking, improves driver's vision.

Intelligent speed control system gathers information on current speed limits from a roadside speed control system and then provides the information through in vehicle devices and warns the driver of a speed violation.

Roadside detectors and warning systems make collision avoidance systems more efficient and reliable.

Technologies specifically designed for commercial vehicles include systems to enhance vehicle stability, vehicle inspection systems, onboard recorders and rear warning systems.

8.10 APPLICATIONS OF ITS IN DIFFERENT COUNTRIES

8.10.1 Japan

Japan used ITS in 1995 is full phase. There was a reduction of accidents by 25 percent fatalities in 10 years by the application of ITS as well as development of road designs.

Japan started working on ITS much earlier then other countries of the world. Following systems were used from 1970s till 2000, for practical use and for improvement of infrastructure. Some systems which provide information and warnings that help drivers to operate their vehicles comfortably are:

CATCS (Comprehensive Automobile Traffic Control System)

RACS (Road/Automobile Communication System)

ARTS (Advanced Road Transportation System)

SSVS (Super Smart Vehicle System)

ASV (Advanced Safety Vehicle)

UTMS (Universal Traffic Management System)

VICS (Vehicle Information and Communication System)

Japan has developed a vehicle equipped with ITS technology called ASV 3. It provides information, warning and control to the drivers and reduces accidents caused by wrong operation of drivers. This is an improvement of ASV-1 and ASV-2.

Japan has objectives to achieve the following in future using ITS technology:

• Cut the number of fatal accidents to half the current number.

• Eliminate traffic congestion

• Reduce vehicle fuel consumption and CO_2 emission by 15 percent each and reduce No_x in urban area by 30 percent for environment.

8.10.2 Australia

Australia uses ITS technologies for safe, efficient and environment friendly transport. Table 8.8 covers some examples:

Table 8.8: ITS technology used in Australia

ITS Technologies	*Use of Technology*
1. Adaptive Traffic Control System	to provide priority for road–based public transport.
2. Freeway Management and Information System	to reduce delays due to traffic incidents
3. Electronic Fare Collection System	to improve the convenience of public transport travel and reduce system costs.
4. In-Vehicle Navigation and Information Systems	to assist drivers and reduce unnecessary travel.
5. Vehicle Location and Scheduling System	to reduce theft, improve roadside service and efficiency of freight movement.
6. Advanced Traveller Information System	to improve user's understanding and efficiency of use of public transport system.
7. Advanced Warning System	to warn drivers of dangerous driving conditions due to an accident, weather or congestion.
8. Crash Avoidance System	to indicate drivers of potential collision and advises appropriate evasive action
9. Enhanced Emergency Response System	to allow isolated drivers or vehicles involved in a serious accident for emergency services.
10. In Car Navigation Systems	to provide visual directions to reduce travel times.
11. Automated Highways	maintain headway appropriate to driving conditions.
12. Automated Enforcement	identifies infringements and offenders remotely.

8.10.3 Europe

ITS application areas and services in Europe are shown in Table 8.9

Projects like "DRIVE" (Dedicated Road Infrastructure for Vehicle Safety in Europe), "PROMETHEUS" (Programme for a European Traffic with Highest Efficiency and Unprecedented Safety) "ROMANCE" (Road Management System for Europe) are advanced technologies, used for improving travel behavior so as to achieve more sustained and efficient transport operation.

Table 8.9: ITS service areas in UK and Europe

Developed Areas	User Services
Automated Debiting (Enabling efficient fee collection)	Motorway tolling Congestion pricing Parking pricing Driver Information and Route Guidance
Control Systems (Making best use of Road space)	Variable Message Signs Ramp metering Motorway automatic incident detection Enforcement of speed limits Tolling
Safety and Environment (Improving Safety on the Network)	Emergency call systems Driver vision enhancement Intelligent cruise control Red light cameras Enviromental monitoring and control
Traffic Control and Information Centers (Relieving traffic congestion)	Computerized Traffic Signals Split Cycle Offset Optimization Technique –SCOOT Automatic Incident Detection Parking Guidance Systems
Public Transport Management (Improving service quality and reliability)	Integrated Ticketing – Smart cards On–line passenger information terminals Bus Priority at traffic signals Real–time information at bus stops Multi–modal systems
Fleet and Freight Management (Improving efficiency and security)	Automatic vehicle location Electronic tagging Real–time logistics Hazardous goods monitoring
Traffic and Travel Information (Providing real time message)	On–line traffic information Traffic message broadcasting In–car route guidance

8.10.4 United States of America

ITS services in U.S.A cover the following:

- Collection and transmission of information on traffic conditions, and transit schedules for travellers before and during the trip.
- Information regarding hazards and delays, so that travelers can change their plans/routes.
- Decrease and early clearing of congestion, re–routing of traffic flow.
- Through automated tracking, improve productivity and safety of people.
- Automatic toll collecting systems.
- Route guidance and path finding to assist drivers in reaching the desired destination.

The technologies in U.S.A are continuously evolving. Some of the systems used are called:

ERGS (Electronic Route Guidance System)

CTSCS (Centralised Traffic Signal Control System)

FMS (Freeway Management System)

With these systems travel times, delays, congestion, and accidents are reduced, while travel speed, and comforts of driving increased. Traffic management and control both at metropolitan as well rural level, became better, providing safety, comfort and convenience in travel by private car or transit systems.

8.10.5 Use of ITS in India

Road improvement programmes and National Highway Development Programmes are aimed to provide better riding surface, better road geometry, time savings and faster and greater movement of traffic. These programmes save vehicle operating costs and fuel, reduce maintenance cost of the road, etc.

India with more than 85,000 fatalities per annum, account for 10 percent of total world's road fatalities (year–2000). Accidents on National Highway are more severe type, compared to other roads, due to high speed. Number of accidents is on increase on yearly basis, resulting in loss of crores of rupees. In India, majority of victims are pedestrians, cyclists, two-wheeler riders, etc. due to mixed nature of fast and slow traffic.

Due to population growth, there is increase in number of vehicle, traffic congestion and pollution due to road traffic in urban and suburban areas. Pollution caused by small particles released from the burning of diesel and petrol along with emission of CO_2, NO_x, Hydrocarbons, are poisonous to health, in metropolitan cities of Delhi, Mumbai, Kolkata. Ahmedabad, etc.

ITS technologies can be very helpful in reduction of vehicular pollution, traffic congestion on roads, and overall delays to traffic, by passing information about driving and road conditions, accidents, etc. Provision of public transport information, traffic restriction information, pedestrian and vehicular route guidance, etc. similar to other countries like Japan, U.S.A, Australia and U.K. The road length in India is very large, and also the number of vehicles as compared to these countries.

ITS technologies can however be tried at least on selected National Highways, and also in urban areas to reduce, congestion, pollution and accidents.

In India, heterogeneous nature of traffic and each road user being treated equally, are some problems, however effort should be made to solve traffic problems in core areas of metropolitan cities by resorting to Intelligent Traffic Management Systems (ITMS) and installing intelligent traffic signals, which are networked into the Area Traffic control (ATC).

Application of ITS to transportation network needs large scale investments and therefore its viability to Indian road network should be studied in terms of costs and benefits. Viable ITS technologies can deliver safe, efficient, comfortable and environmental friendly transportation system in India. We have to use advanced systems where human element is minimized, some examples are:

- A simple use is of a speed sensor which is embedded in pavement. It will catch any violation of speed automatically for all vehicles all the time, unlike police vehicle with radar speed gun, which can check only few vehicles.
- The Global Positioning System (GPS) mounted on vehicles keeps the driver informed and guide him where to turn right or left for reaching the destination at shortest time or cost.
- Geographic Information System (GIS), integrated with GPS, can help in public transport system for providing latest information about routes, services, and schedules.

As management of traffic has strong connection to space and time, the combination of GIS, GPS, and telemetry technologies have enormous possibilities of traffic management and planning in India.

8.11 BASIC ROAD PATTERNS

There are 3 basic types of road patterns. They are grid iron, linear and radial pattern.

8.11.1 Grid-Iron Pattern

It consists of straight lines and rectangular coordinates. It is easy to lay. It is very convenient for traffic as traffic can be spread over a grid. For one–way street system, it is ideal as alternate streets can be made one way in opposite direction. As routes are available in all the four directions, it is easy for through traffic to by-pass central business area.

The disadvantage is that traveling in diagonal direction involves extra distance. This can be overcome by providing diagonal routes upon the grid. Jaipur city in India is planned on this pattern.

8.11.2 Linear Pattern

In the old days, settlements established alongside the long tracks. As towns grew, they began to grow laterally. In the linear system the main traffic flow is channeled into one main street. The disadvantage of this system is that local and through traffic both use the same route. Traffic volumes are thus increased.

8.11.3 Radial Pattern

In this pattern several roads radiate from the city centre. Towns thus develop along the radial routes. Ring roads are developed to divert the traffic.

Ring road is a highway that is circumferential about central area of the city. As traffic grows ring roads that come into being are called inner ring road, outer ring road and intermediate ring road. In reality a ring road may be round, square, elongated or incomplete depending upon local conditions.

Figure 8.14 shows basic road layout patterns. Several combinations of these patterns or even irregular pattern are used.

ALTERNATIVE URBAN TRANSPORTATION SYSTEM

Rail based transportation system has the advantages of higher passenger handling capacity, less intrusion and congestion, higher energy efficiency, minimal pollution, superiority in terms of comfort, reliability, safety and speed, etc.

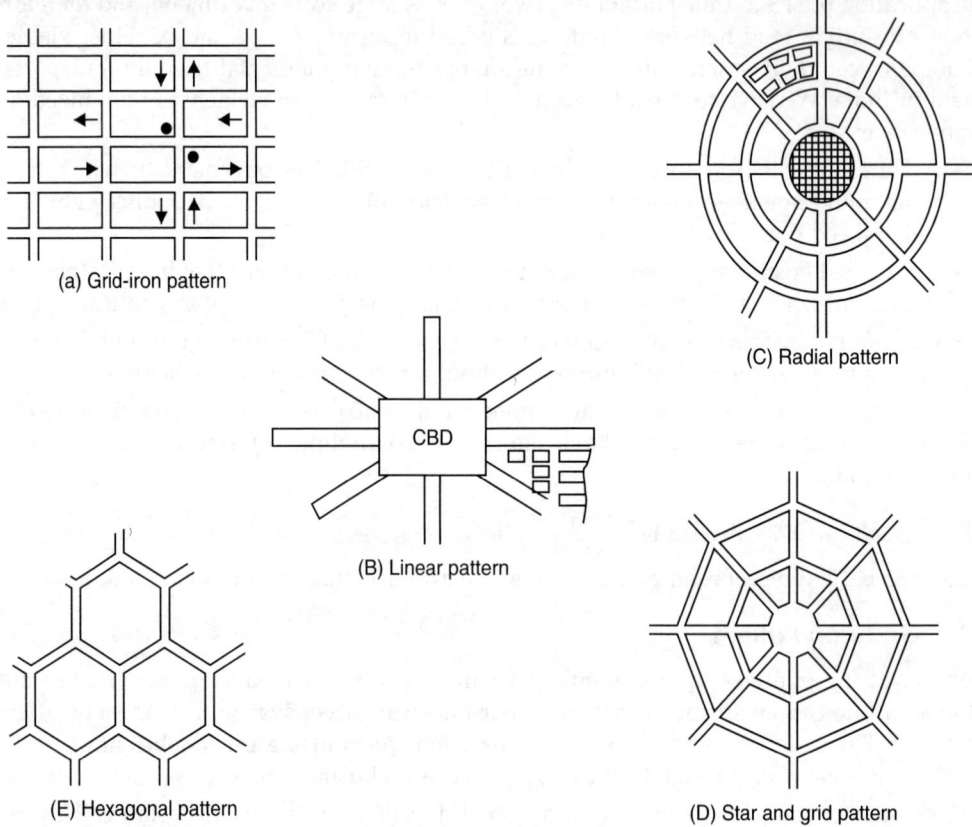

(a) Grid-iron pattern

(C) Radial pattern

(B) Linear pattern

(E) Hexagonal pattern

(D) Star and grid pattern

Fig. 8.14: Basic road patterns

These advantages of rail based transport system should be considered with the technical considerations like topography, terrain, disturbances to adjoining property, geology, sub-soil, availability of technology, ease of maintenance and replacement of parts and scope of incremental augmentation.

Metro rail-services are now being considered as one of the alternative to solve urban traffic problems. Technical and economical consideration, cost-benefit analysis, should be undertaken to justify use of Metro Rail Transport System (MRTS) or Light Rail Transport (LRT), or other rail based system, for urban transport. Merely because other countries or cities posses metros or these systems are symbols of modernity or political visibility, should not dictate the choice.

When cities grows in size and population, the road based transport system of personalized vehicles and buses becomes inadequate to handle the growing traffic. To provide for future needs alternative system of urban transport through railways (MRTS or LRT) is required to be developed. Some examples of railway systems developed in Indian cities for urban transport are the following:

Mumbai Suburban Railway System

Mumbai is the largest city of India, which is well served by suburban railway system, operated by Indian Railways (Western and Central Railways)

The total length of the suburban railway system exceeds more than 300 kms, providing services to Mumbai metropolitan region and nearby areas.

Kolkata Metro

Kolkata Metro transport project was the first railway transportation project of its type. It consists of followung railway lines:

Line 1: North–South axis between Dum Dum and Tollyganj (16.45 km)

Line 2: East–West line between Kolkata and Howrah (14.67 km)

Line 3: Between Joka and BBD Bagh (16.72 km)

Line 4: Between Naopara and Barasat

Line 5: Between New Garia and airport.

It passes through congested parts of the city.

Delhi Metro

Delhi Metro rail project consists of elevated as well as underground system. Three major railway lines for the Metro rail are:

Line 1: Shahdra–Trinagar–Rithala (22.06 km)

Line 2: Vishwa Vidalaya–Central Secretariat (10.84 km)

Line 3: Indraprastha–Barakhamba–Dwarka City (32.10 km)

The system is still developing and expanding to different areas.

Bengaluru Metro

Bengaluru Metro network called 'Mamma Metro' consists of two corridors,. The North–South corridor is named 'Green Line' connecting Nagasandra to Puttenahalli via Yeshwantpur, Rajaji Nagar, Kempegowda, Lal Bagh, Jayanagar, Banashankari, etc. and East–West corridor named 'Purple Line' connecting Baiyappanahalli to Mysore road via Indira Nagar, Gubbon park, Kempegowda, Vijainagar. Two corridors intersect at Kempegowda station which is a two level interchange station. The construction work is still in progress.

Jaipur Metro

Jaipur Metro Rail System (JMRS) is India's fourth, after Kolkata, Delhi and Bengaluru. It has two major lines. The first line called 'Green Line' provides services between Mansarover–on the outskirts of Jaipur to Bari Chaupar via Chandpole, in the walled city. The second phase called 'Orange Line' (North–South corridor) will connect Sitapur Industrial area in South to Ambabari in North via Ajmerigate, MI Road, The track will be partly elevated and partly underground. The work on the first phase is still in progress.

Traffic Flow Theory

9.0 PREAMBLE—BASIC FLOW PARAMETERS

Traffic flow theory is the description of traffic flow characteristics by application of the laws of physics and mathematics, to measure the ability of a roadway to accommodate traffic. The knowledge of the distribution of vehicles in traffic lane is of vital importance in understanding the behaviour of traffic flow and in estimating the required capacity of a given road. The characteristics of traffic flow are influenced by limitations of human behaviour, the vehicle operating characteristics, the need and purpose of the trip and physical parameters of the highway system. The knowledge of the flow of group of vehicles on the road is very important for planning, design, control, operation and management of traffic.

Three traffic flow parameters are of basic importance and they are inter-related:

    i. Speed V

    ii. Volume (Flow) Q

    iii. Density K

Definitions of Flow Parameters

(i) Speed

Speed is the rate of movement of traffic or distance covered in a unit time. It is expressed as distance/time (km/h).

Speed is one of the important characteristics of a traffic stream. Speed adopted by a vehicle depends upon various factors like the vehicle type, road conditions and traffic flow. There are two types of speeds:

    (a) **Time mean speed (V_t):** The mean speed of vehicles over a period of time at a point is called time mean speed. It is spot speed.

    (b) **Space mean speed (V_s):** The mean speed over a space at a given instant is called space mean speed.

(ii) Traffic Volume (Flow) (Q)

Volume is the number of vehicles passing a given point during a specified period of time expressed as number of vehicles/time (vehicles/hour).

(iii) Traffic Density (K)

Density is the number of vehicles occupying a unit length of the moving lane of a roadway at a given instant, expressed as number of vehicles/distance (vehicle/km).

(iv) Space Headway: (Spacing of Vehicles) (S)

Space headway is the distance between the head to head of successive vehicles, expressed as distance/number of vehicles (km/vehicles).

It is thus reciprocal of density. Space Headway = Gap between vehicles + length of rear vehicle. It is measured in metres.

(v) Time Headway (h)

Time headway is the time interval between head to head of successive vehicles passing a given point measured from front to front of the successive vehicles, expressed as time/number of vehicles. It is thus inverse of volume, and is measured in seconds.

By simple definition of the above terms, the relationship between traffic flow (Q), along a road can be related to traffic density (K) and the speed (V_s) as follows:

Volume (vehicles/hr) = Density (vehicles/km) × speed (km/hr)

$$Q = K \times V_s$$

or

$$Vs = \frac{Q}{K}$$

Since density (K) is reciprocal of the space headway (S)

$$V_s = \frac{Q.S}{1000} \cdot$$

or

$$S = \frac{1000 \, V_s}{Q}$$

$$S = \frac{h}{3600} \times V_s \times 1000$$

where h is the time headway.

9.1 RELATIONSHIP BETWEEN TIME MEAN SPEED AND SPACE MEAN SPEED

Speed can be measured in following two ways:

i. Measuring speed at one point using a radar speed meter, which gives immediate reading of the speed of a vehicle. The average speed of several vehicles is obtained by sum of the speeds of each vehicle divided by the number of vehicles observed. This arithmetic mean speed is called "Time Mean Speed" (V_t) and gives distribution of speed of vehicles in time.

ii. Another method of measuring speed is by taking a photograph of a section of a road, at a predetermined time interval (t), by measuring the distance (x), covered by each vehicle and computing speed (v = x/t). This is the speed corresponding to the average of overall travel times (running times) over a specified section of highway. It is called 'Space Mean Speed'. It gives an idea of distribution of speed in space.

These two speeds differ because, greater proportion of slow moving vehicles appear in the photographs as they occupy space for a greater length of time, than faster vehicles.

Time mean speed is therefore higher than the space mean speed. By timing of vehicles over a known distance, both time mean speed (V_t) and space mean speed (V_s), can be obtained using the following two equations:

$$V_t = \frac{x}{N}\left(\frac{1}{t_1} + \frac{1}{t_2} + \ldots\ldots\right) \text{ and}$$

$$V_s = \frac{Nx}{(t_1 + t_2 + \ldots\ldots)}$$

where:
V_t = time mean speed
V_s = space mean speed
x = measured distance
t_1, t_2, t_3 = the times for the passage of vehicles of sample N.
N = number of vehicles.

Space mean speed (V_s) can also be derived by taking the harmonic mean of time mean speed (V_t).

$$V_s = \frac{N}{\sum \dfrac{1}{V_t}}$$

Since it is easier to measure time mean speed (spot speed) in the field, the space mean speed can be calculated as follows:

$$V_s = \frac{N}{\displaystyle\sum_{i=1}^{m}\dfrac{f_i}{V_{ti}}}$$

where:
V_s = space mean speed
V_{ti} = time mean speed
f_i = number of vehicles with speed group
m = number of speed groups
N = Total number of vehicles in the sample $\displaystyle\sum_{i=1}^{m} f_i$

The approximate relationship between time mean speed (V_t) and space mean speed (V_s) is given by:

$$V_t = V_s + \frac{\sigma_s^2}{V_s}$$

Where σ_s^2 is the variance in space mean speed (V_s).

It is often desired to convert spot speed (time mean speed) to space mean speed (V_s). Following approximate relationship can be used for this purpose.

$$V_s = V_t - \frac{\sigma_{v_t}^2}{V_t}$$

where $\sigma_{v_t}^2$ is the variance in the time mean speed (V_t)

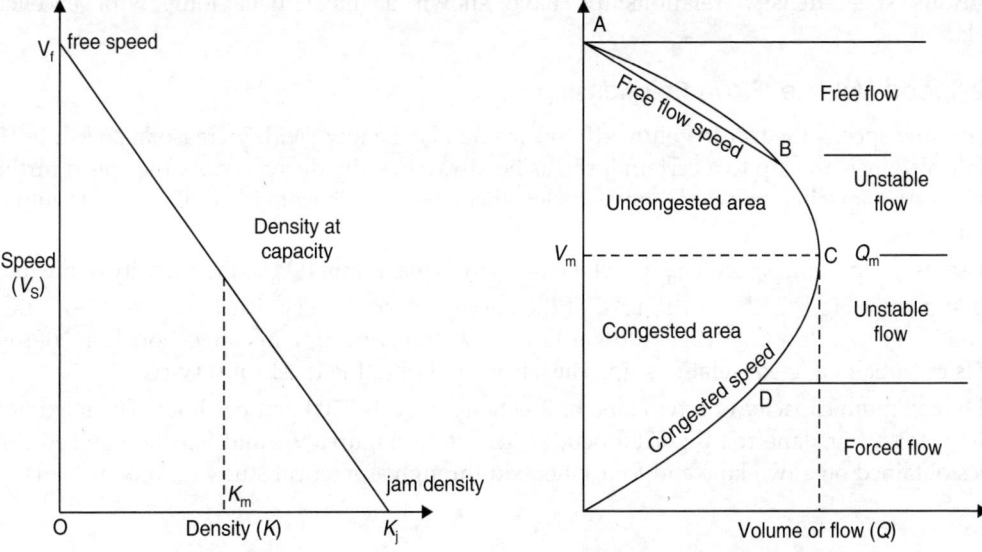

Fig. 9.1: Speed-density relationship

Fig. 9.2: Speed-volume relationship

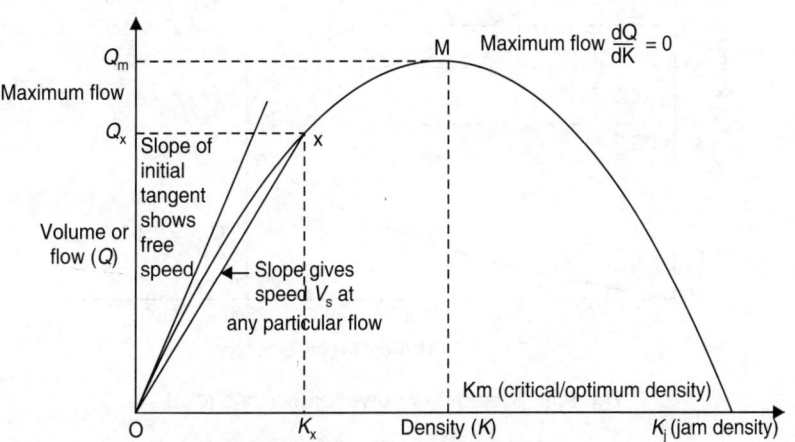

Fig. 9.3: Flow density relationship

9.2 RELATIONSHIP BETWEEN FLOW PARAMETERS

Figures 9.1 to 9.3 show fundamental diagrams of traffic flow parameters, and their relationships, described in following paragraphs.

9.2.1 Speed-Density Relationship

As shown in Figure 9.1 with increase in density, the speed decreases. When there is no vehicle density (K) is zero, the speed is maximum. This speed is called "Free Speed" (V_f).

At very high density (K_j), the congestion is so high that vehicles approach zero speed. This density is called "Jam Density" (K_j). If the average length of cars in a stream is 5.0 m, the jam density is given by:

$$K_j = \frac{1000}{5} = 200 \ VPK.$$

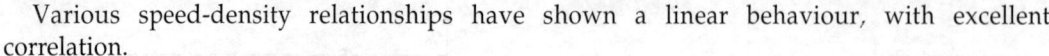

Various speed-density relationships have shown a linear behaviour, with excellent correlation.

9.2.2 Speed-Volume (Flow) Relationship

At very low speeds the traffic volume (flow) would also be low. With increase in speed, traffic volume also increases up to a certain limit, as headway initially decreases. As the speed further increases the spacing between the vehicles increases and becomes so large that volume decreases.

There is an optimum speed (V_m) at which the flow is maximum (Q_m) – the capacity of the road.

As shown in Fig. 9.2, the portion AB of the curve is normally classified as "Free flow", BCD portion is "Forced flow" or "Break-down-flow" condition, resulting in congestion. Zone of free-flow is essentially a linear relationship. This curve describes the levels of service.

The maximum capacity of a two lane bidirectional road is 2500 pcu per hour. The maximum capacity for a four lane road is 6200 pcu/hr/direction. Figures 9.4 and 9.5 show speed flow curves obtained on a two lane and four lane road through a practical study on Indian roads.

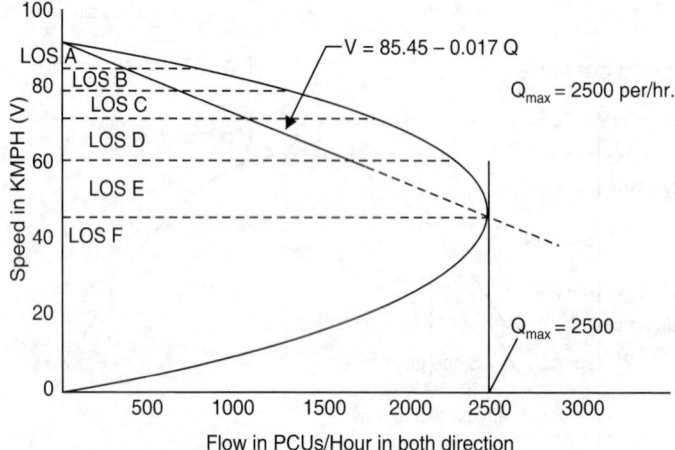

Fig. 9.4: Speed-flow curve for two lane road

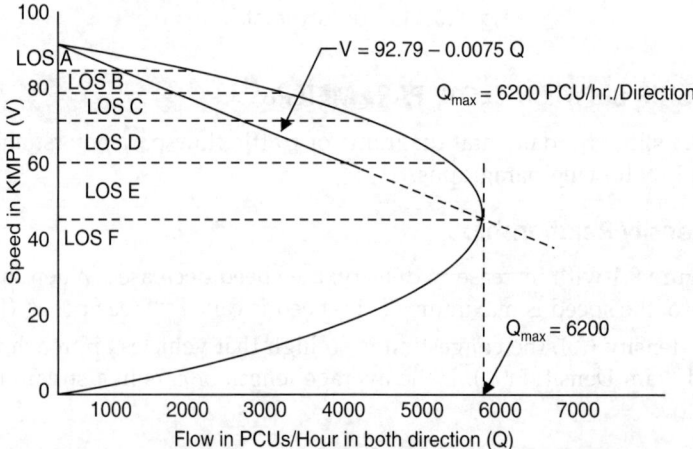

Fig. 9.5: Speed-flow curve for four lane divided carriageway

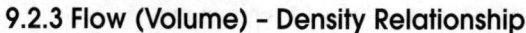

9.2.3 Flow (Volume) – Density Relationship

As shown in Figure 9.3, as the density increases from zero (when there is no vehicle at all), volume increases up to the point of "Critical Density" km–which is the dentistry corresponding to maximum flow Qm. It is also called "Optimum Density". Thereafter the volume decreases as density continues to increase to a maximum value known as "Jam Density" (K), when all vehicles are stopped. As density increases the speed of the vehicles is reduced, reducing the flow, till it reaches jam density when there is no movement ($V = 0$).

If a line is drawn form origin to any point (x) on the curve, them the slope of the line is the space mean speed corresponding to the traffic flow (Qx) represented by the point on the curve, at the density kx.

9.2.4 Density and Spacing Relationship

The density of a traffic stream is a measure of congestion. Density is difficult to quantify in the field, however, it can be obtained by means of aerial photography.

Spacing of vehicles shows the way a driver evaluates traffic stream congestion at any moment. At bumper to bumper packing spacing is zero. Knowledge of density of traffic flow indicate, that at what densities it becomes "Forced Flow" "Unstable Flow" or "Bumper-to-Bumper" condition.

9.2.5. Time headway and Space Headway (Spacing) Relationship

The relationship between headway and spacing is given by:

$$\text{Average headway (seconds)} h = \frac{\text{Average Spacing (m)}}{\text{Space Mean Speed (m/s)}}$$

$$h = \frac{S}{Vs}$$

Headways are fundamental to all traffic operations and control such as overtaking, lane changing, intersection manoeuvers, etc.

9.2.6 Poisson's Distribution

If the vehicles are not subjected to interference each driver operates independently of all other. Under these conditions equal intervals of time or space are equally likely to contain a given number of vehicles. The resulting distribution can then be defined as random, and probability of an event occurring can be determined theoretically by Poisson's distribution given by:

$$P(x) = \frac{e^{-m} m^x}{x!}$$

Where :

$P(x)$ = Probability that x vehicles will arrive in given interval of time.

m = Mean number of vehicles arriving in the given interval of time.

$$\frac{Qt}{3600}$$

Q = traffic volume (veh/hr)

t = given interval of time (seconds)

e = base of Naperian logarithms = 2.718

The probability that no vehicle will arrive in time t, at a volume Q is given by:

$$P(0) = \frac{e^{-m} m^o}{o!} = e^{-m}.$$

9.3. GREENSHIELD'S LINEAR SPEED-DENSITY MODEL

Greenshield analysed data collected on speed and density, in Ohio and found a linear relationship between them.

The basic equation of traffic flow is:

$$Q = Vs \cdot K \qquad \qquad \text{... (i)}$$

Figure 9.6 shows straight line relationship between speed and density, whose equation can be written as:

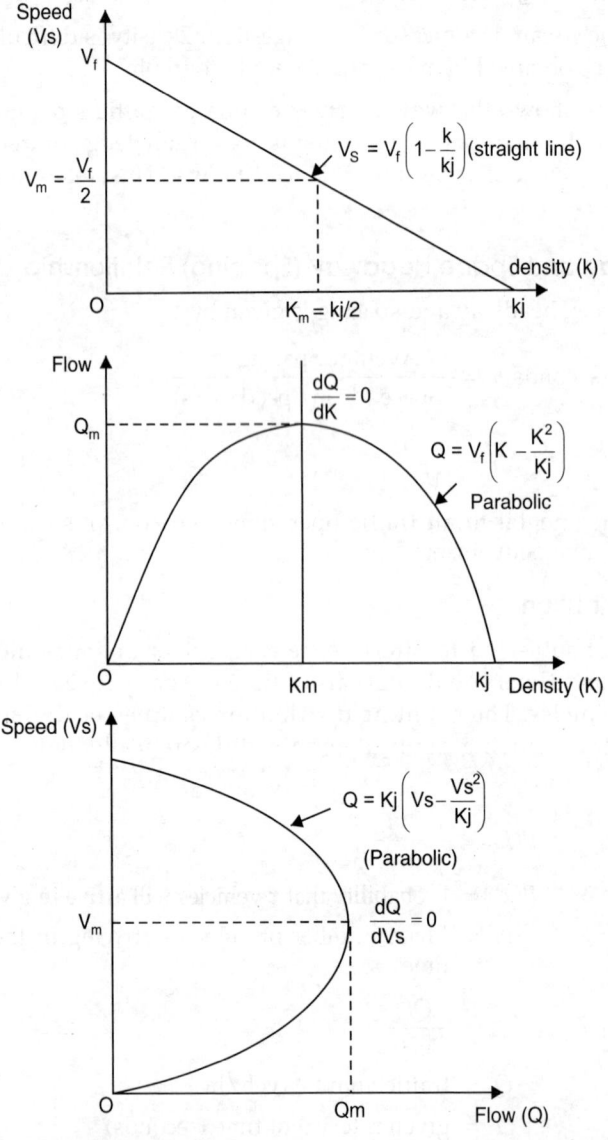

Fig. 9.6 : Linear speed-density, flow characteristics

$$\frac{K}{K_j} + \frac{Vs}{V_f} = 1 \qquad \text{... (ii)}$$

or $$V_s = V_f\left(1 - \frac{k}{k_j}\right)$$

from (i) $$\frac{Q}{K} = V_f\left(1 - \frac{k}{k_j}\right)$$

$$Q = V_f\left(1 - \frac{k}{k_j}\right)K.$$

This model is Greenshield's linear model.

At maximum volume $\dfrac{dQ}{dK} = 0$

Differentiating with respect to K.

$$\frac{dQ}{dK} = \frac{d}{dK}\left(V_f K - \frac{V_f K^2}{K_j}\right)$$

$$= V_f - \frac{V_f - 2K}{K_j}$$

$$\frac{dQ}{dK} = 0 = 1 - \frac{2K}{K_j}$$

or $$\frac{2K}{K_j} = 1, \ K = \frac{K_j}{2}$$

The maximum volume (flow) Q_{max} occurs when density (K) is half of the jam density (Kj). Density at maximum flow is half the jam density.

Equation (ii) is $$\frac{K}{K_j} + \frac{V_s}{V_f} = 1$$

Putting $K = K_j/2$ in above equation for maximum flow.

$$\frac{K_j}{2K_j} + \frac{V_s}{V_f} = 1$$

$$V_s = \frac{V_f}{2} \qquad \text{...(iii)}$$

The space mean speed (*Vs*) is half of the free speed (*Vf*). i.e. speed *Vs* at maximum flow (volume) is half of the free speed.

From equation (i), for maximum flow (volume)

$$Q_{max} = K.Vs = \frac{K_j}{2} \times \frac{V_f}{2} = \frac{K_j V_f}{4} \qquad \text{...(iv)}$$

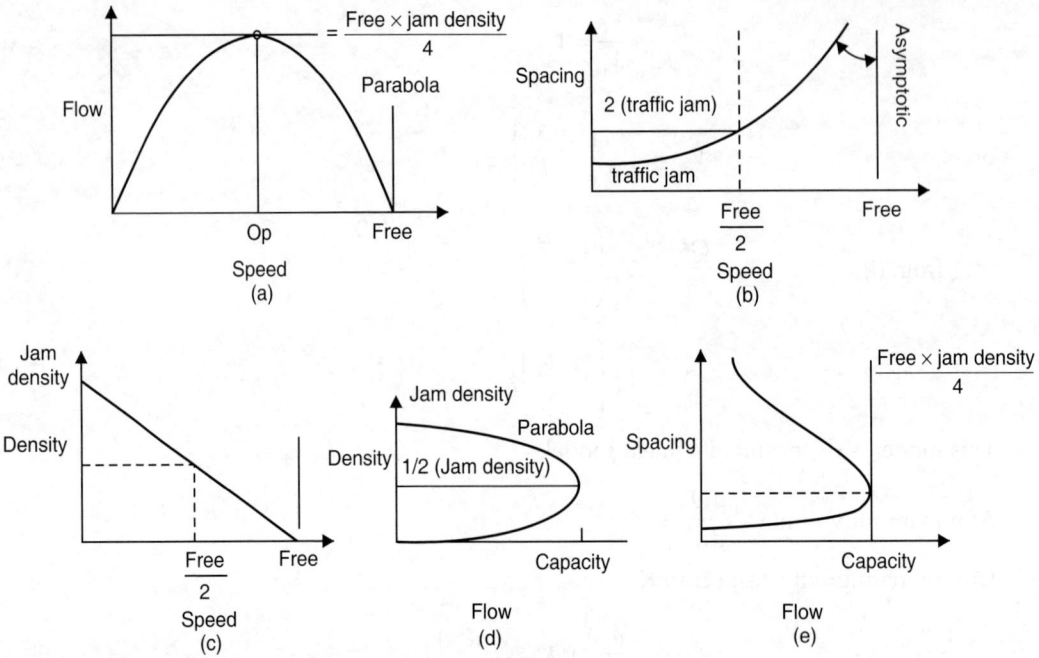

Fig. 9.7: Typical speed-flow-spacing-density relationship based on free speed (liner speed density curve)

Rewriting equation (i)

$$Q = K\,Vs$$

For maximum flow differentiating it w.r. to K and equating to zero.

$$\frac{dQ}{dK} = K\frac{dV_s}{dK} + V_s \times 1$$

$$O = \frac{dV_s}{dK} + V_s$$

$$Vs = K.\frac{dV_s}{dK}, \text{ for point } M, \frac{dQ}{dK} = 0$$

$$K = Km.$$

$$\therefore \quad Vs = -Km.\frac{dV}{dK}$$

Integrating both sides

$$\int \frac{1}{Vs}\,dVs = -\int \frac{1}{Km}.\frac{dV}{dK}$$

$$\log^{Vs} = -\frac{K}{Km} + C \qquad \qquad ...(v)$$

Where C is constant of integration.

As we know.

When $K = 0$, $Vs = V_f$

so $\qquad\qquad \log^{V_f} = 0 + C$ or $C = \log^{V_f}$

Substituting in equation (v)

$$\log^{V_s} = -\frac{K}{K_m} + \log^{V_f}$$

$$\frac{K}{K_m} = \log^{V_f} - \log^{V_s}$$

or $\qquad\qquad K = K_m . \log\dfrac{V_f}{V_s}$ ...(vi)

Various linear speed-density model relationships and equations are shown in Figs. 9.6 and 9.7.

9.4 OTHER THEORIES OF TRAFFIC FLOW

Some of the advanced theories are described in short in the following paragraphs. Detailed description of these theories is beyond the scope of this book.

9.4.1. Analogy with Fluid Flow: (Hydrodynamic Model)

In this theory traffic is described in terms of fluid flow behaviour. Though fluid models have certain shortcomings in treating traffic as a continuous medium, they are found useful in expressing the behaviour of the stream rather than individual cars. Lighthill and Whitham have developed models based on kinematic theory of fluid dynamics.

The behaviour of traffic at bottleneck appears to be acting like a shock-wave in fluids. Several traffic problems are analysed by assuming a system of traffic waves. Shock-waves are formed when traffic approaches a signalised intersection, where vehicles stop in succession or at other locations which brings vehicles successively to halt at bottlenecks. In the region of the halted vehicles the flow is zero and the density is greatest jam density. The region of jam density will be propagated backwards down the road by means of shock transition.

9.4.2. Car Following Theory

In the car following theory, driver-vehicle are considered as one system. In this theory the driver which follows a vehicle, responds to the situation (stimuli) created by the movement of the vehicle moving ahead. The following driver is required to follow the speed, acceleration and spacing etc. according to the vehicle in front. He will allow safe distance between himself and the cars ahead depending upon his own reaction time and the braking capacity of his vehicle.

Empirical speed-flow curves can be predicted using this theory. This theory is based on uniform traffic flow not subjected to any interruptions. Detailed discussion of this theory for uniform and non-uniform flows is beyond the scope of this book.

9.4.3. Queueing Theory

Queues of vehicles are formed at places where demand approaches or exceeds capacity. The cause of queues formation may be either due to temporary increase in demand or a transitory decrease in capacity.

In the absence of overtaking opportunities, the queuing will lead to platoon of vehicles, with slow moving as the platoon leader. Each vehicle in a platoon will adopt the slow lead vehicles speed but its own individual headway. These conditions will be maintained by a continuous process of vehicles catching up and then overtaking within the traffic mix.

The critical smallest gap people will accept in passing through or into the traffic stream will depend upon type of traffic flow, and will vary with drivers. Theoretical models are developed

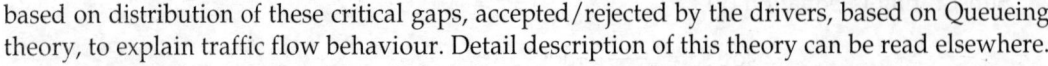

based on distribution of these critical gaps, accepted/rejected by the drivers, based on Queueing theory, to explain traffic flow behaviour. Detail description of this theory can be read elsewhere.

9.5. HIGHWAY CAPACITY

The capacity of a traffic facility is the maximum hourly volume that can reasonably use it under the prevailing or defined road and traffic conditions. The speed at capacity is called optimum speed. Highway capacity is qualitative and quantitative measure to know the service provided by the facility.

Capacity of a facility can be affected by number of factors such as vehicle and roadway characteristics, operation controls, and environmental elements.

9.5.1. Capacity Definitions

(a) **Possible Capacity:** 'Capacity' or 'Possible Capacity' is the maximum number of vehicles that can pass over a given section of a lane or highway in one direction (or both directions for a 2 lane or 3 lane highway) during a given period of time (usually one hour), under prevailing roadway and traffic conditions.

In the above definitions, it is essential to state prevailing roadway and traffic conditions. Prevailing road conditions are those which depend on the physical features of the roadway and can not change unless some reconstruction or improvements are done. Prevailing traffic conditions depend upon the nature of traffic on the roadway and any one of which may change or be changed from hour to hour or during various periods of day.

In addition to these prevailing conditions, conditions of weather (clear, dry, cold, hot, rain, snow, fog, windy etc.) and conditions of visibility during different hours of the day also effect the ability of a roadway to accommodate traffic.

(b) **Basic or Ideal Capacity:** The capacity of lane or roadway under ideal conditions (not prevailing) is called basic or ideal capacity. The ideal conditions are:

– Traffic lanes are 3.5 m wide, with adequate shoulders and no obstructions within 2 m of the pavement edge.

– Horizontal and vertical alignment is satisfactory for average roadway speed of 110 km/hr and overtaking sight distance is not restricted on two-lane and three lane roads.

– Road surfaces is good and smooth.

– There is uninterrupted flow, free from side interferences of vehicles or pedestrians.

– Only passenger cars in the traffic stream.

– Visibility and weather conditions are normal.

Possible capacity is usually less than the basic capacity (ideal capacity) as prevailing roadway and traffic conditions are seldom ideal. In the worst case, when congestion is high, traffic may come to a stand still and possible capacity may approach zero. Thus possible capacity fluctuates from zero capacity to 'Basic Capacity' depending upon prevailing highway and traffic conditions. This gives rise to another definition of practical capacity.

(c) **Practical Capacity:** The maximum number of vehicles that can pass a given point on a lane or roadway during an hour, without traffic density being so large as to cause unreasonable delay, hazard or restriction to drivers freedom to manoeuvre. Service volume can better be termed as practical capacity.

(d) **Design Capacity:** The flow which the designer assumes in order to design all associated traffic facilities. It is usually less than peak volume. It is common to use 30th highest hourly volume as the design capacity.

(e) **Peak Hour Factor:** The traffic volume during peak hour expressed as percentage of the ADT. The peak hour volume is highest hourly volume on the road.

9.5.2. Basic Capacity of a Traffic Lane

Basic or Theoretical capacity, under most ideal conditions as defined above can be arrived as follows:

If V = Speed of vehicles km/h

H = The average headway distance in meters from centre to centre of vehicles.

= Average length of vehicle (L)　+　Clear spacing between Vehicles in the stream (S)

= L + S.

For safe operation and under ideal conditions, the spacing between the two vehicles should be equal to the stopping sight distance.

Stopping sight distance (SSD)　= $0.28\ Vt + 0.1\ V^2$

where t = reaction time of drivers varying from 1 to 2.5 seconds.

V = Speed (design) in km/h

Therefore, Basic capacity $= \dfrac{V \times 1000}{H}$ Vehicles/hr

$= \dfrac{V \times 1000}{L + SSD}$

Highway Capacity Units

Highway capacity is generally expressed in common type of vehicle units in terms of passenger car equivalents. The different vehicles having varying dimensions and speeds are expressed in Passenger Car Units (PCU). The equivalency factors for different types of vehicles under typical geometric and traffic conditions are evaluated.

Typical PCU conversion factors for rural roads are shown in Table 9.1.

Table 9.1: Conversion factors for vehicles in PCU for rural roads (IRC)

Sl. No.	Vehicle type	Equivalency Factor PCU
1.	Passenger car, auto rickshaw, van, etc.	1.0
2.	Motor cycle, scooter, moped	0.50
3.	Cycle	0.50
4.	Light commercial vehicles	1.50
5.	Cycle rickshaw	2.00
6.	Truck, bus, hand cart	3.00
7.	Truck-trailer, tractor	4.50
8.	Horse drawn vehicles	4.00
9.	Bullock cart	6–8

9.5.3 Levels of Service

The concept of levels of service relates to operating conditions encountered by the traffic on a given traffic lane or roadway, depending upon traffic volume, time of the day, interruptions, weather, etc. It is a qualitative measure of the effect of number of factors including:

- Operating speeds and travel time.
- Traffic interruptions and frequency of stops
- Freedom to manoeuvre
- Safety
- Driving comfort and convenience
- Vehicle operating costs, etc.

In practice any given road or its part may operate at a wide range of level of service depending upon time of day, day of a week and period of a year.

Six specific levels of service are designated by letters A to F, defined in terms of particular limiting values of travel speed and ratio of demand (or service) volume to capacity of the road, as described below in brief.

(i) Levels of Service A (Free Flow)

This condition of free flow is accompanied by low volumes of traffic and high speed. Traffic density will be low, with uninterrupted flow speeds controlled by driver desires. There is little or no restriction in maneuverability due to presence of other vehicles. Drivers can maintain desired speeds with little or no delay. Operating speed is more than 95 km./hr. The general level of comfort and convenience to the road user is excellent.

(ii) Level of Service B (Stable flow but speed begins lowering)

This occurs in the zone of stable flow, with operating speeds beginning to be restricted by traffic conditions. Drivers will have reasonable freedom to select their speeds and lane of operation. Reduction in speed is not unreasonable and not much of restriction. Average operating speed ranges between 90-95 km/hr.

(iii) Level of Service C (Stable flow, speeds continue lowering)

C level is still in the zone of stable flow, but speeds and maneuverabilities are more closely controlled by the traffic volumes. Most of the drivers feel restricted in their freedom to select their own speed, changing the lanes or overtake. Operating speeds between 80-90 km/hr.

(iv) Levels of Service D (Unstable flow, speed restricted but tolerable)

This level approaches unstable flow, with tolerable operating speeds. Fluctuations in traffic volume and temporary restrictions to flow may cause substantial drops in operating speed. Drivers have little freedom to manoeuvre and comfort, convenience are reduced. Operating speeds between 60-80 km/hr.

(v) Level of Service E (Speeds Restricted to 45 km/hr or so, unstable flow)

This level is characterized by low speeds, with volumes at or near the capacity of the highway. Flow is unstable and there may be stoppages of momentary durations. Operating speeds are between 45-65 km/hr. Freedom to movement is difficult. Comfort and convenience are very poor, drivers get frustrated. Traffic breakdowns are common.

(vi) Level of Service F (Forced flow, very low speeds, queues formed)

This level is forced flow, at low speeds. In extreme conditions both volume and speed can drop to zero, resulting in queues of vehicles backing up from a restriction down stream. Speeds are reduced substantially and congestion is high. Operating speeds are below 45 km./hr. between 25 to 33 percent of the free flow speed.

Figure 9.8 shows typical relationship between operating speed and volume/capacity ratio.

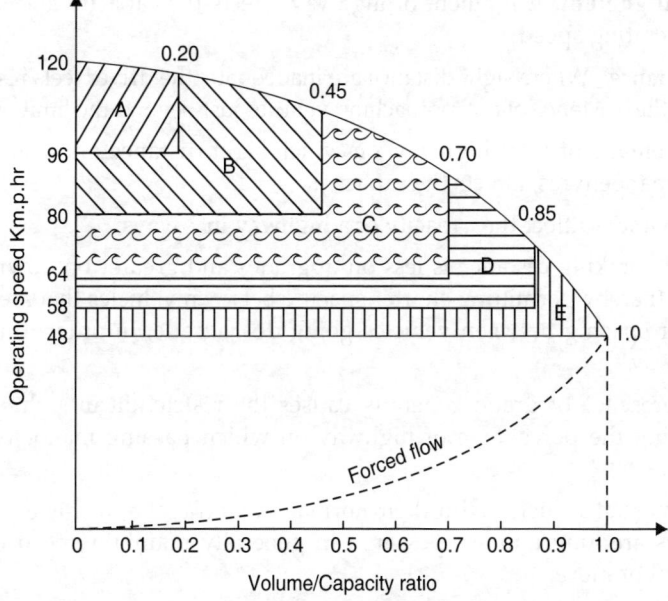

Fig. 9.8: Levels of service representation (Rural highways)

9.5.4. Factors that Affect Capacity and Levels of Service

Roadway and traffic conditions are seldom ideal and therefore capacities are lower than given above. Some adjustment factors are used to determine capacity in such cases, specially in Indian conditions. Following factors affect the highway capacity.

i. **Lane Width:** Width of a lane affects capacity. On narrow lanes capacity is less. 3.5 m wide lanes are considered necessary for mixed traffic volumes that exist these days. A 3 m wide lane reduces capacity up to 70-80% of that of 3.5 m wide lane.

ii. **Restrictive Lateral Clearances:** Side obstructions such as walls, light poles, parked cars adjacent to the edge of the traffic lane reduce the effective width of the lane and the capacity of the road. They also affect driving comfort and accident rates, etc. One lateral restriction within a section of a highway will cause bottleneck and thereby affect the capacity of entire section. This effect increases as the clearance is reduced and is less for continuous obstructions than for intermittent ones.

iii. **Shoulders:** For full capacity utilization, adequate shoulders are necessary. Without a place of refuge out side the traffic lane, one disabled vehicle can reduce the capacity of road to a great extent. Traffic lanes less than 3.5 m wide, a bitumen treated shoulder 1.2 m or more in width, increase the effective lane width by 30 cms. Auxiliary traffic lanes increase the capacity and quality of service.

iv. **Commercial vehicles:** Commercial vehicles reduce capacity of a highway, because they occupy a greater road surface and influence traffic over a larger area, than do passenger cars. They also generally travel at lower speeds.

10% trucks reduce capacity to 91% on level and 77% on rolling terrain.

20% trucks reduce capacity to 83% on level and 63% on rolling terrain.

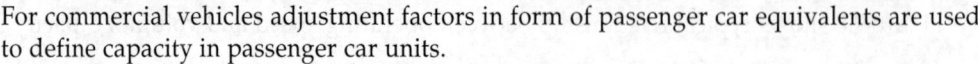

For commercial vehicles adjustment factors in form of passenger car equivalents are used to define capacity in passenger car units.

v. **Imperfect alignment:** Alignment of highway affects the capacity and level of service at different operating speeds:

(a) **Sight distance:** Where sight distance are inadequate, the driver feels restricted, he always assumes the existence of an approaching vehicles just beyond the limit of his sight.

The percentage of total highway, on which sight distance are insufficient, to permit passing manoeuvres, affects the capacity.

(b) **Grades:** Grades affect the capacity of a highway in 3 ways:

1. Vehicle braking distance is less on upgrades and greater on down grades, than on level, thereby permitting shorter spacing between vehicles that are climbing up and longer spacing between vehicles going downgrade, in order to maintain a safety headway.

2. The presence of grade generally causes the restriction in sight distance, thereby affecting the percentage of highway on which passing manoeuvres can be done safety.

3. Commercial vehicles with their normal loads travel at a slower speed upgrade, if grades are long and steep. Cars can generally maintain speed on grads up to 6 percent or more.

   The length and rate of grade are important factors affecting capacity. The usual adjustment to allow for slower speeds on a grade is to provide a climbing lane or passing bay to maintain the level of service for the road. Truck adjustment factors are used on grades.

vi. **Intersections:** Intersections at grade limit the capacity of highway, specially on the city roads. An intersection may be unsignalised or signalised.

(a) **Unsignalised intersections:** On an unsignalised intersection, if vehicles have to reduce speed occasionally, capacity is affected little. As the cross turning volume increases at intersection both the running speed and capacity are reduced.

Stop sign sometimes put on minor roads, but with progressive higher intersection volumes, interference multiplies and uninterrupted flow on the highway becomes impossible and capacity of the highway and intersection is reduced.

Because of cross road traffic interference, the capacity is determined by computing the capacity as if it were operated under traffic signals. This procedure appears to be on safe side, and is suggested as design guide.

The following are the design hour volumes for which signal control should be assumed in geometric design of intersections.

	Minimum	*Volumes per Hour*	
2 lane-through highway	450	500	650
cross road	250	200	100
4 lane-through highway	1000	1500	2000
cross road	100	50	25

Widening, channelisation and signalisation are the devices to adjust traffic volumes at intersections.

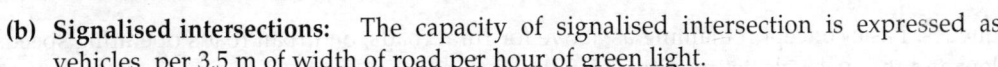
(b) **Signalised intersections:** The capacity of signalised intersection is expressed as vehicles per 3.5 m of width of road per hour of green light.

The capacity varies with the type of road, pavement width, truck percentage, signal timing etc.

Design capacity per lane at signalised intersection generally is in the range of 250 to 600 v.p.h. Additional lanes may therefore be required at intersection if traffic volumes are more, or where large proportion of the signal cycle must be devoted to the movement of traffic on the cross road.

vii. **Ramps:** The capacity and the efficiency of traffic on freeways or expressways, depend directly on the adequacy of interchange ramp and the ramp terminal, where traffic enters or leaves the highway.

The capacity of a ramp may be limited by any of its component parts:

(a) Ramp proper

(b) Entrance terminal (ramp to highway)

(c) Exit terminal (highway to ramp)

(a) **Ramp proper:** The design capacity of a ramp gets affected by lateral restrictions, curvature, gradient and proportion of trucks. Maximum capacity is about 1200 v.p.h.

(b) **Entrance terminal:** At the entrance terminal, the capacity is governed by design of terminal and type of traffic control utilized. Entrances preferably should be designed to permit ramp traffic to merge with through traffic without stop.

In most cases, ramp entrance terminals are designed for single lane entrance, though in special cases, the design may be a 2 lane-entrance. About 1200 pcu/hr. can merge in one lane without reducing speed. Where no acceleration lane is provided the entering traffic is required to stop before merging with through traffic. The capacity of ramp entrance reduces considerably.

(c) **Exit-terminal:** At ramp exit terminals, the number of vehicles that can leave the highway is affected by the volume of through traffic using the lane adjacent to ramp.

The practical capacity, on rural road, of a single lane exit, with adequate deceleration lane is about 1000 p.c./hr. On urban roads, where speeds are lower and densities are high, the capacity is 1200 pcu/hr.

viii. **Weaving sections:** Weaving sections enable one-way traffic streams to cross by merging and diverging maneuvers. They permit crossing of vehicles with least possible interference. They are often provided near to grade separators (interchanges).

Certain weaving traffic volumes, length and width of weaving sections, operate satisfactorily and give proper capacities. The total number of vehicles entering the weaving sections, if all are weaving vehicles, cannot exceed the capacity of a single lane. This is a thumb rule.

For non-weaving vehicles, there should be separate lane of adequate width.

ix. **Traffic interruptions:** Traffic interruptions are created by intersections, toll gates, bridges, and railway crossing, etc. If the demand is high then long queues may develop. Uninterrupted traffic flow on the approach, exceeding the saturation flow, adds to the queue even after the interruption ceases and the capacity of the approach route drops below saturation flow.

9.6. LEVELS OF SERVICE ON URBAN (CITY) ROADS

Urban roads have lower speeds, higher volumes of traffic, many intersections (signalised or non-signalised), roadside developments, mixed traffic conditions etc. resulting in congestion and

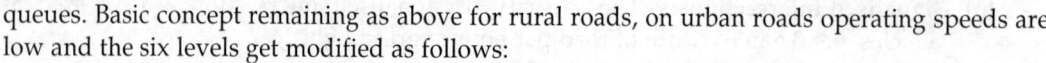

queues. Basic concept remaining as above for rural roads, on urban roads operating speeds are low and the six levels get modified as follows:

Level A   Free flow-with few stops. Average operating speed 40 km/hour or more.

Level B   Stable flow–with reasonable delays. Average operating speed between 30-40 km/hour. (usually 70 percent of free flow speed).

Level C   Stable flow–Significant but acceptable delays. Average operating speed about 25 km/hr. (about 50% of free flow speed).

Level D   Unstable flow conditions approaching still with tolerable delays. Average operating speed between 15 to 25 km./hr. (about 40% of free flow speed).

Level E   Unstable flow, significant delays. Average operating speed 10-15 km/h.

Level F   Forced flow, jam conditions, formation of queues. Very slow movement.

Traffic interruptions and number of intersections have most significant effect on capacity of urban roads.

9.7. PASSENGER CAR UNITS ON URBAN ROADS

Urban roads have mixed traffic and the equivalent PCU of different vehicles change under operating conditions, and proportion of a type in total traffic. IRC recommends value of PCU factors for various types of vehicles as shown in Table 9.2.

Capacity of urban road depends upon the road side conditions such as parking, extent of commercial activities along the road, frontage and crossing road etc. Fringe conditions for different roads are assumed as:

Arterial road have no frontage access, no standing vehicles, very little cross-traffic.

Sub arterial road have some frontage development, side roads, bus stop, no standing vehicles, waiting restrictions.

Collector roads have free frontage access, parked vehicles, bus stops, no waiting restrictions.

Design service volumes for different categories of urban roads are given in Table 9.3, as recommended by IRC. These values may be modified when above conditions change.

Table 9.2: Recommended PCU factors for various types of vehicles on urban roads (ICR)

Vehicle Type	Equivalent PCU factors Percentage composition of vehicle type in traffic stream	
	5%	10% and above
Fast Vehicle		
1. Two wheeler motor cycle or scooter, etc.	0.5	0.75
2. Passenger car, pick-up van	1.0	1.0
3. Autorickshaw	1.2	2.0
4. Light commercial vehicle	1.4	2.0
5. Truck or bus	2.2	3.7
6. Agricultural tractor trailer	4.0	5.0
Slow Vehicles		
7. Cycle	0.4	0.5
8. Cycle rickshaw	1.5	2.0
9. Tonga (horse drawn vehicle)	1.5	2.0
10. Hand cart	2.0	3.0

Table 9.3: Recommended design service volumes (IRC)
(PCUs per hour)

S.No.	Types of carriage way	Total design service volumes for different categories of urban roads		
		Arterial	*Sub-arterial*	*Collector*
1.	2-Lane (one-Way)	2400	1900	1400
2.	2-Lane (two-Way)	1500	1200	900
3.	3-Lane (one-Way)	3600	2900	2200
4.	4-Lane undivided (Two-Way)	3000	240	1800
5.	4-Lane divided (Two-Way)	3600	2900	—
6.	6-Lane undivided (Two-Way)	4800	3800	—
7.	6-Lane divided (Two-Way)	5400	4300	—
8.	8-Lane divided (Two-Way)	7200	—	—

9.8. METHODS OF IMPROVING CAPACITY OF URBAN ROADS

When traffic volume on urban roads increase the capacity, congestion is created reducing travel speed and safety. For increasing capacity, widening of roads is usually not possible due to development on both sides of the road, however certain measures listed below are used to provide increased capacity wherever possible.

- Prohibiting on street curb parking.
- Providing separate space for pedestrian and cyclists.
- Adopting segregating measures of traffic.
- Restricting certain vehicles totally or during peak hours of traffic.
- Controlling land uses and commercial activities along the road.
- Making roads oneway and banning certain turns on intersections.
- Providing medians and banning certain movements and turns on cross roads.
- Providing road markings and imposing lane discipline.
- Providing education to road users on road-use through publicity.
- Imposing strict control measures and fines to defaulters.
- Providing signalisation on intersections, police patrol and strict observance of traffic rules and regulations.
- Staggering of office and school hours to reduce traffic at particular peak hours.

9.9. LEVELS OF SERVICE FOR PEDESTRIANS

Flow of pedestrians, can also be defined according to level of service on walkways in A to F categories as given below and illustrated in Fig. 9.9.

Level A   Pedestrian walks on the desired path without any restrictions by other pedestrians. Pedestrians walking speeds are freely selected. There is no conflict between pedestrians.

Level B   Pedestrian can select desired walking speeds. There is no crossing or passing conflicts. There are other passenger, but no interference between them, and walking paths can be selected freely.

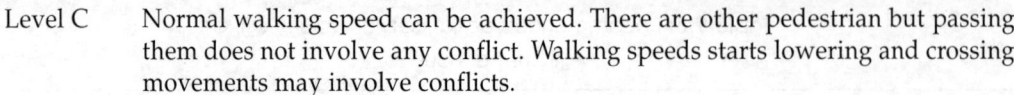

Level C   Normal walking speed can be achieved. There are other pedestrian but passing them does not involve any conflict. Walking speeds starts lowering and crossing movements may involve conflicts.

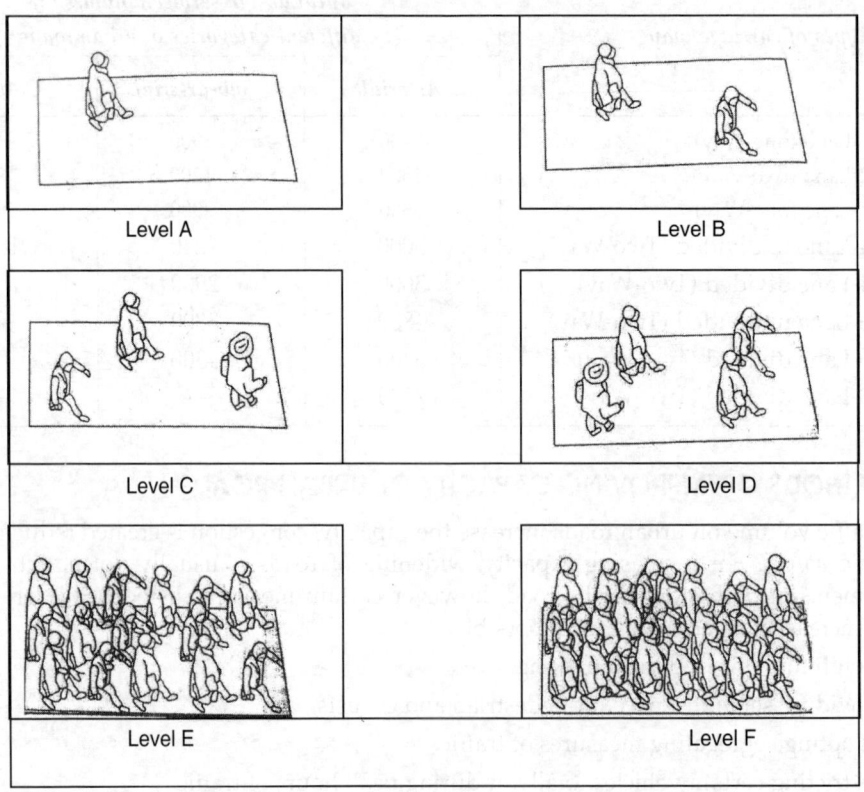

Fig. 9.9: Level of service for pedestrian walkways

Level D   Though speed may be normal, passing other pedestrian may be restricted. Friction and interaction between pedestrians may occur. Pedestrians may require to change speeds frequently.

Level E   The normal speeds are restricted. Pedestrian may have to slow down frequently. Passing is not easy due to speed restriction. Crossing or moving in opposite direction is not easy Walkway capacity is reduced.

Level F   Walking is severely restricted and speed are very slow. There is conflict with other pedestrians crossing and reverse movements are almost impossible Queues may be formed. Flow is unstable.

ILLUSTRATIVE EXAMPLES

Example 1

A two lane, two-way road (major road) is carrying 900 vehicles per hour both direction with a 60% directional distribution. There is a minor road branching off from the heavily trafficked side of the major road. Assuming that vehicle arrival rates at any point along the road follow a random (poisson's) distribution, compute the number of vehicles travelling along the major road in the direction of heavier flow at headway of 4 seconds or more.

If the right turning traffic at this junction is negligible and if traffic from the minor road would accept gaps of 8 seconds or more for merging, what volume from the minor road, would be able to merge into the traffic stream of the major road?

Solution

Poisson's distribution is represented by

$$P(x) = \frac{e^{-m} m^x}{x!}$$

where
$P(x)$ = Probability of the arrival of x vehicles at a given point during a given length of time.

m = mean number of vehicles arriving in the given length of time

$$= \frac{tV}{3600}$$

t = given time gap in seconds

V = hourly volume (VPh)

e = base of naperian log. = 2.71828.

For the probability that no vehicle arrives in a given time interval (i. e. x = 0)]

$$P(0) = e^{-m}$$

The calculated probability given the ratio of vehicles travelling at headway > the selected time interval.

Thus for headway of 4 seconds t : = 4

directional factor

$$m = \frac{tV}{3600} = \frac{4 \times 0.6 \times 900}{3600} = 0.6.$$

$\therefore$
$$P(0) = e^{-0.6} = 0.55$$

55% vehicles are travelling at headway of 4 seconds or more.

For a headway of 8 seconds: t = 8

$$m = \frac{tV}{3600} = \frac{8 \times 0.6 \times 900}{3600} = 1.2.$$

$$P(0) = e^{-1.2} = 0.30$$

30% vehicles are travelling at headway of 8 sec. or more, i.e. 162 vehicles per hour are travelling at headway equal to or more than 8 seconds along heavy trafficked side of major road.

Assuming that all the acceptable gaps can be utilized then 162 vehicles pre hour can merge into the traffic stream of the major road, from the side road.

Example 2

One kilometer long section of a road is travelled by 3 cars in 1 min, 2 min and 3 min respectively.

Calculate the following:

(i) Time mean speed

(ii) Space mean speed

(iii) Standard deviation

(iv) Variance of space distribution of speed.

Solution

(i) Time mean speed (V_t) = Average value of speed of three car.

Speed of cars: (i) 60 km/h (ii) 30 km/h (iii) 20 km/h.

$$\therefore \qquad V_t = \frac{60+30+20}{3} = \textbf{36.7 km/h}$$

(ii) Space mean speed (V_s) = Speed value based on average time of travel

Average time for travelling 1 Km = $\dfrac{1+2+3}{3} = 2\,\text{min}$

1 kms distance covered in 2 min.

$$\therefore \qquad V_s \,(\text{Speed}) = \textbf{30 K.P.h}$$

(iii) $\qquad V_t = V_s + \dfrac{\sigma s^2}{V_s}$ where σs^2 = variance of space mean speed

$$\sigma s^2 = (V_t - V_s)\,V_s = (36.7 - 30)\,30 = \textbf{201}$$

(iv) Standard Deviation σs = $\sigma s = \sqrt{201}$ = **14.17 KPh**

Example 3

On a highway there is a car per minute having a speed of 120 km/h and a truck having a speed of 60 km/h

Calculate :

(a) Time mean speed and

(b) Space mean speed

Solution

(a) Time mean speed (V_t) = Average of two values of speed.

$$= \frac{60+120}{2} = \textbf{90 km/hr.}$$

(b) Space mean speed (V_s) = Based on average time of travel same distance.

Car – 120 kms in 1 hour

Truck– 120 kms in 2 hour

──────────────────

Total 240 kms in 3 hours

$$= \frac{240}{3} = \textbf{80 Km/hr.}$$

Alternatively:

Space mean speed (V_s) = Simple harmonic mean of spot speed

$$= \frac{2}{\left[\dfrac{1}{120} + \dfrac{1}{60}\right]} = \textbf{80 km/hr.}$$

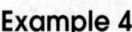

Example 4

Calculate the basic capacity of a traffic lane having a speed of 50 km./h. Average length of the vehicles can be assumed as 6 m and reaction time of drivers as 1.5 second.

Solution

Basic capacity is given by relationship

$$C = \frac{1000 \, V}{L + SSD}$$

SSD stopping sight distance is given by

$$SSD = 0.28 \, Vt + 0.1 \, V^2$$
$$\text{here } V = 50 \text{ KPh}$$
$$L = 6 \text{ m}$$
$$SSD = 0.28 \times 50 \times 1.5 + .01 \, (50)^2$$
$$= 46 \text{ m}$$
$$\text{Basic Capacity } C = \frac{1000 \times 50}{6 + 46}$$
$$= \textbf{960 VPh}$$

Example 5

The average length of vehicles in a traffic stream is 7 m and average distance between the cars is 2 m. Average time headway at maximum flow (Qm) is 1.5 seconds.

Calculate the jam density (Kj), maximum flow (Qmax), optimum density and speed at maximum flow.

Solution

Space headway = Length of average vehicle + Average spacing

$$7m + 2m = 9 \text{ m}$$

(i) K_j = Maximum number of vehicles in given length

$$= \frac{1000 \text{ m}}{9 \text{ m}} = \textbf{111 Veh/Km.}$$

(ii) Qmax as headway is 1.5 sec.

$$Qmax = \frac{60 \times 60}{1.5} = \textbf{2400 Veh/hr.}$$

(iii) $$Kmax = \frac{Kj}{2} = \frac{111}{2} = \textbf{55 Veh/Km.}$$

(iv) $$V_{max}^s = \frac{Qmax}{Kmax} = \frac{2400}{55} = \textbf{44 km/h.}$$

Example 6

The free speed on a highway was found to be 120 km/h and the jam density was 80 vehicles/hr.

(a) What is the maximum flow expected on the highway ?

(b) At what speed maximum flow will occur?

Solution

$$Qmax = \frac{V_f \cdot K_j}{4} = \frac{120 \times 80}{4} = 2400 \text{ Veh/hr.}$$

$$Vmax = \frac{V_f}{2} = \frac{120}{2} = 60 \text{ km/h.}$$

Example 7

Following is the speed-density data, obtained in the field. Analyse this data and obtain various flow parameters.

Density VPK (x)	Speed km/h (y)
10	60
20	50
30	45
40	40
50	35
60	30
70	20
80	15
90	10
100	5

Solution

(1) Plotting a scatter diagram:

Figure shows the plot (straight line) of the above data, and by graphic method, estimated line is drawn.

(2) Determine the regression equation and find multiplying and additive constant:

Regression equation can be determined by least square method.

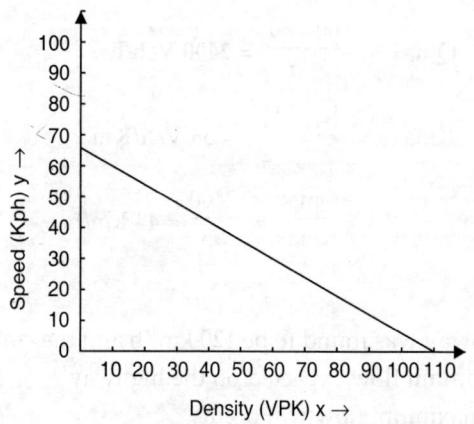

$N = 10$, $\Sigma Y = 310$, $\Sigma X = 550$, $\Sigma Y^2 = 12600$, $\Sigma X^2 = 38500$, $\Sigma XY = 12100$

	Y	X	Y^2	X^2	XY
1.	60	10	3600	100	600
2.	50	20	2500	400	1000
3.	45	30	2025	900	1350
4.	40	40	1600	1600	1600]
5.	35	50	1250	2500	2750
6.	30	60	900	3600	1800
7.	20	70	400	4900	1400
8.	15	80	225	6400	1200
9.	10	90	100	8100	900
10.	5	100	25	100,00	500

Let the equation be Y = a + bx

where
$$Y = \text{Dependent variable (Speed)}$$
$$X = \text{Independent variable (Density)}$$
$$a = \text{is additive constant}$$
$$b = \text{multiplying constant.}$$

The following two statistical equations should be satisfied

$$\Sigma Y = Na + b\Sigma X \qquad \qquad \text{... (i)}$$
$$\Sigma XY = a\Sigma x + b\Sigma x^2 \qquad \qquad \text{... (ii)}$$

So,
$$310 = 10 \times a + b \times 550$$
$$12100 = 550a + b \times 38500.$$

Solving them, we get,
$$b = -0.6 - \text{multiplying constant}$$
$$a = 64 - \text{Additive constant.}$$

The equation is : **Y = 64 − 0.6X.**

(3) Average decrease in speed, for 10 VPK increase in density:

The equation is Y = 64−0.6X

for density x = 10, the decrease is −0.6 × 10 = 6 Km/h

(4) Compute statistical parameter total variation, explained variation and unexplained variation and coefficient of determination:

If
$$Y = \text{Dependent variable}$$
$$\bar{Y} = \text{Average value of dependent variable}$$
$$Y_e = \text{Estimated value of dependent variable.}$$

$$\text{Total variation} = \Sigma(Y - \bar{Y})^2$$

$$\text{Explained variation} = \Sigma(Y_e - \bar{Y})^2$$

$$\text{Coeff. of determination} = \frac{\text{Explained variation}}{\text{Total variation}}$$

$$\text{Coeff. of correlation} = \sqrt{\text{Coeff. of determination.}}$$

V	$\bar{Y}\Sigma y / N$	Y_e from eqn.	$Y-\bar{Y}$	$(Y-\bar{Y})^2$	$(Ye-\bar{Y})$	$(Ye-\bar{Y})^2$	$(Y-Y_e)$	$(Y-Y_e)^2$
60	31	58	+29	841	27	729	2	4
50	31	52	+19	361	21	441	−2	4
45	31	46	+14	196	15	225	−1	1
40	31	40	+9	81	9	81	0	0
35	31	34	+4	16	3	9	+1	1
30	31	28	−1	1	−3	89	2	4
20	31	22	−11	121	−9	81	−2	4
15	31	16	−16	256	−15	225	−1	1
10	31	10	−21	441	−21	441	0	0
5	31	4	−26	676	−27	729	+1	1
				Σ 2990		Σ 2970		

From the above table:

$$\text{Total variation} = (y - \bar{Y})^2 = 2990$$

$$\text{Explained variation} = (Y_e - \bar{Y})^2 = 2970$$

$$\text{Coeff. of determination} = \frac{\text{Explained variation}}{\text{Total Variation}} = \frac{2970}{2990}$$

$$= 0.99$$

$$\text{Coefficient of correlation} = \sqrt{\text{Coeff. of determination}}$$

$$= \sqrt{0.99} = \textbf{0.996}$$

Other method for equation and coefficient of correlation. Coeff. of correlation can also be found directly by

Equation $\qquad Y = \dfrac{\Sigma x_1 y_1}{\sqrt{\Sigma x_1^2\, \Sigma y_1^2}}$ and $\qquad b = \dfrac{\Sigma x_1 y_1}{\Sigma x_1^2}$

$$a = \bar{y} - b\bar{x}$$

where $\qquad \Sigma x_1 y_1 = \Sigma xy - \text{N}.\,\bar{x}.\,\bar{y}$ and $\qquad \bar{y} = \dfrac{\Sigma y}{N}, \bar{x} = \dfrac{\Sigma x}{N}$

$$\Sigma y_1^2 = \Sigma y^2 - N(\bar{y})^2 \qquad\qquad \bar{y} = \frac{310}{10} = 31$$

$$\Sigma x_1^2 = \Sigma x^2 - N(\bar{x})^2 \qquad\qquad \bar{x} = \frac{550}{10} = 55$$

here $\qquad \Sigma x_1 y_1$ $\qquad\qquad\qquad\qquad =12100 - 10 \times 31 \times 55 = -4950$

$$\Sigma y_1^2 = 12600 - 10 \times (31)^2 = 2790$$

$$\Sigma x_1^2 = 38500 - 10\,(55)^2 = 8250$$

$$\therefore \qquad b = \frac{\Sigma x_1 y_1}{\Sigma x_1^2} = \frac{-4950}{8250} = -0.6$$

$$a = \bar{y} - b\bar{x} = 31 - (-0.6 \times 55) = 64$$

equation is $\qquad y = 64 - 0.6x,$ (as got before)

Coeff. of correlation $\qquad = \dfrac{\Sigma x_1 y_1}{\sqrt{\Sigma x_1^2 . \Sigma y_1^2}} = \dfrac{-4950}{\sqrt{2790 \times 8250}}$

$$= -0.996 \quad \text{(as got before)}$$

Significance of coefficient of correlation. Coefficient of correlation (*r*) determines the strength of relationship between dependent and independent variables.

−ve   Signs means that with increase of one variable the other decreases.

r = 1,   means perfect relationship between variables

r = 0,   indicates no relationship between variables

+ve   sign means that with increase of one the other variable also increases.

(5) Convert the speed-density data to speed-volume data:

Volume = Density × speed.

Speed in KPh	Density in vehicles per km	Volume VPh
60	10	600
50	20	1000
45	30	1350
40	40	1600
35	50	1750
30	60	1800
20	70	1400
15	80	1200
10	90	900
5	100	500

(6) Plot of speed volume data : Figure shows the plot of speed volume data:

This curve helps in finding the speed (optimum) at which maximum volume can be had.

(7) Find Equation for speed volume curve: From the linear equation, for speed-density, as obtained before, the volume-speed relationship can be developed:

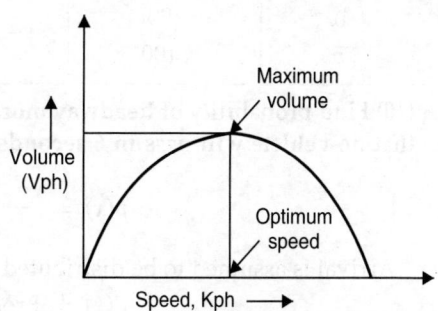

$$y = 64 - 0.6x$$

i.e., $\qquad V = 64 - 0.6 \times K$

multiplying both sides by V (speed), we have

$$V^2 = 64V - 0.6\, V \times K$$

K × V = Volume (Q)

$$\mathbf{V^2 = 64V + 0.6Q}$$

This is the equation for speed volume curve.

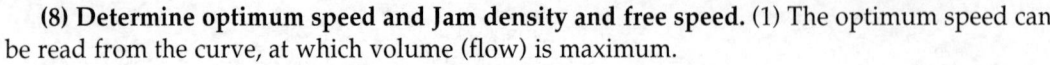

(8) Determine optimum speed and Jam density and free speed. (1) The optimum speed can be read from the curve, at which volume (flow) is maximum.

$$y = 64 - 0.6x$$

(2) At Jam conditions, speed becomes almost zero.

i.e., $y = 0$, $x = \dfrac{64}{0.6} = 107$ vehicles/km

Jam Density = 107 Km

(3) For free flow, density approaches zero.

i.e., $x = 0$. $y = 64$.

Free speed = 64 m.p.h

(9) Convert data for time headway and space headway:

$$\text{Time headway} = \frac{1}{\text{Volume}}$$

$$\text{Space headway} = \frac{1}{\text{Density}}$$

Speed (y) (K ph)	Density (x) (V pk)	Volume (xy) (Vph)	Time headway (1/Vol.) Second	Space Headway (1/Density) (m)
60	10	600	$\dfrac{60 \times 60}{600} = 6$ sec.	$\dfrac{1000}{10} = 100$ m
50	20	1000	3.6	50 m
45	30	1350	2.66	33.3 m
40	40	1600	2.55	25 m
35	50	1750	2.06	20 m
30	60	1800	2.0	16.6 m
20	70	1400	2.57	14.2 m
15	80	1200	3.0	12.5 m
10	90	900	4.0	11.1 m
5	100	500	7.2	10 m

(10) Find probability of headway more than 6 seconds at 600 Vph. or what is the probability that no vehicle will pass in 6 seconds.

$$p(x) = \frac{e^{-m} m^x}{x!}$$

Arrival is assumed to be distributed as Poisson's distribution. So

$P(x)$ = Probability of arrival of x vehicles at a point, during given length of time

m = mean number of vehicles arriving in that period.

e = Base of nap. e = 2.78.

t = 6 sec.

here $m = \dfrac{\text{Volume} \times t}{3600} = \dfrac{600 \times 6}{3600} = 1$ vehicles in 6 seconds.

$$p(o) = \frac{e^{-1} \times (1)^0}{0!} = \frac{1}{2.78} = 0.368$$

Example 8

Assume a linear relationship between speed and density on a length of a highway having the free flow speed of 80 km/hr. and the jam density of 72 vehicles/km.

(a) Determine the speed a maximum flow and the maximum flow expected on this section of the road.

(b) Suppose the observed flow is 75 percent of the capacity, what are the possible flow speed? Under what conditions such speed will occur? Also find out the headway distance in each case.

(e) At what level of service you think road will be operating at above speeds, if it is a rural two-lane highway.

Solution

(a)
$$Q = KVs$$

$$Q_m = K_m V_m = \frac{K_j V_j}{4}.$$

$$V_m = V_f/2 = \frac{80}{2} = 40 \text{ km/h}$$

$$K_m = K_j/2 = \frac{72}{2} = 36 \text{ km/h}$$

∴
$$q_{max} = 40 \times 36 = 1440 \text{ VPh}.$$

(b) The observed flow is 75% of capacity (max. flow)

Observed flow 0.75 × 1440 = **1080 VPh**

The speed-density equation is

$$y = a + bx$$

or
$$\text{Speed } V = a + b \text{ (K-density)}$$

$$\text{For free flow, density} = 0, V = a = \text{free speed} = 80 \text{ km/h}.$$
$$\text{For jam density, speed} = 0, b = 80 + b \times 72$$

∴
$$b = \frac{80}{72} = 1.11$$

$$V = 80 - 1.11 \text{ K}.$$

Multiply both sides by V,

$$V^2 = 80V - 1.11 \text{ (KU)} \quad KV = Q = 1080$$
$$V^2 = 80V + 1.11 \times 1080$$
$$V^2 = 80V + 1200$$

∴
$$V = \frac{80 \pm \sqrt{80^2 - 4 \times 1200}}{2}$$

$$= 40 \pm 20.$$
∴
$$V = 60 \text{ km/h.} \quad \text{or } 20 \text{ Km/h.}$$

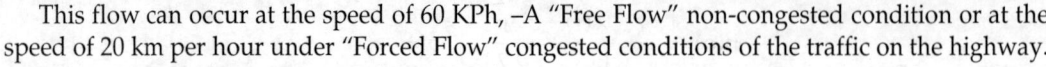

This flow can occur at the speed of 60 KPh, –A "Free Flow" non-congested condition or at the speed of 20 km per hour under "Forced Flow" congested conditions of the traffic on the highway.

$$\text{Corresponding density } (K_1) = \frac{1080}{60} = 18 \text{ VPK}$$

$$\text{and} \qquad (K_2) = \frac{1080}{20} = 54 \text{ VKP}$$

$$\text{Corresponding headway} = \frac{1}{\text{Density}} = \frac{1000}{\text{Density}} \text{ m}$$

$$h_1 = \frac{1000}{18} = \textbf{55.56 m}$$

$$h_2 = \frac{1000}{54} = \textbf{18.52 m}$$

The actual flow of 1080 veh/hr. can occur under free and forced conditions at corresponding speeds of 60 and 20 km/hr, densities of 18 and 54 V/km and headways of 55.56 m and, 18.52 m respectively.

(c) These speeds, as per the description of level of services for two lane rural highway, correspond to

Level of serve D   (60 km/h).
Level of service F   (20 km/h).

Question 9

The maximum flow rate of vehicles is 1000 per hour and flow speed is 80 km/h. At any instant, the traffic is flowing at 70% of the maximum flow rate.

(i) Develop the traffic parameters to draw basic curves of flow and calculate the values.

(ii) Compute the average space and average time headways for the given flow conditions.

Solution

$$Q_{\max} = 1000 \text{ VPh}$$
$$U_f = 80 \text{ km.}$$
$$Q_0 = 700 \text{ VPh}$$

Basic curves are:

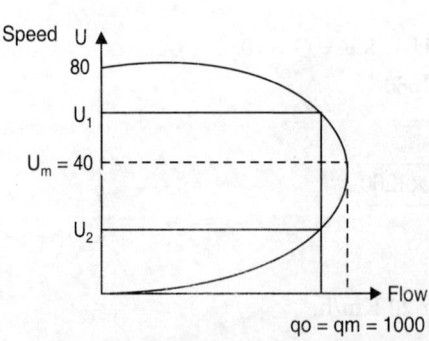

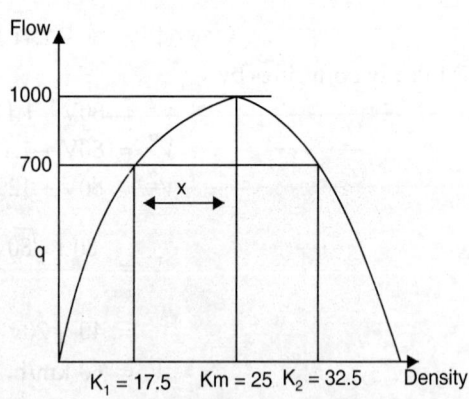

(i) There are two possibilities U_1 and U_2 i.e. uncongested and congested conditions.

$$q_m = U_m K_m$$

or $\qquad\qquad K_m = \dfrac{q_m}{4m} = \dfrac{1000}{40} = 25 \qquad\qquad\qquad \therefore K_j = 50$

By proportioning $\qquad \dfrac{25}{1000} = \dfrac{x}{300} \qquad \therefore \; x = 25 \times 0.3 = 7.5$

$$K_1 = 17.5, \; K_2 = 32.5$$

$$U_1 = \dfrac{700}{17.5} = \textbf{40 km/h} \quad U_2 = \dfrac{700}{32.5} = \textbf{21.5 km/h}$$

(iii) **Space headway: (h)**

when $\qquad\qquad K_1 = 17.5 \quad h_1 = \dfrac{1000}{17.5} = 57.6 \text{ m}$

$$K_2 = 37.5 \quad h_2 = \dfrac{1000}{37.5} = 30.8 \text{ m}$$

Time Headway

$$q = 700 \text{ Uph} \qquad\qquad\qquad ht_1 = \dfrac{3600}{700} = \textbf{5.1 sec.}$$

$$q = 1000 \text{ Uph} \qquad\qquad\qquad ht_2 = \dfrac{3600}{1000} = \textbf{3.6 sec.}$$

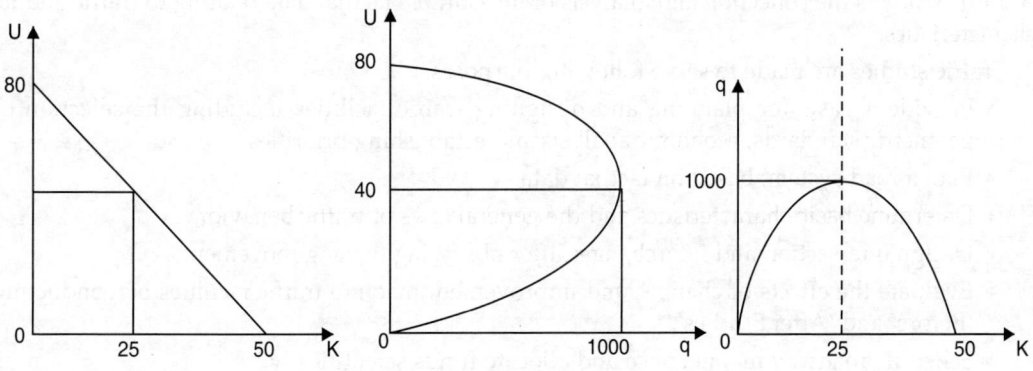

(Basic traffic flow curves)

Traffic Data Collection and Analysis

10 PREAMBLE

A traffic study is the collection and analysis of measurable factual data relating to traffic and its characteristics.

Traffic studies are made to serve following purposes:

- Provide a basis for planning and designing traffic facilities including the selection of geometric standards, economic analysis and establishing priorities.
- Plan a road system, based on factual data.
- Determine basic characteristics and the general laws of traffic behavior.
- Design intersection and interchanges after observing turning movements.
- Evaluate the effects of changes and improvement made to traffic facilities by conducting 'Before' and 'After Studies'.
- Schedule roadway maintenance and allocate funds scientifically.
- Establish the need of traffic control measures such as signs, markings, signals, pedestrian crossing, etc to increase mobility and safety.
- Determine traffic signal timings, type and location of signals.
- Determine trends for future prediction of traffic.
- Select pavement, shoulder and bridge width, as per traffic volumes.
- Know the landuse pattern and socioeconomic factors affecting generation of the traffic.
- Determine accident rates and involvement of different type of vehicles.
- Study and investigate the cause for any specific problem on the road in traffic flow.
- Design pavement thickness as per traffic loadings.

In attempting to improve road transportation, facts must be collected to define accurately, the precise location and extent of the problem. These facts are obtained through conduct and analysis of relevant laboratory and field studies on traffic behavior. Problems that are

approached on the basis of facts can be rationally and economically solved. Successful traffic improvements should be based on reliable facts.

The common type of traffic studies are:

1. Traffic volume (flow) study and pedestrian volume study.
2. Spot speed study
3. Travel time and delay study
4. Origin and destination (O-D) study
5. Parking study
6. Accident study
7. Public transport study

These studies are of vital importance for determining trends and characteristics of traffic on the road.

In addition, many other studies such as axle load survey, traffic conflict studies, vehicle occupancy survey, queue length studies, etc. may be carried out to examine special/specific problems as per the need. Details of these specialized studies can be found in any text on traffic studies.

10.1 TRAFFIC VOLUME (FLOW) STUDY

Traffic volume (flow) study is eventually a counting process, involving counting of different types of vehicles (both slow and fast). The counting periods may be hours, days, weeks, months or year depending upon the purpose of the study and required degree of accuracy. Pedestrian volume also becomes part of the study at certain locations. Traffic volume data is commonly used in planning, design, operation and in dealing with practical traffic problems.

Definitions

(i) **Hourly volumes (HV):** Traffic volume is counted on hourly basis. The highest hourly volume in a day is called "Peak Hour Volume". There is one peak hour volume in morning called" Morning peak hour volume and another in the evening referred as "Evening Peak Hour Volume".

The "30th Highest Hourly Volume" is the volume per hour that is exceeded by 29 hourly volumes during a designated year. The hourly volume which is used for designing the facilities is called" Design Hourly Volume" (DHV).

(ii) **Daily traffic volume (DT):** Traffic volume counted on 24 hour basis is called daily traffic (DT). The total yearly traffic volume divided by number of days in the year is called "Average Annual Daily Traffic (AADT). AADT is 1/365th of total annual traffic flow. The total traffic volume during certain number of days divided by that number of days is called "Average Daily Traffic" (ADT)

10.1.1 Type of Traffic Counts and their Purpose

The traffic volume data to be collected varies according to its use such as:

(a) **Annual total traffic volume used for:**

- Measuring and establishing trends in traffic volume
- Annual travel in vehicle kms used in economic analysis
- Computing accident rate per million vehicle kms
- Estimating highway users revenues, etc.
- Extrapolating present trends into future.

(b) **AADT or ADT volumes used for:**
- Highway planning activities, lane width, shoulder width, etc.
- Measuring the present demands of service.
- Evaluating the present quality of traffic flow on existing highway system, and its capacity.

(c) **Peak hourly volumes used for:**
- Geometric design elements of highway.
- Determining deficiencies in highway capacity.
- Justifying control devices such as signals, markings, etc.
- Developing operational measures and controls such as signals, markings, etc.
- Justifying regulating measures such as parking, turning, stopping, etc.
- Classification of highways/roads.
- Justifying enforcement and its planning.

(d) **Classified volumes used for:**
- Geometric design and turning paths of vehicles.
- Structural design of pavements, bridges, etc.
- Analysis of capacity and effect of commercial vehicles.
- Estimating highway users revenue.

(e) **Short volume counts less than an hour used for:**
- Analyzing maximum rate of flow and variations with in peak hour flows.
- Providing quick means of obtaining volume data.
- Estimating flow variations and capacity limitations.

(f) **Intersectional volume counts used for:**
- Determining total traffic entering the intersection.
- Knowing traffic on different legs of inter sections and turning movements.
- Design of time cycles for traffic signals.

(g) **Cordon volume counts:** Are made to determine the accumulation of vehicles, during a typical time period within the cordon area.

(h) **Screen line volume counts:** Are made at crossing of natural or man-made barriers such as waterways, railways, bridges, etc.

(i) **Pedestrian volume counts:** Are made at locations where pedestrian vehicle conflict or other pedestrian problem occurs. Used for planning pedestrian overpass, subways and pedestrian safety measures.

10.1.2 Methods of Traffic Volume Counts

Traffic volume counts can be carried out manually, by automatic counters, using photographic techniques or by moving observer method.

(1) **Manual method:** Manual traffic counts require simply counting of every vehicle seen to pass a fixed point on a road. Data is recorded on a specially prepared field sheet. It is normally done with pen or pencil, making 'Tally' of marks in groups of five like this ⊞, ⊞ . Field sheets no 1 to 3 show the specially prepared performa for this purpose.

Hand tallies pressing a button every time a vehicle is passed can also be used for manual counting. For each type of vehicle a separate counter is required. Total number of vehicles at the end of each counting period can be thus obtained.

Advantages of manual counting is its simplicity, quickness, flexibility and no special skills requirements. It is cheap and convenient for short and specific purpose counts. Its disadvantages are dependence on capacity and accuracy of observers, and good weather conditions. For long and continuous specially odd hours counting it may be expensive.

Field Data Sheet 1

CLASSIFIED TRAFFIC VOLUME COUNTS

Date and day of the week ....
Direction of traffic

From .... To ....

Road Classification ....
Kilometrage/mileage............
Route number (if any)............
District............State............

Type of vehicles (1) / Hour of count	Cars, jeeps, vans, three wheelers (2)	Buses (3)	Trucks (4)	Motorcycles and scooters (5)	Animal drawn vehicle** (6)	Cycles (7)	Others (specify) (8)	Remarks, including weather conditions (9)
From–Hours* to–Hours								
Hourly Total								
From–Hours* to–Hours								
Hourly Total								
From–Hours* to–Hours								
Hourly Total								
From–Hours* to–Hours								
Hourly Total								

Notes.

1. Record traffic-volume in columns 2 to 8 by making tallies in the form of vertical strokes for first four vehicles and drawing an oblique stroke for every 5th as shown within brackets.

2. Some roads carry appreciable volume of other traffic like cycle rickshaws. Record the volume of such vehicles in column 8 after specifying the vehicle type.

   * The hour of count should be entered before the start of the enumeration. P.M. hours should be recorded after adding 12 to the actual hour for example, 2 P.M. should be recorded as 14.00 hours.

**If felt necessary by Highway Authority, this column could be subdivided into two for recording the volume of pneumatic-tyred and iron-tyred vehicle separately.

Name and signature of enumerator with date

Name and signature of supervisor with date

Field Data Sheet 2

TYPICAL VEHICLE TURNING MOVEMENT COUNTS

(From one leg of the intersection)

Time Hour............Hrs. To ....Hrs

Day ....and Date ....

Weather ....observer ....

Name and location of Intersection ....

Entering	From			Leg A*				Remarks
	Leg B*			Leg C*		Leg D*		
Type	Nos	PCU Equi-valency	PCU	Nos	PCU Equi-valency	Nos	PCU Equi-valency	
	1	2	3 = 1 × 2	1	2	1	2	
Fast Vehicles								
1. Passenger cars, tempos auto rickshaw, tractors, pickup vans		1.00						
2. Motor Cycles, scooters		0.50						
3. Agricultural tractor light Commercial Vehicles		1.50						
4. Trucks, Buses,		3.00						
5. Tractor-Trailer, Truck Trailer units		4.50						
TOTAL FAST								
Slow Vehicles								
6. Cycles		0.50						
7. Cycle Rickshaws		1.50						
8. Hand Cart		3.00						
9. Horse Drawn		4.00						
10. Bullock-Carts		8.00						
TOTAL SLOW								
PEDESTRIAN Nos.								

*legs A, B, C, D, can be specified by names also

Field Data Sheet 3

PRESENTATION OF TYPICAL VEHICLE TURNING MOVEMENT COUNTS

(Four approach intersection)

Intersection location............   Time from............To............

Name of roads............   Date:............Day............

                 Weather............Observer............

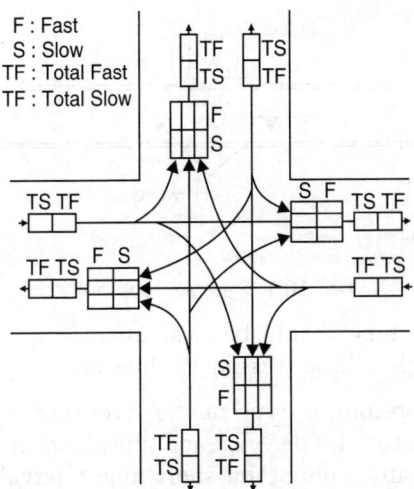

F : Fast
S : Slow
TF : Total Fast
TF : Total Slow

(2) **Mechanical method:** Mechanical counter consists of two elements: a detector and a counter. Counters are fixed permanently for continuous counts, for days, weeks, months or years. For detecting/sensing a vehicle, these counters use any of the following devices:

(i) **Electric Contact Device.** This consists of a sub-surface detector, which provides for a positive electrical contact for each vehicle axle crossing it. A steel base plate over which a moulded and vulcanized rubber pad (which is flush with the road surface) holds suspended a strip of spring steel. Electric contacts are installed in each lane.

(ii) **Photoelectric Device.** Detection is accomplished by the vehicle passing between a source of light and a photo cell. The equipment is mounted above the road surface.

(iii) **Radar Device.** Detection is done by continuously comparing the frequency of a transmitted radio signal with the frequency of the reflected signal. When the moving vehicle intercepts the signal, a frequency difference occurs. The unit is normally mounted above the centre of the lane or lanes, for which detection is desired.

(iv) **Ultrasonic Device.** This is similar in its operation to a radar unit. A beam crosses the road and is broken by the passage of a vehicle.

(v) **Magnetic Device.** Detection is done by a signal or impulse, caused by a moving vehicle (mass of metal) and disturbing a magneting field. The unit is installed in each lane immediately below the road surface. It is more durable but more expensive than pneumatic detector.

(vi) **Infra-red Device.** This device utilizes a pick up cell, which is similar to a photo electric cell, but is sensitive to infrared (heat) radiation, rather than to visible light. The unit is mounted above the road surface on a bridge or sign structure, etc.

(vii) **Pneumatic Detector.** In this device, which is most popular, the wheel of a vehicle, crossing a tube, set up an impulse in the tube. It consists of a thick walled rubber tube (about 10–12 mm diameter with 3 mm thick walls) which is fixed to the road by means of tough canvas straps, nailed across it at about one metre intervals. One end of the tube is connected to the counters and the other is fitted with a plug incorporating a small air release hole, to avoid bounced impulses. As a vehicle wheel crosses the tube it compresses it, causing an air shock-wave to travel along the tube, operating a simple make and break circuit of the counter as shown in Fig. 10.1

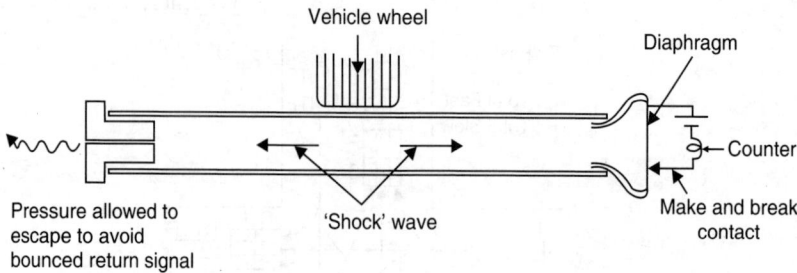

Fig. 10.1: Pressure tube detector

It is important that the tube should be clear of turning path of vehicles, to reduce the multiple counts, due to single vehicle crossing the tube at an angle.

A counter may be non-recording type or the recording type. The non-recorder counter accumulates the number of vehicles detected and must be read at regular intervals as desired. The counters record the traffic volume on short time intervals by printing numbers or by punching holes in a moving paper tape. The punched tapes can be processed by a computer. Advantage of mechanical counting is that they can be used in all weathers in all times, they are reasonably accurate and cheap for long duration counting. The disadvantages are high cost, requirement of skilled and trained labour, and regular and proper maintenance. Mechanical counters can not make classified counts.

(3) **Photographic Method:** This method involves taking of continuous strip photographs, over the area to be studied. Based on the principles of photogrammetry, mathematical relationships are developed to know the volume data, speed data and other traffic flow characteristics from the photographs.

Time lapse photograph (which consists of a device to move the film a single frame at a time, at a given time interval) and a frame numbering device, is also used for providing information on volume, speed and spacing of vehicles.

(4) **Moving Observer Method:** In this method an observer sits in a car. The car is driven at an average speed and observation on number of vehicles overtook by the test car and those which overtook the test car are recorded. This method shall be described in detail on travel time studies.

10.1.3 Periods and Locations for Volume Counts

The periods of volume count should avoid special events, and conditions such as holidays, exhibitions, strikes, etc. Normally counts are made in good weather, on working days, to cover period of study (24, 18, 12, 8 hours or hourly volumes). Isolated counts are made at any location where problem has developed. The usual locations are intersections, high accident locations, major highways or mid-block points.

Short counts are taken sometimes due to shortage of manpower, finance or time. In heavy traffic volumes short term counts of 5–15 minutes, properly planned, can provide dependable information.

Short counts may be expanded, if complete representation data is collected. Expansion factors can be developed for monthly or weekly variations of the traffic and can be used to calculate ADT for the year based on short counts as illustrated in the example

10.1.4 Presentation of Traffic Volume Data

Following techniques are used to present the collected data on traffic volume (flow):

Flow Diagrams: Using the bands proportional to traffic volume, flow maps are prepared to show traffic flow along various routes in an area. The numerical traffic volume are also shown on the band. Volumes shown may be peak hourly, average daily or others. Typical traffic volume data on roads is shown in Fig. 10.2.

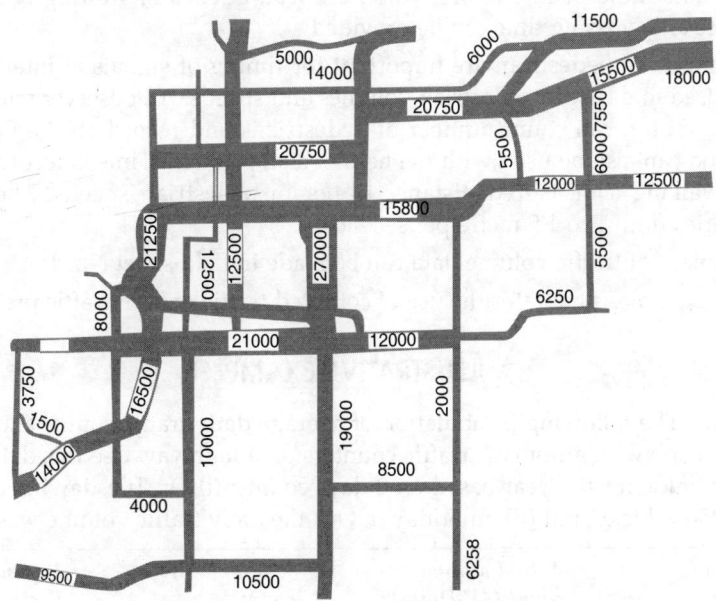

Fig. 10.2: Typical traffic flow diagram on road network

Similarly intersection flow diagrams are drawn showing direction and flow of traffic through an intersection, as shown in Fig. 10.3

Fluctuation Charts: The charts represent hourly, daily, monthly, yearly changes in volume through an area, along a given road. They readily show the peak periods of flow.

Trend Charts: These charts show the volume changes over a period of years.

Summary Tables: Traffic volume data such as ADT, hourly flow, etc. are summarized in the tabular form.

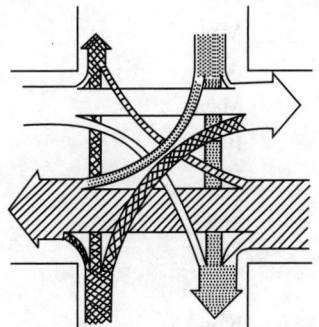

Fig. 10.3: Typical traffic flow diagrams on an intersection

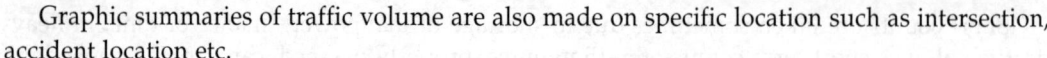

Graphic summaries of traffic volume are also made on specific location such as intersection, accident location etc.

10.1.5 Conversion of Traffic Volume into Passenger Car Units

The traffic consists of various types of vehicles. It is common practice to convert the flow into a single unit of equivalent passenger car unit PCU, using different multiplying factors. PCU takes into account the relative effect of different classes of vehicles, depending upon locality, e.g. urban roads, rural roads, round about or intersections. Tables 9.1 and 9.2 show recommended values for rural and urban roads.

10.1.6 Pedestrian Volume Study

Pedestrian movement and characteristics are studied for providing a design which minimises pedestrian vehicle conflicts, increase pedestrian safety and minimise vehicle delay. It is important to determine peak crossing volumes, peak queues of waiting pedestrians, so that adequate space and crossing time can be provided

Walking speeds of pedestrian are important for timing of signals at intersections. Manual methods are used to measure pedestrian volumes and speeds. An observer with a hand counter at any vantage point, can count number of pedestrians and record on the field sheet. Photographic method can also be used with the help of a stop watch. Time is recorded by observers, for persons walking a measured distance to obtain pedestrian speeds. The mean speed of pedestrian varies from 1 to 1.5 metre per second.

Daily summary of traffic volume data can be made in field sheet No. 4

Following examples illustrates the use of collected traffic data in traffic predictions.

ILLUSTRATIVE EXAMPLES

Example 1.   The following is tabulation of average daily traffic counts and average weekly traffic counts made with automatic traffic counters on a highway. Use this data to calculate the average daily traffic for the year based on 2 days counts: (i) On Tuesday in February the daily counts were 250 vehicles and (ii) on Friday in Oct, the daily traffic volume was 300 vehicles

Months of the year	Actual automatic count (AV-Daily)	% of av daily count	Expansion Factor (calculated)
Jan	1100	53	1.89
Feb.	1200	58	1.72
Mar.	1350	65	1.54
Apr.	1400	67	1.50
May	1600	77	1.3
June	2300	111	0.90
July	3300	159	0.90
August	4200	202	0.50
Sept.	3600	173	0.58
Oct.	1800	87	1.15
Nov.	1600	77	1.30
Dec	1500	72	1.39

Field Data Sheet 4
DAILY SUMMARY OF TRAFFIC VOLUME DATA

From............Hrs on*............To............Hrs on*
Direction of traffic from............To............(UP)
From............To............(Down)

Road Classification............
Kilometrage/milage............
Route number (if any)............
District............State............

Count Hour	Cars, jeeps, vans, three wheelers etc.		Buses		Trucks		Motor cycles and scooters		Total fast			Animal drawn vehicles		Cycles		Others (specify)		Total slow			Remarks
	Up	Down	Up	Down	Up	Down	Up	Down	Up	Down	Total Col.10 and 11	Up	Down	Up	Down	Up	Down	Up	Down	Total Col.19 and 20	
(1)	(2)	(3)	(4)	(5)	(6)	(7)	(8)	(9)	(10)	(11)	(12)	(13)	(14)	(15)	(16)	(17)	(18)	(19)	(20)	(21)	(22)
0600–0700																					
0700–0800																					
0800–0900																					
0900–1000																					
1000–1100																					
1100–1200																					
1200–1300																					
1300–1400																					
1400–1500																					
1500–1600																					
1600–1700																					
1700–1800																					
1800–1900																					
1900–2000																					
2000–2100																					
2100–2200																					
2200–2300																					
2300–2400																					
0000–0100																					
0100–0200																					
0200–0300																					
0300–0400																					
0400–0500																					
0500–0600																					
Total																					
Total Up and Down for vehicle type																					

* Enter Date and Day of week

Name and signature of Supervisor with date ....

Day of the week	Average daily traffic	% of av. Daily traffic	Expansion factor (calculated)
Sun.	500	86	1.16
Mon.	600	103	0.97
Tues.	560	96	1.04
Wed	650	111	0.90
Thur.	550	94	1.06
Fri.	750	128	0.78
Sat.	480	82	1.22

Solution: (*a*) **Using the monthly table:**

Sum of 12 average daily counts for months = 24950

Daily average for a month = 24950/12 = 2079

Compare each month's average to obtain percent average daily traffic.

e.g.   For January = $\dfrac{1100}{2079} \times 100 = 53$

The expansion factor for a month = 100/% average daily count.

e.g   For January = $\dfrac{100}{53} = 1.89$

(Various expansion factors are included in table)

(*b*) **Using the weekly table:**

Sum of 7 days daily counts for the week = 4090

Daily average weekly count = $\dfrac{4090}{7} = 584$

Compare each day of week to the daily average to obtain% of average daily traffic.

e.g   For Sunday = $\dfrac{500}{584} \times 100 = 86$

Calculate the weekly expansion factor by taking inverse of percent of average daily traffic

For Sunday = $\dfrac{100}{86} = 1.16$

(Various expansion factors are gives in the table as calculated)

Average Daily Traffic (ADT) = Coverage count × Seasonal factor × Weekly factor

ADT for February = 250 × 1.72 × 1.04 = 447 vehicles.

ADT for October = 300 × 1.15 × 0.78 = 269 vehicle.

Example 2.   Making use of the above monthly and weekly traffic volume data (example –1 Calculate

(a) The average daily flow if the flows counted on Friday, Saturday and Sunday in July were 8000, 6000 and 5000 vehicles, on 16 hourly count basis (16 hourly counts represent on the average 93% of the 24 hourly counts)

(b) Using this data calculate seven day 16 hourly flow in September.

Solution:

(a) Friday is a normal days of the week (5 days) and Saturday and Sunday are week end days.

Seven day (weekly) volume on 16 hourly basis for July will be

$$= 5 \times 8000 + 6000 + 5000 = 51,000 \text{ vehicles}$$

Seven day (weekly) 24 hourly volume $= \dfrac{51,000}{0.93} = 54839$ vehicles.

Referring to the table above the July traffic is 159% of annual average monthly traffic, therefore annual average seven day traffic on 24 hourly basis will be $= \dfrac{54839}{1.59} = 34490$

$$\text{Average Annual Daily Traffic (AADT)} = \dfrac{34490}{7} = 4927 \text{ vehicles}$$

(b) Seven day July 16 hourly volume is = 51,000 vehicles

September is 173% of annual average flow.

$$\text{Therefore 7 day Sept flow} = \dfrac{51,000 \times 173}{159} = 55490 \text{ vehicles}$$

$$\text{Average daily flow in September} = \dfrac{55,490}{7} = 7927 \text{ vehicles}$$

10.2 SPOT SPEED STUDY

Speed of a vehicle fluctuates from time to time along the road and its value as shown on the speedometer at a particular spot is called the 'Spot Speed.' The traffic police is interested in these speed-checks at problematic locations (spots), while a trip maker is more interested in the total journey time involved in the complete journey, i.e. on the "Journey Speed". It is on the part of the traffic engineer that desirable journey speeds are maintained on the highway system. For maintaining good journey speeds, the delays or involuntary stop due to road congestion should be minimum and vehicle should be running smoothly. This involves the concept of running time and running speed of a vehicle. It will therefore be proper to introduce some basic definitions regarding speeds and times, in the following paragraphs.

DEFINITIONS

Spot speed The speed of a vehicle as it passes a spot or point on a highway.

Average spot speed: The arithmetic mean of the speeds of all traffic, or component thereof, at a specified point. This is also known as time mean speed.

Overall travel speed: The speed over a specified section of highway, being the total distance divided by overall travel time (time of travel including stops and delay).

Running time: The time the vehicle is in motion.

Running speed: The speed over a specified section of highway, that is the distance divided by running time. The average for all traffic or component is the sum of distances divided by the sum of running times

85th percentile speed The speed below which 85% of the vehicles travel and above which 15% travel.

Design speed: It is the selected maximum safe speed, that can be maintained, over a specified section of highway, under favourable conditions of weather, traffic and geometrics of road, to be used in design of highway features such as curvature, superelevation, sight distance etc.

Median speed: The speed represented by a middle value when all speeds are arranged in ascending order. Half the speed values will be above the median and half below. It is 50th percential speed.

Modal speed: The speed value occurring most frequently. Modal speed is the value, which has the highest frequency of observation.

Pace: A given increment of speed that includes the greatest number of observations. It is usually taken as 10 mile (15 km.p.h.) increment.

Operating speed: The highest overall speed, exclusive of stops, at which a driver can travel, under prevailing conditions, without any time exceeding the design speed.

Free-flow operating speed: The operating speed of a passenger car over a section of highway during extremely low traffic densities, i.e. not restricted by any other vehicle.

Time mean speed: It is arithmetic mean, i.e. the speed obtained when sum of all speed values is divided by the number of observation.

Space mean speed: The speed corresponding to the average overall travel time (running time), to cover a specified section (distance) of the road.

Basic desired speed: The speed at which drivers would desire to travel on a road

The curve Fig 10.4. shows that it is not of much use to increase the speed beyond certain optimum value. Also time savings for shorter trips are not very high and therefore there is no much benefit of increasing speeds to too high for short trips within city area.

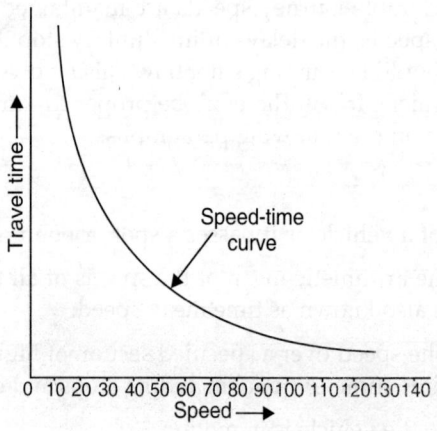

Fig. 10.4: Speed travel time relationship

Factors affecting spot speeds: The various factors that affect spot speed, can be grouped as follows:

1. **The driver:** The distance of his trip, the number of passengers in his vehicle, his sex, age, residence (urban or rural), in-state or out of state, and marital status, etc.

2. **The vehicle:** Its type, age, weight, horse power and manufacturer, etc.

3. **The roadway:** Geographic location, type, curvature, grade, length of grade, sight distance, number of lanes, surface type, lateral clearance, frequency and spacing of intersection, road side development, etc.

4. **The traffic:** Volume, density, passing manoeuvers, opposing traffic, access control, traffic control devices, speed regulations, types of vehicles, etc.

5. **The environment:** Time of the day, month, season, weather, etc.

10.2.1 Purposes of Spot Speed Study

Spot speed data has many applications such as:

- Establishing the trends in operating speeds, by periodic collection of data.
- For checking speeds at problem location where speeds are too high.
- For use in traffic regulation and control such as:

  (a) Establishing speed limits

  (b) Safe speeds on curves and intersections

  (c) Locating traffic signs

  (d) Location and timing of traffic signals

  (e) Establishing speed zones, etc.

- For accident analysis, to determine the relationship of speed to accidents.
- Before and after studies, to evaluate the effect of some change in conditions or control.
- For geometric design features:

  (a) To find out design speed

  (b) For designing curvature, speed change lane, super elevation, sight distance, etc.

- To evaluate capacity in relation to speed. If all vehicles travel at same speed, capacity would be at maximum and overtaking, rear-end collision type accidents would be eliminated
- As an aid in enforcement, to determine its effectiveness.
- Research studies.

10.2.2 Methods of Spot Speed Study

There are two basic methods in use:

1. Measuring of time and distance.

2. Taking advantage of Doppler principle.

The time versus measured distance methods are commonly used. These methods are:

1. Stop watch method:

(a) **Pavement markings:** This method involves marking the pavement at suitable distance apart say 50 m to 100 m, and the observer starts and stops the watch as the vehicle passes the markings.

(b) **Enoscope:** The enoscope (Fig. 10.5) is an L-shaped mirror box (mirror set at 45°), which is used at one or both the ends of the course. This device bends the line of sight of the observer, so that it is perpendicular to the path of vehicle at the limits of the measured course.

If one enoscope is used, the stop watch is started as the vehicle passes the observer and stopped when it passes the enoscope. If two enoscopes are used, the observer stations himself between them (Fig. 10.6).

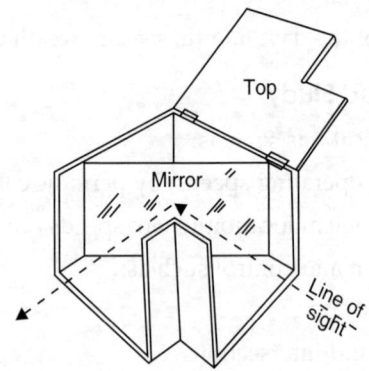

Fig. 10.5: Enoscope with top raised

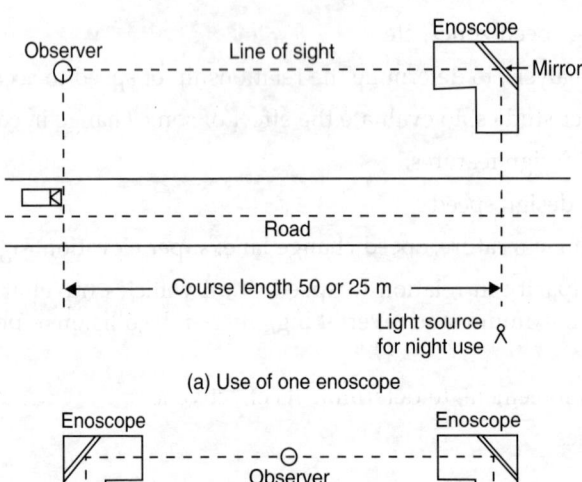

(a) Use of one enoscope

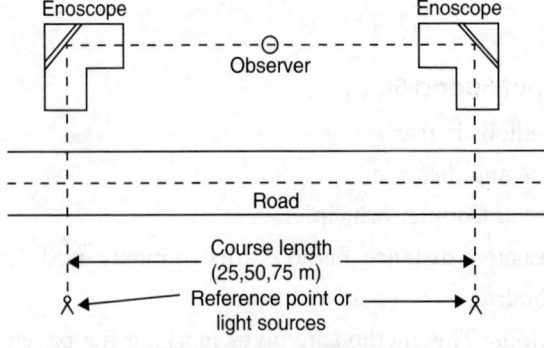

(b) Use of two enoscopes

Fig. 10.6: Use of enoscope for speed measurement

This method eliminates the parallax, in judging the exact time, the vehicle crosses the limits of the course. The method is easy, flexible, low cost and fairly reliable.

The recommended spacing for the enoscopes and the formula for calculating the speed is given in Table 10.1

Table 10.1: Recommended spacing of enoscopes

Average speed Kph	Course length (enoscope spacing m)	Speed formula Kph (t-time in seconds)
0–40	25	90/t
40–65	50	180/t
65 or more	75	270/t

Field sheet 5 shows the data to be collected for spot speed using enoscope method.

Field Data Sheet 5
SPOT SPEED STUDY ENOSCOPIC METHOD

Location............   Date and Day............

Direction............   Time............

Road Surface Condition............   Observer............

Weather............

Seconds	KPH for 50 m	KPH for 75 m		Passenger vehicles No. veh.		Buses No. veh.		Trucks No. veh.		Total
1.0	180	270								
1.25	144	216								
1.50	120	180								
1.75	103	154.5								
2.0	90	135								
2.25	80	120								
2.75	65.5	98.2								
3.0	60	90								
3.25	53.4	80								
3.50	51.4	77								
3.75	48	72								
4.0	45	67.5								
4.25	42.4	63.6								
4.50	40	60								
4.75	37.9	56.8								
5.0	36	54								
5.25	34.3	51.5								
5.50	32.7	49								
5.75	31.3	47								
6.0	30	45								
6.25	28.8	43.2								

(Contd.)

(Contd.)

Seconds	KPH for 50 m	KPH for 75 m		Passenger vehicles		Buses		Trucks		Total
					No. veh.		No. veh.		No. veh.	
6.50	27.7	41.5								
6.75	26.7	40								
7.0	25.7	38.5								
7.25	24.8	37.2								
7.50	24.0	35								
7.75	23.2	34.8								
8.0	22.5	33.8								
8.25	21.8	32.7								
8.50	21.1	31.7								
8.75	20.6	30.9								
9.0	20	30.0								
9.25	19.4	29.1								
9.50	18.9	28.3								
9.75	18.4	27.6								
10.0	18	27								
10.25	17.6	26.4								
10.50	17.1	25.6								
10.75	16.7	25.0								
11.0	16.4	24.6								
11.25	16.0	24.0								
Total Vehicles:										

(2) **Pressure Contact Tubes/Strips Method:** This method utilises either pneumatic tubes or electric type contact strips. The method involves laying two pneumatic tubes across the road at a fixed distance apart. A vehicle crossing the tube compresses the air in the tube, actuating an air impulse switch. The switch operates a timing mechanism, when the second tube is passed, by the same wheels of vehicle, the time is automatically stopped. It is similar to the one used for volume studies

Pneumatic tubes are low in initial cost, simple to install, and easily maintained, but they attract driver's attention, thereby affecting his behavior and speed data. Tubes are subject to traffic hazards. The timing devices of different types can be used with pneumatic tubes, such as milliammeter, photoelectric meter, etc.

(3) **Radar Meter Method:** Radar meter operates on the Doppler principle that the speed of a moving target is proportional to the change in frequency between the radio beam transmitted to the target and the reflected radio beam. The equipment measures this difference and converts it to a direct reading in km/h. The accuracy of speed determin-ations is ±2 kms per hour up to 100 kmph.

In this method, no tubes are required and equipment a 'Black Box' can be kept hidden. It is however difficult to distinguish a single vehicle being observed in heavy traffic. The accuracy of this method may be plus or minus 2 to 3 kmph

(4) **Ultrasonic Meter:** This device also operates on Doppler principle. A high-energy audio tone is directed at oncoming vehicles and the return signal is changed in frequency by the movement of vehicle. This change is proportional to the speed of the vehicle. The

transmitter-receiver is mounted over the center line of a lane and it is directed downward, at a 45° angle toward approaching traffic. The detection zone is very sharp and all vehicles are detected at approximately the some point, on the roadway. Its accuracy and advantages are like that of a radar meter. The data are usually transmitted from the field equipment to a central point by standard telephone circuit. This equipment is permanent and requires telephone lines or other data transmission devices.

(5) **Time Lapse Photography:** In this method, a camera takes photographs at fixed interval of time. (It has a device to move the film a single frame at a time, at a given time interval). When the developed film is projected on to a screen, a vehicle's movement along the roadway is readily determined. The speed equals the distance traversed by the vehicle from frame to frame, divided by the time between the same frames.

This method gives permanent record of flow characteristics, volume, speed, vehicle types, spacing, lateral vehicle manoeuvrs etc. Camera can remain completely hidden and the results are therefore unbiased.

For example a car is seen in 4 frames of the film and frame speed in 88 frames/min. One frame is 3.5 m coverage of road.

88 frames in one minute so 4 frames in 1/22 minutes.

In $\dfrac{1}{22}$ min the distance travelled is $4 \times 3.5 = 14$ m

Therefore $\qquad\qquad$ Speed $= \dfrac{14 \times 22 \times 60}{1000} = 18.5$ kph.

10.2.3 Analysis and Statistical Presentation of Speed Data

The data collected for spot speed is processed, using statistical techniques described in following paragraphs explained through following example.

(1) **Tabular arrangement of speed data:** Large number of observations on speed, are put into speed grouping, and number of vehicle observed in each speed category (group) as shown in Table 10.2

Table 10.2: Typical spot data from a study

	(1)	(2)	(3)	(4)	(5)	(6)	(7)	(8)	(9)
S. No. group	Speed Kph	Mid class x	Frequency (f)	Percent of total observation	Cumulative frequency	$f.x$	Deviation from assumed mean d_1	$f.d_1$	$f.d_1^2$
1	30.0–39.9	35.0	12	4.8	4.8	420	−4	−48	192
2	40.0–49.9	45.0	32	12.8	17.6	1440	−3	−96	288
3	50.0–59.9	55.0	48	19.2	36.8	2640	−2	−96	192
4	60.0–69.9	65.0	60	24.0	60.8	3900	−1	−60	60
5	70.0–79.9	75.0	38	15.2	76.0	2850	0	0	0
6	80.0–89.9	85.0	27	10.8	86.8	2295	1	27	27
7	90.0–99.9	95.0	15	6.0	92.8	1425	2	30	60
8	100–109.9	105.0	8	3.2	96.0	840	3	24	72
9	110–119.9	115.0	5	2.0	98.0	575	4	20	80
10	120–129.9	125.0	5	2.0	100.0	625	5	25	125
	Totals		250	100		17010		−174	1096

(2) **Graphical presentation of speed data:**

(a) **Histogram and frequency curve:** The information from the above table is best represented graphically by histogram and frequency curve as shown in Figs. 10.7 and 10.8.

Histogram can be plotted between column (1) and (4) of table and frequency curve is found by rounding off the histogram (joining mid-points) in such a way that area under the curve is equal to the area of histogram.

(b) **Cumulative frequency curve:** The cumulative frequency curve is used for determining the number of vehicles travel above or below given speed.

It is plotted between cumulative percentage and upper limit of each speed group, i.e., columns (5) and (1) of table as shown in Fig. 10.8. From these curves, the speed characteristics of traffic stream are described by several significant values, which describe the distribution in question, such as given below:

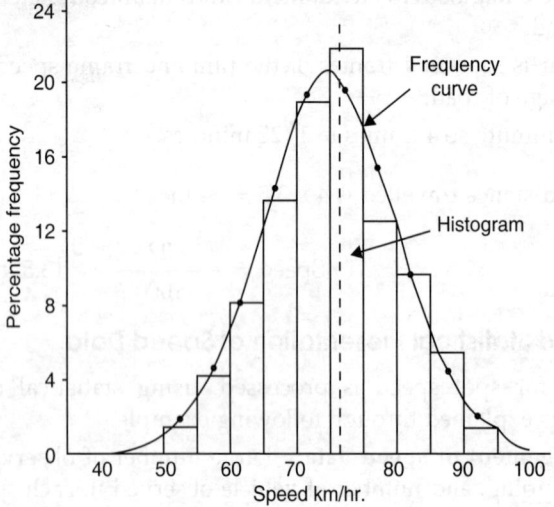

Fig. 10.7: Presentation of spot speed data (Histogram)

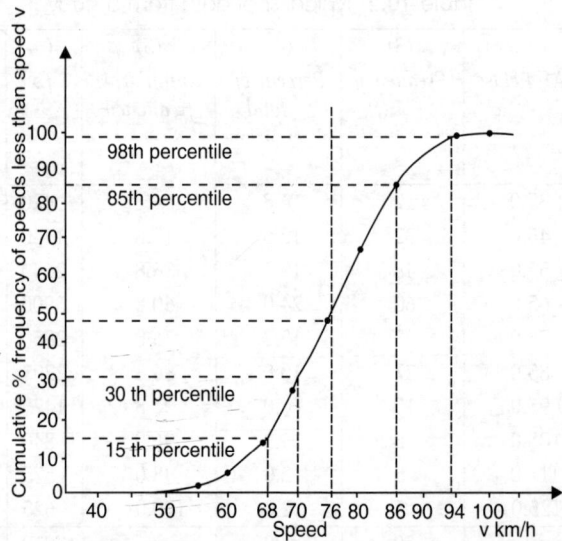

Fig. 10.8: Cumulative frequency distribution curve, for spot speed data

(i) MEASURE OF CENTRAL TENDENCY

(a) **Arithmetic Mean Speed (Time Mean Speed):** Arithmetic mean speed is just the average speed of all observed vehicles.

Arithmetic mean speed

$$\bar{x} = \frac{\Sigma \text{ Mean speed of each group} \times \text{Number of observations in that group}}{\text{total number of observations}}$$

$$\bar{x} = \frac{\Sigma fx}{\Sigma f} = \frac{17010}{250} = 68 \text{ km/hr.}$$

It is rather simple to understand the mean or average speed, but it is possible to have widely different speed conditions on the sections of the road and yet for all conditions have roughly the same average speed. All vehicles may be travelling about the same speed, very near the average speed or they may be travelling at widely varying speeds evenly balanced about mean speed, giving the same speed as average balancing each other. To get a useful and realistic picture of the speed we should know not only the mean speed, but also the spread of the speed variation about the mean.

Though the lowest and the highest observation of the speed value gives a rough idea of spread called 'Range', it does not give a good idea of the actual happening. We therefore measure the spread of a group of readings by a statistical function called 'Standard Deviation' (S). The standard deviation is a measure of the average deviation of readings from their mean. It is the square root of the average of deviations squared.

To standardise the standard deviation, there is a single statistical measure to standardise the spread by expressing the standard deviation as a percentage of mean value. This is known as "Coefficient of Variation" (V).

Coefficient of variation gives a clear impression not only of the spread readings but also of the significance of this spread.

(b) **Median Speed:** The median speed is the middle value. One half of the observed values are higher than the median and one half are lower. It is 50th percentile speed on the cumulative frequency distribution curve, as shown. In this case it is 76 km/hr

(c) **Modal Speed (Mode):** The mode or model speed is the most frequently occurring value in a distribution. It is the speed in which maximum number of vehicles lie. Sometimes it is estimated as the speed corresponding to peak point on the frequency distribution curve.

If the frequency distribution curve is symmetrical around its vertical axis, the mean, median and mode will all be the same. In this case the mode is 76 km/hr.

(d) **Pace:** The pace is defined as 10 kph or 20 kph range in speeds in which highest number of observations were recorded. It is identified by a 20 km/hr speed increment which "Cut Off" the peak of the frequency distribution curve. In this example, the 20 km/hr pace is 50–70 km/hr (in this range highest number of observations lie).

(e) **85th, 98th and 15th Percentile Speeds:** From the cumulative frequency curve 85th, 98th and 15th percentile speed can be determined. 98th percentile speed is used in Geometric design. It is 94 km/hr, 85th percentile speed is used for traffic regulations, it is 86 km/hr. 15th percentile speed shows the slower vehicles whose speed may cause interference with traffic stream, it is 68 km/hr.

(ii) MEASURE OF DISPERSION

(a) **The Standard Deviation:** The standard deviation of a grouped frequency distribution is given by:

$$s = \sqrt{\frac{\Sigma \, fd^2}{\Sigma \, f}}$$

where
s = standard deviation

f = frequency of each class

d = deviation of mid class point from arithmetic mean, *i.e.* $d = x - \bar{x}$

For computations it is often convenient to use an assumed mean as in the following expression:

$$s = \sqrt{\frac{\Sigma \, fd_1^2}{\Sigma \, f} - \left[\frac{\Sigma \, fd_1}{\Sigma \, f}\right]^2} \times i$$

where
d_1 = is the deviation from an assumed mean, expressed as a multiple of class interval.

i = class interval.

In the example: Assumed mean is 75 km/h

Using columns 3, 7, 8 and 9 of the table,

Standard deviation $\quad s = \sqrt{\frac{1096}{250} - \left(\frac{-174}{250}\right)^2} \times 10 = 19.7$ kph.

(b) **Coefficient of Variation (V):** To provide a measure of dispersion related to the mean it is convenient to use the coefficient of variation, V given by

$$V = \frac{100 \, s}{\bar{x}}$$

in this example $\quad V = \dfrac{100 \times 19.7}{68} = 29\%$

A higher value indicates wide scatter about the mean and vice versa.

Sample Accuracy: The determination of spot speed of vehicles on the road is done on a sample basis.

The standard deviation of the distribution of sample means, is known as the standard error and is given by

$$\text{Standard Error} = \text{SE} = \frac{S}{\sqrt{N}}$$

where
SE = Standard error

N = Number of observations in the sample

S = Standard deviation

$$\text{SE} = \frac{19.7}{\sqrt{250}} = 1.25 \text{ km/hr}$$

This means that 95% chances are that true mean of spot speed of vehicles on the road lies within range.

Calculated mean ± standard error

In this case   68 ± 1.25 = 69.25 km/ph to 66.75 km/ph.

This is the property of a normal curve

If it is required to narrow this range sample size must be increased. If standard error is reduced to 0.6, the sample size

$$\text{would be } 0.6 = \frac{19.7}{\sqrt{N}}, \text{ or } N = 1078 \text{ vehicles}$$

Example 2.   The field data collected through a field study is summarized in table below:

Speed class	Number of vehicles observed
10–14.9	3
15–19.9	10
20–24.9	21
25–29.9	31
30–34.9	54
35–39.9	43
40–44.9	21
45–49.9	10
50–54.9	5
55–59.9	2

Making use of the above data, calculate:

(i) Modal speed

(ii) Median speed

(iii) Time mean speed

(iv) Speed used in geometric design of road

(v) Suitable speed for traffic regulation

(vi) Standard deviation and

(vii) Coefficient of variation

Solution:

Table below summarises the calculations made:

Mid class (x)	Frequency (f)	Cumulative	f(x)
12.45	3	1.5	37.35
17.45	10	6.5	174.50
22.45	21	17	471.45
27.45	31	32.5	850.95
32.45	54	59.5	1752.30
37.45	43	81	1610.35
42.45	21	91.5	891.45
47.45	10	96.5	474.50
52.45	5	99	262.25
57.45	2	100	114.90
	Σ 200		Σ 6640

Cumulative frequency curve is drawn as shown.

(i) Modal speed is the one which most frequently occurred

$$= 32.45 \text{ km/h}$$

(ii) Median speed is 50th percentile speed, from graph

$$= 30.65 \text{ km/h}$$

(iii) Time mean speed is the arithmetic mean speed

$$\bar{x} = \frac{\Sigma f x}{f} = \frac{6640}{200} = \textbf{33.2 km/h.}$$

(iv) Speed in geometric design is 98th percentile.

From graph = **49.65 km/h**

(v) Speed for traffic regulation is 85th percentile.

From graph = **39.45 km/h**

(vi) Standard deviation (S) = $\dfrac{P_{85} - P_{15}}{2.07}$

P_{15} from graph = 21.45 km p.h.

$$S = \frac{39.45 - 21.45}{2.07} = \textbf{8.69 km/hr.}$$

(vii) Coefficient of variation

$$V = \frac{100 \times S}{\bar{x}} = \textbf{26.17\%}$$

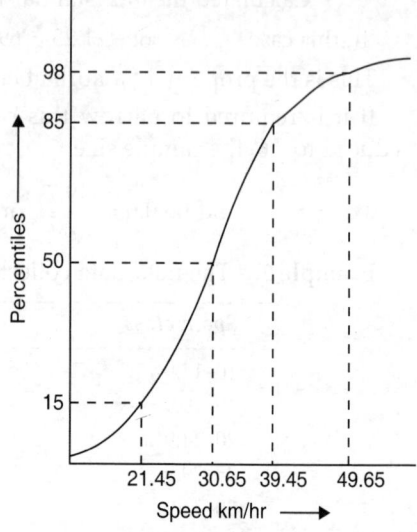

Cumulative Frequency Curve.

10.3 TRAVEL TIME AND DELAY STUDY (SPEED AND DELAY STUDY)

Travel time and delay study measures average travel and running times along sections of a road, while at the same time information regarding location, cause and duration of delay is also collected. Space mean speeds are determined by conducting a travel time and delay study, as mean travel time of several trips over the section is measured.

DEFINITIONS

Delay: This is the time lost while traffic is impeded by some element over which the driver has no control. The delay may be fixed delay or operational daily.

Fixed delay: The delay is caused by traffic control devices. It is the delay to which vehicle is subjected to regardless of amount of traffic volume and interference. It may be caused by traffic signals, stop signs, rail-road crossing etc.

Operational delay: This delay is caused by interference between components of traffic, such as turning vehicles, parking or unparking vehicles, crossing traffic, pedestrians, etc.

A second type of operational delay is caused by interferences, within the traffic stream (internal frictions). This is includes congestion due to high volumes, lack of roadway capacity, merging or weaving manoeuvrs.

Stopped time delay: This is the time period that a vehicle is actually standing still, due to any factor

Travel time delay: This is the delay caused by acceleration and deceleration, in addition to stopped time delay.

10.3.1 Purposes of Travel Time and Delay Study

- Information on amount, location and cause of delays helps in locating 'Congestion' spots, where proper remedial measures can be used.
- Sufficiency ratings, congestion indices or quality indices methods, based on travel time are used to compare different roads and assess the quality.
- Before and after studies, utilizing data on travel time and delay, are used to determine effectiveness of a traffic improvement such as parking restriction, signal timing, new one way street, etc.
- Assignment of traffic is done based on speed delay characteristics.
- Economic studies such as benefit cost analysis use travel time data to evaluate the benefits of time savings.
- Trend studies use travel time data to evaluate the level of service, as it changes with the passage of time.
- To determine the need for traffic signals and to determine optimum timing sequences of the signal cycle.

10.3.2 Methods of Travel Time and Delay Study

Following methods are commonly used:

(i) **License plate method:** Observers are stationed at the entrance and exit of a test section. The timings and vehicle numbers are noted by both the observers of the selected sample. From office computations of travel time of each vehicle is obtained. This method provides information on travel times only. Data is collected as shown in field sheet 6 and 7.

This method requires large number of observers and too much office work. It does not give important details such as cause, duration and location of delays, within the test section.

(2) **Interview Method:** This method involves interviewing selected individuals as to their travel time and delays experienced during the particular trip on the road section under study. Sometimes they are asked in advance to record their experience. With good co-operation the results may be satisfactory and require minimum time.

(3) **Photographic Method:** Photographic methods are suitable for short sections of the road, or intersections. This method provides a means of obtaining a large sample of vehicles and a permanent record, but is very costly and limited to day light and good weather conditions.

(4) **Elevated Observation Method:** In this method observers are stationed at an elevated vantage print. They select certain vehicles at random and record pertinent data, as they progress through study section. This method is limited to short sections, intersections etc. and depends upon availability of suitable elevated observation points.

(5) **Riding a Vehicle Method:** In this method, a test car is driven along the study section according to driver's judgment of the average speed of traffic on the route or at legal speed limit unless impeded by actual traffic conditions. Observers record with a stop

watch the time as the test car passes the pre-determined control points and also notes the location, cause and length of any delay that occurs. Normally 8 to 12 runs are required for reasonably accurate results. Number of runs, however depend upon the degree of confidence, the level of the accuracy required, length of the route and the nature of traffic. If a tape recorder or tachograph is fitted to the vehicle only the driver can collect the data, without the help of a reorder. Care should be exercised in placing the test car in the traffic stream, for proper representation of average conditions.

Field Data Sheet 6

TRAVEL TIME STUDY LICENSE PLATE METHOD

Location on.........................At.........................Weather........................

Date.......................Route Length........................Direction of Traffic.......................

Time of Start.......................Time of End.......................Time Correction (if any).......................

Observor........................Recorder

0	1	2	3	4	5	6	7	8	9
									967(10.1)
					562(11.10)				

Note: Only record last three digits of license plate number and corresponding time in each box, underline and bus or truck with dual rear wheels or heavier vehicles. The entry is made in the indicated column that corresponds with first digit of the last three digits of the license plate number,

(6) **Floating Car Method (Moving Observer Method):** The test vehicle is driven over a given course of travel at approximately the average rate of speed, thus trying to float with the stream. At least 12 trips should be made to adequately measure the average speed and delay for any one direction.

2 or 3 observers in the floating car record the times at various control points and the location, duration and cause of delay.

One observer in the car counts opposing traffic and records the number of vehicles overtaking and overtaken by his vehicle. The other observer records the time at the predetermined control points and notes the location and length of delays that occur. The need for the driver to follow the average speed of the traffic stream is less critical in this method. Sometimes 3 observers are used for the same observations.

Stop watches, suitable field-sheets or voice recording equipment is used to record time and delay. In this method detailed information is obtained concerning all phases of speed and delay. The data collected is analysed by following equations.

The average journey time t in minutes for all vehicles is given by: (in the direction of flow)

$$t = t_w - \frac{y}{q}$$

The traffic volume q per minute, in the direction of the stream being considered is given by:

$$q = \frac{x + y}{t_a + t_w}$$

where, t = Average journey time in minute, for vehicles in a traffic stream in the direction of flow q.

q = Flow of vehicles (volume per min) in one direction of the stream.

x = Average number of vehicles, counted in the q direction when test car was travelling in opposite direction.

y = Number of vehicles overtaking the test vehicle minus the number of overtaken by test car while the test vehicle travelled with the stream (in direction of q).

t_a = Average journey time, in minutes, of the test vehicle travelling against the stream (opposite direction to q).

t_w = Average journey time, in minutes, of the test vehicle, travelling with the stream (direction of q).

Figure 10.9 illustrates the moving observer technique.

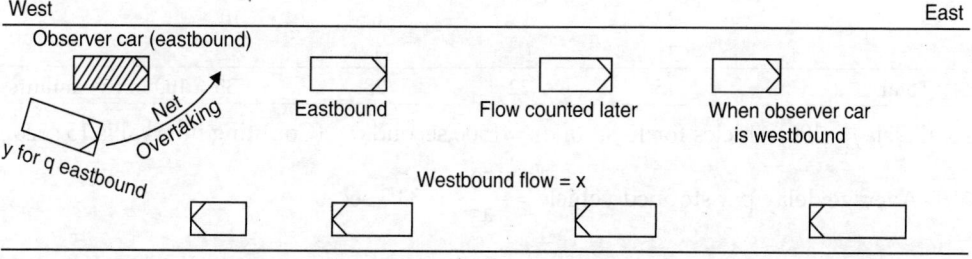

Fig. 10.9: Floating car method of speed-delay study

10.3.3 Intersection Delay Studies

Delay at intersection is the major factor in the analysis of delays. Because junctions are complex areas of traffic interaction, their physical characteristics (number of lanes, gradient, geometric layout, bus stops and pedestrian crossings), traffic use (volume and turning movements, speeds, pedestrian flow), and the form of traffic control (signals, channelisation, turn controls etc.) all influence the nature and amount of delay.

Methods for measuring intersection delay: Methods of study to evaluate delay for isolated intersections are based on travel time studies measured over a length between two points-upstream and downstream of site (intersection)-by Test car, recording registration numbers, elevated point or photographic methods all discussed earlier.

Sampling procedure: A sampling procedure, which involves the counting of vehicles occupying an intersection approach at successive time intervals such as every 15 seconds for a total period of time 5 to 10 minutes.

Each successive count represents an instantaneous density (number of vehicles occupying the length of intersection approach per time interval). During the given time period the number of vehicles leaving the intersection approach is also counted. This count represents the traffic volume.

From these counts, the average travel time for all vehicles traversing the intersection approach can be calculated. In this method a specific vehicle is counted more than once, if it is stopped during more than one observation time. Because of this it is necessary to count the number of vehicles stopping for volume determination. Two observers are needed and can obtain reasonably accurate information very quickly. Taking one approach to the crossing one observer counts all vehicles for five minutes classifying the traffic in 'Stopping' and 'Non Stopping'. Meanwhile the second observer, at fifteen seconds intervals, counts the number of stationary vehicles on the approach. From these readings an average delay per vehicle can be derived. An example below explains the method.

Observer 1.   *APPROACH VOLUME*

	Stop	*Non-stop*	*Total*
	45	40	85

Observer 2.   *STATIONARY VEHICLES*

Min	00	15	30	45	Seconds
0	0	4	10	12	
1	3	2	1	6	
2	7	12	10	5	
3	2	3	6	10	
4	3	1	1	2	
Total:	15	22	28	35 = 100 in five minutes	

Total Delay = 100 vehicles for 15 seconds = 1500 seconds.   (Counting interval is 15 seconds)

$$\text{Average delay per stopped vehicle} = \frac{1500}{45} = 33.3 \text{ sec.}$$

$$\text{Average delay per approach vehicle} = \frac{1500}{85} = 17.6 \text{ sec.}$$

$$\text{Percent of vehicles stopped} = \frac{45}{85} = 52.9\%$$

Data is collected on field sheet No. 8, for intersection delay study. For correlating delays to various road features, road Inventory data sheet N.   ed

Field Data Sheet 7

SPEED AND DELAY STUDY
(License Plate Method)

Route: ....   Date: ....

Observer: ....   Time : ....

S. No.	Vehicle Number	Departure Time in Min.			Travel Time in minute		Speed (KPH)		Remarks
		Station 1	Station 2	Station 3	1 – 2	2 – 3	1 – 2	2 – 3	

Station: 1. ....   Distance:

2. ....   1 – 2........................ Kms

3. ....   2 – 3 .... Kms.

Field Data Sheet 8

INTERSECTION DELAY STUDY

Location........................Approach........................Movement........................

Date........................Weather........................Study No........................Observer........................

Time (Starting at)	Total Number of vehicles stopped in the approach at time				Approach volume	
					Number stopped	Number not stopped
	+ 0 sec	+ 15 sec	+ 30 sec	+ 45 sec		
Sub Total						
Total						

Field Data Sheet 9

ROAD INVENTORY DATA FOR DELAY STUDIES

Name of the Road............

Observer............

S. No.	Sections of the road	Road width in metres	Type of road surface	Width of Side walks		No. of side lane joining on		No. of Intersections			Other geometric characters curve grades etc	Main activity along the section and special remarks
				Left	Right	Left	Right	Uncontrolled	Manually Controlled	Signal ised		

10.3.4 Presentation of Travel Time and Delay Data

(1) **Time Contour Maps:** Contour lines represent minutes of travel time, from some central point. With reference point as a centre, a series of concentric circles are drawn indicating distance from the central reference point, on different roads (A-34, A-38 etc)

These maps shown in Fig. 10.10, are well suited for the comparison of the various routes leading from CBD and for estimating the time that may be saved by planning a new or improved facility.

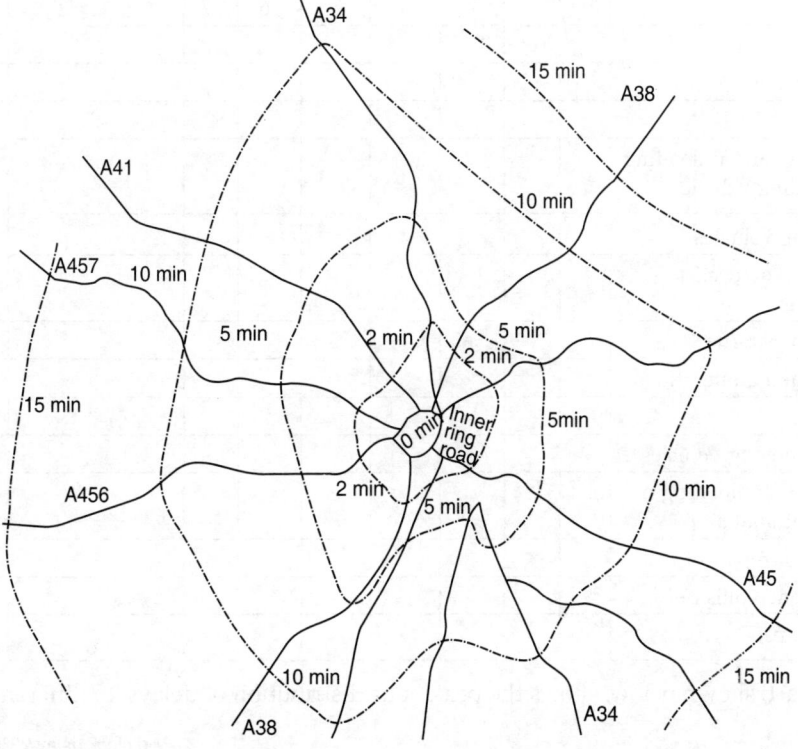

Fig. 10.10: Time contour map

(2) **Bar Charts (Time Charts):** The total delay, caused due to various causes, is summarized to prepare the bar charts. These charts indicate causes and distribution of delays. They show at glance all of the factors contributing to delays. They are useful in comparing before and after results when changes are made.

Field sheet 10 is used to summarise distribution of delay collected through speed and delay study. Bar charts are shown in Figs 10.11 and 10.12

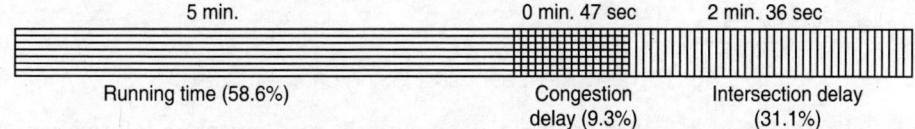

Fig. 10.11: Time spent and delay during peak hour

Field Data Sheet 10
SPEED AND DELAY STUDY
DELAY DISTRIBUTION SUMMARY

Route............   Date............

Observer............   Time............

Major Causes of Delay	DELAY IN SECONDS ON VARIOUS SECTIONS							Average delay in seconds	Maximum Delay in seconds
	1	2	3	4	5	6	7		
1. Traffic signals									
2. Congestion									
3. Parked vehicle									
4. Loading and unloading of loading vehicle									
5. Turning vehicles									
6. Pedestrian/cyclist crossing									
7. Crossing traffic									
8. Overtaking operations									
9. Curves									
10. Slow moving vehicles									
11. Loading/unloading of bus with out stop									
12. Block or repair of road									
13. Accident/collision									
14. Bus stops.									

Bar chart, shown below, shows the peak hour distribution of delays, to different causes.

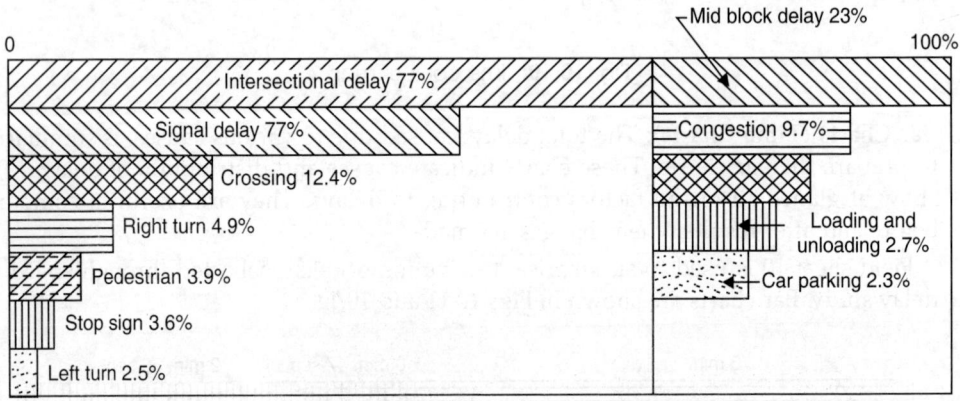

Fig. 10.12: Typical bar chart showing analysis of delay

This delay information is then analysed along with the road inventory information to arrive at remedial measures.

(3) **Speed Profiles:** These maps indicate the variations in average running speed and overall speeds (or travel times) of traffic, along a given route, from section to section or block to block basis.

(4) **Frequency Distribution Curves:** These curves show the number or percentage of vehicles versus the travel time or delay.

(5) **Speed Zone or Delay Zone Maps:** These maps show speed or travel time along different sections of various routes within an area, and are called speed zone maps.

Delay zone maps show the vehicle minutes of delay during different hours, along different sections of various routes within area.

Example 4: A test car was used on a north-south road 0.75 km long, and the following data, for the moving car was collected.

North trips Number (N)	Travel time travelling North (minutes)	Number of vehicles met against stream (x)	Number of vehicles overtaking test car	Number of vehicles over taken by test car
1	2.65	85	1	0
2	2.70	83	3	2
3	2.35	77	0	2
4	3.00	85	2	0
5	2.42	90	1	1
6	2.54	84	2	1
Totals	15.66	504	9	6
Average	2.61	84.0	1.5	1.0

South bound trips (S)	Travel time travel South (minutes)	Number of vehicles met against stream	Number of vehicles overtaking test car	Number of vehicles over taken
1	2.33	112	2	0
2	2.30	113	0	2
3	2.71	119	0	0
4	2.16	120	1	1
5	2.48	105	0	2
6	2.54	100	0	1
Totals:	14.52	669	3	6
Average	2.42	111.5	0.5	1.0

Calculate traffic volume, average travel time and space mean speeds in both directions.

Solution:

$$\text{Traffic Volume: } q = \frac{x + y}{t_a + t_w}$$

in one direction per min. in direction of flow

$$\text{Travel Time: } t = t_w - \frac{y}{q}$$

average journey time in t, in direction of flow (q)

Traffic volume in North direction: $V_N = \dfrac{x + y}{t_a + t_w}$

x = Average number of vehicles counted in direction N (of q) when test car was travelling in opposite direction = 111.5

for V_s = 84.0

y = Number of vehicles overtaking the test car minus the number of vehicles overtaken by test car, while test car travelled in direction N.

= 1.5 − 1.0 = 0.5

for V_s = 0.5 − 1.0 = − 0.5

t_a = Average journey time in minutes of test vehicle travelling against the stream (opposite to q) (S direction)

= 2.42

for V_s = 2.61

t_w = Average travel time of test car travelling with the stream. (N direction)

= 2.61

for V_s = 2.42

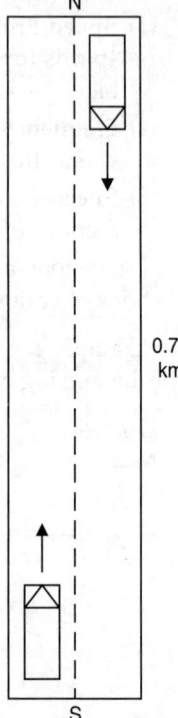

0.75 km

$$V_N = \frac{111.1 + 0.5}{2.42 + 2.61} \times 60 = 1336 \text{ VPh.}$$

$$V_S = \frac{84 - 0.5}{2.61 + 2.42} \times 60 = 996 \text{ VPh.}$$

Average journey time $t_N = 2.61 - \dfrac{0.5}{1336/60} = 2.59$ min.

Average journey time $t_S = 2.42 - \dfrac{(0.5)}{996/60} = 2.45$ min.

Space mean speed in N direction = distance/average time = $\dfrac{0.75}{2.59/60} = 17.5$ KPh.

Space mean speed in S direction = $\dfrac{0.75}{2.45/60} = 18.41$ KPh.

Traffic volume in south direction $V_S = \dfrac{x + y}{t_a + t_w}$, here x = against the stream , $i.e.$, north

= 84

$V_s = \dfrac{84 - 0.5}{2.42 + 2.61} \times 60$ $y = 0.5 - 1 = -0.5$

$t_a = 2.42$

= 996 vehicles per hour. $t_w = 2.61$

Travel time in North direction $= t_N = t_w - \dfrac{y}{q}$

$$= 2.61 - \dfrac{0.5}{1336/60}$$

$$= 2.59 \text{ minutes.}$$

Travel time in South direction $= t_s = t_w - \dfrac{y}{q}$

$$= 2.42 - \dfrac{(-0.5)}{996/60}$$

$$= 2.45 \text{ minutes}$$

Space mean speed in North direction

$$S_N = \dfrac{\text{distance}}{\text{average time}} \qquad\qquad \text{dist} = 0.75 \text{ m}$$

$$= \dfrac{0.75}{2.59/60} = 17.4 \text{ K.P.h}$$

Space mean speed in South direction

$$S_S = \dfrac{0.75}{2.45/60} = 18.4 \text{ K.P.h}$$

Example 5: A test car was used for seven runs, in each direction over a section of the road 1.64 km in length, with the following reading. Calculate traffic volume, time and space mean speed in both directions.

(A) EAST-BOUND TRIPS

Journey time in min-sec	x counts	Vehicles overtaking test car	Vehicles overtaken by test car
1—50	38	1	4
2—02	42	2	3
2—00	40	1	5
1—56	38	1	4
1—58	38	1	4
1—48	35	1	6
2—13	41	2	2
Total: 13—47	272	9	28
Average: 1.97 min	38.8		$\dfrac{9-28}{7} = -2.7$

(B) WEST BOUND TRIPS:

Journey time in min sec	n Count	Vehicles overtaking test car	Vehicles overtaken by test car
2—05	27	3	4
2—00	26	2	4
2—10	28	4	4
1—58	25	2	5
2—15	30	4	3
2—07	28	3	3
2—10	31	2	3
Totals: 14—45	195	20	28
Average: 2.10 min	27.9		$\dfrac{20-28}{7} = -1.1$

Solution:

$$\text{West} \quad \rightarrow \quad \text{East}$$

....

....

Flow in East Direction (q_{east})

t = Average journey time in minutes, for vehicles in East direction of q_{east}.

x = Average number of vehicles counted in direction of East. When test car was travelling West bound

= 195 for 7 trips, and 272 for 7 trips for west and east bound trips

$$\text{Average } \frac{272}{7} = 38.8, \text{ and } \frac{195}{7} = 27.9$$

y = Overtaking the test car—vehicles overtaken by test car while test car is east bound

= 9 – 28 = for seven runs and 20 – 28 for 7 runs for east and west trips

$$= \frac{-19}{7} = -2.7 \text{ and } \frac{-8}{7} = -1.1 \text{ are average values.}$$

t_a = Average journey time when travelling against the flow, *i.e.*, in west bound direction.

= 2.10 min – 1.97 min

t_w = Average journey time of test car while travelling with flow, i.e. in East bound direction.

= 1.97 min – 2.10 min

$$\therefore \quad q_{east} = \frac{x+y}{t_a+t_w} = \frac{27.9+(-2.9)}{2.10+1.97} = \frac{27.9-2.7}{2.10+1.97} \, 60 = 371 \text{ VPh}$$

East bound time $t_{east} = t_w - \dfrac{y}{q} = 1.97 - \dfrac{(-2.9)}{6.18} = 2.41$ minutes

With similar calculations

$$Q_{west} = \frac{x+y}{t_a+t_w} = \frac{38.8+(-1.1)}{1.97+2.10} \text{ VPm}$$

$$= \frac{38.8-1.1}{4.07} \times 60 = 536 \text{ vehicles per hour.}$$

West-bound time $t_{west} = 2.10 - \dfrac{(-11)}{9.26} = 2.22$ min

From these times the speed can be calculated.

East-bound speed $= \dfrac{\text{Distance}}{\text{Average time}} \times 60 = 40.8$ Km/hr.

West-bound speed $= \dfrac{1.64}{2.22} \times 60 = 44$ km/hr.

Example 6: 5 minutes of data, using the sampling procedure, at one approach of an intersection are given below, for calculating total volume and average delays.

Time	Total number of vehicles stopped				Approach volume	
Minutes at starting	in the approach at time				Number	Number
	+0 sec	+15 sec	+30 sec	+45 sec	Stopping	Non-stopping
5.00 pm.	0	4	10	12	9	6
5.01 pm.	3	2	1	6	6	14
5.02 pm.	7	12	10	5	14	2
5.03 pm.	2	3	6	10	12	1
5.04 pm.	3	1	1	2	4	17
Sub-totals:	15	22	28	35	45	40

Totals $\quad\quad\quad\quad\quad\quad\quad\quad\quad\quad\quad$ 100 $\quad\quad\quad\quad\quad\quad\quad\quad\quad\quad\quad$ 85

Solution:

Total volume leaving the intersection approach during the time period of 5 min $\Big\}$ Given in last two columns.

Total delay = Total number of vehicles observed × Observation interval (15 seconds)
$= 100 \times 15 = 1500$ vehicle seconds.

Average delay per stopped vehicle $= \dfrac{\text{Total delay}}{\text{Number of stopping vehicles}}$

$= \dfrac{1500}{45} = 33.3$ seconds.

Average delay per approach vehicle $= \dfrac{\text{Total delay}}{\text{Approach volume}}$

$= \dfrac{1500}{85} = 17.6$ seconds

Percent of vehicles stopped $= \dfrac{\text{Number of stopping vehicles}}{\text{Approach volume}}$

$= \dfrac{45}{85} = 52.9$ percent.

10.4 ORIGIN AND DESTINATION STUDY

Origin and destination study determines the pattern of journey that people make. It is the basic study which provides the information for planning of a transportation facility or system particularly the location, design and programming of a new or improved highway, public transport and parking facility. Data for trip generation, trip distribution, modal split and route assignment analysis, becomes the part of this study. It is often called travel survey fundamental to all transportation studies and consists basically of determining:

- Where travelers are coming from (origin of trip)
- Where trip makers are going to (destination of trip)
- Why people are making trip (purpose of trip)
- By which made do different types of people travel (mode of travel)
- At what time of the day do they travel (time of trip)
- What is the normal length of trip.
- Other socio-economic data of the trip maker.

O-D study may range from a relatively simple survey to collect data for locating a by-pass in the town to a comprehensive transportation study for planning and design of the transportation system of a large metropolitan city. Depending upon its origin and destination, the traffic (trip) going through a town can be classified as:

(i) External to external: Traffic whose origin and destination both lie outside the town study area.

(ii) External to internal: Origin of the traffic is outside and destination inside the town study area.

(iii) Internal to internal: Origin and destination of traffic both inside town/study area

(iv) Internal to external: Origin of traffic inside and destination outside the town/study area

10.4.1 Purposes of Origin and Destination Studies

Origin and destination data is used by the traffic engineer in:

- Determining travel demands on existing and future facilities
- Knowing the adequacy of existing parking and other terminals
- Location of new terminals, bridge or by-pass.
- Judging the adequacy of mass transportation facilities and planning and locating the new services.
- Planning, designing and locating of new facilities or improving the existing system.
- Establishing travel characteristics from various types of land-use.
- Estimating future travel needs, patterns and requirements of transportation facilities.
- Arriving at construction priorities and economic justifications of improvements or new ventures.
- Establishing the socio-economic correlations.

ZONING OF STUDY AREA

Most origin and destination studies studies begin with delineation of the survey area-the boundary of which is called 'External Cordon'. For summarizing origin and destination data, the survey area is subdivided into reasonably small areas called 'Zones'. Normally zones are numbered and all trips with origins and destinations within a zone are assumed to begin or end at the centroid of that zone. A zone should have homogeneous nature of development.

Topographical features often form main boundaries further divided by landuses such as residential, shopping, recreational, industrial, commercial, etc. to determine the size of individual zones.

Zones should also be small enough to reduce any major errors in assuming that the zonal centroid is the centre of the road network. This limits the area to about 1 square kilometer. Smaller zone may give statistically reliable data but corresponding greater computing requirements.

10.4.2 Methods for O and D Studies

Following methods are used for conducting O-D studies:

(1) **Roadside Interview Method:** Drivers are stopped and interviewed at roadside. Data is recorded on prepared forms. The usual information required is:

 (a) Type of vehicle

 (b) Number of persons in vehicle

 (c) Origin and destination of trip

 (d) Purpose of trip

 (e) Parking location and duration

 (f) Intermediate stops

 (g) Routes travelled.

Stopping of the drivers requires the assistance of a police officer.

Advantages

- The interview can be carried on the sample basis.
- It is direct method and accurate information is obtained.
- Well suited where man power available is less.
- Survey can be conducted over extended period of time.

Disadvantages

- Stopping of drivers may create congestion.
- A detailed information, other than the trip, cannot be asked as more time is required.
- Drivers do not like stopping.

(2) **Post Card Method**: Prepaid return post cards are used in two general methods

 (i) Distribution of post cards to vehicle drivers at some locations on their travel route.

 (ii) Mailing the post cards to vehicle owners within the study area. More information can be obtained by this method. The number of questions is usually limited to six or seven and they should be precisely and properly worded. Cards with different colours may be used to distinguish between census stations or different directions of travel.

Advantages

- Requires less money
- Data can be collected in lesser time.
- Fewer trained personnel are required.

Disadvantages

- Since all population is not equally cooperative, results may be biased.

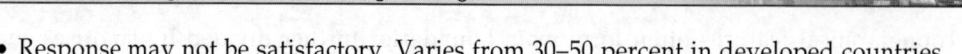

- Response may not be satisfactory. Varies from 30–50 percent in developed countries.
- Many drivers specially of commercial vehicles may be uneducated.
- Information is obtained only on vehicular trips.
- A good programme of advance publicity is required for public cooperation.

(3) **License Plate Method:** Recording of the license plate numbers of moving vehicles or parked vehicles is used, to collect O and D information.

Observers are stationed at various locations in the study area and they record the plate number, time of passage, direction of travel and type of vehicle. In analyzing the data the zone of origin is assumed to be the station where vehicle was first observed, the route is then traced and destination is assumed where the vehicle was last observed.

For parked vehicles, the zone of destination is where the vehicle is seen parked and its origin is assumed to be the address of the vehicle owner. In this method information on trip purpose, route travelled, ultimate destination, can not be obtained.

Advantages

- Simplicity of field organization,
- No interference with moving traffic.
- Good likelihood of having unbiased sample.

Disadvantage

- Extremely difficult in analyzing the data.
- Large number of observers is required.
- Large element of personal error in recording license plate number of moving vehicles.

(4) **Head Light On Method or (Tag On Car Method):** In this method, drivers at the entrance of the study area are requested to turn on the lights until they leave the study area. Observers stationed at all exit points record the number of vehicles which have head lights on. This method is good and simple for tracing vehicle movements along a single route or through an intersection. In this method the personal error in recording the license plate number is eliminated. Sometimes Tags on the vehicles are put in place of 'Light on'.

(5) **License Renewal Method:** The license renewal procedure for drivers requires them to appear in person and so in this method an interview is conducted on O and D.

Advantage

- It is simple, low cost.

Disadvantage

- Duplication may exist, when in some house there are more than one drivers.
- Many unlicensed drivers may not be covered.

(6) **Telephone Method:** An interview, for obtaining O and D information, from a sample of households in the study area, randomly selected, is conducted. The householders are sent pre-survey letter, describing the purpose of the study and kind of questions they will be asked, on telephone.

Advantage

- Low cost.
- High sample can be interviewed.

Disadvantages

- Results may be biased.
- Households without telephone are excluded.

(7) **Photographic Techniques:** Like volume and speed studies, an aerial photograph can record vehicle movements, which can be traced through a given area and recorded for entry and exit point, in the study section.

Advantage

- Good for small area studies.

Disadvantages

- It is costly.
- Data reduction is difficult and time-consuming.
- Limited to small areas only.

(8) **Home Interview Method:** This method provides the most comprehensive procedure for obtaining travel characteristics within a study area. A representative sample of houses is selected, and personal interview is conducted to obtain travel characteristics for all members of the household, by all modes of transportation, some additional socio-economic information may also be collected, if necessary.

For conducting home interviews, the following sample sizes (Table 10.3) are recommended for proper accuracy.

Table 10.3: Sample size for household survey

Population of survey area	Sample size	Percentage
Under 50,000	1 in 5 household	20%
50,000—150,000	1 in 8 household	12.5%
150,000—300,000	1 in 10 household	10%
300,000—500,000	1 in 15 household	6.7%
500,000—1,000,000	1 in 20 household	5%
Over 1,000,000	1 in 25 household	4%

There is some doubt, for justification of a detailed survey of this type, in cities with population less than 50,000. Advance publicity is done to ensure maximum public cooperation.

Typical questions asked in household survey are given in field sheet 10.11.

Advantages

- Completeness and correctness of data.
- The information is obtained directly.

Disadvantages

- The method is costly.
- It is time-consuming and requires large manpower.

(9) **Work Place Interview Method:** In this method O-D information is collected by contacting people at work places such as offices, factories, educational institutions, etc. by a personal interview.

(10) **Transportation Terminal Method:** In this method, the questionnaires are handed over to persons at the terminals such as bus stops, parking lots, etc. This method is specially good, if planning and improvement of terminals is the part of the study.

External Cordon Survey: An external cordon origin-destination survey often provides adequate information for planning of small areas having population of less than 5000 or so. For small urban areas of this size the traffic from outside affects the traffic pattern of the area. Such a cordon survey is useful when predominant flow of traffic is on through routes. Motorists are stopped at the imaginary cordon line and asked questions about their trip. Field sheet no. 12 is used to collect the data

Field Data Sheet 11

TYPICAL QUESTIONNAIRE FOR HOUSEHOLD SURVEY

Name of the colony............   Date and day............

House number............   Enumerator............

Q. 1. How many members are living in your household?

1 2 3 4 5 6 7 8 9 10 11 12 12+

Q.2. Details of family members:

Sr.No.	Name	Relationship to H/Head	Sex	Age	Student	Occupation (profession)	Monthly Income
1							
2							
3							
4							
5							
6							
7							
8							
9							
10							
11							

Q .3. Does anyone in the household possess?

A cycle   A motor cycle   A motorcar

Q. 4. How much does the household spend monthly on:

0	1	2	1 + 2
Food	*Public transport*	*Personal travel modes*	*All transport modes*

Q. 5. Please record all journeys made by household members on.

Name of household member	*Where did the journey*		*Journey purpose*						*Journey time (Overall trip time)*				*Travel mode*										
	Begin	*End*	*For work*	*To work*	*For shopping*	*For personal business*	*To school/college*	*For recreation social/leisure*	*To return home*	*Time start*	*Time finish*	*Journey distance*	*Fare paid*	*Walk*	*Cycle*	*Motor cycle driver*	*Motor cycle pass'er*	*Motor car driver*	*Motor car pass'er*	*Commercial vehicle driver*	*Commercial vehicle pass'er*	*Taxi passenger*	*Auto-rickshaw passenger*	*bus passenger*

Note. All journey made during the 24 hours. of the above day should be recorded.

A journey means a single journey with a single purpose from one place to another.

Commercial vehicle drivers should not include journeys made for employment purpose.

Field Data Sheet 12

CORDON INTERVIEW SURVEY

Sheet No....... Station............ Date............ Time............Inbound Weather
(date) (day of week) (hour beginning)
(month) Outbound

1	2	3	4	5	6	7		8	9
Serial Number	Vehicle type	No. in Vehicle	Where did this trip begin?	Where will this trip end?	Trip purpose	Intermediate		Frequency each trip per week	Type of goods
						stop location	stop purpose		carried by trucks
	*		**	**	***		****	if regular	

*Code for vehicle type	***** Code for trip purpose**	****** Code for purpose**
1. Passenger car	1. Work	1. Work
2. Bus	2. Business	2. Business
3. Taxi	3. Social recreation	3. Social recreation
4. Bicycle	4. Change mode of travel	4. Serve passenger
5. Motorcycle	5. Serve passengers	5. Eat meal
6. Truck		6. Gas or oil
		7. Overnight.

** Specify zone, street and a more detailed location (if possible) if the origin or destination is within the study area. Outside the study area the region and district or village will do.

10.4.3 Presentation of Data

Most of the O and D surveys result in large amounts of data and a number of different methods are used to summarise the data. They include tabulation, graphic and pictorial representation.

(1) **Tabulation for Completeness of Data:** The data is first analysed for its accuracy, for comparisons, with screen line counts. Screen line comparisons are summarized by using separate tables for each type of vehicle (passenger car, truck, taxis, etc.) A table as shown below is prepared to summarise all vehicle trips. Tables 10.4 and 10.5 show typical sample tables.

Table 10.4: Comparison of passenger car trips at screen line
(screen line data summary form)

Location					
	Expanded trip data				
Hours (Period)				*Ground Counts*	*Percent of total ground count*
	Internal	*External*	*Total*		
6.00 a.m to 6.59					
12 noon –12.59					
9 pm –9.59					
Totals					

Tables 10.5: Passenger car trips crossing cordon lines
(cordon line summary form)

Hours (Period)	*Expanded internal*	*Counted external*	*Percent (Ext./int.)*
6.00–6.59 a.m.			
9.00–9.59 p.m			
Totals			

(2) **Comparison Graphs for Completeness of Data:** From the tables, the results are graphically presented. If adjustment of trip data is required, the adjustment factor is applied

(3) **Trip Matrix:** The trip matrixes are then formed for the Zone to Zone flow, for different journey purposes and time periods, as shown in Table 10.6. Sometimes such matrixes are made by vehicle types.

Table 10.6: O and D matrix for passenger cars/business trips

From origin Zone	To (destination) Zones								Σ trips from Zone
	1	2	3	...	—	—	—	N	
1									
2									
3									
N̈									
Σ of trips to Zone →									

(4) **Other Tables.** Table are prepared to summarise the expanded information on zone, number of houses, car owners, total number of persons, total number of trips etc.

Tables also show trips from zone to zone, classified on the basis of purposes, e.g. Work trip, business, medical, school, recreational, shopping, home, etc.

Several other tables for average passenger car occupance by trip purpose, local trips and through trips, economic status vs. trips, time of day vs. trips, etc. can be prepared.

(v) **Desire Lines:** For pictorial representation, the data inter-zonal trip matrixes, are used to prepare desire line charts. The two directional movements between each pair of zone are represented by a straight line, extending between zone centroids, whose width is made proportional to the number of trips between zones. Desire lines show the shortest distance between zones of origins and destination and an ideally located street system will closely correspond to the pattern of the desire lines. Figure 10.13 shows typical desire line and flow diagram.

(vi) **Pie and Bar Charts:** Other methods of illustrating traffic statistics, and economic information is by use of pie or bar charts or columnar representation, as shown in Fig 10.14

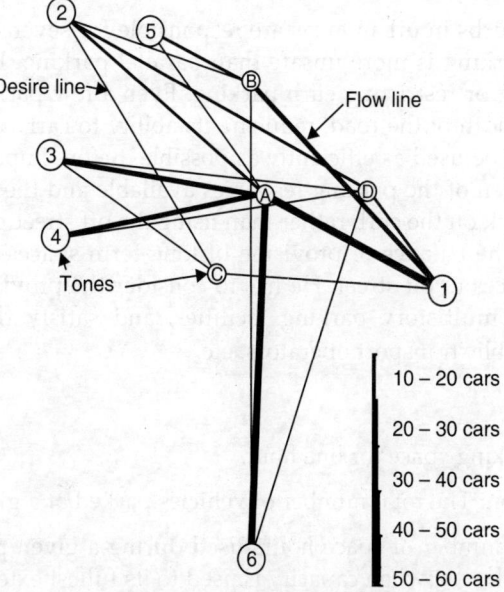

Fig. 10.13: Flow and desire line diagram

Computer oriented automatic methods of plotting the trip data are also used.

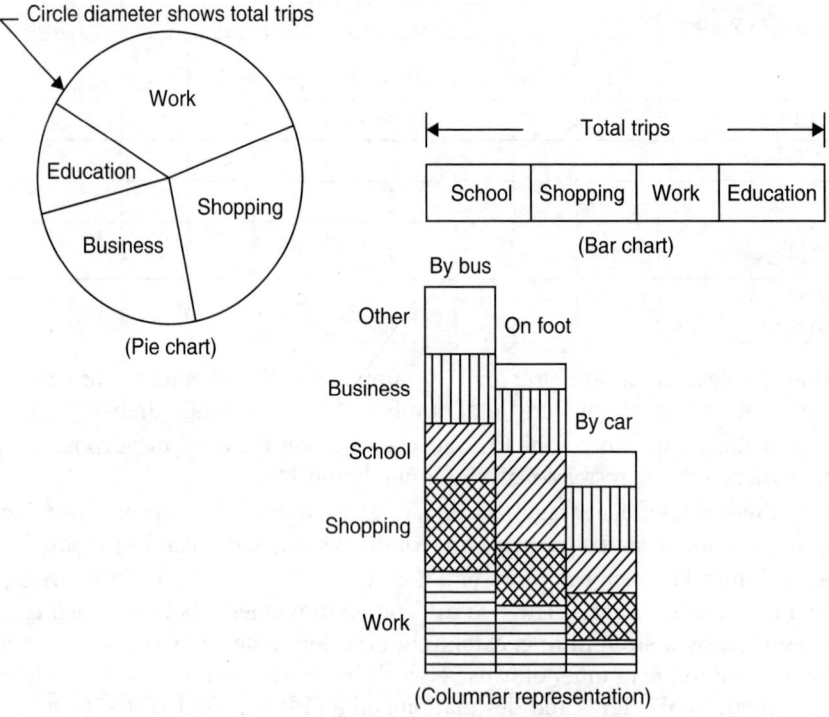

Fig. 10.14: Presentation of total trips data by purposes

10.5 PARKING STUDY

Vehicles parked at the curbs in urban areas are responsible for several accidents while parking or unparking. Angle parking is more unsafe than parallel parking. Roadway capacity can be increased by prohibiting or restricting curb parking. Even a few parked vehicles along a road effectively reduce the width of the road, reducing its ability to carry the traffic flow. The space for curb parking should be used as efficiently as possible. Before imposing the parking control, study must be undertaken of the parking resources available and the present demand. Most of the road users like to park on the curb rather than using the off street parking spaces. The traffic engineer has to weigh the balance of provision of long-term spaces to short-term spaces, the balance of on-street spaces to off-street. He has to consider the provision and need of ground-level, underground or multistory parking facilities, and satisfy the drivers, shopkeepers, commercial vehicles, public transport operators, etc.

DEFINITIONS

Space hour: One parking space for one hour.

Parking accumulation: The total number of vehicles parked at a given time

Parking load: Total number of space hours used during a given period of time. Its peak is reached at peak accumulation, when capacity is used to its fullest extent.

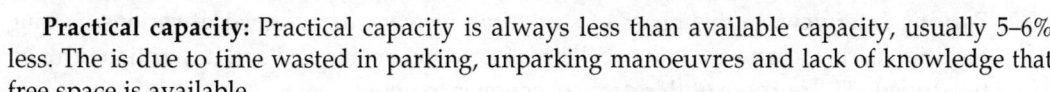

Practical capacity: Practical capacity is always less than available capacity, usually 5–6% less. The is due to time wasted in parking, unparking manoeuvres and lack of knowledge that free space is available.

Turn over: Average number of times a parking space is used by different vehicles during a given period of time, e.g. 150 spaces were used by 1500 vehicles in 10 hours study period: then

$$\text{turnover} = \frac{1500}{150} = 10 \text{ vehicles per space.}$$

Parking duration. The average time spent in a parking space.

N-Number of vehicles that can be parked in a given area is worked out by the formula:

$$N = \frac{(\text{Number of spaces})\,(\text{Period covered in hours})}{\text{Average parking duration (hours per vehicle)}} \times 0.85 \text{ to } 0.95.$$

Parking duration is also dependent on the type of parking facility.

10.5.1 Purposes of Parking Study

The purpose of a parking study is to develop a parking programme which meets the requirements of an area. For a parking study, it is necessary to have the following information:

- The supply and type of parking facilities, both on-street and off-street.
- How and for what purposes parking facilities are used including parking duration and illegal parking?
- The demand for parking space.
- The characteristics of parking demand.
- The location of parking generators.
- Legal, financial and administrative factors associated with parking situation, and the adequacy of the existing enforcement measures.

10.5.2 Methods of Parking Study

Parking studies develop the picture of the parking problem in an area. The study area should include the portions of the town containing most of the business activity and also include those areas that generally cater for the parking.

(1) **Space Inventory:** A physical inventory of existing facilities such as all streets, lanes, the curb and off-street parking spaces are listed with their location and capacity (in numbers and space hours), physical features such as type of building and open areas, operating features, regulation and fee etc. This inventory provides information of existing supply and helps in locating spaces for potential development of additional parking spaces, and also identifies areas where vehicles cannot be parked such as close to intersections, schools, crossings,s etc.

The data collected is summarized in Field Sheets no. 13 and on base maps of the area under study. On street parking facilities must also be investigated and their type and location should be shown on map.

(2) **Parking Interview:** In this phase of study interviewers are assigned to various parking facilities (curb, lots and garage) to collect information on demand and its location through the drivers. The purpose of this data collection is to know:

 (i) The extent to which inventoried parking facilities are utilized, and

 (ii) The extent and location of the demand for parking facilities.

The driver of practically every car parked in the area is interviewed. (100% parkers are interviewed in morning or evening peak hours). The information collected is:

 (a) Trip origin, home address, purpose of stop, type of vehicle.

 (b) Destination to which the driver is going to walk.

 (c) Time of arrival and departure-to know duration of parking.

 (d) Type of parking space-curbed, unrestricted, meter space, etc. Parking interview data is collected on field sheet no. 14

(3) **Parking Usage Study:** The parking space inventory establishes parking facilities both on-street and off-street available in the area. To investigate the demand for that space and the reasons for that demand, the parking usage study is conducted.

The study area is divided into blocks that can be walked around comfortably in half an hour. (About 1200 m of curb length per half an hour). Within each of these parking blocks a parked vehicle count can be made between 8 am and 6 pm, at half hourly interval, by an observer walking and noting down the registration number of every parked vehicle. It is assumed that each vehicle seen once has stayed for half an hour, if seen twice for one hour and so on.

Information collected can be used in plotting a graph of number of vehicles parked by hours through the day, from which peak parking time can be found. By matching the number of the cars, it is possible to calculate the time each vehicle has remained in the area, on half hour basis as shown in Fig. 10.15

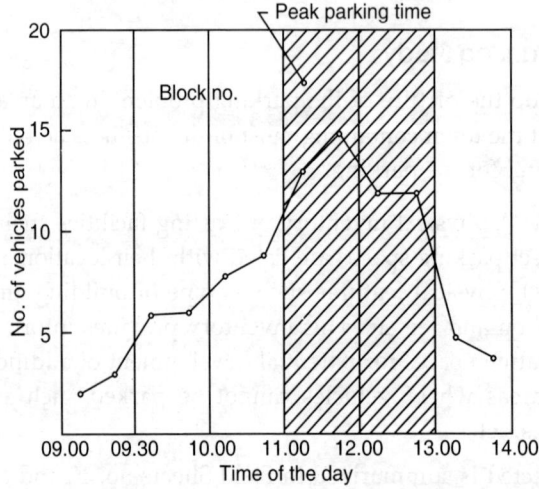

Fig. 10.15: Vehicles parked by hours of the day (on street parking)

Field Data Sheet 13
PARKING INVENTORY SUMMARY SHEET

Area of Inventory ....

Date of Inventory ....

Block	Facility	Street and Alley Stalls						Off-Street Parking		Total Stalls
								Private	*Public*	

Date... Compiled By...

Field Data Sheet 14
PARKING INTERVIEW

Location .... Date and Day ....

Facility ..... Tabulated By .....

Time Parked	Time Leaving	Purpose 1. Work 2. Shop 3. Business 4. Other	Destination (building name, street address, etc.)	Duration		Walking Distance
				hr.	*min.*	

Hence the times parked can be tabulated, and from this peak parking durations can be obtained (so also the most common durations).

Extending the exercise further, it is possible to calculate the total number of parked vehicle hours-a figure which is of utmost importance in case a commercially run, fee paying car park is to be planned.

Parking usage data is collected on field sheet no. 15

Field Data Sheet 15

PARKING USAGE SURVEY

Zone ....Street ....Between ....(Beginning) ....and ....
(End) ....Date ....Recorded by : ....

	TIME																							
	0700		07.30		08.00		08.30		09.00		09.30		10.00		10.30		11.00		11.30		12.00		12.30	
Space and Regulation	L	R	L	R	L	R	L	R	L	R	L	R	L	R	L	R	L	R	L	R	L	R	L	R

*Space and regulation

1. Marked curb
2. Unmarked curb
3. Metered
4. Restricted (No. parking)
5. Driveway
6. Alley

L = Left side of street

R = Right side of street

This data has been developed from the usage survey of on-street parking. Similar data can be collected for off-street parking by a similar survey-walking around at regular intervals recording registration numbers, and can be processed in a similar way.

(4) **Cordon Counts:** A cordon count is a count of all vehicles going out or coming in, on each street, crossing the boundary of the designated study area. The purpose of this phase of study is:

(*i*) To determine the volume and classification of traffic entering and leaving, on each street crossing the boundary of the area under study.

(*ii*) To estimate the total parking load, imposed on the area, during the business day or other selected period and during half hour period throughout the day.

Data is obtained by conducting manual counts and classification of traffic, at all streets crossing the cordon from 7 am to 9 pm. Volume is recorded half-hourly or preferably at 15 minuets interval. Figure 10.16 shows accumulation of vehicles, showing total volume of vehicles, inside the cordon at any particular time.

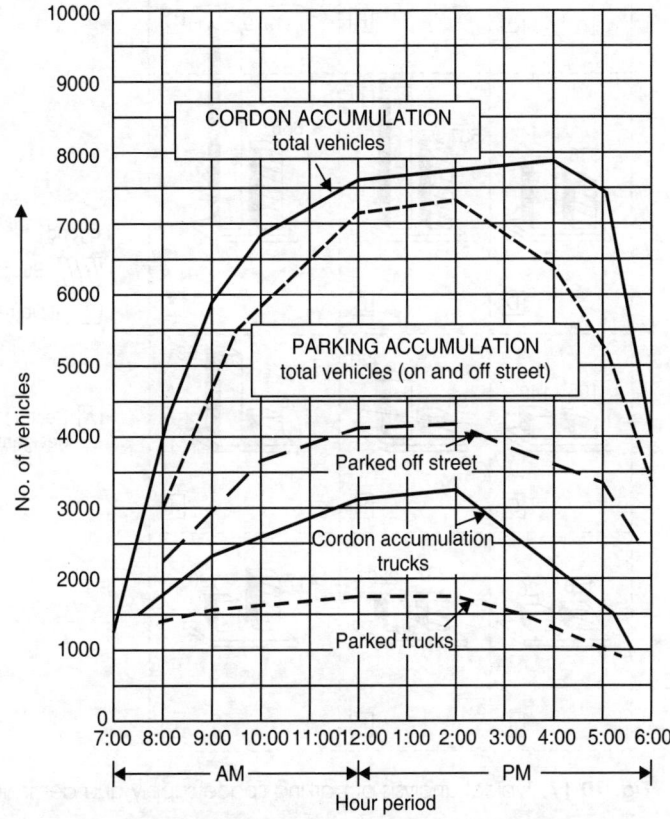

Fig. 10.16: Typical parking cordon accumulation

10.5.3 Analysis of Data

(1) **Demand for Parking Spaces:** Total vehicles parked illegally or legally or just outside study area, when compared with the parking spaces available, will give an idea of priorities and urgency of requirement for additional area (spaces) for parking, and whether the need is for short term parkers or long term parkers. Figure 10.17 shows a typical analysis of supply and demand, for different sectors A, B, C, etc of a city.

(2) **Deficiency in Parking Spaces:** On certain sectors, in the study area deficiency of parking spaces may be large, where tall buildings are located or land values are high. In such cases parking facilities, as near as possible, should be planned at reasonable walking distances. A criteria for walking distances is given in Table 10.7

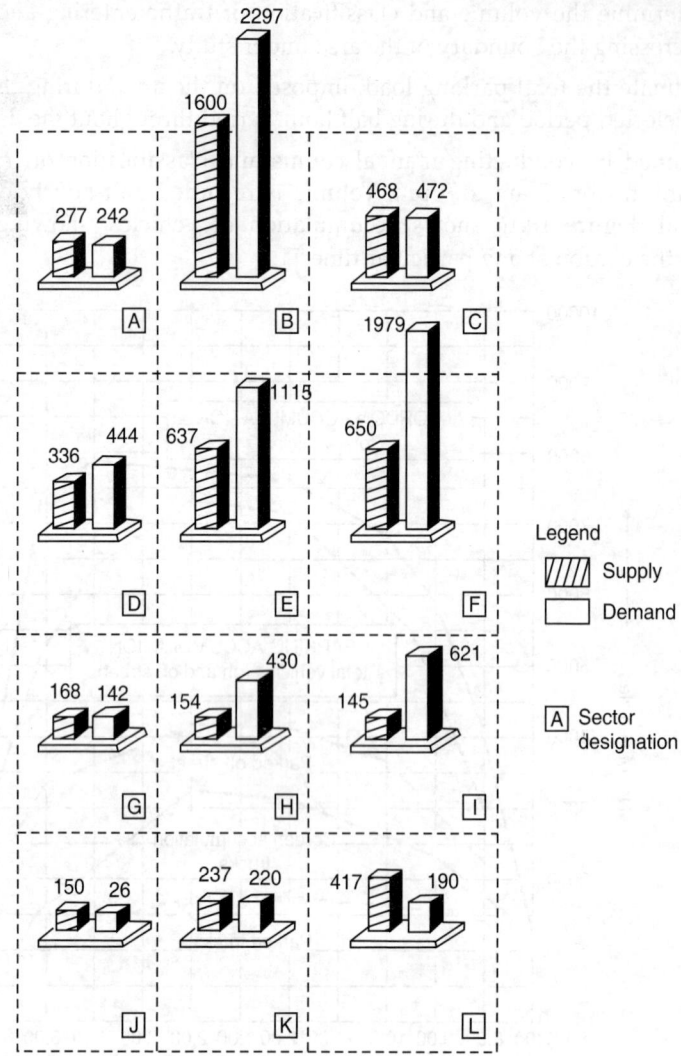

Fig. 10.17: Typical analysis of parking space supply and demand

Table 10.7: Walking distances from parking place

S. No.	Population group of urban area	Reasonable walking distance (m)
1.	under 25,000	90
2.	25,000 – 50,000	105
3.	50,000 – 100,000	150
4.	100,000 – 250,000	162
5.	250,000 – 500,000	225
6.	Over 500,000	230

(c) **Demand pattern:** Based on parking studies a demand pattern with respect to land-use can be developed for future planning of parking spaces.

10.6 ACCIDENT STUDY

Road accidents have grown up in the recent years due to several reasons. A complete realization of the problem and many factors which contribute to the accidents should be known before effective preventive measures are adopted to alleviate the present problem. Analysis of accident data provides causes of the accident so that preventive measures can be taken to improve road safety. Reports and accident statistics are very important to traffic police, engineers and many other individuals and organizations concerned with road safety. Traffic police is usually responsible for making original accident reports, collecting the data and filing it properly. Complete accidents records are required for accident prevention and orienting engineering and enforcement programmes in the most effective manner.

DEFINITIONS

Fatal accident: An accident in which one or more persons loose their life.

Injury accident: An accident in which one or more persons receive injuries such as fractures, crushing, cuts, wounds requiring immediate medical attentions. Such accidents are called grievous injury accidents. Sometimes only minor injuries are sustained by persons, such accidents are called minor injury accidents.

Non-injury accidents (Property damage accidents): Accidents in which no one is killed or injured, but only property is damaged. Such accidents are called property damage accidents.

Comprehensive Insurance Plan: An insurance cover provided to vehicle owner, in which all risks are covered by the insuring company.

Third Party Insurance Plan: An insurance cover provided to vehicle owner in which comprehensive cover is not provided. There are several types of such insurance plans.

Vehicle collisions: Two or more vehicles when they make an accident, there may be collisions of various types such as

Head on Collision: When two vehicles meet an accidents head to head.

Rear-end Collision: When head of one vehicle dashes with tail of the other vehicle.

Side Swipe Collision: When sides of two vehicle meeting an accident collide.

Right Angle Collision: In which two vehicles meeting an accident collide at right angles.

Collision while Right Turning: A vehicle which collides with other vehicle while making a right turn.

10.6.1 Purposes of Accident Data

Detailed and accurate data regarding accidents is very important to different individuals as shown in Table 10.8 and also useful is knowing basic causes of accidents and adopting preventive engineering and enforcement measures.

The number of uses of traffic accident data are:

- Identifying high accident locations for detailed study.
- In locating traffic control devices, such as signals, signs etc.
- Planning regulatory and enforcement measures to increase safety.
- Improving geometrics of intersections and problematic road sections, in relation to accidents.

Table 10.8: Users of accident report and their interests

Users of accident data	Interest/Purposes of data
Traffic police	• Driver identifications • Criminal procedures • Enforcement planning
Vehicle administrators	• Licensing • Inspections • Design standards and improvements • Financial responsibility
Engineers (Highway and Traffic)	• Problem area identification • Design standards • Maintenance standards • Programming improvements
Vehicle manufacturers	• Vehicle design standards • Maintenance practices • Equipment
Public (users of roads)	• Safety promotions • General trends of accidents • Specific problems and interests
Educators	• General traffic safety • Driver training • Specific problem and research
Medical professionals	• Treatment of injured • Physiological and psychological problem of drivers
Insurance agencies	• Claim settlements • Actuarial functions
Public health administrators	• Mortality and morbidity trends • Demographic distribution
Legal and judicial people	• Criminal proceedings • Civil suits
Legislators	• Driver licensing • Traffic laws and administration • Vehicle inspection and control
Motor transport industry	• Driver control and operating policies • Vehicle design and maintenance
Regulatory commissions	• Responsibility • Regulations • Legislation

- Improving geometric design features like curves, grades, widening of roads, etc.
- Justifying provision of dividers, islands etc. on roads/intersections.
- Controlling and guiding pedestrian movement by provision of subways, overhead bridges, pelican crossing, etc.
- Imposing parking restrictions and other restriction, on heavy vehicles, turning, etc.
- Measuring the after effect of an road safety improvement programme by "Before and After" studies.
- Working out financial loss of an accidents by converting the total loss in monetary value.
- Planning driver education programmes for road safety.
- In renewing driving license issuing policy, vehicle inspections periods and other legislative and administrative measures.

10.6.2 Accident Reporting and Recording

The accident reporting system should be such that the data is collected in a standard manner. The data should be factual, collected through a standard and consistent method of reporting. All the three types of the accidents should be reported by the drivers involved and by investigating police officers.

For uniformity, standard forms should be designed for accident reporting including the following information:

- Date, day, time and precise location of accident with sketch.
- Particulars of drivers and vehicles and pedestrians involved and type of accident.
- Persons died, injured and extent of injuries.
- Extent of damage of vehicles and their age and condition.
- Location and description of traffic control devices, etc.
- Regulations in force.
- Roadway lighting, weather conditions and adjacent land-uses.
- Possible violations, alcohol and associated factors.
- Probable causes.
- Diagram of accident (collision diagram).
- Narrative of accident events prior to accident.

Typical accident reporting form used by Indian Road congress is included in field sheet no 16

In order to make comprehensive examination of the accident reports, it is extremely important that traffic accident data be properly compiled and recorded. It is common these days to store relevant data in a computer for rapid and frequent retrieval. The accident record system should include all reportable accidents. The record system should be such that the extraction of accident histories for specified locations can be made without undue delay.

One method of recording accidents for quick visual display is to use large whole-town map, on which location of each accident is indicated by a small circle (or pin) of different colours to indicate the type of accidents. Sometimes a separate map is used for each type. Such maps are called "Accident Spot Maps".

For recording purpose, the classification of accidents to characterize the manner of occurrence may be done in following categories:

- Running off the road type accidents
- Non-collision type accidents
- Overturning and others
- Collisions on the road type accidents

(a) With: – Pedestrians

                        – Another motor vehicle

                        – Parked vehicles

                        – Cyclists

                        – Fixed objects

                        – Animals and

                        – Others

(b) Angle of collision: – Rear end collision (hit from behind)

                        – Side sweep (hit from sides)

                        – Head on collision (hit face to face)

                        – Accidents while backing type

                        – Other types.

Field Data Sheet 16

ACCIDENT REPORTING FORM (IRC)

1. **Identification particulars**
   (i) State
   (ii) District/taluka
   (iii) City/town/village
   (iv) Police zone/station
   (v) Accident number allotted by police station
       FIR Number
       MVA accident number

2. **Location**
   (i) Name of road
      1. National highway
      2. State highway
      3. Other road
   (ii) Location of area
      1. Urban
      2. Rural
   (iii) If intersection, specify the names of roads
   (iv) Details of location (the details under this should be such as to give precise location)
   (v) Type of area
      1. Near school or college
      2. Near or inside a village

       3. Near a factory/industrial area

       4. Near a religious place

       5. Near a recreation place/cinema

       6. In bazaar

       7. Near office complex

       8. Near hospital

       9. Residential area

      10. Open area

      11. Near bus stop

      12. Near petrol pump

      13. At pedestrian crossing

      14. Affected by encroachments.

  (vi) Narrow bridge or culverts

       1. Yes

       2. No

3. Date, Day and Time

  (i) Date   Month   Year

  (ii) Day of the week

 (iii) Time   A.M.   P.M.

  (iv) Holiday   1. Yes   2. No

4. Light Conditions

       1. Daylight

       2. Twilight

       3. Dark hours with good street light

       4. Dark hours with poor street light

       5. Dark hours with no street light

(vi) Type of area	Code No.	Type of area	Code No.
Near school or college	01	Near a recreation place cinema	05
Near or inside a village	02	In bazaar	06
Near a factory/industrial area	03	Near office complex	07
Near a religious place	04	Near hospital	08
		Other built up area	09
		Unbuilt-up area	10

(*ii*) Narrow bridge or culvert

          Yes   1

          No   2

          Not stated   9

1.3. Date, day and time

(*i*) Date (specify)

(ii) Month	Code No.	Month	Code No.	(iii) Day of week	Code No.
January	01	July	07	Monday	1
February	02	August	08	Tuesday	2
March	03	September	09	Wednesday	3
April	04	October	10	Thursday	4
May	05	November	11	Friday	5
June	06	December	12	Saturday	6
				Sunday	7

(*iv*) Time (also to be mentioned in hours and minutes)

Hours of the day	Code No.	Hour of the day	Code No.
6.00 AM—7.00 AM	01	3.00 PM—4.00 PM	10
7.00 AM—8.00 AM	02	4.00 PM—5.00 PM	11
8.00 AM—9.00 AM	03	5.00 PM—6.00 PM	12
9.00 AM—10.00 AM	04	6.00 PM—7.00 PM	13
10.00 AM—11.00 AM	05	7.00 PM—8.00 PM	14
11.00 AM—12.00 Nn	05	8.00 PM—9.00 PM	15
12.00 AM—1.00 PM	07	9.00 PM—10.00 PM	16
1.00 PM—2.00 PM	08	10.00 PM—11.00 PM	17
2.00 PM—3.00 PM	09	11.00 PM—6.00 AM	18
		Not stated	19

(*v*) Holiday or not (If yes, describe)

Holiday	Code No.
Yes	1
No	2

5. Weather conditions

1. Fine
2. Mist/Fog
3. Cloudy
4. Light rain
6. Hail/sleet
7. Snow
8. Strong wind
9. Dust storm
10. Very hot
11. Very cold
12. Other extraordinary weather conditions (specify).

6. Classification of accident

1. Fatal
2. Grievous injury

3. Minor injury
4. Non-injury

7. Type of vehicles with registration numbers and objects involved

7.1. Vehicles involved

	Type of vehicle	Registration number	Year or manufacture
Vehicle 1	...	...	...
Vehicle 2	...	...	...
Vehicle 3	...	...	...

7.2. Pedestrian, animal and other objects involved (specify type)

S. No.

1. ....
2. ....
3. ....

8. Nature of the accident

8.1. Type

1. Overturning
2. Head on collision
3. Rear end collision
4. Collision brush/side swipe
5. Right angled collision
6. Skidding
7. Right turn collision
8. Others (describe)

8.2. Hit and Run   1.Yes   2. No

9. Details of drivers of vehicles involved

9.1. Name, Sex, Age, Education and Address of the Driver(s)

	Name of Driver	Sex M/F	Age	Highest class Passed	Address
Vehicle 1	—	—	—	—	—
Vehicle 2	—	—	—	—	—
Vehicle 3	—	—	—	—	—

9.2. Person driving the vehicle

	Veh. 1	Veh. 2	Veh. 3
1. Owner of private vehicle			
2. Owner of public/commercial vehicle			
3. Paid driver			
4. Other			

9.3. Type of licence

	Veh. 1	Veh. 2	Veh. 3
1. Regular licence			
2. Learner's licence			
3. Without appropriate licence			

9.4. Licence number and other details

	Veh. 1	Veh. 2	Veh. 3
1. Licence number			
2. Date of issue			
3. Date of expiry			
4. Place of issue			

9.5. Type of manoeuvre

	Veh. 1	Veh. 2	Veh. 3
1. Diverging			
2. Merging			
3. Crossing			
4. Stationary			
5. Temporarily held up			
6. Parked			
7. Stopping			
8. Starting from near side			
9. Starting from off side			
10. Turning right			
11. Turning left			
12. Making U-turn			
13. Going ahead, overtaking			
14. Going ahead, not overtaking			
15. Using private entrance			
16. Reversing			

9.6. Responsibility of driver

	Veh. 1	Veh. 2	Veh. 3
1. Consumption of alcohol or drugs			
2. Exceeded lawful speed			
3. Did not give right of way to vehicle			
4. Did not give right of way to pedestrian			
5. Followed too closely			
6. Overtaking on hill			
7. Overtaking on curve			
8. Cut in sharply after overtaking			
9. Other improper overtaking			
10. On wrong side of road			

11. Failed to give signal
12. Gave improper signal
13. Improper turn
14. Disregarded police officer
15. Disregarded traffic light signal
16. Disregarded stop sign
17. Improper starting from parked position
18. Wrong/improper parking
19. Asleep or fatigued or sick
20. Inattentive or attention diverted at the moment
21. Improper use of headlights causing glare
22. Other improper actions (describe)

10. Particulars of vehicle involved

10.1. Particulars of vehicles

	Veh. 1	Veh. 2	Veh 3
(a) Load:			
(i) Overloaded/overcrowded			
(ii) Load protruding			
(b) Left hand drive			
(c) Vehicular defect:			
(i) Defective brakes			
(ii) Defective steering			
(iii) Punctured or burst tyres			
(iv) bald tyre			
(v) Other serious mechanical defect (describe)			

10.2. Certificate of fitness in the case of commercial vehicles

	Veh. 1	Veh. 2	Veh. 3
1. In force			
2. Not in force (expired)			

10.3. Particulars of insurance

	Veh. 1	Veh. 2	Veh. 3
(a) Name of insurance company			
(b) Number of insurance			
(c) Type of insurance			
1. Comprehensive			
2. Third party			
3. Not insured			

11. Details of pedestrian or person other than the driver involved in accident

11.1. Age of person/pedestrian

S. No.	Male/Female	Age
1	...	...
2	...	...
3	...	....
4	...	...

11.2. Particulars of person/pedestrian

1. Intoxicated or drugged
2. School child
3. Pedestrain deaf, blind or otherwise, infirm
4. Crossing at intersection with signal
5. Crossing at intersection-against signal
6. Crossing at intersection-diagonally
7. Crossing at pedestrian crossing
8. Crossing not at inter section/pedestrian crossing
9. Coming from behind parked vehicle
10. Walking on road with traffic, side-walk not available
11. Walking on road with traffic, side-walk available
12. Walking on road-against traffic, side-walk not available
13. Walking on road-against traffic, side-walk available
14. Pushing or working on vehicle
15. Others working on road
16. Playing on road
17. Hanging on the vehicle
18. Sleeping on squatting on the road
19. Dismounting or mounting moving vehicle
20. Dismounting or mounting stationary vehicle
21. Not on road (explain)
22. Other actions (specify)

12. Details of cyclists involved

1. Double riding
2. Overloading
3. Not keeping to the left
4. Without light at night
5. Going on carriageway-cycle track available
6. Cycling in the lane of fast moving vehicle
7. Cutting in the flow of traffic or zigzag moving
8. Turning right carelessly/without giving signal

9. Towing himself with other vehicle

10. Not observing traffic rules

11. Confused by traffic

12. Rider inexperienced

13. Loss of control

14. Skidding on wet/slippery road

15. Others (specify)

13. Type of persons and animals killed or injured

13.1. Type and number of persons killed or injured

Type	No. Killed		No. Injured			
			Grievously		Slightly	
	M.	F.	M.	F.	M.	F.
1. Pedestrian						
2. Bicycles						
(i) Drivers						
(ii) Passengers						
3. Motor cycles						
(i) Drivers						
(ii) Passengers						
4. Scooters						
(i) Drivers						
(ii) Passengers						
5. Mopeds						
(i) Drivers						
(ii) Passengers						
6. Autorickshaws						
(i) Drivers						
(ii) Passengers						
7. Cars, taxis, vans and other light and medium motor vehicles						
(i) Drivers						
(ii) Passengers						
8. Trucks						
(i) Drivers						
(ii) Passengers						
9. Buses						
(i) Drivers						
(ii) Passengers						
10. Other motor vehicles						
(i) Drivers						
(ii) Passengers						

11. Animal driven vehicles
   (i) Drivers
   (ii) Passengers
12. Cycle rickshaws
   (i) Drivers
   (ii) Passengers
13. Hand carts and rickshaws
   (i) Drivers
   (ii) Passengers
14. Other persons

13.2. Number of animals killed/injured
   Killed
   Injured

14. Type of damage to vehicles and property

   Vehicle/Property   *Type of damage (specify)*
   1. Vehicle 1
   2. Vehicle 2
   3. Vehicle 3
   4. Other property

15. Road condition
   (i) Horizontal features of the road
      1. Straight road
      2. Slight curve
      3. Sharp curve
   (ii) Vertical features of the road
      1. Flat road
      2. Gentle incline
      3. Steep incline
      4. Hump
      5. Dip

15.2. Type of surface
      1. Surfaced (black topped/concrete)
      2. Metalled
      3. Kutcha

15.3. Condition of surface
      1. Dry
      2. Wet

15.4. Nature of surface
      1. Good surface
      2. Loose surface

3. Rutted/and or potholed

4. Road under repair/construction

5. Corrugated or wavy road

6. Slippery surface

7. Snowy

8. Muddy

9. Oily

10. Speed breaker

11. Other (specify)

16. **Road features**

(i) Carriageway

1. Single lane

2. Two lanes

3. Three lanes or more without central divider (median)

4. Four lanes or more with central divider

(ii) Cycle track provided   1. Yes   2. No

(iii) Foot path provided   1. Yes   2. No

(iv) Pucca shoulder provided   1. Yes   2. No

17. **Intersection type and control**

17.1. **Type of junction (Refer sec. 21 for figures)**

1. T-junction

2. Y-junction

3. Four arm junction

4. Staggered junction

5. Junction with more than 4 arms

6. Roundabout junction

7. Manned rail crossing

8. Unmanned rail crossing

17.2. **Type of traffic control**

1. Traffic light signal

2. Police controlled

3. Stop sign

4. Flashing signal/blinker

5. Uncontrolled

18. **Traffic regulations**

1. One-way street

2. Entry of heavy vehicles prohibited

3. Entry of slow moving vehicles prohibited

4. Speed restrictions

5. Parking prohibited

6. Any other (specify)

19. **Names of persons (with user's classification such as pedestrian, driver, etc).**

   Name   User Classification   Occupation   Monthly income (Rs)

   Killed

   _____

   _____

   _____

   Grievously injured

   _____

   _____

   _____

   Slightly injured

   _____

20. **Cause of accident**

    1. Fault of driver of motor vehicle
    2. Fault of cyclist
    3. Fault of driver of other vehicle
    4. Fault of pedestrian
    5. Fault of passenger
    6. Defect in mechanical condition of motor vehicle
    7. Poor light condition (including street light)
    8. Defect in road condition
    9. Result of weather conditions
    10. Stray animal
    11. Other causes (specify)

21. **Diagrammatic sketch of accident site and brief description of the accident**

    A computerized system becomes necessary as the number of accidents increases in an area. Such a system can summarise accident data properly. Accident summaries for fatal, injury and property damage types on analysis can provide lot of statistical information such as:

    - Accidents per vehicle km.
    - Accidents per vehicle.
    - Accidents per person.
    - Severity rate of accident as compared to total number of accidents
    - Accidents classified as per locations or spots.
    - Accidents classified as per driver violations, age, sex, types, etc.
    - Trend of accidents from year to year.
    - Accidents classified as 'Rural' and 'Urban' type.
    - Accidents classified on the basis of age, sex and actions of victims.
    - Accidents classified on the basis of day, time, month, weather, etc.
    - Preparation of collision diagrams.

    The computer system also has the advantage of integrating the accident information with other data such as traffic volumes, control devices and road inventories.

10.6.3 Presentation of Accident Data

Traffic data collected for several years can be processed and compiled for statistical purposes in different ways and can be presented in form of tables, charts, histograms, pie charts, etc.

Following are some of the methods of summarizing and analyzing the collected data.

- Different type of accidents, on monthly basis in a particular year.
- Accidents according to the time of the day and night.
- Accidents classified road wise.
- Accidents classified according to different location.
- Accidents classified based on weather conditions.
- Classification as rural, and urban area accident.
- Accidents according to type of vehicles/age of vehicles.
- Number of accidents as per nature of their occurrence collision/manoeuvres.
- Accidents classified as per age/sex of the vehicle drivers.
- Classification as per vehicles involved.
- Accidents classified on the basis of persons/vehicles/objects involved.
- Classifications of accidents according to road condition/road geometry/road type.
- Accidents according to junctions/intersection types and traffic control.

10.7 STUDY OF HIGH ACCIDENT LOCATIONS

Following steps are involved in study of high accident locations:

(1) **Identifications of High Accident Locations:** Four or more accidents at one location in a year is called 'Black Spot' extracted from the accident information already recorded and stored.

   (i) **General:** Date, time, person involved in accidents and their particulars, nature of accident, etc.

   (ii) **Details of vehicles involved:** Registration number, make, description of vehicle loading details, defects, etc

   (iii) **Nature of accident:** Details of collision, pedestrian or objects involved. Injuries sustained

   (iv) **Road and Traffic conditions:** Straight, curved, surface characteristics, type and density of traffic speed.

   (v) **Primary causes and cost of accidents:** Basic probable causes and financial losses due to accidents.

For deciding the high accident locations a more rational method assigns rating number to each accident, taking into account the accident severity as the most important consideration rather than merely the number of accidents. For example 12 points are given to each fatal accident, 3 points to injury accident and 1 point to property damage accident. Acceptable number of points, to classify a 'Black Spot' will vary from location to location and resources of a country.

(2) **Detailed study of dangerous locations (Black-Spots):** For detailed study all the information from accident reports and location files are collected and studied carefully. Based on the information a collision diagram (not to a scale) is prepared to show pattern or similarity of accidents.

(3) **Preparation of a collision Diagrams:** A collision diagram illustrates graphically, by means of directional arrows and symbols, the path and nature of collision of vehicles and pedestrians involved in accidents, at a particular location. Collision diagrams are usually not drawn to scale and the arrows do not show exact path, because they will overlap and become confusing.

On the arrows, representing each accident may be shown the date and time of the day to nearest hour. Unusual conditions should also be noted such as intoxicated driver, or icy pavement, etc. These diagrams are used to study the accident pattern and may suggest possible causes of accidents and improvement measures such as channelization, erection of signals, signs, etc. Figure 10.18 shows a typical collision diagram at an intersection.

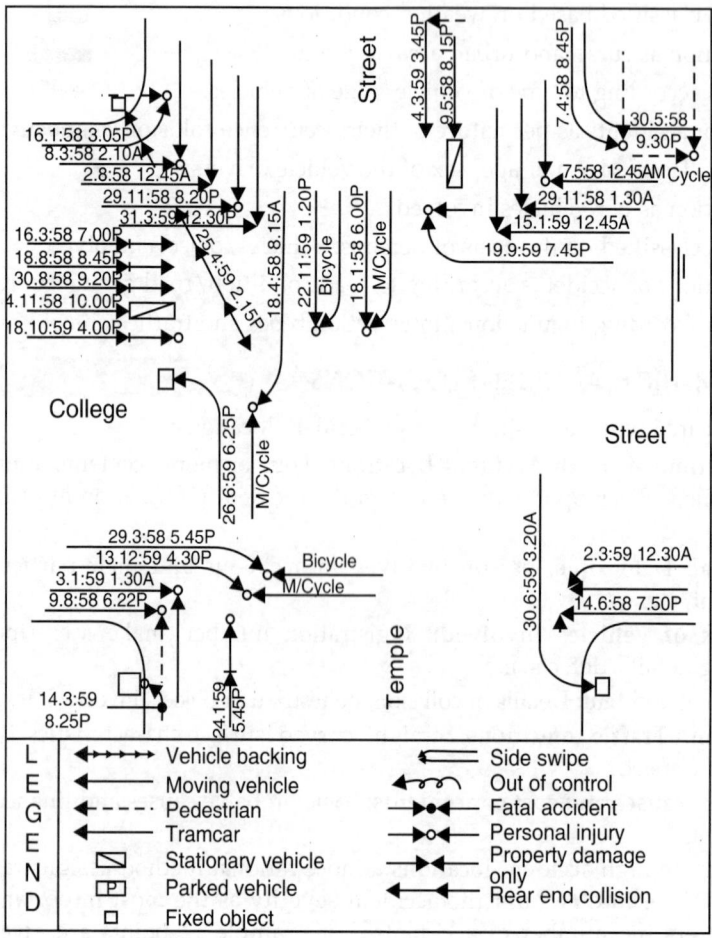

Fig. 10.18: Typical collision diagram

When collision diagrams for "Before" and "After" corrective measures are compared, they show the type of accidents that have been eliminated, and those that have continued.

The pattern of accidents if any is correlated with the features of the site and existing conditions of roads. A condition diagram showing existing conditions is prepared.

(4) **Preparation of a condition Diagram:** A condition diagram is a scaled drawing of the important physical conditions at a high accident frequency location to be studied. It is used as an aid to the interpretation of accident pattern. Following information is usually shown on the condition diagram

• Curb line and roadway limits

- Property lines
- Side walks
- Obstruction to view
- Physical conditions of roads and gradient
- Ditches along the roadway
- Traffic signals, signs, pavement markings
- Street lighting, power poles, trees, etc.
- Type and condition of road surface
- Type of adjacent property
- Bus stops, parking lots, etc.
- Any other specific features of the site.

Figure 10.19 shows a typical condition diagram.

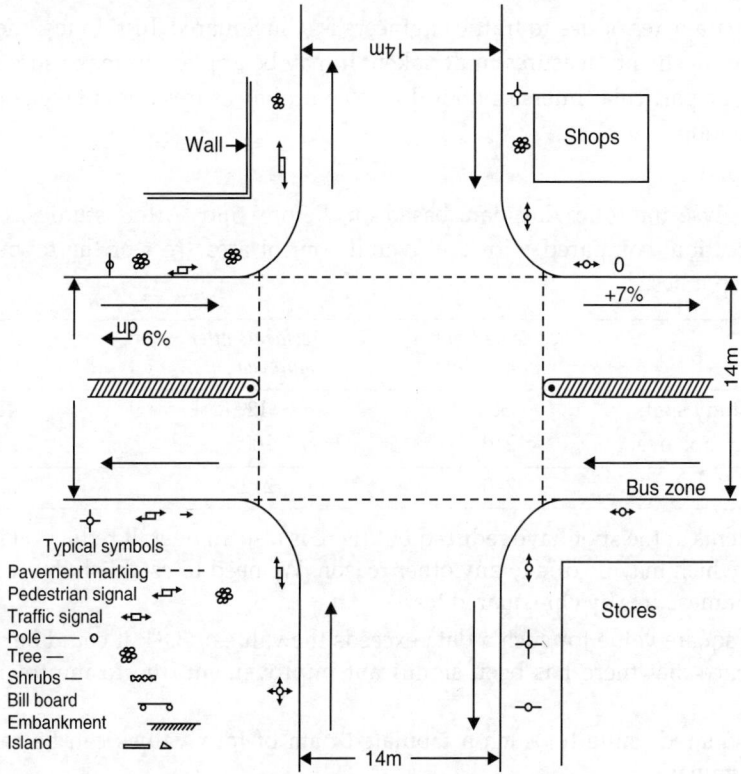

Fig. 10.19: Typical condition diagram

(5) **Collection of other data:** Other data needed is assembled from the existing files or records or collected on site. Data is required on volumes, approach speeds, near by parkings, signal timings etc.

(6) **Alternative remedial measures:** After considering all the factors, possible alternative corrective solutions are worked out using control devices, signs, markings, signals, geometric design improvements, and other controls such as one-way, banning of certain turns etc.

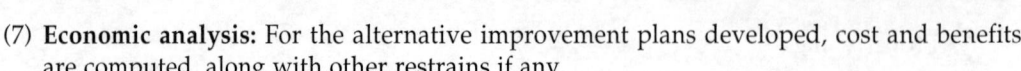

(7) **Economic analysis:** For the alternative improvement plans developed, cost and benefits are computed, along with other restrains if any.

(8) **Selection of best plan:** Based on the economic analysis and feasibility, the best improvement proposal is selected. If there are other locations needing improvements, the level of priority is also worked out.

(9) **Recommend plan and its detailing:** For the improvement plan which is recommended most suitable based on above exercise, detailed plans and drawings are prepared showing the specifications to be adopted in its implementation.

(10) **After Studies:** To check the outcome of the improvement plan implemented, after sometime an 'After' study should be conducted for statistical comparison to 'Before' conditions. A "Chi Squared Test" can be used for this purpose.

THE CHI SQUARED TEST

A statistical parameter of use to traffic engineers is 'Chi Squared Test' to test the significance of the traffic improvement measures undertaken. It may be applied to make sure if the reduction in accidents at a particular intersection is due to control measures under taken or just because of chance fluctuation.

An example:

Let us analyse the following data based on 'Before' and 'After' studies, undertaken at a particular location, compared with the overall control area (rest of the town) in which this problematic spot lies.

	Accidents before improvements	*Accidents after improvements*	*Total*
At the study spot:	30	12	42
In the control area:	210	200	410
Total	240	212	452

The accidents at the spot have reduced but there is also an overall reduction in the rest of the control area which may be due to any other reason. We need to establish the significance of the improvement measures by chi-squared test

If the chi square value for such a data exceeds the value of 3.84, it could be said with 95 per cent confidence that there has been significant improvement due to improvement measures undertaken.

The chi squared value for such a tabulated data of four values can be calculated by the following formula:

$$X^2 = \frac{(\text{Difference of Cross-Products})^2 \times \text{Grand Total}}{\text{Product of all the sub-totals}}$$

Using the above data:

$$X^2 = \frac{(30 \times 200 - 210 \times 12)^2 \times 452}{240 \times 212 \times 42 \times 410}$$

$$= 6.24$$

This value is greater than 3.84 and therefore it can be said with 95 percent of confidence that the improvement measures were effective in reducing the accidents.

The data collected for 'Before' and 'After' studies should be of similar type collected for the same period. There could be several such situations in traffic engineering.

A detailed description of statistical derivations is beyond the scope of this work.

10.8 PUBLIC TRANSPORT STUDY

Public transport is an integrated group of transportation facilities (other than those privately owned for personal use) which enable people or freight to be transported from one place to another. Modes of public transport are buses, railways, minibuses, taxies, autorickshaws, cycle-rickshaws, trams, animal drawn vehicles, etc.

Public transport systems (mass transit system) are important part of urban transportation, because overall transportation is for people rather than vehicles.

Various types of public transport systems, depending upon the service they provide can be classified as:

(a) **Rapid Transit:** Mass transit services, without interference or many stops provided through fast buses, mini buses or railways. Speed usually higher than 30 km/h

(b) **Mass Transit:** Services provided for passengers and their incidental baggage on fixed route, following a fixed schedule of movement and fares. Normally provided by buses, mini buses, or railways.

(c) **Local Transit:** Mass transit services on city roads going through congested residential and market areas, subjected to various types of traffic interference, provided by buses, trams, rickshaws, etc.

(d) **Point to point Transit:** Services provided between two or more points, over short distances in small vehicles like rickshaws, mini buses, escalators, moving belts, etc.

(e) **Railway Transit:** Railway services provided on the periphery of an urban area connected by a system of roads for entering CBD, for mass transportation

(f) **Other Transit:** These services are those types not included above, such as contact vehicles, taxies, rickshaws, etc. which are demand responsive.

10.8.1 Purposes of Public Transport Study

Public transport (mass transit) studies are conducted for one or more of the following purposes

- Assess the quality of existing transport services or proposed new services.
- Establish the extent of usage and determine their suitability.
- Study the operation, problems, location, and adequacy of stops, etc.
- Obtain characteristics of mass transit riders and establish trends for statistical analysis.
- Plan, design and operate suitable mass transit services.
- Scheduling the transit services, special services to schools, CBD, industrial areas, etc.
- Conduct an economical analysis of public transport operations and plan better services or improvements.
- Use in research and further analysis of the system.

10.8.2 Methods of Public Transport Study

The public transport study can be divided in the following categories

(1) **Public Transport Inventory Study:** The basic information about the available public transport services provides the data for analysis and evaluation. Data gathered includes the location of service, frequency of service and hours on each route, travel time between various points, fare schedule, number and types of vehicles used—their capacity, age and conditions.

(2) **Public Transport Origin and Destination Study:** This study gives the information about the origin and destination of the public transport riders. This information is obtained through a comprehensive survey as described earlier for origin and destination study. Questionnaire is handed over to the public transport passengers with the request that they may be returned at the end of the trip.

   The data regarding the destination and purpose of trip collected, through a questionnaire from a transit (bus) stop is summarized in field data sheets no. 17 and 18.

   Return post card method or interview methods are often used to collect O-D information of public transport riders.

(3) **Public Transport Usage Study:** This study provides data on passenger volumes, vehicle occupancy, boarding and alighting, adherence to schedule, etc. which is useful for location of bus stops, planning the operating plans and number and type of vehicles for the transit routes.

   Two types of studies are conducted to collect data on public transport riding characteristics:

   (i) **Transit load checks:** Manual method is used for data collection, at one or more bus stops along the route under study. One or two observers record the vehicle number, time of arrival and departure, and the number of persons alighting, boarding and on board of the vehicle when it leaves. Normally maximum load points-from where maximum number of passengers are carried are selected for transit load checks. If along the same route more than one stop is selected for study, the travel time of the bus between these stops can also be measured.

   (ii) **Riding, boarding and alighting checks:** In this method observers board the bus at one of the transit route and sit at a position from where coming in and going out passengers can be easily seen. At each stop, the observers record the time of arrival at the bus stop, time of departure from the bus stop, location of the bus stop, number of boarding and alighting passengers and any other remarkable observation. The occupancy is computed after adding the number of passengers who came in and subtracting the number of passengers who went out, from the number of bus passengers at the previous stop. A separate field sheet is used for each run of the bus. Typical field sheet no 19 shows the data collected.

(4) **Transit Speed and Delay Study:** Transit speed and delay studies along the bus routes are similar to those described in riding a bus. The time the bus passes the different stops is recorded and also the location, amount and cause of delay is noted. The various cause of delay are given some short notation for quick recording for example B-10 indicates 10 seconds delay due to boarding of passengers.

Field Data Sheet 17

PUBLIC TRANSPORT STUDY

Trip Destination Summary (Passenger's Questionnaire)

Route of Public Transports (Bus): From.................... To....................

Location of Bus stop.................... Date, Day and Time ....

DESTINATIONS ALONG THE TRANSIT (BUS) ROUTE

Arrival Time Interval	Number of users interviewed	Stop 1		Stop 2		Stop 3		Stop 4		Stop 5		Stop 6	
		No. of users	Percentage	No. of users	Percentage	No. of users	Percentage	No. of users	Percentage	No. of users	Percentage	No. of users	Percentage

Field Data Sheet 18

PUBLIC TRANSPORT STUDY

Trip Purpose Summary (Passengers Questionnaire)

Route of Public Transports (Bus): From.................... To....................

Location of Bus stop.................... Date, Day and Time ....

PURPOSES OF TRIP MARKING

Arrival Time Interval	Number of users interviewed	WORK		SHOPPING		ENTERTAINMENT		OFFICE		OTHERS	
		Number	Percentage	Number	Percentage	Number	Percentage	Number	Percentage	Number	Percentage

Field Data Sheet 19

PUBLIC TRANSPORT STUDY

Usage Data (Riding Study)

Route of Public Transport (Bus) From...................... To......................
Vehicle Number...................... Date, Day, Time......................
Seating Capacity...................... Standing Capacity......................

Time		Location of	Passengers			
Arrive	Depart	Stop	Boarding	Alighting	Occupancy	Remarks

10.8.3. Presentation of Study Data

The data collected for public transportation can be presented by tabulation and graphic methods as given below:

Load Profile maps: Load profile maps show the number of passengers in a bus on vertical scale and corresponding route section on horizontal scale.

Boarding and alighting maps: A plot showing the number of passengers boarding and alighting at each bus stop.

Daily passenger load profile: The daily total of bus passengers on a particular stop along the transit rout can be plotted. On the same plot the number of seats available for that day can be superimposed to have a visual indication of demand and supply.

Transit flow maps: The daily passenger loads can be put in form the traffic flow maps–the width of the band along the transit route representing the number of passengers per day on the route.

Time space diagram: These diagrams show the travel time or speed on various sections of the transit route and are similar to speed zone or delay zone maps.

Bar charts (time charts): The charts show the total delay, causes of delay, and their distribution along the public transit route.

11

Road Crossing and Parking Facilities

11.0 PREAMBLE

An at grade intersection is where two or more roads cross or join at the same level. Intersections are important part of road and traffic operation where driver has to choose one out of the available choices. Capacity and safety of road depends upon design of its intersections as accident risk is increased by conflicts. To design or re-design an intersection, considerations should be given to following factors.

- Volume, type and pattern of traffic using the intersection and its future anticipated distribution and growth.

- Topographical and environmental features such as alignment, grades, approach roads, etc.

- The need for and type of traffic control to be adopted and requirements for their install-ation.

- The requirements and provision for road lighting and its installation.

- Possibilities of reducing traffic conflicts, if possible, by eliminating certain movements.

The concepts of safety is most important factor which should be achieved by application of basic principles of design. Intersections are critical spots along a road.

11.01 CONFLICTS AT INTERSECTIONS

The four main types of traffic manoeuvres at an intersection are:

   i Diverging

   ii Merging

   iii Crossing and

   iv Weaving

These are shown in Fig. 11.1. Significance of conflicts depends upon traffic volumes in each direction of flow, the speed and time spacing between the vehicles, number of approaches and their width.

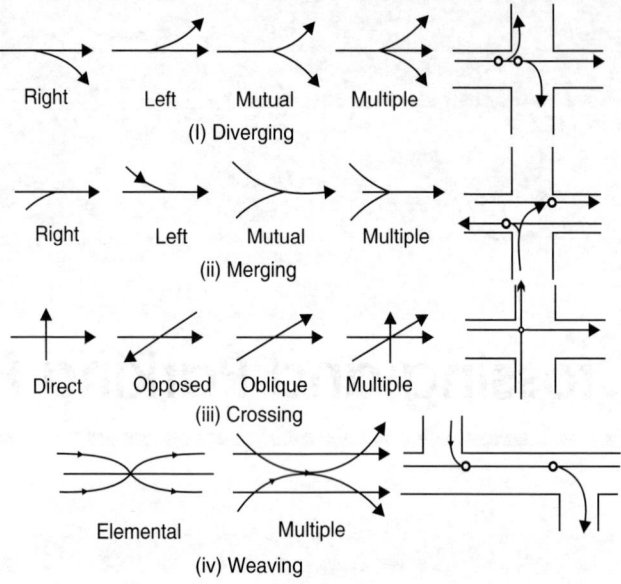

Right   Left   Mutual   Multiple
(I) Diverging

Right   Left   Mutual   Multiple
(ii) Merging

Direct   Opposed   Oblique   Multiple
(iii) Crossing

Elemental   Multiple
(iv) Weaving

Fig. 11.1: Type of traffic manoeuvres

Different types of traffic conflicts, and their numbers at a four legged intersection, for different movements are shown Fig. 11.2.

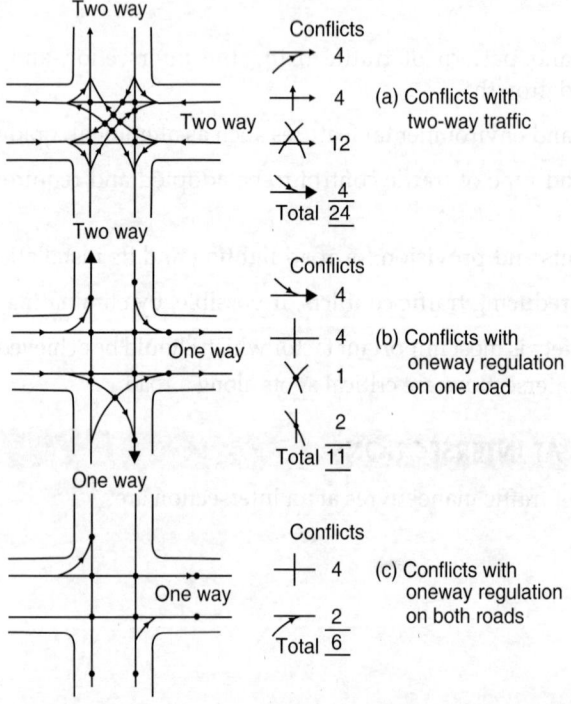

Two way

Two way

Conflicts
4
4   (a) Conflicts with
12   two-way traffic
4
Total 24

Two way

Conflicts
4
4   (b) Conflicts with
1   oneway regulation
2   on one road
Total 11

One way

Conflicts
4   (c) Conflicts with
2   oneway regulation
Total 6   on both roads

One way

. 11.2: Traffic movements and conflicts at four legged intersection

11.02 CLASSIFICATIONS OF INTERSECTIONS AT GRADE

Intersections can be classified in different ways, such as given below:

(a) **According to shape:** At grade intersections are those where roads meet at the same level. They are classified according to the number of intersecting roadways, angles at which they intersect, and the shape they form as shown in Fig. 11.3

   (i) Three-legged intersections: Right angle, 'T', 'Y' or skewed intersection.

   (ii) Four-legged intersections: Right angle, skewed, offset.

   (iii) Multiple-legged intersections with five and more legs.

   (iv) Rotary intersections or roundabouts having a big island in the centre.

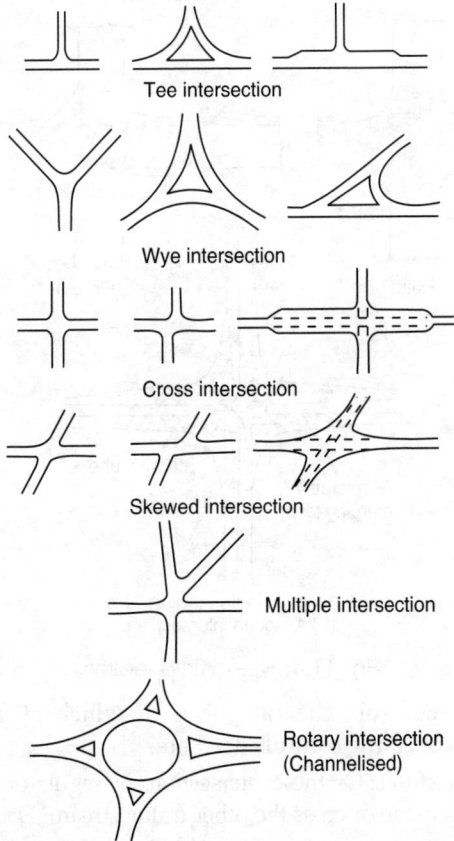

Fig. 11.3: Forms of Intersections according to shape

(b) **According to geometric design features:** Based on the geometric design criterion the at grade intersections may be classified as shown in Fig. 11.4

   (i) **Plain intersections:** They are simplest type, cheap and suitable for low traffic volumes, where minor roads or one major and other minor road meet. They are unchannelised and unflared.

   (ii) **Flared intersections:** They are those where lanes are widened (auxiliary lanes), for easiness of merging or diverging. These additional lanes are called speed change lanes.

(iii) **Channelised intersections:** They are those where traffic islands are used to direct or guide the path of drivers and reduce the area of conflict. Depending upon the number and extent of islands used, it may be called partially channelised or complete channelised.

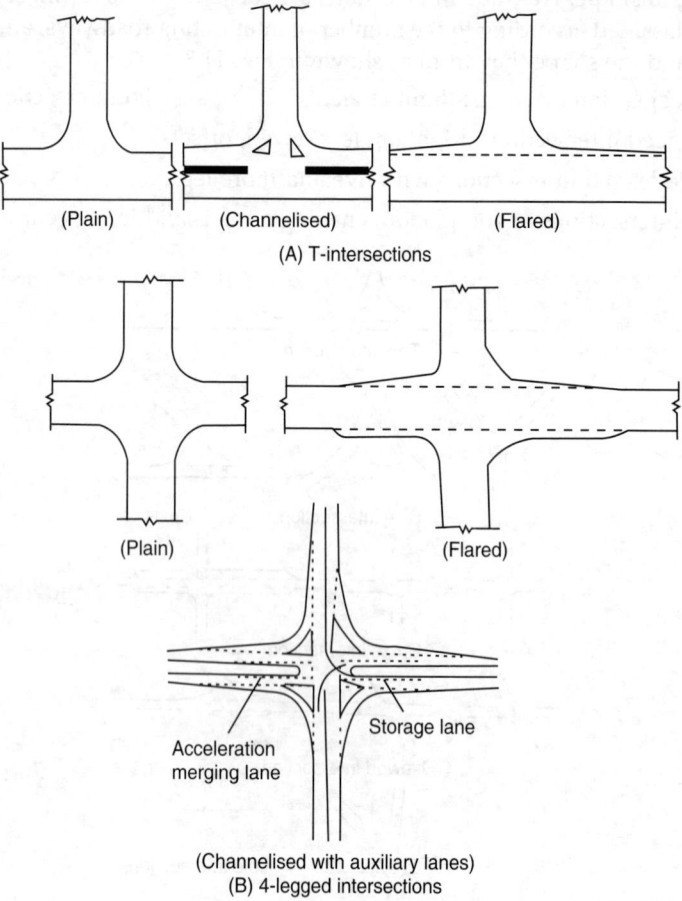

(A) T-intersections

(B) 4-legged intersections

Fig. 11.4: Type of intersections

(c) **According to traffic control:** The categories in which at grade intersections can be classified on the basis of traffic controls used are:

(i) **Uncontrolled intersections:** On these intersections no regulatory devices are used. Driver has to wait for a gap to merge or cross the other traffic stream. Drivers must follow the right of way rules. The roads are more or less of equal importance and there is no established priority.

(ii) **Controlled intersections:** On these intersections traffic is controlled by sign, police or traffic signals. They provide orderly movement and better level of safety. Traffic is controlled by "Give way" or "Stop" regulations.

(iii) **Signalised intersections:** Light signal control is used on these intersections for movement of vehicles. A signalized intersection is justified when major street has a traffic volume of 650 to 800 vph (both directions) and minor road has 200 to 250 vph in one direction only.

(d) **According to levels of crossing:** Intersections are classified at grade intersections and grade separated intersections.

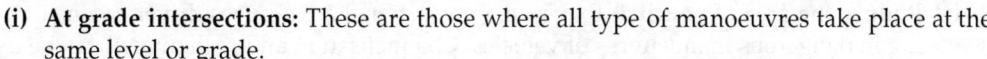

(i) **At grade intersections:** These are those where all type of manoeuvres take place at the same level or grade.

(ii) **Grade separated intersections (Interchanges):** These are those characterized by the separation of one or more crossing conflicts through an overpass or underpass. They are commonly called interchanges or flyovers. A grade separated intersection besides other warrants, is justified when the total traffic of all arms of the intersection is in excess of 10,000 pcu per hour. Figure 11.5 shows criteria for choice of intersection.

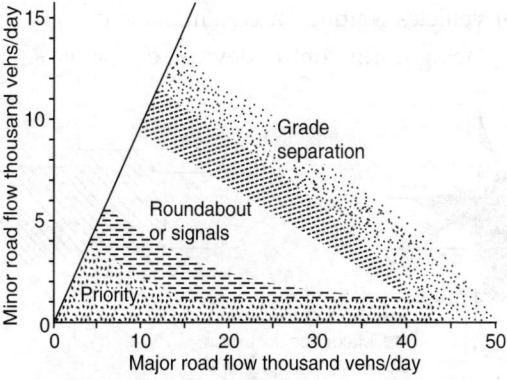

Fig. 11.5 : Criteria for selection of intersection type

11.1 PRINCIPLES OF INTERSECTION DESIGN

Following are the basic principles which should be used in intersection design to provide safety and efficient flow of traffic.

(1) **Reduction of conflict points:** The number of conflict points can be reduced by prohibiting certain movements or by reducing the legs of the intersection. Channelisation helps in separating the conflicts, as shown in Fig. 11.6

(2) **Separate conflict points:** When intersection conflicts are too close together, they result in dangerous conditions. They may be separated through channelisation or by staggering the intersection.

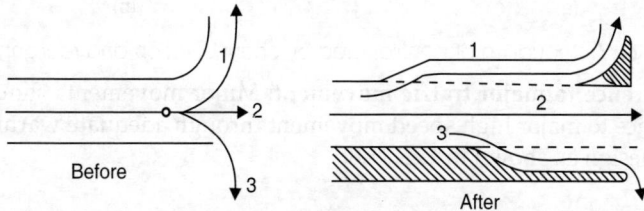

Fig. 11.6 : Separation of conflict through channelisation

Space separation may be provided vertically also be means of grade separator. Time separation is provided by use of turning lanes, wide medians, signals and storage space.

(3) **Control the relative speed:** Relative speed is the speed of convergence of vehicles in the intersection flows. Small difference in speed (upto 25 kph) and small angles (less than 30°) of convergence between paths of vehicles increase the safety of the intersection. Crossing manoeuvres that involve high relative speed should be made at approximately right angles, to allow the driver better judgement. Relative speeds can be reduced by using control devices, signs, traffic island and speed change lanes.

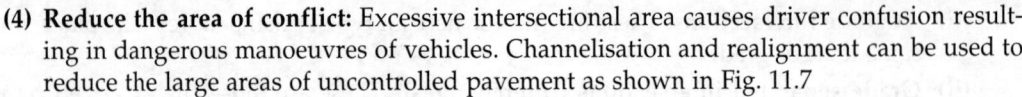

(4) **Reduce the area of conflict:** Excessive intersectional area causes driver confusion resulting in dangerous manoeuvres of vehicles. Channelisation and realignment can be used to reduce the large areas of uncontrolled pavement as shown in Fig. 11.7

Channelisation serves following purposes:

- Controls speed and directs path of vehicles
- Controls area of conflict and separates conflicting streams
- Assists pedestrian to cross
- Provides shelter for vehicles waiting for certain maneuvers
- Provides space for placing traffic control devices, e.g. signs, signals.

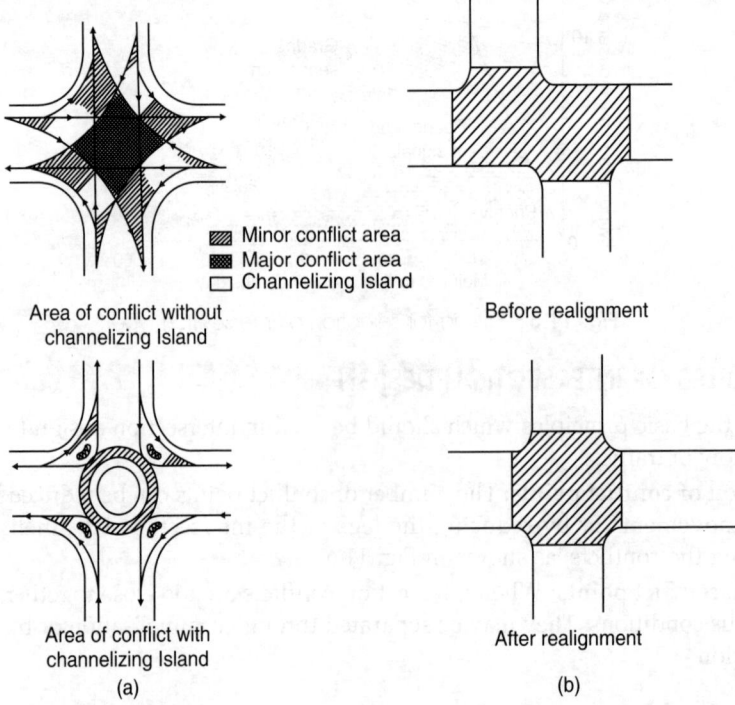

Fig. 11.7: Reduction of conflict area by channelization and realignment

(5) **Allow preference to major traffic movement:** Minor movements should be restricted to give preference to major high speed movement through adequate warning to minor traffic or through design as shown in Fig. 11.8.

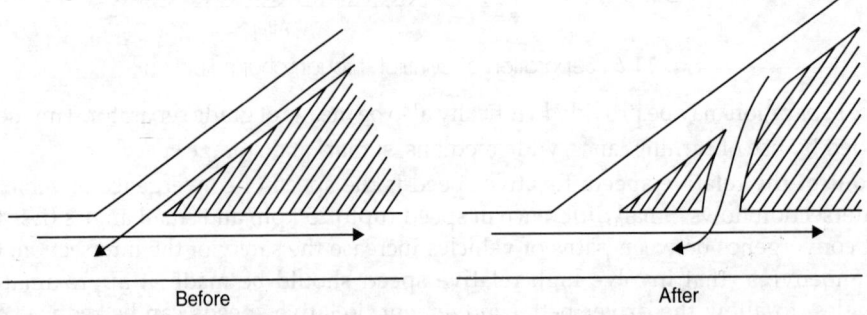

Fig. 11.8 : Restricting minor traffic movement

(6) **Avoid multiple and compound conflicts:** Multiple and compound merging or diverging manoeuvres require complex decisions by the drivers, resulting into unsafe conditions. (Fig. 11.9). They should be avoided.

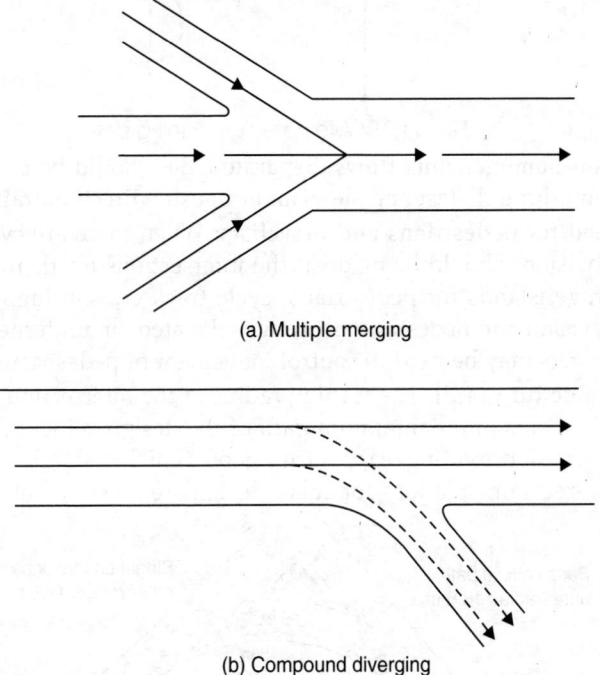

(a) Multiple merging

(b) Compound diverging

Fig. 11.9 : Multiple and compound manoeuvres

(7) **Use proper control devices:** At intersections where manoeuvres take place at high relative speeds, traffic controls such as stop signs, signals, etc. should be used. Signs and markings should be used to warn the drivers. The intersection design and the traffic control system to be used should be planned together. Traffic control should be provided at best locations in the intersection, depending on the actual intersection conditions, width of intersection, etc.

(8) **Use best feasible crossing method:** For crossing of the two streams of traffic one of the following methods can be used:

(i) Uncontrolled crossing at grade

(ii) Traffic sign or signal control

(iii) Weaving-first traffic merges and then diverges.

(iv) Grade separation.

Most suitable of these should be used consistent with the traffic volumes, space available and financial position.

(9) **Consider suitable turning path:** The methods for providing left turn or right turn are many as shown in Fig. 11.10. Separate roadways can be provided for left and right turn, reducing conflict in the intersection area. Direct turns offer shorter paths, less travel time and are easily understood by drivers. When direct path is not possible other most suitable may be chosen.

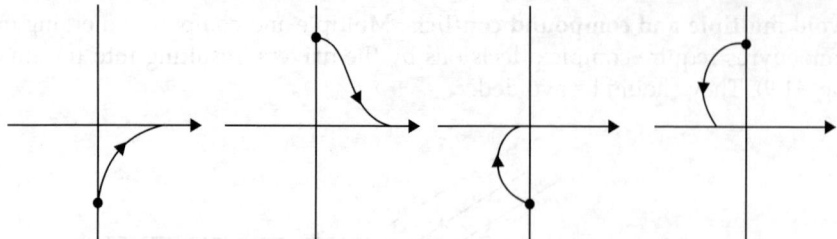

Fig. 11.10: Alternate right turning path

(10) **Segregate non-homogeneous flows:** Separate lanes should be provided at intersections for turning and through, fast and slow, and opposite direction traffic.

(11) **Consider need for pedestrians and bicyclists:** When there are cyclists and pedestrians, adequate provisions should be made at the intersections for their safe manoeuvres. For example, refuge islands for pedestrians, cycle track etc. On high volume intersections, separate provision for pedestrians through elevated or underneath crossings may be required. Railings may be used to control movement of pedestrians.

(12) **Provide suitable turn radii:** The turning radius at the intersection should be provided as per the requirements of the minimum path of the design vehicle. Single radius curves or compound curves providing easy turning of vehicles should be used as shown in Fig 11.11. IRC recommendations for curve design are given in Table 11.1.

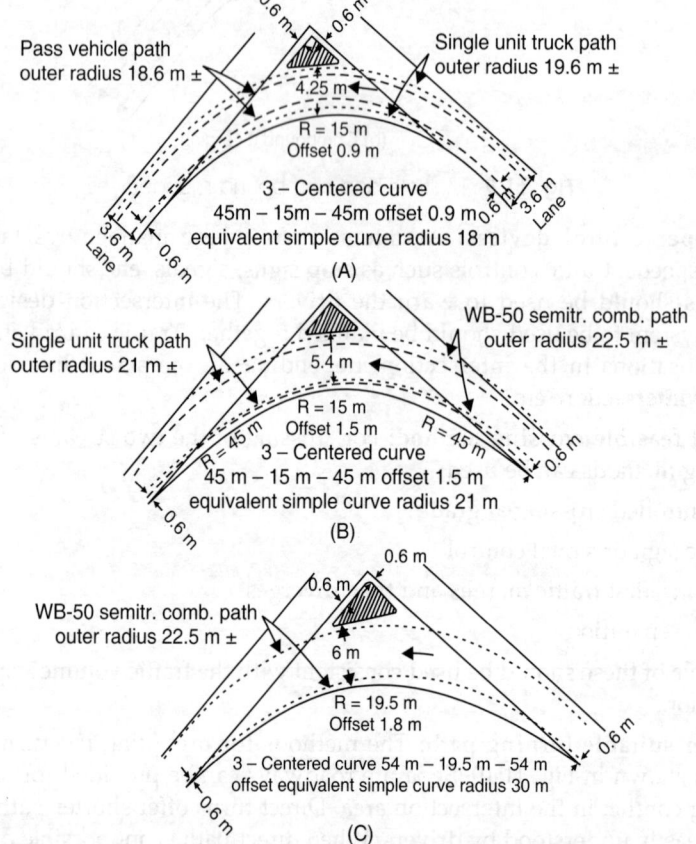

Fig. 11.11: Design for turning roadways with minimum corner island

Table 11.1: Curve design and location of intersection (IRC)

S. No.	Location of Intersection	Curve Design
1.	Rural section	Design for single unit truck is preferred for intersection with local minor roads. Semi-trailer design is preferred for major road intersection where large paved areas result, channelisation also becomes essential.
2.	Suburban arterial section	Designed for semi-trailer with speed change lanes and channelisation. Three-centered compound curves are preferred.
3.	Urban arterial and sub-arterials	Designed for single unit truck
4.	Urban central business districts	Designed for single unit trucks for minimum curve radii with allowance for turning vehicles encroaching on other lanes.
5.	Residential area	Designed for cars with encroachment of tracks into other lanes.

(13) **Adequate design till expected life:** The design provided must take into account the future traffic demands and development of the surrounding area, so that the provisions made are adequate throughout the expected life of the intersection. This may require considerations of 'Stage Construction Technique' before ultimate development of intersection is reached.

(14) **Optimum spacing of intersection:** The number of intersections on a road should be as small as possible in view of the traffic pattern and access to adjacent road and areas. As a guide under ideal conditions, intersections should not be spaced less than 350 to 550 m apart.

(15) **Use proper channelisation:** Layout using islands and channels depends upon traffic pattern, volumes, topography, available land, pedestrian movement, geometry of roads and ultimate development. A proper layout should be worked out for every individual site. Excessive channelisation creates unwanted obstructions, restrictions, maintenance problems and may even lead to confusion among drivers. Types and shapes of islands commonly used are shown in Fig. 11.12.

(16) **Provide adequate lighting:** Safety during night is very critical at the intersections. Adequate visibility should be provided during the night driving through an intersection, and also during bad weather. Lighting should not create glare. For this purpose height of mountings for lights should be at least 9 m.

11.2 ELEMENTS OF INTERSECTION DESIGN

For designing various elements of an intersection, considerations must be given to the following:

- Topography, terrain and environment in which the intersection is located. A location plan is necessary.
- The geometry of the roads meeting at the intersection: A base survey plan may be prepared indicating land boundaries, roadways and physical features in the vicinity.
- Land uses surrounding the intersection: their future control may be required. Landscaping suiting to the location and character of the intersection improves aesthetic and driving pleasure.

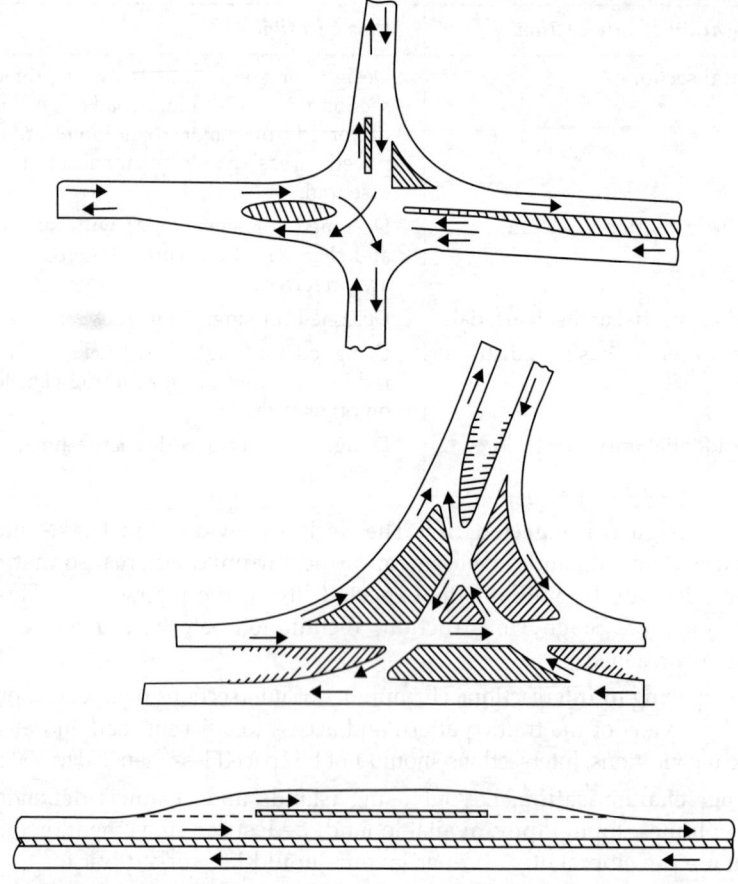

Fig. 11.12: General types and shapes of islands (channelisation)

- Prevailing alignment and grades: longitudinal sectional plan showing ground contour for each leg of intersection at 0.5 m interval be prepared.
- Traffic type, volume and the design vehicle to be used. Peak hour traffic data at each leg, left and right turning expressed in vehicle type is required.
- Future growth of traffic during the design period.
- Volume of pedestrians, cyclists and their tuning movements.
- Approach speed of vehicles on all approaches.
- Type of traffic control devices, signs, pavement markings, etc to be used.
- Drainage plan to be adopted, suitable cross falls and drains to be provided. Hydraulic capacities and locations of drainage facilities should not create risk for traffic. Their design and location should be safe.
- The type of lighting provisions for night use.
- High stability mixes should be used at intersection, because of the turning movements of vehicles.
- Accommodation of utilities like sanitary sewer, water supply pipes, oil, gas, telephone and others should be planned in advance

Important elements of the intersection design are the following:

11.2.1 Sight Distance

(a) **Stopping sight distance:** Using the approach speeds of the particular leg stopping sight distance on each leg should be provided so that the drivers are able to see the intersection markings, islands, etc. to enable them to decide whether to stop, change speed or continue to drive, for avoiding collisions.

(b) **Sight triangle:** Drivers approaching an uncontrolled intersection should have a clear view of the area unobstructed by buildings or other objects across the corners of the intersections, as shown in sight triangle in Fig. 11.13. The minimum visibility triangles should be clear of any obstructions to a height of 1.2 m above the roadway.

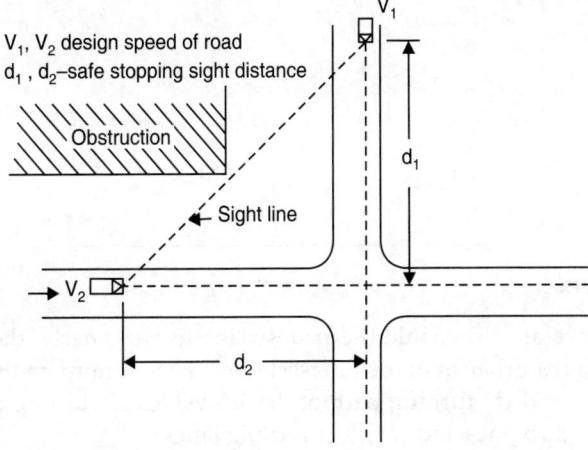

Fig. 11.13: Minimum sight triangle at uncontrolled intersections

(c) **Crossing sight distance.** At minor roads vehicles stop before proceeding into the intersection. The driver of stopped vehicle should be given enough distance on the major road so that he can see the vehicles of the major road, well in advance, before entering the intersection. (Fig. 11.14)

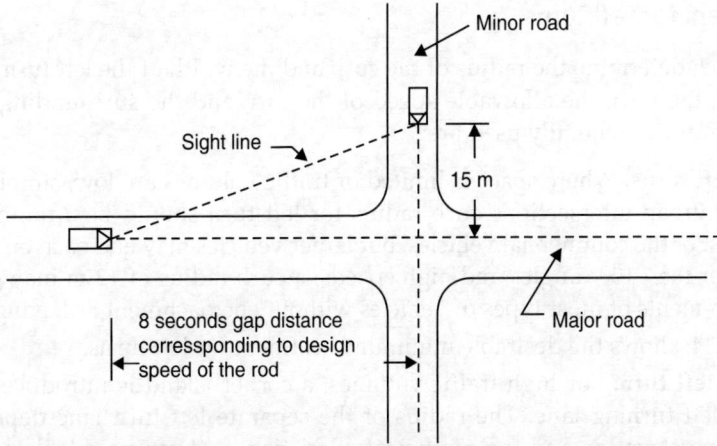

Fig. 11.14: Minimum sight triangle for crossing

When adequate sight distances are not available, as per sight triangle, traffic control measures such as "STOP", "GIVE WAY' sign, etc. should be used.

Safe stopping sight distance at intersections is given in Table 11.2

Table 11.2: Safe stopping sight distance at intersections

Speed	Safe Stopping Sight Distance (m)
20	20
25	25
30	30
40	45
50	60
60	80
65	90
80	130
100	180

11.2.2 Radii of Curves

To economise the space and to enable the pedestrians to cross early, the minimum possible radius is used in design of urban at grade intersections. The minimum radius is governed by the (*a*) speed of operation and (*b*) turning path of design vehicles. Turning paths of the vehicles influence the shape of curb lines and width of turning lanes.

In actual design, curve radius can be determined for each design to fit the shape of the path traced by its inner wheel. The inner curb should be offset 0.6 m from the vehicle path. 3-centred compound curves are often used since they more closely fit to the actual vehicle paths. While providing for the design vehicle a check should be made to ensure that there is sufficient width of pavements available to accommodate occasional higher vehicles.

11.2.3 Treatment for Left Turns

The factors which determine the radius of the curb and the width of the left turn lane are traffic volume making the turn, the allowable speed of the turn and the surrounding development. Two types of left turns generally used are:

(a) Simple left turns: Where space is limited or traffic volumes are low, simple left turns are used. For urban intersections curb radius for left turn should be a minimum of 6m to allow most of the commercial vehicles, but larger vehicles may encroach on opposing lane. For greater than 100° angles and high speeds, a curb radius of 12 m may generally meet the requirements of most types of vehicles without encroachment or having to back up.

Figure 11.14: shows the desirable minimum treatment for left turns.

(b) Separate left turn: For high traffic volumes, a corner island is introduced to provide a separate left turning lane. The radius of the separate left turn lane depends upon the speed, superelevation and the coefficient of friction, and can be calculated by using the super elevation formula:

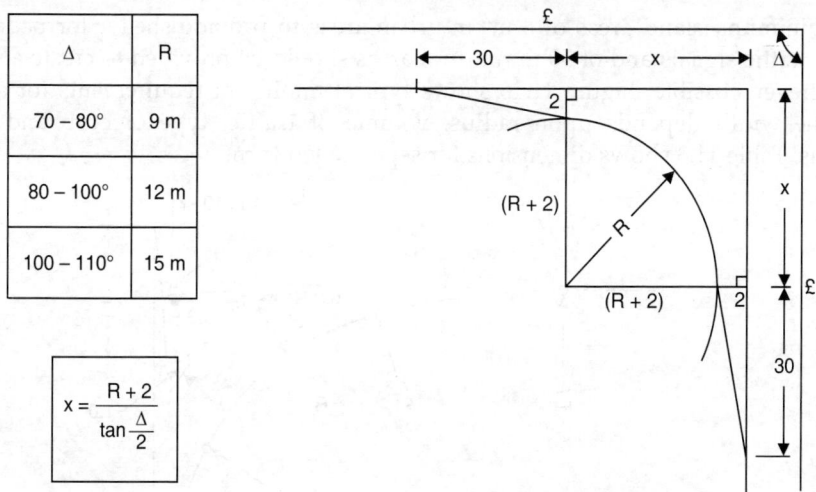

Δ	R
70 – 80°	9 m
80 – 100°	12 m
100 – 110°	15 m

$$x = \frac{R+2}{\tan\frac{\Delta}{2}}$$

Fig. 11.15: Desirable minimum left-turn treatment for rural intersections

$$R = \frac{V^2}{127(e+f)}$$

where

R = Radius of curve in m.

V = Speed in KPh

e = Superelevation (m/m) (desirable values 0.08 in rural area, 0.04 to 0.06 in urban area).

f = Coefficient of friction between vehicle tyres and the pavement surface.

Table 11.3: Dimensions for separate left turn

R_1	W_1	W_2	W_3	S	
12	6.5	7.5	10.3	1.5	where R_1 = 12 m to 30 m
14	6.2	7.2	10.1	1.4	R_2 = 1.5 R_1
16	6.0	7.1	9.9	1.3	R_3 = 3 R_1
18	5.9	6.9	9.7	1.3	where R_1 = 30 m to 45 m
20	5.7	6.8	9.6	1.2	R_2 = 2R_1
22	5.6	6.7	9.5	1.2	R_3 = 2 R_1
24	5.5	6.6	9.4	1.1	where R_1 > 45 m single rad. acceptable.
26	5.4	6.5	9.3	1.1	**Lane Widths**
28	5.4	6.5	9.2	1.1	W_1 = single-lane flow
30	5.3	6.4	9.1	1.0	W_2 = Single-lane flow with provision for
45	50	6.1	8.8	09	passing and stalled vehicle
60	4.8	5.9	8.6	0.9	W_3 = Two-lane flow
90	4.6	5.8	8.4	0.8	(All dimensions in metres)
120	4.5	5.7	8.3	0.7	
150	4.5	5.6	8.2	0.7	

A compound curve of radii 1.5 R, R and 3R, satisfies the requirements for radius between 12–30 m. For radius between 30–45 m, a compound curve of 2R, R and 2R may be satisfactory. For curve beyond 45 m, single radius curves are satisfactory.

The minimum island areas of 8 m² in urban areas to provide shelter for pedestrians and space for traffic signals and of 50 m² in rural areas should be provided to create a separate left turn wherever possible. Figure 11.16 shows typical minimum requirements for separate left turns. The width depends upon radius, volume of traffic, type of curb and other local conditions. Table 11.3 shows dimensions for separate left turn.

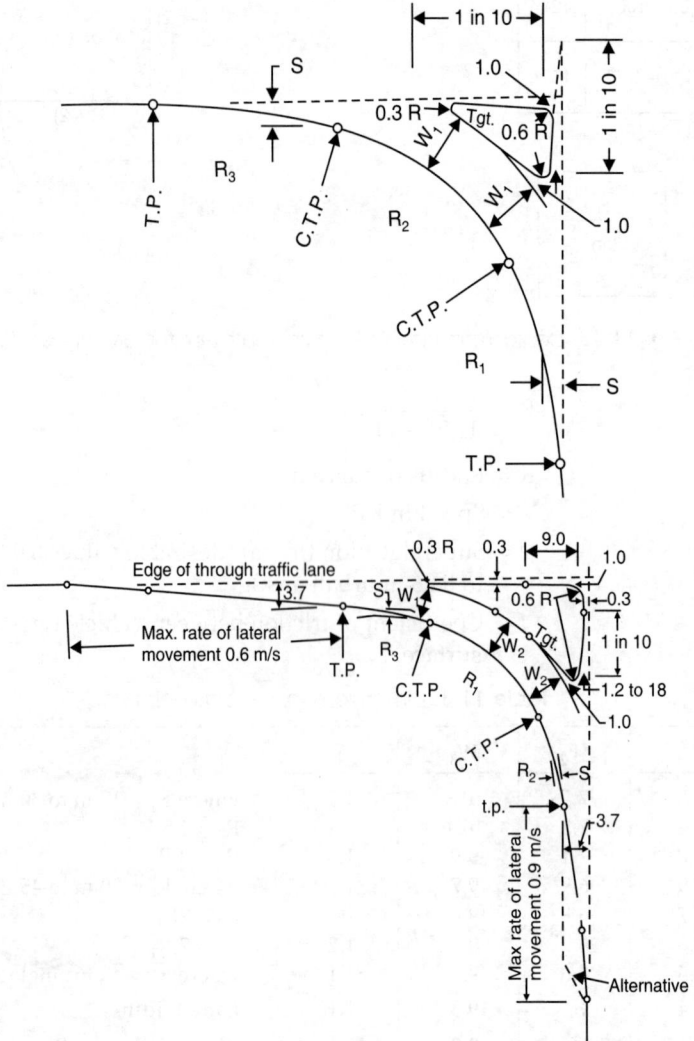

Fig. 11.16: Minimum treatment for separate left turn

11.2.4 Treatment for Right Turns

From the capacity and safety point of view, right turns should be provided:
- At rural intersections for safety
- At undivided urban road intersections, where right tuning traffic volumes are considerable.
- At intersection of divided urban roads with sufficient medians.

Figure 11.17 shows the basic design elements of a right turning lane

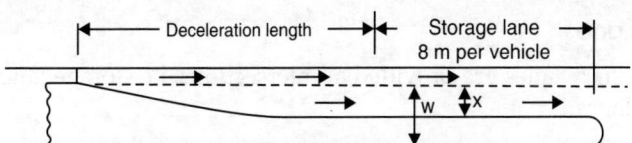

Demension	x(m)	W(m)				V(kph)	Dec. length (m)
		Pedestrain refuge	Signal Pedestal. No Pedestrian	No. signal or Pedestrian			
Desirable	3.7	2.4	2.4	2.4		50	60
						60	80
Minimum	3.1	2.4	1.8	1.2		70	100
Absolute minimum	2.5	2.0	1.5	0.3		80	120
						90	140
						100	170

Fig. 11.17: Elements of a right turning lane

The length of turning lane depends upon (*a*) deceleration length (*b*) storage lane and entering taper lanes. Provision of deceleration clear of through traffic lanes is an important element of high speed arterial streets and should be incorporated into their design whenever feasible. In many cases, the full length of deceleration plus storage and taper cannot be provided. In such case the deceleration is accomplished before entering the tuning lane.

The storage lane should be at least 32 m long to provide for four vehicles.

When two opposing right turns are expected to flow simultaneously, the turning radii and the tangent points should be selected by using the turning paths of design vehicles so that there is a clear width which enables two turning vehicles to pass each other with adequate clearance as shown in Fig. 11.18.

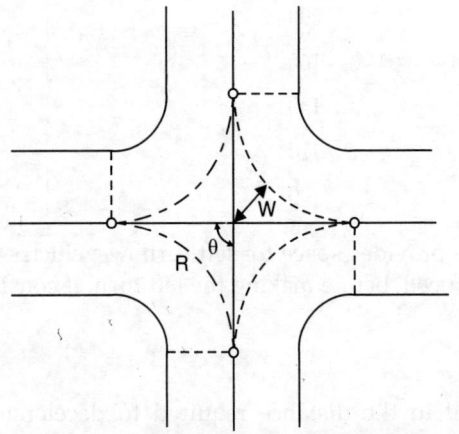

Θ	W(m)	R(m)
70°	7	9
80°	6	10.5
90°	5	12.5
100°	4.5	15.3
110°	4	18.5

*W : min. for opposed turns by S.U. trucks.

Fig. 11.18: Minimum width for turning vehicle

11.2.5 Auxiliary Lanes

Three types of auxiliary lanes are provided at intersections, i.e storage lanes, acceleration lanes and decelerations lanes.

(a) **Storage lane (Storage length).** The design of storage lane for the right turns involves followings steps. It should be sufficient to accommodate the average number of vehicles arriving per cycle.

  • Establish the Peak Hour Factor (PHF), for each approach to the intersection.

  • PHF varies from 0.25 to 1.00. The value of 'Storage Lane Multiplying Factor' is a linear function of PHF as below:

PHF	*Storage lane multiplying factor*
0.25	2.0
1.00	1.5

  • Determine the volume of through traffic and turning traffic per cycle for average design hour

  • Modify this volume in considerations of the number of trucks

  • Each vehicle now requires 8 m storing space.

    (For unsignalised intersections use the number of vehicles wishing to turn in 2 minutes time in place of the cycle time).

Design length of the storage lane is taken as the greater of the two values i.e., between length of right turning traffic storage and length of through traffic storage. The taper at the entering of lane is 8 : 1 for speeds up to 50 K Ph and 15:1, beyond 50 KPh speeds.

In places where not more than one or two vehicles are expected to wait for right turn, as in rural areas, the storage lane may be provided as per Table 11.4

Table 11.4: Length of right turning lane (IRC)

Design Speed (km/h)	*Length of storage lane including 30-45 m Taper (m)*
120	200
100	160
80	130
60	110
50	90

(b) **Deceleration lane.** Deceleration lane provides space for left turning vehicles to separate from through traffic and reduce the speed, before making the left turn. It consists of:

  (i) a length of diverge taper and

  (ii) a length of parallel lane.

The combined length should be equal to the distance required to decelerate from the approach speed of the through road to the design speed of the left turn. The length of deceleration can be found by using Tables 11.5 and 11.6.

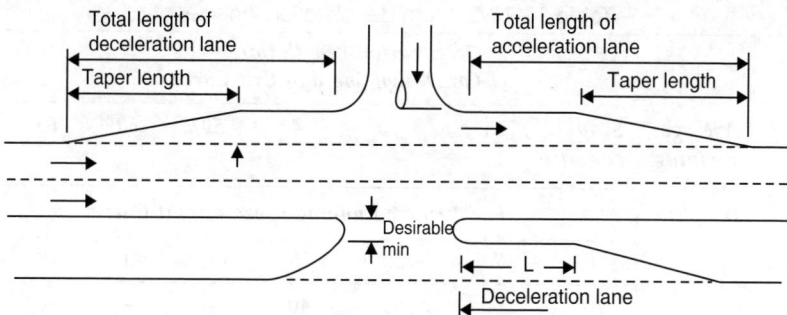

Fig. 11.19: Deceleration and acceleration lanes

(c) Acceleration lane. When the volume of merging traffic is high, the driver of the vehicle may not find any gap immediately available in the through traffic stream to merge. He should therefore continue on a route parallel to the through traffic until he adjusts his speed (accelerates) to create an opportunity to merge. Such lanes are called acceleration lanes. An acceleration lane consists of:

(i) a length parallel to through lane.

(ii) a length of merging taper.

The combined length should be equal to the distance required for a vehicle to accelerate from the design speed of the left turn to the design speed of the through road. Tables 11.5 to 11.7 are used to determine the total length of acceleration lanes.

If acceleration and deceleration lanes are on gradient, a correction factor for its length due to effect of grade should be made using the multiplying factor given in these tables.

Table 11.5: Length of taper-speed change lanes

Highway Design Speed	48 kph	64 kph	80 kph	97 kph	113 kph	129 kph
Length of the Taper	46 m	58 m	70 m	82 m	91 m	107 m

Uniform 50 : 1 taper should be used where lengths of acceleration lanes exceed 1300 ft. (396 m) or where speeds exceed 70 mph (112 kph) or elsewhere if appropriate and space permits.

Table 11.6: Minimum acceleration lane lengths

Highway		*Acceleration Length (m)*								
		for entrance curve design speed (kmph)								
	Stop conditions	*25*	*30*	*40*	*50*	*60*	*65*	*75*	*80*	
Design Speed (kmph)	*Speed Reached (kmph)*	*and initial speed (kmph)*								
		0	*20*	*30*	*35*	*40*	*50*	*60*	*65*	*70*
50	40	60	–	–	–	–	–	–	–	–
65	50	120	100	75	70	40	–	–	–	–
80	60	230	210	190	180	150	100	50	–	–
100	75	360	340	330	300	280	240	160	120	50
110	85	490	470	460	430	400	380	310	250	180

Table 11.7 Minimum deceleration lane Length

Highway Design Speed (kmph)	Average Running Speed (kmph)	Deceleration Length (m) For Design Speed of Exit Curve								
		Stop condition	25	30	40	50	60	65	75	80
		For Average Running Speed of Exit Curve								
		0	20	30	35	40	50	60	65	70
50	45	70	60	50	40	–	–	–	–	–
65	60	95	90	80	70	60	50	–	–	–
80	70	130	120	120	110	100	90	70	50	–
100	85	160	150	150	140	130	125	100	90	70
105	90	175	165	160	150	150	130	120	100	85
110	95	190	180	175	170	160	150	130	120	100

Where acceleration lanes are on a down gradient their length may be reduced to 1-0.08G times the normal length, where G is the gradient expressed as a percentage.

Where deceleration lanes are on an upgrade their length may be reduced to that obtained by multiplying by 1 – 0.03G times, where G is the gradient expressed as a percentage and on downgrade their length may be increased by multiplying 1 + 0.06G

11.2.6 Channelisation

Channelisation involves use of islands at the intersections, to guide and protect the traffic and pedestrians. It is generally used to:

- Reduce the area of conflict.
- Merge traffic streams at small angles.
- Reduce the relative speed.
- Control speed of the traffic entering or crossing an intersection.
- Provide refuge for turning or crossing vehicle.
- Improve the efficiency and layout of the signalized intersection.
- Provide protection to pedestrians.
- Improve and define the alignment of major movements.
- Provide locations for installation on traffic signals and signs.
- Provide reference point within intersection enabling the driver to predict path and speed of other drivers.

Though it is not practicable to standardize the design of channelized layouts, as it depends upon local conditions, following points should be considered in preparing channelized intersection design.

- The islands should fit to the natural path of drivers.
- A few well placed, large islands are better than many small islands.
- Island should be offset 0.6 m or more from the edge of normal travelled way.
- Adequate approach end treatment should be provided to warn the drivers and to permit gradual changes in speed and path.

- The curves used should have sufficient radii and width as per the requirements of design vehicle.
- Islands should be well defined and well illuminated for night use.
- Raised islands should be at least 8 m² in urban area, not less than 2.4 to 3.6 m. on any side after rounding the corners if triangle, and at least 1.2 to 3.6 m, if elongated.
- Islands of at least 50 m² in area should be used in rural areas.

Figure 11.20 shows typical treatment of island noses.

11.2.7 Median Opening

Design of median openings and median ends depends upon the traffic volume and types of the turning vehicles. Paths of the design vehicles are plotted to develop median ends. To ensure that large vehicles can turn right without difficulty to or from a major road, the gap in the median should normally extend 3 m beyond the continuation of both curb lines of the minor road to the edge of the major road as shown in Fig. 11.21 and should also be determined by 12–15 m radius control circles tangential both to the centre line of the minor road and the side of the central verge away from the minor road.

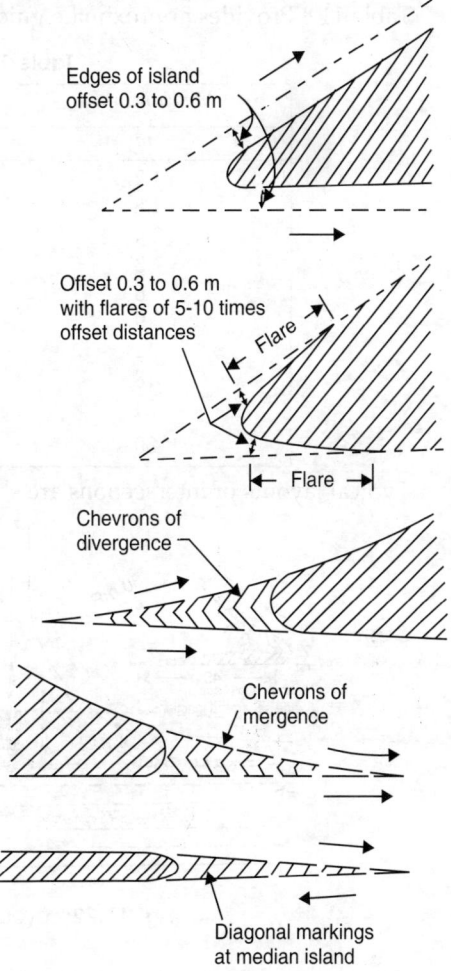

Fig. 11.20: Treatment for island noses

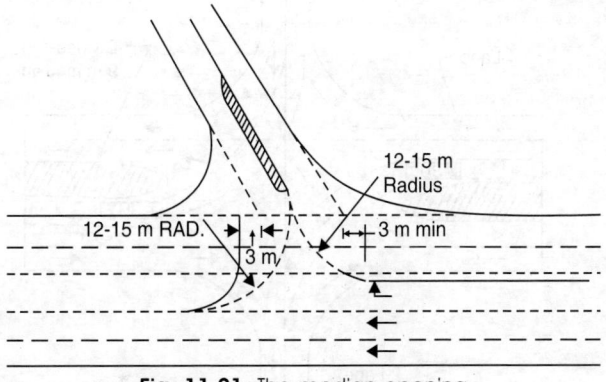

Fig. 11.21: The median opening

For skew intersections the length of median opening shall increase with increase in the skew angle. Where there is insufficient width available to construct a curbed median or small island, a painted median or island may be used.

Table 11.8 Provides approximate guidelines for length of median opening.

Table 11.8: Median openings

Median width	Minimum length of opening
m	m
1.2	42
1.8	43
2.4	42.5
3	42
6	39
9	36
12	30
18	27
24	21
30	15

Typical layouts of intersections are shown in Figs. 11.22 to 11.24.

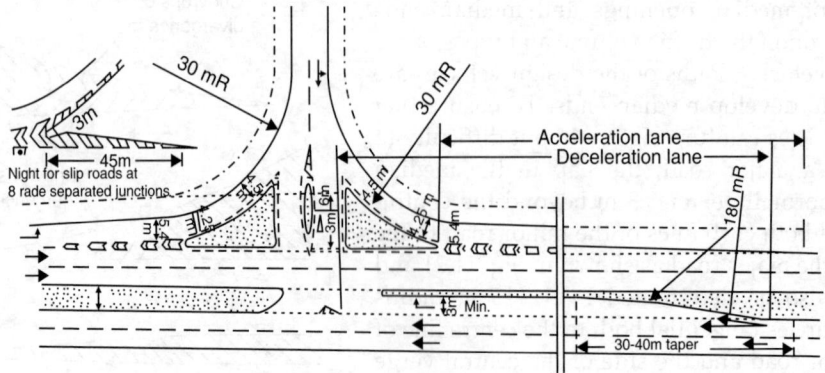

Fig. 11.22: Layout of typical 3 legged intersection

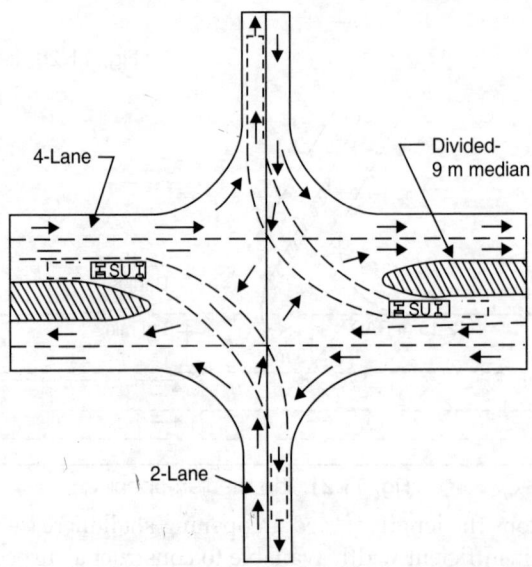

Fig. 11.23: Typical layout of 4 legged intersection

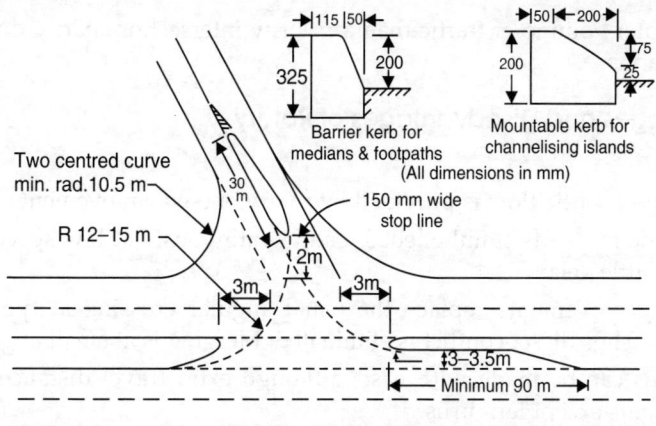

Fig. 11.24: Typical layouts of 3 legged skewed intersection

Detailed guidelines for the design of intersections in rural and urban area in India, are provided in IRC-SP 41 and type design for intersections on National Highways are published by MORT&H.

11.3 PRINCIPLE OF ROTARY OPERATION

A rotary or roundabout is a channelised intersection where traffic moves clockwise around a central island as shown in Fig. 11.25. Roundabouts are commonly used in U.K. where small rotaries have been built with good operational and safety characteristics.

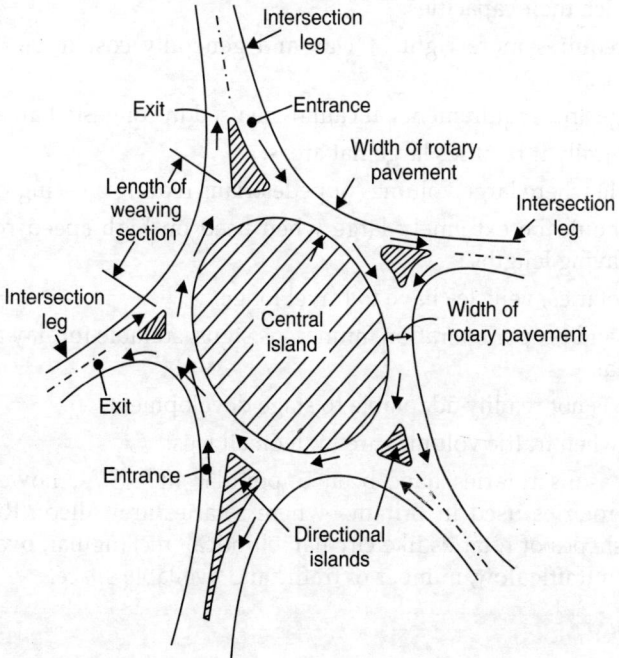

Fig. 11.25: Conventional rotary (design nomenclature)

With the application of the principal "Give Way to Traffic Coming From the Right" a rotary works as series of T-intersections with only left turn in and out. Rotary can accommodate

significantly greater volume of traffic than a priority intersection, partly due to their one-way, circulatory operation.

11.3.1 Advantages and Disadvantages of Rotary

(a) *Advantages*

- An orderly traffic flow is provided by rotary one-way movement.
- All traffic proceeds simultaneously and continuously at low speeds. At low volumes there is little delay.
- Weaving movements replace the usual angular crossing at typical at-grade intersections. Thus direct conflict is eliminated. Merging is at small angels.
- All turns can be made with ease, although extra travel distance is required for all movements except left turns.
- The rotary design is especially suited for intersections with five or more legs.
- It costs less than a grade separator (interchange), though the capacity of rotary is much lower than an interchange.
- It reduces the large area of conflict (if it is there).

(b) *Disadvantages*

- It wastes a lot of space.
- A rotary can accommodate no more traffic than a properly designed channelised layout.
- Does not operate satisfactorily when the traffic volumes on two or more intersection legs approach their capacities.
- A rotary requires more right of way and generally cost more than other at grade intersections.
- Due to large area requirements, it cannot be used in congested areas.
- Topographically it requires large flat areas.
- Not suitable where large volumes of pedestrians require crossing the roads.
- Rotaries should be extremely large when used on high speed roads to provide the proper weaving lengths.
- On large rotaries, vehicles have to travel longer.
- For safety and proper operation, numerous signs, suitable for day and night operation, are essential.
- The rotary is not readily adaptable to stage development.
- Not good when traffic volumes are high on all legs.

Due to these reasons rotaries have been unpopular in U.S.A., however there are three principal types of rotaries used in Britain—where rotaries are called "Roundabouts". There could be different shapes of rotaries like circular, elliptical, rectangular, oval, elongated, square etc. depending upon traffic flow, number of traffic and available space.

11.4 TYPES OF ROTARIES

(1) **Conventional Roundabout:** (Fig. 11.26) consists of one-way circulation around curbed control island usually at least 25 m in diameter (not necessarily circular). If not circular, corners of the island should have a minimum radius of 10 m to accommodate the turning radius of large vehicles.

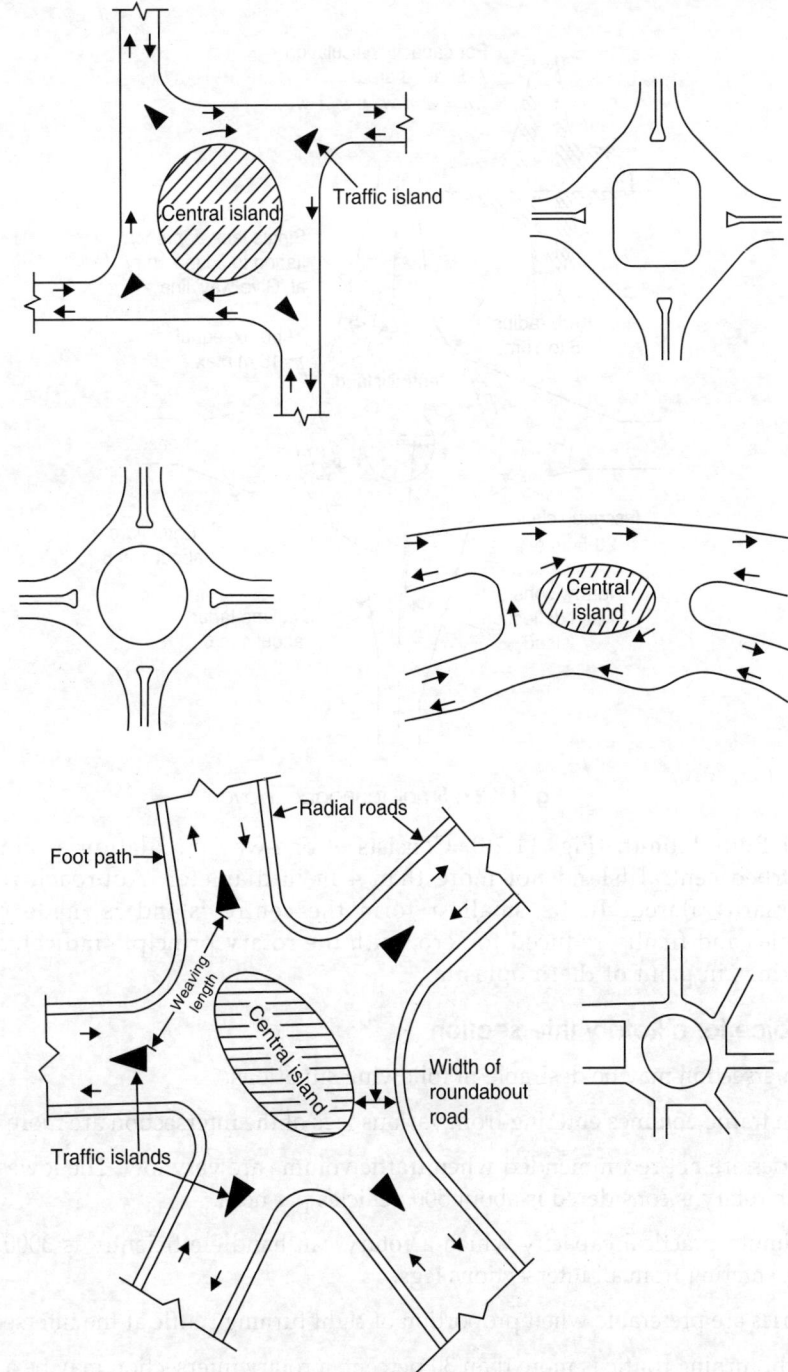

Fig. 11.26: Typical shapes of rotaries

(2) Small Roundabout: (Fig. 11.27) Consists of one-way circulation around a circular, curbed central island at least 4 m in diameter. The approach roads are flared at entry point to facilitate multiple vehicle access.

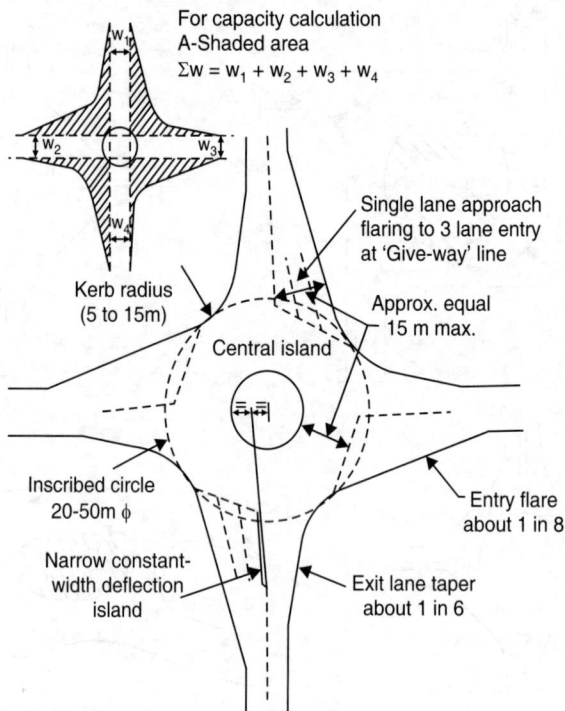

For capacity calculation
A-Shaded area
$\Sigma w = w_1 + w_2 + w_3 + w_4$

Single lane approach
flaring to 3 lane entry
at 'Give-way' line

Kerb radius
(5 to 15m)

Approx. equal
15 m max.

Central island

Inscribed circle
20-50m ϕ

Entry flare
about 1 in 8

Narrow constant-
width deflection
island

Exit lane taper
about 1 in 6

Fig. 11.27: Small roundabout layout

(3) Mini Roundabout: (Fig. 11.28). Consists of one-way circulation around a circular uncurbed central island not more than 4 m in diameter. Approach roads are not necessarily flared. In its smallest form the centre island is made crossable by vehicles and finally reduced to zero, with the rotary principle indicated by circular markings in paint of thermoplastic.

11.4.1 Choice for a Rotary Intersection

A rotary intersection may be desirable in following situations:

- When traffic volumes entering from various legs of the intersection are more or less equal.
- Rotaries are not recommended when traffic volume are very low. The lowest volume for which rotary is considered is about 500 vehicles per hour.
- Maximum practical capacity which a rotary can handle efficiently is 3000 vehicles per hour, entering from all intersections legs.
- Rotaries are preferable when proportion of right turning traffic at the intersection is high.
- If right turning traffic is more than 30 percent, a rotary intersection may be a better choice than a four legged signalized intersection.
- When number of approaches are more than four, a rotary intersection may be considered.
- Layout of the site and availability of space, is also important for considerations of a rotary
- In narrow width of the roads approaching an intersection, a rotary may be a choice.

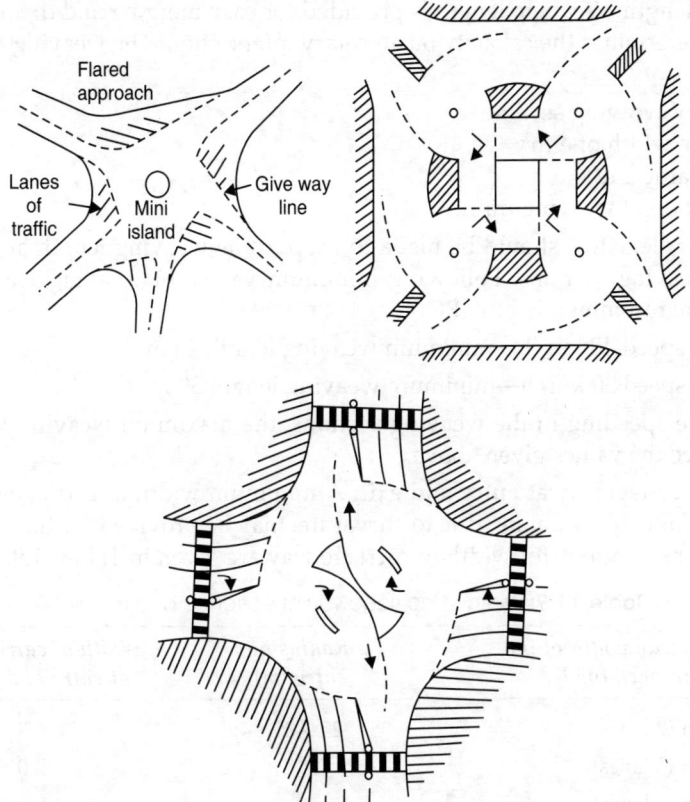

Fig. 11.28: Typical mini roundabouts

11.5 DESIGN OF CONVENTIONAL ROTARY INTERSECTION (ROUNDABOUT)

Following elements of rotary design should be considered

(i) **Design speed:** Design speed is involved in determinations of radius of entry and exist curve, weaving length etc. As per India Road Congress IRC-65 rotary design speeds are:-

   (a) 40 km. p. hour for rural areas and

   (b) 30 km p. hour for urban areas and other restricted locations.

(ii) **Radius of curve at entry:** Radius of curve at the entry depends upon the design speed, amount of super elevation and coefficient of friction. IRC recommends that

   (a) Radius at entry for 40 km p.hr design speed in rural areas = 20 – 35 m.

   (b) Radius at entry for 30 km p. hr design speed in urban restricted areas = 15 – 25 m.

   The lower value is to ensure easy entrance of vehicles into rotary and the higher value to guard any tendency for over speeding.

(iii) **Radius of curve at exit:** The radius of exist curve should be greater than the radius of the rotary island, so as to allow the drivers to pick up speed and clear the rotary rapidly. The radius of curve is therefore kept $1^1/_2$ to 2 times the radius of the entry curve.

   In case number of pedestrian is large across the exit road, to keep the speeds low, the radius of exit curve may be the same as of entry curve.

(iv) **Radius of central island of rotary:** Radius of the central island is kept slightly larger than that of the curve at entry. A value of 1.33 times the radius of entry curve is recommended as a guidance, by IRC.

(v) Weaving length: Weaving length is provided for easy merging and diverging of vehicles, and also determines the capacity on an rotary intersection. The weaving length is decided on the basis of:

- Width of weaving section
- Average width of entry
- Total traffic and
- Proportion of weaving traffic.

As a general rule, effort should be made to keep to the weaving length at least 4 times the width of the weaving sections. Following minimum values of weaving length for different design speeds are recommended by IRC

(a) Design speed 40 km/h—minimum weaving length 45 m

(b) Design speed 30 km/h—minimum weaving length 30 m

To discourage speeding in the weaving sections, the maximum weaving length should be restricted to twice the values given above.

(vi) Width of carriageway at entry and exit: A minimum width of carriageway of 5 m, with provision for extra widening due to curvature, may be provided for the entrance and exit. IRC recommendations for width of carriage way are given in Table 11.9.

Table 11.9: Width of carriageway at entrance and exit (IRC)

Carriageway width of the approach road	Radius at entry (m)	Width of carriageway at entry and exit (m)
7 m (2 lanes)	25 – 35	6.5
10.5 m (3 lanes)		7.0
14 m (4 lanes)		8.0
21 m (6 lanes)		13.0
7 m (2 lanes)	15 – 25	7.0
10.5 m (3 lanes)		7.5
14 m (4 lanes)		10.0
21 m (6 lanes)		15.0

(vii) Width of rotary carriageway: The width of non-weaving section of the rotary should be equal to the widest single entry into the rotary, and should generally be less than the width of the weaving section as shown in Fig. 11.29.

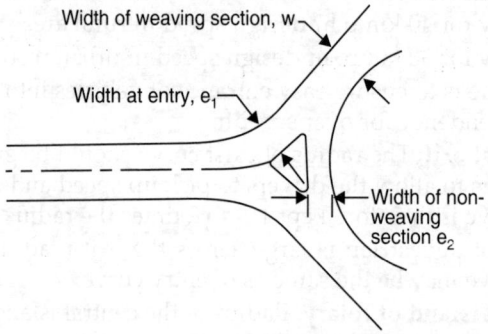

Fig. 11.29: Width of rotary carriageway

The width of the weaving section of the rotary should be one traffic lane (3.5 m) wider than the mean width of weaving section $W = 3.5 + \dfrac{e_1 + e_2}{2}$ m. as shown in figure

(viii) Entry and exit angles: Entry angle should be larger than exit angle and it is desirable that entry angle should be 60° if possible. The exit angle should be small, even tangential. An ideal design is when entry angle is 60° and exit angle is 30°. This condition can only be achieved by staggering the approach roads.

(ix) Capacity of the rotary: The capacity of a rotary is based on the capacity of each weaving section. Capacity of a weaving sections depends upon the geometric layout factors, i.e.

- Width of the weaving section.
- Average width of entry into the rotary.
- Weaving length and.
- Properties of weaving traffic.

Following empirical formula is used for calculating the capacity.

$$Q_p = \frac{280\, w\left(1 + \dfrac{e}{w}\right)\left(1 - \dfrac{P}{3}\right)}{1 + \dfrac{w}{L}}$$

where: Q_p = The practical capacity of the weaving sections of the rotary in PCU per hour.

W = Width of the weaving section in metres (within the range of 6 – 18 m)

e = The average width of entries to the weaving section in metres = $\dfrac{e_1 + e_2}{2}$

$\dfrac{e}{w}$ to be (within the range of 0.4 to 1.00) Fig. 11.30

L = Length of weaving section in metres (w/L to be within range of 0.12 and 0.4)

P = Proportion of weaving traffic i.e, ratio of sum of crossing stream to the total traffic on the weaving section

$p = \dfrac{b + c}{a + b + c + d}$ as in Fig 11.30 being in range of 0.4 to 1.0

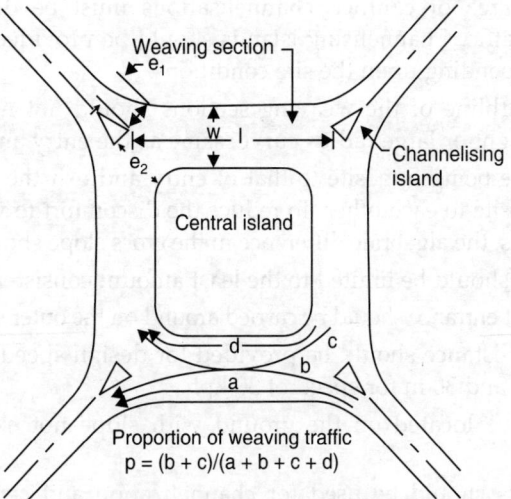

Fig. 11.30: Relevant dimensions of weaving section and proportion of weaving traffic for use in capacity formula for rotaries

The passenger car unit equivalents may be taken as follows:

1. Car and light commercial vehicle including 3 wheelers...................... 1.00
2. Buses and medium and heavy commercial vehicles...................... 2.80
3. Motor cycle and scooter (2 wheelers)...................... 0.75
4. Pedal cycle...................... 0.5
5. Animal drawn vehicles............. 4 to 6

Following adjustments in the capacity calculated by the above formula are suggested.

 (i) Where the entry angle (see Fig. 11.25 for definition) is between 0° and 15°, deduct 5 percent from the capacity of the weaving section.

 (ii) Where the entry angle is between 15° and 30°, deduct $2\frac{1}{2}$ percent from the capacity of the weaving section.

 (iii) Where the exit angle (*see* Fig. 11.25 for definition) is between 60° and 75°, deduct $2\frac{1}{2}$ percent from the capacity of the weaving section.

 (iv) Where the exit angle is greater than 75°, deduct 5 percent from the capacity of the weaving section.

 (v) Where the internal angle (*see* Fig. 11.25 for definition) is greater than 95°, deduct 5 percent from the capacity of the weaving section.

 (vi) Where the pedestrian flow at exit from the roundabout exceeds 300 per hour, an arbitrary deduction of one-sixth should be made in the practical capacity of the preceeding weaving section.

While designing care should be exercised that weaving sections are adequate for the required capacity so that merging and diverging manoeuvres take place smoothly. As a major disadvantage with rotaries is the reduction in speed, the weaving sections should preferably be kept slightly longer than just necessary for capacity, say 33 to 50 percent more.

The capacity of a rotary can be increased above the value given in the above equation by signalizing the rotary intersection or introducing the off-side priority rule.

(x) Other important points for a rotary

• To reduce the area of conflict, channelisations must be done, for safe and orderly movement of traffic. Channelising islands should be provided at entry and exit on each leg of rotary, depending upon the site conditions.

• The external curbline of the weaving sections should not normally be re-entrant, but consist of a straight or large radius curve, same as the entry and exit curves.

• Rotary curvature being opposite to that of entry and exit the super elevation in the two portions is opposite to each other. To reduce the discomfort to vehicles specially buses and trucks due to this, the algebraic difference in the cross slope should be limited to about 0.07.

• Super elevation should be limited to the least amount consistent with design speed.

• The cross-slope at entrance should be carried around on the outer edge of the rotary.

• Stopping sight distance should be provided for design speed of the rotary, i.e. 45 m for speed of 40 kph and 30 m for speed of 30 kph.

• Rotary should be located on the ground with slope not exceeding 1 in 50 with the horizontal.

• Mountable curbs should be used for channelization and central island. To discourage pedestrian from crossing over, a barrier type curb may be used at outer edges of the rotary.

• Proper drainage and lighting facilities for night use should be provided at rotary.

- Pedestrian crossings should be suitably provided and also segregate the cyclists, by providing separate cycle track.
- Adequate signs and markings should be provided for both day and night travel. The standard warning sign indicating 'Rotary Ahead' should be installed to give advance information to traffic.
- A red reflector about 1 m above the road level, should be fixed at the nose of each directional island and on the curb of the central island facing the approach roads. Illumination of the rotary junction at night is important.
- A rotary provides ample space for effective development of the land scape. Landscaping suiting to site may be undertaken, keeping safety of the traffic in view.

11.6 DESIGN OF A SMALL ROUNDABOUT

Small roundabout is an intermediate treatment between the mini roundabout and conventional roundabout. A small roundabout can be accommodated within an inscribed circle of between 20 and 50 metres diameter (up to the lower size limit of the conventional roundabout). The central island is about one-third the diameter of the inscribed circle, within the limits of 4 to 25 metres (central islands of less than 4 m in diameter are classified as mini roundabouts).

The fundamental principles of design of small roundabouts and mini roundabouts are the following:

- The speeds should be low up to about 50 km/h.
- To faster gyratory motion to the left, narrow, curbed deflection islands, constant width and road markings should be used.
- To encourage multi-lane entry to the circulation area, flaring of the approaches is done. Flaring may also be done at exit for easy exit.
- The entry should be taper at about 1 in 3, in order that a single lane approach can provide a three-lane entry each between two and a half and three-and-a-half metres wide at give way line. The exit flare may be less 1 in 6.

The design capacity of a small roundabout is 85 percent of its practical capacity given by the following formula (Fig. 11.31)

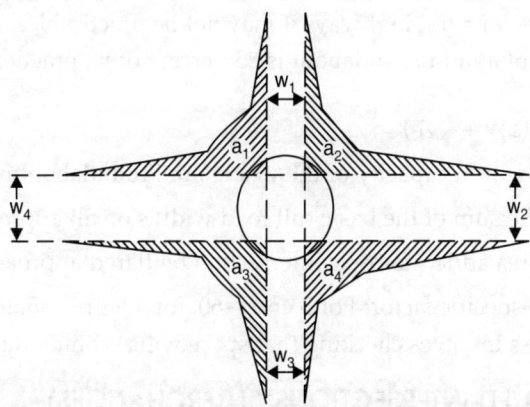

Fig. 11.31: Capacity calculations for a small roundabout

$$Q_p = K(\Sigma W + \sqrt{A})$$

Where Q_p = Practical capacity of the whole small roundabout (vph).

ΣW = The sum of the basic full road widths on all approaches (m).

A = Area added to basic intersection by flared approaches (sq. m).

K = A specific factor: For 3-entry–70, for 4 entry –50, for 5 entry – 45

$Q = K(\Sigma W + \sqrt{A})$

$\Sigma W = w_1 + w_2 + w_3 + w_4$ metres

$A = a_1 + a_2 + a_3 + a_4$ sq. metres

Sometimes a simpler formula is used, as below which gives fairly accurate results;

$$Q_p = KD$$

where Q_p = The saturation capacity in passenger vehicles per hour.

K = 150 – for 3-leg intersections.

140 –for 4-leg intersections.

D = The diameter of the inscribed circle in m. For an oval intersection D is the mean of the major and minor axes.

The design capacity is 85% of this capacity. A rough idea of the inscribed circle diameter for a small roundabout, for the given traffic flow (total design flow through junction) can be made as in Fig. 11.31.

11.6.1 Design of Mini Roundabouts

Where conventional roundabouts are no longer capable of accommodating the demand, and it is not possible to have additional land, mini roundabouts may be used to improve the capacity of the intersection.

The main characteristics of a mini roundabout are the following:

- Its size is small. The island has a diameter of 1 m to 4 m within an inscribed circle of not more than 22 m diameter.
- The centre island is of humped nature (dome shape), raised not more than 125 mm at the centre and virtually flush at the edges. It could merely be marked on the road surface by paint. The characteristic is specifically intended to facilitate over-running. No signs etc. are placed on the island. It is usually painted white all over.
- The approach speed limit for a mini roundabout is 60 km/h. Beyond this speed, it is not used and therefore for rural highways it may not be practicable.

The design capacity of a mini roundabout is 85 percent of its practical capacity, given by the following formula:

$$Q_p = K(\Sigma W + \sqrt{A})$$

where Q_p = Practical capacity of the whole mini roundabout in vph

ΣW = The sum of the basic full road widths on all approaches (m)

A = Area added to basic intersection by flared approaches.

K = A specific factor: For 3 entry–60, for 4 entry –45, for 5 entry –40.

The designing process involves checking that space will accommodate future volumes.

11.7 GRADE SEPARATED INTERSECTIONS (INTERCHANGES)

11.7.1 Types of Interchanges

Grade separated intersections are called interchanges. They involve the use of successive left turns and bridges over or under the main flow. Traffic moves at different levels in space. The classification of interchanges is not simple as they are described by the pattern of the various turning roadways and ramps. Figure 11.32 illustrates the basic types of interchanges i.e., diamond, cloverleaf and the roundabout.

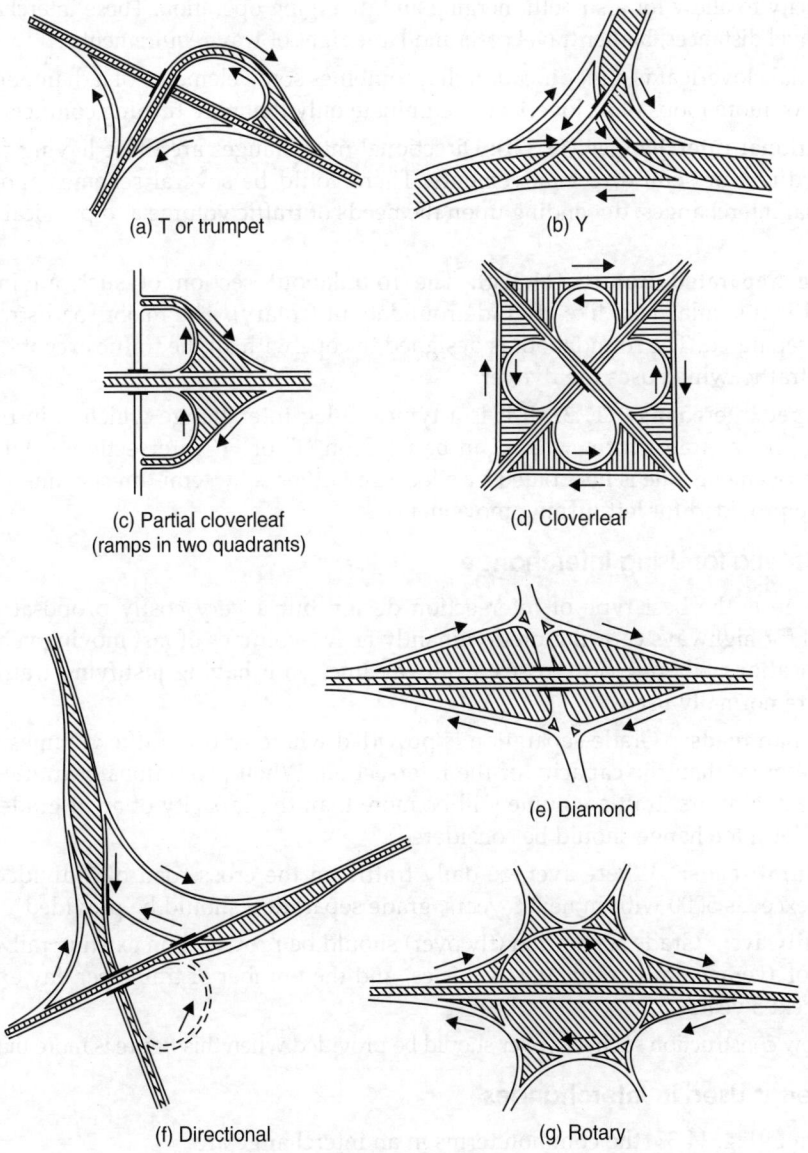

Fig. 11.32: Types of interchanges

Diamond Interchanges (Fig. 11.32e): The diamond interchange is the simplest form of grade separated intersection between two roads. The conflicts between through and crossing traffic are eliminated by a bridge. The diamond interchanges require minimum of land and are economical to construct. The conflict that occurs where ramps meet the grade-separated cross street may require an alternative solution if ramp volumes are high.

Cloverleaf Interchanges (11.32d): The full cloverleaf interchange eliminates all crossing movement conflicts by the use of weaving sections. The weaving section replaces a crossing conflict with a merging, followed some distance further by a diverging conflict. There are two points of entry and exit on each through road. A weaving section is created between the exit and entry points of the structure. This weaving section is a critical element of cloverleaf design. It must have length and capacity to allow for a smooth merging and diverging operation. These interchanges involve longer travel distances, higher travel costs and large right of way requirements.

A partial cloverleaf is a modification that combines some elements of a diamond interchange with one or more loops of a cloverleaf to eliminate only the more turning conflicts.

Directional Interchanges (11.32f): Directional interchanges are those having ramps for one or more direct or semi-direct movements. There could be several schemes and patterns for directional interchanges, depending upon the needs of traffic volume and physical conditions of the site.

Grade Separated Rotary (11.32g): The roundabout section of such an interchange is designed in a similar way like at grade roundabout (rotary). The major road straight through traffic is separated. The roundabout is designed to cope with all the traffic except the major road through traffic, which uses the flyover.

Trumpet Interchange (11.31a): It is a typical 3-leg interchange which is in the shape of a trumpet. This simplest form which can be used on 'T' or 'Y' intersections. Out of two right turning movements one is negotiated by a loop and other is by semi-direct connection. Diagonal ramps are provided for left turning movements.

11.7.2 Criteria for Using Interchange

Interchange is the best type of intersection design but a very costly proposal. It should be provided for highways carrying predominantly heavy volumes of fast moving vehicular traffic, and at locations where road crosses a railway line, both having justifying traffic. Following criteria are normally used.

On urban roads: Grade separation is provided where future traffic volumes within next 5 years, are more than the capacity of the intersection. When projections of traffic indicate that within next 20 years, traffic volume will be more than the capacity of an at grade intersection, the need for interchange should be considered.

On Rural roads: Where average daily traffic on the cross road of a divided rural intersections exceeds 5000 with in next 5 years, grade separation should be provided

On railways: Grade separation (fly over) should be provided on existing railway line if the product of average daily traffic (fast vehicles) and the number of trains per day exceeds 50,000, with in next 5 years.

For new construction such a flyover should be provided when this figure is more than 25,000 only.

11.7.3 Terms Used in Interchanges

As shown in Fig. 11.33, the common terms in an interchange are:

Ramp: An interconnecting roadway of a traffic interchange, or any connection between highways at different level, on which vehicles may enter or leave a designated roadway.

Loop: Oneway turning roadway that curves about 270° to the left to provide a right turning movement.

Outer connection: The ramp provided for traffic for left turning movement from one of the through road separated by a grade separator to the other through roadway.

Direct connection: It is a type of ramp which does not deviate much from the desired direction of travel.

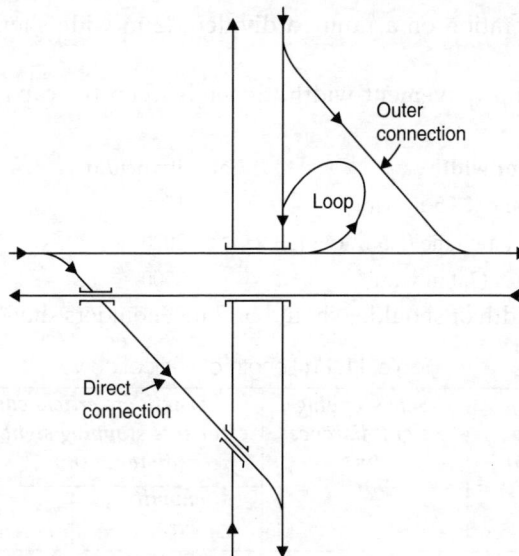

Fig.11.33: Common terms used in interchanges

11.8 SUMMARY OF CONSIDERATION FOR INTERCHANGE DESIGN

Design of an interchange is a structural design problem, however basic principles governing traffic engineering aspect are enumerated below:

- Design speed, radius of horizontal curve and stopping sight distances on the ramp are related to design speed of the highways. Table 11.10 shows these parameters for design speeds of 80 km phr (urban road) and 100 km p. hr (rural roads)
- The gradient on ramp should be as flat as possible. It should be limited to a maximum of 4 percent and should not increase 6 percent in any case.
- The vertical curve at either end of the ramp should be designed to provide for atleast safe stopping sight distance corresponding to design speed of the ramp. Table 11.11 shows length of vertical curves for design speeds 30 to 100 km/h.

Table 11.10: Speed, horizontal curvature and sight distance for ramp design

| Particulars | Design values for major highway designs speed of | | | | For loop ramps | |
| | 80 km/hr | | 100 km/hr | | | |
	Minimum	Desirable	Minimum	Desirable	Minimum	Desirable
Ramp design speed (km/h)	40	50	50	65	30	40
Radius of curvature (n)	60	90	90	55	30	60
Stopping sight distance (m)	45	60	60	90	25	45

Notes: 1. The major highway design speeds of 80 km/h appropriate for highways in urban areas.

2. The radius of curvature values have been worked out for a maximum superelevation of 7 percent.

- For two-way operation on a ramp, a divider 1.2 m wide should be used in the cross-section.
- Width of the ramp (pavement width) depends upon the capacity of the unidirectional flow as below:

Pavement width	Capacity pcu/hr
Single lane (3.75 m wide)	1500
Intermediate lane (5.5 m (wide)	2000
Two lanes (7.0 m wide)	2500

- The minimum width of shoulder should be 2 m. Shoulders should be properly delineated.

Table 11.11: Length of vertical curve

Sl. No.	Design speed (km/h)	Safe stopping sight distance (m)	Length of vertical curve for safe stopping sight distance (m)		Absolute minimum length of vertical curve (m)
			Summit curve	Valley curve	
1	2	3	4	5	6
1.	30	30	2.0A	3.5A	15
2.	40	45	4.6A	6.6A	20
3.	50	60	8.2A	10A	30
4.	65	90	18.4A	17.4A	40
5.	80	120	32.6A	25.3A	50
6.	100	180	73.6A	41.5A	60

Note: 1. 'A' in columns 4 and 5 is the algebraic difference in grades expressed as percentage.

2. Where the length given by columns 4 or 5 is less than that given in column 6 the later value should be adopted.

- Acceleration and deceleration lanes should be suitably provided at entrance and exit respectively. Recommended minimum and desirable lengths including taper are:

Acceleration lane – minimum length 180 m-desirable 250 m

Deceleration lane – minimum length 90 m-desirable 120 m.

- The desirable and minimum lengths of weaving sections are 300 m and 200 m respectively.
- The vertical clearance at underpass should be minimum of 5.5 m in urban areas.
- For converting fast vehicles into PCU's the following equivalency factors may be adopted

(i) Passenger car, tempo, autorickshaw   1.0

(ii) cycle, motor cycle, scooter   0.5

(iii) Truck, bus, agricultural tractor trailer unit   3.0

The type of the interchange to be used depends upon

- Physical conditions of the site
- Available right of way
- Land use and development in area surrounding the roads
- Expected traffic volumes

- Volume of traffic turning and their composition and
- Orientation of intersecting roads etc.

11.9 PARKING FACILITIES (PROVISION FOR STATIONARY VEHICLES)

Parking facilities are basic component of modern transportations system. Besides everyone who drives, parking is of concern to businessmen, public administrators, workers, and shoppers as well. Parking facilities should be provided as per attitudes, values and desires of people. Parking is one of the main traffic problem in cities.

Automobile numbers and usages continue to grow, creating need for more off street parking. As parking demands rise, issues of energy and environment raise important concerns.

New parking garage technology to lift vehicles to different floors, has dramatically changed the state-of-the-art.

The space allocated to motor vehicles may be divide between space for movement and space for vehicle storage/parking. In high density or CBD areas there is a completion for space needs. For example curb parking—a problem how to apportion the space for vehicles in motion and vehicles at rest.

Growing urbanization and increasing car ownership is resulting in greater parking demand both for long term and shot-term parkers in cities. Any vehicle requires a minimum of two places for parking. One at the origin-home and other at destination-work.

11.9.1 Planning for Parking

Parking should be planned and developed as part of an overall parking program to meet the needs of an area, city or major activity centre such as airport and hospitals, etc. It is necessary to determine how much parking should be provided, where it should be located, who will benefit from it, what it will cost and how it will be financed.

Planning and development program calls for:

(i) Parking demands, assessed through different studies described earlier.

(ii) Knowledge of existing transportation and parking policies and

(iii) Formulating and evaluating alternative planning plans.

Contemporary issues of environmental impact, energy conservation etc. should also be considered in individual facility development and overall parking programe.

Parking facility development involves a wide range of interests:-

The motorist want to park near the intended trip destination spending minimum time in walking, full safety of his vehicle at a minimum or no cost.

The Retailer sees parking as an economic necessity as near by parking is pre requisite to retailer.

A pedestrian thinks parking along main shopping street undesirable as it may impede walking path

Among government workers different people think differently. Some like parking in city area, others like central area for parking.

Bus riders need parking close to their working place or outlaying sites from where park and ride facility is available to their destination.

Balancing these conflicting requirements calls for careful study of long term and short term impacts, in providing parking facilities.

Parking studies are pre requisite to planning and developing new facilities or improve/expand existing facilities. Parking studies are described in chapter 10. These studies provide.

(i) Inventory of existing parking spaces.

(ii) Current level of usage of existing parking space (accumulation and space turnover).

(iii) Knowledge of parking characteristics e,g duration, purpose, trip destination and walking distances to destinations.

(iv) Help in quantifying demands, needs and determine.

- How many spaces are needed under present conditions?
- How many spaces will be needed in future?
- Where should additional space be located?
- What type 'Short Term' or 'Long Term' parking are needed and what are their characteristics?
- What parking rates shall be realistic?
- Any other special considerations to be taken into account.

(v) Financial and economic estimates for justifying the overall parking improvement plan through cost benefit analysis.

New parking facilities may attract parkers diverted from other areas, and new parkers who changed their travel mode with improvement in income level.

Planning, location and design of parking facilities involves balancing economic, engineering, environmental and land use requirements. Figure 11.34 shows typical supply-demand analysis.

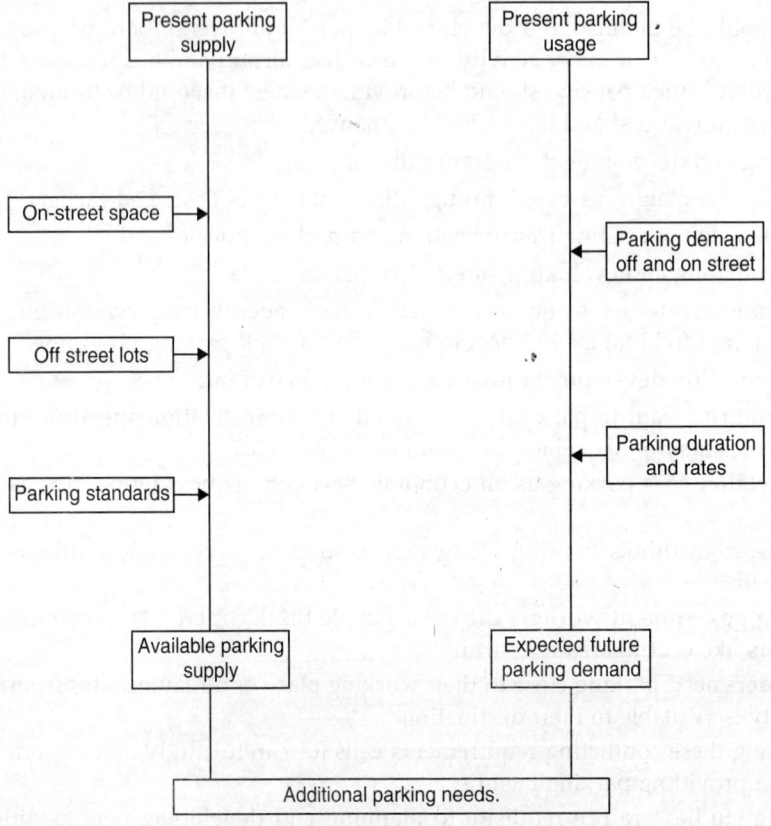

Fig. 11.34: Parking supply–demand analysis

11.9.2 Types of Parking

For supplying the parking space, following facilities are provided:

1. **On street (curb) parking:** Reduces the effective width of the road.

   (a) Unrestricted curb parking—liked by parkers.

   (b) Restricted curb parking—controlled by police or parking meters.

2. **Off street parking facilities:** Vehicles parked away from curb

   (a) Parking garages:

      (i) above ground

     (ii) underground

    (iii) integral

   (b) Surface parking lots

Garages can be classified for inter floor travel as

 (i) Mechanical elevators and

(ii) Ramps

By method of operation parking facilities can be one of the following types:

 (i) Attendant parking: where there are attendant to do or help in parking.

(ii) Self parking : parking is done by vehicle owner himself.

From ownership point of view parking facilities may be:

  (i) Privately owned and operated

 (ii) Publicly owned and privately operated

(iii) Publicly owned and operated

On street (curb parking), if not properly planned and designed results in traffic congestion, spoils aesthetics of the road, and creates unsafe traffic conditions. This reduces the effective capacity of the road when congestion increases. On street parking is removed totally or partially, to meet capacity needs, as traffic increases.

Off-street parking requires vehicles to be parked away from the curb, and does not create any problem on the road. It is therefore safe and economical to provide parking spaces away from the roads.

11.9.3 Parking Standards

Regulations require that new or reconstructed buildings should have adequate parking spaces, within their premises. The desirable parking spaces for different type of land uses as recommended by IRC are given in Table 11.12. Table 11.13 showing parking standards of various metropolitan cities in India, adopted in view of demand and supply of different building activities. The parking demand is changing in nature from time to time and therefore standards vary depending upon density of development and concentration of activities.

Table 11.12: Land used and recommended parking spaces (IRC)

S.No.	Land use activity	Number of parking spaces/area to be provided
1.	**Residential Houses**	
	Plot size 300 to 500 sq. m.	1/3 of open area
	Plot size 50 to 1000 sq. m.	1/4 of open area
	Plot size 100 sq. m. or more	1/6 of open area
2.	**Flats**	
	For every two flats of 50 to 99 sq. m area	1 space
	For every single flat of 100 sq. m or more area	1 space
3.	**Special, costly developed areas (Flats)**	
	Flat area 50 to 100 sq. m	1 space
	Flat area 100 to 150 sq. m	1 1/2 spaces
	Flat area above 150 sq. m	2 spaces
4.	**Multistoried, group housing schemes**	
	For every four dellings	1 space
	Demand may be more in big cities, so provision increased accordingly.	
5.	**Offies**	
	Every 70 sq. m floor area	1 space
6.	**Industrial premises**	
	Up to 200 sq. m floor area for every 200 sq. m or fraction additional area	1 space
7.	**Shops and markets**	
	Every 80 sq. m floor area	1 space
8.	**Restaurant**	
	Every 10 seats	1 space
9.	**Theatres and Cinemas**	
	Every 20 seats	1 space
10.	**Hospitals**	
	Every 10 beds	1 space
11.	**Hotels and motels**	
	Five and four star: every 4 guest room	1 space
	Three star: every 8 guest room	1space
	Two star: every 10 guest room	1 space
	Motels: for each guest room	1 space

Table 11.13: Parking standards (IRC)

Land use activity parking standards in terms of car space	*Land use activity parking standards in terms of car space*
RESIDENTIAL	**COMMERCIAL**
Detached, semi-detached, row house	1. *Offices*
Based on Plot area	For every 70 sqm – 1
• Upto 100 sqm – No private/community parking	2. *Shops/Markets*
• 101–200 sqm – 1 community parking	For every 80 sqm – 1
• 201–300 sqm – 2 community parking	3. *Restaurants*
• 301–500 sqm – min 1/3 of open area earmarked for parking	For every 10 seats – 1
• 501–1000 sqm – min1/4 of open area	4. *Hotels*
• >= 1000 sqm – min 1/6 pf open area	*For 5 & 4 star hotels*
Flats	4 guest rooms – 1
• 50–90 sqm – 1	*For 3 star hotels*
• >–100 sqm – 1	8 guest rooms – 1
	For 2 star hotels
	10 guest rooms – 1
Special, costly developed area	5. *Motels*
• 50–100 sqm – 1	10 guest room – 1
• 100–150 sqm – $1\frac{1}{2}$	6. *Theatres/Cinema*
• > 150 sqm – 2	Every 20 seats – 1
	7. *Hospitals*
	10 beds – 1
Multistoried group housing	8. *Industrial Premises*
• 4 Dwelling units – 1	Up to 200 sqm – 1
	Every 200 sqm – 1 additional

The parking dimensions adopted are:

(i) Car parking space:

3m × 6m = 18 sqm (individual parking space)

2.5 m × 5m – 12.5 sqm (community parking space)

(ii) Truck parking space:

3.75 m × 7.5 m = 28.125 sqm

(loading and unloading berths)

(iii) For loading and unloading berths

For all development except residential warehouses/godowns:

• 500-1500 sqm 1 berth

• For every additional 1000 sqm 1 berth

(iv) *Warehouses and godowns*

• 500–1500 sqm 2 berth

• For every additional 500 sqm 1 additional

11.10 GEOMETRY OF PARKING TERMINALS

(a) Curbs parking

Curb parking is permitted in proper manner. It could be parallel to road parking or parking at an angle. Angle parking is more convenient than parallel parking, but interferes more severely with moving traffic, resulting in higher rate of accidents, and more hazardous situations. Angle parking can accommodate more number of vehicles than parallel parking in given length of the curb as shown in Fig. 11.35.

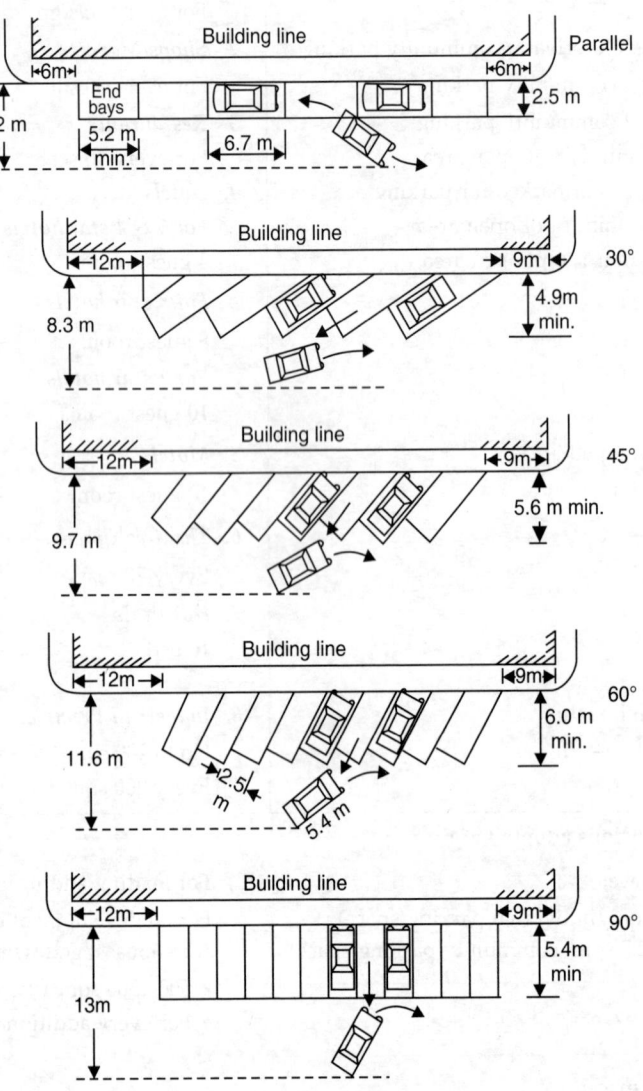

Fig. 11.35: Curb parking layouts

Angle parking should be used for roads which are wide, have good sight distance and carry low volumes of traffic. 45° parking is very commonly used as an angle parking. Parallel parking is used when width of curb parking space and width of road are limited. The right angle parking is allowed only in special situations.

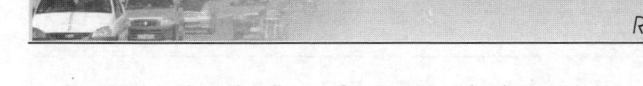

To ensure that the flow of turning vehicle is not impaired, parking near intersections should not be allowed. Curb side parking and its type is decided considering the width of road, volume and type of traffic, development along the road and type of the road. The parking bays, specially angle parking bays should be marked on the road for orderly parking.

Minimum parking space requirement for truck is 3.75 m × 7.5 m and for a car when individually parked is 3 m × 6 m and when parked in community parking lot is 2.5 × 5m.

Separate space should be provided for the taxies to park at convenient locations. A system of main taxi stand and feeder taxi stand may be used.

Buses require to be parked near the curb for loading and unloading of passengers. The minimum desirable length for 40–45 seat passenger bus (11 m long) is about 25 m for the approach side, 19 m for the departure side and 32 m for mid-block in a single bus zone. Bus stands should not be provided too close to an intersection to minimise the obstruction to other traffic. Figure 6.9 shows typical bus bays.

(b) Off street parkings

Site for new off-street parking depends upon the cost, availability of land, site size and shape and accessibility. Parking lots are convenient, where sufficient space is available at comparatively low cost, meeting the parking needs.

The selection of the best parking angle depends primarily on the size and shape of the parking lot. Figure 11.36 shows arrangement of cars at various angles in a parking lot.

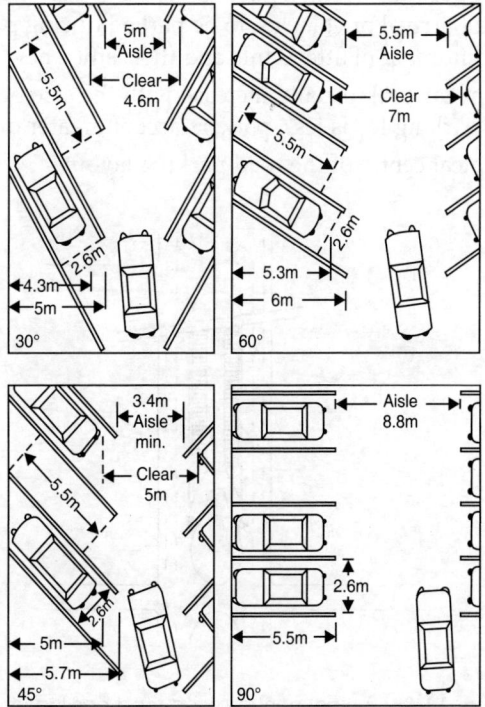

Fig. 11.36: Space and aisle requirement for garage parking at various angles

Multistoried parking garages are used when the floor area space available for parking is less and costly. Ramp or elevator provide interfloor travel facility on a multistoried car park. For ramps, the space requirement is increased, as compared to mechanized garages with elevators. (Fig. 11.37)

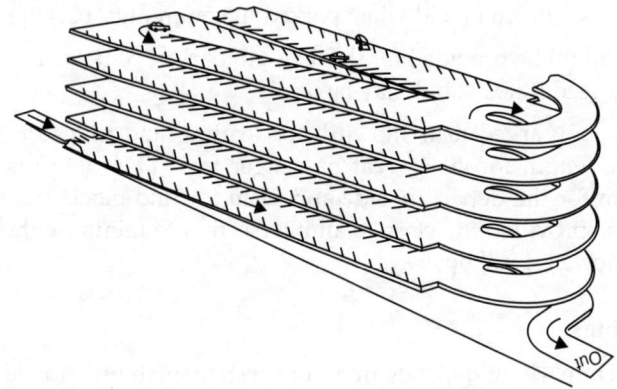

Fig. 11.37: Multistoried car park with ramp

The operations involved in using parking lot or a garage are: entrance, acceptance, storage, delivery and exit. Some space should therefore be available in front of the parking lot or garage for acceptance and exit operations. This space is called "Reservoir Area" the size of which depends on average rate of arrival of vehicles, to be parked during peak hour, the time required to handle one vehicle and number of attendants and their efficiency.

Walking distances from car park usually increase with the size of the city. It varies from 100 to 150 meters or more. If parking fee is less, parkers accept greater distances.

Figure 11.38 shows typical centre of the road parking layout

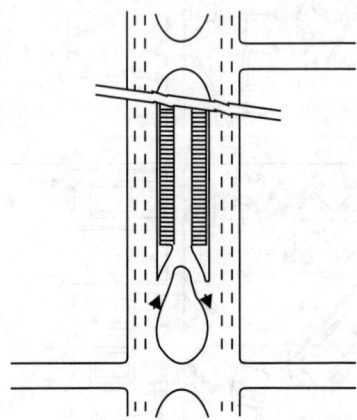

Fig. 11.38: Typical centre of the road parking layout

Typical multiple use parking concepts, used in malls and other commercial places are shown in Fig. 11.39, and a typical mechanical parking system is shown in Fig. 11.40.

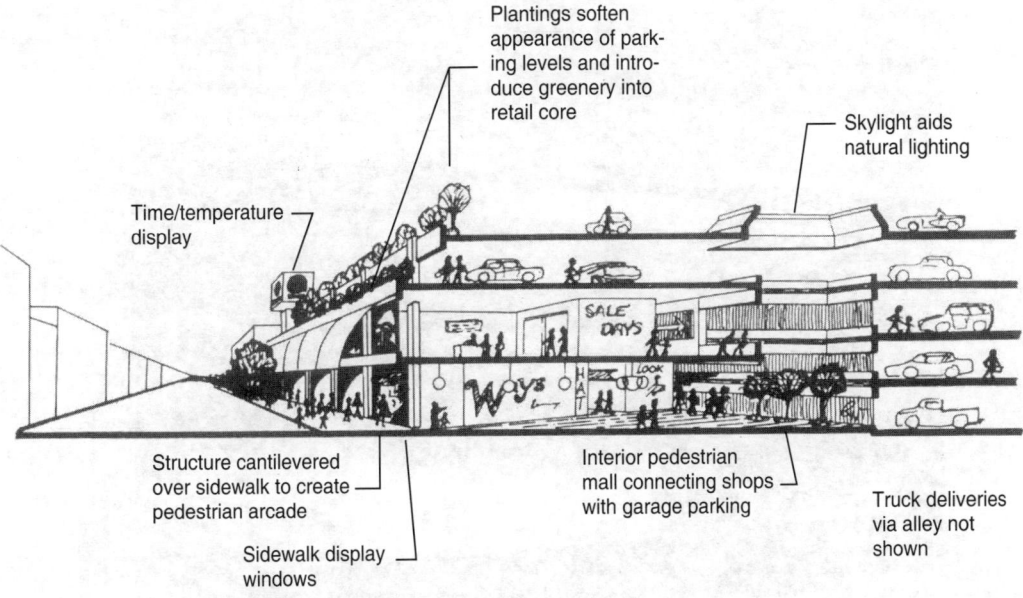

Plantings soften appearance of parking levels and introduce greenery into retail core

Skylight aids natural lighting

Time/temperature display

Structure cantilevered over sidewalk to create pedestrian arcade

Sidewalk display windows

Interior pedestrian mall connecting shops with garage parking

Truck deliveries via alley not shown

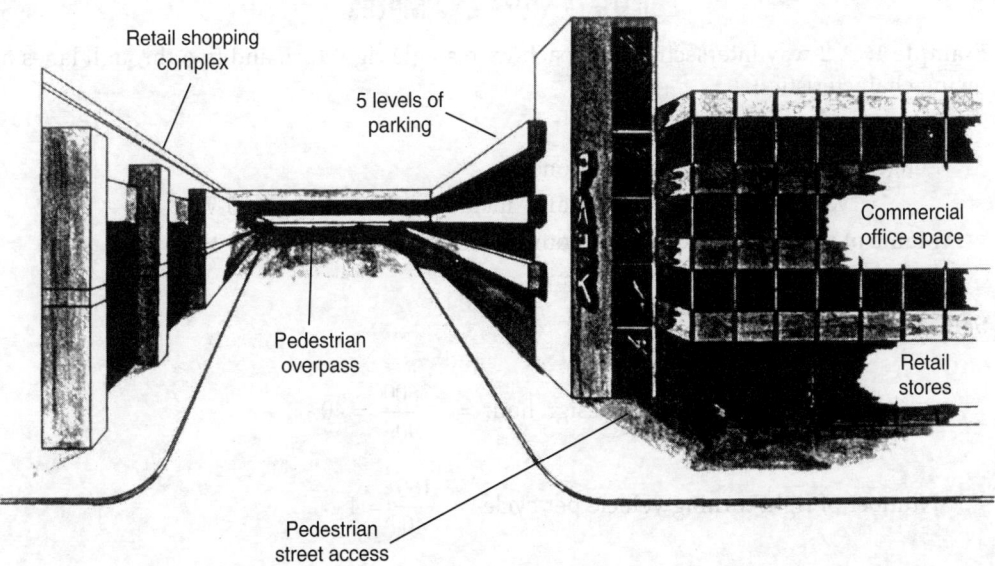

Retail shopping complex

5 levels of parking

Commercial office space

Pedestrian overpass

Retail stores

Pedestrian street access

Fig. 11.39: Typical multiple-use parking design concept

Fig. 11.40: Typical mechanical parking

ILLUSTRATIVE EXAMPLES

Example 1. A 2-way intersection approach with single right turn and two through lanes has following characteristics:

$$\text{Peak hour factor} = 0.85$$

$$\text{Length of the single cycle} = 90 \text{ seconds}$$

$$\text{Volume of right turning traffic including 10\% truck} = 160 \text{ Vph}$$

$$\text{Volume of through traffic in peak hour including 15\% trucks} = 480 \text{ Vph}$$

Design the right turning storage lane.

Solution:

$$\text{Number of cycles in the design hour} = \frac{3600}{90} = 40$$

$$\text{Number of right turning vehicle per cycle} = \frac{160}{40} = 4$$

$$\text{Number of through vehicle per cycle} = \frac{480}{40} = 12$$

i.e. 6 vehicles per lane per cycle.

Storage lane multiplying factor for P.H.F. of 0.85 can be interpolated from

PHF – .25 – multiply factor = 2.0

PHF = – 1.00 – multiply factor = 1.5

for 0.85 multiplying factor = 1.6

Adjusted right hand traffic storage factor for trucks = $1.6 \times 1.1 = 1.76$

Adjusted through traffic storage factor for trucks = $1.6 \times 1.15 = 1.84$

Length of right turn traffic storage (L)

$$= \text{Vehicles/Cycle} \times \text{Storage factor} \times \text{Space for 1 vehicle}$$

$$L = 4 \times 1.76 \times 8 = \mathbf{56.3}$$

Length of the through traffic storage = $6 \times 1.84 \times 8 = \mathbf{88.3} \approx 88$ m

Greater of the above two 88.3 is length of storage lane

Taper length can be taken as 30 m. Total length = 30 + 88 m

$$\approx 118 \text{ m}$$

Example 2. An existing roundabout has a uniform weaving section, having following dimensions.

Width (w) = 15 m

Length (l) = 50 m

Average entry width (e) = 8 m

Future design hourly volume (DHV) = 2400 PCU per hour. Find out if the existing roundabout shall be capable of handling this volume. Proportion of weaving traffic is 70 percent.

Solution:

The practical capacity of the roundabout $Q_p = \dfrac{280w\left(1+\dfrac{e}{w}\right)\left(1-\dfrac{P}{3}\right)}{1+\dfrac{w}{L}}$

$$= \dfrac{280 \times 15\left(1+\dfrac{8}{15}\right)\left(1-\dfrac{.70}{3}\right)}{1+\dfrac{15}{50}}$$

$$= 3806 \text{ pcu}$$

Since future requirement in only 2400, this roundabout shall be able to handle this DHV.

$$\text{Reserved capacity} = \dfrac{\text{Practical Capacity - Required Capacity}}{\text{Required Capacity}}$$

$$= \dfrac{3806-2400}{2400} \times 100$$

$$= \mathbf{58\%}.$$

Traffic Control Devices

12.0 PREAMBLE

Traffic control devices are used to improve the overall traffic conditions in most convenient, safe and economic way. Commonly used traffic control devices are signs, markings, traffic islands, signals, etc. These devices regulate, warn, control and guide the traffic in proper and safe use of the roads and intersections. They indicate traffic rules and regulations such as speed control, parking and other prohibitions, etc. warn of hazards which are not otherwise self-evident, and convey other useful information to road users. Graphic symbols are commonly used to convey messages rather than words, so that drivers from other countries travelling by roads can easily understand their meaning. Efforts are made to adopt international standards as far as possible however, there is no worldwide uniformity of traffic control devices. Uniform and standard control devices are easily recognised and understood by the drivers, enabling them to take the right course of actions. It reduces confusion and decision-making becomes simple. Manufacture, installation and maintenance of standard devices is easy and economical. In India uniformity and standards of traffic control devices are governed by a manual of Indian Road Congress, Delhi.

Most widely followed International document is International Convention on Road signs and signals produced by United Nations in Geneva, in 1968.

12.1 FUNCTIONS AND REQUIREMENTS OF TRAFFIC CONTROL DEVICES

Table 12.1 shows some of the basic functions performed by the traffic control devices, and traffic regulations.

Requirements of Traffic Control Devices

Traffic control devices should be well designed, adequately maintained, to perform the intended functions. The installation of these devices should be such that:

Table 12.1: Main functions of control devices

Control device	Main functions
Traffic signs (such as stop, No parking, etc.)	• Compensate for highway deficiencies • Provide effective environment. • Guide the road users and provide information. • Warn about abnormal driving conditions ahead. • Supplement or modify the basic rules of the road, at specific time or locations. • Help road users in unfamiliar areas.
Pavement markings:	• Channelise traffic in proper lane. • Separate opposing flows of traffic. • Show the points of restricted sight distance. • Define pavement edges, sidewalks, cross walks and stopline to guide the traffic. • Steer the motorist away from obstructions on the road. • Display regulations and warn the traffic. • Promote safety and ensure smooth flow.
Traffic islands:	• Define the path of the driver in complicated locations. • Reduce the area of traffic conflict. • Serve as an area for location of various types of signs and signals. • Direct vehicular traffic to minimise traffic conflicts. • Separate traffic streams. • Prevent undesirable movements and turns. • Provide protection to pedestrians. • Improve the efficiency of operation at intersections.
Traffic signals:	• Provide orderly movement of traffic. • Increase the intersection capacity. • Reduce the number of accidents. • Permit the vehicles and pedestrians to cross in proper manner by controlling conflicts. • Provide continuous movement of traffic at desired speed. • Assign priority of movement and right of way. • Promote driver's confidence, by reducing conflicts. • To warn of possible danger. • To control traffic at railway level crossing and bridges.
Speed controls:	• Control drivers from overspeeding. • Help in safety on roads.
Parking prohibition:	• Improves the efficiency of traffic movement. • Increases capacity of roads and intersections. • Provides safety and uninterrupted flow of traffic. • Provides adequate sight distance at intersections. • Provides additional street space for vehicle movement. • Gives priority to short term parkers by use of time limit.

 (i) Fulfill the need or functional requirements.

 (ii) Draw the attention of the users. Attention depends upon the location, size, colours etc. They should be placed within the cone of clear vision of road user.

 (iii) Convey a simple, clear and only one meaning, at a glance.

 (iv) Road users for which they are meant, respect them. This will be achieved if they are designed as per requirement and need.

 (v) Allow adequate time for proper response. They should be placed adequately ahead of the place where action is required, to ensure proper action with confidence.

 (vi) They are uniform throughout. Similar devices should be used for similar situations, to avoid confusion.

 (vii) Traffic control devices should not conflict, but supplement each other.

 (viii) The unnecessary signs should not be used; only the minimum required devices should be used to avoid any confusion.

 (ix) Uniform system of reflectorisation or illumination should be used for night use.

 (x) Traffic control devices should be put by an authority responsible for traffic rules and regulations.

 (xi) In urban areas, signs should be placed sufficiently high for them to be seen by over parked vehicles or obstructions.

12.2 TRAFFIC SIGNS AND THEIR CLASSIFICATION

Traffic signs are passive, visual traffic control devices mounted on a fixed or portable support whereby specific message is conveyed by means of symbols or words, officially erected for the purpose of regulating, warning or guiding the traffic. Traffic signs communicate the essential instructions which the road user is required to follow, warn about the hazard, and provide information about routes, destinations and points of interests. Signs are read by drivers of moving vehicles therefore they should be simple, large enough, suitably coloured, so that drivers can easily understand the message conveyed by them. It should be located sufficiently in advance to allow a driver to properly detect, read, understand and then act on the message on that sign.

Classification of Signs

Signs can be classified in three main categories described below:

12.2.1 Mandatory (Regulatory or Prohibitory Signs)

These signs inform the road user about prevailing laws and regulations. Violation of such signs is a legal offence. They are also called 'Regulatory' or 'Prohibitory' signs. Prohibitory signs are used to disallow parking, entry, turns, overtaking, etc. IRC specifies circular shape, with a diameter of 60 cms as standard size and 40 cms for the reduced size. These signs have a red border. For speed control the colour of the background is white, for parking restrictions and other signs it is blue. The symbols are in black colour for prohibitory signs and white in colour for direction control signs.

Two important mandatory signs which require to take positive action by the drivers are 'STOP SIGN' and 'YIELD (GIVE WAY) SIGN'. Stop signs mean that all vehicles must stop, before the line, irrespective of the traffic on the other road. As per IRC standards an octagon with a white border and red background is used for stop sign. The side of the octagon should be 90 cm for the standard size and 60 cm for smaller size.

The 'YIELD' or GIVE WAY' sign is used to give right of way to traffic on other road. Vehicle controlled by a give way (yield) sign need to stop only when necessary, to let the other vehicles pass, IRC standards recommend downward pointing equilateral triangle having a red border and white background for these signs. The side of the equilateral triangle is 90 cms long in standard size and 60 cm long in smaller size. It is used in combination with a definition plate written 'GIVE WAY'.

Figures 12.1 and 12.2 show stop and yield signs and Fig. 12.3 shows typical mandatory signs.

Fig. 12.1: Stop sign (IRC)   **Fig. 12.2:** Give way or yield sign (IRC)

Left turn prohibited   Right turn prohibited   U-turn prohibited   Overtaking prohibited

Straight prohibited no entry   Vehicles prohibited both directions   Cycle prohibited   Motor vehicles prohibited

Horn prohibited   No stopping   No parking   Speed limit

Height limit   One way signs   Direction sign

Direction sign   Restriction ends   Cycle track   Load limit   Width limit

Fig. 12.3: Typical mandatory signs (IRC)

12.2.2 Warning (Cautionary Signs)

These signs warn the road user about the existence of certain hazardous condition on or adjacent to the road. Conditions that otherwise would not be immediately apparent. They are very important from safety point of view. Equilateral triangle is used to indicate these signs. As per IRC the side of the triangle is 90 cm for a standard size and 60 cm for a reduced size. These signs have a red border and the indicating symbol in black colour against a white background.

Typical warning signs are shown in Fig. 12.4a and b.

12.2.3 Informatory (Guiding Signs)

These signs are for information and guidance of road users and call attention to facts which might be of interest to them. They guide them to their destinations, and help in identification of different important feature along the route such as rivers, historical places, cities/towns, parks, etc.

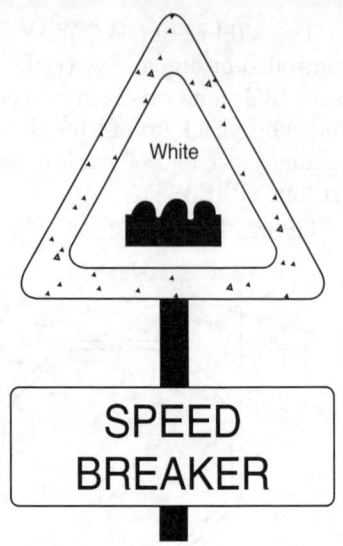

Fig. 12.4: (a) Typical warning sign

Fig. 12.4: (b) Typical warning signs (IRC)

The size of these signs should be big enough so that message can be read easily from the moving vehicles. These signs are rectangular in shape. IRC standards suggest a size of 60 cm × 45 cm with a black symbol, against a white rectangle and blue background.

Informatory signs can be sub classified as indication signs as they show location of hospitals, rest houses, filling station, telephones, etc. Figure 12.5 shows typical information signs.

Taxi stand

Rest house

First aid

Petrol pump

Hospital

Light refreshment

Parking this side

Scooter and motor cycle stand

Cycle stand

No through side road

Parking both sides

No through road

Public telephone

Eating place

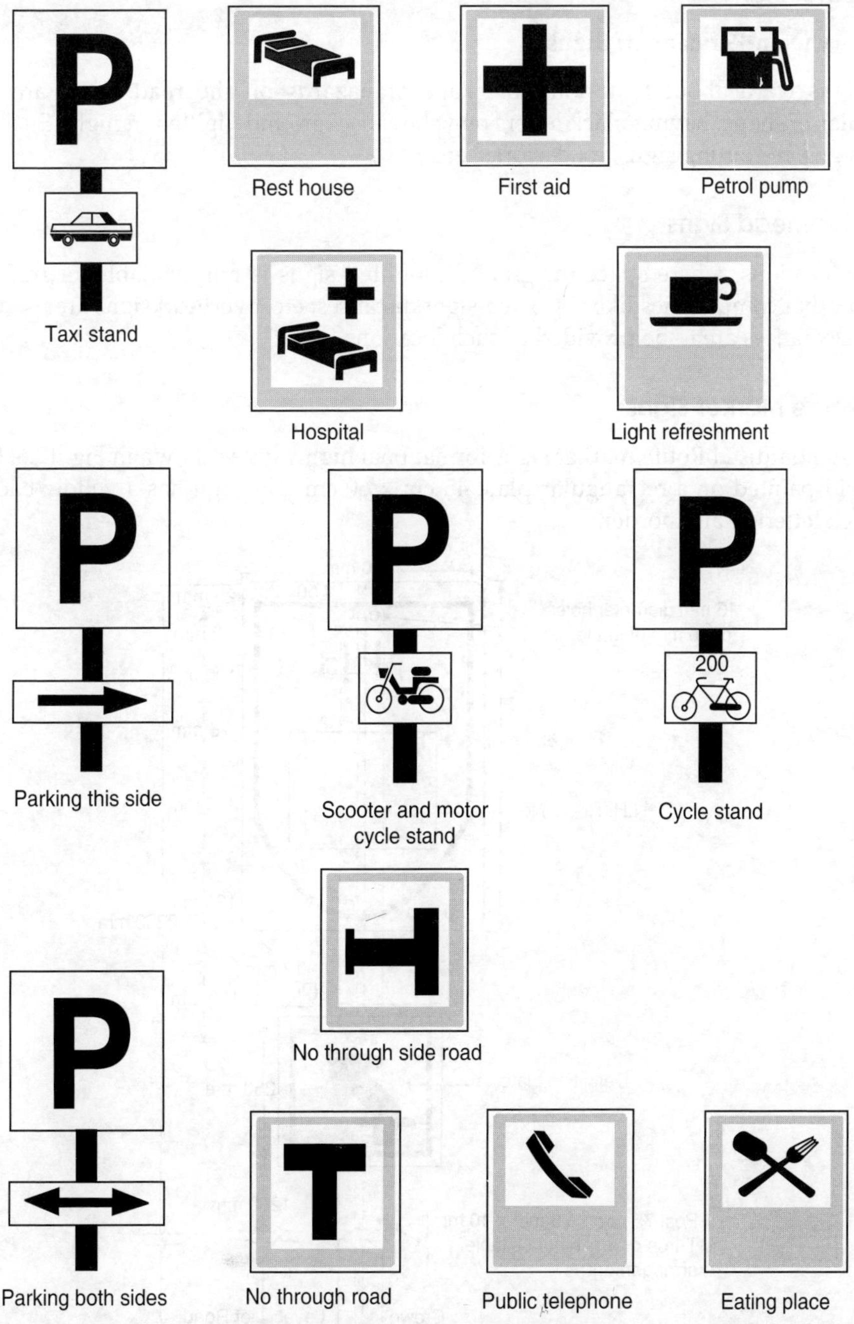

Fig. 12.5: Typical information signs (IRC)

Direction signs provide information about a place or city, its direction and distance away. They are rectangular in shape, an arrow head shows the direction. Distances on such destination signs indicate how far is the place from the sign. At some locations, an advance direction sign may be necessary to indicate intersection, bridge, narrow road, etc. ahead.

12.2.4 Other Signs

Temporary (Non-Permanent) Signs

These signs warn about temporary blockage or hazards on the road. They are generally rectangular in shape, having black letters on yellow background, lighted in night. Such signs are service signs, re-routing signs, road-works, etc.

12.2.5 Overhead Signs

At some locations, where space for ground mounted signs is not available, or traffic is very heavy or other complexities like restricted sight distances, etc. overhead signs are used. Suitable vertical clearances are to be provided at such locations.

12.2.6 Route Marker Signs

IRC has standardised Route marker signs for national highways as shown in Fig. 12.6. It consists of a shield painted on a rectangular plate 45 cm × 60 cm. The sign has a yellow background, with black lettering and border.

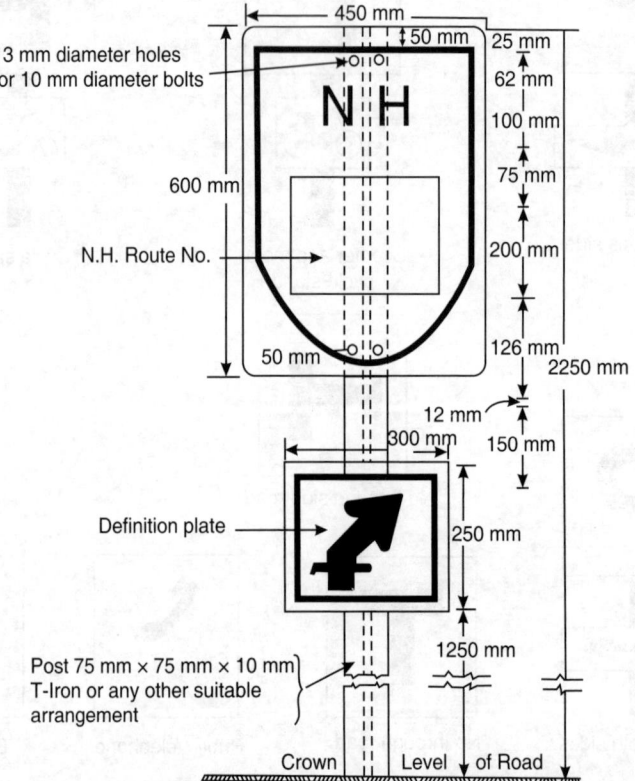

Fig. 12.6: Route marker sign for National Highways (IRC)

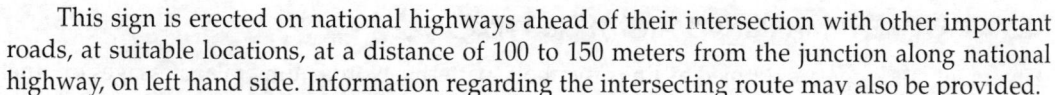

This sign is erected on national highways ahead of their intersection with other important roads, at suitable locations, at a distance of 100 to 150 meters from the junction along national highway, on left hand side. Information regarding the intersecting route may also be provided.

12.3 PLACEMENT, HEIGHT AND MAINTENANCE OF SIGNS

The distance at which a message can be read or recognised is called legibility distance. The distance at which message can be read at once is known as glance distance. The legibility distance is influenced by background colour, height, width, stroke, colour, spacing and number of letters, sign margin and general layout of the sign. The signs should be placed within driver's normal cone of vision (10° in horizontal and 5° in vertical plane). Signs are placed on left side of the road in India. Signs are mounted on posts, or provided overhead. They are repeated on other side of the road if necessary.

As per IRC the signs should be erected not less than 60 cm away from the edge of the curb or at a distance of 2-3 m from the carriageway edge if curbs are not provided. A height of 1.5 m is used for un-curbed roads and 2 m for curbed roads.

Following locations are prescribed for stop, give way and warning signs.

'STOP' signs should be provided 1.5 to 3 m ahead where vehicle is to stop. For pedestrian crossings stop sign should be erected 1.2 m before the pedestrian crossing stop line.

'GIVE WAY' signs should be located 1.5 to 3 m, before the stop line of vehicles. If markings are provided for stop line. 'GIVE WAY' sign should be erected 1.2 m advance of the marking.

'WARNING' signs are to be located at distances in advance of the hazard, as given in Table 12.2.

To compel attention, the signs must be clean, legible, free from obstructions and maintained to high standards. They should be regularly cleaned. Signs should be designed and used in accordance with standards and manuals of the particular country, which are based on several years of experience with traffic operation under conditions normally encountered. Signs should be regularly painted for legibility and damaged signs should be replaced. Maintenance of signs should be the part of routine maintenance of roads. IRC 67, 30, 31, 26, 25 provide design standards, sizes, colours, etc. for signs, marker, and stone indicators followed in India.

Table 12.2: Location distance of warning signs (IRC)

Type of Road	Plain and Rolling terrain (m)	Hilly or Mountainous terrain (m)
In Non-urban areas:		
National highways or state highways (NH) (SH)	120	60
Major district roads (MDR)	90	50
Other district roads (ODR)	60	40
Village roads (VR)	40	30
In urban areas	50	–

12.4 PAVEMENT MARKINGS AND THEIR TYPES

Pavement markings are defined as lines, symbols, patterns, words, numerals, messages or other devices set in the pavement, or applied or attached to the pavement or curb to control, warn, guide or inform the road users. They convey the message without diverting the attention from the road.

White or yellow colours are generally used for markings, to provide strong contrast between pavement and markings. Raised pavement marker or reflective or non-reflective type are also used for markings.

The limitations of the pavement markings are that they may become dusty, invisible on wet surface, wear out by traffic use, and non-usable on unpaved roads. Regular maintenance of markings is therefore essential. They can not be applied on unsealed roads. They can be obscured by traffic.

The major advantage of markings is that they convey continuous information within the driver's direct field of vision. As with traffic signs discussed earlier, the basic requirement for a pavement marking is that drivers should be able to interpret its meaning in sufficient time to properly react to its message. Markings are applied either on the pavement or on the nearby objects such as piers, curbs, islands, abutment, etc.

Types of Pavement Marking

Pavement markings are generally of following types:

12.4.1 Longitudinal Lines

The regulations specify that vehicles may cross a single broken line, single solid line gives warning and double unbroken line should not be crossed, longitudinal lines are used as:

- Lane lines
- Barrier lines—broken or unbroken
- Separation lines—single broken line
- Guidelines
- Continuity lines—broken lines
- Edge lines
- Centre lines

12.4.2 Transverse Lines

These are marked across traffic stream and are generally associated with traffic controls.

Transverse lines are used for:

- Stop lines: Points behind which vehicle must stop.
- Holding lines: Safe position for a vehicle to be held, at a giveway sign, etc.
- Pedestrian crossing marking: Comprising of Zebra crossing, cross-walk, etc.

12.4.3 Other Markings

Other types of markings commonly used are:

Turn lines: Indicating proper course to be followed by turning vehicles.

Give way lines: Broken double lines at intersection entrance.

Zebra crossing: Zig-zag lines on pavements reserved for use of pedestrians.

Directional arrows: Indicating correct direction for lane usage.

Ghost island: Islands painted on the pavement in white (or hatched in white) for minimising conflict, separating traffic streams, prevent undesirable movement, protect pedestrains, protect traffic control devices, show medians, etc.

Worded markings: Such as 'Stop' 'Keep Clear' 'No Parking' and other words and numerals for guiding, warning or regulating the drivers.

Parking control line: Yellow lines marked along edge of the road.

Road Delineators: These are the markings used for safety such as on bridge supports, level railway crossings, culvert walls, etc.

The other type of markings include "No Overtaking" zones, 'Obstruction Approach', "Cycle Crossing" "Bus-Stop", etc.

12.4.5 Common Pavement Markings

(a) Centre lines

(i) Single broken line: Marking the centre of the road.

(ii) Solid and broken line: Permitting crossing of centre line by vehicles from broken line side only if safe to do so.

(iii) Double solid line: Prohibiting any crossing of centre line by vehicles except in emergencies. (Non-overtaking zones).

(iv) Traffic lane lines: Separating the traffic lanes.

(v) Pavement edge lines: Showing edges of the pavement.

(vi) Pavement width reduction lines: Showing that pavement is going to be narrower ahead.

(vii) Obstruction approach marking: Show any obstruction which is ahead.

(b) Transverse (edge) line

(i) Solid lines: Indicating bands or other hazard.

(ii) Wide gap broken lines: Defining edge of the pavement.

(iii) Narrow gap broken lines: At intersections and laybyes.

(iv) Stop lines: Solid double lines at intersection entrance.

(v) Give way lines: Broken double lines at intersection entrance.

(vi) Zebra crossing marking: Zig-zag lines prohibiting parking and overtaking.

(vii) Chevron markings/Ghost islands: Painted white and hatched in white, used at divergings/merging lanes.

(viii) Pedestrian/cycle crossing markings: Where pedestrian/cyclist can cross.

(ix) Intersection approach markings: Near intersection.

(x) Bus stop markings: Indicating locations where buses can stop.

(c) Parking lines

Normally yellow in colour.

(i) Broken lines: Some restrictions-short time parking only.

(ii) Single solid lines: No parking for certain time.

(iii) Double solid line: No parking any time.

A broken line warns, solid line prohibits and double line imposes greater restrictions on parking.

Figures 12.7 and 12.8 show typical pavement markings.

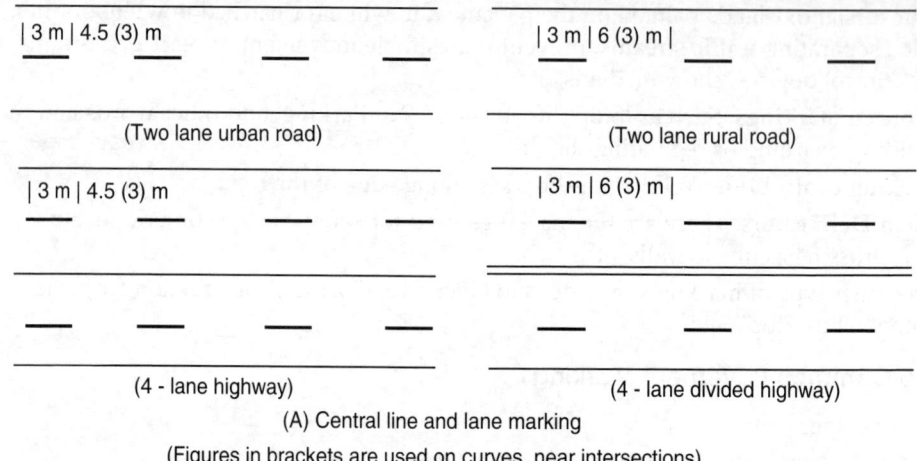

(A) Central line and lane marking

(Figures in brackets are used on curves, near intersections)

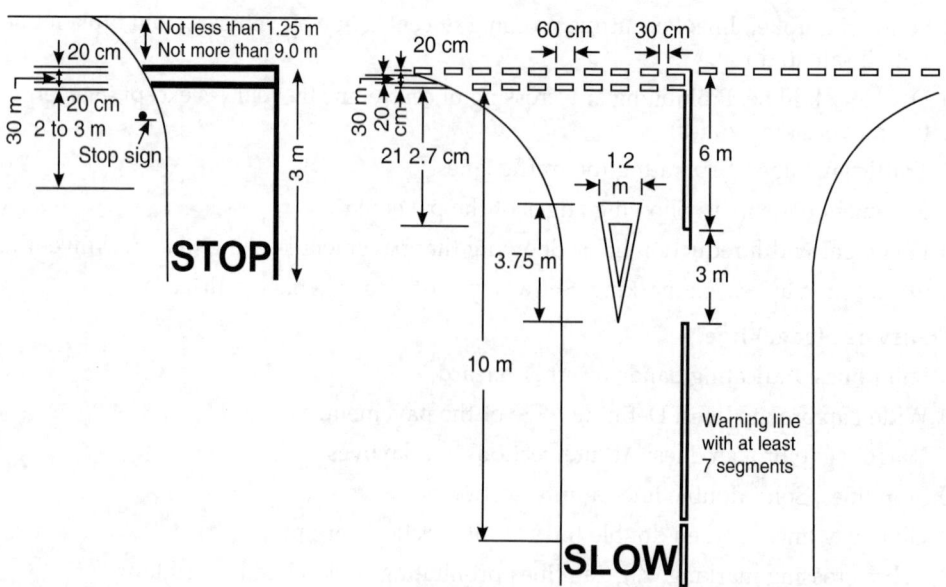

Fig. 12.7: Typical pavement markings

12.5 MATERIALS AND COLOURS FOR PAVEMENT MARKINGS

The materials used for pavement markings are:

(a) Paints-as specified by Indian Standards

(b) Thermoplastics with or without reflective properties

(c) Precut sheeting with or without reflective properties.

(d) Raised pavement markers which may be retroreflective or non-retroreflective. These are studs of plastic, metal set into or stuck to road surface.

Pavement markings are white except that yellow may be used for the unbroken portion of barrier lines and for restrictive parking areas.

Alternative bands of white and black are used on roadside curbs or other objects.

Sizes of pavement lines are specified in code of practice for road markings, Indian Road Congress IRC 35.

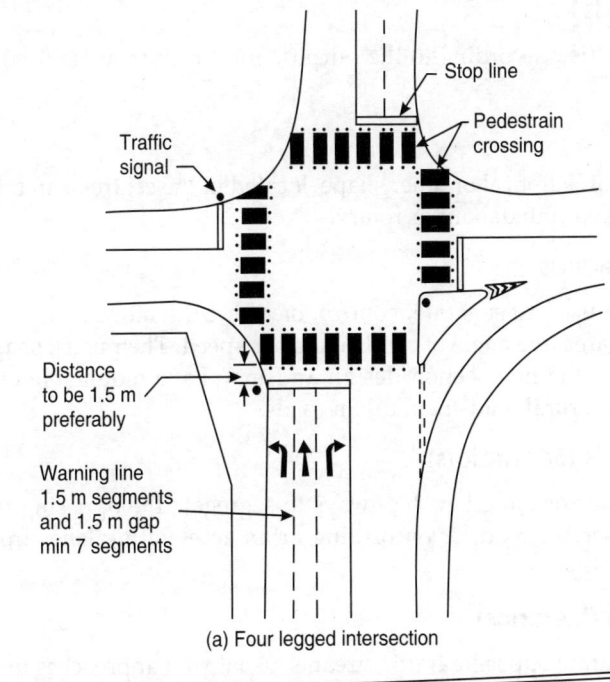

(a) Four legged intersection

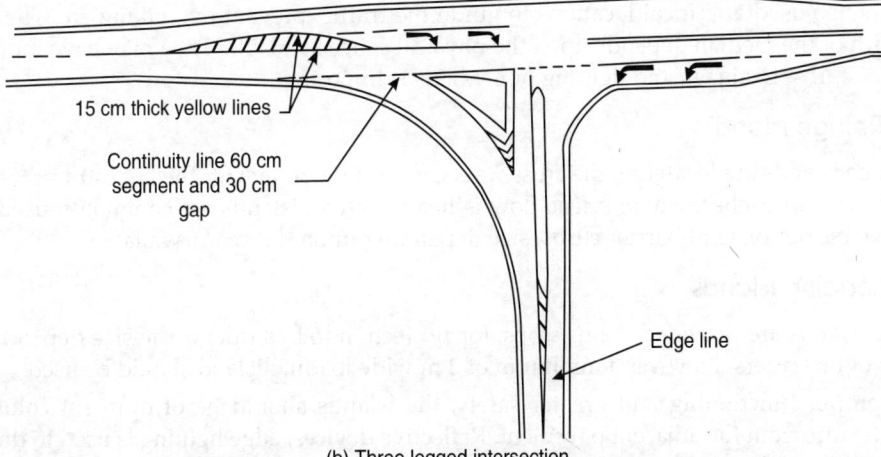

(b) Three legged intersection

Fig. 12.8: Typical markings at intersections

12.6 TRAFFIC ISLANDS AND THEIR CLASSIFICATION

Traffic islands are important form of pavement marking whose functions are to:

  (i) Direct and separate traffic to minimise conflict

 (ii) Separate traffic streams

(iii) Prevent undesirable traffic movement

 (iv) Protect pedestrians

  (v) Protect traffic control devices

They may be constructed by painted markings or alternatively by unsealed areas of pavements, safety bars, coloured pavement material or raised areas. Mountable curbs are used for raised areas except where the island is intended as a pedestrian refuge.

Classification of Islands

Islands are usually classified according to the purpose for which they are used. Following are important types:

12.6.1 Roundabout

A central island is circular, elliptical or other shape, located in the centre of an intersection, used for channelisation, is called roundabout or rotary.

12.6.2 Channelising Islands

Channelised islands are used to separate, control, or direct the movement of traffic streams. They are used to reduce area, the angle of conflict and the speed. Their usual shape is triangular. They are integral part of the intersection design and may have mountable or barrier curbs. Larger islands are used on rural roads than urban roads.

12.6.3 Divisional Islands (Separators)

They are provided on four or more lane highways to segregate the opposing traffic. They are used to separate road, service road, or on turning or an accelerating lanes from the through traffic lane.

12.6.4 Median Islands (Medians)

Medians are used to separate opposing traffic streams, usually on approaches to an intersection. They are also used at critical locations to guide the traffic, prevent overtaking etc. The length and width of the median depends upon the physical space available. They may have mountable or barrier curbs. Drainage and lighting of medians is important.

12.6.5 Refuge Island

The refuge islands are located in the cross walks, to provide refuge for the pedestrains. On wide intersection approaches, where traffic flow is heavy, refuge islands are commonly used. They may have barrier or semi-barrier curbs, size depending upon the space available.

12.6.6 Loading Islands

Loading islands are provided at bus stops, for protection to bus riders. The size depends upon number of bus riders, however a minimum of 2 m wide loading island should be used.

For proper functioning and greater safety, the islands should be of different colour and surface texture from the adjacent pavement. Reflective devices, edge lighting is used to delineate the islands for night use.

12.7 ROAD DELINEATORS

During the night and bad visibility conditions like fog, rain, snow, etc. road delineators provide visual assistance to drivers in knowing road conditions ahead.

Delineators are devices which show the roadway or obstructions and could be raised pavement markings, painted lines, sign posts, post-mounted reflectors or contrasting surfaces, etc. They are retro-reflective type units/devices.

Delineators are roadway indicator, hazard marker or object marker. They are driving aids in addition to signs, markings, barriers, etc. to provide extra alert to drivers.

Normal locations on which delineators are used, include horizontal and vertical curves, narrow bridges/culverts, hilly road, approach to intersections, tunnels, poor visibility sections of road, at hazardous locations, dead ends, etc.

Reflector are made from films, synthetic materials like plastic, glass etc. with stable optical characteristics. The delineators should be so positioned that the reflectorised or painted side is perpendicular to the direction of travel and clearly visible. They should be well maintained and periodically renewed. Figure 12.9 shows typical road delineators.

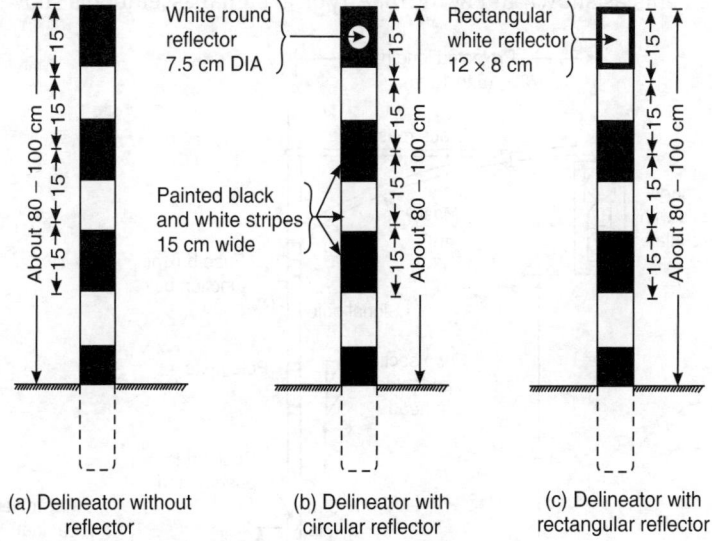

Fig. 12.9: Typical delineators

12.8 TRAFFIC SIGNALS AND THEIR TYPES

The points of conflict at an intersection can be further reduced by the use of traffic signals. Traffic signals are devices which by means of changing coloured lights, regulate the movement of traffic by separating the traffic flows in time. Traffic signals provide for the orderly movement of traffic by alternately allowing one stream of traffic to enter an intersection, then another from

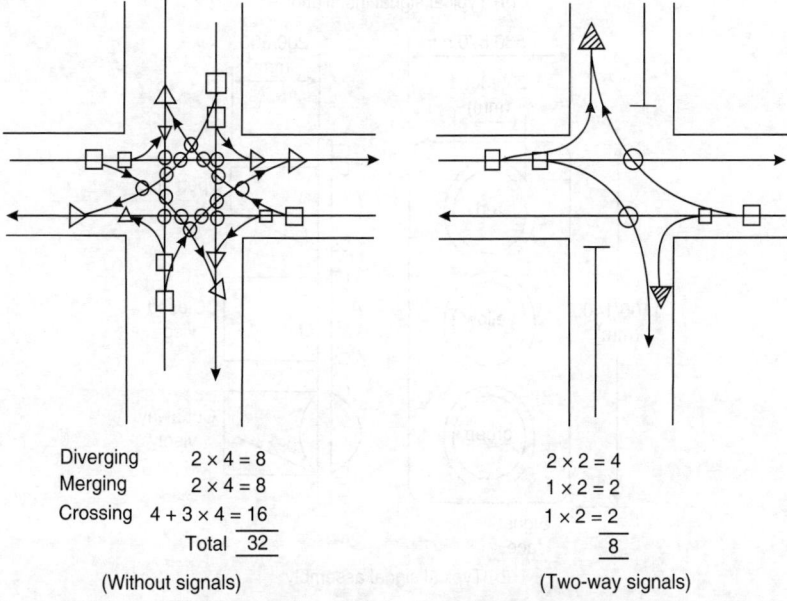

Diverging	$2 \times 4 = 8$	$2 \times 2 = 4$
Merging	$2 \times 4 = 8$	$1 \times 2 = 2$
Crossing	$4 + 3 \times 4 = 16$	$1 \times 2 = 2$
Total	$\underline{32}$	$\underline{8}$

(Without signals)   (Two-way signals)

Fig. 12.10: Reduction of conflicts by signals

an intersecting direction and thus they reduce the right angle collisions. Signals control the speed and provide continuous movement through a series of signals timed for progressive movement along the road. Traffic signals are therefore used at heavily loaded intersections to control traffic, to minimise delay and to provide safety to traffic and pedestrians by reducing the conflicting movements as shown in Fig. 12.10. A typical signal assembly is shown in Fig. 12.11.

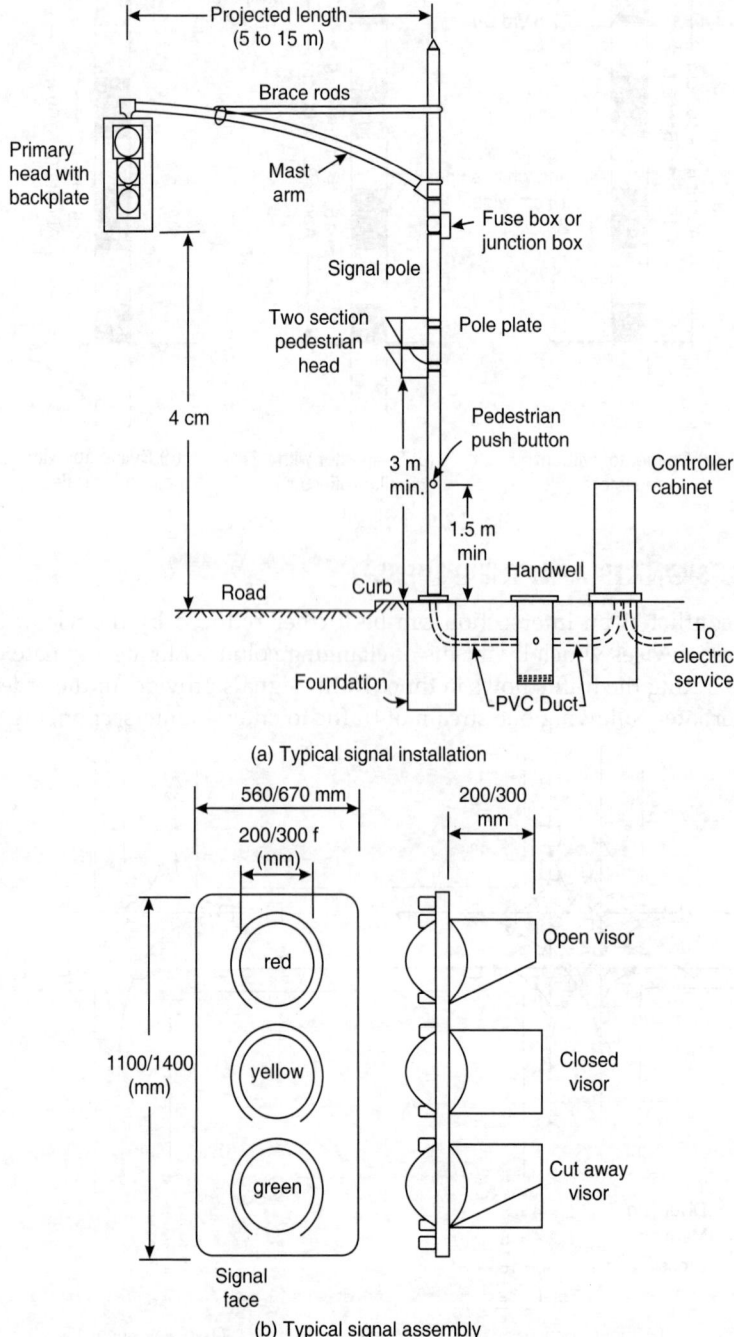

(a) Typical signal installation

(b) Typical signal assembly

Fig. 12.11: Typical signal hardware

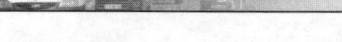

Advantages of Traffic Signals

Traffic signals offer following advantages:

- Alleviate excessive delays at stops.
- Eliminate the problems caused by turning traffic
- Streamline the traffic and provide orderly movement of traffic
- Pedestrian accidents are avoided.
- Capacity of an intersection is fully utilised.
- Angle collisions are reduced or eliminated.
- Manual control by police or traffic personnel is not required.
- Better speed along the route can be obtained through coordinated signals.
- It reduces overall cost and driver's annoyance.

Types of Traffic Signals

12.8.1 Pretimed (Fixed time) Signals

The sequence in which the signals RED, RED-YELLOW, GREEN, YELLOW, as they appear on each approach of an intersection, is fixed and repeated after a fixed interval in seconds. The time period for each signal light is predetermined and fixed in the signal equipment by a timing device. Traffic signals which operate in this way are called pretimed or fixed time signals.

Pretimed signals cannot take into account the fluctuations of traffic flow and so they are satisfactory where fluctuation is small. Different cycle times may be used to take care of morning peak, evening peak and non-peak hour traffic, by using different cycle lengths during those hours. The cycle length varies from 40 seconds to 120 seconds. These signals are cheap and easy to maintain.

In a progressive type of fixed time signal, all the fixed time signals along a route of progressive movement are controlled by a master controller instead of individual timing device at each intersection. The change of cycle can be made from a central control.

12.8.2 Vehicle Actuated Signals

These signals change the length of green signal interval, in accordance with the actual traffic volumes on the particular approach of the intersection. The vehicles on any intersection approach are sensed by a detecting device, e.g. magnetic tape, loops, or pneumatic pads, placed in the road. By recording the vehicles as they cross the detector and by timing the interval between vehicles, the signals are automatically adjusted to give preference to the approach with the heaviest flow.

These signals are useful when traffic flow at intersection fluctuates on a short-term basis. With heavy traffic volumes on all approaches they operate just like pretimed signals. Pedestrian protection through a 'Push Button' system on demand can be incorporated in vehicle actuated signals.

A system of signals may use 'Fixed Time System' as well as 'Vehicle Actuated System' for different approaches and can be called 'Semi-Actuated Signals'.

12.8.3 Pedestrian Actuated Signals

Signals are sometimes provided on busy urban streets for crossing of the pedestrians. They temporarily show a RED light for the traffic to stop, permitting the pedestrians to cross the road. They are justified when the volume of pedestrian traffic is so great as to cause excessive delays to vehicles. A typical hourly volume of pedestrians to cross may vary from 175 to 350 to justify use of pedestrian actuated signals. The setting of traffic signals usually allows for a pedestrian walking speed of 1.2 metres per second, followed by a pedestrian clearance period.

Sometimes additional traffic lights and cycle lengths are provided at intersections, for pedestrians to cross, when vehicles are stopped. Symbol used for such signals are 'WALK', 'WAIT' and 'DON'T WALK'.

12.8.4 Linked Traffic Signals (Co-ordinated Signals)

In congested urban areas, after passing through one intersection, the vehicle may have to wait at next two or three intersections—all within a short length of journey (one kilometre or so), which is an undesirable situation. The overall optimisation of traffic flow, through such nearby intersections, requires linkage or coordination of traffic signals so that green time is available progressively along the street to the approaching vehicles.

The green wave progression depends upon timing of beginning of the cycle at each signals so that at a predetermined speed, a vehicle arrives at each consecutive intersection at the same time at which the green phase occurs.

The signal route green wave concept can be extended to a whole network, using several signal plans, known as 'Area Traffic Control'—ATC, or 'Urban Traffic Control'—UTC, through computer programming and control. Figure 12.12 shows the 'Green Wave' traffic signal linking at four close by intersections. If the cross roads were centred at AA, the situation would not have been ideal as shown by the circle.

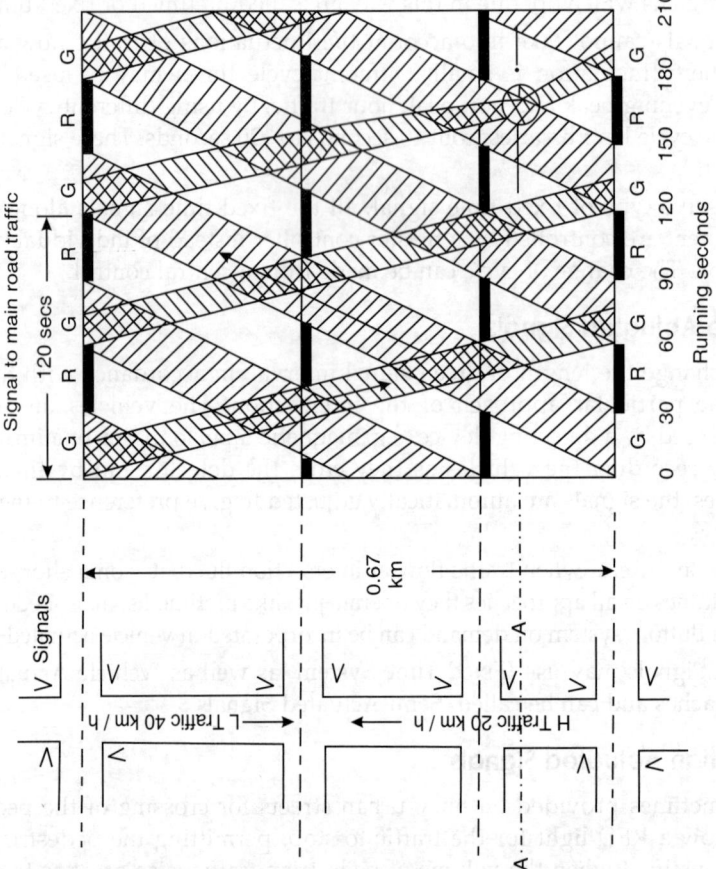

Fig. 12.12: 'Green Wave' traffic signal linking

12.8.5 Flashing Signals

To warn the approaching drivers, sometimes 'Yellow Flashing Signals' on the main road and 'Red Flashing Signals' on the cross road are provided. The drivers are required to slow down and proceed with caution at such locations like sharp horizontal curves, railway crossings or other obstructions on the road. At fixed time traffic intersections, sometimes flashing lights are used during the late night non-peak hours of traffic.

12.8.6 Bus Priority Signals

Such signals are used at special intersections, with the concept that whenever there is a bus in the traffic stream, it should find a green indication of signal immediately so that the total delay to people sitting inside the bus is minimised. The bus on such signals gets priority.

A vehicle unit fitted beneath the bus permits its identification. Detector loops bedded beneath the road surface are connected to an electronic limit for sensing the presence of a relevant bus. A traffic signal controller responds to the message from the detector unit.

The response may be continuation of an existing favourable green phase, or premature curtailment of red phase, to expedite the green phase needed by the waiting bus.

12.9 CRITERIA FOR SIGNALISATION OF AN INTERSECTION (WARRANTS)

The principle function of traffic signals is to provide the time separation of two crossing streams. The major criterion for signal control is the traffic volume entering the intersection from different approaches. Signalisation of the intersection is not needed if traffic is sufficiently light that adequate gaps appear to permit all intersecting traffic to enter and cross the intersection.

There are however other specific factors at a particular location to be considered for providing signals at a location. Some of these warrants are the following:

(i) **Minimum traffic volumes:** The traffic volumes on the two intersecting roads, when they exceed a specified limit, depending upon the traffic and road conditions, indicate the need for signalisation of the intersection. Typical minimum traffic volumes used in America are given in Table 12.3.

Table 12.3: Minimum volume warrants for signals

Number of lanes on each approach		Vehicle per hour on major street (both directions)	Vehicle per hour on minor street (one direction)
Major street	*Minor street*		
1	1	500	150
2 or more	1	600	150
2 or more	2 or more	600	200
1	2 or more	500	200

(ii) **Pedestrian volume:** When pedestrian volumes are heavy a traffic signal may avoid conflicts.

(iii) **School crossings:** Signals may be used where children cross the road near the schools.

(iv) **Accident experience:** When less restrictive remedies have failed to reduce the accident frequency, signals may be installed.

(v) **Progressive movement:** Signals may be necessary at the intersections to secure progressive movement control.

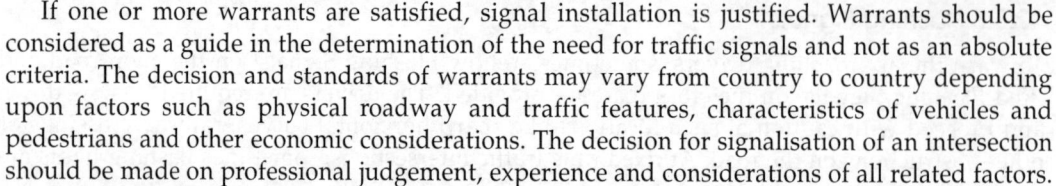

If one or more warrants are satisfied, signal installation is justified. Warrants should be considered as a guide in the determination of the need for traffic signals and not as an absolute criteria. The decision and standards of warrants may vary from country to country depending upon factors such as physical roadway and traffic features, characteristics of vehicles and pedestrians and other economic considerations. The decision for signalisation of an intersection should be made on professional judgement, experience and considerations of all related factors.

12.10 BASIC REQUIREMENTS OF TRAFFIC SIGNALS

The traffic signals shall be effective if the following basic requirements are taken care of:

(i) **Attention:** The signals must attract the attention of its users. Factors that affect the attention are the following:

- Position of placing the signals.
- Height and type of mounting.
- Design of signal head and face.
- The size of lamp lens, (usually 20 cms is used).
- Power of the bulb (usually 40-100 W).
- Background of the lens (generally dark background is required for lighted lens).

(ii) **Meaning:** Each indication of signal (colour or signal) should convey one and only one meaning clearly without any scope of misinterpretation.

Red: Vehicles must STOP.

Green: Vehicle can GO, if safe to do so.

Red Yellow (Red/Amber): Means STOP, but indicates stopped vehicles and pedestrians that light is about to change to GREEN. BE READY—to move.

Yellow (Amber): Warns the coming vehicles that light is going to be RED. It is clearance interval for vehicles and pedestrians within intersection.

The red/amber period is usually fixed at 2 seconds duration and amber period varies from 3 to 6 seconds.

(iii) **Response:** Since it takes time to stop a moving vehicle an immediate change from green to red does not give time for response.

There must be enough time to clear the intersection, before cross traffic is released. YELLOW interval provides this time required for response.

(iv) **Respect:** The signals which are installed at a location should be useful to road users and they must be respected (liked) by them.

Phasing of Traffic Signals

The procedure by which the traffic streams are separated is known as phasing. A phase is defined as the sequence of conditions applied to one or more streams of traffic, which during the signal cycle, receive simultaneous identical signal indications. Signals may have 2, 3 or 4 phases, as shown in Fig. 12.13.

The number of phases required for the proper and efficient operation of a signalized intersection varies with the composition and direction of traffic flows, as well as with the number of intersection approaches and the general intersection layout. In general traffic signals operate on a two-phase cycle, in which right of way is alternately assigned to each of the two cross movements. Intersections having a large and concentrated volume of right turns, or unusually heavy pedestrain movements, and intersections having more than four approaches

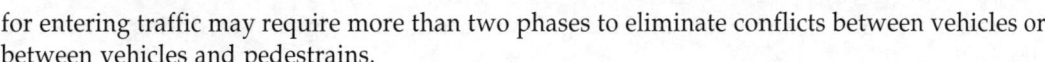

for entering traffic may require more than two phases to eliminate conflicts between vehicles or between vehicles and pedestrains.

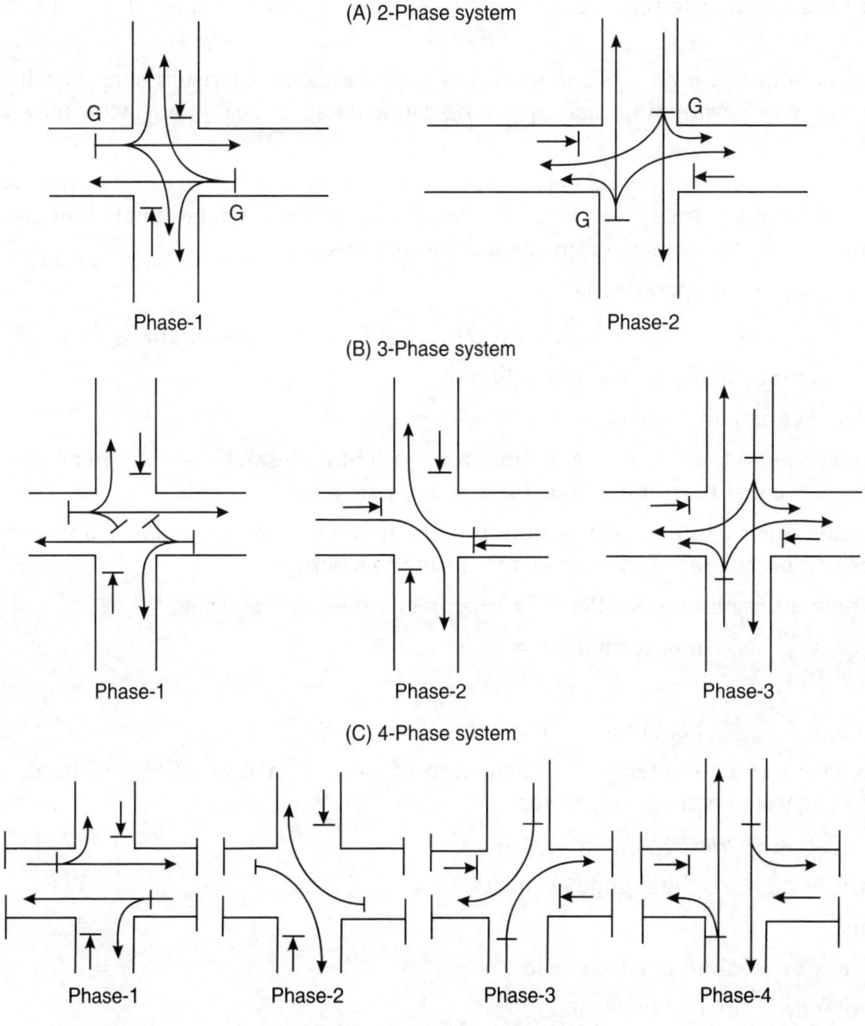

(A) 2-Phase system

Phase-1   Phase-2

(B) 3-Phase system

Phase-1   Phase-2   Phase-3

(C) 4-Phase system

Phase-1   Phase-2   Phase-3   Phase-4

Fig. 12.13: Phases of traffic signals

More than two phases should be avoided, where possible since each additional phase lengthens the overall cycle, thereby increasing delay.

12.11 SIGNAL TERMINOLOGY

Definitions of the common terms, used in design of traffic signals, are given below.

Time cycle or cycle length: The time required for one complete sequence of signal indications (i.e. R—RY—G and Y) is called time cycle or cycle length. It may vary from 35 to 70 seconds depending upon the (i) Volume of traffic on approaches, (ii) Width of the roads and intersection.

Interval: Any one of the divisions of the signal cycle during which signal indications do not change.

Load factor: The load factor is the ratio of the number of green phases that are loaded or fully utilised by traffic to the total number of green phases available for that approach during

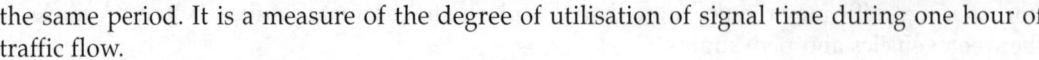

the same period. It is a measure of the degree of utilisation of signal time during one hour of traffic flow.

Peak hour factor: The ratio between the number of vehicles counted during the highest 15 consecutive minutes.

The peak hour factor reflects variations in the peaking characteristics of approach roads and provides means of evaluating their operating characteristics. Peak hour factor may vary from 0.25 to 1.0.

Intergreen period: The period between one phase losing right of way and next phase gaining right of way is called "Intergreen Period". (It is the period between termination of green on one phase and the commencement of green on the next phase).

Sequence of signal operation is:

RED—(RED—AMBER)—GREEN—AMBER

Amber period usually varies from 3 to 6 sec.

Red/Amber usually is taken as 2 sec.

Intergreen periods are kept to a minimum consistent with safety, because in every intergreen period there is a loss of running time, for both approaches.

Lost time during inter-green period. It is the time when no movements are possible from any approach because all signals show red during this period.

Lost time during Intergreen Period = Intergreen time—Amber time

$$\text{e.g. Intergreen time} = 5 \text{ sec.}$$
$$\text{Amber time} = 3 \text{ sec.}$$
$$\text{Lost time } 5 - 3 = 2 \text{ sec.}$$

Passenger Car Equivalency: In calculation of saturation flow PCUs are used, using the following suggested equivalency factors:

Type of vehicle	*PCU equivalents*
1. Private car, taxis, light goods vehicles	= 1.00
2. Bus	= 2.25
3. Heavy or medium goods vehicle	= 1.75
4. Motor cycle, moped or scooter	= 0.33
5. Pedal cycle	= 0.20

Detailed passenger car equivalents are given in Table 6.3.

Effective Green Time: Figure 12.14 "Concept of Effective Green Time" shows that as soon as the green signal is given, the rate of discharge begins to pick up and some time is lost before the flow reaches the maximum value (saturation flow). Similarly at the termination of the green phase, the flow tends to taper off, involving further lost time.

The area under the actual traffic flow curve gives the total volume of traffic (in pcu) passing through the approach in the Green + Amber time of the phase.

At saturation flow rate, the time taken for this volume to pass through the approach is given

by $= \dfrac{\text{Total volume during the phase}}{\text{Saturated flow}} \times (\text{Green time}).$

This time is known as the *effective green time.*

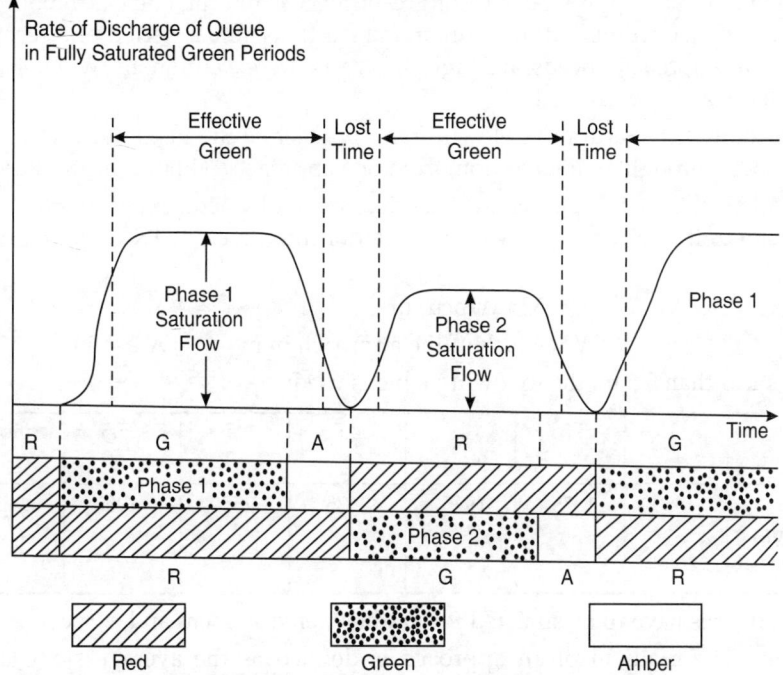

Fig. 12.14: Concept of effective green time

Because of starting delays at the beginning of the GREEN PHASE and the flow reducing to zero towards the end of GREEN PHASE, the GREEN + AMBER times is not fully utilised to clear traffic at saturation flow rate.

The net effect is that there is time lost during the phase due to starting and tapering delay given by:

Delay time = Green time + Amber time – Effective green time.

In actual practice, the lost time due to starting and tapering delays is taken as 2 seconds. So that:

Effective Green Time (EGT) = Actual Green Time + Amber Time – 2

$$EGT = AGT + Amb. T - 2$$

or Actual Green Time (AGT) = EGT + 2 – Amb. Time.

Total Lost Time Per Cycle: The total lost time per cycle (L) is the sum of all the lost times due to starting and tapering delays and lost times (due to All red signal aspects) during intergreen periods. This is given by:

$$L = n\,(l + R)$$

where
L = lost time per cycle

l = lost time due to starting and tapering delay per phase

R = lost time due to all red period per intergreen period

n = Number of phases/cycle.

Vehicle Clearance Interval (Amber time): The purpose of the amber signal following each green interval is to warn moving traffic facing the signal to come to stop and, if possible, to do so with safety. It should provide enough time for vehicles to clear the intersection before cross traffic starts to move. It should, therefore, be sufficient to allow a vehicle to cover a distance of stopping distance plus the cross road width at normal approach speed.

If it is too long it may be used as part of green interval and thus defeat the purpose. If it is too short, it may be hazardous. At most urban intersections amber period of 3 seconds is quite sufficient. Where approach speeds are high or streets are exceptionally wide, 4 to 6 second clearance period may be warranted.

Saturation Flow (S): 'Saturation Flow' of an approach is the maximum flow (in pcu) per hour that can pass through an intersection, from one approach, without impedance by signals.

The saturation flow of an approach is best determined by actual counting; however, if not actually measured, it can be found by a liner relationship given in TRRL, based on approach width.

$$S = 525 \, W \text{ pcu/hr}$$

where W = width of the approach in m $\quad (W \geq 5.5 \text{ m})$

For widths less than 5.5 m, the following table is used:

Approach width (m)	3	3.5	4	4.5	5	5.5	more than 5.5
Saturation Flow (PCU/h)	1850	1875	1975	2175	2550	2900	525 pcu per hour per metre width

The above figures have to be adjusted for gradient, environment and curves.

(i) **Gradient:** The gradient of an approach is defined as the average slope between the stopline and a point 61 m before it.

The saturation flow is decreased (or increased) by 3% for every 1 percent uphill (or downhill) gradient.

(ii) **Environment:** The above figures can be increased or decreased in following situations of the environment:

Very Good Junction

Good visibility, two lanes,   Increase the saturation flow by 20%

no interference by pedestrians,

adequate turning radii

Poor Junction

Low speed, poor visibility,   Reduce the saturation flow by 15%.

poor alignment, some inter-

ference.

(iii) **Curves:** Where a separate right turn phase is provided for vehicles and where vehicles crossing the stopline have to travel immediately around a curve, the rate of discharges around the stop line will be reduced. The saturation flow on curves may be obtained by:

$$S = \frac{1800}{1 + 1.52/r} \text{ pcu/hr for single file streams}$$

or 1600 pcu/hr

$$S = \frac{3000}{1 + 1.52/r} \text{ pcu/hr for double file streams}$$

or 2700 pcu/hr

where r is turning radius in m.

Normal flow (q): The traffic actually using a junction through the hour is called normal flow 'q'. It is normally less than the saturation flow (s). The normal flow is also expressed in pcu.

Y-values (y): The ratio of normal flow (q) to saturation flow s is called the 'y' value of approach.

$$y = \frac{q}{s}$$

q = Normal flow pcu./hr.

s = Saturation flow pcu./hr.

Maximum 'y' values (Y): For a normal four-way junction, two y values of N—S approach and two y values of EW approach are obtained.

The greater of the two values from each pair of approaches is added to give maximum value of y's.

$$Y_{max} = \Sigma y_{max} \quad \begin{Bmatrix} y_N, y_S \text{ Either (NS)} \\ \text{which is high} \\ + \\ y_E, y_W \text{ Either (WE)} \\ \text{which is high} \end{Bmatrix} = \begin{matrix} \text{higher of } y_N \text{ and } y_S \\ + \\ \text{higher of } y_W \text{ and } y_E \end{matrix}$$

This value of Y is a measure of the congestion of the junction.

12.12 DESIGN OF PRE-TIMED TRAFFIC SIGNALS

The cycle time for the traffic signals can be determined by one of the following two method:

(1) Method I (Field method or Trial method). The timing calculations in this method are based on average clearance time of vehicles and pedestrian requirements. Shorter cycles lengths are required for off-peak periods as compared to peak hour requirements. The general steps involved in this method are the following:

(i) Select or calculate the yellow time: There are two methods of finding the yellow time, one based on approach speeds and the other is determined by the dimensions of the intersection and stopping sight distance.

(a) *Based on approach speed:* Following values may be used for different approaches depending upon the approaching speeds of the traffic.

Approach sped (KPh)	Yellow period to be used (Seconds)
≤ 55	3.0
65	3.5
70	4.0
80	4.5
≥ 80	5.0

(b) *Based on dimensions of intersection:* When approach speeds are high and dimensions of the intersection are large, the yellow time can be calculated by the following formula, to provide greater safety and reduce dilemma.

$$Y = \frac{SSD + W + L}{V}$$

where   Y = Yellow time (seconds).

$$SSD = \text{Stopping sight distance (m).}$$
$$= 0.28\ Vt + .01\ V^2.$$
$$t = \text{Reaction time of drivers (sec.).}$$
$$W = \text{Width of intersection (m).}$$
$$L = \text{Average length of vehicle (m).}$$
$$V = \text{Approach speed (KPh).}$$

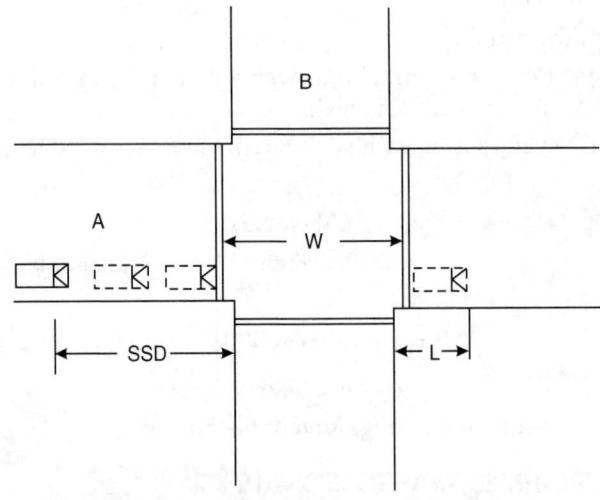

Fig. 12.15: Yellow time based on width of intersection

(*ii*) *Determine the pedestrian clearance time:* Based on the pedestrian speed of 1.2 m per second, determine the timings for the pedestrians to cross.

(*iii*) *Compute the green times:* The minimum green time is equal to the pedestrian clearance time minus the yellow time plus an initial interval, required by the pedestrians to start. Assume a minimum green of not less than 15 seconds and distribute the assumed green time depending upon the number of vehicles and average time spacing between the vehicles using the formula:

$$\frac{T_A\ (\text{Green time at A})}{T_B\ (\text{Green time at B})} = \frac{V_A\ (\text{Volume on approach A})}{V_B\ (\text{Volume on approach B})} \times \frac{t_a\ (\text{Time spacing of vehicle on A})}{t_b\ (\text{Time spacing of vehicle on B})}$$

Adjust the cycle length (sum of all greens and yellows), to ensure that sufficient green time is available, by increasing it by 5 seconds.

(*iv*) *Find red times and draw the phase diagram:* After knowing the yellow and green times, for the approaches, the red times can be calculated.

Green time of A = Red time of B approach.

Green time of B = Red time of A approach.

A phase diagram showing the sequence of signal operation can be drawn as shown below.

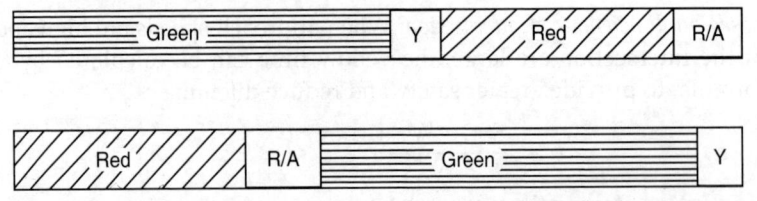

Fig. 12.16: Phase diagram showing sequence of operations of signal cycle

(2) **Method II (British TRRL Method or Minimum Delay Method):** On the basis of extensive research, Webster has given an approximate formula for determining the optimum cycle length, (Co) in terms of minimum delay:

$$Co = \frac{1.5\,L + 5}{1 - \sum Y_{max}} \text{ seconds}$$

L = $nl + R$ = Total time lost during total cycle.

l = Time lost due to starting delays each phase (suggested value is 2 seconds).

R = All red period during the cycle time and is equal to (intergreen-amber) for each phase.

n = Number of phases.

Y_{max} = Design flow/saturation flow for each phase. Maximum value of y, y values (Y_{max}) of various approaches must be considered (added)

The effective green time is distributed to each approach in proportion to respective y values.

$$g_1 : g_2 : g_3 = y_1 : y_2 : y_3$$

Effective green in one cycle = Co – L.

The amber period is taken as 3 seconds and the red/amber as 2 seconds.

12.13 PEDESTRIAN REQUIREMENTS AT SIGNALS

When pedestrian crossing time runs concurrently with the vehicle "Go" period, the total 'Go' period should be long enough to allow a minimum of 5 seconds as pedestrian starting time, to permit the pedestrians to cross the road safely.

Minimum "Go" time = 5 + cross time – vehicle clearance interval (curb to curb)

At some intersections it is the pedestrian crossing time that will decide the "Go" time interval, particularly when the approaches are too wide and vehicles can clear the intersection in a shorter time.

At locations where pedestrian signals are used, a starting time or pedestrian reaction time of 7 seconds is recommended, instead of 5 seconds as used above. IRC-93, provides guidelines for design and installation of signals in India.

ILLUSTRATIVE EXAMPLES

Example 1. On the right-angled crossing of two urban roads A and B, pre-timed traffic signals with pedestrian signals are to be provided using the available data given in table below. Design the traffic signals.

	Road A	*Road B*
Width of road (m)	21	14
Critical volume (VPh)	400	200
Approach speed (KPh)	84	70

Solution:

(i) **Yellow time.** Based on the approach speed of vehicles and using the table, the yellow times are:

y_A (yellow time for approach A) = 5 seconds.

y_B (yellow time for approach B) = 4 seconds.

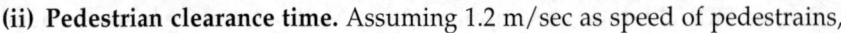

(ii) Pedestrian clearance time. Assuming 1.2 m/sec as speed of pedestrains,

Pedestrian clearance time for approach A $= \dfrac{21}{1.2} = 17.5$ seconds ≈ 18 sec

Pedestrian clearance time for approach B $= \dfrac{14}{1.2} = 11.6$ seconds ≈ 12 sec.

(iii) Minimum red periods. Assuming the pedestrian reaction time of 7 seconds:

Minimum red period for traffic on approach A = 18 + 7 = 25 sec. so that pedestrians may cross, and similarly

Minimum red period on approach B = 12 + 7 = 19 sec.

(iv) Minimum green period

(a) *Pedestrian requirements.* Based on this pedestrian requirements of red time, the minimum green times shall be:

Minimum green time for traffic on approach T_A = 19 – 5 = 14 sec.

Minimum green time for traffic on approach T_B = 25 – 4 = 21 sec.

(b) *Traffic volume requirements.* Based on the approach volumes of the traffic.

$$\dfrac{G_A}{G_B} = \dfrac{V_A}{V_B} \qquad G_A \text{ and } G_B \text{ are green times on A and B.}$$
$$V_A \text{ and } V_B \text{ are approach volumes on A and B.}$$

If green time on B is kept minimum for pedestrians

$$G_B = T_B = 21 \text{ seconds.}$$

Then green time on approach A

$$G_A = G_B \times \dfrac{V_A}{V_B} \; 21 \times \dfrac{400}{200} = 42 \text{ seconds}$$

Total cycle length $= G_A + G_B + Y_A + Y_B$

$= 42 + 21 + 5 + 4 = 72$ seconds.

Red period for lane A $= T_{ra} = 21 + 4 = 25$ seconds.

Red period for lane B $= T_{rb} = 42 + 5 = 47$ seconds.

Pedestrian signal timings

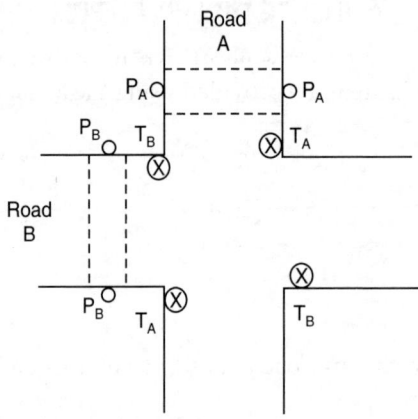

P_A, P_B and T_A, T_B are the locations of pedestrian and traffic signals respectively on the roads A and B as shown in above figure.

"Red" period of traffic signal on road A = 25 sec.

"Red" period of traffic signal on road B = 47 sec.

"Dont walk" period for pedestrian signal on road A = 47 seconds

"Dont walk" period for pedestrian signal on road B = 25 seconds

"Walk" period for pedestrian signal on road A is equal to

Total cycle length – Red time period on B – Pedestrian clearance time on A

Walktime on A = 72 – 47 – 18 = **7 seconds.**

"Walk" period for pedestrian signal on road B is equal to

Total cycle length – Red period at A – Pedestrian clearance time on B

walk time on B = 72 – 25 – 12 = **35 seconds**

Example 2. The dimensions of the approach widths of the four approaches A, B, C and D and the actual volumes of traffic flow in pcu per hour, from each direction are shown in the figure for a right-angled intersection of two roads AB and CD. The intersection is located in normal environment on a flat ground and the turning traffic is not substantial.

For design of a 2-phase traffic signal at this intersection, the integreen period may be taken as 5 seconds on one approach and 6 seconds on another (2 seconds of red/red amber time on one approach and 3 seconds of red/red amber time on other).

Using the above data and TRRL method of design, calculate:

(i) The optimum cycle length.

(ii) The distribution of green time on two roads, and

(iii) Draw the cycle diagram for the two phases.

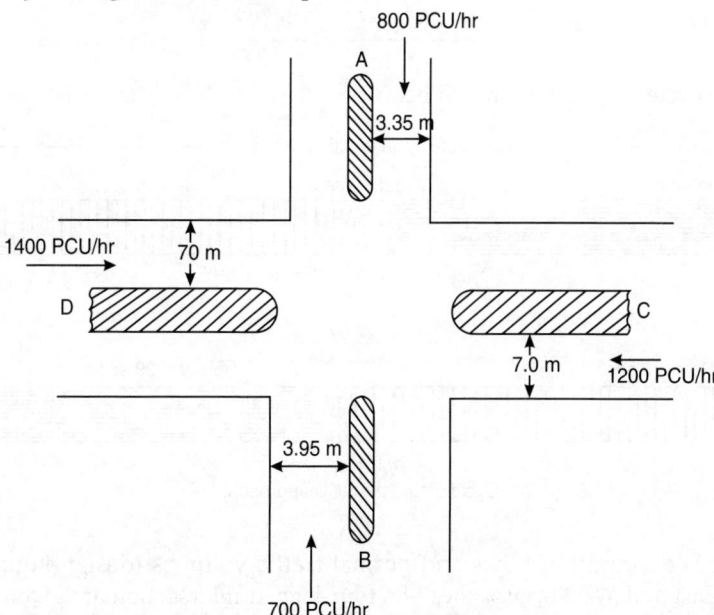

Solution.

(i) Calculate the total delay (L):

Total delay L = 2n + R = 2 × 2 + (2 + 3) = 9 seconds or

L = Σ (Intergreen time – 1) = (5 – 1) + (6 – 1) ≈ 9 seconds.

(ii) Calculate the 'Y' values in tabular form as below:

Particulars	Approaches to Intersection			
	A	B	C	D
Normal flow q (pcu/h) given.	800	700	1200	1400
Saturation flow S (pcu/hr) (calculated from approach widths).	1875	1950	$7 \times 525 = 3675$	$7 \times 525 = 3675$
$Y = q/S$	0.42	0.36	0.33	0.38
Y_{max} (Greater of two Y's)	0.42		0.38	
$Y = \Sigma Y_{max}$	0.8			

(iii) Optimum cycle length $C_0 = \dfrac{1.5\,L + 5}{1 - Y}$

$$= \dfrac{1.5(9) + 5}{1 - 0.8}$$

$$= 92.5 \text{ seconds.}$$

take 92 seconds.

(iv) Calculate the green times (g) in proportion to y values of the approaches:

$$g_{AB} = \dfrac{Y_{AB}\,(C_0 - L)}{Y} = \dfrac{0.42\,(92 - 9)}{0.8} = 44 \text{ seconds}$$

$$g_{CD} = \dfrac{Y_{CD}\,(C_0 - L)}{Y} = \dfrac{0.38\,(92 - 9)}{0.8} = 39 \text{ seconds}$$

(v) Draw the cycle diagram as shown below:

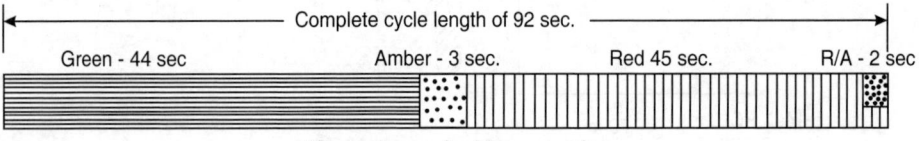

(Cycle diagram for AB approach)

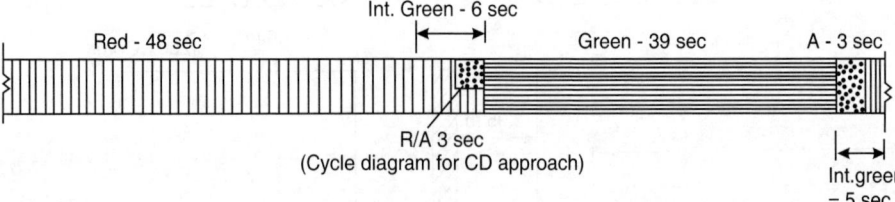

(Cycle diagram for CD approach)

Example 3. The saturation flows and normal traffic volumes (design hourly volumes) on North, South, East and West approaches of a four-legged intersection are given below:

	North	South	East	West
Normal traffic volume pcu/hr	500	400	300	250
Saturation traffic flow pcu/hr	1500	1200	1200	1250

Assume intergreen period as 8 seconds between each phase when red or red-amber shows simultaneously and the lost time delay due to starting on each phase is 2 seconds. Determine the

optimum cycle time for a two phase signal setting and distribution of green time on NS and EW approaches.

Solution:

$$\text{Lost time (L)} = \Sigma(I - a) + \Sigma l$$

where

$$L = \text{Average lost time per phase}$$
$$I = \text{Intergreen period}$$
$$a = \text{Amber time taken as 3 sec.}$$
$$l = \text{Lost time due to starting delay}$$
$$L = (8 - 3) + (8 - 3) + 2 + 2 = 14 \text{ seconds.}$$

	N	S	E	W
Normal flow (q)	500	400	300	250
Saturation flow (s)	1500	1200	1200	1250
$Y = q/s$	0.33	0.33	0.25	0.20
Y_{max} :		0.33		0.25
$Y = \Sigma Y_{max}$:		0.33 + 0.25 = 0.58		

The optimum cycle length C_0 is given by:

$$C_0 = \frac{1.5\,L + 5}{1 - \Sigma\,Y_{max}}$$

$$= \frac{1.5\,(14) + 5}{1 - .58} = 62 \text{ seconds.}$$

The optimum cycle time for these conditions is 62 seconds and the overall delay (i.e. total delay to all vehicles will be minimum when this cycle is used). With this cycle time there will be 14 seconds of the lost time.

So total effective green time in the cycle is equal to 62 – 14 = 48 seconds which should be distributed between phases in proportion to their y-values:

$$\text{Green time on NS} : g_{NS} = \frac{0.33\,(62 - 14)}{0.58} = 27 \text{ seconds}$$

$$\text{Green time on EW} : g_{EW} = \frac{0.25\,(62 - 14)}{0.58} = 21 \text{ seconds.}$$

To convert these effective green times into actual green times, it is necessary to add the portion of lost time and deduct the amber time.

i.e.

$$K = g + l - a$$

where

$$K = \text{actual green time}$$
$$a = \text{amber time (3 seconds)}$$
$$g = \text{effective green time}$$
$$l = \text{lost time due to starting delay (2 seconds)}$$

Actual green time for NS = K(NS) = 27 + 2 – 3 = 26 seconds.
Actual green time for EW = K(EW) = 21 + 2 – 3 = 20 seconds.

Example 4. One approach of a signalised intersection has an effective approach width of 7 m and an overall uphill gradient of 4 percent. The intersection has good geometry with separate pedestrian phases. Traffic flow at the approach consists of 15% of heavy vehicles.

The green signal period at the approach is of 30 seconds and lost time due to starting delay may be assumed as 2 seconds. Calculate the maximum possible number of vehicles discharged from the stop line of this approach per cycle when:

(i) No provision is made for right turning vehicles.

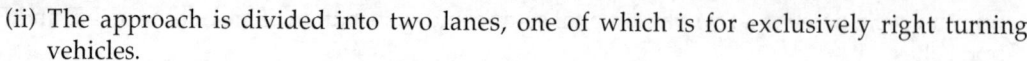

(ii) The approach is divided into two lanes, one of which is for exclusively right turning vehicles.

Solution:

The maximum rate of discharge (saturation flow) at the stop line of an intersection approach depends upon the following considerations:

(a) The approach width.

(b) The geometric design and environment of intersection.

(c) The gradient of the approach.

(d) The composition of traffic.

(e) The length of green period and lost time during green period.

(i) **When no separate provision for right turning is made:** Saturation flow considering:

(a) *Width of approach:* The effective width of the intersection (distance from the curb to central island or to the centre line marking) is given as 7 m.

The saturation flow 'S' for 7 m width = 525 × 7

$$= 3675 \text{ PCU/hr.}$$

(b) *Geometric Design Standard:* For good standards of geometric design and environment, the saturation flow calculated above should be increased by 20%.

Correct saturation flow for geometry (S) = 3675 × 1.2

$$= 4410 \text{ PCU/hr.}$$

(c) *The gradient:* A correction of plus or minus 3 percent in saturation flow for every one percent gradient is used.

Here gradient is uphill of 4 percent, therefore a minus correction of 3 × 4 = 12 percent shall be used.

Corrected saturation flow for gradient (S) = 4410 (1 – 0.12)

$$= 3881 \text{ PCU/hr.}$$

(d) *Composition of traffic:* The traffic consists of 15% heavy vehicles and 85% cars. One heavy vehicle is taken as equal to 1.75 of a passenger car.

$$100 \text{ vehicles} = 85 \text{ PC} + 15 \text{ heavy vehicles}$$
$$= 85 + 15 \times 1.75 \text{ PC}$$
$$= 111.25 \text{ P.C.}$$

Therefore saturation flow in vehicles/hour $= \dfrac{3881}{111.25}$

$$= \textbf{3489 Veh./hr.}$$

(e) *The green period and lost time:* The effective green time is used in calculation of the saturation flow (maximum discharge rate).

Effective green time = Actual green time + Amber period (3 secs)

– The lost time due to acceleration and deceleration at the beginning and end of green period.

Effective greentime ((EGT) = 30 + 3 – 2 = **31 sec.**

Therefore maximum number of vehicles discharged per green cycle time

$$= \frac{3489 \times 31}{3600}$$

$$= \textbf{30 vehicles.}$$

(ii) When exclusive right turning lane is provided: In this case the approach width of 7 m is divided into 2 lanes of 7/2 = 3.5 m each. The saturation flow for the two lanes and the number of vehicles per cycle (green time) should be calculated separately.

(a) *For the straight and left turning lane* : Saturation flow (S) for width of 3.5 m

$$= 1875 \text{ PCU/hr.}$$

Corrected 'S' for good geometry (+ 20%) = 1855 × 1.2

$$= 2250 \text{ PCU/hr.}$$

Corrected 'S' for gradient (– 12%)   = 2250 (1 – 0.12)

$$= 1980 \text{ PCU/hr.}$$

Saturation flow in vehicles/hour $= \dfrac{1980}{1.1125} = 1780$ Veh./hr.

Maximum number of vehicles discharged per cycle length

$$= \frac{1780 \times 31}{3600} = \textbf{15 vehicles.}$$

(b) *For the right turning lane*: For the right turning lane if radius of right turn is not given, a minimum saturation flow value of 1600 PCU/hr. for single file is assumed.

Saturation flow (S) $= 1600 \text{ PCU/hr.}$

Corrected 'S' for good geometry (20% increase)

$$= 1600 \times 1.2$$
$$= 1920 \text{ PCU/hr.}$$

Corrected 'S' for gradient (12% decrease) = 1920 (1 – .12)

$$= 1690 \text{ PCU/hr.}$$

Saturation flow in vehicle/hour $= \dfrac{1690}{1.1125}$

$$= 1520 \text{ Veh./hr.}$$

Maximum number of vehicles discharged per cycle length

$$= \frac{1520 \times 31}{3600}$$
$$= \textbf{13 vehicles}$$

Total number of vehicles discharged per cycle

$$= 15 + 13$$
$$= \textbf{28 vehicles.}$$

Example 5. A three-phase traffic signal is to be installed at right angle crossing of two city streets. The site is "Average" and the approaches are 12 metres wide between kerbs. The approaches are straight and level and parking is prohibited on them. One of the phases is to be a "Pedestrian Only" phase occurring at the end of each cycle. Starting delay may be taken as 2 seconds. An "All-red" period of 4 seconds is to be provided after each vehicle phase to allow clearance of right turning vehicles left over in the crossing. The design hour traffic volumes in PCU/hour are given in the following table:

From	N			E			S			W		
To	E	S	W	S	W	N	W	N	E	N	E	S
PCU/hr	40	800	70	60	500	50	60	600	60	70	680	50

Calculate the optimum cycle time for a fixed time signal installation and calculate the settings required for the controller. Sketch the phasing diagram for each phase. Draw a diagram showing the timings for all three aspects for a complete cycle.

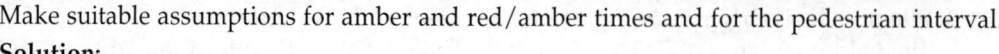

Make suitable assumptions for amber and red/amber times and for the pedestrian interval.

Solution:

The width of roadway being 12 metres, the time taken for pedestrians to cross the street is 10 seconds, with a speed of walk of 1.2 metres per second. Because of the large width of the streets, it is desirable to have a central pedestrian refuge of at least 1 m width. The time that will be needed by a pedestrian to reach the pedestrian refuge from the kerb will then be:

$$\frac{12-1}{2} \times \frac{1}{1.2} = 4.6, \text{ say 5 seconds.}$$

This will be the pedestrian clearance interval during which no signal is displayed to the pedestrians and those who have just left the kerb or the central refuge before the termination of the pedestrian green signal can reach safely the central refuge or the kerb as the case may be. The American practice is to display the flashing DON'T WALK indication for this clearance interval. The pedestrian clearance interval is followed by red/amber of the next vehicular phase and by the Red signal in the pedestrian phase.

A total length of 13 seconds will be provided for the pedestrian phase as follows:

Pedestrain green time = 8 seconds

Pedestrain clearance interval = 5 seconds

The width of the approach road from each direction is

$$\frac{12-1}{2} = 5.5 \text{ m}$$

and this will be used to calculate the saturation flow from the formula:

$$s = 525 \text{ W.}$$

Since the site is "Average" and is level with parking prohibited, no corrections are needed for the saturation flow obtained from the above formula.

The effect of left-turning traffic will be accounted for if it constitutes more than 10 per cent of the traffic by counting each left turner as equivalent to 1.25 straight ahead vehicles. Since no exclusive right turning lanes are provided, the effect of right turning traffic will be accounted for by counting each right turner as equivalent to 1.75 straight ahead vehicle.

The following tabulations indicate the sequence of calculations:

From	N			E			S			W		
To	E	S	W	S	W	N	W	N	E	N	E	S
Given flowPUSs/hr	40	800	80	80	500	52	60	600	60	70	680	60
Correction for left turners	—	—	—	+20	—	—	—	—	—	—	—	—
Correction for right turners	—	—	+60	—	—	+39	—	—	+15	—	—	+15
Total	40	800	140	100	500	91	60	660	75	70	680	75
q		980			691			795			825	
$s = 525$ W		2900			2900			2900			2900	
$y = \dfrac{q}{s}$		0.34			0.24			0.27			0.28	
y (max)												
N—S		0.34									0.28	
E—W												

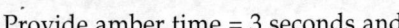

Provide amber time = 3 seconds and

Red/Amber = 2 seconds.

All-red period after each vehicular phase = 4 seconds (given).

This will consist of 2 seconds Red/Amber.

This intergreen period (I) will, therefore, consist of:

$$\text{Amber} = 3 \text{ seconds}$$
$$+$$
$$\underline{\text{All-Red} = 4 \text{ seconds}} \qquad (\text{including 2 seconds Red/Amber})$$
$$= 7 \text{ seconds}$$

Pedestrian phase = 13 seconds

Red/Amber following pedestrian phase = 2 seconds

Starting delay per vehicle phase (1)　　= 2 seconds

Lost time L = (13 + 2) + Σ (I – a) + Σ l

$$= 15 + (7 - 3) + (7 - 3) + 2 + 2 = 27 \text{ seconds.}$$

Optimum cycle time,

$$C_0 = \frac{1.5\,L + 5}{1 - Y} = \frac{1.5 \times 27 + 5}{1 - 0.34 - 0.28} = 120 \text{ seconds}$$

Effective green time per cycle available to the vehicular phases.

$$= C_0 - L = 120 - 27 - 93 \text{ seconds.}$$

This will be apportioned between N – S and E – W phases as follows:

$$g_{NS} = \frac{y_{NS}}{Y} \, (C_0 - L) = \frac{0.34}{0.62} \times 93 = 51 \text{ seconds.}$$

$$g_{EW} = \frac{y_{EW}}{y} \, (C_0 - L) = \frac{0.28}{0.62} \times 93 = 42 \text{ seconds.}$$

Total green times (including amber) are:

$$G_{NS} = 51 + 2 = 53 \text{ seconds (as } l = 2)$$
$$G_{EW} = 42 + 2 = 44 \text{ seconds (as } l = 2)$$

Controller setting for green for N – S phase = 53 – a

$$= 53 - 3 = \textbf{50 seconds.}$$

Controller setting for green for E – W phase = 44 – a

$$= 44 - 3 = \textbf{41 seconds.}$$

Timing Diagrams

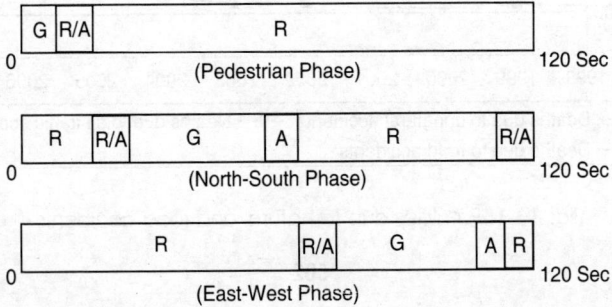

(Pedestrian Phase)

(North-South Phase)

(East-West Phase)

Highway Safety and Design

13.0 PREAMBLE-ACCIDENT SCENARIO

Road transport is a dangerous activity—where many serious and fatal accidents occur world-wide. It is projected that by year 2020, road accidents will be the third leading cause of death, exceeded only by cardiovascular diseases and major depression (WDR–2005). Escalating road safety problem is big social, health and economic disaster for India and other developing countries. According to Road Injury Training Manual, 2006, around 1.2 million people around the world loose their lives annually as a result road accidents with about 20–25 million people being injured or disabled.

In India, road accidents are the biggest cause of unnatural deaths in the country as shown in Table 13.1.

Figure 13.1 shows accident deaths and those due to road accidents, and Fig. 13.2 shows percentage of accidents fatalities on national, state highways and other roads.

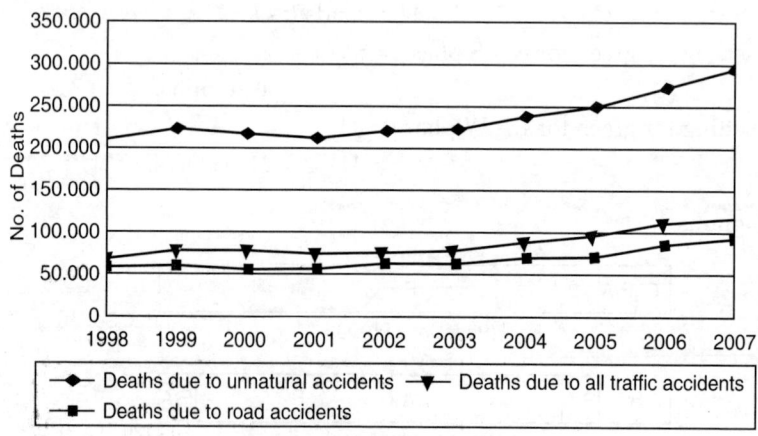

Source: MORT&H website

Fig. 13.1: Fatalities due to natural and road accidents

Table 13.1: Road accidents– a cause of unnatural deaths (Indian Highways)

| *Cause* | *2003* | | *2004* | | *% variation* |
	Fatalities	*%*	*Fatalities*	*%*	*2003–2004*
Causes attributed to nature	14,954	5.8	18,937	6.8	26.6
Rail–Road and other railway accidents	18,521	7.1	20,418	7.3	10.2
Drowning	20,960	8.0	21,190	7.6	1.1
Fire accidents	19,278	7.4	18,445	6.6	–4.3
Sudden deaths	16,749	6.5	17,413	6.3	3.9
By falls	8,800	3.4	8,848	3.2	0.5
By electrocution	6,336	2.4	6,224	2.2	–1.76
Road accident	84,430	32.5	91,376	33	8.2
Poisoning	21,172	8.1	22,035	8.0	4.0
Cause not known	12,405	4.8	14,084	5.0	13.5
By other unnatural causes	36,020	14	38,693	14	7.4
Total	**2,59,625**	**100**	**2,77,263**	**100**	**6.8**

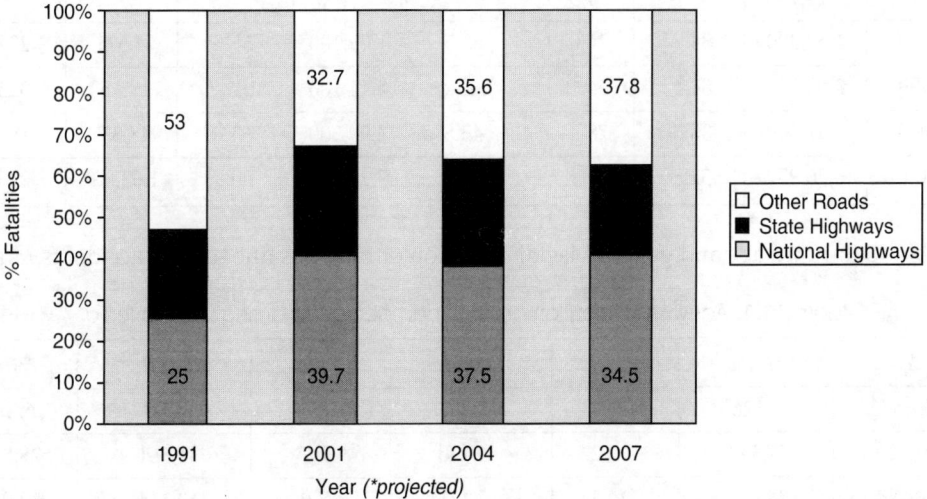

Source: MORT&H website.

Fig. 13.2: Percentage of road fatalities on different roads in India

On an average, about 1,30,000 people loose their lives every year due to road accidents in India, and the economic losses are estimated to be around Rs. 75,000 corers (Financial express, 2009). Half-a-million are injured annually. Figure 13.3 shows road accident fatalities by vehicle user/occupant (2007).

Table 13.2 shows number of road traffic accidents (RTA) decadewise and percent of registered vehicles in India.

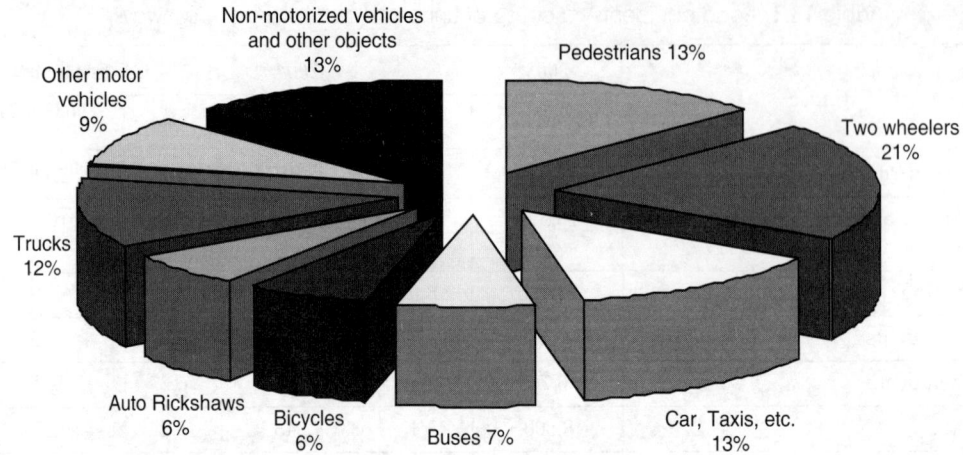

Source: MORT&H website

Fig. 13.3: Fatalities in road accidents by type of motor vehicles use/occupant (2007)

Table 13.2: Road accidents and registered vehicles in India (Indian Highways)

Year	Number of RTA	% growth	RTA fatalities	% Growth	Injured	% Growth	Registered Vehicles 000'	% Growth of registered vehicles
1970	1,14,100		14,500		70,100		1,401	
1980	1,53,200	34.7	24,000	65.5	1,09,100	55.6	4,521	222.7
1990	2,82,600	84.4	54,100	125.4	2,44,100	123.7	19,152	323.6
2000	3,91,449	38.5	78,911	45.8	3,99,300	63.5	48,857	155.1
2010 (P)	5,96,959	52.5	1,41,171	78.9	7,22,333	80.9	1,63,084	233.8

Table 13.3 shows age and genderwise distribution of fatalities due to road accidents in India.

Table 13.3: Agewise distribution of fatal road accidents during 2004 in India

Age	Males	%	Females	%	Total	%
Upto 14	5,126	5	1,807	2.6	6,933	6.2
15–29	27,171	20.2	4,490	7.7	31,661	28.3
30–44	34,673	26.3	5,471	7.0	40,144	40.0
45–59	20,269	17.5	3,636	4.1	23,905	21.4
Above 60	7,389	7.3	1,762	2.3	9,151	8.1
Total	**94,628**	**76.3**	**17,166**	**23.7**	**111,794**	**100**

Above tables show tremendous increase in growth of motor vehicles and subsequent increase in road accident fatalities, and injuries. Highest number of casualities are in the working age groups between 15 to 59 (89.7 percent). Males are most vulnerable victims in all categories, devastating the entire household socially and economically.

Table 13.4 shows road accidents in India during 1970–2004 and Table 13.5 includes world statistics showing that while developing countries lag behind in term of number of motor vehicles on the road, they are ahead in number of accidents. For example, accident rate in India in 1996 was almost 4 times of US. Only Kenya and Sri Lanka were worse. With increasing purchasing power of average Indian, motor vehicle ownership is growing very fast and level of ownership of vehicles in major cities of India is comparable to that of developed countries.

World health organization (WHO) has estimated that fatality per 100,000 population in developing world will grow from 13.3 in 2000 to 19 in 2020, while developed world during the same two decades will have a decline from 11.8 to 7.8. Status of India in 2007 has been 10.1 fatalities per 100,000 population. Many of the accidents with minor injury and property damages

Table 13.4: Road accident statistics of India: 1970–2004 (internet)

Sl. No.	Year	Total no. of road accidents (in numbers)	Total no. of persons killed (in numbers)	Total no. of regd. motor vehicles (in thousands)	No. of accidents per ten thousand vehicles	No. of persons killed per ten thousands vehicles
1	2	3	4	5	6	7
1.	1970	114100	14500	1401	814.42	103.50
2.	1980	153200	24000	4521	338.86	53.09
3.	1990	282600	54100	19152	147.56	28.25
4.	1991	295131	56278	21374	138.08	26.33
5.	1992	275541	60113	23507	117.22	25.57
6.	1993	284646	60380	25505	111.60	23.67
7.	1994	325864	64463	27660	117.81	23.31
8.	1995	351999	70781	30295	116.19	23.36
9.	1996	371204	74665	33786	109.87	22.10
10.	1997	373671	76977	37332	100.09	20.62
11.	1998	385018	79919	41368	93.07	19.32
12.	1999	386456	81966	44875	86.12	18.27
13.	2000	391449	78911	48857	80.12	16.15
14.	2001	405637	80888	54991	73.76	14.71
15.	2002	407497	84674	58924	69.16	14.37
16.	2003	406726	85998	67007	60.70	12.83
17.	2004	429910	92618	72718	59.12	12.74

never got reported. World Bank in its report on Road Traffic Injury Prevention, (based on observations made in many countries) observed that ratio between death, major injury and minor injury accidents is 1: 15: 70. The occurrence of an accident not only causes loss of life or property but also brings agony to the family involved and society at large. Long range vision to reverse the increasing trend of road accidents, through comprehensive measures, is required. In the latest

Table 13.5: Growth of road traffic and accidents in different countries

Country	Motor Vehicles				Passenger Cars	Two-wheelers	Road Traffic	Traffic Accidents (People injured or killed/1000 vehicles)	
	Per 1,000 people		Per kilometer of road		Per 1,000 people	Per 1,000 people	Million vehicle km	1980	1996
	1980	1996	1980	1996	1996	1996	1996		
Argentina	155	154	20	25	127	1	56,590	...	...
Australia	502	604	...	12	485	16	...	5	...
Bangladesh	...	1	...	1	...	1	...	...	30
Brazil	85	79	7	6	84	...	...	3	4
China	2	8	2	7	3	8	165,000	12	22
Cuba	...	5	...	2	2	...	...	...	...
Egypt	...	30	...	28	23	7	6,222	...	16
Germany	399	528	51	68	500	30	563,200	...	12
India	**2**	**7**	**1**	**3**	**4**	**24**	**...**	**...**	**65**
Indonesia	8	22	8	11	11	48	...	59	...
Iran	...	38	...	15	29	...	...	...	15
Japan	323	552	34	60	374	120	689,800	16	14
Kenya	8	13	3	6	10	1	6,200	74	75
S. Korea	14	195	11	106	151	54	56,940	212	42
Malaysia	...	152	...	33	131	164	...	...	17
Mexico	...	140	...	52	92	3	...	...	4
Nigeria	4	12	3	7	7	3	...	123	...
Norway	342	470	17	22	379	46	25,386	8	6
Pakistan	2	7	5	4	5	12	31,950	71	28
Russia	...	158	...	24	92	...	...	...	9
S. Africa	133	134	18	16	106	7	...	25	26
Sri Lanka	...	14	...	3	6	28	8,950	...	75
U. K.	303	399	50	63	359	10	436,470	19	14
U. S.	...	767	25	32	521	14	2,577,620	...	17

Source : World Bank.

statistics published (Times of India, Oct 8, 2012), one in every 10 road deaths across the globe is in India and every sixth road crash in the world happens in India. This is when India is having only 1% of the global vehicle population. The total number of deaths every year due to road accidents

has now passed the 1,35,000 mark according to the latest report of National Crime Record Bureau (NCRB). Trucks and two–wheelers were responsible for over 40 percent of deaths. The total loss from road accidents is estimated to be Rs. 1.5 lakh crore every year.

13.1 UNITED NATION'S DECADE OF ACTION (2011–2020) FOR ROAD SAFETY

To encourage road safety United Nations have started an awareness plan on global scale. During the decade 2011–2020, all the countries are expected to implement changes in policy and encourage actions to reduce economic and social impact of road crashes.

Following are the facts which prompted UN for the action plan:

- 1.3 million people are killed on the world's road each year.
- 50 million people are injured, many disabled.
- 90 percent of these casualties occur in developing countries, which have less than half of the world's motor vehicles.
- Annual deaths are forecast to rise 1.9 million by 2020.
- It is the number one cause of death for young people worldwide, killing 260,000 children under the age of 18 each year.
- The global losses due to traffic injuries total US$ 538 billion a year.
- The economic cost to developing countries is at least US $ 100 billion a year.

UN action plan seeks to stabilize and then reduce the forecast level of road traffic fatalities worldwide by the year 2020. The goal of the action programme is to reduce the forecast level of road deaths by 50%, i.e. from 1.9 million to less than 1 million a year. Achieving the 2020 target could save up to five million lives and prevent 50 million severe injuries in addition to an estimated economic benefit of US $ 3 trillion.

The five pillars on which this programme will focus are road safety management, safer roads, safer vehicles, safer road users and improved crash response. India is a signatory to this programme and is yet to announce its action plan.

13.2 CAUSES OF ROAD ACCIDENTS

Road accident occurrence is a complex phenomenon, essentially due to interaction of vehicle, road user, roadway and environmental conditions. The traffic on the road is increasing rapidly while the capacity of road is not able to meet the demand. Mismatch in interaction of vehicle, road users, and the road environment leads to occurrence of accidents.

Accidents can never be totally eliminated, and certain level of accident risk will always be there and should therefore be tolerated. An understanding of accident causative factors will provide clues to use measures to reduce the accident risk, and the severity of an accident. Usually, the cause of an accident is not a single factor. A combination of events and factors lead to an accident, some of which may be easily identified while others may be indirect and not so easy to predict.

The causes of road accidents are attributed to various factors which can be grouped in following categories:

- Road deficiencies
- Vehicle defects
- Driver's behaviour
- Pedestrian (road–use–human) errors
- Enforcement related (act and rules)
- Environmental factors
- Other factors

The cause of traffic accident are related to country's national condition, regional conditions, economy and character of traffic. Because of these factors traffic accidents have different characteristics from country to country. Based on several studies and research work done in India, the causes of different types of accidents are enumerated in following paragraphs.

13.2.1 Road Deficiencies that Cause Accidents

Highway/road features affect safety aspects in many ways. Inadequate or deficiencies in geometric design of roads could contribute directly or indirectly to road accidents.

The common road features which affect safety of road are enumerated below:

- Inadequate lane width/shoulder width restricts lateral separation.
- Poor design of crest and sag curves reduce the visibility.
- Safety on curves is affected, if super–evaluation is not adequate to counteract lateral force.
- Sign distances for stopping and passing actions, if not property designed result in different type of accidents.
- Plantation and vegetation on sides of the road result in visibility obstructions.
- Surface defects like potholes, corrugations, etc increase accident potential.
- If drainage is not proper, vehicles control in wet weather, gets reduced.
- Raised medians for traffic control are common hazard.
- Too much channelisation at intersection locations creates confusion, resulting in accidents.
- Physical layout of intersections, with unsafe/confusing turning movements, creates hazards.
- Vertical edges of the lanes, pose serious hazard to drivers specially inside of horizontal curves.
- Fixed objects on the roadsides such as signs, barriers, luminaire post, etc. if not properly designed and located result in serious accidents, due to impact with them.
- Side slopes and ditches along the road for draining the water if not properly designed cause vehicles to bounce severely, and overturning may occur.
- Lack of proper road markings and traffic signs to guide drivers, create confusion and hazardous situations.
- Many accidents take place because pedestrian crossings are not marked on the road/ subways are not provided.
- Lack of bus-bays for loading and unloading of passengers along the road, results in many accidents.
- Inadequate/poor light conditions, result in night accidents.
- If footpaths adequate width are not provided, pedestrian walk on road, creating unsafe conditions.
- Sharp bends in horizontal alignment are dangerous. Blind corners result in accidents.
- Advertising hoardings along the road distract driver's attention and result in accidents mainly at intersections
- Lack of guard rails at places like bridges curves and other strategic locations results in run-off type of accidents.
- Bus stands located too close to an intersections have created safety problems for vehicles as well as bus riders

- On many sections of the road poor lighting is responsible for night accidents. Poor visibility increases the risk, because road user are often not seen. A large number of pedestrian and cyclist collision occur around dusk, dawn or at night, because of poor visibility.

- Poor maintenance of roads, bad road surfacing and lack of precautionary measures are major cause of accidents. Road cuts are typical examples.

- If road signs and marking are not properly made and placed or if markings are not clearly visible due to fading or obstacles which interfere with visibility (e.g. vegetation) the risk of accident increases.

The relationship between safety, highway geometrics and other features if well understood by practicing engineers, safety oriented designs, will eliminate many accidents. Road should be constructed so as to prevent the driver from doing a dangerous act.

13.2.2 Vehicle Defects that Cause Accident

There are several types of vehicles slow and fast using the same road. Brakes, lights, tyres, breakdowns, etc are many factors, creating unsafe conditions.

- Conflict between motorized vehicles and non-motorised vehicles plying on the same road gives rise to large number of accidents.

- Road vehicles if not strong, well designed and maintained result in accidents due to failure to perform particular function in time.

- Weak door latches and hinges increase the risk of ejection from the vehicle, with chances of severe injury.

- Failure or non-effectiveness of braking system of vehicles is responsible for some road accidents

- Motor cycles, scooters, etc. offer little protection to rider involved in accidents. Rider frequently makes direct connect with road surface and other vehicles. These vehicles and their riders are at higher risk of an accident than car occupants.

- Tyre failure also plays role in some road accidents, though the number of accidents due to this cause is not very high.

- Other causes leading to accidents due to vehicle factors are:
  - Poor visibility of the driver from driver's seat
  - Presence of protruding objects
  - Absence of seat-belts in the vehicle

Provision and use of seat-belts reduces serious and fatal accidents. The restrained occupant has ten times less chance of being ejected in a crash than does an un-belted occupant. The greatest benefit of a seat-belt is in frontal collision and roll-over accidents.

13.2.3 Driver's Behavior Responsible for Accidents

Drivers are human beings having varying degree of education, intelligence, emotions, while using the road. PIEV time, age, sex, driving habits, education, training, etc. all play role in road safety. Most of the accidents are beings caused by wrong action of drivers at the wheels.

- While overtaking drivers are reckless, not judicious. Their wrong judgment results in many fatal accidents. Overtaking from wrong side is frequent.

- Over-speeding of the vehicles or reckless driving results in accidents as at high speeds controlling and manouvering of the vehicles is difficult. Skidding occurs due to high speeds.

- Drunken driving, during nights is a major cause of all types of accidents. Due to influence of alcohol, the nervous system of drivers gets affected and he loses his mental balance,

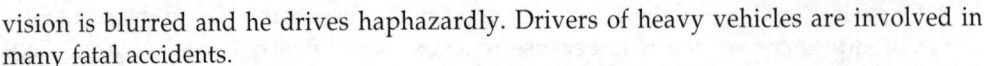

vision is blurred and he drives haphazardly. Drivers of heavy vehicles are involved in many fatal accidents.

- Overloading of a truck or a bus, running at high speeds, results in major losses when accidents is met. Lane discipline is not followed by heavy traffic.
- Illiterate drivers are not able to read traffic signs and warnings. Their failure to do is responsible for many accidents. Driver's behaviour in compliance with traffic rules and regulations is frequently very poor.
- Use of cell phone by the drivers while driving vehicles is also a cause of accidents. 'Talk and Walk' and `Talk and Drive' involves heavy risks on the road.
- Drinking and driving is the major contributing factor, in the young age group of drivers in accidents occurring at nights and during weekend.
- Fatigue of driver is also a contributing factor in night accidents. These drivers get tired and sleepy after long driving hours.

13.2.4 Pedestrian (Human) Errors that Cause Accidents

Pedestrian facilities in India are far from satisfactory and their moving on roads is a big risk.

- Many accidents occur when pedestrians cross the road. Some of them occurs because use of mobile phones, while crossing the road.
- Footpaths are often damaged or occupied by adjoining shop keepers, forcing pedestrian to walk on the road.
- Young male drivers with little driving experience with high level of vehicle usage create more safety problems on roads.

13.2.5 Enforcement related Causes of Accidents

Non-existence or non-observance of legal rules and regulations results in several road accidents. Enforcement authorities role is very important.

- Absence of access control invariably increases accident rate.
- Uncontrolled parking affects road safety adversely. Poor enforcement of parking regulation is often the cause.
- Overtaking rules are never followed and enforcement people never do any checking or punishing.
- Varying traffic regulations, create confusion and result in accidents.
- Speed limit enforcement in India is almost non-existent.
- Due to lack of enforcement and checking by the traffic police to detect drinking and driving, there is no impact on the minds of the people. Roadside bars and liquor shops further encourage this problem.
- Drivers of below the legal age of driving (i.e. 18 years) also drive on the road, due to poor enforcement resulting in unsafe conditions.

13.2.6 Environmental Factors in Accidents

Roads are used in extreme conditions of weather, during rain, fog, snow, etc. These environmental changes affect road safety in several ways.

- During rainy season when roads are slippery many roll over accidents take place.
- 'Hit and Run' accidents are increasing and police is not able to track down vehicles and drivers in most of the cases.

- Generally police personnel who perform traffic controlling jobs are not adequately trained in accident investigation or completion of the accident reporting form.

13.2.7 Other Factors

- Heterogeneous traffic flow consisting of mixed fast and slow type vehicles plying on the same road, competing for space results in accidents.
- Ribbon development (shop and houses) along the road, with increased movement of people, creates dangerous conflicts with vehicular traffic. Roadside garages/repair shops, grain markets create safety problems.
- Stray cattle in many Indian cities has resulted in accidents as drivers while saving the cattle hit other objects/vehicles.
- Advertisers, builders are frequently seen unconcerned about road safety in placing advertising hoardings, construction material on or very close to the road without consulting and informing road authorities.
- Informal markets which spring out on sections of the road, cause traffic safety problems.

In developing countries like India, the nature of road safety problem is different from the industrialized developed countries. Roads in these countries were not planned and designed for the volume and type of traffic which they are now required to carry. In addition unplanned urban growth resulted in incompatible land-uses with high level of vehicle/pedestrian conflicts.

The other problem is movement of people from rural area in large numbers. This brings new urban residents who are unaware of using high traffic roads, thus creating hazards. In general accidents are consequences of high urban density resulting in high traffic concentrations, and congestion.

Highway design standards are generally taken from developed countries without appropriate modifications for needs of the particular country. In these countries emphasis is more on construction rather than maintenance or operational aspects, such as road signs, markings, which are too often left for future. Maintenance of existing roads is ignored as compared to building of new roads.

Accident occurring is so uncertain that someone's accident record has little predictive power, with respect to his future chances of involving in accident. There is no method of accurately predicting the future accident involvement of individual driver.

13.3 REPORTING, RECORDING AND ANALYSIS OF ACCIDENT DATA

For achieving roads safety, it is crucial that good, accurate and comprehensive accident data is available, so that the problem can be properly understood, defined and suitable remedial measures are implemented. It is essential that good accident data systems are established. Accurate accident data is essential for future prevention of accidents, by removing the specific cause through safety measures. A sound scientific approach to accident recording and analysis is essential for planning, designing and engineering improvements, as well as for effective enforcement and educational strategies.

Uniform, complete and accurate report, with rapid retrieval, analysis and compatibility with recording system at national and state levels, can provide information regarding number, type location, time of accident, physical circumstances leading to accidents, people involved, injuries, death and property damage, etc. to plan future safety strategies.

Accident data is required by highway engineers, planners, traffic police, lawyers, research organizations, teachers, politicians, statisticians, insurance companies, doctors, public and other concerned with road safety.

An accident data collecting, storage, retrieval and analysis system should be established for comprehensive programme to identify and improve accident black spots, and ensure safety on roads. The purpose of accident data analysis is to know possible causes of accidents, as related to drivers, vehicles, road, etc. and plan measures to control them and reduce their frequency and severity.

13.3.1 Accident Reporting and Recording in India

Accident data collection is required to be done on the forms suggested by Indian Road Congress (Field data sheet 16, Chapter 10). These forms are filled by untrained and non experienced police personnel, for completion of FIRs (First Information Reports) of an accident. These standard forms are recommended for uniform adoption in country by the police and other traffic agencies who collect and maintain road statistics information. The traffic police is also required to prepare annual summaries of accidents in each state, and report for national road accident statistics. Despite these efforts towards recording and reporting system in India, the existing data base of road accidents is poor, and grossly inadequate for analysis of causes and accident preventive measures. The form recommended by IRC is too complicated, lengthy and cumbersome, for most of the unqualified police personnel to complete scientifically. The interests of police is to record information mainly to determine who is at 'Fault'. This form is quite elaborate having 19 blocks for all types of accidents.

The data gathered by State Government is collected from the individual police stations retrieved from their first information reports (FIRs). Many accidents are not reported or registered. The FIRs are based on statements of persons reporting the case, without proper site investigations, filled by police personnel with a view to fix responsibility for the accident rather than cause for the accidents, relating to road features. The exact location or sketch of the location is not included in FIRs nor the causes like mechanical defect in vehicles, bad weather, poor road conditions, etc. FIR reports do not collect information on socioeconomic background of road users, accident victims, and drivers.

There is a need to have a single, simple, uniform form for recording accident data and it has to be computerised. Trained personnel should be used for completion of these forms using computers and analysis software, etc. The form should be attractive, easy to fill in and contain all the information considered necessary for accident investigation and the evaluation of remedial measures implemented.

13.3.2 Element of Scientific Data System

Accident data base is required by several people involved in road safety. Their requirements may be slightly differing according to the need and reasons for wanting the data, but they all want to acquire as much relevant knowledge as possible from the accident data to serve their purpose and ensure road safety in the future. Complete and accurate data is necessary for road safety improvements.

There are following four basic elements, for assembling and utilising the accident data in a scientific way.

i. Accident reporting and recording system
ii. Accident data storage and retrieval system
iii. Accident data analysis system and
iv. Dissemination of accident data.

(i) Accident reporting and recording system

There should be only one single organisation responsible for collecting and compiling the national data base for accidents. Police may be the most suitable organization for this purpose,

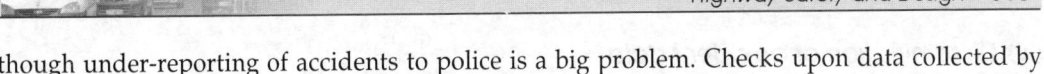

though under-reporting of accidents to police is a big problem. Checks upon data collected by the police should be made for its completeness.

The number of questions in the reporting form should be minimum, containing sufficient information for accident investigators, court requirements and local, regional and national needs. System should be conducive to accurate reporting and improving reporting rates. Details to be recorded for accidents should be requiring minimum guidance in completing the form needed for accident investigation as too long or complicated form will discourage officers filling it accurately. The information required for each accident should be completed at the site of the accident, using the specially designed form for all purposes suitable for filling and computer data entry. The data collection form should provide space for driver, pedestrian and witness statements along with a condition and collision diagram, to make it complete for police purposes, research studies, and establishing causative factors of road accidents.

System of computer software for data base is very useful and reliable. More than one data base can be developed by different organizations like police, hospitals, insurance companies, research agencies, etc. and then it can be cross-checked for its completeness, by integrating, to know complete picture of accident causation. A coordinated safety approach to gain maximum value out of each separate system is best. Road accidents should be managed not by gusts of emotions but run by logic derived from data based on sound research.

(ii) Accident data storage/retrieval system

Accident data may be stored in a manual, micro-based or mainframe based storage software system. It should be easy to store and retrievable as needed. The data structure has to be such that information stored on individual accidents can be retrieved either as a single record or in combination with other records. This permits cross tabulation and more useful annual reports to be produced.

The software package used to enter data, therefore must be standardised and easy to use. It should contain number of standardized logical checks automatically applied on data entry to ensure that the data is accurate from the outset.

It should be possible to produce maps showing location of accidents, and all other information giving details of the accident from the software used. The process of identifying locations with high number of accidents, becomes very quick and simple.

(iii) Analysis of accident data

Accident data collected, should be easy to process. Analysis by hand is time consuming, costly and prone to errors. The rapid development and availability of microcomputers has opened up new possibilities of analyzing road accident data efficiently. Use of microcomputers is relatively cheap, easy, accurate and user friendly.

Trained, experienced personnel are needed to administer and execute the collection of data, to analyse, interpret and present it in a meaningful manner, with a good institute to administer.

Accident investigating team should use the data base through a GIS (Geographical Information System), or other processes to search the pattern for accident occurring in a city or area, to plan for short and long term safety measures.

For investigating the nature and cause of road accidents, tables of frequency and/or percentages can be produced for accidents, casualties and vehicles. For each type, cross tabulation or other type of tabulation should be possible. Accident frequency plots and maps can also be produced, along with other graphical representation of the accident data on a particular road.

(iv) Dissemination of accident data

Better communication and cooperation is needed among different organisations that use the accident data. It is important to ensure that the data is utilised as effectively and widely as possible. Police annual accident statistics reports should be circulated widely and all concerned should use the data. They should be readily accessible to relevant organisations for designing appropriate counter-measures, producing plans, monitoring effectiveness and carrying out research.

With detailed annual accident report, highway authorities are better able to plan and carry out safety improvement programmes and correcting the black spots. Statistical summaries of the accident data by various characteristics, are prepared in order to determine accident trend and plan counter measures to achieve safety. Table 13.6 shows typical accident characteristic categories, for data summation.

13.4 ROAD DESIGN WITH SAFETY CONSIDERATIONS

To ensure safe travel, measures should be taken at various stages and levels. Adoption of safety conscious approach in planning, design, construction and maintenance of new roads and adequate implication of traffic engineering practices, can contribute greatly towards improving traffic safety. By incorporating good design principles from the start, it is possible to avoid many traffic safety problems. Improvement of the existing roads can also be undertaken by introducing safety related traffic, enforcement and engineering measures.

Traffic safety measures are discussed in terms of 5 ES; Engineering and planning, Enforcement, Education, Environment and Economy. 'Engineering' aims to provide the basic design elements to make safe roads. 'Enforcement' provides rules for safe use of the roads and punishment for violators, 'Education' is expected to provide good road sense and feeling of responsibility towards road use, 'Environmental' factors such as climate, wet and dry, hot and cold, day and night, rain, fog etc. should not adversely affect the safety and finally all these measures should require to be implemented in most 'Economic' way. Engineering, enforcement and education has strong relationship and each one should contribute fully in road safety.

A multidisciplinary approach involving various organizations should make efforts to improve road safety. For example engineers to create safer road through planning, design, construction and maintenance, police to influence driver's behaviour through enforcement of traffic rules and regulations and educationalists try to train and inform road users about potential danger on the road and ways to avoid them. Efforts should be coordinated to have maximum effect, by reinforcing each others activity.

For efficient and effective management of road safety adequate funds and responsible organisations capable of carrying out and co-ordinating such activities, should be created so that impact of individual efforts are maximised. Both government and non-government key organisation, should meet regularly to discuss road safety activities.

'National Road Safety Council' (NRSC) is one such effort. It should have greater and dynamic role, with its own operating budget and secretariat to work full time. A strong inter-agency relationship and understanding with clearly defined and shared responsibilities for road safety is required. Clear cut responsibilities between concerned agencies, ministries and departments should be established, checking at various levels that no major aspect affecting road safety is missed. Road safety should be taken as a National mission. Road safety audit has to be made compulsory for all new roads during planning, design, construction, and operation.

13.4.1 Safety Conscious Planning of Roads: (Planning measures to road safety)

Engineers and planners who create a road network for present and future land users can design the roads accommodating human characteristics, and by incorporating good design principles

Table 13.6: Categorywise summation of traffic data

S.No	Summary of Data	Categories of accidents
1.	Nature of accidents	• Left-turn, head on • Right-angle • Sideswipe • Pedestrian-related • Run-off-road • Fixed object • Head-on • Parked vehicle • Others
2.	Severity of accident	• Fatal • Personal injury – Incapacitating – Nonincapacitating – Possible injury • Property damage
3.	Possible causes of accident	• Driving under the influence of alcohol or drugs • Reckless or careless driving • ILL, fatigued, or inattentive driver • Failure to comply with license restrictions • Obscured vision • Defective equipment • Lost control due to shifting load, wind or vacuum • Other road factors
4.	Environmental conditions during accident	• Weather (clear, cloudy, rain, fog, snow) • Ambient light (light, dark, dawn, dusk, street lights) • Roadway surface (dry, wet, snowy, ice)
5.	Time of day of the accident	• 12:00 midnight 6 am • 6 am – 10 am • 10 am – 4 pm • 4 pm – 11 pm • 11:00 pm – 12:00 midnight
6.	Locations of accident	• Intersections - Non-signalised - Signalised - Urban/rural • Mid of the road - National highway - State highway - Urban road - Rural road.

from the start so that many problems can be avoided. Incorporating safety features during planning and design is cheaper than correcting afterwards. Planning and design for traffic needs depends ultimately upon the land uses which they serve. Control of both is vital. Various measures of safety conscious planning and design are:

- The road network should be clearly classified into those meant for through traffic, local traffic, and for local access, in a systematic hierarchy.
- Traffic at the intersection should always be given priority of movement on important road.
- Land uses should be carefully planned and distributed to minimise road traffic, pedestrian conflicts, and total travel.
- Segregation of fast and slow traffic should be considered. Slow-moving traffic and vulnerable road user should be segregated wherever possible from faster motor vehicles.
- The need for travel by vehicle should be minimized by locating shops, schools, etc. within walking distances from home. Self sufficient neighborhoods should be planned.
- The number of direct accesses to main road should be minimized, by bringing traffic to a single T Junction. No access should be permitted at potentially dangerous locations like intersections or bend, etc.
- Land uses should be planned with the aim of minimizing travel and maximizing accessibility to public transport.
- Land uses in urban areas change continuously and therefore consistent control is required to achieve road safety.
- Decision points for the drivers must be simplified and should be non confusing. Drivers should be informed of any adverse characteristics on the road, well in advance to take necessary safety action.
- 'Recovery Zones' wherever possible should be provided so that even if driver makes an error of judgement, there are still opportunities for him to regain control before an accident occurs.
- Signs which are not clearly visible or obstructed should be avoided or splitted into two and more signs used to simplify driver's task.
- Conflict point between pedestrian and vehicles, and other vehicles can be avoided by providing separate lanes for pedestrians (side-walks), cycles and slow moving traffic.
- Residential development should be separated from heavy industry and major commercial uses.
- It is necessary to develop comprehensive set of safety measures, such as speed and parking restrictions, together with measures such as heavy vehicle bans, etc. to achieve overall safety.
- The number, types and spacing of intersection should be consistent with the type and nature of traffic for which road is planned. Intersections should be with same class of road.
- Grid iron pattern layout of roads which does not separate access and movement functions is unsuitable from safety point of view. Existing grid-iron networks should be closed off or restricted.
- When a bypass is provided for through traffic deliberate efforts should be made to reduce speeds of traffic as it passes through built up area. New bypasses should permit very limited access.
- When rehabilitating roads, efforts should be made to ensure that unsafe condition are not created as a result of the increased speeds now possible on the improved road.

- Off road bus-bays should be provided along the roads at regular intervals
- On road parking should be discouraged and displaced to off-road sites.
- Grade separated intersections (interchanges) should be planned for extremely high flows of traffic.
- Vertical re-routing via overbridge or subway is not attractive to pedestrians as compared to at grade facilities. This fact should be considered in planning.
- Area wide traffic management schemes should be planned, to restrict the undesirable effects on road safety, when overall/major improvements are not possible.
- In commercial and retail areas, road safety is a major problem. For safe and efficient movement of both people and goods, adequate facilities should be provided for different activities, speed reducing devices may be needed to protect pedestrians.
- Bus routes should be planned to serve all major needs. Bus-Bays should be provided beyond intersections, and linked with other parts of the general traffic network such as footpaths.

13.4.2 Safety Conscious Designing of Roads
(Engineering/Design measures to road safety)

Accident risks are considerably reduced, if safety is given special attention in initial designing stage or improvement of any road and its intersections. Safety features should be incorporated in the very beginning and carried forward throughout the complete design and construction process. Proper geometry, adequate signs and markings, clear priorities are required for safe and efficient use of the roads by the drivers. Even if drivers makes a mistake, the design should enable him to either recover without accident or minimise severity should an accident occur.

Some of the basic safety considerations, from design stage to final construction of a road are:
- On urban roads accidents are caused by intersections between the conflicting streams of traffic and between different road user groups. Clear segregation and prioritisation is therefore necessary for reducing such conflicts.
- Provision of adequate safe crossing places for pedestrians is required, specially where pedestrian traffic is high such as hospitals, schools, intersections, etc.
- By suitable design, traffic management schemes should incorporate one-way movement systems to reduce conflicts and accidents.
- Geometric design elements should be based on actual speed of vehicles, so that driver's expectations are met. Roads must be designed to induce the required speeds. It is common to use 85th percentile speed for design.
- Adequate sight distance should be available along full length of the road for vehicles to be able to stop safely. There should be a clear signing and marking system to indicate locations where sight distance is inadequate for safe driving. Minimum stopping sight distance should always be available.
- Suitable sight distance may be achieved by increasing the radii of horizontal and vertical curve, widening shoulders, and benching to allow visibility outside the road width.
- Successive short vertical curves should be avoided as they are potentially dangerous. Vertical cures should be comfortable (Shock-free).
- Horizontal alignment and profile should be made as flat as possible at intersections and interchanges where sight distance along both highways is important. Minimum and ruling radii should be used only when site conditions are difficult.
- 'Broken–Back' horizontal curve should be replaced by a signal curve. Adequate super-elevation should be provided on all horizontal curves, to prevent vehicles to go off the road.

- Many safety benefit are obtained by simple careful designing the cross-section profile. Lane width, type and width of shoulders, lateral clearances to road-side must be considered jointly with pavement width, in evaluation of highway safety (2 lane width is 7 m, single lane 3.75 m and intermediate lane 5.5 m)

- A divided highway with wide medians and wide shoulders, result is less number of accidents.

- Accident rate on 3 lane highway is one third higher than accidents on 2 lane or 4 lane roads, for same volumes of traffic. 3 lanes road should therefore be avoided.

- There is considerable variation in a accident rate with sharpness of curves, frequencies of curves and grades. Transition curve should be used for easement.

- Road side objects such as culvert head walls, bridge abutments, utility poles, signs, trees, etc. must be controlled to some degree in design, construction and maintenance of highways to ensure safety.

- Advertising or other roadside signs should not be placed where they distract (divert) the attention of the driver, from driving.

- Increasing percentage of cross-traffic increases the accidents at intersections. Improvement in safety is achieved by channelisation, refuge islands, sidewalks for pedestrians, control devices, signals, etc. These should be considered in geometric design of roads.

- Uniform types and high quality of signs, markings and traffic control devices must be used for better safety.

- Steep side slopes are hazardous. Slopes of 4:1 are considered safe. Steeper slopes should be avoided or guard rails provided for safety, if steeper slopes are used.

- Considerations should be given to adequate lighting arrangements – type of bulb, height of poles, glare, etc. so that night accidents are prevented.

- Control of access, is one of the most important measure which reduces accidents. Depending on the area, it may reduce 40-50% accidents.

- Generally at low speeds accidents are less and non-serious type. High speeds are certainly responsible for more and severe and fatal accidents. Depending upon traffic volumes, weather conditions, road-side developments, cross-traffic and other factors, safe driving speeds should be maintained and used as design speeds.

- Highways should be designed so that a driver needs to make only one decision at a time.

- Open channel drains should be covered, where possible or have some physical barrier to separate them from the carriageway. Deep, steep-sided, drainage channels result in increased danger to vehicles.

- Safety fences should be placed between carriageway and the objects which cause severe accidents, if struck such as bridge supports, on high embankments, etc.

- Reflective delineators posts and marking such as raised pavement markers, centerline, lane and edge markings are especially useful for night time safety on roads, guiding a driver in a potentially dangerous location.

- Roadside obstructions such as trees, or road furniture are dangerous due to potential of collision as well as obstruction to visibility. Great care should be taken in positioning road furniture and obstacles, such as telephone, electricity poles, etc should not be allowed subsequently, so that visibility is maintained

- Median barriers to reduce or eliminate the danger of head on collisions and prevent pedestrian crossing at potentially hazardous location, should be used.

- The design of lighting system should be as per road surface reflection characteristics in order to provide the optimum quality and quantity of illumination. Siting of lamp posts should be carefully done to avoid being an obstruction. Lighting is important for pedestrian and cyclists also. Average level of illumination as per Indian practice should be 30 Lux on important roads carrying fast traffic, on secondary roads carrying light traffic 4 Lux can be used.
- Road (rural) should be planned with dense roadside vegetation to eliminate headlight glare.
- Regular maintenance of lighting system should be done. All installations must be inspected on a regular basic. Sodium lights, particularly at key points are much more efficient than mercury or tungsten lighting.
- Lay-byes and bus stops for allowing vehicles to stop safely without interfering with other traffic, and with less risk to passengers getting on or off should be provided. They are essential elements of road design, for enhancing safety.
- Bus stops should be located beyond pedestrian crossings and away from an intersection for safety.
- Pedestrians are particularly at risk on urban roads. Roads should be designed with raised footpaths, as a part of cross-section in urban areas. Adequate width, maintenance and keeping footpaths unobstructed, especially by street traders and/or parked vehicles is important.
- Inter-urban roads, or rural roads, where people walk many miles, construction of simple foot ways along the road or even wide shoulders should be provided to avoid conflict with high speed traffic. Considerations should be given at the design stage of the highway.
- On bridges, flyovers, special segregated provision should be made for cyclists and pedestrians, in the designs.
- Staggered intersections are preferable to cross-roads on safety grounds if space permits. Multiple lane approaches should be avoided if possible.
- Various safety measure in designing intersections include provision of skid-resistant surfaces, good direction signs, facilities for pedestrians and cyclists, replacement by roundabouts or signals, adequate visibility and sight distances, provision of islands, markings, etc.
- Round abouts are suited for heavy turning movements which cause safety problems with other types of intersections.
- At signalised intersections, the signal heads must be conspicuous in all lighting conditions so that drivers can stop safely. Signals should be so located that they are visible only to the traffic for whom they are intended.
- Signal timings should be updated frequently (each year at least) depending upon current traffic flows, and speeds to ensure that their operation is safe and efficient. Pedestrian phases at signals are particularly effective in road safety.
- As with other elements of design, consistency of signs and marking is important so that drivers can understand warning signs quickly. Symbols are preferred to words for proper understanding even by illiterate and foreign drivers. The location of sign is important as it should not create an obstruction.
- Acceleration and deceleration lanes are useful for safety as they allow traffic to merge and diverge safety, with minimum disruption of other traffic. They should be used wherever possible.

- Badly maintained roads contribute to the growing road safety problems. Greater resources (manpower, equipment and funds) should be made available to ensure that safety-related elements of the road such as road structures, drainage, shoulders, slopes, bridges and control devices are kept properly maintained.
- Guard rails should be installed to reduce the probability of occurrence of an accident, where
  - embankment slopes and height exceed safe values
  - road formation narrows at some bridges/culverts
  - on the outside of sub-standard (sharp) curves.
  - protection of structures and pedestrian is required.
- Use of GPS technology can help in avoiding congested roads by providing advance information about the road, about weather, in alerting police, attending accidents, etc. GPS device can be integrated with traffic monitoring system.

13.4.3 Safety Considerations in Vehicle Design

The fundamental design of vehicles is crucial for road safety. Car designs and manufacturing should be to protect the drivers and co-passengers. The driver and passengers, in some new models are surrounded with high strength steel safety structure which is capable of withstanding enormous stress, thus providing protection to occupants.

Following are some of the basic considerations for safe vehicles on the road. These safety devices are already in use in some cars, and can be easily introduced in all cars in India.

- Energy absorbing structure designed and mounted in front and rear of occupant's cage. Front and rear bumpers mounted on the car are also be designed for aborsbing energy.
- For greater safety new trend is to mount engine in transverse direction to distribute energy across a broad area effectively thus providing better protection against crash.
- Side Impact Protection System (SIPS) helps to protect occupants from side impacts collisions. Reinforced steel profiles provide added strength and keep the energy of impact, away from the occupants.
- The arm rests in the interior of the car are designed to collapse in side on impact and absorb more crash energy, reducing liver injuries.
- For increased protection in side impacts, side airbags may also be used to reduce the risk of serious side impacts.
- Steering column is designed to collapse in stages using an energy absorbing design to direct the steering column downwards away from driver.
- Whiplash protection system uses an advance seating design to diffuse violent energy transfer in rear impact collision.
- Anti spin system prevents the front wheels from spinning even on slippery surface.
- Side marker and day time running lights make vehicles more conspicuous.

13.4.4 Enforcement Measures to Road Safety

Traffic management, better operational controls, and enforcement measures aimed primarily to reduce congestion and improve traffic circulation, help in enhancing road safety. Potential future problems can often be avoided through early enforcement of operational and preventive measures, which lead to safer use of existing roads. Sustained and concentrated efforts on legislation and enforcement are required. Some of these measures are:

- Prohibition of animal drawn vehicles from certain roads, not allowing heavy vehicles in residential areas, etc. help to reduce diversity of traffic and aid in road safety.
- Pedestrian crossing points should be clearly defined and enforced.
- Change of land use and development over time should be controlled for overall preservation of road safety, through strict rules.
- Unauthorised use of facilities, road side developments, hoardings, etc. encroaching onto the road or causing abstractions create danger for road users and need to be prevented, through rigorous enforcement.
- Taxi and para transit vehicles stopping indiscriminately to pickup and discharge passengers should be controlled as they create dangerous conditions.
- Public transport and para transit vehicles should undergo frequent road worthiness checks to ensure that they are safe.
- Public transport operators and drivers should be required to meet minimum criteria before being licensed to operate or drive public transport vehicles.
- Encroachment by other activities into the cross-section should be controlled.
- Obstacles like stalls, structures, shops, etc should not be allowed too close to the road edge.
- Priority rules should be enforced rigidly as their poor enforcement results in high accident rates and inefficiencies in operation.
- Parking and loading control on main traffic routes at least during peak hours, near pedestrian crossing, etc. can improve safety.
- Traffic circulation measures such as ban of certain conflicting movements, road closures, re-routing, etc. can be used to prevent non-essential through or undesirable traffic from entering congested or residential areas.
- One-way system of traffic flow results in reduction of conflict points, and should be adopted wherever possible to improve road safety.
- Effective development control procedures, must be established to prevent unauthorized accesses and encroachments onto the road.
- Due to poor pays of traffic personnel, the chances of corruption are high and many offences remain unreported and unpunished. This needs special attention.
- Overloading of vehicles is dangerous as it causes road damage and create safety problem. It should not be allowed.
- Parked, parking and unparking vehicles, cause obstructing, interference and potential danger to pedestrian and other motorists.
- Police control of unauthorized parking is necessary in central area of the city. Parking bans should be strictly maintained.
- Traffic police personnel should be adequately trained and equipped with modern equipment, to control traffic offenses.
- Random checks on tyres, brakes, and lights at different places and times should be undertaken. Defective vehicles should not be allowed the roads.
- For controlling speeds, speed limit should be practical and signs posted extensively, police should have equipment to measure speeds and training to enforce speed limits. Speed is a common factor to accidents.
- Police enforcement of axle load restriction should be effectively undertaken to save roads from structural damage, and avoid detrimental conditions to road safety.

- Deliveries or collections of goods through heavy goods vehicle (HGV) should be restricted to early morning or late at night only.

- Heavy goods vehicles and vehicles carrying hazardous load should avoid using the roads in residential areas.

- Efforts should be made to attract passengers to public transport (buses), using following techniques:
  - Giving buses priority over other traffic.
  - having good interchange facilities
  - Cost/fares should be reasonable
  - Ensuring safety and comfort by using good quality vehicles, trained drivers, etc.

- By compulsory use of seat–belts, the probability of severe injury, and of head, face or chest injuries are reduced. For serious and fatal injuries seat–belts are single most effective measure.

- Medical checks of the drivers for vision and reaction time, etc. periodically may be beneficial in road safety.

- There are adequate and enough legal provisions to make our roads safe, but the existing laws must be enforced in totality. The nature of punishment does not matter very much, but the fear of being caught is what changes people' s behaviour.

13.4.5 Educational Measures to Road Safety

By changing attitudes and creating awareness through educational programmes and formal education, roads safety can be improved.

- Education, information and training can also play a role by teaching drivers better and safer road behaviour and telling them the meaning of signs, markings, etc. Besides adequate signs and markings placed on the road, it is important that drivers know their meaning to obey them. Driver education will make better and safer drivers and be effective in reducing road accidents.

- Education of pedestrian is necessary to train them to cross the road at specified crossings only for safety. Awareness of traffic rules, publicity of management and enforcement measures can be organized through newspapers, TV and radio advertisements, to educate people and drivers. In school, lessons for safety education could be given, through chapters in their textbooks.

- Special training and refresher in service courses should be devised for traffic police personnel, for using modern equipment, better traffic control, and accident reporting

- Periodic seminars, safety weeks, demonstrating that what should be done and what should not be done, by drivers, and road-users for road safety, may play an important role.

- Our rural population, now rapidly facing new traffic environment, needs to be educated about using the road for their own safety and safety of other.

- It should be realized that carrying out the job of safety enhancement, requires proper professional training. Possessing of formal degree may be a necessary requirement, but in service training and practical experience on road safety aspect, is much more important. Right type of education, training, and experience in road safety engineering, is required for those engineers and staff who are responsible for safety on roads.

- Traffic safety programmes should address different aspects of road user activities like walking, crossing, cycling, parking, using public transport, etc.

13.4.6 Environment and Road Safety

Creating appropriate environment can provide necessary perceptual clues to modify driver's behaviour and have significant beneficial results towards road safety. Road environment should be adequate to guide the driver's safely through the system.

Nights are foggy in certain periods of the year and occurring of accidents is more during fog. Design a benign environment in which people are encouraged to behave sensibly rather than break the rules. Table 13.7 shows common accidents, causing factors and possible counter-measures.

13.5 PEDESTRIAN SAFETY

Pedestrians and cyclists are the most vulnerable of road users who are involved in large number of accidents, mainly while crossing a road. Main problem is the reluctance of people to use the control devices provided. A common cause of accident is sudden emergence of pedestrians from behind stationary vehicles into the motorists line of travel. Children and elderly are also involved in pedestrian-vehicle accidents. Pedestrians often cross the road in an impracticable manner with over-optimistic estimate of a motorist's ability to either stop or avoid impact.

Adults involvement in pedestrian accidents is mainly due to alcohol. Traffic engineers and controllers have an obligation to protect the pedestrian, particularly children and the elderly, from their own inevitable misjudgment. Children are particularly unpredictable and also poor judges of traffic behaviour.

Supervised pedestrian crossings have been successful in reducing accidents, to school children crossing the road. If pedestrian wear light clothing's they will be clearly seen on the road.

Traffic engineering measures to reduce pedestrian accidents include:

- Provision of separate pedestrian facilities, i.e. footpaths
- Provision of fences to marked pedestrian ways
- Provision of median stripes to reduce crossing length.
- Provision of pedestrian signals
- Provision of 'Zebra' markings at pedestrian crossings
- Prohibition of parking near crossings
- Provision of good street lighting
- Use of safety zones at bus, taxi stops
- Provision of over-bridges and under passes wherever possible
- Making exclusive pedestrian malls and shopping area
- Education and training of pedestrians, particularly children, about safe use of the road, and obeying traffic rules.
- Pedestrian side walks should be above the road level.

Among the above measures, which one to adopt depends upon number of pedestrians, type of pedestrians (elderly, children, etc), number and speed of passing vehicles, road geometry, economic consideration, etc. Table 13.8 shows common pedestrian accident patterns, probable causes and general counter measures.

13.6 ROAD SAFETY AUDIT (RSA)

After incorporating safety conscious planning, design, construction and operation of road, road safety audit is an independent assessment of the accident potential and likely safety

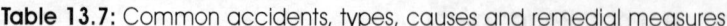

Table 13.7: Common accidents, types, causes and remedial measures

Types of accidents	Probable Causes	General counter measures
1. Collisions at unsignalised intersections	• Restricted sight distance • Large traffic volume • High speed • Pedestrian crossing • Drivers unaware of signal • Large turning traffic • Slippery/poor surface	- Remove sight obstructions - Install stop sign/warning sign - Improve channelistion - Reduce speed limit - Install signals - Install yield sign - Re-route through traffic - Put rumble strips - Re-locate cross walks - Groove pavement - Prohibit certain turns.
2. Collision at signalised intersections	• Poor visibility of signals • Inadequate signal timings • Pedestrian crossing • Slippery surface	- Put advance warning - Install overhead signals - Change signal lenses/heads - Reduce speed on approaches - Adjust amber period - Retime cycle, increase red time - Coordinate with near by signals - Provide 'Pedestrian Walk" phase - Overlay pavement - Groove pavement - Improve drainage
3. Collision with fixed objects or vehicle running of the roadway	• Object near traveled way • Slippery pavement	- Remove obstacle near road - Install barrier curb - Put sign posts/warning signs - Protect object with guard rail - Overlay pavement - Provide adequate drainage - Groove the pavement - Widen lanes - Re-locate islands - Improve/install road side delineators
4. Head-on collisions or side swipe collision vehicle traveling in opposite direction	• Inadequate road design for prevailing traffic	- Improve/install pavement markings - Channelise/divider for lanes - Create one way operation - Remove parked vehicles if any - Widen lanes
5. Side-swipe collision between vehicles traveling in same direction (turning, lane changing)	• Inadequate road design for traffic conditions	- Widen lanes - Channelise intersections - Provide tuning bays - Install signs - Put pavement markings - Remove parking (if any) - Reduce speed limit.
6. Night accidents	• Poor visibility	- Improve/install strect lighting - Improve/put delineation - Put warning signs
7. Accidents on wet pavement	• Slippery pavement	- Overlay with skid resistant surface - Provide adequate drainage - Groove the pavement - Reduce speed limit - Put 'Slippery When Wet' sign

Table 13.8: Common pedestrian accidents pattern, causes and remedies

Type/Pattern	Probable causes	General counter measures
1. Pedestrian accidents at intersections	• Inadequate protection • Unwarranted signals • Large turning volumes • Restricted sight distance • Inadequate signal phasing • School crossing area	- Install pedestrian crossings - Remove signals - Create left or right turn lanes - Prohibit certain turns - Remove sight obstructions - Put pedestrian crossing signs - Put pedestrian islands - Re-route pedestrian paths
2. Pedestrian accidents between intersection	• Pedestrian walking on road • Cross walk far away • Drivers not warned about midblock crossings	- Install side walks - Install warning signs - Put pedestrian barriers - Lower speed of vehicles - Prohibit parking.
3. Pedestrian accidents at intersections right turning collisions.	• Side walk/pedestrian movement too close to travelled way • Large volumes of right turning traffic	- Side walks to be away from the highway. - Provide right turning signal phase - Channelise intersection - Prohibit right turns - Install stop sign - Create one way street - Provide turning guidelines.

performance of the road. Road safety audit is a systematic and rigorous procedure to evaluate the accident potentiality and the performance of the road with respect to safety. It ensures that adverse features are not introduced un-intentionally into the planning, design and construction/improvement of a highway/road. Safety audit may remove such elements or if uneconomic to remove them at least plan appropriate remedial measures to minimise likelihood of accidents. Road safety audit is applicable to any road or traffic scheme whether new construction or an upgradation/improvement of an existing road. Safety audit can be done at each of the successive stages of a highway project, i.e. feasibility, preliminary design, detailed design, during construction before or after opening to traffic, or on existing road audit.

The engineers/planner who were involved in preparing the highway project, are not involved in safety audit, i.e. formal final checking, etc. It should be undertaken by a team of experts in road safety engineering and accident analysis.

The definitions of the safety audit (RSA) are:

Astroid's definition "A formal examination of an existing or future road or traffic project or any project which interact with road users, in which an independent, qualified examiner reports on the project's accident potential and safety performance"

MORT&H defines road safety audit as "A formal procedure for assessing accident potential and safety performance in the provision of new road schemes, the improvement and rehabilitation of existing roads, and in the maintenance of exiting roads".

Details and extent of road audit depends upon funds available for this purpose. It could be for entire road length, covering all features or it could be for key elements only like intersections, geometric features, etc. There is thus lot of flexibility in the implementation of Road Safety Audit (RSA) programme, to make it cost effective.

RSA is a pre-active approach, rather than re-active measure. It helps in better understanding and documentation of road safety aspects, resulting in safety improvements and minimizing risk of crashes, disruption and trauma. RSA is useful for both urban and rural road projects ranging from design/construction of a new road to upgrading/improvements of old roads. RSA is used beneficially for:

- New roads/highways
- Major new rehabilitation/improvement projects
- Maintenance works of roads
- Traffic circulation and management plans in urban area
- Other problematic (high accident) stretches of road or at problematic intersections.

13.6.1 Objectives of Road Safety Audit

The main objectives of road safety audit are:

- To identify potential safety problems – features which threat safety.
- To plan measure to eliminate or reduce those problems/features.
- To emphasis that safety consciousness is exercised throughout from planning to construction and operation of a highway project.
- To minimize the risk of accidents and/or severity of accident, if total elimination is not possible.
- To recognize the importance of safety in every project and in every individual involved in a highway project.
- To realize that unsafe roads are more expensive and it is impossible to correct certain features at a later stage, therefore reduce long term cost by using safety audits.
- In short it is better to "get it done right the first time" is the aim of safety audits.
- RSA is in fact also a check to see if quality is being implemented in road design and construction.
- RSA is an aid to optimum design, rather then following design standards blindly.
- RSA observes poor maintenance of safety features (delineation, barriers etc) to be improved.

13.6.2 Road Safety Check List

A typical check list used during road safety audit compiled under different heading from planning to design, operation of highways, and to the improvement of accident black spots is given below. Database for conducting an audit requires checking the provisions made for safety on site along with plans, drawings.

On-site evaluation through the checklist, will examine a location from the perspective of the road users (pedestrians, cyclists, vehicles — slow and fast). Check list for the audit team ensures that no important safety aspect is overlooked. It helps auditors in evaluation of safety aspect on the road.

(i) General planning considerations

- Is road development plan is conforming to general zonal plan of the area?
- Are essential facilities like hospitals, shopping, schools, etc which create large volume of traffic located within the zone?
- Are bus stops suitably located in the zone and connected by footpaths?

- Are bus stops located beyond the pedestrian crossing and intersections?
- Is there a need for cycleway network? Are they provided? If not are there wide shoulders along the street?
- Unauthorized developments, structures, advertising hoardings, etc. which obstruct visibility are removed?
- Suitable space of auto-rickshaw stands, rickshaw stands, bus stands, etc. are planned?
- Are adequate accesses to roadside properties and car parks atleast 50 m from road intersections have been provided?
- Are roads planned to exclude through traffic, and segregated footpaths/cycle tracks provided?
- Are parkings away from children's play areas, and adequate as per needs.
- Is industrial area physically separated from nearby residential area? If not, specific measures planned to minimize undesirable effects of heavy goods vehicle traffic.
- Are intersections in industrial are wide enough and designed for heavy goods vehicles?
- Has off-road space for parking and unloading of HGV provided?
- Are commercial/business area separated from through traffic by provision of service roads?
- Where pedestrian traffic is significant are there safe facilities for crossing the roads? Are there adequate side walks?
- On recreational roads, special traffic management and parking plans, have been developed?
- Are entry and exit areas for the parking are safe and suitable for volumes of traffic expected?
- Has an adequate signing, marking plan has been prepared for roads in the area.?
- For special events, traffic circulation and direction plans have been prepared to divert traffic suitably.

(ii) Road network considerations

- Have roads been categorized into a hierarchy of the classification?
- Do the roads form primary network for the whole town or region and carry most of the through traffic?
- If two are more lanes are there in each direction are dividers used to separate by means of median or central reserve?
- Do local roads serve only the traffic with in an area or connect these areas with main roads?
- Have access roads been so designed that they are not suitable for through traffic?
- Does each road intersect only with road in the same category?
- All the intersections between two roads channelised, signal-controlled intersections or roundabouts or where very high volumes are involved, grade separated?
- Is intersection spacing on road at least 250 m?
- Is access to local parking areas from access road only?
- Is visibility and signing at intersections such that road user can readily see which road has priority and where they should stop or give way?
- Is vehicle parking controlled or prohibited on roads carrying large volume of traffic?
- Have suitable bus and para transit stopping places been provided at safe locations?

Pedestrian and cyclist Facilities

- On busy roads are pedestrian channeled to safe location where special facilities have been provided for safe crossing?
- Are main footpaths separated from the road wherever possible?
- Do main footpaths always cross roads at well designed, properly signed and where possible, lit pedestrian crossing facilities?
- On the roads is there reservation for a separation strip between the carriageway and the footpath?
- Are all pedestrian crossing on roads are grade separated, or controlled by traffic signals, or designed to have pedestrian refuge such that the pedestrian never needs to cross more than two lanes of traffic at a time, before reaching a safe refuge?
- Are pedestrian crossings on collector roads controlled by traffic signals?
- Have under or overpasses are used by pedestrians or they prefer to cross on the carriageway?
- Have the crossing needs of cyclists been taken into account in detailed intersection design at locations where there are large number of cyclists?
- Are there exclusive bicycle ways at least 2 m in width?
- Are combined bicycle and pedestrian ways of sufficient width?

(iii) Highway design considerations

- Have estimates been made of vehicle speeds likely to occur on the highway section?
- Have estimates of current and future pedestrian usage been made and appropriate facilities provided for pedestrian safety.
- Has geometric design suitable for the prevailing speeds on the road?
- Has stopping sight distance available along the road above the minimum required for the speed?
- Are the radii of horizontal curves with super-elevation where required, above the minimum required for estimated speed?
- Are vertical curves adequate for the estimated speed?
- Is the cross-section adequate for the levels of traffic flow?
- Do the combined geometric design elements produce a consistent and safe alignment?
- Are the combined gradient on the cross-section and longitudinal section sufficient to avoid standing water?
- Does the alignment allow regular overtaking opportunities
- Have combining lanes been introduced where necessary to provide adequate and safe overtaking opportunities?
- Will the road enable safe driving in darkness?
- Have centerline and edge markings been designed which give adequate guidance/control for drivers to position there vehicles and overtake safely?
- Will the design lead to reduced severity in the event of an accident?
- Has adequate provision made for parked and stopped vehicles, including buses, so that they do not pose a danger to other road users?
- Have specific safety provisions been made for non-motorised traffic?
- Does the proposed cross-section include hard or soft shoulders for broken down vehicles, buses etc.

- Are road edge obstructions such as embankments, advertising hoardings, vegetation, building, etc set backed to provide sufficient visibility?
- Are crash barriers on the outside of bends provided where large drops (over 3 m) in levels occur?
- Are crossing facilities with adequate advance signing provided at well sited locations on highways close to villages or agricultural fields where villagers (including animals) frequently cross?
- Are pedestrian and non-motorised traffic discouraged from using main roads or are special provisions made for this traffic.

Intersection Design

- Will the intersections carry the design traffic load with acceptable level of reserve capacity?
- Is the route through the intersection as simple and as clear to all users as possible?
- Is the presence of the intersection clearly evident at a distance to approaching vehicles from all directions?
- Are warning and information signs placed sufficiently in advance of the intersection for a driver to take appropriate and safe action?
- On the approach to the intersection, is the driver clearly made aware of the actions necessary to negotiate the intersection safely?
- Are the different turning movements sufficiently segregated for capacity and simplicity of action by the drivers?
- Are lane widths adequate for all vehicle movements and all vehicle types?
- Do the decisions which need to be made by a driver follow a simple, logical and clear sequence?
- Are the drainage features sufficient to avoid presence of standing water?
- Is the level of lighting adequate to identify the intersection at night?
- Is the level of lighting adequate to silhouette pedestrian and other movements?
- Are sight lines sufficient and clear of obstructions, including parked and stopped vehicles?
- Are accesses prohibited within 50 m of the intersection?
- Have adequate facilities for cyclist and pedestrian (footpaths, refuges, crossings, etc) provided?
- Is the design of intersection consistent with the road type and adjacent intersections?
- Are the gaps in central islands of sufficient size to store waiting/turning traffic?

Check list for road safety audit as above covers information basically regarding:

- General planning
- Road location and network
- Pedestrian/cyclist provisions
- Road geometrics, surface condition and its consistency
- Signs, markings, delineations
- Intersections design and approaches
- Traffic volumes and speeds
- Provisions of bus bays and refuge areas

- Railway crossings
- Adequacy of lighting
- Provision of parking facilities

There will be different check lists for different stages of road project for which RSA is required. Check lists are means to an end and not an end in themselves as professional wisdom and experience of auditing team in safety engineering is vital. Figure 13.4 shows a typical procedure and share of responsibilities for conducting RSA.

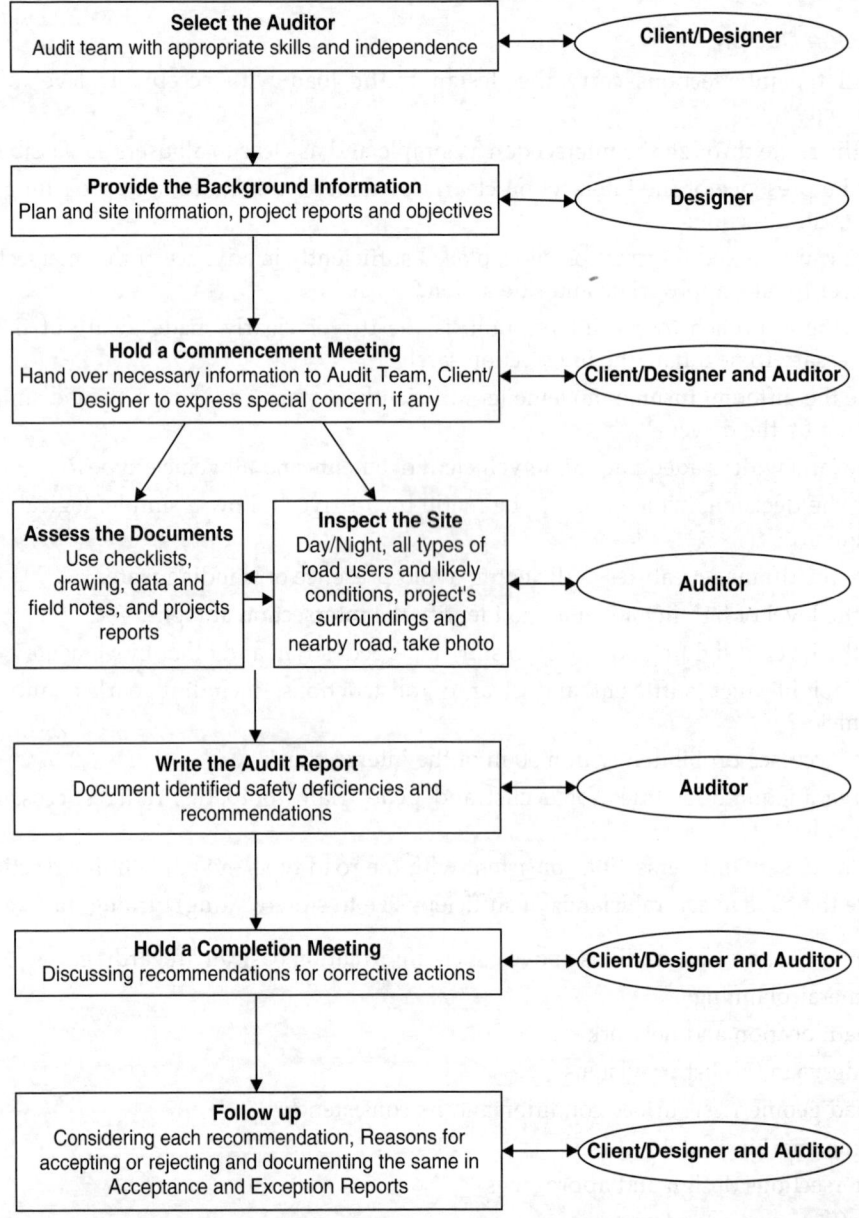

Fig. 13.4: Typical RSA procedure and share of responsibilies (Indian highways, Mar, 2005)

13.6.3 Safety Audit Report

The main contents of a RSA reports should be:

(a) Features of the project which involve safety concerns and

(b) Recommendations as to what corrective actions are required.

Based on the recommendation on what needs to be done the highway engineers/designers will take step to revise and improve the situation in best possible way.

RSA report should be brief, to the point, mentioning remedial treatments from road safety point of view, precisely location wise.

The basic contents of the report are project details, series of findings and recommendations and concluding remarks.

13.7 TRAFFIC CALMING

Speeds of motor vehicles are increasing rapidly. High speeds in the urban areas create unsafe conditions and increase number of accidents. Residents want that speeds of vehicles should be lowered down in their locality as incidence of overspeeding accidents are increasing in urban areas, as young drivers drive at high speeds. High speed is a factor in every accident – lower speeds reduce both the likelihood of the accident and severity if accident does occur.

Realizing this need of local residents, safety requirements and social considerations, efforts are made to mange/reduce speeds by engineering measures called "Traffic Calming" or "Speed Management by Design". In countries like Netherlands, Denmark, Japan, and other European countries, traffic calming measures to reduce speed and volume of vehicular traffic in residential area are becoming very common. Speeds of 15 to 20 km/hour are considered safe speed of travel. Traffic calming involves changes in road/street alignment, installation of barriers and other physical measures to reduce traffic speeds and volumes. Traffic calming measures are intended to be self-enforcing. These measures/techniques use various means like physical, psychological, visual, social, regulatory and enforcement, etc.

13.7.1 Objectives of Traffic Calming

Objectives of traffic calming, besides safety can be summarized as:

   i. Improve safety and reduce accidents
  ii. Environmental improvement
 iii. Developing appropriate driving behaviour
  iv. Reduced need for police enforcement
   v. Increase in all modes of traffic
  vi. Better quality of urban life

It is therefore important, that roads passing through towns and villages, should be calmed down, in the interest of safety and meeting other objectives described above.

Provision of warning signs, road markings are used to show the beginning and end of calming sections, to ensure speed reduction for vehicles entering the urban area from the fast highways/expressways.

13.7.2 Benefits of Traffic Calming

Traffic calming engineering measures, result into following basic benefits:

(a) **Reduction in Number and Severity of accidents:** Reduction of speed to 15-20 km/hr, result in reduced number of pedestrian collisions. Even if in such incidents severity is highly reduced. Studies in other countries have revealed that 5 to 10 km/hr reduction in speed, reduces fatal accidents by 30 to 20 percent, e.g. at 30 km/hr 50% of pedestrian hit will die compared to 85% at 65 km/hr.

(b) **Improved Facilities for Non-motorised modes:** Walking, cycling, rickshaw movement becomes comfortable, convenient and safer, as speeds and volumes of traffic are reduced- Elderly and children shall be able to walk easily.

(c) **Reduction in Noise and Air-pollution:** Due to reduced speeds vehicle emission and noise is less. It improves the environment. Traffic calming, encourages more people to visit those area, green spaces are increased, people feel relaxed, giving several indirect environmental and health benefits.

(d) **Reduction in Crimes and Community Interaction:** As speeds are low, more people visit, streets become more lively encouraging community interation. Due to this crowd of people, does not allow criminals to involve in crime, as they would be easily caught, and can not run away with speeds.

(e) **Property values are increased:** Due to better accessibility, pedestrian amenities, commercial activities and customer grow up, result in increased values of the properties.

(f) **Reduced Suburban Sprawls:** Due to pleasant urban environment and greater use of non-motorised travel, traffic calming can help in reducing "Suburban Sprawl" which imposes economic, social and environmental costs on the society.

13.7.3 Techniques of Traffic Calming

Traffic calming measures need to be applied in a formal structured way. Some of the techniques are:

(i) Advance warning to drivers for speed reduction

There are several techniques to warn the drivers in advance to reduce speed or do not enter. They could be

(a) Visual warnings, like signs, markings, gates, etc and

(b) Physical warning, like humps, rumble strips, narrowing, etc

Humps: Humps are effective in speed reduction. The profile, height, gradient, length and material used in making of the hump, affect the utility of a hump. Commonly used at intersections, near schools, etc. They should not cause inconvenience:

Rumble Strips: Rumble strips of contrasting material are laid across the travel path of vehicles. Vehicles make tyre noise while passing over them. Commonly used at accident prone locations, sharp curves, etc.

Narrowing: Road width is narrowed, so that drivers are forced to reduce the speed. This method is used mostly on two lane roads.

All these technique including speed limit signs are to let the driver know in advance that speed reduction is required for journey ahead.

(ii) Restrictive techniques influencing vehicles

The accelerating capabilities of the vehicles are restricted by these techniques. Staggering of lane, which involves merging in small area reduces the speed. Narrowing of the lane also helps in reducing the speed.

Use of roundabout also makes vehicles to reduce the speed. Careful planning of staggering and narrowing is essential due to mixed nature of traffic on our roads.

Speed control is also implemented with rough textured road surface of coarse aggregates, etc.

Though free speeds are wanted by every driver, for the safety requirements, traffic calming measures suiting to particular situation of urban development, road geometrics, and intended purposes, should be properly planned and monitored. Traffic calming is used in many advanced countries. In India also there is need and scope for this technique. Ministry of Road Transport and Highways has started research on "Development of Guidelines for Traffic Calming".

Role and requirements of traffic engineer/planner

With the alarming increase in traffic congestion and road accident in cities, traffic engineers are required to play an vital role in solving different traffic and transport problems. For solving growing traffic problems traffic engineer has to deal with various departments and also with public. He has to posses overall skills, attitudes and different approaches. In changing situation it is not only technical knowledge but an integrated awareness of various disciplines/subjects, makes a successful traffic/transport engineer. He should possess capabilities such as:

- Ability to communicate
- Command over language and knowledge of local language.
- Explaining technical aspects in simple and understandable way, using audio-visual and other techniques.
- An open minded approach in listening and responding to people.
- Welcoming suggestions and developing a sense of participation.
- Accepting useful suggestions without defending.
- Establish credibility and trust by not accepting an unsuitable proposal under pressure.
- Beaware of possible resistance and disagreement and plan himself accordingly.
- In many situations, controversies are inevitable, ability to handle them with assertiveness.
- Should have sense of humor and control over emotions and anger.
- Aware of techniques of conflict management and dispute settlement.
- Sound technical knowledge, is most important.
- Patience, maturity and sincerity are basic requirements, besides engineering skills.

Structural Design of Highway Pavements

14.0 PREAMBLE

The natural soil is seldom strong enough to support repeated applications of wheel loads without significant deformation. It is therefore necessary to interpose between the wheels and the soil a structure to support loads. This structure is called 'Pavement'.

Highway pavements provide adequate support for loads imposed by heavy vehicles, and produce a firm, stable, smooth, all year, all weather surface, free from dust or other particles. In addition it must possess sufficient inherent stability to withstand, without damage the abrasive action of traffic, adverse weather conditions, and other damaging elements.

The pavement structure is composed of a number of horizontal courses of materials (unbound or bound), whose primary function is to distribute the applied vehicle load to the sub-grade. The courses are sub-divided into layers. A layer is that part of a course which is placed and compacted as one entity.

The pavement structure should:

- Ensure that the transmitted stresses are reduced to an extent that they will not exceed the supporting capacity of the sub-grade, and pavement courses
- Provide a firm surface with good acceptable riding quality, with adequate skid resistance and do not deform excessively.
- Have favourable light reflecting characteristics and low noise generation.
- Possess sufficient stability to support without damage, the abrasive action of traffic
- Withstand adverse weather and environmental conditions and other damaging elements.
- Aim at a balance between construction, road user and maintenance costs.

To serve these requirements, high quality materials are used in the surface course, with a steady decrease in quality towards the sub-grade.

14.01 Type of Pavements

There are two type of pavements 'Flexible' and 'Rigid'. A flexible pavement consists of sub-base, base and bituminous surfacing on the top and a rigid (concrete) pavement normally consists of a concrete slab laid on a base or sub-base, which rests on sub-grade. Figure 14.1 shows typical flexible and rigid pavements.

This classification is an attempt to distinguish between cement concrete pavements as rigid and all others (bituminous or unbound) as flexible. The relatively stiff rigid pavement produces a uniform distribution of stress on the sub-grade, whereas in flexible pavements deformation is more as shown in Figure 14.2.

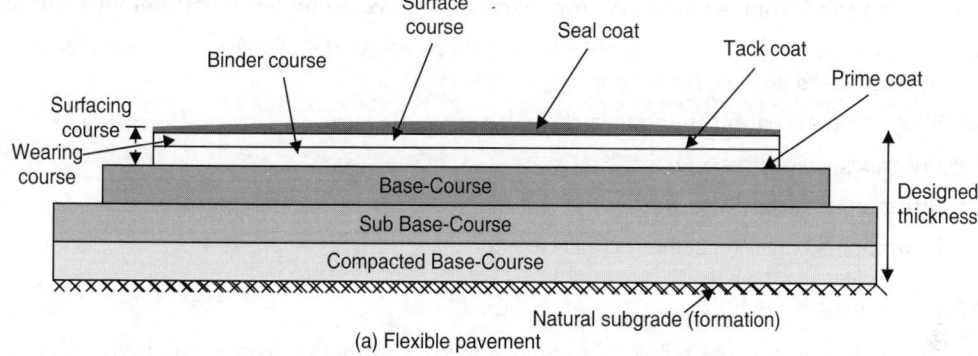

(a) Flexible pavement

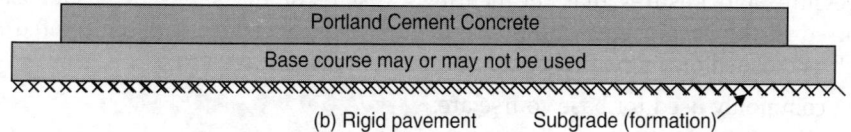

(b) Rigid pavement   Subgrade (formation)

Fig. 14.1: Typical section through (a) Flexible and (b) rigid pavement

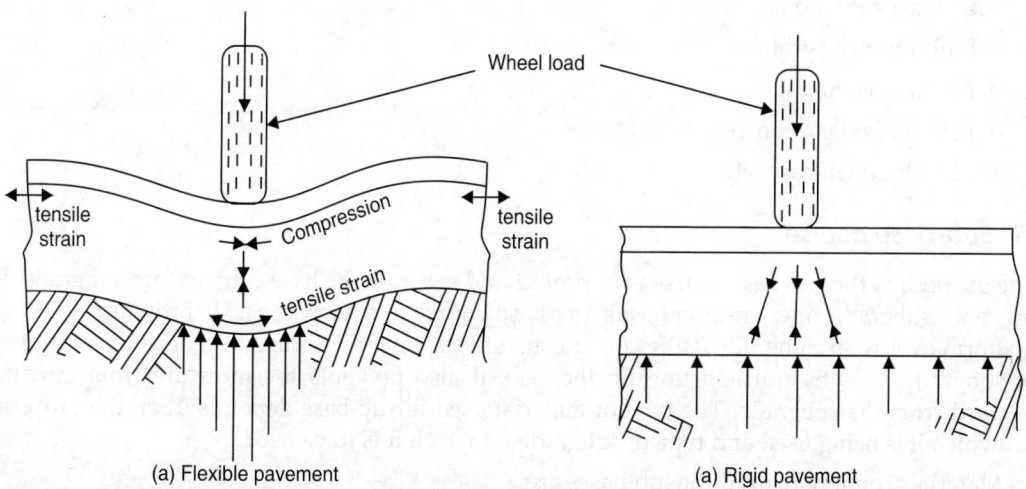

(a) Flexible pavement   (a) Rigid pavement

Fig. 14.2: Behaviosur of 'Flexible' and 'Rigid' pavements under wheel loads

14.02 Components of a Pavement

Modern flexible pavements consist of three main courses, i.e bituminous surfacing, road base and sub-base, over natural soil called sub-grade generally referred as formation. Pavement courses are known by their location and function within the pavement structure:

(i) Surface course (wearing course)

It is the upper most layer of a pavement. The main purpose of the surface course is to carry the wheel loads and withstand atmospheric conditions. It should prevent penetration of surface water to the base course. It provides a smooth, well bound surface free from loose particles. It resists shear stresses, provides a non skid surface without causing undue wear on tires. The material for surface course may vary from earth and gravel to high type bituminous surface or concrete.

Commonly used surface courses are:

- Single or two coat bituminous surface dressing
- Pre-mix surface dressing
- Bituminous carpet
- Bituminous concrete, cement concrete, etc.

(ii) Base course

The base course is the main load carrying course within the pavement. It distributes the imposed wheel loads to the subgrade and withstand stresses produced in the base. It resists vertical pressure, and ensures that the bearing capacity of the subgrade is not exceeded. Materials used in the base course should be of high quality, which may be treated or non-treated such as crushed or uncrushed aggregates, or stabilised soil.

Material, commonly used for base course are:

- WMM (wet mix macadam)
- WBM (water bound macadam)
- Cemented material as lean concrete
- Soil-cement base
- Built up spray grout
- Bituminous macadam
- Lime pozzolana concrete
- Sand-bitumen base, etc.

(iii) Sub-base course

The purpose of the sub-base course is to protect and separate the base course from subgrade. It prevents subgrade fines from entering the base course and vice versa. It provides working platform over weak subgrade (CBR <6). In concrete pavement it is used to reduce the pumping of subgrade fines by traffic action on the slab. It also prevents the moisture from moving upward from the subgrade. The type of material used in sub-base depends upon the purpose for which it is being used and type of subgrade on which it is to be used.

Materials commonly used for sub-bases are:

- Stone soling

- Brick soling
- WBM sub-base
- Stabilized soils
- Sand-bitumen mix
- Soil/gravel/ moorum sub base, etc.

(iv) Subgrade

The subgrade is the natural or filled prepared surface (formation) on which the pavement structure is constructed. It is subjected to lower stresses than sub-base, base and surface course. It should resist shear deformation. Intersection of the subgrade and the pavement is known as "Formation". Subgrade soil ultimately provides support for the pavement and imposed loads.

In a flexible pavement the pressure is transmitted to the subgrade through the lateral distribution of the applied load with depth, while in a rigid pavement pressure is supported by beam and slab action, as with a concrete slab. Due to rigid structure the concrete slab is able to bridge over the localized inadequate support of the subgrade, and therefore quality and strength of subgrade does not affect the thickness requirements of the slab, if the minimum requirements are met. Strength of the subgrade is the main criterion in thickness requirement of flexible pavement, as when subgrade deflects the overlying flexible pavement should deform to a similar shape and extent.

The load carrying capacity of a flexible pavement depends upon distribution characteristics of various layers. The total thickness of pavement depends upon strength of the subgrade.

Rigid pavements are made of Portland cement concrete and may or may not have a base course. Base course is provided

(a) to control pumping

(b) to control frost action (if any)

(c) to have effective drainage

(d) to control swelling and shrinkage of clayey-soils of subgrade.

(e) for quick construction.

The rigid pavement distributes load over a relatively large area of subgrade soil. This is due to rigidity and high modulus of elasticity of cement concrete slab. The major factor considered in design is the strength of cement concrete. For this reason, minor variations in subgrade strength have little or no effect on the structural capacity of the pavement.

Table 14.1 shows a comparison between rigid and flexible pavements. Basic principles of design of highway and airport pavements are common however several differences exist. Table 14.2 shows a comparison between highway and airport pavement.

14.1 STRESSES IN FLEXIBLE PAVEMENTS

The behavior of pavements subjected to wheel load applications has been studied theoretically and through full-scale experiments, to justify different models applicable to stresses developed in pavements and to arrive at a rational design approach. There is no theory which can accurately predict pavement behaviour under different loading and environmental conditions, which can be adopted universally.

Elastic theories are the only best practical approach towards analysing flexible pavements, and defining stresses and strains.

Table 14.1: Comparison of rigid and flexible pavements

S. No.	Point of comparison	Rigid Pavement	Flexible Pavement
1.	Load distribution	It distributes the surface load to a relatively wide area of subgrade soil. This is due to the rigidity or flexural strength or high E value of CC slab	Pavement structure maintains close contact with subgrade. The load distribution depends upon the property (cohesion, internal friction, aggregate interlocking) of pavement structure.
2.	Pavement structure	It consists of CC slab and may or may not have base course over the subgrade.	It consists of wearing course, base course, over compacted subgrade. The various layers are placed in such a way that quality of material in each layer is higher than its underlying layer.
3.	Deformation characteristics	Since it is strong in bending it is able to bridge over localized failure of subgrade.	Since it is able to resist only small tensile stress, any permanent deformation of lower layers and the subgrade is reflected on the road surface.
4.	Effect of subgrade on pavement thickness	Due to high flexural rigidity the major portion of the sturctural capacity is supplied by the pavement itself. In design of rigid pavement flexural strength of CC is the main factor.	The load distribution is through many layers which gradually reduce the stress reaching the subgrade level so as to be within safe bearing capacity. The thickness design is mainly influenced by subgrade strength.
5.	Design basis	Various mathematical stress formulae are used.	Empirical or semiempirical approach for design is adopted.
6.	Cost	High initial cost, but low maintenance cost.	Low initial cost but higher maintenance cost.
7.	Life	May be 25-30 years	5 to 15 years.
8.	Effect of temperature	Heavy temperature stresses are produced	There are no temperature stresses.
9.	Traffic	Good for high volume and heavy traffic	Traffic carrying capacity is lower.
10.	Penetration of water	Low	More
11.	Night Driving	Light coloured surface so better driving	Dark coloured, not so safe in nights
12.	Noise Pollution	Low	More
13.	Extreme weather	Not affected	Affected

Pavements layer thickness disperses the concentration of stress, acceptable to the various materials used according to their capacity. High quality materials are therefore used in surface course, decreasing quality wards subgrade. Figure 14.3. shows the dispersion of wheel load.

Table 14.2: Comparison of highway and airport pavement

S. No.	Point of comparison	Highway Pavement	Airport Pavement
1.	Paved width	Normally 3.5 m to 7.0 m	15 m to 65 m
2.	Thickness of pavement (geometric section)	Normally uniform thickness is used	Runway and Taxiway ends may have thicker sections
3.	Wheel load	4500 kg on duals	20,000 kg on each wheel
4.	Tyre pressure	4 to 7 kg/cm^2	5 to 15 kg/cm^2
5.	Load repititions	Very high may be for 1000 to 2000 vehicles per day	Low may be 20,000 to 40,000 coverage for the whole life.
6.	Lateral placement of Traffic	Wheel load is applied whithin 1 m of the edge of pavement	Concentrated in the 10 m portion
7.	Impact of loads	Low	Very high
8.	Drainage requirements	Normal	High attention and elaborate requirements.

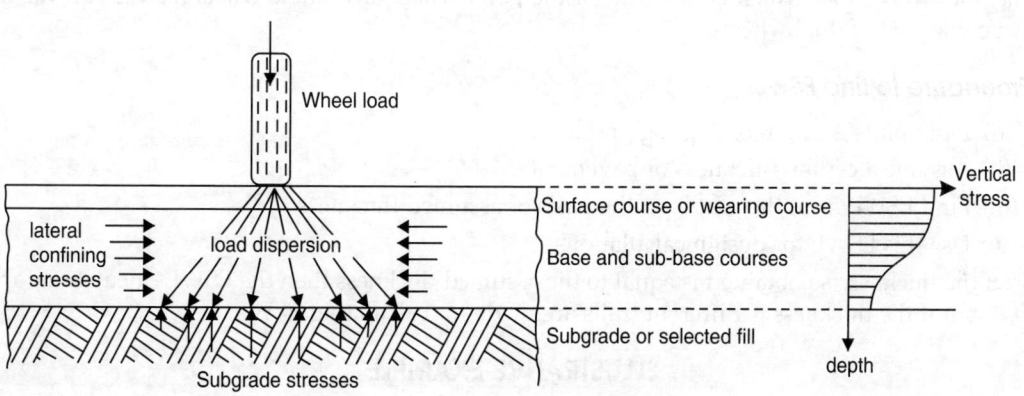

Fig. 14.3: Dispersions of wheel load through different courses

Pavement ultimately distributes the traffic loads to the subgrade over an area larger than that of the tire contact area. The greater the thickness of the pavement, the larger is the area over which the load on the subgrade is distributed. Weaker or unstable subgrade will require larger area of load distribution and consequently the thicker will be the required pavement.

Figure 14.4 shows dispersion of vertical stresses under a wheel gear assembly. The maximum depth at which each tyre of a dual tyred assembly acts as an independent unit is equal to half the distance 'd' between the inner faces of the two tyres.

At depths greater than 'd/2' the pressure by the two wheels begins to overlap. At an approximate depth of 2S (Twice the distance between the centre lines of the two tyres) the dual

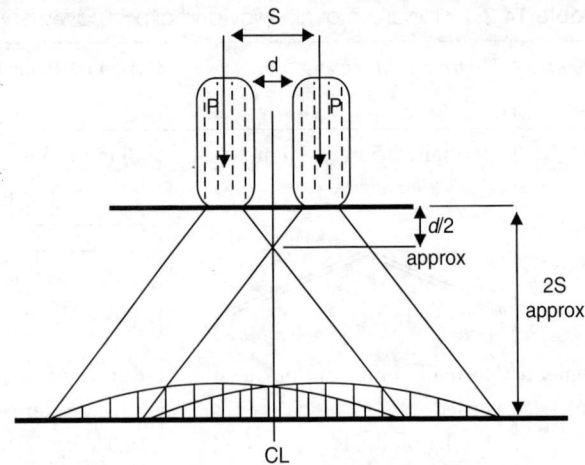

Fig. 14.4: Dispersion of wheel load stresses for twin wheel assembly

wheel assembly acts as a single unit carrying a load equal to $2P$. This observation was made by US Corps of Engineers, when conducting investigations on airport pavements.

Thus the value of equivalent single wheel load (ESWL) will be equal to P up to a depth $d/2$ and P to $2P$ between depths between $d/2$ and $2S$. The percentage to overlap of stress beyond depth $2S$ is so large that ESWL can be taken as $2P$, although the value is slightly less.

Two points A and B are plotted on log-log graph with coordinates $A(d/2, P)$ and $B(2S, 2P)$. The line AB is a plot which is the locus of the point where any single wheel load is equivalent to a certain set of dual wheels.

Procedure to find ESWL

   (i) Plot points A and B on log-log graph.

  (ii) Assume a certain thickness of pavement.

 (iii) Find ESWL from the graph on the basis of assumed thickness

 (iv) Use this ESWL for design calculations.

If the thickness so obtained is equal to the assumed thickness then the ESWL calculations are O.K. but if the thickness is different trails are made for calculating ESWL.

ILLUSTRATIVE EXAMPLE

Example 1. Assuming a cement concrete pavement has to carry the maximum loaded weight on the rear axle of a truck = 8200 kg. There are two pair of dual wheels attached to the axle, one pair on each side of the axle.

The centre to centre and clear space in the dual wheels are 30 cms and 10 cms respectively. If the thickness of the pavement slab is 20 cms, Find E.S.W.L.

Solution: Here
$$d = 10 \text{ cms}$$
$$S = 30 \text{ cms}$$

$$P = \frac{8200}{4} = 2050 \text{ kg, on each wheel}$$

Point $\qquad A\,(d/2, P) = (10/2, 2050)$

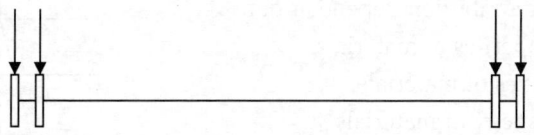

Point $\qquad B\ (2S,\ 2P) = (2 \times 30,\ 4100)$

For thickness of 20 cms the ESWL from line $AB = 3000$ Kg.

Example 2. The gross weight of a truck is 10 Ton. The front axle has single wheel on each side, whereas it carries a pair of dual wheels on each side of the rear axle. The centre to centre spacing of tires and clear span between wheels of dual assembly are 30 cms and 10 cms, respectively.

Find the wheel load, for the design of a road pavement assuming its thickness as

    (i) 8 cms.

    (ii) 20 cms.

    (iii) 40 cms.

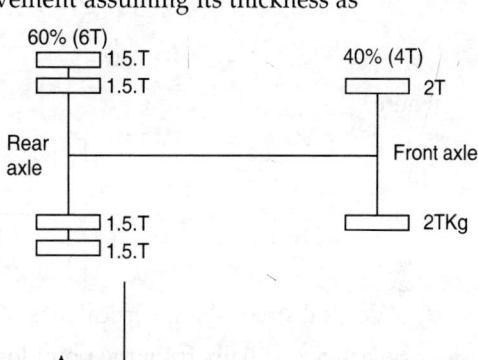

Solution: $\qquad$ Total weight $= 10T$

$\qquad\qquad d = 10$ cms $d/2 = 5$ cm.

$\qquad\qquad\qquad S = 30$ cms.

$\qquad\qquad\qquad 2S = 60$ cm.

$\qquad$ Load on rear axle $= 6T$

$\qquad$ load on front axle $= 4T$

$$P = \frac{6T}{4} = 1.5\ T = 1500 \text{ kg.}$$

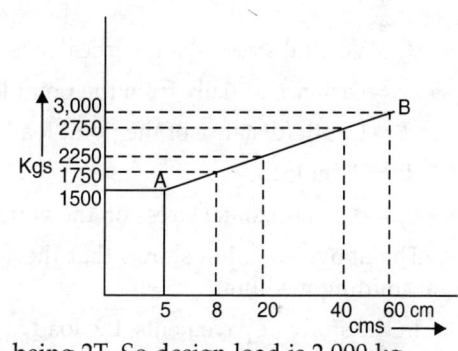

Coordinates of $A = d/2,\ P = 5$ cm, 1500 kg.

Coordinates of $B = 2S,\ 2P = 60$ cm, 3000 kg.

    (*i*) **For thickness of 8 cms.** From Graph

$\qquad\qquad\qquad$ ESWL $= 1750$ kg

The load on the front axle is more than 1750 kg. being 2T. So design load is 2,000 kg.

    (*ii*) **For thickness of 20 cms.** From Graph

$\qquad\qquad\qquad$ ESWL $= 2250$ kg.

$\therefore \qquad\qquad$ Design load $= 2250$ kg

    (*iii*) **For thickness of 40 cms.** From Graph

$\qquad\qquad\qquad$ ESWL $= 2750$ kg.

$\qquad\qquad$ Design load $= 2750$ kg.

There are two simple assumptions for load distribution through pavement structure.

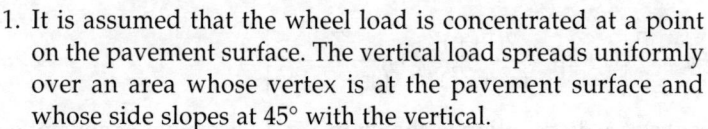

1. It is assumed that the wheel load is concentrated at a point on the pavement surface. The vertical load spreads uniformly over an area whose vertex is at the pavement surface and whose side slopes at 45° with the vertical.

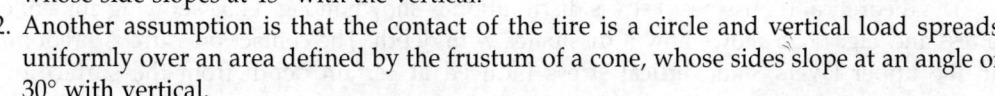

2. Another assumption is that the contact of the tire is a circle and vertical load spreads uniformly over an area defined by the frustum of a cone, whose sides slope at an angle of 30° with vertical.

The angle of load distribution depends upon
  (i) interlocking properties of materials.
 (ii) Physical properties of materials.
(iii) Modulus of elasticity of materials.

These properties widely vary with different materials used in road construction.

14.1.1 Theories of Stress Distribution

1. One-layer system (Boussinesq's Elastic Theory)

According to Boussinesq's Elastic Theory, for an isotropic, homogenous, elastic and infinite soil mass the stresses at varying depth, for a uniformly distributed circular load are given as follows:

$$\sigma_Z = k.\frac{P}{Z^2}$$

where,

$$k = \frac{3}{2\pi}\left[\frac{1}{1+\left(\dfrac{r}{Z}\right)^2}\right]^{5/2}$$

σ_Z = Vertical stress along vertical axis of loading at depth Z.

r = Distance radially from the point load.

Z = Depth (distance of the point load from surface)

P = Point load.

$\sigma_x = \sigma_y$ = horizontal stress on the vertical axis of loading.

The above equation shows that the vertical stress is independent of the properties of the transmitting medium.

In the study of pavements, the load at the surface is not a point load, but a distributed load. For a circular area of contact, stress on a vertical plane passing through the centre of the area is obtained by integration of the above equation as below:

$$\sigma_Z = P\left\{1-\frac{Z^3}{(a^2+Z^2)^{3/2}}\right\}$$

where

a = Radius of loaded area

b = Unit stress on circular plate (contact pressure)

μ = Poisson's ratio

$$\sigma x = 6y = \frac{P}{2}\left\{(1+2\mu-\frac{2(1+\mu)Z}{(a^2+Z^2)^{1/2}}+\frac{Z^3}{(a^2+Z^2)^{3/2}}\right\}$$

The theoretical (Boussinesq) stress distribution is shown in Fig. 14.5a showing zone of equal stress and Fig. 14.5b shows how it dissipates with depth. The contact pressure is predominant in the upper layers, and vertical stress induced at certain depth from the surface of any pavement is practically negligible.

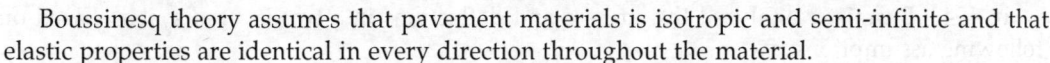

Boussinesq theory assumes that pavement materials is isotropic and semi-infinite and that elastic properties are identical in every direction throughout the material.

The thickness of the pavement layer can be chosen so that the shear stress at the depth of the pavement – subgrade interface does not exceed the shear strength of the subgrade.

Alternatively, surface deflection at the centre of the loaded area may be used as limiting factor (using the modulus of elasticity of the material,) assuming an incompressible pavement layer on top of a compressible subgrade.

Full scale experiments have shown that actual stresses below bituminous pavements are similar or slightly more than those computed using Boussinesq theory.

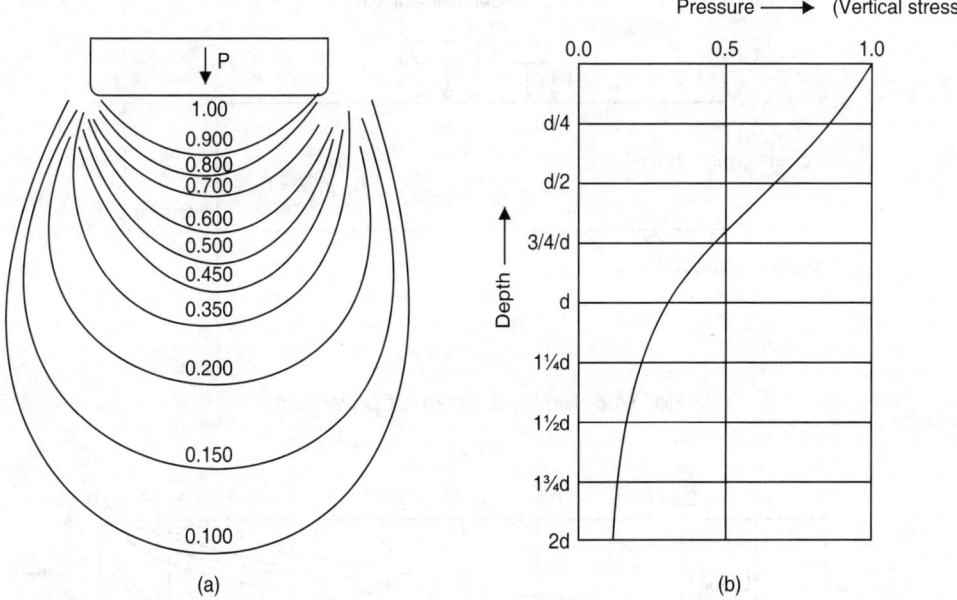

Fig. 14.5: Vertial stress distribution beneath a single wheel

The vertical deflection at surface ($Z = 0$) at the centre of loaded area is given by

$$\Delta = 1.5 \frac{Pa}{E} \text{ for flexible pavement}$$

and $\Delta = 1.18 \dfrac{Pa}{E}$ for rigid pavement

where E is the modulus of elasticity of soil.

The limitations of this approach reveal from the assumptions that:

– Soil is elastic and homogenous

– Load is uniformly distributed and

– Pavement layers have a single property

2. Two layers system (Burmister's theory)

The assumptions made by Boussinesq are not true is case of flexible pavement as it is made of layers. The modulus of elasticity (E) decreases with the depth of pavement. As a result the stresses and deflections are less than those obtained by Boussinesq equation.

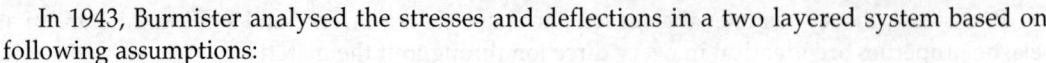

In 1943, Burmister analysed the stresses and deflections in a two layered system based on following assumptions:

   (i) The top elastic layer has higher modulus of elasticity, than its lower layer.

  (ii) The top layer is assumed to be infinite is horizontal direction only. The bottom layer is assumed to be infinite in depth and horizontal direction.

 (iii) The materials is each layer are homogenous, isotropic and elastic.

 (iv) There is a complete continuity across the interface, between two layers.

  (v) For the two layers $\mu_1 = \mu_2 = 0.5$.

Figure 14.6 shows as two layer system of pavement.

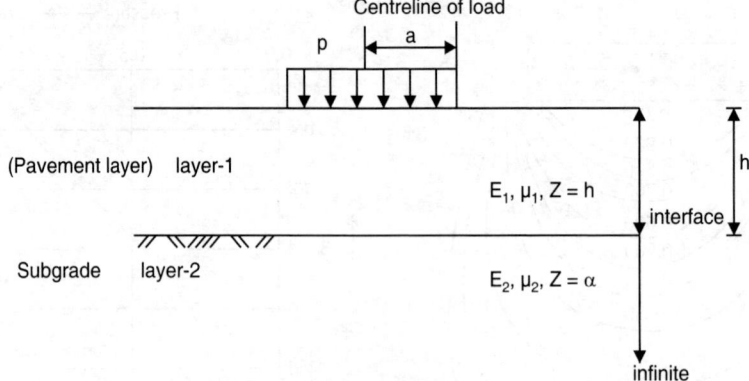

Fig. 14.6: Two layer system of pavement

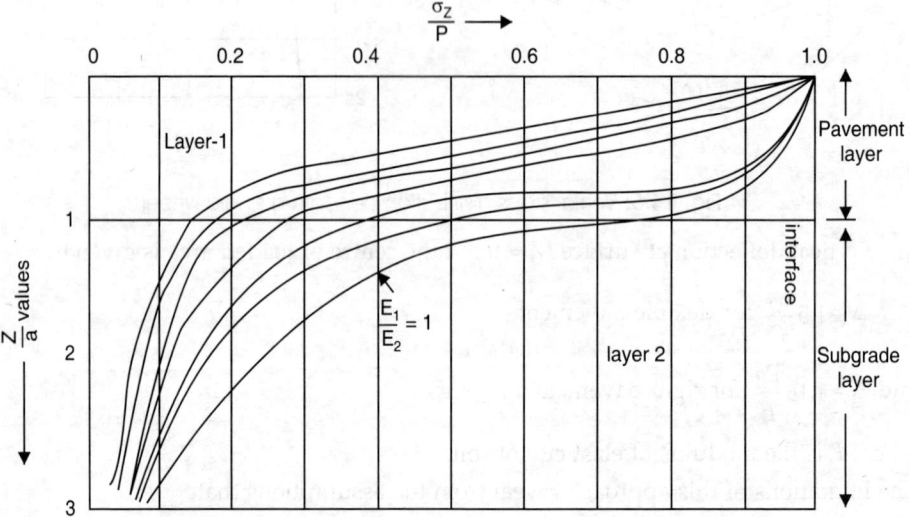

Fig. 14.7: Typical Burmister's two layer stress influence curves

(A) STRESS:

From the above curves (Fig. 14.7) Boussinesq $\left(\dfrac{E_1}{E_2} = 1\right)$ vertical stress at the interface is about 70% of the applied pressure, but for $\dfrac{E_1}{E_2} = 100$, the vertical stress is only 10% of the applied pressure.

At great depths $\left(\dfrac{z}{a} \geq 3\right)$, the two analysis approach a common level, whereas at the interface they give very much different values. This shows that in addition to depth, the quality of base and surface is also important in stress distribution.

(B) DEFLECTION

For a two layer system, Burmister's deflection equations are:

For flexible plate: $\qquad D = 1.5\dfrac{pa}{E_2}.F_2$

For rigid plate: $\qquad D = 1.18\dfrac{pa}{E_2}F_2$

where

p = Unit load on circular plate

a = Radius of circular plate

E_2 = Modulus of elasticity of lower layers

F_2 = Deflection factor, depends upon E_1/E_2 and $\dfrac{h}{a}$.

Burmister calculated the vertical displacement at the surface and at the centre of the applied load for various ratios of moduli of elasticity for the two layers, for various ratios of depth of the upper layer to the radius of the applied load, and for a uniform contact pressure over a circular area and a perfectly rough interface between two layers.

14.1.3 Multilayer System

A multilayer approach in analysis of pavement stresses is possible if elastic or stiffness modulus and Poisson's ratio for each layer is known, using a computer programme. With availability of computer techniques an analysis to determine stresses and strains at critical points in the pavement structure under a load is possible. The stresses and strains calculated can be compared with the bearing capacities of the materials, used in a particular layer and design thicknesses can be worked out accordingly.

The finite element method, which divides the pavement structure in finite elements can be useful for multilayer analysis. The solution to complete system can be obtained by assembling the behavior of elements.

The theoretical approach supports elastic theories in pavement design but it suffers with certain limitation and assumptions which are not really true. For example stress/strain relationship for pavement materials in non-linear, and depends upon other factors like loading time and temperature. Repetition of wheel load also decreases strength of materials.

However, it is hoped that theoretical analysis, with the use of powerful computers and studies on behavior of materials will be able to provide rational approach to pavement design in near future. The practical highway design adopts methods which are largely based on experience, as theoretical methods have not yet reached the stage when they can be used as a complete design method, mainly because:

(i) Complexities of the calculations involved.

(ii) Difficulties in evaluating effect of non-elastic behaviour of pavement and subgrade under repetitive loading.

(iii) Assumptions made in development of the theory that stress/strain relationship for road pavement materials is linear, which may not be true.

14.2   CONSIDERATIONS FOR FLEXIBLE PAVEMENT DESIGN

Pavement design aims in providing adequate cover to the subgrade so that stresses at the subgrade are low enough to prevent excessive deformation and also use pavement materials which are strong enough to resist stresses and strains imposed by wheel loads. The complete design should also ensure that the pavement structure is adequately drained.

Factors affecting Pavement Design: The thickness design of highway pavement requires the following complex factors to be considered:

- Magnitude and method of application of wheel loads.
- Number of repetition of applied wheel loads.
- Contact area between the tyre carrying the load and the road surface (pavement)
- Stiffness, stability, durability, elastic and plastic deformation and resistance to fatigue of different pavement layers.
- Strength and volumetric changes in subgrade due to climatic factors.
- Deformation characteristics of subgrade under load and ability of the pavement layers to reduce the stress imposed by wheel loads.
- Temperature and moisture conditions, because it
  - Results in weakening the subgrade
  - Affects modulus of elasticity
  - Results in warping stresses in concrete
- Fatigue failure: All materials tend to fail under repeated loads.

14.2.1 Fundamental Considerations for Flexible Pavement Design

The thickness design of flexible highway pavements require the following factors to be considered for satisfactory design.

1. Concerning traffic
   (a) Wheel loads
   (b) Contact area and contact pressure
   (c) Initial traffic (number of commercial vehicles/day)
   (d) Future traffic prediction for the design year.
2. Design life of the road
3. Vehicle damaging factor
4. Distribution of wheel load on carriageway
5. Design traffic
6. Subgrade strength
7. Stiffness and characteristics of pavement layer material
8. Climatic conditions and drainage provisions

Due to complexity of so many factors involved in practical highway pavement design, many thickness design methods are based on practical field experience, gained over years.

i. Wheel loads

The road pavement is normally loaded by pneumatic tyred wheels of varying sizes and inflation pressure. Load from the wheels of the vehicles is transmitted to pavement layers. To control pavement damage from heavy commercial vehicles limits are imposed on the total load which can be carried by a single axle.

The single axle load in India and USA is 8160 kg (18,000 lbs or 8.16 tonnes). For dual-wheel assembly, normally used in commercial vehicles, the wheel load shall be half of the axle load. With multiple wheels, the combined effect varies according to the wheel load spacing in relation to the pavement thickness and others factors, but can never exceed that for a single wheel load carrying the same load. Figure 14.8 shows typical axle configurations and codes.

Maximum single axle load have been incorporated in national laws and proposed at an international level by the Geneva Convention in 1949 (8 tonns) in order to control the damage of road pavement.

IRC recommends standard axle load of 8160 kgs (8.16 tonnes)

Contact area and pressure: The contact pressure between the wheel and the road surface depends on the tyre inflation and the stiffness of the tyre walls. Although the pressure distribution of a pneumatic tyre on a flexible pavement is not uniform, the assumption of an uniform distribution over a circular area is sufficiently accurate for most analytical purposes used in theoretical concept. Tyre pressure of commercial vehicles is about 0.5 MN/m^2 (70 psi)

Present traffic (initial traffic): Present day (initial) traffic after construction is calculated in commercial vehicles per day (CVPD). For design purpose only the commercial vehicles of gross vehicle weight of 3 tonnes or more, and their axle-loading is considered. As per IRC guidelines initial daily average traffic flow for any road normally be based on at least 7 day, 24 hour classified traffic counts. In case of a new road traffic on a similar existing road in the locality can be taken as a guide.

Future traffic: Future traffic is calculated based on prediction of attracted, generated, and developed traffic which shall occur within the design life of the road, considering socioeconomic, and other factors like growth in traffic and population, agricultural and industrial growth, petrol consumption estimates, per capita income, NDP, etc. as described in chapter 6.

In situations where adequate data is not available, IRC recommends that an average annual growth rate of 7.5 percent may be adopted, for design life of the road.

Only commercial vehicles (more than 3 tonnes) are counted in design, light vehicles like passenger cars, pick-up, motor cycles, etc can be eliminated.

ii. Design life of the road

Design life is defined as the cumulative number of axles which a road can carry, before strengthening of the pavement is necessary. Design life is adopted as follows:

For national and state highways — 15 years

For expressways and urban roads — 20 years

Other category of roads— 10 to 15 years.

It is usual to take design life of 10 years in first instance. Stage construction may be planned to meet traffic requirements beyond this period.

iii. Vehicle damage factor (VDF)

Road tests have shown that the damaging power of an axle load depends very much on its magnitude. Vehicle damage factor is a factor which should be used to convert commercial vehicles of different axle configurations, to equivalent number of standard axle load of 8.16 tonnes (8160 kg). Equivalency factors have been derived which reveal the damaging power of different axle loads compared with a standard axle load of 8.16 tonnes (8160 kg).

VDF is obtained by large scale road tests to study damaging by different axles on road sections under varying conditions of traffic mix, commodities carried, time and season of the year, type and conditions of road and enforcement measures used. Realistic VDF values should be used, after conducting axle load survey for major road projects. Where project does not demand conducting of an exhaustive axle load survey indicative values recommended by IRC can be used as in Table 14.3.

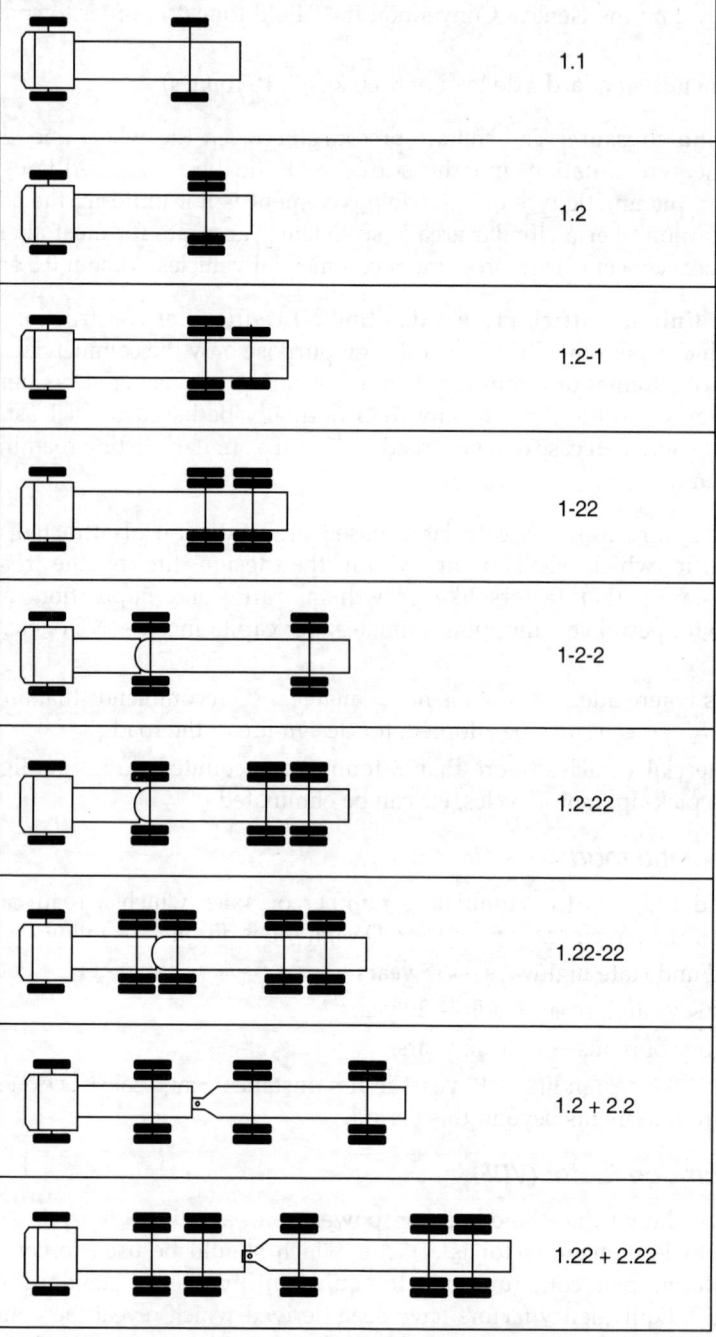

Fig. 14.8: Typical axle configuration and codes

Table 14.3: Indicative VDF value (IRC)

Initial traffic volume *(number of commercial vehicles/day)*	*VDF value* *Plain/Rolling terrain*	*VDF value* *hill terrain*
0–150	1.5	0.5
150-1500	3.5	1.5
more than 1500	4.5	2.5

The design should be based on traffic in most heavily traffic lane and used for whole pavement width.

Besides the magnitude of traffic loads, the number of repititions of axle loads influence the damaging effect of traffic on road pavement.

Table 14.4 shows equivalency factors and damaging power of different axle loads as recommended by IRC.

Following formula may be used for converting axle loads in to equivalent standard axle load.

(*i*) Single axle load: Equivalency factor = $\dfrac{(\text{axle load in kg})^4}{8160}$

(*ii*) Tandon axle load: Equivalency factor = $\left(\dfrac{\text{axle load in kg}}{14968}\right)^4$

iv. Distribution of wheel load on carriageway

On dual carriageway roads or single carriageway roads with more lanes, it is assumed that the slow traffic lanes carry all the commercial vehicles. The design thickness for the slow traffic lanes would normally be used over the whole carriageway.

On single lane roads (3.75 m width) traffic tends to be more channelised than on 2 lane road, however, a realistic assessment of commercial vehicles by direction and by lane is necessary as it directly, affects the total equivalent standard axle load applications for the design. IRC has recommended following distribution for design, till more reliable data is available for Indian roads.

(i) **Single-lane roads (3.75 m width):** To allow for concentration of wheel load repititions, the design should be based on total number of commercial vehicles in both direction.

(ii) **Two lane single carriageway roads:** The design should be based on 75 percent of the total number of commercial vehicles in both direction.

(iii) **Four-lane single carriageway roads:** The design should be based on 40 percent of the total number of commercial vehicles, in both direction.

(iv) **Dual carriageway roads:** The design of dual two lane carriageway road should be based on 75% of the number of commercial vehicles in each direction. For dual three lane carriageway and dual four lane carriageway the distribution factor will be 60 percent and 45 percent respectively.

Where significant difference in traffic flow in the two direction is observed, more heavily trafficked lane should be used for design. Same design will apply to whole carriageway width.

Table 14.4: Equivalency factor of different axle loads (IRC)

Gross Axle Weight Kg.	Load Equivalency Factors	
	Single Axle	Tandem Axle
900	0.0002	0.0000
1810	0.002	0.0002
2720	0.009	0.001
3630	0.031	0.003
4540	0.08	0.006
5440	0.176	0.013
6350	0.35	0.024
7260	0.61	0.043
8160	1.00	0.070
9070	1.55	0.110
9930	2.30	0.166
10890	3.27	0.242
11790	4.48	0.342
12700	5.98	0.470
13610	7.8	0.633
14520	10.0	0.834
15420	12.5	1.08
16320	15.5	1.38
17230	19.0	1.73
18140	23.0	2.14
19051	27.7	2.61
19958	33.0	3.16
20865	39.3	3.16
21772	46.5	4.49
22680	55.0	5.28
23587	-	6.17
24494	-	7.15
25401	-	8.20
26308	-	9.4
27216	-	10.7
28123	-	12.1
29030	-	13.7
29937	-	15.4
39844	-	17.2
31752	-	19.2
32660	-	21.3
33566	-	23.6
34473	-	26.1
35380	-	28.8
36288	-	31.7

V. Design traffic

The design traffic in term of the cumulative number of standard axles in lane carrying maximum traffic, to be carried during the design life of the road can be calculated by following formula:

$$N = \frac{365 \times \left[(1+r)^n - 1 \right]}{r} \times A \times D \times F$$

Where

N = The cumulative number of standard axles to be catered for in the design in terms of Million Standard Axles (MSA)

A = Initial traffic in the year of completion of construction in terms of number of commercial vehicles CV/day.

D = Lane distribution factor

F = Vehicle damage factor

n = Design life in years

r = Annual growth rate of commercial vehicles assumed 7.5% (r = .075) when data for actual rate is not available.

The traffic (A) in the year of completion is estimated by formula:

$$A = P (1 + r)^x$$

where

P = Number of commercial vehicles as per last count

x = Number of years between last count and the year of completion of construction of road.

Example 3: Calculate the cumulative standard axles, for designing a two lane road using following data:

Present day traffic – 1500 CV per day

Design period – 15 years

Time required in construction – 5 years

Vehicle damage factor – 2.5

Overall traffic growth – 10 percent.

Solution: Traffic on completion of road in 5 years is given by

$$A = P(1 + r)^x$$
$$= 1500 (1 + 0.1)^5$$
$$= 2416 \text{ commercial vehicles/day}$$

Cumulative standard axles can be calculated by

$$N = \frac{365 \times \left[(1+r)^n - 1 \right]}{r} \times A \times D \times F$$

$$= \frac{365 \times \left[(1+0.1)^{15} - 1 \right] \times 2416 \times 0.75 \times 2.5}{0.1 \times 10^6}$$

$$= \textbf{52.46 msa}$$

vi. Subgrade strength

The assessment of subgrade strength is of major importance in design of flexible pavement as it provides support to the higher quality material in a road pavement (sub-base, base and bituminous surfacing) and has to bear the stresses induced by traffic.

The strength of subgrade is governed by

- The type of soil, which is determined by the location of the road. Every effort should be made to locate road on soils, having good bearing capacity.

- The density of soil, which can be controlled during construction by compaction, subgrade should be well compacted to utilize its strength and economise on total thickness required.
- The moisture content which is dictated by climatic conditions in the area and particularly by the location of water table. Subgrade strength is assessed in most critical moisture conditions, likely to occur.

A commonly used indicator of subgrade strength, is (CBR) California Bearing Ratio test which includes above governing factors. CBR test can be conducted in the laboratory or in-situ. Laboratory tests are usually carried out for subgrade soils which can be and shall be compacted during construction, and in-situ test are used on subgrade soil that can not be improved by further compaction (i.e. heavy clays) or when no further compaction of the subgrade is expected during the course of construction..

Pavement design methods, are based on standard laboratory tests. Standard procedure for conducting CBR test in laboratory is described is chapter 5, the test must be performed on remoulded samples of soils in the laboratory. In-situ tests do not reflect critical conditions and therefore not commonly used to determine design CBR. Laboratory test conditions should reproduce as closely as possible the weakest conditions likely to occur under the road after construction. To simulate the worst condition it is recommended that sample of subgrade should be soaked in water for a period of 4 days prior to testing. Four days soaking may be unrealistic in many situations, giving unduly conservative designs and therefore in such cases CBR test may be performed on samples prepared at the natural moisture content of the soil at subgrade depth immediately after recession of the monsoon.

Design CBR: In the laboratory at least three samples at the same density and moisture content are tested for each type of soil, to obtain a reliable average value. Permissible maximum variation within CBR values from the 3 specimen is given in Table 14.5.

If variation is more than above limits, the design CBR value should be average of six samples and not three.

Table 14.5: Permissible variation is CBR value

CBR (percent)	Maximum variation is CBR value
5	±1
5-10	±2
11-30	± 3
31 and above	± 5

The lowest value of CBR on the subgrade soil should be used for designing the pavement thickness. Pavement thickness may be modified at intervals as CBR value changes, but frequent changes in design are not possible for practical considerations. Sometimes it may be more economical to remove and replace a weak layer than designing on it. Local conditions dictate the best treatment for improving the weak subgrade.

vii. Stiffness and characteristics of pavement layer material

The load spreading properties of a road pavement are related to the elastic module of the respective layer, and therefore use of a material of higher modulus of elasticity should allow a decrease to be made in the overall pavement thickness.

The elastic module of material to be used in pavement layer can be measured in laboratory, but since these are likely to change during the life of the road as a result of compaction, breaking up by traffic, or due to temperature changes for bituminous materials, it is necessary to observe the performance of different materials in actual pavement structure constructed under different conditions. Experiments like AASHO road test in USA are conducted to obtain information on the performance of various pavement materials. Design charts have been developed which allow to read the required thickness of pavement layers for the respective materials for given traffic loads and subgrade strength.

Materials generally used in base and sub-base course are gravel, crushed stone, stabilised soils, etc. The variations in the stiffness between these materials are not large and it is assumed that they have equal stiffness, i.e. equal load spreading properties. This assumption is justified by the very small difference in design thickness that a more complicated analysis, would make possible.

viii. Climatic conditions and drainage provisions

Design of flexible pavements must consider climatic conditions, rainfall, and provision of drainage system, during its construction and service life. Heavy rainfall, frost, etc need extra attention.

Figure 14.9 shows a flow diagram for structural design of flexible pavements.

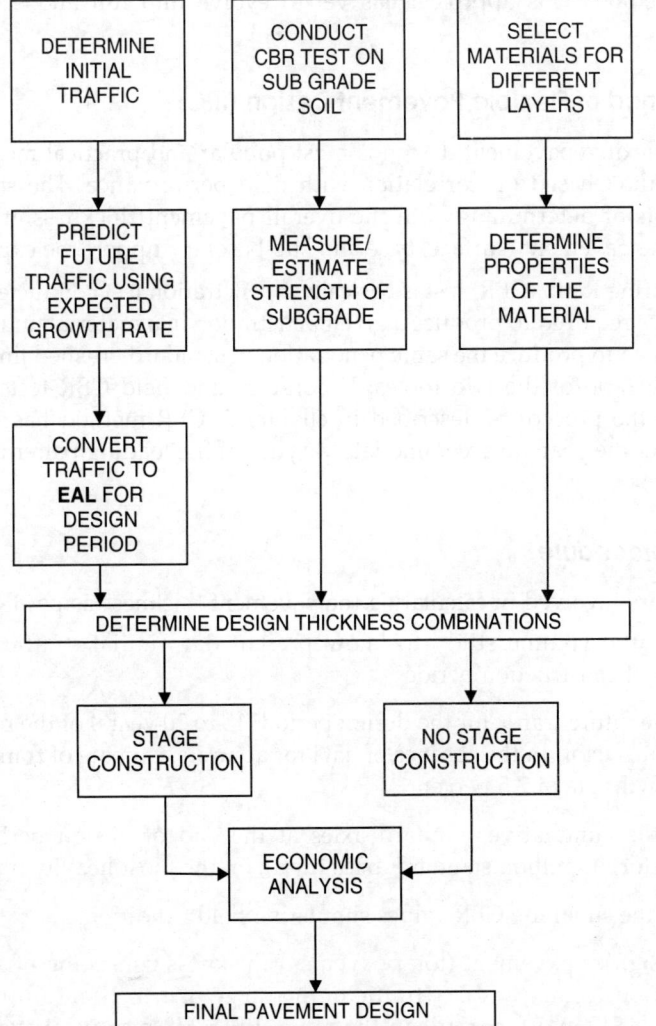

Fig. 14.9: Flow diagram for structural design of flexible pavements

14.3 METHODS OF FLEXIBLE PAVEMENT DESIGN

Many design methods have been developed to suit different climatic and traffic loading condition. Basically the methods can be classified as:

(a) **Empirical and Semi-empirical:** Based on long term performance of existing pavement structures, for specific traffic loading and environments conditions. They are satisfactory so long as the material and conditions of loading for which they are developed do not change, like:

- CBR method
- Group index method

(b) **Theoretical and Semi-theoretical concepts (Analytical methods):** These methods are based on the stress-strains behaviour in pavement and subgrade. They can provide better understanding of empirical results and help in further research and development of a rational method. This approach has yet to evolve into full and easily usable design method.

14.3.1 CBR Method of Flexible Pavement Design (IRC)

CBR method of flexible pavement design is most popular and practical method. It is basically an empirical method based on correlation with field performance. The structural design of pavement consists of determining both the overall pavement thickness and the thickness of components of the pavement (surface, base and sub-base) to support the expected traffic.

California Bearing Ratio (CBR) test is basically a penetration test conducted at a uniform rate of strain. The force required to produce a given penetration in a material under test is compared to the force required to produce the same penetration in standard crushed lime stone. The result is expressed as a ratio of the two forces. Laboratory and field CBR tests are performed in accordance with the procedure described in chapter 5. CBR method has undergone several modifications over the years to accommodate varying traffic loading patterns and environment conditions.

14.3.2 Design procedure

Following steps are involved in calculating the pavement thickness as per IRC: 37-2001.

(i) Conduct traffic volume study to know present day (initial) traffic and project it for completion of construction period.

(ii) Estimate the future traffic for the design period (15 to 20 years) of the road by establishing a growth rate factor. In the absence of data for accurate estimate of traffic, assume average annual growth rate of 7.5 percent.

(iii) Calculate the cumulative standard axles at the end of design period, using suitable damage factor, in million standard axles (msa), on the most heavily trafficked lane.

(iv) Determine the subgrade CBR values simulating field condition.

(v) For the design of pavement (total pavement thickness consisting of sub-base, base and bituminous surfacing) to carry traffic in the range of 1 to 10 msa use pavement design chart (Fig. 14.10) and for traffic in the range 10 to 150 msa use pavement design chart (Fig. 14.11).

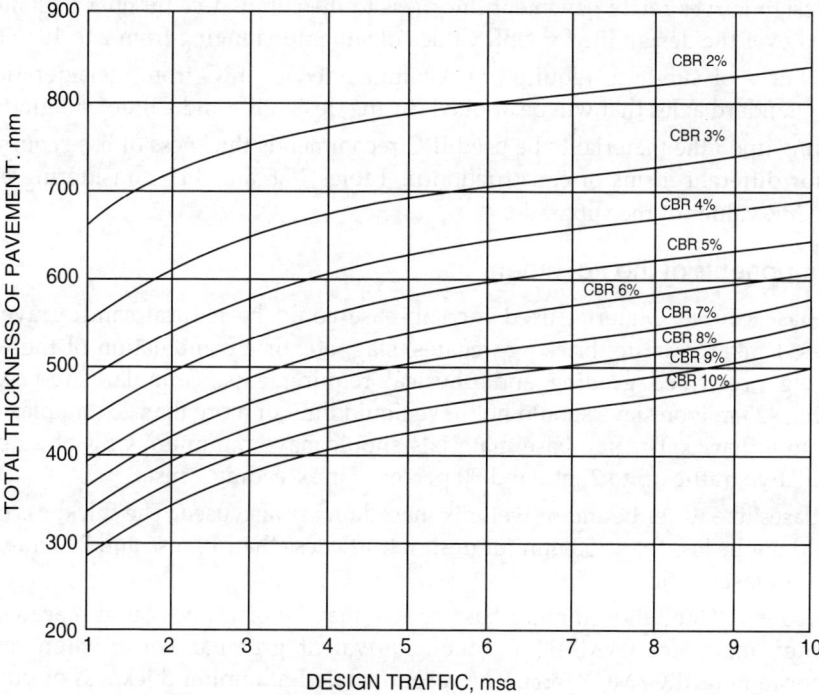

Fig. 14.10: Pavement thickness design chart for traffic 1-10 msa (IRC: 37-2001)

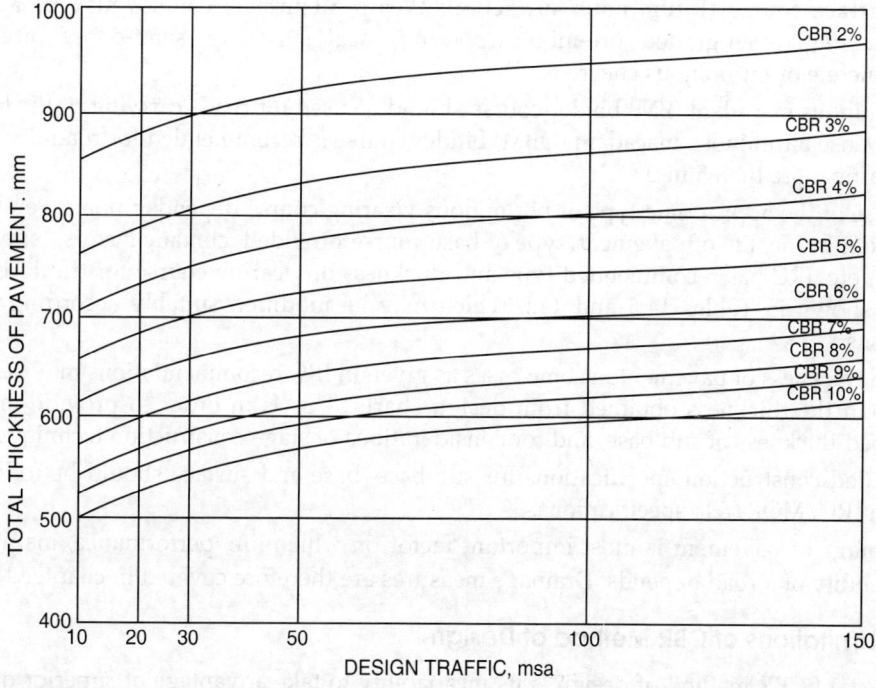

Fig. 14.11: Pavement thickness design chart of traffic 10-150 masa (IRC: 37-2001)

These design curves relate pavement thickness to the cumulative number of standard axles to be carried over the design life for CBR values of subgrade ranging from 2 to 10 percent.

The thickness of sub-base required is recommended by IRC from considerations of the cumulative standard axles that will be applied during design life and CBR of subgrade material.

Depending upon the material to be used IRC recommends thickness of base course and also surfacing for different forms of construction in Tables 14.6 and 14.7 considering cumulative traffic and CBR values of the subgrade.

14.3.3 Components of the Pavement

(a) **Sub-base course:** Material used for sub-base could be natural sand, gravel, crushed stones, kankar, laterite, brick aggregates, slag, etc. or a combination of these materials meeting prescribed grading and physical requirements. Granular sub-base material passing 425 micron sieve should not have liquid limit of more than 25 and plasticity index not more than 6. The sub-base materials should have minimum CBR of 20 percent for cumulative traffic up to 2 msa and 30 percent for exceeding 2 msa.

Sub-bases like water bound or wet mix macadam are also used. The thickness of sub-base should not be less than 150 mm for design traffic less than 10 msa and 200 mm for design traffic 10 msa or more.

(b) **Base course:** Unbound granular base course may be of water bound macadam (WBM), wet mix macadam (WMM), or other equivalent granular construction meeting the specification of IRC/MORT&H. The recommended minimum thickness of granular base is 225 mm for traffic up to 2 msa and 250 mm for traffic exceeding 2 msa.

IRC has suggested thickness of various components for varying CBR values from 2 to 10 percent and cumulative traffic of 1 to 150 msa.

(c) **Surface course (Bituminous surfacing):** Wearing (surface) courses may be a surface dressing, open-graded pre-mix carpet, mix seal surfacing, semi-dense bituminous concrete or bituminous concrete.

Bituminous macadam (BM) binder course should be used for roads carrying traffic less than 5 msa. Dense bituminous macadam (DBM) binder course is recommended for road designed to carry traffic more than 5 msa.

Choice of the appropriate type of bituminous wearing course depends upon several factors like design traffic, life of pavement, type of base course provided, climatic factors, temperature rain fall, etc. IRC has recommended type and thickness of wearing course for traffic 10 to 150 msa as shown in Tables 14.6 and 14.7 which may be modified suitably according to local conditions.

Total thickness of pavement in some cases as given in IRC recommendations may be slightly more than the thickness obtained from design charts. This is in order to provide minimum prescribed thickness of sub-base, and for considerations of stage construction technique.

Detailed construction specifications for sub-base, base and surface should be followed as given in IRC/MORT&H specifications.

Drainage of pavement is most important factor on which life, performance, maintenance, serviceability of a road depends. Drainage measures are therefore covered in chapter 18.

14.3.4 Limitations of CBR Method of Design

Drawback of CBR method of design is its incapability to take advantage of superior quality of material used in layers, i.e. method of design gives total thickness requirement above the

Table 14.6: Recommended designs for traffic range 1-10 msa (IRC: 37-2001)

Subrade CBR (%)	Cumulative Traffic (msa)	Total Pavement Thickness (mm)	PAVEMENT COMPOSITION			
			Bituminous Surfacing		Granular Base (mm)	Granular Sub-base (mm)
			Wearing Course (mm)	Binder Course (mm)		
2	1	660	20 PC	–	225	435
	2	715	20PC	50BM	225	440
	3	750	20PC	60 BM	250	440
	5	795	25 SDBC	70DBM	250	450
	10	850	40BC	100 DBM	250	460
3	1	550	20 PC		225	435
	2	610	20 PC	50 BM	225	335
	3	645	20 PC	60 BM	250	335
	5	690	25 SDBC	60DBM	250	335
	10	760	40 BC	90DBM	250	380
4	1	480	20PC		225	255
	2	540	20 PC	50 BM	225	265
	3	580	20 PC	50 BM	250	280
	5	620	25 SDBC	60DBM	250	285
	10	700	40 BC	80 DBM	250	330
5	1	430	20 PC		225	205
	2	490	20 PC	50 BM	225	215
	3	530	20 PC	50 BM	250	230
	5	580	25 SDBC	55 DBM	250	250
	10	660	40 BC	70 DBM	250	300
6	1	390	20 PC		225	165
	2	450	20 PC	50 BM	225	175
	3	490	20 PC	50 BM	250	190
	5	535	25 SDBC	50 DBM	250	210
	10	615	40 BC	65 DBM	250	260
7	1	375	20 PC		225	150
	2	425	20 PC	50 BM	225	150
	3	460	20 PC	50 BM	250	160
	5	505	25 SDBC	50 DMB	250	180
	10	580	40 BC	60 DBM	250	230
8	1	375	20 PC		225	150
	2	425	20 PC	50 BM	225	150
	3	450	20 PC	50 BM	250	150
	5	475	25SDBC	50 DBM	250	150
	10	550	40 BC	60 DBM	250	200
9 and 10	1	375	20 PC		225	150
	2	425	20 PC	50 BM	225	150
	3	450	20 PC	50 BM	250	150
	5	475	25 SDBC	50 DBM	250	150
	10	540	40 BC	50 DBM	250	200

Table 14.7: Recommended designs for traffic range 10-150 msa (IRC: 37-2001)

Subgrade CBR %	Cumulative Traffic (mm)	Total Pavement Thickness (mm)	Pavement composition		Granular Base and Sub-base (mm)
			Bituminous Surfacing		
			BC (mm)	DBM (mm)	
2	10	850	40	100	
	20	880	40	130	
	30	900	40	150	Base = 250
	50	925	40	175	
	100	955	50	195	Sub-base= 460
	150	975	50	215	
3	10	760	40	90	
	20	790	40	120	
	30	810	40	140	Base = 250
	50	830	40	160	
	100	860	50	180	Sub-base = 380
	150	890	50	210	
4	10	700	40	80	
	20	730	40	110	
	30	750	40	130	Base = 250
	50	780	40	160	
	100	800	50	170	Sub-base = 330
	150	820	50	190	
5	10	660	40	70	
	20	690	40	100	
	30	710	40	120	Base = 250
	50	730	40	140	
	100	750	50	150	Sub-base = 300
	150	770	50	170	
6	10	615	40	65	
	20	640	40	90	
	30	655	40	105	Base = 250
	50	675	40	125	
	100	700	50	140	Sub-base = 260
	150	720	50	160	
7	10	580	40	60	
	20	610	40	90	
	30	630	40	110	Base = 250
	50	650	40	130	
	100	675	50	145	Sub-base = 230
	150	695	50	165	

(Contd.)

Table 14.7: Recommended designs for traffic range 10-150 msa (IRC: 37-2001) *(Contd..)*

Subgrade CBR %	Cumulative Traffic (mm)	Total Pavement Thickness (mm)	Pavement composition		
			Bituminous Surfacing		Granular Base and Sub-base (mm)
			BC (mm)	DBM (mm)	
8	10	550	40	60	
	20	575	40	85	
	30	590	40	100	Base = 250
	50	610	40	120	
	100	640	50	140	Sub-base = 200
	150	660	50	160	
9	10	540	40	50	
	20	570	40	80	
	30	585	40	95	Base = 250
	50	605	40	115	
	100	635	50	135	Sub-base = 200
	150	655	50	155	
10	10	540	40	50	
	20	565	40	75	
	30	580	40	90	Base = 250
	50	600	40	110	
	100	630	50	130	Sub-base = 200
	150	650	50	150	

subgrade and this thickness remains the same irrespective of the quality of material used in the component layers. There is no consideration about the strength characteristics, shrinkage, particle size distributions, drainage, etc. of the different materials used in the construction of layers (courses) of the road.

- The CBR is generally evaluated on the soaked samples which are kept submerged in water for 96 hours. This situation seldom occurs at site and therefore designs by CBR method are oversafe.

- With the new materials available for road construction like modified bitumen, paving fabrics, geosynthetics, etc. CBR method should be modified to accommodate the use of these materials with reduced overall thickness of the pavement. Structural equivalency factor concept should be developed for various types of pavement courses.

- Another limitation associated with design curves is that they do not take into account the sensitivity of a pavement to damage under wheel loads which exceed certain assumed design value. Vehicle damage factors may change during the life of a pavement, creating higher damage than assumed.

- CBR may not be very reliable method of design in case of semi-rigid materials, when CBR value exceeds 60. In such cases the compressive strength test of the material would provide more reliable strength parameter for design of flexible pavements.

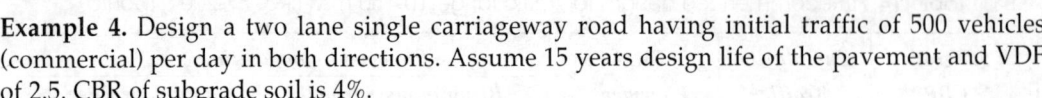

Example 4. Design a two lane single carriageway road having initial traffic of 500 vehicles (commercial) per day in both directions. Assume 15 years design life of the pavement and VDF of 2.5. CBR of subgrade soil is 4%.

Solution:

(a) Distribution factor for 2 lane road = 0.75

(b) Cumulative number of standard

axles to be catered for in the life $= \dfrac{365\left[(1+0.75)^{15}-1\right]}{.075} \times 500 \times .75 \times 2.5$

$= 900,0000$

$= \mathbf{9\ msa}$

(c) Total thickness required for
CBR value 4%, traffic 9 msa $= 690$ mm
(Fig. 14.10) say 700 mm (70 cm)

(d) Composition of pavement layers
for CBR (Table 14.6) $= 4$

(i) Bituminous surface courses : 40 mm BC wearing course
+ 80 mm DBM binder course

(ii) Thickness of WBM Base course: 250 mm WBM

(iii) Thickness of granular
sub-base (CBR 30%): 330

Total thickness = **700 mm (70 cm)**

Example 5. Making use of the following data calculate the cumulative number of standard axles for design of a flexible pavement:

(i) Initial traffic in each direction after construction $= 5000$ com/veh/day

(ii) Design life $= 15$ years

(iii) Traffic growth rate as predicted $= 8\%$

(iv) Vehicle damage factor as determined by axle load survey $= 4.5$

(v) Traffic distribution factor $= 0.75$

Solution: Cumulative number of standard axles in lane carrying maximum traffic

$$N = 365 \times \left[\dfrac{(1+r)^{N}-1}{r}\right] \times A \times D \times F$$

here $A = 5000$

$D = 0.75$

$F = 4.5$

$N = 15$

$r = 8$

Number of standard axles $= 365\left[\dfrac{(1+.08)^{15}-1}{0.08}\right] \times 5000 \times 0.75 \times 4.5$

$= \mathbf{165.7\ MSA}$

Example 6. Design a 4 lane divided highway having total cumulative number of standard axles for the design life as 100 msa. The CBR value of the subgrade is 6%.

Solution: Referring to Table 14.7, for CBR value 6, and cumulative traffic of 100 msa.

The total pavement thickness	= 700 mm
(a) Thickness of bituminous sufficing	= 50 mm BC + 160 mm DBM layer
(b) Thickness of base course	= 250 mm (wet mix madam)
(c) Thickness of sub-base course	= 260 mm (granular materials CBR not less than 30)
Provide the total thickness	= **720 mm**

14.3.5 Group Index Method of Flexible Pavement Design

This is an empirical design method, based on the particle size distribution and plasticity of the subgrade material. The group index of soil is determined from the following formula.

$$GI = 0.2a + 0.005ac + 0.01\ bd$$

where

GI = Group Index;

a = that portion of the percentage passing the no. 200 sieve which is greater than 35 and which does not exceed 75, expressed as a positive whole number (0 to 40);

b = that portion of the percentage passing the no. 200 sieve which is greater than 15 and which does not exceed 55, expressed as a positive whole number (0 to 40);

c = that portion of the numerical liquid limit which is greater than 40 and which does not exceed 60, expressed as a positive whole number (0 to 20); and

d = that portion of the numerical plasticity index which is greater than 10 and which does not exceed 30, expressed as a positive whole number (0 to 20).

GI ranges between 0 and 20; GI = 0 implies very good material (high bearing capacity), and GI = 20 implies very poor material (low bearing capacity)

Design method considers traffic volume on road as light, medium and heavy.

Light traffic < 50 commercial vehicles

medium traffic 50 – 300 commercial vehicles

Heavy traffic > 300 commercial vehicles.

Design procedure

Following steps are involved is designing by this method.

(i) Conduct the sieve analysis and find Atterberg limits of the subgrade soil (P.L. and L.L)

(ii) Calculate Group Index (GI) of the subgrade using above formula.

(iii) Estimate the expected traffic volume for the design year (light, heavy, medium)

(iv) Find the pavement thickness, using the proper curve as shown in Fig. 14.12, for the calculated group index value and design traffic volume.

Drawback of the group index method is, that it is based on physical properties of the soil. Quality of base and sub-base course material are not considered. The assumption made is that soil with identical group index give identical strength after compaction on field, so it is essential that compaction and constructional specifications are followed rigidly.

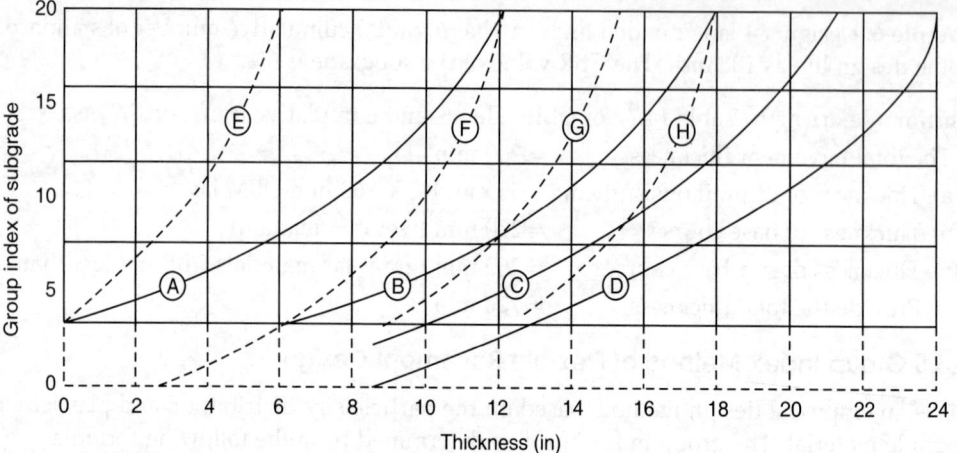

Curve *A* – thickness of selected material sub-base only
Curve *B* – combined thickness of surface, base and selected material sub-base (light traffic)
Curve *C* – combined thickness of surface, base and selected material sub-base (medium traffic)
Curve *D* – combined thickness of surface, base and selected material sub-base (heavy traffic)
Curve *E* – thickness of additional base which may be substituted for sub-base of curve A
Curve *F* – combined thickness of surface and base (no sub-base, light traffic)
Curve *G* – combined thickness of surface and base (no sub-base, medium traffic)
Curve *H* – combined thickness of surface and base (no sub-base, heavy traffic)

Fig. 14.12: Group Index method – design curves

Example 7. A subgrade soil sample has the following properties:

Soil passing sieve no 200 = 60%

Liquid limit = 65%

Plastic limit = 45%

Design pavement for 400 commercial vehicles per day, using Group Index method.

Solution: $\qquad$ G.I. $= 0.2a + 0.005\ ac + 0.01\ bd$

where,

$$a = 60 - 35 = 25$$
$$b = 55 - 15 = 40$$
$$c = 60 - 40 = 20$$

Plast. index $= 65 - 45 = 20$

$$d = 20 - 10 = 10$$

$$\text{G.I.} = 0.2 \times 25 + 0.005 \times 25 \times 20 + 0.01 \times 40 \times 10$$
$$= 11.5 = 12$$

(i) Using curve A, (Fig. 14.12)

$\qquad$ Thickness of subbase for GI = 12 is 23 cm

(ii) Using curve D, for heavy traffic,

$\qquad$ Thickness of surface + base + Sub base = 54 cm.

Thickness of surface and base = 54 – 23 $\qquad$ = 31 cm.

Assume surface thickness for heavy traffic $\qquad$ = 8 cm

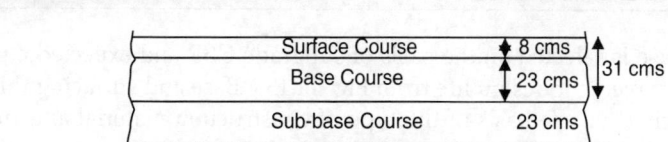

Surface Course	8 cms	
Base Course	23 cms	31 cms
Sub-base Course	23 cms	

subgrade

Example. 8 A soil sample from a proposed highway site was tested in laboratory and following parameters were obtained:

Soil passing sieve No. 200 (B.S) = 55%

Liquid limit = 40%

Plasticity index = 20%

It is estimated that traffic volume consisting of 200 trucks and buses per day will be using this facility when it is opened. Determine thickness using Group Index method of design.

Solution:

$$\text{Group Index (GI)} = 0.2a + .005 \, ac + .01b \, d$$

here
$$a = 55 - 35 = 20$$
$$b = 55 - 15 = 40$$
$$c = 40 - 40 = 0$$
$$d = 20 - 10 = 10$$

∴
$$\text{GI} = 0.2 \times 20 + .005 \times 20 \times 0 + .01 \times 40 \times 10 = 8.$$

Using group index design chart (Fig. 14.12) and daily traffic of 200 cv/day.

Total thickness = 37.5, which can be distributed as :

Bituminous top layer thickness = 7.5 cm

Granular base thickness = 10 cm

Selected material for sub-base thickness = 20 cm

Total thickness of pavement = 37.5 cm

14.4 FLEXIBLE PAVEMENT DESIGN PROCEDURES IN OTHER COUNTRIES

The following is a brief description of a number of design methods developed for use in some countries, to let the readers know their peculiar features. Further details will be available in the technical manuals of the particular country, as design procedures will keep changing/ modifying with the development of new knowledge and experience. It is not possible to cover entire field of structural design of highway pavement in a book of this size. A review of some methods for flexible pavement design is covered.

Due to some degree of correlations with pavement performance, which is influenced by environmental and soil conditions there can not be an entirely satisfactory method of design in one country, which can suit to another country.

14.4.1 British Method of Pavement Design

Design standards followed in United Kingdom are contained in Road note 29, based on full scale experimental pavement test carried out by Transport and Road Research Laboratory (TRRL).

In this method, cumulative number of commercial vehicles passing over the highway on the most heavily trafficked lane during the design life is obtained either by calculations or graphs prepared for this purpose. Design life is normally taken as 20 years for flexible pavement.

The thickness of sub-base is selected on the basis of subgrade CBR and expected cumulative standard axles during the pavement design life from Fig. 14.13a. Base and surfacing thicknesses are determined from the chart on the basis of the type of construction material and the design life of the pavement. Figures 14.13b and 14.13c show design charts for the dense macadam, wet mix and dry-bound macadam road bases.

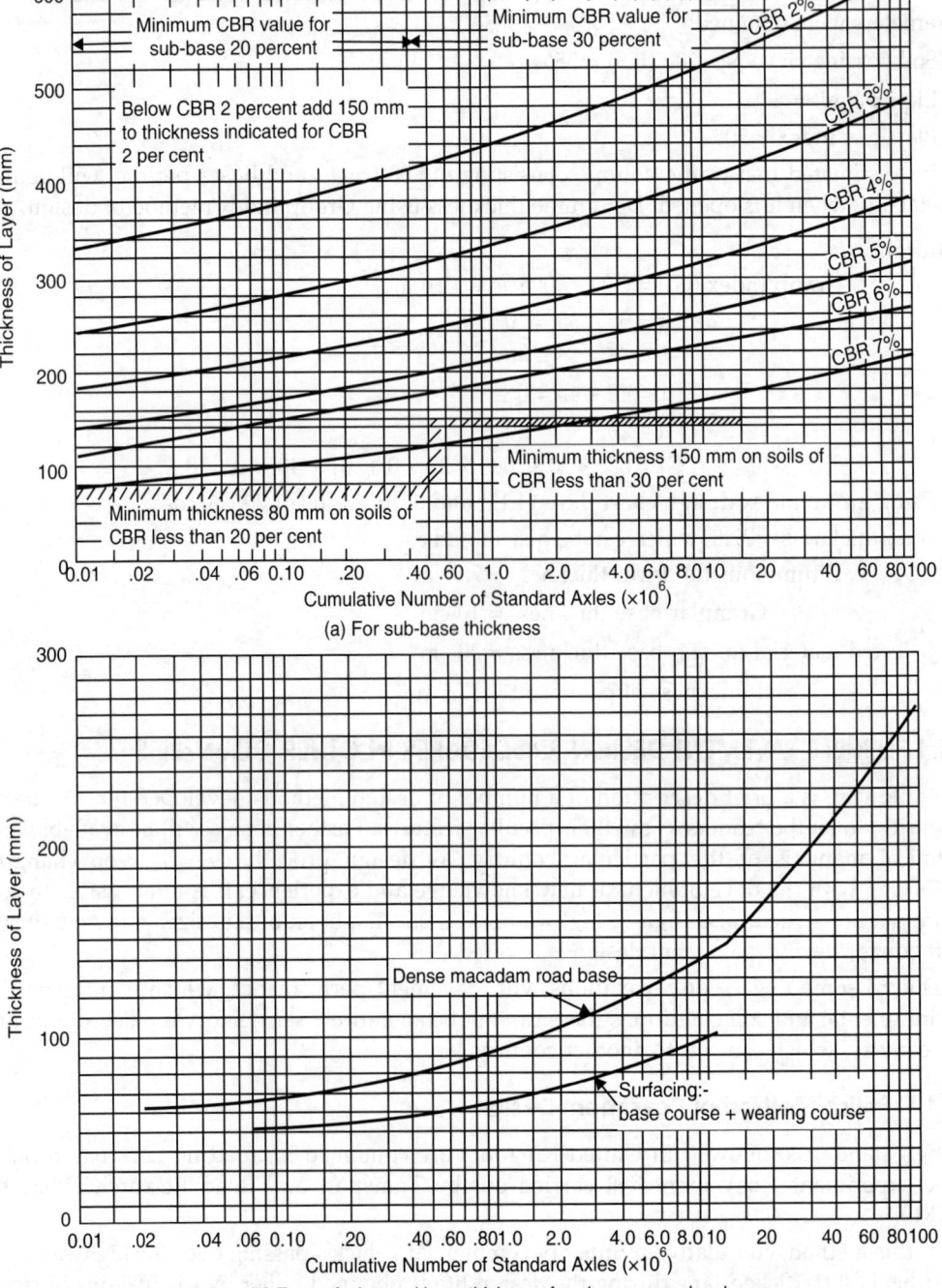

(a) For sub-base thickness

(b) For surfacing and base thickness for dense macadam bases

Figs 14.13a and b: Design charts for pavement component thickness (RN 29)

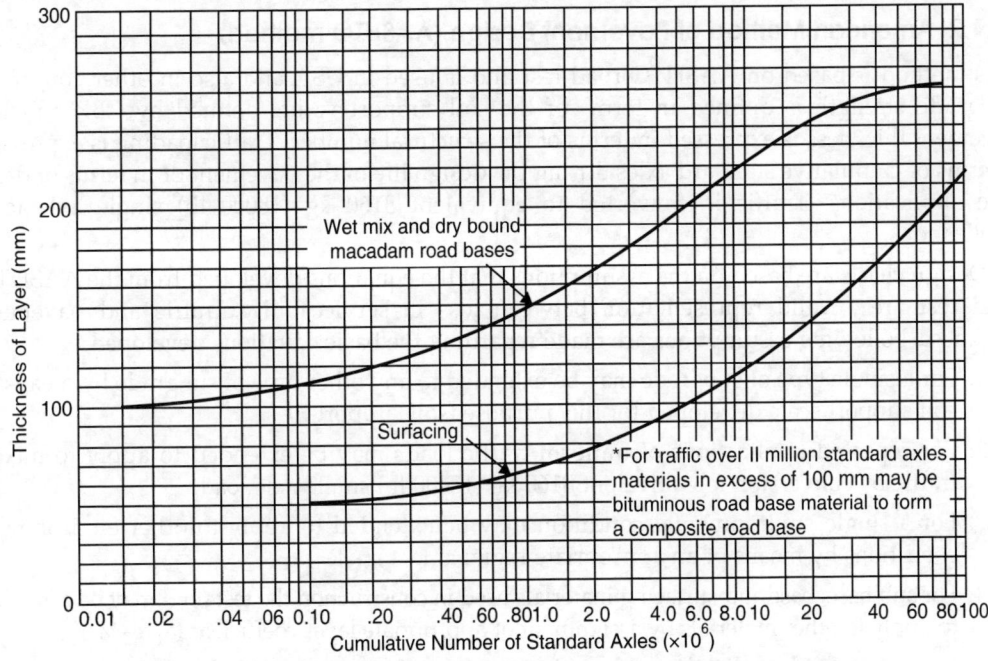

(c) For surfacing and base thickness for wet mix and dry bound macadam bases

Figs 14.13c: Design charts for pavement component thickness (RN 29)

Road note 31 considers the traffic loading in terms of cumulative number of standard axles on the basis of which the type of surfacing, and thickness of the base and sub-base are selected. Selection of the sub-base thickness is also based on the bearing strength (CBR) of the subgrade as shown in Fig. 14.14.

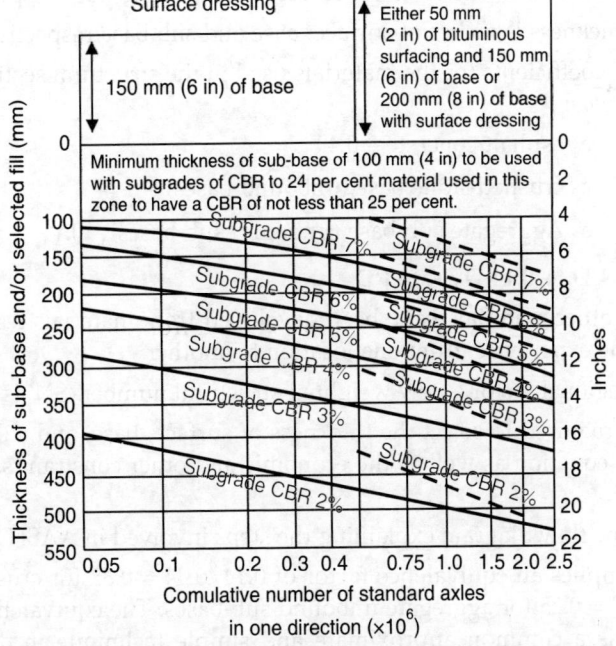

Fig. 14.14: Pavement design chart for flexible pavement (RN31)

14.4.2 American Method of Pavement Design (AASHTO method)

This method is based on AASHTO road test and is used in USA and also in other countries. Subgrade strength is defined in terms of the soil support value, defined by CBR and the pavement thickness is expressed in terms of the structural number. Traffic loading is expressed is terms of cumulative standard axles during the design life of the pavement or in terms of daily axle applications. Traffic is converted to equivalent 8160 kg (18000 lb) single axle load applications.

Design charts are based on the assumptions that the equations developed from the AASHTO road test are a valid representation between loss in serviceability, traffic and pavement thickness. Following assumptions are made regarding the basic equations developed.

1. For a single type of subgrade may be extended to any other type of subgrade by means of soil support scale developed for this purpose (soil support-S)

2. For repeated applications of uniform traffic loads may be extended to apply to mixed traffic by conversion to equivalent 8160 kg (18000 lb) single axle load.

3. For a single environmental condition may be extended to apply to other environment conditions by means of an appropriate regional factor - R.

4. For sub-base, base and surfacing materials used in construction the test road may be extended to apply to other materials by assignment of appropriate layer coefficient (a_1, a_2, a_3)

5. For accelerated applications of traffic during the 2 year test period may be extended to apply to repetitions of traffic during an extended period of time (up to 20 years)

For a specific terminal serviceability the design equations can be expressed by structural number as a functions of S, R, number of equivalent single axle loads. The overall strength of the pavement is presented as a structural number (SN).

Structural Number: SN (Thickness Index) = $a_1 D_1 + a_2 D_2 + a_3 D_3$

where a_1, a_2, a_3 are layer coefficients of surface, base and sub-base respectively.

D_1, D_2, D_3 are thickness is inches of surface, base and sub-base respectively.

Values of the 'a' coefficients for the materials used in the structural sections of the AASHTO test road are as follows:

a_1 asphalt concrete – 0.44

a_2 crushed stone base (unbound bases) – 0.14

a_3 Aggregate sub-base (unbound sub-bases) – 0.11

So that **SN = 0.44 D_1 + 0.14 D_2 + 0.11 D_3**

Coefficients for other materials have been developed from materials tests and modified the values in light of the experience, from one location to another.

By applying a regional factor a new weighted structural number 'SN' is obtained.

The designer is required to select the thickness of surface, base, and sub-base course which satisfies the design equation as well as the economic and other constraints. The design chart is shown in Fig. 14.15.

Figure 14.16 is the flow diagram explaining the steps involved in AASHTO design approach.

This equation implies an equivalence factor of 0.14/0.44 = 0.32 for crushed stone unbound bases and 0.11/0.44 = 0.25 for aggregate unbound sub-bases. The equivalent thickness factor or layer equivalency is a common approximate and simple technique in design of multilayer pavements where the layers (courses) are of different materials.

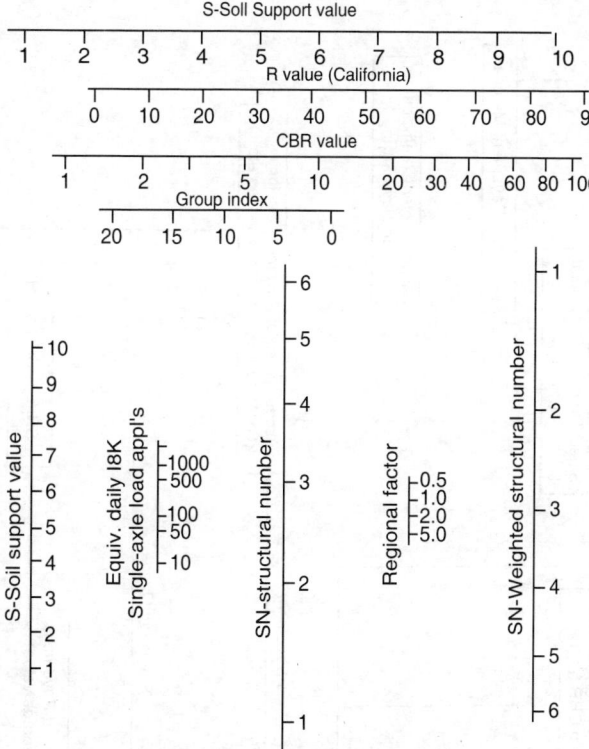

Fig. 14.15: AASHTO design chart

It is normal to use asphaltic (bituminous) concrete as the standard material with an equivalence factor of 1.00.

The factor for a layer of another material is the amount that its layer thickness would need to be multiplied in order that an asphaltic concrete layer of this new thickness would perform identically to the real thickness of the material layer in question.

A layer coefficient converts the structural number to actual pavement thickness and is a measure of the relative ability of the material to function as a structural component of the pavement.

Use of design chart

The solutions of the design equation for flexible pavement for terminal service is done, using the straight edge. First the soil support value of the subgrade and the total equivalent standard axle loading are used to obtain the unweighted structural number (SN). This latter value is then used with the regional factor to obtain the weighted structural number.

Suitable designs are then those whose material types and thicknesses satisfy the general equation.

$$SN = a_1 D_1 + a_2 D_2 + a_3 D_3$$

Minimum thickness of 50 mm are specified for wearing courses and 100 mm for road bases and sub-bases.

Layer coefficient values, can be developed for the locally available materials. For example in USA several states developed following layer coefficients.

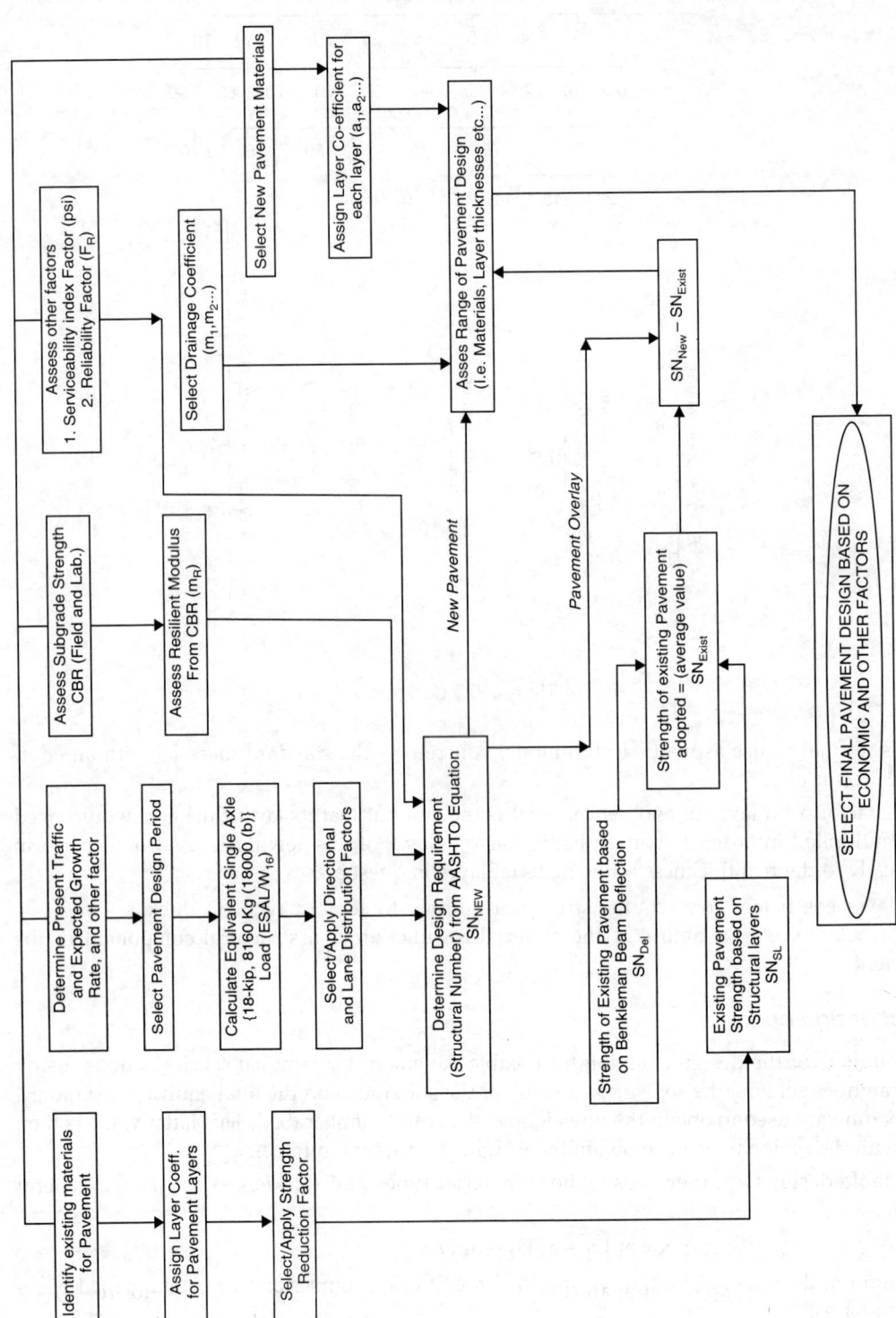

Fig. 14.16: AASHTO design approach for pavement design

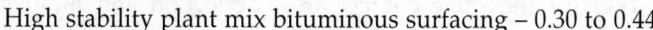

High stability plant mix bituminous surfacing – 0.30 to 0.44

Untreated crushed stone road bases – 0.10 to 0.14

Crushed – stabilized soil road bases – 0.20

Bituminous road base materials – 0.30 to 0.34

The design chart for determining the thickness of flexible pavement requires the following data:

   i. Soil support of the road bed soil.

   ii. Total or daily equivalent 18000 lb single axle load estimated for the design lane for design life (usually 20 years)

   iii. The regional factor applications to the site.

   iv. Unweighted structural number can be found from the chart which later on can be modified with appropriate region factor to give weighted structural number.

   v. Knowing the structural number, various combinations of materials and thickness, satisfying the general equation SN = $a_1 D_1 + a_2 D_2 + a_3 D_3$ can be selected.

14.4.3 French Method (CEBTP Method)

Centre Experimental de Recherches et d' Etudes du Batiment e des Travaux Publish (CEBTP), developed a design method in French speaking countries. In this method subgrade strength is determined by CBR and traffic is classified in four categories. CBR method has been modified according to the local physical and environmental conditions. Four basic pavement structures are worked out and designing involves selection of the most suitable one, modified as per local conditions.

Design of a flexible pavement involves two stages.

   i. Collection of data

     • Traffic data (load and movements)

     • Characteristics of the natural soil.

   ii. Determination of equivalent pavement thickness and selection of pavement structure which provides an equivalent thickness.

CBR value is used to determine bearing strength of subgrade. Graphs are then used to determine the pavement thickness.

14.4.4 Canadian Method (Mcleod method)

Mcleod through Canadian department of transport conducted the plate bearing test on the surface, base and subgrade at a large number of test locations, in addition to cone bearing tests. CBR tests and tri-axial compression tests were also performed on subgrade soils at each test location, to correlate with the plate bearing test results.

Based on these investigations, following equation was developed for thickness design of pavements.

$$T = K \frac{P}{Log_{10}^S}$$

where

   T = The required thickness of base course - cm

   P = Gross wheel load (ESWL)

S = Total subgrade support (kg) for the same contact area, deflection and number of repititions of load as for the applied load.

K = Base course constant.

From the experiments, it was found that the value of base course constant (K), changes with the size of the bearing plate, as shown in Fig. 14.17.

The subgrade support (S), for highway pavement is calculated from the plate bearing test using 30 diameter plate at 0.5 cm deflection and 10 repetitions of the load.

The relation of subgrade support 'S' and perimeter area ratio (P/A) is shown is Figure 14.18.

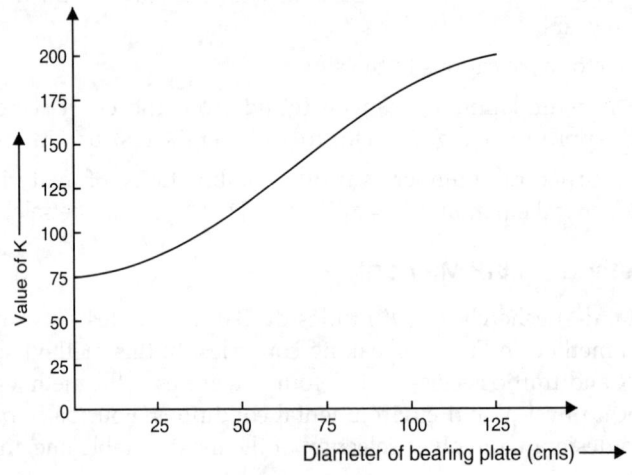

Fig. 14.17: Base course constant (K) Vs diameter of load bearing plate

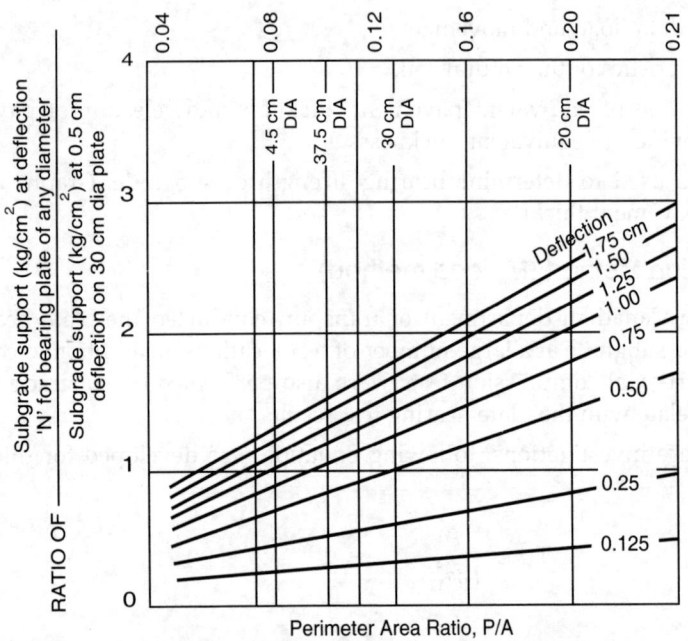

Fig. 14.18: Subgrade support (s) vs perimeter area ratio (P/A)

Example 9: Find the thickness of granular base course of a highway pavement for the following conditions.

The maximum weight of the truck = 12,000 kg

The truck has dual wheels on the rear axle and single wheel on the front axle.

Clear span in duals = 10 cm

Centre to centre spacing in duals = 27 cm

Tire pressure = 4kg/cm^2

Factor of rigidity = 1.25

The plate bearing test conducted on subgrade soil using 30 cm diameter plate yielded a pressure of 3kg/cm^2 after 10 repetitions.

Solution:

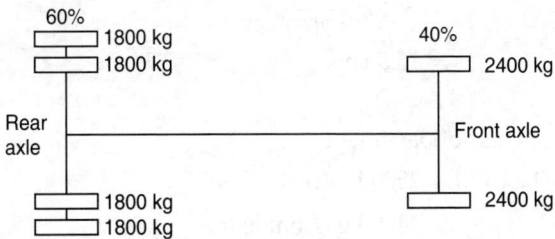

Assuming 60% of total load on rear axle.

Load on rear axle = $0.6 \times 12000 = 7200$ kg

$$\text{load on each wheel of rear axle} = \frac{72,00}{4} = 1800 \text{ kg}$$

Front axle load = $0.4 \times 12000 = 4800$ kg

$$\text{Load on each wheel of front axle} = \frac{4800}{2} = 2400 \text{ kg}$$

I Trial:

Assuming pavement thickness = 30 cms

plot graph between (d/2, P) and (2S, 2P) line AB

(5, 1800) and (54, 3600)

From the graph for 30 cm thickness

$$\text{ES WL} = 3100 \text{ kg}$$

$$\text{For front wheel} = 2400$$

Therefore design wheel load = 3100 kg

Given that: Tire pressure = 4kg/cm^2 and Rigidity factor 1.25

$$\text{Contact pressure} = 4 \times 1.25 = 5 \text{ kg/cm}^2$$

$$\therefore \quad \text{Area of contact} = \frac{3100}{5} = 620 \text{ cm2} = \pi a^2$$

$$\text{Radius of contact } a = \sqrt{\frac{620}{\pi}} = 14 \text{cm}$$

∴   Diameter of contact = 2 × 14 = 28 cm

Value of K, corresponding to diameter 28 cm

$$K = 85 \text{ (from the curve for highway pavement)}$$

$$\text{Ratio} = 1.1 = \frac{\text{Subgrade support any condition}}{3}$$

Unit subgrade support for 28 cms = 1.1 × 3 = 3.3 kg/cm^2

∴ $S = 620 \times 3.3 = 2031$

Using Mcleod equation

$$T = K \log_{10} \frac{P}{S}$$

$$= 85 \log_{10} \frac{3100}{2031}$$

$$= 15.5 \text{ cms}$$

II Trial:

Assuming 15 cm thickness of pavement

$$\text{E.S. W.L} = 2500 \text{ kg (rear wheel)}$$

$$= 2400 \text{ kg (front axle)}$$

∴   Design wheel load = 2500 kg

$$\text{Contact area} = \frac{2500}{5} = 500 \text{ cm}^2 = \pi a^2$$

∴ $a = 12.62$ cm and dia = 25.24 cms

Ratio for 25.24 diameter = 1.25

Unit support for 25.24 cm dia plate = 1.25 × 3

$$= 3.75 \text{ kg/cm}^2$$

∴ $S = 3.75 \times 500 = 1875$ kg

K for 25.24 cm dia plate (from curve) = 83

∴ $$T = K \log_{10} \frac{P}{s}$$

$$= 83 \log_{10} \frac{2500}{1875}$$

$$= 10.3 \text{ cms}$$

III Trial:

Assume thickness = 10 cm., and repeat the steps as in trial I and II.

14.5 ANALYTICAL METHODS OF FLEXIBLE PAVEMENT DESIGN

In general, analytical pavement design involves the assumption of a pavement structure as a system. The strength characteristics (in terms of modulus and poisson's ratio) for each layer, may be established or can be assumed. Traffic loading is then introduced and structural analysis carried out to determine stresses and strains at critical points in the structure. The values of

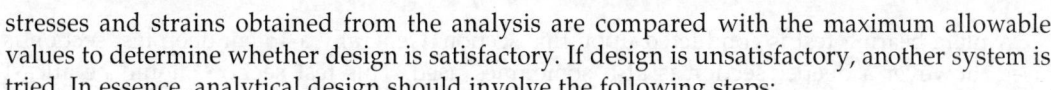

stresses and strains obtained from the analysis are compared with the maximum allowable values to determine whether design is satisfactory. If design is unsatisfactory, another system is tried. In essence, analytical design should involve the following steps:

- Development of mathematical models to represent pavement behaviour.
- Selection of appropriate solution techniques to the equations developed above and thereby compute stresses, strains and deformation at critical points of the pavement structure.
- Material characterization for the various layers of the pavement structure for a given environment
- Setting out design criteria such as in terms of allowable stresses or strains in certain critical points in the structure.

Several computer programmes have been developed to carry out pavement analysis and design and are commercially available.

14.5.1 Burmister's Method of Flexible Pavement Design

This method based on two layered theory was developed by Mr. Burmister. The flexible pavement section consists of number of layers. The elastic modulus of the layers decreases with increase in depth.

Boussinesq single layer analysis can be considered as a special case of Burmister's two layer theory. Thus

As per Bossinesq theory $E_1 = E_2$ and as per Burmister theory $E_1 > E_2$

where E_1 = elasticity modulus for base course and E_2 = elasticity modulus of subgrade soil. The deflection equations given by Burmister are:

Deflection for flexible plate $\Delta = 1.5\dfrac{Pa}{E_2}F_2$ (conditions obtained by rubber tyred wheel loading)

Deflection for rigid plate $\Delta = 1.18\dfrac{Pa}{E_2}F_2$ (condition obtained by using staked steel plates).

14.5.2 Design Steps

Following steps are involved in design of pavements:

(i) Determination of the value of E_2 (modulus of elasticity of subgrade)

A plate bearing test on subgrade soil is conducted and contact pressure for a deflection of 0.5 cm or 0.25 cm is obtained.

Since the soil, below subgrade can be considered as one layer, the value of $F_2 = 1$

using the equation.

$$\Delta = 1.18\frac{Pa}{E_2}F_2$$

the value of E_2 can be calculated for $\Delta = 0.5$ or 0.25, $F_2 = 1$, load (P) and area (a) are known.

(ii) Determination of value of E_1 (modulus of elasticity of base course)

After E_2 is known, a test section consisting of base course materials, that will be used in actual construction is built over the subgrade.

A plate bearing test is conducted upon this section. Generally a 15 cm deep test section is used, however a deeper section is also sometimes used. This test section should be atleast 4.5 m × 4.5 m square or larger. The plate bearing test on the test section, should be made in same manner, that on the subgrade.

Using stress-deflection data, obtained from plate load test, on the base course, the value of F_2 for base course-subgrade (Two layer system) can be worked out as P and a, at $\Delta = 0.5$ cm or 0.25 cm, can be found out from the test data as value of E_2 is already known in step (1).

Knowing the value of F_2 for two layered system the $\dfrac{E_1}{E_2}$ is found from curve as h/a is also known (h is thickness of the test section)

As E_2 is known, E1 can be calculated.

(iii) Determination of pavement thickness

After working out the ratio of $\dfrac{E_1}{E_2}$, the value of F_2 is calculated using the eqn.

$$\Delta = 1.5\frac{Pa}{E_2}F_2$$

For any condition of load on the pavement.

From the curves, given by Burmister for displacement factor F_2, using the value E_1/E_2, the value of pavement thickness can be calculated.

Example 10: The plate bearing tests were conducted with 30 cm dia plate on soil subgrade and over 15 cms base course. The pressure yielded at 0.25 cm deflection is 1.25 kg/cm^2 and 5 kg/cm^2.

Design the pavement thickness for 4080 kgs wheel load with tire pressure of 4.5kg/cm^2 Rigidity factor = 1.11

Solution:

(1) Calculation for E_2

For single layer condition $F_2 = 1$ (subgrade)

$$\Delta = 1.18\frac{Pa}{E_2}\times 1 \qquad\qquad (a = 30/2 = 15)$$

or $$E_2 = \frac{1.18Pa}{\Delta} = \frac{1.18\times 1.25\times 15}{0.25} = 88.5\text{kg/cm}^2$$

(2) Calculation of E_1

Two layer theory will be applicable

$$\Delta = 1.18\frac{Pa}{E_2}\times F_2$$

$$0.25 = 1.18\frac{5\times 15}{88.5}\times F_2$$

$$F_2 = 0.25$$

$$\text{here } \frac{h}{a} = \frac{15}{15} = 1$$

From Burmister deflection curve $\dfrac{E_1}{E_2} = 85$

or $E_1 = 85\ E_2 = 85 \times 88.5 = 7520\ \text{kg/cm}^2$

(3) Design thickness of pavement

$$P = 4080\ \text{kg}$$

$$p = \text{tire pressure} \times \text{Rigidity factor}$$

$$= 4.5 \times 1.11 = 5\ \text{kg/cm}^2$$

$$A = \frac{P}{p} = \frac{4080}{5} = 816\ \text{cm}^2 = \pi a^2$$

$$a = \sqrt{\frac{816}{\pi}} = 16\text{cms}$$

For rubber tyred vehicle, condition of flexible plate

$$\Delta = 1.5 \frac{Pa}{E_2} F2$$

$$0.25 = 1.5 \frac{5 \times 16}{88.5} \times F_2$$

$F_2 = 0.184$, $\dfrac{E_1}{E_2} = 85$, for this condition from curve $h/a = 1.5$

∴ $h = 1.5a = 1.5 \times 16 = 24$ cms Ans.

Example 11: Using Burmister's theory calculate the expected pavement surface deflection for a 300 mm stabilised gravel pavement structure lying on a murram subgrade of infinite depth. Assume the average modulus of elasticity of stabilised gravel is 344, 500 kN/m² and that of the murram subgrade to be 68900 kN/m². Also assume uniform surface loading of 689 kN/m², distributed over a circular area of radius 150 mm.

Solution:

$$E_1 = 344{,}500\ \text{kN/m}^2$$

$$E_2 = 68{,}900\ \text{kN/m}^2$$

$$E_1/E_2 = 5$$

$$h = 300\ \text{mm} = 2a = \text{thickness of pavement}$$

$$a = \text{radius of loaded area} = 150\ \text{mm}$$

$$p = 689\ \text{kN/m}^2.$$

From Burmister's charts

$$F = 0.41$$

∴ Δ = vertical displacement on the road surface

$$\Delta = \frac{1.5pa}{E_2} \times F$$

$$= \frac{1.5 \times 689 \times 150}{68{,}900} \times 0.41$$

$$= \textbf{0.92 mm}$$

14.5.3 Drawbacks of Burmister's Method

(i) The pavement thickness calculated by Burmister's method is objectionable as the value of E_1 and E_2, have no physical meaning and can be used only as an estimate. To get the true pavement thickness trial section can be built and tested for critical deflections.

(ii) The two layer theory applies only to single wheel loads. Therefore for design load, for dual wheel ESWL must be calculated.

(iii) While conducting the plate bearing test at natural moisture content, if the test is made during summer month, when the subgrade is relatively dry, the strength value obtained will not present the strength of the soil in its weakest condition, subsequent saturation of subgrade resulting from surface infiltration, freezing and thawing and other climatic factors will try to increase soil moisture content. This will result into decreased stability. Thus the plate bearing test may yield different test values than the critical one.

(iv) It is to be noted that no consideration of the type and depth of wearing course is given. However the structural qualities of wearing course material are always better and hence a certain thickness of material can be substituted for the base course material.

14.6 RIGID (CONCRETE) PAVEMENTS

Rigid pavement are composed of a concrete slab which functions as a beam and is the major load spreading component. The concrete slab is supported by a sub-base, of granular or cement bound material. This element of the pavement structure distributes the wheel load to the subgrade, ensures an adequate depth of material to prevent frost, moisture penetration and provides a working base for construction plant. Between the concrete slab and the sub-base a separation membrane of usually a thermoplastic sheeting is put to reduce the restraint of the sub-base upon the slab movement and also prevent moisture movement between the slab and the sub-base.

Purpose of sub-base in rigid pavement

Material which supports a rigid pavement is termed as sub-base. Granular/treated material is placed over the existing soil (subgrade) as sub-base to serve following purposes:

- It prevents mud pumping (ejection of water near joints and cracks)
- Adverse action due to swell and shrinkage in soils is avoided, specially in clayey, silty subgrades.
- Supporting capacity of existing soil (subgrade) is improved.
- Continuous, level, smooth and stable platform for concrete slab is provided.
- In forest areas, damage of pavement is prevented, by insulating layer.

Sub-base of portland cement or bitumen are used for heavy trafficked highways.

Temperature changes cause concrete slab to expand and contract as a result of which stresses are set up when contraction or expansion is prevented.

Movement of slab could be wholly or partially prevented by friction between the slab and the subgrade thereby leading to tensile stresses developing in the slab with a drop in the temperature. The design aims at providing a concrete section capable of resisting the stresses developed and if necessary reinforcement may be provided.

Joints are provided in concrete pavements for a variety of reasons. The contractions joints are provided so as to relieve tensile stresses resulting from contraction and warping of the slab. Dowel bars are used for load transfer across the joint. Expansion joints are provided in form of clear breaks in the concrete slab to allow for expansion. Dowel bars are used for load transfer across the joints. Construction joints make the end of a days construction and are normally of

the butt type with dowel bars provided for load transfer. Hinge and warping joints are used to control cracking along the centre line of the concrete slab.

Pumping and blowing are major problems of concrete pavements. Pumping involves the ejection of water and subsoil through joints, cracks and along the edges of pavement due to downward slab movement caused by passage of heavy axle loads over the slab. Extensive deformation of the concrete slab leads to transverse cracking due to cantilever action. Blowing is a form of pumping in which base or subbase under concrete slab leads to longitudinal cracking. Mud jacking and joint sealing are used to correct these defects.

14.6.1 Type of Rigid Pavement

Rigid pavement can be of following types:

   (i) Cement concrete pavement.

   (ii) Reinforced cement concrete pavement.

  (iii) Continuously reinforced cement concrete pavement.

  (iv) Prestressed cement concrete pavement.

14.7 FACTORS THAT GOVERN DESIGN OF RIGID PAVEMENTS

The design of rigid pavements is affected by following factors:

1. Wheel loads

Higher the wheel loads on the road, higher will be stresses. Axle loads of vehicles, vary widely on the roads, and on heavy traffic roads the commercial vehicles often exceed the legal limits. The legal axle load limits in India are:

   Single axle – 10.2 Tonnes

   Tandem axle – 19.0 Tonnes

   Tridem axles – 24.0 Tonnes

Design load is the maximum wheel load of the heavy vehicle which uses the highway frequently. For single axle vehicles the design wheel load is 5100 kg.

2. Tire-pressure and contact area of wheels

Wheel load are transmitted to pavement over the contact area. Stresses is rigid pavement also depend upon the tire inflation pressures of commercial vehicles. The tire imprint is assumed circular. The tire pressure for commercial vehicles ranges from 0.7 to 1.0 MPa. A tire pressure of 0.8 mPa is adopted in design.

3. Repetition of wheel loads

The deflection of pavement or subgrade due to one application of wheel load may be small, but due to repeated application of loads, there would be increased elastic and plastic deformations. The accumulated plastic deformations may cause pavement failure. Design procedure takes this factor into account.

4. Moving loads and impact factor

Moving loads cause shocks and thereby result into additional deflections and stresses. Presence of transverse joints in rigid pavement cause impact. The amount of impact depends upon.

   • The amount concrete slab is depressed at the joint relative to adjoining slab and

   • The speed of the vehicle

Generally an impact allowance of 20 to 25% is used for rigid pavement design.

5. Position of wheel load on the concrete slab

Loading positions affect the intensity of stresses. The three critical positions are:

Load at the interior: Tensile stresses are produced at the bottom of the concrete slab.

Load at the edge: Produces tensile stresses at the bottom of the concrete slab parallel to the edge and at the top of the slab, right angles to the edge.

Load at the corner: Produces tensile stresses at the top of the slab parallel to the bisector of the corner angle. In Westergaard formulae for stress calculation these loading positions are analyzed.

6. Design period

Design life for a rigid pavement is taken from 20 to 30 years. In India, a 30 year design period is normally adopted, for rigid pavements.

7. Design traffic

Design traffic in terms of cumulative number of axles during the design period is calculated by the following formula

$$C = \frac{365 \times A(1+r)^N - 1}{r}$$

Where

C = Cumulative number of axles during design period.

A = Initial number of axles per day in the year after construction is over.

 r = Annual rate of growth of commercial traffic (To be determined from actual data and various correlation techniques as correctly as possible. In the absence of such reliable data, IRC recommends that a growth rate of 7.5% may be assumed.

N = Design period of road in years.

8. Quality of subgrade/sub base (subgrade modulus 'K')

The quality of subgrade/subbase (foundation for concrete slab) is determined by the modulus of subgrade reaction (K) which is pressure per unit deflection, determined by the plate bearing test as defined earlier. K value – the subgrade modulus, indicates the bearing value of the supporting material.

Strength of the subgrade/sub-base foundation is reduced due to presence of moisture and therefore the minimum value under worst moisture conditions encountered during rainy season must be used in design.

A dry lean concrete (DLC) sub-base is generally used in modern concrete pavement, specially if traffic intensity is high.

9. Separation layer between sub-base and concrete slab

To reduce inter layer friction a foundation layer below concrete slab is provided using a smooth membrane of minimum thickness of 125 micron of polythene to separate the slab from foundation.

10. Characteristics of concrete

(a) **Flexural strength of concrete:** Crushing strength of concrete is important in design of the pavement, but concrete rarely fails in compression as it has very high crushing strength.

The flexural strength is more important strength parameter to be considered in design. The flexural strength is evaluated as the modulus of rupture. IRC specifies a minimum modulus of rupture of 40 kg/cm^2 (4MN/m^2) This can be attained by a concrete of 28-day strength.

(b) Modulus of elasticity (E): Modulus of elasticity (E) of concrete determines relative stiffness of the slab and therefore should be considered in design. Value of modulus of elasticity increases with increase in strength of the concrete. E value of concrete is 3×10^5 kg/cm^2 for concrete having flexural strength in the range of 38-42 kg/cm^2 (3.8 - 4.2 MN/m^2)

(c) Poisson's ratio (μ): In calculation of stresses in the rigid pavement Poisson's ratio is a parameter which is used. Its value is taken as μ = 0.15. Poisson's ratio decreases with increase in the modulus of elasticity.

(d) Coefficient of thermal expansion: Coefficient of thermal expansion of concrete varies with type of aggregates. For design of slab it is taken as 10×10^{-6} per °C.

(e) Shrinkage properties of concrete: Initially concrete expands slightly due to hydration of cement, but on drying it shrinks. The shrinking causes stresses. Change in moisture content also cause shrinkage or expansion.

These additional stresses due to shrinkage and expansion need consideration in design.

(f) Temperature stresses: Due to temperature changes stress are caused by concrete slab, due to:

(i) Change in temperature gradient through the slab, causing expansion and contraction.

(ii) Expansion and contraction of slab due to temperature changes is restrained by friction between subgrade and the slab.

(g) Friction between slab and sub-base: The friction between slab and sub-base imposes restrain on expansion and contraction due to temperature changes.

(h) Arrangement of joints: For allowing expansion, contraction and warping, joints are used in concrete slab. The type, spacing, and arrangement of joints is important in design of rigid pavements.

(i) Reinforcement: Rigid pavement can be reinforced or unreinforced concrete slab. The reinforced pavement and continuously reinforced pavements completely change the design approach.

11. Drainage layer

Drainage is important for subgrade strength. A drainage layer should be provided beneath the pavement throughout the road width, to facilitate quick disposal of water that is likely to enter the subgrade.

14.8 STRESSES IN RIGID PAVEMENT

Stresses in rigid pavement (concrete slab) are caused by following:

i. Wheel load stresses: Higher the wheel loads, greater shall be the flexural stresses in pavement.

ii. Temperature stresses

(a) Warping stresses

(b) Frictional stresses

Due to temperature variations slab expands and contracts resulting in frictional stresses (when expansion and contraction is prevented) and difference is top and bottom temperature of the slab results in warping stresses. The design aims in providing a concrete section that resists developed stresses.

iii. **Stresses due to moisture variation:** Moisture changes result is plastic cracks due to rapid absorption of moisture. Excessive water results in muddy conditions known as 'Crazing' resulting in cracks. Stresses due to moisture changes are however very small and are opposed by temperature stresses.

iv. **Combined stresses:** The stresses by above factors get combined to produce final stress in the concrete pavement. The maximum combined stresses in the 3 regions of the slab will be when effect of temperature is additive to the wheel load stresses.

14.8.1 Wheel Load Stresses

For theoretical calculations of the stresses in the concrete slab, Westergaard developed formulae making following assumptions.

(i) The contact area is circular for corner and interior loading and semi-circular for edge loading as shown in Fig. 14.19.

(ii) The slab acts as a homogenous, isotropic, elastic solid in equilibrium.

(iii) The reaction of the subgrade is vertical only and proportional to deflection.

(iv) The thickness of slab is uniform.

(v) Concrete is homogeneous and uniformly elastic.

Westergaard's formulae are the simplest to use. The tensile stress occurs to the bottom of the slab and of the same magnitude in all direction for

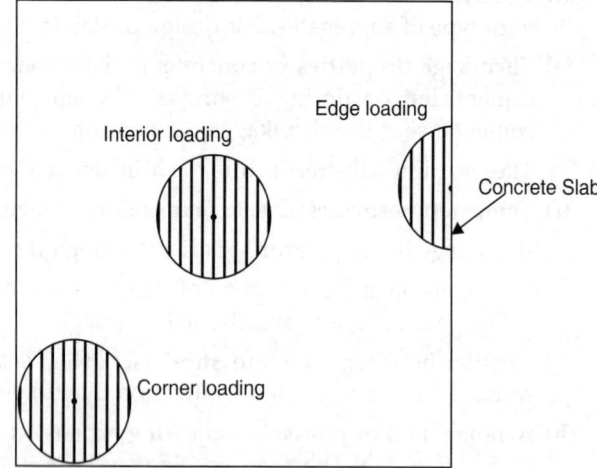

Fig. 14.19: The three positions of loading considered by Westergaard

the in-terior loading. When loading it at the edge the tensile stress occurs at the bottom of the slab parallel to the edge. Some tensile stress will occur at the top of the slab at right angles to the edge. For loading at the corner tensile stress occurs at the top of the slab parallel to the bisector of the corner edge.

Westergaard formulae

Westergaard formulae for stresses in rigid pavements uses terms, which are defined as below:

(a) Modulus of subgrade reaction "K"

It is a measure of pressure required to produce unit deflection in a subgrade. Expressed as load/length^3 (Kg/cm^3) Modulus is calculated for 0.125 cm deflection

$$K = \frac{P}{0.125}$$ where P = load intensity kg/cm^2 for 0.125 cm deformation.

(b) Radius of relative stiffness "L"

It is a measure of stiffness relationship between slab and subgrade. Westergaard suggested:

$$L = \left[\frac{Eh^3}{12(1 - \mu^2)K} \right]^{1/4}$$

L = Radius of relative stiffness cms.

E = Modulus of elasticity of concrete 0.21×106 kg/cm^2

h = Slab thickness cm

k = Subgrade modulus kg/cm^2

μ = Poisson's ratio, of lateral extension to longitudinal contraction for concrete usually taken as 0.15.

Physical concept of Westergaard's 'Radius of Relative Stiffness' is illustrated in Fig. 14.20.

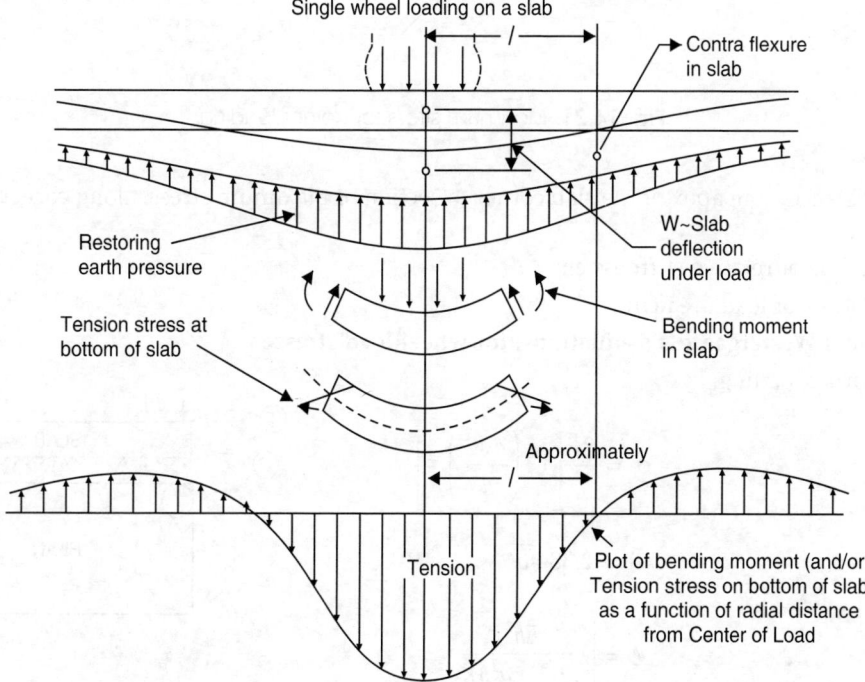

Fig. 14.20: Physical concepts of Westergaard's radius of relative stiffness (FAA)

(c) Equivalent radius of resisting section "b"

It is equivalent radius of a portion of the slab which is effective in resisting the moment caused due to wheel load.

$$b = \sqrt{1.6a^2 + h^2} - 0.675h \quad \text{when a} < 1.724 \text{ h}$$
$$= a \text{ when a} > 1.724 \text{ h}$$

b = Equivalent radius of resisting section cm

a = Radius of wheel load distribution cm.

h = Slab thickness cm

(d) Distance for maximum stress for corner loading "d"

The maximum stress does not occur around the load when loaded in corner, but it exists at a distance 'd' along the corner bisector as shown in Fig. 14.21.

Distance d is given by

$$d = 2.38\sqrt{al}$$

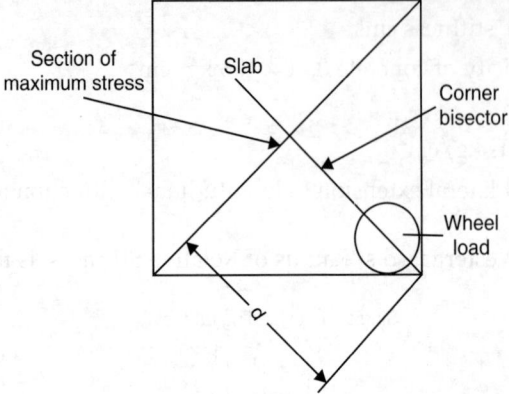

Fig. 14.21: Maximum stress for corner loading

where

d = distance from apex of the slab corner to section of maximum stress, along corner bisector cm.

l = Radius of relative stiffness cm

a = Radius of load area cm

Original Westergaard's's equations, for wheel load stresses.

(a) Corner loading:

$$\sigma_c = \frac{3P}{h^2}\left[1-\left(\frac{a\sqrt{2}}{l}\right)^{0.6}\right]$$

$$d = 2.38\sqrt{al}$$

$$\ell = \left[\frac{Eh^3}{12\left(1-\mu^2\right)K}\right]^{1/4}$$

$$b = a \qquad\qquad \text{when } a > 1.724\ h$$

$$b = \sqrt{1.6a^2+h^2}-0.675\ h \qquad\qquad \text{when } a < 1.724\ h$$

$$K = \text{modulus of subgrade reaction} = \frac{p}{\Delta} = \frac{p}{0.125}$$

(b) Interior loading:

$$\sigma i = \frac{0.3162P}{h^2}\left[4\log_{10}^{\left(\frac{\ell}{b}\right)}+1.069\right]$$

(c) Edge loading

   (i) Slab warped down

$$\sigma e = \frac{0.572\ P}{h^2}\left[4\log_{10}^{\frac{\ell}{b}}+0.359\right]$$

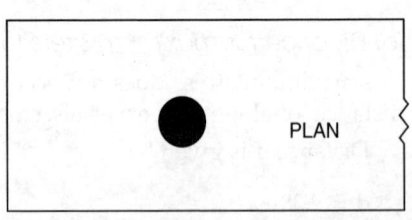

(ii) Slab warped up

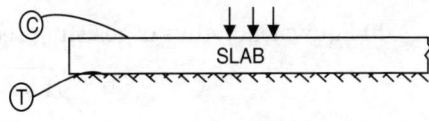

$$\sigma e = \frac{0.572\,P}{h^2}\left[4\log_{10}^{\left(\frac{\ell}{b}\right)} + \log_{10}^{b}\right]$$

where

P = Design wheel load (ESWL) Kg.

σc = Stress due to corner loading (kg/cm^2)

d = Distance of the position of maximum stress
   from the corner loading along bisector

ℓ = Radius of relative stiffness cm

b = Equivalent radius of resisting section cm.

K = Modulus of subgrade reaction, (kg/csm^3)

a = Radius of the area of contact, cm

h = Pavement thickness cms

σe = Maximum stress due to edge loading kg/cm^2

σi = Maximum stress due to interior loading kg/cm^2

Various modifications in the above stress formulae were suggested by various research workers like Older Goldbeck, Bradbury, Kelly, Spanglar, Picket and Ray, Teller and Sutherland from time to time.

Indian Road Congress in its design method uses Westergaard equations modified by Teller and Sutherland for load stresses in critical edge region vide IRC : 58-2002.

The modified formula for stress due to edge loading is:

$\sigma_e = 0.529\,P/h_2\,(1 + 0.54\,\mu)\,[4\log_{10}^{(l/b)} + \log_{10}^{b} - 0.4048]$ kg/cm^2

all symbol as described earlier.

Example 12. Calculate the stresses at interior and corner regions of a concrete pavement using Westergaard's original equations on the basis of following data:

wheel load = 4100 kg

modulus of elasticity of concrete = 3.0×10^5 kg/cm2

pavement thickness = 15 cm

Poisson's ratio = 0.15

modulus of subgrade reaction = 3.0 kg/cm^2

radius of contact area = 15 cm

Solution:

(a) Radius of relative stiffness $\ell = \left[\dfrac{Eh^3}{12(1-\mu^2)K}\right]^{1/4}$

here

$E = 3.0 \times 10^5$

$h = 15$

$\mu = 0.15$

$k = 3$

$\qquad = \left[\dfrac{3 \times 10^5 \times 15^3}{12(1-.15^2)3}\right]^{1/4}$

$\qquad = 73.25$ cm

(b) Equivalent radius of resisting section b

$$b = \sqrt{1.6a^2 + h^2} - 0.675h$$

$$= \sqrt{1.6 \times 15^2 + 15^2} - 0.675 \times 15$$

$$= 14.05 \text{ cm}$$

(c) Stress at the interior σ_i

$$\sigma_i = \frac{0.3162P}{h^2}\left[4\log_{10}^{\left(\frac{\ell}{b}\right)} + 1.069\right]$$

$$= \frac{0.3162 \times 4100}{15^2}\left[4\log_{10}^{\frac{73.25}{14.05}} + 1.069\right]$$

$$= \mathbf{22.67 \ kg/cm^2}$$

(d) Stress at the corner $\sigma_e = \dfrac{3P}{h^2}\left[1 - \left(\dfrac{a\sqrt{2}}{\ell}\right)^{0.6}\right]$

$$= \frac{3 \times 4100}{15^2}\left[1 - \left(\frac{15\sqrt{2}}{73.25}\right)^{0.6}\right] = \mathbf{28.70 \ kg/cm^2}$$

14.8.2 Temperature Stresses

Stresses in a concrete slab due to temperature can be calculated from the general equations by allowing the degree of restraint. If there is adequate space for expansion, there will be no stress at the extreme end of the slab. Near the end, internal stress will occur, plus some stress due to restraint at warping which will increase from zero at the extreme end to a maximum value at the point where warping is completely restrained by the weight of the slab, usually 2 to 5 m from the end.

If adequate space is allowed for expansion of the slab, longitudinal (or end) restrain will be small in short slabs, but will increase with slab length.

Changes in temperature (gradient likely to occur during day and night temperature) of a slab causes the slab to expand or contract and this movement is usually partially resisted by restraint of the subgrade and joints.

On the underside of the slab the stresses due to subgrade restraint will be compressive during expansion and tension during contraction. Seasonal expansion and contraction of a slab takes place by a series of daily cycles.

Where the expansion of a slab is prevented by the absence or complete closure of joints or by frictional restraint of the subgrade, high compressive stresses are developed in hot weather. If slab is also restrained from warping, stresses will be higher at the top of the slab than bottom.

Following equations are recommended for calculation temperature of stresses.

(A) Warping stress due to temperature gradient

Day time: Warped downwards, slab convex upwards

Night time: Warped upwards slab concave upwards

1. EDGE STRESSES

$$\sigma et = \frac{C_x \cdot E \cdot \alpha \cdot T}{2} \text{ or } \frac{C_y \cdot E \cdot \alpha \cdot T}{2}$$

2. INTERIOR STRESSES

$$\sigma it = \frac{E\alpha T}{2} \left[\frac{C_x + \mu C_y}{1 - \mu^2} \right]$$

3. CORNER STRESSES

$$\sigma ct = \frac{E\alpha T}{3(1 - \mu^2)} \sqrt{\frac{a}{\ell}}$$

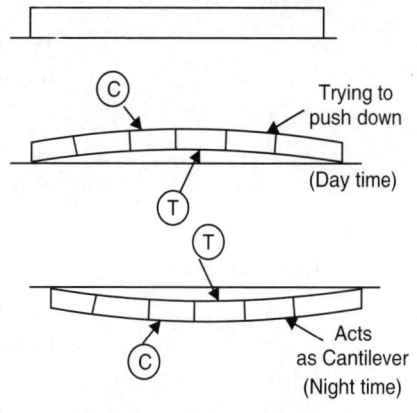

Trying to push down

(Day time)

Acts as Cantilever (Night time)

(Warping during day and night)

Figure 14.22 shows the typical curve for temperature stress coefficients, and slab dimensions.

In above equations

C_x = Coefficient based on length (L_x) of the slab in x direction

C_y = Coefficient based on width (L_y) of the slab in y direction

t = temperature differential between top and bottom of the slab in °C.

α = Coefficient of thermal expansion of concrete per °C.

If temperature t is not given use a 'Thumb - Rule'

– During day time: Temp gradient = 1°C per cm thickness of slab.

– During night time: 1°C for each 2.5 cm thickness of slab.

14.8.3 (B) Frictional stress

$$\sigma_f = \frac{WLf}{2} \times 10^{-4} \text{kg / cm}^2$$

where:

σ_f = frictional stress

W = unit weight of concrete = 2400 kg/m³

L = length of C.C. slab in meter

f = coefficient friction (assumed = 1.5)

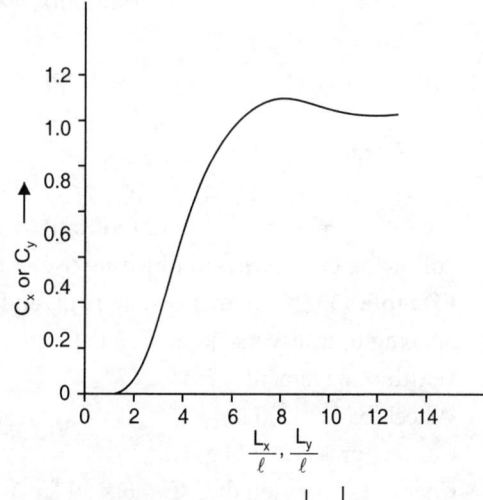

Fig. 14.22: Typical curve $\frac{L_x}{\ell}, \frac{L_y}{\ell}$, Vs stress coefficient

14.8.3 (C) Combined Load, Temperature and Friction Stresses

1. Slab warped upwards:

(A) Expanding: Positive (+) represents compression and Negative (–) represents Tension.

Combined stresses in:

Interior:

$$\sigma_{ic} \text{ (top): } = +\sigma_i + \sigma_{it} + 0 \qquad\qquad (\sigma_f = 0)$$
$$\sigma_{ic} \text{ (bottom): } = -\sigma_i + \sigma_{it} + \sigma_f$$

Edge:

$$\sigma_{ec} \text{ (top): } = +\sigma_e + \sigma_{et} + 0 \qquad (\sigma_f = 0)$$
$$\sigma_{ec} \text{ (bottom): } = -\sigma_e + \sigma_{et} + \sigma_f$$

Corner:

$$\sigma_{cc} \text{ (top): } = -\sigma_c - \sigma_{ct} + 0 \qquad (\sigma_f = 0)$$
$$\sigma_{cc} \text{ (bottom): } = +\sigma_c + \sigma_{ct} + \sigma_f$$

(B) Contracting:

Interior:

$$\sigma_{ic} \text{ (top): } = +\sigma_i - \sigma_{it} + 0 \qquad (\sigma_f = 0)$$
$$\sigma_{ic} \text{ (bottom): } = -\sigma_i + \sigma_{it} - \sigma_f$$

Edge:

$$\sigma_{ec} \text{ (top): } = +\sigma_e - \sigma_{et} + 0 \qquad (\sigma_f = 0)$$
$$\sigma_{ec} \text{ (bottom): } = -\sigma_e + \sigma_{et} - \sigma_f$$

Corner:

$$\sigma_{cc} \text{ (top): } = -\sigma_c - \sigma_{ct} + 0 \qquad (\sigma_f = 0)$$
$$\sigma_{cc} \text{ (bottom): } = +\sigma_c + \sigma_{ct} - \sigma_f$$

2. Slab warped downwards:

Interior:

$$\sigma_{ic} \text{ (top): } = +\sigma_i + \sigma_{it} + 0 \qquad (\sigma_f = 0)$$
$$\sigma_{ic} \text{ (bottom): } = -\sigma_i - \sigma_{it} + \sigma_f$$

Edge:

$$\sigma_{ec} \text{ (top): } = +\sigma_e + \sigma_{et} + 0 \qquad (\sigma_f = 0)$$
$$\sigma_{ec} \text{ (bottom): } = -\sigma_e - \sigma_{et} + \sigma_f$$

Corner:

$$\sigma_{cc} \text{ (top): } = -\sigma_c + \sigma_{ct} + 0 \qquad (\sigma_f = 0)$$
$$\sigma_{cc} \text{ (bottom): } = +\sigma_c - \sigma_{ct} + \sigma_f$$

Following examples will illustrate use of these equations

Example 13. The following data is given for a cement concrete pavement:

Spacing of transverse joint = 10 m

Width of pavement = 7m

Wheel load = 4000 kg

Contact pressure = 5kg/cm^2

Coeff. of expansion of C.C. = 8×10^{-6} per °C

Modulus of elasticity for C.C = 3×10^5 kg/cm^2

Modulus of subgrade reaction = 4 kg/cm^3

Maximum difference of temperature between top and bottom of C.C. pavement = 20 °C

Find the nature and value of combined stresses at the top and bottom of the C.C. pavement 25 cm. thick, warped up and contracting.

Solution:
$$\ell = \left[\frac{Eh^3}{12\left(1-\mu^2\right)K} \right]^{1/4} = \left[\frac{3 \times 10^5 \times 25^3}{12(1-0.15^2)4} \right]^{1/4} = 99.2 \text{ cms}$$

$$a = \frac{P}{p} = \frac{4000}{5} = 800 = \pi a^2 \therefore a = \sqrt{\frac{800}{\pi}}$$

$$= 16 \text{ cms}$$

$$b = \sqrt{1.6a^2 + h^2} - 0.675h$$

$$= \sqrt{1.6 \times 16^2 + 25^2} - 0.675 \times 25$$

$$= 15.3 \text{ cms}$$

$$K = 4 \text{kg/cm}^3 \text{ given}$$

1. WHEEL LOAD STRESSES

$$6_c = \frac{3P}{h^2}\left[1 - \left(\frac{a\sqrt{2}}{\ell}\right)^{0.6}\right] = \frac{3 \times 4000}{25^2}\left[1 - \left(\frac{16\sqrt{2}}{99.2}\right)^{0.6}\right] = \textbf{11.32 kg/cm}^2$$

$$\sigma_i = \frac{0.3162\,P}{h^2}\left[4\log_{10}^{\frac{\ell}{b}} + 1.069\right] = \frac{0.3162 \times 4000}{25^2}\left[4\log_{10}^{\frac{99.2}{15.3}} + 1.069\right] = \textbf{8.8kg/cm}^2$$

$$\sigma_e = \frac{0.572P}{h^2}\left[4\log_{10}^{b} + \log_{10}^{b}\right] = \frac{0.572 \times 4000}{25^2}\left[4\log_{10}^{\frac{99.2}{15.3}} + \log_{10}^{15.3}\right] = \textbf{16.25 kg/cm}^2$$

2. TEMPERATURE STRESSES

a. Frictional stress

$$\sigma_f = \frac{WLF}{2} \times 10^{-4} = \frac{2400 \times 10 \times 1.5}{2} \times 10^{-4}$$

$$= 1.80 \text{kg/cm}^2$$

b. Warping stress

$$\ell_x = 10\text{m} = 1000 \text{ cms } Lx/\ell = \frac{1000}{99.2} = 10 \qquad\qquad Cx = 1.04$$

$$L_y = \frac{7}{2} \times 100 = 350,\ Ly/\ell = \frac{350}{99.2} = 3.5 \qquad\qquad Cy = 0.25$$

$$t = 20°C$$

$$\sigma_{et} = \frac{E\alpha t}{2}C_x = \frac{3 \times 10^5 \times 8 \times 10^{-6} \times 20}{2} \times 1.04 = \textbf{25kg/cm}^2$$

$$\sigma_{it} = \frac{E\alpha t}{2}\left[\frac{C_x + \mu C_y}{1 - \mu^2}\right] = \textbf{26 kg/cm}^2$$

$$\sigma_{ct} = \frac{E\alpha t}{3(1 - \mu^2)}\sqrt{\frac{a}{\ell}} = \textbf{7.53 kg/cm}^2$$

3. COMBINED STRESSES

Warped up and contracting (+ comp, – tension)

a. Corner region

$$\sigma_{cc} \text{ (top)} = -\sigma_c - \sigma_{ct} + 0$$
$$= -11.32 - 7.53 = -18.85 \text{ kg/cm}^2$$
$$\sigma_{cc} \text{ (Bottom)} = 6_c + 6_{ct} - 6_f$$
$$= 11.32 + 7.53 - 1.8 = 17.05 \text{ kg/cm}^2$$

b. Edge region

$$\sigma_{ec} \text{ (top)} = 6e - 6et + 0$$
$$= 16.25 - 25 = -8.75 \text{ kgcm}^2$$
$$6_{ee} \text{ (Bottom)} = -6_e + 6_{et} - 6_f$$
$$= -16.25 + 25 - 1.8 = +6.95 \text{ kg/cm2}$$

c. Interior region

$$\sigma_{ic} \text{ (top)} = +6_i - 6_{it} + 0$$
$$= 8.8 - 26 = -17.2 \text{ kg/cm}^2$$
$$\sigma_{ic} \text{ (Bottom)} = -6_i + 6ct - 6_f$$
$$= -8.8 + 26 - 1.8 = +15.4 \text{ kg/cm}^2$$

maximum tension = **18.85 kg/cm²**

maximum compression = **17.05 kg/cm²**

Example 14. Calculate the combined stresses for a cement concrete pavement, given the following data. Assume single tire assembly:

Spacing of transverse joint = 15 m

Width of c.c slab = 3.5 m

$a = 12$ cm, $\alpha = 8 \times 10^{-6}$ per ºC, $k = 2.5$ kg/cm³

$E = 3 \times 10^5$ kg/cm², p = 4100 kg, h = 20 cm

Assume the slab is warped down and expanding.

Solution: $\ell = \left[\dfrac{Eh^3}{12(1 - \mu^2)k} \right]^{1/4} = 94.5$ cm

$$1.724 \, h = 1.724 \times 20 = 34.48 \text{ cm}$$

$$b = \sqrt{1.6a^2 + h^2} - 0.675 \, h = 11.7 \text{ cm}$$

$$k = 2.5 \text{ kg/cm}^3$$

LOAD STRESSES

$$\sigma_i = \frac{0.3162P}{h^2}\left(4\log_{10}^{\frac{\ell}{b}} + 1.069 \right) = \textbf{14.5 kg/cm}^2$$

$$\sigma_e = \frac{0.572P}{h^2}\left(4\log_{10}^{\frac{\ell}{b}} + 0.359 \right) = \textbf{22.2 kg/cm}^2$$

$$\sigma c = \frac{3p}{h^2}\left[1-\left(\frac{a\sqrt{2}}{\ell}\right)^{0.6}\right] = 20 \text{ kg/cm}^2$$

TEMPERATURE STRESSES

$$L_x = 1500, \quad \frac{L_x}{\ell} = \frac{1500}{94.5} = 15.9, \qquad\qquad C_x = 1.0$$

$$L_y = 350, \quad L_y/\ell = \frac{350}{94.5} \; 3.7, \qquad\qquad C_y = 0.445$$

$$\sigma_{et} = \frac{C_x \cdot E \cdot \alpha \cdot t}{2} = \frac{1\times3\times10^5 \times 8 \times 10^{-6} \times 20}{2} = 24 \text{kg/cm}^2$$

$$\sigma_{it} = \frac{E\alpha t}{2}\left[\frac{C_x + \mu C_y}{1-\mu^2}\right] = 26.2 \text{ kg/cm}^2$$

$$\sigma_{ct} = \frac{E\alpha t}{3(1-\mu^2)}\sqrt{\frac{a}{\ell}} = 6.7 \text{ kg/cm}^2$$

Frictional stress

$$\sigma_f = \frac{WLf}{2}\times10^{-4} = 2.7 \text{ kg/cm}^2$$

COMBINED STRESSES

a. Interior stresses

σ_{ci} (Top) = 14.5 + 26.2 + 0 = 40.7 kg/cm^2 (C)

σ_{ci} (Bottom) = $-$ 14.5 $-$ 26.2 + 2.7 = $-$ 38 kg/cm^2 (T)

b. Edge stresses

σ_{ce} (Top) = 22.4 + 254 + 0 = + 46.2 kg/cm^2 (C)

σ_{ce} (Bottom) = $-$ 22.2 $-$ 24 + 2.7 = 43.5 kg/cm^2 (T)

c. Corner stresses

σ_{cc} (Top) = $-$ 20 + 6.7 + 0 = $-$ 13.3 kg/cm^2 (T)

σ_{cc} (Bottom) = 20 $-$ 6.7 + 2.7 = 16 kg/cm^2 (C)

14.9 DESIGN OF RIGID PAVEMENTS

Different design procedures are utilized to determine the pavement thickness, using different assumptions and modifications to the basic Westergaard analysis, in view of available materials, conditions of loading, climatic parameters and existing soils by different countries to meet the challenges of heavy traffic. Suitable design models, curves, based on computer solutions of stresses, strains and deformation are used.

The basic procedure of designing rigid pavements (concrete slab) involves following steps:

(i) Assuming a suitable thickness for the concrete slab.

(ii) Calculating the total combined stresses in critical regions of the slab, using suitable equations and computer programs or curves.

(iii) Making sure that stresses are lesser than the allowable stresses in the concrete with a reasonable factor of safety.

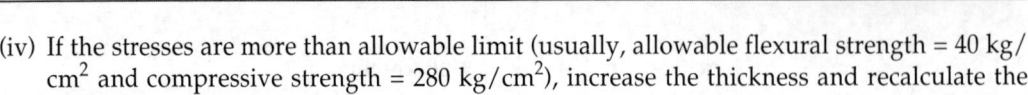

(iv) If the stresses are more than allowable limit (usually, allowable flexural strength = 40 kg/cm^2 and compressive strength = 280 kg/cm^2), increase the thickness and recalculate the stresses

(v) Continue the trials, till a most suitable thickness for the concrete slab is determined.

IRC method of rigid pavement design is described in following paragraphs.

14.9.1 IRC Method of Rigid Pavement Design

Indian Road Congress has provided revised guidelines for the design of plain jointed rigid pavements vide IRC: 58-2002. In these guidelines flexural stresses due to traffic load and stresses due to temperature differential are considered. The maximum flexural stress is caused when the tyre touches the longitudinal edge, both for single as well as tandem axles.

Tandem axles carrying twice the load of a single axle cause flexural stresses which are about 20 percent lower than that of the single axle load because of super position of negative bending moment due to one dual wheel load over the other.

The maximum combined tensile stress occurs during the day in case of interior and edge regions, when there is maximum temperature differential in the slab. The concrete slab is designed to withstand the maximum stresses due to warping and wheel load which are caused at the edge region.

Stresses at the edge are calculated due to

(a) Wheel loads using Westergaard's modified picket and Ray's chart/computer programme (IITRIGID)

(b) Temperature stresses using Bradbury's coefficient, as per Westergaard analysis.

(A) Wheel load stresses: (corner stress)

In its earlier edition IRC-58-1988, Westergaard's equation modified by Teller and Sutherland were adopted, but due to certain limitation regarding wheel configuration the new approach (IRC-58-2002) uses Picket and Ray's modification of Westergaard's theory and stresses are computed by a computer programme named IITRIGID as it is developed by IIT Kharagpur.

The corner stress due to wheel load is calculated as per Westergaard's equation modified by Kelly, given below:

$$S_C = \frac{3P}{h^2}\left(1 - \left(\frac{a\sqrt{2}}{1}\right)^{1.2}\right)$$

Where,

Sc = load stress in the corner region, kg/cm^2

P = wheel load, kg

a = radius of equivalent circular contact area, cm

Temperature stresses in corner region are negligible as corner are free to warp. Ready to use charts for calculation of load stresses in the edge region of rigid pavement slab for single axle loads (6, 8, 10, 12, 14, 16, 18, 20, 22, 24 tons), and Tandem axle loads (12, 16, 20, 24, 28, 32, 40, 42, 44 tons) for different magnitude of k values of sub-base (6, 8, 10, 15 and 30 kg/cm^3) have been developed. Typical set of charts for single and tandem axle is shown in Figs 23 to 25. A computer programme is also available for calculating the stresses at the edges.

14.9.2 Temperature Stresses

The temperature stresses are calculated by the following equation.

$$S_{te} = \frac{E \propto t C}{2}$$

Where

S_{te} = stress due to temperature in the edge region kg/cm^2

E = modulus of elasticity of concrete kg/cm^2

t = maximum temperature differential during day between top and bottom of the slab °C

$\propto$ = coefficient of thermal expansion of cement concrete per °C.

C = Bradbury's coefficient which can be taken from Bradbury's chart against values of L/ℓ and W/ℓ

(chart is shown in Fig. 14.26 and Table 14.8.

L = length of the slab or spacing between consecutive construction joints, cm

l = radius of relative stiffness, cm

$$= \sqrt[4]{\frac{Eh^3}{12(1-\mu^2)K}}$$

W = width of the slab or spacing between longitudinal joints.

μ = Poisson's ratio

h = thickness of concrete slab, cm

K = modulus of subgrade reaction kg/cm^3.

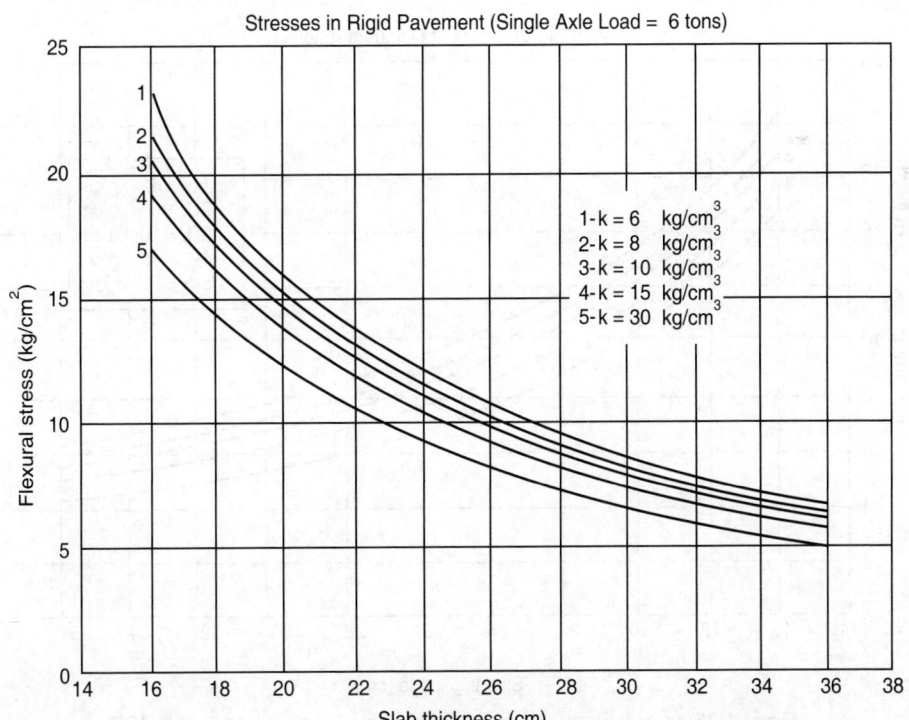

Fig. 14.23: Load stresses in corner region by single axle load (6T)

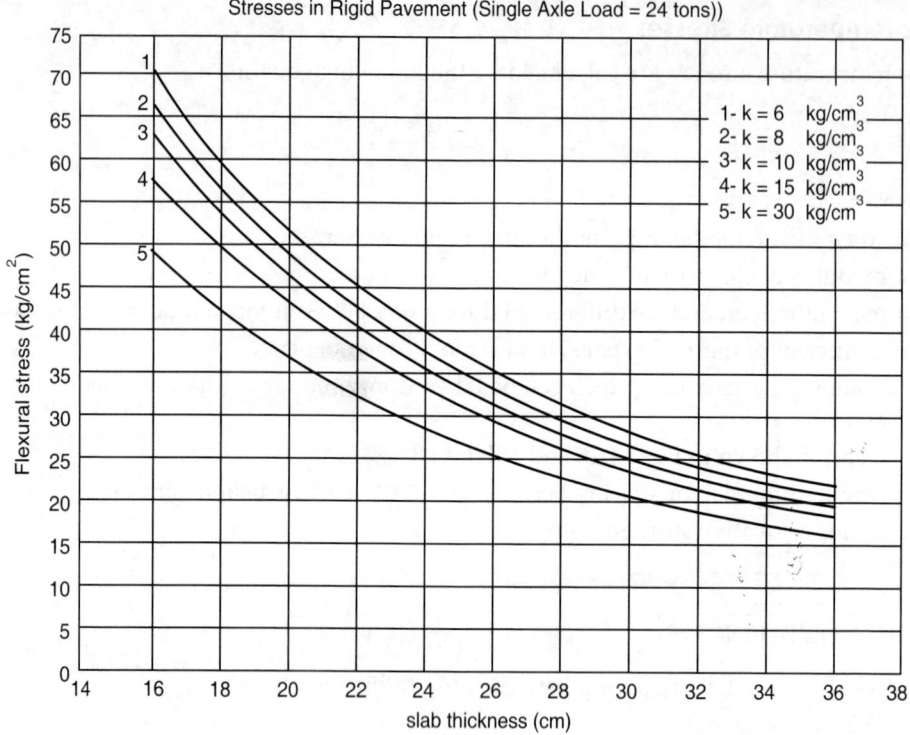

Fig. 14.24: Load stresses in corner region by single axle load (24T)

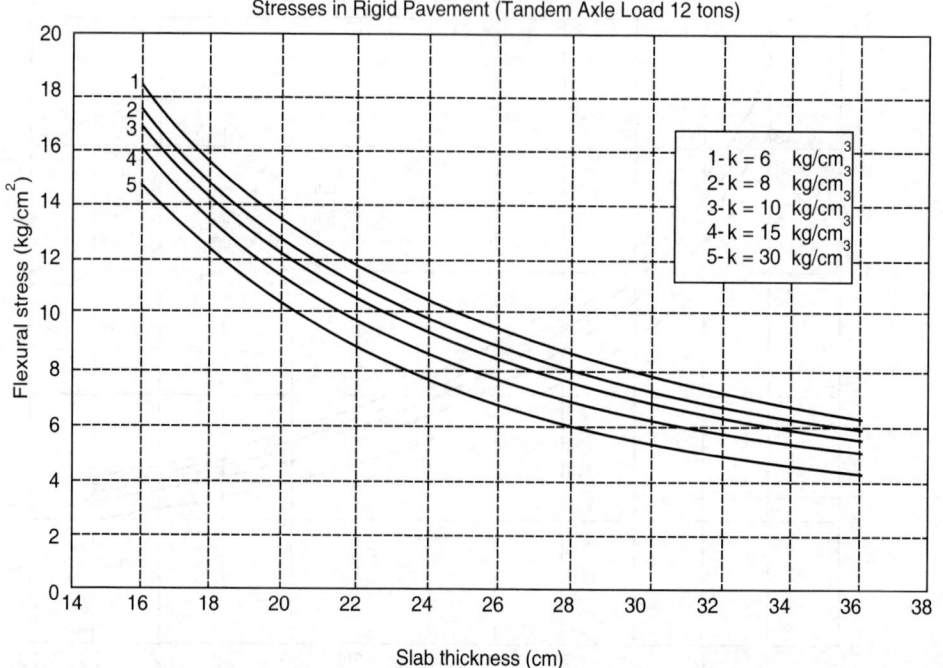

Fig. 14.25: Load stresses in corner region by Tandem axle load (12T)

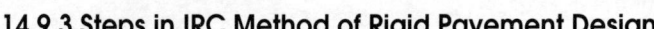

14.9.3 Steps in IRC Method of Rigid Pavement Design

   i. Arrive at the design parameters like wheel load, subgrade modulus (K), type, properties of concrete, and design life of pavement.
  ii. Decide the width and length of a slab, type and spacing of joints.
 iii. Assume a suitable trial design thickness of the pavement slab.
  iv. Compute the repetitions of axle loads of different magnitude during the design life.
   v. Calculate the stresses due to single and tandem axle loads and determine cumulative fatigue damage (CFD)
  vi. If the cumulative fatigue damage is more than 1, select higher thickness than assumed and repeat steps (i) to (v).
 vii. Calculate the temperature stress at the edge. Add temperature stress and flexural stress due to highest wheel load. If the sum is greater than the modulus of rupture, select a higher thickness and repeat step (i) to (vi).
viii. Pavement thickness should be designed/adopted on the basis of corner stress if no dowel bars are used and there is no load transfer due to lack of aggregate interlock (The worst case).

An example illustrates the above design procedure.

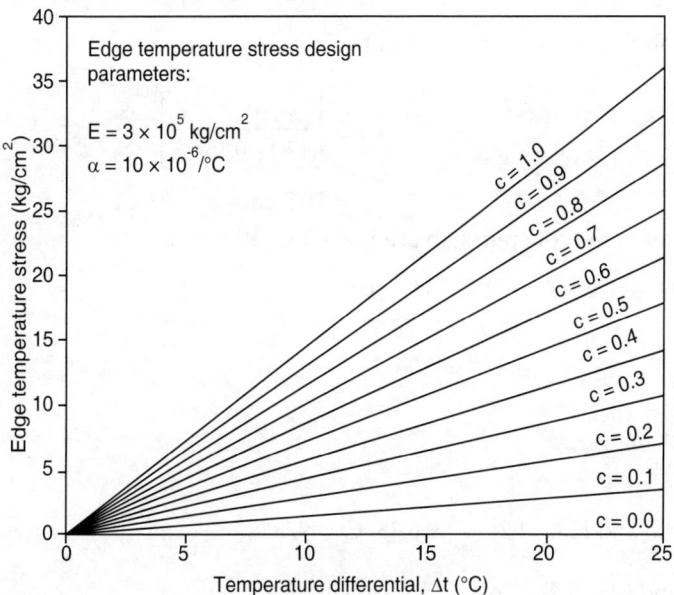

Fig 14.26: Chart for determination of edge temperature stresses

Table 14.8: Determination of coefficient C

L/ℓ or B/ℓ	C	L/ℓ or B/ℓ	C
1	0.000	7	1.030
2	0.040	8	1.007
3	0.175	9	1.080
4	0.440	10	1.075
5	0.720	11	1.050
6	0.920	12	1.000

Example 15. Design a cement concrete road having lane width of 3.5 m for 20 years design life. The present traffic as per studies is 500 vehicles per day (commercial vehicles). The design wheel load is 4100 kg and growth of traffic can be assumed as 7.5 percent. Following values of other parameters may be used in design.

Modulus of subgrade reaction (k) = 6 kg/cm^3

Flexural strength of concrete = 40 kg/cm^2

Modulus of elasticity of concrete (E) = 3×10^5 kg/cm^2

Poisson's ratio of concrete (μ) = 0.15

Temperature differential of slab = 12°C

Coefficient of thermal expansion = 10×10^{-6} per°C

Spacing of joints = 4.5 cm

Solution:

Future traffic (design hour volume) = $500(1+0.075)^{20}$ = 2125

$$\text{Modulus of relative stiffness } \ell = \left[\frac{Eh^3}{12(1-\mu^2)K} \right]^{1/4}$$

Assuming a slab thickness of 18 cms as a trial

$$\ell = \left[\frac{3 \times 10^5 \times 18^3}{12(1-0.15^2)6} \right]^{1/4}$$

$$= 70.5 \text{ cm}$$

Here dimension of the concrete slab are L = 4.5 m, W = 3.5m

$$\frac{L}{\ell} = \frac{4.5 \times 100}{70.5} \qquad\qquad \frac{W}{\ell} = \frac{3.5 \times 100}{70.5}$$

$$= 6.38 \qquad\qquad\qquad = 4.96$$

$\dfrac{L}{\ell}$ is critical than $\dfrac{W}{\ell}$

From Table 14.8 and Fig. 14.26 $\dfrac{L}{\ell}$ = 6.38, C = .978

For t = 12 °c and C = .978 $\quad \sigma_{et}$ = 15.0 kg/cm^2

Residual strength of concrete = 40 – 15 = 25 kg/cm^2

For corner loading (using figure, for h = 18, k = 6,) σ_c = 14 kg/cm. So S.F. = $\dfrac{25}{14}$ = 1.8

Assumed 18 cms. thick slab is safe, and can be adopted.

14.10 DESIGN OF JOINTS IN CONCRETE PAVEMENTS

Variation in temperature and moisture content causes volume changes and slab warping results in significant stresses. Joints are therefore placed in concrete pavements to permit expansion and contraction of the pavement, to relieve stresses due to curling and friction and facilitate construction.

There are three types of transverse joints as shown in Fig. 14.27.

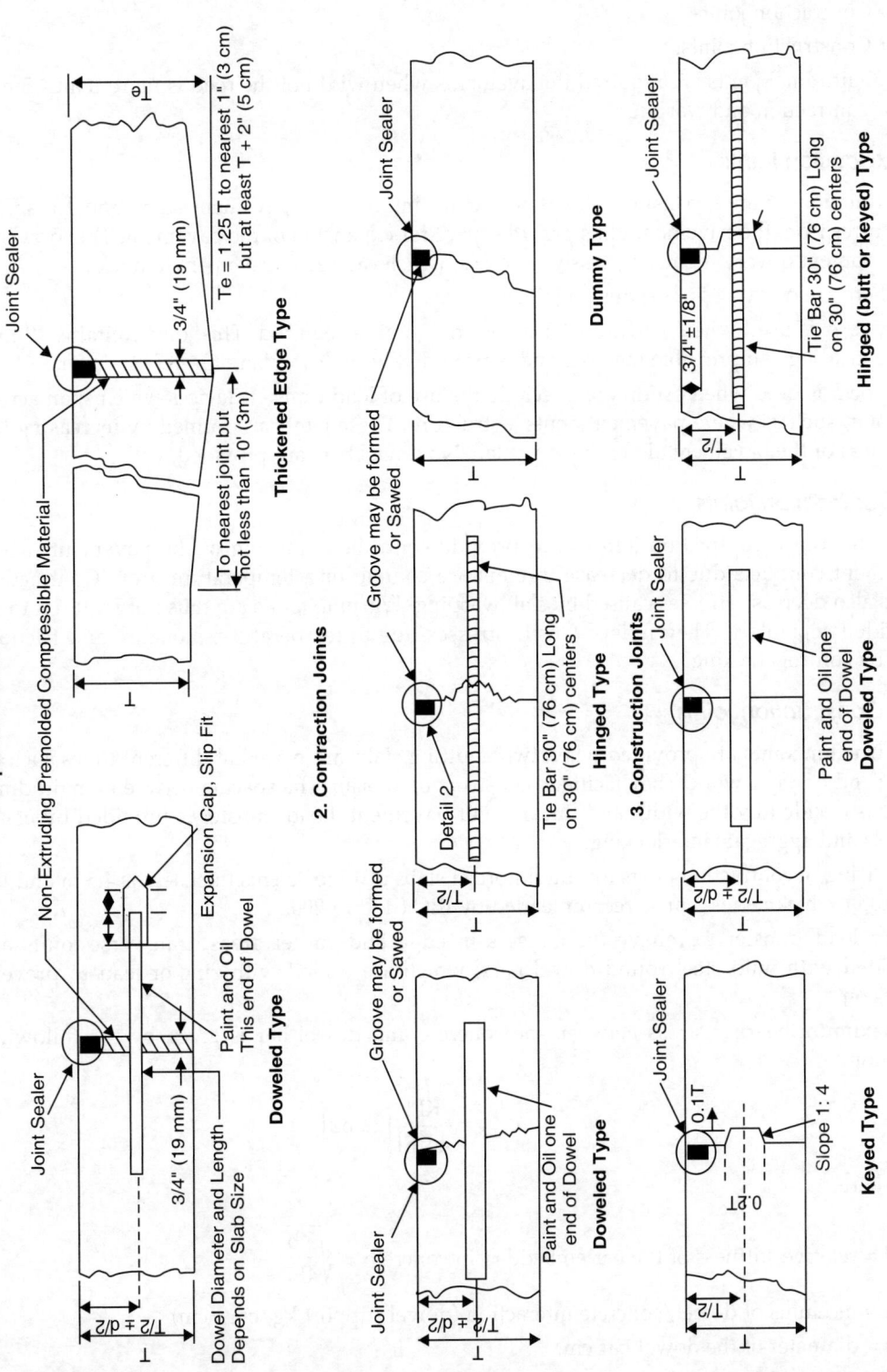

Fig. 14.27: Typical joints in concrete pavements

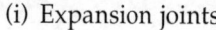

   (i) Expansion joints

  (ii) Contraction joints

 (iii) Construction joints.

Longitudinal joints are required in pavements when width of the road is more than 4.5m to allow contraction and warping.

(i) Expansion joints

The function of an expansion joint is to isolate intersecting pavement slabs and to isolate structures from the pavement, thus provide space for expansion of the pavement. This prevents development of very high compressive stresses, which can cause pavement to buckle.

There are two types of expansion joints:

Type A is used when load transfer across the joint is required. This joint contains 19 mm non-extruding compressible material and is provided with dowel bars for load transfer.

Type B is used when conditions preclude the use of load transfer devices which span across the joint, such as where pavement abuts a structure. These joints are formed by increasing the thickness of the pavement along the edge slab. No dowel bars are provided.

(ii) Contraction joints

The function of contraction joints is to provide controlled cracking of the pavement, when pavement contracts due to decrease in moisture content or a temperature drop. Contraction joints also decrease stresses caused by slab warping. Warping joints are thus not required to be provided separately. They relieve tensile stresses due to temperature, moisture and friction, thereby control cracking.

(iii) Construction joints

Construction joints are provided when two abutting slabs are placed at different times such as at the end of day's work. They facilitate construction of slab. The spacing between longitudinal joints is decided by the width and thickness of pavement. Load transfer is provided by using dowels and aggregate interlocking.

Spacing of contraction joints for unreinforced slabs of 15 to 25 cms thickness is 4.5 m and for 30 to 35 m thickness is 5 m as recommended by IRC (IRC 15-2002).

For load transfer to relieve the stresses in edge and corner areas, transverse joints are provided with mild steel round dowel bars, which are coated with zinc or lead to prevent corrosion.

Maximum bearing stress between the concrete and dowel bar is given by the following equation.

$$\sigma_{max} = \frac{KPt}{4\beta^3 EI}[2+\beta z]$$

Where:

β = relative stiffness of the bar embedded in concrete = $\sqrt[4]{\dfrac{Kb}{4EI}}$

K = modulus of dowel/concrete interaction (dowel support kg/cm^2/cm)

b = diameter of the dowel bar cm.

Z = joint width, cm

E = modulus of the elasticity of the dowel, kg/cm^2

I = moment of inertia of the dowel cm^4

Pt = load transfer by a dowel bar.

Each dowel bar should transfer load that is less than maximum bearing pressure. Allowable bearing stress on the concrete can be calculated by:

$$F_b = \frac{(10.16 - b)f_{ck}}{9.525}$$

Where F_b = Allowable bearing stress kg/cm^2

b = Dowel diameter, cm

f_{CK} = Ultimate compression strength of the concrete kg/cm2 (400 kg/cm^2 for m 40 concrete)

Suitably placed dowel bar is assumed to transfer 40 percent of the wheel load.

IRC has recommended diameter and length of dowel bars as given in Table 14.9

Table 14.9: Recommended dimension of dowel bars (IRC)

Thickness of concrete slab cm	Dowel bar detail		
	Diameter mm	Length mm	Spacing mm
20	25	500	200
25	25	500	300
30	32	500	300
35	32	500	300

For less than 15 cm thickness dowel bars are not necessary. The actual value of dowel bar size, length and spacing should be calculated as per the axle load used in design.

14.10.1 Design of Tie Bars

For longitudinal joints, when traffic is heavy or the subgrade is weak or expansive tie bars may be designed and used. The area of steel required per meter length of joint may be computed using the following formula given by IRC.

$$As = \frac{bfw}{S}$$

Where

As = area of steel in cm^2, required per meter length of joint.

b = lane width in meters

f = coefficient of friction between pavement and the sub-base/base (taken as 1.5)

w = weight of slab in kg/m^2

S = allowable working stress of steel in kg/cm^2

The length of tie bar should be at least twice that required to develop a bond strength equal to the working stress of steel, calculated by the formula.

$$L = \frac{2SA}{B'P}$$

Where

L = length of tie bar (cm)

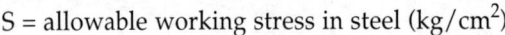

S = allowable working stress in steel (kg/cm^2)

A = cross – sectional area of the tie bar (cm^2)

P = perimeter of tie bar, (cm)

B′= permissible bond stress of concrete

For (i) plain tie bars – 17.5 kg/cm^2

      (ii) deformed tie bars – 24.6 kg/cm^2

The maximum diameter of the tie bar is limited to 20 mm to permit warping. To avoid concentration of tensile stresses they should not be spaced more than 75 cm apart. To take care of inaccuracy in placing, length of tie bar is increased 5 to 8 cms.

14.11 REINFORCED CONCRETE PAVEMENTS

The reinforcement in cement concrete pavement is provided, under following conditions:

1. When the joints are provided at large intervals, e.g. more than 6 m, because in longer slabs the temperature stresses will be high, resulting into cracks. Therefore to care for this reinforcement is provided.

2. When no base course is used. Small diameter wire fabric or bar mat reinforcement is used in rigid pavement, for control of temperature cracks. In the design it is assumed that reinforcement does not add to the structural strength of the pavement. The only function of reinforcement is to hold the cracked slab portions together and thus not to allow them to open up any more.

Amount of reinforcement

The amount of steel required is obtained assuming that a crack will develop and the resistance to movement will be offered by the tension in steel and is computed by the formula:

$$A_s = \frac{LfW}{2S}$$

where:

A_s = area of steel in cm^2 required per m width or length of slab

L = distance in m between free transverse joints (for longitudinal steel) or free longitudinal joints (for transverse steel),

f = coefficient of friction between pavement and sub-base/base (usually taken as 1.5),

W = weight of slab in kg/m^2 and

S = allowable working stress in steel in kg/cm^2 (usually taken as 50 to 60 percent of the minimum yield stress of steel)

From the above equation, it is clear that the amount of steel required depends upon the length of slab. Therefore in short length the amount of steel can be reduced practically to zero. Due to shorter slabs in airfield pavement, steel reinforcement is generally not provided.

Highways with transverse joint spacing from 6 m to 30m, and with lanes 3.7m wide require heavier longitudinal steel in the transverse direction. The transverse wire acts mainly as spacers.

14.11.1 Considerations In Design and Construction of Reinforced Pavements

– In general, closely spaced, small diameter bars are more effective, than large diameter bars at greater intervals.

– For longitudinal steel – use 10 to 32 mm dia bars at spacing of 10 to 20 cms.

For Transverse steel – use 8 to 20 mm dia bars at spacing of 15 to 40 cms.

– Position of reinforcement in mid depth or slightly above.

– At full depth joints, the reinforcement is kept at least 5 cm away from the face of the joint.

– If it is necessary to lap welded wire fabric, then the lap length should not be less than 30 times the diameter of wire.

Reinforcement is placed 5 cm below the surface.

Example 16. A cement concrete pavement 3.75 m wide has transverse joints at a spacing of 16m. Assume allowable tensile stress in steel = 1400 kg/cm^2 and weight of the slab as 450 kg/m^2. Design the reinforcement for the pavement.

Solution: Area of longitudinal steel $A_L = \dfrac{LfW}{2S}$

$$= \frac{16 \times 1.5 \times 450}{2 \times 1400}$$

$$= 3.86 \text{ cm}^2$$

Provided 10 mm dia bars a 25 cm c/c giving an area of 3.9 cm^2.

Area of transverse steel $= A_T = \dfrac{LfW}{2S}$

$$= \frac{3.75 \times 1.5 \times 450}{2 \times 1400}$$

$$= 0.904$$

Use 6.25 mm dia bars a 30 cm c/c.

14.12 CONTINUOUSLY REINFORCED CONCRETE PAVEMENTS (CRCP)

The temperature reinforcement used in RCC pavement may not be enough to keep the cracks completely tight for very long slabs, e.g. slabs without transverse joints except construction and very few expansion joints. Such long slabs (120 m to 200 m) are known as continuously reinforced pavements. In these pavements, cracks may develop due to shrinkage, contraction, warping and load.

Some agencies do not permit a reduction in thickness of continuously reinforced pavements, compared to plain cement concrete pavement.

The usual practice is to use 0.6% steel for hot climatic areas and 0.7 to 1.0% for cold climatic areas. This is due to fact that the shrinkage stress will be high, if the temperature falls soon after the slab is laid.

If higher percentage of steel is used, the cracks occur at close interval, but these crack remain held together. But if lower percentage of steel is used, the cracks occur at large intervals, but they are wide and the water and grit may enter through them. This causes many detrimental affects on the pavement.

Advantages of continuously reinforced concrete pavements are:

– Less joints therefore (a) smooth riding (b) low maintenance cost.

– Better performance if properly designed and constructed.

14.13 PRESTRESSED CONCRETE PAVEMENTS

A prestressed concrete pavement is one in which high tensile steel is stretched and anchored to the ends of the slab, causing compression stress in cement concrete slab, before a prestressed pavement cracks. This compressive stress must be neutralised and then enough tensile stress must be produced to exceed the tensile resistance of cement concrete. Even after cracking on the removal of load, the cracks close due to a prestressing compressive force.

Length of Slab

The slab up to 200 m, length have been successfully used. Wide joints are provided between the adjoining slabs to accommodate prestressing devices.

Thickness

Generally 12.5 cm to 17.5 cm thick pavements have been used.

Amount of Prestressing

It depends upon the length of the pavement. 20 to 40 kg/cm^2 of prestress have been used in longitudinal direction and 3-5 kg/cm^2 in transverse direction, if it is necessary.

Tendons

Tendon is high steel wire, group of wires, cable, or bar used for prestressing concrete.

Wire of 1 mm diameter and with ultimate tensile strength of 14000 to 17500 kg/cm^2 are used. In case of airport, due to heavy wheel loads, thick plain C.C. pavement are used, whereas for highways, thinner pavements (about 20 cms) are used and also the problem of curves on vertical and horizontal alignment are more.

Therefore such type of pavements are particularly beneficial in places where construction materials are in short supply.

Subgrade Restraint:

The loss in pre-stressing due to subgrade restraint is a major problem. The subgrade restraint reduces the effect of pre-stressing force. To reduce this a layer of sand about 5 cm thick is spread, smoothed and covered with waterproof paper. This is done to reduce the sliding frictional force between the bottom of the pavement slab and top of the subgrade.

Advantagess

(i) Less material required

(ii) Thinner pavement means lesser temperature gradient and so less warping stress.

(iii) Due to practically no joints:

    (a) A good riding quality

    (b) Low maintenance cost

    (c) Good performance.

(vi) There is no or very little crack formation.

Disadvantages

(i) Highly skilled labour, good quality control and strict supervision is required.

(ii) If due to some reasons, cable inside breaks, pavement fails.

14.14 RIGID PAVEMENT DISTRESS

Pavement distress may occur due to one or more of the following reasons:

1. Pavement pumping
2. More cracking
3. Spalling of joints.

Overlay

For successful maintenance of pavement the engineer has to ensure that its thickness is adequate for

(i) Traffic intensity and volume

(ii) Wheel load and load repetitions.

Due to unexpected economic development in a region the above conditions may change and become move severe, then the pavement rapidly undergoes the distress and no amount of periodic maintenance can help it. In such a case, the alternatives are:

(a) Either to divert traffic to some adjacent road or

(b) To strengthen the existing pavement.

Strengthening means to provide extra thickness of the pavement layer, over the existing pavement, i.e to provide an overlay. If the pavement is completely deteriorated, then an overlay will not serve the purpose and the solution will be to remove the existing pavement structure and reconstruct the new one. The engineer therefore should take the decision, in time, for providing an overlay as and when needed. Refer chapter 16, for overlays and their design.

Highway Construction and Stabilisation

15.0 PREAMBLE

Performance of a highway, besides other factors, also depends upon the quality and standard of construction including the materials used in highway construction, the technology of construction, management and cost control. Highway construction covers new construction, re-construction or rehabilitation and improvements of an existing road, overlays, spot improvements, etc. starting from earthwork and finishing with top bituminous or concrete surface.

For cost optimisation, and to ensure overall economy, the materials selected for highway construction should be available locally requiring minimum haul and those which will require no or little treatment to improve their strengths. Ideally, the most economic road construction material would be which is available closed to proposed road alignment. When such materials do not satisfy strength requirements it becomes necessary to improve strength by stabilisation or other treatments. The specifications for subgrade preparation, laying of sub-base, base and surface courses should be appropriate and realistic to local conditions and environment.

Highway pavement is a structure composed of superimposed horizontal courses. The courses used in a pavement are themselves sub-divided into layers of selected materials whose primary function is to distribute the applied load of vehicles so that transmitted stresses are reduced and will not exceed the supporting capacity of these courses. Functions of these components are:

(i) Subgrade: Subgrade in a road pavement structure is the natural or improved ground on which the road structure is constructed. Road may be in cut or on a fill or it may be on an embankment. Subgrade may be the existing material or borrowed material to be prepared and compacted before pavement structure is constructed over it. During construction it is important to check that specifications for preparing the subgrade are met.

Subgrade preparation involves checking of levels of the formation, checking the horizontal alignment, and the compaction required (by measuring in-situ density and moisture content). Suitable equipment should be used for excavation, grading and compaction.

(ii) Sub-base course: Sub-base is an intermediate layer between subgrade and granular base course. Sub-base layer is provided in a pavement structure, depending upon the bearing capacity of the subgrade so that, when the subgrade is strong, there is no need for a sub-base layer. Specifications for the sub-base course are stipulated based on the standard of the road and intensity of the traffic. The prime purpose of sub-base is to protect and separate the base course from the subgrade and vice-versa. In rigid pavements sub-base reduces the pumping of subgrade fines.

When locally available materials do not meet the specifications, it may be required to stabilise such materials either mechanically (by improving grading) or using additives such as cement, lime or bitumen.

Materials for construction of sub-base should be tested for their strength characteristics and specifications should specify the construction techniques and equipment to be used. Common materials for sub-base construction include granular materials like coarse sand, gravels, crushed stones, cement or lime stabilised soils, etc. Sometimes locally available waste materials like coal ash, slags, marble waste, etc. are also used as sub-base. Materials for construction of sub-base and base course layer are normally obtained from quarries and should be tested at the design stage to establish their strength characteristics, and to make sure that it meets the requirements of the specifications.

(iii) Base course: Base course is the main structural component of a pavement structure. Materials used in base courses include graded crushed stone, water bound macadam, bitumen bound macadam, wet-mix macadam, stabilised material, lean concrete, etc.

Sub-base layer and base course layer should be compatible, meeting the requirements of the specifications regarding strength characteristics, based on laboratory and field testing. An on site laboratory is useful for testing the materials.

(iv) Surface course: Surface course should provide acceptable riding quality and skid resistance, favourable light reflecting properties, and low noise, and be able to withstand wheel load and atmospheric conditions. Most common surfacing material used in flexible pavements are bituminous materials, which may be applied in different thicknesses varying from 2.5 cm to 10 cm thick. Such courses include surface dressing, gap-graded bituminous course, asphaltic concrete, emulsion slurry seal, etc.

Selection of materials used in preparation of bituminous mixes, i.e. coarse and fine aggregate, filler component and the bindser (bitumen), should be done properly. The mixing process, laying and compacting should be rigidly as per specifications, and rigid quality control should be exercised, at each level. Correct temperature should be maintained, during compaction and proper equipment should be used to achieve the specified compaction.

With variety of materials available, need of the traffic, climate and cost considerations, many methods and processes of road construction, have been developed, varying from earth roads to high type cement concrete or bituminous concrete roads. Engineers decide which of the many possible choices best meets the requirement of lowest overall cost for particular needs. Each component of the road, from earthwork to high quality surface course, has to be constructed as per specifications regarding materials, technology and equipment.

An engineer should adopt all possible ways and means to reduce the cost of the highway project to a minimum, without unduly reducing the service which the project is to furnish. There are possibilities of achieving savings on a highway project by considering:

• Proper planning of construction
• Selection of suitable method of construction

- Suitable plant, machinery and equipment
- Supervision by properly trained, skilled, and honest supervisors.

15.1 CONSTRUCTION TECHNOLOGY

The alternative technologies for construction of roads can be labour intensive on one extreme and machine intensive on the other. Labour intensive method, where output and quality are determined solely by labour productivity. With the labour intensive works, equipment means hand-held tools like hoes, spades, shovels, pickaxes, hammers, wheel-barrows, rollers, etc. labour intensive methods are useful where employment of labour is beyond the optimum requirement relative to equipment.

At the other extreme, machine-intensive method where output and quality is determined by the productivity of the equispment. Such equipment would include excavators, serapers, bulldozers, graders, rollers, mixers, dump-trucks, etc. equipment intensive methods are useful where equipment is used beyond optimum requirements relative to labour.

Between the two extremes there are many man-machine (labour-equipment) combinations that might be employed, with varying degree of labour and equipment productivity towards overall output and quality. An optimum mix of labour and equipment assuring cost effectiveness should be worked out, with the objective of maximising the productivity. Therefore in reality, a choice of technology is a choice of equipment on the one hand and labour on the other and various combination of labour and equipment based on output and cost, varying from situation to situation.

The technology adopted affects the design of highways and therefore considerations of technology are important during design and writing specifications. For example a labour intensive technology would seriously affect the geometric design standards achievable especially in rolling, or in hilly terrain. Hauling distance will be limited by the wheel-barrow, and low maximum cuts and fills would have to be adopted. In developing countries, the highway design process is based on technology availability, especially with respect to availability and skills of labour, and to the availability and type of equipment.

Conventional methods of construction in developing countries are labour intensive, most of the works being done manually, making use of local material. These methods generate employment to the local people. Because labour is available cheaper and in large numbers. Labour based methods have been largely used for rural road construction in India. Trend is now changing towards manual cum mechanical techniques using equipments for paving, rolling, transporting and grading.

Recently mechanisation is introduced in construction of National Highways to achieve greater productivity, better quality and time savings in construction to cover greater kilometers in short time. Selection of machines for particular construction work should be done after study of available types of equipment, for selecting the best suited to concerned work. In this regard publication MOST "Handbook on Road Construction Machinery" provides necessary guidelines.

15.2 CONSTRUCTION MANAGEMENT AND PLANNING

The object of construction of any road is to achieve the best facility at minimum initial cost of construction, subsequent maintenance and vehicle operating costs. Management of construction is the efficient utilisation of Manpower, Material and Machines (equipment) and Money—the four "Ms" for maximum output of the highest quality.

Proper management should utilise the full capacity, for example lack of incentive, improper site organisation, lack of supervision, inadequate skills etc. may be the reasons for lower output from labour, and for equipment it might be low operator skills, poor instructions, mechanical defects etc. Whatever may be the reason, they reflect ineffectiveness in managing the resources adequately.

Having decided the construction work, by appropriate technology the next aspect is that of planning and programming which provides means for monitoring the progress from commencement to the completion of the construction. Any deviation from the initial programme should be detected at an early stage and action should be taken to bring the work back on schedule. Planning and programming gives overall picture of the progress and also points out problems requiring attention. It also ensures that many problems are identified in advance and resolved in time.

Planning and programming mainly involves:

- Estimating the resources needed to undertake the construction project with adequate quality control of materials and final output.
- Programming the work and flow of associated resources.
- Forecasting future costs and resources needed on regular basis.
- Recording and comparing actual costs and outputs.
- Replanning and reprogramming work, as per changing needs.
- Continuously review and update the information on cost occurred, progress of work, and scheduling of construction resources.

Techniques like 'Network Planning', Critical Path Method (CPM), Programme Evaluation and Review Techniques (PERT), 'Stage Construction', etc. are useful tools for programming and planning of construction projects.

15.3 QUALITY CONTROL IN HIGHWAY CONSTRUCTION

Quality control is important for highway construction to ensure long lasting roads in view of increasing demands of heavy traffic and better service levels with reduced vehicle operation and maintenance costs.

Quality control can be ensured by preparing proper specifications and estimates for effective quality control measures to be undertaken and regular checks during construction for their implementation. The site engineer for ensuring quality control should be adequately trained and familiar with laboratory and field checks, to ensure quality of work on regular basis.

Basically there are two types of quality control 'Process Type' and "End Result Type" measures to be used at site. The 'Process Type' in which site engineer makes sure that work is being done according to laid down procedures and specifications using desired equipment and material. In the 'End Result Type' to ensure through regular tests that the final road meets all the requirements of quality.

For proper quality control during the construction standard and specifications of Indian Road Congress for various items of construction works should be rigidly followed, by performing tests and checks as specified, using standard procedures and equipment. IRC: Special Publication 11, "Handbook of Quality Control for Construction of Roads and Runways" gives details of essential requirements and procedures for quality control.

The pre-requisite for achieving the quality in a construction project are:

- Construction specifications and estimates should provide effective quality control measures and checks.

- Need of quality control should be emphasised and how it has to be translated into reality.
- Use of right technology, apt materials, and relevant processes and methods, for each activity, should be adopted.
- Systematising the measurements and accuracy should be followed.
- Total commitment for quality work, from top to bottom of the organisation personnel is necessary.
- Adequate training of the staff on the concept and significance of quality first, should be imparted through education and on the job training.
- Commitment of the concerned staff involved in quality control should be firm.
- Avenues of quality are known through comparision of performance standards with global best as bench marks.
- Periodic appraisal of quality control measures and their effectiveness, for incorporating further improvements, based on review of the quality control criteria used.

15.3.1 Total Quality Management Activities Involved in Highway Construction

The main activities involved in any highway construction project affecting quality of work can be classified as:

(a) Pre-design activities

(b) Design and project preparation activities

(c) Construction activities

(a) Pre-design activities: The first step is a need-based assessment of the requirement of the project, to be determined by the highest authority/client. Adjustments and modifications (if any) will be in line with the project requirement.

The next important step is realistic time allotment for the completion of the project along with a judicious provision of in built margin arising out of contingencies for each component of work. The assessment of the requirement of time in realising each of the component and fitting in all the activities with in overall time frame is very important in achieving quality.

The organization and personnel responsible for the project with elaborate description of each one's responsibility, authority and inter-relationship of functions has to be worked out without any ambiguity. The required qualification and experience of the personnel to be deployed for the project work should match the requirements of the project and a comprehensive selection criteria is to be worked out, for each category of staff. Past exposure and experience of personnel play the vital role.

The type and degree of control desired in each activity of the construction project critical to quality has to be identified before hand and shall have to be documented, for inclusion with the **Quality Assurance Plan** (QAP). Documentation and records testifying to satisfactory exe-cution of the activity with required quality control is to be maintained. Any corrective action proposed on the basis of quality record is to be clearly stated and incorporated in an overall surveillance and review plan.

(b) Design activities: The quality of engineering design and specifications have direct relationship to quality of work. Quality design depends upon reliability and accuracy of data made available to designer.

The manner in which survey and investigation work is undertaken has major impact on the project quality. Design based on faulty surveys and investigations will not lead to the quality even if all quality requirements are satisfied during construction. Consultants responsible for the design of the project play important role in quality of the work. Design consultants must have adequate knowledge and experience on similar works. Errors in the detailed design report (DPR) can embarrass the construction contractors and supervisory staff.

Technical specifications describing in detail the materials, equipment and construction method, should be properly written making use of best type of equipment, new materials, procedure of construction and good workmanship. The specifications dealing with performance should allow the workers and his supervisor to utilise their expertise to achieve quality results. Good Quality work is achieved when specification are clear and unambiguous. It should provide for variability giving tolerances, sampling procedure and acceptance criteria.

Construction proceeds on the basis of working drawings. Designs therefore must be translated into working drawings, giving minute details. Working drawings must show all important quality requirement so that nothing is left to chance.

Costing of the project is as important as its design and specifications. Cost estimates must be realistic so that contractor can do a quality work without suffering financial loss. Faulty estimates can lead to major problems affecting quality. Conditions of contract must be fair to the employer and the contractor, otherwise hindrances are created in smooth operation of the contract leading to stoppage of work, deterioration of quality and prolonged litigation. Bill of quantities should cover all items of work without any scope for extra items or extra quantity of work, so that additional burden on contractor is not involved.

(c) Construction activities: Implementation of a successful project depends considerably on the selection of a contractor having expertise and resources for the specified quality and quantity of work. To ensure quality in construction it is important to procure the services of proven, competent and experienced organizations, agencies, and personnel, right from the beginning Engaging a competent construction supervision consultant may be a practical alternatives.

A complete and detailed programme of work using CPM analysis of various activities from commencement to completion should be finalized between the contractor, the client and the supervising consultants and depicted in form of a bar chart alongwith supporting details for mobilization and deployment of required material, labour and equipment, including those already in possession, proposed to be purchased/hired. Such programme finalized at initial stage may need reviewing through periodic meetings.

Schedules for each activity has to be merged in a detailed master schedule integrated to show design, procurement, construction and other connected activities. Scheduled construction plan assures quality construction due to orderly and planned pace of progress in works, and the quality control units may better plan their control process and control tests. Figure 15.1 shows overall components and activities of TQM System for a highway construction project.

Along with a quality assurance plan an overall **Quality Control Plan** is also required to be prepared. The quality control plan has two components. One describes in details the materials, apparatus, equipment and methods to be adopted and the other lays down the essential required condition of the final structure. All the professional employed on the project are required to be given induction training, detailed instructions to ensure quality on different items of work.

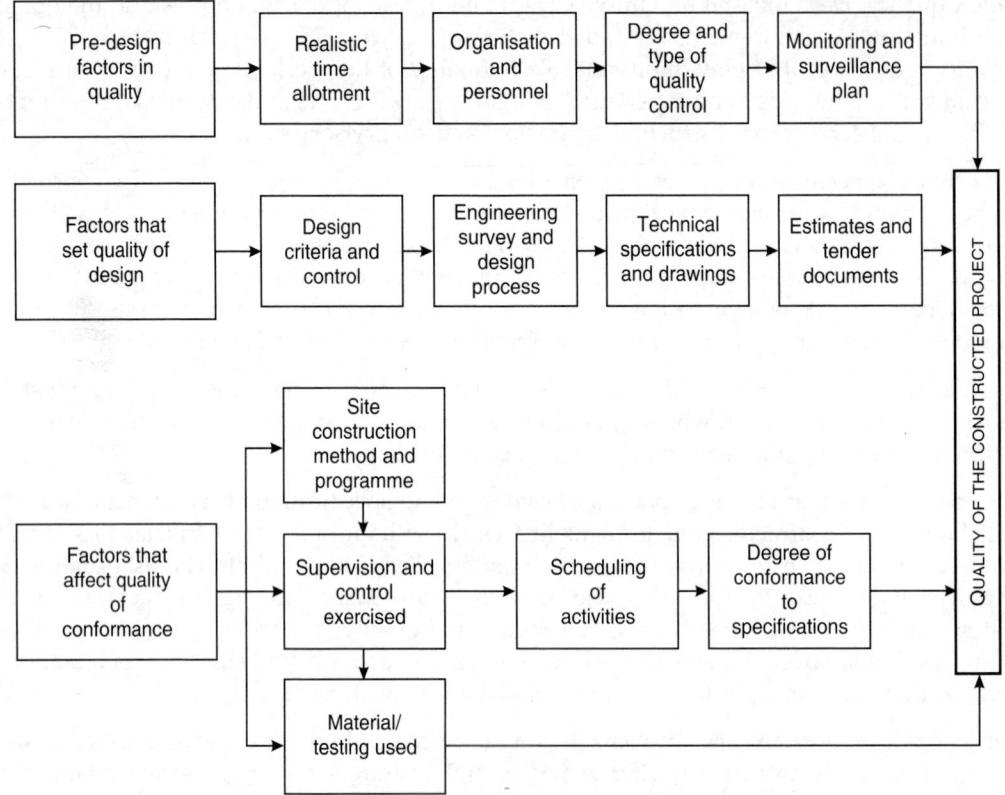

Fig. 15.1: Total quality assurance/management system for a highway construction project

15.3.2 Management of Quality Assurance

Overall **Quality Assurance Plan** is devised either by the owner or the consultant for the owner who furnishes overall quality requirements and the basic quality control tests to be carried out. It is contractor's prime responsibility to achieve desired quality of work. Quality Management/ Quality control provisions are to be made inflexible and mandatory. They have to be enforced rigidly.

For ensuring quality control/management, two essential basic inputs are:

(1) Technical supervisions and fair play by competent staff, and

(2) Control and test checks through laboratories and field equipment.

(i) Technical supervision: For quality control measures it is important to have well qualified and experienced supervisory team which becomes integral part of the quality control culture. Technical supervision is the prime mover. Its vision, conviction, decision and support is vital for quality adherence.

(ii) Control and test check: Total quality of construction work depends upon the materials used. Testing is required for compliance with specifications and to get technical information about suitability of materials and construction techniques used. Figure 15.2 shows steps in monitoring the quality.

IRC-SP-11, "Handbook of Quality Control for Construction of Roads and Runways" provides details of tests and their frequency for particular type of works.

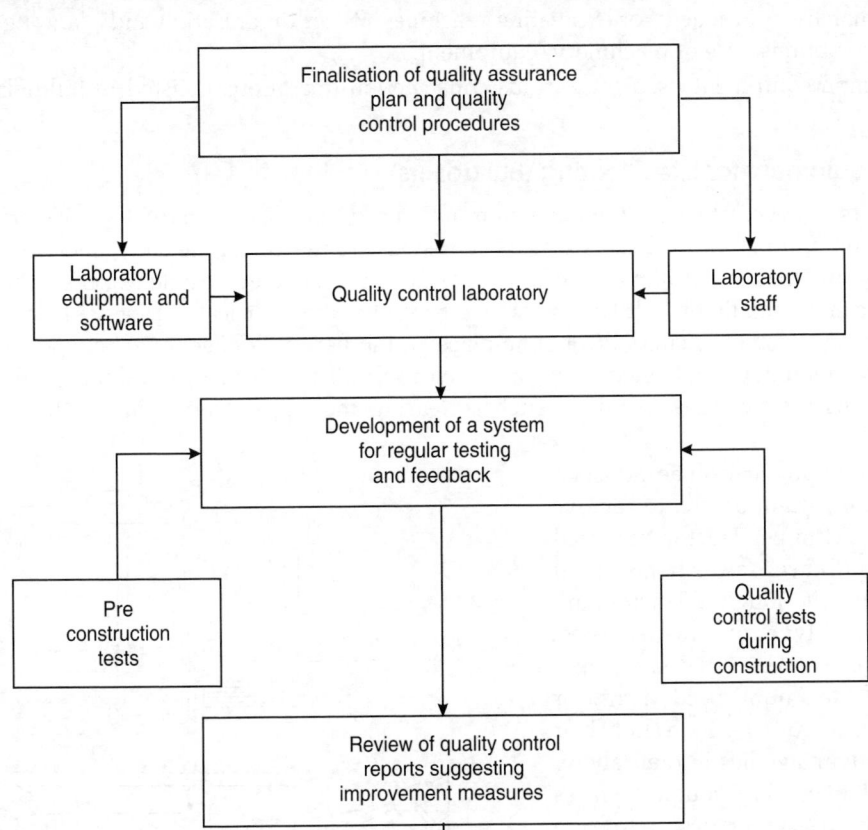

Fig. 15.2: Steps in quality control management for a highway construction project

15.4 CONSTRUCTION EQUIPMENT (HIGHWAY MAKING MACHINES)

After deciding the type, size, and sequence of all construction operations, the next step is to determine the type, size and number of equipment/plant needed for the purpose. In selection of equipment choice is made between type of equipment. Large and expensive, inflexible equipment may complete the job quickly while small, cheap and flexible equipment may be easily available, but may increase construction period. Equipment selected should have lesser breakdowns, lower maintenance cost and capable of working in the climate and environment of the project site.

Use of machines/equipment in road projects offers following benefits:

- Construction time is reduced and cost of construction may be reduced.
- Quality of construction is improved and uniformity is achieved.
- Due to proper construction, maintenance work is reduced.
- Vehicle operation cost and maintenance costs are reduced, because of good quality surface for driving.
- Life of the road is increased.

In selection of the equipment, another important factor to be kept in view is that capacities of handling, related to operations should match. For example capacity of earth-moving plant should match with capacity of excavating machines. Better organisation and planning work is needed to optimise use of machinery/equipment.

Common equipment used for various construction operations is listed in following paragraphs.

15.4.1 Equipment for Site Clearing (Bulldozers)

Bulldozers are versatile equipment commonly used, in road construction. Tractors with horizontally front mounted blades which is transverse to the line of the body, and which can be raised or lowered in vertical plane, also work as dozers. It is called "Angle Dozer', its blades can be angled horizontally or 'Tilt Dozer' its blades can be tilted vertically. Heavy blades attached to the tractor push the material from one place to another. Bulldozers may be wheel mounted or crawler mounted, controlled by cable or hydraulically controlled. Bulldozers are versatile machines used for various operations such as clearing, moving and spreading earth, backfilling, etc.

This equipment can be attached with rooter, stumper, etc. to remove trees and stumps. Tractor mounted winch and power saws are also used in clearing the jungles. Tractors are also wheel type or crawler type. Figure 15.3 shows a bulldozer (crawler type). Maximum speed of crawler tractors is about 10–12 km/hr. Their special advantage lies in their ability to travel over very rough surfaces and to climb steep grades up to 25 to 30%, at a speed of 2 to 3 km/hr. Wheel tractors travel with higher speeds (30–40 km/hr) and are used for jobs requiring travel over long distances.

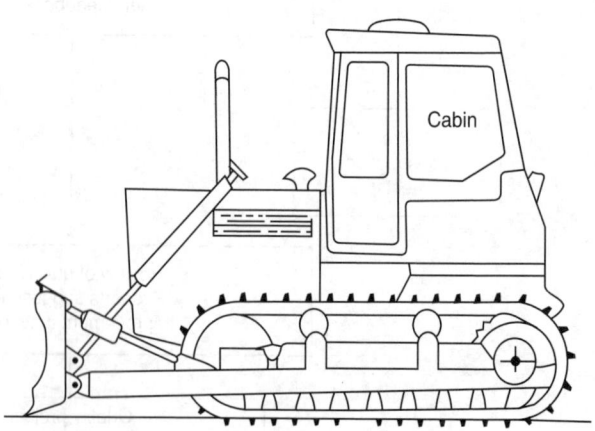

Fig. 15.3: Crawler type bulldozer

15.4.2 Earth Moving Equipment

The earth moving bulldozer consists of a heavy blade of concave profile. They are used in difficult ground and for short distances. For longer distances scrapers loading into the trucks or elevating graders are used.

Crawler dozer is appropriate for working on soft and rock formations and wheel dozers having better travel speeds are used for longer hauls. In hard materials, not rocks, hydraulic ripper attachment can be used with tractor before dozing.

Construction of Formation (Scrapers, Graders): For formation construction, graders, scrapers and bulldozers are used. Scrapers are very useful for excavation, loading, hauling and spreading by itself without help of any other equipment except a pusher tractor. Motor graders, pay loaders, crawler tractors are also used. A typical scraper is shown in Fig. 15.4.

A scraper is loaded by lowering the front end of the bowl, till the cutting edge enters the ground and at the same time raising the front apron to provide an open slot through which the earth flows into the bowl. As the scraper is pulled forward a strip of earth is forced into the bowl. The cutting edge is raised and the apron is lowered to prevent spillage during hauling.

The dumping operation consists of lowering the cutting edge to the desired height above the fill, raising the apron and forcing the earth out.

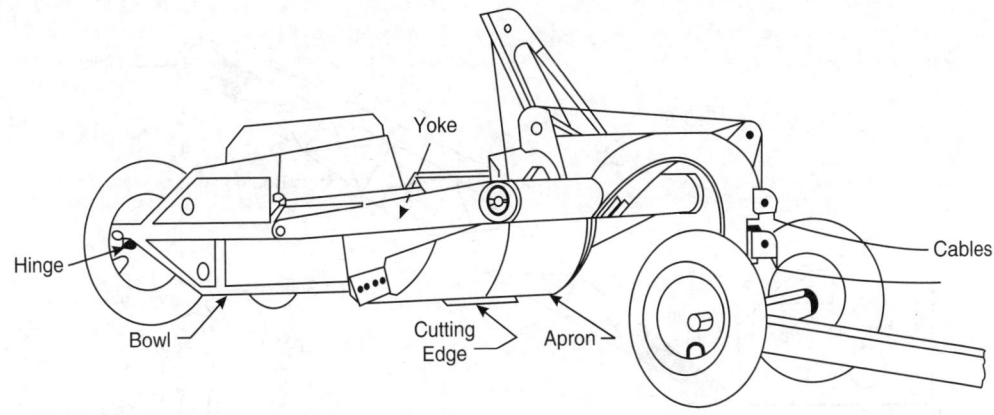

Fig. 15.4: A scraper

The graders are able to steer both forward and backward. Graders are very useful in embankment construction.

Grader has a ploughing blade to cut the earth. It is suited for ploughable materials. Tractor-scraper outfit is adaptable to great variety of earth moving and formation jobs and requirements.

15.4.3 Earth Digging in Trenches (Draglines, Shovels, Hoe)

For earth digging, power shovels are used above ground while, back actors, back hoes are used for trench digging and draglines, grabs, clam shells are used for digging below ground. Draglines can dig and dump over larger distances than shovels.

Hoe is used in firm and hard ground, for digging below the operating level. It can do excavation very precisely.

Power shovels are used for excavating and loading all type of soils except rock. They can be mounted on crawlers for low speeds. The shovel with its sharp edges is suitable where large working space is available. It is one of the basic equipment in construction of roads. A dragline, hoe and power shovel are shown in Figs 15.5 to Fig. 15.7 respectively.

Power shovels may be single engine self propelled, operated from a cabin or non-self propelled mounted on rear of trucks. Size of power shovel and draglines is indicated by size of dipper bucket expressed in cubic meters.

Shovel is put in correct position near the earth to be excavated and the dipper is lowered with teeth pointing into the face. Force is then applied through point line to pull the dipper bucket up.

Draglines are also used to excavate earth and load it into hauling units such as trucks, tractor-pulled wagons, etc. or deposit it on banks.

Power shovels can be converted into dragline by replacing the boom of the shovel with a crane boom and substituting the drag line bucket for the shovel dipper bucket. Both dragline and shovel can be used for performing same job. Draglines are however better for excavating trenches. Draglines can be crawler mounted, wheel mounted self propelled, truck mounted or walking draglines.

The term Hoe is applied to an excavating machine of the power shovel group. Other names of hoe are: back hoe, back shovel, pull shovel, etc.

Hoes are used to excavated below the natural surface of the ground on which machine rests, as in truches, pits etc. It can also be used for general grading work. Hoe can exert greater pressure than shovels or draglines.

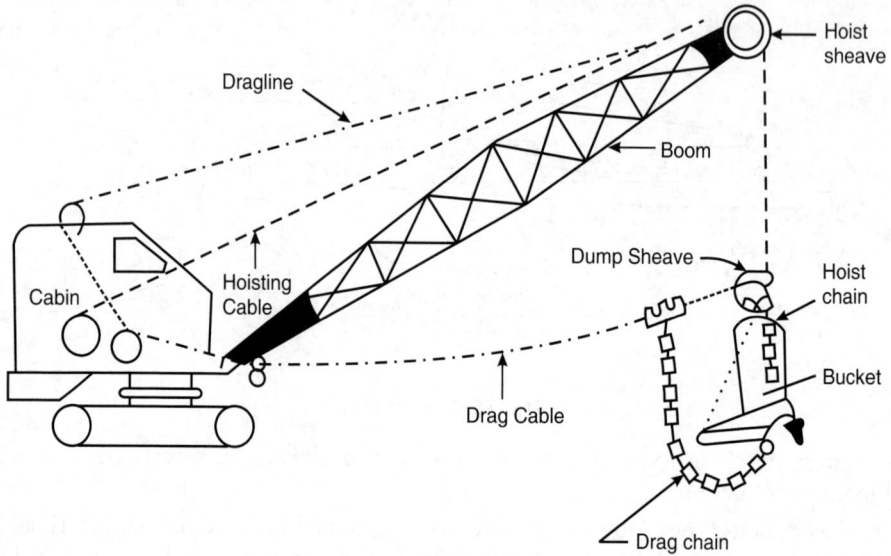

Fig. 15.5: A dragline

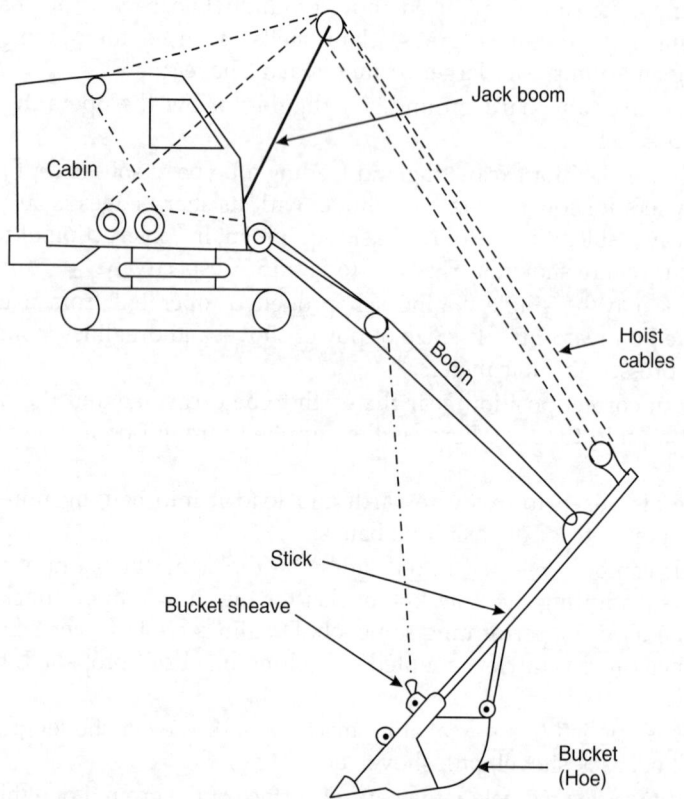

Fig. 15.6: Hoe-excavating machine

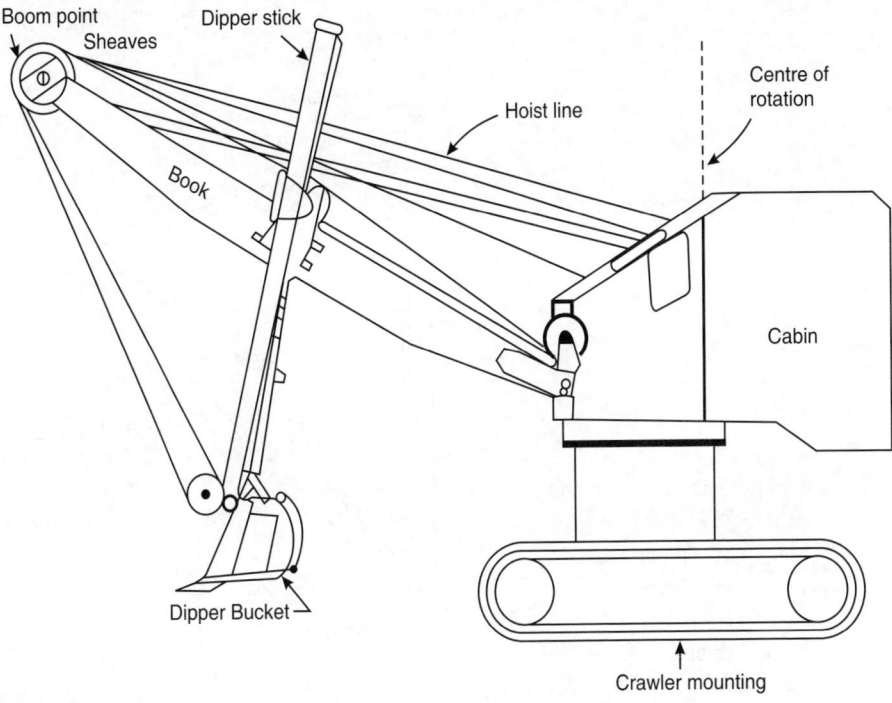

Fig. 15.7: Power shovel

15.4.4 Rock Drilling

Rock drilling may be done by hand-held drills or machine mounted drills. Usually compressed air is used in drilling. Drilling may be followed by use of explosives and blasting. Wagon drills, jack hammers, with accessories are commonly used. Various types of drilling equipment are used to drill holes. The type is selected depending upon the size of the project, nature of the terrain, the kind of rock, the depth of rock etc. Common types of drilling equipment are:

(a) Percussion drills: These drills break the rock by impact from repeated blows. The popular types are Jack hammers, Sinkers, Tripod drills, Stop hammers, Drifters, Piston drills, Churn drills, Wagon drills etc.

(b) Abrasion drills: These drills grind rock into small particles through abrasive effect of a bit that rotates in the holes. Common types are: Blast hole drills, Shot drills, Diamond drills, etc.

Ripping: When excavation becomes difficult, ripping is used. Ripping attachment is a plough—like device with one or more blades and is usually mounted on bulldozers. Ripping is scraping the surface by conventional earth moving equipment to remove soft rock from in-situ. It avoids blasting and drilling operation.

15.4.5 Hoisting Equipment (Cranes)

Hoisting is lifting of weight from one place and moving it to another location for dumping. Hoisting equipment includes jacks, winches, chains and cranes. Cranes are most commonly used for safety, speed and convenience.

Derrick crane consists of a mast, boom, and bull wheel on which boom rotates about vertical axis. A typical crane is shown in Fig. 15 8.

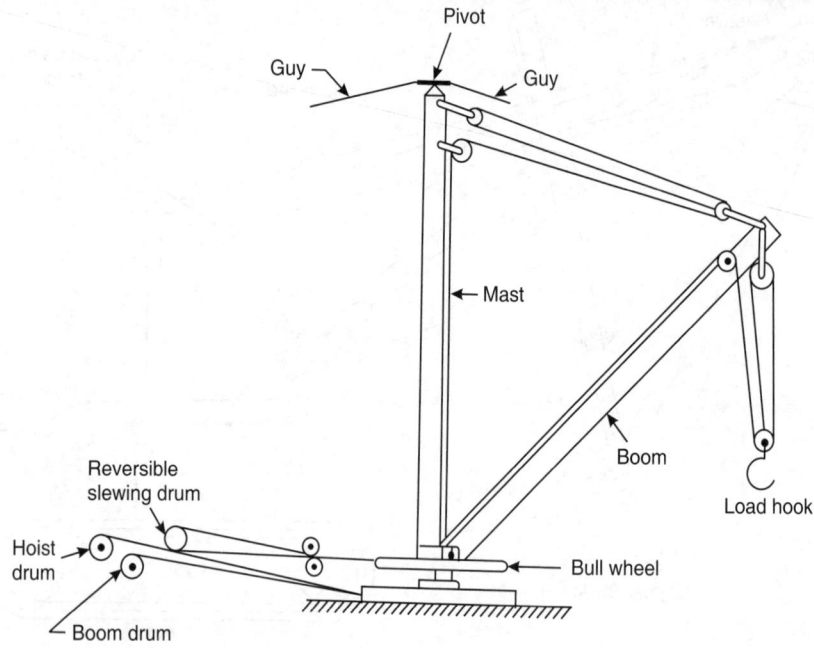

Fig. 15.8: A Derrick crane

15.4.6 Compaction Equipment

To increase density and strength soil is compacted at suitable moisture content.

(a) Grading equipment

- For grading material into individual sizes devices such as grizzlys (fixed sloping screens), or power screens (rotating or vibrating screens) are used. Mixing of material is done in place by using motorised graders.

(b) Rolling equipment (compacting equipment)

- For rolling, used for compacting process, following equipment may be used:

(i) **Smooth steel wheel roller:** Shown in Figs 15.9 and 15.10 self-propelled rollers with suitable diameter to minimise horizontal displacement of soil are used. They are versatile and suit to many situations. Usually they have two axles driving two steel rolls on either side of the driving wheel. It is also called a three-wheel roller. This provides compaction over the whole width of the road.

(ii) **Grid rollers:** It consists of two drums of open mesh grid with self cleaning device to prevent clogging. It is puled at towing speed of 25 km/hr to create impact forces to crush large pieces of rock below surface. The drums work by both direct pressure and a kneading action and are well suited for compaction of gravelly soils.

(iii) **Pneumatic tyred roller:** For material with low cohesion, this roller produces a more uniform compaction than smooth steel wheel roller. The tyre pressure should be kept high. It compacts the material through kneading action. It may be towed or self-propelled, and can be of different weights. 8-10 tonner roller is commonly used.

(iv) **Sheep foot and tamping foot roller:** This roller has steel drum fitted with feet. It produces high contact pressures and is useful in compacting cohesive soil and clays. Sheep foot rollers can also be used successfully on rough or steep areas inaccessible to other rollers. It can be towed or self-propelled. Their weight may vary from 2 to 20 tonnes.

Power rammers are used when space is restricted.

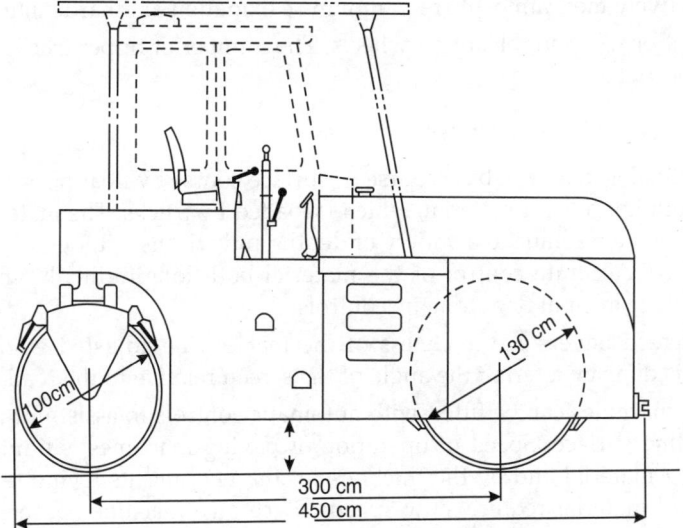

Fig. 15.9: Two-axle steel wheel roller

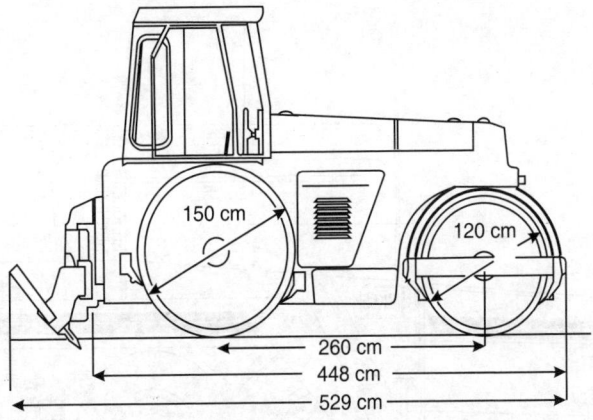

Fig. 15.10: Three wheel roller

(v) Vibrating rollers: Vibration raises the effective compactive efforts, particularly by causing momentary drops in internal friction. Low frequency, high amplitude rollers are used for unbound courses and high frequency, low amplitude rollers are used for bituminous concrete. High impact pressures are often effective for clays. The vibrations imparted by the roller cause the particles of the material to shift themselves into tightest arrangement, thus achieving high degree of compaction.

15.4.7 Bituminous Mixing Plants

Following are main type of plants common in use:

(i) Batch plants: Batch sizes of different outputs can be used. The plant can also operate intermitently. Aggregates are dried by heating to temperatures of between 150 and 190°C, which reduces moisture. All ingredients are separately proportioned on scales. The adequate mixing time is used to coat all aggregates with bitumen.

(ii) Continuous plants: In this plant, aggregates are dried, heated, and mixed in one operation. Large quantities of single mix are produced. The heating and mixing are accomplished in a relatively inert atmosphere to minimise oxidation and hardening.

Plants are stationary, portable, or on wheels. The number of tipper trucks must be matched with the plant output.

15.4.8 Bituminous Paving Machines

Bituminous material is placed in base course or surface course by floating–screed pavers. These machines use both hot or cold material, which is loaded by trucks. The material is then spread and compacted by the machine to a variety of depths and widths without segregation to form a hard level surface. Accurate control of the material both longitudinally and transversely is possible using either manual or automatic controls.

In floating-screed pavers the thickness of the mat can be adjusted while the machine is travelling. This is done by altering the angle of the screen relative to the road surface.

Most paving machines can be fitted with automatic controls to assist in the production of a high quality riding surface. Speed of operation of paving machines is limited by the rate of supply of paving material and by the thickness of the mat and its laying temperature. Thick mats and/or cool material require slow speeds. Excessive machine speed results in uneven surface of poor riding quality. Typical speeds are 2 to 3 m/s for base course materials and 3 to 5 m/sec for wearing course materials. Figure 15.11 shows a typical paving machine.

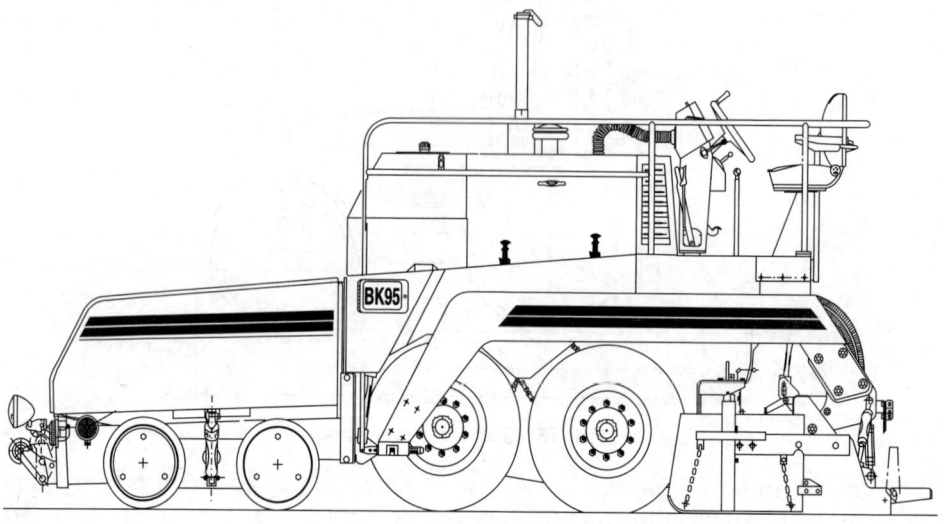

Fig. 15.11: Typical paving machine

15.4.9 Other Equipment

(i) Air compressor

Air compressors operating a single jack hammer are useful for exploratory work on hill road construction. It can be transported to site by manual labour or animals. Air compressors with two jack hammers are also available for other development works.

(ii) Stone Crushing Plants (Production of Aggregates)

For production of aggregates, stone crushers of required size are used. Capacity of the plant may vary from 5 to 100 tonnes per hour, depending upon the size of the project and requirement of various sizes of aggregates.

Production of crushed stone aggregates involves

(a) Quarrying operations: i.e. drilling, blasting, loading and transporting.

(b) Reduction operations: i.e. crushing, screening, handling, and storing.

Different types of crushers are used. They are classified as primary crusher, secondary crusher and tertiary crushers.

Primary crushers are Jaw crushers, Gyratory crushers, Hammer mill, etc. The secondary crushers are Conc crushers, Roll crushers, and Hammer mills, etc. and the teritary crushers are Roll crushers, Rod mills, Ball mills, etc.

The choice of a crusher depends upon type of stone, maximum size of stone, method of feeding the crusher, required size or sizes of aggregates and their quantity.

Figure 15.12 shows flow diagram of a typical aggregate processing plant.

Screening is done before feeding the material to a primary crusher. This process called scalping is required to prevent oversize stones entering the crusher, removing the dirt, mud etc. Number of widely spaced bars called "Grizzly" are used for scalping purpose.

Different screens of specified sizes are used to separate aggregates of various size ranges. Different type of screens used for this purpose may be (a) revolving screens or (b) vibratory screens.

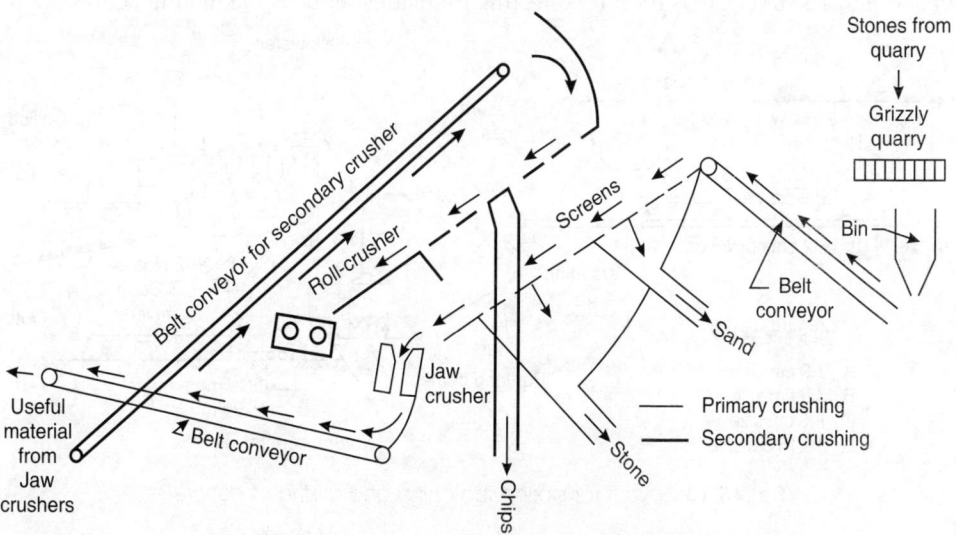

Fig. 15.12: Flow diagram of an aggregate processing plant

Vibrating screens are multideck units, permitting installation of one screen over the other. Vibration is obtained by means of an eccentric shaft or electromagnet. They are most widely used. In a multiple deck unit the sizes of the openings will be progressively smaller for each lower deck.

15.4.10 Equipment for Concrete Road Construction

Basically concrete is obtained by mixing cement, aggregate and water. Operations involved in construction of concrete roads requiring equipment/machinery are:

(a) Batching the Aggregates (Batchers)

Batching is usually done by weight. Specifications may require concrete to be prepared from 2 to 5 or more different size ranges of aggregates. It is the function of the batching equipment to perform the measuring of quantities, for each size range.

For batching proportioning is generally done by weight. An elevated storage bin equipped with a weighing batcher is used.

For large projects storing, batching, and mixing of aggregates is done continuously, as shown in Fig. 15.13.

(b) Mixing of Concrete (Mixers)

Concrete mixers are available in different sizes from 0.375 to 3 cub. m. capacities, suiting to the requirements of the project. Two types of mixers are common.

Tilting type mixers and non-tilting type mixers. On large works tilting type are suitable, while non-tilting are suitable for small works.

Paving mixers are used to mix and place concrete for highways and streets. They are mounted on crawler trucks so that they move along with placing of concrete. They are single drum or double drum units. In the double drum unit, aggregate is charged in the first compartment where it is pre-mixed, then transferred to second compartment, as soon as it is empty. This operation permits substantial output of double drum mixers.

Central mixing plants are established to mix concrete for large projects. Suitable vibratory drums mounted on trucks are used, where large volumes of concrete are to be transported to site. In modern paving mixers, mixing time is controlled automatically. Blades in the mixing drum should be kept clean as mortar collecting on blades reduces the mixing action.

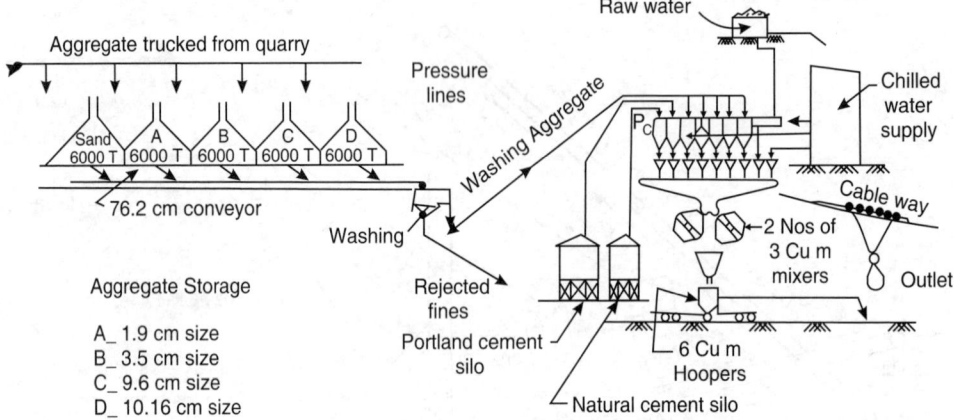

Fig. 15.13: Layout for storing, batching and mixing of concrete

Transit mixer is a truck on which concrete mixer is mounted. Aggregate and cement are charged into the mixer at a central batching station and mixing is done on route to the site.

(c) Consolidation and Finishing Machine

After concrete is laid on subgrade and spread to the desired thickness, it is consolidated and finished. This is accomplished by mechanical finishing machines. Finishing machines have two screeds that not only give the concrete its proper contour but also consolidate it by pressure. The front screed is higher to allow for consolidation by the rear screed. Other machine adjustments include those for variation in crown, forward speed of the machines, and front to back screed tilt. Proper adjustments of these features is important. The finishing quality of a wide range of concrete mixtures can be handled by proper use of these adjustments.

Small vibrators attached to hand operated strike-off screeds are also used. Vibration if required is obtained by either a pan-type vibrator mounted on the concrete spreader or by vibrating the front screed of the finishing machine.

Finishing is often followed by a mechanical longitudinal float. The float is operated traversely across the pavement with its longitudinal axis approximately parallel to the centreline, and successive passes lapped at least one-half the length of the float. The finishing machine and longitudinal float are so adjusted that later acts as a float and does not have to cut the surface to the required elevation. Excess mortar is carried ahead of the float. If mechanical finishing is not required, a manually operated longitudinal float is used. It is made of a plank of wood 75 cm by 10 cm, 5 to 7.5 cm thick, about 3 m in length, provided with handles on each end. The longitudinal float is handled by two persons who stand on bridges spanning the pavement. It provides smooth riding surface.

Each type of machinery is characterised by certain advantages and disadvantages that must be weighed when selecting the equipment/machinery for a particular road project. Manually operated, semi-automatic and fully automatic equipment is available for selection.

15.5 ACTIVITIES IN CONSTRUCTION OF HIGHWAY PAVEMENTS

Construction of highway pavement is providing a firm surface which could bear the traffic loads and stresses. Two basic type of pavement surfaces are flexible pavements and rigid pavements (concrete slabs).

The different activities involved in construction of a new road may be grouped as follows:

(a) Flexible pavements

(1) Clearing of site

(2) Earthwork (preparation, compaction and shaping of subgrade in cut and fill)

(3) Construction of sub-base and base course (road bases)

(4) Surface courses of bitumen

(5) Construction of shoulders

(6) Protection of embankments and cut slopes (drainage provisions)

(7) Painting/providing of markings

(8) Installation of signs

The above activities, can further be broken into sub-activities. Some activities may not be required at some locations, while other activities may be added at certain type of constructions. For example an earth road will not require activities 4, 5 and 7.

(b) Rigid pavements

For rigid pavements (concrete slab) activities can be slightly modified as given below:

(1) Clearing the site

(2) Earthwork (Preparation, compaction, shaping of subgrade)

(3) Form laying

(4) Mixing and placing of concrete

(5) Compacting and finishing

(6) Curing

(7) Painting/providing markings

(8) Installation of signs

15.6 CONSTRUCTION OF FLEXIBLE PAVEMENTS

Following activities are involved in construction of a flexible pavement.

15.6.1 Clearing of Site

The first step in construction of flexible or rigid pavement is to clear the site by removing the natural waste material. In urban area clearing may also require shifting of public utilities, removing of structures, buildings, footpaths, etc.

During clearing from the right of way of all trees, tree stumps, vegetation, and rubbish material is removed. The top soil is removed till all the vegetation is removed. All holes created by removal of the stumps should be filled and compacted prior to the compacting of subgrade as a whole.

In urban areas, removal of utilities, poles, power lines, water mains, etc. should be carefully planned, keeping in view the future use.

Site clearing work involves earthwork which may include excavation, loading and hauling of excavated material, unloading, spreading and compacting. Bull dozers, rooter, stumpers type equipment is used in clearing operation, depending upon nature of the site.

15.6.2 Earthwork, Preparation, Compaction and Shaping of Subgrade

Construction of embankment and preparation of subgrade are part of earthwork. Subgrade should be prepared to required stability and density so that it is able to withstand the load (stresses) transmitted to it by the pavement. It has therefore to be well compacted, to a state stipulated in design, using suitable type of rolling equipment, for the construction site, type of soil and climate. When the subgrade passes from cut to fill, care should be taken that there is no change in the degree and uniformity of compaction.

Embankments are necessary depending upon the flooding conditions of the area. Height of the road embankment should be 0.6 to 1 m above the highest flood level. Embankments are made of suitable height by compacting thin layers of earth by suitable rollers, to achieve desired density, using optimum moisture. IRC: 36, IRC: 56 and IRC: 75 provide detailed specifications and guidelines for construction of embankments. Chapter 17 describes embankment design and construction in details.

After compaction of subgrade, in level, embankment or cutting the next step is to shape the rough surface to the final shape of carriageway. By doing this any moisture intruding into the subgrade will drain away rapidly and pavement will be constructed to its proper shape. Shaping is done by blade-grader and any irregularity is removed by light rolling with smooth wheeled or pneumatic-tyred rollers.

15.6.3 Construction of Sub-Base and Base Courses (Road-Bases)

After preparation, compaction and shaping of the subgrade, sub-base (if necessary) and base course is constructed. These courses can be made of different materials like stones, aggregate, stablised soils and waste materials from industry like clinkers, quarry waste, shales, slag etc. where available.

Techniques used in construction of bases/sub-base are:

(1) Hand pitched base
(2) Macadam bases
(3) Stablised bases (stabilisation)

(1) Hand pitched bases (stone soling)

Hand packing of rectangular stone, side by side on the prepared subgrade used to be done, in olden days. Each stone was firmly settled and smaller stones were hand-wedged into the gaps,

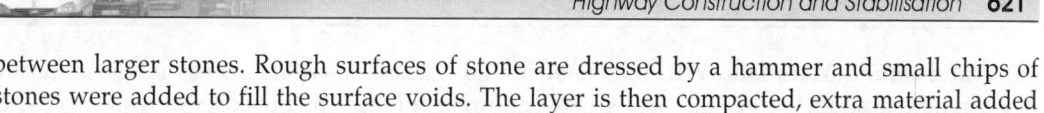

between larger stones. Rough surfaces of stone are dressed by a hammer and small chips of stones were added to fill the surface voids. The layer is then compacted, extra material added to surface till smooth base course having no depressions or hole is obtained. Camber is provided to the surface.

It is labour intensive, slow technique, still common where labour is cheap or equipment is not available. Thickness of layer should not be too much as compaction will then be inadequate, allowing ingress of water.

(2) Macadam type bases

Macadam construction is an open graded pavement course of angular aggregates to which smaller size stones are added. It is named after its inventor who took advantage of the natural interlock developed between the stone particles. Macadam type construction of bases are most common these days suiting to present traffic requirements, and equipment. With mechanical means it requires lesser labour, construction speed is higher and so initial cost is low, time is saved and quality of work is high. Stable road surface without deformations is produced, permitting bituminous surface course to be laid concurrently. Macadam type of construction depends upon the interlocking and frictional characteristics of the constituent material.

Different types of Macadam pavements are used in highway construction of bases such as:

(i) Dry Bound Macadam (DBM)

(ii) Water Bound Macadam (WBM)

(iii) Crushers Run Macadam (CRM)

(iv) Wet Mix Macadam (WMM)

(v) Bituminous Coated Macadam Bases

(vi) Cement Bound (Stabilised) Bases

All type of Macadam construction basically rely for their strength and resistance to deformation, on the interlocking of the individual crushed stone particles and on friction generated between the rough surfaces in contact. Some techniques also use water or bituminous binder to cement stone particles together.

(i) Dry-Bound Macadam: The first step in 'Dry Bound Macadam' type of construction is to compact the underlying layer on which Macadam construction will be done, to provide stable formation. Silty or clayey soils may require an insulating layer of stones chips or sand of suitable thickness, to provide a granular sub-base, on the top of which Macadam construction is laid.

Secondly the edges of the roadway should be supported to provide lateral stability, to resist the lateral thrust of aggregates when they are compacted. Steel or timber forms laid along the edges provide this stability and prevent flying away of the material laterally. Compaction is normally begun from the edges in a direction parallel to the centre line and is extended progressively towards the centre of the road.

In Dry Bound Macadam base crushed aggregates are laid of two or more different sizes and rolled dry. The first layer is of coarse aggregates is laid and compacted. After removing the depressions and shaping of the first layer, subsequent layer of finer aggregates is laid so that voids in the larger particles can be filled to attain interlocking and internal friction. The sizes of the material, thickness and number of layers depend upon the requirements of total thickness of the road, as per design of pavement.

For obtaining the final dense compact surface heavy smooth wheel roller is used, in the final stage when all the inter spaces are filled. After this rolling the surface is broomed to remove excess fines and surface course may be laid.

(ii) Water Bound Macadam (WBM): 'Water Bound Macadam' type construction is similar to dry-bound macadam, except that water is used during rolling. The purposes of water spraying during compaction are:

(i) Act as a lubricate for the aggregates allowing them to slip easily.

(ii) Dust of the aggregates when wet gives better binding property.

In WBM also, dry coarse aggregates layer is laid, compacted and surface irregularities removed. Over it a layer of fine material is laid on the coarse layer and compacted, loose material is removed from the surface and irregularities are removed.

In the next step, compacted surface is to be sprayed with water, and rolled, water is spread by mechanical sprayer and rolling is carried out. More material may be added to fill voids and depressions if necessary. Water spraying and rolling is continued till a wave of grout appears ahead of the roller, indicating that all voids are full. Compaction is then completed.

Different gradings are recommended for W.B.M. construction by Indian Road Congress as given in Table 15.1. In WBM construction each layer is given time for moisture to escape and dry thoroughly before next layer is laid.

Coarse aggregates for water bound macadam should meet physical requirements as given in Table 15.2.

Table 15.1: Size and grading requirement of coarse aggregates and screening for WBM (IRC)

A. Coarse Aggregates:	Range of sizes	Sieve designation	Percent by weight passing the sieve
S.No.			
1.	90 mm to 45 m	125 mm	100
		90 mm	90–100
		63 mm	25–60
		43 mm	0–15
		22.4 mm	0–5
2.	63 mm to 45 mm	90 mm	100
		63 mm	90–100
		53 mm	25–75
		45 mm	0–15
		22.4 mm	0–5
3.	53 mm to 22.4 mm	63 mm	100
		53 mm	95–100
		45 mm	65–90
		22.4 mm	0–10
		11.4 mm	0–5
B. Screenings			
1.	13.2 mm	13.2 mm	100
		11.2 mm	95–100
		5.6 mm	15–35
		180 micron	0–10
2.	11.2 mm	11.2 mm	100
		5.6 mm	90–100
		180 micron	15–35

Table 15.2: Requirement of coarse aggregates for WBM (IRC)

S.No.	Construction Course	TEST	Requirements (Test Value)
1.	Sub-base	Los Angles Abrasion or Aggregate impact	Max. 60% Max. 50%
2.	Base course with bituminous surfacing	(a) Los Angles Abrasion or Aggregate Impact (b) Flakiness index	Max. 50% Max. 40% Max. 15%
3.	Surface course	(a) Los Angles Abrasion (b) Aggregate impact (c) Flakiness Index	Max. 40% Max. 30% Max: 15%

Detailed specifications for Water Bound Macadam construction are given in IRC: 19-2005. Following steps are involved in construction of a W.B.M. road:

- Preparation of the subgrade, sub-base or base to receive the water bound madam course.
- Provision of lateral confinement of aggregates.
- Spreading of coarse aggregates on prepared base
- Compaction of the aggregates, by rolling with suitable roller, progressively from edge to the centre.
- Application of screenings to fill the interstices, followed by dry rolling.
- Sprinkling of water and brooming the surface to sweep into the voids.
- If required, binding material shall be spread and sprinkled with water.
- Surface to be left for setting and drying for a day, before traffic is allowed.
- Bituminous surface if to be provided on WBM it should be done after surface is completely dried and before allowing traffic on it.

Construction of water bound macadam if done manually using lot of water, segregation of aggregates is likely to occur, which may result in non-uniform surface.

(iii) Crusher-Run-Macadam: Crusher-run aggregates may be natural or artificial which give graded mixture, such that fine particles completely fill the voids between the coarse particles, thus giving the required stability on compaction. If only one material can give the required combination this type of construction becomes cheap. It is spread easily and required compaction is obtained easily.

This method has limited use, depending upon availability of such material.

(iv) Wet-Mix Macadam (WMM): An improvement over WBM is WMM construction. In this method crushed graded aggregates are mechanically mixed with controlled amount of water (2 to 5 percent by weight). They are thoroughly mixed in a mixing plant, before transporting to the site. Because water is added in the beginning it is called 'Wet-Mix'. Addition of water facilitates laying of the mix with good speed in desired shape by automatic pavers. Compaction is done soon after laying.

Amount of water to be added is very important in this technique of construction. Too much water will result in segregation and cohesive bond will not be formed. Just optimum amount of water is to be added. Laying of W.M.M. in rainy weather is avoided, as moisture affects compaction.

WMM can be used in sub-base and base course and laid in many layers as per required thickness, and type of equipment available for compaction. Physical requirement and grading of aggregates recommended by IRC are given in Tables 15.3 and Table 15.4 Guide lines for wet mix macadam construction are given in IRC: 109–1997. Construction operations of WMM are similar to WBM, except that addition of screening etc. is not required.

(v) Bituminous Macadam Base: In bituminous bases, well graded aggregates are premixed with bitumen, before placing in the pavement. Bituminous bound road-bases can be of lesser thickness than other macadams, having increased stability and flexibility.

The bituminous mixture is laid by machines in layers. It protects sub-base and subgrade in bad weather. Quality and bitumen type is selected according to climatic conditions. Tar can also be used for this purpose. Bituminous macadam is not laid during rainy weather. The base or subgrade should not be damp for placing bituminous macadam.

Bituminous binder is applied uniformly, using a tack/prime coat where necessary, using sprayers or manually. Mechanical mixers for mixing aggregates give better mix than hand mixing drums. Mix should be spread immediately after mixing by mechanical pavers, spreaders or graders in required thickness and rolled with 8 to 10 tonne rollers, commencing from edges towards centre.

Table 15.3: Requirement of coarse aggregates for wet-mix macadam (IRC)

S.No.	Test	Requirement (Test Value)
1.	Log Angles Abrasion or Aggregate Impact	Max: 40% Max: 30%
2.	Combined Flakiness and Elongation indices (Total)	Max: 30% (for bases) Max: 35% for (sub-bases)

Table 15.4: Grading requirements of aggregates for wet mix macadam (IRC)

IS Sieve Designation	Percent by Weight Passing Sieve	
	Grading 1	Grading 2
53.00 mm	100	—
45.00 mm	95–100	—
26.50 mm	—	100
22.40 mm	60–80	50–100
11.20 mm	40–60	—
4.75 mm	25–40	35–55
2.36 mm	15-30	—
600 micron	8–22	10–30
75 micron	0.8	2.9

For material finer than 425 micron plasticity index should be less than 6.

(vi) Cement Bound Bases: Lean concrete mixes can be used as base and sub-base materials, on high traffic major roads. Cement-Bound material are considered in flexible pavement category because the lean surface cracks upon setting and hardening of cement after construction.

Water content is critical in lean concrete. If the mix is too dry the surface may crack (shear) under roller during compaction and if mix is too wet, it will stick to roller wheels or roller wheels will sink into it. Thickness of the lean concrete in one layer for compaction depends upon the quality of subgrade or sub-base on which it is to be put and the performance of the compaction equipment. It could be put in layers also if total thickness required is more than the maximum recommended for good compaction.

Top layer should be as thick as possible consistent with obtaining good compaction. Spreading of concrete can be done by hand, machines or grader etc. Forms are used to resist vertical and lateral displacement.

Generally cement bound road bases/sub-bases are laid without joints, expect day work joints. Expansion joints are not required. Lean concrete bases required curing to prevent evaporation of moisture.

15.7 SURFACE COURSES OF BITUMEN

Most of the roads having flexible pavements are provided with bituminous surfacings of various types and thicknesses. Bituminous surfacing serves following basic functions:

- Protect underlying base and subgrade from damage due to weather, and moisture.
- Provide smooth, quiet, not-slippery, and impermeable running surfaces for vehicles.
- They are highly resistant to surface wear and deformation.
- Provide adequate resistance to skidding, by locking the aggregate particles in position.
- Bituminous surfaces are dust and mud free.
- They help in transmitting applied loads in such a way that the layers underneath are not overstressed.
- The maintenance cost of bituminous roads are relatively low.

Bituminous surfaces can be divided into following types:

- Surface treatments: Use low viscosity binders
- Medium textured or open textured surfaces: Binder range from very viscous materials to very fluid ones such as cut-back bitumen or low viscosity tar.
- Closed textured (dense) surfaces: Use very viscous binders such as penetration grade bitumen or high viscosity tar.

Surface treatments are not in themselves surfacings but they help meeting the above purposes, except transmitting load to underneath layers. In surface treatments aggregates are usually laid cold in the field while the binders may or may not be heated, before application.

15.7.1 Types of Bituminous Surfaces

Bituminous surfaced roads are most common type of roads in India. Applications of the various bituminous materials in road construction make use of mixture of bitumen/tar and aggregates, involving anyone of the two basic techniques:

(i) Spreading of bitumen/tar and aggregates and compacting or

(ii) Mixing of bitumen/tar and aggregates, then spreading of the mix and compacting.

Bitumen could be in form of paving bitumen, cut-back bitumen or bitumen emulsion. Paving bitumens require heating to certain temperature to make them viscous, while cut-backs can be used in cold condition and emulsions can be used with wet aggregates, even in raining weather.

Bituminous construction is used both for base course and surface course. Different types of construction using bituminous material, can be closed-textured (Dense), medium-textured, or open-textured surfaces or surface treatments.

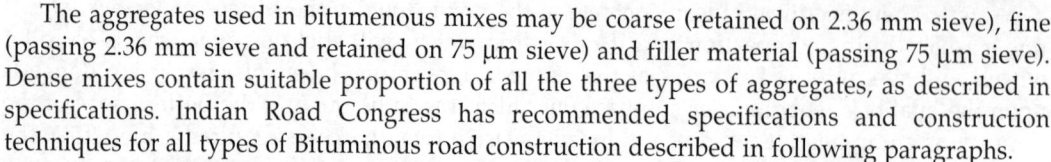

The aggregates used in bitumenous mixes may be coarse (retained on 2.36 mm sieve), fine (passing 2.36 mm sieve and retained on 75 μm sieve) and filler material (passing 75 μm sieve). Dense mixes contain suitable proportion of all the three types of aggregates, as described in specifications. Indian Road Congress has recommended specifications and construction techniques for all types of Bituminous road construction described in following paragraphs.

15.7.2 Techniques Involving Spreading of Aggregates and Bitumen and Compacting

(i) Tack Coats: Application of a bituminous material is done on an existing old bituminous road, to ensure proper bond between it and superimposed layer. Tack coat is not a surface treatment. Its purpose is to bind the old and new layers. Bituminous binders a cutback bitumen, emulsion or low viscosity tar is used as a single application for tack coat, on old bituminous surface or cement concrete surface.

(ii) Prime Coats: Application of bituminous binder on previously untreated porous layer such as WBM, earth or gravel road etc, is done so that it penetrates the top of the layer and fills voids is called a prime coat. It is a special form of tack coat, used on porous surfaces initially before placing the bituminous wearing surface. Prime coat helps in providing bonding between the base and surface course, and also binds dust and loose particles. It also acts as a membrane to prevent capillary rise of moisture into pavement from underneath.

Low viscosity cut-back bitumens are used for prime coats. Prime coats also provide a firm surface for placing pre-mix carpet. Detailed specifications for prime and tack coat are given in IRC: 16–2008.

(iii) Seal Coat: A seal coat is a thin layer of bituminous binder applied on an existing or new bituminous surface, with the object of closing (sealing) the voids and rendering it water proof. Emulsions are useful as seal coat material. Seal coat is not the wearing surface itself.

Seal coats develops skid resistance for smooth and slippery surfaces, it rectifies the defects on the surface and improves riding quality and night visibility. Seal coat also seals entry of water and moisture and improves (re-news) the appearance of the road. It provides a dust and mud free surface.

(iv) Surface Dressing: Surface dressing consists of applying single or two coats of bituminous binder and aggregates over prepared base. A surface dressing is simply a thin layer of aggregates held by the binder and has no structural strength.

In a single coat of surface dressing a thin layer of bitumen binder is applied on the prepared surface to receive surface dressing. Over it aggregates are spread and rolled. In double coat surface dressing the same process is repeated second time. Purposes of surface dressing are:

- Prevent entry of water from the top, thus make it water proof.
- Prevent disintegration of the base and provide dust free surface.
- Improve non-skid property of the surface.
- Provide demarcation between the carriageway and shoulders.
- Protect the base course.

Following operations are involved in single coat surface dressing:

(i) Preparation of road surface

(ii) Application of bituminous binder

(iii) Application of cover aggregates

(iv) Rolling of the cover aggregates

(v) Finishing the surface to conform to cross-section and grades.

(vi) Opening to traffic

For two coat surface dressing, steps (i) to (iv) are repeated before steps (v) and (vi), mentioned above.

In place of conventional single or double coat surface dressing, another similar technique of aggregates pre coated with bitumen is adopted where better adhesion between aggregates and bitumen is required for heavy traffic roads. Detailed specifications for good construction practices for the three types of surface dressing are given in IRC: 17–1965, IRC: 23–1966 and IRC: 48–1972 and IRC: 96–87. IRC: 100–1988, gives specifications for use of catonic emulsion in single and double coat surface dressings. IRC: 110–205 covers design and construction of surface dressing. Table 15.5 shows gradings for surface dressings.

Table 15.5: Grading for surface dressing (IRC)

Seive size (mm)	Percent Passing	
	First coat (19 mm)	Second coat (13 mm)
26.5	100	—
19.0	85–100	100
13.2	0–40	85–100
9.5	0–7	0–40
6.3	—	0–7
2.36	0–2	0–2
75 micron	0–1	0–1

(iv) Penetration Macadam: Penetration macadam is the type of construction in which bitumen is poured after the aggregates are already spread, on the road. It is also called 'Grouting' when binder penetrates full depth it is called 'Full-Grout Macadam' and when binder penetrates half depth, it is called 'Semi Grout Macadam'. Thickness for full grout is about 50 to 75 cm and semi grout is about 50 cm or less.

Following steps are involved in penetration macadam:

- Preparation of existing road surface
- Spreading of coarse aggregates and compacting
- Application of bitumen
- Opening to traffic

Depending upon the climate paving bitumen 30/40 to 80/100 or cut-back bitumen MC-3, are used for penetration macadam. Detailed specifications for penetration macadam are given in IRC-20–1966.

(v) Built-up Spray-grout: The built-up spray-grout construction involves a technique in which dry aggregates are spread on the road and rolled after applying a layer of bituminous material as a prime coat or track coat. Bituminous layer is spread again on the layer of compacted aggregates and followed by another layer of coarse aggregates, and finished with key aggregates.

This method is used for strengthening an existing bituminous pavement. Following steps are involved in Built-up spray-grout.

- Preparation of surface
- Spreading and compacting one layer of coarse aggregates
- Application of bitumen
- Spreading and compacting of second layer of aggregates
- Application of second layer of bitumen
- Finishing with key aggregates

Total thickness of such construction is about 75 cms and grade of bitumen to be used depends upon the climate. Detailed specification for this type of construction are given in IRC-47-1972.

15.7.3 Techniques Involving Premixing of Bitumen and Aggregates

(vi) Pre-Mix Carpet: Pre-mix methods of construction involve preparing bituminous pre-mix by coating aggregates with binder in special mixers/plants before spreadings on the road surface. In bituminous pre-mixing process aggregates of specified size are mixed with hot bitumen.

Pre-mix carpet involves preparing bituminous pre-mix, laying over previously prepared base and compacting it to required thickness-about 2 cm in a single course. The premix carpet is immediately covered with a seal coat. Seal coat used may be one of the two types: Viz. liquid-seal coat and premix seal coat. In heavy rain fall area liquid-seal coal is used. In liquid seal coat hot bitumen is spread over pre-mix carpet, covered with coarse aggregates and compacted.

Following steps are involved in pre-mix carpet:
- Heating of aggregate and bitumen
- Mixing of aggregates and bitumen
- Spreading of hot-mix as a layer
- Compacting the layer
- Application of liquid seal coat or pre-mix seal coat.

Detailed specifications for pre-mix carpet are given in IRC: 14–2004.

(vii) Bituminous Macadam: Bituminous macadam type construction has already been described. Bitumen macadam mix can be open graded or dense graded. An open-graded type construction is used for base-courses and dense graded is used in binder course.

Construction operations involved are:
- Heating of aggregates and bitumen
- Mixing of the two
- Laying of hot mix
- Compacting the layer

Detailed specifications of open graded bituminous macadam are given in IRC: 27–1967 and for dense bitumen macadam in IRC: 94–1986.

(viii) Sand Bitumen Mix and Sheet Asphalt: Sand bitumen base courses are used in desert areas, where no other construction material except sand is economically available. Mix is properly designed based on stability requirements using Hubbard test. Type of bitumen used depends upon type of sand available and atmospheric conditions. Water is not easily available in desert areas.

Sand bitumen construction consists of following steps:
- Preparation of road surface
- Heating of sand and bitumen
- Mixing and laying of hot mix
- Compacting the layer and finishing

Wearing course 2 cm in thickness is used to protect the base and rectify the ruts and undulations. The specification details are given in IRC: 55–1974.

Sheet asphalt is a mixture of well graded sand, bitumen and mineral filler, used in balanced proportion, mixed and laid hot to be used as surface/wearing course. It provides smooth, dense and impervious surface using locally available sand.

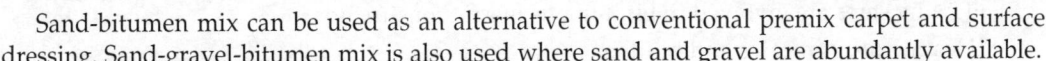

Sand-bitumen mix can be used as an alternative to conventional premix carpet and surface dressing. Sand-gravel-bitumen mix is also used where sand and gravel are abundantly available.

(x) Bituminous Concrete (Asphalt Concrete): Bituminous concrete consists of construction with a hot mix of crushed aggregates, stone dust, filler and bitumen and laid and compacted immediately after mixing. The filler material used is material whole of which passing 600 micron sieve, at least 90% passing 150 micron sieve and not less than 70% passing 75 micron sieve. Lime-stone dust, cement, hydrated lime, stone dust, fly-ash etc. can be used as filler material.

Bituminous concrete is suitable for heavy traffic roads. The construction involves following steps:

- Preparation of the existing base course layer.
- Heating of aggregate and bitumen
- Mixing hot
- Spreading the hot-mix as a layer
- Compacting the layer and finishing

The bitumen type is selected as per requirements of traffic and climate. 30/40 and 40/50 grades of bitumen are used for very heavy traffic and 60/7 and 80/100 for lighter traffic. The detailed specifications and construction technique is covered in IRC: 29–1988. The thickness of bituminous concrete, viz. 25 mm, 40 mm, 50 mm is decided on its functional requirement for the traffic on the road.

Advantages of bituminous concrete are:

- It can withstand high stresses and is good for heavy duty vehicles and large traffic intensities.
- It has good riding quality, is durable, strong, comparable to a concrete pavement.
- It provides dust free, resilient and smooth surface.
- It is best type of bituminous road construction and thickness requirement is comparatively less. Bituminous mix is designed for bituminous concrete as described article 15.8.

(xi) Semi Dense Bituminous Concrete: Semi dense bituminous concrete consists of construction of one course with a hot mix of crushed aggregates, stone dust/natural sand, and filler, premixed with bitumen, laid and compacted, immediately after mixing. It uses less fine aggregates as compared to dense bituminous concrete and thus have more voids. It is coarse graded type construction, good for heavy traffic roads in low rainfall areas. Following steps are involved in its construction.

- Heating of aggregates and bitumen
- Mixing the aggregates and bitumen
- Spreading the hot mix as a layer
- Compacting the layer

Bitumen is selected as per traffic intensity and climate of the region. Detailed specifications of Semi Dense Bituminous Concrete are given in IRC: 95–1987.

(ix) Mastic Asphalt: Mastic asphalt is a mixture of bitumen, fine aggregates and filler material, in suitable proportions to give a voidless, coherent, impermeable mass, solid or semi-solid under normal temperature conditions.

The bitumen mastic can absorb shocks, vibrations, deflections, it is voidless, posses the quality of self-healing of cracks and prevents bleeding. When cooled mastic asphalt becomes hard, stable, providing a durable layer to withstand heavy traffic. It is therefore used as a surfacing material specially for bridge decks, bus stops, roundabout, etc.

Following steps are involved in bituminous mastic construction:
- Heating of aggregates and bitumen
- Mixing and cooling of ingredients for 2 to 3 hours
- Spreading the mastic as a layer

Detailed specifications for mastic construction for road wearing courses are given in IRC: 107–1992 and for bridge decking and roads are given in IS: 5317–1969.

15.8 BITUMINOUS MIX DESIGN (MARSHALL METHOD)

Design of bituminous mixes is basically selecting and proportioning aggregates and bitumen to obtain desired properties in the constructed road.

The desired properties of the mix are durability, flexibility, stability, tensile strength, imperviousness, resistance to skidding, fracturing, and workability during construction. To achieve these properties Marshall method of mix design is commonly used. The procedure of performing this test is already described in Chapter 5. It is an unconfined compressive strength test.

The design procedure based on Marshall test involves following steps:

Step (i) Choosing the desired test values for the purpose for which mix is designed (Design Criteria)

For dense bituminous concrete used for surface course in roads, following are required values:

(a) No of blows on each face of specimen: 50
(b) Marshall stability value, in kg: Min: 340
(c) Percent voids in the mix: 3–5
(d) Percent voids in mineral aggregates filled with bitumen   75–85
(e) Marshall flow value, in mm.   2–4

Step (ii) Selection of grading of aggregates to be used in mix

For different purpose different gradings are used depending upon the availability of materials and requirements of the mix. IRC in its detailed specification for various bituminous constructions has suggested gradings of individual types of aggregates as shown in Table 15.6. Generally, aggregates of nominal size 20 mm, 10 mm, stone dust or screenings and filler are blended in suitable proportion to obtain a grading.

Table 15.6: Gradings of aggregates for bituminous mix

Sieve size	A	Grading of Aggregates B	C	D	Suggested for Dense Bit. concrete limits	Mean
20 mm	100	100	—	—	100	100
12.5 mm	20	94	—	—	80–100	90
10 mm	3	60	—	—	70–90	80
4.75 mm	—	22	—	—	50–70	60
2.36 mm	—	2	100	—	35–50	42.5
600 micron	—	—	50	—	18–29	23.5
300 micron	—	—	27	—	13–23	18
150 micron	—	—	14	—	8–16	12
75 micron	—	—	4	100	4–10	7

For dense bituminous concrete IRC recommendation percent passing, and mean values are shown in the table.

Step (iii) Producing of the selected grading to be used in mix

The proportioning of various sizes of aggregates is done by trial and error method.

Figure 15.4 shows the combined grading of aggregates for selected proportion of the bituminous concrete. Blending proportion of combined aggregates using approximations will be the following:

Group and percentage	Specific gravity
A–10	2.65
B–45	2.66
C–40	2.71
D–5	3.10

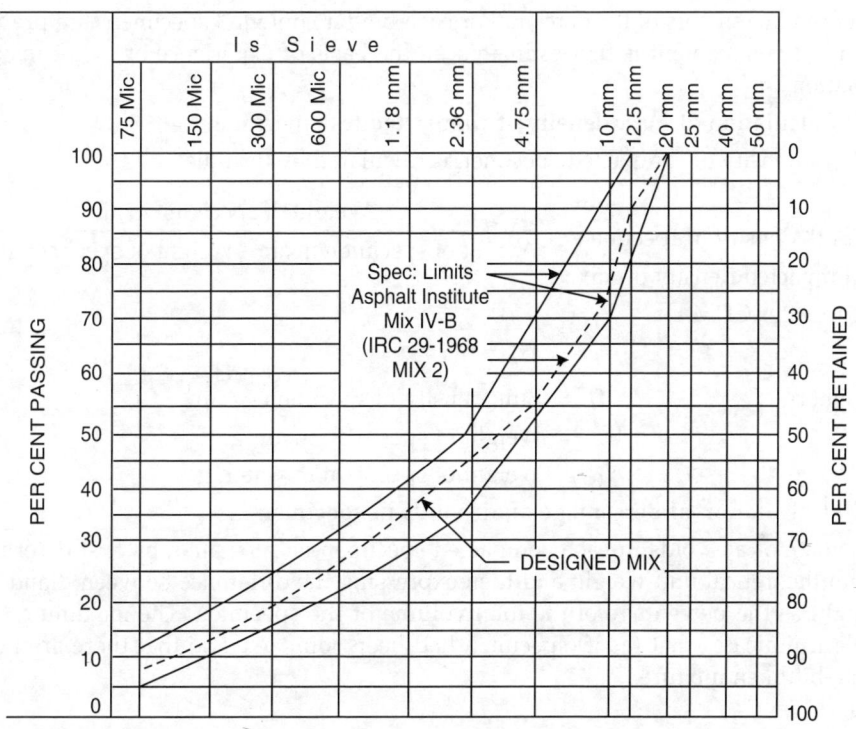

Fig. 15.14: Combined grading for bituminous concrete mix

Step (iv) Determination of specific gravity of combined aggregates

For computation of voids in mineral aggregates, average specific gravity of combined aggregates is determined as follows:

Average specific gravity of combined aggregates (ASG) = (Specific gravity of mix aggregates SGMA)

$$\text{ASG} = \frac{100}{\dfrac{\text{Percentage by Weight of A}(P_a)}{\text{Specific gravity of A (Spg. A)}} + \dfrac{\text{Percentage by Weight of B}(P_b)}{\text{Specific gravity of B (Spg. B)}} + \dfrac{P_c}{(\text{Spg.C})} + \dfrac{P_d}{(\text{Spg.D})}}$$

$$= \frac{100}{\dfrac{P_a}{\text{Spg. A}} + \dfrac{P_b}{\text{Spg. B}} + \dfrac{P_c}{\text{Spg. C}} + \dfrac{P_d}{\text{Spg. D}}}$$

Where: P_a P_b P_c P_d are percentages of aggregates A, B, C and D and Spg. are their respective specific gravities.

$$= \frac{100}{\dfrac{10}{2.65} + \dfrac{45}{2.66} + \dfrac{40}{2.71} + \dfrac{5}{3.1}}$$

$$= 2.699$$

Step (v) Preparation of specimen for Marshall test

Test specimens are prepared using different percentages of bitumen content so as to plot desired curves. The percentage of bitumen is by weight of dry aggregates varying from 4.5 to 7.5 percent in increments of 0.5 percent. For each binder content 3 specimens are prepared. The optimum bitumen content is first estimated on the basis of experience or by means of surface area equation.

Step (vi) Calculation of bulk density of compacted test specimen

The bulk density of compacted specimen is calculated by formula:

Bulk density of specimen = $\dfrac{\text{Weight of specimen in air}}{\text{Weight of specimen in air} - \text{Weight of specimen in water}}$
(Compacted density of mix aggregates CDMA)

$$D = \frac{W_A}{W_A - W_W}$$

where
$\quad D$ = Bulk density of specimen g/cm^3
$\quad W_A$ = Weight of specimen in air, g.
$\quad W_W$ = Weight of specimen in water, g.

Step (vii) Calculation of percentage of air-voids in specimen

Percentage of air voids in each compacted specimen is calculated, by first determining the maximum theoretical unit weight and then expressing the difference between it and the actual unit weight as the percentage of the total volume of the specimen. The maximum theoretical unit weight indicates that (as if) specimen had been compacted so that there are no voids in aggregate-bitumen mixture.

Thus

$$W_{tu} = \frac{W_A}{V_b + V_c + V_f + V_{mf}}$$

$$= \frac{W_A}{\dfrac{W_b}{G_b} + \dfrac{W_c}{G_c} + \dfrac{W_f}{G_f} + \dfrac{W_{mf}}{G_{mf}}}$$

where
$\quad W_{tu}$ = Maximum theoretical unit weight (g/cm^3)
$\quad W_A$ = Weight of specimen (g)
$\quad V_b, W_b, G_b$ = Volume (cm^3), weight (g) and apparent specific gravity of bitumen in specimen.

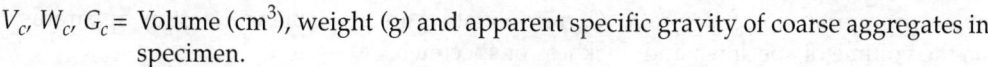

V_c, W_c, G_c = Volume (cm³), weight (g) and apparent specific gravity of coarse aggregates in specimen.

V_f, W_f, G_f = Volume (cm³), weight (g) and apparent specific gravity of fine aggregate in specimen.

V_{mf}, W_{mf}, G_{mf} = Volume (cm³), weight (g), and apparent specific gravity of mineral filler in specimen.

here W_b W_c W_f W_{mf} are weight of each component fraction of specimen determined by assuming each to be a proportional amount of the total weight of specimen and specific gravity values of the aggregates are apparent specific gravity and not the bulk specific gravity.

Percentage air voids in the compacted specimen are given by:

$$\% \text{ V.T.M.} = \frac{W_{tu} - d}{W_{tu}} \times 100$$

where

%V.T.M. = Voids in the specimen (%)

W_{tu} = Theoretical maximum unit weight (g/cm³)

d = Bulk unit weight (g/cm³)

Step (viii) Calculation of percentage voids filled with bitumen

For calculation of percentage voids in the compacted specimen filled by bitumen, first amount of voids in the aggregates (V.M.A.) is to be calculated.

V.M.A. is obtained by substracting the volume occupied by aggregates in the compacted specimen from the bulk volume of the compacted specimen, i.e., this is the volume of voids which is theoretically available for filling with binder.

$$V.M.A. = V - V_c - V_f - V_{mf}$$

$$= \frac{W}{d} - \frac{W_c}{G_c} - \frac{W_f}{G_f} - \frac{W_{mf}}{G_{mf}}$$

where

$$V.M.A. = \text{Voids in the mineral aggregates (cm}^3\text{)}$$
$$V = \text{Total volume of specimen (cm}^3\text{)}$$

percent V.M.A. will be expressed as

$$\% \text{ V.M.A.} = \frac{V.M.A.}{V} \times 100$$

The percentage of voids in aggregates which is filled with bitumen is given by:

$$\% \text{ voids filled with bitumen} = \frac{V_b \times 100}{V.M.A.}$$

Step (ix) Determination of Marshall stability and flow values

For each specimen marshall stability is determined by testing in the compression testing machine, where load is applied at a constant rate of strain of 50.8 mm/min. The maximum load in kg at which the test specimen fails is recorded as "Stability Value", while the stability test is in progress a dial gauge is used to measure vertical deformation of specimen. The deformation read at the load failure point is the "Flow Value" of the specimen.

Step (x) Correction for measured stability values

Measured stability values are corrected to those which would have been obtained if the specimen was exactly 63.5 mm high. This is done by multiplying each measured stability value

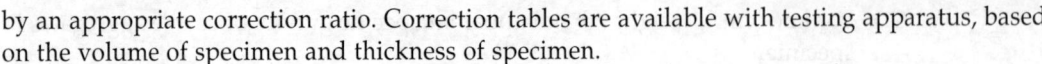

by an appropriate correction ratio. Correction tables are available with testing apparatus, based on the volume of specimen and thickness of specimen.

Step (xi) Plotting of curves and determination of optimum bitumen content

Following graphical plots are made:

(a) Bitumen content vs corrected Marshall stability

(b) Bitumen content vs Marshall flow values.

(c) Bitumen content vs percentage of voids in the total mix

(d) Bitumen content vs percentage of voids in the mineral aggregate filled with bitumen

(e) Bitumen content vs unit weight

Points are joined with a smooth curve, to fit in the experimental data. From these plots, values of bitumen contents required for following are read:

(i) Maximum unit weight

(ii) Maximum stability

(iii) 4% air voids

(iv) 8% voids in mineral aggregates filled with bitumen. Optimum bitumen content is average value of above four, binder contents. Typical relationship between Marshall test parameters as above are shown in Fig. 5.29 and described in following paragraphs.

Step (xii)

Check that optimum bitumen content gives a mix which will meet the Marshall design criteria. An example on the next page shall explain the calculations mentioned above.

The desired properties of the mix at optimum bitumen content are:

Criteria	Properties at optimum bitumen content
1. Marshall stability (kg)	885
2. % air voids	3.6
3. Percent voids in mineral aggregates filled with bitumen	79.5
4. Flow value (mm)	3.4

Marshall test results show following characteristics:

- The unit weight of the mix increases with increasing bitumen content until a maximum value is obtained, after which unit weight decreases. At first bitumen acts as a lubricant and helps the aggregates particles to slide over each other. After optimum, it starts displacing the particles. For dense mixes therefore it is important that amount of binder should not exceed the optimum for the compaction used.

- The stability value of the mix also increases with increasing bitumen content until a maximum value is obtained, after which stability decreases. At maximum stability best internal friction, interlocking of aggregates and cohesion is provided.

- The flow value increases as bitumen content increases the rate of deformation change is slow at low bitumen contents, but increases rapidly at higher bitumen content. Surfacings with low flows and high stability will not deform easily under traffic.

- The percentage of voids in the total mix decreases with increasing bitumen content until a value is reached at which it begins to level off. When air void content is too low, the surfacing may 'Bleed' under traffic.

- The percentage of aggregate voids filled with bitumen increases with increasing bitumen content, until a maximum value is reached. Rate of increase is fastest at low bitumen content and levels off at high contents. When surfacing is deficient in bitumen content, it may become brittle, and crack early and ravel under traffic.

Example: A bituminous mixture for road surfacing project has been prepared using the following aggregates:

Coarse aggregate   percent by weight 60%   specific gravity = 2.6
Fine aggregate   percent by weight 30%   specific gravity = 2.65
Filler material   percent by weight 10%   specific gravity = 2.70

The bitument content used is 7.5% and specific gravity of bitumen is 1.04.

A field sample cored from the compacted layer weighted 3680 gms in air and 2060 gms in water.

Evaluate the mixing and compaction achieved in the field.

Solution:

Following calculations are to be done:

(1) Specific Gravity of Mixed Aggregate (SGMA)

$$= \frac{100}{\left(\dfrac{W_1}{G_1} + \dfrac{W_2}{G_2} + \dfrac{W_3}{G_3}\right)}$$

$$= \frac{100}{\left(\dfrac{60}{2.60} + \dfrac{30}{2.65} + \dfrac{10}{2.7}\right)}$$

$$= \frac{100}{37.9} = \mathbf{2.62}$$

(2) Specific Gravity of Mix (SGM)

$$= \frac{100 + \text{Bitument Content (\%)}}{\dfrac{100}{SGMA} + \dfrac{\text{Bitument Content}}{\text{Sp. gr. of Bitumen}}}$$

$$= \frac{100 + 7.5}{\dfrac{100}{2.62} + \dfrac{7.5}{1.04}}$$

$$= \frac{107.5}{45.38} = 2.37$$

(3) Compacted Density of Mix (CDM)

$$= \frac{\text{Weight of specimen in air}}{\text{Weight in air} - \text{Weight in water}}$$

$$= \frac{3680}{3680 - 2060}$$

$$= \mathbf{2.27}$$

(4) Compacted Density of Mix Aggregates (CDMA)

$$= \frac{CDM}{1 + \dfrac{\text{Bitumen content}}{100}}$$

$$= \frac{2.27}{1 + \dfrac{7.5}{100}}$$

$$= \frac{2.27}{1.075} = \mathbf{2.11}$$

(5) Voids in Compacted Mix (VIM)

$$= \frac{SGM - CDM}{SGM} \times 100$$

$$= \frac{2.37 - 2.27}{2.37} \times 100$$

$$= \frac{0.1}{2.37} \times 100 = \mathbf{4.2\%}$$

(6) Voids in Mixed Aggregates (VMA)

$$= \frac{SGMA - CDMA}{SGMA} \times 100$$

$$= \frac{2.62 - 2.11}{2.62} \times 100$$

$$= \frac{0.51}{2.62} \times 100 = \mathbf{19.5\%}$$

(7) Voids filled with Bitumen (VFB)

$$= \frac{VMA - VIM}{VMA} \times 100$$

$$= \frac{19.5 - 4.2}{19.5} \times 100$$

$$= \frac{15.3}{19.5} \times 100 = \mathbf{78.5\%}$$

Since voids filled with bitumen are 78.5%, mixing done is satisfactory.

(8)

$$\frac{CDM}{SGM} = \frac{2.27}{2.37} \times 100 = \mathbf{95.8\%}$$

Since relative compaction achieved in field is more than 95% the degree of compaction is satisfactory.

15.9 LOW VOLUME LOW COST ROADS (RURAL ROADS)

For roads of low traffic volume, surfaces of untreated soil mixture like earth roads, gravel roads, crushed rock roads are commonly used. They consist largely of locally available materials such as lime rock, shells, crushed rock combined with clay etc. Fine material is used to bind the particles together. In rural areas and where traffic volume is low (possibly in range 100 to 250 vehicles per day), low cost earth and gravel roads making use of locally available materials are used for village connectivity. As traffic volumes increase surface pitching, formation of corrugations, dust formation occurs and cost of rectifying these mitigate against their use. Low cost means that road should have low construction and low maintenance cost. The word 'Rural' and 'Low Cost' are synonyms.

Requisites of untreated soil-mixed low cost roads are:

(i) Stability—i.e. they must support the superimposed loads without deformation.

(ii) They must stand abrasive action of traffic.

(iii) They should be able to stand rains which falls on the surface, maintaining stability of surface or softening of sub grade.

(iv) They should posses capillary properties to replace moisture lost by surface evaporation.

(v) The size should not be large, so that blading and dragging operations can be undertaken.

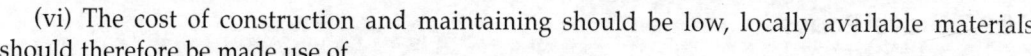

(vi) The cost of construction and maintaining should be low, locally available materials should therefore be made use of.

(vii) Using their geometry it should be possible to upgrade/strengthen these roads in future with increase in traffic.

Specifications and materials, for construction of low cost roads should meet above requirements keeping in view the availability of man, material and equipment. Quality control measures that can be effectively used should be followed.

Sound engineering principles should be adopted in the design and construction of low cost roads in order to secure long term economy. Standards adopted for low cost road should be such that it may be possible to upgrade the roads to higher standard and specifications without involving too much wastage. Stage construction techniques should be profitably adopted. At early stage pavement may be made entirely using locally available material, but in the future it should be possible to use the road as a part of sub base or base course for higher type of pavement, designed to carry heavier traffic.

Common types low cost roads are:

(1) Earth roads

(2) Gravel roads

(3) Kankar roads

(4) Soil stabilised roads

(5) Traffic bound Macadam road

15.9.1 Earth Roads

Earth roads are constructed by shaping the native soils to the cross-section of the finished road. Locally available soils can be mechanically stabilised for better quality of earth roads in rural areas.

The operations involved in construction of earth roads are:

(i) Clearing the site

(ii) Grading of the earth (filling or excavating)

(iii) Compacting with 8 to 10 tonne roller at 95% density at optimum moisture content.

(iv) Drying of surface and checking camber etc. Earth roads require regrading and reshaping every year after rainy season. An earth road is the lowest and cheapest type of road.

The material used for making earth roads should be free from logs, stumps, roots, rubbish or any other ingredient likely to affect its stability. Highly expansive clays exhibiting marked swelling and shrinking properties should not be used. The quality and character of clay or other fines is important as it serves both a binder and a moisture retainer. Table 15.7 shows physical properties of an earth material suitable for making earth roads.

Table 15.7: Physical properties of soil for earth roads

Physical property	Base course material (percent)	Wearing course material (percent)
Sand content	60-80%	65-80%
Silt content	9-32%	5-15%
Clay content	> 5%	10-18%
Liquid Limit (L.L.)	25%	35%
Plasticity Index (P.I.)	< 6%	4-10%

15.9.2 Gravel Roads

Gravel roads can cater for about 100 tonnes of pneumatic tire vehicles and an average daily traffic of about 350 to 400 vehicles/day.

Gravel is a mixture of stones, pebbles, sand, and fine sized particles which can be used in specified gradings for road constructions as sub-base or base. When required grading is not available naturally, blending of the materials in required proportion is done. Gravel should be hard, tough and durable.

Gravel roads require proper selection of material and their grading, adequate drainage, regular maintenance and lateral support from the shoulders. Proper compaction of subgrade is important for gravel roads, so that when traffic demands call for upgradation or strengthening of the existing road, it can be done on the same foundation (sub grade).

There are two methods of construction for a gravel road:

(i) Trench method: In which gravel is put on the prepared subgrade, in a shallow trench. Figure 15.15 shows the trench type construction method. Shoulder are constructed of embankment material. Draining of water is a problem in this method because of side of the trenches, which however provide better confinement for gravel.

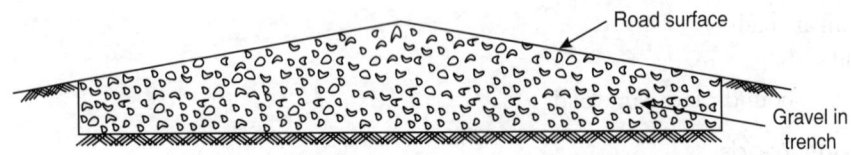

Fig. 15.15: Trench method of gravel road construction

(ii) Feather edge method: In this method gravel is spread on the prepared subgrade, thick at the centre and thin at the shoulders/edges. This type of construction is better from drainage point of view as surface water is easily drained out. Figure 15.16 shows feather edge type method of construction of gravel roads. In two layer type of construction it is better to lay first layer in the trench and second by feather edge type of construction.

Fig. 15.16: Feather edge method of gravel road construction

Gravel roads are generally built in two courses (layers).

Following steps are involved in construction of the gravel roads:

(i) Clearing of the site.

(ii) Preparation of the subgrade, and compacted to 100% maximum dry density by Proctor test.

(iii) Spreading of the gravel on prepared subgrade and compacting by rollers of light weight, from edges towards centre.

(iv) Second layer of the gravel is put after rolling the first one, water may be sprinkled during rolling for better consolidation.

(v) Camber should be checked as specified. A layer of fine aggregates passing ISS 4.75 mm is provided before opening to traffic.

(vi) Thin bituminous surface as dust palliative may be used for dust control.

Gravel roads become dusty in dry weather and during rainy season. Proper drainage is required to keep them usable. Grading and reshaping of gravel road by removing pot holes, depressions and ruts is regularly done, to provide desired level of serviceability.

15.9.3 Kankar Roads

Kankar which is used in low cost roads should be hard, tough and free from sand or earth. Like gravel kankars are spread in 15 cm thick layers over the prepared subgrade or brick or kankar soiling. To confine the metal and preventing spreading under traffic, mud/clay walls may be made on the edges of the road. Kankar is available as a natural deposit either in form of nodules or block. It is an impure form of lime stone.

After spreading Kankar, water is flooded and ramming is done until the surface is thoroughly compacted, light weight rollers can be used for compacting and better consolidation.

Surface dressing on Kankar roads with tar or bitumen is also done, to increase the life of road and handle heavier traffic.

15.9.4 Soil Stabilised Roads

Different soils are mixed in a desired proportions so that resultant soil is more resistant to weathering, and traffic loads. Stability of soil is increased by proper grading. Cement, lime, bituminous materials can be used as stabilisers or binders for improving engineering properties of soils. Stabilisation of soils is described separately in detail.

15.9.5 Traffic Bound Macadam Roads

Just like gravel roads, stone aggregate are put on the prepared subgrade in suitable thickness, and rolling or compacting is done by the traffic moving on it. Use of rollers for compacting is avoided in this type of construction. Gravels, lime stone, and other granular material like slag can be used for these roads.

In this type of construction the first course of material is driven into subgrade which provides the internal friction necessary to support the traffic load. Subsequent layers are compacted by traffic giving good surface. This road is suitable for light to moderate traffic.

Generally lime stone, with following grading is used:

Sieve designation	Percent passing (by weight)
25 mm	100
20 mm	90–100
10 mm	20–55
No: 4	0–10
No: 8	0–5

As road traffic increases, rural roads are upgraded to modern type roads, described earlier, or using soil stabilisation techniques, by effective utilisation of local soils and suitable stabilising agents. Soil stabilisation deals with physical, physico-chemical, and chemical methods to make the soil serve its purpose as a pavement material.

15.10 STABILISATION

When natural occurring soil and other materials on which pavement is to be constructed are poor in quality, either local material should be replaced by better material brought from other places or the existing material should be improved in quality by certain treatments, to enhance its ability to perform its pavement functions. A variety of techniques are available for this purpose through soil stabilisation. Any treatment used to improve the strength characteristics of a soil by reducing its susceptibility to influence of water and traffic, can be called "Soil Stabilisation". Soil stabilisation implies improvement of soil so that it can be used for sub bases, bases, and subgrade.

Soil stabilisation can be achieved by:

(i) Proportioning various locally available soils and aggregates suitably and compacting the mixture to serve desired objective called "Mechanical Stabilisation".

(ii) By addition of suitable cementing agent like cement, lime and lime-flyash to the soil.

(iii) By adding a stabiliser like lime in small proportion to modify the undesirable characteristics in certain soils.

(iv) By adding water proofing or repelling agents to mitigate the effect of water, on the soil.

(v) By adding retarding agents like calcium chloride to a non-cohesive soil.

(vi) Heat treatment called 'Thermal Stabilisation' has useful effect on some clayey soils.

(vii) There are several chemicals, which when added single or in combination, may impart useful changes in certain type of soils.

Considerable investigations and laboratory tests are needed before selecting a particular method of stabilisation.

Soil Stabilisation Methods

Methods of soil stabilisation can be classified as:

1. Mechanical stabilisation
2. Cement stabilisation
3. Lime stabilisation
4. Bituminous stabilisation
5. Chemical stabilisation
6. Other methods of stabilisation
   (a) Electrical stabilisation
   (b) Thermal stabilisation
   (c) Complex (Two-stage stabilisation)

After substandard material is upgraded to 'Normal' through any method of soil stabilisation, then use of conventional methods of design and construction of pavement are used. Overall cost to obtain satisfactory improvements in soil, by treatments to stabilise is considered in selecting the method of stabilisation.

Figure 15.17 shows areas of maximum efficiency for soils and stabilisers.

15.10.1 Mechanical Stabilisation

There are two basic types of soils: Granular soils and cohesive soils. Properties of these soils are:

Granular soils	**Cohesive soils**
Like Sand, moorum, gravel,	Like Clay
(4.75 mm to .075 mm)	(1.075 mm to .002 mm and less)
They have:	They have:
• Internal friction	• Cohesion
• Strength	• Better moisture holding but weak, in strength
• Hardness	• Better binding property

- High unit weight
- Excellent drainage properties
- Low capillarity
- Do not undergo large changes when wet

- Low unit weight
- Poor drainage property
- High capillarity
- Have too much shrinkage, swell with moisture change

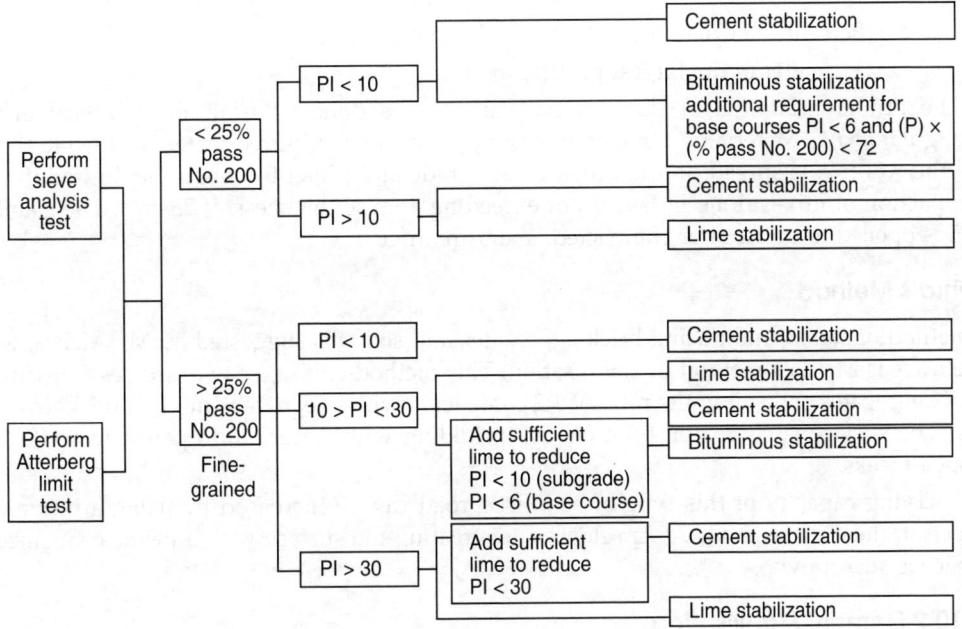

Fig. 15.17: Soil characteristics and stabilisation

Strength of a soil, depends upon cohesion and internal friction. Both the above type of soils independently are incapable of carrying loads as granular soils like sands lack cohesion and are spread laterally under vertical loads, while clays when wet swell, become muddy and lack strength.

Proper combination of these two soils in specific proportion (grading) gives the required qualities of strength and cohesion. This process is called "Mechanical Stabilisation". In mechanical stabilisation cohesion is provided by clay or silt and internal friction is achieved by coarser particles. A proper blend of these two fractions provides the property of resisting the lateral displacement under load.

Additives like chlorides, lignin, molasses are also used in stabilisation. Freezing or heating the soil to improve its properties is also a type of mechanical stabilisation.

Mechanical stabilisation has been successfully used for sub-base and base courses in roads. In low cost roads it is also used as a surface course, when traffic volume is not high as in villages, and where rainfall is low.

Following are the recommended values of the liquid limit and plasticity index for the material passing 425 micron sieve, to be used for mechanical stabilisation.

	Base course	Surface course
Liquid limit (LL)	25%	35%
Plasticity Index (PI)	6% max.	5 to 10%

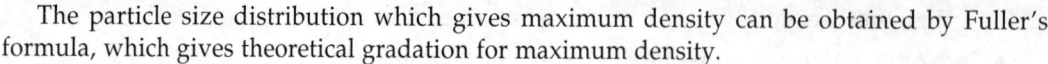

The particle size distribution which gives maximum density can be obtained by Fuller's formula, which gives theoretical gradation for maximum density.

Fuller's formula is:

$$P = 100 \left(\frac{d}{D}\right)^{1/2}$$

where

P = Percent finer than diameter 'd' in the material and

D = Diameter of the largest particle in mm

The proper gradation of the material can also be done by trials or graphical method (triangular chart). Materials are mixed in desired proportion, required water is added to the mix and the wet mix is spread on the prepared subgrade and rolled to obtain the desired density. Compaction of mix is done in layers not exceeding a loose thickness of 25 cms. The stabilised road is opened to traffic after compacted layers are dried.

Mehra's Method

A method of stabilisation using brick aggregates and soil was suggested by Mr. Mehra, which is known as Mehra's method of stabilisation. This method envisages two courses consisting of brick aggregates and soil in the ratio of 1:2 compacted with adequate water to a thickness of 7.5 cms. The surface is cured for 4-5 days by sprinkling with water, and opened to traffic after about 2 weeks.

Carrying capacity of this type of stabilised road can be increased by using a bituminous carpet on the surface about 2.5 cm thick, after treating the surface with a prime coat since the surface is adsorptive.

15.10.2 Cement Stabilisation

Stabilisation of soil with cement consists of adding portland cement to a pulverised soil and permitting the mixture to harden.

In cement stabilisation hydration of cement occurs due to cement-soil interaction. The hydrated cement fills the voids in the soil. When water is added to cement, hydration produces calcium silicate, hydrated calcium aluminate, and hydrated lime.

Cement stabilisation is most popular method of stabilising soils due to following reasons:

(i) It can be used for almost all type of soils, by using enough amount of cement, water and proper compaction and curing for both the sub-base and base course.

(ii) Cement is available easily as it is manufactured in plenty.

(iii) Relatively cost of cement is not very high.

(iv) Lesser quality control and care is required with cement stabilisation as compared to other methods.

(v) Behaviour of soil-cement mixture can be easily known through tests in the laboratory and field.

The strength developed depends upon the quantity of cement and intimacy with which soil particles are mixed with cement. As it has poor resistance to abrasion and impact, it can not be used as surface course. A bituminous wearing course is necessary over the base course.

Soil-cement stabilisation is affected by the following factors:

(a) Nature and type of soil

(b) Quantity and type of cement

(c) Water content

(d) Mixing and compaction

(e) Curing

(f) Additives (if any)

(a) Nature and Type of Soil: Soil stabilisation with cement can be used with any type of soil, but required cement content increases with silt and clay content in soil, making its use uneconomical in heavy clays. Poorly graded soils also require high quantity of cement. A well graded soil, is well suited for cement stabilised construction. Organic matter present on the surface, should be completely removed for stabilisation with cement. Most of the coarse-grained soils and many fine grained soils with clay content not exceeding about 30% can be stabilised successfully by cement. Peaty soils, heavy clays, highly organic soils and soils containing salts of sulphate can not be successfully stabilised with cement.

Table 15.8 shows grading of materials for stabilisation with cement. Gravelly, sandy, clayey type of soils can be stabilised using cement. It is the only method used in water logged and high rainfall areas.

(b) Quantity and type of cement: Portland cement is commonly used for soil stabilisation, though any type of cement can be used. Minimum quantity of cement depends upon the minimum compressive strength required at some minimum compaction. Cement should comply with the requirements of IS–269, 455 and IS: 456–2000.

High cement content of the order of 7 to 10% produces hard mass, and smaller percentage of 2–3% improves strength and C.B.R. value of the soil.

For gravelly or sandy soils 6 to 10% of cement, for well graded sandy clays 9 to 10% of cement, for silty soil 8 to 12% of cement and for clayey soils 10 to 14% of cement is usually required. Higher the clay content, more cement is needed.

Table 15.8: Gradation of material for stabilisation with cement (IRC)

IS Seive Designation	Percentage by weight passing	
	Sub-Base (Finer than)	*Base (Within the range)*
53 mm	100	100
37.5 mm	95	95–100
19.0 mm	45	45–100
9.50 mm	35	30-100
4.75 mm	25	25-100
600 micron	8	8–65
300 micron	5	50–40
75 micron	0	0–10

The cement requirement for satisfactory stabilisation of soils increases with specific surface area of the soil. Well graded soils require less cement than poorly graded or fine grained soils, with high clay content.

Cement should comply with Indian standard requirements. The quantity of cement is determined on the basis of unconfined compressive strength and durability test under 12 cycles of wet-dry conditions.

(c) Water Content: Water is needed in soil cement mixture to hydrate the cement, improve workability and to facilitate compaction. Water used should be clean and free from alkalies, acids, organic and deleterious matter. Potable water is generally used.

The amount of water present in soil cement mixture has influence on its strength and unit weight. If mixture is not fully hydrated the strength will be reduced. Optimum moisture content for maximum density at given compactive effort for a particular type of soil (Sandy, clayey, etc) should be used. It is determined by laboratory experiments.

(d) Mixing and Compacting: Depending upon the type of the soil, it should be specified that how long water will be added to the soil-cement mixture and when compaction will be completed. Increasing the water mixing time and delaying compaction after water mixing, reduces the durability and unconfined compressive strength. The soil is watered during mixing as required. Water should be well distributed throughout the soil, specially during hand mixing so that there will be no dry places where cement can not hydrate. The success of stabilisation depends upon degree of pulverisation and mixing in the field.

Compaction is done little above the optimum moisture content which gives highest density. Tests should be conducted to arrive at the quantity of water to be used. Soil cement should be placed in position soon after mixing for compacting. Good compaction is necessary for high strength.

(e) Curing: Soil-cement must be cured during the initial stage with water enough for hydration of cement in the mixture. Requirement of curing will depend upon the environmental temperature and the weather, for preventing drying of the surface during initial period of development of the strength. Curing can be done by covering it by wet soil, straws or gunny bags and keeping them moistened for at least 7 days after which water sprinkling is done for 28 days. The mixture sets in 7 to 28 days to give a hard and resistant surface.

(f) Additives: (if any) Additives are sometimes used in specified amounts of soil, cement and water, for certain reasons. For example hydrated lime or calcium chloride are added to remove detrimental effects in organic soils. Small quantity of lime may be added to highly plastic soils to facilitate pulverisation. Certain chemicals are added to soil-cement mix to increase strengths, or reduce required quantity of cement. Flyash is often used for this purpose.

Soil-cement mix may be designed based on compressive strength of soil-cement specimen, as required for heavy traffic (28 to 35 kg/cm^2) to light traffic (17.5 to 20 kg/cm^2).

Steps in Cement/Lime Type Stabilisation of Soils

Following steps are used in mix-in place type of construction.

(i) Pre-preparation: Before beginning the actual construction works, subgrade or sub-base should be prepared by compacting and bringing it to proper shape.

(ii) Pulverization: Scarifying and pulverizing of the soil may be necessary in certain soil types. When loose soil is in suitable condition, it is bladed and required shape, crown and grade is provided.

(iii) Spreading cement: Cement can be spread mechanically or by hand. Machine spreading is uniform and quick.

(iv) Dry mixing: Soil and cement are mixed dry to distribute the cement through out the soil mass. Cement balls are not formed when water is added due to this dry mixing. Dry mixing is continued until the mixture has a uniform colour.

(v) Addition of water and wet mixing: Water in required amount is now added as per design and material is remixed as required, to obtain uniform colour and texture.

(vi) Compacting: After ending of the mixing, compaction should start as early as possible. After addition of the cement it should be completed within a short time. It should be ensured that at edges of the pavement and transverse and longitudinal construction joints, full compaction is achieved. Self propelled smooth tandem rollers provide a smooth surface after rolling.

(vii) Finishing: After completion of the compaction, the finishing process involves the removal of all surface irregularities. Final shape is given to the compacted layer.

(viii) Curing: The finished surface should be cured for the specified period, by any curing method.

Soil-cement roads perform well for moderate to heavy traffic. Bituminous surface can be used on a soil-cement road. IRC: 50–1973 describes use of cement modified soil in road construction.

During the entire construction operation, proper quality control is to be kept at various stages described above.

(ix) Surfacing: A priming coat is provided before the final surfacing is laid.

15.10.3 Lime Stabilisation

Stabilisation with lime refers to the use of hydrated or slaked lime, calcium hydroxide or calcium oxide (quick lime). Lime stabilisation is used in clayey soils to reduce its moisture sensitivity, plasticity and provide strength making it more friable, and increasing CBR. As unlike cement, lime does not stabilise rapidly and, delayed compaction is not as big a problem as in cement stabilisation.

Flyash a waste product of thermal power plant when added to lime, it helps in reducing the cost and providing higher strengths in lesser time. It expedites the construction.

When lime or a mixture of lime and fly ash is added to a soil, following reactions take place:

(i) Lime and moist clayey soil, when mixed in loose condition, a flocculation effect occurs, which lowers liquid limit of soil and raises plastic and shrinkage limit. This alteration in soil characteristics occurs, because of a cation-exchange reaction between lime and clay particles, resulting in soil aggregation.

(ii) A chemical reaction occurs when lime is added to moist cohesive soils, which results in slow and long-term cementing together of the soil particles. This may be due to pozzolanic reaction between lime and alumina and silica present in soil, which form hydrous calcium aluminate and silicate. Soils such as coarse silt and sand, which contain low silica and alumina, react very little with lime.

IRC-51–1992, gives detail, specifications for soil-lime mixes. IRC: 88-1984 describes construction of lime-flyash stabilised buses. Pozzolanic reaction depends upon type of soil and quantity and type of lime used. There is an optimum amount of particular type of lime to be used for the given soil, depending upon amount of clay present in soil.

(iii) Carbon dioxide from the air and rain water converts into calcium and magnesium oxides and hydroxides back to their respective carbonates. This reaction is called carbonation. Strength due to this reaction is very low.

Soils having medium plasticity index between 5 and 20 and some clays which do not respond to lime alone are improved by lime-flyash stabilisation method. The procedure of construction of lime stabilised surface is similar to soil-cement stabilisation as described earlier except that in lime stabilisation reaction is relatively slow and the time allowed between the addition of cementing agent and the compaction of the mixture is not critical with lime-soil and lime-soil-flyash as it is with soil-cement. IRC: 49–1973 and IRC: 51 provide guidelines for use of lime in stabilised road construction.

Lime stabilisation is used for silty clays and blacks cotton soil. In soils having a plasticity index of greater than 8 lime stabilisation is recommended. Good quality and pure lime (IS: 1514–1990) having calcium hydroxide content of more than 70% is normally used for stabilisation. The quantity of lime depends upon the type of the soil and is added as percentage by weight of the dry soil. The optimum quantity to be used is decided by tests in the laboratory and desired test value. Medium and heavy clays having PI of at least 10 and containing at least 15% of material finer than 425 micron are suitable for lime stabilisation. Proper quantity of lime may be determined based on required stability.

15.10.4 Bitumen Stabilisation

Various bituminous materials such as bitumen, cut-back, road oil, low viscosity bitumen, emulsion, etc. can be used to stabilise a wide range of soil types. The basic principles in bituminous stabilisation is water proofing and binding.

Bitumen stabilisation is usually used with more granular materials where mixing is easy. Bitumen stabiliser act predominantly as glue like cohesion agents in granular soils and as water proofers in clayey soils. Combination of cementing of soil particles by gluing (binding) and water proofing, occurs together in stabilising the soil.

Sand-bitumen mixtures may use penetration grade bitumen, tars, cut-backs, or emulsions for stabilisation. Soils to be stabilised should not contain organic matter or acidic matter as it will be detrimental to the stability of a soil-bituminous mixture. Bituminous materials in liquid form are suitable for stabilising cohesive soils. Best results are obtained by medium and slow curing cut-backs. Earth roads, are improved by spraying the dry soil surface with a bituminous stabiliser like a cutback, oil, emulsion or tar, in small quantities.

The bituminous stabilised layer may be used as a sub-base, or base-course for roads with low traffic. It can be used as surface course also. The various factors on which the properties of the stabilised surface depend are soil type, proportion of bituminous material, mixing, compaction during construction and additives (if any).

15.10.5 Chemical Stabililization

When materials of suitable type are not available or their transportation is too costly chemical stabilisers, for improving engineering properties of soils are used:

(a) Sodium silicate as stabiliser: Sodium silicate is available in various forms. The use of sodium silicate as a soil stabilising agent is due to its ability to react with soluble calcium salts in water to form cementing agents composed of insoluble gelatinous calcium silicates.

Various sodium silicates are used as secondary additives to improve the strength and durability of non-plastic soils stabilised with cement.

In desert areas where soil is sandy 2 to 3 percent of sodium silicate is used for improving soil qualities.

(b) Ammonium, Calcium and Sodium Chloride as stabiliser: When Ammonium chloride is added to a soil a rapid cation exchange reaction takes place between the organic cationic compounds and the inorganic cations on the clay surfaces. This tends to flocculate the clay particles and lower the plasticity index, lessing the ability of the clay to take up moisture.

Calcium chloride ($CaCl_2$) and Sodium chloride ($NaCl$) absorb moisture from the atmosphere. This quality is used in sandy soils as rate of evaporation of water is reduced and compaction becomes easy. Calcium chloride is a by-product of the manufacture of sodium carbonate by the ammonia-soda process.

Calcium chloride comes in flake, pellet and granular forms. In areas of high relative humidity calcium chloride acts as dust palliative as it holds the moisture to bind the road surface together. Calcium chloride as a dust palliative is most effective where the soil is clayey rather than silty or sandy. Calcium chloride is mixed with surfacing material on the road, by about 1 to 2% of weight of aggregates. Mixture is then easily compacted by conventional rollers.

Calcium chloride increases the surface tension, viscosity and lubricating properties. Sodium chloride (NaCl) is hygroscopic. It absorbs and retains moisture. It is also used for stabilisation of base and surface courses like calcium chloride.

(c) Resinous stabilisers: Common plastics are composed of synthetic resins and filler materials. 'Vinsol Resin' and 'Rosin' are two natural resinous agents which have water proofing characteristics.

Vinsol resin ($C_{27}H_{30}O_4$) is made from the residues obtained after distillation of pine tree stumps in the manufacture of turpentine. Resins are supposed to waterproof the soil, and help to preserve the stability and strength. There is an optimum amount of admixture for a given soil.

'Calcium Acrylate' is a powder which has the advantages of mixing with soil and dispersed throughout the soil. This is 30 per cent soluble in water. when a catalyst (ammonium persulphate) and an activator (sodium thiosulphate) are added to the solution, polymerization takes place, resulting in the formation of water insoluble gel which imparts rubber like properties to the soil. When dried, hard rigid stabilised soil is formed. This is a costly process and lasts for short time only.

(d) Geotextiles as stabiliser: Synthetic material in form of sheets provides water tightness and is used to improve quality of soils. Jute fabrics, coir mats are type of geotextiles used for this purpose. Geotextiles may be of different forms: Woven, non-woven, knitted, composite, etc.

Geotextile when interposed between two layer of soils it prevents intermixing of the two layers which may be of dissimilar materials, and also prevents pumping of soil particles. It is therefore commonly used over soft soils in base and sub-base. Geotextile material prevents subgrade material from penetrating into the aggregate base. The permeable nature of geotextile material removes the water from the road and prevents pumping action.

15.10.6 Other Methods of Soil Stabilisation

Some new methods of soil stabilisation are on experimental and trial stage. Some of them are:

(a) Electrical stabilisation: By passing electric current properties of soil are improved. This technique is also called electro-osmosis. The electrodes get decomposed by passing of the current and get deposited in soil pores, resulting in dense mass, and increase in shear strength.

Electric current results in a process of exchange of ions, which re-arranges soil particles to give dense and strong surface.

(b) Thermal stabilisation: By heating the soil, its properties are improved. In this method soil to be stabilised is sintered by gaseous products of fuel combustion. Heating of the soil is done to a temperature high enough to cause the necessary changes in the soil. Soil characteristics changes are achieved mainly by infiltration of compressed heated air.

Thermal stabilisation is a costly technique and is restricted to conditions in which other methods of stabilisation are not possible. Use of this techniques is done in Russia for stabilisation of road embankment.

(c) Complex stabilisation: When more than one method or material is used for improving the qualities of soil, it is called complex stabilisation. For example in water logged areas a combination of lime and bitumen or cement and bitumen method of stabilisation is used.

In heavy clays, the clay is first treated with lime to reduce plasticity and to facilitate pulverisation. In second stage, the resulting soil is stabilised with cement, bitumen, lime or lime-flyash as usual.

15.11 CONSTRUCTION OF RIGID (CEMENT CONCRETE) PAVEMENTS

15.11.1 Materials

Ordinary cement which is called Portland cement, (because of its resemblance to a rock in British island of Portland) is used for making cement concrete roads. Cement mixed with water forms a paste and when used with aggregates to form concrete, it hardens and binds the particles together, to form a hard, durable mass. Concrete is a mixture of cement, aggregates and water.

(a) Cement: Cement used for construction of roads should be of good quality, meeting the requirement as per IS: 456-2000, based on following tests:

- Compressive strength test
- Soundness test
- Setting time and
- Fineness

There are following three grades of ordinary portland cement, based on 28 days compressive strength in MPa, i.e. Grade 33, Grade 43, and Pozzolana cement Grade I. For cement grade 33 and grade 43, properties are:

	Cement Grade 33	Cement Grade 43
1. **Compressive strength of cement sand mortor not less than**		
72 ± 1 hour	16 MPa	23 MPa
168 ± 2 hour	22 MPa	33 MPa
672 ± 4 hour	33 MPa	43 MPa
2. Fineness (sq.cm/gm) not less than	2250	2250

For Portland Pozzolana Cement (PPC) Grade I (Fly-ash based) properties are:

	Portland Pozzolana cement I
1. **Compressive strength of cement sand mortar not less than**	
168 ± 2 hrs	22 MPa
672 ± 4 hrs	33 MPa
2. Fineness sq. cm/gm not less than	3000
3. Flyash/calcined clay	10-25 percent

Cement used for construction of concrete roads should meet standards as above.

(b) Aggregates: Mineral aggregates form about 75% of the volume or about 80% of the weight of normal concrete used for pavements. Aggregates used in cement concrete roads are classified as fine (passing 4.75 mm sieve) and coarse (retained on 4.75 mm sieve). Fine aggregates are generally natural sand from river beds, or obtained by crushing stone and gravel. Coarse aggregates are either gravel or crushed stone. Indian Road Congress specifies that maximum size of aggregates should not exceed 1/4th of the pavement thickness, and should not contain more than 0.5 percent of sulphate as SO_2 and should not absorb water more than 2 percent of their own weight.

Aggregates used for construction of concrete roads should be hard, durable, clean, and free from dirt, alkali, vegetable matter, and deleterious substances. The maximum size of aggregates commonly used is 20-25 mm for concrete roads, satisfying following strength requirements:

Property	Used in wearing course	Other than wearing course
Aggregate impact value	$\not> 30$	$\not> 45$
Los Angles abrasion value	$\not> 30$	$\not> 50$

Rounded or cubical aggregates are better than flat or elongated ones. Sand should be clean, sound, properly graded and free from organic matter.

(c) Water: Water for making cement concrete should be free from harmful acids, oils, alkalis, salts and organic material. Its pH value should be between 6 to 8. Potable water is suitable for mixing and curing of concrete.

The materials used for concrete roads should conform to the requirements of Bureau of Indian Standards and IRC: 15 and IRC: SP-49. Concrete mix using the selected materials is designed as per IRC: 44 "Guidelines for Cement Concrete Mix Design".

(d) Admixtures: Many admixtures (substances) are added to change the properties/ characteristics of the concrete. 'Air Entraining' admixture is very commonly used. Air entrainment is the entrapment of air in concrete in the form of well distributed bubbles. Air entraining improves workability and reduces bleeding in fresh concrete. It allows surface finishing soon after the concrete is placed, which saves time and cost. Depending upon other factors like grading, percentage of air bubbles, and their distribution, etc. the durability of concrete is also increased. Air-entraining admixtures are commercially available in market, to be added to the batch at the time of mixing. Air entraining cements are also manufactured. Air entraining cements increase resistance of concrete pavements to freezing and thawing.

Other admixtures used are water reducer, pozzolans, super plasticizers, etc.

Polymers have also been used alone or in combination with portland cement to bind aggregates together in concrete. Polymers impart higher strengths but it is costly, limiting its use to bridge decks, joints, etc. only.

(e) Concrete for Pavement: Concrete used for pavements must have durability, strength and resistance to wear. Each of these requirement should be considered in designing the mix. These properties are closely related. Air-entrained concrete is used in certain locations where climatic conditions, require its use.

Compressive strength and flexural strength of the concrete govern the pavement design. The water-cement ratio plays an important role in mix design. Optimum amount of water gives maximum strength.

15.11.2 Construction of Concrete Roads
Following steps are involved in construction of rigid/concrete pavements.

(i) Preparation of subgrade, Sub-base and compaction

On prepared subgrade cement concrete slab is laid. In preparation of the subgrade/sub-base it should be ensured that foundation is compacted to a smooth hard surface. The subgrade should be dressed to conform to the required line, profile, and cross-section. It should provide uniform support. The type of equipment to be used for compaction depends upon the material to be compacted. Thorough compaction is absolutely essential. Preparation of subgrade and sub-base is same as discussed for flexible pavements. Sub-base is provided for weak subgrades (CBR less than 2). A layer of natural coarse sand, moorum, gravel, open graded crushed stones, etc. can be placed as drainage layer over subgrade compacted to about 15 cm, if subsoil water is a problem, and to prevent any absorption of water by the sub-soil or sub-grade.

Subgrade should have uniform strength over its entire width, weak spots if any should be strengthened by placing new material. When subgrade is of good strength sub-base is not

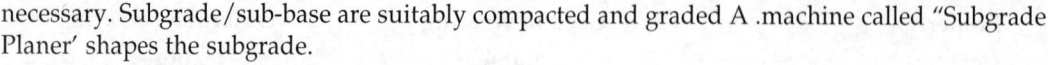

necessary. Subgrade/sub-base are suitably compacted and graded A .machine called "Subgrade Planer' shapes the subgrade.

(ii) Laying of forms and sprinkling the subgrade

After grading and compaction of subgrade and sub-base, the next step is to lay the side-forms for the concrete. Timber or in major works steel forms are used these days. Steel forms can be used on different sites and are thus economical and do not warp like wooden after contact with water.

After completion of form laying, the subgrade/subbase is checked that elevation and shape of the surface is exactly as desired. To prevent water being absorbed from the concrete, when it is placed it is common to wet the surface just before the concrete is placed. Sometimes a desirable method is to cover the surface with a polythene sheeting or water proof bitumen-bonded paper. Sometimes instead the foundation surface is sealed with bituminous material. When insulating water-proof paper is placed, the moistening of the surface prior to placing of concrete is not required.

Forms should be well set, true to line and grade, on a thoroughly compacted subgrade with uniform bearing throughout their entire length.

(iii) Mixing of concrete

The ingradients to concrete are mixed in proper proportion to secure uniform distribution.

Mixing and placing of the concrete can be done by any one of the following methods:

(a) Batching of the material at central plant and transporting dry batched material to site, for mixing and placing.

(b) Batching and wet mixing of the material at central plant and transporting concrete to site for placing.

(c) Batching of material at central plant and then mixing and transporting the material at the same time on way to site in a transit mixer.

In the first method freshly mixed concrete is available close to the place where it has to be placed. There is good control over the mixing time, proportioning of ingradients and quality of the concrete. Material can be used as per requirement and unforeseen situations can be easily controlled. This method may require moving of mixing plant from site to site, during construction of a road.

The second method eliminates the necessity of moving the mixer continuously in course of construction. The other advantage of this method is that concrete can be produced in huge quantity and can be supplied to several sites at the same time. Good degree of quality control can be achieved.

The disadvantage of this method is that the site should be within the reasonable distance. If site is far away concrete will begin to stiff, affecting its workability and consistency. During the haul sometimes segregating of aggregates may take place, if road is bumpy or rough. Another disadvantage is that if placing is delayed or stopped due to breakdown of equipment, whole concrete will be wasted.

The third method is the one which takes care of the disadvantages of the first two methods. It is useful, when central plant is located for away from the site. Mixing and addition of water can be done, when mixer is at a particular distance from the site as per time required for mixing. Concrete costs are higher in this method and quantity of concrete in a single batch of mixing is lesser than those supplied by central plant or produced by on site mixers. Thus mixing operation tends to be less efficient.

In the modern paving mixers, mixing time is controlled automatically.

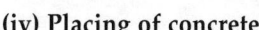

(iv) Placing of concrete

Placing of the concrete should be done in such a way that minimum segregation occurs and even depth of spread is obtained. Placing is done through machines to obtain uniform spreading and depth of concrete. Concrete is usually spread to depth of 15 to 25% greater than the required depth of the compacted slab, depending upon the workability of the concrete.

The Hopper spreading machines, the blade and screw spreader machines are commonly used for this purpose. Concrete should be put in 20 minutes and compacted within 60 minutes in summer and 75 minutes in winter. Usually concreting should not be done when atmospheric temperature is below 5°C and above 40°C.

There are two methods of placing the concrete:

(1) Alternate Bay method and

(2) Continuous method

(1) Alternate bay method: In this method, concrete is placed in bays 3.5 to 4.5 m long and width equal to width of a lane. Alternate bay is filled at one time. The remaining adjacent bay is filled after 1 week for ordinary cement and after 2 days for rapid hardening cement.

For double lane roads, the placing of concrete is done in odd bays of one lane and even bays of other lanes as shown in Fig. 15.18.

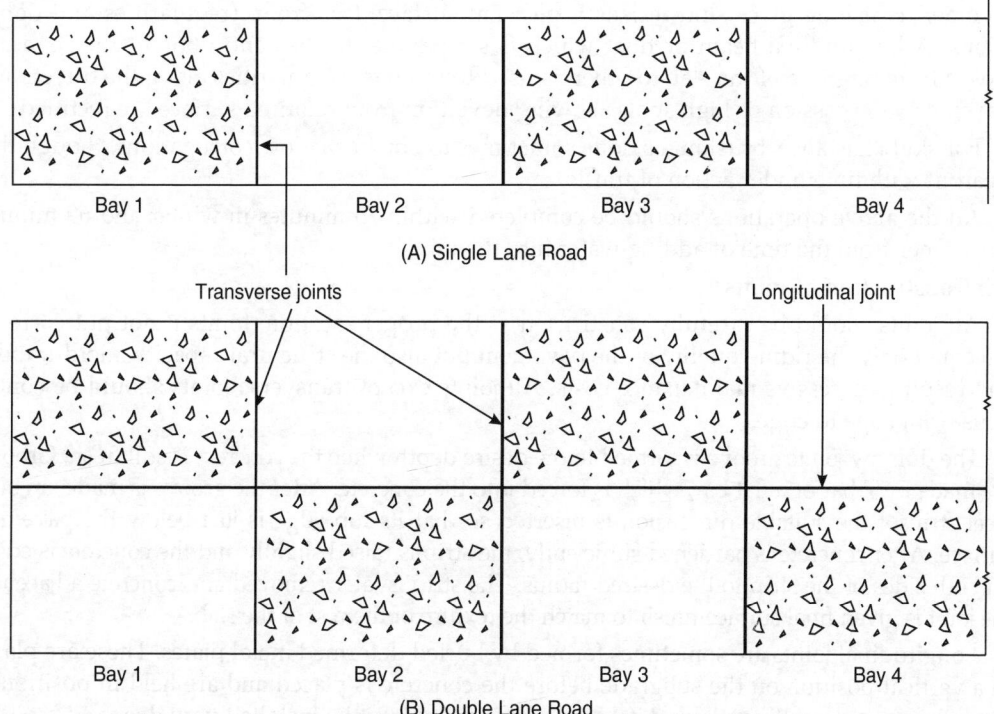

Fig. 15.18: Alternate bay method of placing concrete

In this method, construction work is spread over the whole road width, and number of transverse joints is increased. Due to this more time is required in construction. Cost is increased and due to many joints, riding quality of the surface is not too good.

(2) Continuous bay method: In this method, all bays are filled with concrete continuously without break. It is also called full width method. Width of the slab is one traffic lane and the entire length is constructed continuously from one end to another.

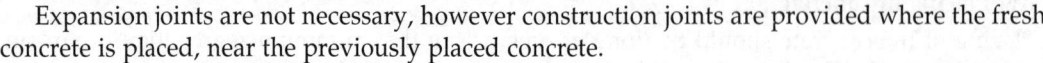

Expansion joints are not necessary, however construction joints are provided where the fresh concrete is placed, near the previously placed concrete.

Concrete along the forms or joints should be spaded to draw mortar to the edge of the slab and insure a face free from honeycomb. A vibrator may be used for this purpose.

When reinforcement is used it is placed at least 5 cm below the surface. A layer of concrete is spread to the desired depth, and reinforcement is placed on it. The remaining concrete is placed and finished over it.

(v) Compacting and finishing

After spreading and placing the concrete to the desired depth it is to be compacted. Finishing then gives the exact surface texture to the completed slab. Finishing provides skid resistant and smooth riding surface.

Through compaction a dense homogeneous slab, free from voids is obtained. Different type of compactors are available for compacting such as power propelled vibrating machines or hand-operated ones. Vibrators are moved longitudinally as compaction is carried out. Water in the fresh concrete should not be in excess of the stipulated quantity to prevent cracking on drying.

Finishing operation is done soon after completion of compaction with screed, a float and a broomer. Finishing gives skid-resistant, pleasant surface, free from irregularities and depressions. A longitudinal beam or float which is set parallel to the centre line of the road and moves from one side of the slab to other, is used to remove irregularities along the direction of travel. Any depression or high spots showing deviation from required surface are rectified.

For skid resistance, brushing is done sometimes to obtain shallow corrugations, though they wearout with time under action of traffic.

All the above operations should be completed within 75 minutes in winter and 60 minutes in summer from the time of adding water into the mix.

(vi) Construction of joints

All joints should be carefully constructed at the proper location. Joints if not put correctly affect not only the riding quality of the pavement but also the structural capacity and durability and result in excessive maintenance costs. All joints except transverse joints should be continuous from edge to edge.

The dummy joints groove is formed to the desire depth while the concrete is still fresh. Grooves are made by T-bar or a flat bar, which is forced into the concrete. After the groove is made, an oiled steel strip of the suitable dimension is inserted so that its top edge is just below the pavement surface. After concrete is hardened sufficiently, the strip is raised slightly and the concrete is edged on both sides of the strip to the desired radius. The strip is later removed and concrete adjacent to the joint is given final surface finish to match the texture of the rest of the slab.

Longitudinal joints are sometimes formed by buried deformed metal plates. These are placed in a vertical position on the subgrade before the concrete is placed and are held in position by pins. To prevent spalling, these deformed plates should not be installed very deep.

Expansion joints consist of nonextruding filler which is placed in position on the subgrade prior to placing concrete. This filler, which is shaped to fit the cross-section of the pavement, must be securely staked to the subgrade to maintain proper alignment during placing and finishing. When the concrete is hardened sufficiently, the space above the filler should be cleaned and the joint edged. No concrete should be left in the joint space.

At construction joints, both longitudinal and transverse care should be taken that no concrete from the last slab placed overhangs any portion of concrete already in piece. This may be insured by making a small groove at the top of the last slab placed.

(vi) Curing

To prevent evaporation of water, curing of concrete slab is done. If curing is not done adequately soon after finishing, the slab may crack in long run, due to shrinkage and hardening of the concrete during weather changes, and temperature fluctuations.

Curing is usually done for 14 days, soon after placing and finishing in hot climate, for ordinary cement. For rapid-hardening cement this period may be reduced to 7 days. This period may be increased in cold and frosty weather.

There are several methods of curing the concrete such as:

(a) Covering with damp cloth: Damp cotton mating, jute, etc. is placed on the top of the concrete. This cover is sprinkled with water and kept wet throughout the curing period.

(b) Spraying water proof membranes: This method involves spraying a white-pigmented insoluble resinous compound on to the concrete slab immediately after finishing. This prevents moisture from evaporating by the impervious cover, and the white pigment reflects heat of the sun. The advantage is that no further treatment is needed.

(c) Covering with damp sand: Often concrete slab, after initial hardening is covered with about 5 cm thick layer of wet sand. This sand is kept wet throughout the curing period. This is quite effective and convenient method, except that constant wetting of sand is required and it should be removed at the end of curing.

(d) Covering with water proof papers: In this method a layer of water proof paper is laid on the top of the concrete slab and kept in close contact with the surface so that wind does not penetrate and free circulation of air is prevented. This allows droplets of condensed water to collect on the underside of the paper and concrete is kept moist and thus cured.

(e) Water ponding: Small clay dams are made on the slab and ponded with water. This is though most effective method of curing, is not commonly used these days.

Pavement can be opened to the traffic after 28 days of curing of concrete slab.

15.12 JOINTS IN CEMENT CONCRETE ROADS

Joints are provided in concrete roads to allow for expansion, contraction and warping of the slabs due to changes in the temperature, and to allow for the break in construction at the end of the day's work. The number of joints should be kept minimum, to reduce extra work, cost and interference to concreting work. Closely spaced joint spoil the looks of the road and main-tenance cost is also increased.

15.12.1 Purpose of Joints

In concrete roads, joints are provided to serve the following purposes:

- To bear expansion and contraction of concrete slabs due to variation in temperature, so that cracking is avoided.
- Top of the slab, becomes hot during the day, while bottom is at lower temperature. This results in warping of the slab. Joints are necessary to control adverse effects of warping.
- After end of days work of constructing the concrete road, when work starts next day, suitable joint will have to be provided to maintain continuity.
- On some soils joints are required to allow for differential shrinkage and swelling of the soil between the edges and centre of the road.

Joints between two slabs, are weaker in carrying the load than rest of the slab and it is this lower structural strength that determines the load-carrying capacity of the pavement. Joints therefore should be strong, to transmit full load across the joints to the adjacent slab.

15.12.2 Requirement of Joints

A joint to be satisfactory, for providing good performance and construction, must fulfil the following requirements:

   (i) The joint must permit slabs to move without restraint, so that stresses are not developed in slab.
  (ii) The joint must not unduly weaken the road structurally.
 (iii) The joint must be effectively sealed so as to not allow the water, grit, stones to infiltrate.
  (iv) The joint must not detract from riding quality of the road.
   (v) The construction joint must not interfere with laying of the concrete.
  (vi) The joint must not protrude or come out of the road level, to cause inconvenience to vehicles.
 (vii) It must be easy to maintain the joint.

15.12.3 Type of Joints

Joints are broadly classified as transverse and longitudinal joints:

(a) Transverse Joints

   Joints which are provided perpendicular to the centreline of the road are called transverse joints. Transverse joints are classified into four groups:

   (i) Expansion joints
  (ii) Contraction joints
 (iii) Warping joints
  (iv) Construction joints

(b) Longitudinal Joints

   Longitudinal joints are required in concrete roads more than 4.5 m wide, in order to allow for transverse warping and for uneven settlement of the subgrade. They also help in constructing the road in convenient widths.

   If due to some reason it is desired not to use the longitudinal joints, reinforcement with heavier transverse bars must be provided.

   Figure 15.19 shows typical layout of transverse and longitudinal joints.

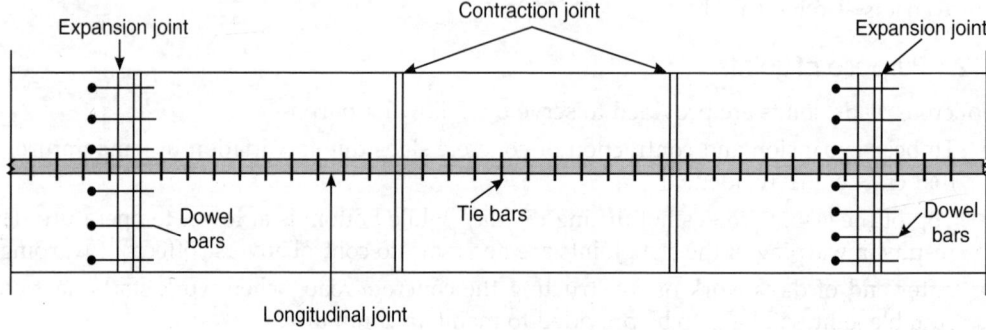

Fig. 15.19: Layout plan of transverse and longitudinal joints

   (i) Expansion joints: Expansion joints provide gaps in the concrete slab to allow for expansion of the slab when temperature rises above the temperature at which the concrete was laid. The gap is filled with a compressible material.

The spacing of expansion joints, may vary from 12.5 m to 45 m depending upon the temperature during placing the concrete, thermal expansion coefficient of concrete, thickness of slab, amount of reinforcement and method of construction used. The width of the expansion joint is about 20 mm to 25 mm extending to full depth of slab.

Figure 15.20 shows a typical expansion joint. Expansion joints should be strengthened structurally to avoid cracking in the region of joints and then spread into the slab.

Load-transfer devices are used for this purpose, which also help in preserving the alignment of the slabs and reduce the risk of mud pumping and spalling by traffic. Dowel bar joint is the simplest and most effective device for expansion joints.

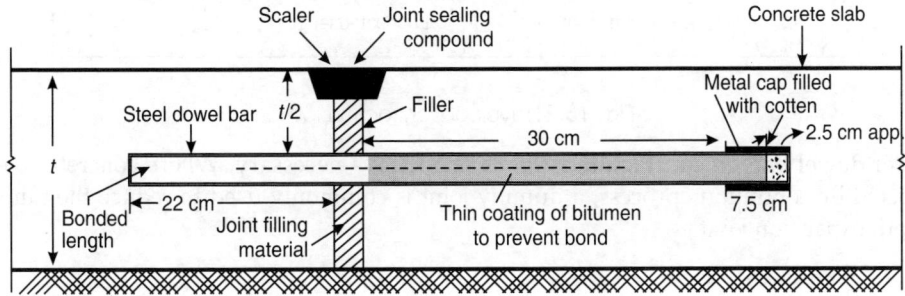

Fig. 15.20: Dowel-bar expansion joint

In dowel-bar joint the load is transferred across the joint by a series of steel dowel-bars, each bar being bonded into one slab and sliding in the other, as shown in Fig. 15.20. Dowel bars are designed for sizes in various conditions.

The spacing of the dowel-bars should not be so close as to cause interference with the construction and not so wide to permit flexure of the concrete between the dowel bars. The dowel bar may 1.5 cm to 1.8 cm in diameter and they are placed centre to centre spacing of about 40 cm to 50 cm. The length of dowel bar may bar 40 cms to 70 cms.

One end of dowel bar is fixed in concrete of the slab and the other end is painted with bitumen to move in a metal cap partly filled with cotten, and is fixed in the adjoining slab. The joint is filled with compressible material made from bitumen, cork and rubber. During expansion of the slab, in hot weather, the end of the dowel bar in metal cap moves and reduces thickness of the joint.

A sealing material is put on the top of the expansion joint after concrete slabs are laid in alternate bays to prevent ingress of water, dirt etc. The depth of sealer is about 25 mm.

(ii) Contraction joint: Contraction joints are breaks in the structural continuity of the concrete, permitting it to contract when the temperature falls below the temperature of laying. As concrete is weaker in tension than in compression, contraction joints are spaced closer than expansion joints. Contraction joints also allow the slab to warp under influence of temperature gradient.

Close spacing of contraction joints is used because:

- Concrete is weaker to take tension
- The cracks will not open up
- Warping stresses are considerably reduced

The type of contraction joint used depends upon the whether alternate-bay or continuous construction is to be used. Butt contraction joints are most suitable for alternate bay method of construction. They are formed by painting the ends of the first slab laid with a bituminous material, after the forms are removed. This prevents bond with the adjacent slabs. Sealing compound is used to seal this joint against water, etc. They are either full depth breaks in continuity of slab or break only 1/3rd to 1/4th of the slab, to prevent cracking called 'Dummy Joint' as shown in Fig. 15.21.

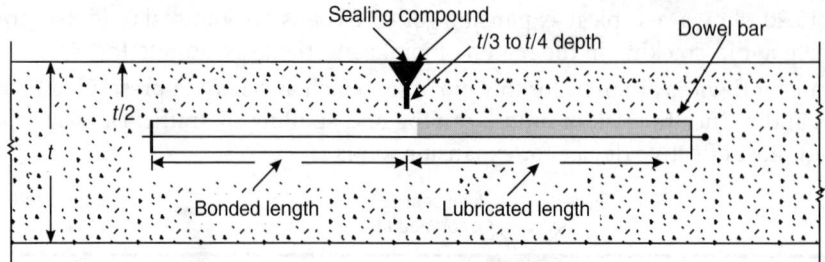

Fig. 15.21: Typical contraction joint

Use of dowel bars in contraction joints is not always necessary. Where concrete roads are constructed by continuous process, a dummy joint is commonly used to reduce the number of full depth expansion joints.

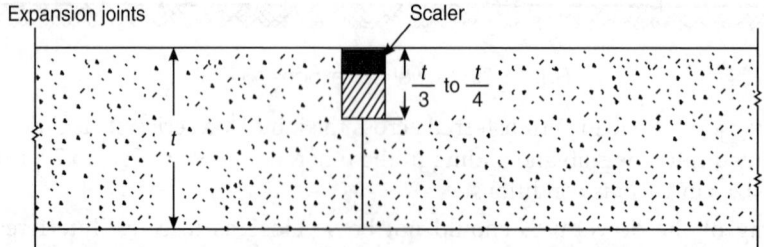

Fig. 15.22: Dummy contraction joint with groove at surface of slab

Dummy joints consist of a groove in the slab to form a plane of weakness, as shown in Fig. 15.22.

(iii) Warping joint: Warping joints are break in the concrete pavement. These joints allow small amount of angular movement to occur between the slabs and so prevent high stresses due to restrained warping. The opening of joints is prevented by tie-bars or reinforcement. If expansion and contraction joints are suitably designed to take care of warping stresses, warping joint may not be necessary.

The use of dummy warping joint instead of dummy contraction joints are used to reduce amount of progressive opening of the joints, due to expansion and contraction of single unit.

Warping joints are formed using tie bars or continuing reinforcement across a contraction joint of the dummy joint as shown in Fig. 15.23. The opening of the joint is prevented, but there will be sufficient flexibility to allow the slab to warp.

(iv) Construction joints: When construction of concrete road is suspended for 30 minutes or more and when work of placing the concrete starts next day, construction joint is provided with suitable tie bars and additional reinforcement across the joint.

Normally construction should be so planned that the day's work ends at transverse or contraction joint, to avoid the need of construction joint. The joint when provided is tied together either by continuing the reinforcement across the joint or by using tie-bars.

Construction joints are provided with a groove for sealing against the ingress of water or grit, etc.

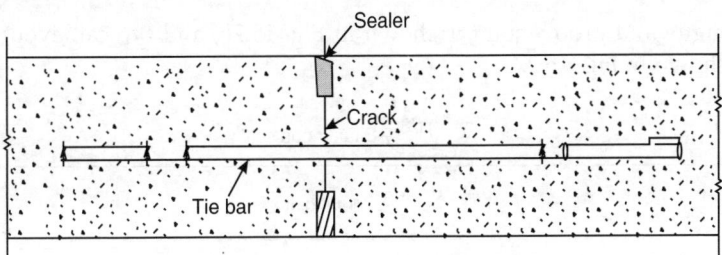

Fig. 15.23: Transverse dummy warping joint

Longitudinal Joints

The longitudinal joint is required to act as a hinge and also to construction the road in convenient widths.

Load-transfer device are used in longitudinal joints to reduce deflection and prevent the development of high tensile stresses in the concrete, and also to maintain the alignment of the slab.

Commonly used longitudinal joints are:

(i) Butt joint: A butt joint is the simplest longitudinal joint and is formed by painting the joint face with bitumen, after forms are removed from the first line of the slab concreted. When second line of the slab is concreted (completing the joint) a groove should be formed at the surface for sealing the joint, as shown in Fig. 15.24.

Tie bars are used sometimes for load transfer as shown in Fig. 15.25. Diameter of tie bar varies from 10 mm to 25 mm depending upon the traffic and the length of tie bar is 70 cm to 100 cms spaced about 50–80 cms centre to centre.

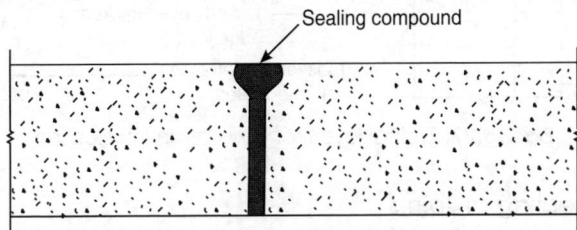

Fig. 15.24: Plain butt joint

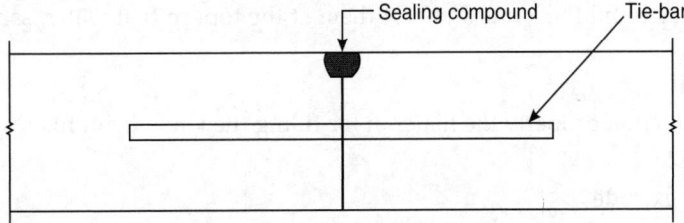

Fig. 15.25: Butt joint with tie bar

(ii) Tongue and groove joint: In tongue and groove type of longitudinal joint the transfer of load is provided by the interlocking tongue and groove. This joint is not suitable for slabs thinner than 20 cm. Thorough compaction of the concrete in the tongue and groove is essential. It is useful in poor subgrades to help in preventing the differential uplift between the slabs.

A typical tongue and groove joint is shown in Fig. 15.26, and typical layout of joints at an intersection is shown in Fig. 15.27.

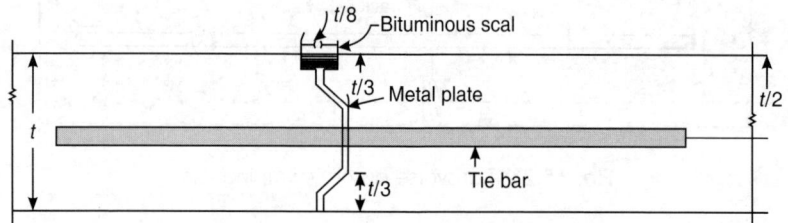

Fig. 15.26: Tongue and groove type longitudinal joint

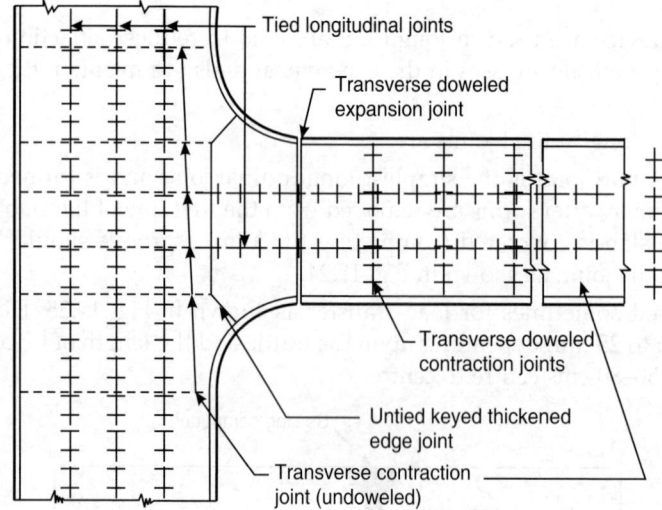

Fig. 15.27: Typical layout of joints at an intersection

15.12.4 Filling and Sealing of Joints

Joints create a gap in concrete slab, and water and grit, etc. are liable to enter them, causing failure of the pavement and preventing expansion of the concrete, resulting in blow ups. It is therefore necessary to fill the joints and seal them at the top, so that water, sand, grit, etc. may not enter in.

Filler Material

To fulfill its function properly the material for filling the joints (joint filler) must be:

- Compressible
- Should not extrude
- Elastic
- Strong
- Rot-proof
- Easy to cut
- Durable
- Sufficiently rigid to support during construction

The materials suitable, meeting above requirements, commonly used as joint fillers are the following:

- Soft wood or fibre wood (creosted to prevent rotting)
- Cork, which is available in several types
- Bituminous compounds
- Rubber and rubber compounds
- Resins (synthetics)

15.12.5 Sealing Material

To seal the joint effectively, joint sealing compounds should posses following properties:

- Good adhesion to cement concrete
- Extensible without cracking
- Should not themselves flow along the sides of joints in hot weather
- Not too hard nor too soft
- Durable against climate
- Resistance to the ingress of sand, grit, etc.
- Easy to apply
- Resistive to abrasive action of traffic

Sealing compounds which are commonly used on the concrete pavements are the following:

- Straight-run bitumen
- Air blown bitumen used alone or with plasticising oils and/or fillers. This compound is considered the best
- Resinous compounds
- Rubber-bitumen compounds

15.13 SPECIAL ROAD PAVEMENTS

15.13.1 Concrete Block Pavements (CBP)

Stone blocks, bricks, concrete blocks have been used in pavements in urban areas. Concrete blocks are made from high strength concrete, in various sizes, shapes and thicknesses. (Usually 8 to 10 cm thick). These blocks must have good compressive strength, durability and abrasion resistance.

Blocks are laid on a prepared bed, usually of sand. During compaction bedding sand fills in the gaps between the blocks. Interlocking between the blocks develops due to friction between adjacent block faces transmitted by sand in the joints and by wedging and interlocking between adjacent blocks.

Some of the common shapes of the concrete blocks are shown in Fig. 15.28. Paving quality concrete is used for production of precast blocks, as per IS: 383–1970.

Blocks are generally laid in herring-bone or stretcher bond pattern. Herringbone pattern can be easily used on curved sections of the road. Fig. 15.29 shows type of bonds.

Concrete block pavements are used for low speed only as riding quality is not so smooth and they are noisy.

15.13.2 Brick Pavements

When traffic is less as on town streets, bricks on edges are also used for pavements. Brick pavements use a layer of well burn clay bricks laid on edges over a thin bedding of a sand layer, provided over the prepared and compacted subgrade or sub-base. The compressive strength of bricks used for paving should not be less than 7 MPa. They can be set on a bed of cement mortar.

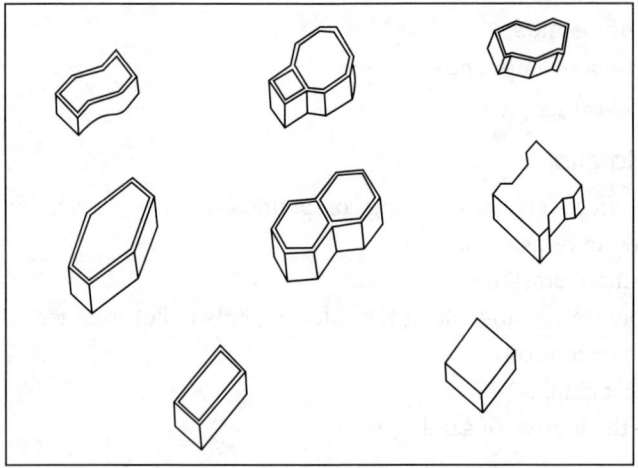

Fig. 15.28: Typical shapes of concrete blocks

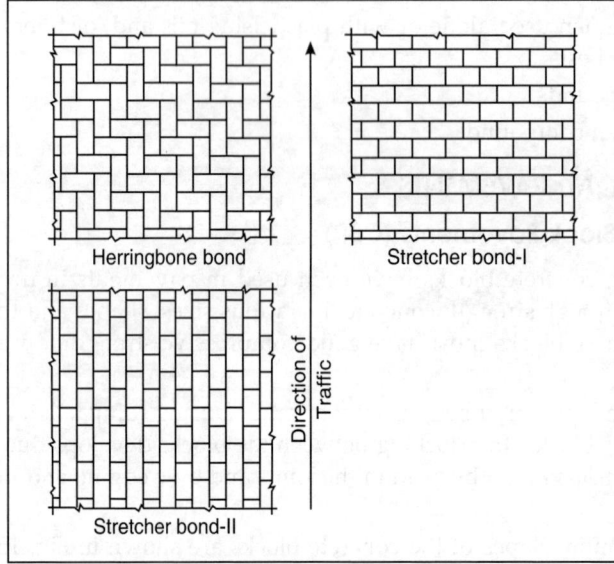

Fig. 15.29: Type of bonds used in concrete block pavements

Different patterns of brick laying can be used for proper bonding, like concrete block pavements. Good quality bricks, having high strength and durability are used for making pavements. Depending upon quality of bricks, these pavements may have a life of 15 to 20 years, when traffic intensity is low. Sometime two layers of bricks are also used for better performance. The gaps in the joints are filled with sand, bitumen, or cement mortar.

To give smooth riding surface a layer of pre-mix bituminous carpet or a coating of bitumen or tar with chips, can be given over the brick pavements.

15.13.3 Stone Pavements

Stones are set on a bedding of sand layer provided over compacted subgrade. Stones used for making pavements should be hard and tough. Where good and hard stones are available locally, they can be used as pavement material. Irregular stones obtained from the near by quarries can be cut to regular sizes and shapes and used for constructing pavements. Type of stones suitable for making roads are granite, sandstones, basalt, and lime stones, etc.

Stone-set pavements, depending upon the quality of stones may have a life of 15 to 20 years in low and slow traffic conditions.

Stones may be laid in herringbone or stretcher bond pattern. They can also be set on a layer of sand-bitumen mix. The joints are filled with sand, stone dust or sand-cement mortar or bitumen mix.

The operations involved in construction of concrete block, brick, and stone pavements are the following:

(1) Preparation and compaction of subgrade

(2) Putting a bed of sand and screening

(3) Laying of the blocks/bricks/stones, in proper bond.

(4) Compacting

(5) Joint filling

(6) Opening to traffic

Pavement Maintenance, Evaluation and Strengthening

16.0 PREAMBLE

Once the road is constructed and opened to traffic, immediately then, the necessity of maintenance arise due to:

(i) Strength of road goes on reducing and surface starts deteriorating by movement of high speed vehicles, heavy vehicle loads and repetition of loads.

(ii) Pavement is also subjected to other deteriorating effects like rain water, action of sun and wind.

(iii) Pavement structure has to bear too many other effects of curvature, gradient, etc.

Maintenance is the process to conserve, as nearly as possible the originally designed condition of paved and unpaved roads, and of traffic signs, signals, markings and associated facilities, to minimise vehicle operating costs and accident costs.

Maintenance is necessary to lessen the rate of deterioration of a road and restore the service levels to earlier standard. It includes all works which are required for upkeep of a road and its associated structures to prevent the deterioration of quality and efficiency, which pertained immediately after construction. Maintenance of a road is different from rehabilitation which implies strengthening or changes in alignment/width, etc.

Highway maintenance involves correcting deficiencies in highway developed due to various reasons (certain defects may be due to faulty design, improper quality of construction or material), and prevent further development of deficiencies, so that road is safe and provide smooth movement of traffic. Poor maintenance may be a cause of large number of accidents and high vehicle operating costs.

Formation of cracks, patches, waves, corrugations, ruts, unevenness, etc. are the causes of deterioration in flexible pavements. Cracks due to temperature stresses, uneven joints, mud pumping, etc. are common cause of deterioration in rigid pavements. If immediate attention is not paid to these defects in pavements, they may result in failure of pavements in serving the intended purposes.

As shown in Fig. 16.1, the rate of deterioration of a pavement accelerates with time. By adopting a maintenance strategy, overall costs can be minimised.

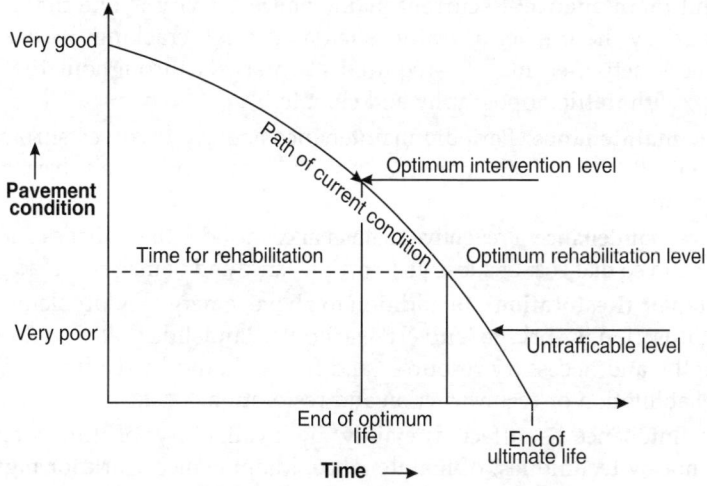

Fig. 16.1: Pavement condition levels with time

16.01 OBJECTIVE OF MAINTENANCE OF HIGHWAYS

Objects of maintenance can be listed broadly as:

- To prolong the life of the pavement.
- To provide a defined minimum level of surface quality (good serviceability)
- To preserve the design level of surface quality
- To provide a defined minimum level of passability
- To keep signs in an easily recognisable condition
- To minimise incidences of signal malfunctions
- To keep markings visible
- To minimise total cost of maintenance and road user's (vehicle operating costs)
- To maximise utilisation of available manpower.
- To maximise ratio of quality of maintenance (output) to value of resources (input)
- To defer large capital costs.

16.02 ADVANTAGES OF GOOD MAINTENANCE

- The life of both pavement and vehicles increases by good and proper maintenance.
- Well maintained road imparts smooth riding surface for vehicles which result in comfort to passengers and safety.
- There is also savings in vehicle operating costs as fuel consumption is less due to reduced friction and speeds can be maintained.

16.1 TYPES OF MAINTENANCE

Maintenance of roads can be classified in following categories:

(i) Routine maintenance: Routine maintenance activity is expected to be performed on a regular basis and is usually seasonal. It does not depend upon traffic usage. It consists of drain cleaning, grass cutting, painting signs, markings, grading shoulders, collecting litter, maintenance of drainage structures such as bridges and culverts, etc. Routine maintenance is not amenable to detailed planning, which can be assessed on the basis of accepted levels of service.

(ii) Recurrent maintenance: Recurrent maintenance activity is one that is necessitated by problems induced by the impact of traffic, such as rutting, cracking. It consists of patching, grading etc. These activities may be required at intervals throughout the year but whose frequency varies with traffic, topography and climate.

(iii) Periodic maintenance: Periodic maintenance activity involves surface dressing, over laying etc. periodically at intervals more than a year. These works are subjected to planning for a specific maintenance plan.

(iv) Preventive maintenance: Preventive maintenance includes the activities and measures which are taken beforehand so that cost of subsequent maintenance and damages to road are reduced.

(v) Special repair (Restoration): In addition to above, emergency problems, though difficult to predict, such as floods, bridge/culvert washouts, landslides, etc. on highways must be attended promptly, and necessary resources and funds should be readily available. Such cases may involve rehabilitation or reconstruction and restoration.

Highway maintenance is affected mainly by availability of funds and management problems, and not by techniques, of maintenance. Maintenance work for highways is carried out either manually or by use of mechanical appliances or by combination of both, i.e. machine and labour, using hand tools. In India, machine and labour method is commonly used, unlike America and other developed countries where mechanical methods are largely used.

16.1.1 Maintenance of Roads in India

Roads in India are maintained by government agencies. The responsibility of development and maintenance of the national highways and expressways is vested with government of India, while that of state highways (SH) and major district roads (MDR) are vested with state governments and ordinary district roads (ODR) and village roads (VR) are vested with the respective zila panchayats. The municipalities maintain all the roads in their respective jurisdiction, in urban areas.

In the organisational structure and the mechanism for executing maintenance work, departments engage casual labour, who are not provided with proper equipment and machinery, nor they have knowledge/training of techniques and trends in maintenance. The output therefore is not commensurate with the expenditures and required quality.

Recently maintenance by departmental agency is being switched over to contract system. A detailed list of various maintenance activities, specifications to be adopted, minimum skilled labour to be employed, could be indicated for allotting contracts. The periodicity, the quality expected to be achieved and various milestones in maintenance could also be indicated in the bid document.

16.2 DEFECTS IN FLEXIBLE PAVEMENTS

Deflect/distress is the condition of the pavement structure that reduces its serviceability due to degenerating. Several types of defects develop in flexible and rigid pavements which ultimately lead to failure of the road to perform its intended purposes. Cause of defects in a flexible pavement may be due to:

  (i) Poor quality of material used in construction.
 (ii) Road not properly designed structurally.
(iii) Poor construction due to inadequate quality control.
 (iv) Faulty or inadequate drainage system.
  (v) Wheel loads higher than those for which road was designed.
 (vi) Poor compaction resulting in settlement of fill material in embankments.

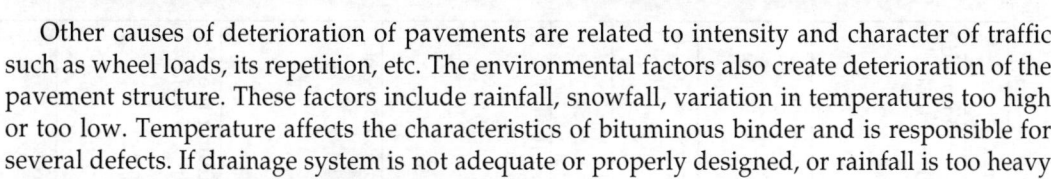

Other causes of deterioration of pavements are related to intensity and character of traffic such as wheel loads, its repetition, etc. The environmental factors also create deterioration of the pavement structure. These factors include rainfall, snowfall, variation in temperatures too high or too low. Temperature affects the characteristics of bituminous binder and is responsible for several defects. If drainage system is not adequate or properly designed, or rainfall is too heavy it causes erosion and other defects in the pavement.

Repeated loads on pavements, combined with heavy traffic load and adverse climatic condition, result in fatigue failures of the road.

16.2.1 Fatigue Failures of Pavements

A pavement subjected to repeated loads will result in both plastic and elastic deformations, and the relative magnitudes of each depend upon the number of load applications. Accumulated plastic deformation may be of such a magnitude as to cause pavement failure, and therefore the fatigue failures resulting from repeated load may be significant. The behaviour of most materials under repeated loads is extremely complex, and matter becomes more complicated when considering the base course structure.

Concrete exhibits the property of fatigue. Bituminous concrete may also fail by fatigue. Alligator cracks in asphalt concrete surfaces may result from repeated load applications. There is a linear relationship between log of repitition and the stress which causes failure. If base or subgrade are highly plastic, permanent deformation in these layers may cause failure before the pavement fails in fatigue.

Three principal effects of repeated load application upon base course material are:

1. Some aggregates breakdown when subjected to high repeated stress conditions. Particularly true of soft particles like shale.
2. Subgrade intrusion into the base course is greatly affected by number of load applications.
3. Blowing of base course material is a function of the number of load applications.

Repeated loads give excessive plastic deformations in the soils. In general the amount of deformation under repeated loads varies directly with the log of the load applications.

On flexible pavement the early spring period are most critical for fatigue failures. Seasonal variations and frost have great influence on the behaviour of the pavement concerning deformation under the repeated loads. On Rigid pavements distress becomes more in several ways like pavement pumping, blowing, etc. Uphill traffic lanes generally show more distress than the down hill traffic lanes.

Pavement distress under moving loads is greatest when the loads travel close to the pavement edge. Distribution of vehicle over pavement is determined by the width of the pavement. Flexible pavements show greater distress in the outer-wheel path than inner wheel path for reasons given below:

(i) Shear failures and lateral shoving are likely to occur through the shoulder.

(ii) Subgrade soil near the edge of the pavement is subjected to severe moisture conditions.

Most of the road failures are the fatigue failures with limited amount of destruction and pavement cracks. Usually the failure condition will be reached by a process of cumulative deterioration, therefore defects in pavements should be immediately rectified through proper maintenance and rehabilitation.

Tables 16.1 to 16.3 show common defects in flexible pavements, the probable causing factors and remedial measures. These defects are removed or repaired through regular maintenance work for keeping the road in good condition. Figure 16.2 shows some common defects in flexible pavements.

Table 16.1: Deformation related defects in flexible pavements (A)

Defect/Distress	Description	Possible causes	Remedial measures
A 1. Settlement	General lowering of the road surface	• Due to change in subgrade moisture. • Inadequate compaction during construction • Heavy axle loads.	Defective portion should be removed and strengthening should be done.
A 2. Subsidence	Localised/abrupt lowering of road surface	• Poor drainage • Compressible soil • Underground cavities • Inadequately compacted pockets	Filling with pre-mix bituminous material.
A 3. Bumps (Heaving)	Localised upward displacements of the pavement (upheaval)	• Moisture in swelling soils • Load induced plastic deformation pushing upward.	Repair with deep patches.
A 4. Waves	Waves formed on the pavement surface with crust few cms apart (60 cm or so)	• Deep differential subgrade deformation.	Levelling the surface with bituminous layer.
A 5. Corrugations (Rippling)	Closely spaced waves (less than 60 cm) on the pavement surface, transverse to the traffic direction like ripples across the bituminous pavement. Regular undulations	• Poorly laid surface coarse • Low stability bituminous mix • Braking, accelerating, turning of vehicles. • Mix too rich in bitumen or fine aggregates.	Scarify the surface and apply seal coat or plant mix.
A 6. Depressions	Dish type localised deformation, 2.5-3 cm below grade. Water collects in them	• Localised subgrade failure base or surface failure. • Poor compaction and stability of layers • Heavier loads resulting excessive stress.	Grading and then apply tack coat and bituminous plant mix.
A 7. Rutting (Channels))	Longitudinal wheel track depressions in bituminous pavements.	• Plastic deformation of bituminous layer due to high temperature • Depression of surface layer • Poor consolidation of layers • Deformation of pavement layers • Mix without enough stability.	Levelling pavement by filling the channels with hot plant mix overlay.
A 8. Distortion	Irregular deformation of the pavement	• Differential settlement of lower pavement layers.	Adopt levelling procedures

Contd...

Table 16.1: Deformation related defects in flexible pavements (A) (contd...)

Defect/Distress	Description	Possible causes	Remedial measures
A 9. Creep (Shoving)	- Horizontal displacement of surface - A series of shallow transverse depress-ions	• Low stability/stiffness of the mix • Lack of bond between surface and layer below • Horizontal wheel forces	Level the pavement by applying bituminous mix.
A 10. Edge failures	• Broken edges of sealed pavements	• Narrow width of pavement traffic using edges • Scouring of shoulders by water	Control by preventive maintenance cutting affected area and rebuilding.
A 11. Utility cut Depressions	• Depressions in pavement resulting from cut made for utility repair or installation.	• Lack of adequate compaction of backfill.	Same as for grade depressions.
A 12. Roughness	• Unevenness of a road, an extreme manifestation of which is corrugations	• Poor surface quality/mix	Make the surface even by seal
A 13. Frost heaving	Localised heaving up due to formation of ice-lense.	• Ground water table near the subgrade. • Poor drainage (sub-surface)	Improve drainage and methods of lowering water table.

Table 16.2: Cracking related defects in flexible pavements (B)

Defect/Distress	Description	Possible Cause	Remedial measures
B 1. Alligator cracks (Crocodile cracks)	Small blocks in the pavement similar to alligator skin, interconnected forming small blocks.	• Shrinkage of hardened bitumen at low temperatures • Excessive deflection of the surface over unstable base • Fatigue failure of base • Overloading of surface courses • Repeated load applications.	- Deep patching - Improve drainage - Use seal coat/slurry seal as a temporary measure
B 2. Reflection cracks	Cracks in bituminous overlays, longitudinal or transverse, reflect the cracks pattern.	• Cracks in base course/subgrade reflected through surface course due to poorly laid overlays. • Horizontal movement of under laying layer below overlay.	- Sealing with emulsion/liquid asphalt plus sand to prevent pickup by traffic
B 3. Edge cracks	Cracks (longitudinal) near the pavement edge.	• Lack of side or shoulder support for pavement • Inadequate pavement width • Very flexible surface course • Bitumen hardening.	- Fill the cracks with emulsion slurry or liquid asphalt, mixed with sand - Improvement of drainage.

Contd...

Table 16.2: Cracking related defects in flexible pavements (B) (contd...)

Defect/Distress	Description	Possible cause	Remedial measures
B 4. *Edge Joint Cracks (longi-tudinal cracks)*	Separation of joint between pavement and shoulder. Cracks near the edges. cracks developed in longitudinal direction throughout the thickness.	• Poor joints between layers • Wetting and drying in joints • Differential volume changes in subgrade • Settlement or shrinkage. • Inadequate pavement width.	• Improve drainage • Repair like reflection cracks.
B 5. *Lane joint cracks*	Longitudinal separation along the seam between two paving lanes.	• Weak joint between adjoining spreads in courses of the pavement.	Same as for reflection cracks.
B 6. *Wheel path cracks*	Cracks along the wheel path.	• Fatigue or excessive settlement • Water infiltration from shoulder.	• Improve drainage.
B 7. *Meandering cracks*	Cracks on the surface	• Due to poor construction leading to settlement of fill or embankment.	Fill the cracks with bituminous coat.
B 8. *Shrinkage cracks*	Inter-connected cracks forming a series of large blocks usually with sharp corners or angles.	• Volume changes in bituminous mix or in base or subgrade • Temperature shrinkage, contraction of surface layer.	Fill cracks with emulsion slurry followed by surface treatment.
B 9. *Slippage Cracks (crescent cracks)*	Crescent shaped cracks pointing in direction of the thrust of wheels on pavement surface. (Traffic-direction)	• Lack of good bond between the surface layer and the course below. • Slippage of wearing course under horizontal wheel loads.	• Remove the entire slipped area apply tack coat and place hot mix bitumen layer.
B 10. *Hair line cracks*	Small, fine, closely spaced, irregular cracks.	• Low bitumen content. • Poor compaction procedure • Excessive filter	Cracks filled with low viscosity binder or slurry seal, fog seal.
B11. *Widening cracks*	Longitudinal reflection cracks shown up in bituminous overlays above the joint between old and new sections of pavement widening	• Cracks in base/subgrade due to poorly laid overlays • Horizontal movement under laying layer below overlay.	• Sealing with emulsion or liquid asphalt with sand.
B 12. *Shear cracks*	Upheaval of pavement resulting in fracture or cracking	• Excessive loading • Weak material in pavement mix.	Fill cracks with emulsion slurry.

Table 16.3: Surface related defects in flexible pavement (c)

Defect/Distress	Description	Possible cause	Remedial measures
C-1. Pot Holes	Bowl-shaped holes of various sizes in the pavement	• Localised disintegration • Inadequate bond of between layers • Weakness in pavement • Too little bitumen • Poor drainage • Fatigue failure.	Skin patching as temporary repair and deep patching for permanent repair.
C-2. Ravelling	Progressive separation of aggregates in pavement from surface downward or from edges inward. - Loss of aggregates. Binder unable to hold aggregates	• Aging of the binder • Lack of compaction during construction • Weathering of aggregates • Too little bitumen or overheated use of bitumen.	Surface treatment/dressing or fog seal or slurry seal.
C-3. Bleeding	Film of bitumen on pavement-sleek pavement	• High temperatures • Heavy loads • Improper grade of bitumen • Exces sive bitumen content.	• Resurfacing with open graded mix • Spreading of sand • Seal coat
C-4. Polished aggregates	Aggregate particles in the surface of the pavement have been polished smooth	• Aging of the pavement • Loose of cover aggregates • Heavy vehicular traffic	• Cover with skid resistant surface-hot plant mix or sand seal or surface dressing.
C-5. Scoring (wheel imprints)	Localised marks of the wheels on the surface.	• Prolonged parking • Low stability bituminous concrete • Poor construction.	Resurface with bituminous mix.
C-6. Stripping	Stripping of small surface from the bitumen film.	• Poor mix design • Presence of water • Use of hydrophilic aggregates	• Surface treatment/dressing • Replacement with fresh bituminous mix.
C-7. Weathering	Wearing surface becomes rough in texture due to partial loss of both aggregates and binder.	• Ageing of the binder • Weathering of aggregates	Surface dressing and seat coat.
C-8. Glazing	The surface becomes smooth, hard, shiny, and slippery. Low skid resistance.	• Poor aggregate grading • Loss of ductility with age.	Use skid resistant surface treatment.

Table 16.3: Surface related defects in flexible pavement (c) (*contd....*)

Defect/Distress	Description	Possible Causes	Remedial measures
C-9. Peeling	The wearing of surface loosens and breaks away from underlying unbound base course layer.	• Inadequate thickness • Poor adhesion of layers.	Put new surface dressing course.
C-10. Longitudinal streaking	Alternate lean and heavy lines of bitumen running to the centreline of the road.	• Spray bars not correct in height or angle. • Wrong pump pressure or speed.	Hot course of sand and bitumen over affected area. Laying of new surface.
C-11. Transverse streaking	Alternate lean and heavy lines of bitumen running across the road	• Spurts in bitumen spray from distributor bars.	Same as above.
C-12. Erosion Gullies	Gullies form on the surface of unsealed roads.	• Poor drainage • Heavy rainfall, water runoff.	Improve drainage.

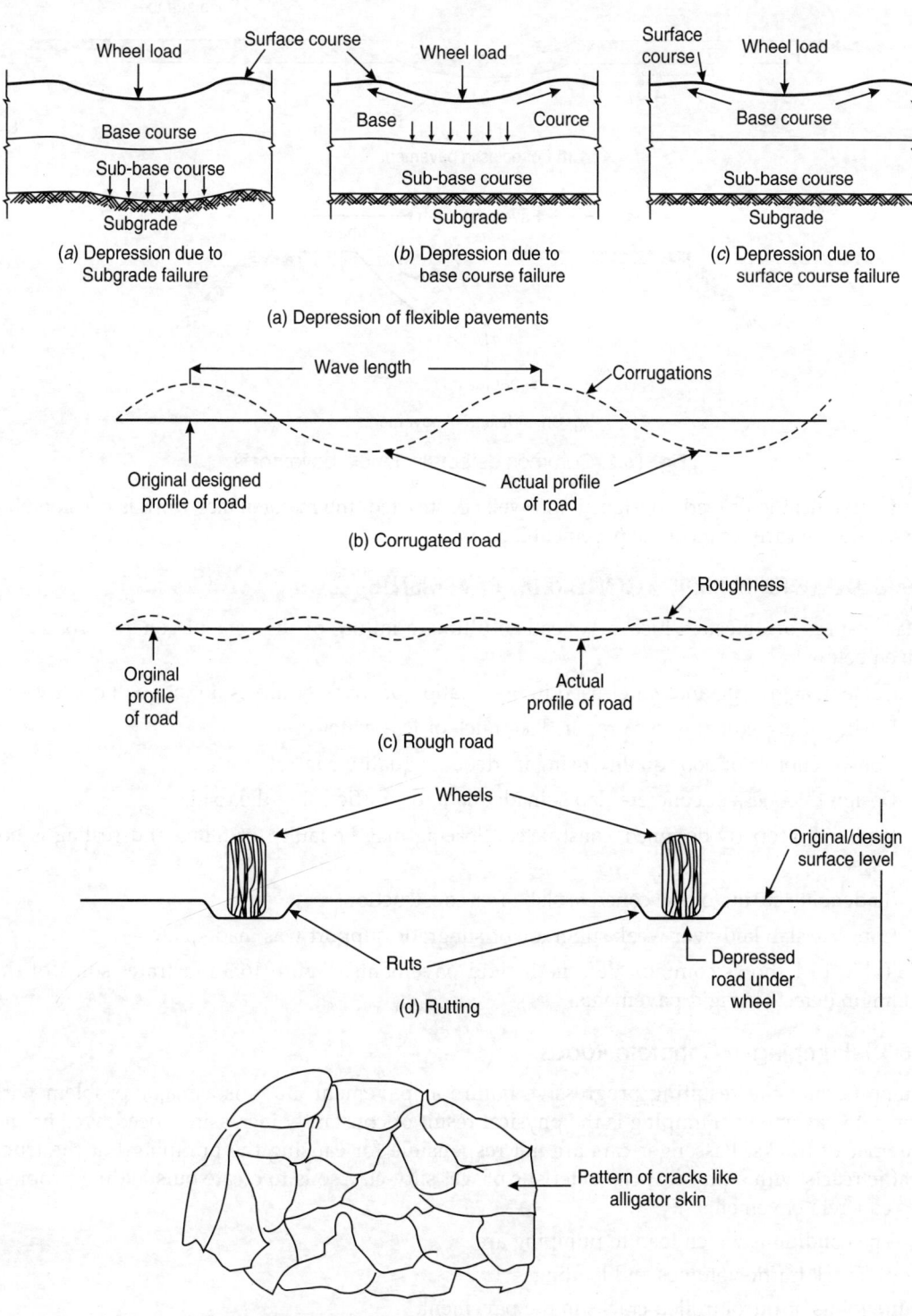

(a) Depression due to Subgrade failure

(b) Depression due to base course failure

(c) Depression due to surface course failure

(a) Depression of flexible pavements

(b) Corrugated road

(c) Rough road

(d) Rutting

(e) Alligator cracks on road surface

Fig. 16.2: Common defects in flexible pavements (*contd...*)

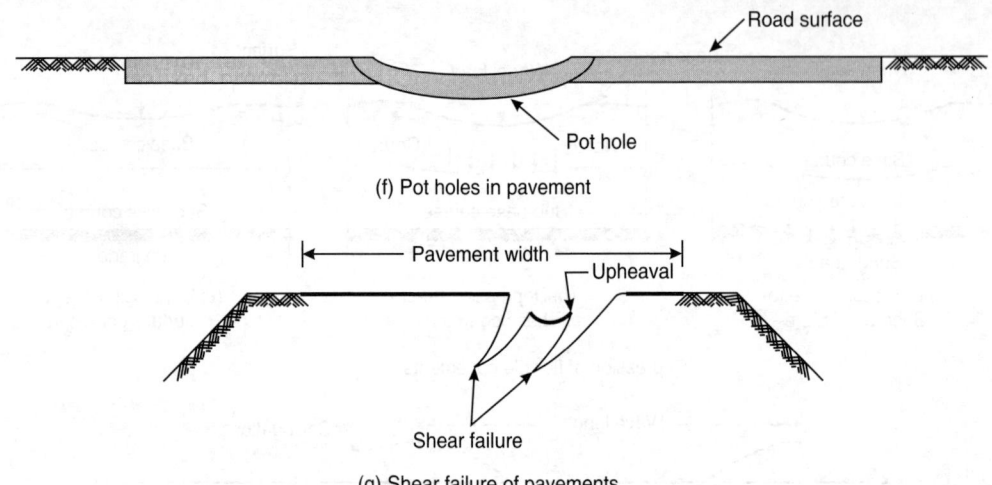

(f) Pot holes in pavement

(g) Shear failure of pavements

Fig. 16.2: Common defects in flexible pavements

If pavement is properly designed and well constructed, the maintenance work is considerable lesser, during early years of the pavement life.

16.3 DEFECTS IN RIGID (CONCRETE) PAVEMENTS

Main causes of defects which may lead to ultimate failure of the cement concrete roads are given below:

- Cement, aggregate and water used in preparation of concrete mix is not of good quality.
- Mixing of ingredient is not proper. Too much or less water used.
- Construction is of poor quality using inadequate quality control.
- Design thickness of concrete slab is inadequate for traffic using the road.
- Joints not properly designed/constructed. Spacing may be faulty, or filling and sealing is not correct.
- Inadequate curing of the concrete slab after construction.
- Concrete slab laid over weak subgrade or subgrade support was inadequate.

Table 16.4 shows common defects in rigid pavements. Figure 16.3 illustrates some of the common defects in rigid pavements.

16.3.1 Pumping in Concrete Roads

Pumping and the resulting progressive failure at pavement joints is a major problem with concrete pavements. Pumping is the physical result of constantly increasing speed, weight and number of trucks. Passenger cars are not responsible for causing the pumping but the truck traffic reacts with a physical characteristic of wet silty-clay soils to create outstanding problem for concrete pavements.

The conditions which load to pumping are

(i) Truck traffic volumes and loading.

(ii) Joints or uncontrolled cracks in the pavement.

(iii) Silty clay subgrade.

(iv) Access of moisture in the subgrade.

Table 16.4: Defects in rigid pavements

	Defect/Distress	Description	Possible Cause	Remedial measures
1.	Pumping-mud	Process by which vertical movement of the base course pumps to the surface usually through joints or cracks. Water and sand are pumped out	• Moisture between base course and subgrade. • Heavy loads • Lack of subgrade support.	Sealing with bitumen or cement mortar. Overlay of bituminous material.
2.	Faulting	Relative movement of slabs of concrete pavement. Difference in elevation at joints.	• Differential deformation • Inadequate load transfer device between slabs. • Pumping.	Faulted slabs be brought to the original level, under sealing.
3.	Spalling	Pieces of concrete spall off cement concrete pavement, at joints, cracks, edges.	• Poor construction/material • Relative movement at cracks, improper joints	After cleaning put dense graded bitumen carpet.
4.	Blow-ups	Localised buckling or shattering of a rigid pavement usually at transverse joints or cracks.	• Longitudinal movement of slabs (Excessive expansion) during hot weather.	Remove the damaged portion of slab level, apply tack coat and dense bitumen carpet.
5.	Scaling	It is peeling away of the surface of concrete road. Surface becomes rough.	• Chemical reaction of salts • Improper curing or mixing of concrete. • Poor mix design.	• Apply emulsion slurry seal or • Overlay.
6.	Excessive Joint Seal	Sand or pebbles lodged in the joint seal. Blow up of joint occurs.	• Expansion of seal or spalling • Oxidation/hardening of filler.	Resealing of joints after removing or cutting off the joint seal.
7.	Coner cracks	Diagonal cracks cornering a triangle with longitudinal, edge or transverse joints	• Traffic loads on unsupported corners. High traffic • Weak subgrade	Remove and Patch
8.	Diagonal cracks	Cracks diagonal to the centre line of the pavement.	• Traffic loads on unsupported corners • Sub soil pumps out.	Cleaning cracks and filling the voids.
9.	Longitudinal cracks (Shrinkage cracks)	Cracks parallel to centre line of the road.	• Shrinkage of concrete • Expansive sub-base or subgrade	Clean and fill with rubberised bituminous seal.

Contd....

Table 16.4: Defects in rigid pavements (contd....)

Defect/Distress	Description	Possible Cause	Remedial measures
10. Restraint cracks	Cracks near the outside edge of pavement (within 1 m)	• Foreign matter lodged deep and restraining slab from expansion.	• Joints plowed out and resealed.
11. Transverse cracks	Cracks approximately at right angles to the centreline of pavement.	• Overloads • Repeated bending of pumping slabs. • Soft foundation.	• Clean and fill the cracks with rubberised bitumen sealar.
12. Polished aggregates	Aggregate particles in the surface of pavement that have been polished smooth.	• Due to traffic movement.	• Cover the pavement with a skid resistance treatment.
13. Crow foot cracks	Cracks parallel to pavement edge or at an angle	• Failure of expansion joint filler to extend. • Infiltration of foreign matter into the joints.	• Change/put new filler material. • Improve the joint.
14. Crazing	Series of fine shallow cracks which extend only in to the upper surface of the slab.	• Restrained shrinkage • Common in hot weather concreting	• Thin hot mix overlay of bituminous material. • Thin bonded topping.
15. Opening of longitudinal joints	Longitudinal joints open up, panel can open with or without faulting.	• Poor construction. • Absence of tie bar.	• Install the bar by cutting slots across the joints or cracks.
16. Inadequate surface texture	Surface texture of pavement falls below the acceptable limit.	• Passage of heavy traffic.	• Lay bonded concrete overlay.

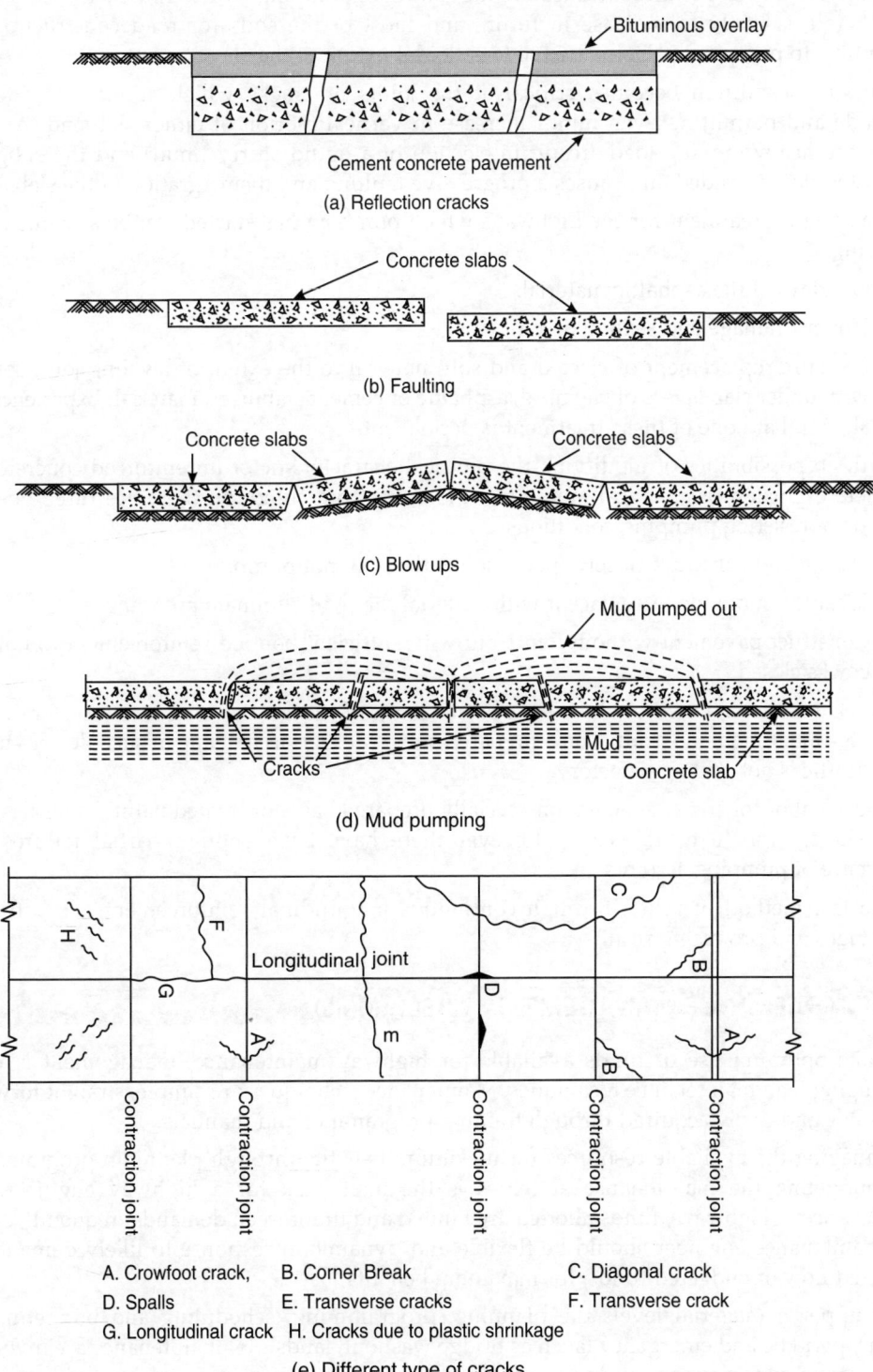

(a) Reflection cracks

(b) Faulting

(c) Blow ups

(d) Mud pumping

A. Crowfoot crack,   B. Corner Break   C. Diagonal crack
D. Spalls   E. Transverse cracks   F. Transverse crack
G. Longitudinal crack H. Cracks due to plastic shrinkage

(e) Different type of cracks

Fig. 16.3: Defects in rigid pavements

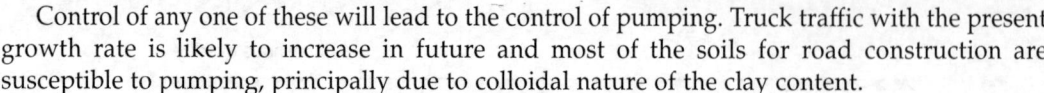

Control of any one of these will lead to the control of pumping. Truck traffic with the present growth rate is likely to increase in future and most of the soils for road construction are susceptible to pumping, principally due to colloidal nature of the clay content.

Pumping condition becomes critical at the joints or cracks which admit water to the subgrade and permitting even minor increase in vertical motion of either slab end. At such points the heavy moving loads result in ejection of a liquid slurry (mud) and the subgrade support is slowly eroded and causes a progressive faulting and disintegration of the slab.

Treatment: Treatment for the highways where pumping has started can be any one of the following:

(i) Overlays of the asphaltic material.

(ii) Under drainage and sealing of joint

(iii) Pressure replacement of ejected and soft material to the extent of leveling joint corners with under slab layers of chemical, asphaltic or cement stabilized material. Experience has shown that none of these treatment is permanent.

With no possibilities of modifying the physical characteristic of unreinforced concrete and complete exclusion of a critical amount of moisture from subgrade there are three possible methods of resisting pumping conditions.

1. Design and construct closely spaced joints that will not pump.

2. Treat the subgrade or replace it with material that will eliminate pumping.

3. Construct pavement without joints but with sufficient bonded reinforcement to hold all cracks closed.

As regards the first method there have been a range of success by using load transfer devices for joints, under moderate traffic, on a prepared subgrade. Serviceable life of such devices under heavy traffic is not quite satisfactory.

Placement of concrete pavement on specially prepared bases or treated natural soils has, had some success for 10 to 15 years. However there have been enough partial failures and occurrence of pumping at joints.

The last method of relief, through continuous longitudinal reinforcement has sufficient advantages and promising results.

16.4 MAINTENANCE MANAGEMENT SYSTEM (MMS)

To make optimum use of funds available for highway maintenance, management is more challenging than maintenance techniques. Maintenance techniques are simple, straight forward, repeatable and easily acquired through training programmes and manuals.

Managing the available resources for maximum benefits through planning, programming and budgeting the maintenance activities is the main task for a highway engineer. For maintenance of highways, funds allotted are limited and unforeseen demands frequently occur. The maintenance engineer should be flexible and dynamic in response to likely demands or restricted flow of budgeted funds, for maximum benefits/use.

At upper managerial levels the planning, programming, scheduling and budgeting for routine, periodic and emergency (such as bridge washout, landslide) maintenance are involved, while at on-site managerial level, construction management by effective utilisation of manpower, material, equipment, time allocated, funds is required. The efficient highway maintenance is an extremely challenging management problem.

Management in maintenance planning is to ensure consistent and adequate standards, methods, procedures, specification, and then to develop means of objectively preparing maintenance programmes related to needs and constraints.

Figure 16.4 shows a general maintenance management flow chart. Various steps are described briefly as below.

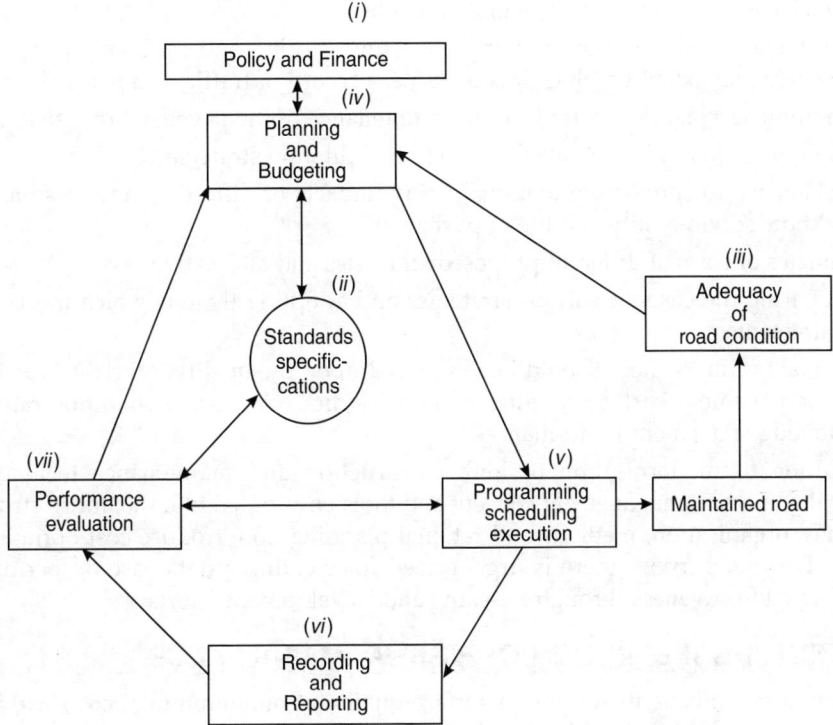

Fig. 16.4: Typical chart of maintenance management

(i) Policy and finance related decision are to be taken by higher authorities.

(ii) Standard and specifications are established to promote consistent and uniform maintenance procedures for different type of roads and standards for the same. Indian Road congress takes care of these for Indian roads. It suggests what action to be taken at certain defects (Pothole, broken edge, etc), and specify procedure, equipment and material. The output and costs should also be related to works.

(iii) Monitor the condition of the existing network and compare with (*ii*) to assess needs. A system should be established to assess the needs, for maintenance of different roads.

(iv) For meeting the needs as in (*iii*) planning budgeting and procuring, funds, is required. Priority programmes and schedules should be established for each year. This step involves political decision which may take time. Advance planning is therefore necessary.

(v) Within the plan and budget established in (*iv*) above, planning programme and executing the works should be done at district/area level. The engineer and his staff should workout the schedule of maintenance for the whole year, and allocate priorities. This however is not a simple task.

(vi) Measuring, recording, reporting costing and other relevant data is required for future planning and expenditure monitoring. This step is necessary for local short term monitoring and cost control.

(vii) Evaluation should be done whether the above road needs in (*iii*) above were met. If not or to refine and improve going back to step (*v*) and revising the whole steps for better output, is to be done.

An effective management system should include:

- A systematic technique of regularly collecting, storing and retrieving condition of the pavement and extent to which it is used, i.e. a record of traffic data using the pavement.
- Minimum acceptable standard for the maintenance of the pavement of different types.
- Developing alternative maintenance and rehabilitation strategies.
- Mechanisms for predicting and evaluating impacts of different strategies on pavement condition, serviceability and life of pavement.
- Estimates of costs of different proposed strategies and alternatives.
- Comparing the costs of various strategies and adopting the one which meets the stated requirements.

A pavement rating system should be developed upon the quality, severity, type of distress affecting the pavement surface condition. The periodic collection of condition-rating data is essential to judge pavement performance.

In addition to modernisation of high construction and maintenance techniques, it is necessary that modern management concept and tools be adopted in India, for re-structuring of maintenance organisation, methods and rational planning, to introduce cost optimisation and efficiency. For this purpose there is urgent need for creating a data-base on performance of roads and cost effectiveness through Research and Development efforts.

16.5 MODELLING THE PROBLEM OF HIGHWAY MAINTENANCE

For maximum social benefits (optimum surface quality at minimum total cost) within allotted funds, optimum strategies should be worked out. Figure 16.5 shows a typical model for surface deterioration as a function of climate, road type, traffic and maintenance. Road roughness as a measure of surface quality which is a major link between maintenance and operational costs, because it affects speed, and fuel consumption. Speed also depends on vehicle loads and types, road geometry (curvature and gradient etc.)

Several models are developed for optimising these variables. Two pioneer models are: one developed by Transport and Road Research Laboratory (U.K.) "Road Investment Model (RTIM 2)" and other developed by World Bank "Highway Design and Maintenance Standard Model (HDM III)". These two models are refined and modified by different countries from time to time to suit their local conditions. In India also, use of such models has started recently.

16.5.1 The TRRL Road Investment Model (RTIM-2)

RTIM-2 is designed to study various aspects of road investment project, optimising maintenance standards, strategies and related benefits of selecting road types (earth, gravel, bituminous pavements, etc), stage construction etc. A flow diagram of the model is shown in Fig. 16.6.

Lot of data on various aspects of road, traffic, costs, road failures, defects, speeds, geometry of road, etc. is required for computer based analysis in this model. Further details on RTIM-2 can be obtained by Transport and Road Research Laboratory, U.K.

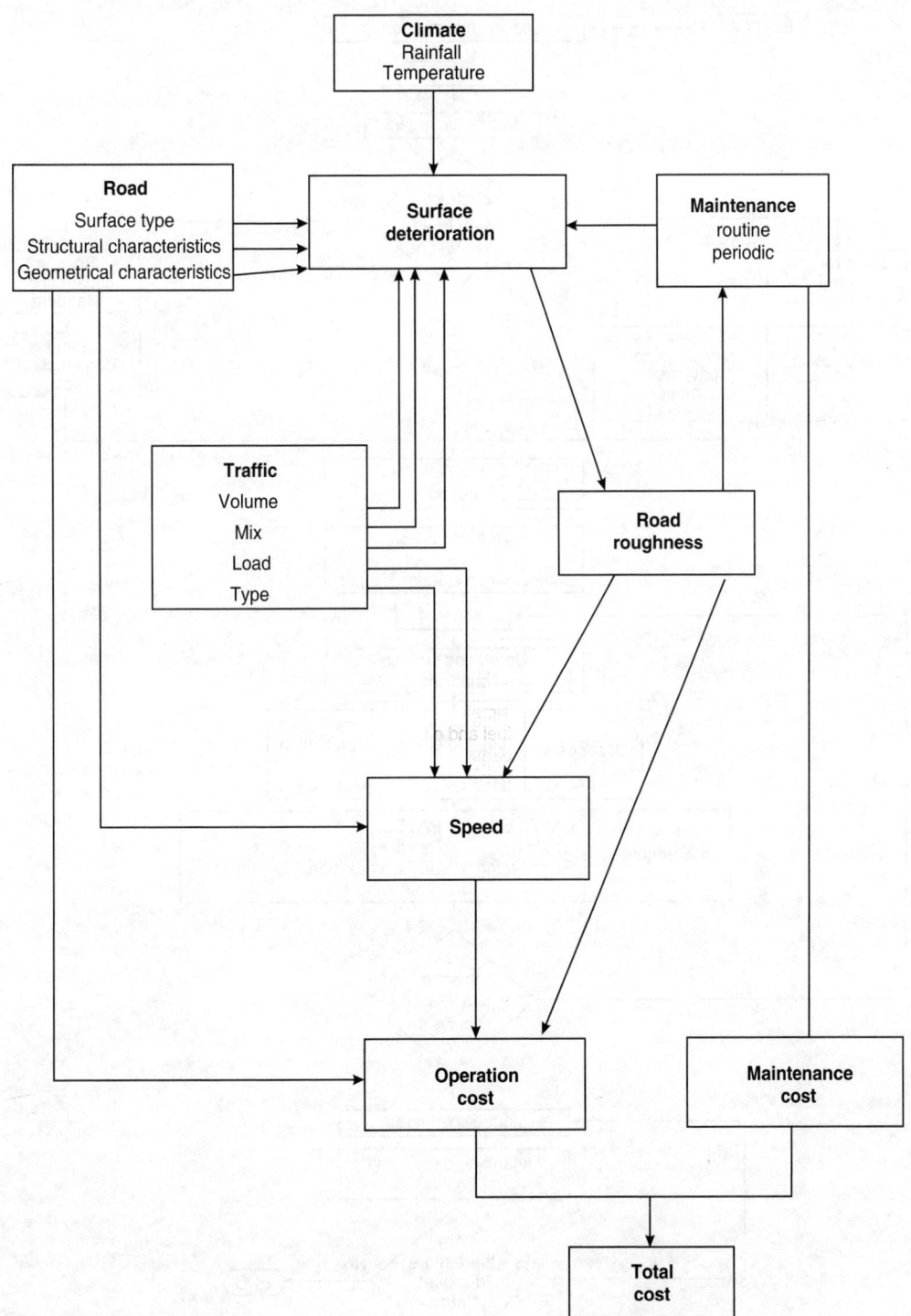

Fig. 16.5: Typical model showing interaction between surface deterioration, maintenance and vehicle operating costs

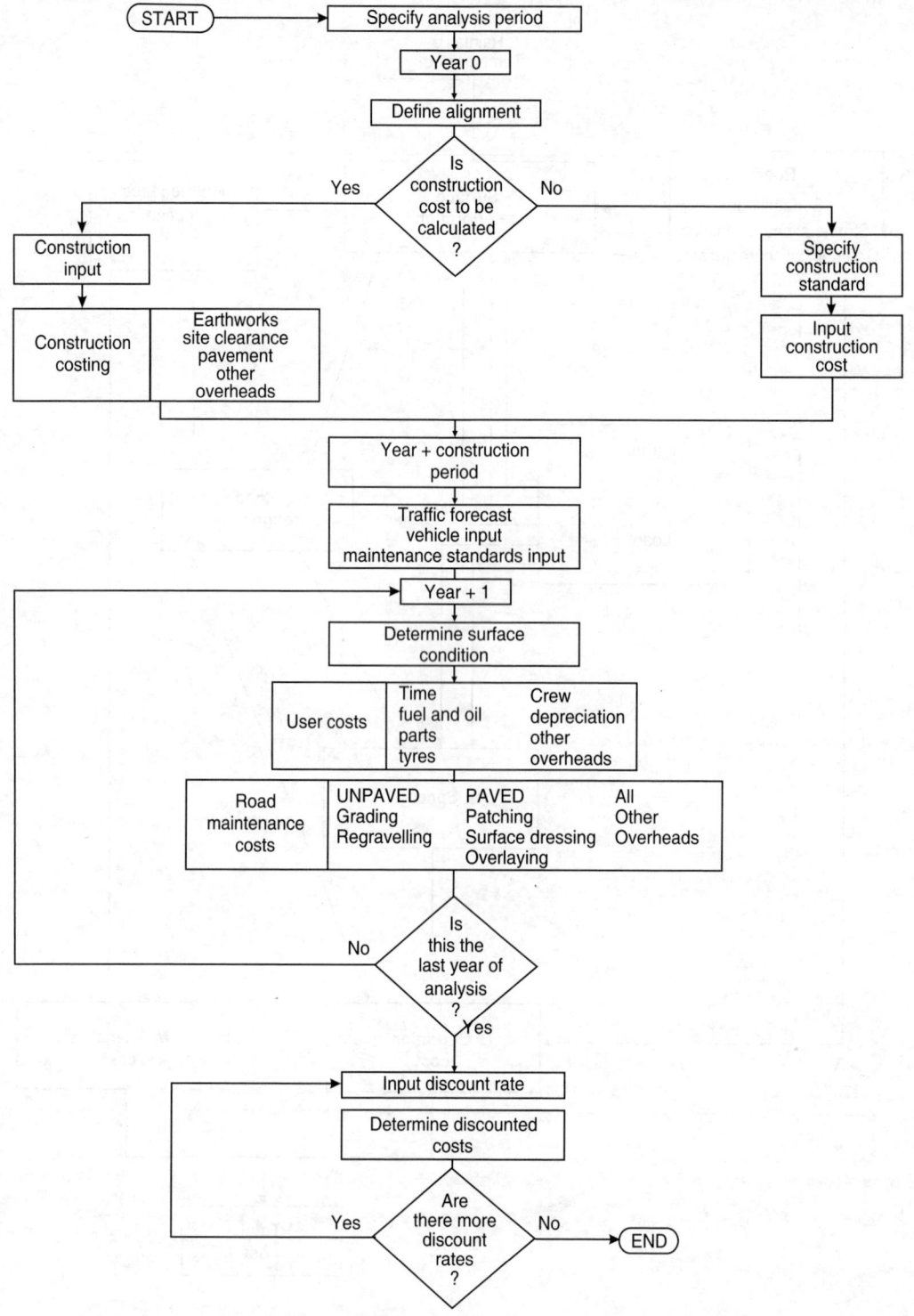

Fig. 16.6: Flow diagram of TRRL model RTIM-2

16.5.2 HDM Model (World Bank's Model)

Objectives of HDM model are almost same as that of RTIM-2.

The HDM is an analytical tool for engineering and economic assessment of:

- road investments;
- maintenance strategies;
- transport pricing and regulation;
- network programme optimization; and
- budget strategy analysis

It is based on physical and economic relationships derived from extensive research on road deterioration, the effects of maintenance activities and road user costs. The interaction of construction, maintenance and road user costs in HDM model is shown in Fig. 16.7. The flow diagram is shown in Fig. 16.8.

The HDM computes:

- deterioration of paved and unpaved roads for a set of specified road agency strategies;
- road user costs as a function of the roadway and vehicle characteristics; and
- time-streams of road agency and user costs for the specified strategies.

It compares these strategies by presenting relevant economic indicators and provides analytical support for selecting the best strategy. The HDM provides/defines:

- analytical support to justify funding of road works;
- provides optimum maintenance strategy under different budgetary constraints;
- economic priorities amongst projects;
- financial and physical needs for preserving the road network; and
- information for policy formulation studies (pricing of road user charges, optimum axle loading, fleet modernization, etc.)

The operation of the HDM takes place in three phases:

1. data input and diagnostics phase, in which input data are generated and examined;
2. simulation phase, in which traffic flows and changes in road conditions from initial construction through annual cycles of use, deterioration and maintenance are analyzed; and
3. economic analysis and comparison phase, during which alternative construction and maintenance policies are analyzed and compared to the base case for selected groups of road links.

HDM-III can evaluate up to 20 different road links in a single computer run, each link may have up to 10 sections of different design standards and environmental conditions. Reports are generated to give differences between financial, economic and foreign exchange costs for pairs of alternatives, comparing them in terms of net present values at various discount rates, internal rates of return, incremental benefit-cost ratio and first year benefits. Flow diagrams for road deterioration and maintenance sub model for paved road is shown in Fig. 16.9 and for unpaved road in Fig. 16.10.

The model capabilities/applications include:

- Congestion analysis ("Q"-Model);
- Project formulation;
- project evaluation;
- Project prioritization;

- Network programme optimization;
- Budget strategy analysis;
- Policy formulation; and
- Expenditure budgeting model for the highway sector.

For more details of these models, World Bank's latest publications should be followed.

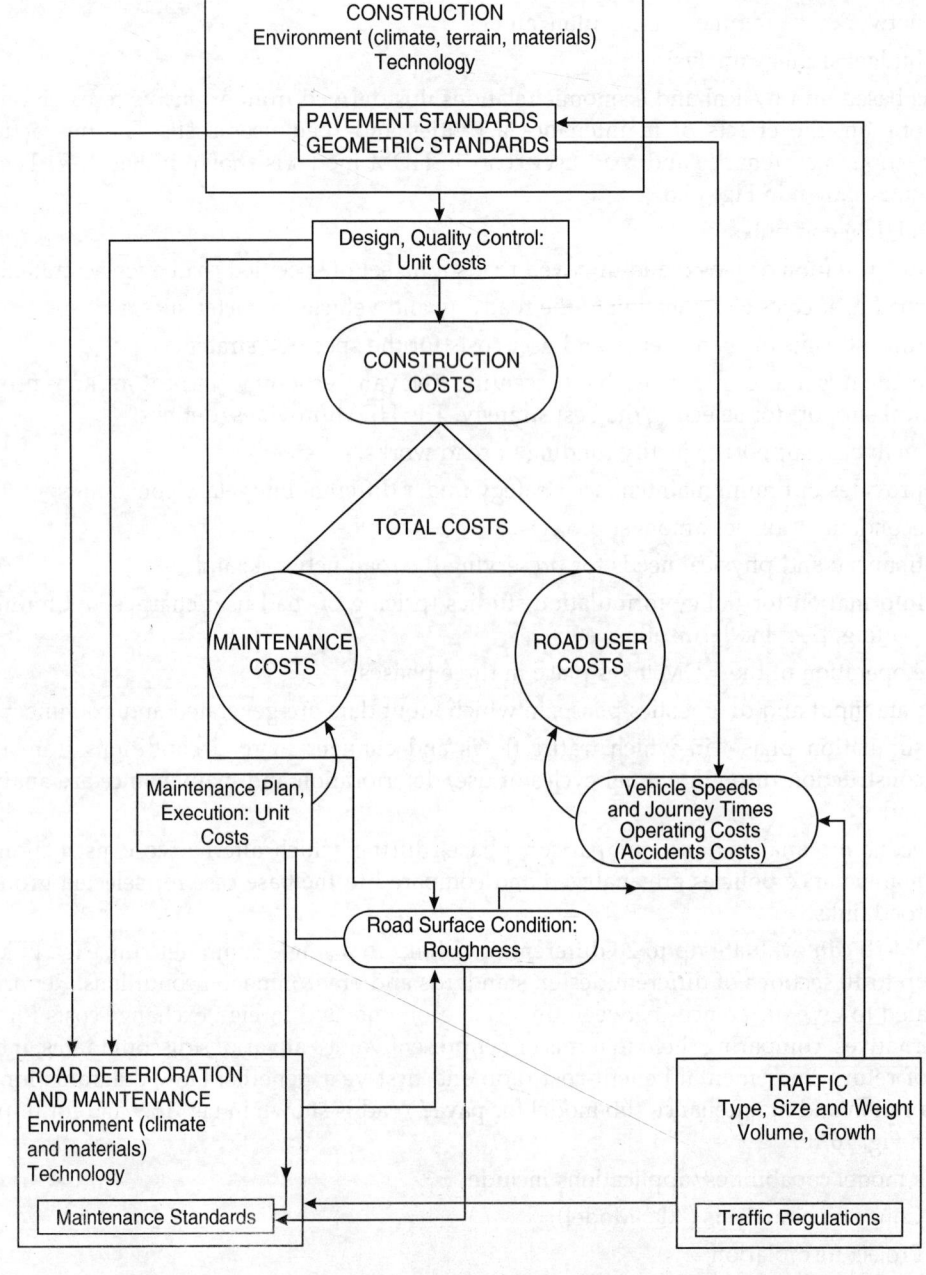

Fig. 16.7: The HDM Model: Interaction between costs of road construction, maintenance, and road users

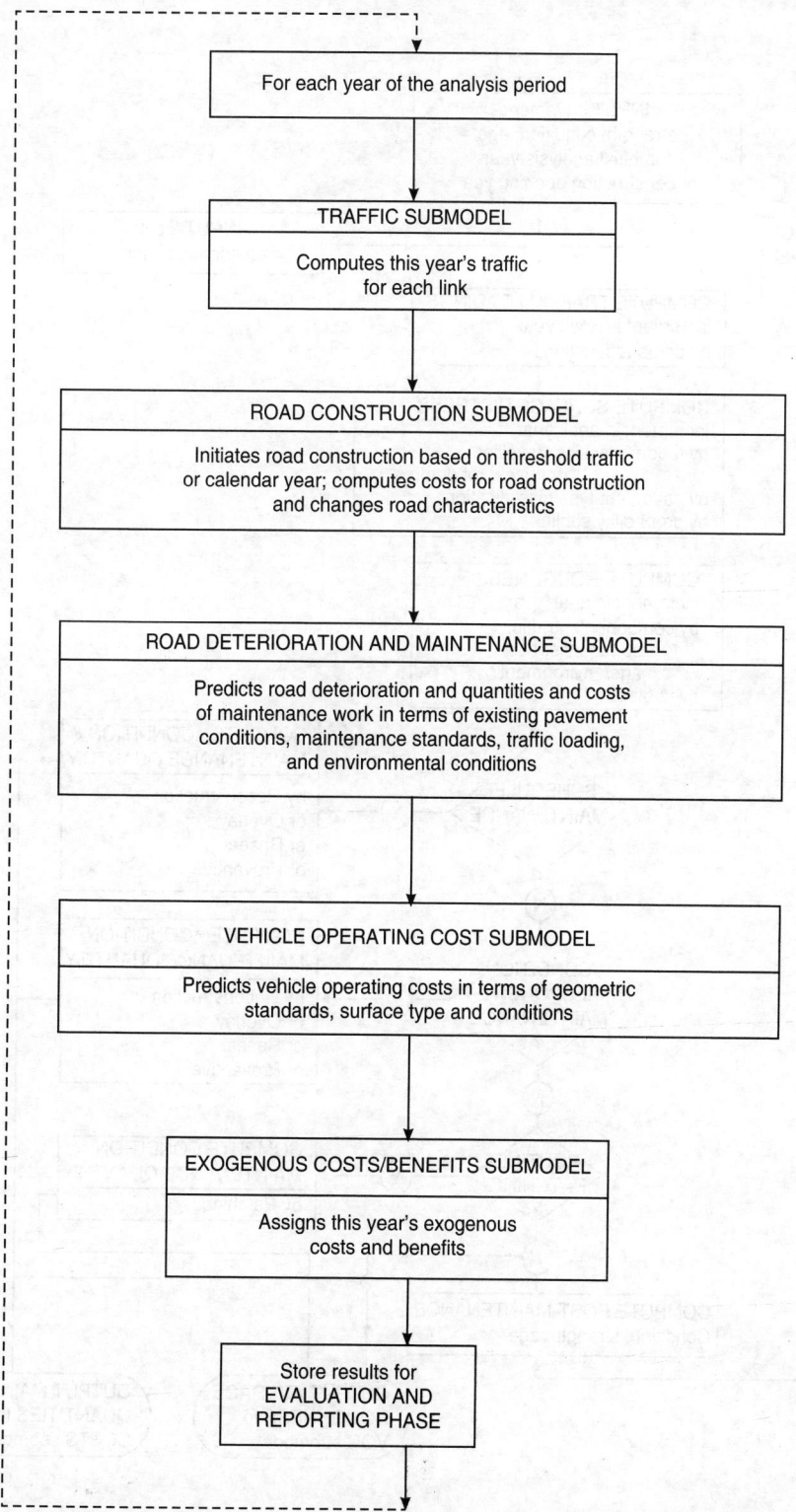

Fig. 16.8: The HDM-III model-flow diagram, showing sub-models

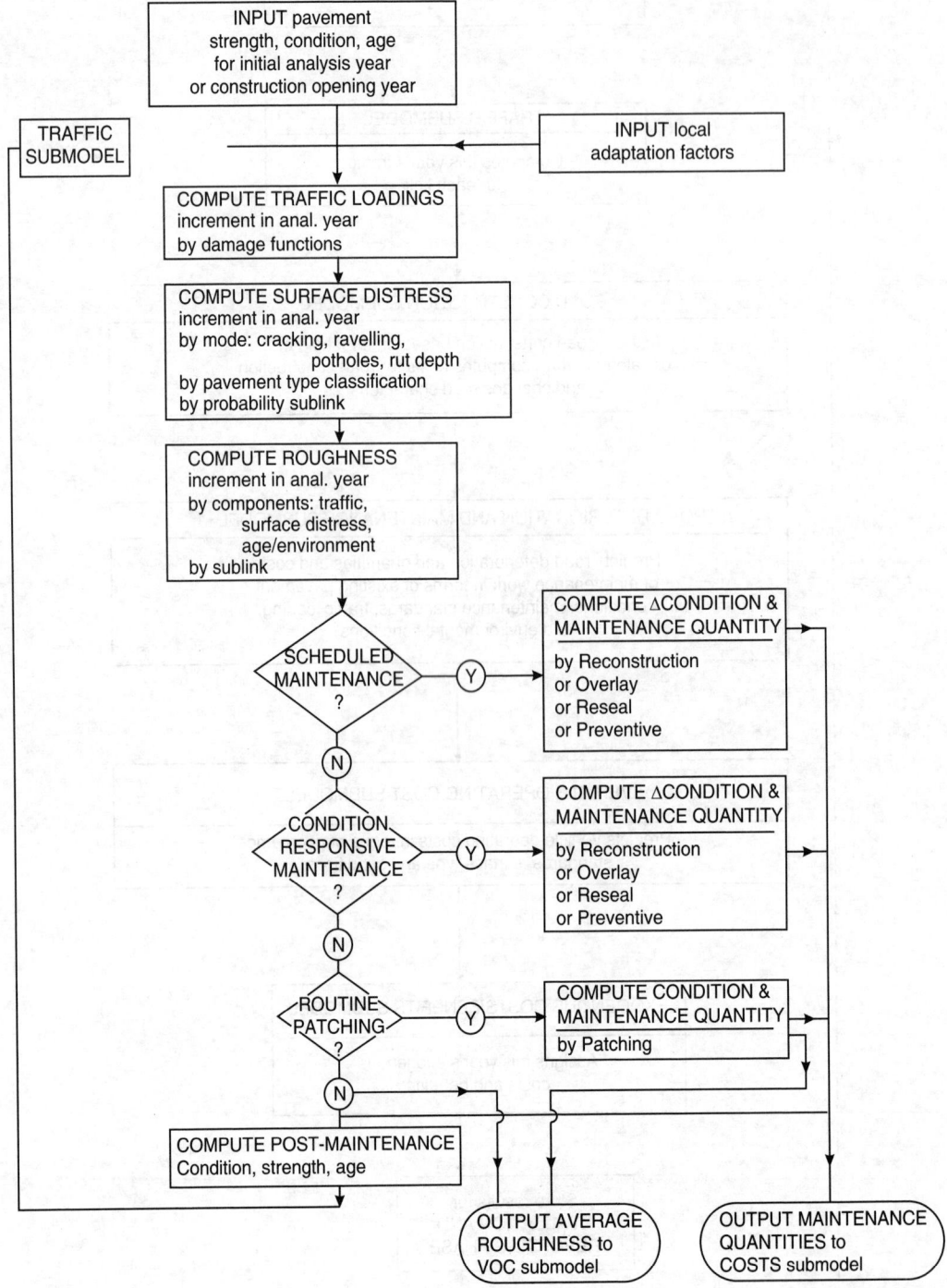

Fig. 16.9: HDM Sub model for paved roads (Flow-diagram) (road deterioration and maintenance)

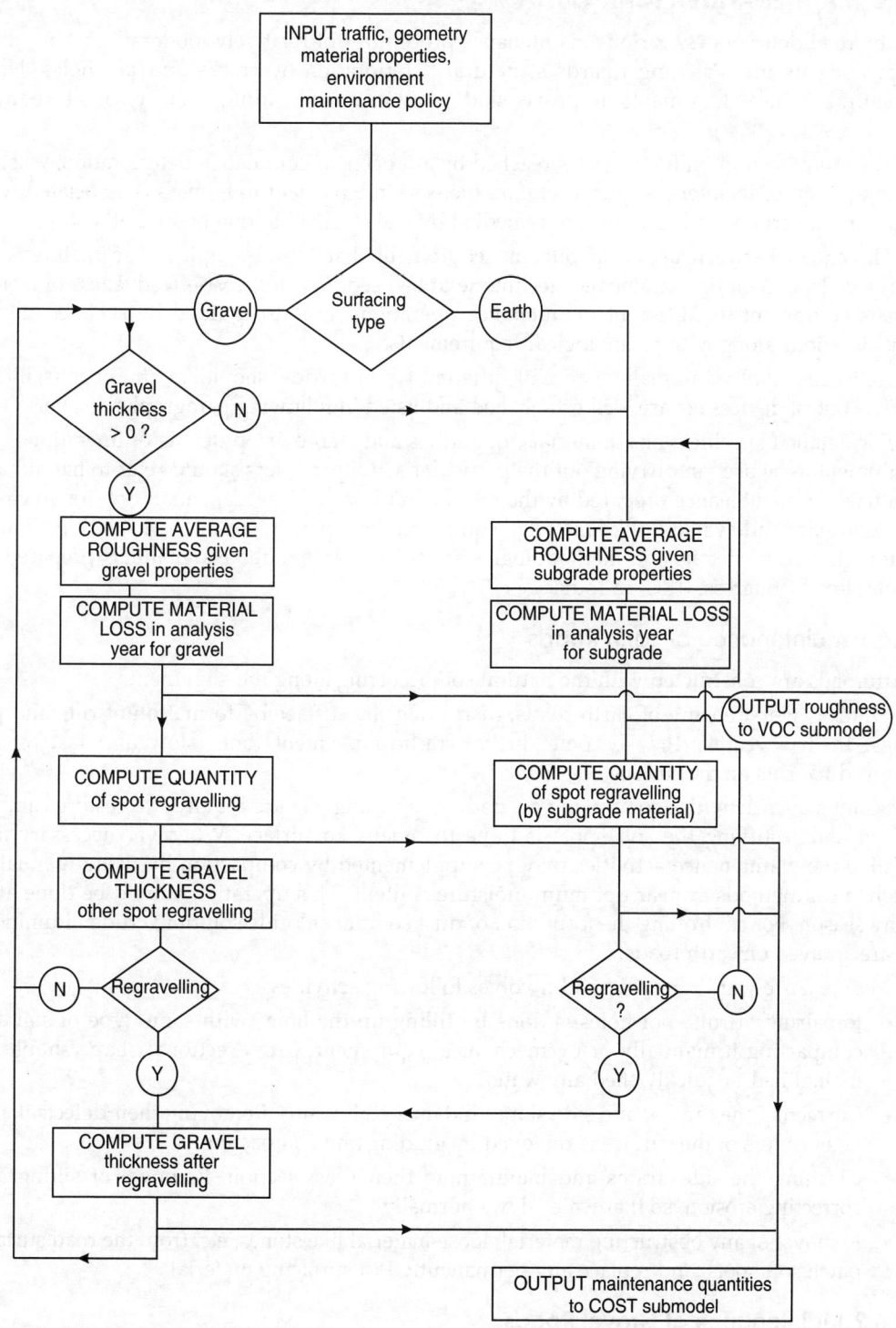

Fig. 16.10: HDM sub-model for unpaved roads (flow diagram)
(road deterioration and maintenance)

16.6 MAINTENANCE TECHNIQUES

As the road deteriorates, various maintenance procedures of relatively moderate cost are used to prolong its life. Patching retards structural deterioration by cracks and pot holes. Non-structural surface treatments improve skid resistance and riding quality, as it reduces roughness.

The failure condition for a road is reached by a process of cumulative deterioration, when it no longer serves its intended purpose. If an increase in pavement roughness is associated with structural deterioration it can only be remedied by major rehabilitation or strengthening.

The choice between use of labour and equipment that can be employed in maintenance works is dependent on maintenance technique to be used. Use and cost-effectiveness of labour versus equipment for different maintenance operations, are also guided by socio-economic considerations along with technological requirements.

Activities involved in maintenance of different type of roads, shoulders, drainage facilities, traffic control devices etc are well established and listed in following paragraphs, in short.

For detailed specifications, materials, quantities and step by step details of operations and equipment to be used in carrying out the particular activity, readers should refer to handbook/manuals for maintenance prepared by the relevant authorities. These manuals are drawn based on experience and availability of material, equipment, manpower, etc. and considering climatic factors. Extensive details and methodologies are covered in handbooks/manuals prepared by Indian Road Congress, (IRC) in India.

16.6.1 Maintenance of Earth Roads

Earth roads are constructed with the natural soil occurring along the alignment.

Water is worst enemy of earth roads, disrupting the surface by formation of ruts and pot holes. In dry weather they become dusty. Traffic movement (both slow and fast) create longitudinal ruts on the surface.

Maintenance activities comprising of grading, patching, etc are required to restore camber. Filling and rectifying the soft spots is done to repair the surface. Wherever necessary and feasible the maintenance activities may be supplemented by compaction by using manual or mechanical methods at near optimum moisture content. This operation should be done after rainy season. For controlling the formation of dust sodium chloride solution, crude bituminous oil are sprayed on earth roads.

Maintenance work for earth road involves following activities:

- Repairing of ruts, pot holes is done by filling up the holes with same type of soil and compacting it manually or by mechanical equipment. Cross-sectional shape should be maintained to quickly shed any water.
- Correcting the camber and side slopes if damaged by rains. Removing their defects if any. Unevenness of the surface is removed by grading and compacting.
- Cleaning the side drains and maintaining their cross-section. Removal of silting and correcting erosion, so that water flows normally.
- Removal of any obstructing material, loose material like stones, etc. from the road surface, patch soft spots, and replace any permanently lost surfacing material.

16.6.2 Maintenance of Gravel Roads

Gravel roads are surfaced with a layer of gravel or moorum which is mechanically better than the natural soil in earth roads.

The main objective of maintenance of gravel roads is to keep the water away from the road by immediately disposing it away. It may cause ruts or pot holes.

Regravelling, and grading is done regularly. Gravel is restored from shoulders, pot holes and corrugations are removed. Gravel road requires regular maintenance to keep the surface smooth, free from pot holes, ruts and maintain the camber.

Normal activities for maintaining the gravel roads are:

- Repairing of pot holes, depressions, corrugations, etc. as soon as possible.
- Checking of camber and side slopes and maintaining them to normal requirements so that water does not remain standing on the road surface.
- If badly damaged by rains, renewal may be done for the whole road, using fresh material in a way similar to its construction as new.
- To prevent formation of dust on gravel roads, calcium chloride solution is applied at the rate of 0.25 to 1.25 kg/sq.m, depending upon climatic conditions.

16.6.2.1 Patch repairing procedure

Steps involved in patching work of road surface are:

(a) Kankar/Gravel Roads

- Patches should be repaired as soon as they appear and should not be allowed to develop into pot-holes.
- The patch should be first marked out on the ground inform of series of rectangles, encompassing the patches.
- These rectangles are then dug cutting the edges vertically to a depth of 5 to 7 cm.
- The periphery of the bottom of the whole should be made slightly deeper say 2 cm in a width of 7 cm.
- The hole should then be saturated with water for some time.
- The hole is then hand packed placing large size material in the bottom, then medium size and small size on the top.
- Consolidation by rammer is done starting from edge towards centre using plenty of water keeping finished surface slightly higher say 5 mm than road surface.
- The patch is then left for a day protected by the traffic using it.
- The next day patch is rammed again, before opening to traffic. During this operation, the patch comes in line and level with the road.

(b) WBM Roads

- Patches are prepared and repaired as above, except that second day ramming is not done. Rolling by rollers may be done in place of ramming if repair work is extensive.

16.6.2.2 Rut filling procedure

- Rut filling is done when renewal coat is not to be used, immediately. Following procedure is used:
  - A continuous trough is excavated over each rut with side slope of about 60°.
  - The old rut trough surface is spread with new or washed and cleaned old material. It is then hand arranged true to camber and level.
  - Consolidation is done as usual, using water and rollers.
  - By this type of filling of ruts, road performance is improved and renewal coat can be postponed by an year or two.

16.6.3 Maintenance of Water Bound Macadam Roads

Water-bound macadam roads develop defects like ruts, pot holes, corrugations, ravelling, damaged edges, grinding of stone, etc. due to use by traffic, climatic factors, or flaws in construction or materials. The normal maintenance procedure for water bound Macadam roads-which are very common in India involves:

- Repairing/renewal of ruts by clearing them and filling with fresh metal and screenings and compacting using water.
- Correcting the pot-holes by patching, and filling with new material and compacting. Excavated material can also be used after cleaning.
- Corrugations are scarified, surface is broomed and a new layer of WBM may be put to correct the surface irregularities.
- The ravelling phenomenon can be corrected by using a thin building layer on the surface. Screening and blindage material are mixed, watered and compacted.
- Repairing of damaged edges, shoulders is done by putting fresh material, and compacting it.
- The surface of water bound macadam road is regularly cleaned, by sweeping.
- WBM surfaces deteriorate very fast and renewal is done every 2-3 years. Renewal process is same as that of new WBM layer.

16.6.4 Maintenance of Bituminous Roads/Surfaces

Maintenance of bituminous roads basically involves three activities: general repairs, surface dressing and overlays. Maintenance of bituminous roads involves following works.

- Sealing of cracks and small areas showing loss of aggregates.
- Sanding in cases of bleeding. Coarse sand or small chippings of stone work as blotting materials.
- If there is any subsidence, re-shaping is done.
- Patching of potholes, and removal of surface irregularities such as ruts or depressions. Mixture of sand, aggregates and bitumen is used for surface repairs. Process involves cutting, digging, filling the holes, ramming and sand covering.
- Surface dressing is done to arrest surface deterioration, provide impermeable surface and improve skid resistance. Surface dressing is done after general repairs have been done to the surface, specially in defects like bleeding, cracking, glazing, fretting or ravelling, streaking, local deformation, etc.
- Overlays are used where there is structural failure or traffic has increased demanding upgrading of the road. In some case when deformation is progressive, it is to be arrested.

  Overlays consist of hot mixes of bitumen and aggregates, meeting the specifications and laid like a new construction.

IRC-82 "Code of Practice for Maintenance of Bituminous Surfaces" provides details of treatments for different type of distresses in bituminous surfaces.

16.6.5 Maintenance of Concrete Roads

Concrete roads have the main advantage that their maintenance costs are not high, if construction and design is properly done. However due to weathering and traffic movement defects do appear on concrete road such as cracks, mud pumping, etc.

Maintenance of concrete roads involve following main operations.

- Cracks are filled with bituminous sealing material, after cleaning. For deep cracks bituminous material can be forced in by pressure blower for filling. Filled cracks are topped up by sand to prevent damage by passing traffic.

- Joints are considered weakest, in cement concrete pavement and they should be regularly checked and maintained. The filler and seal get damaged with time due to expansion and contraction of concrete slab. Damaged joints are cleaned and fresh filler material is placed and top is sealed with sealer material. Winter days are best time to re-make joints.

- Surface of the road can develop defects like spalling, depressions, irregularity, scaling, etc. These are removed by patching the defective areas, by bituminous pre-mixes or by cement grout using epoxy resins in mortar.

- Blowing or mud pumping from the joints or cracks in concrete slab is a major problem. The defective joints and cracks which blow-off are refilled and sealed. A bituminous material is pumped underneath the slab to prevent re-occurrence of pumping.

16.6.7 Maintenance of Shoulders and Slopes

The objective of shoulder maintenance is to allow the shoulders to continue giving adequate lateral support to the pavement, and to allow their use by vehicles, e.g. overtaking, parking, etc.

Maintenance activities for unpaved shoulders involve retaining the cross-section and grade by grass-cutting, shoulder grading and dressing, water should not remain standing on the shoulders. It should be drained off from the roadway surface towards the side drains and ditches.

Slopes should be maintained to protect from damaging effect of water so that they retain their shape and stability, cross-section and camber for successful pavement performance. During the rainy season, unpaved shoulder are damaged due to rains and pits may be formed. These pits should be filled and compacted soon after the rain. Side slopes are protected by pitching or growing suitable grass on them (turfing).

Following items should to be checked by maintenance engineers during the maintenance of shoulders and slopes:

- There are no obstructions on the shoulders.
- Shoulder surface is at lower level than carriagway surface.
- Rut and depressions are removed.
- Vegetation on shoulders is not too high on shoulders and slopes.
- Slopes are not eroded by erosion. Slipped slopes are rectified.
- Shoulders if they are made of gravel, WBM, or bituminous surfacing, their maintenance is done similar to the surface of the road.
- Slopes which are protected by pitching by stones require attention during maintenance as damaged or loose stones should be removed and pitching should be redone properly.

16.6.8 Maintenance of Traffic Control Devices

Traffic control devices such as signs, marking, signals, islands, etc. get spoiled, hardly visible or dull, and correcting these defects and maintaining in good condition is required for safety and convinience of road and drivers.

The normal maintenance works for traffic control devices include:

- Painting of signs and markings for proper visibility when required.
- Traffic signals should be checked for their cycle times (Green-Yellow-Red) in view of changing volumes of traffic periodically.
- The signal mountings and lights should be cleaned as they become dusty after some time.

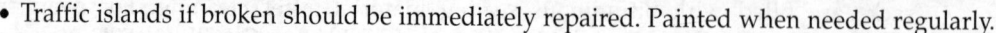

- Traffic islands if broken should be immediately repaired. Painted when needed regularly.
- When a traffic signal fails, it is desirable to alert users to the fact that it is no longer operating normally for safety in such situations.

16.6.9 Maintenance of Drainage Structures

Proper and efficient drainage is the most important feature in the maintenance of roads. Adequate drainage keeps the road in good condition, free from several defects. Water erodes the slopes, weakens the pavement, destroys shoulders and slopes, and washes culverts, embankments and even bridges.

The objective of highway drainage system is to ensure that components of the drainage system, viz ditches, drains, culverts, causeways, manholes, underdrains, culverts, remain free from obstructions, retain their intended cross-sections and slopes.

In maintenance operations of drainage structures following items should be checked and repaired/corrected.

(a) *Ditches and drains*
- Ditch cross-section is not destroyed.
- There is no ponding in ditch and shoulder.
- There is no silting in the ditch or drain.
- There is no erosion of sides.
- Ditch lining is not damaged.
- There is no obstruction to flow of water like vegetation/grass in the ditches.

(b) *Slopes*
- Slopes are not eroded.
- Slopes are protected by vegetarian, correctly. Pitching (if any) is not disturbed.
- Slipping has not occurred.

(c) *Causeways*
- There are no pot holes in unpaved surfaces.
- Cracks are not there in paved surfaces.
- Guide post at causeway are not missing and are well painted for visibility.
- There are no loose material or obstruction.

(d) *Manholes and underdrainage*
- Manhole covers are intact and not broken.
- Soil or vegetation is removed from man-hole covers.
- Water is not flowing out of manhole cover at the surface.

(e) *Culverts*
- Silting, sanding, or blockage by debris, etc. is removed.
- Erosion at inlet or outlet is corrected.
- Settlement cracks, rut, damages are repaired.

(f) *Bridges*
- Condition of all components such as foundation, wing walls, approaches, sub-structure, floor, etc. is kept alright.

- Bearings are greased and there is no sign of erosion.
- There is no settlements.
- Highest flood levels are not reached.
- Steel components are protected by painting when needed.
- Repairing of joints, in masonry works and other defects should be removed.
- Any obstructions like vegetation, which clogs flow of water should be removed.
- Scouring of the foundation should be measured and corrected by construction or dumping stone boulders, concrete blocks, etc. as required.
- Joints should be periodically renewed.

16.7 EVALUATION OF HIGHWAY PAVEMENTS

The performance of all roads should be consistently evaluated to assess the ability of existing pavement to support traffic volumes and loads. The purposes of highway evaluation are:

- To check if the required pavement function and performance objectives are being achieved.
- To establish load carrying capacity of existing pavements.
- To provide guidance for planning, maintenance and rehabilitation.
- To judge the adequacy of existing design, construction and maintenance procedures and where they require improvements.
- To establish road performance data for use in future planning, design and research. Data is also required for economic analysis.
- To asses the condition of pavement, for devicing maintenance strategies for its up-keep.
- To asses the need for strengthening or structural overlays for the distressed pavements, to meet expected demands.
- To help in need-based budgetary allocations.
- Residual life of the existing pavement can be assessed through evaluation.

16.7.1 Indicative Parameters

The objective assessment of the present condition of a road requires rating of the individual components that make up that condition. The usual parameters considered are :

(a) *Riding quality:* The riding quality measured by roughness is not affected by either rut depth or surface texture. Roughness is related to riding comfort.

(b) *Skidding:* When the road surface becomes slippery and is not able to provide adequate resistance to skidding.

(c) *Pavement deterioration:* When pavement surface is distressed by having pot holes, ravelling, corrugations, cracks, etc.

(d) *Structural strength:* This is the condition when structural support/strengthening becomes necessary to carry the heavy traffic loads.

Other factors may be travel speed, level of service, delays and accidents. Considering these parameters a single performance indicator that will reflect overall effect of all parameters is evaluated for comparisons of various roads/or sections of the same road.

16.8 TECHNIQUES OF PAVEMENT EVALUATION

Procedures used for evaluation of pavements range from visual techniques to sophisticated electro mechanical non-destructive methods, as described below:

(i) Pavement condition survey (visual rating).

(ii) Roughness measurement

(iii) Serviceability index for pavement

(iv) Measurement of deflection

(a) Rolling wheel: This technique measures pavement surface deflection when a standard moving wheel load is applied like Bankleman Beam, Deflectograph.

(b) Stationary loading: This technique measures pavement surface deflection when a standard weight is allowed to fall on to a circular plate of standard dimensions like falling weight deflectometer, plate bearing test.

(i) Pavement condition survey (visual rating system)

Method I Straight edge

There are different methods of visual rating system followed in different countries. This procedure involves visual inspection to identify pavement distress features (such as cracks, potholes, etc.) pavement distortions (such as rutting and corrugations, etc) and edge failures, etc. This is supplemented by quantitative simple techniques, for measuring rutting, cracks and roughness.

The technique to measure rutting and other pavement distortions uses a straight edge. The straight edge is laid across the top of the distortion and the distance to lowest level is measured from the top bottom of the straight edge, as shown in Fig. 16.11. The value of the measurement is affected by the length of the straight edge which is normally of standard length - 2.0 m to 3 m

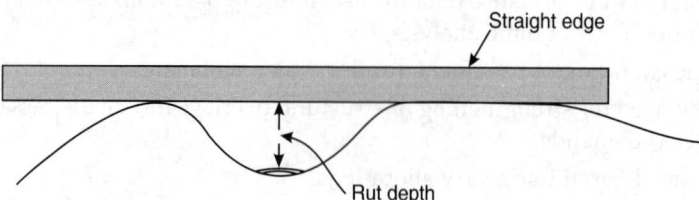

Fig. 16.11: Straight edge technique for measuring rutting

The cracking measurement is done by simply placing a shallow box that is open ended top and bottom, having inside dimensions of 1 m^2 on the pavement surface and measuring the total length of cracks within that area (Fig. 16.12).

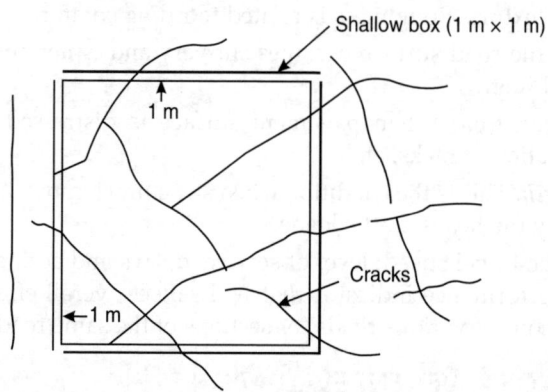

Fig. 16.12: The crack measurement technique

Method II Visual Rating Method

Another method of visual rating was developed in Texas, USA. The team of experts would inspect the road about 1 km in each inspection. The rating of each deficiency in terms of both the severity and extent is undertaken and points are deducted indicating the importance of distress. Each inspection begins with a value of 100. The defects or distress considered are: corrugations, cracking, rutting, patching, revalling, pavement failure, edge failure, etc. Greater points are deducted for severity of a particular defect from slight, moderate to severe.

After deduction of the points for each defeciency/defect out of 100, an overall rating score for the particular inspected length of the pavement is obtained.

Method III Australian Method

On the lines similar to the Texas system, Australia deviced a system producing condition indices for 6 categories, i.e. wearing course, pavement, shoulders, drainage, roadside and traffic devices. Each index begins with a value of 100 and points are then deducted for each defeciency as prescribed. The final score for condition indices of the roads is used as shown in Table 16.5.

Most of the systems of visual rating followed, have 3 parts:

1. The selection of items on which data is to be collected, e.g. the distress types (patching, rutting)
2. The rating of each item: e.g. 5% of the road length is patched, with fair quality of patching.
3. The combination of each individual item rating into a single total rating.

Table 16.5: Index scale for condition indices (NAASRA, Australia)

Score Range	Management considerations	Treatment considerations
Negative – 0	Sign posting required to alert driver to abnormally deteriorated road condition	Road fails to provide safety and comfort level for the class of road. Reconstruction is required salvage will be nil.
0 – 20	Sign posting should be considered depending upon alignment and weather. Regular (weekly) inspection may be advisable.	As above Reconstruction is required
20 – 40	The section should be on current program of works	Rehabilitation work is critical in order to avoid loss of value of existing asset.
40 – 60	These sections should be considered for incoming programme. The rate of deterioration should be considered in assigning work for action.	Normal maintenance to rehabilitation will arrest deterioration in the value of existing pavement.
60 – 80	These sections should be monitored for deterioration rate in order to timely maintenance action	Preventive maintenance in the form of minor repair. Regular supervision for maximum effectiveness.
80 – 100	Routine monitoring of condition.	Deterioration since construction is barely visible.

(ii) Roughness measurements

Roughness is the unevenness of a road, and is measured in millimeters per kilometer (mm/km), as the sum of the maximum deviations of the actual profile from designed profile. The smaller this number is more even is the surface. International standard measure of roughness recommended by

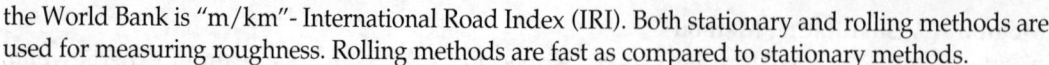

the World Bank is "m/km"- International Road Index (IRI). Both stationary and rolling methods are used for measuring roughness. Rolling methods are fast as compared to stationary methods.

Method I: Use of levels

The simplest technique is to measure the profile of the road by measuring its longitudinal profile by a levelling measure instrument, at suitable intervals. This is a time consuming and tedious method but quite accurate in results.

Method II: Bump integrator

Another method is use of 'Bump Integrator', which measures the vertical movement of a single wheel relative to its mounting frame as it travels over the road at a uniform speed. The speed of 32 km per hour is used. The roughness measured is recorded in mm/km and can also be plotted on electromagnetic tape.

It consists of a trailer of single wheel with a pneumatic tyre mounted on a chasis over which integrating device is fitted. The vehicle is usually a Jeep which towes the machine at tyre pressure of 2.1 kg/cm.

The unevenness index value is calculated by dividing the bump integator counter values (in cms) by the distance travelled in kms.

$$\text{The unevenness index} = \frac{\text{Integrator counter value (cms)}}{\text{Distance travelled (kms)}}$$

Method III: Abay's beam

Another method is called 'Abay Beam' named after its inventor. It consists of a beam, which is 3 m long, along which a fixture is made to slide. The transducers fixed in the slider detect the vertical profile of the road along the beam.

Method IV: (Car mounted unit TRRL/or NAASRA roughness meter)

Like the bump integrator, car mounted integrator unit is placed directly above the rear axle of a car and a flexible steel cable taken from the axle is wound around a pulley in the integrating unit. Distance along the road is recorded by any distance measuring device. The roughness is given in mm, and recorded by a mechanical counter. Travel speed is normally 50 km/hr in urban and 80 km/hr in rural areas.

It is a laser profilometer fitted with laser-based measurement system that records the measured longitudinal profile of the road.

Recommended values for maximum unevenness for different types of roads surfaces are given in Table 16.6. The values are to be used in conjunction with the maximum permissible frequency values given in Table 16.7.

Table 16.6: Maximum permissible surface unevenness for roads (IRC)

Type of Surfacing	Maximum Permissible Surface unevenness	
	Longitudinal Profile 3m straight edge (mm)	Transvers Profile camber template (mm)
1. Surface dressing	8	8
2. Open graded premix carpet*	8	6
3. Mix seal surfacing	8	6
4. Semi dense bituminous concrete	6	4
5. Bituminous concrete	5	4
6. Cement concrete	5	4

* These values are for mechanised construction. For manual construction the tolerance may be increased by 2 mm.

Table 16.7: Maximum permissible frequency of surface unevenness in 300 m length in longitudinal profile (IRC)

S.No.	Type of surfacing	Unevenness (mm)	Max'm Number of surface unevenness NH & SH	Max'm number of surface unevenness MDR and other lower roads
1.	Surface dressing	8 – 10	20	40
2.	Open graded premix carpet	6 – 8	20	40
3.	Mix seal surfacing	6 – 8	20	40
4.	Semi-Dense bituminous mix	4 – 6	20	40
5.	Bituminous concrete	3 – 5	15	30
6.	Cement concrete	4 – 5	15	30

It is important to determine that roughness is associated with structural deterioration of pavement. If it is so, deflection tests are required and major rehabilitation of the road will be needed.

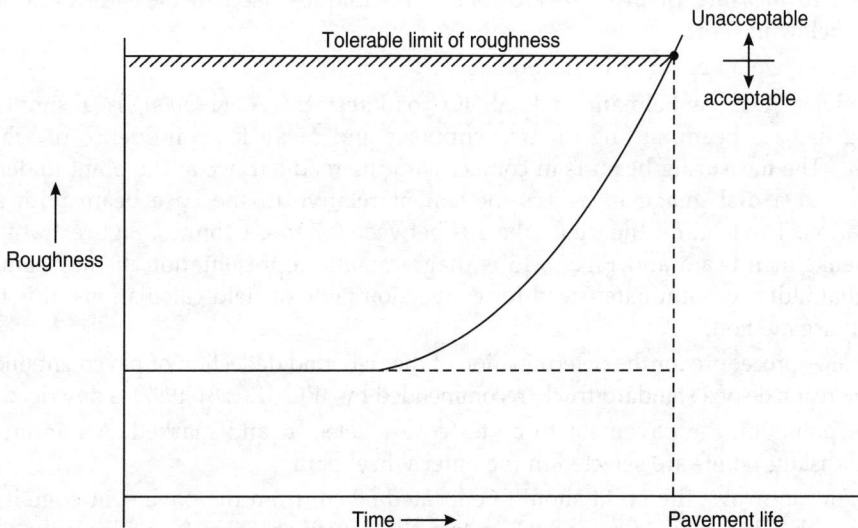

Fig. 16.13: Roughness change with time

Figure 16.13 shows the roughness of a road over a period of time. It happens partly with traffic and partly with time dependent environmental deterioration.

Table 16.8 shows maximum permissible values of roughness (mm/km) for road surfaces.

Table 16.8: Maximum permissible values of roughness (IRC)

Type of surface	Condition of Road surface		
	Good (mm/km)	average (mm/km)	Poor (mm/km)
1. Surface dressing	< 3500	3500 – 4500	> 4500
2. Open graded premix carpet	< 3000	3000 – 4000	> 4000
3. Mix seal surfacing	< 3000	3000 – 4000	> 4000
4. Semi dense bituminous concrete	< 2500	2500 – 3500	> 3500
5. Bituminous concrete	< 2000	2000 – 3000	> 3000
6. Cement concrete	< 2000	2000 – 3000	> 3000

(iii) Pavement serviceability index [present serviceability index (PSI)]

The serviceability of a pavement is the extent to which it meets the requirements of safety and comfortable ride for its users. Pavement serviceability index method developed by AASHO road road test is a numerical rating of the suitability of a road for high speed, high volume mixed traffic. It depends upon

- wheel path slope variance
- cracked area
- ruts and their depth.

Because it measures the riding quality, the basic present serviceability rating (PSR) is obtained from a panel of drivers, driving on the road. The drivers are asked to rate the road on its ability to serve the traffic at its operating speed on a linear seale from 0 (extremely poor) to 5 (extremely good). For development of data, individuals (the rater) ride the pavement and assign a PSR value. A relation is developed between PSR and measurable parameters of pavement distress. The value determined by this relationship is called "Present Serviceability Index".

(iv) Measurement of deflection

Deflections caused by fully loaded truck travelling at high speeds are realistic parameters to obtain, but to measure them is very difficult. Techniques used to measure deflections are described below in short.

Method I: Benkelman beam or Deflection beam

Developed by A.C. Benkelman, in WASHO road test U.S.A. (1950's), is a simple device consisting of base beam, sitting on two supports and away from influence of wheel load deflections. The measuring beam is in contact with the road surface at the point under test. At the other end a dial guage measures movement relative to the base beam with a 2 to 1 magnification. The load on the dual wheel is between 2.7 to 4.1 tonnes. Figure 16.14 shows a typical Benkelman beam and Fig. 16.15 is diagrammatic representation of the beam. Direct-reading dial indicator eliminates need for conversion table or field calculations. It is therefore an easy to use method.

The CGRA procedure for the determination of the re-bound deflection of pavement under static load of the rear axle of a standard truck, recommended by IRC (IRC: 81-1997) is described below:

(i) The points on the pavement to be tested are selected and marked. A minimum of 10 equidistant points are selected in the outer wheel path.
  (a) For highways, the point should be located 60 cm from the pavement edge if the lane width is less than 3.5 m and 90 cm from the pavement edge for wider lanes.
  (b) For divided four lane highway, the measurement point should be 1.5 m from the pavement edge.
(ii) The dual wheels of the truck are centred above the selected point.

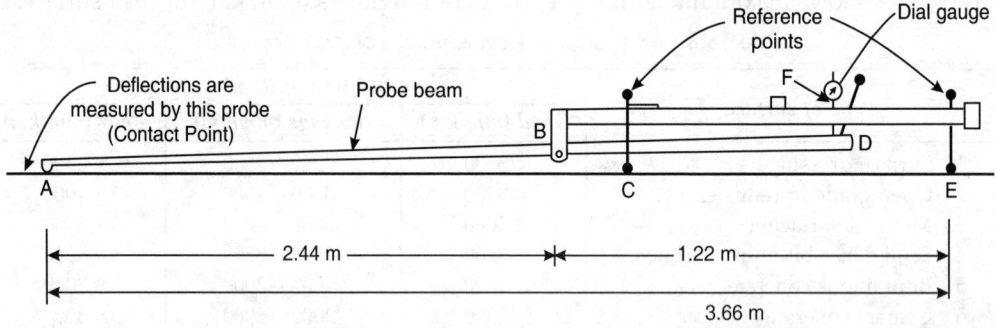

Fig. 16.14: A Benkelman beam

The Benkelman beam consists of a slender beam 3.66 m long, having following dimensions:

(i) Length of probe arm from pivot to probe point — 2.44 m
(ii) Length of measurement arm from pivot to dial — 1.22 m
(iii) Distance from pivot to rear legs — 1.66 m
(iv) Distance from pivot to fornt legs — .25 m
(v) Lateral spacing of front support legs — .33 m

It has a mandril for making 4.5 cm deep hole in the pavement for temperature measurement. The diameter of the hole at the surface shall be 1.25 cm and at bottom 1 cm.

By suitably placing the probe between the dual wheels of a loaded truck, rebound (elastic) and residual (plastic) deformations of the pavement structure are measured. Rebound deflections are used in overlay design.

For measuring pavement deflection either the CGRA procedure, which is based on testing under static load or the WASHO procedure based on creep load test may be used. In both these methods a standard truck transmitting a load of 8170 kg and equipped with dual tyres inflated to a pressure of 5.6 kg/cm is used.

(i) The probe of the Benkelman beam is inserted between the duals and placed on the selected points.
(ii) The locking pin is removed from the beam and the edges are adjusted so that the plunger of the beam is in contact with the stem of the dial guage.
(iii) The beam pivot arms are checked for free movement.
(iv) The dial gauge is set at approximately 1 cm. The initial reading is recorded when the rate of deformation of the pavement is equal or less than 0.025 mm/minute.
(v) The truck is slowly driven a distance of 270 cms and stopped.
(vi) An intermediate reading is recorded when the rate of recovery of the pavement is equal to or less than 0.025 mm per minute.
(vii) The truck is driven forward a further 9 m.
(viii) The final reading is recorded when the rate of recovery of pavement is equal to or less than 0.025 mm per minute.
(ix) Pavement temperature is recorded at least every hour, inserting thermometer in the standard hole and filling up the hole with glycerol.
(x) The tire pressure is checked at two or three hours interval during the day and adjusted to the standard if necessary.

The calculations of the rebound deflection are done in the following steps:

(i) Substract the final dial gauge reading from the initial dial reading.
(ii) Substract the intermediate dial gauge reading from the initial reading.
(iii) If the differential readings obtained compare within 0.025 mm, the actual pavement deflection is twice the final differential reading.
(iv) If the differential reading obtained do not compare to 0.025 mm, twice the final differential dial reading represent "Apparent Pavement Deflection".
(v) Apparent pavement deflections are corrected by means of the following formula:

$$X_T = X_A + 2.91 \, Y$$

where
X_T = True pavement deflection
X_A = Apparent pavement deflection
Y = Vertical movement of the front legs, i.e. twice the difference between the final and intermediate dial reading.

(vi) The rebound deflection percentage, shall be twice of the X_T value.

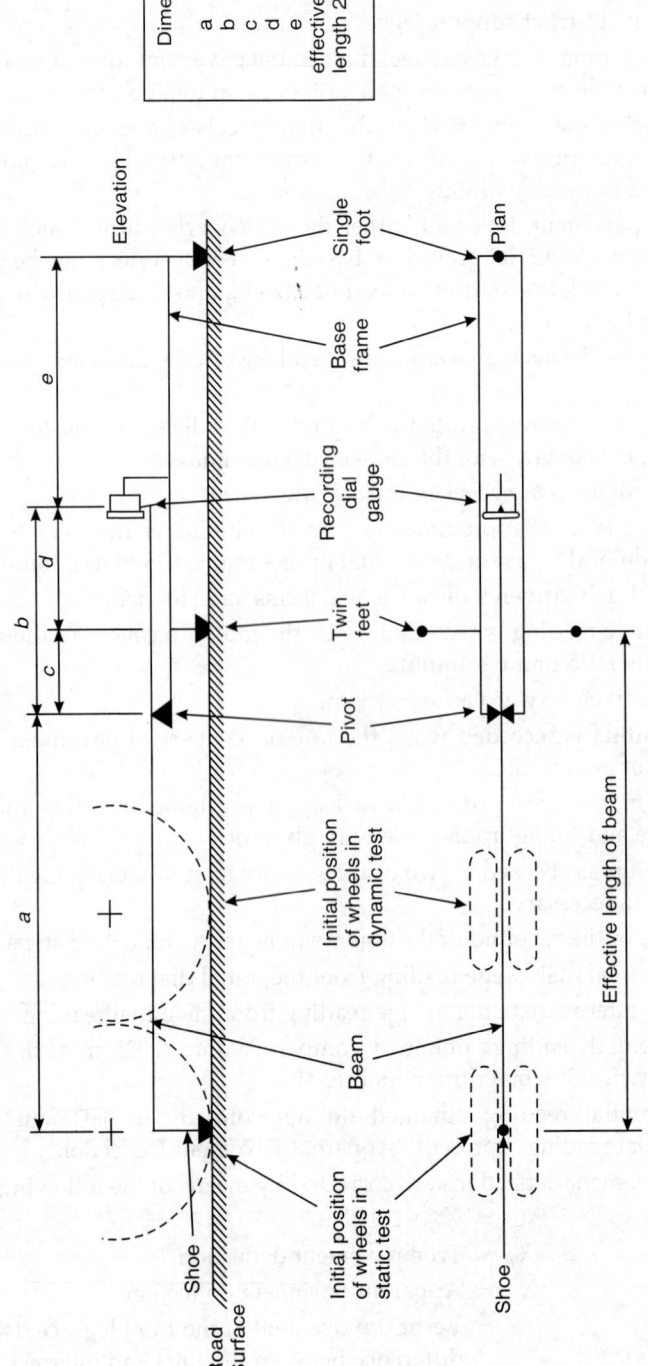

Dimensions	
a	2.44 m
b	1.22 m
c	0.30 m
d	0.91 m
e	0.61 m
effective length	2.74 m

Fig. 16.15: Diagrammatic representatioin of the deflection beam

Example: The dial gauge reading at the selected point on the pavement is 0.51 mm, while the reading behind the selected point and infront of the selected point are 0.07 mm and 0.09 mm respectively in a Benkelman beam test for measuring the deflection.

Calculate the deflections.

Solution.

$$\text{Maximum deflection} = 2 \text{ (Reading at selected point – initial reading)}$$
$$= 2 \text{ (0.51 – 0.07)} = 0.88 \text{ mm.}$$
$$\text{Rebound deflection} = 2 \text{ (Reading at selected point – final reading)}$$
$$= 2 \text{ (0.51 – 0.09)} = \mathbf{0.84 \text{ mm}}.$$
$$\text{Residual deflection} = \text{Maxm deflection – rebound deflection}$$
$$= 0.88 – 0.84 = \mathbf{0.04 \text{ mm}}.$$

The WASHO (Western Association of State Highway Organisation) test based on creep load, is similar to Benkelman beam method for measuring pavement deflection. The truck in this method is moved forward at as creeping speed of 2 km/hr. The maximum reading shall be when the wheels are in the line with the probe arm. The probe arm is located 1.2 m infront of the wheel. This reading is initial and the final reading is when the dial needle comes to rest; after some time.

The rebound deflection is = 2 (Maximum reading – initial reading)

The residual deflection is = 2 (Final reading – initial reading)

Method II: The Lacroix Deflectograph

Based on the Benkelman concept, partially automated deflectometers have been developed. The Lacroix deflectograph is a loaded truck with a 8.2 tonne rear axle and a deflection-measuring undercarriage. It travels at a constant speed of 2 km/hr. Actual measurement is taken at about every 4 m. The observations can be recorded on tapes and discs and analysed on computers. The data obtained is used for the analysis and assessment of the load bearing capacity of the road.

Figure 16.16 shows a typical Lacroix deflectograph.

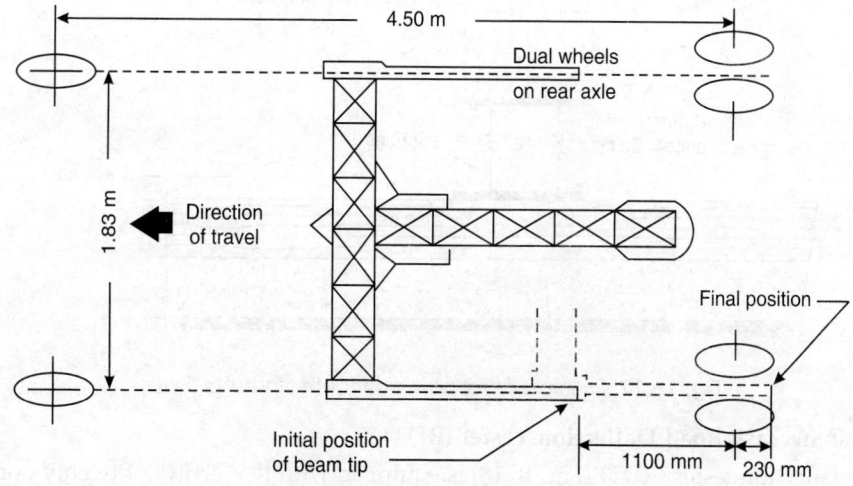

Fig. 16.16: Lacroix deflectograph

Method III: Dynaflect Dynamic Deflection

Dynaflect dynamic deflection system, determines the deflection of a pavement under a known dynamic load. It consists of a force generator and live deflection sensor (geophones) housed in a small trailer towed by a light vehicle. The force is generated by two eccentric vertically rotating masses. The unit can be operated from driver's seat.

The system provides rapid and precise measurement of road deflections, which in this test are caused by forces generated by unbalanced flywheels rotating in opposite directions. A vertical force of 1000 Lbs is produced at the loading wheels, and deflections are measured at five points on the pavement surface located 1 ft. apart.

Dynaflect system is designed to be operated behind any vehicle and is self contained with a control box so that the operation can be performed by one man. It is used to determine overlay requirements for pavements.

Method IV: Falling Weight Deflectometer (FWD)

Falling weight deflectometer (Fig. 16.17) is another device, which uses realistic loading levels and times. The load is applied by a mass falling onto a set of springs that are mounted on a rigid circular plate resting on the pavement surface. The equipment is carried on a single axle trailer. Force impulses created by falling load more closely resemble the pulse created by a moving load.

Variations in applied load may be achieved by either altering the magnitude of the mass or the height of drop.

Vertical peak deflections are measured by the FWD in the centre of the loading plate and at varying distances away from the plate to draw "Deflection Basins".

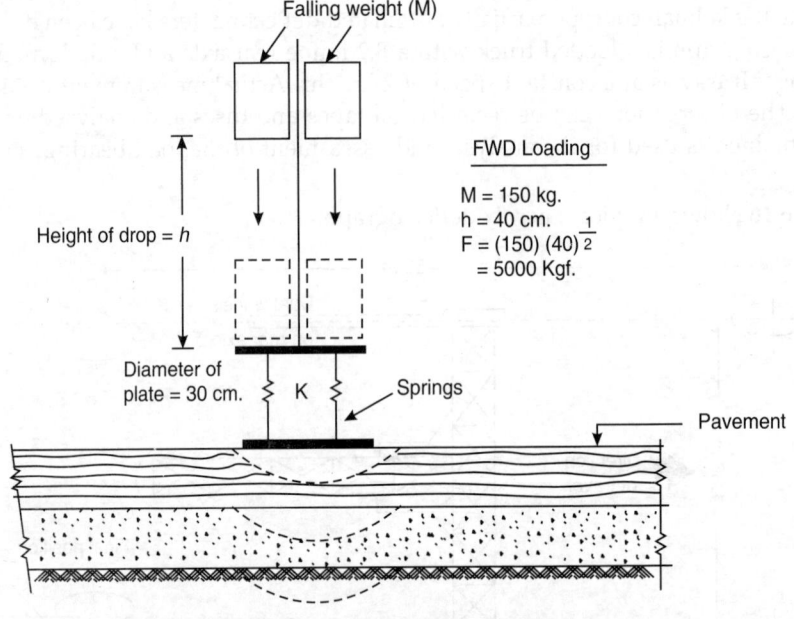

FWD Loading

$M = 150$ kg.
$h = 40$ cm.
$F = (150)(40)^{\frac{1}{2}}$
$= 5000$ Kgf.

Fig. 16.17: Principle of falling weight deflectometer (FWD)

Method V: Swedish Road Deflection Tester (RDT)

The road deflection tester (RDT) (Fig. 16.18) is equipped with two arrays of twenty non-contact sensors that collect transversal surface profiles at normal traffic speeds. One profile placed between the wheel axles, constitutes an unloaded case. The other profile just behind the rear axle of the vehicle, constitutes the loaded case. By substracting the front cross profile from the corresponding rear one, the "Deflection profile" is assessed.

In order to use large loads on the rear wheels, the engine was mounted in the back of the vehicle, slightly behind the rear axle. In testing mode the rear axle force is approximately 112 KN and the front axle force is about 30 KN. An incremental wheel pulse transducer, two force transducers and two accelerometers, an optical speed meter and a gyroscope are also mounted on RDT.

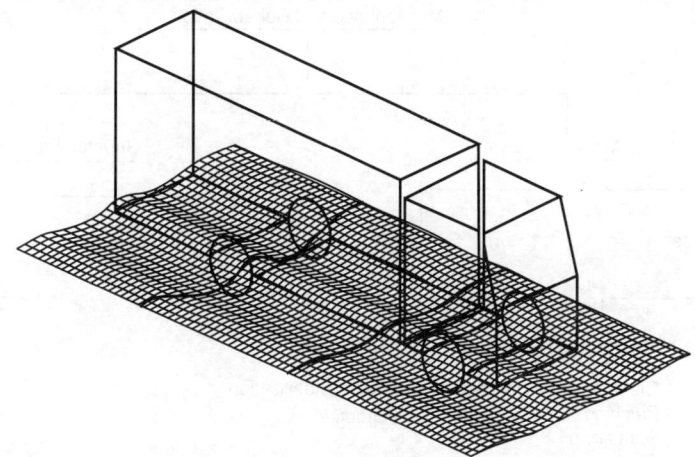

Fig. 16.18 The working principle of RDT vehicle. (Two arrays of non-contact laser sensors acquire the transversal deflection profile)

Method VI: Laser Method (Road Surface Profiler)

Laser based system can measure road profile and deflection accurately while travelling at normal traffic speed. Laser based road surface profiler (RSP) is designed to provide an advanced, automated, high quality pavement roughness and related measurements.

The RSP system consists of two primary components:

(i) A transducer unit carrying laser sensor and accelerometers.

(ii) A PC with expansion slot capability.

The accelerometer is used to obtain vertical vehicle body movement and a laser is used for measuring the displacement between the vehicle body and the pavement. The profiler can be provided on any vehicle. This method is non-destructive.

Rolling Weight Deflectometer (RWD) is laser based system for measuring deflection. Using multiple lasers, a spot on the road can be measured before and during its deflection. A semi-trailer carries a 7.8 m aluminium beam specially designed to house 4 lasers spaced 2.6 m apart. A computer is located inside the trailer to record the data.

Results of the pavement evaluation can be used for planning maintenance and rehabilitation of the highways and also predict pavement conditions in future years. When considerable data is collected overtime, the results can be used to develop mathematical models that relate pavement condition to age of pavement. Prediction models help in inspection, maintenance and rehabilitation activities. Figure 16.19 shows alternative uses of the collected data.

Benkleman Beam is used for designing the overlays for pavements, which are badly deteriorated and need strengthening.

Pavement evaluation is in fact inversion of design process. To design a pavement, one selects, a design method to determine the thickness and acceptable characteristics of materials for each layer and the wearing course, taking into account the subgrade on which pavement will rest and intensity of traffic (loading) the pavement must support. For evaluation, the process is reversed since the pavement is already existing. Character of subgrade and thickness and character of each structural layer must be established, from which the maximum allowable magnitude of traffic loading can be determined, by using the choosen design method in reverse. The elements of the design method in the existing pavement must be evaluated in accordance with the selected design method.

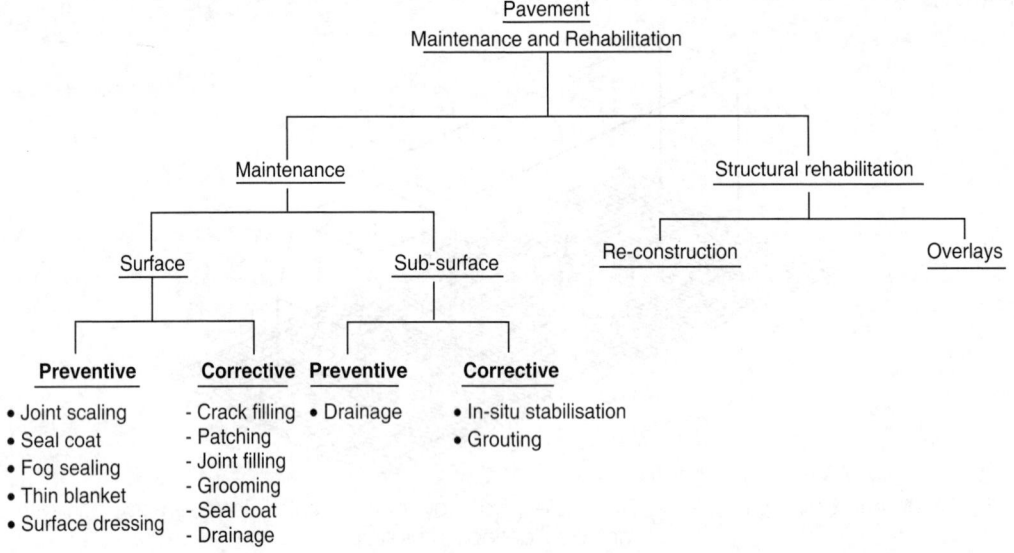

Fig. 16.19: Alternatives for maintenance and rehabilitation

16.9 STRENGTHENING OF PAVEMENTS (OVERLAYS)

Overlays means putting a new layer over the top of existing pavement. Overlay may be required due to variety of reasons such as:

- A pavement may have been damaged by overloading by increased or heavy traffic and overlay is required to maintain serviceable level.
- A pavement may require strengthening by overlaying to serve heavier traffic than that for which pavement was originally designed.
- A pavement which is worn-out after serving its design life, and may require an overlay.
- An overlay is required sometimes to merely improve the riding quality of the pavement.
- Overlay may also be used to improve skid resistance and provide better safety.

By conducting condition surveys, the pavement condition is assessed either by individual ratings or roughness, or by the pavement serviceability index. The condition surveys or evaluation surveys, determine the structural adequacy of the pavement. Thus it is determined that whether surface maintenance is adequate or overlays are necessary in view of expected future traffic. To provide strength or stiffness to pavement, a new layer of material of design thickness is put over the existing surface. This added layer is called an 'Overlay'.

Type of overlays

Pavement overlays can be of following types:

(i) Bituminous overlays: Bituminous pavement placed on an existing pavement (flexible or rigid)

(ii) Concrete overlays: Cement concrete pavement placed on an existing pavement (flexible or rigid)

(iii) Bonded concrete overlays: An overlay which is bonded to existing rigid pavement.

Figure 16.20 shows typical overlay types

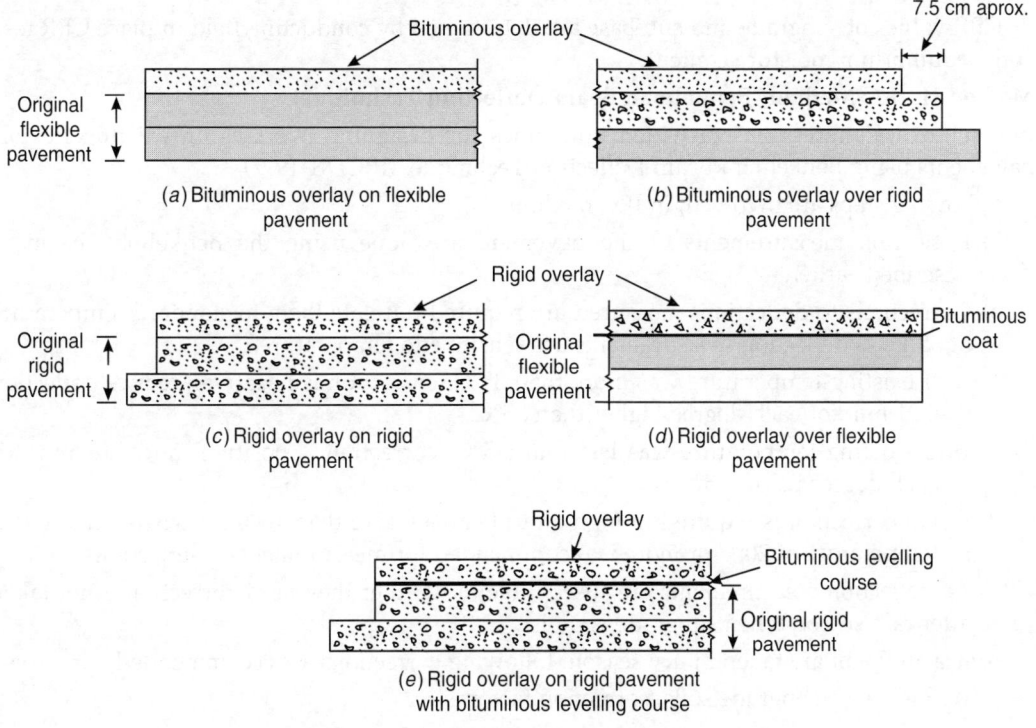

Fig. 16.20: Typical overlays on pavement

Rigid pavement constructed with bituminous overlay is called a 'Composite Pavement': It reduces noise of the concrete slab in driving. Concrete overlay on top of an existing bituminous surface is termed as 'White Topping'.

16.10 DESIGN OF OVERLAYS

16.10.1 Design of Bituminous Overlays

Bituminous overlays can be used on either flexible or rigid pavements.

(A) Bituminous overlay on existing flexible pavements

Method I: FAA Method

Following steps are used in FAA method:

(a) Use the appropriate basic flexible pavement design method to determine the thickness requirement for a flexible pavement for the design traffic, using CBR value of subgrade and sub-base, assuming that existing pavement does not exist. Thickness of all pavement layers must be determined.

(b) The thickness of pavement required over the subgrade and sub-base must be compared with the existing pavement to determine overlay requirements. The thickness of the bituminous overlay is equal to the difference between computed thickness and the thickness of the existing pavement.

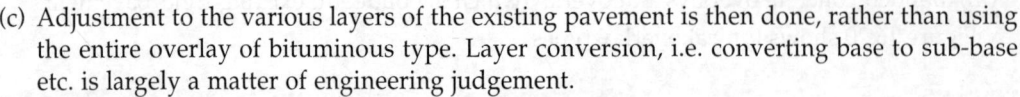

(c) Adjustment to the various layers of the existing pavement is then done, rather than using the entire overlay of bituminous type. Layer conversion, i.e. converting base to sub-base etc. is largely a matter of engineering judgement.

CBR values of subgrade and sub-base are determined by conducting field in place CBR test, at the equilibrium moisture content.

Method II: IRC Method (Benkelman Beam Deflection Technique)

Indian Road Congress has provided guidelines for designing overlays on existing flexible pavements using Benkelman Beam Deflection Technique. (IRC: 81-1997)

Following steps are involved, in IRC method:

(a) Deflection measurements of the pavement are done using the Benkelman beam as described earlier.

(b) Corrections to deflections measured are required to relate them to standard temperature of 35°C, as deflection measurements are sensitive to temperature.

   (i) If existing temperature was more than 35°C (correction is negative)–reduce deflection 0.01 mm for each degree higher than 35°C.

   (ii) If existing temperature was less than 35°C (correction is positive)–add 0.01 mm for each degree below 35°C.

   (iii) No correction is required in high altitude area (more than 1000 m) where an ambient temperature of 20°C or more is recommended for measurement of deflections.

(c) As deflection measurements are susceptible to weather they need correction if not taken after rain season when subgrade is moist.

If measurement are taken in dry season, following corrections are recommended:

   (i) For clayey subgrade soils, correction factor is = 2.

   (ii) For sandy subgrade soil, correction factor is = 1.2 to 1.3.

Subgrade type and the field moisture content is determined, by digging a pit, adjacent to pavement edge, in the prescribed manner. For different types of subgrade, field moisture content and annual rainfall curves are provided to select the correction factors.

The deflection value corrected for temperature is multiplied by the appropriate value of seasonal correction factor to obtain corrected value of deflection.

(d) The design traffic in terms of million standard axle (MSA) is calculated and vehicle damage factor (VDF) is arrived at as is done for design of flexible pavements. (Refer chapter 14).

(e) Number of deflection observations say n are taken and corrected for temperature and seasonal variation.

The characteristic deflection for design of overlays is taken by statistical analysis of the data as given below:

(i) Mean deflection = $\bar{x} = \dfrac{\sum x}{n}$

(ii) Standard deviation = $\sigma = \sqrt{\dfrac{\sum (x - \bar{x})^2}{n-1}}$

(iii) Characteristic deflection (mm) $D_C = \bar{x} + 2\sigma$ for major roads like NH and SH.

(iv) Characteristic deflection (mm) $D_C = \bar{x} + \sigma$ for all other roads.

Where x = Individual deflection observations mm

$\bar{x}$ = Mean deflection mm

n = Number of deflection measurement observations

σ = Standard deviation mm

D_C = Characteristic deflection mm to be used in design of overlays.

(f) The design curve shown in Fig. 16.21 is used to get the overlay thickness. The curve is between cummulative number of standard axles to be carried over design life of the pavement and the characteristic deflection (D_C) as calculated in step (e).

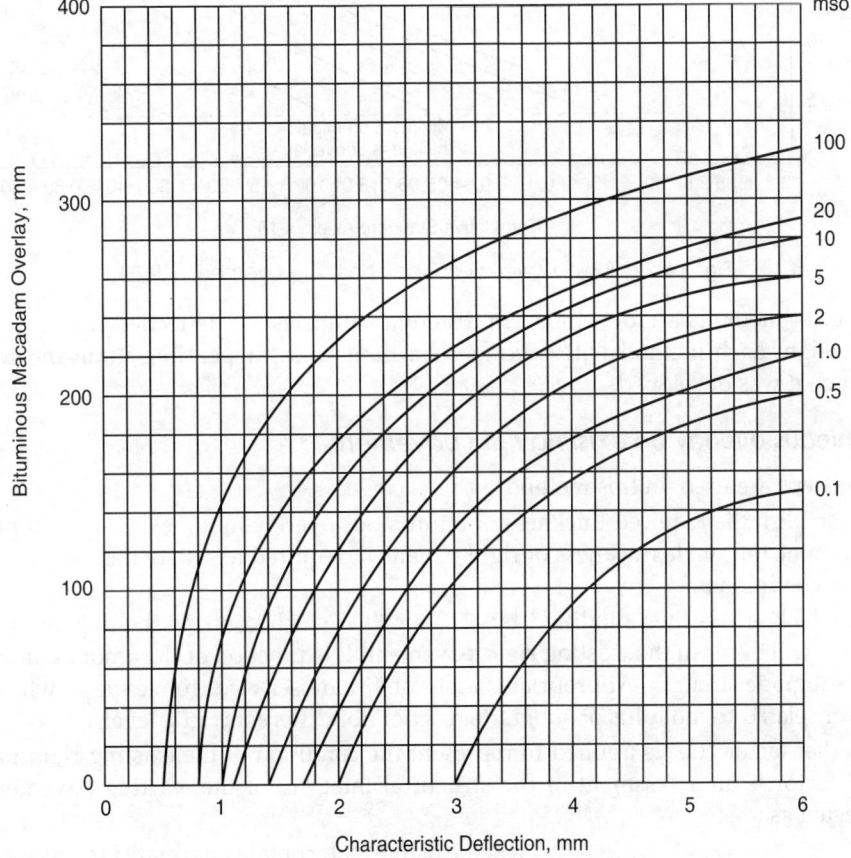

Fig. 16.21: Overlay thickness design curves.

The design thickness deduced from the design curves is in terms of bituminous macadam construction. Equivalency factors suggested are:

1 cm of bituminous macadam = 1.5 cm of WBM/WMM/BUSG or

= 0.7 cm of DBM/AC/SDC

Recommended minimum bituminous overlay thickness is 50 mm (from structural considerations) of bituminous macadam, with an additional surfacing course of 50 mm DBM or 40 mm bituminous concrete.

Method III: TRRL Method

Transport and Road Research laboratory has developed a method and chart for use in the design of overlays. The designer is expected to measure the pavement surface deflection in order to use the design chart and then to determine the required overlay thickness to reduce the deflection to a given level, consistent with the performance under future traffic forecast (Cumulative Standard Axles). The chart is shown in Fig. 16.22.

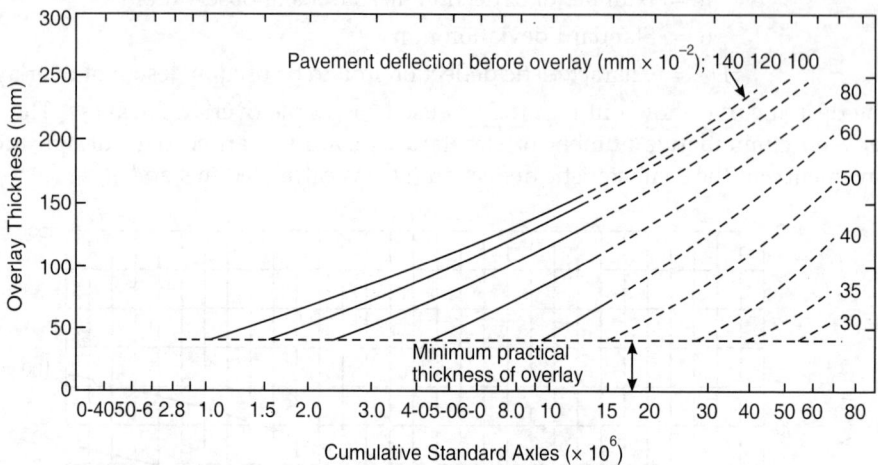

Fig. 16.22: Typical overlay design chart for pavements (TRRL)

Number of methods are developed in different countries for deflection measurement and overlay design. Each method will have its own limitations, depending upon the conditions under which it was developed.

(B) Bituminous overlay on existing rigid pavement

Following steps are used, in this method:

(i) To establish the required thickness of bituminous overlay for an existing rigid pavement, determine the single thickness of rigid pavement required to satisfy the design traffic and other conditions.

(ii) This thickness is then modified by a factor "F" which controls the degree of cracking which will occur in the existing rigid pavement. It is function of the amount of traffic and the subgrade strength. Appropriate values of "F" are selected from a graph which gives F factor related to modulus of subgrade reaction for different traffic levels.

(iii) Another factor "C_b" is applied to represent the condition of the existing rigid pavement. 'C_b' factor is an assessment of the structural integrity of the existing pavement and is selected as :

$$C_b = 1 \text{ when existing slab contains normal cracking}$$
$$C_b = 0.75 \text{ when slab contains multiple cracking}$$

(iv) After knowing the 'F' factor (from graph) and condition factor 'C_b' the thickness of the bituminous overlay is worked out from the formula:

$$t = 2.5 (Fh - C_b h_e)$$

Where

t = thickness of bituminous overlay in cm.

F = factor which controls the degree of cracking in the base pavement.

h = single thickness of rigid pavement required for design conditions (traffic) in cm.

C_b = condition factor for base pavement ranging from 1.0 to 0.75.

h_e = thickness of existing rigid pavement in cms.

For an existing bituminous overlay on a rigid pavement the procedure is to assume as if existing overlay was not present. This is illustrated in following example.

Example: An existing pavement consists of 25 cm rigid pavement with 7.5 cms bituminous overlay. The existing pavement is to be strengthened to be equivalent to a single rigid pavement of 36 cms in view of increase traffic loads. Assume a F factor of 0.9 and C_b of 0.9 for existing pavement.

Solution. (*i*) Calculate the required thickness of bituminous overlay as if existing 7.5 cm overlay was not there.

$$t = 2.5\,(Fh - C_b\,h_e)$$
$$= 2.5\,(0.9 \times 36 - 0.9 \times 25)$$
$$= 23 \text{ cms.}$$

(*ii*) This overlay would have been required if there was no existing overlay, however since overlay of 7.5 cm thickness is existing, an allowance will be made.

The effective thickness of existing overlay can be assumed only 6 cms, rather than 7.5. It is a matter of engineering judgement.

The required overlay thickness is = 23 – 6

$$= \mathbf{17 \text{ cms.}}$$

16.10.2. Design of Concrete Overlays

Concrete overlays are of following three type:

  (i) Bonded overlays in which thin overlay slab is bonded on the existing rigid pavement. Minimum thickness of 25 mm is commonly used, where slab is not badly cracked.

 (ii) Partially bonded in which overlay slab is placed directly over the existing slab. They are thicker than bonded overlays (120 mm or more) used if slab is not badly failed.

(iii) Unbounded in which thick overlay slab is put over the existing slab. The overlay slab acts independently, and is infact a replacement of badly damaged rigid pavement. Thickness may be 150 mm or more.

(a) Concrete overlay on existing flexible pavement

For design of concrete overlay, the existing flexible pavement is considered a foundation for the overlay slab. The existing flexible pavement is assigned a "K" (modulus of subgrade reaction) value either

  (i) Using the chart or

 (ii) By conducting a plate bearing test on the existing pavement.

K value assigned should not exceed 500 (14 kg/cm^3). The overlay is then designed like rigid pavement design as per IRC 58-2002.

(b) Concrete overlay on rigid pavements

Following steps are involved.

  (i) The rigid pavement design is done to find out the thickness of concrete slab, required to satisfy the design load and other conditions, for a single thickness of concrete pavement. The K value to be assigned to existing foundation is determined by field plate bearing tests conducted in the test pit cut through the existing rigid pavement or may be estimated from construction records of the existing pavement.

 (ii) Structural integrity of the existing rigid pavement is assessed on the basis of engineering judgement or through non-destructive testing (NDT). Following values are adopted to provide uniform assessment of condition factor C_r.

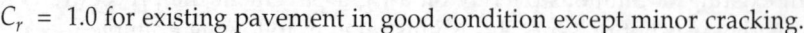

C_r = 1.0 for existing pavement in good condition except minor cracking.

C_r = 0.75 for existing pavement having initial corner cracks due to loading but not progressive

C_r = 0.35 for existing pavement in poor existing condition – badly cracked.

Within these ranges, intermediate values of C_r may be adopted.

(iii) There may be two situations for design of overlays:

1. Concrete overlay without levelling course.

2. Concrete overlay with levelling course

Different formulae are used, based on research work.

1. Concrete overlay without levelling course

The thickness of concrete overlay slab, applied directly over the existing rigid pavement is computed by the formula:

$$h_c = (h^{1.4} - C_r h_e^{1.4})^{1/1.4}$$

$$= 1.4\sqrt{h^{1.4} - C_r h_e^{1.4}}$$

Where

h_c = required thickness of concrete overlay

h = required single slab thickness determined from design

h_e = thickness of existing rigid pavement

C_r = condition factor. Varying from 1 to .35

Due to inconvenience of exponents in the above formula graphic solutions are provided for different C_r values.

2. Concrete overlay with levelling course

Sometimes it may be necessary to apply a levelling course of bituminous concrete on an existing rigid pavement, prior to application of concrete overlay.

When the existing pavement and overlay pavement are separated, the slab acts more independently than when slabs are in contact with each other.

Following formula is used for the thickness of overlay slab, when levelling course is used.

$$h_c = \sqrt{h^2 - C_r h_e^2}$$

Where

h_c = required thickness of concrete overlay

h = required single slab thickness determined by design

h_e = thickness of existing rigid pavement

C_r = condition factor.

Levelling course must be constructed of highly stable bituminous course.

Graphic solutions for the above equation, for different condition factors are developed.

3. Fully bonded overlays

In some situations overlays are bonded to existing rigid pavements. In such cases the new section behaves as a monolithic slab.

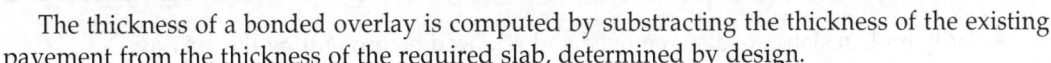

The thickness of a bonded overlay is computed by substracting the thickness of the existing pavement from the thickness of the required slab, determined by design.

$$h_c = h - h_e$$

Where

h_c = required thickness of concrete overlay

h = required single slab thickness determined from design

h_e = thickness of existing rigid pavement.

Bonded overlays should be used when existing rigid pavement is in good condition. Adequate bond should be achieved through elaborate surface preparation and precise construction techniques.

16.11 PREPARATION OF SURFACE FOR THE OVERLAYS

Before constructing an overlay, all defective areas in the existing surface, base, sub-base and subgrade should be corrected to have satisfactory overlay pavement, otherwise deficiencies in the base pavement may reflect in the overlay.

(a) Flexible pavements

Pavement breakups, patches, surface irregularities and depressions, etc in the existing flexible pavement should be removed, before putting the overlay.

- Broken pavement areas are repaired by improving the subgrade or drainage facilities. After correction of subgrade condition, the sub-base, base and surface courses of required thickness should be placed and thoroughly compacted layer by layer.
- Surface irregularities and depressions like shoving, rutting should be levelled by rolling and or by filling with suitable hot mix bituminous mixtures.
- Bleeding surface, i.e. hot mix bituminous material accumulated at one place on the surface of the pavement is bladed off/or blotted up.
- Old joints or cracks are removed by putting the crack filler–a lean mixture of sand and liquid bituminous material. The mixture is well tamped in place, levelled with the pavement surface and any excess removed. Material placed should be dry and become hard, before placing the overlay.
- Pot holes should be thoroughly cleaned and filled with suitable bituminous mixture and tamped in place.

For hot mix overlay, a light tack coat is put immediately on the pavement after cleaning. The overlay should not extend to the edge of the pavement, but should be cut-off approximately 7.5 cms from the edge.

In recent years, new pavement overlay materials such as fibrous concrete, roller compacted concrete and rubberised asphalt are used with varying degree of success.

(b) Rigid pavements

In rigid pavements, narrow transverse, longitudinal and corner cracks will generally need no special attention unless there is displacement and faulting between the separate slabs. If the subgrade is stable and no pumping has occurred, no corrective measures are required as overlay will take care of such defects.

If subgrade support is not stable and pumping has occurred, pumping cement grout or bituminous cement is used to fill the voids, to provide even support for the overlay.

- Slab may be removed and replaced if the pavement is broken and subject to racking because of uneven bearing on the subgrade.

- Badly broken slabs are also removed and replaced before putting the overlay.
- If the existing pavement is rough due to slab distortion, faulting or settlement, a provision should be made for a levelling course of hot bituminous mix concrete, before overlay is placed.
- Cracks and joints 1 cm or more in width should be filled with a lean mixture of sand and liquid bituminous material and tamped firmly in place and levelled with pavement surface.
- After repairing and prior to placing of the overlay, the surface should be swept clean of all dirt, dust and foreign material. Any extruding joint sealing material should be trimmed from the rigid pavement.
- Bonded concrete overlays require special attention to ensure bond with the existing pavement. Bond can be achieved by placing a neat cement grout on the prepared surface, before placing the overlay. Shot peening or mechanical texturing by cold milling are two other techniques which have been used to provide a surface which will allow bonding.

IRC-SP-17, provides recommendations about overlays on cement concrete pavements.

17

Highway Embankments

17.0 PREAMBLE

Vertical alignment for highways inevitably require embankment structures to bring natural topography to an acceptable design grade. An embankment is a structure of soil, soil-aggregate mix between the existing ground and the subgrade. Highway embankments should provide a stable surface on which the pavement structure may be constructed. The embankment should be stable, not prone to any type of failure and settlement. Settlement (if any) should occur within a specific time period and should be within tolerable limits.

Highway embankments should be constructed with good soil, adequately compacted and properly designed keeping in view the economic aspects. Embankments should be designed to protect the subgrade from capillary rise of water and ingress of water from surface flow.

Field and laboratory explorations on soil profile, soil strength, depth of ground water table are done for stability analysis of embankments and working out a factor of safety. The internal embankment stability problems generally result from the selection of poor quality embankment material and/or improper placement of fills or poor compaction. Internal stability can be insured through proper specifications using granular materials, its grading and compaction requirements.

Natural slopes that have been stable for many years, may suddenly fail because of changes in topography, seismicity, ground water flows, loss of strength, stress changes and weathering. Planning, design, construction, improvement, rehabilitation and maintenance are the aspect of embankment which require analysis in the beginning and sometimes throughout its life.

17.1 DIMENSIONS OF EMBANKMENT

(a) Width: Width of embankment should be sufficient to accommodate pavement, median (if any), shoulders, as prescribed for a particular class of road, as shown in Fig. 17.1. Road width standards are covered in chapter 6.

(b) Side slopes: Height of the embankment, type of soil, and economic considerations are important in fixing the side slopes of an embankment. In general slopes of $1\frac{1}{2}:1$ to $2:1$ are used for un-inundated conditions and 2 : 1 to 3 : 1 for inundated conditions. Flatter slopes are better for safety but they are costly in construction.

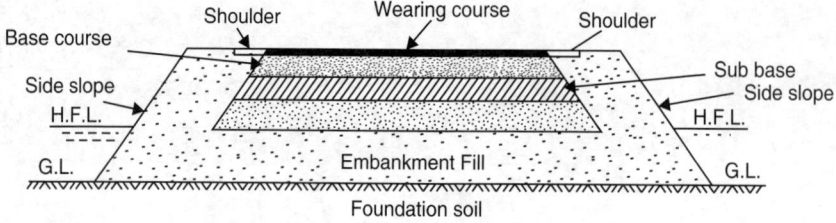

Fig. 17.1: Embankment with pavement layers

Slopes recommended by IRC are shown in Fig. 17.2.

(c) Height: Height of the embankment depends upon the site requirement and vertical alignment of the road. The lower most layer of the pavement (subgrade) should be at least 0.6 to 1 m above the highest flood level/level of water table.

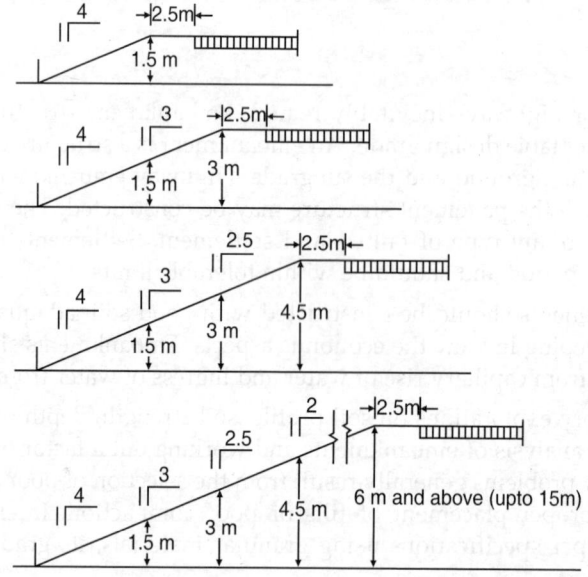

Fig. 17.2: Side-slopes for road embankments (IRC)

17.2 STABILITY OF EMBANKMENTS

Stability of an embankment depends upon:
- The geometry of the embankment.
- The gravitational and seepage forces.
- Foundation soil.

Bearing capacity and slope stability failure are the dominating failures for an embankment. Embankment failures are basically of two types:

1. Rotational (Slope) failures and sliding failures of the embankment. It may be circular or non-circular path. Normally circular slips are formed in homogeneous, isotropic soils and non-circular slips formed with non-homogeneous soils (slope failures).

2. Bearing capacity and plastic failures (squeezing of the soil) of the foundation. (settlement failures).

Figure 17.3 shows typical embankment failures.

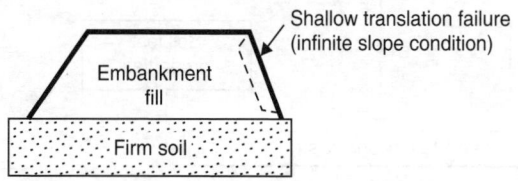

(a) Infinite slope failure in embankment fill

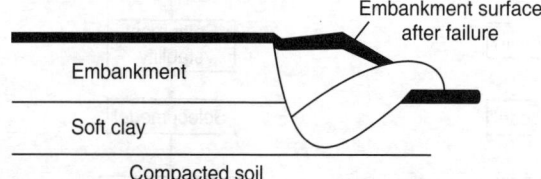

(b) Circular arc failure in embankment fill and foundation

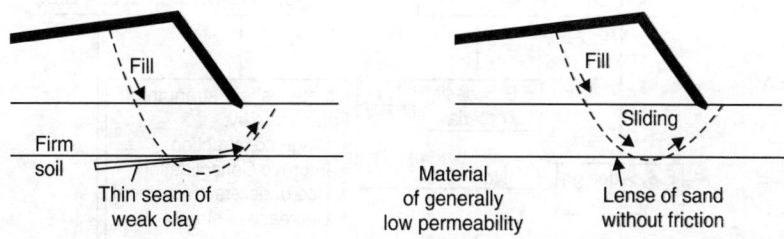

(c) Sliding block failure in embankment fill and foundation soil

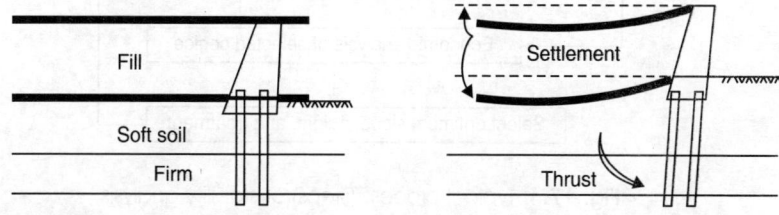

(d) Lateral sqeeze of foundation soil.

Fig. 17.3: Typical embankment failures

An embankment is said to be stable if it can maintain its original shape during its design life and can withstand against bearing capacity of foundation. A stable embankment does not penetrate into the foundation and should maintain its original geometry.

To ensure that an embankment is stable it is usual to adopt a minimum factor of safety, against failure. A factor of safety is defined as the ratio of the available shear strength of the soil to the shear stress developed along a critical failure surface.

Stability is usually specified on the basis of short and long term behaviour of the embankment.

- *Short term stability* requires the embankment to be stable during construction period.
- *The long term stability* is associated with the period after construction and up to the end of its design life.

Figure 17.4 show steps involved in the stability analysis of embankments.

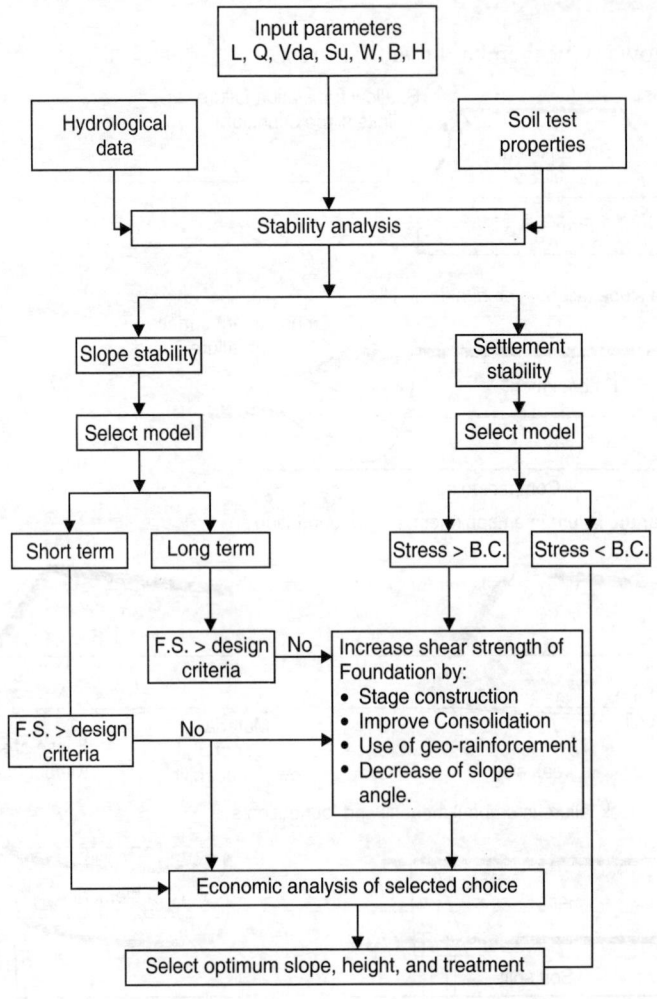

Fig. 17.4: Typical process (steps) of stability analysis

17.2.1 Procedure of Estimating Stability

There are 3 steps in estimating stability:

1. *Estimate Disturbing Forces:*
   - Gravity acting on body of soil
   - Superimposed loads (if any)
   - Seepage forces due to water flow (if any)
   - Earthquake forces.
2. *Shearing Resistance of Soil:*
   - Determine the number, thickness and average strength parameter of each soil layer.
   - Find soil strength using equations:

$$\text{Total stress} = C_u + \sigma_N \tan \theta$$

   (shearing Resistance at failure)

- Include a factor of safety, to limit the maximum mobilised shearing resistance on a failure plane.

3. *Select Appropriate Method of Analysis (Model)*

There are two approaches, used by different scientists:

(a) *Limit State equilibrium.* A circular arc is chosen because it is the simplest to analyse and sufficiently accurate.

(b) *Stress Analysis.* Uses principles of elasticity to evaluate stresses and strain throughout the slope. Finite element analysis is common technique.

17.3 SLOPE STABILITY ANALYSIS

The degradation of slopes in embankments is a result of mass movement which occur mainly due to action of gravitational forces.

The objective of slope stability analysis is to contribute to safe and economic design of embankments for highways. Slope stability evaluation is concerned with identifying critical geological, material, environmental and economic parameters that will affect the stability. The analysis of slopes takes into account topography, geology, material and soil properties, climatic condition, etc. with following objectives:

(i) To understand natural slopes, their development and form.

(ii) To assess the stability of slopes on short term (during construction) and long term basis.

(iii) to assess the possibility of land slides due to natural or man-made causes.

(iv) To analyse the land slides and causes of failure, their mechanism and effect of environmental factors.

(v) To enable the re-design of failed slopes, and adopt preventive and remedial. measures.

(vi) To study the effect of seismic loading on slopes and embankments.

17.3.1 Methods of Slope Analysis

Slope may fail by rotational and or translation type of failure, and the failure mechanism is analysed by limit equilibrium methods which assume some limiting conditions for maintaining slope equilibrium. Figure 17.5 shows forces acting on a slip circle.

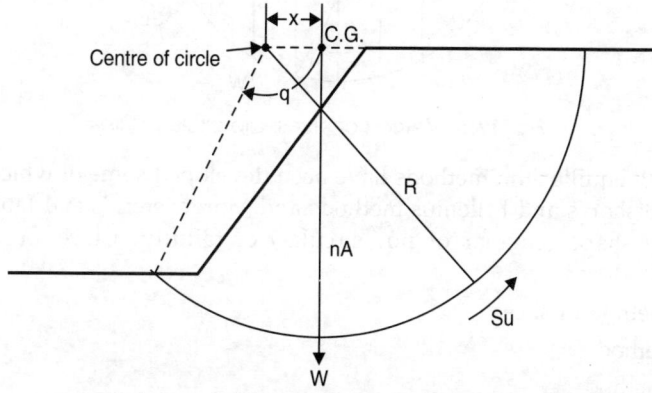

R = Radius of the circle
q = Angle defined as above in radians
x = Distance of centre of gravity of soil mass from centre of circle
W = Weight of soil mass = nA
n = Unit weight of soil
A = Area of soil mass
Su = Soil strength

Fig. 17.5: Forces acting on a slip circle

In the limit equilibrium method, a known trial failure surface is conceived and available shear strength along the surface is compared with the required shear strength to prevent the failure. The ratio of the two values is known as the factor of safety.

In ordinary method of slices, the sliding surface is considered as circular to simplify the calculations. The factor of safety (F) for the short term or total stress analysis of stability is given by the ratio of restoring moment to the disturbing moment.

$$\text{S.F.} = \frac{\text{Re-storing moment}}{\text{Disturbing moment}}$$

In Fig. 17.5

$$\text{Restoring moment} = S_u R^2 \theta \quad \text{and}$$
$$\text{Disturbing moment} = Wx$$

$$\therefore \qquad \text{S.F.} = \frac{S_u R^2 \theta}{Wx}$$

As the shear strength of soil varies with the depth, the soil mass above the circular arc can be divided into a number of slices, and factor of safety in this case is given by

$$\text{S.F.} = \frac{\sum_n S_u L}{\sum_i W_i x}$$

where L is length of each slice along failure,

$$n = \text{number of slices, as shown in Fig. 17.6}$$
$$W_i = \text{Weight of each slice.}$$

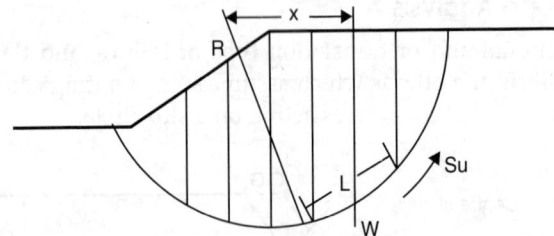

Fig. 17.6: Method of slices: Slip circle failure

Number of limit equilibrium methods have been developed some of which are applicable to circular slip like Bishop's and Fellenius methods and Morgenstern's and Janbau's methods are applicable to any shape (circular or non-circular) of failure surface. Some of the popular methods are:

– Ordinary method of slices
– Bishop's method
– Spencer's method
– Bishop and Morgenstern's method
– Janabu's generalised method
– Low's method
– TRL method, and several other methods.

Detailed description of these methods can be found in any text on soil mechanic as covering them here is beyond the scope of this book.

17.3.2 General Slope Stability Concept

Figure 17.7 shows variation of various parameters of slope stability on a clayey foundation.

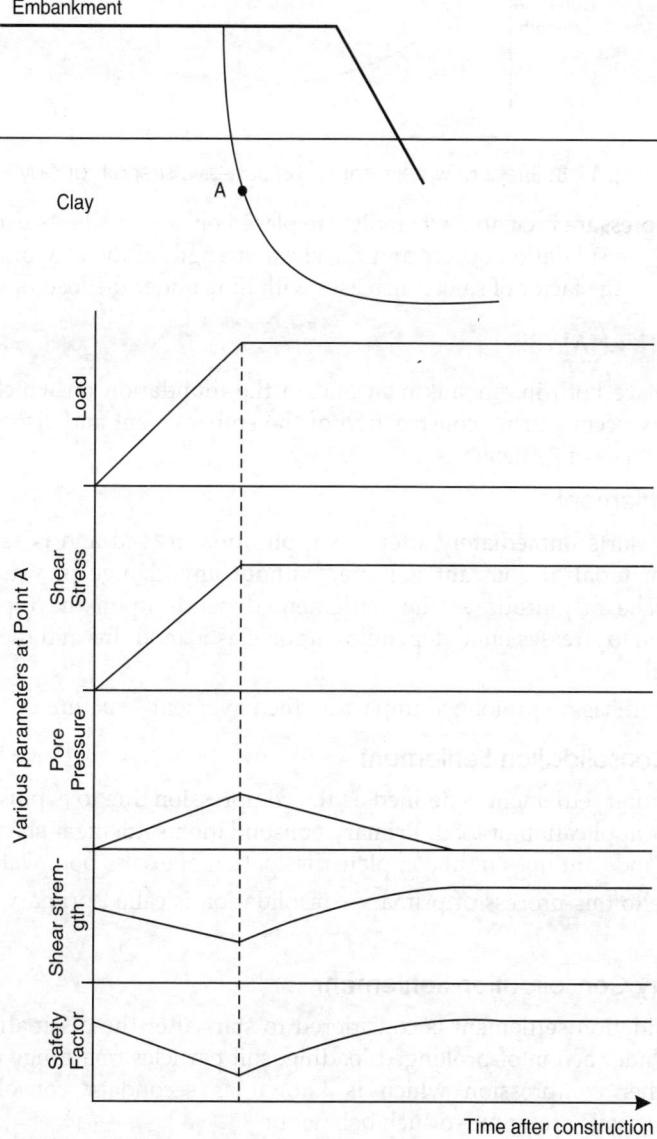

Fig. 17.7: Stability condition of an embankment slope over clay foundation

17.3.3 Effect of Water on Slope Stability

Next to gravitational forces, water is the most important factor in slope stability. In cohesionless soils, water does not affect the angle of internal friction (ϕ), however in cohesive soils clay minerals do react with water and cause volume changes of the clay mass. An increase in moisture is a major factor in the decrease in strength of cohesive soils, as water absorbed by clay minerals cause increased water content that decreases cohesion of clay soil. This effect increases with type of clay/expansive soil, as shown in Fig. 17.8.

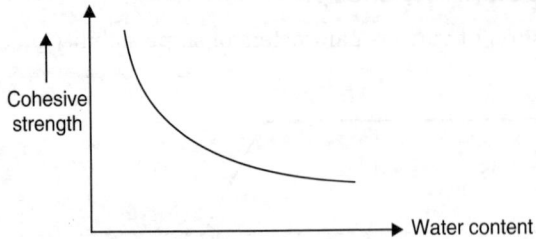

Fig. 17.8: Effect of water content on cohesive strength of clay

Excessive pore pressure is created when fills are placed on clay or silt. As excessive pore water pressure disciples, consolidation occurs and the shear strength of the clay or silt increases with time. For this reason, the factor of safety increases with time under the load of the fill.

17.4 SETTLEMENT ANALYSIS

Settlement take place both in embankment and in the foundation on which embankment is constructed. It may occur during construction of the embankment and also afterwards. There may be following types of settlements.

17.4.1 Elastic Settlement

Elastic settlement starts immediately after the application of load and is associated with the shearing of the material at constant volume, without any change in water content of the foundation soil. The magnitude of the settlement depends upon the types of underlying material in addition to stresses on it, depending upon elastic modulus and the Poisson's ratio of the foundation soil.

Elastic settlement has no prolonged impact on the pavement structure.

17.4.2 Primary Consolidation Settlement

Primary consolidation settlement is defined as the compression due to expulsion of water from pore spaces under application of load. Primary consolidation settlement starts when the soil is settled elastically and continues until complete dissipation of excess pore water pressure.

Settlement due to this process of primary consolidation is called primary settlement due to consolidation.

17.4.3 Secondary Consolidation Settlement

Secondary consolidation settlement is considered to start after the dissipation of excess pore water pressure. Under action of prolonged loading, soil particles re-arrange and re-orientation resulting in further compression which is known as secondary consolidation or creep settlement. Organic soils are prone to such behaviour.

17.4.4 Long term Settlement

Due to traffic loads and self weight of embankment long term settlement occurs within the embankment it self. This is due to:

(i) the deformation of soil particles

(ii) arrangement of soil particles

(iii) dissipation of air and water from the interparticle void spaces.

The settlement of the foundation on which an embankment is built occurs due to embankment and traffic loading. Traffic load induced foundation settlement is negligible for embankment height more than 3 m, although it may be significant for low embankments.

The consolidation settlements are considered in the design process of the embankments to achieve design requirements. To keep the consolidation settlement within desired limits measures are adopted to accelerate the settlement process. To prevent failures, it is usual to specify limit of the amount of time taken for the foundation to settle a given amount during construction.

The two main methods of achieving the accelerated consolidation are:

- Pre-compression and
- Use of vertical drains.

When soil is highly compressible, such as peat soil a surcharge load can be applied on top of the embankment before the pavement is constructed. Thus pre-compresssion is achieved.

Vertical drains contribute to sub soil drainage by adding radial drainage paths in addition to vertical ones. Two types of vertical drains commonly used to accelerate the consolidation settlement process are:

- Sand drains and
- Prefabricated vertical drains (PVD)

Sand drains are constructed by making holes and filling them with sand, in layer(s), to be consolidated.

Prefabricated drains are manufactured from synthetic polymers. PVD are faster to install than sand drains, and are becoming more popular.

17.4.5 Methods of Settlement Analysis

Number of methods are available in literature on soil mechanics for determining the total settlement of an embankment constructed on soft soil: Some of the basic ones are:

- Standard method
- Finite element method
- Finite difference method
- Empirical methods.

Standard method uses Taylor's and Terzaghi's theory of finite element and is based on computer software. The finite difference method is also a computer method for which software is available. It determines pore water pressures in foundation soil, and effective stresses using Terzaghi's one-dimensional consolidation equation. There are number of empirical methods used in different countries.

Design charts for embankments are developed using finite element methods of analysis. These charts are useful and supplement to conventional stability analysis.

Embankment fills over the soft clay foundation are usually stronger and stiffer than their foundation. Embankments may crack as foundations deform and settle under its own weight. This may lead to progressive failure because of stress-strain incompatibility between the embankment and its foundation, as shown in Fig. 17.9. The use of geosynthetic reinforcement in the fill may prevent the initiation of cracking and subsequent failure.

Peak strength of the embankment and the foundation soil can not be mobilised simultaneously because of stress-strain incompatibility. Stability analysis should be performed using soil strengths which are smaller than the peak value as using peak strengths of soils would overestimate the factor of safety.

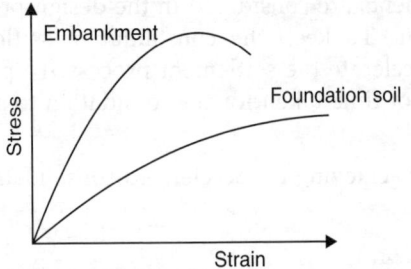

Fig. 17.9: Typical stress-strain imcompatibility

17.5 DESIGN OF EMBANKMENTS

Design of highway embankment based on analysis of slope stability and settlement is done to satisfy the following requirements :

- Limit settlement of embankment to an acceptable amount
- Limit the time for consolidation
- Provide adequate stability
- Provide good surface for pavement.

There are several methods of design and analysis of stability and settlement developed by Bishop. Janabu, Spencer, Low and Fellenius, etc. Following criteria may be used in selecting the method for design and analysis of embankment:

- Accuracy-model that accurately represents site, sub surface conditions
- Ease to use
- Availability and cost of software
- Input parameters should be possible through field and laboratory testing
- Previous field performance, of structure designed compared to predicted results.

17.5.1 Design Factor of Safety

In design of embankments it is usual to specify factor of safety as a means of taking into account uncertainty. A factor of safety is defined as the ratio of the available shear strength of soil to the shear stress developed along a critical failure surface.

The selection of factor of safety depends upon:

- Method of stability analysis
- Method of determining shear stress
- Degree of confidence in the reliability of subgrade data
- The consequence of failure
- Properties of foundation and fill material
- Geometry of the embankment and its height
- Water table level.

A minimum factor of safety as low as 1.25 is used for highway embankments. This value of safety factor is increased depending upon situation and above factors.

17.6 MATERIALS FOR EMBANKMENTS

In highway embankments, soil (earth) is compacted in fill slopes. Engineering properties of materials used in the embankments are controlled by the grain size distribution, method of construction and the degree of compaction.

Embankment slopes are designed using shear strength parameters, by performing tests on the samples of the proposed material. Suitable materials are pre-selected and embankments are constructed in layers, and compacted. Compaction of soil increase their strength. Different types of compactors, ranging from large sheep's foot and vibratory rollers to small hand-operated tempers have been used to compact the soil.

Materials for embankment fill consist of :

(i) Cohesionless soils such as sands or gravel

(ii) Cohesive soils such as silts and clays

(iii) A mixture of above two types (earth-rock mixture).

Organic soils, soft clays and silts are generally avoided. The range of particle size distribution for embankment fills depends upon the availability of the material from nearby borrow areas and economic considerations.

Cohesionless fills consisting of clean sand and gravel remain pervious when compacted. These soils represent soil groups: SW, SP, GW, GP, of unified classification system of soils.

During compaction, cohesionless soils are not affected significantly by water content, as they are relatively pervious. Their compactness is usually evaluated by their relative density, given by formula

$$D_r = \frac{e_{max} - e}{e_{max} - e_{min}} \qquad \text{(i)}$$

Where

$$D_r = \text{Relative density expressed as a percentage}$$
$$e_{max} = \text{Void ratio of the soil in its loose state}$$
$$e = \text{Void ratio of the soil being tested}$$
$$e_{min} = \text{Void ratio of the soil in its densest state.}$$

As void ratio is related to dry density the above equation (i) can be written in terms of specific gravity as :

$$D_r = \frac{d_{max}(d - d_{min})}{d(d_{max} - d_{min})} \qquad \text{(ii)}$$

Where

$$D_r = \text{Relative density expressed as percentage}$$
$$d_{max} = \text{Dry density of the soil in its densest state}$$
$$d_{min} = \text{Dry density of the soil in its loose state}$$
$$d = \text{Dry density of the soil being tested.}$$

The loose or dense state of cohesionless soil is usually judged by relative density as defined by Terzaghi. The typical ranges of relative density of sand are :

$$0 < D_r < \frac{1}{3} - \text{Loose sand}$$

$$\frac{1}{3} < D_r < \frac{2}{3} - \text{medium dense and}$$

$$\frac{2}{3} < D_r < 1 - \text{Dense sand.}$$

Terzaghi further defined compactibility (F) as:

$$F = \frac{e_{max} - e_{min}}{e_{min}}$$

Where F is very large for well graded cohesionless soils such as SW and GW soils. Coarse grained granular material make a very good fill material, specially on soft foundation soils.

Cohesive fills for embankments contain sufficient quantities of silt and clay to render the soil mass relatively impermeable when properly compacted. Unlike compacted cohesionless soils, whose properties are generally improved by compaction to the maximum dry unit density, the physical properties of cohesive soils are not necessarily improved by compaction to maximum unit density. In fact, strength of silty clay decreases with increasing molding water content. By changing the water content of compacted clays, a pronounced change in their engineering properties result. This type of soil has low permeability and develops high excess pore water pressure under increased loading.

Earth-rock mixtures used as embankment fill for high embankments, produce a heterogeneous mixture of particles, that may range from large boulders to clay. Large particles enhance the workability of the fill in the field and increase the overall strength of the soil.

The strength will increase with certain amount of rock (depending upon its quality), for example 50 to 70 percent for sand-gravel mix. Further increase in gravel content produces little or no increase in strength.

Selection of the material to be used in construction of embankment is done after judging its suitability through field and laboratory tests. Commonly used tests are
- Sieve analysis (particle size distribution)
- Atterberg limits (LL and PL)
- Proctor test for establishing moisture-density relationship
- Chemical content test (if any)
- Shear and consolidation test.

For designing the pavement over the embankment CBR test is conducted on the top 50 cm sample of soil, in the laboratory, after soaking the sample for 4 days. Pavement is designed based on this CBR.

Based on the tests on the materials, following points are kept in view for selecting the suitable fill for satisfactory performance of embankments :
(i) For good stability and strength, a well graded soil is better.
(ii) Coarse material is preferred to fine. Sand and gravel are better than clay and silt.
(iii) Soil which contain organic matter or chemicals like sodium sulphate should be avoided.
(iv) Cohesion less sands are good from drainage point of view.
(v) It should be possible to compact the soil to the acceptable degree.
(vi) Must not contain muck, roots, sod, and any other deleterious materials.
(vii) Soils having laboratory maximum dry density of less than 1.44 gm/cc are considered unsuitable for use in embankments.

Table 17.1 provides general guidelines for selecting the embankment soils as recommended by Indian Road Congress.

Table 17.1: General guide to the selection of soils on basis of anticipated embankment performance

P.R.A Classification	Comparable soil groups in Indian Standard Soil Classification System		Visual description	Max. density range gm per c.c.	Optimum moisture content range per cent	Anticipated embankment performance
	Most probable	*Possible*				
1	2	3	4	5	6	7
A-1	GW, GP, GB, GM, SW, SP, SB, SM	— —	Granular material	1.84.2.28 (115-142)	7-15	Good to excellent
A-2	GM, GC, SM, SC	— —	Granular materials with soil	1.76.2.16 (110-135)	9-18	Fair to excellent
A-3	SP	—	Sand	1.76-1.84 (110-115)	9-15	Fair to good
A-4	ML, MH, OL, OH	CL, SM, SB, SC	Sandy silts and silts	1.52-2.08 (95-130)	10-20	Poor to good
A-5	MH, OH	—	Elastic silts and silts	1.36-1.60 (85-100)	20-35	Unsatis-factory
A-6	CL, CI	MH, OH, SC	Silt clay	1.52-1.92 (95-120)	10-30	Poor to good
A-7	MH, CI, CH, OH	SC	Clay	1.36-1.84 (85-115)	15-35	Poor to fair

Specialised materials like tires, commercial products, flyash and slag, cellular concrete/foam concrete have also being used as fill in embankments, with varying degree of success.

17.7 EMBANKMENT CONSTRUCTION

Overall performance of a pavement structure depends upon the proper construction. Embankment construction should start as early a possible to avoid degradation of the prepared foundation. Embankment should be raised at a uniform rate. The fill should be spread in layers and fully compacted. Each layer should be placed and compacted along the entire length of the embankment in a continuous process, any openings for access should be kept minimum.

Prior to further placing of a layer, the previous layer should be ripped to key new material into the previously placed layer. Fill should not be placed as a sloping layer on the side of the embankment as this will subsequently tend to soften and may lead to slope instability.

Fill is usually compacted by mechanical equipment, and rolling with vibration is done for granular material and simple rolling for cohesive soils. There is a particular moisture content depending on the nature of the fill material and compaction equipment called optimum moisture content, at which material should be compacted for maximum density. Optimum moisture content (omc) is determined through laboratory testing.

The quality of the compaction should be monitored during the construction, by in-situ nuclear density meters or by sand replacement test.

The density to which a particular soil can be compacted depends upon :

• Moisture content
• Type of compaction equipment

- Thickness of layer to be compacted
- Number of passes of the roller
- Roller speed.

Sheep foot rollers are normally used on relatively soft cohesive soils in conjunction with a bulldozer blade to spread, mix and compact the soil. Granular soils if not well graded require vibratory compaction.

Following are the steps involved in construction of an embankment.

- *Clearing and grabbing:* Clearing is removal of trees, bush, debris and other large matter and grabbing is the removal of stumps roots and top soil. After grabbing a firm foundation is prepared.
- *Erosion control:* Erosion and siltation controls must be installed before construction to protect surrounding land and waterways from effects of erosion and siltation.
- *Grading:* To protect against rainfall during construction the top of earthwork should be shaped to permit run-off of rain water, to intercept run off water and prevent damage to earth slopes by erosion.
- *Placement of layers:* Successive layers should be uniform layers parallel to finished grade. Because of large amount of soil in an embankment, it may not be feasible to blend it so that the entire embankment is homogenous. The method used should be able to maintain uniform thickness of layer, proper mixing of soil and/or aggregates in layer, and uniform compactive effort. Uniform thickness of layer or lift thickness is essential for achieving proper compaction. The layer should not exceed 25 cm in loose thickness.
- *The Compaction of layers:* Layers must be compacted with suitable equipment. The moisture content at the time of compaction should be the optimum moisture content. Proctor tests are good guide for field control of moisture/density.

Nuclear testing equipment also gives quick values of moisture/density, comparable to proctor's test and is widely used these days. Conventional sand replacement method is also used and is comparable to proctor's test.

The rolling equipment for compaction is selected for different types of soils according to Table 17.2 recommended by Indian Road Congress.

Table 17.2: General guide to the selection of compaction plant for different types of soil (IRC)

| Type of compaction | Suitability of compaction plant for different types of soil | | | Remarks |
	Cohesive soil	Well graded granular and dry cohesive soils	Uniformly graded materials	
1	2	3	4	5
1. Smooth-wheeled roller	Suitable	Suitable	Suitable only if the roller is towed by tractors and the load per cm width of the roller is less than 55 kg.	
2. Pneumatic-tyred roller	Suitable	Suitable when load on each wheel is more 2 tonnes (tons).	Suitable only if the roller is towed by tractors and the load on each wheel is less than $1\frac{1}{2}$ tonnes (tons).	

(Contd.)

Table 17.2: General guide to the selection of compaction plant for different types of soil (IRC) (*Contd.*)

Type of compaction	Suitability of compaction plant for different types of soil			Remarks
	Cohesive soil	*Well graded granular and dry cohesive soils*	*Uniformly graded materials*	
1	*2*	*3*	*4*	*5*
3. Vibratory roller	Suitable only when the static load per cm width of the vibrator roller is more than 7 kg	Suitable	Suitable; but when the static load per cm width of the vibratory roller is more than 12 kg the roller should be towed by tractors.	
4. Sheepsfoot roller	Suitable	Unsuitable	Unsuitable	
5. Power rammer*	Suitable	Suitable	Unsuitable	* Normally used only when space is restricted such as behind abutments.

Note: For the purpose of this table, soil are grouped as follows:

(i) Cohesive soil includes clays with upto 20 per cent of gravel and having a moisture content not less than the value of the plastic limit minus 4;

(ii) 'Well-graded granular and dry cohesive soil' include clays containing more than 20 per cent of gravel and or having a moisture content less than the value of the plastic limit minus 4; well graded sands and gravels with a uniformity co-efficient exceeding 10 and all shales and clinker-ash.

(iii) 'Uniformly-graded' material includes sands and gravels with a uniformity coefficient of 10 or less and all silts and pulverised fuel ashes. Any soil containing 80 per cent or more of material in the particle size range 0.06-0.002 mm will be regarded as silt for this purpose.

Densities to be aimed in compaction should be chosen with regards to soil type, height of embankment, drainage conditions, type of compaction equipment. Tables 17.3 and 17.4 provide guidelines in this direction. Number of passes of the compacting equipment to achieve desired density is determined by a trial section of the road.

Table 17.3: Specification requirements for embankment soil compaction (IRC)

Condition I		Condition II	
Fills 3 m or less in height and not subject to extensive floods		*Fills exceeding 3 m in height or fills of any height subject to long periods of flooding*	
Laboratory max. dry density, gm per c.c.	*Min. field compaction requirements, per cent laboratory max. dry density*	*Max. laboratory dry density gm per c.c.*	*Min. field compaction requirements per cent laboratory max. dry density*
Less than 1.44 (90.0)	Φ	Less than 1.52 (95.0)	w
1.44 to 1.64 (90—102.9)	100	1.52 to 1.64 (95.0—102.9)	102*
1.65 to 1.75		1.65 to 1.75	

(*Contd.*)

Table 17.3: Specification requirements for embankment soil compaction (IRC) (*Contd.*)

Condition I		Condition II	
Fills 3 m or less in height and not subject to extensive floods		Fills exceeding 3 m in height or fills of any height subject to long periods of flooding	
Laboratory max. dry density, gm per c.c.	Min. field compaction requirements, per cent laboratory max. dry density	Max. laboratory dry density gm per c.c.	Min. field compaction requirements per cent laboratory max. dry density
(103.0—109.0) 1.76 to 1.91	98	(103.0—109.9) 1.76 to 1.91	100
(110.0—119.9) 1.92	96	(110.0—119.9) 1.92	98
(120.0) and more	95	(120.0) and more	96

Φ Soils having maximum dry densities of less than 1.44 gm per c.c. are ordinarily considered unsuitable and shall not be used in embankments as far as possible.

w Soils having maximum dry densities of less than 1.52 gm per c.c. are ordinary considered unsuitable and shall not be used in embankments.

* If not attainable in any soil, then at least 100 per cent.

Table 17.4: Specification requirements for minimum subgrade soil compaction (IRC)

Laboratory Max. Dry Density, gm per c.c.	Min. Subgrade Compaction Requirements, (per cent of Laboratory Max. Dry Density)
Less than 1.65 (103.0*)	102**
1.65 to 1.74 (103—109.0)	102**
1.76 to 1.90 (110—119.0)	100
1.92 (120.0) and more	98

* Soils with a maximum dry density of less than 1.65 gm per c.c. are considered unsuitable for use in the top 50 cm soil layer immediately below the surface of the subgrade and shall be replaced with suitable soil or granular material.

** If not attainable in any soil, then at least 100 per cent.

Highly expansive soils (such as black cotton soil) if used, should be compacted at moisture content 3 to 4% above the optimum to a density not exceeding 90% of the laboratory standard proctor dry density. IRC–36, "Recommended practice for the construction of embankment" is followed for detailed specifications and procedures.

Construction of Embankment on Hill Side: (Benching)

To ensure stability, on hill side slopes, we must provide for a foundation and a bond. The foundation is called a 'Bench'. Benching can be used for new construction and also for repairs of failed slopes.

Simply constructing the new embankment directly on top of the existing one is not correct. In addition to compaction, two conditions must be met to ensure the new embankment is secured to the existing slope.

(i) The existing slope must be benched to provide a foundation for the new embankment. Benches are series of horizontal cuts beginning at the intersection with original ground and continuing at each vertical intersection with the previous cut.

(ii) Secondly, the existing slope or hillside is to be continuously blended with the fill material to provide a bond between the old and new material. Figure 17.10. illustrates the concept of benching. Transverse benches as shown will be required where proposed grade intersects existing ground.

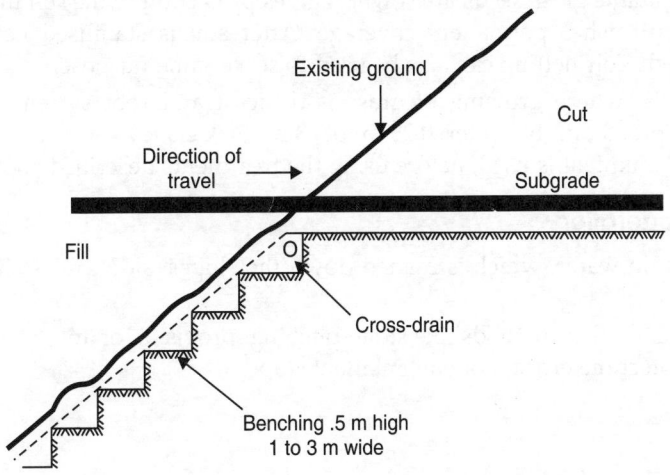

Fig. 17.10: Typical cross-section of benching at cut/fill transition

Embankments on Weak Foundations

Sometimes embankments are built on weak foundation materials, resulting in sinking, spreading and piping failures irrespective of the stability of the new overlying material used in embankment construction.

In such case considerations for embankment-foundation system, rather than just embankment may be necessary.

17.8 EROSION CONTROL OF EMBANKMENT SLOPES

Embankments for roads are made with different type of soils. Slopes of the embankments may be damaged by due to erosion from rain and wind. Lack of vegetation cover or other techniques to protect embankment slopes, result in formation of rills and rain cuts, which may lead to surfacial slide or to an undermining of the edges of the road structure.

It is therefore necessary to protect embankment slopes, from erosion, by any one of the following techniques (IRC-56)

- *Vegetative turfing:* This method is simplest using vegetation over embankment slopes. Depending upon the altitudes above sea level different grasses and shrubs, are recommended for plain and hilly areas.
- *Transplantation of ready-made turfs of grass:* In this method called 'Sodding' grass can be transplanted from original site, with its root to the side slopes of the embankment, to be treated.
- *Straw with cowdung or wood shavings/saw dust*: If height of embankment is less than 3 m and severity of erosion problem is not much, use of straw/wood shavings/saw dust, mixed with cow dung, cover 25 mm thick may be used to prevent erosion.
- *Asphalt mulch as an aid to vegetative turfing*: Asphaltic emulsion (mulch) is sprayed to protect the roots of plants. Most promising type of grass locally available is dibbled on the prepared beds. The asphaltic film gradually disintegrates and its place is taken by the

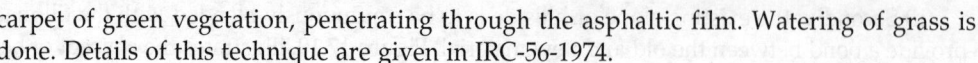

carpet of green vegetation, penetrating through the asphaltic film. Watering of grass is done. Details of this technique are given in IRC-56-1974.

- *Vegetative turfing using jute netting:* Heavy mesh of jute fabric is firmly laid on loose earth and on it suitable grass seeds are sown. It gives protection to the soil until the grass takes roots and furnishes permanent coverage. After soil is stabilised nettings of fabric is decomposed. Coir netting can also be used to serve same purpose.

On sandy soils, where growing of grass is difficult and root system are susceptible to 'washed out' or 'piped out' by water, 0.25 m to 0.3 m thick clayey soil as a blanket covering the slopes of the embankment is used, before using the treatments described above.

Drainage Consideration

To shed-off the rain water, which is carried down the slopes, side drains should be placed at proper interval.

There are several other methods like stone-pitching, provision of aprons, use of stabilisation techniques to protect the erosion of embankment slopes.

Highway Drainage

18.0 PREAMBLE

Water is the worst enemy of any road, be it surface water or sub-surface (groundwater) because it erodes slopes, weakens pavements, destroys shoulders and slopes, washes embankments and even culverts. The objective of highway drainage is to intercept and remove the water from over and under the road (from road surface and also from subgrade).

If drainage is not proper, roads develop cracking, pot holes, corrugations, and ruts, affecting riding quality and strength of the pavements. Inadequate drainage causes more severe damages to the roads, heavy traffic intensify results in premature failure of the pavement, besides causing traffic management problems. Poorly drained pavements are responsible for reduction in road capacity and traffic hazards to the users and they also increase the maintenance/ rehabilitation costs. As per an American study poorly drained pavements are twice as costly as well drained ones.

An efficient drainage system will enhance life of pavement, provide facility for free flow of traffic, reduce yearly maintenance cost, perform better and receive appreciations from the users, Drainage is very important for safe movement of traffic as friction level drops significantly from dry to a wet pavement.

To get the best out of a highway, various drainage elements should be properly designed, constructed, periodically inspected and maintained. They should be promptly modified or corrected when required by environmental changes. IRC special publication No. 42, provides guidelines on road drainage.

18.1 SOURCES OF WATER ENTRY IN ROAD STRUCTURE

There are several ways in which water can enter the road structure and subgrade including the following:

(a) Seepage from higher ground, which may be:
  (i) Transverse or
  (ii) Longitudinal
(b) Ponding at the edge of the formation in
  (i) Cutting or
  (ii) Embankment
(c) Flooding of the surface either
  (i) Pavement or
  (ii) Shoulder
(d) Infiltration of rain water through
  (i) Pavement or
  (ii) Shoulder
(e) Rise in permanent water table
  This is illustrated in Fig. 18.1

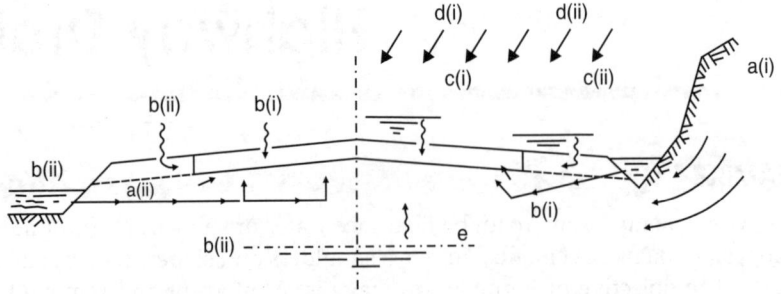

Fig. 18.1: Sources of water entry in pavement

The rate of water entry from one of the above sources varies, depending upon quality of road and intensity of rain fall. Road surface could be quite permeable, if poorly compacted, inadequately maintained. Infiltration through unsealed shoulders may be a major cause.

Moisture entry also depends upon overall climate of the region and the surrounding topography, in addition to above local factors. Moisture content under road pavement will vary seasonally, annually and over long periods.

Most serious damages from water occur regardless of the thickness of pavement or stability of base course because it is caused by pore pressure and movement of moisture contained in the road structure.

The influence of water in different type of soils can generally be as following:
  (a) Sandy soils—little change in volume or strength
  (b) Clayey soils—little change in strength, but large volume changes may occur
  (c) Silty soils—little change in volume, but large changes in strength due to pore pressure.

Volume changes are minimized if material is compacted at a moisture content which represents most likely service value (equilibrium moisture content).

The adverse effect of water into pavement are well established and it is most important to design the pavement in such a way that the water infiltrating into it is drained away quickly, through a proper drainage system, at the design stage itself.

Figure 18.2 shows how water infiltrates in a pavement through unsealed (permeable) shoulders.

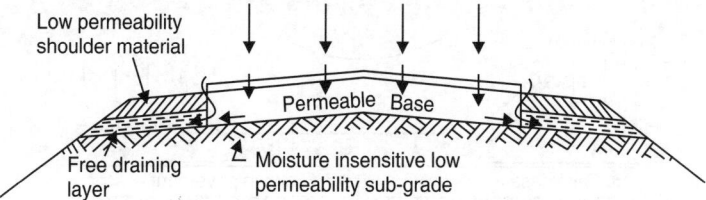

Fig. 18.2: Infiltration through permeable (unsealed) shoulders

18.2 CONSEQUENCES OF POOR DRAINAGE SYSTEM

Water should be removed from the road structure as early as possible after rainfall through a well designed and regularly maintained drainage system. Many pavement failures are direct result of water entering the pavement courses and/or subgrade.

Poor drainage system results in water entry leading to

(a) Reduction in strength, elastic and plastic stiffness of the pavement material and subgrade.

(b) Increase in rate of traffic induced deterioration of the road in time.

(c) Rutting, pavement edge failures, and other forms of surface deformation like waves, corrugations occur due to loss of strength and stiffness.

(d) Shear strength reductions resulting from increasing moisture content can lead to land-slides in cuttings and embankments.

(e) Performance of the surface course is affected by causing layer separation between courses, stripping of bitumen and pothole formation, as shown in Fig. 18.3

(f) Pavement edges are most failure prone regions when moisture conditions are severe. Shoulders and edges get damaged.

(g) In expansive soils, volume changes due to moisture result in major pavement failures.

(h) Due to poor drainage, erosion of soil occurs on unsurfaced roads, gullies may be formed along the road side.

(i) Moisture in subgrade, reduces the bearing capacity leading to failures due to excessive hydrostatic pressure caused by application of dynamic loads of traffic.

(j) In rigid pavements, 'Pumping' occurs, in which the pressure and movement induced by traffic, pumps water and fine particles through cracks or joints to the pavement surface.

(k) Stability of slopes in embankments, due to entry of water is affected adversely, because of increased weight and reduction of shear strength.

(l) In snow fall areas, ice lenses are formed due to water present in pores of the soil, which result in freezing and thawing, damaging the pavement.

18.3 BASIC PRINCIPLES OF GOOD DRAINAGE

Proper and efficient drainage is most important for highways. The main objective of the drainage for highways is to ensure that surface water and sub-surface water are removed from the highway so that the road pavement is not adversely affected by it. This can be achieved by considering certain requirements for good drainage while designing and constructing the highways. Some of the basic requirements are listed below:

(i) Fast dispersal of precipitation on the surface by proper geometric design of the road, e.g. using proper cross-slopes, gradients for pavement and shoulders.

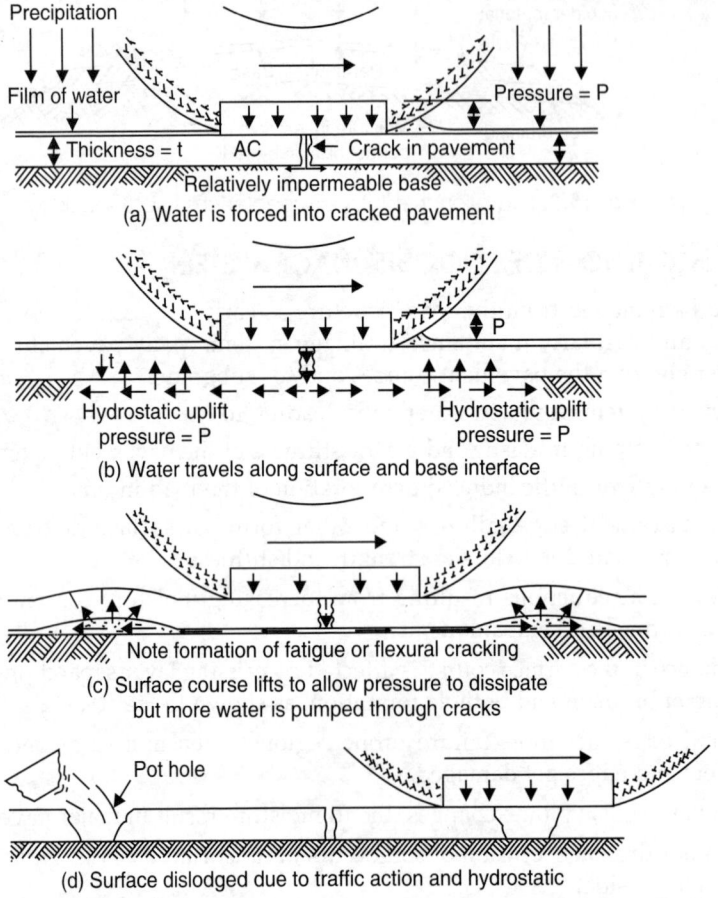

Precipitation

Film of water

Pressure = P

Thickness = t   AC   ← Crack in pavement

Relatively impermeable base

(a) Water is forced into cracked pavement

P

↓t

Hydrostatic uplift
pressure = P

Hydrostatic uplift
pressure = P

(b) Water travels along surface and base interface

Note formation of fatigue or flexural cracking

(c) Surface course lifts to allow pressure to dissipate
but more water is pumped through cracks

Pot hole

(d) Surface dislodged due to traffic action and hydrostatic

Fig. 18.3: Layer separation and formation of potholes

(ii) Water from the road and the surrounding area should be intercepted and led away to natural outfalls. This can be accomplished by suitable surface drains, shallow ditches by the side, catch drains on hill roads, etc.

(iii) Adequate cross-drainage structure should be built at river crossings and minor streams.

(iv) The surface water from nearby adjoining lands should be prevented to enter the road structure.

(v) Drains used for removal of water should have proper longitudinal slope and adequate capacity.

(vi) Level of the groundwater table should be maintained at least 1.2 m below, the subgrade.

(vii) The carriageway should be provided higher than the highest flood level of the surrounding area.

(viii) Sub-surface drainage system suitable to the site requirement should be used to drain of the water. A rational basis to calculate probable inflow quantities and to determine required outflow should be used.

(ix) In water logged areas, special care should be taken according to site requirements.

18.4 DRAINAGE SYSTEMS: SURFACE WATER DRAINAGE

The design of drainage system should aim at providing an efficient disposal system which is consistent with the cost and importance of the highway. The highway alignment is selected in such a way that drainage costs are not high.

Drainage system is broadly classified into three categories:

(1) Surface water drainage

(2) Sub-surface water drainage and

(3) Cross water drainage

Surface water drainage includes:

(a) Removal of rain or flood water from the carriageway, road formation and right of way as early as possible and

(b) Interception of surface water flowing towards the road formation and disposal of the same, to natural or artificial streams, valley or water course.

At the initial stage, for drainage of the carriageway the cross falls (camber) at the appropriate rates for different types of pavements should be provided, along with satisfactory shoulder drainage. Shoulder usually have same slope as the pavement as they are built flush with the top edge of pavement.

For collecting the surface water, longitudinal side drains or ditches are laid, for disposal of water to nearest stream. The side drains may be 'V' shaped or trapezoidal, consistent with drainage requirements. They may be lined or unlined depending upon their location, and economy. Roadside drainage channels should have adequate capacity. For design of drains, two steps are:

(1) To calculate maximum rate of run-off water, "Hydrologic Analysis", for estimation of the amount of surface water expected.

(2) To find size and shape of the drains, "Hydraulic Analysis" and design for a facility capable of disposing that quantity of water without causing damage to adjacent property or overflowing into the road shoulders or road margin.

18.4.1 Hydrological Analysis

The factors which affect run-off or discharge from a catchment area are:

- Size, shape, soil type, geology, land use, surface infiltration of the catchment area.
- Characteristics of the stream channels such as their geometry, capacity, and configuration.
- Loss of rainfall in filling the small depressions in ground surface.
- Flood area characteristics such as geometry, vegetation, configuration, etc.
- The rainfall intensity and characteristics like dizzle, storms, their frequency and duration.

All the rainfall which falls on the catchment area does not reach the place where drainage facility is being planned.

There are number of methods to estimate the amount of run-off from the catchment area, but widely used method is 'Rational Formula' which is empirical given by following expression:

$$Q = .028 \, PAI_C.$$

where:

Q = Maximum run-off in cub.m/sec

P = Coefficient of run-off for catchment characteristics

A = Area of the catchment in hectares

I_C = Critical intensity of rainfall in cm per hour occurring during the time of concentration.

Suggested values of coefficient of run off (P) are given in Table 18.1

Table 18.1: Suggested values of coefficient of run-off (IRC)

S.N.	Description of surface	Coefficient of Run off (P)
1.	Steep bare rock and water tight pavement, concrete or bitumen	0.90
2.	Steep rock with some vegetative cover	0.80
3.	Plateau areas with light vegetative cover	0.70
4.	Bare stiff clayey soils (impervious soils)	0.60
5.	Stiffy clayey soil with vegetative cover	0.50
6.	Loamy lightly cultivated or covered and macadam or gravel road	0.40
7.	Loamy largely cultivated or turfed	0.30
8.	Sandy soil, light growth, parks, lawns, meadows	0.20
9.	Sandy soil covered with heavy bush or wooded/forested areas	0.10

18.4.2 Hydraulic Design

The estimated discharge by the Rational formula should be accommodated, in a drainage channel of suitable cross-section

Manning's formula given below is commonly used for this purpose.

$$Q = \frac{AR^{2/3}S^{1/2}}{n} = A.V$$

$$V = \frac{R^{2/3}S^{1/2}}{n}$$

where:

Q = Discharge (capacity) in cub. m/sec (quantity of water to be carried by the pipe or drain

V = Velocity of flow (m/sec)

n = Manning's roughness coefficient, which depends upon type of channel or pipe.

R = Mean hydraulic radius in m which is area of flow cross-section of drain/pipe divided by wetted parameter of drain/pipe-A/P

P = Wetted parameter

S = Longitudinal slope-which is taken as slope of the drain bed

Generally, the designer uses charts that are available for selecting the size of the chosen facility (drain size, pipe diameter) for a given longitudinal gradient which also satisfies the velocity requirements in terms of scour and self cleansing. Slope and velocity are kept below the critical level, by selecting and redesigning channel section. Values of '*n*' in some typical type of drains/pipes are given in Table 18.2

Table 18.2: Typical values of Manning's coefficient of roughness '*n*' (FAA)

	Type of drain	*n*
(a)	**Clay and concrete pipes**	
	Good alignment, smooth joint, smooth transitions	0.013
	Less flow favourable conditions	0.015
(b)	**Corrugated metal pipes**	
	100% of periphery smoothly lined	0.013
	Paved invert, 50% of periphery paved	0.018
	Paved invert, 25% of periphery paved	0.021
	Unpaved, bituminous-coated or uncoated	0.024
(c)	**Open channels paved**	0.015–0.020
(d)	**Unpaved channels**	
	Bare earth, shallow flow	0.020–0.025
	Bare earth depth of flow over 3 m	0.015–0.020
	Turf depth of flow over 0.3 m	0.040–0.060

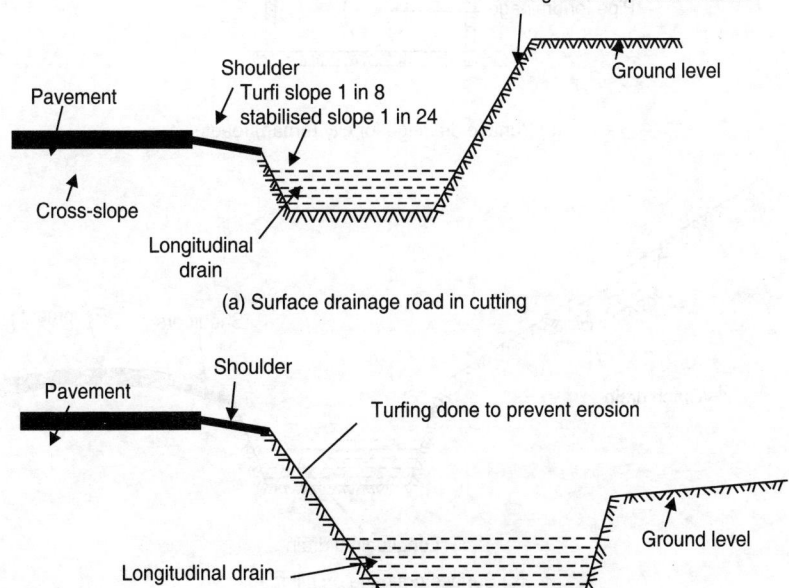

(a) Surface drainage road in cutting

(b) Surface drainage road on embankment

Fig. 18.4 A: Typical cross-sections for surface drainage drains (*Contd.*)

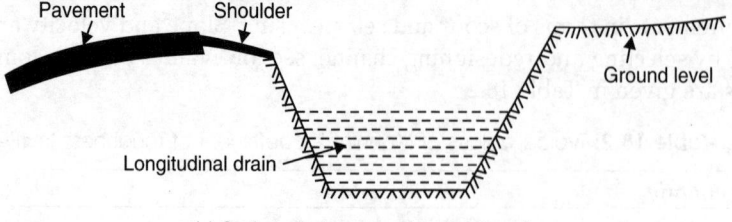

(c) Surface drainage road on level road

Fig. 18.4 A: Typical cross-sections for surface drainage drains

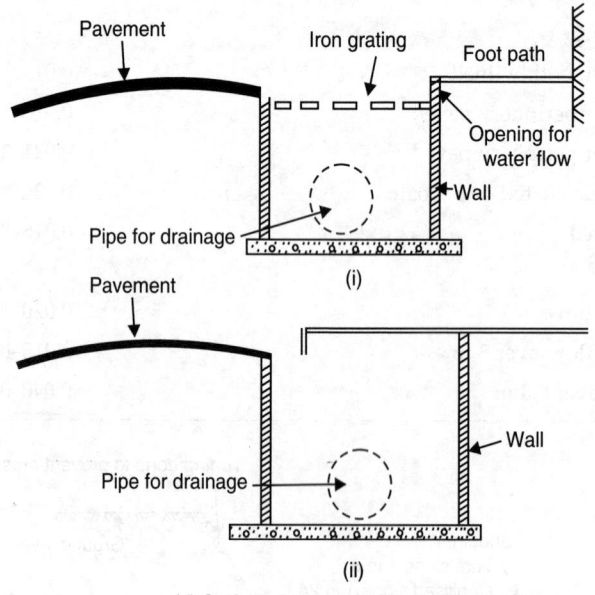

(d) Surface drainage for city (urban) roads

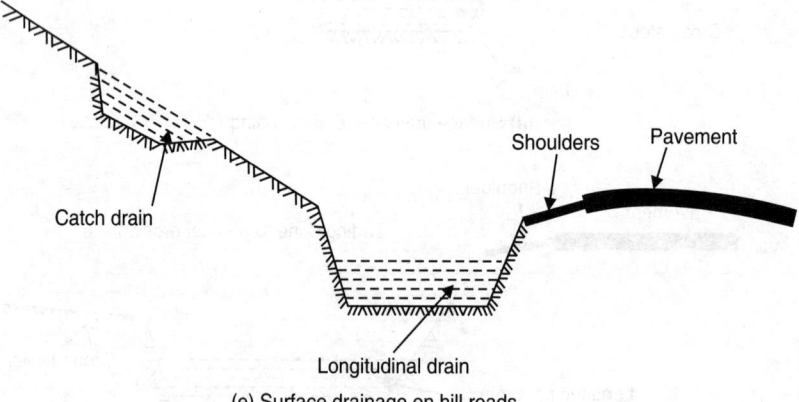

(e) Surface drainage on hill roads

Fig. 18.4 B: Typical cross-sections for surface drainage drains

For hill roads, due to practical considerations 'V' shaped drains having section large enough to carry the peak hour run-off without overflow are used. Fig. 18.4 (a) and (b) show typical cross sections of drains for surface drainage.

There are four basic types of draining structures which are commonly used for highway drainage.

(1) Earth channel or side ditches: They are cheap to construct, but costly to maintain. They are commonly used in rural areas.

(2) Lined open channels: These are commonly used in urban areas. They may be pre cast units, masonry walled rectangular or trapezoidal in shape.

(3) Pipes: They are used mainly where land is costly in urban centres for laying channels. Precast concrete pipes of various diameter are commercially produced and used (Fig. 18.5).

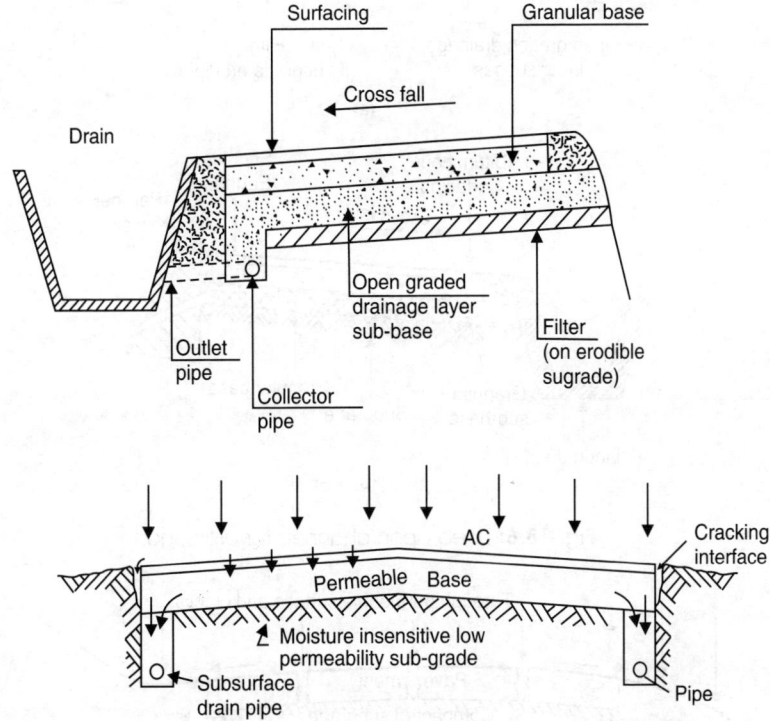

Fig. 18.5: Drainage by pipes

(4) Culverts: Culverts are used to provide cross-drainage. Different types of culverts are used. They may be circular, box, rectangular, arched in shape.

Typical examples of drainage by lined open channels are shown in Fig. 18.6.

Drainage channels should have gentle side/slopes slightly rounded bottom so that vehicles forced off the road can run down into the channel without overturning.

Catch drains of adequate width and depth are constructed above cut slopes to prevent water from the hill washing down the slope. Catch drains finally connect to cutverts.

18.5 SUB-SURFACE DRAINAGE SYSTEM

Figure 18.7 shows the causes through which water may enter the sub-surface in certain locations. Sub-surface drainage is concerned with draining the water in the ground from flowing or fluctuations in groundwater table.

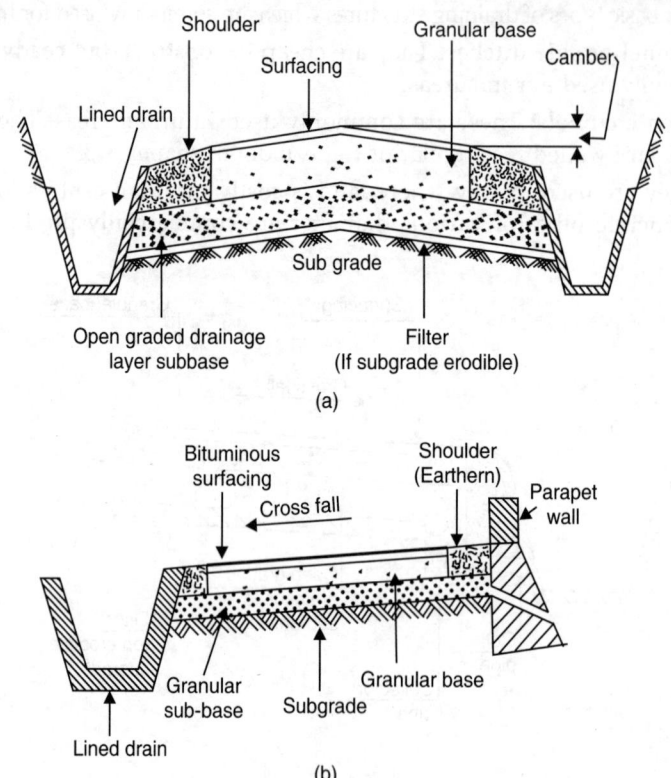

(a)

(b)

Fig. 18.6: Lined open channels for drainage

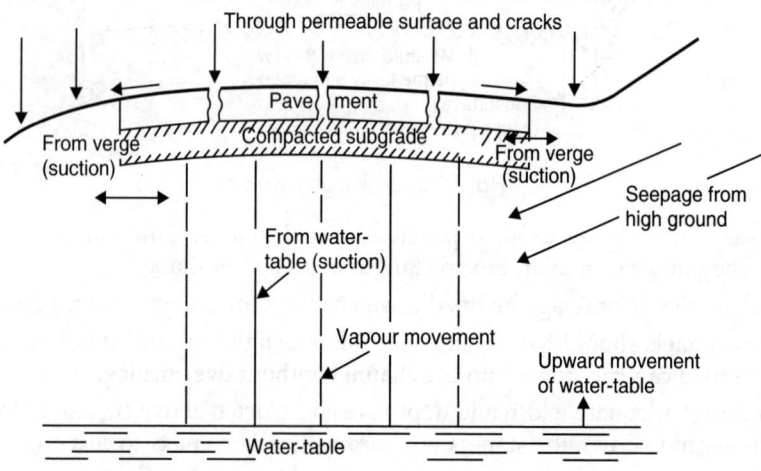

Fig. 18.7: Sources of sub-surface water entry

Purposes of sub-surface drainage are

(i) To intercept or drain under groundwater, and keeping the water away from subgrade, sub-base and base.

(ii) To lower the level of water table

The sub-surface drainage must be capable of removing any water that enters fast enough to prevent harmful accumulation of free water within the structure.

Some materials used in pavement construction like sand and gravel are free draining, while clayey elements are difficult to drain. It is better to design base and sub-base as free draining layer to maintain their bearing capacity and to protect them from changes in moisture, and thus providing horizontal sub-surface drainage. The water collecting from such a layer will drain into the side drains.

Where sub-grade consists of very poorly drain soil or is in marshy area sub-surface drainage may be necessary through vertical and/or horizontal sand drains. Intercepting drains for intercepting ground water may also be used, when sub-surface waters from adjacent areas is seeping towards the pavement.

Sub-surface drainage is mainly required where:

(i) Groundwater table is likely to rise to the base course level, within 1.2 m below the road subgrade.

(ii) Subgrade is highly impervious (low permeability)

(iii) There are chances of water seepage from side slopes, hill, etc.

(iv) Frost action takes place in subgrade, beneath the pavement

(v) There is likelihood of rise of capillary water to the subgrade

Hydraulic design of sub-surface drains is difficult, but now-a-days monographs have been developed to determine the size of collector pipe and the outlet spacing.

18.5.1 Methods of Providing Sub-surface Drainage

The techniques used for sub-surface drainage are:

(1) **Lowering the water table:** Water table can be lowered, by constructing longitudinal pipe drains and filter as shown in Fig. 18.8, on each side of the road. Drainage pipes used are of perforated metal/vitrified clay/porous concrete, etc. laid with open joints to allow the water to enter. Perforated pipes are preferred as they are less liable to silting.

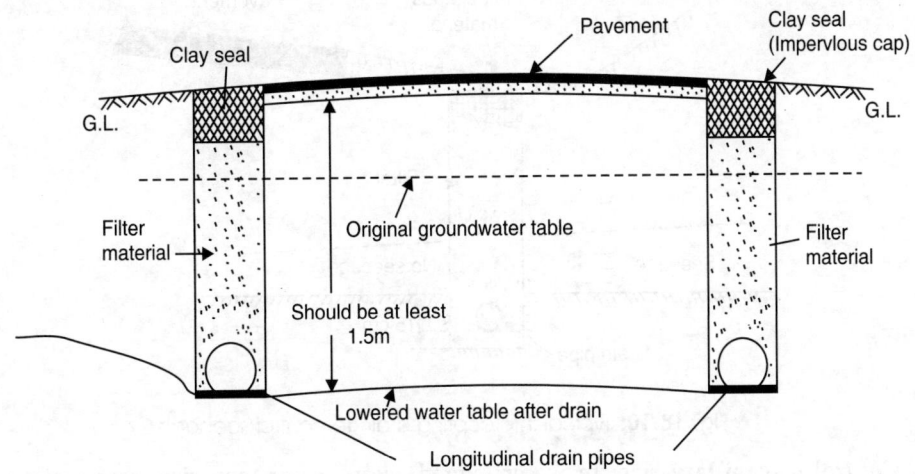

Fig. 18.8: Lowering water table by longitudinal drains

This system may not be effective if subgrade is of fine grained soils such as clay. In such case a combination of longitudinal and transverse drain is provided for lowering the water table as shown in Fig. 18.9. Perforated flexible plastic drainage pipes are becoming more popular these days due to light weight, long lengths and flexibility.

The transverse drains are inclined at about 60° and staggered. They are laid at suitable slope and discharge water into longitudinal drains.

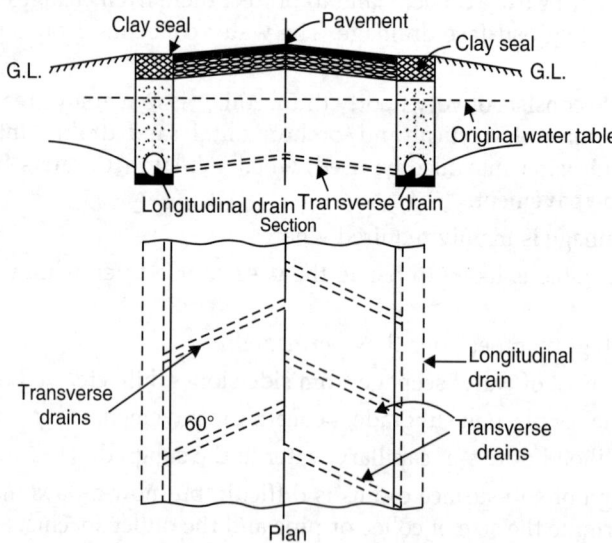

Fig. 18.9: Lowering water table by longitudinal and transverse drains

(2) **Controlling seepage flow:** Intercepting drainage is provided by means of open ditches well beyond the pavement areas, to prevent water flowing towards pavement. Figure 18.10 shows a typical intercepting drain to control seepage flow.

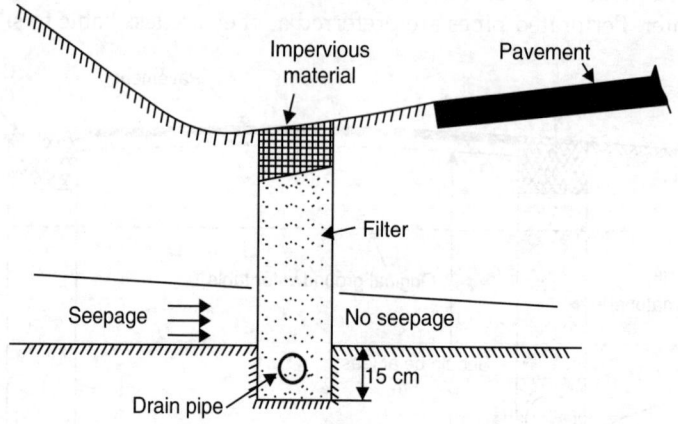

Fig. 18.10: Typical intercepting drain to control seepage

(3) **Control of capillary rise:** In water logged areas, water may rise from water table by capillary action and soften the subgrade. In such cases capillary cut-off layers are provided to control rise of water.

Commonly used capillary rise cut-off layers are:

- Sand blanket
- Bituminous prime treatment
- Heavy duty tar treatment
- Polythene enveloping
- Geotechnical fabric layer is easy to put.

Figure 18.11 shows typical capillary cut-off layer

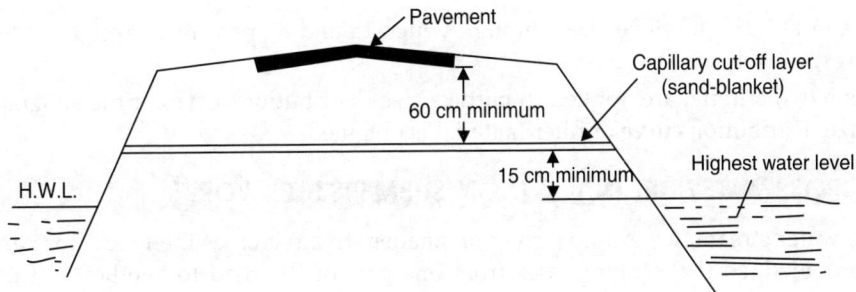

Fig. 18.11: Capillary cut-off sand blanket

The capillary cut off should be placed at least 15 cm above the water level and should be at least 60 cm away from the subgrade as shown in Fig. 18.11.

Figure 18.12 shows free draining sub-base layer to provide sub-soil drainage.

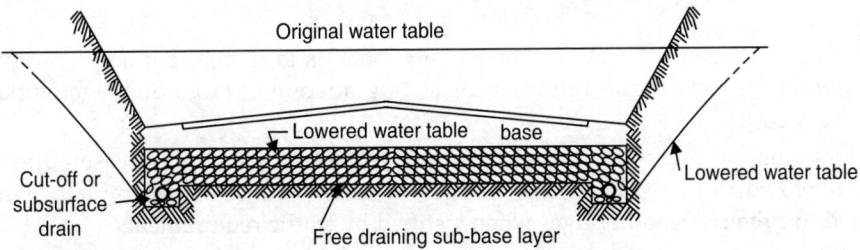

Fig. 18.12: Free draining layer with sub-surface drains

18.5.2 Design of Filter

The granular filter material used in draining system should be coarse enough so that it does not migrate into the drainage layer and fine enough so that the subgrade soils do not migrate into it. It must be permeable enough to drain out water.

Following two requirement should be met.

(i) Sufficiently permeable to allow quick drainage of water.

For this purpose permeability ratio should be greater than 5.

$$\text{Permeability ratio} = \frac{15 \text{ percent of the filter material (coarse layer)}}{15 \text{ percent size of the subgrade material (fine layer)}}$$

$$= \frac{D_{15}\text{ (Filter)}}{D_{15}\text{ (Subgrade)}} \geq 5$$

(ii) Should be fine enough to prevent intrusion of fine soil particles, i.e. silt and fine particle of the subgrade material should not be washed into the filter. For this, piping ratio should be less than 5.

$$\text{Piping ratio} = \frac{15\% \text{ size of the filter material (coarse layer)}}{85\% \text{ size of the subgrade material (fine layer)}}$$

$$= \frac{D_{15} \text{ (Filter)}}{D_{85} \text{ (Subgrade)}} < = 5$$

D_{15} and D_{85} are the sieve sizes through which 15 and 85 percent by weight of the material passes respectively.

Above two criteria, are applied to particle size distribution curve for the subgrade soil and grain size distribution curve of filter material is obtained.

18.6 CROSS WATER DRAINAGE (NON-SUBMERSIBLE) WORKS

When a water stream, for example river or smaller stream crosses the road, cross drainage has to be provided, so that water passes from one side of the road to another . At the point of crossing masonry structure known as "Cross-Drainage Works" or " Drainage Structures" are constructed.

Non-Submersible structures: The commonly used non-submersible drainage structures are:

(a) Culverts: used up to 6 meter height, to provide cross drainage

(b) Small bridges: If height is more than 6 m

For small stream and nallas culverts are used and for large streams or rivers bridges are constructed.

Cost of cross-drainage works is a major component and therefore it needs careful attention in calculating the run-off and future needs as upgradation of cross-drainage works at a later stage is very costly.

On lesser important roads, where traffic interruptions during rainy season due to flooding can be tolerated, fords, submersible bridges or causeways are used initially. They can be upgraded to culverts/small bridges when justified by traffic requirements.

(a) Culverts: Culvert and bridges have much in common, but in general when the span of the structure is less than 6 m it is considered to be a culvert and when span is greater than 6 m the structure is considered a bridge (small or large).

For run-off calculations for culvert, the Rational formula described earlier is used, while Manning's formula is commonly used to establish the size of the culvert. Structural design of culverts requires an assessment of the loading that will be applied on the culvert by traffic and earth pressure. The structural design is normally carried out by a structural engineer and highway engineer is concerned with selection of type of structure and its location. Culverts commonly used are pipe culverts, slab culverts or arch culverts. Figure 18.13 shows different type of culverts.

Scuppers: Scupper is a cheap type of culvert or cross drain of rubble dry masonry. Typical sections of two types of scuppers are shown in Fig. 18.14. In hill roads scuppers are quite common. It is made to cut the road from hill-face and to open up the road thereafter for traffic. Temporary stone scuppers are constructed after the road is cut. This may be replaced by permanent structures eventually.

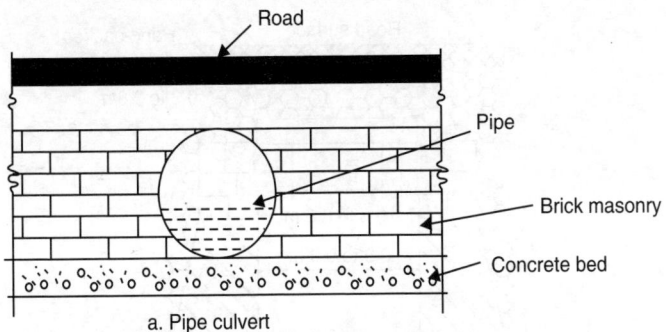

a. Pipe culvert

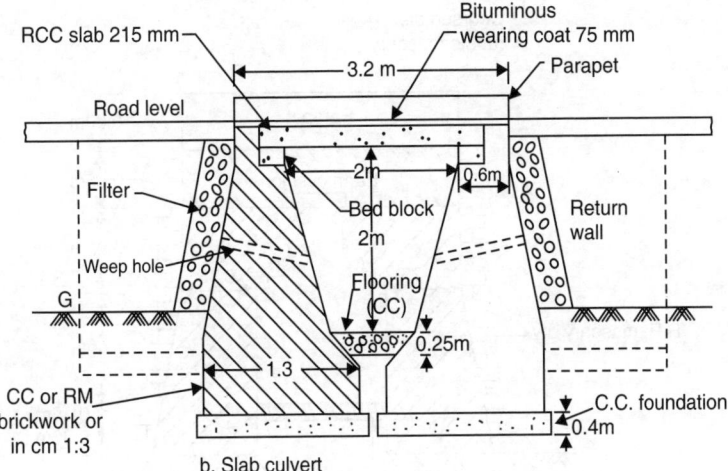

b. Slab culvert

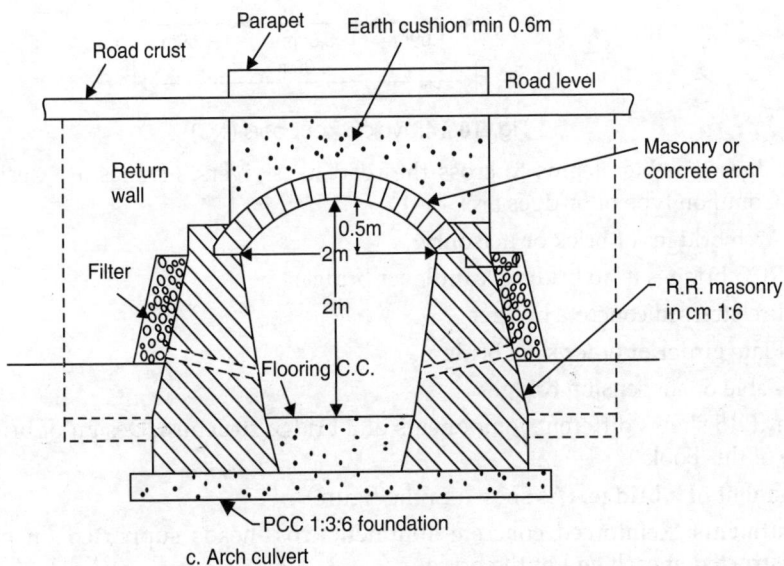

c. Arch culvert

Fig. 18.13: Typical box, slab and arch culverts (IRC)

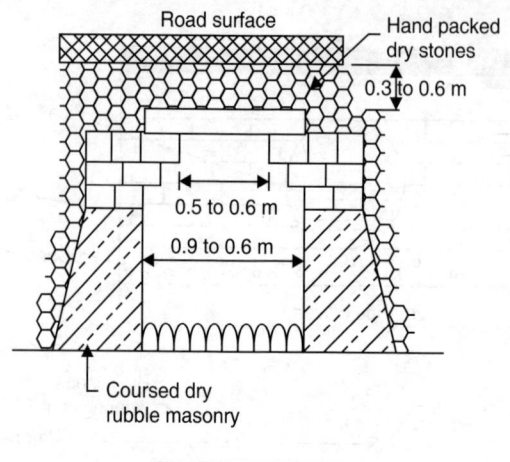

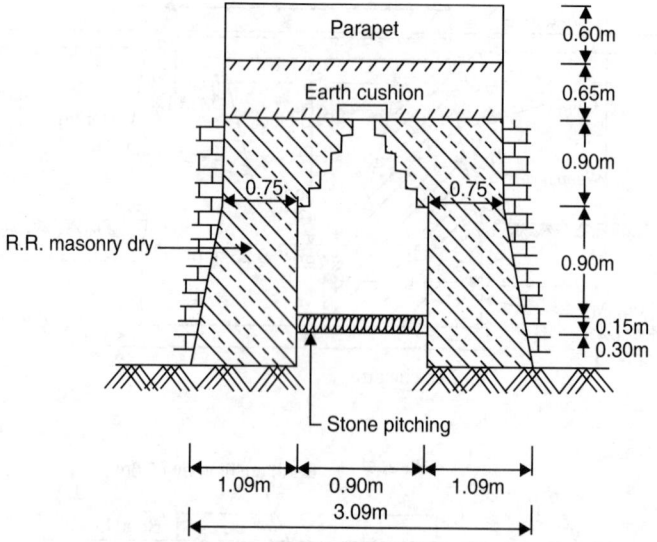

Fig. 18.14: Typical scuppers (IRC)

(b) Bridges: For big steams to cross the road likes rivers, bridges are designed and constructed. Common type of bridges are:

(a) Arch bridges of brick or masonry

(b) RCC bridges (slab bridges, cantilever bridges)

(c) Pre-stressed concrete bridges

(d) Plate girder or other steel bridges.

(e) Cable or suspension bridges

Figure 18.15 shows different components of a bridge structure. Design of bridges is beyond the scope of this book.

Component of a bridge: (As shown in the figure)

1. **Abutments:** Reinforced concrete abutment cross-heads supported on piles which are constructed at each end of the bridge.

2. **Approach slab:** A slab of reinforced concrete placed on the approaches to the bridge as a transition from road to bridge.

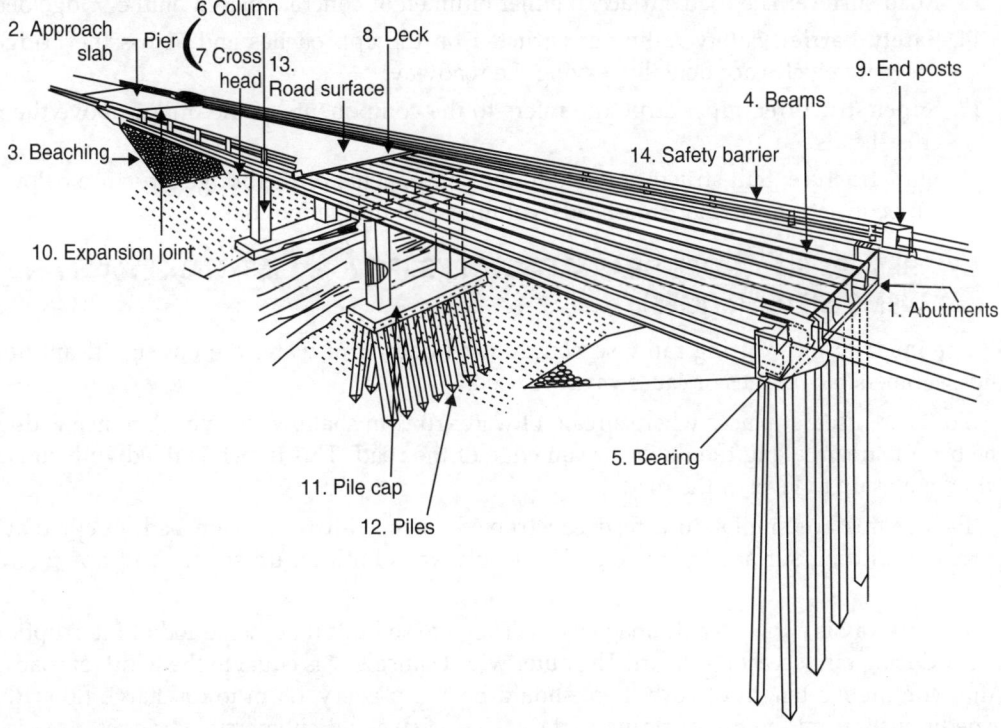

Fig. 18.15: Typical components of a bridge (CRB)

3. **Beaching:** A stone or concrete surface placed on the embankment under the bridge to prevent erosion.

4. **Beams:** Precast, reinforced or prestressed concrete beams supported on bearings which rest on the pier and abutment cross-heads.

5. **Bearings:** Pads of natural or synthetic rubber placed under the beams so that the beams can move as the bridge expands and contracts. Most common bearings are elastomeric composed on interleaved laminated plates of steel and of natural rubber.

6. **Columns:** Reinforced concrete columns constructed on top of the pile caps.

7. **Cross-heads:** Reinforced concrete cross-heads constructed above the columns to complete the piers and to provide support for the bridge beams and deck.

8. **Deck:** Reinforced concrete cast over the beams to form the bridge deck and the curbs. It is like base course in a pavement structure. They may be precast, reinforced or prestressed concrete.

9. **End Posts:** Reinforced concrete end post blocks constructed to support the rails on each side of the bridge

10. **Expansion joints:** Metal or synthetic expansion joints placed across the width of the deck to allow for expansion and contraction.

11. **Pile cap:** A reinforced concrete cap cast on top of the piles to serve function to spread footing.

12. **Piles:** Pre-cast reinforced concrete piles driven into the ground to form the foundation of the bridge

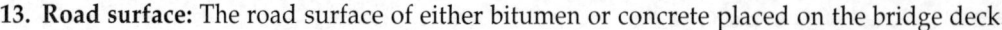

13. **Road surface:** The road surface of either bitumen or concrete placed on the bridge deck.

14. **Safety Barrier:** Safety railing constructed on the approaches and across the bridge to prevent vehicles accidentally leaving the roadway.

15. **Super structure:** Super-structure refers to the components of the bridge above the pier cross-heads

16. **Sub-structure:** Sub-structure refers to the components of the bridge structure below the super-structure, i.e. pier crossheads, columns and bridge foundation.

18.7 SUBMERSIBLE CROSS-DRAINAGE WORKS (FORDS, DIPS, CAUSEWAYS AND SUBMERSIBLE BRIDGES)

Where interruptions during rainy season due to flooding of water are not significant, fords, dips, submersible bridges or causeways are used.

Fords are used at places where stream of water runs in shallow channel. A trench is dug on the bed of stream along the down stream edge of the road. This trench is filled with stones in wired crates up to level of the road.

Paved dips are similar to a ford constructed flush with the stream bed except that the permanent riding surface is protected by cut-off walls both on up stream and down stream sides.

A causeway is a low cost drainage work. They are so built that the period of interruption to traffic during rainy season is short. The outer width causeway is equal to the width of roadway. After continuous length of curb there should be a gap every 1.5 m to discharge flood water expeditiously from the top of riding surface. Causeways and submersible bridges, are dips in the road profile to allow water flow across the road surface during floods. They get submerged during rainy season in high floods of short duration, and are available for use by the traffic rest of the times. On roads which are not of major importance submersible bridges/causeways are used.

Causeways can be of flush, low-level, or high-level type.

Flush causeway is a paved dip or roadway, built to cross a shallow water-course. The top level of the road shall be the same as that of the bed of the water course.

A low level causeway is a structure provided with a few openings made by pipes, short span slabs or small arches, etc. with a raised road top level to a height of up to 1.2 m to 1.5 m.

A high level causeway is a submersible structure provided with larger openings comprising of a simply supported or continuous RCC slab or multiple arches or boxes and a raised road top level to a reasonable height up to 1.5 to 3 m.

Road surface of a causeway should be such that it does not get damaged by passing of water over it. Figure 18.16 shows a typical submersible pipe causeway.

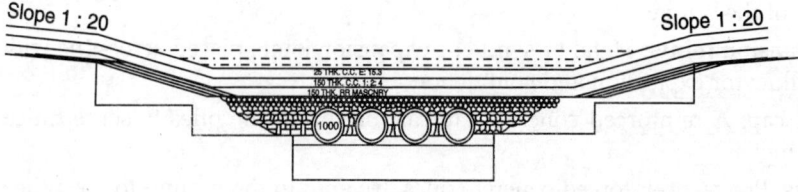

Fig. 18.16: Typical submersible vented pipe causeway

18.8 SPECIAL CONSIDERATION FOR DRAINAGE OF URBAN ROADS

In urban areas, capacity of drains, can not be increased due to land problem and therefore initial drains should be designed to cater for increased discharge. Following points need special attention in urban areas for drainage system.

- Good velocity of flow of water should be ensured so that silt and other materials do not collect in the bed of the drain. Self cleaning velocities should be achieved.
- The water collected in the manhole should be taken away through a drain pipe to the side drain.
- All drains should preferably be covered so that they are not used as dustbins and blocked for flow of water.
- On narrow roads in urban areas, water may be allowed to flow in the kerb channels, which is led to main drainage system, as it may be difficult to provide separate space for drains.
- Gratings (iron, precast, fibre reinforced) for surface run-off to the drain, is important in urban roads for safety.
- In some urban locations like underpasses, subways, flyovers, bridges, intersections, rotaries etc. it may be necessary to remove rain water through pumping after collecting it into a sump tank to nearest drain.

18.9 SPECIAL CONSIDERATION FOR DRAINAGE IN WATER LOGGED AREAS

In water logged areas, water remains standing in sub-soil (subgrade) for prolonged periods and groundwater table is within 1.5 m. Due to capillary action from high water table, water migrates towards pavement, resulting in loss of subgrade strength.

Due to flooding for long time, wearing course is disintegrated loosing its water proofing quality as stripping of binder occurs. Entry of water from shoulders is another factor increasing the problem. Water entry reduces the strength and the elastic and plastic stiffness of the pavement material and subgrade, leading to ultimate failure of the pavement. The problem is further complicated if salts like sulphates of calcium, magnesium or sodium and sodium carbonate are present in subgrade soil or in the groundwater

To drain the water in waterlogged areas elaborate measure are used including

- Providing drainage channels 1.5 to 2 m deep and connect them with suitable outfalls.
- Measures (already described) to lower the groundwater table could be used to prevent subgrade from capillary water.
- High embankments may be provided if drainage channels involves high cost.
- Provision of capillary cut off layer could be used where economically justified.

Road design (structural thickness) in such cases should be based on saturated CBR value of the subgrade to simulate worst situation in the field and strengthening of pavement should be done if not properly designed. Measures for drainage to be adopted depend upon the location and available funds (cost-analysis).

Hill Roads

19.0 PREAMBLE

Hill road is a road which passes through a terrain which has a cross slope of 25 percent or more. It is characterized by steep slopes, differing elevations, gorges, variety of climate, unique geological and ecological characters. Difficult terrain conditions pose constraints in ensuring adherence to design, and construction norms. Roads in hilly regions, in general, are more accident prone due to inadequate sight distances, sharp curves, steep gradients and lack of visibility in poor weather conditions. Hill roads therefore require special attention in planning, design, construction and maintenance.

For economic and industrial development it is necessary to build hill roads. Hill roads at the borders of the country are important for strategic reasons during the war, for moving the army and equipment. Hill stations attract tourists to enjoy climate and scenic beauty. For development of tourism, good system of roads in hilly area is the basic requirement.

19.1 CLASSIFICATION OF HILL ROADS

Border road organization is responsible for hill roads in the border region, and they have four categories of roads given below:

(a) **Border road classification:**

    (1) National Highways

    (2) Class 9: Roads which are 6 m wide for 3 tonne vehicles.

    (3) Class 5: Roads which are 4.9 m wild for 1 tonne vehicles

    (4) Class 3: Roads which are 2.45 m to 3.65 m wide for jeeps.

(b) **General classification:** Hill roads like other roads in the country are classified as:

    (1) National Highways (NH)

    (2) State Highways (SH)

    (3) Major District Roads (MDR)

    (4) Other District Roads (ODR)

    (5) Village Roads (VR)

(c) **Classification based on uses:** According to the use of a hill road, they are classified as:

(1) Motor roads: Roads which are used by motor vehicles. They are properly designed, from engineering considerations.

(2) Bridle paths: These paths are used by pedestrians and animals. They are feeder to motor roads. They may be 2 to 3.5 m wide, having gradient of 1 in 10 or 1 in 7.5. They have side drains, proper camber and gradient. They may be earth, gravel or natural surface.

(3) Village tracks: They connect small villages and pedestrian and cattle moves on them. Width varies from 1 m to 1.5 m.

19.2 PLANNING OF HILL ROADS

Construction of hill roads is costly because of rock cutting, drainage works, construction of retaining and breast wall, measures for slope stability and erosion control, etc. Besides the normal principles of planning of roads, following basic principals should be kept in view for hill roads. Villages in hilly areas are scattered in valleys and over slopes, hill tops and can not be connected by straight roads as in plains.

(i) **Development of total plan:** To avoid changes at heavy cost, an overall road development plan should be developed for the hilly area, for which planning is done. Out of the total master plan, priorities can be assigned for construction of roads. Roads in hilly areas are shorter in length and connecting villages of low population requires clustering them together.

(ii) **Widening as per traffic needs:** In view of economy, it is better to plain for jeepable roads in the beginning and widen them to full width as traffic increases, in future years, following the standards of geometric design and construction according to the category of the road.

(iii) **Stage construction technique:** The construction work of the hill road can be completed in stages, as per requirements of traffic in number of years, rather than completing all works requiring more funds. Stage construction should therefore be planned.

(iv) **Hydrological parameters and climatic conditions:** Hilly regions pose unique problems of climate and hydrological nature, which require considerations in planning, and route location. Snowfalls and heavy rains need considerations in fixing the alignments. Drainage and cross drainage works need greater attention. Reduction in cost is achieved if cross-drainage works are less.

Heavy winds should also be avoided in locating the route.

(v) **Slope failure and landslides:** Slope failures are major issue for the safety and stability of hill roads. Slope failures and landslides occur in humid climates during unusually heavy precipitation or during heavy concentration of rain fall. Slope stability is an important element of planning of hill roads. Geologically unstable and fissured areas prone to landslide and erosion should be avoided.

(vi) **Conservation of forests and ecological balance:** While planning the hill roads, and during its construction all considerations should be given for maintaining the ecological balance and conservation of forests. Deforestation is responsible for many landslides in the hills. Road construction activities should be minimum on reserve forests, in order to avoid any disturbance to natural wealth and ecology.

(vii) Safety features: Adequate safety features such as parapet/guide walls, safety barriers, retaining and breast walls, signs and markings, etc. should be planned and provided. High accident prone approaches should have properly placed traffic signage, delineators and markings. Laminating paints on signs and markings help in foggy weather.

(viii) Other Measures:
- Balanced cut and fill with a retaining wall, use of reinforced soil to minimize earth work, and provision of adequate erosion control measures should be planned.
- When trucks negotiate a grade, there is a fall in speed, depending upon initial speed, type of commercial vehicle, its tractive effort and steepness of the grade. To avoid interference, with movement of other vehicles, separate truck climbing lanes should be planned on steep grades.

19.3 ALIGNMENT OF HILL ROADS

For efficient and safe operation of vehicles through the hilly terrain, special considerations are required for fixing the alignment, some special considerations are:

(i) The alignment should be such that ruling gradient is attained in maximum length, minimizing steep gradient, hairpin bends and unnecessary rise and fall.

(ii) In hilly areas, road should be aligned through the side of the hill that is stable, where there are no or less chances of landslide.

(iii) Road should be aligned in such a way that number of cross drainage structures is minimum. This will reduce construction cost.

(iv) In cases where alignment passes through landslide prone areas, advance geological studies should be conducted and protection works like breast walls, retaining walls, toe walls should be constructed.

(v) Alignment should be such that hairpin bends are minimum

(vi) Deep cuttings and steep terrain should be avoided as they increase chances of landslide.

(vii) Tunnels though they reduce travel lengths should be avoided, to save the cost.

(viii) The alignment towards the side exposed to sun should be preferred.

(ix) In hill roads, obligatory points, in view of river crossing, valleys, location of tunnels, bridges should be identified before deciding the alignment, along with those points through which alignment should not pass.

(x) The alignment should cross ridges at the lowest elevation.

(xi) As geologic conditions differ from one side of a valley or mountain slope to other, even where the material are similar, the alignment should be selected towards stable conditions, where there is stability against geological disturbances.

(xii) Destruction of forest area should be avoided.

For alignment surveys of hill roads use of aerial survey methods and air-borne laser terrain mapping are useful. Aerial surveys help in final decision in the selection of the alignment, for which detail ground survey should be undertaken. Details of survey works to be undertaking and drawings to be prepared are given in Chapter 4.

19.4 GEOMETRIC DESIGN OF HILL ROADS

Construction cost increases with the gradient in the hilly terrain. The geometric design standards are different for hilly terrain to keep the cost of construction and time of construction under control . Sharper curves and steeper gradients are adopted for hill roads.

Geometric design standards used in mountain and steep terrain for basic features are given below:

(i) **Building line and control line:** Distance between the building line and road boundary in rural area and urban areas, for different type of roads is given in Table 19.1

(ii) **Roadway width and carriageway width:** Table 19.2 shows standards for roadway widths and carriageway widths in mountainous and stop terrain for different type of roads. In heavy snowfall areas roadway widths may be increased by 1.5 m for MDR, ODR and VR.

(iii) **Shoulder width:** Shoulder width recommended is 1.25 m for National and State highways on each side of the roadway, for single lane and 0.9 m for double lane roads. For MDR, ODR and VR the shoulder width of 0.5 m on either side is recommended. In hard rocks, shoulder width may be reduced by 40 cm on either side for 2 lane roads and 20 cm in other cases.

Table 19.1: Standards for building lines and control lines in mountainous and steep terrain

		Distance between building line and road boundary (meters)			
		Rural areas		Urban areas	
S. N.	Class of Road	Normal	Exceptional	Normal	Exceptional
1.	National and state highways (NH and SH)	5	3	5	3
2.	Major district road (MDR)	5	3	5	3
3.	Other district road (ODR)	5	3	5	3
4.	Village road (VR)	5	3	5	3

Table 19.2: Standards for roadway width in mountainous and steep terrain

S. N.	Class of road	Width of roadway (m)	Carriageway width (m)				
			Single lane	Intermediate lane	Double lane	Double lane with curb	Multi lane
1.	National and state highway						
	(a) Single lane	6.25	3.5	5.5	7.0	7.5	3.5 m per lane for multi-lanes
	(b) Double lane	8.80	7.0	–	–	–	
2.	Major district road (single lane or two lane)	4.75	3.5 7.0	5.5	7.0	7.5	
3.	Other district road	4.75	3.5	5.5	7.0	7.5	
	(a) Single lane	–	7.01	–	–	–	
	(b) Double lane	–	–	–	–	–	
4.	Village road (Single lane)	4.00	3.5	5.5	7.0	7.5	

(iv) **Right of way widths:** Table 19.3 shows the recommended right of way width for different roads in mountainous and steep terrain.

Table 19.3: Right of way widths for different roads in mountain and steep terrain (IRC)

S.N	Class of Road	Right of way width in meters			
		Rural areas		Urban areas	
		Normal	Range	Normal	Range
1.	National and state highways (NH and SH)	24	18	20	18
2.	Major district road (MDR)	10	15	15	12
3.	Other district road (ODR)	15	12	12	9
4.	Village road (VR)	9	9	9	9

(v) Cross slope (camber): Cross slopes are recommended, in view of the rainfall and type of surface. For hill roads following values are recommended:

(a) Bituminous concerts, cement concrete and other high type bituminous surfaces — 2 percent (1 in 50)

(b) Thin bituminous surfaces — 2 to 2.5 percent (1 in 50 to 1 in 40)

(c) Gravel or water bound macadam surface — 2.5 to 3 percent (1 in 40 to 1 in 33)

(d) Earth roads, shoulders, sub-grades, etc. — 3 to 4 percent (1 in 33 to 1 in 25)

Flatter camber is used, if gradient is more than 1 in 20. Camber can be made to slope either on one side or on both sides.

(vi) Design speed: The design speed for different types of hill roads recommended by IRC are given in Table 19.4

Table 19.4: Design speeds for hill roads (IRC) (km/hr)

S.N.	Road type	Mountainous terrain		Steep terrain	
		Ruling km/hr	Minimum km/hr	Ruling km/hr	Minimum km/hr
1.	National and state highways	50	40	40	30
2.	Major district road	40	30	30	20
3.	Other district road	30	25	25	20
4.	Village road	25	20	25	20

(vii) Sight distances: Sight distances both stopping sight distance and overtaking sight distance are calculated by the formula given in Chapter 6. The calculated values recommended for hill roads are given in Table 19.5. Passing (overtaking) places are required to facilitate crossing of vehicles approaching from opposite directions.

Table 19.5: Sight distances for hill roads

S. N.	Design speed	Stopping sight distance m	Passing right distance m
1.	30	35	90
2.	40	50	145
3.	50	70	210

As overtaking sight distance is not always available on hill roads, minimum setback distance from centre line of the inner side of horizontal curve, equal to stopping sight distance should be provided for design speed.

(viii) Gradients: Table 19.6 shows recommended ruling, limiting and exceptional gradients for different types of terrain.

Table 19.6: Gradients for hilly and steep terrain (IRC)

S.N.	Terrain	Ruling gradient %	Limiting gradient %	Exceptional gradient %
1.	Mountain	5 (1 in 20)	6 (1 in 16.71) for a distance	7 (1 in 14.3) not more than 100 m at a stretch.
2.	Steep (i) up to 3000 m height above msl	5 (1in 20)	6 (1 in 16.7)	7 (1 in 14.3) for a distance not more than 100 m at a stretch
	(ii) Above 3000 m height above mean sea level	6 (1 in 16.7)	7 (1 in 14.3)	8 (1 in 12.5) for a distance not in more 100 m at a stretch

(ix) Design service volume: For single, intermediate and two lane roads, design service volume in pcu/day, for hill roads as recommended by IRC are given in Table 19.7.

Table 19.7: Suggested design service volume for hill roads (IRC)

S.N.	Type of road	Design service volume pcu/day	
		Curvature (Degree/km) Low (0 – 200)	Curvature (Degree/km High (above 200)
(1)	Two lane roads (7. 0 m)	7,000	5,000
(2)	Intermediate lane roads (5.5 m)	5,200	4,500
(3)	Single lane roads (3.75 m)	1,800	1,400

(x) Super-elevation: IRC recommends that super-elevation on hill roads should not exceed 7 percent on road which are in snowfall area and should not exceed 10 percent on other roads.

(xi) Radius of horizontal curve: Table 19.8 shows minimum radii of horizontal curve for different types of roads.

Table 19.8: Minimum radii of curve on hill roads (IRC)

S. N.	Classification of road	Minimum Radius of horizontal curve			
		Mountainous terrain		Steep terrain	
		Snow free area	Snow bound area	Snow free area	Snow bound area
1.	National and state highway (NH and SH)	30	60	30	33
2.	Major district road (MDR)	30	33	14	15
3.	Other district road (ODR)	20	23	14	15
4.	Village road (VR)	14	15	14	15

(xii) Transition curves: The minimum length of transition curve recommended by IRC is given in Table 19.9.

Table 19.9: Minimum length of transition curves (IRC)

S.N.	Curve Radius (m)	Design speed Km/h				
		50	40	30	25	20
1.	15				NA	30
2.	20				35	20
3.	25			NA	25	20
4.	30			30	25	15
5.	40		NA	25	20	15
6.	50		40	20	15	15
7.	55		40	20	15	15
8.	70	NA	30	15	15	15
9.	80	55	25	15	15	NR
10.	90	45	25	15	15	
11.	100	45	20	15	15	
12.	125	35	15	15	NR	
13.	150	30	15	15		
14.	170	25	15	NR		
15.	200	20	15			
16.	300	15	NR			
17.	400	15				
18.	500	NR				

NA-Not applicable   NR-not required

(xiii) Extra widening on curves: Road requires widening on curves due to psychological and mechanical reasons, as desesibed in chapter 6. For hilly roads, extra width to be provided on horizontal curves is given in Table 19.10.

Table 19.10: Widening of pavements on horizontal curves

S.N.	Extra width(m)	Radius of curve (m)				
		upto 20 m	21–40	41–60	61–100	101 to 300
1.	Two lanes	1.5	1.5	1.2	0.9	0.6
2.	Single lane	0.9	0.9	0.6	Nil	Nil

No extra widening is required for curve of radius more than 300 m.

(xiv) Grade compensation on curves: At horizontal curves, the gradient should be easy to offset the extra tractive effort involved on curves. For gradients flatter than 4% compensation is not required.

(xv) Vertical curves: For smooth transition of grade change, vertical curves (summit curves, sag curves) are introduced. Length given in Table 19.11 is recommended.

Table 19.11: Minimum length of vertical curves

Design speed (km.hr)	Maximum grade change (%) not requiring a vertical curve	Minimum length of vertical curve (m)
Up to 35	1.5	15
40	1.2	20
50	1.0	20

(xvi) Hairpin bend: It is a sharp compound curve, which turns 180°, on the same side down the hill. Its shape resembles a hairpin and so it is named as hairpin bend. It helps in attaining the height, without long horizontal travel. Figure 19.1 shows a typical hairpin bend. Hairpin bend should be located at sites where there is no danger of landslides or ground-water and provided with long arms.

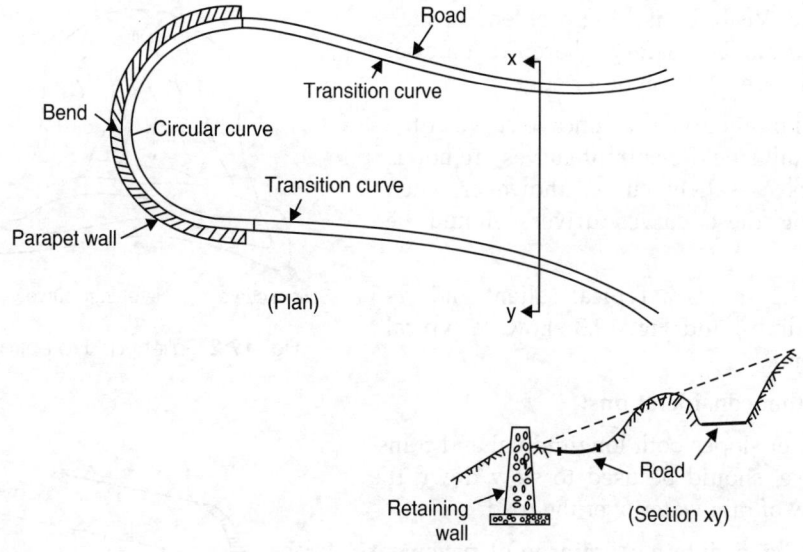

Fig. 19.1: A typical hairpin bend

The number of hairpin bends should be minimum possible. The minimum distance between the successive bends should be 60 m.

A hairpin bend is designed as a circular curve with transition following IRC standards given below:

(i) Minimum design speed = 20 km/h.

(ii) Minimum width of carriage way at apex of the curve

    (a) NH and SH

        Double lane   = 11.5 m

        Single lane   = 9.0 m

    (b) MDR and ODR   = 7.5 m

    (c) VR   = 6.5

(iii) Minimum radius of inner curve = 14 m

(iv) Minimum length of transsition curve = 15 m

(v) Gradient

    Maximum   = 1 in 40 (2.5%)

    Minimum   = 1 in 200 (0.5%)

(vi) Super elevation in circular portion = 1 in 10 (10%)

(vii) Minimum straight length between two successive hairpin bends = 60 m

(viii) The approach gradient should not be steeper than 5 percent for = 40 m

Proper sight distance should be available at hairpin bend for visibility. They should be located on stable and flat slopes. Valley bends should be avoided.

(xvii) Salient and Re-entrant curves: Salient and re-entrant curves are provided if there are ridges and valleys in the hill. Salient curves have convexity outwards on the outerside of the road at the ridge of hill side. The bend caused by salient curve is called corner bend. Visibility is a big problem on such location. For safety, therefore parapet walls are used, on outer edge.

A re-entrant curve is a concave curve at the valley of hill side. Re-entrant curves are not as dangerous as salient curves, however when negotiating these curves drivers should be careful.

Fig. 19.2 shows a typical salient and re-entrant curves, and Fig. 19.3 shows a typical corner bend.

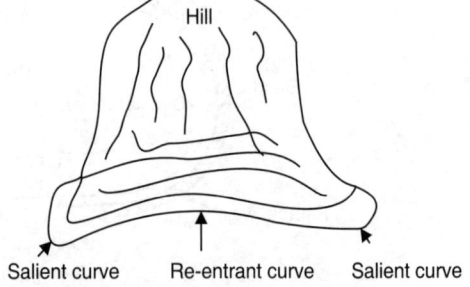

Fig. 19.2: Salient and re-entrant curves

(xviii) Other considerations:

- Flatter slopes both longitudinal and transverse, should be used to slow down the flow of rain water over the road.

- For better internal drainage of pavement layers, especially of granular material, a slight longitudinal gradient may be preferable say 0.3% or so.

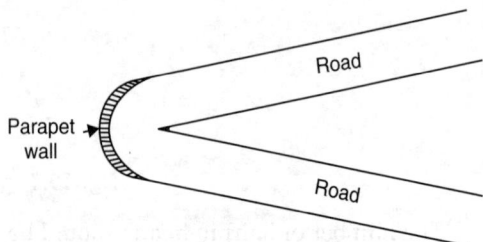

Fig. 19.3: Corner bend on hill roads

- Paved hard shoulders are better than earth shoulder. Shoulder level should not be higher than road surface, otherwise water will accumulate on the top of it.

- For disabled vehicles, parking places should be available at least 2 per kilometer to enable other vehicles to pass.

- Adequate vertical and lateral clearance should be provided on hill roads, for safety. Vertical clearance in tunnels should not be less than 5 m.

- Passing places and lay-byes should be provided frequently to facilitate crossing vehicles and tow aside a disabled vehicle.

Geometric features for bridle roads and bridle paths are shown in Table 19.12.

Table 19.12: Geometric features for bridle roads and paths (IRC)

S.No.	Geometric feature	Bridle road	Bridle path (Border/Village track)
1.	Road land width in open areas	6 m	3.0 m
2.	Formation width		
	a. Normal	2 m	1.0 m
	b. Exceptional	1.7 m	0.8
3.	Radius of curves (minimum)	5 m	5 m
4.	Widening at sharp curves upto 3 m radius	1.0 m	0.3 m
5.	Inside slope (cross fall)/camber	3 to 4%	3 to 4%
6.	Minimum radius at H.P. bends	3.0 m	1.0 m
7.	Gradient		
	Ruling	17%	17%
	Limiting	–	25%
	Exceptional (not more than 30 m length)	25%	30%
8.	Drains	0.30 m	0.2 m
9.	Scuppers	1 m span 3 to 10 nos per km	0.6 m span 3 to 5 nos per km
10.	Bridges and culverts		
	a. Design load	400 kg/sqm	400 kg/sqm
	b. Clear roadway between kerbs	2.0 m	1.0 m

19.5 DRAINAGE OF HILL ROADS

One of the major cause of failure of slopes and road formation in hills is poorly designed, constructed and maintained drainage system. Road drainage (surface, sub-surface or cross drainage) should be of enough capacity, not very deep for safety of vehicles and pedestrians. Gentle slopes and shallow depths are preferred for safe traffic movements.

Natural drainage pattern of the area should not be disturbed. If not taken care it may result in floods. Rains fall heavily on hills and water flows very fast.

Large quantity of excavated material during construction if not disposed off properly, moves towards the hill slope and accumulates in dams and reservoirs.

Along the entire side of the road, a suitable drainage system (surface and sub-surface) must be provided so as to avoid any flash flood, soil erosion and damage to vegetation, etc. In hilly areas where rainfall is heavy, it is better to use frequent and suitable type of cross drainage works, with proper disposal. Drains on valley side are to be avoided as far as possible. Fig. 19.4 shows typical shapes of drains, used on in hill roads.

Catch drains (intercepting drains) are provided on up hill side, above side drains, to prevent over loading of side drains. Design principles of drainage system are the same as described in chapter 18.

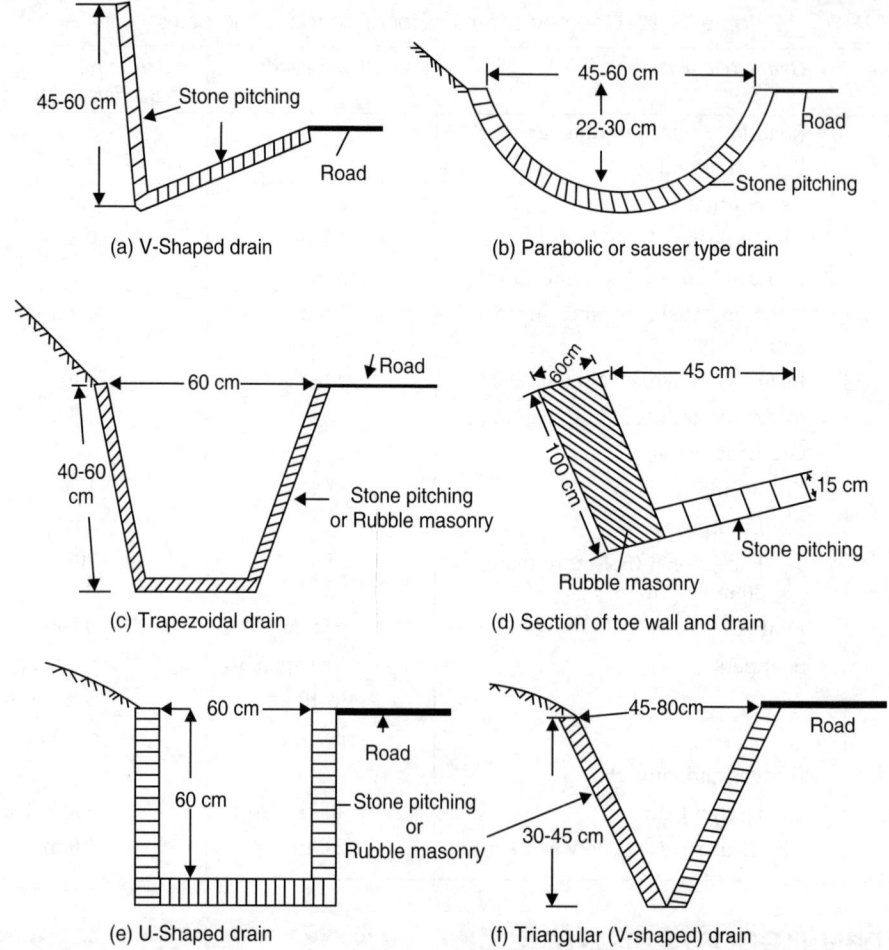

Fig. 19.4: Typical side drain shapes for hill roads

19.6 PROTECTIVE STRUCTURES IN HILL ROADS

Different structures are used in hill roads for preventing landslides, floods, etc. and providing safety. Commonly used structures are:

19.6.1 Retaining walls

Retaining walls are most important structures in hill road construction to provide adequate stability to the roadway and to the slope. Retaining walls are constructed on the valley side of the road and also on cut hill side to prevent landslide towards the road. Retaining walls, retain the back filling and common locations for their use are:

  (i) Where road cross-section passes partly in cutting and partly in embankment
 (ii) On re-entrant curves
(iii) For cross-drainage works

Retaining walls are built in stone masonry, brick masonry, cement concrete, etc. depending upon the pressures of earth fill and economics, as construction of retaining walls is costly. The near side of a retaining wall is kept vertical. If there is water accumulation, behind the retaining walls, weep holes are provided for draining the water.

Figure 19.5 shows different type of retaining walls.

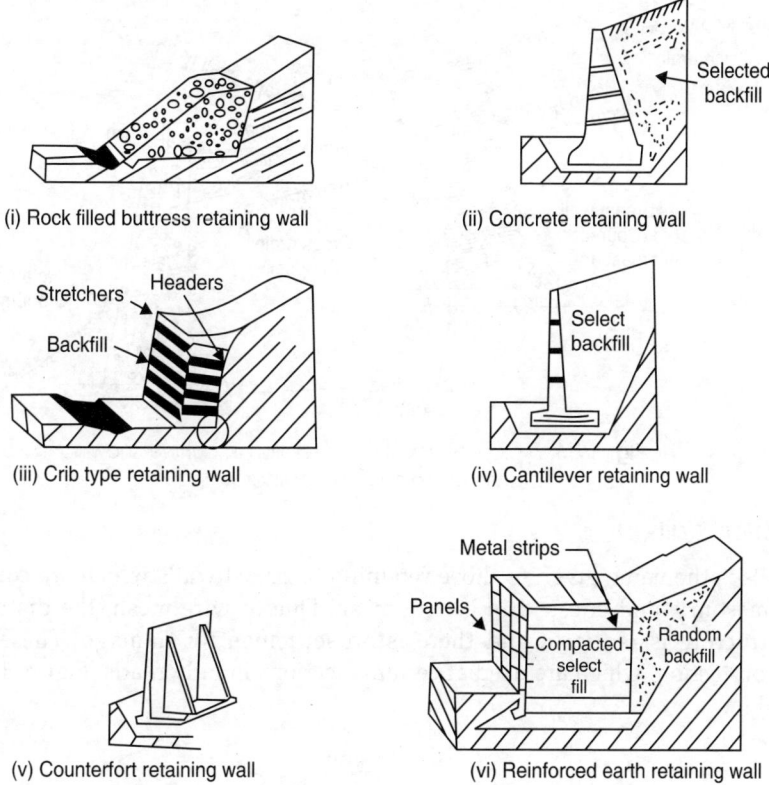

(i) Rock filled buttress retaining wall

(ii) Concrete retaining wall

(iii) Crib type retaining wall

(iv) Cantilever retaining wall

(v) Counterfort retaining wall

(vi) Reinforced earth retaining wall

Fig. 19.5: Different types of retaining walls

19.6.2 Breast Wall

Breast wall is a wall which is constructed on the uphill side of the road to prevent cut portion of rock from slippage. Breast walls may be battered from one side or both the sides. They should be able to withstand the pressure from the back and retain the slope. They may be of rubble masonry, bricks masonry or concrete. The cross-section and height of the retaining wall varies from site to site. Number of weep holes may be provided to remove water from the back of the wall.

Figure 19.6 shows a typical breast wall. The height of breast wall is 1.5 to 3.5 m. Front batter is usually kept 1 : 3

19.6.3 Parapet Wall

The wall constructed above the formation level, on one side or both sides of a road to provide psychological and physical protection to the drivers, on hilly roads is called a parapet wall. It is usually provided on the down hill side, and provide protection on roads with steep valley side. They are .45 m thick of stone masonry.

These walls need not be continuous. Usually they are 2 to 6 m long with gaps of 0.6 m to 1 m. Figure 19.6 shows typical location of a parapet wall. The height of the wall is about 0.6 m Parapet walls should be white washed and maintained in good condition.

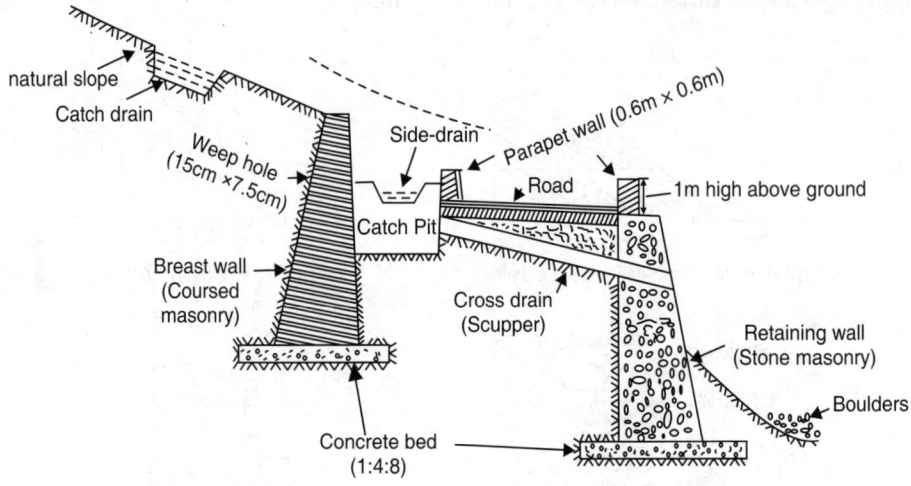

Fig. 19.6: Typical hill road cross-section in cutting showing breast, retaining and parapet walls

19.6.4 Gabion Walls

Gabion walls is the name given to those retaining or breast walls which are constructed with dry stone masonry, and encased in a wire mesh. Due to wire mesh, the disturbance to the masonry structure is reduced, and there is no settlement or damage. These are a sort of additional protection, which are used at certain locations in hilly roads. Figure 19.7 is a typical gabion wall.

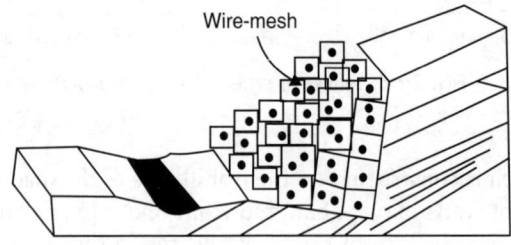

Fig. 19.7: A typical gabion wall

19.6.5 Catch Drains

Catch drains are provided high up on the hill slope, to catch the water before it flows towards the road. These are placed parallel to roadway to intercept the water from higher ground at suitable intervals and divert it away from the road.

Typical catch water drains are shown in Fig. 19.8

19.6.6 Toe/check walls

Locations where culverts/scuppers are constructed, water falling from a considerable height on the valley side results in erosion at the toe of the retaining walls. In order to prevent this erosion, one or more toe wall are constructed to break the water force.

In some places check wall are made in the beds on upstream of nallas, to reduce the flow of debris which may block the road. Figure 19.9 shows typical toe and check wall.

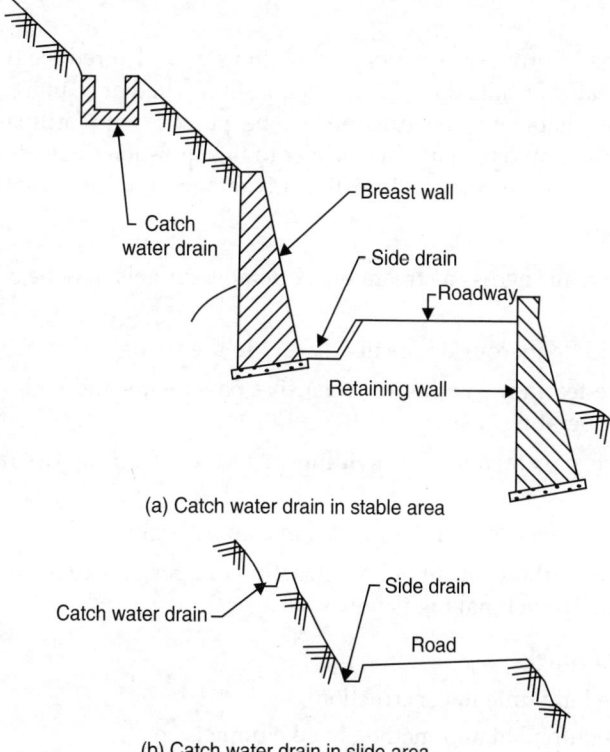

(a) Catch water drain in stable area

(b) Catch water drain in slide area

Fig. 19.8: Typical catch water drains

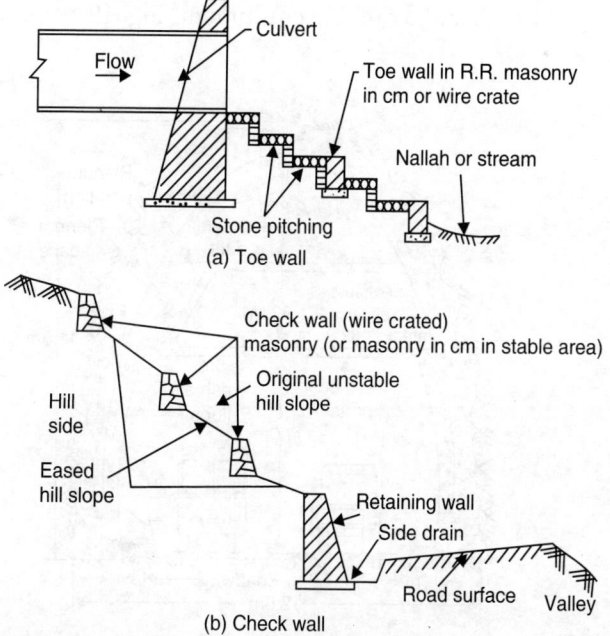

(a) Toe wall

(b) Check wall

Fig. 19.9: Typical toe and check walls (IRC)

19.6.7 Tunnels

Tunnels though costly provide eco-friendly solutions and also reduce the distances for travel. For roads in mountains or hills there is often a question whether to make a tunnel or make such a detour that will enable the obstructions to be passed with ordinary surface or cut the mountain/hill to make an open cut. The answer to this question depends upon the comparative cost of construction, maintenance and relative advantages and disadvantages.

Advantages of tunnels

- For carrying roads across a stream or mountains tunnels may be cheaper than bridges or open-cut
- Tunnels avoid dangerous open-cut adjacent to the roads.
- Tunnels save tearing/wearing of expensive pavements and so lesser maintenance costs and vehicle operating costs.
- Tunnels save pavements by providing protection from snow, rain and other natural influences.
- During warfare and bombing, tunnels have an intangible value.
- With modern methods of tunnel construction danger of settlement of overlaying ground is eliminated. Tunnel making is now safe.

Disadvantages of tunnels

- Tunnels take long time in construction
- Specialised equipment and methods make tunnels costlier.

It is generally assumed as a thumb rule that when the cut required have a vertical depth of more than 20 m it is less expensive to built a tunnel, unless excavated material is needed for a nearby embankment or fill. This rule is not absolute and varies according to local conditions, type of rock and available equipment. Fig. 19.10 shows a typical two-lane highway tunnel.

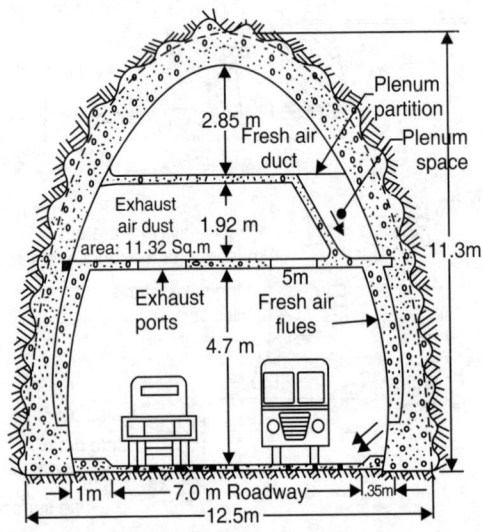

Fig. 19.10: Two lane highway tunnel

19.7 LANDSLIDES AND REMEDIAL MEASURES

Landslides or slope failures occur due to imbalance of forces, i.e when driving forces exceed resisting forces. Slope stability and prevention of landslides is an important aspect of highway safety and stability on hill roads.

A landslide occurs if the shearing stresses in a soil mass or rock exceed the magnitude that the soil or rock is able to resist. Reasons for increase in active shearing forces are:

- Release of large quantity of impounded water from upper hill, due to heavy rainfall.
- Construction of the road on the slope.
- Increase in the weight of soil mass, due to rain, snow, etc.
- Undermining due to excavation or erosion.
- External additional load due to traffic.
- Removal of lateral support, due to cutting or quarrying.
- Removal of soil particles by external agents such as wind or water.
- Shock and vibration due to blasting or earthquakes.
- Entry of water into the road bed through fissures, faults or voids.
- Increase in pore water pressure due to rising groundwater table.
- Higher gradients/slope brought about by construction activity like cut or fill.
- Due to decomposition of the geologic materials with time.

The reduction in resisting forces (strength) in soil or rock may be due to:

- Formation of fissures and faults in the beds.
- Climatic factors like heat, heavy rains, snow, etc. change the characteristics of the soil mass or the rock and decrease shear strength.
- Increase in water content resulting in swelling and pore water pressure.
- Due to deforestation and removal of vegetation
- Seepage pressure of water within the soil mass is increased due to fissures or cracks.
- Development of cracks in the rock or soil due to swelling and shrinking

19.7.1 Stability of Slopes

Stresses are usually highest at the toe of the slope and failure therefore begins from toe and progresses up slope. Stability of slope generally depends upon the following variables.

(i) Topography — slope inclination and height

(ii) Geology — material structure and strength.

(iii) Weather — rainfall, seepage forces and run-off quantity and velocity of flow.

(iv) Seismic activity — it affects internal and seepage forces.

In evaluation of instability of slopes and landslides following basic factors need considerations.

- Type and distribution of geologic materials in the slope.
- The geologic structure-weak and heterogenous composition.
- Existing groundwater conditions.
- Possibilities of future rise in seepage pressure during rainy season, inclination and height of slopes.
- Excavation for cuts or the placement of fill which might change above factors, such as natural slope.

- Removal of vegetation decreases slope stability.
- Earthquake force or fatigue (decomposition) of material over period of time.
- Discontinuities in rock like, joints, cracks, faults, etc. due to weathering action.
- Lack of sufficient drains for removal of water or faulty design.

19.7.2 Remedial Measures for Landslides

There are several remedial techniques adopted to correct landslides depending upon the site, magnitude of the problem and economic considerations. Some of the remedial measures used alone or in combination for correcting landslides are described below:

(i) Vegetation in an important slope stabilizer. Planting of thick native vegetation on slopes serves to strengthen the shallow soils with root system, and prevents erosion

(ii) Seeding or soding can also be used for slope treatment. Asphalt mulch technique can be used in which the slopes are prepared into vast seed beds. Asphalt mulch is spread by sprayer, which gradually disintegrates resulting in carpet, of grass, with deep roots.

(iii) Effective drainage measures (surface, sub-surface and cross-drainage) should be undertaken to intercept and divert the water. Water is a contributing factor in all landslides. Improvement in drainage system with catch drains, new drains, and culverts may be used.

(iv) Construction of buttress walls at toe, breast walls, and retaining walls. Retaining walls can be designed sufficiently large to withstand any landslide. However, a long wall may be beyond economics of the project. Rock filled buttresses are used when good foundation is available at toe in shallow or deep soil.

(v) Gabion walls are also used to provide toe support. They are free draining and retention is obtained from the stone weight and its interlocking. Typical heights are about 5 to 6 m.

(vi) Cementation of loose material. Cement grout is injected into the moving area to achieve stability. Cracks, joints, fissures are filled with grout. It produces higher shear resistance.

(vii) Surface erosion can be controlled by laying geogrids or jute netting.

(viii) Treatment of slopes to improve the stability and prevent erosion, like chemical treatment, or vegetation

(ix) Slopes should not be used for dumping of garbage, etc. or activities like agriculture

(x) Piling can be used in shallow soils to hold the slide mass temporarily.

(xi) For retaining rock slopes, rock bolts, wire mesh and shortening can be used. Rock bolts can retain foliating slabs and other loose blocks. Shot creating stabilises highly fractured rocks.

(xii) Grazing by animals should be prevented and afforestation should be resorted. Indiscriminate cutting of trees should be avoided.

(xiii) To ensure quick and efficient drainage of water sufficient number of weep holes of adequate size and design should be provided at various levels in the wall.

Slope stability and settlement analysis is discussed in detail in chapter 17

19.8 CONSTRUCTION OF HILL ROADS

Due to varying climatic conditions, geological features, and hydrological parameters in hilly regions, unique problems are posed in highway construction, though the techniques of construction are the same as in flat areas.

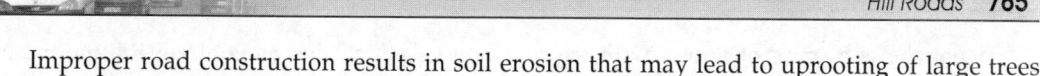

Improper road construction results in soil erosion that may lead to uprooting of large trees and degeneration of lower plants, leading to serious ecological imbalances and landslides.

Operations like blasting, excavation, clipping of mountain slopes create geological disturbances in the mountain body, causing movement of slip zones, cracks, fissures and weak planes. Blasting may also distract the flora and fauna, and wild life in the area.

Some of the points to be considered during hill road construction are:

- Retaining walls when required should be constructed before hill cutting so that cut soil can be used as back fill.
- During the construction earthwork and drainage works should proceed side by side.
- Cross-fall should be on hill side to control run-off over the valley side.
- During the construction work, excavation of formation, and the rock cutting precaution should be taken that conditions are not created for landslides.
- Blasting operations should be minimum. All visible fissers and cracks should be sealed before start of construction.
- The cut and throw approach to road construction should be avoided, as it may cause huge waste and landslides.
- Heating of bitumen should be done properly to protect the environment, during the construction works.

Techniques of construction of roads in hilly area are same as described in chapter 15.

19.9 MAINTENANCE OF HILL ROADS

Some problems which occur on hill roads, needing immediate and careful attention in their maintenance are different from those associated with roads in plain terrain. Maintenance of pavements remain more or less the same as in plain terrain but in hill roads maintenance and protection of slopes, their stabilization, snow removal, maintenance of drainage facilities and cross drainage works, require more funds and timely action through regular and prompt maintenance.

19.9.1 Stability of Slopes and Landslides

Slides are common phenomenon on hill roads, which often block the road. The maintenance of road involves removal of these obstructions with suitable equipment as early as possible, so that traffic flow is resumed in shortest time with minimum delay. Removal of the cause for slides to re-occur such as filling of cracks, grouting, vegetation, etc. should be undertaken wherever possible.

The effort in maintenance should be to ensure that original slope is maintained and stability is retained. Methods of slope protection and maintenance are described in chapter 17.

19.9.2 Snow Removal on Hills

On high altitudes snowfall is common during winter. Annual snowfall varies from heavy exceeding 4 m to average (1.5 to 4 m) from place to place. In certain areas, snowfall is so heavy that roads are closed during winter, while in other areas snow removal is done to keep roads open to traffic.

After winter ends, the snow starts melting due to rise in temperature, resulting in release of water, which should be removed immediately to prevent formation of ice lenses. Snow clearance is done by machines like ploughs, bulldozers, scrapers, graders, etc. For a road located in snowfall area, snow clearance is serious maintenance problem in winter.

Avalanche is a mass of snow, with ice and debris, which descends along the mountain slopes with high momentum. Avalanches are caused due to winds, temperature differential, gravitational forces, earth tremors, snow storms, etc. Frequency of occurring, type and intensity of an avalanche is unpredictable in most cases. Detailed advance planning of labour and equipment during snowfall period is required, keeping the men on alert all the time for undertaking snow clearance immediately, on occurrence.

Damage done by snow and avalanches to the drainage system should be immediately corrected and road formation and surface should be repaired to make them traffic worthy.

19.9.3 Maintenance of Drainage System

Maintenance of surface drains, catch drains, cross-drainage works like culverts, causeways, etc. is a major task on hill roads as their blockage may result in slope failures, slips and landslides. Regular inspections and cleaning of debris and other obstruction which prevent water flow should be undertaken.

An efficient system of monitoring maintenance work is required for quick and proper management of maintenance operations. The techniques of maintenance for pavements, markings, signs, and other structures are the same as described in chapter 16.

'Hill Road Manual' published by Indian Road Congress, provides detailed specification and guidelines for hill roads in India.

Highway Lighting

20.0 PREAMBLE

The number and rate of highway accidents and fatalities that occur during night driving is several times higher than that during the day driving. Night time illumination of a roadway is very important in promoting safety and operational efficiency. In general, lighting is used more extensively for urban than rural roads. In addition to furthering highway safety, lighting in urban area promotes safety to pedestrians. In rural areas, lighting is generally applied in critical areas such as intersections, interchanges, rail-road crossings, narrow or long bridges, tunnels, sharp curves and areas where roadside interference is likely. The purpose of highway lighting is to provide a safe and comfortable environment for driving in the night time. The objectives of lighting are:

(a) Promote safety at night by providing quick, accurate and comfortable visibility for drivers and pedestrians.

(b) Improvement of traffic flow at night by providing light beyond that provided by vehicle lights, which aids drivers in orienting themselves, delineating roadway geometrics, and obstructions and judging opportunities of overtaking.

(c) Illumination in long underpasses and tunnels during the day to permit drivers entering such structures from the day light to have adequate visibility for safe vehicle operation.

(d) Reduction in street crime after dark in nights. Anti-social activities are discouraged due to lighting.

(e) Enhancement of commercial activities by attracting evening/night shoppers, audiences and other users.

(f) Police petrol and performing of its functions becomes easy due to lighting system.

Road lighting should be able to ensure that the road environment and objects are sufficiently visible to ensure that drivers are able to drive efficiently and safely. The lighting must be able to display the carriageway, its surroundings and permitting the detection of presence, position and movement of other road users (pedestrians), parked vehicles, traffic control devices, changes in road conditions, etc. In driving task it is necessary to detect relative motion between objects as well as the object themselves. The safety (reduction in accidents) justifies the install-ation of an effective light system on highways.

Anything is visible when light being reflected from it, enters the eye of the viewer. The two sources of light on the road are vehicle lights and road lighting.

Road lighting makes the carriageway bright. Dark objects are thus seen as a contrast against a bright background. Vehicle lighting, illuminates objects against a dark background. Contrast factor is one which can be influenced by street lighting. Contrast begins as soon as an object can just be detected separately from its background.

Ambient light level falls in night time conditions and the human visual system loses sensitivity. Discrimination of colour and details are impaired and object are seen mainly by contrast. The broad principles of highway lighting are described in following articles so that national or regional standards, according to site conditions can be developed. Practices in lighting and standards may differ from country to country.

20.1 GENERAL PRINCIPLES OF VISIBILITY ON ROADS

The visibility of an object on the road from a vehicle depends upon the combination of the following factors:

- The difference in luminance (contrast) between the objects and their background
- Brightness of the object
- Size of the object
- Reflection properties of the object
- The luminance of the background against which it is seen
- Movement of eyes
- The duration of the observation (time factor in response)
- Glare
- Driver's vision
- Condition of windshield

The performance of the lighting system of a highway is measured by:

(i) **Illuminance:** The amount of light from the installation source incident upon the surface of interest (visibility target) in the highway environment and

(ii) **Luminance:** The amount of the light reflected to the driver's eye from the visibility target.

Visibility is extremely important in the design of highway lighting. The lighting system should not only provide the amount of light required but also its distribution on the pavement and the amount of glare experienced by the drivers. Light distribution affects the contrast of targets viewed by drivers.

The design of highway lighting system is an interplay of many factors which should be taken into account such as:

- Different combination of lamp types
- The housing for lamps (for glare control)
- The height of light source (luminaire)
- Spacing of light source (luminaire)

These factors contribute to overall system output and jointly determine the visibility of the target at specified location on the road.

In road lighting it is important that driver is able to discern clearly the presence and movement of any object on or adjacent the road which may be a potential hazard. To achieve this, proper and even illumination on the road and its surroundings is to be provided.

20.2 LIGHTING SYSTEM AND ROAD ACCIDENTS

Studies and investigations conducted around the world have indicated that, although estimates vary, the savings can be enough to pay for a lighting installation on highways in a few years. Road accidents at night are much higher in number and severals compared to day time accidents.

Some of the finding from accident studies in different countries are summarised below:

- Lighting can reduce the ratio of night to day accidents by as much as 14% of total accidents. Box (1989).
- Poor visibility involving pedestrian and cyclists resulted in more fatal accidents. Owen and Sivak (1993).
- Elvick (1995) observed after study of number of accidents that roadway lighting has large safety benefits for pedestrians and at intersections. Fatal accidents were reduced by 65% due to lighting and night time injury and property damage accidents were reduced by 30% and 15% respectively.
- Simultaneous introduction of channelisation and illumination at locations experiencing high number of accidents (intersections, etc.) improves safety to a great extent. Lipinki and Wortman (1976)
- Based on research conducted in USA, AASHTO (1984) recommended that lighting of spot locations in rural areas should be considered whenever the driver is required to pass through a section of road with complex geometry and/or raised channelisation.
- Several research studies on freeways, have indicated that lighting has beneficial safety effects. It was found that the average night/day accident rate was 66% greater on unlighted freeway than on lighted freeways. It was noted that best results were obtained when illumination was combined with roadside delineation.
- For all volumes of traffic, vehicle to vehicle accidents, vehicle to fixed object accidents, property damage accidents rate decreased with increasing light levels. This was found by Gramza, et al. (1980), in a study to predict frequency of occurring of each type of accident per year, in various levels of illumination on certain interchanges.
- Accident studies that have investigated age-related benefits of highway lighting, suggest that night time accidents involvement rate is much higher for younger drivers relative to day time accidents.
- Analysis of accident studies in UK suggests that up to half of the night road accidents might be avoided if adequate lighting was provided at the following critical areas of driver's decision.
  - Entrances and exits
  - Interchanges and intersections
  - Bridges overpasses and viaducts
  - Tunnels and underpasses
  - Guide sign locations
  - Dangerous hills and curves
  - Heavily travelled sections in urban and rural areas
  - Rest areas and connecting roads

– Rail road crossings
– Elevated and depressed roadways

20.3 DEFINITION OF TERMS USED IN LIGHTING SYSTEM

Some important terms used in highway lighting design system are defined below and illustrated in Fig. 20.1.

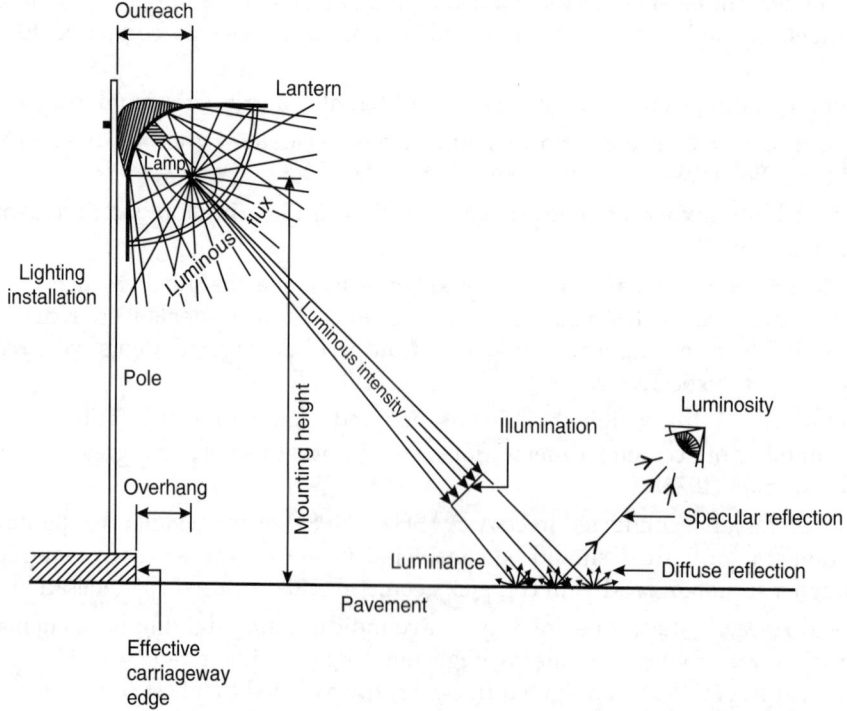

Fig. 20.1: Terms used in highway lighting design

Average initial illumination: The average level of horizontal illuminance on the pavement area, at the time when lighting system was installed and lamps were new and clean expressed in average foot candle (FC)/(lux) for the pavement area.

Glare: Luminance greater than that to which the eye is accustomed. Glare may be sub-divided as

'Disability glare' which affects visibility but may pass un-noticed. It is caused by scattering of light within the eye and its effect is like passing an external luminous veil over the field of view.

'Discomfort glare' relates to levels at which physical reaction is experienced.

'Blinding glare' is so intense that for an appreciable length of time no object can be seen.

Diffuse reflection: Reflection such that light incident from one direction is reflected from the surface in many directions.

Candela (Cd): A measure of the luminous intensity of a light source as seen by eye. Light sources have different luminous intensities and it varies with the angle at which it is viewed, i.e. direction. The unit is candela and a candela is equivalent to a lumen/steradian (lm/sr).

Light: Radiant or luminous energy in the visible spectrum (wave lengths of 380 to 780 nm)

Candle power (CP): The luminous intensity in a specific direction, expressed in candles (cd). It is not an indication of total light output.

Luminous flux: The light being emitted by a light source, e.g. by a lantern or received by a pavement. It is measured in lumen (lm). It is radiant energy per unit time emitted by a radiation source.

Coefficient of utilisation: The ratio of luminous flux (lm) from a luminaire received on the surface of the roadway to the lumens emitted by the luminaire's lamp.

Lantern: A housing for one or more lamps, together with a reflector, or refractor, or diffuser, etc. associated with lamp.

Illumination (illuminance): The unit flux incident on a surface per unit area (e.g. pavement). Its unit is the lux (x) and is equivalent to lm/m^2 or Cd/m^2.

Luminous intensity being flux per solid angle is thus the lantern measure and illumination being flux per surface area, is the pavement measure.

Luminance: The luminous intensity per unit area measured at a point on a surface and in a given direction. It is measured in candela per unit area, i.e. Cd/m^2 and is a measure of brightness.

Brightness levels range from -6 to -1.5 log Cd/m^2 at night and from $+1.5$ to $+5.5$ log Cd/m^2 in the day.

Foot candle (FC): The unit of illumination or luminance when the foot is taken as the unit of length. It is the illumination on a surface one square foot in area on which there is uniformly distributed flux of one lumen.

Luminance contrast: The ratio of the luminance increment of an object with respect to its background, divided by the background luminance.

Luminosity (Brightness): The visual sensation related to a surface appearing to emit or reflect light. Brightness is usually taken as the subjective equivalent of luminance.

Light pole (Standard): A pole (standard) provided with necessary internal attachment for wiring and the external attachments for the bracket and luminaire.

Luminaire dirt depreciation factor (LDD): A depreciation factor that indicates the expected reduction of a lamp's initial lumen output due to the accumulation of dirt on or within the luminaire over time.

Luminous efficacy (lm/W): The quotient of the luminous flux (lumen) emitted by the total lamp power input (watt). It is expressed as lumens per watt (lm/W).

Specular reflection: Reflection, as in a mirror, from a flat surface according to the law of optics. It is reflection from wet road surface.

Mounting height: The vertical distance between the roadway surface and the centre of the light source in the luminaire.

Overhang: The horizontal distance between a vertical line nadir of a luminaire and the edge of travelled way or edge of the area to be illuminated.

Nadir: The vertical axis which passes through the centre of the luminaire light source.

Spacing: For roadway lighting the distance between successive lighting units, measured along the centre line of the road.

20.4 WARRANTS OF HIGHWAY LIGHTING

Prior to actual design of highway lighting system, it should be determined that lighting at particular section, location or area is actually warranted. Providing lighting for all highway facilities is not practical nor cost effective.

Accident data of the location should be collected to determine the night to day ratio of accidents. This ratio is one of the most dominating factors, for justifying the lighting requirement. Sound engineering judgement and criteria is required for justifying lighting.

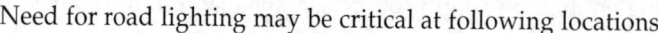

Need for road lighting may be critical at following locations:

(i) Continuous highway/freeway lighting.

(ii) Complete intersection/interchange lighting.

(iii) Partial intersection/interchange lighting.

(iv) Lighting for ramps, acceleration lanes, side walks, curves, grades, etc.

(v) Lighting of bridges and tunnels.

(vi) Lighting at bike paths, pedestrian walk ways and overpasses.

(vii) Lights where roadway reconstruction is undertaken.

The main considerations for analysing highway lighting needs are:

(i) Vehicular traffic volume and pedestrian volume.

(ii) Intersection/interchange spacing, turning movements, signalisation.

(iii) Land development.

(iv) Lighting conditions in the area surrounding the road.

(v) Night to day accident ratio.

(vi) Varying geometric features, etc.

The effect of above factors on driver's visibility should be considered. Besides impacts of local conditions like fog, rain, snow, roadway geometry, ambient lighting, sight distance, signing, etc. should also be considered while analysing highway lighting needs.

20.4.1 Lighting at Intersections

At intersections specially in rural areas following factors should be considered to justify lighting.

- Accident frequency and severity during night time.
- Substantial volume of pedestrian, specially during the night time.
- Alignment of any intersection leg/geometrics, is inadequate or less than desirable.
- Complex turning movements are involved at the intersection due to its peculiar geometry.
- Commercial activity around the intersection and its vicinity generates high volumes of traffic during night time.
- Site restrictions distract illumination due to land development.
- Fog or industrial smog in the area, occurs several times, affecting safety of traffic using the road.

Lighting should also be provided at high conflict locations where pedestrian movement, commercial activity, residential development, etc. to improve safety and efficiency of such locations.

Lighting is also required at roadside picnic areas, rest areas, kiosks, toll tax sites, weigh stations, underpasses, bridges, tunnels, parking places, cycle tracks, footpaths, certain highway signs (most are reflective paint type, illuminated by vehicle lights), areas where reconstruction is taking place, etc.

20.5 TYPES AND SELECTION OF HIGHWAY LIGHTING SYSTEM

A typical highway lighting installation system consists of an aluminium or steel pole on top of which is mounted a luminaire. The lighting fixture comprises of a lamp, its housing and a lens. Like other roadside elements, lighting poles are susceptible to vehicle impact and therefore should be placed outside the roadway clear zone. Figure 20.2 shows typical lighting installation.

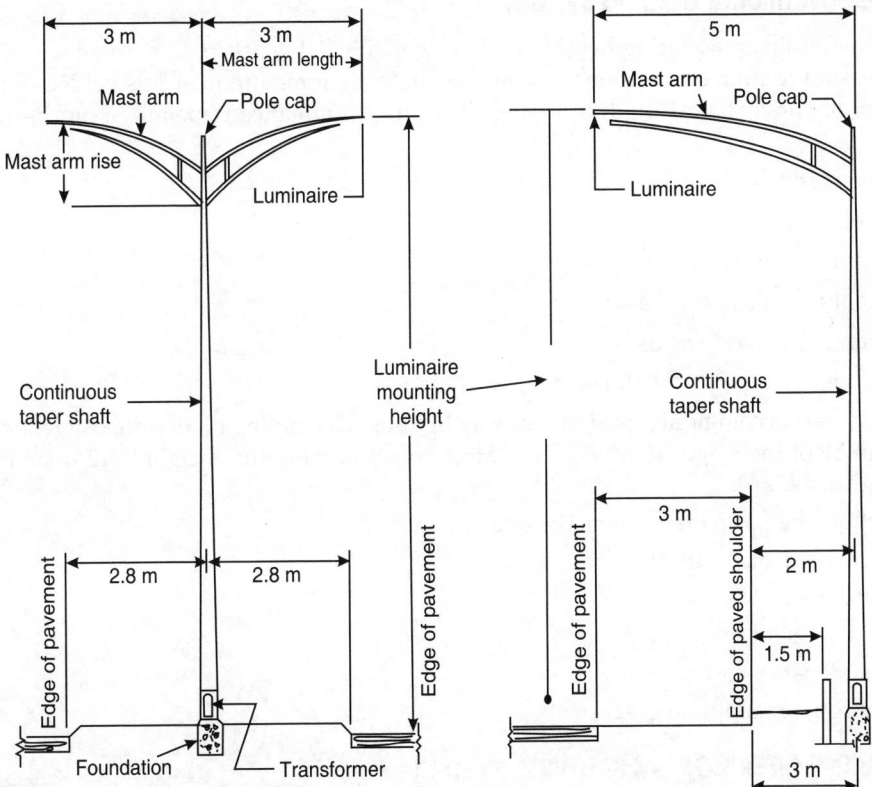

Fig. 20.2: Typical highway lighting structure

There are basically two types of highway lighting systems:

1. High mast lighting system (overhead lighting system)
2. Conventional lighting system.

One type or combination of the two can be used, depending upon site.

20.5.1 High Mast Lighting System (Overhead Lighting System)

This is the system which utilises mounting height of about 24 to 80 m, with a cluster of a maximum of eight 400 watt high pressure sodium luminaires. Required illuminance depends upon the fallowing factors:

- The area to be lighted
- Set-back of lighting assembly from pavement edge
- Type and wattage of luminaires
- Mounting height

This system is used for intersections, interchanges, toll plazas, etc. Luminaires in this system are mounted on tapered steel poles or triangular steel towers. A minimum assembly set-back of 10 m measured from the face of curb or edge of pavement to centreline of high mast lighting system is required. Lesser set-back in certain situations can also be used. This system is good where large area requires lighting. High mast lighting and higher wattage lamps reduce the number of poles and enhance illumination uniformity.

20.5.2 Conventional Lighting System

This system utilises mounting height of 8 m or so with 150 watts or 12 m or so with 250 watts high pressure sodium conventional luminaires (bulbs). Luminaire may be shielded to minimise the glare of conventional lighting system. Required luminance depends upon the following factors:

- The road width
- Set-back from edge of pavement or curb
- Luminaire mounting height
- Lighting pole bracket arm
- Luminaire overhang as required
- Luminaire type and wattage

This system is commonly used in highway lighting. Depending on design, conventional light poles can be of the height (6 m to 15 m). Most common mounting height is 12 m or 13.5 m on wide highways.

Selection of a particular system depends upon:

(i) Environmental impact on residents

(ii) Public opinion

(iii) Initial cost of installation

(iv) Maintenance cost

(v) Energy consumption cost

20.6 PARAMETERS OF A LIGHTING SYSTEM

The setting of the light, technical parameters are important for driver's visual performance and comfort. The following parameters are acceptable for visual environment.

(i) Level of illuminance: Highways and other important roads are designed to provide an average maintained horizontal illuminance of at least 1 Cd/m^2 and a range of 2 to 0.5 Cd/m^2, depending on the importance of the road. This ensures the silhouettes of objects seen with sufficient contrast to provide visibility.

(ii) Uniformity of illuminance: The uniformity of illuminance on various highways/roads should produce a uniformity ratio of 3 : 1 to 4 : 1. The ratio is defined as the average minimum illuminance. It is intended that no dark areas should extend more than 25 m longitudinally when seen from 75 m, thus concealment from a moving driver would be only momentary. The minimum/maximum ratio along centre lines of at least 0.25 should be aimed.

(iii) Spacing and location of lighting poles: Lighting poles spacing and offsets should be as uniform as possible. Varying in spacing or offset if it is necessary, it should be done gradually.

Lighting poles are located along outside lanes, spaced opposite or staggered to suit the geometry of the highway and to provide the best lighting uniformity. The height may vary from 9 m to 20 m or so. To facilitate maintenance on ramps, etc. it is better to locate the lighting along the inside curve. A set back of 2 to 3 m may be required.

If it is not possible to place poles outside the roadway clear zone, they should be installed along stretches of roadway where vehicles should be travelling at relatively low speeds, so that damage to vehicles striking a pole will not be severe.

On a divided highway, lighting poles may be placed either in the median or on left side of the road. Poles may be fabricated as tapered or straight, made of aluminium, steel, fibre glass, concrete or wood.

(iv) Glare control: Light sources of considerably greater luminance than rest of the visual field, may accentuate glare. It should be minimised by specifying the light distribution characteristics of the lanterns, which provide for a choice between cut-off and semi-cutoff distributions.

(v) Other consideration:

- Size of luminaires (bulb) should be proper for required level and uniformity of illumination.
- Length of bracket arm should be selected such that maximum efficiency and uniformity of lighting is provided.
- For obtaining foot candle values, contribution from all luminaires which have effect on the area should be considered.
- Lighting system should not obstruct driver's view.

(vi) Mast arm: Mast arms allow placement of the light source near the edge of the travelled way. The use of long arms is better, through initial cost may be higher. Longer arms allow poles to be placed farther away from the travelled way, thus providing greater safety. Typical luminaire mast arm lengths are 3 m and 4.5 m.

(vii) Light sources: (lamp) A complete lighting unit (luminaire) consists of a lamp or lamps, designed to distribute the light. High intensity light sources which are used for highway lighting are:

(a) High pressure sodium (HPS): Due to their excellent luminous efficiency, power usage and long life, high pressure sodium is commonly used in installation of conventional and high mast highway lighting.

The HPS lamps produce a soft, pinkish-yellow light by passing an electric current through a combination of sodium and mercury vapour.

(b) Low pressure sodium (LPS): Low pressure sodium is most efficient light source, but its disadvantage is that it requires very long tubes and has poor colour quality. Use of LPS however is common on less important roads. LPS lamps produce a yellow light by passing an electrical current through a sodium vapour.

(c) Mercury vapour (MV): Before introduction of high pressure sodium (HPS) lamp, mercury vapour was most commonly used light source. Mercury vapour lamps produce a bluish-white light and are still in use in many locations.

(d) Metal halide: Light in metal halide lamps is produced by passing a current through a combination of metallic vapours. They have better light efficiency than mercury lamps, but their life expectancy is shorter than the HPS or MV.

They are commonly used for sport areas, stadium, high-mast highway lighting, parks, etc.

The lamps commonly used in highway lighting are 200 W, 250 W, and 400 W.

(viii) Optical system: The optical system consists of a light source, a reflector and usually a refractor.

Light source are the type of lamp described above. The reflector is the device used in optical control to change the direction of the light rays. Its purpose is to take that portion of light emitted by the lamp that otherwise would be lost or poorly utilised, and redirect it to a more desirable distribution pattern. Reflector are designed to work either alone or with a refractor. Reflectors are of two types: Specular and diffused. Specular reflectors are made from a glossy material that provides a mirror like surface. Diffuse reflectors are used where light is to spread over a wide area.

Refractor is another means in optical control to change the direction of the light. Refractors are made of a transparent, clear material, usually high strength glass or plastic. Glass refractors are commonly used.

(ix) Glare shield: Glare shield should be used, where additional glare control is necessary. The glare shield has a conical shape and is attached under the luminaire housing for control of light direction and distribution.

(x) Housing units: Luminaire housing and reflector holders are typically made of aluminum with a weather proof finish. They should be accessible from the road side to allow adjustments to the lamp. The unit should be sealed to prevent dust, moisture and insects to enter the inside of luminaire. The mounting attachment is adjustable to allow for directing the light.

Electric components, fuses, wiring, conduits, connections, circuit breakers, etc. should be properly done, in consultation with electric engineers.

20.7 STEPS IN HIGHWAY LIGHTING DESIGN PROCESS

When designing the highway lighting system there are number of factors to be considered regarding illuminance, and luminance. The basic factors and steps involved in highway lighting design process are the following:

   (i) Road in urban/rural location

  (ii) Type of pavement

 (iii) Lighting will be continuous. What extent of lighting is warranted.

 (iv) Assess the facility to be lit and determine minimum foot candle levels.

  (v) Selection of luminaire type: 250 W or 400 W, HPS or LPS.

 (vi) Selection of parameters of conventional lighting: Types of poles, height of poles, mast-arm length and mounting etc.

(vii) Spacing and arrangement of poles: Staggered, opposite, or same side, etc.

(viii) Check the accuracy for above arrangement

 (ix) Determine source of power and plan layout of the lighting system using electrical services for wiring, circuits, etc.

  (x) Calculate voltage drops and see that it is within allowable limits of 5 percent.

 (xi) Iterate steps (v) to (vii), if necessary.

Highway lighting design process is an iterative process that is quite effectively implemented by computer. If criteria is not initially satisfied, it will be necessary to change design parameters (e.g. pole spacing, mounting height, luminaire wattage etc.) until an acceptable alternative is found and design is optimised. Figures 20.3 to 20.5 show typical lightings on roads and intersections. There are computer software package for lighting design process.

20.8 SOLAR HIGHWAY LIGHTING SYSTEM

Automatic solar street light control system is a simple technique which uses transistor as a switch. It automatically switches 'ON' lights when the sunlight goes below the visible region of our eyes. This is done by a sensor called light dependant resistor (LDR) which senses the light like our eyes. It automatically switches 'OFF' lights whenever the sunlight comes, visible to eyes.

By using this system energy consumption is reduced and manual work is removed, for 'ON' and 'OFF' operations. LDR and transistor are the main components of this system. The LDR is connected as biasing resistor of the transistor. This transistor switches the relay to switch on/off the lights.

A crystal based solar panel is used to charge a rechargeable battery of 6 volts. Additional battery charger circuit is provided for emergency applications. This charger uses regulated 6V, 750 mA power supply.

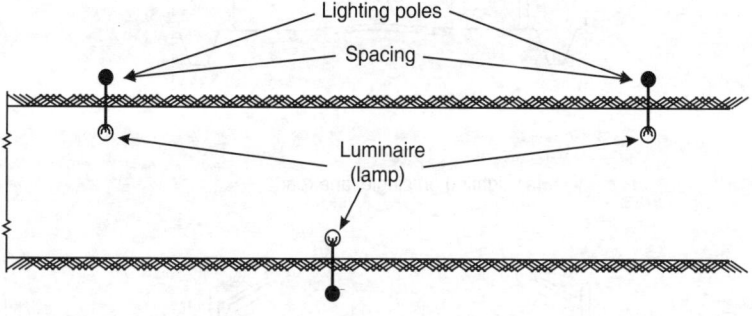

(a) Staggered-both side lighting arrangement

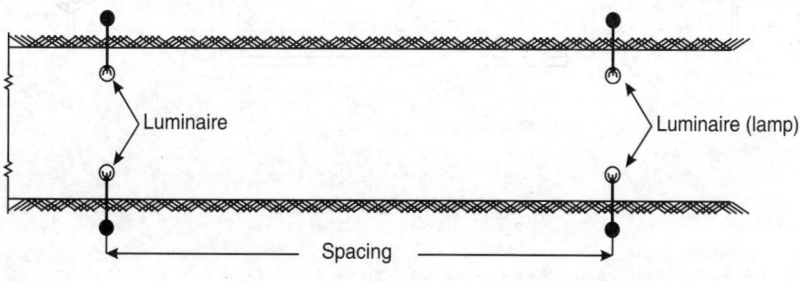

(b) Opposite - both side lighting arrangement

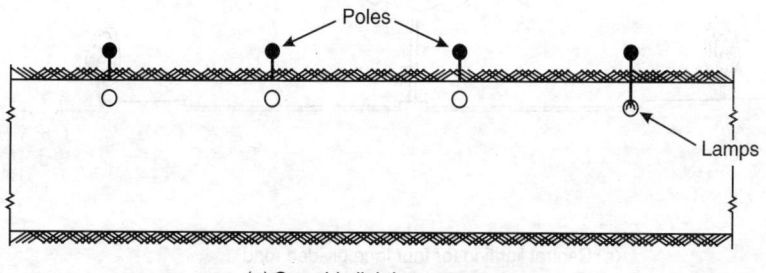

(c) One side lighting arrangement

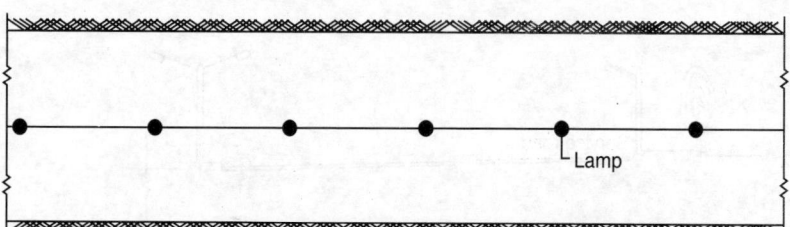

(d) Central-lighting arrangement

Fig. 20.3: Arrangement of lights on highways (plan)

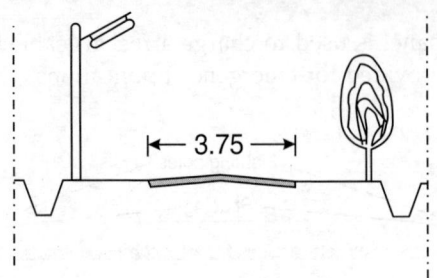

(a) Lighting on single lane road

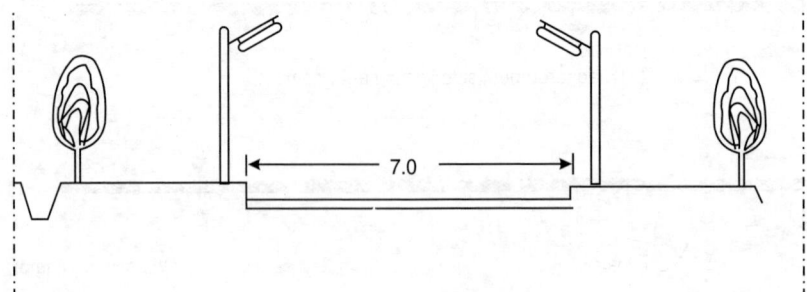

(b) Lighting on double lane road

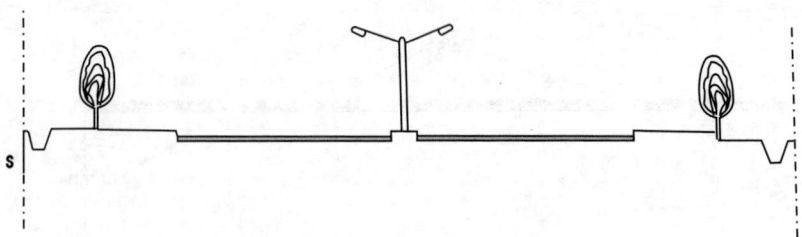

(c) Central lighting for four lane divided road

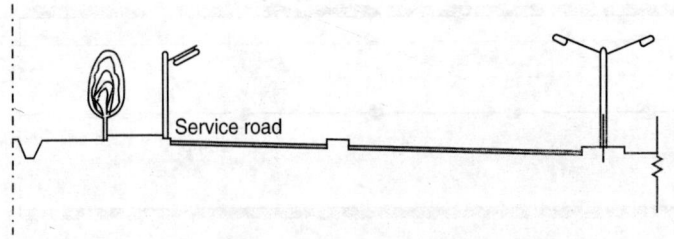

(d) Lighting for six lane divided road with service road

Fig 20.4: Typical lighting for different types (widths) of roads

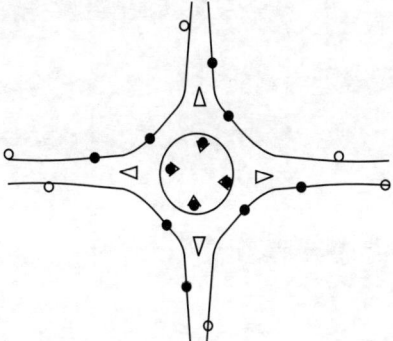

(a) Light arrangement at a roundabout

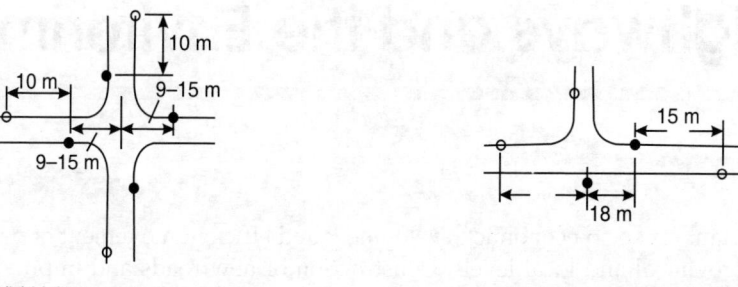

(b) Light arrangement at four
legged intersection

(c) Light arrangement at three
legged intersection

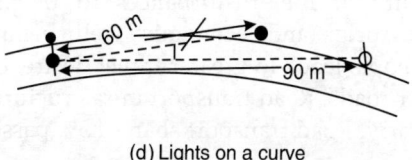

(d) Lights on a curve

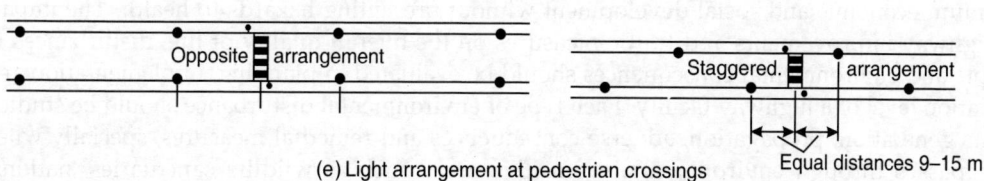

(e) Light arrangement at pedestrian crossings

Fig. 20.5: Typical light arrangements at critical locations

Highways and the Environment

21.0 PREAMBLE

Highways are important for socio-economic development and efficient movement of people and goods at national, provincial and local levels. Construction of new roads and improvement of existing roads is undertaken, under National Highway Development Programme (NHDP). A large number of highways are being upgraded, realigned and constructed in urban and rural areas. Environmental deterioration seems to be an inevitable consequence of rapid and convenient transportation system. Drastic effects of highway improvements, are the impact of traffic movement on overall quality of life, disturbances to people and environment, interference with health, mental, social and economic well being of human beings. Improvements in economic life of people leads to greater travel desire, creating more vehicles on the road, needing more and better roads. Road transport infrastructure is one of the priority areas of the development in the country. Road transport shares 85% passenger and 70% freight traffic.

The problem, then is to determine the level of environmental degradation that permits optimum economic and social development without presenting hazards to health. The impact of highway improvements and traffic measures on the overall quality of life, disturbances to people and environmental consequences should be evaluated in planning, implementation and operation level of a highway facility. Each type of environmental disturbance should be studied for its generation, propagation, adverse consequences and remedial measures, specially when road passes through environmentally sensitive areas such as wildlife sanctuaries, national parks, coastal area, etc.

The changes in the environment and their extent call for a detailed study and careful evaluation of all alternatives in view of community benefits, attitudes, responses, to ensure that highway and traffic project will adequately protect the environment. Development of sustainable road transport is necessary for meeting the growing demands of traffic and challenges of providing safe, environment friendly and cost effective movement of passengers and goods without negative impacts on natural environment.

Careful planning of road alignment and design can minimise the adverse effects of highways and traffic on the environment. Urban planning and control of land uses through improved transport infrastructure, improved vehicle technology, and strict traffic management can mitigate environmental problems. Some of the techniques commonly used to reduce adverse effects of traffic on urban environment are:

Banning of certain vehicles, priority of high occupancy vehicles (buses), coordinated traffic signals, speed controls, traffic restraints, road pricing, parking controls, grade separated facilities, improvements in traffic flow and obviate frequent stopping and starting, thus reducing air pollution.

Road development and good environment, both are related to quality of life of the people and therefore the ideal approach lies in striking a balance between the two, by adopting suitable mitigation measures.

In view of the growing awareness of the community and the government, to preserve and enhance the environmental values, highway engineers should plan, construct and maintain roads/highways, with environmental requirements in view. The issue for all types of roads is, how we can protect the environment and the traffic functions, in a sustainable way. Beyond 'Traffic Capacity' we should concentrate on 'Environmental Capacity' in which safety, air quality and noise level are fundamental starting points, along with many other elements to be considered.

Few Definitions

Sustainable development: The sustainable development involves eco-friendly growth and human activity, allocation and conservation of resources over space and time in a sustainable manner. It takes future consequences into considerations. Sustainable development must meet the needs of the present generation without compromising the ability of future generation to meet their own needs and aspirations.

Biodiversity: Biodiversity is the totality of genes, species and ecosystems in any region. It means the variability among living organism from all sources including interalia, terrestrial, marine and other aquatic ecosystems and the ecological complexes of which they are part.

Environmental overview: Environmental overview (EO) is an assessment tool that forms the basis for environmental management strategy (EMS). It aims to provide basic information on the present environmental situation of the region/project is being designed/implemented within an environmentally sound and sustainable approach.

Environmental impact assessment: Environmental Impact Assessment (EIA) is an analytical process that systematically examines the possible consequences/impacts of implementation of projects, policies and programmes on the environment.

Environmental management plans/strategies: An environmental management plan/ strategy (EMP/EMS) is an action plan prepared both for an ongoing and proposed project after identifying the environmental objectives, issues involved, arriving at operational strategy including suggested alternatives for project designs, implementation and monitoring.

Environmental audit: Environmental audit (EA) is a management tool for systematic, periodic and objective evaluation of environmental practices adopted including meeting the regulatory needs.

Carrying capacity of ecosystem: Ecosystem carrying capacity provides the physical limit to economic and technological development and may be defined as maximum rate of resource consumption and waste discharge that can be sustained indefinitely in a defined planning region without progressively impairing bio-productivity and ecological integrity.

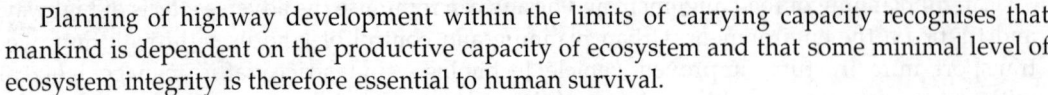

Planning of highway development within the limits of carrying capacity recognises that mankind is dependent on the productive capacity of ecosystem and that some minimal level of ecosystem integrity is therefore essential to human survival.

21.1 ENVIRONMENTAL CONSEQUENCES OF HIGHWAY DEVELOPMENT

Highways provide positive impacts such as accessibility, connect remote and isolated areas, generate employment opportunities, enhance and encourage local industry, help in rural development, improve economic life of people, develop tourism, reduce traffic congestion, improve road safety, better quality of road reduces pollution, provide even distribution of products and offer several other social, and economic benefits nevertheless there are some negative impacts of highways, specially on the environment, which spoil the ecology by its effect on plant and animal life, human settlements, geo-hydrology, etc. Highway projects require land for the right of way, quarries, borrow areas, road side amenities, interchanges, flyovers, etc. Such land may be agricultural, forest, residential or industrial, or in tribal area, etc. Acquisition of such lands and construction of highway through different ecological regions, cause environmental impacts.

Environmental impacts due to roads are caused as:

(a) **Road related effects:** These are due to location, design, construction and maintenance of roads/highways. Physical presence of road creates these consequences, such as visual intrusion, severance, social impacts, etc.

(b) **Traffic related effects:** These are due to movement and characteristics of traffic on the road such as air pollution, noise, vibration, dust, safety, etc.

(c) **Administrative effects:** These are due to administrative procedures for planning of roads/highways in an area such as compensation for land, relocation of people, etc.

(d) **Construction effects:** These are impacts during the construction of roads/highways such as noise, dust, air pollution etc. They are temporary impacts.

These effects may also be grouped as

• Physical environmental consequences, e.g. noise, vibration, pollution.

• Social consequences, e.g. community severance, accidents.

• Ecological consequences, e.g. water pollution, air pollution.

• Economic consequences, e.g. land value changes.

Table 21.1 shows environmental impacts of a highway projects, which are likely to occur during the planning and design, construction and operation (traffic) phase of the project.

A brief description of these impacts is covered in following paragraphs:

21.1.1 Air Pollution

Undesirable changes in physical, chemical and biological characteristics of air are called "Air pollution".

Air pollution is created by automobiles plying on highway. As traffic increases pollution becomes more noticeable. At lower levels an unpleasant smell in the environment creates fatigue. The heavy traffic maintains smelly emissions, and particulate matter in the air at levels which are in excess of desirable. Air pollution levels in urban centres generally exceed the National Ambient Air Quality Standards (NAAQS).

Air pollutants caused by road traffic include pollutants directly emitted by vehicle exhaust such as Carbon mono-oxide (CO), Carbon dioxide (CO_2), oxides of nitrogen, sulphur dioxide,

hydrocarbons (HC), etc. and secondary pollutants formed by chemical reaction in atmosphere such as photo-chemicals, dangerous lead and other compounds, together with unburnt petrol, smog and aldehydes. Exhaust from diesel have more of smoke, but lesser pollutants. Pollution output is more in stop-go type driving situation on congested roads. In old and improperly maintained automobiles emissions are more. Traffic queuing at signalised intersections produces the highest vehicular air pollution.

Table 21.1: Environmental impacts of highway projects during different phases

Environmental, Social and Community Factor which may be affected by highway projects	Positive/Negative impacts during		
	Planning and Design Phase	Construction Phase	Operation Phase
(a) Environmental factors:			
• Air pollution		●	●
• Noise		●	●
• Vibrations		●	●
• Dirt and dust		●	●
• Water pollution		●	●
• Soil erosion and sedimentation		●	●
• Ecological impacts		●	●
• Aesthetic effects		●	●
• Geological disturbances		●	●
• Natural drainage		●	●
• Global warming			●
(b) Social environmental factors:			
• Neighbourhood and social life pattern	●	●	●
• Relocation of residents	●	●	●
• Relocation of employments	●	●	●
• Land compensation	●	●	
• Number and severity of accidents			●
(c) Community factors:			
• Public open space		●	●
• National parks forestry and wild life		●	●
• Historical and architectural features	●	●	
• Community facility and services	●	●	●

Prolonged exposure to these pollutants causes irritation to eyes leading to reduced visual activity, affects respiratory system, creates headaches, lung damage, anaesthetic effects, reduction in defence mechanism against infections. Lead compounds are most injurious as they are taken up in biological chain causing long term harms. The impact of pollution on human health is due to the fact that man is completely and constantly exposed to atmosphere from which he breathes air. The increasing air pollution in metropolitan cities is considered to be responsible for various health related problems faced by public.

Dust from the roads and shoulders also results in air pollution. Temperature and relative humidity influences the formulation of secondary pollutants. In winter months the concentration of pollutants is highest. NO and CO concentration is maximum during evening and morning peak hours of traffic on congested city roads. Table 21.2 shows average emissions from petrol and diesel engines in kg per 1000 litres of fuel burnt.

Table 21.2: Components of exhaust gases (kg/1000 litres)

S.No.	Components	Petrol Engine	Diesel Engine
1.	Carbon monoxide (CO)	27.40	07.10
2.	Hydrocarbons (HC)	24.00	16.40
3.	Nitrogen dioxide (NO_2)	13.50	26.40
4.	Sulphur dioxide (SO_2)	01.10	04.80
5.	Organic acid	00.50	03.70
6.	Aldehydes	00.50	01.20
7.	Solid particles	01.40	13.20
8.	Lead (Pb) (mg/cu of exhaust gases)	0.5.30	–

The highest emission rates occur during motor idling, deceleration and at slower speeds as shown in Table 21.3.

Table 21.3: Emission from vehicle operation

Operation mode	CO	HC	NO_X
Idling	Very high	High	Low
Deceleration	Very high	High	Very low
Low speed	Moderate	Moderate	Low
Acceleration	Low	Low	High
High speed	Very low	Very low	Very low
Uniform speed	Minimum	Minimum	Minimum

In India, exhaustive standards for emission for light and heavy vehicles plying on our highways, should be prepared for various categories, and ages of vehicles based on field studies. This will help in working out the impact and control of air quality in the area.

The relationship between amount of travel and air pollution emissions and concentration is complex. Besides air pollution, the pollutants emitted from motor vehicles are responsible for various regional and global problems such global warming, acid rain, ozone depletion, etc.

Better engine of automobiles with better combustion and exhaust system, can reduce harmful auto-exhaust emissions. The legal restrictions on auto exhaust from automobiles encourage in development of new techniques to design engines to reduce harmful emissions.

Another measure to reduce the pollution is to reduce congestion, stop-go operation of vehicles on urban roads. Through proper urban planning the home and working places should be brought closer to reduce the length of auto journeys and thereby reduce environmental pollution. As per Central Pollution Control Board the concentration for the following pollutants shall be 95 percent of the time, within limits prescribed in Table 21.4. Vehicle emission norms are given in appendix-C.

Table 21.4: National ambient air quality standards
(CPCB notification, Delhi 11/4/90 and 14/10/98)

Pollutants	Time Weighted Average	Concentration in Ambient Air			Method of Measurement
		Sensitive Area	Industrial Area	Residential Rural and Other Areas	
Sulphur Dioxide (SO$_2$)	Annual[#]	15 µg/m^3	80 µg/m^3	60 µg/m^3	- Improved West and Gaeke method
	24 Hours[##]	30 µg/m^3	120 µg/m^3	80 µg/m^3	-UV Fluorescence
Oxides of Nitrogen (NO$_X$)	Annual[#]	15 µg/m^3	80 µg/m^3	60 µg/m^3	-Jacob and Hochheiser Modified (Na Arsenite) Method
	24 Hours[##]	30 µg/m^3	120 µg/m^3	80 µg/m^3	-Gas phase Chemiluminescence
Suspended Particulate Matter (SPM)	Annual[#]	70 µg/m^3	360 µg/m^3	140 µg/m^3	- High Volume Sampler (HVS)
	24 Hours[##]	100 µg/m^3	500 µg/m^3	200 µg/m^3	(Average flow Rate not less than 1.1 m^3/min
Respirable Suspended Particulate Matter (RSPM)	Annual[#]	50 µg/m^3	120 µg/m^3	60 µg/m^3	- Respirable Dust Sampler (RDS)
	24 Hours[##]	75 mg/m^3	150 mg/m^3	100 mg/m^3	
Lead (Pb)	Annual[#]	0.50 µg/m^3	1.00 µg/m^3	3.00 µg/m^3	- AAS method after sampling using EPM 2000 or equivalent filter paper
	24 Hours[##]	0.75 µg/m^3	1.50 µg/m^3	1.00 µg/m^3	
Carbon Monoxide (CO)	Annual[#]	1.0 µg/m^3	5.0 µg/m^3	3.0 µg/m^3	- Non dispersive IR Spectroscopy
	24 Hours[##]	2.0 µg/m^3	10.0 µg/m^3	4.0 µg/m^3	

[#] Annual arithmetic mean of minimum 1.4 measurements in a year taken twice a week 24 hourly at uniform interval

[##] 24-hourly/8-hourly values should be met 98% of the time in a year. However, 2% of the time, it may exceed but not on two consecutive days.

21.1.2 Noise Pollution

Noise is unwanted sound, and one of the environmental pollutant from motor vehicles specially diesel trucks which causes problems to daily life in the adjoining areas. Some mechanical plants used for highway work also produce noise. Noise interferes with conversation, television watching, sleep and congenial life of people living in nearby houses, schools, libraries and hospitals. Continuous and sustained exposure to noise beyond tolerable levels may permanently affect the hearing power of human beings. Sound waves cause an oscillation of the ear drum, affecting sensitivity of the ear.

The noise level depends upon traffic intensity, type and power of vehicles, condition and age of vehicles, acceleration/deceleration characteristics, traffic speed, along with other factors such as texture of road surface, congestion, etc. Roads with lower volume/capacity ratio, and smoother surface will produce lesser noises. Grades, air pressure and direction, temperature, humidity, time of the day also affect propagation of sound waves.

The noise from motor vehicles is measured in decibels dB(A), on the A scale of a sound level meter. The normal tolerance level of noise in residential areas is about 70 dB(A). A level of 150 dB may result in deafness. The measurement of noise and its correlation with human annoyance is a very complex matter. Tolerance of noise depends upon age, sex, frequency of noise, occupation, social status, neighbourhood type, time of the day, etc. During night time the acceptable levels are low as compared to day hours.

The approaches to reduce the noise disturbances are:

(i) Better and quieter design of motor vehicles to reduce engine noise and exhaust through stricter legislation and to ensure that vehicles maintain allowable noise levels. Vibrations of external surfaces of engine must be controlled.

(ii) Traffic management and control techniques to reduce noise levels such as smoother flow of traffic, without stop-start operation, lesser volumes and speeds of vehicles, etc.

(iii) Planning and design of residential buildings with adequate sound insulation so that noise intrusion is controlled. Road with smoother surfaces, providing noise barriers, through buffer strips, set-backing of buildings, etc. are other techniques to reduce noise.

(iv) The tyre noise is reduced by Radial-Play tyres and quiet tread pattern. Some people become quickly accustomed to road noise and therefore it becomes difficult to put a value on noise pollution. The evaluation of noise level is therefore subjective. Steady exposure to noise of 90 dB(A) can cause irreversible damage.

Noise levels acceptable in residential areas as given by Indian Standard IS: 4954-1968 are as given, in Table 21.5.

Table 21.5: Acceptable noise levels in residential areas (IS)

S.No.	Location	Acceptable outdoor noise level in residential areas dB(A)
1.	Rural	25–35
2.	Suburban	30–40
3.	Residential (urban)	35–45
4.	Urban (residential and business)	40-50
5.	City	45–50
6.	Industrial area	50–60

It is important to predict probable highway noise level and their effects on users of adjacent property, so that noise may be considered in the design of highway features and reduced through legislations, enforcement of vehicles and design changes of vehicles. Considerations of noise at planning and design stages of highways are essential. The green belt development along the road corridor (s) reduces the air and noise pollution and also improves aesthetics and overall environment of the region.

Figure 21.1 shows traffic noise and its effects on people and Fig. 21.2 shows equipment for measuring noise and recording it.

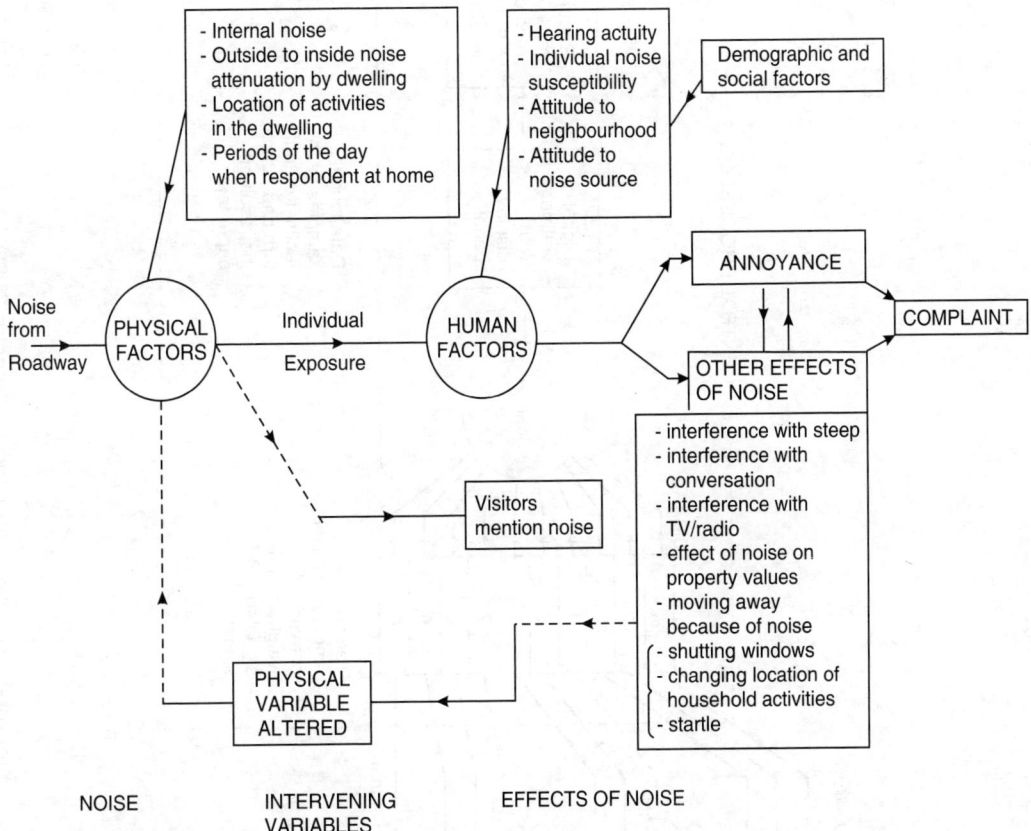

Fig. 21.1: Road traffic noise and its effect on human beings

21.1.3 Vibrations

Highway traffic creates vibrations, but there amplitude and frequency is usually low to cause structural damages to nearby buildings. Traffic vibrations caused by the load transmitted to the road, by the vehicle's suspension and by impact with surface irregularities are transmitted through the road pavement directly to the supporting soil mass. Such ground vibrations generally do not result in damages, except where resonance occurs.

Traffic however may cause unpleasant vibrations and worrying feelings as experienced when standing on a foot bridge, crossing a major traffic road etc. Many people get bothered by traffic vibration as they do by traffic noise and therefore prolonged vibrations are also a source of nuisance and must be controlled.

A smooth surface of roads do not produce vibrations annoying to people or causing damages to structures nearby. The effect of vibrations created by traffic are rattling of doors and window glasses, cracking of plasters and creating resonance in buildings. Vibrations may cause structural damage to bridges by overstressing and other susceptible buildings nearby.

Vibrations can be controlled by providing smoother road surfaces, reducing number of commercial vehicles and keeping minimum desirable space between road and adjacent buildings.

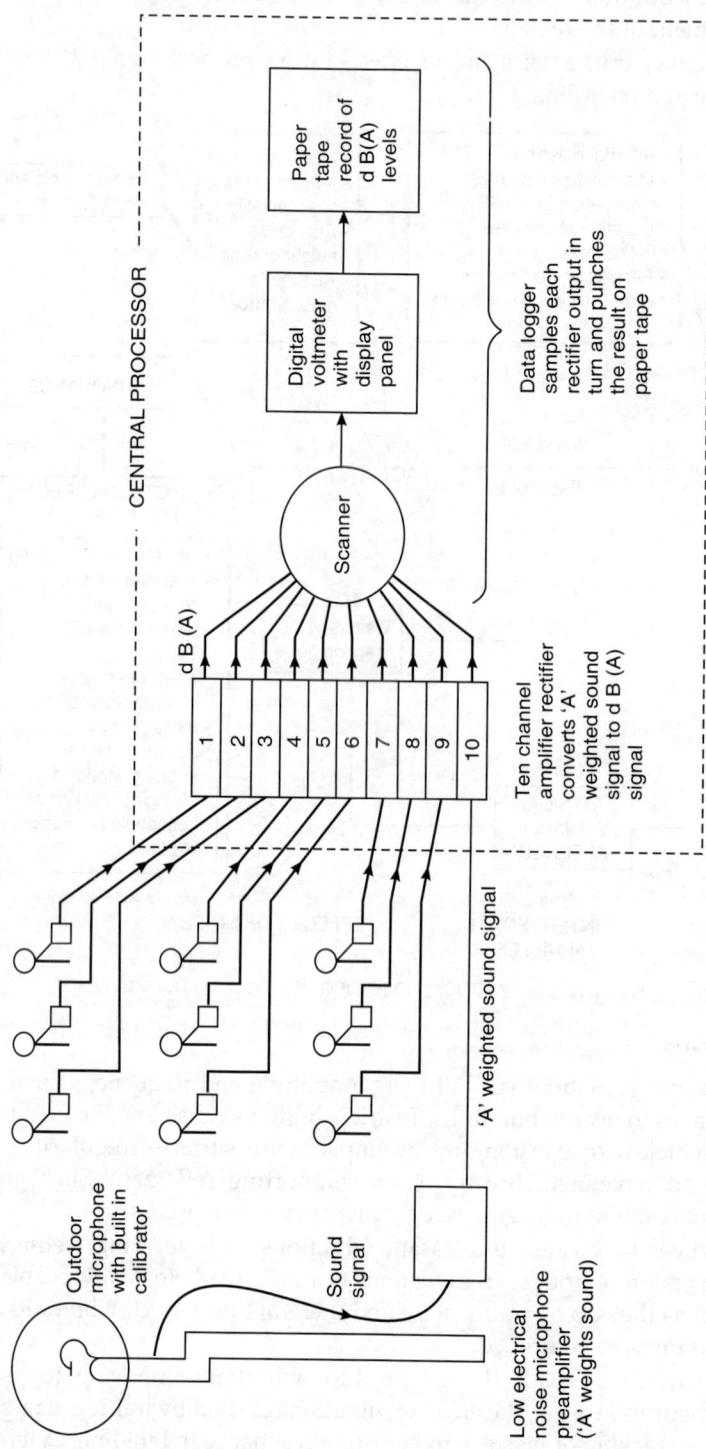

Fig. 21.2: Equipment for measuring and recording traffic noise (TRRL)

21.1.4 Dust Pollution

Unpaved roads are affected by dust which is generated by the interaction between vehicular tyres and the unbound surface. Dust and smoke are also created by the shrubs and trees cleared from the right of way of road and burnt in a haphazard and uncontrolled manner. During dry season dust is created by bare surfaces of embankments, cut sections, gravel, earth and village roads of low category. In addition to the effects of traffic, dust is also produced by the climatic influences of wind and rainfall.

Clouds of dust present a safety problem by reducing the visibility on the roads thereby contributing to traffic accidents. Furthermore, because it pollutes the atmosphere close to the road, dust is also an environmental problem and health hazard to vehicle operators, road users, construction workers and those living in the vicinity. Roadside plants and crops are hindered in their growth due to dust. Dust can reduce the value of pasture land and lower crop production.

The emission of fine particles from the road surface arises from loosening of the soil structure and reduction in the cohesion of the gravel/earth wearing course by the action of traffic and climate. This loss of material increases the permeability of the surface layer and results in development of pot holes, and surface irregularity, which in turn give rise to increased vehicle operating costs.

Unpaved roads have much higher rate per vehicle kilometer of accidents than paved roads, primarily due to the effect of dust in reducing visibility in the region.

The major remedial measures to deal with dust problem on unpaved roads is to use dust palliatives, or application of bituminous seal. Turfing of earth shoulders, embankments is used to prevent dust nuisance.

21.1.5 Water Pollution

Surface water as well as groundwater may be polluted due to inflatration of water through cracks, fissures and other openings by the use of fuel and lubricants in vehicles and machines, during construction and maintenance of roads.

Polluted water transmits a large number of diseases such as cholera, typhoid, dysentery, jaundice, bacterial or viral diseases. The water having foul smell, or water containing oil, grease, floating material etc. are examples of polluted water. Pollution is caused mainly due to organic material getting into water streams, rivers or lakes. Hot effluents to river water removes oxygen content rendering it unsuitable for living creatures in water.

Petrol and oil washed from roads, along with illegal discharge of engine oil, leads to pollution of hydro carbons in fresh water ecosystems. The water soluble components of crude oils and refined products may prove toxic to fresh water animals. Emulsifiers and dispersion agents used to clean up oil spillage themselves are highly toxic.

In planning, aligning and construction of roads near a water body, care should be taken so that pollution of water is minimised.

21.1.6 Soil Erosion, Sedimentation of Lakes and Reservoirs (Chemical Pollution)

Erosion of soil from freshly exposed and distributed surfaces with the consequent transportation of sediments to streams by surface flow of water results in silting of lakes and reservoirs. Channelisation of meandering streams during highway construction may result in reduced channel length and changed gradients. New equilibrium is established by up-stream scouring and deposition in the downstream reaches.

Use of stream bed for highway material sources produces large and deep excavations which may lead to upstream erosion or to stagnant pools. Surface erosion also increases the tendency of water to seep through and render the slopes unstable.

Soil erosion results in leaching and transportation of soil nutrients, toxicants, etc. into the acquatic system, resulting in chemical pollution. Chemical pollution also results from lead salt particulates from the exhausts systems and various petroleum products used in the auto-mobiles.

Attention should be paid in selection of road alignment to avoid unstable and erosion prone areas. Measures to control carriage of sediments from highway land to property and streams nearby should be undertaken, and incorporated in the contract, for strict implementation during highway works. In hilly areas, landslide study should be carried out. Road alignment should be taken sufficiently above the zone of influence of erosion.

21.1.7 Visual Intrusion and Social Effects (Severance)

The desired life style and social pattern of the community changes due to a new road in the vicinity. The visual intrusion by road and traffic is disturbed life by the highway structure, loss of privacy due to vehicle movement, restriction of people outside the house, as well as privacy inside the house. The neighbourhood may be cut out, creating psychological disturbances, change in social pattern, life style and day to day living pattern, causing severance of land and uprooting a settled community.

Building conditions, adjacent development, overhead utilities, posters, signs, blockages, crowded corridors, intermingled vehicles and pedestrains, etc. are the factors which affect the visual qualities in urban areas. Such visual pollution, may contribute to confusion and accidents. Flicker of vehicle lights at night is another source of visual disturbance and discomfort. Highway planners should avoid such situations.

21.1.8 Disturbance to Eco System

Each organism has its own preference with regard to depth of water, velocity of flow, water temperature, availability of light, etc. If acquatic environment conditions are varied acquatic biotic will also be varied. The process of construction and operation of highways interferes with the hydrological process. Removal of vegetation for construction of highways affects the process of transpiration and surface flow. Embankments compress acquifers and consequently reduce flow/stop flow through acquifer. Highway projects may result in changes in acquatic system principally in the physical form and to some extent in the chemical and biological forms.

Another problem created is alteration of physical equilibrium of the system. Removal of vegetation sets off a chain of events starting from erosion, increased stream turbidity due to sediment and consequent reduction in light penetration. This results in process of photo synthesis which reduces food to acquatic organism and ultimate reduction in the number of these organism in the aquatic eco system.

Removal of vegetation might result in temperature risk, killing those lives which survive at lower temperature.

21.1.9 Loss of Forestry and Vegetation Affecting Wild Life

Along the proposed alignment of road, trees, forests, vegetations, etc. are cut, resulting in serious imbalance of ecology, adversely affecting land pattern.

Destruction of forests, blasting operations, noise of vehicle and machineries during construction of roads, displaces the animals and birds destroying and damaging their natural habitat.

21.1.10 Aesthetic Degrations

At some places, the basic environmental assets like woodland, recreational spots, historical and cultural sites, natural springs, etc. may be destroyed. A historical site forms a significant aesthetic appeal for an area.

21.1.11 Geological Disturbances

Due to blasting, excavation, deforestation operations during construction of highways, geological disturbances occur in the area. Landslides occur in unprotected slopes, during rainy season. Sliding of debris may cause destruction of valuable cultivating land, spoiling growth of vegetations and causing imbalance of ecosystem.

21.1.12 Interruption of Natural Drainage System

Natural system of flow of water in the area is disturbed due to construction of highways. Natural course of flow is blocked due to excavated debris, resulting in formation of water pools/lakes near by and beneath the roadway.

21.1.13 Global Warming

Motor vehicles emit both CO_2 and CFC_S along with some quantities of CH_4 and N_2O. These gases are strong green house gases. The increasing emission of these gases from motor vehicles throughout the world, results in global warming effect, which can have significant environmental consequences.

As per the framework convention on climate change (FCCC) 15% of the total worldwide CHG emission is contributed by transportation sector only. CHG emission is likely to increase due to vehicular growth and greater fuel demands in various Indian cities.

21.1.14 Impact of Mixed Traffic

The mixed traffic involving cars, trucks, animal drawn vehicles, man pulled vehicles, on the same road in urban areas contributes towards poor visual impacts, reduced speeds, greater interference of traffic, resulting into dangerous conflicts, congestions, increased pollution and damaging the environment.

Segregation of traffic, control of commercial vehicles, distinguishing vehicles carrying dangerous loads, etc. are the essential measures to be used in urban areas to protect environment and safety.

21.1.15 Land Development Impacts and Reduced Safety

Highways affects the social-economic environment status of society by the impact they make on land development. High intensity development results in form of major and tall buildings, interchanges, congestion, etc. creating a negative visual and social impact, such as neighbour-hood, cohesion and identity, school access, access to recreational facilities, community services and zoning etc.

Safety of road users is reduced due to increasing number of vehicles, concentration of activities, intense development along the roads.

21.1.16 Impacts During Construction of Highways

When a highway facility is constructed, the worst environmental conditions are developed creating noise, dust, vibrations, visual intrusions and severance.

The deteriorative impacts should be minimised by proper planning, timing of operations, and using preventive measures.

Regulations on control of pollution should specify detailed measures for contractors and construction agencies/engineers, to be followed while construction works are undertaken in urban areas. Contract conditions should stipulate that the equipment to be used for highway construction work is provided with pollution control devices, and engineer incharge should ensure this.

To prevent dust nuisance while loading, unloading and transporting materials such as soils, sand, morrum it is advisable to slightly moist the material at source before loading, and/or to cover the loaded vehicle with tarpaulin etc.

During actual construction temporary screens should be erected to prevent dust, noise nuisance to nearby areas.

21.2 ENVIRONMENTAL IMPACT ASSESSMENT (EIA)

Environmental impact assessment (EIA) study identifies the potential effects of highway activity on environmental system. EIA study involves comparing of all feasible alternatives and determining which alternative represents an optimum mix of environment and economic aspects of the highway project. Steps can then be taken to mitigate potential environmental problems through project planning and design. It provides necessary information for making decision regarding acceptability of a highway project from environmental point of view. EIA prevents environmental degradation in two ways. First some projects may not be viable after EIA report revealing an adverse environmental impact. Secondly, many proposals may be modified or abandoned in view of EIA.

At present, during planning and design of a highway project in India, considerations are given to economic and traffic aspects, ignoring/overlooking adverse environmental impacts. If environmental impacts can be identified during planning stage measure can be undertaken during design and construction to mitigate them as far as possible. Though not easy, if a system can be developed to assign monetary value to these impacts, they can be included in cost-benefit analysis, for appropriate weightage in decision making. EIA involves carrying out environmental and socio-economic studies for a project in parallel with analysis of engineering and socio-economic feasibility before any project is approved and implemented.

Monetary valuation of all environmental effects is difficult, so valuation by road users is commonly used. The 'Delphi Technique'–a procedure for eliciting and processing the opinions of a group of people, is involved. The impact that can be reliably measured in monetary terms, is only a small part of total environmental and social impacts of a highway projects. Attempts are being made to evolve techniques of putting monetary values to certain impacts such as accidents, air-pollution, etc.

21.2.1 Goals of EIA

EIA has developed due to increased public awareness of the harmful environmental and social effects of a highway project.

The prime objectives and scope of an environmental impact assessment (EIA) can be summarised as:

- To maintain the long term ability of natural resources to support human, animal and plant life.
- To conserve the board activity of plants, animals and ecosystems.
- To preserve the natural processes such as recycling of air, water and soil nutrients.
- To minimise irreversible environmental damage.
- To protect social, historical and cultural values of people and their communities.
- To avoid adverse effects of a highway/road project on basic needs of the people for food, clear water, clean air, shelter, health and sanitation.
- To support environmentally sustainable development.
- To provide plans for monitoring and managing environmental impacts.

- To ensure that to the extent possible, the local people be no worse of than they were before the highway/road project is implemented.

In preparation of EIA, it is essential to have open-mindness, must not have erroneous data, should have clarity of information and must also suggest the post-development impact audits, as there may be significant discrepancies between predicted and actual impacts.

21.2.2 Guidelines for Environmental Impact Assessment

Experience over the years has shown that EIAs are always conducted under severe limitations of time, manpower, financial resources and data. In India no reliable comprehensive environmental information base exists and need for extensive data collection makes EIA an extremely cumbersome and time-consuming exercise.

The public acceptance of impact assessment is not fully practised in India, and these are taken up, if at all, by a few sensitive groups. Public participation needs to be incorporated in the review process not only to enable considerations of local knowledge and preferences in highway project planning but also for avoidance of conflicts, and identification of issues of public concern.

Planning for sustainable development in context of ecosystem carrying capacity, requires systematic identification, quantification and management of cumulative environmental variables of a region.

Functional planning needs are to be identified in a region based on ecological criteria such as climate, vegetation pattern, soil classification, water shed boundaries, wild life and forestry, etc. in context of sustainable development. EIA procedure based on key indicators, and a method of evaluation and valuation of a highway project for sustainable development, based on the Indian Road Congress (IRC) guidelines is described in following paragraphs.

IRC guidelines assist in:

(i) providing information about potential adverse environmental effects of the highway project.

(ii) evaluating and arranging the different available alternatives from environmental point of view.

(iii) Identifying mitigation measures to be used in project proposal to eliminate/reduce environmental problems. "Do Nothing" alternative as one the solutions should also be considered.

EIA is prepared in consultation with experts of departments of environment and forest. Steps involved in a EIA are shown in Fig. 21.3. These steps are as per the EIA notification dated Sept. 14, 2006, by Ministry of Environment and Forest (MoEF), Govt. of India.

21.2.3 Procedure for Environmental Impact Assessment (EIA)

Preparation of EIA, for a highway project involves following steps.

1. General description of project: This describes the project broadly including description of terrain, topography, climatic conditions, existing alignment and its condition, improvement proposal, its objectives and restraints.

2. Existing conditions: Existing information regarding the following aspects is covered:

(a) **Road factors**

(i) Land width available (m)

(ii) Geometrics-curvature, gradient, roadway/pavement widths, etc.

(iii) Structural condition of road and road structures

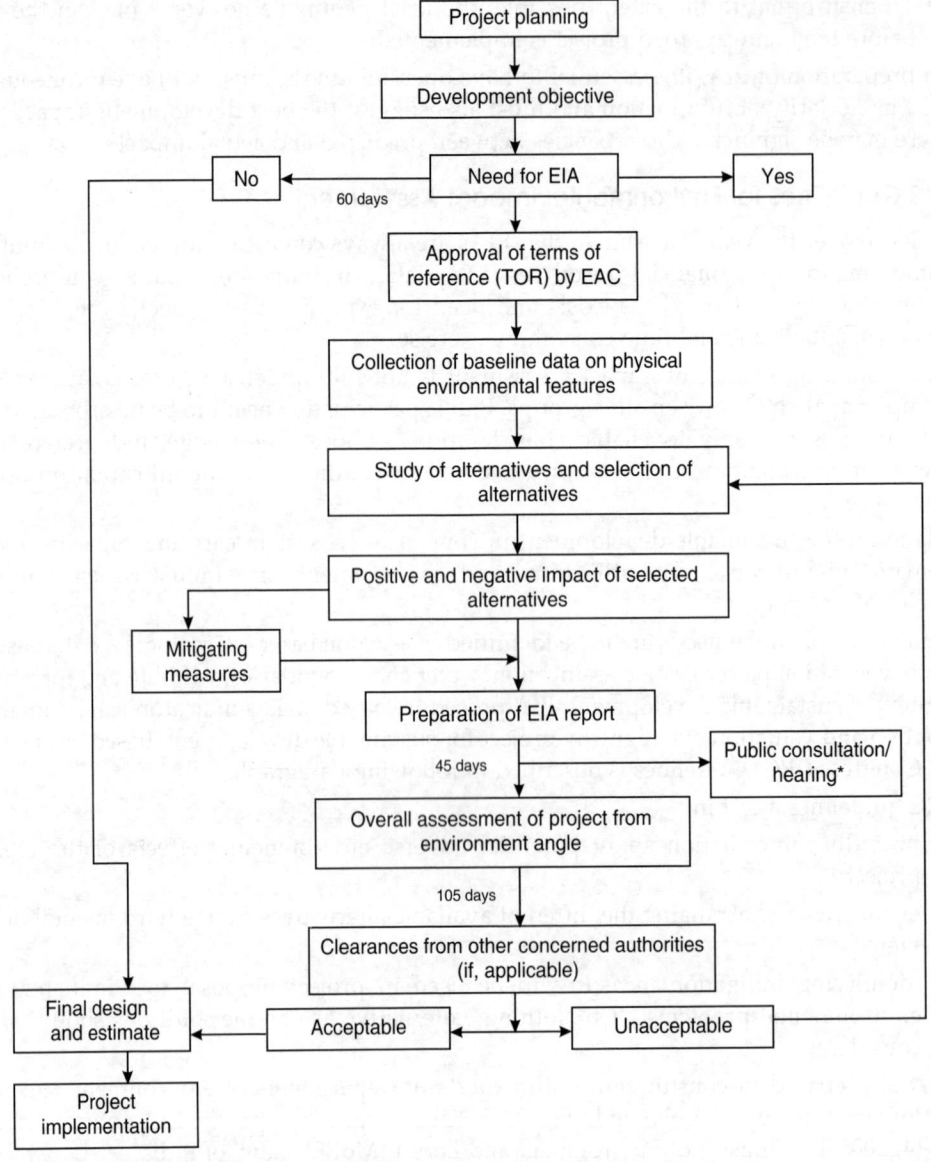

Total procedure takes 210 days as per time frame set by EIA Notification, 2006

Fig. 21.3: Flow diagram for environment assessment (EIA) for highway projects

(b) Terrain

  (i) Type of terrain (plain, rolling, hilly)

(c) Traffic factors

  (i) Traffic volume (veh/day)

  (ii) Traffic composition

  (iii) Average speed of travel

  (iv) Time delays at railway level crossing, etc.

  (v) Presence of road intersection (Nos./km)

 (vi) Access control

 (vii) Accidents (fatal and injury accidents per year)

(d) Land use

 (i) Type of area (urban, semi-urban, rural)

 (ii) Number of towns/villages traversed enroute with population figures

 (iii) Location of major land-use types such as commercial, residential, industrial, recreational, agricultural, etc.

(e) Environmental factors

 (i) Climate (annual rainfall (mm)), snowfall, maximum and minimum temperatures (°c)

 (ii) Vegetation

 (iii) Ribbon development and encroachments

 (iv) Roadside facilities

 (v) Air pollution level (very high, high, moderate, low)

 (vi) Noise level (very high, moderate, low)

 (vii) Wildlife (any endangered species)

For a new road information regarding likely traffic on the road after completion is also included.

3. Purpose/need of the proposed project: Following points are to be covered regarding the new project.

 (a) Transportation demand —present demand, projection for design year.

 (b) Access —present position about access to the area and the importance.

 (c) Capacity —the highway width and type needed to cater to design traffic

 (d) If "No action alternative" is to be chosen (i.e. the proposed action is not implemented), discuss the anticipated adverse effects on the following:

 (i) Traffic convenience —free traffic movement/congestion, delays, safety.

 (ii) Environmental —air quality, noise, vibration, general aesthetic quality, etc.

 (iii) Economic —costs of vehicle operation, road maintenance/improvement, accidents, etc.

4. Proposed projects and their detail: In this part, physical and environmental feature of the different alternative projects under investigation, is recorded in a tabular form for all alternatives (Table 21.6).

Table 21.6: Proforma for recording the physical and environmental features of alternatives investigated (IRC)

Particulars	Selected alternative	Alternative A	Alternative B
1	*2*	*3*	*4*
1. Length (km)			
—New construction			
—Improvement of existing road			
2. Terrain (plain/rolling/hilly)			
3. Land width proposed (m)			

(Contd.)

Table 21.6: Proforma for recording the physical and environmental features of alternatives investigated (IRC) (*Contd.*)

Particulars	Selected alternative	Alternative A	Alternative B
1	2	3	4

4. Category of land proposed to be acquired (ha)
   —Forest land
   —Agricultural land
   --Waste land
   —Swampy land
5. Displacement of households (Nos.)
6. Cut sections
   —Length in cut (km)
   —Max. depth of cut (m)
7. Fill sections
   —Length in fill (km)
   —Max. height of fill (m)
8. Vegetation
   —No. of trees exceeding 60 cm in girth to be cut
9. Flood hazard (encroachment on flood plain)
10. Erosion potential
11. Landslide potential
12. Stretch in geologically unstable areas
13. Drainage and adverse impact on water flow
14. Number of major river crossings (exceeding 60 m)
15. No. of road intersections
16. No. of railway crossings
17. Schools, colleges, hospitals falling enroute
18. Number and type of utilities requiring relocation
19. Possibility of providing wayside amenities
20. Air quality (very poor, poor, fair, good)
21. Noise level
22. Estimated cost

Considering above factors, the best alternative selected is described in detail regarding environmental aspects. Its probable impact on environment and proposed mitigation measures are worked out.

5. Probable environmental impacts and proposed mitigation measures of selected alternative: Both harmful and beneficial impacts of the proposed plan are considered and measures are suggested to mitigate adverse impacts. This information is summarised in tabular form as included in Table 21.7.

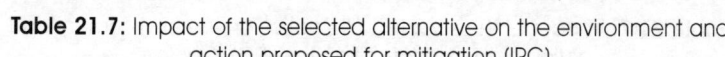

Table 21.7: Impact of the selected alternative on the environment and
action proposed for mitigation (IRC)

S.No.	Particulars	Probable impact in brief/ proposed action
1	2	3

1. **Land acquisition**
   (a) Is acquisition of forest land involved?
   (b) If so, whether discussions have been held with Forest Deptt. Also state as to what action has been taken to get clearance from forest angle.
   (c) Is acquisition of wet land/swampy land/mangroves, wildlife habitat involved?

2. **Highway location**
   (a) Is the road to traverse any unstable area, avalanche area, marsh, etc? if so, have necessary remedial measures been planned?
   (b) Have geological maps been studied or local Geological Deptt. consulted to avoid unstable strata?

3. **Highway alignment**
   (a) Does the alignment follow level of the land and avoid large scale cutting?
   (b) Is any section susceptible to damage/ erosion by streams and torrents? If so, have protection measures been planned?

4. **Highway cross-section**
   (a) Does the road cross-section involve a lot of disturbance to the natural ground?
   (b) For sections in cut, is the half cut and half fill type of cross-section which involves least disturbance to the natural ground being adopted?
   (c) Are the proposed cut slopes stable for the strata?
   (d) Are slope stabilising structures like breast walls, pitching, etc. required and being proposed?
   (e) Does the cut hill face require any special treatment to prevent slips. If so, are such measures being proposed?

5. **Erosion control**
   (a) Has erosion potential been considered for the alignment?
   (b) Are erosion control measures before start of work and between successive construction stages required? If so, have these been worked out?

(Contd.)

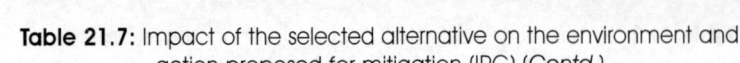

Table 21.7: Impact of the selected alternative on the environment and action proposed for mitigation (IRC) (*Contd.*)

S.No.	Particulars	Probable impact in brief/ proposed action
1	2	3

(c) Have location and alignment of culverts been chosen to avoid severe erosion at outlets and siltation at inlets?

(d) Are necessary erosion control measures proposed at outfall of culverts?

6. **Drainage**

(a) Does the project provide for necessary cross drainage structures so as not to obstruct the natural drainage of the area?

(b) Does the project provide for necessary side drains, catchwater drains, etc. for safe disposal of surface water?

(c) Will the road cause undue increase in the HFL or create ponding situation for long periods?

7. **Vegetation**

(a) Does the project provide for sodding grassing all embankment/cut slopes and other bared areas?

(b) Does the project provide for planting trees/plants on the roadside at the appropriate location.

8. **Traffic movement**

(a) Does the road affect the traffic circulation in the area? If so, have necessary measures been taken to provide suitable access to crossing roads?

(b) Will the proposed highway improve traffic movement, in terms of speed, convenience and safety?

(c) Are school children, hospitals, etc. affected by the highway? If so, are necessary traffic control measures taken into account?

(d) Are road user facilities like fuel-filling station, rest areas, truck-parks, etc. planned along the highway?

9. **Construction**

(a) Has the proper disposal of surplus excavated material been thought of and provided for?

Table 21.7: Impact of the selected alternative on the environment and action proposed for mitigation (IRC) (*Contd.*)

S.No.	Particulars	Probable impact in brief/ proposed action
1	2	3

(b) Have the type of equipment to be used for construction been identified? Will it be specified that these equipment are provided with pollution control devices?

(c) Have quarry/material/sources/borrow areas been identified? Is opening up of new quarry/material sources involved? If so, will this affect the inhabitants in the nearby areas?

(d) What measures have been planned to control dust from construction site?

(e) Will the proposed borrow areas affect the environment by way of soil erosion, water ponding, projecting a shabby look, etc. If so, have the necessary remedial measures been planned?

10. **Air quality**
(a) What is he estimated number of motorised vehicles expected on the highway in he design year
—petrol vehicles (nos.)
—diesel vehicles (nos.)

(b) Volume-capacity ratio and traffic flow condition expected on the proposed highway in the design year.

(c) Will the highway improve or deteriorate the air quality in the population centres enroute?

11. **Traffic noise and vibrations**
(a) What is the existing noise levels and noise level expected on completion of project?

(b) What is the type of surfacing proposed for the highway?

(c) Are residential, institutional areas located within a distance of 100 m from the centreline of the highway?

(d) If so, have noise—abatement measures like noise screens/screen plantation, etc. been included in the project?

12. **Water quality**
Is any pollution expected to affect water quality on completion of the project?

13. Any other point?

6. Implementation of suggested mitigation measures: The proposed mitigation measures as listed in Table 21.7, to mitigate negative impacts of the final plan, are considered to decide whether the project is acceptable. If not acceptable other modifications are done, till it is acceptable for implementation.

Field studies/surveys for EIA have been adequately described in detail in various IRC code of practices/manual such as IRC–104, EIA integrates the objectives of environmental management and a tool for achieving sustainable development. It establishes a procedure for environmental protection and ensuring that natural resources are utilised within the carrying capacity of the system.

EIA is necessary for environment clearance for development of road and highway project and is also a pre-requisite for consideration for funding of a project by international funding agencies like World Bank, UNDP, Asian Development Bank (ADB), etc. As per the existing guidelines, environmental clearance is required for highway projects except projects relating to improvement works including widening and strengthening of roads with marginal land acquisition (not exceeding 20 m on either side put together) along the existing alignment provided they do not pass through ecologically sensitive areas such as national parks, reserve forests etc, as per MoEF notification dated Oct, 15, 1999). Byepass project would require environmental clearance, if project exceeds Rs: 100 crores.

The concept of 'Social Impact Assessment' (SIA) along with environmental considerations into decision making/project implementation, process has also been introduced in highway projects specially those funded by Asian Development Bank (ADB) and World Bank (WB), as per their mandatory requirements.

21.3 ENVIRONMENT MANAGEMENT PLAN (EMP)

Environment management plan (EMP) is an implementation plan for mitigation, protection or enhancement measures for the environment. It is not always possible to remove all the environmental impacts totally, however the problems could be minimised, by proper understanding, and attention taken from stage of planning, surveys, alignment selection, project formulation, construction and maintenance. EMP is to ensure that highway construction schemes will adequately protect the environment, the relevant laws and regulations will be implemented and the proposed project shall be environmentally acceptable to the community at large and so also to politicians. The project should also be acceptable in its planning, economics and engineering design.

Adverse environmental impacts can be reduced through improved project selection, more responsible project planning and design. In order to mitigate adverse impacts on the environment, it is imperative to consider the environmental impacts of highway projects at all stages including the planning stage. This necessitates identification of activities of highway construction/improvement programme, which can adversely affect the environment and take necessary steps to minimise detrimental effects. Specifications and techniques stipulated in contractor's documents for highway project must include mandatory provisions for preservation of environment.

21.4 MANAGEMENT AND MITIGATION MEASURES FOR SUSTAINABLE ENVIRONMENT

Sustainable road transport system should meet need of today, taking care of future generations in providing safe, efficient, environmental friendly and cost effective system of roads. To minimise the environmental and ecological degradation and meet the demands of sustainable environment, several remedial measures need to be taken at the planning, construction, operation and maintenance stages of a road project. The important measures which should be

considered for minimising adverse impacts during planning and design, construction, operation and maintenance stage of a project are briefly described below:

21.4.1 Measures During Planning

- Selected road alignment should avoid landslide prone and geological unstable areas, forests, agricultural land, sensitive ecosystems and important cultural and religious sites.
- Selected road alignment should avoid large scale cutting and filling and should be as per mass balancing.
- Road alignment should cause least disturbance to natural surroundings.
- Considerations should be given for provision of suitable drainage facilities at the planning stage itself, utilising discharge to natural drainage channels.
- Quarry sites selected for the project roads should be away from populated areas and natural drainage system, in structurally stable areas.
- Cultural heritage and religious sites should be avoided in deciding road alignment during the planning stage.
- Noise mitigation measures should be introduced to reduce impacts to acceptable levels.
- No noise sensitive land uses such as schools, hospitals should be permitted within 200 m from the roads.
- Adequate compensatory afforestation should be ensured. Adequate provision for tree plantation by road side should be included in the project.
- Cattle-passes across the highway should be provided for safe movement of the cattle and wild life.
- Project affected people (PAP) should be rehabilitated properly, close to their displaced places.

21.4.2 Measures During Design Stage

- The removal of trees and other useful vegetation should be discouraged by proper engineering design.
- The design should specify the bio engineering techniques for stabilising exposed slopes and other bare areas.
- Use of locally available materials should be encouraged in design wherever possible.
- During the design, quantities of waste material should be estimated and safe disposal of spoil material should be worked out.
- In the design, environmental mitigation measures should be clearly worked out, so that they may be included in contract document during construction of roads/highways.
- Environmental mitigation costs should be included in tender document for proper management and control of environmental problems. 'Environmental Management Plan' will then be satisfactorily implemented, and monitored. Costs of mitigation the impacts should be a part of the project costs, to be recovered from beneficiaries, i.e. the road users.
- Highways should be designed to minimise congestion and improve road safety. A proper disaster management plan should be evolved for the project.

21.4.3 Measures During Construction

- Removal of trees, shrubs, from the land should be done properly without damaging pipeline, sewers, etc. adjacent to proposed highway.
- To prevent soil erosion large portions of exposed land by clearing and grubbling should be avoided. Exposed cut and fill slopes should be turfed using bioengineering techniques.
- Storing of material should not disrupt the natural water flow and should prevent erosion and migration of soil particles into surface water.

- Undesirable ponding of water should not be allowed. Temporary drains may be used for diverting such water to natural drainage channels.
- Site should be restored to near natural conditions soon after the construction is over, without steep slopes and standing water, etc.
- Depth of material should be limited and extraction areas should be selected where there is little fine material to be carried down stream. Surface area of borrow pits should be minimised.
- Spoil material should be disposed off in abandoned quarries or borrow pits.
- Bitumen should not be applied during winds and rainy day. Bitumen material should not be discharged into side drains.
- Hazardous materials should not be stored near surface water. A plastic sheeting should be placed under hazardous material to collect and retain leaks and spills.
- The blasting operations for excavation in rocks should be well planned and properly monitored.
- All equipment should be fitted with dust controlling devices and be operated in day time only. Noise dampening and air pollution devices should operate properly, all the time.
- Stockpiled sand and soil should be slightly moist for stock piling and before loading specially in windy conditions to prevent dusting. While transporting such material, vehicle must be covered with tarpaulin.
- Suitable drainage facilities should be provided utilising discharge to natural drainage channels in conjunctions with erosion protection.
- Construction of roads in the areas where disturbances to wildlife, acquatic fauna, or pollution of water bodies is likely, should be carefully done to minimise disturbances and avoid spilling of construction material into water bodies.
- Adequate noise barriers should be provided where road passes through inhabited localities.
- Labour camps should be provided with sanitary facilities and fuel supplies to avoid pollution of water and indiscreet tree cutting for fuel.
- An environmental monitoring mechanism should be established to ensure various mitigation measures during construction of the project.

21.4.4 Measures During Operation

It should be recognised that walking and cycling are unavoidable modes for movement of people in a cost-effective way and sustainable road transport must provide adequate facilities for these non-motorised modes of transport (NMT).

Following operational measures help in developing pollution free environment:
- Phasing out of old vehicles and lead from petrol.
- Compulsory provision of catalytic converters in all petrol driven four wheelers.
- Introduce CNG and LPG in place of petrol in vehicles through better engine technology.
- Check quality of fuel supplied to motor vehicles and fuel adulteration.
- Improve flow of traffic without stops and interruptions.
- Provide acceleration and deceleration lanes where required.
- Remove sight distance restriction and obstacles.
- Restrict speed limits, conflict points, turning movements.
- Install/improve signs, signals, rumble strips, markings, etc.
- Use coordinated and vehicle actuated traffic signals at possible locations.
- Provide one-way streets and re-route through traffic.

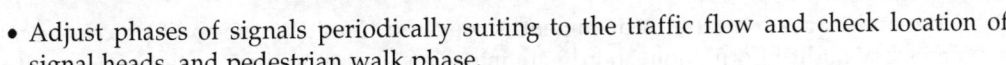

- Adjust phases of signals periodically suiting to the traffic flow and check location of signal heads, and pedestrian walk phase.
- Overlay pavements with grooves-skid resistant surface at critical locations and intersections.
- Channelise intersections, provide pedestrian refuge islands.
- Relocate pedestrian cross walks.
- Protect objects with guard rails.
- Provide adequate drainage and street lighting.
- Enforce strict anti-pollution laws regarding manufacture and maintenance of vehicles.

21.4.5 Measures During Maintenance

- Exposed areas and slopes should be planted with suitable vegetation, as soon as possible, after construction.
- All drainage along roadside should be kept clean during rainy season.
- Eroded areas should be immediately treated and protective measures like turfing should be used.
- Water seepage and water blackage should be immediately attended and rectified.

21.5 ENVIRONMENTAL AUDIT (POST PROJECT MONITORING)

There is a need for a feedback mechanism in environmental impact assessment (EIA), which involves the knowledge of the actual environmental effects of a highway project, rather than simply the predicted consequences. This feedback mechanism is provided by the post-project monitoring of environmental quality through environmental audit (EA).

United States Environmental Protection Agency's (USEPA) definition states that "EA is a systematic, documented, periodic and objective review by regulated entities of facility, operations and practices related to meeting environmental requirements."

Auditing—the activity of verification, is comparison of outcomes against expectations, where standards of expectations do not exist, the EA becomes an assessment. Environmental audits and assessments are evaluation activities, to provide quality control over ongoing operations associated with environmental compliance. With such a feedback remedial actions can be undertaken.

An EA programme provides the following benefits:
- Assessment of environmental risks
- Prioritisation of pollution control activities
- Verification of the correct operation of compliance assurance activities
- Evaluation of compliance management activities
- Assurance that proper actions are taken to minimise loss associated with pollution control
- Proper budgeting to address environmental problems
- Minimising waste and cost savings.

There is no significant disadvantages to the adoption of an EA programme. Designed and implemented conscientiously, an EA programme can only enhance an environmental performance.

Procedures for a logical and comprehensive environmental audit has three major components:

1. Environmental audit guideline: Since most audits are long and complex, written guidelines are needed in order to ensure consistency of approach. The guidelines ensure that interviews, site inspections, and filing and reporting of results are standardised.

2. Pre-audit questionnaire: A document is prepared to initiate the audit process to collect the necessary audit information, coordinate interviews, arrange the logistics of an audit visit and prepare for the process. This coordination is vital in maximising the efficiency of on-site audit as it allows to identify the major environmental issues which will be addressed.

3. Audit work book: A note-book is usually used by the auditors in a step-by-step process of conducting what is essentially an environmental inspection and scrutiny of procedures.

Considerable research was conducted on procedural and methodological issues related to environmental impact assessment (EIA) and environmental audit (EA), in the past. An acceptable standard practices for EIA and EA are now beginning to emerge and proving useful and effective. Table 21.8 shows typical check-list for environmental monitoring.

Table 21.8: Typical checklist for environmental monitoring

Part-I

(A) General Information

1. Location: District: Mandal:
                   Gram Panchayat: Road No.:
2. Project Name/Identification:
3. The road work connects: Width   Length   km
4. Type of Road: National/State/District/Feeder/Local   Surface type: Asphalt/gravel
                                                          Existing   Final
5. Physical dimensions: Embankment Base width (m):
                        Pavement (m):
                        Average Height/depth (m):
                        ROW width (m):
6. Attach a map of road, if available

(B) Ecological Impact

1. Does the road pass through?

Landuse	Forest Reserve	Wetland	Dryland	Wildlife Sanctuaries
Total Length (km):				

2. Are any natural water resources interrupted or reduced in waterway area?   Yes/No
   If yes, attach list of bridges/culverts (location, length, size) and stream width.
3. Numbers of trees (over 3 inch diameter) to be removed:

                                                             Species:
   Number to be planned   Species:
4. Is the road embankment likely to interfere with fisheries production or mitigation?   Yes/No
5. Are there any animals or vegetation unique to this area or known to be endangered?   Yes/No
6. Is it necessary to open borrow pits outside the road alignment?   Yes/No

(C) Impact upon Physiochemical Environment

1. Is the construction and soil to be used likely to lead to siltation downstream?   Yes/No
   If yes, is better protection around waterways included?   Yes/No
2. Is the embankment likely to obstruct natural drainage?   Yes/No
3. Does the area regularly flood?   Yes/No
   Both sides?
   If yes, to what depth above ground level (m)?
   Frequency (times per year)?
4. Are there urban or industrial area along the road alignment?   Yes/No
   If yes, attach details of sites, distance off centreline and type of activity.

5. Are the proposed construction camp sites away from water resources and settlement? Yes/No

    If no, has provision been made for safe disposal of wastes and septic?   Yes/No

(D) Socioeconomic Impacts

1. Have the local people been informed when construction is to commence?   Yes/No

2. Is land acquisition and compensation finalised?   Yes/No

3. Number of residences to be removed or shifted?

4. Are there any mosques, temples, graveyards or historical sites on the road reserve? Yes/No

Part-II (Complete at hand over of works from Contractor)

(E) Record of Compliance

1. Were there any complaints from the local people during construction?   Yes/No

2. Has there been any erosion or damage to the embankment?   Yes/No

3. Were there any spillage of chemicals or bitumen?   Yes/No

4. Have all borrow pits, campus sites and roadsides been restored?   Yes/No

5. Pavement structure used Macadam/Gravel/FCR.

6. Have all replacement trees been planted.   Yes/No

    Environmental guidelines for highway project are also given in Appendix H.

21.6 ENVIRONMENTAL REGULATION RELATED TO ROADS/HIGHWAYS

Various rules and regulations, acts/legislations are prescribed by Central Govt./State Govt. (s) to be strictly followed by the project proponent like National Highway Authority of India (NHAI), CPWD or other agencies, for highway projects and route alignment. They must obtain 'Clearances' from various concerned agencies/govt. departments, before undertaking the project. The purpose of guidelines, acts and regulations is to avoid any kind of alteration in vulnerable ecosystems and indirectly on human beings and wild life or if not possible to fully avoid them, minimise any adverse impact on them by implementing suitable mitigating measures in the form of Environmental Management Plan (EMP). There are various laws and acts which cover wide range of rules, regulations, and/or provisions related to ambient air quality standards, vehicle emission norms for different category of vehicles, specification for fuels, guidelines for EIA for highway and road projects, but their proper enforcement is still lacking.

Govt. of India's Notification dated 14th Sept. 2006 and 19th Jan., 2009, on Road/Highway projects, for decentralisation of various projects/activities between centre and state level, established new criteria for these projects requiring environmental clearance either at centre level or state level. According to this notification, road and highway projects have been divided into 8 major heads requiring 'Environmental Clearance' (EC) either from Central Government or State Government. This notification also changed the purview of EIA procedure for roads and highway projects. Road and highway projects are appraised both by centre as well as state level, depending upon the location and length of the highway, and its distance from the boundaries of:

(i) Protected areas notified under the wildlife protection act, 1972

(ii) Critically polluted areas notified by central pollution control board from time to time.

(iii) Notified eco-sensitive areas and

(iv) Inter-state boundaries and international boundaries.

Salient features of the various acts concerning highway and road projects are given below.

The Wild Life Protection Act 1972

This act was introduced to save endangered species, flora and fauna as well as those which are on the verge of extinction. Under this act areas were declared as "Protected Areas", "Wildlife Sanctuary" and "National Parks" and "Closed Areas" to protect the wild life.

Central Government in the year 2003, constituted "National Board for Wild Life (NBWL)" under wild life protection Amendment Act, which is headed by Prime Minister of India, to promote conservation of forests and wildlife and to carry out assessment of various projects/activities on wildlife and its habitat.

Road and highway projects, require "Wildlife Clearance," if it is proposed to be located in or within 10 km of any "Wildlife Sanctuary" or "National Park." It is desired that new proposed road and highway project is aligned in such a way that it does not pass through any of the protected areas like national parks or wildlife sanctuaries. Even upgradation projects acquiring additional land are discouraged. Permission for upgradation within existing right of way is granted under special circumstances.

The Forest Conservation Act 1980, Amended in 1988

Road and highway projects which pass through any reserved or protected forest, require special permission from the appropriate authority. Permission for such projects is usually conditional or provided with certain strict conditions for compliance. Ministry of Environment and Forests (MoEF), has issued guidelines under Forest Conservation Act, 1980 and Forest Conservation Rules-2003, to grant permission and tackles these problems, for conservations, protection and management of forests, and alteration in forest ecosystem, by stipulating strict conditions and ensuring minimum adverse impact on forests.

Forest clearance as per the specified procedure is required for upgradation or widening of lanes requiring adjoining land.

The Air Prevention and Control of Pollution Act, 1981

Under this act, the Central Pollution Control Board has been authorised, in the Motor Vehicle Act (MVA) to lay down the emission standards for automobiles, and overall ambient air quality standards. This act is to take appropriate steps for the preservation of the quality of air and control of air pollution. The act exclusively deals with the preservation of air quality and control of air pollution.

Section 20 of the act prescribes standards for emission of pollutants from automobiles. Accordingly, the state governments are empowered to give such instructions as may be deemed necessary to the concerned authority in charge of registration of motor vehicles under motor vehicle act.

The Environment (Protection) Act, 1986, Amended up to 1992

Under this act central government can enact and formulate various rules and regulations related to vehicular pollution and can constitute authority/expert committee to perform and exercise powers to control the vehicular pollution.

Specifications for petrol and diesel are proposed under this act, to improve the fuel quality, including removing of lead from petrol and reducing sulphur dioxide from diesel in order to reduce vehicular emissions.

It is a comprehensive legislation for the protection and improvement of environment and for matters, connected therewith. The scope and definition of 'Environment' is expanded to include water, air and land and the inter-relationship which exists among and between water, air and land and human beings, other living creatures, plants, micro organisms and property.

The objectives of the act are to meet the standards of air, water or soil in various areas, regulate the maximum allowable concentration limits of environmental pollutants (including noise) for different areas, to regulate the procedure and safeguard for the handling of hazardous substances, prohibition and restriction on the location of industries, safeguards for the prevention of accidents which may cause environmental pollution and providing for remedial measures for such accidents.

The Motor Vehicle Act 1988 and Central Motor Vehicle Rules 1989

This act contains various provisions for regulation and control of vehicle emission both for vehicles at manufacturing stage as well as for in-use vehicles.

Important provisions of MVA, relate to vehicular emissions, alternation in motor vehicles, suspension and cancellation of registration, grant of certificate of fitness, age limit of vehicles, provisions regarding construction and maintenance of vehicles, emission standards, punishment for offences for using the vehicle in unsafe conditions and for non-compliance of various vehicular emission as prescribed. Rules for vehicle emission are made after consultation with Ministry of Environment and Forest (MoEF).

All vehicle must obtain a certificate of fitness (CoF) periodically as specified in the act.

The Central Motor Vehicles Rule (CMVR), is more specific about manner in which exhaust gases should be discharged, specifies location and position of the exhaust pipe, and covers extensively the procedure for testing and quantification of emissions from motor vehicles. It also authorises persons who can carry out vehicle emission test, and also the type of instruments to be used.

Section of the rules also relates to general provision of punishment of offences applicable to vehicles, vehicle emissions related offences and penalty for using vehicles in unsafe conditions. Provisions if strictly followed and implemented pollution should be controlled effectively. Details of motor vehicle act are covered in Chapter 3.

The Water Preservation and Control of Pollution Act, 1974 as Amended up to 1988

This act provides for preservation and control of water pollution and maintaining/restoring the wholesomeness of water. Central Pollution Control Board (CPCB) performs duties and functions enumerated under the water act. It promotes cleanliness of streams and wells in different areas of the states, prevention and control of water pollution. The objective of this act are:

(i) To provide for the prevention and control of water pollution and maintaining or restoring of wholesomeness of water (in the streams or wells or sewer or on land).

(ii) To establish central and state boards with a view to carrying out the above purposes and

(iii) For conferring on, and assigning to such boards powers and functions relating thereto and for matters connected therewith.

The Noise Pollution (Regulation and Control) Rules 2000

The noise standards have been designated for different type of land-use, i.e. residential, commercial, industrial areas, and silence zone, under these rules.

In India, there is no provision in the motor vehicle act, exclusively about control of noise pollution. In factory act, section 11, there is a provision that factories to provide adequate measures for control of noise to specified levels.

No exposure in excess of 140 dB peak sound pressure level is permitted.

Other rules and acts include the following:

- The hazardous waste (Management and Handling) Rules 1989
- The public liability Act 1991.
- The prevention of cruelty to animals act, 1960
- The Coastal Zone Regulations (CZR) Notifications 1991 and 1994.

Also schemes such as "Paryavaran Vahini' to ensure people's participation in environmental protection are also being under taken/implemented by the ministry.

Legislation is essential to set the goal of environment we desire to achieve. A legislation needs to be successfully enforced to get results. In respect of environmental management, it has to be clearly understood that this process is complex, time consuming, and costly. Unless the objectives are crystal clear and so defined, the approach is gradual, phased, practical, integrated and very well coordinated with all the agencies concerned and obstacles and hindrances first clearly understood and then overcome, the environment is bound to suffer.

21.7 LANDSCAPING, AND AESTHETICS

Landscape is external environment, which enhances the beauty and provides aesthetic effects for road users and surrounding residents. Landscaping is an effective way of utilising the land portions left from land resumption activities.

Landscaping should be planned on the basis of vegetation survey of ecological areas in the vicinity of the road. The stability, interdependency, sensitivity, vulnerability and resistance to change of each type of vegetation needs to be considered along with nature of the existing vegetation.

Natural features for planning landscaping that require evaluation include soil, topography, rainfall, climate, environmentally sensitive areas, existing land usage and landscape, etc. Existing trees can also be used as part of the overall planting pattern.

Landscaping design includes both, the view from the road and view of an observer away from the road. Longitudinal, vertical alignment provides better and greater visual impact than horizontal alignment. Planting for landscaping should be planned to minimise erosion, headlight glare, and side winds. Shrubs and trees used for landscape are useful for delineating the road layout, absorbing traffic noise, and work as crash barriers. Trees should not be planted on steep slopes, or on locations where they obstruct drains or reduce sight distance.

The basic principles for landscaping along a highway/road to be kept in view are:

- To preserve existing features of aesthetic nature like ponds, lakes, falls, streams, monuments etc.
- To protect plans and existing vegetation.
- Road alignment should be in harmony with the natural terrain as far as possible.
- Intersections can be made beautiful in appearance by planting flowering plants on islands, medians and rotary circles.
- Bridges, flyovers, culverts and other structures can be made good in appearance by architectural design and painting parapet wall colourfully.
- Borrow pits along the road give ugly look. They should be properly treated.
- Turfing of slopes and shoulders adds to the aesthetics along the roads and is pleasing to eyes.
- Any ugly looking spot, along the road should be made good by proper vegetations.
- At locations where soil erosion is likely to occur; turfing and planting could be used to prevent it.

21.8 ROADSIDE ARBORICULTURE

Roadside planting of trees is termed as 'Arboriculture'. Trees are planted along the road for providing shade, preventing from heat, cutting off ugly views, reducing soil erosion, creating pleasant look, breaking winds, reducing temperature on road surface, stabilising the formation, intercepting sound waves and fumes from vehicles, preventing dust, etc. The worst enemy against environment is pollution and best friends for preservation and protection of the environment are the trees.

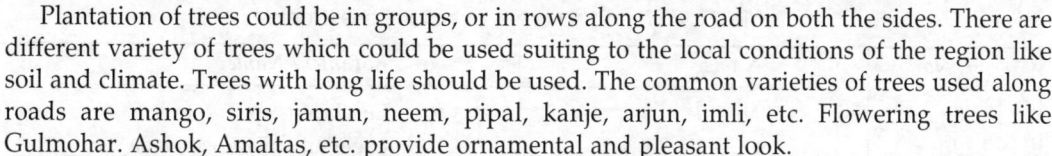

Plantation of trees could be in groups, or in rows along the road on both the sides. There are different variety of trees which could be used suiting to the local conditions of the region like soil and climate. Trees with long life should be used. The common varieties of trees used along roads are mango, siris, jamun, neem, pipal, kanje, arjun, imli, etc. Flowering trees like Gulmohar. Ashok, Amaltas, etc. provide ornamental and pleasant look.

The spacing of trees and distance from the road depends upon the type of trees and provisions for future widening of roads. Spacing of 10-15 m is commonly used. Trees should be at least 1.8 m away from the edge of the roadway so that they do not obstruct the traffic. When trees are planned on both sides of the road, trees on one side should be staggered with those on other side to provide better shade on the road.

The operation of planting trees involves preparation of seedlings, transplanting the seedlings to the excavated pits, protection from cattles, by wiring or providing wall along the pit, and watering regularly. Trees when they are grown are often numbered to keep an updated list.

Trees require to be maintained regularly, protection against pests, pruning and lopping of branches and removal of dead and fallen trees.

Table 21.9 shows the botanical names of some common trees, used for planting on roadsides.

Table 21.9: Botanical names of Indian trees

Sr. No.	Name	Botanical Name
1.	Neem	Azadirachta indica
2.	Pipal	Ficus religiosa
3.	Sisoo	Dalbergia sisoo
4.	Kanji	Pongamia glabra
5.	Vad	Ficus bengalensis
6.	Peltophorum	Peltophorum ferrugineum
7.	Gulmohor	Delonix regia
8.	Garamaro	Cassia fistula
9.	Kasid	C. siamia
10.	Pink cassia	C. renigera and other varieties
11.	Piludi	
12.	Saru	Casuarina equisetifolia
13.	Bangli baval	
14.	Gliricidia	Gliricidia maculata
15.	Keshudo	Butea monosperma
16.	Bougainvillaea	Bougainvillaea cultivars
17.	Bauhinia	1. Bauhinia purpurea
		2. B. variegata
18.	Melitia	Milletia ovalifolia
19.	Arduso	Ailanthus excelsa
20.	Asopalav	Polyathia longifolia
21.	Rain tree	Somania soman
22.	Jaman	Eugenia jambolana
23.	Badam	Terminalia catappa
24.	Kigellia	Kigellia pinnatta
25.	Bakul	Mimusops elengi
26.	Bottle brush	Callistemon lanccollus
27.	Siris	Albizzia lebbek

(Contd.)

Table 21.9: Botanical names of Indian trees (*Contd.*)

Sr. No.	Name	Botanical Name
28.	Nilgiri	Eucalyptus sp.
29.	Queen's flower	Lagerstroemia flosregiane
30.	Teak	Tectona grandis
31.	Khair	Acacia catechu
32.	Cashew (Kaju)	Anacardium occidentales
33.	Bamboo	Dendrocalamus strictus

Selection of species to be planted is done considering type of soil, weather conditions, type of tree which gives dense shadow, and is strong enough against wind, storms, etc. Roadside nursery may be developed for easy and convenient planting of trees along the road and their regular maintenance and replacement.

Table 21.10 shows suitable trees, shrubs and grasses for different soil types.

Table 21.10: Suitable plants for specific soils

Soil Type	Plant Species		
	Tree	*Shrubs*	*Grasses*
Sandy	Acacia arabica	Aerva javanica	Aristida spp.
	Acacia cynophylola	Aerva pseudoto-menosa	Cenchrus prieurii
	Acacia modesta	Capparis aphylla	Cenchrus setigerus
	Acacia senegal	Calotropis procera	Erianthus munja
	Acacia sieberriana	Calligonum polygonoides	Eragrostis supp.
	Acacia tortilis		Eragrostis tremula
	Ailanthus excelsa	Cassia auriculata	Eragrostis ciliaris
	Albizzia amara	Clerodendron phloemoides	Lasiurus hirsutus
	Ailanthus gregii		Panicum antidotale
	Albizzia lebbek	Crotolaria burhia	Panicum turgidum
	Azadirachta indica	Euphoria bivula	Schima nervosum
	Balaniles aegyptiaca	Grewia tenex	
	Butea monospherma	Indigofera argentea	
	Cordia rothii	Leptadenia pyrotechniea	
	Dalbergia sissoo		
	Eucalyptus melonophloia	Zizyphus spp.	
	Ruvslypud pspusns		
	Eucalyptus populifolia		
	Eucalyptus oleosa		
	vargalauca		
	Eucalyptus tesseleris		
	Eucalyptus terminalis		
	Holoptelia integrifolia		
	Parkinsonia aculeata		
	Pongammia pinnata		
	Prospis juliflor		
	Prospis spicigera		
	Salvadora oleoides		

(Contd.)

Table 21.10: Suitable plants for specific soils (*Contd.*)

Soil Type	Plant Species		
	Tree	*Shrubs*	*Grasses*
Shallow Rocky	Tamarix articulata Tecomella undulata Acacia catechu Acacia leucophloea Acacia senegal Anogeissus pendula Azadirachta indica Boswellia serrata Dichrostrachys cin-erea Prosopis juliflora Salvadora oleoides Tecoma stans	Acacia jacouemontii Aerva pseudotomen-tosa Aerva tomentosa Barleria acanthoides Commiphora mukul Capparis aphylla Cassia auriculata Euphorbia nerifolia Euphorbia royleana Grewia tenax Indigofera argentea Zizyphus argentea	Aristida spp.
Gravelly	Acacia senegal Acacia catechu Anogeissus pendula Azadirachta indica Boswella serata Cassia siamea Dalbergia melono-xylon Hardwickia binata Prosopis spicigera Prosopis juliflora	Boehmeria diffusa Cassia auriculata Euphorbia royleana Tribulus terrestis Zizyphus nummularia	Aristidfa spp. Aristida mutabilis Dactyloctenium sindicum Elusine compressa Eleusine artistata Elesine aegyptica
Saline and Alkaline	Azardirachta indica Albizzia lebbeck Albizzia procera Acacia modesta Acacia tortilis Acacia arabica Butea monosperma Eucalyptus teretecor-nis Eucalyptus gomphocephaea Zizyphus spp. Eucalyptus robusta Parkinsonia aculeata Prosopis juliflora Pongammia pinnata Salvadora oleoides Tamarix articulata	Clerodendron phloemoides Calotropis procera Capparis decidua Leptadenia pyrotechnica Salsola foetida Scaevola putascens Scaevola koeniquii Sporobolus pallidus	Aristida spp. Cenchrus ciliaris Cenchrus setigerus Chloris montana Cynodon dactylon Dicanthium annu-latum Eragrostis spp. Panicum antidotale Saccharum munja

Source: Rajasthan State Highway Project

Economic Analysis of Highway Projects

22.0 PREAMBLE

Government provides highways on which private automobiles, trucks, buses, other private and public vehicles travel. Expenditures on highways are made to raise the level of entire economy by providing access to different places and transportation of goods, to make easy and quick provision for police, medical, fire, school and other community services, to assist in national defense and open opportunities for recreation and travel. Due to location of highways land values are increased, travel time is saved, accidents and vehicle operating costs are reduced, comfort and ease of travel is increased. On the other hand new highways or improvement of old roads consumes resources and land which could be used for other productive purposes.

In view of scarcity of resources and competing demands from various sectors it is necessary to allocate resources in most beneficial manner among various sectors and with in highway sector among various schemes.

Highway expenditures can be justified only if net sum of the consequences are favourable, i.e benefits from highway improvement exceed the costs including some allowance for return on the money invested. However, it should be realized that in public policy economic studies are not the final answer, it only aids in decision making by determining the economic resources gained or lost if given alternative, including that of 'Do Nothing' is chosen. The final decision of the project should be made in the best overall public interest by taking into consideration the need for safe, fast, and efficient transportation and cost of eliminating the adverse effects such as:

(i) air, noise and water pollution

(ii) destruction or disruption of man-made and natural resources, aesthetic values, community cohesion and availability of public facilities and services.

(iii) adverse effects on property, values and employment

(iv) dislocation of people, business, commerce and agriculture and disruption of community and regional growth.

Economic assessment and application of economic concepts to justify highway projects/improvements establishes priority in expenditure of limited funds. There are always number of alternative plans to achieve desired goal and objective for a highway improvement, involving different costs and benefits is to select one.

A minimum of expense is highly desirable, but the road which is truly cheapest is not the one which has least cost, but the one which makes the most profitable returns in proportion to the amount which is spent on it. However highways can not be justified strictly on the economic consideration, but must be qualified by its overall service to the community, by taking into account such factor as traffic service, effect on land uses, impact on life of the community, environmental considerations, national objectives, etc. The economic feasibility of a proposed public work as the overall benefit and development of the society must be properly evaluated.

Table 22.1: Objectives for highway projects

Objectives related to	*Objective purposes*

(a) Highway/road users	• To minimize motor vehicle operating costs
	    – Motor vehicle travel time
	    – Accidents and their cost
	• To maximize travel comfort and convenience

(b) Highway design	• To minimize cost of highway construction, maintenance and operation • To maximize the quality of travel service • To provide higher capacity for demands of traffic.
(c) Community benefits	• To maximize coordination and integration of total transportation system of all modes. • To provide adequate transportation to all areas and land uses • To decrease congestion on roads through better public transportation
(d) Community social factors	• To minimize adverse effects due to land acquisition and route location • To preserve historical areas, temples, etc. • To preserve neighbourhood and social characteristics of the area • To minimize noise and air pollution • To preserve/enhance aesthetics • Improve opportunities for recreation and tourists. • To achieve overall regional development • To preserve/enhance community goals. • To conserve open spaces, parks, etc.
(e) Economic factors	• To minimize adverse impacts to business, industry, commerce before, during and after construction of roads. • To preserve land values • To maximize conservation of resources • To promote land use according to economic development plans.

The resources available are always limited and not sufficient as compared to demands of development for any country, more so for developing countries. Economic and technical soundness of a project/scheme, to provide greatest benefits from the resources available is very important. Costs of various project suiting to the desired objectives may be different and so also the benefits which are likely to accrue from them. To select the best scheme which provides maximum benefits from available resources, economic analysis is done.

In the mid 1950s the economic analysis approach was first applied to public sector highway projects. Highway projects are undertaken to meet the following possible objectives, for which best alternative plan is selected through economic evaluation (Table 22.1).

22.1 ROLE OF ECONOMIC EVALUATION

Economic evaluation provides a decision making approach to expenditure of public funds, for judging overall economic viability.

The analysis of economic benefits of a highway project is related to the costs and is done to meets the following broad objects:

1. To determine whether a scheme is economically justified
2. To assist in decision making whether to invest fund in the proposed project or not
3. To establish priority of one scheme to another
4. To arrive at the best alternative out of several plans which are drawn with proper consideration to future requirements, and stated objectives and standards.
5. To develop information which will help in evaluating particular improvement as against other proposals in public works or community projects.
6. To decide which type of structure to be adopted where and when.
7. To aid in the selection of engineering design features of the project, such as location, structural design, specifications, etc. that will provide maximum economy consistent with safety and reliability.
8. To determine the value of the returns, by the proposed work in comparison to cost of producing it.
9. To know how well the proposed project will serve general social objectives and economic activities of the society.
10. To plan phasing of a programme over a period of time in view of availability of funds.

22.2 BASIC PRINCIPLES OF ECONOMICS

Highway economics is concerned with all measurable and non-measurable costs of constructing and operating highways and road users (vehicle and pedestrian) costs. Highway projects bring benefits to society and road users, by connecting places and providing cheaper, efficient, and safe travel to different places. Evaluation of costs and benefits of a highway project in terms of money and compare them, is the basic principle of economic analysis.

Economic evaluation involves quantification of the consequences of each alternative in terms of a single common monetary measure. This is a ranking procedure which converts performance measures to monetary cost and benefits and summed in order to obtain best alternative. In economic evaluation, efforts are made to include not only the direct immediate costs and benefits to road users but also indirect costs and benefits accruing to society at large. Economic evaluation techniques try to produce a single measure of effectiveness upon which, plan selection may be made and this common measure is monetary value.

The impacts of a highway project are divided in two groups costs and benefits. The costs are the capital cost of investment and operating costs. The benefits are those received by users of the facility such as value of travel time in rupees per hour. All the benefits to road users are

measured in economic terms. Through the use of appropriate interest rates and other factors, costs and benefits are made comparable. Then each alternative project/plan is evaluated by determining the total costs and the total benefits for the project and computing a benefit cost ratio. Projects are compared on the basis of these ratios. Classes of benefits usually used are savings in travel time of users, savings in operating costs of users and savings in accident costs to the community. Changes in vehicle operating costs include savings in fuel consumption, tyre, wear, brake wear, engine and other maintenance converted into monetary units. The accident savings are estimates in terms of less accidents, medical expenses, property damage and administrative expenses, however it is not possible to place money value on factors such as personal suffering, etc. Savings in working time are calculated by referring to average wage rates. To put value on non-working time saved are difficult. Another aspect of time saving that should be considered is reduction in vehicle fleet because saving in time of vehicles, the same volume of commodity could be carried in fewer vehicles.

Following guidelines (principles) are followed in conducting any economic analysis:

   (i) Forecasting of future consequences of possible investments should be done accurately and carefully. Economic analysis is not concerned with past investment.

  (ii) Each alternative among which choices are to be made must be fully and clearly understood and spelled out 'Do Nothing' alternative should be the starting point to measure costs and benefits.

 (iii) For public works, broad approach, weighing all consequences of the improvement proposal should be included not ignoring social, economic and community impacts of the new facility.

 (iv) Economic (use of resources), financial (getting and use of money) and political and administrative factors should be considered in final decision making.

  (v) When considering number of alternatives, it is desirable to consider the marginal differences in costs and benefits, between two alternatives and carry out incremental analysis to see if increment in cost yield justified incremental benefits.

 (vi) All costs and benefits should be discounted to same time period, spread uniformly over analysis period.

(vii) Taxes should not be included in economic studies for public projects. Method of financing should be ignored in economic analysis.

(viii) Double counting of costs and benefits should be avoided.

 (ix) Cost of existing facility should not be counted in economic analysis. Benefits and costs which can be converted into money and those which can not be converted into money value should both be considered, in decision making. For example the time saved in a journey may not be a gain as it may go for leisure or recreation, while a life saved by accident reduction has economic value which can be measured through productive capacity of the individual. Economic analysis helps in decision making and is not a rigid criteria. Its purpose is to know how well the proposed highway improvement and construction project will serve the society.

Economic analysis is not concerned with the sources of financing, the availability and the allocation of funds.

22.3 COST CONSIDERATIONS

The costs of highway construction/improvement consist of total initial capital, cost of construction, the cost of any delays during the period of construction and cost of maintaining the facility for its life and road user costs. Road costs are commonly expressed as per unit length in economic analysis

Following are the three items that make up total highway costs.

22.3.1 Construction Costs

(i) **Land cost:** This includes total cost of land including cost of legal fees, transfer documents and cost of rehousing the displaced people (if any). The land is often purchased much before a highway construction begins, however the current value of the land should be used in economic analysis. Land costs are more in urban areas than in rural areas. Land costs are not more in widening of roads projects.

(ii) **Cost of survey, investigation and design:** This includes survey and investigation work and design of road and other items.

(iii) **Costs of walling/fencing:** Many times some sort of borderline in form of walling or fencing is required adjoining the road to separate the properties of highway authority and land owners. Such costs are to be included in initial costs.

(iv) **Removal/relaying of services:** If removal of services like water mains, electric poles, cables, etc. and relaying it else where away from the highway land, is involved. Costs of undertaking such works should be included in initial costs.

(v) **Costs for preparation of site:** The first part of this cost includes clearing of shrubs, trees, etc from the site and demolition of structure (if any). The second part is the earthworks undertaking to prepare a smooth surface for laying construction courses, i.e. preparation of subgrade.

(vi) **Cost of construction of carriageway and shoulders:** This includes cost of laying the sub-base course, base course and surface course. Cost of pavement layers depends upon the cost of aggregates, binder, hire charges of equipment and labour. It may be different from place to place depending upon specifications and type of construction.

(vii) **Cost of curbs, channels, retaining walls, etc:** This includes costs of items along the road like curbs, channels, retaining walls, cycle tracks, or any fencing required.

(viii) **Cost of drainage works:** This includes cost of side drains, culverts, longitudinal drains, channels for draining the water.

(ix) **Cost of structures:** Bridges, subways, etc are costly items. They are major items and often a big component of total highway costs

(x) **Miscellaneous and overheads:** Guard rails, traffic control devices, signals, markings, signs etc are miscellaneous items to be included in highways costs. Overhead costs include administration, supervision, quality control, etc.

For overall construction of a highway the total costs occur in the following sequence.

1. Right of way:
   - acquisition of land
   - fencing
   - alteration to services and utilities
2. Survey and plan preparation
3. Clearing
   - vegetation
   - buildings

4. Drainage
   - culverts
   - parapet walls
   - erosion protective
   - channels
5. Formation:
   - excavation
   - earth work
   - reshaping
   - trenches
   - drains
6. Kerbs and paths:
7. Pavements:
   - various courses
   - sealing
   - shoulders
8. Roadside features:
   - topsoil and seeding
   - erosion protection
   - parking facilities
   - guard rail
   - guide post
   - sign, signals
   - markings
9. Structures:
   - bridges
   - retaining walls
   - others
10. Surface treatments:
    - prime coat
    - surface course
    - seal coat
11. General
    - administration
    - supervision
    - work signs
    - setting out
    - delays due to weather
    - site establishment and plant.

The convenient way of costing is to estimate the quantities of each individual work activity and multiply it with the unit rate for that activity.

22.3.2 Maintenance Costs

The road is a capital investment which deteriorates with time and its use. Maintenance is necessary to lessen the rate of deterioration and also restore to those service levels which pertained soon after construction, as far as possible.

Maintenance costs to upkeep the highways and associated works in good condition are expressed on yearly basis, though maintenance expenses are incurred at frequent intervals and are of recurrent nature. Cost of maintenance of roads/highways depends upon several factors such as cost of labour and equipment, volume, type and intensity of traffic, type of wearing surface, climatic conditions, acceptable serviceability level, etc. Annual maintenance cost for good quality roads such as cement or bituminous concrete roads are significantly lower, than other roads. Cost of maintenance includes repairs of pot holes, patch holes, dressing of surface periodically, emergency repairs, maintaining of signs, markings, supervision and administration, etc.

In economic analysis, the maintenance costs of existing and reduced maintenance costs of the improved road are considered. Future maintenance cost of the facilities are discounted to the base year for economic evaluation.

Total highway cost is the sum of the capital cost of all items expressed on an annual basis, plus annual cost of maintenance. Total annual cost is given by equation:

$$H = C_1K_1 + C_2K_2 + C_3K_3 \ldots + M$$

where   H = total annual cost

C$_1$, C$_2$, C$_3$ ... initial capital costs of the items mentioned above.

K$_1$, K$_2$, K$_3$ .... the capital recovery factors for a known (given) rate of interest, and amortization of total cost of each of the above items based on its average life. (such tables are readily available)

M = Annual cost for maintenance of highway including signs, lighting, etc. This must be adjusted to include all costs anticipated within the weighted average life, such as resurfacing work, a number of years hence.

Construction cost are broken down as used year to year during the period of construction. For maintenance of roads costs year-by-year are to be identified. During construction period there are no maintenance costs. During initial years after construction maintenance costs are low.

22.3.3 Road User Costs

Road user costs include vehicle operating costs, accident cost and time cost. Two types of costs a vehicle operator bears can be classified as (a) vehicle operating costs and (b) vehicle ownership costs.

Vehicle operating costs depend upon type of vehicle, speed, type of area in which vehicle is operated (urban, rural), type of the road/highway, quality of road surface, geometry of road (horizontal and vertical alignment), etc. These are the costs of operating the vehicle on the road such as cost of fuel, cost of engine oil, tyres, repair and maintenance of vehicles. They depend upon the length of the road, on which vehicle is driven. They are used in economic analysis and referred as operating costs. The unit values of operating costs are calculated for different type of vehicles, under different conditions of operation. Summary tables of such costs are developed so that they can be used directly in the economic analysis. These table are based on:

(i) Type of vehicle: Passenger cars, trucks, others vehicles, their type, age, H.P, make, etc. and weight/power ratio

(ii) Type of area: rural, urban

(iii) Type of highway: two lane, divided, pavement width

(iv) Type of operation: free, normal, restricted traffic volume and composition congestion, level of service.

(v) Running speed: likely values

(vi) Gradient: 0 to 3, 3 to 5, 5 to 7, 7 to 9 percent and vertical profiles

(vii) Type of surface: paved, loose, unsurfaced and riding quality

(viii) Alignment : curved, tangent and horizontal geometry

(ix) Number of junctions per kilometer.

Such realistic information should be collected through research and studies to be used in economic analysis and presented in form of charts and table.

The ownership costs do not depend upon running of the motor vehicles. They are the costs of owing the vehicle such as license fee, registration free, insurance fee, rent of garage/parking, etc. These are not included in cost-benefit analysis.

22.3.4 Other Costs

Major improvements of roads/highways increase noise, air pollution and other ecological impacts to neighbouring areas. To mitigate these adverse impacts, additional costs may be involved, e.g. noise barriers.

Displacement of people and their rehabilitation may also involve additional costs in some projects (locations) for widening and improving the existing roads.

22.4 CONSIDERATIONS OF BENEFITS

Improvement of roads/new roads bring several benefits to road users, non road user and to the community in many ways. Benefits to road users are due to the fact that operation costs for vehicles on new facility are lesser, journey is faster, safer and cheaper. Some of the direct benefits to road users are:

(i) Reduction in the cost of vehicle operation

(ii) Reduction in traffic accident and accident costs

(iii) Savings in travel time

(iv) Reduction in maintenance cost of vehicle.

These are those primary or tangible benefits which can be converted into money value. Non road user benefits of highway improvements on social life, include the following:

(i) Increase in land values, adjacent to roads

(ii) Better land uses and economic growth

(iii) Easy movement and social contacts, which improves health and education

(iv) Increased employment potential

(v) Reduction in noise from roads

(vi) Reduced fumes and air pollution by vehicles

(vii) Improved aesthetics

These are intangible or indirect benefits which are difficult to quantify.

From economic evaluation point of view it is easy to estimate changes in traffic volumes, speeds, savings in time, accidents reduction, etc. and translate these tangible benefits into monetary terms, but it is difficult to put money value to intangible benefits like comforts, convenience, aesthetics, increased amenity, and land values. The effects of intangible consequences are identifiable and can be analysed to offer assistance in decision making, but can not be easily priced in monetary terms. They are subjective and vary from project to project. Table 22.2 summarizes favourable and unfavourable consequences of highway improvements.

Table 22.2: Consequences of highway improvements

S.N.	Related to	Consequences
1.	Road users	• Vehicle operating costs reduced • Delays reduced • Traffic accidents reduced • Comfort and convenience increased • Travel distance and travel time saved • Disruption to traffic during construction
2.	Community/society	• Coordination of various transport modes • Transfer of trips between modes • Park and open spaces • Land uses and townscape • Utility services, health, educations • Population movements • Growth of urban area, CBD, etc. • Relocation of households and social facilities and welfare activities. • Changes in living conditions and environment • Cost of living, housing supply and demand • Relocation of places of employment • Noise pollution
3.	Economic factors	• Relocation of business and after effects • Disruption during construction • Land use and land values • Competition for business • Environmental impacts • Land severance effects • Construction, maintenance and operating budgets. • Traffic pricing • Overall economic growth

Disadvantages due to new or improved/widened roads at some locations may be increased noise, visual intrusion, vibrations, severance, air pollution, etc. to be considered for all traffic (generated, diverted, developed) on the new road.

22.4.1 Savings (Benefits) in Vehicle Operating Costs (VOC)

Vehicle operating costs are a major portion of the road transportation. Reduction in these costs by improvement of an existing road can easily be quantified as savings in monetary terms. These benefits should be counted for all the normal, diverted, generated and induced traffic.

Benefits to normal traffic and diverted traffic are evaluated by multiplying the savings (changes) in user costs, by the change in traffic, however for generated traffic-which is not certain, the benefits could be found by multiplying half the changes in user costs by the change in traffic, as a safety factor.

Components of vehicle operating costs (VOC) are the costs of

(i) fuel

(ii) lubricant

(iii) tyres

(iv) spare parts

(v) maintenance

(vi) depreciation

(vii) wages of crew

(viii) fixed costs-like overheads, administration, interest on borrowed capital, taxes, parking, insurance, etc.

Tables, charts, equations are developed for various components of vehicle operating costs (VOC), for different type of vehicles, for any country, based on research and investigations. For Indian conditions such information is available in India Road Congress (IRC) publication: IRC : SP 30 – 2009. Vehicle operating costs depend upon vehicle types, road characteristics, volume and speed of vehicles and environmental factors like weather, altitude, etc.

22.4.2 Reduction in Accident Costs

Road accidents not only cause suffering and misery, but waste huge economic resources. Reduction in road accidents is a positive benefit to road users and society . Accident benefits are of two types—those which can be assigned money value and others like pain and sufferings which can not be measured in money terms. Money value can be assigned to death, injury/property damage and medical expenses, etc.

Road improvements reduce/prevent the number of accidents and provide increased safety. Benefits are reduction of those costs which would have occurred, if accidents were not prevented by improvement of roads. Accident cost estimates are necessary in benefit-cost analysis for highway improvements. Accident occurrence on a road depends upon design of geometry of road and intersections, sight distance and traffic control measures, etc. Type of accidents, their frequency, type of vehicles involved, seriousness of accident, etc. all are considered in economic analysis.

Changes in accident frequency and number due to improvement of roads are estimated by two different methods:

(i) Using 'Before and After' improvement studies on the improved road which gives the indication of percentage decrease in accidents.

(ii) Estimation changes in accident occurrence by using the known data regarding the accidents on similar type of road in the area.

For assigning the cost, accidents are classified in different categories, e.g. fatal, non-fatal, injury and property (vehicle) damage.

Cost of an accident should include the summation of some or all of the following direct costs, depending upon the nature of the accident. These costs are calculated on the basis of the consequences of the events that occurred in an accident.

- Vehicle damage
- Public property damage
- Private property damage
- Output loss due to death or injury

- Medical
- Ambulance
- Police
- Legal
- Insurance administration
- Lost income
- Hospital
- Court
- Traffic delays
- Sick leave
- Lost use of vehicle
- Dependent's pension
- Worker's compensation
- Rehabilitation

Indirect costs are such as bereavement and sufferings, anxiety for those not directly involved, etc. to which money value cannot be assigned.

Tables 22.3 and 22.4 show the typical economic cost of different types of road accidents and quantum of vehicle damage due to accidents in India. These costs vary from city to city and year to year, depending upon economic status, vehicle types and several other factors.

Table 22.3: Economic cost for different type of accidents (IRC)

S.N.	Type of accident	Economic cost of accident (in RS)
1	Fatal	8, 64,350
2	Serious injury	3,91,800
3	Major injury	1,72,650
4	Minor injury	30,450

Table 22.4: Quantum of vehicle damage due to accidents (IRC)

S.N.	Type of accident	Economic cost of accident (in RS)
1	Cars	26,150
2	Two wheelers	6,650
3	Three wheelers	7,650
4	Buses	76,050
5	HCV	8,600
6	MAV	1,340,400

22.4.3 Savings in Travel Time

Improved roads (widening, constructing flyovers, bridges, subways, etc.) result in better speeds and reduce journey time significantly. Vehicle time savings, results in greater usage of the vehicle. Benefits accrue from increased use of the vehicle. For commercial vehicles savings in time may result in extra output, but for work trips and pleasure trips savings in time are difficult to convert in monetary terms. Person-hour and vehicle-hour saved are not easily convertible in money value for use in economic analysis, as it varies from person to person, vehicle to vehicle. There is however a portion of savings in the wages of drivers and other worker

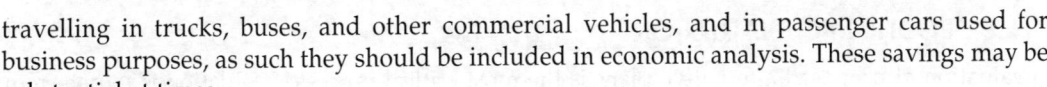

travelling in trucks, buses, and other commercial vehicles, and in passenger cars used for business purposes, as such they should be included in economic analysis. These savings may be substantial at times.

In general it is argued that such small increments of savings in time can not be put to productive use. However cumulative effects of these small savings of time from road improvements may be significant. Time savings though not directly convertible in money terms, should be reported separately to help decision makers. "Willingness to pay more" if time is saved concept is now developing in travelling, by which it may be possible to assign some value to savings in time, in economic analysis.

Congestion increases travel time and also the cost of fuel, lubricant, tyre, spare-part and maintenance. Value of time in wages of crew and passenger and commodity time saving can be studied through research and statistical investigations based on volume-capacity (V/C) ratios, for different road and traffic conditions. Congestion occurs severely during peak hours. Tables and equations are developed for Indian roads by Indian Road Congress and reported in IRC: SP: 30.

22.4.4 Increased Comfort and Convenience

Most people prefer comfort and convenience, by driving on a new and improved road, even though greater distances and time is involved. Driving smoothly without interference and stop-start operation, in a relaxed manner helps in conservation of health. It is therefore possible to place money value on comfort and convenience, as users should be willing to pay more for use of good and better roads. Assignment of money values for different degree of comfort and convenience is not easy. At present in economic studies costs of comfort and convenience are not included.

Many drivers are willing to use toll roads as they can reach their destination in lesser time avoiding congested road. This shows that drivers put money value to the comfort and convenience of driving provided by improved/new road.

22.4.5 Reduced Environmental Pollution

Road improvements are planned with a view to reduce environmental degradation. Due to lesser congestion, faster speeds with reduced stop-start operations, exhausts from vehicles are reduced. Benefits due to clearer air, and better environment, lesser dust pollution etc. should be considered as a benefit of improving an existing road. Some of these benefits are convertible in money value while others are difficult to quantify.

22.4.6 Increase in Land Values

Highway improvements increase land values as accessibility in the area is enhanced. This increases demand for various type of land uses to develop and people move from other areas. New/improved road attracts all types of land uses such as industrial, commercial, residential etc. resulting in competing demand and increased value of the land.

It is however not easy to predict on long-term basis how land values will increase and account them in highway economic analysis as different land uses are affected differently.

Highway improvements also improve access to cultural and historical places, parks, recreational facilities, sight scenes etc, however such benefits are non quantifiable. They are useful in decision making to undertake a road improvement project.

22.4.7 Other Benefits

Other benefits are those which can not be easily quantified or converted into monetary value. Improved roads attract tourists from different areas who spend for food, lodging and other activities bringing economic advantages and employment benefits to the society.

22.5 ECONOMIC EVALUATION

Evaluation of proposed alternative plans is the most critical process. Evaluation is to determine the best required out of several alternative plans, which are drawn with proper considerations to future requirements of traffic and stated objectives and standards. The objective of evaluation is to select and implement that alternative which provides maximum benefits for required financial outlay, fulfils traffic demands without adverse environmental impacts. In other words "which of the proposals is best". This is answered by finding whether the increment of investment between cheaper and more expensive alternative plan also appears attractive, by successively eliminating those plans that fail either the first or second of these tests. The best of the lot is selected.

Highway economic evaluation or highway project appraisal produces a single measure of effectiveness upon which plan selection is made. This common measure is monetary value. Various methods quantify all consequences (cost and benefits) of each proposed alternative plan in monetary terms over a selected period of time. It involves comparing the benefits that will accrue from the expenditure of funds for different alternative highway improvement plans.

22.5.1 Period for Economic Analysis

The period for economic analysis should be the one for which reliable forecast can be made. Normally 15 to 20 years period, after completion of the project is considered for highway projects. For expressway projects a period of up to 25 years is recommended.

22.5.2 Interest/Discount Rate on Capital Cost of Construction

Highway projects are financed by the governments as well as through private funds. On government funds interest on initial cost of construction is not included, in economic analysis, whereas for private funded projects (BOT) it is necessary to include interest charges.

The rate of interest (or discount rate) is one of the most important elements in economic analysis. The choice of the interest rate is governed by number of complex factors depending upon the future availability of finance and opportunity for its use.

The discount rate should always be more than the rate of borrowing or lending by the government or market rate of interest.

A rate of 10–12% is often selected when evaluating highway schemes, by different methods.

22.5.3 Compound Interest Calculations

Economic evaluation is based on the fact that money earns income over a period of time. Costs such as re-surfacing, annual cost of maintenance, and other costs are incurred at different times during the life of the road. For economic studies the money value for the different element/ alternatives, must be accumulated in a manner that makes them comparable, irrespective of the method used. Future benefits and costs are devaluated to the present time in order to determine their present worth. Following formulae are useful in economic analysis.

Compound interest

When interest is paid on the original investment as well as on the interest earned, the process is known as compound interest.

If an initial sum P, is invested at an interest rate i, for a period of N years. The sum of future money will be given by:

$$F = P(1 + i)^N \qquad \qquad \text{...(i)}$$

$$P = \frac{F}{(1 + i)^N} \qquad \qquad \text{...(ii)}$$

where   P = Present sum of money (cost of particular element of road improvement)
                 i = interest rate per interest period. (years)
                 N = number of interest period (years)
                 F = Sum of money at future date (N years)

Factor $(1 + i)^N$ is known as *Compound Amount Factor* (CAF)

Factor $\dfrac{1}{(1+i)^N}$ is known as *Present Worth Factor* (PWF)

CAF answers what future sum (F) will accumulate assuming a given amount of money (P), invested at an interest (i) for N years. PWF tells us what amount P should be invested today at interest (i) to recover a sum of (F) at the end of N years.

If it is required to find the receipt in a uniform series continuing for the coming N years, i.e. A

then $$A = F \cdot \dfrac{i}{(1+i)^N - 1} \qquad \qquad \dots \text{(iii)}$$

or $$F = \dfrac{(1+i)^N - 1}{i} \cdot A \qquad \qquad \dots \text{(iv)}$$

where:

        A = Receipt in a uniform series continuing for the coming N years.
        (The entire series equivalent to P at interest rate *i*)

        $\dfrac{1}{(1+i)^N - 1}$ is known as *Sinking Fund Factor* (SFF)

and $\dfrac{(1+i)^N - 1}{i}$ is known as *Compound Amount Factor of Uniform Series* (SCA)

The first cost of capital investment (P) may be converted into equivalent uniform annual costs (A), using interst i and period N, by combing formula (iii) and (i).

                $$A = F \cdot \dfrac{i}{(1+i)^N - 1} \qquad \qquad \dots \text{(iii)}$$

From (i) $$F = P(1 + i)^N$$

∴ $$A = P \cdot \left[\dfrac{i(1+i)^N}{(1+i)^N - 1} \right] \qquad \qquad \dots \text{(v)}$$

or $$P = \left[\dfrac{(1+i)^N - 1}{i(1+i)^N} \right] A \qquad \qquad \dots \text{(vi)}$$

$\dfrac{i(1+i)^N}{(1+i)^N - 1}$ is a constant for given value of C, and N. It is called *Capital Recovery Factor* (CRF)

and $\dfrac{(1+i)^N - 1}{i(1+i)^N}$ is called *Present Worth Factor of Uniform Series* (SPW)

To simplify the calculations, handy ready made tables are available for capital recovery factors, single payment present worth factors, etc. for various time periods and interest rates.

SFF indicates "how much money should be invested at the end of each year at interest rate i for N years to accumulate a stipulated sum of money".

CRF answers "If an amount of money is invested today at an interest i, what sum can be secured at the end of each year for N years such that the initial investment is just depleted".

Another way to consider CRF is "If a sum of money is borrowed today at an interest rate of i, for N years, what sum must be paid at the end of each year to pay off the loan in N years.

SFF is the reciprocal of CAF, and

CRF is the reciprocal of PWF.

i.e.
$$SFF = \frac{1}{CAF}$$

$$CRF = \frac{1}{PWF}$$

Tables 22.5 and 22.6 show typical value of CRF and PWF for different time periods and interest rates.

Table 22.5: Capital recovery factors (CRF) for various life and interest rates

Life	Interest Rate						
(yr)	0%	3%	6%	8%	10%	12%	15%
5	0.20000	0.21835	0.23740	0.25046	0.26380	0.27741	0.29832
10	0.10000	0.11723	0.13587	0.14903	0.16275	0.17698	0.19925
15	0.06667	0.08377	0.10296	0.11683	0.13147	0.14682	0.17102
20	0.05000	0.06722	0.08718	0.10185	0.11746	0.13388	0.15976
25	0.04000	0.05743	0.07823	0.09368	0.11017	0.12750	0.15470
30	0.03333	0.05102	0.07265	0.08883	0.10608	0.12414	0.15230
40	0.02500	0.04326	0.06646	0.08386	0.10226	0.12130	0.15056
50	0.02000	0.03887	0.06344	0.08174	0.10086	0.12042	0.15014
60	0.01667	0.03613	0.06188	0.08080	0.10033	–	–
80	0.01250	0.03311	0.06057	0.08017	0.10005	–	–
100	0.01000	0.03165	0.06018	0.08004	0.10001	–	–

Table 22.6: Single-payment present-worth factors (PWF) for various life and interest rates

Period	Interest Rate						
(yr)	0%	3%	6%	8%	10%	12%	15%
1	1.000	0.9709	0.9434	0.9259	0.9091	0.8929	0.8696
2	1.000	0.9426	0.8900	0.8573	0.8264	0.7972	0.7561
3	1.000	0.9151	0.8396	0.7938	0.7513	0.7118	0.6575
5	1.000	0.8626	0.7473	0.6806	0.6209	0.5674	0.4972
10	1.000	0.7441	0.5584	0.4632	0.3855	0.3220	0.2472
15	1.000	0.6419	0.4173	0.3152	0.2394	0.1827	0.1229
20	1.000	0.5537	0.3118	0.2145	0.1486	0.1037	0.0611
25	1.000	0.4776	0.2330	0.1460	0.0923	0.0588	0.0304
30	1.000	0.4120	0.1741	0.0994	0.0573	0.0334	0.0151
40	1.000	0.3066	0.0972	0.0460	0.0221	0.0107	0.0037

22.6 METHODS OF ECONOMIC EVALUATION

Having known costs and benefits of a scheme/project, a method is used to arrive at justification in economic terms. The following are the important methods:

(1) Annual cost method

(2) Short term rate of return method

(3) Internal rate of return method

(4) Nett present worth (NPW) method

(5) Benefit-cost ratio (BCR) method

22.6.1 Annual Cost Method

Annual cost of an element of capital investment is found by multiplying its first cost by appropriate capital recovery factor (CRF). The amount so found if charged at the end of each year for the assumed life of the facility (Road), will exactly repay the initial investment, with interest.

Total annual cost of a particular highway improvement is sum of all the annual costs of capital recovery, plus any other annual periodic expenditures plus annual maintenance cost and the road user costs. Annual costs are calculated for the existing road and for each of the proposal for improvement. That alternative which has the smallest 'Total Annual Cost' is considered best alternative.

Results of annual cost method are highly affected by the interest rate. Low interest rates favours those alternatives that combine large capital investments with low maintenance or road user costs, whereas high interest rates favour low capital investment with high maintenance or road user costs.

22.6.2 Short Term Rate of Return Method

In this method the rate of return is obtained by using the current traffic which will use the highway/road facility soon after opening. The first year savings in labour, vehicle time and operating costs over the road to be improved are calculated. The expected annual savings in accidents cost and average annual maintenance costs are related by the formula.

$$R = \frac{O + A - M}{C} \times 100$$

where

R = Percent rate of return (first year)

O = Savings in vehicle time and operating costs

M = Maintenance costs per year

A = Annual savings in cost of accidents

C = Capital cost of improvement

This simple method compares benefits accruing in the first year alone with capital cost of construction.

The alternative which would promise the highest rate of return on the investment will be the best choice. The priority of different alternatives can also be decided on the basis of first year rate of return.

22.6.3 Internal Rate of Return Method

This method produces a long term rate of return which is that rate which discounts all future benefits to equal the initial investment. It overcomes some of the limitation of the short term rate-of return method described above, by recognizing that future must be taken into account.

The capital cost of improvement is first determined as done before. Next the benefits (vehicle-time savings, operating cost and accident cost savings, etc.) occurring in each year in the future are estimated. The maintenance costs likely to occur during the same years are then worked out.

The rate of return is calculated by the following formula:

$$C_0 = \frac{B_1 - M_1}{1 + r} + \frac{B_2 - M_2}{(1 + r)^2} + \ldots \ldots \frac{B_N - M_N}{(1 + r)^N}$$

where

C_0 = Capital cost of the investment in the year zero

B = Value of the benefits which occur in a particular year

M = Maintenance costs incurred in that particular year.

N = Number of years into the future for which the rate of return is to be calculated.

r = Internal rate of return per year

In order that a project be economically feasible, the resulting 'Rate of Return' (r) should be more than the prevailing opportunity cost, i.e. rate of interest obtainable by investing the capital cost in the open market.

The above equation is solved by trial and error. Now a days computer softwares are developed for this purpose. This method is popular when lending of funds from agencies like World bank, IMF, etc is involved.

22.6.4 Net Present Worth (NPW) Method

In this method, the best solution is that alternative which shows the highest positive difference between the present worth of benefits and the present worth of costs. The procedure requires net present value of the improvement, to be calculated. This method is based on discounted cash flow.

Cost and benefits associated with an improvement alternative, over an extended period of time are calculated and discounted at a selected discount rate to give present value.

Selection of discounting rate is a complex/difficult question which depends upon future socio-economic opportunities and growth. As a guidance therefore this rate should not be less than the rate of borrowing or lending money by the government or the market interest. A rate of 12% is generally used in India.

The net present value is expressed algebraically as:

$$NPV_0 = (B_0 - C_0) + \frac{B_1 - C_1}{(1 + r)} + \frac{B_2 - C_2}{(1 + r)^2} + \ldots \ldots \frac{B_N - C_N}{(1 + r)^N}$$

where

NPV_0 = Net present value in the year o

B_1, B_2, B_3 = Value of the benefits which occur in year 1, 2, 3, ... N

$$C_1, C_2, C_3 = \text{Costs which occur in the year 1, 2, 3, ... N}$$
$$r = \text{Discount rate per annum (taken as 12\%)}$$
$$N = \text{Number of years for which the return is to be calculated.}$$

Benefits in this method are positive and costs are negative. The net present value is found every year. Any project with positive net present value is considered acceptable. If more than one projects are to be analysed the one giving highest NPV should be selected.

If the capital expenditure is to take place in more than one year, then it has to be discounted back to its present value before being used in this method.

A positive NPV indicates that the project is able to repay the interest charges implied in the selected discount rate and also yield some surplus. If the NPV is negative the project is unacceptable and should be postponed or rejected. This method is also based on the assumed rate of discount which is not easy to determine.

22.6.5 Benefit-cost Ratio (BCR) Method

Benefit-cost ratio method which is favoured by most of the highway planners/engineers, expresses comparative worth of projects (alternatives) by the ratio of annual benefits to annual costs. The benefit cost ratio is given by:

$$\text{BCR} = \frac{\text{Annual benefits from improvement}}{\text{Annual costs of improvement}} = \frac{R_0 - R_1}{H_1 - H_0}$$

where

$$\text{BCR} = \text{Benefit cost ratio}$$
$$R_0 = \text{Total annual road user cost for the basic condition or existing road}$$
$$R_1 = \text{Total annual road user cost for a proposed improvement}$$
$$H_0 = \text{Total annual highway cost for the basic condition or existing road}$$
$$H_1 = \text{Total annual highway cost for a proposed improvement.}$$

First benefit-cost ratio between each alternative improvement proposal and basic (existing) condition are calculated. Those plans which fail to reach minimum desired BCR are rejected. Then the benefit cost ratio (BCR), for each increment of added investment is computed, each plan being compared against the preceding acceptable plan. The alternative that reaches prescribed BCR (usually 1.0) on both total and increment of investment is considered acceptable on the basis of the assumed interest rate.

Benefit-cost ratio value equal to 1, indicates that annual savings in road-user costs are just equal to the cost of improvement, maintenance and operation of the highway.

Benefit-cost ratio value of less than 1, indicates that in the road user benefit sense, the basic condition is to be preferred over the proposed alternative of improvement, such an improvement scheme is not economically justified.

In the simplest form one alternative proposed for improvement is compared with basic condition, usually the existing highway or system. Similarly other alternatives can be compared with existing road to know which gives highest BCR.

Incremental comparing can be done by arranging alternatives in ascending order of total cost and then comparing alternative 1 and 2. If BCR is 1 alternative 2 is preferred and is then compared with alternative 3. The process is continued till the last alternative is reached and the best alternative is selected.

BCR method depends heavily on the discount interest rate adopted which depends on opportunity costs of capital which is uncertain. This method may not fit many projects like drainage and others where user benefits are either not affected or affected very little. If BCR difference is very small say 1.1 to 1.25 between two alternatives, it becomes difficult to appreciate such small differences.

Sometimes it is not easy to distinguish certain items as cost or benefits to be placed in denominator or numerator of the BCR, specially when alternatives of substantially different character are being compared.

22.7 LIMITATIONS OF ECONOMIC ANALYSIS

Some of the limitations of economic analysis are summarized below:

(i) Economic analysis of highways is focused on an individual project of improvement, yet improvement of one road affects the traffic network in the large area. Highway improvements brings benefits not only to local people living in the vicinity but also to road users who live far way outside the local area. It is important that the geographical cut off area should be specified for recipient of benefits. System-wide effects are not attempted.

(ii) Alternative investment decision in the public sector are not easy as there is no price system and consumer response which makes it difficult to place monetary value to intangible benefits

(iii) The goals and objectives of highway improvement programme and the evaluation process considers primarily the physical goals, while several social and economic goals, on which quality of urban life depends are ignored.

(iv) Comparative analysis of alternative plans depends upon the ability to reduce all costs and benefits to monetary terms. There are many non-quantifiable costs and benefits. Even for quantification value judgments are required which are subjective involving arbitrary assumptions regarding growth of traffic, service life, interest rate, etc.

(v) Discounting of cost and benefits over a period of time is done at an appropriate rate of interest. Deciding and correctness of this rate is not easy, as it depends upon future which is always uncertain

(vi) It is difficult to forecast future traffic accurately, specially for long periods of time 15 to 20 years.

(vii) Economic analysis of highways is focused on locations where there is already access by motor vehicles so that principal benefits in motor vehicle operating costs are calculated. Many roads in rural areas which are just tracks, passable in good weather only, such an economic analysis may not work, and a quite different approach for analysis and decision making is required.

(viii) Distributional problem associated with economic analysis is that those who gain benefits (e.g passing motorist) are different group of people from those who pay for the costs. Benefits may go to people considered relatively undeserving in the eyes of the community.

22.8 STEPS IN ECONOMIC EVALUATION

Economic analysis of highways is to assist in decision making whether or not to invest in proposal under study, and to decide how much funds should be spent considering various alternatives, and returns from them on the investment, consistent with safety and objectives.

Sequential steps involved in a common type of economic evaluation process are summarized below:

1. Study the project and define its scope.
2. Collect available economic data from various sources.
3. Conduct traffic field studies required for the project such as ADT, O-D speed and delay.
4. Select design life, discount rate or minimum IRR which shall be used for the project.
5. Investigate sources of material, labour and equipment, etc.
6. Conduct inventory of the existing road in terms of its structural, geometric and operational characteristics which will help in arriving at design and specifications
7. Estimate future traffic using different methods, to help in pavement design, geometric design, and road user costs.
8. Work out engineering design for the proposed alternatives which can be considered meeting the stated objectives.
9. Estimate the costs involved for the proposed alternatives.
10. Analyse traffic on the existing road and estimate all type of traffic on new/proposed road.
11. Estimate road user benefits from improved road.
12. Conduct economic analysis.

Finally it must be ascertained whether sufficient resources exist within the overall allocation of funds to permit the needed and economically justified project to be undertaken.

ILLUSTRATIVE EXAMPLES

Example 1: A road 60 kms long needs improvements, involving widening, re-surfacing, etc. Following are the details of the project.

(i) Cost of total improvements = 10 lacs per km.
(ii) Vehicle operation cost = Rs. 1.5 per km per vehicle on existing road
(iii) Vehicle operation cost = Rs. 1.0 per km per vehicle on improved road
(iv) Traffic volume = 2000 vehicles/day
(v) Cost of maintenance on the existing road is Rs. 6000/km which shall be Rs. 8000 per km on improved road.

Show by economic analysis if, this project is worth undertaking assuming analysis period of 20 years and interest rate of 10 percent.

Solution:

Capital recovery factor $(n = 20, i = 10)$ CRF $= \dfrac{i(1+i)^N}{(1+i)^N - 1}$ or taken from tables

$$= 0.11746$$

Costs:

Total cost of improvement $= 60 \times 10 =$ Rs 600 Lacs.

(i) Annual cost of improvement

Present value = Total cost × CRF

$$= 600 \times .11746$$

$$= \text{Rs } 70.476 \text{ lacs/year}$$

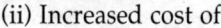

(ii) Increased cost of

maintenance per year $= 60 \times (8,000 - 6000)$

$$= 60 \times 2000 = 12000$$

$$= 1.2 \text{ lacs}$$

Total cost/year $= 70.476 + 1.2$

$$= 71.676 \text{ lacs}$$

Benefits:

Road user costs on existing road per year $= 365 \times 60 \times 2000 \times 1.5$

$$= \text{Rs. } 65700000$$

$$= \text{Rs. } 657 \text{ lacs}$$

Road user costs on improved road per year $= 365 \times 60 \times 2000 \times 1$

$$= \text{Rs. } 43800000$$

$$= \text{Rs. } 438 \text{ lacs}$$

Net benefits/year $= 657 - 438 = 219 \text{ lacs}$

$$\text{Benefit cost ratio} = \text{BCR} = \frac{219}{71.676} = 3.05$$

As benefits are more than 1 (equal to 3), the project is worth for investments.

Example 2: An existing road link of length 21 kms, has bad alignments, inadequate width and poor surface conditions.

There are three proposals to improve the situation.

Alt A: Increase width of road and re-surface the existing road (21 kms)

Alt B: Have a new alignment 17.5 m long

Alt C: Have an alternative alignment 16.7 m long

The traffic for the design period, on the average shall be 600 vehicles per day.

The life and estimated details of the cost are given below:

Element	Estimated useful life (years)	Cost in hundreds of rupees		
		proposal A	Proposal B	Proposal C
(a) Right of way	100	0	270	310
(b) Grading	50	150	290	330
(c) Structures	50	160	250	290
(d) Pavement	10	310	1550	1450

Rate of interest may be taken as 8% and annual maintenance cost of Rs: 500 per km. Other particulars are given below:

Alternative	Speed	Length	Vehicle operating cost at these speeds
A	50 km/h	21 km	Rs. .25 per vehicle km
B	65 km/h	17.5 km	Rs. .1 per vehicle km
C	65 km/h	16.7 km	Rs. .1 per vehicle km

Analyse the alternatives by benefit cost ratio method.

Solution:

Highway costs for various alternatives

Using the life in years and interest rate of 8%, the capital recovery factors (CRF) can be taken from the tables. Total highway cost = cost of elements + maintenance cost

Total highway costs for alt. A = $100 \times 0 + 150{,}00 \times .8174 + 160{,}00 \times .8174 + 310{,}00 \times .14903$
$$+ 500 \times 21$$
$$= \text{Rs. } \mathbf{17{,}652}$$

Total highway costs for alt. B = $270{,}00 \times .08004 + 290{,}00 \times .08174 + 250{,}00 \times .8174 + 1550{,}00$
$$\times .14903 + 500 \times 17.5$$
$$= \text{Rs. } \mathbf{38{,}190}$$

Total highway costs for alt. C = $310{,}00 \times .08004 + 330{,}00 \times .8174 + 290{,}00 \times .08174 +\_1450{,}00$
$$\times .14903 + 500 \times 16.7$$
$$= \mathbf{37.510}$$

Road user costs for alternative = $365 \times$ operation cost/km. $\times$ length $\times$ traffic

Alternative A = $365 \times .25 \times 21 \times 600$
$$= \mathbf{114975}$$

Alternative B = $365 \times .01 \times 17.5 \times 600$
$$= \mathbf{38325}$$

Alternative C = $365 \times .01 \times 16.7 \times 600$
$$= \mathbf{36573}$$

Comparing alternative B to A

$$\text{BCR} = \frac{114975 - 38325}{38190 - 17652}$$

$$= 3.71$$

Comparing alternative C to A

$$\text{BCR} = \frac{114975 - 36573}{37510 - 17652}$$

$$= \frac{78402}{19858}$$

$$= 3.94$$

As BCR for alternative C is higher than B. Alternative C is better.

Computer Applications in Highway (Transportation) Engineering

23.0 PREAMBLE

In the modern scientific and technological age, computers and softwares play an increasingly important role. Computer can be used to generate models of fundamental physical processes, which can be solved using numerical methods. The application of computers in highway (civil) engineering requires engineers to be intelligent and cogent user of computers in order to derive physically sound design and analysis.

There are many challenges and opportunities to utilise computer applications. Computer technology facilitates the assembly and analysis of information and permits a high level of automation. The broad areas of application of computer technology are :

23.01 GIS Applications

Global Information System is an effective means to generate constraint maps and identify potential corridors for highway routes. Extensive environmental data and range of corridor alternatives can be evaluated to select the preferred option. GIS provides a thorough and quantitative measurement of the environment impact associated with each highway corridor option. The GIS data base effectively documents all the factors used in the evaluation.

23.02 Engineering Applications

Planning in digital environment allows to assess the cost of highway alternatives early in the process. Use of computer software (CADD, Digital Terrain Modelling-DTM, etc.) permits the generation of alternatives considering alignment and grades. Impact and cost assessment (using GIS and highway design software tools), are performed to an extent not possible using manual methods.

23.03 Transportation Planning Process

Computers and information technology have been applied to highway and transportation planning for some time. Advanced applications are developed to support specific analysis

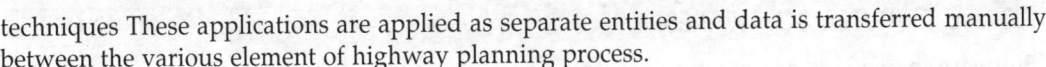

techniques These applications are applied as separate entities and data is transferred manually between the various element of highway planning process.

Total 'Digital Planning' is still a way off and requires refinement of the planning methodology techniques, mainly in terms of application integration. Use of computers and information technology allows for automation, digital documentation and a high level of traceability of the planning process.

23.04 Evaluation Process

A spread sheet evaluation model is the tool developed to assess corridor options. The model computes scores for each alternative based on a set criteria and a scoring system derived from public values. The model processes the information developed in previous stages. The automated process allows for interactive evaluation to support the decision making process.

23.05 Public Participation

Computer technology facilitates communication (newsletter, public forum, survey opinion, etc) using data-base management techniques in public participation and communication between highway planners, local residents and interested groups. Electronic publishing and presentation help in public participation.

23.06 Monitoring of Decay

Decay of an structure can be monitored, with suitable hardware support, which can take into account multicriterias. Probes like deflection gauges, can be implemented in the structure and their output can be fed into the computer. The designer can modify designs accordingly to any hazard in future.

23.07 Satellite Surveys

Satellite surveying is conducted for map preparation and highway alignment. Most of the work is done on computers including the camera in satellite is completely controlled by the computer.

The photographs taken by a satellite are converted into digital images and sent to earth. At the ground station, there is a computer, which interprets the data and provides the real image. The photograph taken can be used as a base map. Without the aid of satellite survey, it will take many weeks and huge manpower to do the work, and that too with many errors.

23.1 ADVANTAGES OF USING COMPUTERS IN HIGHWAY AND TRAFFIC ENGINEERING

- Elements of any process are integrated to streamline the flow of information.
- An effective data record management methodology is developed.
- The efficiency of work (automation, avoidance of duplication, interchangeability of data) is enhanced.
- Large number of data can be effectively managed and analysed.
- The multidisciplinary work activities can be integrated into a single approach.
- Digitilisation of documents, allows for traceability for further study.
- The decision making process becomes transparent.
- Using the digital tools, public participation becomes effective and easy.
- Using effective graphic material, expanded presentation capabilities are produced.
- Large number of alternatives can be analysed, to determine most cost effective solution considering various cross-section and alignment options.

23.2 COMPUTER SOFTWARES

Software, nowadays are mostly used to solve and suggest remedial measures on problems related to many aspects of transportation. Computer-aided Designs (CAD) are extensively used to give the clients an idea of how a facility will look like and how the traffic will move on this facility once implemented? Applications of this type of works include design of facilities like highway design, rail track design, airport and air strip design, and operation of facilities and traffic which may include a mix of vehicles or specific lanes like Bus Rapid Transit (BRT). It is not limited to only the above mentioned aspects but may also include computer-aided dispatch of taxi, metro bus and handicapped van, dispatch and control of bus, train, boat and airplane schedules, design of traffic signals and so on. The software can also be used to create traffic operation scenarios which can be used to examine the effect of change in a variable (defining scenario) on traffic operations and to come out with solutions. Such applications of the software can be used for both visualization and modeling of concerned aspects of transportation.

23.2.1 Classification of Software

Broad-classification of transportation related software can be done based on the intended purpose like planning, designing and operations. These may be on-line for the real-time control of processes or off-line for the pre or post implementation analysis. The classification is shown in Figure 23.1. An overview of these softwares is given in subsequent paragraphs.

23.3 PLANNING SOFTWARE

Planning software automates the four-step-process of trip generation, trip distribution, modal choice and trip assignment. A large number of inputs are usually required for these types of software, such as a full description of the network, the existing traffic and transit volumes and origin-destination (O-D) tables by zones. As depicted in Fig. 23.1 these can be Geographical Information System (GIS) based or non-GIS based.

Geographic Information Systems (GIS)

GIS is defined as "an information technology composed of hardware, software and data used to gather, store, edit, display and analyze the geographic information." Transportation systems have a strong geographical aspect, hence GIS is the very important part for transportation engineering providing various input data. Many Non-GIS transportation software provide linkages to these types of GIS software for the purpose of sharing data displaying the results of their analysis.

23.3.1 Non GIS Based Planning Software

The major advantage of all the planning software is that after inputting the data, complex analysis can be done and large number of alternative can be evaluated in a short time. At a grand scale these software permit the analysis of whole metropolitan areas and the evaluation of proposed new installations or alterations. By the end of the 1990s most of these software offered linkage to GIS software as well.

EMME/2

The major feature of this software package is the incorporation of multimodal equilibrium. It is used for both automobiles and transit related characteristics. This property does not only offer the ability to assess the impact of transit service on road networks, but also it aids in the identi-

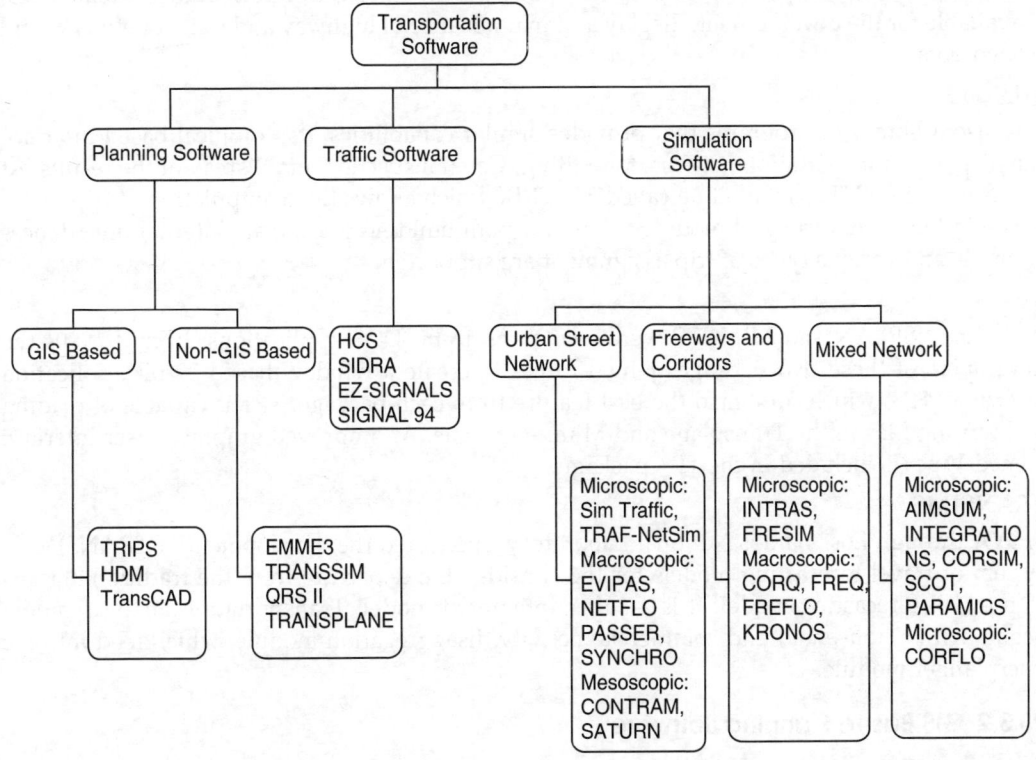

Fig. 23.1: Classification of transportation software

fication of more efficient routes for transit service. The inputs for this software require a network representation either in the form of coordinates or can be digitized directly from maps. Zone characteristics such as demand, socioeconomic variables and travel impedance are inputs. Traffic surveys, accident statistics, pavement characteristics and other custom information, can be incorporated with user defined attributes. The main output of EMME/2 is the overall network equilibrium assignment and the presentation of comprehensive results (most in graphic, interactive way).

QRS II

The Quick Response System was developed in the 1970s as a set of manual travel demand analysis techniques. In 1981 EHWA released a microcomputer version of the system as QRSI. It was capable of handling larger problem than those that could be analyzed by the earlier manual methods but it proved to be awkward to use. QRS II was subsequently developed and upgrades at the center for the Urban Transportation Studies of the University of Wisconsin at Milwaukee. It features a powerful interactive graphics, general network editor (GNE) which can be used to draw and quickly modify highway and transit networks on the computer screen and to display and plot network and the results of travel demand analysis. It is capable of performing "All-Or-Nothing" assignment with interactive capacity restrained conditions and through a feature that averages the results of successive interactions for a true equilibrium traffic assignment.

TRANSPLANE

TRANSPLANE Software is a tool box of more than 40 "Functions" that are grouped in the following categories, namely trip generation, model choice, networks, paths, loading, matrix utilities, reporting and plotting. A graphical package, the "Network Information System" (NIS) is available for the development, display and maintenance of highway and transit networks and related data.

MINUTP

It is also a library of programs that provides similar capabilities. Its graphical based interface can display a transportation network for editing, visual inspection and display of the results. At the heart of MINUTP is a module called MATRIX which allows the manipulation of trip tables and skin tables. A variety of modules allow for path building to estimate internal impedances by mode and various types of trip assignment are supported.

TP+

Both TRANSPLAN and MINUTP were written as 16-bit DOS applications. In mid 1990s the developers of these two packages joined forces to create a window-based 32-bit application known as TP+, which combined the best feature from each package. TP+ is capable of reading and writing files in the Transplane and Minute formats. An improved graphical user interface named Viper is included in the TP+ package.

TRANSIMS

FHWA through Los Alamos National Laboratory, sponsored the development of TRANSIMS a system of travel forecasting models for the considerable departure from the traditional, four-step travel forecasting model. It is a super integrated model. Transportation analysis model begins with a household and commercial activity disaggregation module, which feed into the interplanner module.

23.3.2 GIS Based Planning Software

Computerized methods are general and geographically well suited to the management and sharing of transportation related data. The quick and accurate determination of geographical positioning offered by the GPS can enhance the development of GIS applications, which can greatly enhance the task of data collection relating to real world objects as well as more advanced applications and automatic vehicle location, emergency vehicle dispatching, transit system schedule adherence analysis, travel time surveys and so on.

TRIPS

TRIPS is an interactive graphic multimodel urban transportation planning software and is based on viper system with model development tools such as application manager and scenario manager and GIS related features. It contains various modules such as demand modeling module, matrix estimation module, highway assignment module and public transport module. Its other important features are environmental analysis, cost benefit analysis, registration number matching and survey analysis.

HDM-4

HDM-4 can be used in a wide range of environments and is a powerful system for the analysis of road management alternatives, with different application tools such as project analysis, program analysis, strategy analysis and research and policy studies. There are four modules built in HDM-4 to calculate and analyze the input data in each application, which are:

1. Road Deterioration (RD)
2. Work Effects (WE)
3. Road User Effects (RUE)
4. Social and Environmental Effects (SEE)

All these modules depend on GIS system for getting the required data.

TRANSCAD

TRANSCAD is the best tool for working with transportation data by using the GIS functions to prepare, visualize, analyze and to use the application modules to solve routing, logistics and other transportation problems with greater ease and efficiency than any other product. TRANSCAD provides a comprehensive solution for many types of transportation applications, such as,

1. Network Analysis
2. Transportation Planning and Travel Demand Modeling
3. Vehicle Routing and Logistics
4. Districting and Location Modeling.

23.4 TRAFFIC SIMULATION SOFTWARE

Computer simulation is important for the analysis of freeway and urban street systems. Through simulation transportation, specialists can study the formation and dissipation of congestion on roadways, assess the impacts of control strategies and compare alternative geometric configurations.

Simulation models have different characteristics: Static, dynamic or stochastic, microscopic or macroscopic. These are presented below for urban areas and for freeway corridors.

23.4.1 For Urban Areas

Microscopic Simulation Software: SIM TRAFFIC, TRAF-NETSIM

Sim Traffic

It is a microscopic simulation and animation software that produces the animations which requires a lesser user effort. Its capabilities have been upgraded to include simple freeway modeling, which can be combined with an arterial network. The interaction between the two networks is limited because each freeway ramp is modeled as an unsignalized intersection requiring turning movement inputs.

NETSIM

Being a microscopic simulation model, NETSIM requires a considerable amount of inputs such as topology of the roadway network, characteristics of each roadway link, traffic control system, traffic demand, traffic composition, bus operations, etc. Outputs of NETSIM includes travel times, total and stopped delays, timing data, queue length, signal phase failures, vehicle occupancies, fuel consumption, pollutant emissions and so on.

Macroscopic Simulation Software: EVIPAS, NETFLO, PASSER, SYNCHRO, TRANSYT

EVIPAS

The Enhanced Value Interaction Process Actuated Signals (EVIPAS) software optimize actuated controller setting for isolated signalized intersections operation with a type of 170 controller on the bases of delay, fuel consumption, etc.

NETFLO

The NETwork Traffic FLOw simulation model can simulate the traffic flows at two levels: NETFLO I and NETFLO II. NETFLO I is a stochastic, event-based model. It moves each vehicle intermittently according to events and moves each vehicle as far downstream as possible in a single move. NETFLO II is a deterministic, interval based model. The traffic stream is represented in the form of movement specific statistical histograms.

PASSER

The Progression Analysis and Signal System Evaluation Routine (PASSER)-II 90 performs traffic signal optimization on a single arterial street based on band width maximization. PASSER II-90 and III-98 are separate software designed to perform signal optimization on dual-signal diamond interchanges. All PASSER software are developed by the Taxes Transportation Institution for the Taxes DOT.

SYNCHRO

SYNCHRO is a traffic signal software designed to generate optional signal timings (cycles, splits and offsets). A secondary product of the analysis is capacity and performance estimations similar to those in HCM. Its unique feature is the choice between Webster's delay formula (appears in HCM) or the percentile delay formula.

TRANSYT

Traffic Network Study Tool original version was developed by Dennis Robertson at the TRR Laboratories UK in 1967. There is no representation of individual vehicles in TRANSYT and all calculations are made on the basis of the average flow rates, turning movements and queues. TRANSYT can perform plain simulation which results in the performance of the existing network without any alternations.

Mesocopic Simulation Software: CONTRAM, SATURN

CONTRAM

The CONtinuous TRaffic Assignment Model is a traffic assignment and simulation model that treats a group of vehicles as a single entity (called a packet) CONTRAM determines time-varying link flows and route costs in terms of given time-varying route inflows in a dynamic setting. Traffic assignment equilibrium is achieved through interactions in which each packet is removed from the network and reassigned to a new minimum path.

SATURN

The Simulation and Assignment of Traffic in Urban Road Networks model is a traffic assignment model based on the incorporation of two phases. First is the simulation phase of intersection delays which determines flow-delay curves based on a given set of turning movements and feeds them to the II phase which is the assignment phase which uses these curves to determine route choice and updated turning movements. These interactions continue until the turning movements reach reasonably stable values.

23.4.2 For Freeways and Freeway Corridors

Microscopic Software: INTRAS, FRESIM, CARSIM, WEAVSIM, FREESIM

INTRAS

The INtegrated TRAffic Simulation model is a stochastic model, which uses vehicle-specific; time-stepping, detailed lane-changing and car-following logic to represent traffic flow and control of a freeway including the surrounding surface street network.

FRESIM

FRESIM is the revised reprogrammed model of INTRAS, associates structure design techniques and is more user-friendly. FRESIM can simulate complex freeway geometric such as inclusion of auxiliary lanes and variation of slopes, super-elevations and radius of curvature.

CARSIM

CARSIM is a new car-following model, based on INTRAS, offers additional realistic features and capabilities for simulation of car-following behaviour on freeways.

WEAVSIM

WEAVSIM is also based on INTRAS and was developed specifically for the study of the dynamics of traffic flow at weaving sections.

FREESIM

FREESIM is a stochastic model whose logic is based on a rational description of the behavior of the drivers in a freeways lane-close situation. A set of algorithms were established to simulate driver car-following/lane-changing behaviour in response to advance MUTCD warning signs.

Macroscopic Software: CORQ, FREQ, FREFLO, KRONOS

CORQ

The corridor Queuing model consists of a directional freeway, its ramps, major cross streets and any competing alternative surface streets. Traffic flows are approximated as fluids and travels times are calculated as simple step functions for both free flowing and congested conditions. A key element of CORQ is the dynamics assignment technique for allocating time-slice OD demands to a time dependent traffic network.

FREQ

FREQ is a deterministic simulation model for a directional freeway corridor. FREQ 10 is its new version contains an entry model (FREE10PE) for analyzing ramp metering and a freeway priority model (FREQ10PL) for analyzing HOV (High Occupancy Vehicle) facilities.

FREFLO

FREFLO simulates traffic flow on freeways using a formulation of aggregate variables based on suitably modified analogies of fluid flows.

KRONOS

KRONOS is a freeway simulation model that uses a simple continuous model to represent traffic flow. It has been applied for evaluating the effectiveness of different freeway design/ operational alternatives.

23.4.3 For Mixed Networks

Several composite, synthetic or fully integrated simulation models have been developed in this field since 1980s. Most of the models in this category are able to model complex networks in considerable detail. Consequently a common disadvantage is the extensive requirement for input data and calibrations.

Microscopic Software: AIMSUN, CORSIM, INTEGRATION, PARAMICS, SCOT, WATSIM

AIMSUM

The Advanced Interactive Microscopic Simulator for Urban and Non-urban Network is the analytic part of the GETRAM, whose input processor is TEDI, AIMSUM deals with all types of streets and accommodated traffic flows in either turning movement or OD matrix form. When turning movements are used, traffic is distributed over the network stochastically. When OD data are used, vehicles are assigned to specific routes. This sophisticated software has been of little use in the United States.

CORSIM

CORSIM is a COMbination of NETSIM and FRESIM. The model is capable of simultaneously simulating traffic operations on surface streets as well as on freeways in an integrated fashion. A window version of TSIS was released by the FHWA to provide an integrated, user-friendly,

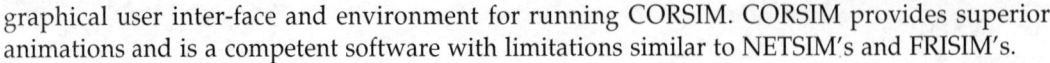

graphical user inter-face and environment for running CORSIM. CORSIM provides superior animations and is a competent software with limitations similar to NETSIM's and FRISIM's.

INTEGRATION

In INTEGRATION software only the aggregate speed-volume interactions of traffic are considered and not the details of a vehicles lane-changing and car-following behaviours. INTEGRATION is routing-based; that is a vehicle trip origin, destination and departures times are specified external to the model. Important additional features of INTEGRATION are user specified detector location for data collection as well as basic signal optimization at user defined intervals.

PARAMICS

PARAMICS is a network-level microsimulator having one base and three optional components. The modeler is the Core simulation component and processor, analyzer and programmer are optional components. PARAMICS's modeling approach makes it suitable for the assessment of detailed effects such as: response to signing, temporary lane closure (e.g. for incident and road constructions), loop detectors linked to variable speed signs and different speed limits for specific types of vehicles.

SCOT

The Simulation of Corridor Traffic is the earliest model for integrated networks. SCOT may also be classified as a mesoscopic model since it is the synthesis of UTCS-1, which is the precursor of NETSIM and DAFT, which is a mesoscopic simulation model for freeway ramps and arterials, in which vehicles are grouped into platoons.

WATSIM

The Wide Area Traffic SIMulation model is a stochastic, integrated network simulation model that extends the functionally of TRAF-NETSIM to incorporate both freeway and ramp operations with surface street traffic Its operational features include those in TRAF-NETSIM plus HOV configurations. It also includes an interface with a traffic assignment model.

Macroscopic Software: CORFLO

CORFLO

CORFLO is a combination of NETFLO I, NETFLO II and FREFLO models integrated within the TRAF/TSIS operation environment. Two important enhancements to CORFLO are the addition of new logic for user optional traffic assignment based on simulated link travel time and the introduction of capacity for enroute diversion modeling.

Other than the above-mentioned software some capacity related software are listed below.

23.5 TRAFFIC CAPACITY SOFTWARE

The capacity analysis software are computational models not the simulation models. All these software analyze three and four leg intersections. Only SIDRA can analyze intersections with five or more legs.

HCS

The Highway Capacity Software is a precise replication of the Highway Capacity Manual (HCM) on a personal computer platform. Most chapters of the manual including freeway or intersectionally analyst are included in the HCS. A full windows version of HCS became available before the turn of the century.

SIDRA

The software was developed by the Australian Road Research Board. Signalize and unsignalized intersection analysis is based on the HCM. SIDRA is one of only a few models

available that can analyze round about as well as unsignalized intersections. It can do the analysis for both left-and right-hand driving.

EZ-SIGNALS

EZ-SIGNALS was developed by Viggen corporation. It is a windows application of signalized intersection chapter of the HCM. It became available in 1991 and later was incorporated into the HCS.

HCM/Cinema

HCM/Cinema was developed by KID Associates. It is also a replication of HCM's signalized intersection chapter, with the additional option of executing NETSIM at the intersectional level, which produces a wealth of measure of effectiveness as well as animation.

SIGNAL 94

SIGNAL 94 was developed by strong concepts. It too replicates HCM's signalized intersection analysis along with signal timing optimization and extensive ability to analyze intersection geometry and control alternatives.

23.6 ROAD DESIGN SOFTWARE

CIRCLY

CIRCLY is the leading pavement design software for roads and highways. CIRCLY implements a rigorous flexible mechanistic pavement design methodology that incorporates state-of-the-art pavement material properties and performance models. CIRCLY is an integral component of the AUSTROADS Australian Pavement Design Guide.

CIRCLY offers benefits to all involved in designing, constructing, owning and using pavement. Designers benefit from the inclusion of automated functions and faster analyses. Material suppliers will make significant saving because the costs and dangers of over-specification or under-specification due to inaccurate information are significantly reduced. The facility owner benefits from more efficient use of materials and will reap long term benefits because changing traffic patterns on proposed designs can be considered and planned for.

AnalDeltaTessera

Tessera is a multifunctional road design and construction program. It is very easy to define the initial road layout, which can later be refined by using the program's advanced road planning functions. Tessera features a complete set of transition curves which can be freely combined to produce the actual corridor layout. The roadway surface is constantly recalculated and visible during the design stage.

The 3D environment is a native part of the Tessera CAD platform. It can be quickly and easily employed at any stage of the design.

Bentley MXROAD

Bentley MXROAD is an advanced, string-based modeling tool that enables the rapid and accurate design of all road types. With MXROAD V8i Edition, one can quickly create design alternatives to achieve the "Ideal" road system. Upon selection of the final design alternative, MXROAD automates much of the design detailing process, saving the users' time and money. MXROAD includes features like Data Interoperability, Alignment Design, Road Design, Interactive Design Editing, Google Earth Integration, and Earthworks Design.

ARD for Road Designers

Advanced Road Designs (ARD) has been developed to provide Australian based road design and outputs directly inside AutoCAD Civil 3D. The philosophy of combining template and string based design, as well as automating common design elements such as intersections and roundabouts, provides a familiar and complete set of tools for rapid creation and output of road subdivision, reconstruction, rural, highway and other design projects. The dynamic interaction between elements allow the mixing of cross section and string elements, and the ability to build own models by using any collection of strings and codes for designed road elements. With ARD, one can completely automate road subdivision design processes using intelligent objects to handle common design elements such as intersections, kerb returns, roundabouts and knuckles. Advanced cross section design tools allow managing cross section easily and the dynamic cross section viewer allows reviewing the edits as they are made.

23.7 SIGNAL OR INTERSECTION DESIGN SOFTWARE

Transmodeler

Transmodeler is a powerful and versatile traffic simulation package applicable to a wide array of traffic planning and modeling tasks. Transmodeler can simulate all kinds of road networks, from freeways to downtown areas, and can analyze wide area multimodal networks in great detail and with high fidelity. You can model and visualize the behavior of complex traffic systems in a 2-dimensional or 3-dimensional GIS environment to illustrate and evaluate traffic flow dynamics, traffic signal and ITS operations, and overall network performance.

Dynamic Network Assignment-Simulation Model for Advanced Roadway Telematics

DYNSMART-P, released by the Federal Highway Administration through McTrans, provides many new features and enhancements. These include:

- Convenient import of network and demand data from other planning models.
- Click-and-drag network/control creation and editing interface based on background imagery.
- Easy conversion of baseline static OD matrix to time-dependent OD matrices.
- Generation of default movements and signal control data for fast deployment.
- Improved loading and display speed for large-scale network datasets.
- Calculation and display of link and network-level toll revenue for HOT or tolling applications.
- Easy redistribution of user classes within vehicle and path files. This allows for flexible implementation of various planning scenarios.
- Capability of modeling large networks.

SOAP

Signal Operation Analysis Package (SOAP) can be used to determine signal timing plans for pre-timed controllers and limited capabilities for actuated controllers. Although SOAP is still used by several agencies, it has been largely overshadowed by more advanced and broader programs. Its main appeal is its inclusion in the Wizard of Helpful Intersection Control Hints (WHICH) package, so it serves many users as a timing plan optimizer for use in conjunction with WHICH-supported tools. The original version (still separately available) can perform multiperiod analysis, but the WHICH version is a single-period tool, since all other programs in the WHICH suite are limited to single periods. The program optimizes cycle lengths and splits based delay and stops.

SYNCHRO

Synchro is a macroscopic traffic signal timing tool that can be used to optimize timing parameters for isolated intersections, generate coordinated traffic signal timing plans for arteries and

networks, and also develop time-space and platoon dispersion diagrams for interactive fine-tuning. SYNCHRO can analyze fully actuated coordinated signal systems. The program has no limitations on the number of links and nodes. It can analyze multi-legged signalized intersections with up to six approaches per intersection. SYNCHRO does not, however, analyze sign-controlled intersections. SYNCHRO is designed to optimize cycle lengths, splits, offsets, and phase orders. The program also optimizes multiple cycle lengths and performs coordination analysis. When performing coordination analysis, SYNCHRO determines which intersections should be coordinated and those that should run free. The decision process is based on an analysis of each pair of adjacent intersections to determine the "Coordinability Factor" for the links between them. Using SYNCHRO the user can optimize the entire network or groups of arteries and intersections in a single run and determine the control boundaries of the different arterial groups, based on the program's selection of the cycle lengths.

PASSER IV

Using hourly traffic volumes, user-defined saturation flow rates and optional minimum green times, PASSER IV can optimize the progression bands for main arteries as well as coordinated crossing arteries by computing the optimum cycle length, splits, phase sequences and subsequently adjusting the offsets for a maximum of 20 arteries and 35 intersections.

The user interface for PASSER IV is very friendly. Its features include file management functions, context-sensitive help, a powerful output view/print capability, ability to display a network map and three data entry/edit capabilities. The basic data structure consists of a set of fully labeled screens arranged in hierarchical order. Data can be entered using the hierarchical data entry process, by directly working with the input data file using a built-in text editor, or by choosing entities from the network map display. The user interface also allows the user to run TRANSYT-7F for performing bandwidth-constrained disutility optimization. The inputs include link lengths, saturation flow rates, traffic volumes, average travel speeds, minimum green splits and the cycle length range.

TSDWIN

TSDWIN is a Windows-based graphical tool, designed to assist analysts responsible for fine tuning signal timing plans. The purpose of the program is to provide a quick and easy method to achieve graphical representation of time-space diagrams for either a single artery or a group of arteries, based on existing or proposed signal timing data (cycle length, splits, offsets). TSDWIN organize, intersection into arteries and arterial groups. The program also allows the user to select a double-cycle option for any intersection. Data inputs required for TSDWIN include spacing and travel speeds between intersections, cycle length, splits, and offsets for all intersections. Traffic demands are not required. The program's outputs include graphic displays of the green band and flows. The green band is colour-coded and measured in seconds. A green band for both directions of traffic movements is shown if one can be calculated, based on the timing data entered by the user. If a continuous green band cannot be calculated, the link-to-link green band is presented. This allows the user to evaluate how offsets might be adjusted to achieve a continuous green band. The directional flow displays show the calculated green band, if one exists, together with a yellow band and red band. The yellow band indicates vehicles outside the green band, but that will clear the next intersection, and the red band represents vehicles that will not clear the next intersection. TSDWIN allows the user to vary the speeds between the intersections and determine the associated impact on existing or proposed progression bands.

WHICH

WHICH (Wizard of Helpful Intersection Control Hints) is an integrator tool that integrates traffic control design parameters for isolated intersections and arteries into one common data set. By doing so, the user can run several programs without getting into their data editors. It interacts with several component programs such as SOAP, HCS, TRAF-NETSIM and several other non-signalized programs. WHICH can also be linked with the AAP. Users should note that WHICH has the capability to run TRAF-NETSIM only for isolated intersections. TRAF-NETSIM, and HCS are not distributed with WHICH.

23.8 OTHER SOFTWARES

RONET

It is designed to assess the current characteristics of road network and their future performance depending on different levels of interventions to the network. It was used for sub- Sahara Africa Transport Project.

ROCKS

Rocks provides an international knowledge system of road work unit cost for road preservation and development activities. It is an institutional memory (data base) that collects historical data on road works. Costs per kilometer or per Km^2 that could ultimately improve the reliability of new cost estimates and mitigate the risk generated by cost over runs.

RED

RED aims at improving the decision making process for the development and maintenance of low volume roads. The model performs an economic evaluation of road investment option using the consumer surplus approach and is customised to the characteristics of low volume roads.

Red computes benefits for normal, generated, induced and diverted traffic and takes into account changes in road length, conditions on the dry and wet seasons, geometry, surface type and accident rates.

HDM-III

HDM-III is a computer programme for analysis the total transport cost of alternative road improvement and maintenance strategies through life-cycle economic evaluation. The programme provides detail modelling of pavement deterioration and maintenance effects and calculates annual costs of road construction, maintenance, vehicle operation and travel time etc. needed to perform economic evaluation of the alternatives being considered.

HDM-VOC

HDM-VOC is a program that estimates vehicle operating costs using the HDM-III relationship, for ten vehicle types as a function of the vehicle characteristics, vehicle utilisation, vehicle unit costs and the road characteristics.

AUTO ROADS

AUTO-ROADS is a software, which automates and precise the steps involved in road design starting from surveys to project report preparation of different types of roads. It is capable of identifying and processing various formates of data collected from modern or conventional instruments. It can identify survey co-ordinates and prepare plan of the road and fix the horizontal and vertical alignment.

HDM-PRD

HDM-PRD is a program that estimates the road deterioration of paved roads, using HDM-III relationship, as a function of the road characteristics, traffic, loading, and maintenance operations. The program is suited for road deterioration prediction, remaining life prediction, sensitivity analysis, road design analysis, road network deterioration, etc.

EBM-32

EBM-32, is an enhanced version of the Expenditure Model for the Highway Sector (EBM-HS) released by the World Bank in 1995. EBM-32 model is an analytical tool for optimising multiyear programme of expenditures under budgetary constraints and it is customised to analyse highway investments, (maintenance, construction/re-construction projects), that are subjected to budgetary constraints: Capital and Re-current expenditures.

RUC

RUC estimates road user charges required to ensure that for a particular country the costs of operating and maintaining all roads are fully funded, and that each vehicle class covers its variable costs. The model is an Excel Work that :

   (i) estimates annualised maintenance costs needed to maintain a stable road network.
  (ii) define countrywide annual recurrent expenditure, annual investment needs, and source of financing.
 (iii) estimates road user revenues from annual fees, fuel levies, tolls, etc.
  (iv) analyses the allocation of road user revenues and optimises road user charges.
   (v) computes externalities and summary macro indicators.

FLEXPAVE

FLEXPAVE is developed by ASHTO (American Association of Highway and Transportation Officials), for design of flexible pavements structures. The system evaluates required thickness for initial pavement design and the subsequent overlays. A sensitivity analysis allows to investigate the required precision for design inputs.

CARROLS

CARROLS is an optimisation model which selects well defined option from a near-infinity of alternative routes between two points on the basis of cost. The use has the choice of using construction, vehicle operation or maintenance costs, either singly or in various combinations of the three.

AUTO-CIVIL

AUTO-CIVIL uses auto-CAD in Civil Engineering field. The usual areas of application are :
 • Digital terrain modelling
 • Roadway design
 • Drainage/hydraulics.

The digital terrain modelling can be done with Auto-Counter and Auto-DTM in which generation of the contour map, point interpolation, 3D residual cut and mapping, and even volume calculations can be done.

The highway design can be done with great ease with Auto-roads which is developed only for this purpose.

STAAD

STAAD is a popular structural software product for 3D model generation, analysis, and multi-material designs. The software is fully compatable with all window operating systems.

ANSYS

ANSYS programme has many finite-element analysis capabilities, ranging from a simple linear, static analysis to a complex, non-linear, transient dynamic analysis. ANSYS provides specific procedures to perform analysis for different engineering disciplines.

Very rapid progress have been made in the last few years in development of engineering software that are very efficient in design process, economic analysis and delivering engineering solutions efficiently and promptly, thus saving time and money.

With recent advances in computer speed, data storage and process capacity, animation software, etc. the engineers are able to optimize the material utilisations, and construction cost, earthquake resistance of structures, and other components of a highway. Computer data handling and analysis capacity have increased manifolds and they are used in design, optimisation and as an innovation tool. Information processing and calculations are done by computers at a very high speed. It can execute million of instructions in a fraction of a second. A single engineer with software knowledge can do large computations, in relatively lesser time.

Typical Questions for Practice

A. Theoretical and Numerical

Q.1. "Transportation plays an important role in development of a Nation". Justify this statement, with specific reference to highway transportation.

Q.2. How transportation contributes to the development of a city?

Q.3. Compare highway as a mode of transportation with air and rail transportations. What are special merits of road transportation.

Q.4. What do you understand by coordination of transportation, explain giving example? Also mention advantages and disadvantages of coordination.

Q.5. Mention the salient features of Nagpur Road Development Plan. To what extent they were met?

Q.6. What were the objective of road development plan 1981–2001? Explain the rationale for deciding road lengths to be achieved.

Q.7. What are the functions of National Highway Authority of India (NHAI). Mention the major projects undertaken by NHAI, for National Highway Development in India.

Q.8. The present average daily traffic on a road is 15,000 vehicles. Studies for future indicate normal traffic growth of 60%. The generated traffic is likely to be 15% and development traffic is estimated to be 25%, of present traffic in 20 year life period of the road. If the directional distribution is 60–40 on this road and the design hourly volume is 15% of ADT, calculate the number of lanes required in one direction assuming capacity of a lane is 1000 VPh.

Q.9. As a highway engineer, you are required to prepare a 'Master Plan' for future development of highways in a country.

(a) What plan period will you select? Give reasons for your answer.

(b) Set out the objectives of the plan.

(c) Name different studies and data you would require for planning.

(d) Suggest a method for phasing of the road development programme.

Q.10. It is proposed to construct a new two lane bituminous road connecting two industrial towns. Mention the factors you will consider for selecting the best alignment for the road. What surveys will be required.

Q.11. Explain how would you develop a suitable alignment for a highway to minimise the total cost of construction, maintenance and vehicle operation.

Q.12. State the various factors you consider necessary for fixing the alignment of a highway connecting two industrial towns. What are the requirements of an ideal alignment?

Q.13. Explain how roads are classified? Name various elements of highway cross-section.

Q.14. Write short notes on:

(i) Piggyback System

(ii) Central Road Fund

(iii) Pradhan Mantri Gram Sadak Yojna (PMGSY)

(iv) NTPC

Q.15. How transportation technologies are classified? What are special merits of highway transportation?

Q.16. Describe with a neat sketch the penetration test for determining hardness of bitumen.

Q.17. Describe the basic tests that are carried out to establish the suitability of stone aggregates to be used in road construction.

Q.18. What are the control tests necessary for judging bitumen suitability for road projects?

Q.19. Explain the procedure of determining modulus of subgrade reaction 'K', using plate bearing test.

Q.20. Name the various tests carried out on bitumen mentioning their uses.

Q.21. What are the requirements of a good bitumen mix? Describe briefly the Marshall method of mix design.

Q.22. Explain the following terms:

(i) Viscosity

(ii) Penetration grade bitumen

(iii) Cut-back

(iv) Emulsion

(v) Modified bitumen

(vi) Tar

Q.23. What is gradient of a road? How is it expressed? Explain different types of gradient.

Q.24. Design a suitable curve for an alignment having rising grade of 1 in 75 followed by a rising gradient of 1 in 80. Draw the typical sketch.

Q.25. Describe the standards you would adopt for the geometric design of a 2 lane road.

Q.26. Derive an expression for calculating the superelevation on highway curves, in terms of speed of vehicles and radius of the curve.

Q.27. Explain the need for pavement widening, on curves. Describe how you would obtain such widening on circular curves.

Q.28. Calculate the design speed of a vehicle on a horizontal curve having radius of 90 m. The permissible superelevation is 7%. Assume the coefficient of friction as 0.15.

Q.29. What are the purposes of transition curves in horizontal alignment?

Q.30. The design speed of the highway is 100 km/h Calculate the stopping sight distance assuming reaction time of drivers as 2.5 seconds and the coefficient of friction as 0.26, in following cases:

    (i) When highway is on level ground

    (ii) When highway descends a slope of 5%

    (iii) Calculate absolute minimum radius of curve using equilibrium superelevation of 1 in 10

    (iv) How much widening shall be required on this curve? (Wheel base = 7 m)

Q.31. Derive the formula for calculating overtaking sight distance on a 2 lane—2 way road. What additional considerations are suggested by AASHTO.

Q.32. The speed of overtaking and overtaken vehicles are 70 and 40 km/hr respectively, on a 2 lane road. If the acceleration of the overtaking vehicle is 3.5 km/hr/sec. Calculate

    (i) Safe overtaking sight distance for

        (a) one-way traffic

        (b) Two-way traffic

    (ii) Draw a sketch showing the overtaking zone and position of signs.

Q.33. Calculate the following geometric design features of a 2 lane highway having design speed of 100 km/hr.

    (a) Passing (overtaking) sight distance

    (b) Absolute minimum radius of the curve

    (c) Extra widening and width of the road on above curve. Make suitable assumptions.

    (d) Theoretical capacity of the lane.

Q.34. A falling gradient of 1 in 25 meets a rising gradient of 1 in 50. Design length of the valley curve using following data:

Design speed of highway—70 km/hr

Stopping sight distance—150 m.

Q.35. Define following terms used in highway engineering.

    (i) Auxiliary lanes

    (ii) Articulated vehicle

    (iii) Boulevard

    (iv) Causeways

    (v) Candela

    (vi) Dowel

    (vii) PERT

    (viii) Para transit

    (ix) Traffic calming

    (x) Weaving of traffic

Q.36. Explain the PIEV theory of human response.

Q.37. A driver travelling at 50 km/hr behind a car decides to overtake it and presses the accelerator. Assume that the accelerating behaviour of vehicle is given by the equation:

$$\frac{dV}{dt} = 1.22 - 0.015 \, V$$

where $\qquad V =$ speed in m/s

$t =$ time in seconds.

Calculate:

(i) The rate at which the vehicle shall be accelerating after 4 seconds.

(ii) The time which takes the vehicle to reach a speed of 120 km per hour.

(iii) The maximum speed which vehicle can attain.

(iv) The possible maximum rate of a acceleration for the vehicle.

Q.38. Define 'Design Speed'. On what factors it depends? The speed data on a road is given in table below:

Speed Range (km/hr)	Number of vehicles
10–14.9	3
15–19.9	10
20–24.9	21
25–29.9	31
30–34.9	54
35–39.9	43
40–44.9	21
45–49.9	10
50–54.9	5
55–55.9	2

Using the above data calculate:

(a) Speed to be used in (a) geometric design and (b) traffic regulations.

(b) Capacity of one lane of the road. Average length of vehicle may be taken as 6 m and reaction time of drivers as 1.5 sec.

Q.39. Assume linear relationship exists between speed and density. If on a length of highway it is noted that the free speed is 80 km/hr and the jam density is 72 vehicles/km.

(a) Determine the speed at maximum flow and the maximum flow expected on this section of the highway.

(b) Suppose that flow is 75% of capacity, what are the possible flow speeds? Also calculate the distance headway in each case.

Q.40. Assuming linear relationship between speed-density show that:

(a) $u = u_f \left(1 - \dfrac{k}{k_j} \right)$ 　　　　　　(b) $q = u_f \left(k - \dfrac{k^2}{k_j} \right)$

(c) $q = k_j \left(u - \dfrac{u^2}{u_f} \right)$

where

$\qquad u =$ speed

$\qquad k =$ density

$\qquad q =$ flow

$\qquad u_f =$ free speed

$\qquad k_j =$ jam density

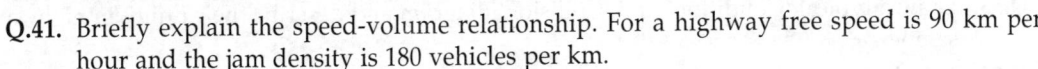

Q.41. Briefly explain the speed-volume relationship. For a highway free speed is 90 km per hour and the jam density is 180 vehicles per km.

    (i) Determine speed-volume relationship.

    (ii) Calculate the speed when density is 60 V.P.K.

    (iii) Determine time and space headway at 80 km/hr speed.

Q.42. What is volumetric count? What are the uses of Traffic Volume data?

Q43. What is a random sample? How would you obtain random sample in actual surveys?

Q.44. Write the purposes of conducting origin destination survey. Which method of O-D survey is best suited for India? Give reasons for your answer.

Q.45. Describe floating car method of travel time and flow studies.

Q.46. Differentiate between the following:

    (i) Space mean speed and time mean speed.

    (ii) Median speed and modal speed.

    (iii) Condition diagram and collision diagram.

    (iv) Fixed delay and operational delay.

    (v) Attracted, generated and developed traffic.

    (vi) Design speed and running speed.

Q.47. What are the various design elements in a roundabout (rotary). Give typical minimum requirement of these elements.

Q.48. Discuss advantages and disadvantages of a rotary intersection as compared with those of a signal controlled intersection.

Q.49. For designing an urban intersection, explain with the help of neat sketches the various principles you will keep in view.

Q.50. What are the advantages of using traffic signals? Explain the significance of "Yellow Period" in traffic signal design. How it is decided?

Q.51. A four legged intersection on level ground and with normal environment has peak hourly volumes (p.c.u./hr) as shown below.

North			South			East			West		
ST	LT	RT	ST	LT	RT	ST	LT	RT	ST	LT	RT
550	50	100	370	30	300	430	40	150	365	35	100

The north-south approaches are 10.5 m wide and the east-west approaches are 9 m wide. The average turning radius from all approaches is to be taken as 30 m. A three phase signal system is to be designed with intergreen periods of 7 secs.

(a) Determine the optimum cycle time

(b) Determine the effective and actual green times for each phase.

(c) Show in a neat timing diagram all signal indications.

Make and state explicitly any assumptions you need to make.

Q.52. On right angled crossings of two urban roads A and B, pre timed traffic signals with pedestrian signals are to be provided, using the following data:

	Road A	Road B
Width of road (m)	21	14
Critical volume (VPh)	400	200
Approach speed (KPh)	85	70

Design the traffic signal.

Q.53. Draw a typical 'Condition' and 'Collision' diagram and describe their utility in highway safety.

Q.54. Mention advantage and disadvantages of the following

(i) Rotary intersections

(ii) One-way streets

(iii) Pedestrian subways

(iv) On-road parking

Q.55. What are joint filler materials? Explain properties of satisfactory filler material.

Q.56. Discuss the factors that affect the properties of cement stabilised soil and lime-stabilised soil, used in road-base construction.

Q.57. Discuss the principal difference between penetration grade, cut-back and emulsion bitumen. Mention typical applications in highway construction for each type.

Q.58. How stability, flexibility and durability requirements of a bituminous mix are affected by following:

(i) bitumen binder

(ii) binder type

(iii) aggregate shape and surface texture

(iv) aggregate type and grading

Q.59. Describe in detail, the basic properties and functions of the following in a bituminous mix:

(a) Coarse aggregates

(b) Fine aggregates

(c) Filler material and

(d) Binder

Q.60. For a bituminous mix following grading of the aggregates is to be used. The specific gravity of bitumen is 1.04 and the Compacted Density of Mixed Aggregates (CDMA) is 2.20.

Sieve Size	Percent Passing
20 mm	100
14 mm	95
10 mm	80
6 mm	60
2 mm	45
1 mm	35
0.425 mm	20
0.075 mm	8

Specific gravity of coarse, fine aggregates and filler material is 2.70, 2.75 and 2.75 respectively. Calculate

(i) the specific gravity of mixed aggregates (SGMA)

(ii) the optimum bitumen content for 4% voids in compacted mix.

(iii) percentage of voids in mixed aggregates (VMA)

(iv) specific gravity of mix, based on value of the optimum bitumen content.

Q.61. Describe how Boussinesq and Burmister's theories can be applied in the evaluation of stresses and deformation in flexible pavements. What are limitation of these theories in design of flexible pavements?

Q.62. Outline the IRC procedure of designing flexible pavements and the selection of sub-base, base and surfacing thickness.

Q.63. Drive the expression for stresses due to temperature changes in a concrete slab of a rigid pavement in which no cracking is allowed.

Q.64. Compare the functions of a sub-base in a flexible pavement with those of a sub-base in a rigid pavement.

Q.65. On a subgrade, a pressure of 3.125 N/cm^2 produces a deflection of 0.125 cms. On this subgrade a cement concrete pavement of 15 cm thickness is constructed. Taking contact pressure of 90 N/cm^2, E = 2.8 × 10^6 N/cm^2, poisson ratio of 0.15 and wheel load of 40500 N. Calculate

(i) The radius of relative stiffness

(ii) The maximum wheel load stresses for interior, corner and edge loading using Westergaard analysis.

Q.66. Describe procedure for conducting a CBR test in laboratory. Why correction to load-penetration curve is required and how it is done?

Q.67. Define the term equivalent standard axles with a sketch showing load distribution through a pavement for twin wheel assembly.

Q.68. Following are the test results obtained by a CBR test on the subgrade:

Penetration (mm)	Load (N)	Penetration (mm)	Load (N)
0.00	00	3.00	580
0.50	50	4.00	700
1.00	170	5.00	780
1.50	290	7.50	920
2.00	420	10.0	102
2.50	500	12.50	108

Design the pavement structure using IRC method of designing flexible pavements. The traffic survey revealed that the present ADT of commercial vehicles is 1200. The annual rate of growth of traffic is found to be 8%. The life of the pavement may be taken as 10 years and the construction period of road is 3 years.

Draw a neat sketch showing the different components of the designed pavement.

Q.69. Describe the Benkelman method for measuring the deflection, for evaluation of roads.

Q.70. What are the major types of distress in flexible pavements? How you would correct those defects.

Q.71. Describe two methods of assessing the structural suitability or performance of an existing flexible pavement.

Q.72. Describe the procedure you would adopt to determine the thickness of a flexible overlay for strengthening an existing pavement.

Q.73. Why grading and regravelling are carried out on earth and gravel roads?

Q.74. Explain why it is necessary to maintain a road? How preventive maintenance differs from periodic maintenance?

Q.75. What are the main causes of defects in concrete pavements. How pumping can be controlled?

Q.76. Why maintenance management system is necessary for Indian roads? What are the components of the system?

Q.77. Write short notes on:

  (i) HDM model

  (ii) The Lacroix deflectograph

  (iii) Mud pumping in concrete road

  (iv) Fatigue failure of pavements

  (v) Maintenance management system

Q.78. Differentiate between structural and functional failure of pavements. State the factors which would influence pavement distress in both rigid and flexible pavements.

Q.79. Explain the terms pumping and blowing as applied to rigid (concrete) pavement.

Q.80. Explain what techniques you would use to evaluate the pavement strength and the physical condition of the road.

Q.81. Discuss the importance of both surface and sub-surface drainage for highways.

Q.82. With the aid of sketches show potential sources of subsurface moisture under a road pavement. How this moisture can be controlled?

Q.83. Explain briefly the principal requirements of a good highway drainage system.

Q.84. Describe with help of sketches, the methods used to control capillary rise of water and seepage flow for a pavement.

Q.85. What is the necessity of cross-drainage? What are the various types of highway culverts? Show them through sketches.

Q.86. What additional measures you would adopt in providing drainage system in water logged areas?

Q.87. What are principal functions of a highway drainage system?

Q.88. Draw a neat sketch showing sub-surface drainage system with longitudinal and cross-drains.

Q.89. What are the sources of sub-surface moisture under road pavement? Why it is important to control this moisture?

Show with the aid of sketch how a higher water table under a road pavement may be substantially lowered. Show also the essential features of a typical drain.

Q.90. Mention the objectives of highway lighting system. On what factors visibility of an object depends.

Q.91. What factors need consideration for highway lighting system? At what locations high-mast system is justified? Give reasons for your answers.

Q.92. With the help of a neat sketch, show the components of a typical highway lighting structure. Name different components.

Q.93. Explain how various environmental factors affect highway engineering decisions?

Q.94. Why the roads in hilly areas require special considerations in geometric design standards? Mention the need for hairpin bends.

Q.95. Mention the purpose of retaining and parapet walls in hill roads. Why weep holes are provided?

Q.96. Draw a typical cross-section of a hill road partly in cutting and partly in fill. Name the various elements.

Q.97. With the help of neat sketches, explain the functions of the following in hill roads:
  (i) Hairpin bend
  (ii) Salient and re-entrant curves
  (iii) Gabion wall

Q.98. What are the causes of landslides on hill roads? Describe the preventive measures to be adopted.

Q.99. Why embankments are necessary? Name factors on which stability of an embankment depends.

Q.100. Write short notes on:
  (a) Slip circle method of slope stability
  (b) Protection of highway slopes
  (c) Material suitable for embankments
  (d) Alignment of hill roads

B. Objective Type Multiple Choice

Q.1. Indian Road Congress came into existence in the year:
  (a) 1947   (b) 1934
  (c) 1940   (d) 1939

Q.2. According to Bombay plan, the period for long term planning for road development is:
  (a) 10 years   (b) 25 years
  (c) 15 years   (d) 20 years

Q.3. Minimum recommended width of village road as per Nagpur plan was:
  (a) 3 m   (b) 3.45 m
  (c) 2.45 m   (d) 4 m

Q.4. Who were the pioneers in road construction?
  (a) Romans   (b) French
  (c) Dutch   (d) British

Q.5. Duration of Nagpur plan was:
  (a) 1941–1961   (b) 1943–1963
  (c) 1930–1950   (d) 1951–1971

Q.6. According to second 20-year development plan (1961–81), the maximum distance of any place in developed area, from a metalled road should be:
  (a) 3.6 kms   (b) 4.6 km
  (c) 6.4 kms   (d) 6.3 km

Q.7. The highway connecting Jaipur to Delhi is:
  (a) NH–2   (b) NH–5
  (c) NH–8   (d) NH–11

Q.8. The Nagpur road plan formulae were developed for a road pattern which is:
  (a) Star and block   (b) Block
  (c) Star and grid   (d) Star and circular

Q.9. The alignment of a road means:
  (a) Its width   (b) Superelevation
  (c) Camber   (d) Centre line on ground

Q.10. The main road which connects important towns and cities within a province is called:

(a) State Highway   (b) National Highway

(c) District Highway   (d) Capital Highway

Q.11. The rate of rise and fall of a road along its length is called:

(a) Ramp   (b) Slope

(c) Curve   (d) Camber

Q.12. The aggregates having affinity with water are:

(a) Hydrophilic   (b) Hydrophobic

(c) Hydraulic   (d) Coarse

Q.13. Aggregate crushing test is done for:

(a) Toughness   (b) Hardness

(c) Strength   (d) Durability

Q.14. Los Angles abrasion testing machine is rotated at a speed of:

(a) 5–10 rpm   (b) 10–15 rpm

(c) 30–33 rpm   (d) 40–45 rpm

Q.15. Chemical used in soundness test of aggregates is:

(a) Na_2SO_4 (b) NaCl

(c) NA_3PO_4 (d) NaL

Q.16. In flakiness index dimension/thickness is less than:

(a) $\frac{3}{5}$ th   (b) $\frac{2}{5}$ th

(c) $\frac{4}{5}$ th   (d) $\frac{7}{8}$ th

of the mean dimension.

Q.17. Angularity number is defined as:

(a) 60% of solid volume   (b) 67% of solid volume

(c) 65% of solid volume   (d) 75% of solid volume

Q.18. Total weight of needle assembly in penetration test of bitumen is:

(a) 30 gm   (b) 50 gm

(c) 80 gm   (d) 100 gm

Q.19. Most fluid cutback bitumen is:

(a) RC–0   (b) MC–2

(c) RC–1   (d) SC–3

Q.20. The instrument used for determining specific gravity of bitumen is called:

(a) Lactometer   (b) Pycnometer

(c) Viscometer   (d) Hydrometer

Q.21. When a volatile material is added to bitumen for changing its viscosity, the final product is:

(a) Emulsion   (b) Asphalt

(c) Cut back   (d) Tar

Q.22. The wet analysis to determine soil type, clay or sand, is based on:

(a) Burmister's law   (b) Hazen's law

(c) Pascal's law   (d) Stoke's law

Q.23. The plastic limit of soil is determined using sieve no:
    (a) 425 micron   (b) 300 micron
    (c) 200 micron   (d) 200 micron

Q.24. The maximum width of road vehicles in India is:
    (a) 2.44 m   (b) 2.8 m
    (c) 2.25 m   (d) 3 m

Q.25. The maximum permissible load per axle in India is:
    (a) 10.5 tonnes   (b) 8.16 tonnes
    (c) 11.6 tonnes   (d) 15.6 tonnes

Q.26. In overtaking, time taken is most affected by:
    (a) Vehicle width   (b) Vehicle length
    (c) Road surface   (d) Power/weight ratio

Q.27. Vehicle stopping distance is most affected by:
    (a) Driver's strength   (b) Thickness of tyre tread
    (c) Type of brakes   (d) Road quality

Q.28. Superelevation is need to:
    (a) Stop vehicle quickly   (b) Remove rain water
    (c) Improve visibility   (d) Reduce sideway forces

Q.29. An equipment to determine spot speed of vehicles is called:
    (a) Enoscope   (b) Total station
    (c) Epidiascope   (d) Speedometer

Q.30. As per IRC the height of the eye level of the driver for measuring sight distances is:
    (a) 2.44 m   (b) 1.22 m
    (c) 1.44 m   (d) 2.02 m

Q.31. Total extra widening on horizontal curve is given by:

    (a) $\dfrac{NL^2}{2R} + \dfrac{V}{9.1\sqrt{R}}$ (b) $\dfrac{NL^2}{2R} + \dfrac{V}{9.5\sqrt{R}}$

    (c) $\dfrac{NL^2}{2R} + \dfrac{V}{9.3\sqrt{R}}$ (d) $\dfrac{NL^2}{R} + \dfrac{V}{9.5\sqrt{R}}$

Q.32. For a transition curve, the shape recommended by IRC is:
    (a) Circular   (b) Spiral
    (c) Parabola   (d) Lemniscate

Q.33. The length of the overtaking zone provided on roads is:
    (a) 2 (OSD)   (b) 3 (OSD)
    (c) 4 (OSD)   (d) 5 (OSD)

Q.34. A test used to determine modulus of subgrade is called:
    (a) CBR test   (b) Consolidation test
    (c) Plate bearing test   (d) Flexural strength test

Q.35. The "Level of Service" of a road means
    (a) Assumed design speed   (b) Permitted capacity
    (c) Peak hour vehicle concentration   (d) General flow-speed characteristics

Q.36. In PIEV theory 'P' stands for:
(a) Presence of mind
(b) Physiological treats
(c) Perception
(d) Practical

Q.37. ITS stands for
(a) Indian Transport System
(b) Intelligent Transport System
(c) International Transport System
(d) Internal Transport System

Q.38. Lowest level of service, for urban roads is:
(a) 'D' level
(b) 'E' level
(c) 'F' level
(d) 'H' level

Q.39. Design hourly volume is taken as:
(a) 100th hourly highest volume
(b) 30th hourly highest volume
(c) 90th hourly highest volume
(d) 10th hourly highest volume

Q.40. Turning radius of a passenger car is:
(a) 5–6 m
(b) 7 to 10 m
(c) 10–15 m
(d) 12–15 m

Q.41. Overtaking sight distance for a speed of 80 km/hr, is about:
(a) 100–20 m
(b) 200–300 m
(c) 450–500 m
(d) More than 600 m

Q.42. Vehicle stopping sight distance is most affected by:
(a) Driver's strength
(b) Thickness of tyres
(c) Adhesion with road
(d) Type of brakes

Q.43. Superelevation is needed to:
(a) To stop vehicle quickly
(b) To improve visibility
(c) To reduce sideway forces
(d) To remove rain water

Q.44. Providing a transverse slope throughout the length of a horizontal curve is called:
(a) Superelevation
(b) Camber
(c) Cross-slope
(d) Gradient

Q.45. The curve provided to change the gradient is called:
(a) Horizontal curve
(b) Vertical curve
(c) Reverse curve
(d) Transition curve

Q.46. Structural design of a highway pavement involves:
(a) Finding the total space requirement
(b) Designing best geometry for the road
(c) Finding thickness of the pavement
(d) Fixing the alignment

Q.47. Rigid pavement consists of:
(a) Hard stone pavement
(b) Full depth bituminous pavement
(c) Pavement with high quality material
(d) Cement concrete pavement

Q.48. A test used to determine modulus of subgrade reaction is:
(a) Plate bearing test
(b) CBR test
(c) Consolidation test
(d) North Dakota Cone test

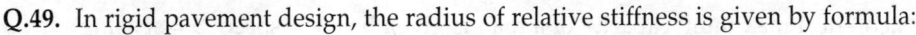

Q.49. In rigid pavement design, the radius of relative stiffness is given by formula:

(a) $4\sqrt{\dfrac{1000\,Eh^2}{12\,(1-\mu^2)\,k}}$

(b) $4\sqrt{\dfrac{3000\,Eh^3}{12\,(1-\mu^2)\,k}}$

(c) $3\sqrt{\dfrac{1000\,Eh^3}{12\,(1-\mu^2)\,k}}$

(d) $2\sqrt{\dfrac{1000\,Eh^3}{12\,(1-\mu^2)\,k}}$

Q.50. The purpose of pavement overlay is:
(a) To strengthen the existing pavement
(b) To build a new pavement replacing the old one
(c) To change flexible pavement to rigid
(d) To change rigid pavement to flexible

Q.51. The thickness of flexible pavement using IRC method is based on:
(a) Marshall test
(b) CBR test
(c) Impact test
(c) RCB test

Q.52. Burmister's method for analysis of stresses in pavement is based on:
(a) One-layer theory
(b) Two-layer theory
(c) Multilayer theory
(d) Rigid theory

Q.53. The base course in flexible pavement is:
(a) A concrete slab
(b) A protective layer to withstand wear and tear
(c) A layer to provide smooth and water proof surface
(d) A prepared material of hard core whose function is to act as a weight distribution layer.

Q.54. A sign showing a Railway Crossing is:
(a) Regulatory sign
(b) Warming sign
(c) Informatory sign
(d) Mandatory sign

Q.55. A drain provided in the slope of cutting to intercept the water flowing down the cut slope is:
(a) Catch drain
(b) Side ditch
(c) Slope drain
(d) Gradient

Q.56. In CBR method of flexible pavement design following information is not required:
(a) CBR value of subgrade and sub-base material
(b) Modulus of subgrade reaction
(c) Wheel load
(d) Number of commercial vehicles, 3T and above

Q.57. Number of vehicles per unit length of a road at a given time is called:
(a) Traffic volume
(b) Capacity
(c) Congestion
(d) Traffic density

Q.58. A diagram showing physical layout of the location where an accident took place is called:
(a) Spot map
(b) Collision diagram
(c) Condition diagram
(d) Site diagram

Q.59. The theoretical maximum capacity of a road is given by: (V = speed, S = spacing)

(a) $\dfrac{1000\ V}{S}$ (b) $\dfrac{1000\ S}{V}$

(c) $\dfrac{100\ V}{S}$ (d) $\dfrac{100\ S}{V}$

Q.60. As per IRC warning signs, should be of the shape-on the top:

(a) Circle   (b) Triangle

(c) Rectangle   (d) Square

Q.61. Wave like deformations on the road surface are called:

(a) Corrugations   (b) Wave defects

(c) Blow holes   (d) Patches

Q.62. A sharp curve which turns 180° on the same side down the hill is a:

(a) Summit curve   (b) Hairpin bend

(c) Salient curve   (d) Briddle curve

Q.63. A curve having convexity outwards on the outer side of the road, at the ridge of hill side is known as:

(a) Transition curve   (b) Re-entrant curve

(c) Salient curve   (d) Summit curve

Q.64. A road sign "No Parking" is which type of sign:

(a) Warning sign   (b) Prohibitory sign

(c) Mandatory sign   (d) Informatory sign

Q.65. For parking, greater number of vehicles can be parked at the angle of:

(a) 90°   (b) 30°

(c) 45°   (d) 60°

Q.66. The wall used at the formation level on hill road is:

(a) Parapet wall   (b) Retaining wall

(c) Gabion wall   (d) Breast wall

Q.67. The wall constructed towards up slope side of a hill road to prevent sliding is called:

(a) Retaining wall   (b) Parapet wall

(c) Breast wall   (d) Side drain

Q.68. The drain used to drain water across the road is:

(a) Catch drain   (b) Cross drain

(c) Longitudinal drain   (d) Side drain

Q.69. Stability of slope in hill roads depends upon:

(a) Weather   (b) Seismic activity

(c) Geological structure   (d) All of the above

Q.70. Highway embankments are:

(a) Summit curves   (b) Road in hilly terrain

(c) Raised structures   (d) Natural slopes

Q.71. Stability of embankments depend upon:

(a) Gravitational forces   (b) Seepage forces

(c) None of above   (d) Both of above

Q.72. Soil that is good as embankment fills:

(a) Sands and gravel   (b) Clays

(c) Silts   (d) Silty clays

Q.73. A test used in laboratory to establish moisture-density relationship is:

(a) Procter's test   (b) Compression test

(c) Consolidation test   (d) Shear test

Q.74. Benching is used, to ensure stability of:

(a) Slopes in cutting   (b) Hill side slopes

(c) Very high embankments   (d) None of the above

Q.75. The basic cause of subgrade failure is

(a) Poor stability   (b) Excessive loads

(c) Low bearing capacity   (d) Poor drainage

(e) All of the above

Q.76. Which of these is not a failure of flexible pavements?

(a) Pot holes   (b) Corrugations

(c) Ruts   (d) Pumping

Q.77. Road maintenance work involves maintenance of:

(a) Pavement   (b) Shoulders

(c) Traffic control devices   (d) Drainage system

(e) All of the above

Q.78. Longitudinal depression formed in a flexible pavement are called:

(a) Pot holes   (b) Cracks

(c) Ruts   (d) Heaving

Q.79. Which of the following is not a rigid pavement failure?

(a) Corrugations   (b) Shrinkage cracks

(c) Mud Pumping   (d) Warping cracks

Q.80. A measure of luminous intensity of light source is:

(a) Candela   (b) Radian

(c) Watts   (d) Ohm

Q.81. The system of collecting and disposal of rain water from the road is called:

(a) Sub surface drainage   (b) Cross-drainage

(c) Surface drainage   (d) Control of seepage flow

Q.82. Causeways are the type of a:

(a) Side drain   (b) Cross drainage structure

(c) Sub-surface structure   (d) None of the above

Q.83. The disposal of water across the road is called:

(a) Cross-drainage   (b) Intercepting drainage

(c) Surface drainage   (d) Sub-surface drainage

Q.84. Horizontal illuminance provided on road should be at least in the range of:

(a) $10\text{–}15 \ Cd/m^2$ (b) $0.5 \text{ to } 2 \ Cd/m^2$

(c) More than $15 \ Cd/m^2$ (d) Less than $0.5 \ Cd/m^2$

Q.85. The average illumination on important roads as per Indian practise is:

(a) 10 lux   (b) 20 lux

(c) 30 lux   (d) 40 lux

Q.86. An analytical process which examines the possible consequences of a highway project on environment is:

(a) Environmental impact assessment

(b) Environmental mitigation

(c) Aesthetical degradation

(d) Protection of environment

Q.87. Roadside arboriculture is:

(a) Keeping road clear of obstruction

(b) Removal of road side trees

(c) Plant of trees along the road

(d) Clearing of side drains

Q.88. Economic analysis for highway project is done for:

(a) Justifying a highway project

(b) Providing safety on road

(c) Saving the environment from degradation

(d) Pleasing the public

Q.89. Cause of environmental pollution is:

(a) Planting of trees

(b) Heavy rains

(c) Automobiles

(d) Flora and fauna

Q.90. Which of the following gas is released by automobiles?

(a) Carbon monoxide   (b) Oxygen

(c) Ammonia   (d) Sulphur dioxide

Q.91. Benefit-cost ratio in economic analysis is:

(a) $\dfrac{\text{Operation cost} + \text{Construction cost}}{\text{Benefit}}$

(b) $\dfrac{\text{Benefits}}{\text{Total cost (annual)}}$

(c) $\dfrac{\text{Operation cost}}{\text{Benefits} - \text{Maintenance}}$

(d) $\dfrac{\text{Benefits} - \text{Operation cost}}{\text{Total cost (annual)}}$

Q.92. Control of access on highways means:

(a) Give preference to through traffic

(b) Control of heavy vehicles

(c) Control of high speed vehicles

(d) Collect tax for use

Q.93. The objective of quality control of highways is:
  (a) To study specifications   (b) To suggest corrective measures
  (c) To ensure adherence to quality   (d) To inspect the road

Q.94. The overlay thickness of highway pavement is decided by:
  (a) Benkleman beam test
  (b) PCUs on highway
  (c) Number of slow moving vehicles
  (d) Commercial vehicles in traffic stream

Q.95. The pavement width of a highway is decided by:
  (a) Number of commercial vehicle
  (b) Total PCUs on the highway
  (c) Length of the largest vehicle
  (d) Number of fast moving vehicles

Q.96. Safe bearing capacity of soil is determined by:
  (a) Benkleman beam test   (b) Marshall stability test
  (c) Proctor's test   (d) Plate load test

Q.97. Arithmetic mean of the speed observations on a road is called:
  (a) Space mean speed   (b) Running speed
  (c) Design speed   (d) Time mean speed

Q.98. Road delineators are used for:
  (a) Visibility during bad weather conditions/night
  (b) Guiding pedestrians
  (c) Providing safety during day time
  (d) Eliminating use of markings

Q.99. Traffic calming measures involve:
  (a) Stopping vehicles on the road
  (b) Roads used for pedestrian only
  (c) Lowering the speed of vehicles in urban area
  (d) Keeping commercial vehicles away

Q.100. Maintenance of highways means:
  (a) Conserve the road in original condition
  (b) Make a better road than designed
  (c) Strengthening and widening the road
  (d) Rehabilitate the road

ANSWERS

1. (b)	**2.** (d)	**3.** (c)	**4.** (a)	**5.** (b)	**6.** (c)
7. (c)	**8.** (b)	**9.** (d)	**10.** (a)	**11.** (b)	**12.** (a)
13. (d)	**14.** (c)	**15.** (a)	**16.** (a)	**17.** (b)	**18.** (d)
19. (a)	**20.** (b)	**21.** (c)	**22.** (d)	**23.** (a)	**24.** (a)
25. (b)	**26.** (d)	**27.** (c)	**28.** (d)	**29.** (a)	**30.** (b)
31. (b)	**32.** (b)	**33.** (d)	**34.** (c)	**35.** (d)	**36.** (c)

37. (b)	**38.** (c)	**39.** (b)	**40.** (b)	**41.** (c)	**42.** (d)
43. (c)	**44.** (a)	**45.** (b)	**46.** (c)	**47.** (d)	**48.** (a)
49. (a)	**50.** (a)	**51.** (b)	**52.** (b)	**53.** (d)	**54.** (c)
55. (a)	**56.** (b)	**57.** (d)	**58.** (c)	**59.** (a)	**60.** (b)
61. (a)	**62.** (b)	**63.** (c)	**64.** (b)	**65.** (a)	**66.** (a)
67. (c)	**68.** (b)	**69.** (d)	**70.** (c)	**71.** (d)	**72.** (a)
73. (a)	**74.** (b)	**75.** (e)	**76.** (d)	**77.** (e)	**78.** (c)
79. (a)	**80.** (a)	**81.** (c)	**82.** (b)	**83.** (a)	**84.** (b)
85. (c)	**86.** (a)	**87.** (c)	**88.** (a)	**89.** (c)	**90.** (a)
91. (b)	**92.** (a)	**93.** (c)	**94.** (a)	**95.** (b)	**96.** (d)
97. (d)	**98.** (a)	**99.** (c)	**100.** (a)		

Appendices

	DISTRIBUTION OF DENSITY OF NATIONAL HIGHWAY LENGTHS					
Sl. No.	Name of State/UT	Total NH length in km	Area in 1000 sq km	Length of NH in km/1000 sq km	Population in lakhs as per 2001 census	Length of NH in km/lakh popul-ation
1	Andaman & Nicobar Islands	300	8.249	36.4	3.56	84.3
2	Andhra Pradesh	4472	275.068	16.3	762.10	5.9
3	Arunachal Pradesh	392	83.743	4.7	10.97	35.7
4	Assam	2836	78.438	36.2	266.55	10.6
5	Bihar	3642	94.163	38.7	829.98	4.4
6	Chandigarh	24	0.114	210.5	9.00	2.7
7	Chhattisgarh	2184	135.194	16.2	208.33	10.5
8	Dadar & Nagar Haveli	0	0.491	0.0	2.20	0.0
9	Daman & Diu	0	3.814	0.0	1.58	0.0
10	Delhi	72	1.483	48.6	138.50	0.5
11	Goa	269	3.814	70.5	13.47	20.0
12	Gujarat	3245	196.024	16.6	506.71	6.4
13	Haryana	1512	44.212	34.2	211.44	7.2
14	Himachal Pradesh	1208	55.673	21.7	60.77	19.9
15	Jammu & Kashmir	1245	222.236	5.6	101.43	12.3
16	Jharkhand	1805	79.714	22.6	269.45	6.7

(Contd.)

(Contd.)

SI. No.	Name of State/UT	Total NH length in km	Area in 1000 sq km	Length of NH in km/1000 sq km	Population in lakhs as per 2001 census	Length of NH in km/lakh population
17	Karnataka	3843	191.791	20.0	528.50	7.3
18	Kerala	1457	38.863	37.5	318.41	4.6
19	Lakshadeep island	0	0.032	0.0	0.61	0.0
20	Madhya Pradesh	4670	308.252	15.1	603.48	7.7
21	Maharashtra	4176	307.69	13.6	968.78	4.3
22	Manipur	959	22.327	43.0	22.93	41.8
23	Meghalaya	810	22.429	36.1	23.18	34.9
24	Mizoram	927	21.081	44.0	8.88	104.4
25	Nagaland	494	16.579	29.8	19.90	24.8
26	Orissa	3704	155.707	23.8	368.04	10.1
27	Pudducherry	53	0.495	107.1	9.74	5.4
28	Punjab	1557	50.362	30.9	243.58	6.4
29	Rajasthan	5585	342.239	16.3	565.07	9.9
30	Sikkim	62	7.096	8.7	5.40	11.5
31	Tamil Nadu	4462	130.258	34.3	624.05	7.2
32	Tripura	400	10.486	38.1	31.99	12.5
33	Uttarakhand	1991	55.845	35.7	84.89	23.5
34	Uttar Pradesh	5874	238.566	24.6	1661.97	3.5
35	West Bengal	2524	88.752	28.4	801.76	3.1
	TOTAL	**66754**	**3291.08**	**20.3**	**10287.20**	**6.5**

APPENDIX B

STATEWISE LENGTH OF NATIONAL HIGHWAYS IN INDIA

Sl. No.	NH No.	Route	Length (km.)
Andhra Pradesh			
1	4	Karnataka border-Palmaner-Chittoor-Naraharipeta-upto Tamil Nadu border	83
2	5	From Orissa border-Ichchapuram-Narasannapeta-Srikakulam-Bhimunipatnam-Visakhapatnam-Prattipadu-Rajahmundry-Eluru-Vijaywada-Guntur-Ongal-Nellor-Gudur-upto Tamil Nadu border	1000
3	7	From Maharashtra border-Adilabad-Nirmal-Ramayampet-Hyderabad-Kurnool-Gooty-Anantpur-Penukonda-Karnataka border	753
4	9	From Karnataka border-Zahirabad-Hyderabad-Suriapet-Vijaywada-Machillipatnam	430
5	16	Niambabad-Armur-Jagtial-Chinnur upto Maharashtra border	220
6	18	Kurnool-Nandyal-Cuddapah-Rayachot-Chittor	369
7	43	From Orissa border-Ramabhadrapuram-Vizianagaram-Jn. with NH-5 near Natavalasa	83
8	63	From Karnataka border-Guntakal-Gooty	62
9	202	Hyderabad-Warangal-Venkatpuram upto Chhattisgarh border	244
10	205	Ananthapur-Kadiri-Madanapalle-Renigunta-upto TN border	360
11	214	Kathipudi-Razole-Kakinada-Narasapur-Pamurru	270
12	214A	The highway starting from the junction of NH-214 near Digamarru connecting Narasapur-Machilipatnam-Challapalle-Avanigadda-Repalle Bapatla-Chirala and terminating at its junction with NH-5 near Ongole	255
13	219	Madnapalli-Punganuru-Palmaner-Kuppam upto Tamil Nadu border	128
14	221	The highways starting from the junction of NH-9 near Vijayawada connecting Kondapalli-Mailvaram-Tiruvuru-Penuballi-Kottagudam-Paloncha-Bhadrachalam-Nellipaka-Chinturu-Konta upto Chhattisgarh border	155
15	222	From Maharashtra border to junction with NH-7 near Nirmal	60
		Sub Total	**4472**
Arunachal Pradesh			
1	52	From Assam border-Pasighat-Dambuk-Roing-Paya-Tezu-Wakro-Namsai-upto Assam border	310
2	52A	From Assam border-Itanagar-upto Assam border	42
3	153	From Assam border-Myanmar border (Still Well road)	40
		Sub Total	**392**

(Contd.)

(*Contd.*)

		Assam	
1	31	From WB border-Gouripur-North Salmara-Bijni-Charaliamingaon Junction with NH No. 37	322
2	31B	North Salmaria-Abhayapuri-Junction with NH No 37 near Jogighopa	19
3	31C	From WB border-Kochugaon-Sidli Ju. With NH-31 Near Bijni	93
4	36	Nagaon-Dabaka-Amlakhi-Nagaland border	167
5	37	Junction with NH No. 31B near Goalpara-Paikan-Guwahati-Dispur-Nowgong-Numaligrah-Jorhat-Jhanzi-Dibrugarh-Tinsukia-Makum-Saikhoghat	680
6	37A	Kuwari Tal-Junction with NH No. 52 near Tezpur	23
7	38	Makum-Ledo-Likhapani	54
8	39	Numaligarh-Naojan-Bokajan-upto Nagaland border	115
9	44	From Meghalaya border-Badarpur-Karimgant-Patharkandi-upto Tripura border	111
10	51	Paikan-upto Meghalaya border	22
11	52	Boihata-Charali-Mangaldai-Dhekiajuli-Tezpur-Gohpur-Bander Dewa-North Lakhimpur-Dhemaji-Kulajan-Arunachal Border-Junction with NH No. 37 near Saikhoaghat	540
12	52A	Gohpur-AP border-Bander Dewa	15
13	52B	Kulajan-Dibrugarh	31
14	53	Junction with NH-44 near Badarpur-Silchar-Lakhipur-upto Manipur border	100
15	54	Dabaka-Lumding-Langting-Haplong-Silchar-Dwarband upto Mizoram border	335
16	61	Jhanzi-Amguri-Nagaland border	20
17	62	Dudhani-Damara-upto Meghalaya border	5
18	151	Karimganj-Bangladesh border	14
19	152	Patacharkuchi-Hajua-Bhutan border	40
20	153	Ledo-Lekhapani-Arunachal Pradesh border	20
21	154	Dhaleshwar (Badarpur)-Bhairabhi-Mizoram border	110
		Sub Total	**2836**
		Bihar	
1	2	From UP border-Mohania-Jahanabad-Sasaram-Dehri-Aurangabad Madanpur-Dobhi-Barachati-Jharkhand border	202
2	2C	Dehri-Akbarpur-Jadunthpur-Bihar/UP border	105
3	19	From UP border-Manjhi-Chhapra-Sonpur-Hajipur-Patna	120
4	28	Barauchi-Bachiwara-Tajpur-Muzaffarpur-Mehsi-Chakia-Gopalganj-up to UP border.	259
5	28A	Junction with National Highway No. 28 near Pipra Kothi-Sagauli-Raxaul Indo/Nepal border	68

(*Contd.*)

		Bihar (*Contd.*)	
6	**28B**	Chapwa-Bettiah-Lauriya-Bagaha-Chhitauni Rail-cum-Road Road Bridge upto UP border	121
7	**30**	Junction with NH-2 near Mohania-Kochas-Dinara-Bikramganj-Piro-Ara Danapur-Patna-Phatuha-Bakhitiyarpur	230
8	**30A**	Phatuha-Chandi-Harnaut-Barh	65
9	**31**	From Jharkhand border-Rajauli-Nawada-Bihar Sharif-Bakhtiyar Barh-Mokoma-Barauni-Begusarai-Balia-Khagaria-Bihpur-Kursela-Purnia-Baisi-WB border-Kishanganj-upto WB border	393
10	**57**	Muzaffarpur-Darbhanga-Jhanjharpur-Narahia-Narpatganj-Forbesganj-Araria-Purnia	310
11	**57A**	The highway starting from the junction of NH-57 near Forbesganj and Terminating at Jogbani	15
12	**77**	Hajipur-Muzaffarpur-Sitamarhi-Sonbarsa	142
13	**80**	Mokamah-Luckeesarai-Munger-Bhagalpur-Kahalgaon-upto Jharkhand border	200
14	**81**	Kora-Katihar-upto WB border	45
15	**82**	Gaya-Hisua-Rajgir-Bar Bigha-Mokama	130
16	83	Patna-Jahanabad-Bela-Gaya-Dobhi	130
17	84	Ara-Buzar	60
18	85	Chhapra-Ekma-Siwan-Gopalganj	95
19	98	Patna-Arwal-Dadnagar-Aurangabad-Amba upto Jharkhand border	157
20	99	Dobhi-Hardawan-upto Jharkhand border	10
21	101	Chhapra-Baniapur-Mohamadpur	60
22	102	Chhapra-Rewaghat-Muzaffarpur	80
23	103	Hajipur-Hazrat Jandaha-Mushrigharari	55
24	104	Chakia-Madhuban-Shivhar-Sitamarhi-Sursand-Jaynagar-Narahia	160
25	105	Darhanga-Keotiranway-Aunsi-Jaynagar	66
26	106	Birpur-Pipra-Madhepura-Kishanganj-Bihpur	130
27	107	Maheshkund-Sonbarsa Raj-Simribakhtiarpur-Bariahi-Saharsa-Madhepura-Banmankhi-Purnia	145
28	110	The highway starting from its junction with NH-98 from Arwal connecting Jahanabad-Bandhuganj-kako-Ekangarsari and terminating at its junction with NH-31 Biharsharif	89
		Sub Total	**3642**
		Chandigarh	
1	21	Junction with NH-22 near Chandigarh-Ropar-Bilaspur-Mondi-Kulu-Manali	24
		Sub Total	**24**

(*Contd.*)

Chhattisgarh

1	6	From Maharashtra border-Baghnadi-Chichola-Rajnandgaon-Durg-Bhilai-Raipur-Arang-Pithora-Basna-Saraipali-up to Orissa border	314
2	12A	From MP border-Chilpi-Kawardha-Pipariya-Bemetara-Simga.	128
3	16	From Maharashtra border-Bhopalpatnam-Bijapur-Bhairamgarh-Gidam-Jagdalpur	210
4	43	Raipur-Marod-Dhamtari-Charama-Kanker-Keskal-Parasgaon-Kondagaon-Jagdalpur-upto Orissa border	316
5	78	From MP border-Mahendragarh-Baikunthpur-Surajpur-Ambikapur-Kunkuri-Pathalgaon-Paikera-Jashpurnagar-Rupsera-Jharkhand border	356
6	111	Bilaspur-Ratanpur-Katghore-Kendai-Laxmanpur-Amikapur	200
7	200	Raipur-Simga-Baitalpur-Bilaspur-Ramgarh-Champa-Sakti-Uravmiti-Raigarh-upto Orissa border	300
8	202	Bhopalpatnam-Bhadrakali-Kotturu-upto AP border	36
9	216	Raigarh-Sarangarh-Saraipali	80
10	217	Raipur-Mahasamund-Suarmar-upto Orissa border	70
11	211	From AP border Konta-Sukma-Kukanar-Darba-Sosanpal-Terminating junction with NH-16 near Jagdalpur	174
		Sub Total	**2184**

Delhi

1	1	Outer Ring Road/Transport Nagar-Haryana border	22
2	2	NH-2/Ring Road-Delhi-Haryana border	12
3	8	Ring Road-Haryana border	13
4	10	Outer Ring Road-Mundka-Haryana border	18
5	24	Nizamuddin Road-UP border	7
		Sub Total	**72**

Goa

1	4A	From Karnataka border-Darabandora-Ponda-Bhoma-Banastari-Panaji	71
2	17	From Maharashtra border-Pernem-Mapca-Panaji-Cortalim-Verna-Margao-Cunoclim-Chauri (Chauri)-Polem upto Karnataka border	139
3	17A	Cortalim (kortali)-Sancoale-Chicalim-Murmugao	19
4	17B	Ponda-Verna-Vasco da Gama	40
		Sub Total	**269**

Gujarat

1	NE1	Ahmedabad-Vadodara Expressway	93
2	6	Hajira-Surat-Bardoli-Vyara-Songadh-upto Maharashtra border	177

(Contd.)

Gujarat (*Contd.*)			
3	8	From Rajasthan border-Himatnagar-Ahmadabad-Nadiad-Vadodara-Karjan Bharuch-Ankleshwar Navsari-Valsad-Vapi-Maharashtra border	498
4	8A	Ahmadabad-Bagodra-Limbdi-Bamenbore-Morvi-Samakhiali-Kandla-Mandvi	618
5	8B	Bamanbor-Rajkot-Gondal-Jetpur-Dhoraji-Kutiyana-Porbandar	206
6	8C	Chiloda-Gandhinagar-Sarkhej	46
7	8D	Jetpur-Junagadh-Maliya-Somnath	127
8	8E	Dwarka-Porbandar-Somnath-Kodinagar-Mahuva-Talaja-Bhavnagar	445
9	14	From Rajasthan border-Palanpur-Deesa-Sihori-Radhanpur	140
10	15	Samakhiyali-Santalpur-Radhanpur-Bhaghar-Tharad-upto Rajasthan border	270
11	59	Ahmedabad-Kathua-Gudhra-Dahod-upto MP border	211
12	113	Dahod-Limdi-Zalod-Rajasthan border	40
	228	Ahmedabad-Dandi route (Dandi heritage route)	374
		Sub Total	**3245**
Haryana			
1	1	From Delhi border-Kundli-Murthal-Samalkha-Panipat-Karnal-Pipli-Shahbad-Ambala-upto Punjab border	180
2	2	From Delhi-Faridabad-Ballabgarh-Palwal-Rundhi-Hodal-UP border	74
3	8	From Delhi border-Gurgaon-Dharuhera-Bawal-Rajasthan border	101
4	10	From Delhi border-Bahadurgarh-Rohtak-Maham-Hansi-Hissar-Agroha-Bodopal-Fatehabad-Sirsa-Odhan-Dabwali-Punjab border	313
5	21A	Pinjaur-Karapur upto HP border	16
6	22	Ambala-Panchkula-Chandi Mandir-Pinjaur-Kalka-HP border	30
7	64	Dabwali-Punjab border	0.5
8	65	Ambala-Pehowa-Kaithal-Narwana-Barwala-Hisar-Siwani-upto Rajasthan border	240
9	71	From Punjab border-Narwana-Jind-Julana-Rohtak-Dighal-Jhajjar-Guraora-Rewari-Rajasthan border	177
10	71A	Rohtak-Gohana-Israna-Panipat	72
11	71B	Rewari-Dharuhera-Taoru-Sohna-Palwal	69
12	72	Ambala-Shahzadpur-Narayangarh-Kala Amb-upto HP border	45.5
13	73	From UP border-Yamnanagar-Mulana-Saha-Raipur-Panchkula	108
14	73A	The highway starting from junction of NH 73 near Yamuna Nagar in state of Haryana and connecting Jagadhri Chowk (Jn with NH-73, via Chhachhrauli, Tajewala, Khizrabad, Kalesar, Lal Dang in Haryana	42
15	NE2	**Eastern Peripheral Expressway around in UP and Haryana** (UNDER CONSTRUCTION)	44
		Sub Total	**1512**

(*Contd.*)

(Contd.)

Himachal Pradesh			
1	1A	From Punjab border-Damtal-upto Punjab border	14
2	20	Mandi-Jogindernagar-Baijnath-Palampur-Bagwan-Nagrota-Kotla-Nurpur-upto Punjab border	210
3	21	From Punjab border-Swarghat-Bilaspur. Sunder Nagar-Mandi-Pandoh-Aut-Bajaura-Kullu-Ralsan-Manali	232
4	21A	Swarghat-Kundlu-Nalagarh-upto Haryana border	49
5	22	From Haryana border-Parwnoo-Dharampur-Barog-Solan-Kandaghat-Shimla-Kufri-theog-Narkanda-Kingal-Rampur-Wangtu-Puh-Namgya-Indo China border near Shipkila	398
6	70	Mandi-Dharampur-Sarkaghat-Awadevi-Hamirpur-Naduan-Amb-Mubarakpur-Gagret-Punjab border	120
7	72	From Haryana border-Kala Amb Nahan-Kolar-Majra-Uttranchal border	50
8	73A	From Haryana border and terminating at Bata Chowk (Junction with NH 72 near Paonta-Sahib) in Himachal Pradesh	20
9	88	Shimla-Sallaghat-Bilaspur-Ghumarwain-Hamirpur-Naduan-Jawalamukhi-Kangra-Mataur	115
		Sub Total	**1208**
Jammu and Kashmir			
1	1A	From Punjab border-Kathua-Samba-Nagnota-Udhampur-Batot-Ramban-Khanabal-Awantipur-Pampore-Srinagar-Pattan-Baramula-Uri	541
2	1B	Batote-Doda-Kistwar-Symthanpass-Kanabal	274
3	1C	Domel-Katra	8
4	1D	Srinagar-Kargil-Leh	422
		Sub Total	**1245**
Jharkhand			
1	2	From Bihar border-Chauparan-Barhi-Barakatha-Bagodar-Dumri-Topchanchi-Gobindpur-Nisa-upto West Bengal border	190
2	6	From Orissa border-Baharagora-upto WB border	22
3	23	Chas-Gola Ramgarh-Omanjhi-Ranchi-Bero-Sisai-Gumla-Palkot-Kolebira-Simdega-Thethaitanagar-Orissa border	250
4	31	Jn. With NH-2 near Barhi-Kodarama upto Bihar border	44
5	32	Junction with NH-2 near Govindpur-Dhanbad Chas-West Bengal border-Chandil-Jamshedpur	107
6	33	Junction with NH-2 near Barhi-Hazaribag-Ramgarh-Ranchi-Bundu-Chandil-Mahulia-junction with NH-6 near Baharagora	352
7	75	From UP border-Nagar Untari-Garhwa-Daltenganj-Latehar-Chandwa-Kuru-Mandar-Ranchi	447
8	78	From Chhattisgarh border-Silam-Gumla	25

(Contd.)

Jharkhand *(Contd.)*			
9	80	From Bihar border-Sahibganj-Talihari-Tinpahar-Rajmahal-Barharwa-upto West Bengal border	100
10	98	From Bihar-border-Hariharganj-Chhatarpur terminating near Rajhara at NH-75	50
11	99	Chandwa-Balumath-Chatra-Hunterganj-upto Bihar border	100
12	100	Chatra-Tutilawa-Hazaribagh-Meru-Daru-Kharika-Bagodar	118
		Sub Total	**1805**
Karnataka			
1	4	From Maharashtra-border-Sankeshwar-Belgaum-Dharwad-Hubli-Haveri-Davangere-Chitradurga-Sira-Tumkur-Nelamangala-Bangalore-Hoskote-Kolar-Mulbagal-Andhra Pradesh border	658
2	4A	Belgaum-Khauapur-Gunji-Goa border	82
3	7	From Andhra Pradesh border-Chik Ballapur-Devannalli-Bangalore-Chandapura-Attibele-Tamil Nadu border	125
4	9	From Maharashtra border-Rajeshvar-Homnabad-Mangalgi-Andhra Pradesh border.	75
5	13	From Maharashtra border-Horti-Bijapur-Hungund-Kushtagi-Hospet-Jagalur-Chitradurga-Holalkere-Bhadravati-Shimoga-Tirthahalli-Karkal-Mangalore	648
6	17	From Goa border-Karwar-Ankola-Honavar-Bhatkal-Baindur-Kundapura-Udupi-Mangalore-Kerala border.	280
7	48	Bangalore-Nelamangala-Kunigal-Channarayapatna-Hassan-Alur-Sakleshpur-Uppinangadi-Mangalore	328
8	63	Ankola-Yellapur-Hubli-Gadag-Lakkundi-Bhanapur-Koppal-Munirabad-Torangallu-Kudatini-Bellary-Hagari-Karnataka/Andhra Pradesh border.	370
9	67	Gundlupet-Hangala-Bandipur-Karnataka/Tamil Nadu border	50
10	206	Tumkur-Nittur-Kibbanahalli-Tiptur-Arsikere-Banavar-Birur-Bhadravati-Shimoga-Anandapuram-Sagar-Telgapp-Gersoppa-Honavar	363
11	207	Hosur-Kodugadi-Devanhalli-Dod Ballapur-Gadigarpalya-Nelamangala	135
12	209	Karnataka/Tamil Nadu border-Punjur-Chamrajnagar-Agra-Sattengala-Malavalli-Sathnur-Bangalore	170
13	212	Kerala/Karnataka border-Maddur-Gundlpupet-Begur-Mysore-Kollegal	160
14	218	Homnabad-Kinhi-Kamalpur-Gulbarga-Firozabad-Jevargi-Moratagi-Sindgi-kannolli-Halagali-Bijapur	399
		Sub Total	**3843**
Kerala			
1	17	Kerala/Karnataka border-Manjeshwar-Kumbla-Kesaragod-Paniyal Mordrug-Charuvattur-Kokkanisseri-Talipparamba-Pappinisseri-Valapattanam-Kannur (Cannon)-Edakkad-Vadakara-Payyoli-Tikkodi-Quilandi-Elattur-Kozhikode (Calicut)-Ferokh-Valancheri-Ponnani-Manattala-K	368

(Contd.)

Kerala (*Contd.*)

2	47	Kerala/Tamil Nadu border-Palakkad (Palghat) Kulalmannam-Alattur-Vadakkancheri-Pattikad-Trichur-Nellayi-Karukurti-Angamali-Chovvara-Aluva-Edappali-Ernakulam	416
3	47A	Willingdon Island terminating at Cochin on NH-47 bypass	6
4	47C	The highway starting from NH-47 near Klmassery, crossing NH-17 and terminating at vallapadam	17
5	49	Kerala/Tamil Nadu border-Devikulam-Pallivagal-Kotamangalam-Cochin	150
6	208	Kollam-Kottarakara-Tenmalai upto Tamil Nadu border	70
7	212	Kozhikode-Kalpatta-Sultan Battery-upto Karnataka border 90	90
8	213	Palghat-Olavakod-Mundur-Mannarkkad-Alanallur-Melattur-Pandikkad-Manjeri-Kondotti-Ferokh	130
9	220	Kolam-Kottarakara-Adu-Kottayam-Kanjirapalli-Vendiperyar	210
		Sub Total	**1457**

Madhya Pradesh

1	3	Rajasthan/MP border-Morena-Gwalior-Ghatigaon-Shipuri-Luckwara-Badarwas-Bhadaura-Guna-Binaganj-Penchi-Biaora-Karaswar-Sarangpur Shajapur-Dewas-Indore-Mhow-Thikri-Julwania-Sendhwo-MP/Maharashtra border	712
2	7	UP/MP border-Mauganj-Mangawan-Rewa-Amarpatan-Murwara-Katni-Sihora-Jabalpur-Bargi-Hulki-Dhuma-Lakhnadon-Seoni-Gopalganj-Khawara	504
3	12	Jabalpur-Shahpura-Deori-Bareli-Bari-Obaidullaganj-Bhopal-Duraha-Shampur-Narsinghgarh-Biaora-Raigarh-Khilchipur-MP/Rajasthan border	486
4	12A	UP/MP border-Orchha-Pithipur-Tikamgarh-Shahgarh-Hirapur-Batigarh-Damoh-Tendukheda-Jabalpur-Mandla-Motinala-MP/Chhattisgarh border.	482
5	25	Shivpuri-Karera-MP/UP border	82
6	26	UP/MP border-Barodh-Bandra-Sagar-Gourjhama-Deori-Maharaipur-Kareli-Nirsimhapur-Lakhandon	268
7	26A	The highway starting from its junction with NH-86 near Sagar-connecting Jeruwakhera-Khurai and terminating at Bina	75
8	27	UP/MP border-Sohagi-Katra-Mangawan	50
9	59	Gujarat/MP border-Rama-Raigarh-Bhandheri-Dhar-Ghat-Bilod-Betma-Indore	139
10	59A	Indore-Chapra-Kannod-Khategaon-Nemawar-Handia-Harda-Timurni Muafi-Sodalpur-Bori-Chirapatla-Chicholi-Betul	264
11	69	Obaidullaganj-Barkhera-Hoshangabad-Itarri-Kesla-Chaukipura-Shahpur-Nimpani-Batul-Multai-Tigaon-Pandhuma-Chicholi-MP/Maharashtra border	295
12	75	UP/MP border-Alipura-Nowgaon-Chhatarpur-Ganj-Panna-Baroura Nagod-Satna-Madhogarh-Connecting on NH-7 near Rewa	600
13	76	Rajasthan/MP border-Kota-Shivpuri	60
14	78	Katni-Umaria-Pali-Shahdol-Burhar-Anupur-MP/Chhattisgarh	178

(Contd.)

		Madhya Pradesh (*Contd.*)	
15	86	MP/UP border-Malhara-Chhatarpur-Gulganj-Shahgarh-Rurawan-Dalpatpur-Banda-Sagar-Rahatgarh-Vidisa-Sanchi-Raisen-Bhopal	379
16	92	MP/UP border-Phup Kolan-Bhind-Mahgawan-Gwalior	96
		Sub Total	**4670**
Maharashtra			
1	3	MP/Maharashtra border-Sangvi-Hadakhed-Dahibad-Amode-Nardana-Songir-Dhule-Arvi-Malagaon-Saundane-Umbrane-Chandvad-Ojhar-Nasik-Padli-Gatpuri-Shahapur-Padghe-Bhiwandi-Thane-Mulund-Mumbai	391
2	4	Thane-Mumbra-Panvel_Chauk-Khalapur-Pune-Khed-Bhatgaon-Surul-Limb-Satara-Valase-Borgaon- Umbraj-Karad-Itakare-Wadgaon-Kolhapur-Kagal-Maharashtra/Karnataka border	371
3	4B	Urban-Chimer and connecting on NH-4 near Chauk	20
4	4C	Km. 16.687 (NH-4B) to Kalamloi on NH-4	7
5	6	Maharashtra/Gujarat border-Visarwadi-Kondaibari-Sakri-Shevali-Ner-Kusumbe-Dhule-Phagne-Mahasva-Erandol-Varad-Jalgaon-Edalabad-Malkapur-Nandura-Khamgaon-Balapur-Akola-Badners-Amravati-Nandgaon-Panjara-Nagpur-Bhandara-Lakhni-Sakoli-Duggupar-Deori-Mah	813
6	7	Maharashtra/MP border-Bandra-Mansar-Nagapur-Gumgaon-Sonegaon-Jamp-Hinganghat-Wadner-Pohna-Wadki-Kinhi-Andhar Kawada-Wajri Bori-Maharashtra/Andhra Pradesh border.	232
7	8	Maharashtra/Gujarat border-Amgaon-Talasar-Karakhu-Mandvi-Thane-Boriyali-Malad-Andheri-Greater Mumbai.	128
8	9	Pune-Loni Kalbhor-Yevat-Bhigvan-Loni-Indapur-Tembhurni-Varawadi-Modnimb-Mohol-Solapur-Naldurg-Yenugur-Umarga-Maharashtra/Andhra Pradesh border.	336
9	13	Solapur-Hattur-Nanandi-Maharashtra/Andhra Pradesh border.	43
10	16	Andhra Pradesh/Maharashtra border-Sirancha-Kopela-Pathagudam-Maharashtra/Chhattisgarh border	30
11	17	Pavel-Pen-Negothane-Kolad-Mangaon-Desgaon-Mahad-Ambavli-Poladpur-Pratapganj-Khed-Asurda-Ankhali-Udgi-Lanja-Vaked-Raipura-Wargaon-Talera-Nandgaon-Kankavli-Kasat-Kudal-Vengurla-Ajgaon-Maharashtra/Goa border.	482
12	50	Nasik-Sinnar-Sangamner-Dolasne-Ghargaon-Bote-Pimpalwandi-Narayangaon-Kalamb-Manchar-Peth-Khed-Chakan-Pune	192
13	69	Nagpur-Koradi-Saoner-Maharashtra/MP border	55
14	204	Ratnagiri-Tink-Pali-Sakharpa-Malkapur-Shahuwadi-Kolhapur	126
15	211	Solapur-Tuljapur-Bav-Badgaon-Osmanabad-Terkhed-Samarkundi-Balsepargaon-Pali-Beed-Pachegaon-Gevrai-Warigodri-Adul-Chetegaon-Aurangabad-Daulatabad-Khuldabad-Ellora-Kannad-Bhamarvadi-Chalisgaon-Mehunbare-Vinchur-Borvihir-Dhule.	400
16	222	From the junction of NH-3 near Kalyan and connecting Ahmednagar-Prabhani-Nanded upto Andhra Pradesh border	550
		Sub Total	**4176**

(*Contd.*)

(Contd.)

		Manipur	
1	39	Manipur-Nagaland border-Maosongsang-Maram-Karong-Kangpokpi-Imphal-Thoubal-Wangling-Palel-Sibong-Indo/Myanmar border.	211
2	53	Manipur/Assam border-Oinamlong-Nungba-Imphal	220
3	150	Manipur/Mizoram border-Parbung-Thanlong-Phaiphengmum-Churachandpur-Moirang-Bishnupur-Imphal-Humpum-Ukhrul-Kuiri-Manipur/Nagaland border	523
4	155	Passam to Manipur/Nagaland border	5
		Sub Total	**959**
		Meghalaya	
1	40	Meghalaya/Assam-Barni Hat-Nongpoh-Umsning-Barapani-Shillong-Meghalaya-Indo/Bangladesh border.	216
2	44	Nongstoin-Shillong-Meghalaya/Assam border	277
3	51	Meghalaya/Assam-Bajengdoda-Tura-Kherapara-Burengapara	127
4	62	Damra-Dambu-Baghmara-Burengapara.	190
		Sub Total	**810**
		Mizoram	
1	44A	Mizoram/Tripura border-Tukkalh-Mamiti-Sairang-Aizawl	165
2	54	Mizoram/Assam border-Chhimlung-Bilkhawthr-Kolasis-Bualpui-Mualvum-Alzawl-Zobawk-Pangzawl-Leite-Zobawk-Sairep-Saiha Kaladan-Tuipang	515
3	54A	Lunglei-and connecting on NH-54 near Zowawk	9
4	54B	Saiha	27
5	150	Mizoram/Manipur border-Thingsa-Ratan-Darlawn-Phaileng-Seling	141
6	154	Meghalaya/Assam border-Connecting on NH-54 near Bualpuri	70
		Sub Total	**927**
		Nagaland	
1	36	Nagaland/Assam border-Dimapur	3
2	39	Dimapur-Cichuguard-Kohima-Viswema	110
3	61	Kohima-Narhema Tseminya-Wokhal-Mokokchung-Chantongia-MerangKong-Nagaland/Assam border	220
4	150	Kohima-Chizami-Nagaland/Manipur border	36
5	155	Mokokchung-Tuensang-Sampurre-Akhegwo-Meluri upto Manipur border	125
		Sub Total	**494**
		Orissa	
1	5	Jharpokharia-Buramara-Kuliana-Baripada-Bentnoti-Baisinga-Balashwar-Bhadrak-Bhandarpokhari-Jagatpur-Cuttack-Bhubaneswar-Chhatarpur-Brahmpur-Golantra-Orissa/Andhra Pradesh border	488

(Contd.)

Orissa *(Contd.)*			
2	5A	Dhanmandal-Patharajpur-Marshaghai-Paradweep Port	77
3	6	Orissa/Chhattisgarh border-Lobarchatti-Bargarh-Attbina-Sambalpur-Deogarh-Barakot-Govindpur-Kunar-Kendujhargarh-Jashipur-Manda-Bangriposhi and upto Orissa/West Bengal border	462
4	23	Orissa/.... border-Birmitrapur-Raiboga-Panposh-Banki-Darjing-Banel-Pala laharha-khamar-talcher-Jn. With NH-42	209
5	42	Sambalpur-Mundher-Jujumura-Charamal-Redhakhol-Bamur-Angul-Dhenkanal-Chaudwar-Jn. With NH-5	261
6	43	Dhanpunji –Kotapad-Nuagan-Bariguna-Rondapolli-Jaypur-Koraput-Dumuriput-Similigurha-Pottangi-Orissa/Andhra Pradesh border.	152
7	60	Orissa/West Bengal border-Jaleswar-Amarda-Basta-Rupsa-Haldipada	57
8	75	Orissa/Jharkhand-Champua-Parsora	18
9	200	Orissa/Chhattisgarh-Lakhanpur-Jharsuguda-Kuchinda-Bhojpur-Deograh-Gogua-Bajrakot-Talcher-Kualo-Kamakyanagar-Bhuban-Sukinda-Chandikhol.	440
10	201	Boriguma-Nabarangapur-Poppada Landi-Maidalpur-Ampani-Koksara-Moter-Junagarh-Bhawanipatna-Dadpur-Utkela-Kesinga-Kusrupara-Belgan-Saintala-Balangir-Luisinga-Jogisuruda-Dungripali-Barpali-Bargarh	310
11	203	Bhubaneswar-Dhauli-Pipili-Puri-Baligahi-Konark	97
12	203A	The highway starting from its junction with NH-203 at Puri, connecting Bhamhagiri and terminating at Satpada	49
13	215	Rajamundra-Bimlagarh-Kora-Parsora-Palasponda-Dhenkikot-Ghatgan-Similia-Anadapur-Ramachandrapur and connecting on NH-5	348
14	217	Orissa/Chhattisgarh-Kharhial-Nauparha-Taraborh-Komana-Khariar-Bongomunda-Tilagarh-Belgan-Ramapur-Baligurha-Mahasingha-Sirtiguda-Simanbadi-Pippalapanka-Sorada-Asika-Pukkundakhandi Brahmapur-Varendrapur-Gopalpur.	438
15	224	Khordha-Begunia-Bolagarh-Nayagarh-Nuagan-Dashapalla-Purunakata-Bauda-Sonapur-Torabha-Balangir.	298
		Sub Total	**3704**
Punjab			
1	1	Punjab/Haryana border-Raipura-Khanna-Ludhiana-Lodhowal-Phillaur-Goraya-Phagwara-Jalandhar-Sara Nussi-Kartarpur-Beas-Butari-Tangra-Amritsar-Atari-Punjab/Pak border.	254
2	1A	Jalandhra-Sanaura-Dasuya-Bhangala-Pathankot-Punjab/J and K border	108
3	10	Punjab/Haryana border-Mandi Dabwali-Lambi-Abul Kharana-Malaut-Abohar-Nihalkhera-Fazilka-Indo/Pak border.	72
4	15	Amritsar-Gohtwar Varpal-Taran Taran–Sirhali-Makhu-Zira-Faridkot-Kot-Lambwali-Bhatinda-Fakarsar-Abohar-Bakayanwala-Punjab/Rajasthan border	350
5	20	Pathankot and upto HP border	10
6	21	Punjab/Chandigarh border-Kharar-Kurali-Rupangar-Ghanauli-Nirmohgarh-Punjab/HP border	67

(Contd.)

Punjab (*Contd.*)

7	22	Punjab/Haryana border-Lalru-Basi upto Haryana border.	31
8	64	Punjab/Haryana border-Banur-Rajpura-Patiala-Sangrur-Dhanaula-Barnala-Tapa-Rampura Phul-Bathinda-Punjab/Rajasthan border.	255.5
9	70	Jalandhar-Adampur-Nasrala-Hoshiarpur-Punjab/HP border	50
10	71	Jalandhra-Nakodar-Mahatpur-Moga-Dala-Barnala-Dhanaula-Sangrur-Dirba-Dogal-Punjab/Haryana border.	130
11	72	Punjab/Haryana border and upto Punjab/Haryana border.	4.5
12	95	Kharar-Marinda-Khamnon-Samrala-Ludhiana-Jagraon-Moga-Ferozpur-Husainiwala Cantt.	225
		Sub Total	**1557**

Pudducherry

1	45A	Villuppuram-Pondicherry-Chidambaram-Nagappattinam	43
2	66	Pondicherry-Tindivanam-Kirishnagiri	10
		Sub Total	**53**

Rajasthan

1	3	Rajasthan/UP border-Majiyan-Rajasthan/MP border	32
2	8	Rajasthan/UP border-Ajarka-Behror-Kotputli-Pragpura-Shahpura-Manoharpur-Chandwali-Dhand-Amer-Jaipur-Bagru-Mahlan-Dadu-Kishangarh-Ajmer-Kharwa-Bayawar-Bali-Bhim-Barar-Bagar-Dewair-Kelwa-Rajnagar-Nathdwara-Dalwara-Eklingji-Chirwa-Udaipur-Passad-Khairwara-Bechiwara-Ratanpur.	635
3	11	Rajasthan/UP border-Luharu-Halena-Mahwa-Manpur-Bhankri-Dausa-Jatwara-Kanota-Jaipur-Chomu-Ringas-Palsana-Goria-Sikar-Lachhmangarh-Harsawa-Fatehpur-Rol-Ratangarh-Lachharsar-Sridungargarh-Benisar-Seruna-Benisar-Naurangdesar-Bikaner	531
4	11A	Manoharpur-Partapgarh-Dausa-Lalsot and terminating at Kothum	145
5	11B	The highway starting from its junction with NH 11 near Lalsot connecting Mandaori-Gangapur-Kurgaon-Karauli-Sri Muthra-Barauli-Anjali-Bari- Dhaulpur.	180
6	11C	Old alignment of NH no. 8 passing through Jaipur from Km 220 to 273.50	53
7	12	Rajasthan/MP border-Ghatoli-Aklera-Ameta-Jhalawar-Khemai-Darrah-Mandara-Kota-Talera-Bundi-Sathur-Hindoli-Umar-Devli-Mendwas-Tonk-Baroni-Newai-Chaksu-Sheodaspura-Sanganer-Jaipur.	400
8	14	Gujarat/Rajasthan border-Mawal-Abu Raod-Swarupgarh-Banas-Pindwara-Sirohi-Palri-Posaliya-Sanderav-Gondoj-Pali-Jadan-Khamal-Sojat-Chandawal-Raipur-Bayawar.	310
9	15	Rajasthan/Punjab border-Ganganagar-Mahiyanwali-Ganeshgarh-Suratgarh-Rajiyasar-Mokalsor-Lunkaransar-Jagdevwala-Bikaner-Gajner-Nokhra-Bap-Phalodi-Kalra-Khara-Lathi-Chandan-Jaisalmer-Devikot-Khoral-Gunga-Shiv-Bharka-Barmer-Dhogimanna-Kabuli-Chitalwana-Sanch	906
10	65	Rajasthan/Punjab border-Sadulpur-Budwa Khare-Ratannagar-Ramgarh-Kaymsar-Fatehpur-Mugluna-Salasar-Kasumbi-Ladnun-Nimbi Jodhan-Karnaota-Surpalia-Borwa-Deh-Nagaur-Bhakrod-Tankla-Khimser-Soila-Kherapa-Bawari-Daijar-Jodhpur.	450

(*Contd.*)

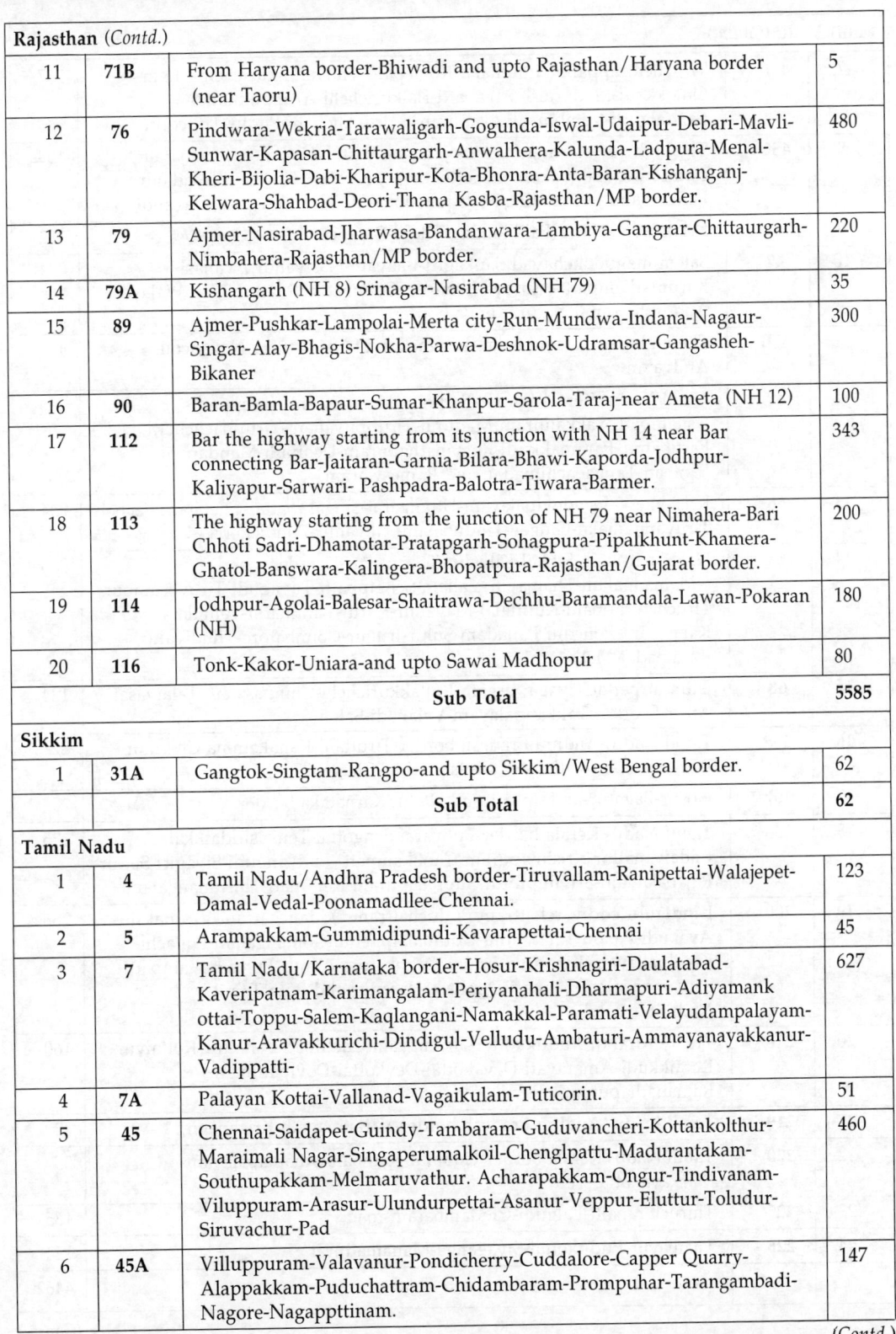

Rajasthan (*Contd.*)

11	71B	From Haryana border-Bhiwadi and upto Rajasthan/Haryana border (near Taoru)	5
12	76	Pindwara-Wekria-Tarawaligarh-Gogunda-Iswal-Udaipur-Debari-Mavli-Sunwar-Kapasan-Chittaurgarh-Anwalhera-Kalunda-Ladpura-Menal-Kheri-Bijolia-Dabi-Kharipur-Kota-Bhonra-Anta-Baran-Kishanganj-Kelwara-Shahbad-Deori-Thana Kasba-Rajasthan/MP border.	480
13	79	Ajmer-Nasirabad-Jharwasa-Bandanwara-Lambiya-Gangrar-Chittaurgarh-Nimbahera-Rajasthan/MP border.	220
14	79A	Kishangarh (NH 8) Srinagar-Nasirabad (NH 79)	35
15	89	Ajmer-Pushkar-Lampolai-Merta city-Run-Mundwa-Indana-Nagaur-Singar-Alay-Bhagis-Nokha-Parwa-Deshnok-Udramsar-Gangasheh-Bikaner	300
16	90	Baran-Bamla-Bapaur-Sumar-Khanpur-Sarola-Taraj-near Ameta (NH 12)	100
17	112	Bar the highway starting from its junction with NH 14 near Bar connecting Bar-Jaitaran-Garnia-Bilara-Bhawi-Kaporda-Jodhpur-Kaliyapur-Sarwari- Pashpadra-Balotra-Tiwara-Barmer.	343
18	113	The highway starting from the junction of NH 79 near Nimahera-Bari Chhoti Sadri-Dhamotar-Pratapgarh-Sohagpura-Pipalkhunt-Khamera-Ghatol-Banswara-Kalingera-Bhopatpura-Rajasthan/Gujarat border.	200
19	114	Jodhpur-Agolai-Balesar-Shaitrawa-Dechhu-Baramandala-Lawan-Pokaran (NH)	180
20	116	Tonk-Kakor-Uniara-and upto Sawai Madhopur	80
		Sub Total	**5585**

Sikkim

1	31A	Gangtok-Singtam-Rangpo-and upto Sikkim/West Bengal border.	62
		Sub Total	**62**

Tamil Nadu

1	4	Tamil Nadu/Andhra Pradesh border-Tiruvallam-Ranipettai-Walajepet-Damal-Vedal-Poonamadllee-Chennai.	123
2	5	Arampakkam-Gummidipundi-Kavarapettai-Chennai	45
3	7	Tamil Nadu/Karnataka border-Hosur-Krishnagiri-Daulatabad-Kaveripatnam-Karimangalam-Periyanahali-Dharmapuri-Adiyamankottai-Toppu-Salem-Kaqlangani-Namakkal-Paramati-Velayudampalayam-Kanur-Aravakkurichi-Dindigul-Velludu-Ambaturi-Ammayanayakkanur-Vadippatti-	627
4	7A	Palayan Kottai-Vallanad-Vagaikulam-Tuticorin.	51
5	45	Chennai-Saidapet-Guindy-Tambaram-Guduvancheri-Kottankolthur-Maraimali Nagar-Singaperumalkoil-Chenglpattu-Madurantakam-Southupakkam-Melmaruvathur. Acharapakkam-Ongur-Tindivanam-Viluppuram-Arasur-Ulundurpettai-Asanur-Veppur-Eluttur-Toludur-Siruvachur-Pad	460
6	45A	Villuppuram-Valavanur-Pondicherry-Cuddalore-Capper Quarry-Alappakkam-Puduchattram-Chidambaram-Prompuhar-Tarangambadi-Nagore-Nagappttinam.	147

(*Contd.*)

Tamil Nadu (*Contd.*)

7	45B	Tiruchchirappalli-Viralimalai-Koviepatti-Thuvarankurichchi-Melur-Ottaikkadi-Madurai-Kariyapatti-Kalkurichchi-Aruppukkottai-Pandalgudi-Vembur-Ettaiyapuram-Kurukkuchabi-Pudur-Tuticorin.	257
8	45C	The highway starting from its junction with NH 67 near Thanjavur connecting Kumbakonam, Sethiathope, Neyveli Township, Vadalur, Panruti and terminating at its junction with NH 45 near Vikravandi.	159
9	46	Krishnagiri-Natra-Pallikonda-Vellore-Ranipet.	132
10	47	Salem-magudanchavad-Sankagiri-Bhavani-Nasiyanur-Avanashi-Karumathampatti-Arasur-Nilampur-Coimbatore-Madukkarai-Walayer And upto Tamil Nadu/Kerala border.	224
11	47B	The highway starting from the junction of NH 47 near Nagencoil-Aralvaym.	45
12	49	Tamil Nadu/Kerala border-Bodinayakkanur-Teni-Andippatti-Usilampatti-Sakkanurani-Madurai-Tiruppuvanam-Tiruppachchetti-Partibanur-Paramakkudi-Ramanathapuram-Uchipuli-Mandapam-Pamban-Tangachchimadan and Rameswaram.	290
13	66	Krishnagiri-Jegadevipalayam-Uttangarai-Singarapattai-Chengam-Pachel-Tiruvannamalai-Gingee-Tindivanam-Kiliyanur-Pondicherry.	234
14	67	Nagappattinam-Sikkal-Kilvelur-Thiruvarur-Koradachcheri-Nidamangalam-Thanjavur-Vallam-Sengippatti-Tuvagudi-Tiruverumbur-Tiruchchirappalli-Kulittalai-Mayanur-Karur-Paramathi-Tennilai-Kangayam-Pongalu-Palladam-Sulur-Irugur-Coimbatore-Thudiyalur-Karamadai-Met	505
15	68	Ulundurpettai-Tiyagai urgam-Kallakkurichchi-Chinnasalem-Talaivasal-Attur-Peddanayakkanpalayam-Valapadi-Salem.	134
16	205	Tamil Nadu/Andhra Pradesh border-Tiruttani-Kanakamma Chatram-Ramanjeri-Tiruvallur-Avadi-Ambathur-Chennai.	82
17	207	Hosur-Bagatur-and upto Tamil Nadu/Karnataka border.	20
18	208	Tamil Nadu/Kerala border-Puliyarai-Sengottai-Tenkasi-Idaikkal-Kadaiyanallur-Krishnapuram-Chokkampatti-Puliyangudi-Sivagiri-Settur-Rajapalaiyam-Srivilliputtur-Suppalapuram-Kellupati-Thirumangalam.	125
19	209	Dindigul-Reddiyarchattram-Puduchattram-Oddanchatram-Virupakshi-Ayakudi-Palani-Talaiyuthu-Udumalaippettai-Gomangalam-Pollachi-Kovilpallayam-Kinattukkadavu-Coimbatore-Annur-Punjaibuliampatti-Satyamangalam-Bannari-Dimbam-Hasanur-and upto Tamil Nadu/Karnataka	286
20	210	Trichy-Kiranur-Pudukkottai-Tirumayam-Chettinad-Pallattur-Kottaiyur-Karaikkudi-Amaravati-Devakottai-Devkottai-Dovipattinam-Ramanathapuram	160
21	219	Tamil Nadu/Andhra Pradesh border-Varatanapalli-Krishnagiri.	22
22	220	Tamil Nadu/Kerala border-Gudalur-Kamban-Uthamapalayam-Markayakottai-Teni	55
23	227	Thiruchirapalli-Lalgudi-Chidambaram road	135
24	226	Thanjavur-Pudukottai-Sivaganga-Manamadurai	144
		Sub Total	**4462**

(*Contd.*)

(*Contd.*)

Tripura			
1	44	Tripura/Assam border-Ambasa-Chandrasadhubari-Barjala-Udaipur-Sabrum.	335
2	44A	Tripura/Mizoram border-Sakhan-Manu	65
		Sub Total	**400**
Uttar Pradesh			
1	2	UP/Haryana border-Kosi-Chhata-Mathura-Farah-Agra-Firozabad-Shikohabad-Sirsaganj-Jaswantnagar-Etawah-Sarai-Muradganj-Sikandra-Rasdhan-Bara-Sachendi-Kanpur-Moharajganj-Aung-Fatehpur-Haswa-Sat Narain-Khaga-Palhana-Kaushambi-Allahabad-Saidabad-Hardia-Gopiganj	752
2	2A	Sikandra-Raipur-Bhognipur	25
3	3	Agra-UP/MP border	26
4	7	Varanasi-Mirzapur-Lalganj-UP/MP border	128
5	11	Agra-Kiraoli-UP/Rajasthan border	51
6	12A	MP/border upto junction with NH-26 near Jhansi.	7
7	19	Ghazipur-Ballia-Rudrapur-Bakutha-UP/Bihar border	120
8	24	Delhi/UP border-Ghaziabad-Rajabpur-Bibauli-Pakbara-Moradabad-Mirgang-Bareilly-Banthra-Uncholia-Neri-Mohli-Sitapur-Lucknow.	431
9	24A	Badshi-Ka-Talab-Chenhat (NH 28)	17
10	24B	Lucknow-Rai Bareily-Allahabad road	185
11	25	UP/MP border-Jhansi-Baragaon-Ghirgaon-Amargarh-Moth-Pirauna-Orai-Usargaon-Kalpi-Bara-Kanpur-Unnao-Aigain-Lucknow.	270
12	25A	Km19 (NH 25)-Bakshi-Ka-Talab	31
13	26	UP/MP border-Karari-Jhansi-Babina-Talbahqt-Bansi-Lalitpur-Birdha-Gona-UP/MP border.	128
14	27	Allahabad-Jasra-UP-MP border.	43
15	28	Lucknow-BaraBanki-Ramsanehighat-Faizabad-Haraiya-Khalilabad-Piprauli-Hata-Kasia-Fazilnagar-Pawanagar-Tamkuhi-UP/Bihar border.	311
16	28B	UP/Bihar border-Nibua Raiganj-Paiganj-Padrauna-Kasia.	29
17	28C	Bara Banki-Ramnagar-Jarwal-Krisarganj-Fakharpur-Bahraich-Matera Bazar-Nanpara-Babaganj-Rupidiha-Nepalganj.	140
18	29	Sonauli-Kolhu-Pharenda-Rawatganj-Gorakhpur-Bhaurapur-Kauriram-Ghasi-Mardah-Ghazipur-Zamania-Chandauli-Varanasi.	306
19	56	Lucknow-Gosainganj-Amethi-Bhetwa-Haidargarh-Inhauna-Jagdishpur-Musafir Khana-Hasanpur-Sitapur-Singramau-Badlapur-Bakhsha-Junpur-Phulpur-Varanasi.	285
20	56A	Chehat Km. 16 of NH 56	13
21	56B	Km. 16 on NH-56 to Km 19 of NH-25	19
22	58	UP/Delhi border-Noida-Muradnagar-Modi-Nagar-Muhiuddinpur-Meerut-Mujjafar Nagar-UP/Uttaranchal border.	165

		Uttar Pradesh (*Contd.*)	
23	72A	UP/Uttaranchal border-Chhutmalpur-Biharigarh and UP/Uttaranchal border.	30
24	73	UP/Haryana border-Sarsawa-Pilkhani-Saharanpur.	60
25	74	UP/Uttaranchal border-Najibabad-Nagina-Afzalgarh-Rehar and UP/Uttaranchal border.	147
26	75	UP/MP border-Dudhinagar-Wyndhamganj	110
27	76	UP/MP border-Srinagar-Mahoba-Banda-Khuhand-Attarra-Badausa-Karwi-Raipura-Mau-Shankargarh-Bara-Jasra-Allahabad-Naini-Astabhuja Mirzapur.	587
28	86	Kanpur-Ghatampur-Sajet-Hamirpur-Sumerpur-Maudeha-Khanna-Kabrai-Mahoba-Srinager-UP/MP border.	180
29	87	Rampur-Bilaspur-UP/Uttaranchal border.	32
30	91	Ghaziabad-Dadri-Sikanderabad-Bulandsharh-Khurja-Amiya-Aligarh-Pilwa-Etah-Kurawali-Sultanganj-Bewar-Nabigaon-Chhibramau-Gurusahayganj-Kannauj-Araul-Bilhaur-Kanpur.	405
31	91A	The highway starting from its junction with NH 2 near Etawah connecting Bharthana-Bidhuna-Bela-Mundarwagaj and terminating at its junction with NH 91 near Kaunauj.	126
32	92	UP/MP border-Udi-Etawah-Chaubia-Kusmara-Bewar.	75
33	93	Agra-Khandauli-Sadabad-Halhras-Mandrak-Daud Khan-Aligarh-Danpur-Dibal-Babrala-Bahjoi-Chandausi-Bilari-Moradbad.	220
34	96	Faizabad-Bilharghat-Bikapur-Sultanpur-Bhada-Piparpur-Kohadaur-Bela-Soraon-Allahabad.	160
35	97	Ghazipur-Zamania-Said Raja.	45
36	119	The highway starting from its junction with NH 58 near Meerut connecting Mawana-Bahsuma-Bijnor-Kiratpur-Najibabad and upto UP/Uttaranchal border.	125
37	NE2	**Eastern Peripheral Expressway around in UP and Haryana (UNDER CONSTRUCTION)**	90
		Sub Total	**5874**
Uttarakhand			
1	58	Uttranchal/UP border-Manglaur-Roorkee-Hariswar-Motichur-Rishikesh-Shivpuri-Bhuint-Srinagar-Khankra-Pudraprayag-Nagrasu-Nandaprayag-Chamoli-Bhimtalla-Mayasur-Belakuchi-Langsi-Helang-Joshimath-Vishnupravag-Govindghat-Hanuman Chatti-Badarinath-Mana.	373
2	72	Uttaranchal/HP border-Dhalipur-Sahaspur-Jhajra-Dehradun-Bullawala-Kansrao and connecting with NH 58 near Motichur.	100
3	72A	Uttaranchal/UP border-Majra-Dehradun	15
4	73	Roorkee-Bhagwanpur-Uttranchal/UP border	20
5	74	Haridwar-Jaspur-Kashipur-Barakhera-Rudrapur-Kichha-Sitarganj and upto Uttaranchal/UP border.	153
6	87	Uttranchal/UP border-Rudrapur-Pantnagar-Jitpur-Kathgodam-Nainital-Bhowali-Kwarab-Almora-Majkhali-Ranikhet-Dwrahat-Mehalchauri-Adbadri and connecting with NH 58 (near Karnaprayag)	284

(*Contd.*)

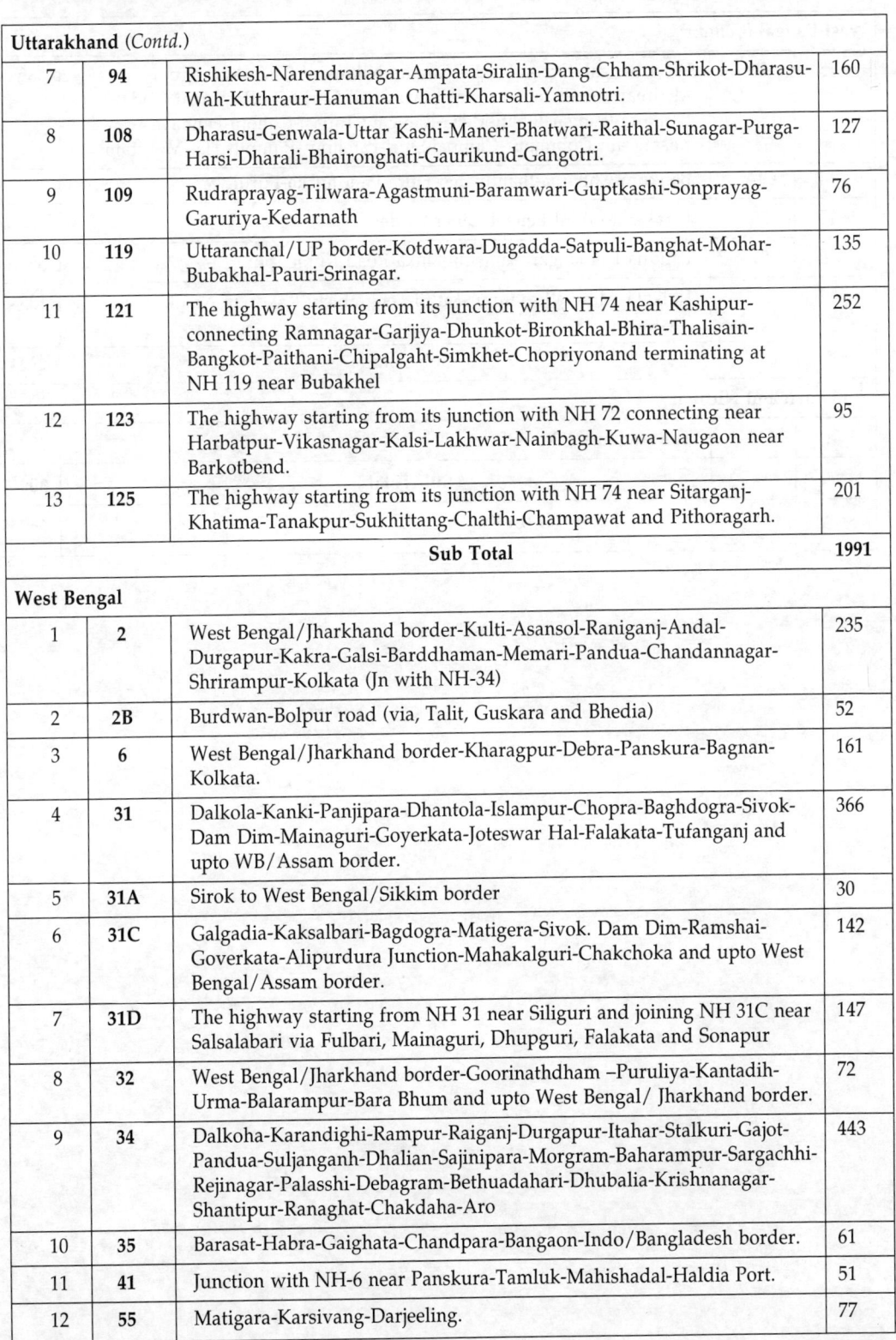

		Uttarakhand (*Contd.*)	
7	94	Rishikesh-Narendranagar-Ampata-Siralin-Dang-Chham-Shrikot-Dharasu-Wah-Kuthraur-Hanuman Chatti-Kharsali-Yamnotri.	160
8	108	Dharasu-Genwala-Uttar Kashi-Maneri-Bhatwari-Raithal-Sunagar-Purga-Harsi-Dharali-Bhaironghati-Gaurikund-Gangotri.	127
9	109	Rudraprayag-Tilwara-Agastmuni-Baramwari-Guptkashi-Sonprayag-Garuriya-Kedarnath	76
10	119	Uttaranchal/UP border-Kotdwara-Dugadda-Satpuli-Banghat-Mohar-Bubakhal-Pauri-Srinagar.	135
11	121	The highway starting from its junction with NH 74 near Kashipur-connecting Ramnagar-Garjiya-Dhunkot-Bironkhal-Bhira-Thalisain-Bangkot-Paithani-Chipalgaht-Simkhet-Chopriyonand terminating at NH 119 near Bubakhel	252
12	123	The highway starting from its junction with NH 72 connecting near Harbatpur-Vikasnagar-Kalsi-Lakhwar-Nainbagh-Kuwa-Naugaon near Barkotbend.	95
13	125	The highway starting from its junction with NH 74 near Sitarganj-Khatima-Tanakpur-Sukhittang-Chalthi-Champawat and Pithoragarh.	201
		Sub Total	**1991**
West Bengal			
1	2	West Bengal/Jharkhand border-Kulti-Asansol-Raniganj-Andal-Durgapur-Kakra-Galsi-Barddhaman-Memari-Pandua-Chandannagar-Shrirampur-Kolkata (Jn with NH-34)	235
2	2B	Burdwan-Bolpur road (via, Talit, Guskara and Bhedia)	52
3	6	West Bengal/Jharkhand border-Kharagpur-Debra-Panskura-Bagnan-Kolkata.	161
4	31	Dalkola-Kanki-Panjipara-Dhantola-Islampur-Chopra-Baghdogra-Sivok-Dam Dim-Mainaguri-Goyerkata-Joteswar Hal-Falakata-Tufanganj and upto WB/Assam border.	366
5	31A	Sirok to West Bengal/Sikkim border	30
6	31C	Galgadia-Kaksalbari-Bagdogra-Matigera-Sivok. Dam Dim-Ramshai-Goverkata-Alipurdura Junction-Mahakalguri-Chakchoka and upto West Bengal/Assam border.	142
7	31D	The highway starting from NH 31 near Siliguri and joining NH 31C near Salsalabari via Fulbari, Mainaguri, Dhupguri, Falakata and Sonapur	147
8	32	West Bengal/Jharkhand border-Goorinathdham –Puruliya-Kantadih-Urma-Balarampur-Bara Bhum and upto West Bengal/ Jharkhand border.	72
9	34	Dalkoha-Karandighi-Rampur-Raiganj-Durgapur-Itahar-Stalkuri-Gajot-Pandua-Suljanganh-Dhalian-Sajinipara-Morgram-Baharampur-Sargachhi-Rejinagar-Palasshi-Debagram-Bethuadahari-Dhubalia-Krishnanagar-Shantipur-Ranaghat-Chakdaha-Aro	443
10	35	Barasat-Habra-Gaighata-Chandpara-Bangaon-Indo/Bangladesh border.	61
11	41	Junction with NH-6 near Panskura-Tamluk-Mahishadal-Haldia Port.	51
12	55	Matigara-Karsivang-Darjeeling.	77

(*Contd.*)

West Bengal (*Contd.*)

13	60	West Bengal/Orissa border-Dantan-Nikursini-Belda-Kharagpur-Medinapur-Salbani-Chandra Kona Road-Garhbeta-Ramsagar-Onda-Bankura-Gangajalghati-Mejia-Raniganj Tapasi-Chhora-Pandaveswar-Kastagram-Dubrajpur-Chinpai-Siuri-Fatehpur-Rampur Hat-Morgram.	389
14	60A	Bankura-Chhatna-Jhantipahai-Hura-Landhurka-Puruliya.	100
15	80	Farakka to West Bengal/Bihar border.	10
16	81	Pandua-Kumangarj-Samsi-Bhaluka Road-Kumedpur.	55
17	117	Kolkata-Alipur-Bishnupur-Kulpi-Tengrabichi-Kakdwip-Namkhana-Bokkhali.	133
		Sub Total	**2524**
Andaman and Nicobar			
1	223	The Andaman Trunk Road.	300
		Sub Total	**300**
		TOTAL LENGTH (in Km.)	**66754**

APPENDIX C

VEHICULAR EMISSION NORMS

- The first State emission norms came into force in 1991 for petrol vehicles and in 1992 for diesel vehicles.
- From April 1995, fitting of catalytic converters in new petrol driven passenger cars was mandated and unleaded petrol was also introduced in the four Metros.
- From April 2000, unleaded petrol is made available in the country.
- In developed countries lead was phased out from petrol over a period of more than 10 years, while in India this was achieved just in 6 years.
- The comparative statement of emission norms as under, indicates that the time gap between the introduction of norms in Europe and our country is narrowing:

	Euro I	Euro I	Euro II	Euro III
European Norms	1983	1992	1996-97	2000-2001
Indian Norms	1996	1.4.2000	*	**

* Bharat Stage-II norms, which are akin to Euro-II norms have been introduced in National Capital Region (NCR) for passenger vehicles upto GVW 3.5T from 1-4-2000 and for heavier vehicles form 24-10-2001 in National Capital Territory (NCT) or Delhi.

** Bharat Stage-III emission norms have been introduced with effect from 1-4- 2005 in respect of Four Wheeled vehicles manufactured on and from 1st April, 2005 in the National Capital Region and the cities of Mumbai, Kolkata, Chennai, Bangalore, Hyderabad including Secundrabad, Ahmedabad, Pune, Surat, Kanpur and Agra except for four wheeled transport vehicles plying on Inter-State Permits or National Permits or All India Tourist Permits within the jurisdiction of these cities.

- In case of Mumbai, these have been extended from 1-1-2001 to 31-10-2001 respectively.
- For both Chennai and Kolkata, the corresponding dates are 1-7-2001 and 31-10-2001 respectively.
- These norms have further been extended to Agra, Ahmedabad, Bangalore, Hyderabad / Secundrabad, Kanpur, Pune and Surat from 1-4-2003, and Lucknow and Sholapur from 1-6-2004 for all categories of vehicles.
- The transport vehicles plying on inter-State permits or on National Permits or on All India tourist Permits or plying from these cities to the other regions of the respective States have been exempted.
- Bharat Stage-II emission norms have been extended to entire country vide Gazette Notification No. G.S.R. 927 (E) dated 5-12-2003 and these have become effective from 1-4-2005.
- Basically, Bharat Stage-II norms involve supply of petrol and diesel with 0.05% sulphur content.
- In rest of the country, petrol has a sulphur content of 0.1% with effect from 1-4-2000 as against 0.05% in these cities and NCR of Delhi.
- Similarly, sulphur content in diesel has been reduced in the country, from a level of 1.0% maximum in 1996 to 0.25% on 1-4-2000.
- In respect of NCT of Delhi and the above mentioned cities the sulphur content in diesel is similarly 0.05%.

- In addition to petrol and diesel, CNG and LPG are permitted to be used as auto fuels. Alternative fuels like di-methyl ether, bio-diesel, hydrogen, electric and fuel cell vehicles etc. are at various stages of experimentation.

- The emission norms for tractors were first notified in the year 1999. The next generation norms have been laid down. While Bharat (Trem) Stage II norms have come into force from 01-06-2003, the Bharat (Trem) Stage III norms has come into force from 01-10-2005.

- Next generation emission norms for two-wheelers and three-wheelers manufactured on and after 1-4-2005 have been notified.

APPENDIX D

HIGH SECURITY REGISTRATION PLATES

On the basis of the recommendations made by the Technical Standing Committee on Central Motor Vehicles Rules, the Central Government had amended rule 50 of the Central Motor Vehicles Rules, 1989, mandating introduction of new High Security Registration Plates, both in respect of new and in-use motor vehicles throughout the country. The relevant Gazette Notifications are G.S.R. 221 (E) dated 28-03-2001, S.O. 814 (E) dated 22-08-2001 and S.O. 1041 (E) dated 16-10-2001.

Features of high security registration plates are:

(i) Chromium hologram.

(ii) A retro-reflective film, bearing a verification inscription 'India' at 45 degree inclination.

(iii) Laser numbering, which is unique in nature containing alpha-numeric identification of both testing agencies and the manufacturers.

(iv) The registration numbers to be embossed on the plates.

(v) In case of rear registration plate, same to be fitted with a non-reusable snap lock to make it tamper proof.

(vi) A chromium based third registration plate in the form of sticker is to be attached to the wind shield, wherein the number of engine and chassis are indicated along with the name of registering authority. If tampered with, itself destructs.

(vii) In front and rear registration plates, letter IND in blue color is hot stamped.

(viii) Letters 'IND' in blue colour on extreme left centre of the plates.

APPENDIX E

SMART CARD BASED DRIVING LICENSE AND REGISTRATION CERTIFICATE

- In order to usher in transparency and e-governance through induction of IT in Transport Sector, Department of Road Transport and Highways, in consultation with National Informatics Centre (NIC), State/UT Governments and Smart Card Industry, have developed a standardized software.

- The software has been made available to States and UTs free of cost.

- The software covers both back-end automation of RTOs and front-end computerization to enable issue of:
  - Smart Card based driving licenses,
  - Registration Certificates/permits for transport vehicles.

- Under an MOU, National Informatics Centre has been remitted to oversee smooth implementation of the programme throughout the country.

- Smart Card was defined in the Gazette Notification No. G.S.R. 400 (E), dated 31st May, 2002.

- This Gazette Notification also empowered the States to prescribe the date of introduction of Smart Card based documents.

- Further, precise specifications to be adopted for smart cards, hand held terminals, printers and smart card readers were also circulated to States/UTs.

- By a recent Gazette Notification No. G.S.R. 153 (E) issued on 10-08-2004, these specification have been embodied in the Central Motor Vehicles Rules, 1989 to ensure their uniform applicability throughout the country.

- The notification lays down the specifications for the microprocessor chip based smart card of minimum 4KB memory capacity which must conform to specifications of International organization for Standardization (ISO)/International Electro Technical Commission (IEO).

- The note to the Notification clarifies that the microprocessor chip of the smart card shall contain the information prescribed under Central Motor Vehicles Rules and shall not contain any other information not prescribed.

- In case the States/Union Territories require extra information to be stored, they may use any other additional information storage media or technological media, outside the microprocessor chip embodied in the smart card so long as such media do not conflict with the prescribed smart card or its operation.

- The symmetric key infrastructure for Smart Card based driving license and Registration Certificate have established at the Central level. State level operations are being commenced by Governments of Delhi, Jharkhand, Maharashtra, Tripura and West Bengal. Several other States are in various stages of progress on this account.

APPENDIX F

MISCELLANEOUS SCHEMES

Public Private Partnership (PPP)

Traditionally, the road projects where financed only out of the budgetary grants and were controlled/supervised by the Government. The road sector has attracted little private sector participation in the past. The traditional system of financing road projects through budgetary allocation has proved to be inadequate to meet the growing requirements of this sector. To encourage private sector participation, several initiatives have been taken by the Union Government, which include:

- Provision of capital subsidy up to 40% of the project cost to make projects commercially viable.
- 100% tax exemption in any consecutive 10 years out of the first 20 years of a project.
- Provision of encumbrance free site for work, i.e. the Government shall meet all expenses relating to land and other pre-construction activities.
- Foreign Direct Investment up to 100% in road sector.
- Higher concession period, (up to 30 years).
- Right to collect and retain toll.

National Highway Accident Relief Service Scheme (NHARSS)

In order to provide immediate medical assistance to road accidents victims and also to remove damaged vehicles from the site, cranes and ambulances are provided under this scheme to NGOs/State authorities. The recurring expenditure on account of operation and maintenance cost is to be borne by the beneficiary State Government/NGO.

Public Grievance Redressal Machinery (PGRM)

The public Grievance Redressal Machinery is headed by Joint Secretary (Admn. and Public Grievances), who is assisted by Under Secretary (O and M) for public grievances and by Director (Admn.) and Staff Grievances Officer for staff grievances. All public grievance petitions sent to the Department are acknowledged and their disposal is closely monitored. The time-limit of 60 days from the date of receipt of the grievance has been laid down for setting the grievance in the Department, in consultation with the concerned office/organization. A wooden box in which members of public can place their grievance petitions without having to enter the security zone has been provided near the Reception on the Ground Floor of the Building (Transport Bhawan).

APPENDIX G

SPECIFIC INFORMATION FROM ROAD DEVELOPMENT PLAN 2001-2021

(ii) There are 11 major ports, and 140 minor ports. The length of railway network is 63000 km.

(iii) The share of public sector investment in transport is 12-13 percent.

(iv) The share of private sector in bus fleet ownership and operation is 77 percent.

(v) Population growth rate computed between 2001 and 2006 was 1.6 percent per year and is projected as 1.57 percent between 2006 and 2011 and 1.50 percent afterwards. The share of urban population by 2021 would be 35 percent.

(vi) In case of automobiles, the growth of automobile industry is 28 percent per annum, vehicle population growth is 5 percent per annum, heavy vehicle growth is 10.6 percent per annum, and light commercial vehicle growth is 19 percent per annum.

(vii) Around 77 percent of transport operators own small fleet size of less than or equal to 5 trucks, whereas, 6 percent own fleet size of greater than or equal to 20 trucks.

(viii) The tax burden in the price of vehicle is 31 percent and in operating cost is 56.5 percent.

(ix) Trucks are responsible for 30 percent of total fatalities though their share is 5.2 percent of vehicle population. Head-on collisions followed by rear-end collisions account for 70 percent of truck accidents. Almost 50 percent truck accidents occur near road junctions and inhabited areas. One-third accidents occur during normal sleeping hours. Sixty-five percent drivers are involved in rash and negligent driving.

(x) The size of urban road network is estimated as 230000 km.

(xi) It is estimated that an investment of Rs. 1000000 in rural roads will bring 165 persons above the poverty line.

APPENDIX H

ENVIRONMENT GUIDELINES FOR HIGHWAY PROJECTS

Purpose of Guidelines

These assist the project authorities in planning and carrying out EIA and developing Environmental Management Plan. These guidelines apply to both new project and upgrading of existing facilities.

Environmental Impact Assessment (EIA)

EIA Procedure identifies the possible positive and negative impacts resulting from a proposed project. The parameters are classified under four categories, namely natural physical resources, natural biological resources, Human development resources, and Quality of life values.

Environmental Impact Statement (EIS)

EIA should cover brief description of the project, description of existing environment within influence area of proposal, adverse or beneficial or reversible or irreversible impacts, mitigation, protection and enhancement measures and alternatives.

Environmental Management Plan (EMP)

EMP is an implementation plan including mitigation, protection and/or enhancement measures, defined by objectives, work plan, implementation schedule, manpower requirements and monitoring.

Identification of Impacts

Impacts are identified based on listed parameters under categories mentioned in EIA. These are impacts on hydrology, surface water quality, air quality, soils and noise impacts; impacts on fisheries, forestry, wild life, and ecosystem, impacts on navigation, flood control and land-use, and impacts on socio-economic aspects, resettlement issues, public health, aesthetics, historical values, etc.

Measures for Mitigation of Adverse Impacts

Measures to deal with the various impacts are as under:

(a) Air quality: Mobile source emissions-
  (i) Construction during off-peak hours in heavy traffic areas
  (ii) Use of low emission construction vehicles
  (iii) Periodical check on all vehicles for emission control
  (iv) Use of lead free gasoline
(b) Air quality: Fixed source emissions-
  (i) Stationary equipment to be located as far away as possible from the receptors
  (ii) Areas prone to dust emissions be sprinkled with water
  (iii) Dust covers over the beds of trucks
  (iv) Low emission equipment for construction
(c) Noise: Construction period –
  (i) Specify permissible standards for noise for construction equipment
  (ii) Specify maximum permissible noise levels in residential, commercial and institutional areas

(iii) Specify time restrictions in sensitive areas such as schools, hospitals

(iv) Describe methods of enforcement for the above

(d) Noise: Operation phase-

(i) Right-of-way to have buffer strip on each side of road and where possible tree belts be planted

(ii) Noise insulation including noise barriers in certain areas such as schools/hospitals

(iii) Rerouting heavy traffic

(iv) Changing speed limits

(v) Changing alignment

(e) Vibrations

Appropriate construction technology should be used to reduce damage due to vibrations during construction.

(f) Relocation

(i) Adequate time be given to relocate

(ii) Public relations through media

(iii) Advance payment to relocates

(iv) Compensation for land property based on fair market value

(v) Cost of reestablishment

(vi) Special low cost housing

(vii) Special section to look into the problems of relocates.

Environmental Monitoring

Monitoring should be done at different levels and of different types like air monitoring, noise monitoring and water quality monitoring.

Management Considerations

Following may be the considerations for the management of impacts:

(i) Cut and fill technology

(ii) Treatment of unstable areas

(iii) Vegetative cover on slopes

(iv) Erosion control measures

(v) Drainage needs

(vi) Channel training and erosion control works of culverts

(vii) Controlling blasting of rocks

(viii) Dumping of excavated material

(ix) Provision of adequate protective works

(x) Adequate provision of water supply, power and sanitation facilities.

References

1. AASHTO *Road user Benefit Analysis for Highway Improvements* National Press Building. Washington, D. C.

2. Agarwal M. K., *Urban Transportation–Some Aspects and Prospects* Indian Highways, June 1996.

3. Agarwal P. D., *Maintenance of Unsurfaced roads in Uttar Pradesh* Indian Highways, Sept. 1980.

4. Anderson H.L., *The Traffic Engineer and Highway Safety* Journal of Institute of Engineers, U.S.A. Nov. 1979

5. Babu K.V.G, Asakura Y and Katsuhiko K., *ITS around the world and need for ITS on Indian Roads* Indian Highways, Feb. 2005.

6. Baerwald J.E., *Traffic Engineering Handbook* Institute of Traffic Engineers, Washington D. C., U. S. A.

7. Basu D, *Total Road Connectivity to Rural Bengal (West Bengal) in the Perspective of Pradhan Mantri Gram Sadak Yojna (PMGSY)* Indian Highways, Jan. 2002.

8. Bhattacharya, *Traffic Calming Measures–Some Issues*, Indian Highways, Jan. 2002.

9. Bhattacharya et.al. *Development of Comprehensive Highway Noise Model for Indian Conditions* Paper No: 481, Journal of Indian Road Congress, Dec. 2001.

10. Bindra H.A, Mookerjee A.K., *Sub-surface Drainage of Flexible Pavements* Indian Highways, Jan. 1999.

11. Bose R, Sarkar P.K. and Mattri V., *Cycle-Rickshaws–A Sustainable Mode of Transport: Case Study-walled city, Delhi,* Indian Highways, Oct. 2004.

12. Box and Oppenlander, *Manual of Traffic Engineering Studies* Institute of transportation Engineers, U.S.A., March 1981.

13. Chakraborty S. S. *Sustainable Road Transport for India–A Need of Today and Challenge for Tomorrow* Indian Highways, Nov. 2009.

14. Dasgupta D and Taneja R. C. *Design and Planning of Urban Roads, Case Studies* Indian Highways, April 2007.

15. Deb S.K. *Environmental Landuse Transport (ETL) Approach for Urban-Rural Continuum.* Indian Highways, July 2009.

16. Deshpande G. K., Jadhav G. B and Shinde *Overview of Environmental Considerations in Highway Projects and Procedure for obtaining clearance from Ministry of Environment and Forest,* Indian Highways, April 2002.

17. Dharmraj S. J. P, *Rubberised Bitumen for Roads* Indian Highways, Oct. 2002.

18. Dhir, Merani and Sarna *Planning and Development of Road Systems for Metropolition Areas*, Paper No. 370, Journal of Indian Road Congress, Vol. 46, 1985.

19. EkSe and Hennes *Fundamentals of Transportation Engineering* McGrawHill Book Company, New York, U.S.A.

20. Eno foundation *Parking Garage Planning and Operation*, Connecticut, U.S.A.

21. Eno Foundation *Statistics with Application to Highway Traffic Analysis*, Connecticut, U.S.A.

22. Faik Abid *Speed-Flow-Density Relationship on Sulaimania Streets* A Master of Engineering Thesis, University of Sulaimania, Iraq.

23. Gerlough D. *Statistics with Application to Highway Traffic Analysis Eno* foundation for Transportation, Westport, U.S.A.

24. Gichaga F.J. and Parker N.A. *Essentials of Highway Engineering* Macmillan Publisher, 1988.

25. Goldstein Gary *Electronic Traffic Management Civil Engineering ASCE*, Nov. 1984.

26. Gupta AK, Nigam S.P and Hansi J. S. *A Study on Traffic Noise for Various land uses for mixed Traffic Flow* Indian Highways, Feb. 1986.

27. Gupta A. K. *Waste Road Construction Material–A Remedy to Environmental Pollution* Indian Highways, Jan. 1997.

28. Gupta AK, Jani S.S, Srivastava JB and Jain J.K. *Traffic Flow Behaviour and Environmental Impact Analysis of a Highway Corridor* Indian Highways.

29. Gupta D.P. *Rural Roads–An Entry Point for Poverty Alleviation and Employment Generation* Indian Highways, June 2006.

30. Guram M.S. *Road Development and Environment* Indian Highways, Jan. 1997.

31. Handa, D. N. *Pradhan Mantri Gram Sadak Yojna in Himachal Pradesh—Some Suggestions for Hilly Areas*, Indian Highways, January 2003.

32. Hay W.W. *An Introduction to Transportation Engineering* John Wiley and Sons, Inc, New York, U.S.A.

33. Highway Research Board *Highway Capacity Manual* Special Report No. 87, National Research Council, U.S.A.

34. Hobbs F.D. *Traffic Planning and Engineering* Pergamon Press, 1979.

35. Homburger and Kell *Fundamentals of Traffic Engineering* Institute of Transportation Studies, University of California, U.S.A.

36. Hurd M.S. *Traffic Engineering*, McgrawHill Book Company, New York, U. S. A.

37. Hutchinson B.G. *Principles of Urban Transport Systems Planning* Scripta Book Company, Washington, D.C.

38. John W.D. Metropolitan Transportation Planning Scripta Book Co. Washington, D.C. U.S.A.

39. Kadiyali LR, Lal NB et.at. *Speed Flow Characteristics on Indian Highways* Paper No. 406, IRC. Nov. 1991.

40. Kadilyali LR *Traffic Engineering and Transport Planning–a book* 6th edition 1997.

41. Kadiyali LR *Principle and Practices of Highway Engineering* Khanna Publishers, Delhi.

42. Kand C. V. *Utility Services along Urban Roads* Indian Highways, Oct. 2008.

43. Kapila KK, Sikdar P.K *Road Safety Scenario and Proposed Action Plan* Indian Highways, Nov. 2009.

44. Karjinni V, Sahu M.K, Bhatia O.P. *Calibration of Gravity Models for an Intermediate City-A Case Problem*, Indian Highways, Sept. 2003.

45. Kaushik S. *Intelligent Transport System (ITS) and its Applications in Public Transportation System* 49th NTCP Congress, Hyderabad.

46. Krishnarao K. V, Maitra B, Sikdar P.K and Dhingra S. L. *Estimation of Traffic and Toll Sensitivity for a Proposed Major Urban Transport Infrastructure—A Case Study* of BWSL, Mumbai, Journal of Indian Road Congress, New Delhi.

47. Kumar K. Urban *Transportation–Problems and Solutions* Indian Highways, June 2002.

48. Kumar S. V. *Online Management, Monitoring and Accounting System for the Pradhan Mantri Gram Sadak Yojna* Indian Highways, June 2005.

49. Kumar, P. *Progress of Privatisation in Highways* Indian Highways, June 2000.

50. Lall B.K. and Khisty CJ. *Transportation Engineering—An Introduction* 3rd edition. Prentice-Hall of India Pvt. Ltd., New Delhi.

51. Lay M. G. *Source book for Australian Roads* Australian Road Research Board, 1985.

52. Logavinayagam K.S, Ramdas C. *Improvements to Anna Salai in Madras City* Indian Highways, May 1982.

53. Logavinayagam K.S. *Drainage Aspects in Road Maintenance* Indian Highways, Sept. 1982.

54. Mahayni R. G. *Passive and Dynamic Concepts of Transportation Planning in Developing Countries* University of Washington, Washington, U.S.A.

55. Murugesan R. *Noise Nuisance by Road Traffic Assessment and Control* Indian Highways, July 1997.

56. NAASRA. *Guide to Traffic Engineering Practice* Sydeny, 1982.

57. Narain A.D. *Road Sector Institutional Initiatives,* Indian Highways, Nov. 2005.

58. National Research Council USA. *Traffic Flow Theory a Monograph,* special report No: 165.

59. Nishal S, Sangita, Sharma B.M., Sengupta J. B. A *Laboratory Study of Efficacy of Conventional and Modified Bituminous Binders in Construction of Roads* Indian Highways, Jan. 2012.

60. O'Flaher C. A. *Highway Engineering* published by Edward Arnold, London.

61. Oglesby and Hewes. *Highway Engineering* John Wiley and Sons, New York.

62. Panday D.K. *Major Environmental Mitigation concerns in Road Projects* Indian Highways, January 2007.

63. Pandey I.K. *Economic Evaluation of the Environmental Impacts of Highway Projects (with special reference to accidents).* Indian Highways, Feb. 97.

64. Pandit U.B. *A New Track in Safety* Indian Highways, May 2003.

65. Paquette, Ashford and Wright. *Transportation Engineering Planning and Design* The Ronald Press Company, New York, U.S.A.

66. Parker N.A. *Highway Maintenance–Techniques or Management* The Tanzania Engineer Journal, March 1987.

67. Parmar CM. *Pedestrian-Most Neglected Major Road User* Indian Highways, Feb. 2003.

68. Pawar A.B. *Community Participation in Highway Projects* Indian Highways, June 2002.

69. Pignatro L.J. *Traffic Engineering Theory and Practice* Prentice Hall Inc. Englewood Cliffs, New Jersy, U.S.A.

70. Puri, A. K. *Problems of Highway Sector in India* Indian Highways, May 2001.

71. Rajoria K.B, Sharma V.K. *Perspective Planning for the Road Infrastructure of Delhi—Futuristic Action Plan,* Indian Highways, June 2003.

72. Rajoria. Singh P and Singh A. P. *Pedestrian Subways in Urban Areas with special reference to Delhi* Indian Highways, Aug. 1999.

73. Ram S, Singh J.B., Singh A.K. *Environmental Impact Assessment and Evaluation for Road Construction–A Case Study* Indian Highways, Sept. 1999.

74. Ramanjee M. *Some Design Related Aspects of Flexible Pavements* Indian Highways, March 2005.

75. Ramesh C.R. *Maintenance of roads a new approach* Indian Highways, Jan. 2000.

76. Rao N.T. *Derivation of Transportation Implications from a Proposed Urban Master Plan* Indian Highways, Aug. 2003.

77. Rastogi R. *A Study of Accidents in and around Kota City* Indian Highways, Apr. 2006.

78. Rathore S.S. *Private Sector Participation in Gujarat* Indian Highways, June 2002.

79. Rathore S.S, Andrew M.B., Patel K.A. Urban *Improvements along the Gujarat Highway Projects issues and solutions* IRC paper No. 515, Oct. 2005.

80. Recommendations about the Alignment Survey and Geometric Design of Hill Roads IRC-52-1973.

81. Recommendations for Road Construction in Water Logged Areas IRC-34-1970.
82. Recommended Practice for 2 cm Thick Bitumen and Tar Carpet IRC: 14(1977).
83. Recommended Practice for Road Delineators IRC-79-81.
84. Recommended Practice for Sand-Bitumen Base Courses IRC: 55(1974).
85. Recommended Practice for the Construction of Earth Embankment for Road Works. IRC-36-1970.
86. Recommended Practice for Traffic Rotaries IRC 65-1985.
87. Recommended Practice for Traffic Rotaries IRC: 65-1976.
88. Recommended Practice for Treatment of Embankment Slopes for Erosion Control IRC-56-1974.
89. Reddy T.S. Road Safety Audit of Indian National Highways—A Need of the Day Indian Highways, Jan. 2002.
90. Salter R.J. *Highway Traffic Analysis and Design* Addison Wesley Publishing Company. Reading U.S.A.
91. Salter, R.J. *Highway Design and Construction* University of Bradford, Macmillan Press Ltd. 1979.
92. Sarin S. M and Bhaita N. L. *Urban Transportation Planning and Operation* Indian Highways, Oct. 2000.
93. Sarin S. M. Sharma N, Sharmak. Singh A. *An Overview of the Veluicular Emission Legislations in India* Indian Highways, Dec. 1999.
94. Sarin S.M., Mitlal N. *Road safety audit? Frequently asked questions* Indian Highways, March 2005.
95. Sarkar P. K, Bose S and Ghosh P. *A Critical Appraisal of Traffic and Transportation Sector in Delhi and possible solutions* Paper No. 532, Indain Road Congress Journal, Sept. 2007
96. Satish Chandra, Devraj *Role of Shoulders in Traffic Operation* Indian Highways, Nov. 1999.
97. Saxena S. C. *A course in Traffic Planning and Design* Dhanpat Rai and Sons, New Delhi 1989.
98. Saxena S. C. *A Text Book of Railway Engineering* Published by Dhanpat Rai and Sons, Delhi 1973.
99. Saxena S. C. *Airport Engineering Planning and Design* a book Published by C.B.S. publishers, Delhi 2008.
100. Saxena S. C. *Atmospheric Pollution from Automobile Emissions.* HRD Times, April. 1998.
101. Saxena S. C. *Master Plan for Traffic and Transportation of Jaipur City*—a report under Rajasthan Urban Improvement Plan of Govt. of Rajasthan, Apr. 2002
102. Saxena S. C. and Arora SP *Structural Failures of Flexible Pavements* Roorkee University Research Journal, 1972.
103. Saxena S. C. and Khalid FF *Study on Iraqi Lime Stone Aggregates in Bituminous Mixes* International Symposium on Aggregates, International Association of Engineering Geology, Nice, France, May 1984.
104. Saxena S. C. Azmim *Rice Husk Ash as Filler in Bituminous Mixes* Journal of Institution of Highway Engineers London, U.K., June 1984.
105. Saxena S. C. Jain Renu *Need for an Environmental Impact Assessment Policy* HRD Times, June 1997.
106. Saxena S. C. *Complex Nature of Environmental Management Problems* American Society of Civil Engineering, Journal of India Section, Feb. 2001.
107. Saxena S. C. *Non-Destructive Method for Soil Compaction Testing* First International Conference of the Engineering Association of Asia and Australia (EAAA) held in Bangkok, Thailand, 1976.
108. Saxena S. C. *Traffic Flow Characteristics in Dar ES Salaam* Journal of Kenya National Academy of Sciences, Nairobi, Kenya, 1990.
109. Saxena S. C. *Transportation Modes and Environmental Hazards* National Seminar on Traffic Enforcement and Environment in Metropolitan Cities Madras, 1972.
110. Saxena S. C. and Ismail Y.I. *Marshall Stiffness Criteria for Bituminous Mix Design* Journal of Institution of Engineers, Malaysia, 1988.
111. Saxena S. C. and Ramprasad C. *Economic Analysis is for Hired Vehicles in Urban Transportation* Road Research Bulletin No. 17, Indian Road Congress, New Delhi, 1974.

112. Saxena S. C. *Layer System Analysis for Varying Values of Poisson Ratio* Journal Indian Road Congress, New Delhi, May 1970.

113. Saxena S. C. *Tunnel Engineering* Book Published by Dahnpat Rai and Sons, 1994.

114. Schwarts S. I. *Towards 2000: Challenges and Strategies for Transportation Profession* Institute of Transportation Engineers, Jan. 1981.

115. Sharama N, Gangopadhyay S and Singh A *Environmental Impact of Road Transport an Indian Scenario* Indian Highways, Jan. 2008.

116. Sharama S. C. *Trends in Construction and Maintenance of Road and Bridges* Indian High·vay, Jan. 2002.

117. Sharma N, Dhyani R and Gangopadhy *Review of Environmental Laws and their Applicability to Road/Highway Projects.* Journal of the Indian Road Congress, July-Sept 2009.

118. Sharma R. S and Khinda G.S. *S.I. Units, their Symbols and Numeration* Indian Highways, June 2005.

119. Sheoran R. R. *Control and Protection of Highways* Journal of Indian Road Congress, New Delhi.

120. Sheoran R. R. *Road Safety Audit* Indian Highways, Jan. 2004.

121. Shukla AK, Jain S. S. Parida M and Sharma S.P. *Assessment of Traffic Impact on Ambient Air Quality of Lucknow City* Indian Highways, May 2006.

122. Shukla R. S., Pundhir N.K.S. and Harit M. C. *Use of Natural Rubber in Bituminous Road Construction* C.R.R.I. Research Publication.

123. Sikdar P.K. *Future Traffic Management The dot com Approach* Nov. 2000.

124. Sikdar P.K. *Pradhan Mantri Gram Sadak Yojna-A mission for Rural Connectivity by All weather Roads* Indian Highways, May 2001.

125. Singh R.S. J. *Urban Transportation Planning in India—Alarming Issues* Indian Highways, Dec. 2006.

126. Sinha N. K. *Road Development –New Perspective* Indian Highways, June 2002.

127. Sivaguru N *Environmental Protection Measures in Highway Construction and Transportation* Indian Highways, Dec. 1983.

128. Srinivasan N. S. *Coordinated Efforts and Positive Approach for Successful Implementation of Urban Traffic Improvement Schemes* Indian Highways, June 1998.

129. Srinivasan N. S. *Critical Review of Thrust Areas in Road Safety Action Plan* Indian Highways, Dec. 2011.

130. Srinivasan N. S. *Traffic and Transportation Improvement Plan for Trivandrum* Indian Highways Feb. 1986.

131. Srinivasan P. P. *Transportation Infrastructure in Pondicherry Past, Present and Future Perspective* Indian Highway, June 2003.

132. Srivastava R. K. Jalota AV and Singh RB *Environmental Impact Assessment and Evaluation for Highway Projects* Indian Highways, Apr. 1991.

133. Subhani S. M. and Balakrishna S. *Upgradation of Transportation Infrastructure in Class I Cities of Andhra Pradesh*—a Methodological approach. 49th NTCP Congress, Hydrabad.

134. Thamizh V, Koshy R. Z. Simulation of Heterogeneous traffic to derive capacity and service volume standards for urban roads Paper No. 500, Indian Road Congress, New Delhi, Dec. 2004.

135. Trivedi R. N Transportation-A Systems Approach Indian Highways, Sept. 2000.

136. TRRL, London *A Road Investment Model for Developing Countries* RTIM-2, TRRL Report No. LR 1057.

137. Weigelt, Gatz and Weiss City traffic Van Nastrend Reinhold Company.

138. Westerman H. L. *Road and Environments* Austration Road Research, Dec. 1990.

139. Winfrey R *Economic Analysis for Highways* International text book company Scarnton.

140. World Bank *Highway Design and Maintenance Standard Model*—HDM III, 1985.

141. Wright, P. H. Ashord N. J, *Transportation Engineering*—Planning and Design, 3rd edition, John wiley and sons, Newyork.

142. Yoder E.J. *Principles of Pavement Design* John Wiley and Sons, New York, 1964.

Indian Road Congress Publications:

(i)   Dimensions and Weights of Road Design Vehicles IRC-3-1984.

(ii)   Geometric Design of Urban Roads IRC 86-1983.

(iii)   Geometric Design Standards for Rural (Non-urban) Roads IRC 73-1980.

(iv)   Guidelines on Regulation and Control of Mixed Traffic in Urban Areas IRC: 70-1977.

(v)   Guidelines for Capacity of Urban Roads in Plain Areas IRC: 106-1990.

(vi)   Guidelines for Control of Access on Highways IRC-62-1976.

(vii)   Guidelines for Design of Horizontal Curves for Highways IRC-38-1988.

(viii)   Guidelines for Environmetal Impact Assessment of Highway Projects IRC-104-1988.

(ix)   Guidelines for Pedestrian Facilities IRC-103-1988.

(x)   Guidelines for the Design of At Grade Intersections in Rural and Urban Areas special publication 41, 1994.

(xi)   Guidelines for the Design of Interchanges in Urban Area IRC: 92-1985.

(xii)   Guidelines for Traffic Prediction on Rural Roads IRC: 108-1996.

(xiii)   Guidelines for Wet Mix Macadam IRC:109-1997.

(xiv)   Guidelines on Accommodation of Utility Services on Roads in Urban Areas IRC : 98-1997.

(xv)   Guidelines on Road Drainage IRC-SP-42-1994.

(xvi)   Guidelines on the Choice and Planning of Appropriate Technology in Road Construction IRC: special publication 24

(xvii)   Guidelines on Urban Drainage IRC SP-50

(xviii) Guidelines on use of polymer and rubber modified Bitumen in Road Construction IRC: SP-53-2000.

(xix)   Handbook of Quality Control for Construction of Roads and Runways. IRC-special publication 11.

(xx)   Hill Road Manual IRC: SP-48-1998.

(xxi)   Lateral and Vertical Clearances at Underpasses for Vehicular Traffic IRC-54-1974.

(xxii)   Nagraj BN., Chandrasekhar BP and Raghvachari S. *Delineation of Regions for Transportation Planning* Paper No. 344, Indian Road Congress, New Delhi, Dec. 1981.

(xxiii) Route Marker Signs for National Highways IRC: 2-1968.

(xxiv)   Rural Road Manual IRC: SP-20-2002.

(xxv)   Space Standards for Roads in Urban Areas IRC: 69-1977

(xxvi)   Specification for Dense Bitumious Macadam IRC: 94-1986

(xxvii) Specifications for Bituminous concrete (Asphaltic concrete) IRC: 29(1988)

(xxviii) Standard Specifications and Code of Practice for Road Bridges IRC-5-1970.

(xxix)   Tentative Recommendations on the Provision of Parking Spaces for Urban Areas IRC : SP-12.

(xxx)   Tentative specification for Built-up spray grout IRC: 47(1972)

(xxxi)   Tentative specifications for Bituminous macadam (Base and Binder course) IRC: 27(1967)

(xxxii) Tentative specifications for Bituminous surface dressing using precoated aggregates IRC: 48(1972)

(xxxiii) Tentative specifications for single coat bituminous surface dressing IRC:17(1965)

(xxxiv) Tentative specifications for single coat surface dressing using Catonic Bitumen emulsion IRC 100 (1988)

(xxxv)   Tentative specifications for the coat surface dressing using Catonic Bitumen emulsion IRC: 96(1987)

(xxxvi) Tentative specifications for two coat Bituminous surface dressing IRC: 23(1966)

(xxxvii) Traffic Studies for Planning Bypasses around Towns IRC-102-1988.

(xxxviii) Vertical Curves for Highways IRC Publication No. 23

Transport and Road Research Laboratory (U.K) Publications:

a. Laboratory Report 638 Speed/Flow relations on recreational roads
b. LR 108 *Public Transport Study in Indian Cities*
c. LR 1110 *Dust mission from unpaved roads in Kenya*
d. LR 479 *A data logging system for the measurement of road traffic noise*
e. LR 670 *Observations of pedestrian behavior at four sites*
f. LR 676 *Road accident data collection and Analysis in developing countries*
g. LR 775 *A study of road accidents in selected urban areas in developing countries*
h. LR 798 *Exposure of drivers to carbon monoxide*
i. LR 896 *Road surfaces and traffic Noise*
j. SR 610 *Review of published research into formation of corrugations on paved roads*
k. Supplementary Report 277 *Road accidents as a cause of death in developing countries*
l. Supplementary Report 332 *A Controlled study of the role of alcohol in fatal pedestrian accidents*

Index

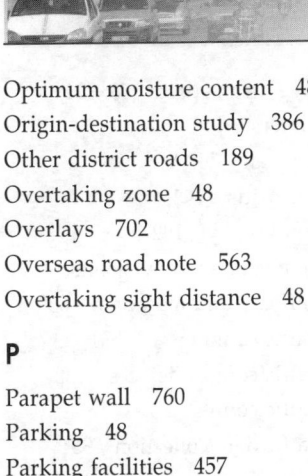